Daily Values for Food Labels

The Daily Values are standard values developed by the Food and Drug Administration (FDA) for use on food [labels], based on 2000 kcalories a day for adults and children over 4 years old.

Proteins, Vitamins, and Minerals

Nutrient	Amount
Protein[a]	50 g
Thiamin	1.5 mg
Riboflavin	1.7 mg
Niacin	20 mg NE
Biotin	300 µg
Pantothenic acid	10 mg
Vitamin B$_6$	2 mg
Folate	400 µg
Vitamin B$_{12}$	6 µg
Vitamin C	60 mg
Vitamin A	5000 IU[b]
Vitamin D	400 IU[b]
Vitamin E	30 IU[b]
Vitamin K	80 µg
Calcium	1000 mg
Iron	18 mg
Zinc	15 mg
Iodine	150 µg
Copper	2 mg
Chromium	120 µg
Selenium	70 µg
Molybdenum	75 µg
Manganese	2 mg
Chloride	3400 mg
Magnesium	400 mg
Phosphorus	1000 mg

[a]The Daily Values for protein vary for different groups of people: pregnant women, 60 g; nursing mothers, 65 g; infants under 1 year, 14 g; children 1 to 4 years, 16 g.

[b]Equivalent values for nutrients expressed as IU are: vitamin A, 1500 RAE (assumes a mixture of 40% retinol and 60% beta-carotene); vitamin D, 10 µg; vitamin E, 20 mg.

Nutrients and Food Components

Food Component	Amount	Calculation Factors
Fat	65 g	30% of kcalories
Saturated fat	20 g	10% of kcalories
Cholesterol	300 mg	Same regardless of kcalories
Carbohydrate (total)	300 g	60% of kcalories
Fiber	25 g	11.5 g per 1000 kcalories
Protein	50 g	10% of kcalories
Sodium	2400 mg	Same regardless of kcalories
Potassium	3500 mg	Same regardless of kcalories

GLOSSARY OF NUTRIENT MEASURES

kcal: kcalories; a unit by which energy is measured (Chapter 1 provides more details).

g: grams; a unit of weight equivalent to about 0.03 ounces.

mg: milligrams; one-thousandth of a gram.

µg: micrograms; one-millionth of a gram.

IU: international units; an old measure of vitamin activity determined by biological methods (as opposed to new measures that are determined by direct chemical analyses). Many fortified foods and supplements use IU on their labels.
- For vitamin A, 1 IU = 0.3 µg retinol, 3.6 µg β-carotene, or 7.2 µg other vitamin A carotenoids.
- For vitamin D, 1 IU = 0.025 µg cholecalciferol.
- For vitamin E, 1 IU = 0.67 natural α-tocopherol (other conversion factors are used for different forms of vitamin E).

mg NE: milligrams niacin equivalents; a measure of niacin activity (Chapter 10 provides more details).
- 1 NE = 1 mg niacin.
  = 60 mg tryptophan (an amino acid).

µg DFE: micrograms dietary folate equivalents; a measure of folate activity (Chapter 10 provides more details).
- 1 µg DFE = 1 µg food folate.
  = 0.6 µg fortified food or supplement folate.
  = 0.5 µg supplement folate taken on an empty stomach.

µg RAE: micrograms retinol activity equivalents; a measure of vitamin A activity (Chapter 11 provides more details).
- 1 µg RAE = 1 µg retinol.
  = 12 µg β-carotene.
  = 24 µg other vitamin A carotenoids.

mmol: millimoles; one-thousanth of a mole, the molecular weight of a substance. To convert mmol to mg, multiply by the atomic weight of the substance.
- For sodium, mmol × 23 = mg Na.
- For chloride, mmol × 35.5 = mg Cl.
- For sodium chloride, mmol × 58.5 = mg NaCl.

About the Author

JUDITH E. BROWN is Professor Emerita of Nutrition at the School of Public Health, and of the Department of Obstetrics and Gynecology at the University of Minnesota. She received her Ph.D. in human nutrition from Florida State University and her M.P.H. in public health nutrition from the University of Michigan. Dr. Brown has received competitively funded research grants from the National Institutes of Health, the Centers for Disease Control and Prevention, and the Maternal and Child Health Bureau and has over 100 publications in the scientific literature including the New England Journal of Medicine, the Journal of the American Medical Association, and the Journal of the American Dietetic Association. A recipient of the Agnes Higgins Award in Maternal Nutrition from the March of Dimes, Dr. Brown is a registered dietitian and the successful author of Everywoman's Guide to Nutrition, Nutrition for Your Pregnancy, and What to Eat Before, During, and After Pregnancy.

EDITION

6 | NUTRITION NOW

JUDITH E. BROWN

University of Minnesota

WADSWORTH
CENGAGE Learning™

Australia • Brazil • Japan • Korea • Mexico • Singapore • Spain • United Kingdom • United States

Nutrition Now, **Sixth Edition**
Judith E. Brown

Nutrition Editor: Peggy Williams

Developmental Editor: Nedah Rose

Assistant Editor: Elesha Feldman

Editorial Assistant: Alexis Glubka

Media Editor: Miriam Myers

Marketing Manager: Laura McGinn

Marketing Assistant: Elizabeth Wong

Marketing Communications Manager:
Belinda Krohmer

Content Project Managers: Trudy Brown,
Jerilyn Emori

Creative Director: Rob Hugel

Art Director: John Walker

Print Buyer: Rebecca Cross

Rights Acquisitions Account Manager, Text:
Mardell Glinski Schultz

Rights Acquisitions Account Manager, Image:
Robyn Young

Production Service: Eric Arima / Elm Street
Publishing Services

Text Designer: Diane Beasley

Photo Researcher: Sarah Evertson / Image
Quest

Cover Designer: Brian Salisbury

Cover Image: Getty Images / Purestock

Compositor: Integra Software Services Pvt. Ltd.

Library of Congress Control Number: 2009940217

Student Edition:

ISBN-13: 978-1-4390-4903-7

ISBN-10: 1-4390-4903-3

Wadsworth
20 Davis Drive
Belmont, CA 94002-3098
USA

Cengage Learning is a leading provider of customized learning solutions with office locations around the globe, including Singapore, the United Kingdom, Australia, Mexico, Brazil, and Japan. Locate your local office at **www.cengage.com/global**.

Cengage Learning products are represented in Canada by Nelson Education, Ltd.

To learn more about Wadsworth, visit **www.cengage.com/Wadsworth**

Purchase any of our products at your local college store or at our preferred online store **www.ichapters.com**.

Printed in Canada
1 2 3 4 5 6 7 13 12 11 10 09

Contents in Brief

Contents

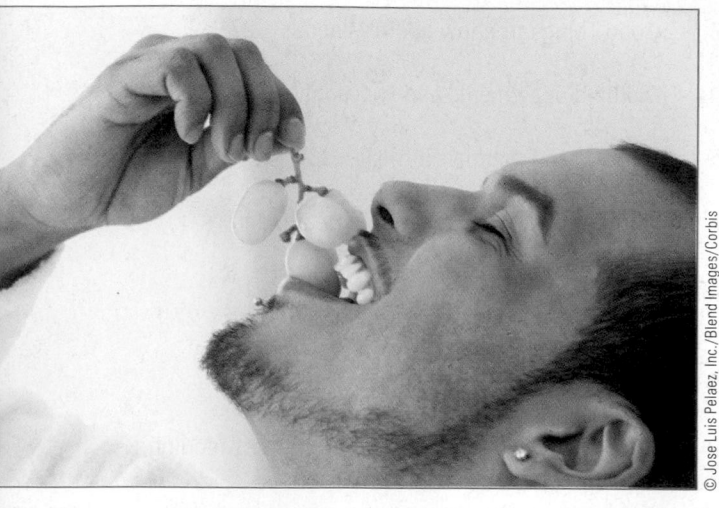

© Jose Luis Pelaez, Inc./Blend Images/Corbis

NUTRITION Timeline →

1621
First Thanksgiving feast
at Plymouth colony

PhotoDisc

1702
First coffeehouse in America
opens in Philadelphia

PhotoDisc

1734
Scurvy recognized

1744
First record of ice cream in America
at Maryland colony

PhotoDisc

PhotoDisc

1747

Lind publishes "Treatise on Scurvy," citrus identified as cure

PhotoDisc

1750

Ojibway and Sioux war over control of wild rice stands

1762

Sandwich invented by the Earl of Sandwich

PhotoDisc

1771

Potato heralded as famine food

1774

Americans drink more coffee in protest over Britain's tea tax

PhotoDisc

NUTRITION Timeline →

1775
Lavoisier ("the father of the science of nutrition") discovers the energy- producing property of food

Stefano Bianchetti/CORBIS

1816
Protein and amino acids identified, followed by carbohydrates and fats in the mid-1800s

1833
Beaumont's experiments on a wounded man's stomach greatly expand knowledge about digestion

1862
U.S. Department of Agriculture founded by authorization of President Lincoln

1871
Proteins, carbohydrates, and fats determined to be insufficient to support life; there are other "essential" components

© Ilene MacDonald/ Alamy

1895
First milk station providing children with un-contaminated milk opens in New York City

Bettman/CORBIS

1896
Atwater publishes Proximate Composition of Food Materials

1906
Pure Food and Drug Act passed by President Theodore Roosevelt to protect consumers against contaminated foods

Bettman/CORBIS

1910
Pasteurized milk introduced

PhotoDisc

1912
Funk suggests scurvy, beriberi, and pellagra caused by deficiency of "vitamines" in the diet

AP Images/Mandatory Credit: Lauren Greenfield / VII

NUTRITION Timeline →

1913
First vitamin discovered (vitamin A)

PhotoDisc

1914
Goldberger identifies the cause of pellagra (niacin deficiency) in poor children to be a missing component of the diet rather than a germ as others believed

1916
First dietary guidance material produced for the public released; title is "Food for Young Children"

1917
First food groups published the Five Food Groups: Milk and Meat; Vegetables and Fruits; Cereals; Fats and Fat Foods; Sugars and Sugary Foods

1921
First fortified food produced: iodized salt, needed to prevent widespread iodine-deficiency goiter in many parts of the United States

Morton Salt Co.

© Felicia Martinez/PhotoEdit

1928
American Society for Nutritional Sciences and the Journal of Nutrition founded

1929
Essential fatty acids identified

PhotoDisc

1930s
Vitamin C identified in 1932, followed by pantothenic acid and riboflavin in 1933 and vitamin K in 1934

1937
Pellagra found to be due to a deficiency of niacin

PhotoDisc

1938
Health Canada issues nutrient intake standards

1941
First refined grain enrichment standards developed

© MB Pictures/Getty Images/UpperCut Images

NUTRITION Timeline →

1941
First Recommended Dietary Allowances (RDAs) announced by President Franklin Roosevelt on radio

FDR Library

1946
National School Lunch Act passed

PhotoDisc

1947
Vitamin B$_{12}$ identified

1953
Double helix structure of DNA discovered

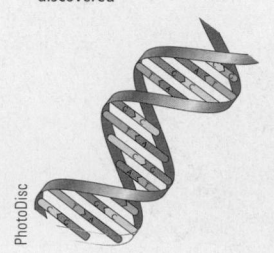
PhotoDisc

1956
Basic Four Food Groups released by the U.S. Department of Agriculture

Richard Anderson

1965
Food Stamp Act passed, Food Stamp program established

1966
Child Nutrition Act adds school breakfast to the National School Lunch Program

PhotoDisc

1968
First national nutrition survey in U.S. launched (The Ten State Nutrition Survey)

1970
First Canadian national nutrition survey launched (Nutrition Canada National Survey)

1972
Special Supplemental Food and Nutrition Program for Women, Infants, and Children (WIC) established

© James And James/Getty Images/FoodPix

NUTRITION Timeline ⟶

1977	1978	1989	1992	1997
Dietary Goals for the U.S. issued	First Health Objectives for the Nation released	First national scientific consensus report on diet and chronic disease published	The Food Guide pyramid is released by the USDA	RDAs expanded to Dietary Reference Intakes (DRIs)

PhotoDisc

PhotoDisc

1998

Folic acid fortification of refined
grain products begins

2003

Sequencing of DNA in the
human genome completed;
marks beginning of new era
of research in nutrient–gene
interactions

2009

Global epidemics of obesity and
diabetes threaten gains in life
expectancy

© Living the dream/Getty Images/Flickr

© Paul Bradbury/Getty Images/The Image Bank

© Digital Vision/Alamy

PhotoDisc

Richard Anderson

Preface

"Everything should be made as simple as possible. But not simpler."
—ALBERT EINSTEIN

IT IS WITH GREAT PLEASURE that we present to you the sixth edition of *Nutrition Now*. Athough the principles of the science of nutrition as developed for and presented in this text have not changed, much else has. Knowledge gains in nutrition are advancing at an incredibly high rate, and the implications of the advances to human health and well-being are impressive. Growth in our understanding of the role of antioxidants and anti-inflammatory components of food on disorders ranging from cancer to Alzheimer's disease is changing dietary guidance. Advances in knowledge about health effects of nutrient-gene interactions are changing fundamental concepts about the origins and prevention of disease. Knowledge about the interrelated activities of antioxidants and other nutrients in foods (versus supplements) is refining recommendations for nutrient intake and supplement use. The expanding epidemics of obesity and type 2 diabetes are reaching into younger age groups, and much more public attention is being paid to the roles of diet and physical activity in their prevention and management.

Advances in the science of nutrition are increasingly being brought to us by a deeper level of understanding of the roles of nutrients and other components of food at the cellular level. To understand nutrition and nutrition and health relationships, introductory courses in nutrition now must include topics such as visceral fat, insulin resistance, fatty liver disease, endothelial function, chronic inflammation, and nutrigenomics. Behavioral changes in diet, weight, and physical activity accomplished through "small steps" programs are leading the way to population-based lifestyles that improve health. This sixth edition covers these and other emerging topics directly related to the state of the science of nutrition now. It attempts to introduce these topics to students in a straightforward and clear way that keeps coverage as simple as possible, but no simpler than that.

Nutrition Now continues to be oriented toward helping students build a firm foundation of scientific knowledge about nutrition that will serve them well throughout life. Units are concise, focused on key facts and concepts, and provide ample real-life examples intended to enhance students' understanding of the material presented. Students are asked to apply their newly gained knowledge about nutrition in decision-making activities and exercises incorporated throughout the units.

Pedagogical Features

In designing the pedagogical features for this text, particular attention has been paid to the development of interactive learning opportunities for students. New and updated features include:

- **Take Action** features that ask students to select small step options aimed at, for example, increasing vegetable and fruit intake, stabilizing weight, and increasing the intensity of physical activity.

- Review Questions, and Answers to Review Questions for each Unit.

- Revised and updated Health Action, Reality Check, and Nutrition Up Close features.

- Broadly updated Web links to instructional videos, PowerPoint™ presentations, and other interactive learning activities.

Features familiar from earlier editions include:

- **Reality Check**. Reality Check presents brief, real-life scenarios and asks students to give a thumbs up or down to the optional solutions posed.

- **Nutrition Scoreboard**. Each unit begins with a three- to five-question pretest. Answers to the questions are given on the second page of the units.

- **Key Concepts and Facts**. Each unit begins with a listing of key concepts and facts related to the central topics covered.

- **On the Side**. Boxed inserts containing interesting facts related to nutrition topics covered in the units are sprinkled throughout the text.

- **Health Action**. Every unit has at least one Health Action box that relates to the personal application of information covered.

- **Margin definitions**. Unfamiliar terms are highlighted in bold in the text, and defined in nearby margins. Pronunciation guides are provided as needed.

- **Review questions and answers**. Students are invited to review their knowledge of key points made in each unit by answering eight to twelve review questions presented at the end of each unit.

- **Nutrition Up Close**. Each unit closes with an activity that gives students an opportunity to relate nutrition knowledge gained to their daily lives and experiences.

- **Media Menu**. Internet sources of reliable nutrition information are listed at the end of each unit. These have been thoroughly updated in the sixth edition and include interactive links.

- **Appended materials**. Resources included in the appendixes have been updated for the sixth edition of *Nutrition Now*: The Food Composition Table, the Reliable Sources of Nutrition Information list, the Food Exchange System, and Canada's Food Guide.

- **Glossary**. Terms defined in the margins are listed in the Glossary near the end of the text. Approximately 20 new terms have been added to the glossary of the sixth edition.

- **Nutrition Timeline**. A Nutrition Timeline bannered across the bottom of the table of contents highlights major developments in the science of nutrition.

- **Daily Values for Food Labels and Glossary of Nutrient Measures**. These appear on the last book page.

What's New in the Sixth Edition

This edition of *Nutrition Now* contains hundreds of small, factual updates and many extensive revisions to previous content. A comprehensive listing of those changes follows.

General changes include:

- In-depth coverage of new MyPyramid and DASH Eating Plan tools and resources in Unit 6: "Healthful Diets: My Pyramid, the Dietary Guidelines, and More."

- New or revised dietary recommendations for heart disease, hypertension, diabetes, and cancer; and for physical activity in people of all ages.

- Integrated coverage of the roles of diet, food components, body fat, and physical activity in oxidation reactions and the development of chronic inflammation-related diseases.

- Major changes in nutrition and physical performance content (from amino acid supplements and muscular recovery and strength to pre-event, event, and recovery foods and fluids).

- A revised Instructor's Activity Manual.

Specific changes, by unit, include:

Unit 1: Key Nutrition Concepts and Terms

- Changed Nutrition Scoreboard questions

- Added content on "Energy Density" including a table of the energy density of foods

- Added content on the DRI revision process

- Updated content related to nutrient functions at the cellular level

- Modified content on "good" foods and "bad" foods concept

Unit 2: The Inside Story about Nutrition and Health

- Modified margin definitions, added definitions of cirrhosis and Alzheimer's disease

- Updated content on nutrition and the current leading causes of death

- Expanded content on chronic inflammation, oxidative stress, and related dietary factors

- Added a vitamin joke as an On the Side feature to lighten student stress

- Added a table titled "Types of food associated with decreased or increased inflammation, oxidative stress, or both"

- Expanded coverage of MyPyramid food types and discretionary calories

- Added table of descriptions of basic foods included (e.g. lean meats, no-fat milk)

- Replaced the fifth edition's Nutrition Scoreboard activity with one that focuses on the types of foods that fit into the MyPyramid five basic food groups and those that characterize Western-type diets.

Unit 3: Ways of Knowing about Nutrition

- Updated Reality Check feature to address current concerns about bisphenol A

- Updated methods to be used in "The Research Design: Gathering the Right Information" section of the Unit.

- Replaced several of the Review Questions at the end of the Unit

Unit 4: Understanding Food and Nutrition Labels

- Condensed coverage of nutrition content claim definitions and deleted coverage of nutrition labeling in other countries

- Revised section on "Key Elements of Nutrition Labeling Standards"

- Updated table on "Examples of claims not approved by the FDA for use on food labels"

- Updated table on FDA approved health claims

- Added coverage of FDA's current efforts to enforce nutrition labeling standards and to crack-down on fraudulent claims and hazardous products

- Added coverage of the now required country-of-origin label (the COOL rule)

- Added a presentation on the "The Nutrition Labeling Transition" that includes the New Wave of Nutrition Labels (e.g. Nutrition at Glance, Smart Choices Made Easy, and the Smart Choices Program™)

- Added a section titled "Calories on Display" that covers recent legislation related to calorie labeling of fast foods

- Added content on new labeling systems and requirements that are being instituted by local and state governments

- Expanded content on organic foods and updated regulations

Unit 5: Nutrition, Attitudes, and Behaviors

- Modified content in table covering examples of ways in which diet may affect behavior

- Revised section titled "Food Additives, Sugar, and Hyperactivity"

- Added detail to the table "Changing food choices for the better"

- Deleted content on carbohydrates, neurotransmitters, and behavior

Unit 6: Healthy Diets, Dietary Guidelines, MyPyramid, and More

- Added content on variety as a core feature of healthful diets

- Updated sections on the ways in which the U.S. diet is out of balance

- Added and updated content on health promotion and chronic disease prevention effects of the Mediterranean diet, the MyPyramid Food Guide, and the DASH Eating Plan

- Increased discussion, added a table on calorie, fat, fiber, and vegetable and fruit content of fast food choices

- Expanded coverage to include the new MyPyramid web resources

- Abbreviated discussion of food guides in other countries

Unit 7: How the Body Uses Food: Digestion and Absorption

- Added content on the absorption of alcohol

- Expanded coverage of diet and constipation, diet and heartburn, and added table on myths related to constipation and on diet and other factors related to heartburn

- Added content on probiotics and ulcers

- Expanded section on irritable bowel syndrome

- Replaced the fifth edition Nutrition Up Close exercise with an activity that focuses on the products of digestion

Unit 8: Calories! Food, Energy, and Energy Balance

- Added a section on, and margin definition of energy density
- Added a Health Action feature titled "Moving toward foods lower in energy density"
- Added a section titled "Changing the Environment" that addresses environmental improvements and weight loss
- Revised content on genetic influences on the development of obesity
- Updated content on weight loss drugs

Unit 9: Obesity to Underweight: The Highs and Lows of Weight Status

- Expanded discussion of health effects of obesity to include metabolic disturbances (e.g. metabolic syndrome, insulin resistance, fatty liver disease), markers of inflammation (e.g. C reactive protein), importance of visceral fat and waist circumference

Unit 10: Weight Control: The Myths and Realities

- Updated illustration on examples of bogus nutrition products
- Modified content on organized weight loss programs, added section on Internet weight loss frauds
- Added a table that describes specific fad diets
- Added a section on physical activity and weight control and a table containing the new American College of Sports Medicine's recommendations
- Modified and updated the section on weight loss benefits
- Added a new Take Action feature related to taking small behavioral change steps that would improve diet and physical activity
- Added a section and illustration on lap band surgery
- Updated, modified information and illustration on body contouring surgery
- Deleted the existing Health Action feature
- Replaced existing Nutrition Up Close activity with one that gives students practice in setting small behavior change goals for weight management

Unit 11: Disordered Eating: Anorexia Nervosa, Bulimia, and Pica

- Added information on ipecac use and its adverse health effects
- Expanded presentation on body shape/appearance concerns
- Added information on night-time eating syndrome

Unit 12: Useful Facts about Sugars, Starches, and Fiber

- Expanded presentation of the structure and function of carbohydrates
- Modified presentation of the relationship between carbohydrates and alcohol

- Added illustration showing the chemical structure of glucose, fructose, xylitol, and ethanol
- Added a table on the monosaccharides and the disaccharides they form
- Deleted tables on added sugar names and the simple sugar content of breakfast cereals
- Modified the discussion and illustration related to labeling the sugar content of packaged foods
- Added a section on simple sugar intake and health effects of sucrose and high fructose corn syrup
- Completely modified the table related to artificial sweeteners (they are called non-nutritive sweeteners in the sixth edition)
- Added sections on neotame and rebiana (stevia)
- Expanded content on soluble fiber
- Updated content of the calorie value of fiber
- Updated section on diet and dental health

Unit 13: Diabetes Now

- Added content on The American Diabetes Association's recommendations for physical activity and dietary choices for individuals with prediabetes
- Added a section on insulin resistance that includes the topic of fatty liver disease
- Updated content on type 1 diabetes to include information on autoimmune diseases
- Updated information and illustration related to insulin pump technology, added section on insulin and new technologies in the management of type 1 diabetes
- Completely revised information on the management of type 1 diabetes
- Added an illustration that shows projected increases in rates of type 2 diabetes
- Revised and expanded content on hypoglycemia

Unit 14: Alcohol: The Positives and Negatives

- Added content of alcohol and chronic inflammation, steatohepatitis, and cirrhosis
- Added a new section labeled "Help for Alcohol Dependence"

Unit 15: Proteins and Amino Acids

- Added content on albumin, the "tramp steamer"
- Added an On the Side feature that presents information about beef grades
- Revised content on amino acid supplements and their use and hazards
- Updated content related to protein and amino acids and muscle mass and recovery
- Deleted section on tryptophan, serotonin, and melatonin supplements
- Completely revised and updated section of protein deficiency, kwashiorkor, and protein calorie malnutrition

Unit 16: Vegetarian Diets

- Added new information about availability of vitamin D fortified foods for vegans

Unit 17: Food Allergies and Intolerances

- Expanded coverage of anaphylactic shock and immunoglobulin E
- Added content on allergies and autoimmune disease
- Increased presentation about celiac disease and added a table on gluten-free foods
- Added the new recommendations related to food allergies in infants released by the American Academy of Pediatrics.
- Added a Take Action feature on the prevention of anaphylaxis
- Replaced Nutrition Up Close activity with one that focuses on how to select gluten-free meals and snacks

Unit 18: Fats and Cholesterol in Health

- Expanded coverage of functions of fat
- Revised and updated table on DHA and EPA content of fish and fortified foods
- Added section on increasing intake of omega-3 fatty acids
- Added a Take Action feature that offers choices for getting enough EPA and DHA from foods other than fish
- Revised table of trans fatty acid content of foods, updated regulations regarding trans fats in fast foods

Unit 19: Nutrition and Heart Disease

- Expanded discussion of chronic inflammation, diet, and inflammation markers
- Added content on anticipated increase in heart disease due to child and adolescent obesity

Unit 20: Vitamins and Your Health

- Updated and added choline to vitamin summary table
- Added a section on vitamin D, inflammation, and related diseases such as osteoporosis
- Added a section on recommendations for vitamin D intake, and on the current controversy about how much is enough
- Added a Take Action feature on personal options for increasing vitamin D status
- Added food sources of choline to table of food sources of vitamins
- Added a section and table on preserving the vitamin content of foods

Unit 21: Phytochemicals and Genetically Modified Food

- Updated table of rich plant sources of phytochemicals
- Added a Take Action feature on the selection of antioxidant-rich vegetables and fruits

- Added a section on animal clones and expanded discussion of genetically modified meats

Unit 22: Diet and Cancer

- Expanded depth of discussion on nutrient-gene interactions in susceptibility to cancer
- Expanded presentation of roles of antioxidants and anti-inflammatory food components on the development of cancer

Unit 23: Good Things to Know about Minerals

- Added a discussion and table related to preserving the mineral content of foods
- Added a Take Action feature aimed at making small changes in behavior to reduce salt intake
- Replaced DASH Eating Plan educational materials with newly released information

Unit 24: Dietary Supplements and Functional Foods

- Updated federal regulatory efforts related to dietary supplements
- Replaced definitions of prebiotics and probiotics with newly developed definitions
- Added a presentation on the potentially harmful effects of probiotics in some people

Unit 25: Water Is an Essential Nutrient

- Expanded presentation of functions of water
- Added a Reality Check on the effects of water consumption with meals
- Added a presentation on the safety of water bottles containing bisphenol A

Unit 26: Nutrient-Gene Interactions in Health and Disease

- Inserted presentation on non-protein coding segments of DNA, individual uniqueness in genetic makeup, and susceptibility to specific environmental exposures
- Expanded presentation on genetic risks for the development of obesity in the current food and physical activity environment
- Added presentation on lifestyle changes that reduce genetic tendencies toward the development of obesity

Unit 27: Nutrition and Physical Fitness for Everyone

- Updated section on population-based physical activity recommendations with content on the 2008 Physical Activity Guidelines for Americans
- Added a new focus on exercise intensity
- Added a Take Action feature on options for ramping-up exercise intensity levels
- Added new information on physical activity of Americans

Unit 28: Nutrition and Physical Performance

- Updated coverage of the effects of protein and amino acids on muscle recovery and strength
- Replaced content of carbohydrate loading with the new, simplified methods for increasing glycogen stores
- Expanded the table on grams of carbohydrate corresponding to 60 and 70% of total calories to include higher calorie need levels
- Revised and updated content on protein needs of athletes
- Added new sections on pre-event, event, and recovery foods and fluids, deleted previous coverage
- Added new Illustration on examples of pre-event, event, and post-event foods and fluids
- Added a Reality Check on the issue of whether athletes should avoid high fructose corn syrup
- Added a table on carbohydrate and sodium content of fluids.
- Added content on the World Anti-Doping Agency's rules for ergogenic aid use and its "Prohibited List" of aids
- Added energy drinks to the table on ergogenic acids: claims and evidence

Unit 29: Good Nutrition for Life: Pregnancy, Breast-Feeding, and Infancy

- Added information on ethnic/racial disparities on birth and health outcomes
- Updated content on DHA, EPA, and the course and outcome of pregnancy
- Updated pregnancy weight gain recommendations for obese women based on the 2009 Institute of Medicine report
- Revised table and illustration related to the prevalence of breastfeeding in the U.S.
- Updated infant feeding recommendations to reflect current American Academy of Pediatrics advice

Unit 30: Nutrition for the Growing Years: Childhood through Adolescence

- Added emphasis on school food and other environmental conditions influencing the development of childhood and adolescent obesity
- Replaced content on physical activity recommendations with the 2008 Physical Activity Guidelines for Children and Adolescents
- Updated information on recommended intake and roles of dietary fat in the diets of youths
- Updated information on the status of dietary intake of U.S. children and adolescents, changes needed, and ways to facilitate needed behavioral changes
- Highlighted the role of vitamin D in health promotion in youth

Unit 31: Nutrition and Health Maintenance for Adults of All Ages

- Updated information on physical activity and longevity

- Added content on functions of EPA and DHA and the development of Alzheimer's disease, cognitive and muscular strength decline, dementia, and depression

- Added an illustration that contains the new MyPyramid tool: "Modified MyPyramid for Older Adults"

Unit 32: The Multiple Dimensions of Food Safety

- Replaced headlines addressing foodborne illness outbreaks with contemporary examples

- Added a Take Action exercise on small steps that limit your exposure to potential sources of foodborne illness

- Added a section on the crisis in food safety in the U.S. and steps planned for its correction

Unit 33: Aspects of Global Nutrition

- Updated profile presented for the Population map: State of the village report

- Added content on the dual public health problems of increasing rates of overweight as well as underweight in many developing countries

- Updated information on protein-calorie malnutrition as it now exists and its causes

- Modified definitions of marasmus and kwashiorkor

Glossary

- Updated food composition tables

- Inserted Health Canada's new Food Guide illustrations and information

- Inserted updated Food Exchange lists

- Modified the Glossary to reflect additions and revisions made in the sixth edition of Nutrition Now

Resources for the Instructor

- **Interactive Learning Guide.** Activities for students included in the Instructor's Activity Manual are intended to get students involved in the topics being covered in class through hands-on and interactive experiences. The activities can be undertaken in classes ranging in size from tens to hundreds. Activities include taste testing to identify genetically determined sensitivity to bitterness, developing a dietary behavioral change plan, anthropometry lab, designing fraudulent nutrition products, a physical activity assessment, and an assessment of three days of dietary intake. The Instructor's Activity Manual, as well as the forms used to conduct and submit activities, may be accessed by Instructors using *Nutrition Now* at the Web address http://www.cengage.com/login

    Although a number of elements have changed in this new edition, many of the basic tenets of the text's approach have stayed the same. The text remains focused on meeting the needs of instructors offering introductory nutrition courses to students from a variety of majors. The 33 units in the text stand alone and can be covered in the order of the instructor's choosing. Instructors may choose to customize their selection of units to be included in the text. The text remains heavily illustrated, and updated and revised figures, tables, and photographs have been added to the new edition.

- **Instructor's Manual and Test Bank.** This features lecture outlines, suggested classroom activities, Web resources, discussion questions, transparency masters, and chapter-by-chapter test questions linked to the ExamView Computerized Testing program.

- **ExamView® Computerized Testing.** An easy-to-use assessment and tutorial system facilitates creation, delivery, and customizing of tests and study guides, both print and online.

- **Multimedia Manager CD-ROM.** Book-specific Microsoft®, PowerPoint®, lecture slides with teaching points, graphics from the book, electronic versions of the Instructor's Manual and Test Bank, ABC video clips, and links to nutrition resources on the Web are included on this disk.

- **CNN Today: Nutrition Videos.** Three volumes of engaging video clips are available for launching lectures.

- **Instructor's Activity Book.** Classroom activities can be used by varying sizes of student groups.

- **Transparency Acetates.** This includes 80 full-color transparencies of key illustrations in the text.

- **Diet Analysis Plus 6.1.** This software enables students to track and assess their food choices and create personal nutrition and activity profiles. It is available on CD-ROM or online and may be packaged with the text.

Acknowledgments

This edition introduces Peggy Williams as the new Executive Editor for Nutrition Now, and Nedah Rose as the Senior Developmental Editor. Both have invested themselves fully in making Nutrition Now and its pedagogical aids effective teaching and learning tools. The results of their efforts are reflected in the new design, photographs, and features presented in this edition of Nutrition Now.

My applause goes out to the sales representatives who are fans of *Nutrition Now* and who work hard to introduce faculty to its contents and features. You do a terrific job of communicating with instructors, as well as with me when you send instructors' thoughts my way.

It is said that instructors adopt a specific textbook but that students play a major role in instructors' decision to keep it. I am honored that you chose to adopt *Nutrition Now* and deeply pleased with the thought that students are helping you decide to keep it.

Reviewers' feedback is the lifeline of text writing, and the reviewers for the new edition conveyed very useful advice. The advice led me to some very interesting places on specific topics that changed my thinking and writing. May you remain students of *Nutrition Now* the textbook and keep your comments coming.

JUDITH E. BROWN

Key Nutrition Concepts and Terms

NUTRITION SCOREBOARD

	TRUE	FALSE
1 Calories are a component of food.		
2 Nutrients are substances in food that are used by the body for growth and health.		
3 Inadequate intakes of vitamins and minerals can harm health, but high intakes do not.		
4 The consumption of energy-dense diets is related to the development of overweight and diabetes.		

Key Concepts and Facts

- At the core of the science of nutrition are concepts that represent basic "truths" and serve as the foundation of our understanding about normal nutrition. (They are listed in Table 1.4 on p. 1-19.)

- Most nutrition concepts relate to nutrients.

Answers to **NUTRITION SCOREBOARD**	TRUE	FALSE
1 Calories are a measure of the amount of energy supplied by food. They're a property of food, not a substance present in food.		✔
2 That's the definition of nutrients.	✔	
3 Excessive as well as inadequate intake levels of vitamins and minerals can be harmful to health.		✔
4 Energy-dense diets are related to the development of overweight and diabetes.[1]	✔	

To be surprised, to wonder, is to begin to understand.

—José Ortega y Gasset

The Meaning of Nutrition

What is nutrition? It can be explained by situations captured in photographs as well as by words. This introduction presents a photographic tour of real-life situations that depict aspects of the study of nutrition.

Before the tour begins, take a moment to make yourself comfortable and clear your mind of clutter. Take a careful look at the photographs, pausing to mentally describe in two or three sentences what each photograph shows.

© Gary Conner/PhotoEdit

© David Frazier/Corbis

© Phil Schermeister/Corbis

© Royalty free/CORBIS

© Jupiterimages/Getty Images/FoodPix

Key Nutrition Concepts and Terms

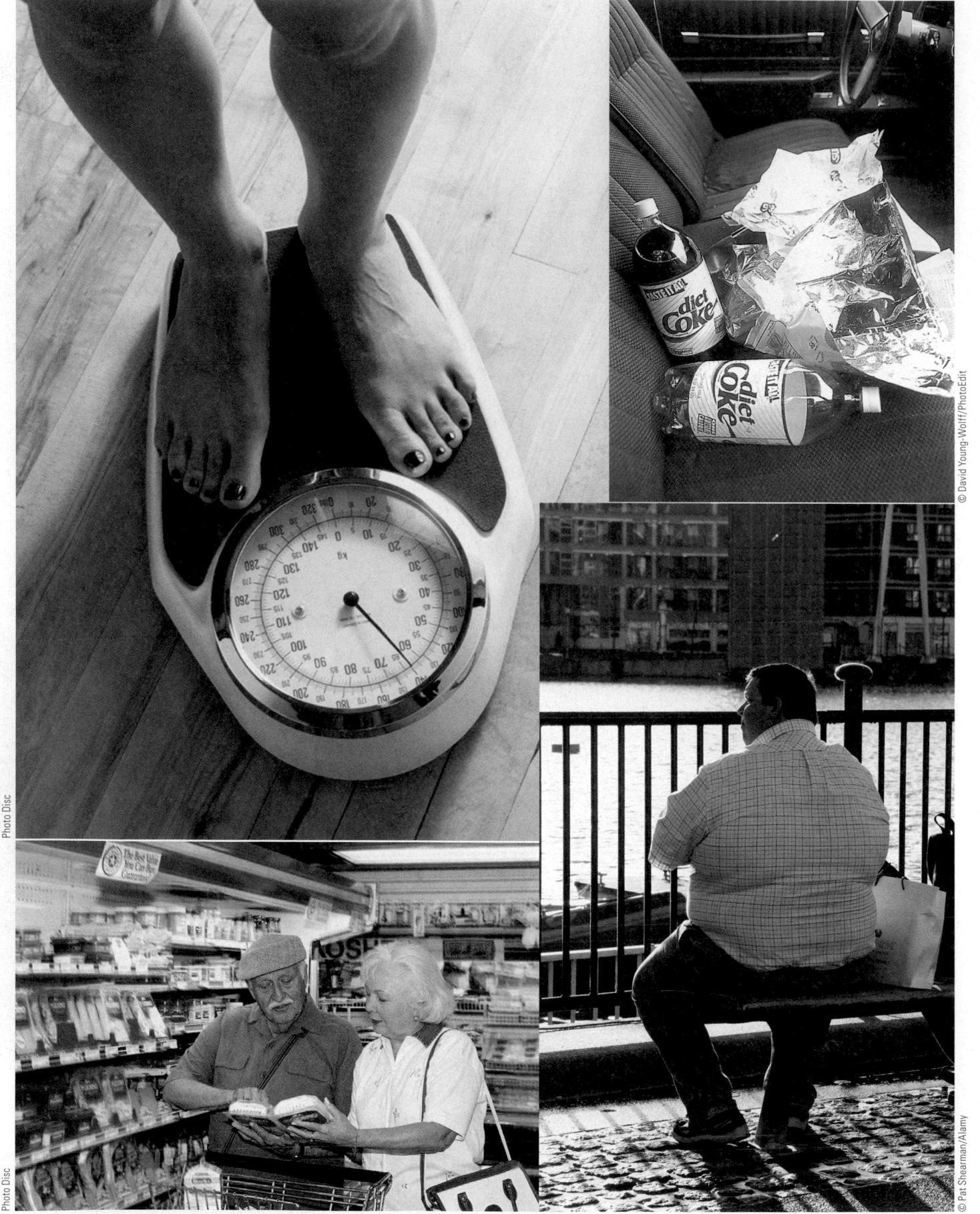

Photo Disc

© David Young-Wolff/PhotoEdit

Photo Disc

© Pat Shearman/Alamy

© image100/Corbis

Royalty-Free/CORBIS

AP Images

Not everyone who looks at the photographs will describe them in the same way. Reactions will vary somewhat due to personal experiences, interests, attitudes, and beliefs. An individual trying to gain weight will probably react differently to the photograph of the person on the scale than someone who is trying to lose weight. If you grew up in a family that farmed for a living, the picture of pesticides being sprayed on a crop may mean increased food production to you; another person may see the photograph and think about how pesticide residues on foods are harmful to health. Although knowledge about nutrition is generated by impersonal and objective methods, it can be a very personal subject.

Nutrition Defined

nutrition
The study of foods, their nutrients and other chemical constituents, and the effects that food constituents have on health.

In a nutshell, **nutrition** is the study of foods and health. It is a science that focuses on foods, their nutrient and other chemical constituents, and the effects of food constituents on body processes and health. The scope of nutrition extends from food choices to the effects that specific components of foods have on health.

Nutrition Is a "Melting Pot" Science   The broad scope of nutrition makes it an interdisciplinary science. Knowledge provided by the behavioral and social sciences, for example, is needed in studies that examine how food preferences develop and how they may be changed. Information generated by the biological, chemical, physical, and food sciences is required to propose and explain diet and disease relationships. Methods developed by quantitative scientists such as mathematicians and statisticians are needed to guide decision making about the significance of results produced by nutrition research. The study of nutrition will bring you into contact with information from a variety of disciplines (Illustration 1.1).

Nutrition Knowledge Is Applicable   As you study the science of nutrition, you will discover answers to a number of questions about your own diet, health, and eating behaviors. Is obesity primarily due to eating behaviors, physical inactivity, or your genes? How do you know whether new information you hear about nutrition is true? Can sugar harm more than your teeth? Can the right diet or

Illustration 1.1   Nutrition is an interdisciplinary science.

© Scott Goodwin Photography

supplement give you a competitive edge? What is a healthful diet and how do you know if you have one? If improvements seem warranted, what's the best way to go about changing your diet for the better? These are just a few of the questions that will be addressed during the course of your study of nutrition. You will take from this learning experience not only knowledge about nutrition and health, but skills that will keep the information and insights working to your advantage for a long time to come.

Foundation Knowledge for Thinking about Nutrition

You don't have to be a bona fide nutritionist to think like one. What you need is a grasp of the language and basic concepts of the science. The purpose of this unit is to give you this background. The essential topics covered here are explored in greater depth in units to come, and they build on this foundation of knowledge. With a working knowledge of nutrition terms and concepts, you will have an uncommonly good sense of nutrition.

NUTRITION CONCEPT #1

Food is a basic need of humans.

Humans need enough food to live, and they need the right assortment of foods for optimal health. In the best of all worlds, the need for food is combined with the condition of **food security**. People who experience food security have access at all times to a sufficient supply of the safe, nutritious foods that are needed for an active and healthy life. They are able to acquire acceptable foods in socially acceptable ways; they do not have to scavenge or steal food in order to survive or to feed their families. **Food insecurity** exists whenever the availability of safe, nutritious foods—or the ability to acquire them in socially acceptable ways—is limited or uncertain (Illustration 1.2).[2]

Adults who live in food-insecure households are more likely to have poor quality diets, to be overweight, and to have heart disease or diabetes than adults who are poor but food secure.[3] Poor nutritional quality of food consumed, the absence of local supermarkets, lack of exercise, and episodic food shortages may be partly responsible for the higher rates of overweight and chronic disease among food-insecure adults.[4] High-calorie, filling, and inexpensive fast and convenience foods are often the foods of choice when money is low. Although some children living in food-insecure households are nourished adequately and successful in school, and develop high levels of social skills, as a group they are at higher risk of poor school performance and social and behavioral problems.[5]

Common sense requires a common knowledge.

THOMAS PAINE, 1737–1809

If we could give individuals the right amount of nourishment and exercise, we would have found the safest way to health.

—HIPPOCRATES

food security
Access at all times to a sufficient supply of safe, nutritious foods.

food insecurity
Limited or uncertain availability of safe, nutritious foods—or the ability to acquire them in socially acceptable ways.

Illustration 1.2 "It is possible to go an entire lifetime without knowing about people's experiences with hunger."

—Meghan LeCates, Capitol Area Food Bank

Ricin, a deadly nerve toxin, comes from the seeds of the beautiful castor bean plant. The oil from the seeds (sometimes called beans) is used commercially in castor oil, motor oil, nylon, and paint. Castor plants are becoming increasingly abundant in southwestern parts of the United States, although their cultivation is discouraged. Ricin poisoning occurs when seeds with broken coatings or seed contents are ingested. 70 micrograms—the amount of a grain of salt—is enough to kill an adult. The U.S. Senate offices were closed for several days in 2004 when powdered ricin was discovered in a letter sent to Senator Bill Frist. No one was injured. Periodic reports of ricin exposure continue.[8]

calorie
A unit of measure of the amount of energy supplied by food. (Also known as a kilocalorie, or the "large Calorie" with a capital *C*.)

nutrients
Chemical substances in food that are used by the body for growth and health. The six categories of nutrients are carbohydrates, proteins, fats, vitamins, minerals, and water.

Illustration 1.3 Foods provide nutrients. "Please pass the complex carbohydrates, thiamin, and niacin … I mean, the bread!"

Food insecurity exists in 12.6% of U.S. and 7.0% of Canadian households.[6] It is most likely to occur in poor, female-headed households with young children living in inner-city areas. The rate in the United States is over twice as high as the target of 6% established as a national health goal.[7]

Food Terrorism The term *food security* now means more than it used to. Food is a potential weapon of bioterrorism. Although public health concerns related to bioterrorism center on disease threats such as smallpox and anthrax, food and water could also be used to intentionally spread illness. Simply recalling the latest announcement about a foodborne illness outbreak makes it easy to imagine the panic and health consequences that could result from intentional contamination of food or water supplies.

Toxic substances that could be introduced into food supplies include botulism toxin, ricin, radioactive particles, and microorganisms such as *Salmonella, E. coli 0517:H7,* and *Shigella.* Botulism toxin is the single most poisonous substance known. Most bacteria are killed when heated above 160°F or when municipal water supplies are treated with chlorine[9,10]; however, it takes about 15 minutes of boiling to destroy botulism toxin. Ricin (pronounced rye-sin) is a widely available toxin found in the seeds of the castor plant, which is also widely available in tropical and warm areas of the world.[11] Castor plant seeds are sufficiently interesting to warrant being featured in the "On the Side."

NUTRITION CONCEPT #2

Foods provide energy (calories), nutrients, and other substances needed for growth and health.

People eat foods for many different reasons. The most compelling reason is that we need the **calories, nutrients,** and other substances supplied by foods for growth and health.

Calories

A calorie is a unit of measure of the amount of energy in a food—and of how much energy will be transferred to the person who eats it. Although we often refer to the number of calories in this food, or that one, calories are not a substance present in food. And, because calories are a unit of measure, they do not qualify as a nutrient.

Nutrients

Nutrients are chemical substances present in food that are used by the body to sustain growth and health (Illustration 1.3). Essentially everything that's in our body was once a component of the food we consumed.

There are six categories of nutrients (Illustration 1.4). Each category (except water) consists of a number of different substances that are used by the body for growth and health. The carbohydrate category includes simple sugars and complex carbohydrates (starches and dietary fiber). The protein category includes 20 amino acids, the chemical units that serve as the "building blocks" for protein. Several different types of fat are included in the fat category. Of primary concern are the saturated fats, unsaturated fats, essential fatty acids, *trans fats*, and cholesterol (read more about them in Illustration 1.4). The vitamin category consists of 14 vitamins, and the mineral category includes 15 minerals. Water makes up a nutrient category by itself.

Carbohydrates, proteins, and fats supply calories and are called the "energy nutrients." Although each of these three types of nutrients performs a variety of functions, they share the property of being the body's only sources of fuel. Vitamins, minerals, and water are chemicals that the body needs for converting carbohydrates, proteins, and fats into energy and for building and maintaining muscles, blood components, bones, and other parts of the body.

Illustration 1.4 Nutrients are grouped into six categories. Here the major types of nutrients in each category are listed, along with a description of each.

SIX CATEGORIES OF NUTRIENTS

1. CARBOHYDRATES are substances in food that consist of a single sugar molecule, or of multiple sugar molecules in various forms. They provide the body with energy.

Simple sugars are the most basic type of carbohydrates. Examples include glucose (blood sugar), sucrose (table sugar), and lactose (milk sugar). **Starches** are complex carbohydrates consisting primarily of long, interlocking chains of glucose units. **Dietary fiber** consists of complex carbohydrates found principally in plant cell walls. Dietary fiber cannot be broken down by human digestive enzymes.

2. PROTEINS are substances in food that are composed of amino acids. Amino acids are specific chemical substances from which proteins are made. Of the 20 amino acids, 9 are "essential," or a required part of our diet.

3. FATS are substances in food that are soluble in fat, not water.

Saturated fats are found primarily in animal products, such as meat, butter, and cheese, and in palm and coconut oils. Diets high in saturated fat may elevate blood cholesterol levels. **Unsaturated fats** are found primarily in plant products, such as vegetable oil, nuts, and seeds, and in fish. Unsaturated fats tend to lower blood cholesterol levels. **Essential fatty acids** are two specific types of unsaturated fats that are required in the diet. **Trans fats** are a type of unsaturated fat present in hydrogenated oil, margarine, shortening, pastries, and some cooking oils that increase the risk of heart disease. **Cholesterol** is a fat-soluble, colorless liquid primarily found in animals. It can be manufactured by the liver.

4. VITAMINS are chemical substances found in food that perform specific functions in the body. Humans require 13 different vitamins in their diet.

5. MINERALS are chemical substances that make up the "ash" that remains when food is completely burned. Humans require 15 different minerals in their diet.

6. WATER is essential for life. Most adults need about 11–15 cups of water each day from food and fluids.

Other Substances in Food

Food also contains many other substances, some of which are biologically active in the body. One major type of these substances are the **phytochemicals** and there are thousands of them in plants. Illustration 1.5 presents examples of plant foods that are particularly rich sources of phytochemicals. Phytochemicals provide plants with color, give them flavor, foster their growth, and protect them from insects and diseases. A specific example of how one phytochemical works is described in the "On the Side" box on the next page. In humans, consumption of phytochemicals in diets is strongly related to a reduced risk of developing certain types of cancer, heart disease, infections, and other disorders.[13]

Specific phytochemicals have names that are often hard to pronounce and difficult to remember. Nevertheless, here are a few examples. Plant pigments, such as lycopene (like-o-peen), which help make tomatoes red, anthocyanins (an-tho-sigh-an-ins), which give blueberries their characteristic blue color, and beta-carotene (bay-tah-kar-o-teen), which imparts a dark yellow color to carrots, are phytochemicals that act as **antioxidants.**

phytochemicals (phyto = plant)
Chemical substances in plants. Some phytochemicals perform important functions in the human body. They give plants color and flavor, participate in processes that enable plants to grow, and protect plants against insects and diseases. Also called phytonutrients.

antioxidants
Chemical substances that prevent or repair damage to cells caused by exposure to oxidizing agents such as environmental pollutants, smoke, ozone, and oxygen. Oxidation reactions are a normal part of cellular processes.

Illustration 1.5 Examples of good food sources of phytochemicals.

They protect plant cells—and in some cases, human cells, too—from damage that can make them susceptible to disease. Various types of sulfur-containing phytochemicals are present in cabbage, broccoli, cauliflower, brussels sprouts, and other vegetables of the same family. These substances may help prevent a number of different types of cancer.[14]

essential nutrients

Substances required for normal growth and health that the body can generally not produce, or not produce in sufficient amounts. Essential nutrients must be obtained in the diet.

nonessential nutrients

Nutrients required for normal growth and health that the body can manufacture in sufficient quantities from other components of the diet. We do not require a dietary source of nonessential nutrients.

Some Nutrients Must Be Provided by the Diet   Many nutrients are required for growth and health. The body can manufacture some of these from raw materials supplied by food, but others must come assembled. Nutrients that the body generally produce in sufficient quantity are referred to as **essential nutrients**. Here "essential" means "required in the diet." Vitamin A, iron, and calcium are examples of essential nutrients. Table 1.1 lists all the known essential nutrients.

Nutrients used for growth and health that can be manufactured by the body from components of food in our diet are considered nonessential. Cholesterol, creatine, and glucose are examples of nonessential nutrients. **Nonessential nutrients** are present in food and used by the body, but they are not required parts of our diet because we can produce them ourselves.

Both essential and nonessential nutrients are required for growth and health. The difference between them is whether or not we need to obtain the nutrient from a dietary source. A dietary deficiency of an essential nutrient will cause a specific deficiency disease, but a dietary lack of a nonessential nutrient will not. People develop scurvy (the vitamin C–deficiency disease), for example, if they do not consume enough vitamin C. But you could have zero cholesterol in your diet and not become "cholesterol deficient," because your liver produces cholesterol.

On the Side

The distinctive odor garlic produces when chopped or crushed is actually this plant's defense against insect predators. Garlic cloves contain an odorless, sulfur-containing phytochemical called "alliin." When the clove is disrupted, alliin is released and reacts with an enzyme located in neighboring cells that converts the alliin to the odoriferous "allicin." Allicin is garlic's bug repellant—and the "people repellant" that makes many individuals shy about eating it.[9]

Our Requirements for Essential Nutrients   The amount of essential nutrients humans need each day varies a great deal, from amounts measured in cups to micrograms. (See Table 1.2 to get a notion of the amount represented by a gram, milligram, and other measures.) Generally speaking, adults need 11 to 15 cups of water from fluids and foods, 9 tablespoons of protein, one-fourth teaspoon of calcium, and only one-thousandth teaspoon (a 30-microgram speck) of vitamin B_{12} each day.

We all need the same nutrients, but not always in the same amounts. The amounts needed vary among people based on:

Table 1.1

Essential nutrients for humans: A reference table

Energy Nutrients	Vitamins	Minerals	Water
Carbohydrates	Biotin	Calcium	Water
Fats[a]	Folate	Chloride	
Proteins[b]	Niacin (B$_3$)	Chromium	
	Pantothenic acid	Copper	
	Riboflavin (B$_2$)	Fluoride	
	Thiamin (B$_1$)	Iodine	
	Vitamin A	Iron	
	Vitamin B$_6$ (pyroxidine)	Magnesium	
	Vitamin B$_{12}$	Manganese	
	Vitamin C (ascorbic acid)	Molybdenum	
	Vitamin D	Phosphorus	
	Vitamin E	Potassium	
	Vitamin K	Selenium	
	Choline[c]	Sodium	
		Zinc	

[a]Fats supply the essential nutrients linoleic and alpha-linolenic acid.
[b]Proteins are the source of 9 "essential amino acids": histidine, isoleucine, leucine, lysine, methionine, phenylalanine, threonine, tryptophan, and valine. The other 11 amino acids are not a required part of our diet; they are considered "nonessential."
[c]A dietary source of choline may not be required during all stages of the life cycle.[15]

Table 1.2

Units of measure commonly employed in nutrition

Measure	Abbreviation	Equivalents
Kilogram	kg	1 kg = 2.2 lb = 1000 grams
Pound	lb	1 lb = 16 oz = 454 grams = 2 cups (liquid)
Ounce	oz	1 oz = 28 grams = 2 tablespoons (liquid)
Gram	g	1 g = 1/28 oz = 1000 milligrams
Milligram	mg	1 mg = 1/28,000 oz = 1000 micrograms
Microgram	mcg, µg	1 mcg = 1/28,000,000 oz

1 egg = 50 grams or 1³/₄ oz; 212 milligrams (0.2 grams) of cholesterol in yolk

1 slice of bread = 1 oz = 28 grams

1 nickel = 5 grams

1 teaspoon of sugar = 4 grams, 1 grain of sugar = 200 micrograms

- Age

- Sex

- Growth status

- Body size

- Genetic traits

- Disease states

and the presence of conditions such as:

- Pregnancy

- Breastfeeding

- Illnesses

- Drug/medication use

- Exposure to environmental contaminants

Each of these factors, and others, can influence nutrient requirements. General diet recommendations usually make allowances for major factors that influence the level of nutrient need, but they cannot allow for all of the factors.

Nutrient Intake Standards   Recommendations for levels of essential nutrient intake were first developed in the United States in 1943 and have been updated periodically since then. Called the Recommended Dietary Allowances (RDAs), these standards were established in response to the high rejection rate of World War II recruits, many of whom were underweight and had nutrient deficiencies. The recommended levels of nutrient intake provided are based on age, gender, and condition (pregnant or breastfeeding have). The most recent update, completed in 2004, was a major one. Developed by nutrition scientists from the United States and Canada, the updated nutrient intake standards apply to people in both countries. The RDAs have a whole new look and a new name—Dietary Reference Intakes (Illustration 1.6).

◼ **Dietary Reference Intakes (DRIs)**   Previous editions of the RDAs were based on levels of essential nutrients that protected people from deficiency diseases such as scurvy and rickets. As the science of nutrition advances, it is becoming abundantly clear that other components of food besides essential nutrients affect health. It is also becoming apparent that levels of nutrient intake associated with deficiency disease prevention may be too low to help prevent cancer, heart disease, osteoporosis and other diseases and disorders. Excessively high intakes of nutrients from fortified foods and supplements are another problem that was not considered when previous editions of the RDAs were prepared.

The DRIs provide recommended intake levels of essential nutrients and safe upper levels of intake. The recommended daily levels of intake not only meet the nutrient needs of almost all healthy people (97 to 98%) but also promote health and help reduce the risk of chronic disease.[13] Table 1.3 provides examples of endpoints aimed at health promotion and disease prevention used by the DRI committees to estimate recommended levels of nutrient intake. Tables showing the DRIs begin on the inside front cover of this book.

◼ **Adequate Intakes and Estimated Average Requirements**   Recommended intakes of some nutrients, such as calcium, vitamin D, and fluoride, for which too little conclusive evidence on disease prevention exists, are represented by an "Adequate Intake," or AI, level (Table 1.4). Regardless of which label is attached, the RDAs and AIs represent the best estimates of nutrient intake levels that promote optimal health. These values are represented in the

Illustration 1.6   One of the six reports on the DRIs.

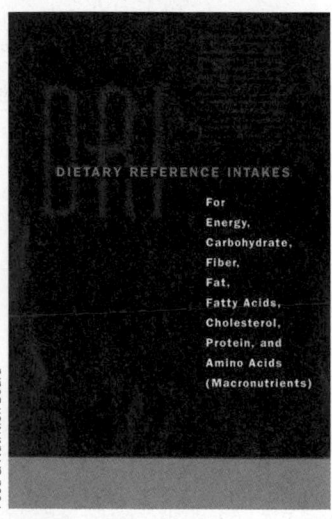

Food & Nutrition Board

Table 1.3

Examples of primary endpoints used to estimate DRIs[15]

Carbohydrate: Amount needed to supply optimal levels of energy to the brain.

Total Fiber: Amount shown to provide the greatest protection against heart disease.

Folate: Amount that maintains normal red blood cell folate concentration.

Iodine: Amount that corresponds to optimal functioning of the thyroid gland.

Selenium: Amount that maximizes its function in protecting cells from damage.

Table 1.4

Terms and abbreviations used in the DRIs and a graphic representation of their meaning[15]

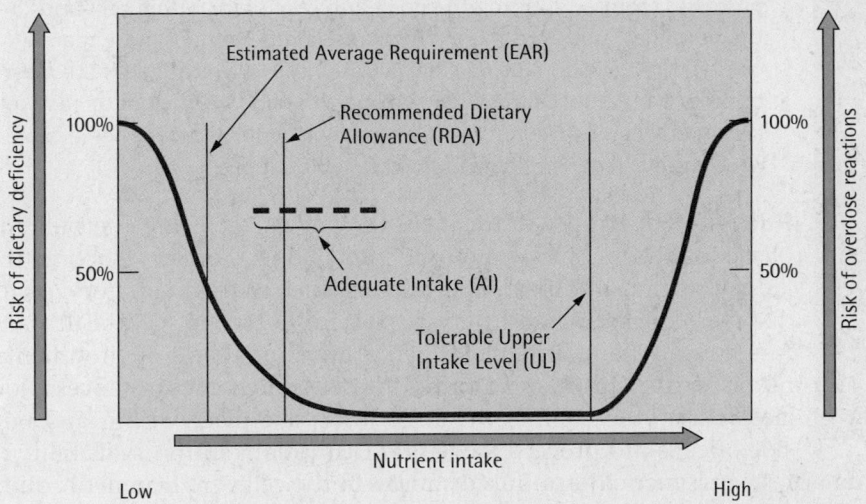

- **Dietary Reference Intakes (DRIs).** This is the general term used for the new nutrient intake standards for healthy people.
- **Recommended Dietary Allowances (RDAs).** These are levels of essential nutrient intake judged to be adequate to meet the known nutrient needs of practically all healthy persons while decreasing the risk of certain chronic diseases.
- **Adequate Intakes (AIs).** These are "tentative" RDAs. AIs are based on less conclusive scientific information than are the RDAs.
- **Estimated Average Requirements (EARs).** These are nutrient intake values that are estimated to meet the requirements of half the healthy individuals in a group. The EARs are used to assess adequacy of intakes of population groups.
- **Tolerable Upper Levels of Intake (ULs).** These are upper limits of nutrient intake compatible with health. The ULs do not reflect desired levels of intake. Rather, they represent total, daily levels of nutrient intake from food, fortified foods, and supplements that should not be exceeded.

main DRI tables presented on the inside front cover of this book. The term Estimated Average Requirement (EAR) represents the intake level of a nutrient that is estimated to meet the requirement of 50% of the individuals within a group.

■ **Tolerable Upper Intake Levels** Estimates of safe upper limits of nutrient intake are called Tolerable Upper Intake Levels, abbreviated ULs. The ULs do not represent recommended or desired levels of intake. Instead, they represent total daily levels of nutrient intake from food, fortified food products, and supplements that should not be exceeded due to the potential for toxicity reactions.

■ **Revising the DRIs** Nutrition and health is an intensively studied area. New results from scientific studies emerge all the time. Some of the results change our understanding of what constitutes optimal levels of nutrient intake. Such advances in nutrition science are reflected in new DRIs for nutrients. Although all nutrients in the DRI table are examined about every ten years, special, scientific committees can be formed periodically to review and update intake standards for specific nutrients. The latest nutrients to receive this attention are vitamin D and calcium.

Illustration 1.7 Schematic representation of the structure of a human cell.

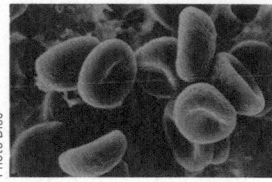

What's the most common cell in the human body? The red blood cell. The human body contains more than 30 billion of them.

On the Side

NUTRITION CONCEPT #3

Health problems related to nutrition originate within cells.

Cells are the main employers of nutrients. (Illustration 1.7). All body processes required for growth and health take place within cells and the fluid that surrounds them. The human body contains more than one hundred trillion (100,000,000,000,000) cells. (Which type is the most common? See the "On the Side" box on this page.) Functions of each cell are maintained by the nutrients it receives. Problems arise when a cell's need for nutrients differs from the available supply.

Nutrient Functions at the Cellular Level Cells are the building blocks of tissues (such as muscles and bones), organs (such as the kidneys, heart, and liver), and systems (such as the respiratory, reproductive, circulatory, and nervous systems). Normal cell health and functioning are maintained when a nutritional and environmental utopia exists within and around the cells. Such circumstances allow **metabolism**—the chemical changes that take place within and outside of cells—to proceed flawlessly. Disruptions in the availability of nutrients—or the presence of harmful substances in the cell's environment—initiate diseases and disorders that eventually affect tissues, organs, and systems. Here are two examples of how cell functions can be disrupted by the presence of low or high concentrations of nutrients:

- Folate, a B vitamin, is required for protein synthesis within cells. When too little folate is available, cells produce proteins with abnormal shapes and functions. Abnormalities in the shape of red blood cell proteins, for example, lead to functional changes that produce loss of appetite, weakness, and irritability.[17]

- When too much iron is present in cells, the excess reacts with and damages cell components. If cellular levels of iron remain high, the damage spreads, impairing the functions of organs such as the liver, pancreas, and heart.[18] Health problems in general begin with disruptions in the normal activity of cells. Humans are only as healthy as their cells.

Maximum health and lifespan require metabolic harmony.

—B. Ames[16]

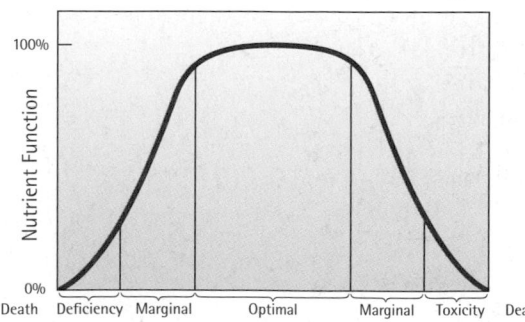

Illustration 1.8 For every nutrient, there is a range of optimal intake that corresponds to the optimal functioning of that nutrient in the body.

NUTRITION CONCEPT #4

Poor nutrition can result from both inadequate and excessive levels of nutrient intake.

For each nutrient, every individual has a range of optimal intake that produces the best level for cell and body functions. On either side of the optimal range are levels of intake associated with impaired body functions.[12] This concept is presented in Illustration 1.8. Inadequate essential nutrient intake, if prolonged, results in obvious deficiency diseases. Marginally deficient nutrient intakes generally produce subtle changes in behavior or physical condition. If the optimal intake range is exceeded, mild to severe changes in mental and physical functions occur, depending on the amount of the excess and the nutrient. Overt vitamin C deficiency, for example, produces bleeding gums, pain on being touched, and a failure of bone to grow. A marginal deficiency may cause delayed wound healing. On the excessive side, high intakes of vitamin C cause diarrhea.[15] Nearly all cases of vitamin and mineral overdose result from the excessive use of supplements or errors made in the level of nutrient fortification of food products. They are almost never caused by foods. For nutrients, "enough is as good as a feast."

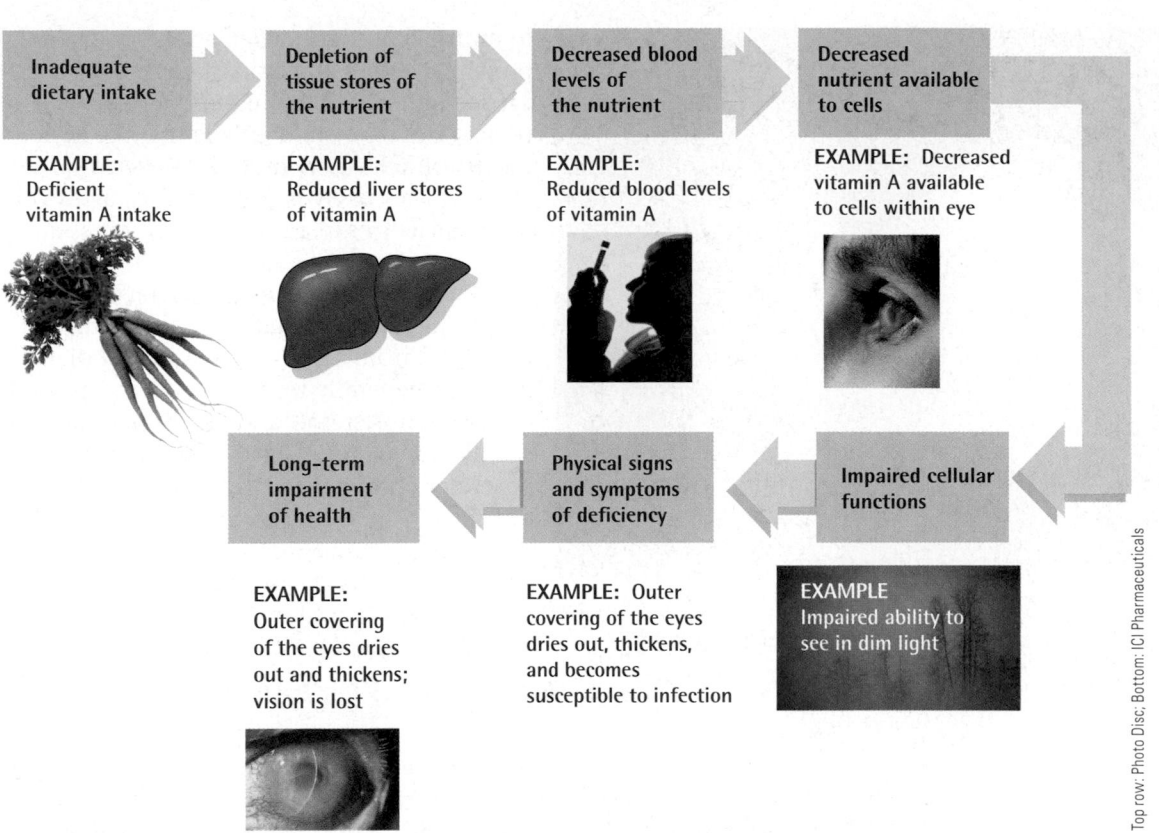

Top row: Photo Disc; Bottom: ICI Pharmaceuticals

Illustration 1.9
The usual sequence of events in the development of a nutrient deficiency and an example of how vitamin A deficiency develops.[13,19]

Steps in the Development of Nutrient Deficiencies and Toxicities Poor nutrition due to inadequate diet generally develops in the stages outlined in Illustration 1.9. To help explain the stages, this illustration includes an example of how vitamin A deficiency develops.

After a period of deficient intake of an essential nutrient, the body's tissue reserves of the nutrient become depleted. Blood levels of the nutrient then decrease because there are no reserves left to replenish the blood supply. Without an adequate supply of the nutrient in the blood, cells get shortchanged. They no longer have the supply of nutrients needed to maintain normal function. If the dietary deficiency is prolonged, the malfunctioning cells cause sufficient impairment to produce physically obvious signs of a deficiency disease. Eventually, some of the problems produced by the deficiency may no longer be repairable, and permanent changes in health and function may occur. In most cases, the problems resulting from the deficiency can be reversed if the nutrient is supplied before this final stage occurs.

Excessively high intakes of many nutrients such as vitamin A and selenium produce toxicity diseases. The vitamin A toxicity disease is called "hypervitaminosis A," and the disease for selenium toxicity is called "selenosis." Signs of the toxicity disease stem from increased levels of the nutrient in the blood and the subsequent oversupply of the nutrient to the cells. The high nutrient load upsets the balance needed for normal cell function. The changes in cell functions lead to the signs and symptoms of the toxicity disease.

For both deficiency and toxicity diseases, the best time to correct the problem is usually at the level of dietary intake, before tissue stores are adversely affected. In that case, no harmful effects on health and cell function occur—they are prevented.[20]

■ **Nutrient Deficiencies Are Often Multiple** Most foods contain many nutrients, so poor diets will affect the intake level of more than one nutrient (Illustration 1.10). Inadequate diets generally produce a spectrum of signs and symptoms related

metabolism
The chemical changes that take place in the body. The formation of energy from carbohydrates is an example of a metabolic process.

to multiple nutrient deficiencies. For example, protein, vitamin B_{12}, iron, and zinc are packaged together in many high-protein foods. The protein-deficient children you may see in news reports on television are rarely deficient just in protein. They likely have iron, zinc, and B vitamin deficiencies in addition to protein.

■ **The "Ripple Effect"** Dietary changes affect the level of intake of many nutrients. Switching from a high-fat to a low-fat diet, for instance, generally results in a lower intake of calories, cholesterol, and vitamin E as well. So, dietary changes introduced for the purpose of improving the intake level of a particular nutrient produce a ripple effect on the intake of other nutrients.

Illustration 1.10 This woman has iron deficiency anemia. She also has suboptimal blood levels of vitamin A, zinc, and iron. These nutrients are primarily found in animal products.

NUTRITION CONCEPT #5

Humans have adaptive mechanisms for managing fluctuations in nutrient intake.

Healthy humans are equipped with a number of adaptive mechanisms that partially protect the body from poor health due to fluctuations in dietary intake. In the context of nutrition, adaptive mechanisms act to conserve nutrients when dietary supply is low and to eliminate them when they are present in excessively high amounts. Dietary surpluses of energy and some nutrients—such as iron, calcium, vitamin A, and vitamin B_{12}—are stored within tissues for later use. In the case of iron, copper, and calcium, the body regulates the amounts absorbed in response to its need for them. The body has a low storage capacity for other nutrients, for instance vitamin C and water, and eliminates any excesses through urine or stools.

Here are some examples of how the body adapts to changes in dietary intake:

- When caloric intake is reduced by fasting, starvation, or dieting, the body adapts to the decreased supply by lowering energy expenditure. Declines in body temperature and the capacity to do physical work also act to decrease the body's need for calories. When caloric intake exceeds the body's need for energy, the excess is stored as fat for energy needs in the future.

- The ability of the gastrointestinal tract to absorb dietary iron increases when the body's stores of iron are low. To protect the body from iron overdose, the mechanisms that facilitate iron absorption in times of need shut down when enough iron has been stored.

- The body protects itself from excessively high levels of vitamin C from supplements by excreting the excess in the urine.

Although these built-in mechanisms do not protect humans from all the consequences of poor diets, they do provide an important buffer against the development of nutrient-related health problems.

NUTRITION CONCEPT #6

Malnutrition can result from poor diets and from disease states, genetic factors, or combinations of these factors.

Malnutrition means "poor" nutrition and results from both inadequate and excessive availability of calories and nutrients in the body. Vitamin A toxicity, obesity, vitamin C deficiency (scurvy), and underweight are examples of malnutrition.

malnutrition
Poor nutrition resulting from an excess or lack of calories or nutrients.

Malnutrition can result from poor diets and also from diseases that interfere with the body's ability to use the nutrients consumed. Diarrhea, alcoholism, cancer, bleeding ulcers, and HIV/AIDS, for example, may be primarily responsible for the development of malnutrition in people with these disorders.

In addition, a percentage of the population is susceptible to malnutrition and increased disease risk due to genetic factors. For example, people may be born with a genetic tendency to produce excessive amounts of cholesterol, absorb high levels of iron, or use folate poorly. Some cases of obesity, diabetes, heart disease, and cancer are related to a combination of genetic predisposition and dietary factors.[22]

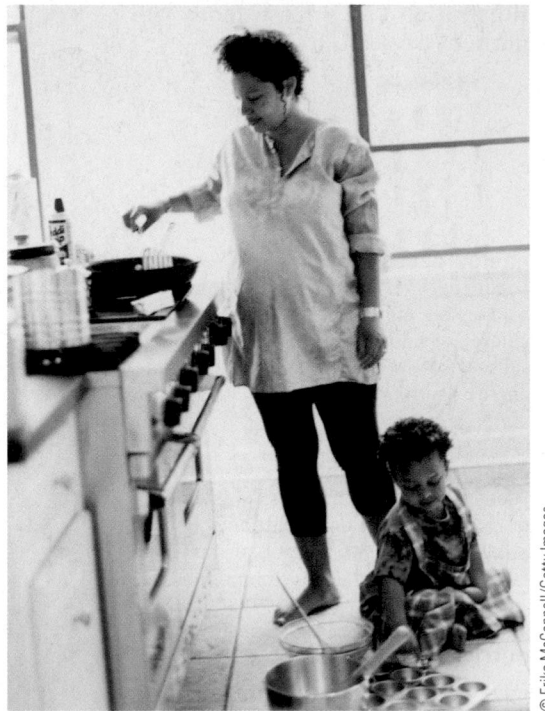

NUTRITION CONCEPT #7

Some groups of people are at higher risk of becoming inadequately nourished than others.

Women who are pregnant or breastfeeding, infants, growing children, the frail elderly, the ill, and those recovering from illness have a greater need for nutrients than other people. As a result, they are at higher risk of becoming inadequately nourished than other people (Illustration 1.11). In cases of widespread food shortages, such as those induced by natural disasters or war, the health of these nutritionally vulnerable groups is compromised the soonest and the most.

Within the nutritionally vulnerable groups, certain people and families are at particularly high risk of malnutrition. These are people and families who are poor and least able to secure food, shelter, and high-quality medical services. The risk of malnutrition is not shared equally among all persons within a population.

Illustration 1.11 Women who are pregnant or breastfeeding and infants are among the people who are at a higher risk of becoming inadequately nourished.

NUTRITION CONCEPT #8

Poor nutrition can influence the development of certain chronic diseases.

Poor nutrition does not result only in nutrient deficiency or toxicity diseases. Faulty diets play important roles in the development of heart disease, hypertension, cancer, osteoporosis, and other **chronic diseases**. Diets high in animal fat, for example, are related to the development of heart disease; those low in vegetables and fruit to cancer; low-calcium diets and poor vitamin D status to osteoporosis; and high-sugar diets to tooth decay. It may take years to understand the harmful effect negative dietary practices have on the development of certain diseases and disorders.

chronic diseases
Slow-developing, long-lasting diseases that are not contagious (for example, heart disease, diabetes, and cancer). They can be treated but not always cured.

NUTRITION CONCEPT #9

Adequacy, variety, and balance are key characteristics of a healthful diet.

Diets that promote growth, development, and the maintenance of health provide:

- An adequate amount of essential nutrients from foods while delivering a level of calorie intake that corresponds to a healthy weight. Adequate amounts of essential nutrients approximate the RDA (Recommended Dietary Allowance) or the AI (Adequate Intake) levels cited in the DRI (Dietary Reference Intake) tables.

- A variety of foods from each of the basic food groups. A variety of foods is needed to obtain the wide assortment of nutrients and beneficial phytochemicals we need for the optimal functioning of our bodies. No one food (except breast milk for young infants) contains them all, and most single foods don't come close.

  Many combinations of foods can supply the variety of nutrients and phytochemicals needed for a healthy diet, so no specific foods are required. Healthy diets could include apples, snails, sea cucumbers, ants, and burdock root. Diets that include a variety of foods in the amounts recommended by the MyPyramid Food

Illustration 1.12 "Basic foods" are nutrient dense.

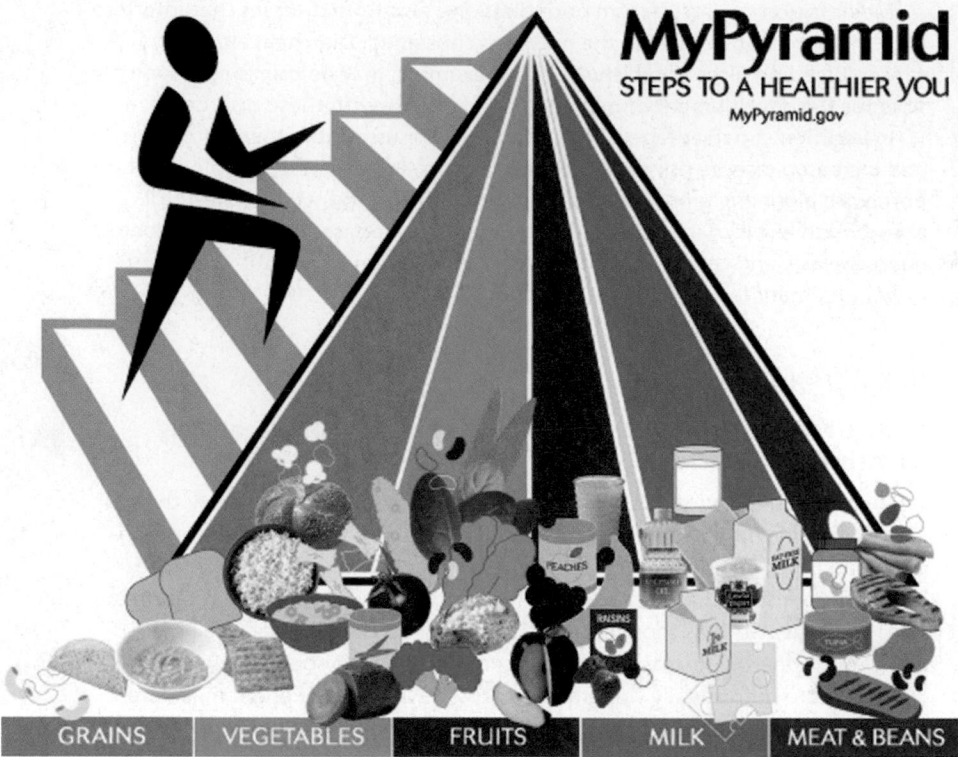

energy-dense foods
Foods that provide relatively high levels of calories per unit weight of the food. Fried chicken; cheeseburgers; a biscuit, egg, and sausage sandwich; and potato chips are energy-dense foods.

empty-calorie foods
Foods that provide an excess of energy or calories in relation to nutrients. Soft drinks, candy, sugar, alcohol, and animal fats are considered empty-calorie foods.

nutrient-dense foods
Foods that contain relatively high amounts of nutrients compared to their calorie value. Broccoli, collards, bread, cantaloupe, and lean meats are examples of nutrient-dense foods.

Guide (Illustration 1.12) are most likely to supply the body with adequate amounts of nutrients and other beneficial substances.[22]

- A balanced selection of food types and amounts from the MyPyramid basic food groups. Regular consumption of fried foods, high fat meats, and sweets, and infrequent intake of whole grains and colorful vegetables, for example, can knock a diet out of balance.

To bring diets into balance, Americans are being urged to consume more vegetables, fruits, whole grains, dried beans, and low-fat meats and dairy products; and to consume less sugar, animal fat, and saturated fat than is the case now.[22]

Energy- and Nutrient Density   Most Americans consume more calories than needed, become overweight as a result, *and* consume inadequate diets.[23] This situation is partly due to over-consumption of **energy-dense foods** such as processed and high fat meats, chips, candy, desserts, and full-fat dairy products. Energy-dense foods have relatively high calorie values per unit weight of the food. Intake of energy-dense diets is related to the consumption of excess calories and to the development of overweight and diabetes.[1,24]

Many energy-dense foods are nutrient-poor, or contain low levels of nutrients given their caloric value. These foods are sometimes referred to as **empty-calorie foods** and include products like soft drinks, sherbet, hard candy, alcohol, and cheese twists. Excess intake of energy-dense and empty-calorie foods increases the likelihood that calorie needs will be met or exceeded before nutrients needs are met.[23] Diets most likely to meet nutrient requirements without exceeding calorie need contain primarily **nutrient-dense foods**, or foods with high levels of nutrients and relatively low calorie value. Nutrient-dense foods such as non-fat milk and dairy products, lean meat, dried beans, vegetables, and fruits provide relatively high amounts of nutrients compared to their calorie value.[25] Illustration 1.13 shows a comparison of the calorie and nutrient content of an empty-calorie and a nutrient-dense food.

NUTRITION CONCEPT #10

There are no "good" or "bad" foods.

People tend to classify foods as being "good" or "bad," but such opinions over-simplify each food's potential contribution to a diet.[26] Typically hot dogs, ice cream, candy, bacon, and french fries are judged to be bad, whereas vegetables, fruits, and whole grain products are given the "good" stamp. Unless we're talking about spoiled stew, poison mushrooms or something similar, however, no food can be accurately labeled as "good" or "bad." Ice cream can be a "good" food for physically active, normal weight individuals with high calorie need who have otherwise met their nutrient requirements by consuming nutrient-dense foods. Some people who only eat what they consider to be "good" foods like vegetables, fruits, whole grains, and tofu may still miss the healthful diet mark due to inadequate consumption of essential fatty acids and certain vitamins and minerals.

All foods can fit into a healthful diet as long as nutrient needs are met at calorie intake levels that maintain a healthy body weight.[27] If nutrient needs are not being met and calorie intake levels are too high, then the diet likely includes too many energy-dense or empty-calorie foods. A greater intake of nutrient-dense foods would be the healthy solution.[28]

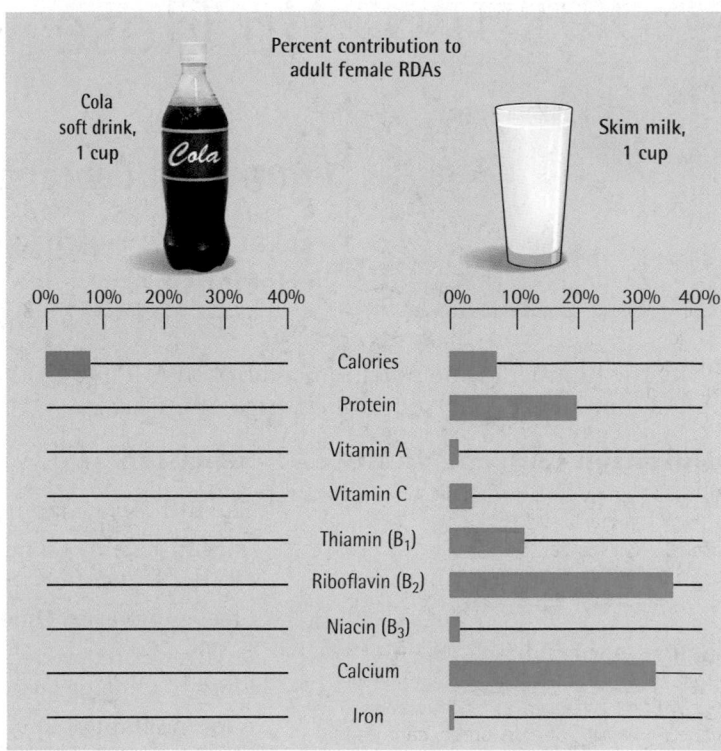

Illustration 1.13 Calorie and nutrient content of an empty-calorie. Calorie and a nutrient-dense food. Percentages given represent percent contributions to adult female RDAs.

All things in nutrient are good or bad relatively.

—HIPPOCRATES

The basic nutrition concepts presented here are listed in Table 1.4. It may help you to remember the concepts and to start "thinking like a bona fide nutritionist," if you go back over each concept and give several examples related to it. If you understand these concepts, you will have gained a good deal of insight into nutrition.

Table 1.4

Nutrition concepts

1. Food is a basic need of humans.

2. Foods provide energy (calories), nutrients, and other substances needed for growth and health.

3. Health problems related to nutrition originate within cells.

4. Poor nutrition can result from both inadequate and excessive levels of nutrient intake.

5. Humans have adaptive mechanisms for managing fluctuations in nutrient intake.

6. Malnutrition can result from poor diets and from disease states, genetic factors, or combinations of these factors.

7. Some groups of people are at higher risk of becoming inadequately nourished than others.

8. Poor nutrition can influence the development of certain chronic diseases.

9. Adequacy, variety, and balance are key characteristics of a healthy diet.

10. There are no "good" or "bad" foods.

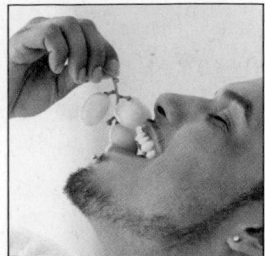

Nutrition Concepts Review

Focal Point: Nutrition concepts apply to diet and health relationships.

Write the number of the nutrition concept from Table 1.4 that applies to the situation described. Use each concept and do not repeat concept numbers in your responses.

Nutrition Concept Number	Situation
1. _____	The Irish potato famine caused thousands of deaths.
2. _____ and _____	Otis mistakenly thought that as long as he consumed enough calories, protein, vitamins, and minerals, he would stay healthy no matter what he ate.
3. _____	I feel guilty every time I eat potato chips. I wish they weren't bad for me.
4. _____	Phyllis was relieved to learn that her chronic diarrhea was due to the high level of vitamin C supplements she had been taking.
5. _____	A low amount of iron in Tawana's red blood cells was the reason for her loss of appetite and low energy level.
6. _____	Far more young children than soldiers died as a result of the 10-year civil war in Sudan.
7. _____	For the past 20 years, Don's idea of dinner was a big steak and potatoes. His recent heart attack changed his view of what's for dinner.
8. _____	During the two weeks they were backpacking in the Netherlands, Tomás and Ozzie ate very few vegetables and fruits. Their health remained robust, however.
9. _____	Zhang wasn't aware that he had the inherited condition hemochromatosis until he began taking iron supplements and developed iron overload symptoms.

FEEDBACK (answers to these questions) can be found at the end of Unit 1.

[Key Terms

antioxidants, page 1-9
calorie, page 1-8
chronic diseases, page 1-17
empty-calorie foods, page 1-18
energy-dense foods 1-18

essential nutrients, page 1-10
food insecurity, page 1-7
food security, page 1-7
malnutrition, page 1-16
metabolism, page 1-15

nonessential nutrients, page 1-10
nutrients, page 1-8
nutrient-dense foods, page 1-18
nutrition, page 1-6
phytochemicals, page 1-9

Review Questions

Answers can be found on the last page of this unit.

TRUE FALSE

1. The word *nonessential* as in *nonessential nutrient* means that the nutrient is *not* required for growth and health. ☐ ☐

2. Nutrition is defined as "the study of foods, their nutrients, and other chemical constituents and the effects of food contituents on health." ☐ ☐

3. Food insecurity is a problem in developing countries, but it is *not* a problem in the United States or Canada. ☐ ☐

4. Water is an essential nutrient.

5. The development of standards for nutrient intake levels was prompted in

TRUE FALSE

part by the high rejection rate of World War II recruits due to underweight and nutrient deficiencies. ☐ ☐

6. Tissue stores of nutrients decline after blood levels of the nutrients decline. ☐ ☐

7. To maintain health, all essential nutrients must be consumed at the recommended level daily. ☐ ☐

8. An individual's genetic traits play a role in how nutrient intake effects disease risk. ☐ ☐

9. Nutritionally speaking, potato chips can be accurately described as a "bad" food and broccoli as a "good" food. ☐ ☐

Media Menu

www.fns.usda.gov
The Food and Nutrition Service's main site provides information in English and Spanish on federal food and nutrition programs such as WIC and Food Stamps (including income eligibility and how to apply), responses to FAQs, links to the Dietary Guidelines, and nutrition education materials.

www.nutrition.gov
Go here for your guide to nutrition and health information. It includes information on food facts, food safety, life-cycle nutrition, food assistance, and nutrient databases.

fnic.nal.usda.gov/nal_display/index. php?info_center=4&tax_level=1
This terrific site from the Food and Nutrition Center at the National Agricultural Library is stuffed with information on public food and nutrition programs, nutrition labeling, the Dietary Guidelines and MyPyramid Food Guide, and much more. To avoid the long address, search the term "USDA Food and Nutrition Center" and select this site.

WebMD.com
WebMD is a science-based health and medicine site that offers extensive nutrition

information. Resources range from a healthy eating and diet center to the regular postings of updates on developments in nutrition and health.

www.iom.edu/CMS/3708.aspx
The Food and Nutrition home page of the Institute of Medicine is located here. It presents updated information on obesity, nutrient intake standards, and other nutrition topics.

www.feedingamerica.org
The former Second Harvest home page, this is a leading site for information on food insecurity.

Notes

1. Wang J et al. Dietary energy density predicts the risk of incident type 2 diabetes: the European Prospective Investigation of Cancer (EPIC)-Norfolk Study. Diabetes Care, 2008;31:2120–5

2. Anderson SA. Core indicators of nutritional state for difficult-to-sample populations. J Nutr 1990;120:15–8.

3. Scheier LM. What is the hunger-obesity paradox? J Am Diet Assoc 2005;105:883–6.

4. Larsen Nl et al. Access to healthy foods varies by neighborhood. Am J Preventive Med Jan 9, 2009,

available at www.medscape.com/viewarticle/585965.

5. Cook JT et al. A brief indicator of household energy security: associations with food security, child health, and child development in U.S. infants and toddlers. Pediatrics 2008;122:e867–75.

6. A comparison of household food security in Canada and the United States, Jan. 2009, www.ers.usda.gov/WhatsNew, accessed 1/3/09.

7. Wilde PE, Peterman JN. Individual weight change is associated with household food security status. J Nutr 2006;1395–1400.

8. CDC alert on ricin, http://:emergency.gov.cdc.gov./agents/ricin/han_022008.asp, accessed 11/17/08.

9. Bruemmer B. Food biosecurity. J Am Diet Assoc 2003:688–92.

10. Meinhardt PL. Water and bioterrorism: preparing for the potential threat to U.S. water supplies and public health. Ann Rev Public Health 2005;26:213–37.

11. Ricin toxin from the castor bean plant. www.ansci.cornell.edu/plants/toxicagents, and waynesword.palomar.edu, accessed 6/03.

12. Herber D. The stinking rose: oregano-sulfur compounds and cancer. Am J Clin Nutr 1997;66:425–6.

13. Hayes DP. Adverse effects of nutritional inadequancy and excess: a hormetic model. Am J Clin Nutr 2008;88(Suppl):578S–81S.

14. Hayes JD et al. The cancer chemopreventive actions of phytochemicals derived from glucosinolates, Eur J Nutr 2008:47(Suppl)2:73–88.

15 The Dietary Reference Intakes: The Essential Guide to Nutrient Requirements, National Academies Press: Washington, DC, 2006

16. Ames BN. The metabolic tune-up: metabolic harmony and disease prevention. J Nutr 2003;133:1544S–48S.

17. Herbert V. Chapter 26. Folic acid. In: Modern Nutrition in Health and Disease, 9th ed. Shils ME et al eds. Baltimore, MD:Lippincott Williams Wilkins, 1999, pp.433–58.

18. Fairbanks VF. Chapter 10. Iron in medicine and Nutrition. In: Modern Nutrition in Health and Disease, 9th ed. Shils ME et al eds. Baltimore, MD:Lippincott Williams Wilkins, 1999, pp. 193–221

19. Goodman, AS, Vitamin A and retinols in health and disease. N Engl J Med 1984;310:1023–31.

20. Mahoney DH Jr. Anemia in at-risk populations—what should be our focus? Am J Clin Nutr 2008;88:1457–8.

21. Trujillo E, Milner J.Nutrigenomics, proteomics, metabolomics, and the practice of dietetics. J Am Diet Assoc 2006;106:403–13.

22. Reedy J et al. Comparison of food-based recommendations and nutrient values of three food guides:USDA's MyPyramid, NHLBI's Dietary Approaches to stop Hypertension Eating Plan, and Harvard's Healthy Eating Pyramid, J Am Diet Assoc 2008;108:522–8.

23. Discretionary Calories, www.health. gov/DIETARYGUIDELINES/dga2005/ report/HTML/D3_DiscCalories. htm#tabled31, accessed 11/08.

24. Kant AK et al. Association of breakfast energy density with diet quality and body mass index in American adults: NHANES, 1999–2004, Am J Clin Nutr 2008;88:1396–404.

25. Leahy KE et al. Reducing the energy density of multiple meals deceases the energy intake of preschool-age children. Am J Clin Nutr 2008;88:1459–68.

26. Nutrition and you: Trends 2008,available at www.eatright.org/ trends2008, released 10/26/08.

27. Position of the American Dietetic Association: total diet approach to communicating food and nutrition information. J Am Diet Assoc 2002;102:100.

28. Drewnowski A. Concept of a nutritious food: toward a nutrient density score. Am J Clin Nutr 2005;82:721–32.

Answers to Review Questions

1. False, see page 1-10
2. True, see page 1-6
3. False, see page 1-8
4. True, see page 1-11, Table 1.1
5. True, see page 1-12
6. False, see page 1-15 and Illustration 1.9
7. False, see page 1-16
8. True, see page 1-17
9. False, see page 1-19

NUTRITION | Up Close

Nutrition Concepts Review

Feedback for Unit 1

The nutrition concepts apply to the situations as follows:

1. 1

2. 2, 9

3. 10

4. 4

5. 3

6. 7

7. 8

8. 5

9. 6

The Inside Story about Nutrition and Health

NUTRITION SCOREBOARD

	TRUE	FALSE
1 How long people live and how healthy they are depends on four factors: lifestyle behaviors, the environment to which people are exposed, genetic makeup, and access to quality health care.		
2 Diet is related to the top two causes of death in the United States.		
3 The body of modern humans was designed over 40,000 years ago.		

Key Concepts and Facts

- Health and longevity are affected by diet. Other lifestyle behaviors, genetic traits, the environment to which we are exposed, and access to quality health care also affect health and longevity.

- Dramatic changes in the types of foods consumed by modern humans compared with early humans are related in some ways to the development of today's leading health problems.

- The diets and health of Americans are periodically eveluted by national studies.

- The health status of a population changes for the better or worse as diets change for the better or worse.

Answers to **NUTRITION** SCOREBOARD	TRUE	FALSE
1 There are no secrets to a long, healthy life.	✔	
2 Diet is associated with the development of heart disease and cancer (which cause about half of all deaths in the United States).[1]	✔	
3 Hairstyles may be different, but our bodies are the same as they were 40,000 years ago.	✔	

Nutrition in the Context of Overall Health

Think of your body as a machine. How well this machine performs depends on a number of related factors: the quality of its design and construction, the appropriateness of the materials used to produce it, and how well it is maintained.

A machine designed to produce 10,000 copies a day will break down sooner if it is used to make 20,000 copies a day. The repair call will, in all probability, come earlier if the machine is overused *and* poorly maintained or if it has a part that doesn't work well. On the other hand, chances are good the copy machine will function at full capacity if it is free from design flaws, skillfully constructed from appropriate materials, properly used, and kept in good shape through regular maintenance.

Although much more complex and sophisticated, the human body is like a machine in some important ways. How well the body works and how long it lasts depend on a variety of interrelated factors. The health and fitness of the human machine depend on genetic traits (the design part of the machine), the quality of the materials used in its construction (your diet), and regular maintenance (your diet, other lifestyle factors, and health care).

Lifestyles exert the strongest overall influence on health and longevity (Illustration 2.1).[2] Behaviors that constitute our lifestyle—such as diet, smoking habits, illicit drug use or excessive drinking, level of physical activity or psychological stress, and the amount of sleep we get—largely determine whether we are promoting health or disease. Of the lifestyle factors that affect health, our diet is one of the most important.[3] In a sense, it is fortunate that diet is related to disease development and prevention. Unlike age, gender, and genetic makeup, our diets are within our control.

People have an intimate relationship with food—each year we put over a thousand pounds of it into our bodies! Food supplies the raw materials the body needs for growth and health; these, in turn, are affected by the types of food we usually eat. The diet we feed the human machine can hasten, delay,

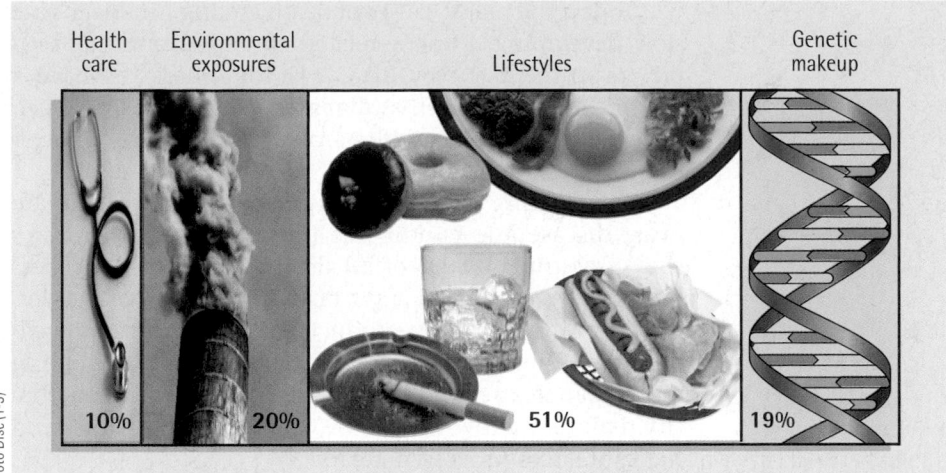

Health care | Environmental exposures | Lifestyles | Genetic makeup

10% | 20% | 51% | 19%

Illustration 2.1 Conditions that contribute to death among adults under the age of 75 years in the United States.
Health care refers to access to quality care; environmental exposures include the safety of one's surroundings and the presence of toxins and disease-causing organisms in the environment; lifestyle factors include diet, exercise, obesity, smoking, genetic traits, and alcohol and drug use.[4]

or prevent the onset of an impressive group of today's most common health problems.

The Nutritional State of the Nation

Since early in the twentieth century, researchers have known that what we eat is related to the development of vitamin and mineral deficiency diseases, to compromised growth and impaired mental development in children, and to the body's ability to fight off infectious diseases. Seventy years ago in the United States, widespread vitamin deficiency diseases filled children's hospital wards and contributed to serious illness and death in adults (Illustration 2.2). Now, however, dietary excesses are filling hospital beds and reducing the quality of life for millions of Americans.

Whosoever was the father of a disease, an ill diet was the mother.
—George Herbert, 1660

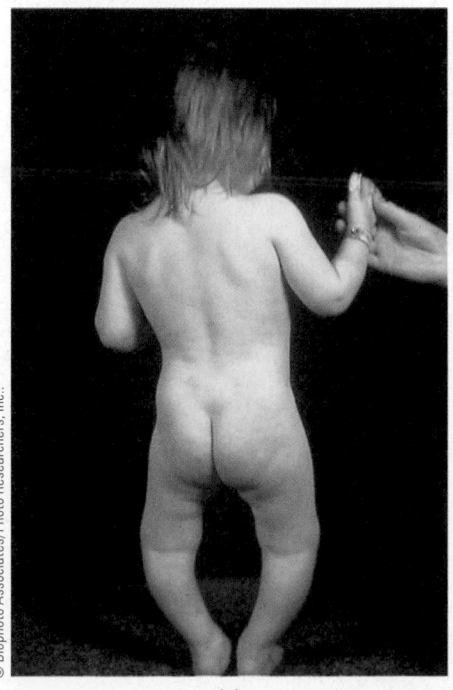

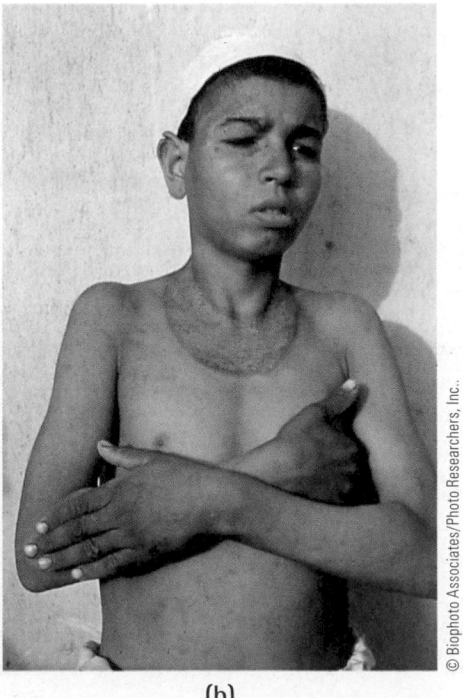

(a)

(b)

Illustration 2.2 Vitamin D deficiency (rickets shown on the left) and niacin deficiency (pellagra pictured on the right) were leading causes of hospitalization of children in the United States in the 1930s.

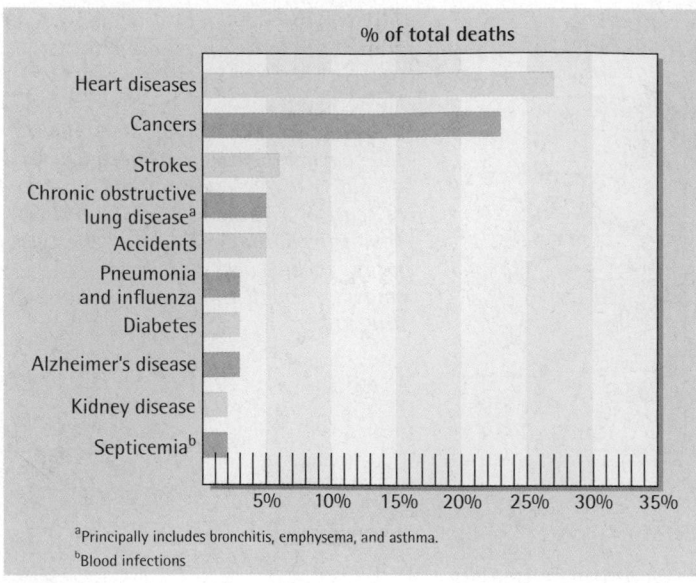

% of total deaths

Heart diseases
Cancers
Strokes
Chronic obstructive lung disease[a]
Accidents
Pneumonia and influenza
Diabetes
Alzheimer's disease
Kidney disease
Septicemia[b]

5% 10% 15% 20% 25% 30% 35%

[a]Principally includes bronchitis, emphysema, and asthma.
[b]Blood infections

Illustration 2.3 Percentage of total deaths for the top ten leading causes of death in the United States, 2005.[1]

chronic diseases
Slow-developing, long-lasting diseases that are not contagious (for example, heart disease, cancer, diabetes). They can be treated but not always cured.

diabetes
A disease characterized by abnormal utilization of glucose by the body and elevated blood glucose levels. There are three main types of diabetes: type 1, type 2, and gestational diabetes. The word *diabetes* in this text refers to type 2 diabetes, by far the most common. Diabetes is short for the term "diabetes mellitus."

hypertension
High blood pressure. It is defined as blood pressure exerted inside of blood vessel walls that typically exceeds 140/90 mm Hg (or, millimeters of mercury).

stroke
The event that occurs when a blood vessel in the brain suddenly ruptures or becomes blocked, cutting off blood supply to a portion of the brain. Stroke is often associated with "hardening of the arteries" in the brain. (Also called a *cerebral vascular accident*.)

Alzheimer's disease
A brain disease that represents the most common form of dementia. It is characterized by memory loss for recent events that expands to more distant memories over the course of five to ten years. It eventually produces profound intellectual decline characterized by dementia and personal helplessness.

Today, the major causes of death among Americans are slow developing, lifestyle-related **chronic diseases.** Based on government survey data, 44% of Americans have a chronic condition such as **diabetes**, heart disease, cancer, **hypertension**, or high cholesterol levels; 13% have three or more of these conditions.[5]

The leading causes of death among Americans are heart disease and cancer (see Illustration 2.3). Together they account for 50% of all deaths. Western-type diets high in saturated and *trans* fats, and low in vegetables, fruits, and whole grain products are linked to the development of heart disease.[6] Six types of cancer—including colon, pancreatic, and breast cancer—are related to obesity, habitually low intakes of vegetables and fruits, and high intake of processed meats.[7]

Diet is related to three other leading causes of death: diabetes, **stroke**, and **Alzheimer's disease.** Relationships between these, other diseases and disorders, and diet are overviewed in Table 2.1.

Shared Dietary Risk Factors A number of the diseases and disorders listed in Table 2.1 share the common dietary risk factors of low intakes of vegetables, fruits, and whole grains; excess calorie intake and body fat, and high animal fat intake. These risk factors are associated with the development of **chronic inflammation** and **oxidative stress**, conditions that are strongly related to the development of heart

Table 2.1

Examples of diseases and disorders linked to diet.[6-14]

Disease or Disorder	Dietary Connections
Heart disease	High saturated and trans fat, and cholesterol intakes; low vegetable, fruit and whole grain intakes; excessive body fat
Cancer	Low vegetable and fruit intakes; excessive body fat and alcohol intake, regular consumption of processed meats
Stroke	Low vegetable and fruit intake; excessive alcohol intake, high animal fat diets
Diabetes (type 2)	Excessive body fat; low vegetable and fruit intake; high saturated fat and energy-dense, nutrient-poor food intake
Cirrhosis of the liver	Excessive alcohol consumption; poor overall diet
Hypertension	Excessive sodium (salt) and low potassium intake, excess alcohol intake; low vegetable and fruit intake; excessive levels of body fat
Iron-deficiency anemia	Low iron intake
Tooth decay and gum disease	Excessive and frequent sugar consumption; inadequate fluoride intake
Osteoporosis	Inadequate calcium and vitamin D, low intakes of vegetables and fruits
Obesity	Excessive calorie intake, over consumption of energy-dense, nutrient-poor foods.
Chronic inflammation and oxidative stress	Excessive calorie intake, excessive body fat, high animal fat diets, low intake of whole grains, vegetables, fruit and fish, poor vitamin D status

disease, diabetes, **osteoporosis**, Alzheimer's disease, cancer, and other chronic diseases.[17]

Chronic Inflammation and Oxidative Stress Inflammation is a normal response of the body to the presence of infectious agents, toxins, or irritants. It neutralizes these threats, in part, by triggering the release of oxidizing agents such as **free radicals** that destroy the offending substances. In the process, however, free radicals also oxidize lipids and DNA, and damage cells and tissues. In the short-term, damage induced by oxidation reactions can generally be repaired by **antioxidants** produced by the body and consumed in vegetables, fruits, whole grain products, and other plant foods.[16,18]

Inflammation is related to chronic disease development when it is present at a low level for a long time, or is "chronic." Chronic inflammation and the resulting oxidative stress are sustained by irritants continually present in the body. Excess body fat and habitually high intakes of saturated and trans fats are examples of such irritants. If not countered by a sufficient supply of antioxidants, chronic inflammatory processes and oxidative stress impair the normal functioning of cells and tissues in ways that promote disease development.[19]

Adverse effects of chronic inflammation and oxidative stress can be diminished by loss of excess body fat, diets low in saturated and trans fats, and intake of omega-3 fatty acids from fish and seafood. Adequate intake of antioxidant-rich vegetables, fruits, whole grain products, and other plant foods can reduce the damage done to cells and tissues by oxidative stress.[6,12,16,19] Table 2.2 lists types of foods that decrease and increase inflammation, oxidative stress, or both.

Nutrient–Gene Interactions and Health Some diseases are promoted by interactions between nutrients and genes. Here are a few examples:

- Cancer-related genes can be activated or deactivated by certain components of food. Sulforaphane (pronounced sul-four-ah-phane) found in cabbage, broccoli, and brussel sprouts and helps prevent several types of cancer. Sulforaphane inactivates a gene that produces a substance that encourages cancer development.[20]

- About half of the U.S. population is genetically susceptible to cholesterol in the diet. For them, blood cholesterol levels increase with high cholesterol intake. People who lack the trait are considered "cholesterol nonresponders" because their blood cholestrol levels do not increase in response to a high cholesterol intake.[21]

- Fish and seafood rich in omega-3 fatty acids improve mental functioning in adults who are genetically susceptible to dementia. Initial evidence indicates that omega-3 fatty acids may lower the risk of Alzheimer's disease in genetically susceptible adults.[22]

Knowledge of nutrient–gene interactions in health and disease is expanding rapidly and is greatly enhancing our understanding of the relationship between diet and health.

The Importance of Food Choices

People are not born with a compass that directs them to select a healthy diet—and it shows. If given access to a food supply like that available in the United States, people show a marked tendency to choose a diet that is high in energy-dense, nutrient-poor foods[23] (Illustration 2.4). Such a diet tends to include processed foods high in saturated fat, salt, or sugar and low in fiber, vegetables, and fruits. This type of diet poses the greatest risks to the health of Americans.

Table 2.2

Types of food associated with decreased or increased inflammation, oxidative stress, or both. [6-15]

1. Decreased
Colorful fruits and vegetables
Dried beans
Whole grains
Fish and seafood; fish oils
Red wines
Dark chocolates
Extra virgin olive oils
Nuts
Coffee

2. Increased
Processed and high-fat meats
High-fat dairy products
Baked products, snack foods with *trans* fats
Soft drinks, other high-sugar beverages

chronic inflammation
Low-grade inflammation that lasts weeks, months, or years. Inflammation is the first response of the body's immune system to infectious agents, toxins, or irritants. It triggers the release of biologically active substances that promote oxidation and other reactions to counteract the infection, toxin, or irritant. A side-effect of chronic inflammation is that it also damages lipids, cells, and tissues.

oxidative stress
A condition that occurs when cells are exposed to more oxidizing molecules (such as free radicals) than to antioxidant molecules that neutralize them. Over time, oxidative stress causes damage to lipids, DNA, cells and tissues. It increases the risk of heart disease, type 2 diabetes, cancer, and other diseases.

osteoporosis
A condition in which bones become fragile and susceptible to fracture due to a loss of calcium and other minerals.

free radicals
Chemical substances (often oxygen-based) that are missing electrons. The absence of electrons makes the chemical substance reactive and prone to oxidizing nearby molecules by stealing electrons from them. Free radicals can damage lipids, proteins, DNA (genetic material contained in cells), cells, and tissues by altering their chemical structure and functions.

Illustration 2.4 Lopsided, all-American food choices.

(a)   (b)

Photo Disc

(c)   (d)

antioxidants
Chemical substances that prevent or repair damage to cells caused by oxidizing agents such as pollutants, ozone, smoke, and reactive oxygen. Oxidation reactions are a normal part of cellular processes. Vitamins C and E and certain phytochemicals function as antioxidants.

Diet and Diseases of Western Civilization

Why is the U.S. diet—a "Western" style of eating—hazardous to our health in so many ways? What is it about a diet that is high in animal fat, salt, and sugar and low in vegetables, fruits, and fiber that promotes certain chronic diseases? A good deal of evidence indicates that the chronic diseases now prevalent in the United States and other Westernized countries have roots in dietary changes that have taken place over centuries.

Our Bodies Haven't Changed

The biological processes that control what the human body does with food were developed more than 40,000 years ago. These hardwired, evolution-driven processes exist today because they are firmly linked to the genetic makeup of humans, and genes change very little over great spans of time.[24]

Then . . . For the first 200 centuries of their existence, humans survived by hunting and gathering (Illustration 2.5). They were constantly on the move, pursuing wild game or following the seasonal maturation of fruits and vegetables. Meat, berries, and many other plant products obtained from successful hunting and gathering journeys spoiled quickly, so they had to be consumed in a short time. Feasts would be followed by famines that lasted until the next successful hunt or harvest.[24]

. . . and Now The bodies of modern humans, adapted to exist on a diet of wild game, fish, fruits, nuts, seeds, roots, vegetables, and grubs; to survive periods of famine; and to sustain a physically demanding lifestyle are now exposed to a different set of circumstances. The foods we eat bear little resemblance to the foods available to our early ancestors (Illustration 2.6). Sugar, salt, alcohol,

(a)　　　　　　　　　　　　　　　　　(b)

Illustration 2.5 Hunter-gatherers still exist in the world, but their numbers are diminishing. It is estimated that hunter-gatherers consume approximately 3000 calories daily due to their physically demanding way of life.

food additives, oils, margarine, dairy products, cereal grains, and refined and processed foods were not a part of their diets. These ingredients and foods came with Western civilization.[24] Furthermore, we do not have to engage in strenuous physical activity to obtain food, and our feasts are no longer followed by famines.

Illustration 2.6 The disconnect between high animal fat, high salt, high sugar, and processed foods in Western-type diets (right) and wild plants and animal foods consumed by our early ancestors (left). *Foods consumed by hunter-gatherers shown in the photograph include bird's eggs, wild cucumbers, roots, nut, and berries. Not shown are grubs and other insects, which might be consumed as quickly as they are discovered.*

(a)　　　　　　　　　　　　　　　　　(b)

Table 2.3

Life expectancy at birth for countries with high life expectancies, 2005[26]

Country	Life expectancy (years)
Japan	82.1
Switzerland	81.3
Australia	80.9
Sweden	80.6
Canada	80.4
Italy	80.4
France	80.3
Norway	80.1
Spain	80.1
New Zealand	79.6
Austria	79.5
Ireland	79.5
Netherlands	79.4
Greece	79.3
Germany	79.0
United Kingdom	79.0
Finland	78.9
Belgium	78.7
Portugal	78.2
Denmark	77.9
United States	77.8
Mexico	75.5

Illustration 2.7 Typical Japanese foods. Compare these to the "all-American" food choices shown in Illustration 2.4.

The human body developed other survival mechanisms that are not the assets they used to be. Mechanisms that stimulate hunger in the presence of excess body fat stores, conserve the body's supply of sodium, and confer an innate preference for sweet-tasting foods—as well as a digestive system that functions best on a high-fiber diet—were advantages for early humans. They are not advantageous for modern humans, however, because our diets and lifestyles are now vastly different.

Although the human body has a remarkable ability to adapt to changes in diet, health problems of modern civilization such as heart disease, cancer, hypertension, and diabetes are thought to result, in part, from diets that are greatly different from those of our early ancestors. The human body was built to function best on a diet that is low in sugar and sodium, contains lean sources of protein, and is high in fiber, vegetables, and fruits (Illustration 2.6).[25] Strong evidence for this conclusion is provided by studies that track how disease rates change as people adopt a Western style of eating.

Changing Diets, Changing Disease Rates

Many countries are adopting the Western diet and the pattern of disease that accompanies it. People in Japan, for example, live longer than anyone else in the world—until they move to the United States (Table 2.3). In Japan, the traditional diet consists mainly of rice, vegetables, fish, shellfish, and meat (Illustration 2.7).

(a) (b)

(c) (d)

Photo Disc

When Japanese people move to the United States, their diets change to include, on average, more meat and fat, and less fiber. Japanese living in the United States are much more likely to develop diabetes (Illustration 2.8), heart disease, breast cancer, and colon cancer than people who remain in Japan.[27]

Dietary habits in Japan are rapidly becoming similar to those in the United States. Hamburgers, fries, steak, ice cream, and other high-fat foods are gaining in popularity. Rates of diabetes, heart disease, and cancer of the breast and colon are on the rise in Japan.[27] Similarly, the "diseases of Western civilization" are occurring at increasing rates in Russia, Greece, Israel, and other countries that are adopting the Western diet.[14,28]

Today's food supply makes it a bit challenging to eat more like our early ancestors did. What types of foods would you choose if you wanted to shape a diet that is closer to that of hunter-gathers? Join Beth and Shandra in making these dietary decisions in the "Reality Check."

Illustration 2.8 An increased rate of diabetes in Japanese men immigrating to Seattle corresponds to dietary changes.

Source: CH Tsunehara, DL Leonetti, and WY Fujimoto, Diet of second-generation Japanese-American men with and without non-insulin-dependent diabetes (Am J Clin Nutr 1990;52:731– 8).

The Power of Prevention

Heart disease, cancer, and other chronic diseases are not the inevitable consequence of Westernization. High-animal-fat diets and lifestyle behaviors (such as smoking) that promote chronic disease can be avoided or changed. Although heart disease is still the leading cause of death in the United States, its rate has declined by 50% in the last 25 years. About half of this decline is related to improvements in risk factors such as smoking, hypertension, and elevated blood cholesterol levels. (To read about an underappreciated aspect of heart health promotion, visit the nearby On the Side feature.) the other half is due to medical interventions for people with heart disease.[31] The American Heart Association warns that declines in heart disease deaths may not continue. The prevalence of risk factors for heart disease are increasing due to the obesity and diabetes epidemics, and that may lead to increased rates of heart disease in the future.[32]

Now it is taken for granted that the quality of the diet is at least as important as the quantity . . .
ROBERT HEANEY[11].

Improving the American Diet

Many efforts are under way to improve the diet and health status of Americans. Like some other countries with high rates of "Western" diseases, the United States has set national health goals and implemented programs aimed at improving health. Goals for changes in health status in the United States are presented in the report *Healthy People 2010: Objectives for the Nation.* Its objectives for improvements in nutrition are outlined in Table 2.4.

What Should We Eat?

The Dietary Guidelines recommendations for disease prevention and health promotion are translated into food choices in the MyPyramid Food Guide (Table 2.5). This guide recommends that people consume basic foods in their most nutrient-dense forms. Meat choices from the meat and beans group, for example, consists of lean meats, fish, and dried beans that are not prepared with fat. It is assumed that vegetables have no added butter or margarine, fruits no added sugar, and milk no fat. The guide encourages consumption of dark-green- and orange-colored vegetables, whole grains, and whole grain products.[3] Fats and sweets are included in the MyPyramid food guide under the heading

On the Side

Laughter

Janie comes home from school to find her aunt visiting. "What do you want to be when you grow up?" the aunt asks. Janie smiles and says "A vitamin!" "A vitamin!" her aunt replies. "Why?" "Because," Janie said, "I walked by the pharmacy on the way home today and I saw a sign that said "Vitamin B-1."

It turns out that laughter, like a nutritious diet, exercise, and happiness, is good for your heart's health.[29,30]

Table 2.4

A summary of nutrition objectives for the nation[33]

Improvements in Health Status	Improvements in Health Practices	Improvements in Food Services and Programs
Reduce rates of: • Overweight • Heart disease and stroke • Hypertension • Cancer • Osteoporosis • Diabetes • Congenital (inborn) abnormalities • Growth retardation • Iron deficiency • Baby bottle tooth decay • Food allergy deaths	Increase rates of: • Breast-feeding • Safe and effective weight-loss practices (diet and exercise) • Healthy weight gain in pregnancy Reduce dietary intake of: • Fat • Saturated fat • Sodium Increase dietary intake of: • Vegetables • Fruits • Grain and grain products (especially whole grains) • Calcium	• Improve food safety practices • Increase food security • Increase nutrition education in schools • Improve nutritional quality of school meals and snacks • Increase work site programs in nutrition and weight management • Increase nutrition counseling for patients with heart disease, diabetes, and other nutrition- and diet-related conditions

"Discretionary Calories" and should only be included in the diet if nutrient needs are met from basic foods and calorie need allows. Diets that conform to this healthy eating pattern are related to a reduced risk of a variety of chronic diseases.[34]

Nutrition Surveys: Tracking the American Diet

The food choices people make and the quality of the American diet are regularly evaluated by national surveys. Table 2.6 summarizes the major surveys conducted by the federal government. The first survey was started in 1936. It was conducted in conjunction with the original national program aimed at reducing hunger, poor growth in children, and vitamin and mineral deficiency diseases in the United States. Results of nutrition surveys are used to identify problem areas within the food supply, the characteristics of diets consumed by people in the United States, and the nutritional health of the population. They provide information ranging from the amount of lead and pesticides in certain foods to the adequacy of diets of low-income families. Together with the results of studies conducted by university researchers and others, they provide the information needed to give direction to food and nutrition programs and efforts aimed at improving the availability and quality of the food supply.

Reality Check

Getting back to the basics—but how?
Beth and Shandra have been roommates for a year and usually shop for groceries together. For the next trip to the grocery store, they decide that each of them will make up a shopping list that includes foods that resemble those that their early ancestors might have eaten. Here are the results:

Which list do you think comes closest to matching the basic foods consumed by our early ancestors?
Answers on page 2–11

Beth:
rice, yogurt, pork, honey, olive oil

Shandra:
carrots, nuts, asparagus, fish, blueberries

Table 2.5

MyPyramid basic food groups

MyPyramid.gov
STEPS TO A HEALTHIER YOU

Food Group	Example foods
Bread, Cereal, Rice, Pasta	Oatmeal, whole-wheat raisin English muffin, cornflakes, brown or white rice, macaroni
Fruits	Grapes, bananas, oranges, melons, fruit juices, dried fruits, berries, papayas
Vegetables	Broccoli, carrots, tomatoes, spinach, turnip greens, radishes, sweet potatoes, cabbage, vegetable juice
Meats, Poultry, Fish, Dried Beans and Peas (legumes), Eggs, and Nuts	Lean steak, baked chicken or turkey without skin, broiled and canned fish, hard boiled egg, cooked beans, peanut butter, walnuts
Milk, Yogurt, Cheese	Skim milk, buttermilk, low-fat cottage cheese, yogurt

Table 2.6

Periodic, national surveys of diet and health in the United States

Survey	Purpose
1. **National Health and Nutrition Examination Survey (NHANES)**	• Assesses dietary intake, health, and nutritional status in a sample of adults and children in the United States on a continual basis
2. **Nationwide Food Consumption Survey (NFCS)**	• Performs regular surveys of food and nutrient intake and understanding of diet and health relationships among a national sample of individuals in the United States
3. **Total Diet Study** (sometimes called Market Basket Study)	• Ongoing studies that determine the levels of various pesticide residues, contaminants, and nutrients in foods and diets

ANSWERS TO *Reality Check*

Getting back to the basics—but how?

Shandra's list contains unprocessed plant foods and fish, which would have been available to our early ancestors.

Beth:

Shandra:

NUTRITION

Up Close

Nutrition Scoreboard: Food Types for Healthful Diets

Focal Point: Identifying the types of foods that fit into the MyPyramid five basic food groups and those that characterize Western-type diets.

The MyPyramid five basic food groups consist of food choices that make up a healthful, disease-preventing diet. The typical Western-type diet, on the other hand, is related to the development of a number of diseases and includes many types of food that are not part of the MyPyramid basic food groups. To which dietary pattern do the following types of food most appropriately belong?

	Dietary pattern	
Food types	**MyPyramid**	**Western**
Mixed vegetables	_____	_____
Cold cuts (ham, bologna, salami)	_____	_____
Broiled fish and seafoods	_____	_____
Whole grain breads	_____	_____
Fruit jams and jellies	_____	_____
Potato, tortilla, and other snack chips	_____	_____
Dried beans	_____	_____
Skinless poultry	_____	_____
Fruit juices	_____	_____
Vegetables with margarine	_____	_____
Fatty meats	_____	_____
Ice cream	_____	_____
Regular soft drinks	_____	_____
Salad dressing, mayonnaise	_____	_____
Skim milk	_____	_____
Cake, cookies, pastries	_____	_____

FEEDBACK (answers to these questions) can be found at the end of this Unit.

Key Terms

Alzheimer's disease, page 2-4
chronic diseases, page 2-4
chronic inflammation, page 2-5
antioxidants, page 2-6

cirrhosis of the liver, page 2-4
diabetes, page 2-4
hypertension, page 2-4
free radicals, page 2-5

osteoporosis, page 2-5
oxidative stress, page 2-5
stroke, page 2-4

Review Questions

	TRUE	FALSE
1. Vitamin deficiency diseases remain a major health problem of poor children in the United States.	☐	☐
2. Our diets are related to the majority of the top ten causes of death.	☐	☐
3. Low intake of vegetables and fruits is related to the development of heart disease, cancer, hypertension, and osteoporosis.	☐	☐

	TRUE	FALSE
4. Our genetic makeup changes as our diets change.	☐	☐
5. Although individuals cannot change their genetic makeup, they can change their risk for chronic disease development by making healthful eating and lifestyle changes.	☐	☐

6. The incidence of chronic diseases such as diabetes, cancer, and heart disease increases as countries adopt a Western style of eating. ☐ ☐

7. National surveys in the United States assess the nutritional health of the population and the safety of the food supply. ☐ ☐

8. In the United States, the *Basic Four Food Groups* is the most recently published guide to selection of a healthful diet. ☐ ☐

[Media Menu

www.healthfinder.gov
A gateway for nutrition and health information can be found at this site.

www.nutrition.gov
This site provides one-stop shopping for information on government-sponsored food and nutrition programs, health statistics, and diet and health relationships. You can go directly from this site to information on MyPyramid Food Guide, the Dietary Guidelines for Americans, nutrition and health issues, and Federal Nutrition Assistance programs.

cdc.gov/nutritionreport
Additional information about national nutrition surveys can be found at this site.

www.healthypeople.gov
This site presents the Healthy People Objectives for the Nation.

www.hc-sc.gc.ca
Search the word *nutrition* and gain access to diet and health information, food and nutrition programs, and resources in Canada.

www.mayoclinic.com
Head to the Food and Nutrition center to obtain a wealth of information on nutrition, wellness, and disease.

www.nal.usda.gov/fnic/consumersite/justforyou.htm
This site presents interactive tools for dietary assessment and provides tools for making informed decisions about food, nutrition, and supplement use.

www.nlm.nih.gov/medlineplus
This site provides excellent information on diet–disease relationships.

[Notes

1. Leading causes of death, www.cdc.gov. nchs/FASTSTATS, accessed 1/09.

2. Gold, M. HALY'S: Measuring lifestyle-related factors that contribute to premature death and disability. In: Estimating the Contributions of Lifestyle-Related Factors to Preventable Death: A Workshop Summary (2005). Washington, DC: National Academies Press, 2005:1–54.

3. Krebs-Smith SM et al. How does MyPyramid compare to other population-based recommendations for controlling chronic disease? J Am Diet Assoc 2007;107:830–7.

4. Contributors to death among individuals under the age of 75 years in the United States, www.cdc.gov/nchs, accessed 6/06.

5. Paez K. More Americans getting multiple chronic illnesses, www. medscape.com/view article/586363, accessed 1/09.

6. Van Horn L et al. The evidence for dietary prevention and treatment of cardiovascular disease, J Am Diet Assoc 2008;108:287–331.

7. World Cancer Research Fund/American Institute for Cancer Research. *Food, Nutrition, Physical Activity, and the Prevention of Cancer. A Global Perspective*, Washington, DC: AIRC;2007.

8. Utsugi MT et al. Fruit and vegetable consumption and the risk of hypertension determined by self measurement of blood pressure at home: the Ohasama study. Hypertens Res. 2008;3:1435–43.

9. Hitz MF et al. Bone mineral density and bone markers in patients with a recent low-energy fracture: effect of 1 y of treatment with calcium and Vitamin D. Am J Clin Nutr 2007;86:251–9.

10. Adrogue HJ et al. Sodium and potassium in the pathogenesis of hypertension. N Engl J Med 2007;356:1966–78.

11. Heaney RP. Nutrients, endpoints, and the problem of proof. J Nutr 1008;138:1591–6.

12. O'Keffe JH et al. Dietary and lifestyle strategies for improving postprandial glucose, lipid profile, markers of inflammation, and cardiovascular health, Am Coll Cardiol. 2008;51: 249–255.

13. Scarmeas N et al. Adherence to the Mediterranean diet and the risk of developing Alzheimer's disease. Ann Neurol, posted online 4/18/06, www. medscape.com/viewarticle/530121, accessed 4/20/06.

14. Hossain P et al. Obesity and diabetes in the developing world—a growing challenge, N Engl J Med 2007;365:213–5.

15. Jimenez-Gomez Y et al. Olive oil and walnut breakfast reduce the postprandial inflammatory response in mononuclear cells compared with a butter breakfast in healthy men. Atherosclerosis 2008, Sept 17, available at www.nutraingrediants-usa.com/content/view/230758, accessed 1/09.

16. Valtuena S et al. Food selection based on total antioxidant capacity can modify antioxidant intake, systemic inflammation, and liver function without altering markers of oxidative stress. Am J Clin Nutr 2008;87:1290–7.

17. Esmaillzadeh A et al. Home use of vegetable oil markers of systemic inflammation, and endothelial dysfunction among women. Am J Clin Nutr 2008;88:913–21.

18. Dai J et al. Association between adherence to the Mediterranean diet and oxidative stress. Am J Clin Nutr 2008;88:1346–70.

19. McDade TW et al. Adiposity and pathogen exposure predict C-reactive protein in Filipino women. J Nutr 2008;1138:2442–7.

20. Hayes JD et al. The cancer chemo-preventive actions of phytochemicals derived from glucosinolates, Eur J Nutr 2008;47(Suppl) 2:73–88.

21. Greene CM et al, Maintenance of the LDL cholesterol: HDL cholesterol ratio in an elderly population given a dietary cholesterol challenge. J Nutr.2005 Dec;135 (12):2793–8

22. Whalley LJ et al. n-3 fatty acid erythrocyte membrane content, APOEe4, and cognitive variation: an observational follow-up study in late adulthood. Am J Clin Nutr 2008;87:449–54.

23. Bachman JL et al. Sources of food group intake among the US population. J Am Diet Assoc 2008;108:804–14.

24. Cordain L et al. Origins and evolution of the Western diet: health implications for the 21st century. Am J Clin Nutr 2005:81:341–54.

25. Lindeberg S et al. A Paleolithic diet improves glucose tolerance more than a Mediterranean-like diet in individuals with ischaemic heart disease, Diabetologia, 2007;50:1795–807.

26. U.S. Census Bureau, International Statistics,www.census.gov/compendia/ statab/cats/international_statistics.html, accessed 1/09.

27. Kato H, Haybuchi H. Study of the epidemiology of health and dietary habits of Japanese and Japanese Americans. Japan Journal of Nutrition 1989;47:121–30.

28. Kontogianni MD et al. Adherence rates to the Mediterranean diet are low in a representative sample of Greek children and adolescents, J Nutr 2008;138:1951–6.

29. Bennett MP et al. Humor and Laughter May Influence Health: III. Laughter and Health Outcomes, Evid Based Complement Alternat Med, 2008;5:37–40.

30. Miller M et al. Music, like laughter, benefits heart health. American Heart Association 2008 Scientific Session, Abs no, 5132, presented 11/11/08, accessed at www.medscape.com/viewarticle/583554.

31. Ford ES et al. Explaining the decrease in U.S. deaths from coronary disease, 1980-2000. N Engl J Med 2007;356:2388–98.

32. Lloyd-Jones D et al. Heart disease and stroke statistics: an update. A report from the American Heart Association Statistics Committee and Stroke Statistics Committee, Circulation 2008;circ:ahajournals.org, accessed 1/09.

33. Healthy People 2010-Objectives for the Nation, www.nutrition.gov, accessed 1/09.

34. Brunner EJ et al. Dietary patterns and 15-y risk of major coronary events, diabetes, and mortality. Am J Clin Nutr 2008;87:1414–21.

Answers to Review Questions

1. False, see page 2-3
2. True, see page 2-4
3. True, see pages 2-4
4. False, see page 2-6
5. True, see page 2-9
6. True, see pages 2-9
7. True, see page 2-10
8. False, see page 2-11

NUTRITION | Up Close

Food Types and Healthful Diets

Feedback for Unit 2

To reduce the selection of energy-dense, nutrient-poor foods and excessive calorie intake levels, the MyPyramid dietary pattern contains foods in their basic form. That means, for example, that fruits in the fruit group have no added sugar, vegetables no added fat, and meats are lean.

Food types	Dietary pattern MyPyramid	Western
Mixed vegetables	X	
Cold cuts (ham, bologna, salami)		X
Broiled fish and seafoods	X	
Whole grain breads	X	
Fruit jams and jellies		X
Potato, tortilla, and other snack chips		X
Dried beans	X	
Skinless poultry	X	
Fruit juices	X	
Vegetables with margarine		X
Fatty meats		X
Ice cream		X
Regular soft drinks		X
Salad dressing, mayonnaise		X
Skim milk	X	
Cakes, cookies, pastries		X

Ways of Knowing about Nutrition

NUTRITION SCOREBOARD

	TRUE	FALSE
1 It is illegal to convey false or misleading information about nutrition in magazine and newspaper articles and on television.		
2 Knowledge about nutrition is gained by scientific studies.		
3 "Double-blind" studies are used to diminish the "placebo effect."		

Key Concepts and Facts

- For the most part, nutrition information offered to the public does not have to be true or even likely true.

- Nutrition information offered to the public ranges in quality from sound and beneficial to outrageous and harmful.

- Science is knowledge gained by systematic study. Reliable information about nutrition and health is generated by scientific studies.

- Misleading and fraudulent nutrition information exists primarily because of financial interests and personal beliefs and convictions.

Answers to **NUTRITION SCOREBOARD**	TRUE	FALSE
1 Freedom of speech applies to information about nutrition in articles, speeches, pamphlets, and broadcasts. However, it is illegal to make false or misleading claims about nutrition in advertisements or on product labels and packaging.		✔
2 The "way of knowing" for the field of nutrition is science.	✔	
3 In a double-blind study, neither the research staff nor the research subjects know which subjects are getting the real treatment and which are receiving the fake treatment. This reduces the placebo ("sugar pill") effect, or changes in health that are due to the expectation that a specific treatment will have a particular impact on health.		✔

Maria, a college student: "You really ought to try this new Herbal Melt Down Diet I found on the Internet. I used the herbs for a week and lost five pounds!"

Jessie, a premed student: "If I were you, I'd treat that urinary tract infection with cranberry juice."

Newspaper headline: "Eating Cauliflower Daily Prevents Breast Cancer!"

Infomercial: "Lose fat, gain muscle and energy! Eat Complete Cereal Pro bars every day!"

Photo Disc

How Do I Know if What I Read or Hear about Nutrition Is True?

How do you know if what you read or hear about nutrition is true? With only the information given in these examples, you couldn't know. In reality, however, this is how much of the information we receive about nutrition comes to us—in bits and pieces. The nutrition information offered to the public is a mix of truths, half-truths, and gossip. The information does not have to meet any standard of truth before it can be represented as true in books, magazines, newspapers, TV and radio reports and interviews, pamphlets, the Internet, and speeches.

Opinions expressed about nutrition are protected by the freedom of speech provisions of the U.S. Constitution. Although it is misleading and fraudulent for a tabloid article or an Internet site to announce that 19 foods have negative calories, it is not illegal. The promotion of nutritional remedies that are not known to work, such as amino acid supplements for hair growth and beef extract for the treatment of cancer, is likewise protected by freedom of speech. It is illegal, however, to put false or misleading information about a product or service on a product label, in a product insert, or in an advertisement. In addition, the U.S. and Canadian mail systems cannot be used to send or to receive payments for products that are fraudulent.

With so much misinformation available, it is difficult to know what to do when we hear or read something about nutrition that may benefit us personally or perhaps help a friend or relative. Why does such a mix of nutrition sense and nonsense exist? How can you separate the sound information from the highly questionable? Where does nutrition information you can trust come from? These questions are addressed in this unit.

Why Is There So Much Nutrition Misinformation?

Consumers are bombarded with nutrition misinformation and ineffective or untested products and services. These reasons for this can be grouped into two categories: profit, and personal beliefs and convictions. The first and the more important reason, not surprisingly, is the profit motive.

Motivation for Nutrition Misinformation #1: Profit As long as consumers seek quick and easy ways to lose weight, build muscle, slow aging, and reduce stress—goals that cannot be achieved quickly or easily—there will continue to be a huge market for nutritional products and services that offer assistance. The financial incentives are in place.

Not everyone who is in the business of nutrition has the goal of maintaining or improving people's health. Many seek to make money from people who are willing to believe their advertisements and buy products or services just in case they work.

People believe nutrition nonsense and buy fraudulent nutrition products for many reasons:

- Many people believe what they want to hear.

- People tend to believe what they see in print.

- Promotional materials sound scientific and true.

- The products offer solutions to important problems that have few or no solutions in orthodox health care.

- Promotional materials often appeal to people who are disenchanted with traditional medical care, fear the side effects of medications, or want a "natural" remedy.

- Laws that govern truth in advertising are often not enforced.

- New bogus products and services find a ready market because existing bogus products and services don't work.

Instead of evidence and facts, profit-oriented companies use paid testimonials ("It worked for me, it will work for you!"), "medical experts," and Hollywood stars and sports heroes to promote their products. Their advertisements promote ideas like "a wonder to science" or "miraculous" that appeal to some people's inclinations toward the mysterious or the divine. Real remedies aren't advertised in such terms. Nor are they portrayed as being effective because they come from Europe, the Ecuadoran highlands, the ancient Orient, or organic algae ponds.

Illustration 3.1 offers a formula for developing and marketing a fraudulent nutrition product. Try following the steps and make up your own miraculous nutritional cure. Once you've devised your own product, you'll find it easier to detect fraudulent marketing.

■ **Controlling Profit-Motivated Nutrition Frauds** The Federal Trade Commission (FTC) has the authority to remove advertisements that make false claims from the airwaves and Internet. In the past, the FTC has exerted its authority by removing blatantly false and misleading advertisements, including those for the "European Weight Loss Patch," colon defoxifiers, juices with herbal extracts, coral calcium, male enhancement dietary supplements, and bee pollen. Nevertheless, misleading and inaccurate advertisements still appear because enforcement efforts are weak and are concentrated on very dangerous products. The FTC and other federal agencies do respond when several consumers register complaints about a nutritional product or advertisement. You can learn how to do that in the On the Side box.

...[join] a special company with 22nd century breakthrough nutritional products and unequaled compensation plan!
—USA TODAY BUSINESS ADVERTISEMENT

Teaming up nutritional science with network marketing makes an awesome twosome!
—PROMOTIONAL ARTICLE FOR "NUTRITIONAL SYSTEM" IN AN ADVERTISING MAGAZINE

Max, a 50-year-old engineer: "If I hadn't been following a low-fat diet for the last 10 years, I would have had a heart attack by now—just like my dad did when he was 48."

Photo Disc

Illustration 3.1 Create Your Own Fraudulent Nutrition Product.

of Tennessee made $7,000 in commission. No Gimmicks – No competition – Big money every week! For more information call
(800) 555-1234

HAVE YOU EVER SEEN A FAT PLANT?

Scientists in our laboratory have discovered the secret. For a lifetime of being reed thin, send $29.95 to:

Obesity Research Center
P.O. Box 281752,
Miami, FL

HOT! HOT! HOT!

Banks & Financial Institutions made HUGE PROFITS trading foreign currencies & so can you! 10K invested

A Step-by-Step Guide

1. Identify a common problem people really want fixed that cannot easily or quickly be fixed another way.
   EXAMPLES: Obesity, low energy, weak muscles.

2. Make up a nutritional remedy and connect it to a biological process in the body. Try to think of a remedy that probably won't harm anyone. While you are at it, create a catchy name for your product.
   EXAMPLES: An herb that burns fat by speeding up metabolism, vitamins that boost the body's supply of energy, protein supplements that go directly to your muscles.

3. Develop a scientific-sounding explanation for the effect the product has on the body. Refer to the results of scientific studies.
   EXAMPLES: "Research has shown that the combination of herbs in this product stimulates the production of chemicals that trigger the energy-production cycle in the body." "As nutritional scientists have known for a century, the body requires B vitamins to form energy. The more you consume of this special formulation of B vitamins, the more energy you can produce and the more energy you will have!" "The unique combination of amino acids in this product is the same as that found in the jaw muscles of African lions. It is well known that an African lion can lift a 600-pound animal in its teeth."

4. Dream up testimonials from bogus previous users of the product, before-and-after photos, or "expert" opinions to quote. Use terms like "magical," "miraculous," "suppressed by traditional medicine," "secret," or "natural" as much as possible.
   EXAMPLES: Jane Fondu, Indianapolis: "I didn't think this product was going to work. The other products I tried didn't. My doctor couldn't help me lose weight. Thank God for [insert name of product]! I miraculously lost 20 pounds a week while eating everything I wanted." Dr. J. R. Whatsit, DM, NtD, Director of Nutritional Research: "We discovered this secret herb in our laboratories after years of looking for the substance in plants that keeps them from getting fat. Voilà! We found it. This discovery may win us a Nobel Prize." Or, show photos of a skinny person and a beefed-up person. (The photos don't have to be of the same person; the people only need to look similar.)

5. Offer customers a money-back guarantee.

On the Side

■ **Science for Sale** Here's a true story:

Friday, 2:00 P.M. A call came into the Nutrition Department from a man who wanted to talk to a nutrition expert. He had heard on last night's news that zinc lozenges were good for treating cold symptoms. Since he had a cold and didn't think the zinc would hurt, he purchased and consumed a whole roll of lozenges. Now he had a horrible, metallic taste in his mouth that he couldn't get rid of, no matter how often he brushed his teeth or gargled with mouthwash. He was worried that the taste was going to stay in his mouth forever. The nutrition expert assured him that it wouldn't. He had overdosed on zinc, and the taste would go away slowly.

This man was not the only one who heard the newscast about zinc and consumed too many lozenges the next day. Sales of zinc lozenges skyrocketed after the newscast, and as it happened, the author of the research study profited handsomely from the increased sales. After completing the study, the author bought shares of stock in the company that made the lozenges.[1] Such a financial tie should have been reported in the article. It is possible that a financial incentive could have influenced how the author presented the study's results.

Many researchers have a vested interest in their research results (Illustration 3.2). A study of 1000 Massachusetts scientists who had published research articles found that one-third held a patent for the product tested, were paid industry consultants, or had another form of financial stake in the research.[1] Another study found that authors of research supportive of a new artificial fat were four times more likely to have financial affiliations with the company producing the product than were authors with no such company ties.[2]

Illustration 3.2 Researchers may have a financial interest in their research.

HEALTH

Study links researchers to money

By Tim Friend
USA TODAY

— A study of 1,000 scientists who published ... major scientific journals shows at least ... in their research.

ing affiliated with a biotechnology company involved in commercializing it. Although the study looked only at Massachusetts researchers, Krimsky says the situation is comparable to other stat... cal research and ... " No one ...

Researcher Made Profit After Study

Invested in Maker of Cold Treatment

By PHILIP J. HILTS

A Cleveland researcher whose study showed that zinc lozenges might be an effective cold treatme... said yesterday that he had invest... in the company that makes the... enges before the paper announci... the results was published. Once ... paper appeared, the company'... stock soared, and he sold

| RESEARCH AND PRACTICE |

Authors' Financial Relationships With the Food and Beverage Industry and Their Published Positions on the Fat Substitute Olestra

| Jane Levine, EdD, Joan Dye Gussow, EdD, Diane Hastings, EdD, and Amy Eccher, MS

The Procter and Gamble (P&G) indigestible fat substitute olestra was approved as a food additive by the Food and Drug Administration (FDA) in 1996 with the proviso that olestra-containing products carry a warning statement about the additive's potential nega-

Objectives. This study examined the association between the authors' published positions on the safety and efficacy in assisting with weight loss of the Procter & Gamble (P&G) fat substitute olestra and their financial relationships with the food and beverage industry.

Methods. Jouranal articles about olestra, and their authors, were classified as supportive, critical, or neutral with respect to its use. Authors not known to have industry

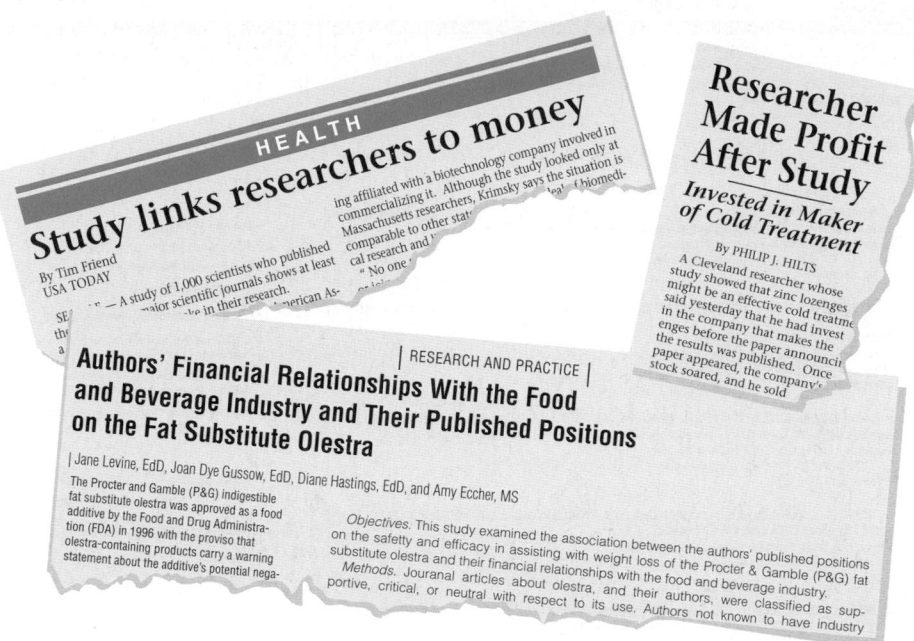

■ **Who Is Conducting the Research?** *Nutrition research is often conducted by people or companies that have a stake in the results.* Although this doesn't mean the research results are invalid, it does mean that the study design should be carefully scrutinized before it is published in a scientific journal and broadcast on the news. People and companies with vested interests in the results of studies tend to not report findings that reflect negatively on a product. Also, their promotional materials may neglect to mention studies that produced different results. For these reasons, it is important to consider whether the results of the study and other claims may be tainted by a financial motive.

■ **A Checklist for Identifying Nutrition Misinformation** Illustration 3.3 lists some common features of fraudulent information for nutrition products and services. If you find any of these characteristics in articles, advertisements, Web sites, or pamphlets, or hear them on infomercials or TV and radio interviews, beware of the product or service. People offering legitimate information are cautious and don't exaggerate nutritional benefits to health. They usually aren't selling anything but rather are trying to inform people about new findings so they can make better decisions about optimal nutrition.

■ **The Business of Nutrition News** Newspapers, TV stations, magazines, books, and other sources of information make money when the number of viewers or readers is high. To increase viewers or readers, the media attempts to present information that will pique people's interest. Nutrition breakthroughs tend to do that. The media have access to a large number of nutrition studies; more than ten thousand nutrition-related research articles are published each year (Illustration 3.4).

Illustration 3.3 If a red flag comes up as you read about a product or service, beware.

A checklist for identifying nutrition misinformation

1. Is something being sold? ____ Yes ____ No
2. Does the product or service offer a new remedy for problems that are not easily or simply solved (for example, obesity, cellulite, arthritis, poor immunity, weak muscles, low energy, hair loss, wrinkles, aging, stress)? ____ Yes ____ No
3. Are such terms as "miraculous," "magical," "secret," "detoxify," "energy restoring," "suppressed by organized medicine," "immune boosting," or "studies prove" used? ____ Yes ____ No
4. Are testimonials, before-and-after photos, or expert endorsements used? ____ Yes ____ No
5. Does the information sound too good to be true? ____ Yes ____ No
6. Is a money-back guarantee offered? ____ Yes ____ No

Illustration 3.4 The search term "nutrition" on the National Library of Medicine's PubMed research site indicates that at least 11,410 nutrition research papers were published in 2008. That's about 220 every week.

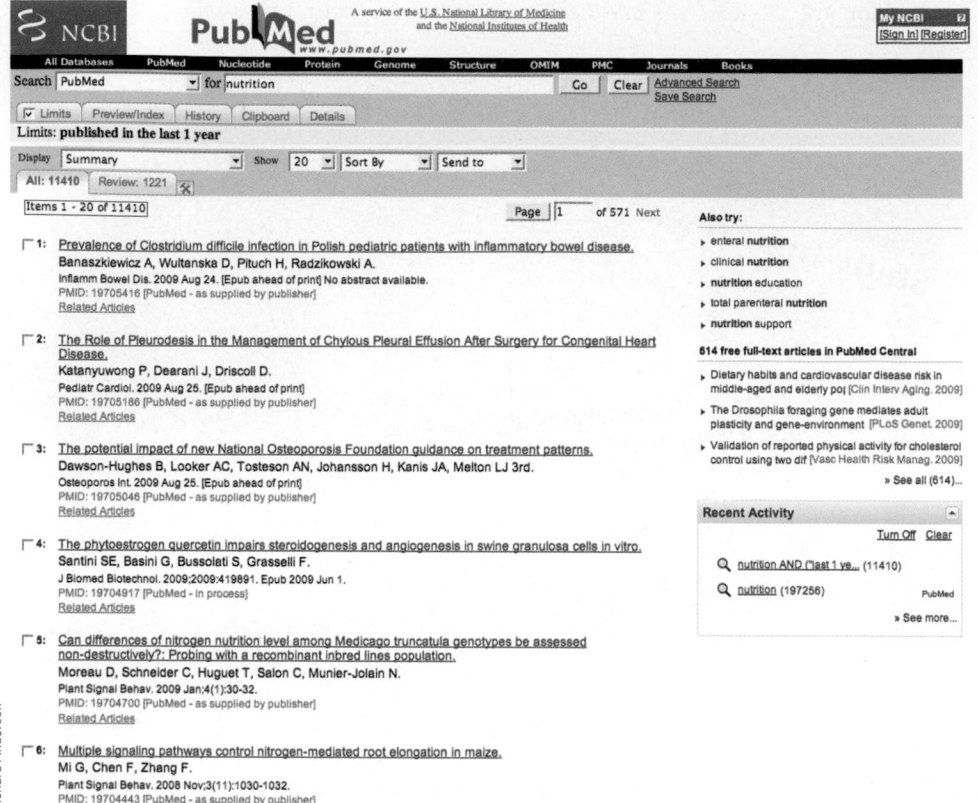

Richard Anderson

Studies reporting new results related to obesity, cancer, diabetes, heart disease, vitamins, and food safety are particularly hot topics. To keep people interested, the media may sensationalize and oversimplify nutrition-related stories. They may report on one study one day and describe another study with opposite results the next. A classic example of headlines about nutrition that confused the public is shown in Illustration 3.5. Such study-by-study coverage of nutrition news leaves consumers not knowing what to believe or what to do. A Nutrition Trends Survey by the American Dietetic Association uncovered a strong consumer preference (81% of those sampled) for learning about the latest nutrition breakthroughs only *after* they had been generally accepted by nutrition and health professionals.

Nutrition studies are complex, and the results of one study are almost never enough to prove a point. Decisions about personal nutrition should be based on accumulated evidence that is broadly supported by nutritionists and other scientists.

Motivation for Nutrition Misinformation #2: Personal Beliefs and Convictions
Numerous alternative health practitioners (such as nutripaths, irridologists, electrotherapists, scientologists, and faith healers) use nutrition remedies. Although

Illustration 3.5 How to confuse the public: a lesson delivered by the headlines about vitamin E and heart disease.

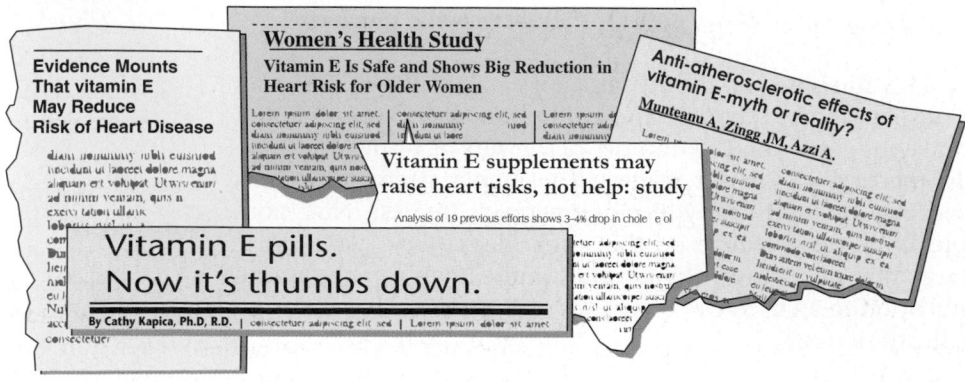

Richard Anderson

Illustration 3.6 Supplements that may be used to treat a variety of health problems. **The safety and effectiveness of many of these products are yet to be proved.**

their approaches to and philosophies about health may vary, these practitioners often have strong beliefs in the benefits of what they do. These firm beliefs may be enough to "talk" some people out of their problems—problems that might have gone away anyway.

There's no way to say whether most of the remedies used by alternative nutrition practitioners work. (Some of the unproven remedies employed are shown in Illustration 3.6.) Strong belief in these remedies, rather than proof that they work or a desire for profits, may rule these practitioners' distribution of nutrition advice and products. The nutritional cures offered are not based on science, and they have not been scientifically evaluated. Until they are evaluated, the logical conclusion is that they are neither safe nor effective.

■ **Professionals with Embedded Beliefs** Professionals who work in health care and research are not immune to the pull of deeply rooted convictions about diet and health relationships. Although they are scientists, they sometimes remain wedded to theories even after they have been disproved. The nurse or doctor who holds onto the belief that salt intake should be restricted in pregnancy and the university professor who is convinced that pesticides on foods pose no risk to health are two examples of professional sources of misinformation. The failure of such professionals to give up erroneous convictions about diet and health adds to the flow of nutrition misinformation as well as to consumer confusion and increased health risk.

How to Identify Nutrition Truths

Where does accurate nutrition information come from? There is only one way to identify sound nutrition information: put it to the test and see if it survives the dispassionate, systematic examination dictated by science.

Science is a unique and powerful explanatory system. It produces information based on facts and evidence. Our understanding of nutrition is based on scientifically determined facts and evidence obtained from laboratory, animal, and human studies. These studies provide information that qualifies for use when developing public policies about nutrition and health (Illustration 3.7). Science delivers information eligible for inclusion in textbooks about nutrition. (Many faculty members review the information in nutrition textbooks for scientific accuracy.)

The true method of knowledge is experiment.
—WILLIAM BLAKE, 1788

Microwaves and Plastic   You've got mail: DO NOT microwave foods in plastic! Hot plastic releases highly toxic dioxins that can cause cancer …

No source for the facts stated is given in the e-mail. Cyndi's and Scott's responses:

Who gets the "thumbs up?"
Answers on next page

Cyndi:
I'll check it out at the WebMD.com

Scott:
I was afraid of that. Now I know microwaving food in plastic containers is dangerous.

Illustration 3.7   This evidence-based tool for assessing poor nutritional health in seniors was developed as part of the Nutrition Screening Initiative by the American Academy of Family Physicians, the American Dietetics Association, and the National Council on the Aging.

The Warning Signs of poor nutritional health are often overlooked. Use this checklist to find out if you or someone you know is at nutritional risk.

DETERMINE YOUR NUTRITIONAL HEALTH

Read the statements below. Circle the number in the yes column for those that apply to you or someone you know. For each yes answer, score the number in the box. Total your nutritional score.

	YES
I have an illness or condition that made me change the kind and/or amount of food I eat.	2
I eat fewer than two meals per day.	3
I eat few fruits or vegetables, or milk products.	2
I have three or more drinks of beer, liquor, or wine almost every day.	2
I have tooth or mouth problems that make it hard for me to eat.	2
I don't always have enough money to buy the food I need.	4
I eat alone most of the time.	1
I take three or more different prescribed or over-the-counter drugs a day.	1
Without wanting to, I have lost or gained 10 pounds in the last 6 months.	2
I am not always physically able to shop, cook and/or feed myself.	2
TOTAL	

Total Your Nutritional Score. If it's . . .

0–2   **Good!** Recheck your nutritional score in 6 months.

3–5   **You are at moderate nutritional risk.** See what can be done to improve your eating habits and lifestyle. Your office on aging, senior nutrition program, senior citizens center, or health department can help. Recheck your nutritional score in 3 months.

6 or more   **You are at high nutritional risk.** Bring this checklist the next time you see your doctor, dietitian, or other qualified health or social service professional. Talk with them about any problems you may have. Ask for help to improve your nutritional health.

These materials developed and distributed by the Nutrition Screening Initiative, a project of:

AMERICAN ACADEMY OF FAMILY PHYSICIANS   THE AMERICAN DIETETIC ASSOCIATION   NATIONAL COUNCIL OF THE AGING

Remember that warning signs suggest risk, but do not represent diagnosis of any condition.

Table 3.1

Reliable sources of nutrition information[a]

Source of Nutrition Information	Examples
Nonprofit, professional health organizations	American Heart Association American Cancer Society American Dietetic Association American Diabetes Association
Scientific organizations	National Academy of Sciences American Society for Clinical Nutrition American Society for Nutrition
Government publications: nutrition, diet, and health reports	National Institutes of Health Surgeon General Food and Drug Administration Centers for Disease Control U.S. Department of Agriculture
Registered dietitians	Hospitals Public health departments Extension service Universities
Nutrition textbooks	College and university nutrition courses Nutrition faculty of accredited universities

[a]See Appendix B for more details about reliable sources of nutrition information.

Sources of Reliable Nutrition Information

Reliable sources of information about nutrition meet the standards of proof required by science. They report decisions about nutrition and health relationships that are based on multiple studies and achieve "scientific consensus." These decisions represent the majority opinion of scientists who are knowledgeable about a particular nutrition topic.

Nutrition recommendations that are made to the public, such as the Dietary Guidelines for Americans and the Dietary Reference Intakes, are based on the consensus of scientific opinion. The information is made available to the public not to sell a product or to pass on an ideology, but to inform consumers honestly about nutrition and to help people use the information to maintain or improve their health. Organizations and individuals offering reliable nutrition information are listed in Table 3.1.

ANSWERS TO
Reality Check

Microwaves and Plastic Good thinking, Cyndi. Information gained from WebMD.com indicates that plastic containers may contain bisphenol A, a softening compound that makes plastic flexible. Heat increases its release from plastic. Preliminary evidence indicates that regular intake of bisphenol A may be related to an increased risk of heart disease, diabetes, and reproductive problems. Canada has banned the substance, but the FDA is awaiting the arrival of stronger evidence. If concerned, only use glass and other containers marked "microwave safe" and don't use plastic wrap in the microwave.[4,5]

Cyndi: 👍

Scott: 👎

Nutrition Information on the Web   Web-based search engines are transforming the way we get information about foods, diets, supplements, and health. Eight in ten Internet users, including scientists and consumers, look for health and nutrition information online.[3] As is the case for many other media sources, the accuracy of nutrition information made available on the Web varies considerably. In general, the most reliable sites are those developed by government health agencies (Web addresses that end in or include .gov), educational institutions (.edu), and professional health and science organizations (.org).[6] Specific Web sites that offer accurate nutrition information you can count on are given on page 3-19 under "Media Menu."

Who Are Qualified Nutrition Professionals?   Many people refer to themselves as "nutritionists," but only some of them are qualified based on education and experience. These individuals are registered, licensed, or certified dietitians or nutritionists who meet qualifications established by national and state regulations. They have demonstrated a mastery of knowledge about the science of nutrition and appropriate clinical practices. Table 3.2 below describes the qualifications of those who legitimately use the title *dietitian* or *nutritionist*. The specific titles vary somewhat depending on state regulations and laws governing the practice of nutrition and dietetics.

It is difficult to make sound decisions about every nutrition question or issue that arises. When trying to make sense out of the information you read or hear, don't hesitate to get help. Visit the Web sites at the end of this unit, check out the index of this book for the relevant topic, or call your local health department or area Food and Drug Administration office. The more solid your information, the better your decisions will be.

The Methods of Science

"Supposing is good. Finding out is better."

—MARK TWAIN

Does the term *scientific method* conjure up an image of beady-eyed, white-coated scientists working diligently in windowless basement laboratories? Although we tend to assume that scientists are busy advancing our knowledge of nutrition and many other fields, for most people the methods of science are a mystery. They should not be (Illustration 3.8). Consumers use a lot of nutrition information. To make sound judgments about nutrition and health, people need to know how to distinguish results produced by scientific studies from those generated by personal opinion, product promotions, and bogus studies.

Table 3.2	
Who's who in nutrition and dietetics: Laws and regulations governing practice.[7,8]	
Title	**Qualifications**
Registered dietitian	Individual who has acquired the knowledge and skills necessary to pass a national registration examination and participates in continuing professional education. Qualification is conferred by the Commission of Dietetic Registration.
Licensed dietitian/nutritionist	Individual (usually a registered dietitian) who is qualified to practice nutrition counseling based on education and experience. Nonlicensed individuals who practice nutrition counseling can be prosecuted for practicing without a license. Licenses are conferred by state law in 46 states.
Certified dietitian/nutritionist	Individual who meets certain educational and experience qualifications. Unqualified individuals may practice nutrition counseling but have to call themselves something else (such as "nutritional counselor"). Certification is established by state regulation.

There are many established methods of science. The specific methods employed vary from study to study depending on the type of research conducted. Nevertheless, all types of scientific studies have one feature in common: They are painstakingly planned. Planning is the most important, and often the most time-consuming, part of the entire research process.

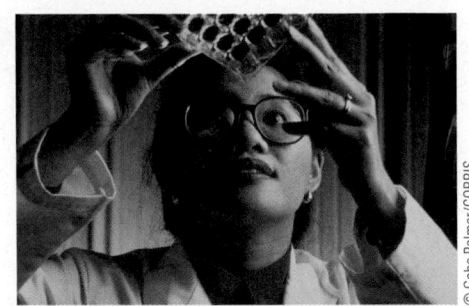

Illustration 3.8 The scientific method is not mysterious at all, but a carefully planned process for answering a specific question.

Developing the Plan

The first part of the planning process entails clearly stating the question to be addressed—and, one hopes, answered—by the research. For the purposes of illustration, let's say our research will address the effects of vitamin *X* supplements on hair loss (Table 3.3 and Illustration 3.9). The idea for the research came from a study that hinted that users of vitamin *X* supplements had increased hair loss. In that study, adults were given 25,000 international units (IU) of vitamin *X* for three months to test its safety. Although the study found that vitamin *X* had no adverse effects on health and offered some benefits, several subjects who had received the supplements complained of hair loss. Accordingly, our study poses the question, "Do vitamin *X* supplements increase hair loss?"

The Hypothesis: Making the Question Testable

The question is then transformed into an explicit hypothesis that can be proved or disproved by the research. (**Hypothesis** and other research terms used in this

Illustration 3.9 Do high doses of vitamin *X* increase hair loss?

Table 3.3

Overview of a human nutrition research study: Vitamin *X* supplements and hair loss[a]

A. **Pose a clear question:** "Do vitamin *X* supplements increase hair loss?"

B. **State the hypothesis to be tested:** "Vitamin *X* supplements of 25,000 IU per day taken for three months increase hair loss in healthy adults."

C. **Design the research:**

   1. *What type of research design should be used?* "In this study, a clinical trial will be used. The supplement and placebo will be allocated by a double-blind procedure."

   2. *Who should the research subjects be?* "The study will exclude subjects who may be losing or gaining hair due to balding, hair treatments, medications, or the current use of vitamin *X* supplements."

   3. *How many subjects are needed in the study?* "The required sample size is calculated to be 20 experimental and 20 control subjects."

   4. *What information needs to be collected?* "Information on hair loss, conditions occurring that might affect hair loss, and the use of supplements and placebos will be collected."

   5. *What are accurate ways to collect the needed information?* "The study will use the 'measure the hairs in a square inch of scalp' technique."

   6. *What statistical tests should be used to analyze the results?* "Appropriate tests identified."

D. **Obtain approval for the study from the committee on the use of humans in research:** "Approval obtained."

E. **Implement the study design:** "Implemented."

F. **Evaluate the findings:** "Subjects receiving the 25,000 IU of vitamin *X* for three months lost significantly more hair than subjects receiving the placebo. Hypothesized relationship found to be true."

G. **Submit paper on the research for publication in a scientific journal or other document.**

[a]This is a fictitious study used for illustrative purposes only.

Table 3.4

A short glossary of research terms

Association	The finding that one condition is correlated with, or related to another condition, such as a disease or disorder. For example, diets low in vegetables are associated with breast cancer. Associations do not prove that one condition (such as a diet low in vegetables) causes an event (such as breast cancer). They indicate that a statistically significant relationship between a condition and an event exists.
Cause and effect	A finding that demonstrates that a condition causes a particular event. For example, vitamin C deficiency causes the deficiency disease scurvy.
Clinical trial	A study design in which one group of randomly assigned subjects (or subjects selected by the "luck of the draw") receives an active treatment and another group receives an inactive treatment, or "sugar pill," called the placebo.
Control group	Subjects in a study who do not receive the active treatment or who do not have the condition under investigation. Control periods, or times when subjects are not receiving the treatment, are sometimes used instead of a control group.
Double blind	A study in which neither the subjects participating in the research nor the scientists performing the research know which subjects are receiving the treatment and which are getting the placebo. Both subjects and investigators are "blind" to the treatment administered.
Epidemiological studies	Research that seeks to identify conditions related to particular events within a population. This type of research does not identify cause-and-effect relationships. For example, much of the information known about diet and cancer is based on epidemiological studies that have found that diets low in vegetables and fruits are associated with the development of heart disease.
Experimental group	Subjects in a study who receive the treatment being tested or have the condition that is being investigated.
Hypothesis	A statement made prior to initiating a study of the relationship sought to be tested by the research.
Meta-analysis	An analysis of data from multiple studies. Results are based on larger samples than the individual studies and are therefore more reliable. Differences in methods and subjects among the studies may bias the results of meta-analyses.
Peer review	Evaluation of the scientific merit of research or scientific reports by experts in the area under review. Studies published in scientific journals have gone through peer review prior to being accepted for publication.
Placebo	A "sugar pill," an imitation treatment given to subjects in research.
Placebo effect	Changes in health or perceived health that result from expectations that a "treatment" will produce an effect on health.
Statistically significant	Research findings that likely represent a true or actual result and not one due to chance.

unit are defined in Table 3.4.) The results of the research must provide a true or false response to the hypothesis and not an explanatory sentence or paragraph. So, in this example, "vitamin X increases hair loss" wouldn't do as a hypothesis because it leaves too many questions about the relationship unanswered. The effect of vitamin X on hair loss may depend on the amount of vitamin X given; how long it is taken; or on the research subjects' health, age, and other characteristics. The hypothesis "vitamin X supplements of 25,000 IU per day taken for three months by healthy adults increase hair loss" is concrete enough to be addressed by research.

The Research Design: Gathering the Right Information

Poor research design, or the lack of a solid plan on how the research will be conducted, is often a weak link that renders studies useless. It's the "oops, we forgot to get this critical piece of information!" or the "how did all the measurements come out wrong?" at the end of a study that can ruin months or even years of work. Each step in the research process must be thoroughly planned. In research, there are no miracles. If something can go wrong because of incomplete planning, it probably will.

Research designs are often based on the answers to the following questions:

- What type of research design should be used?

- Who should the research subjects be?

- How many subjects are needed in the study?

- What information needs to be collected?

- What are accurate ways to collect the needed information?

- What statistical tests should be used to analyze the findings?

What Type of Research Design Should Be Used?   Several different research designs can be used to test hypotheses. We could use an **epidemiological study** design to determine if hair loss is more common among people who take vitamin X supplements than among people who do not. Researchers commonly use this type of study to identify conditions that are related to specific health events in humans. To provide preliminary evidence, we could use animal studies that determine hair loss in supplemented and unsupplemented animals. Or we could use another design, such as a **clinical trial**. Since this design would work well for the proposed hypothesis about vitamin X supplements, we'll follow the rules for conducting a clinical trial in this example. (You can see a map of the research design used for this hypothetical study in Illustration 3.10.)

The purpose of clinical trials is to test the effects of a treatment or intervention on a specific biological event. In our example, we will test the effect of vitamin X supplements on hair loss.

■ **The Experimental and Control Groups**   Clinical trials, as well as other research studies that address questions about nutrition, require an **experimental group** (the group of subjects who receive vitamin X in this example) and a **control group** (the comparison group that receives a **placebo** and not vitamin X). Never trust the results of a study that didn't employ both, and here's why. How do we know the effect of a certain treatment isn't due to something *other* than the treatment? What if we gave a group of adults vitamin X supplements and they lost hair? Does this mean the vitamin X caused the hair loss? Could it be that the subjects lost no more hair during the treatment period than they would have lost without the vitamin X? We can't know whether vitamin X supplements produce hair loss if we don't know how much hair is lost without vitamin X.

After measuring their usual hair loss, we will randomly assign (by the "luck of the draw") people in the study to serve in either the experimental or the control group. Individuals assigned to the control group will be given pills that look, taste, and feel like the vitamin X supplements but have no effect on hair loss or gain. Neither the research staff nor the people in the study will know who is getting vitamin X and who is getting the placebo (Illustration 3.11). Scientists use this **double-blind procedure** because knowing which group is which may affect people's expectations and change the results.

■ **The "Placebo Effect"**   The **placebo effect** can cause a good deal of confusion in research. That's because people tend to have expectations about what a treatment will do, and those expectations can influence what happens.

Illustration 3.10 Map of the design of the vitamin *X* supplement study.

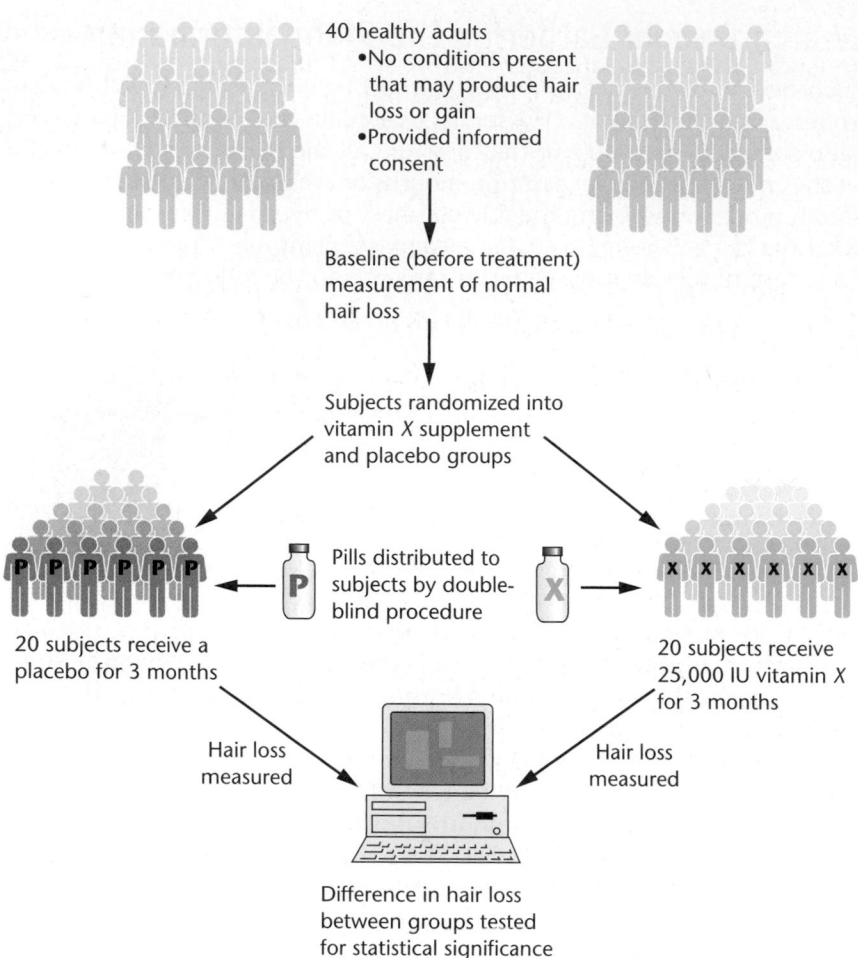

40 healthy adults
- No conditions present that may produce hair loss or gain
- Provided informed consent

Baseline (before treatment) measurement of normal hair loss

Subjects randomized into vitamin *X* supplement and placebo groups

Pills distributed to subjects by double-blind procedure

20 subjects receive a placebo for 3 months

20 subjects receive 25,000 IU vitamin *X* for 3 months

Hair loss measured

Hair loss measured

Difference in hair loss between groups tested for statistical significance

Illustration 3.11 Which is the vitamin *X* supplement? Neither the subject nor the investigator should know which is vitamin *X* and which is the placebo.

A good example of the placebo effect occurred in a study that tested the effectiveness of a medication meant to reduce binge eating among people with bulimia.[9] After the usual number of binge-eating episodes was determined, 22 women with bulimia were given either the medication or a placebo in a double-blind fashion. The number of binge-eating episodes was then reassessed. At the end of the study period, a whopping 78% reduction in binge-eating episodes was found among women taking the medication. But binge-eating episodes dropped by 70% in the control group. Was the medication effective? No. Was the expectation that a medication would help effective? Yes, at least for the time covered by the study. Due to the placebo effect, a reduction of greater than 90% in episodes of binge eating would have been needed to conclude that the medication had a real effect on binge eating.

Who Should the Research Subjects Be?

Although expedient, it is not enough to say, "Well, let's use ten faculty members in the study." The type and number of subjects employed by the research are important considerations.

The type of subjects involved in research is important, because we need to exclude people who have conditions that might produce the problem the research is examining. For example, hair loss may result from periodic balding, a bad perm, the use of mousse or hair spray, or illnesses or medications. If we include people with these conditions and circumstances in the study, it will be difficult to determine whether hair loss is due to vitamin *X* or something else. In addition, it is important to exclude from the study people who already take vitamin *X* supplements or are bald. An inappropriate or biased selection of subjects could make the results useless or, in the worst case, make the study's results come out in a predetermined way.

How Many Subjects Are Needed in the Study?   The number of subjects needed for a study is a mathematical question, and we won't go into the formulas here. The number of subjects is based upon the number needed to separate true differences between the experimental and control groups from differences that are due to chance (Illustration 3.12). Suppose that hair loss normally varies from around 50 to 150 strands per day.[10,11] We'll need to include enough subjects in the study to make sure differences in hair loss between the groups are not due to normal fluctuations in hair loss. This point is important because studies employing too few subjects lead to inconclusive results and a great deal of controversy about nutrition. One or a few subjects are never enough to prove a point about nutrition.

Let's assume that the mathematical formulas applied to the vitamin *X* study show that 20 people are needed in both the experimental group and the control group.

What Information Needs to Be Collected?   Information recorded by researchers must represent an evenhanded approach to discovering the facts. We must identify not only the findings that may support the hypothesis, but *those that may refute it*. In order to know if vitamin *X* supplements increase hair loss, we need to know how much hair is lost, whether the subjects faithfully took the supplement, and whether something came up during the study (such as a bad perm) that might alter hair loss. The presence of such conditions should be determined by methods known to provide accurate results.

© Herbert Spichtinger/Zefa/Corbis

Photo Disc

Illustration 3.12   How many subjects are needed in the study? It's important to get the number right.

What Are Accurate Ways to Collect the Needed Information? "Garbage in—garbage out." If the information obtained on and from subjects is inaccurate, then the study is worthless. To avoid this problem, good research employs methods of collecting information known to produce accurate results.

Let's consider the problem of measuring hair loss. How can we do that accurately? First, we look for a method that has already been demonstrated by research to be accurate. We find one called "the 60 second hair count"[11] and employ it.

What Statistical Tests Should Be Used to Analyze the Findings? Before collecting the first piece of information, we'll select appropriate tests for identifying **statistically significant results** of the research. Statistical tests are used to identify significant differences between the findings from the experimental and control groups. Without such tests, it is often very hard to decide what the findings mean. What if the study shows the group that took vitamin X lost an average of 5% more hair than the control group? Is that 5% difference due to the vitamin X or to something else, such as chance fluctuations in normal hair loss? Well-chosen statistical tests will tell us whether the differences between groups are in all probability real or due to chance occurrences or coincidence.

Obtaining Approval to Study Human Subjects

An important step in the planning process is applying for approval to conduct the proposed research on human subjects. Universities and other institutions that conduct research have formal committees, called institutional review boards, that scrutinize plans to make sure proposed studies follow the rules governing research on human subjects. For this study, we will have to show to the committee's satisfaction that the level of vitamin X employed is safe. As a part of this process, the committee usually requires that human subjects consent, in writing, to participate in the study.

Implementing the Study

With the design in place and the appropriate approvals obtained, it is time to implement the study. Assume that, due to the study's solid design and importance, we have been awarded a grant to fund the research. We can now recruit subjects for the study and enroll them if they are eligible and consent to participate. Measurements of hair loss are performed according to schedule, use of the pills by subjects is monitored, results are checked for errors and entered into a computerized data file, and statistical tests are applied. This process usually ends with computer printouts that exhibit the findings.

Making Sense of the Results

Let's say subjects who received the supplement lost 30% more hair than the control subjects did. Having applied the appropriate statistical test, we find that 30% is a highly significant difference. Does that mean that vitamin X supplements *cause* hair loss? *Cause* is a strong word in research terms. It implies that a **cause-and-effect relationship** (that the vitamin X supplements caused the hair loss) exists. But many factors can contribute to hair loss, and since many causes may be unknown or not measured by the study, it is difficult to conclude with absolute certainty that vitamin X by itself caused the hair loss. Assume 4 of the 20 subjects in the experimental group who took the vitamin X supplement lost less hair than usual. The vitamin X supplements *didn't* cause them to lose hair.

We could conclude from this research (if, as a reminder, there was such a thing as vitamin X) that vitamin X supplements, given to healthy adults at a dose of 25,000 IU per day for three months, are strongly *associated* with hair loss. The term *associated* as used here means that the vitamin X supplements were strongly *related* to hair loss but may not have *caused* the hair loss. The hypothesized relationship between supplemental vitamin X and hair loss would be found to be true. Although the research strongly indicates a cause-and-effect relationship, additional studies would be needed to prove the relationship exists in other groups of people and at different doses of vitamin X.

After determining the results, we can write a paper describing the research and its conclusions and submit it for publication in a scientific journal. If judged to be acceptable after review by other scientists, the paper is published.

Complete coverage of how nutrition questions are addressed through research would take a three-course sequence. Nevertheless, it is hoped that the information presented here makes it easier to judge the likely accuracy of reports about nutrition studies in the popular media. There is a lot more to a "Vitamin X Supplements Linked to Hair Loss" story than meets the eye.

Science and Personal Decisions about Nutrition   Science is based on facts and evidence. The grounding ethic of scientists is that facts and evidence are more sacred than any other consideration. These characteristics of science and scientists are strong assets for the job of identifying truths.

Although imperfect because they are undertaken by humans in an environment of multiple constraints, only scientific studies produce information about nutrition you can count on. Evidence is the single best ingredient for decision making about nutrition and your health.

NUTRITION

Up Close

Checking Out a Fat-Loss Product

Focal Point: How to make an informed decision about the truthfulness of an advertisement for a fat-loss product.

Read the accompanying advertisement for a fat-loss product. Then, check out the information using the checklist for identifying nutrition misinformation.

Obesity Research Ce...
P.O. Box 281752,
Miami, FL

Lose Over 2 Inches In 3 Weeks

A Scientifically Proven Fat Reduction Cream That Actually Works!

A scientific, double-blind, placebo-controlled research study of both sexes demonstrated that LIPID MELT fat reduction cream reduces inches from the thigh area. Subjects lost an average of 2 inches from each thigh in only 3 weeks!

Subjects from a preliminary pilot study reported that LIPID MELT cream reduced inches from their abdomen. Both studies reported no skin rashes, discomfort or sensitivity. Formulated with patented liposome technology, LIPID MELT contains no drugs, only natural active ingredients. Initially sold only in salons and spas.

Six-week supply: $25 + $4.95 s/h. We guarantee that you will lose inches or your money will be refunded.

Obesity Research

1. Is something being sold?

   Yes   No

2. Is a new remedy for problems that are not easily or simply solved being offered?

   Yes   No

3. Are terms such as *miraculous, magical, secret, detoxify, energy restoring, suppressed by organized medicine, immune boosting,* or *studies prove* used?

   Yes   No

4. Are testimonials, before-and-after photos, or expert endorsements used?

   Yes   No

5. Does the information sound too good to be true?

   Yes   No

6. Is a money-back guarantee offered?

   Yes   No

Optional: Repeat the activity using a nutrition-related advertisement from the Internet.

FEEDBACK (answers to these questions) can be found at the end of Unit 3.

[Key Terms

association, page 3-12
cause-and-effect relationship, page 3-12
clinical trial, page 3-12
control group, page 3-12
double-blind procedure, page 3-12

epidemiological study, page 3-12
experimental group, page 3-12
hypothesis, page 3-12
meta-analysis, page 3-12
peer review, page 3-12

placebo, page 3-12
placebo effect, page 3-12
statistically significant results, page 3-12

Review Questions

1. By law, claims about nutrition that are presented on product labels and packaging must be scientifically accurate. ☐ ☐

2. The leading motivation underlying the presentation of nutrition misinformation is the profit motive. ☐ ☐

3. The Food and Drug Administration (FDA) has the authority to remove advertisements making false claims about nutrition from the airwaves and the Internet. ☐ ☐

4. Individuals can help combat the proliferation of bogus nutrition advertisements and products by notifying the Federal Trade Commission (FTC) of their existence. ☐ ☐

5. Nutrition products and services that offer a "money back guarantee" are most likely to work as advertised. ☐ ☐

6. Nutrition information offered by a scientist or a medical doctor can be counted on as being accurate. ☐ ☐

7. The term *registered dietitian* means that individuals with that title have acquired the knowledge and skills necessary to pass a national examination on nutrition and participate in continuing education. ☐ ☐

8. An advantage of clinical trials over other study designs is that they consistently identify cause and effect relationships. ☐ ☐

9. Research results may be unreliable if too few subjects are used in studies. ☐ ☐

10. Research articles published in scientific journals undergo peer review, a process by which experts in the areas under investigation evaluate articles prior to their acceptance for publication. ☐ ☐

Media Menu

pubmedcentral.nih.gov
PubMed Central is the National Institutes of Health's source of free, full-text journal articles for the biomedical and life sciences.

www.ncbi.nlm.nih.gov/pubmed
Home of "PubMed," the National Library of Medicine's site for identifying scientific journal articles on specific nutrition and other science topics. Abstracts of articles are usually available, free full text of articles may not be.

www.ftc.gov/bcp/consumer.shtm
Use the federal Trade Commission's Consumer Information site to report on fraudulent nutrition advertisements. The second top bar on the opening page contains the "File a Complaint" choice.

www.fda.gov
The home page of the Food and Drug Administration provides a listing in the right-hand column called "Report a Problem." Use it to report adverse reactions to foods or dietary supplements and false and misleading label information.

WebMD.com
A long-time leading and reliable health information site, WebMD offers an interactive diet and nutrition center, resources on healthy eating, and a food and fitness planner. Search nutrition and health terms of interest and find scientifically based articles, video presentations by experts, and other resources at this site.

Medscape.com
Offered through WebMD, Medscape supplies interested readers with health, nutrition, and medical updates on a regular basis via email subscription. Registration is required and free.

RealAge.com
EverydayHealth.com
These popular health sites offer practical advice on nutrition, beauty, stress, weight loss, and other topics. Although the site may contain behavioral tips that some may find helpful, there are lots of advertisements, expert advice pages by authors of less-than scientifically based diet and nutrition books, and misleading and hyped-up nutrition reports.

www.hc-sc.ga.ca/dhp-mps/medeff/report-declaration/index-eng.php
This is Health Canada's site for adverse health product and drug reporting. You can also access reports of adverse effects from this page.

scholar.google.com
Google Scholar provides a simple way to search for peer-reviewed papers and articles from academic publishers, professional societies, and other scholarly organizations.

site:gov, site:nih,gov, site:edu, and site:org
Using these terms in Google searches limits results to those from the government (gov), the National Institute of Health (nih), academic institutions (edu), or nonprofit organizations (org).

www.senseaboutscience.org.uk
Developed to respond to the misrepresentation of science and scientific evidence on issues that matter to society, this site covers and updates topics ranging from scares about plastic bottles to fluoride. The site helps people query the status of science reported on the internet.

www.eatright.org
The American Dietetic Association's Web site is an excellent source of information about diet, health, and qualified dietetics personnel in your area.

www.nlm.nih.gov/medlineplus
Maintained by the National Library of Medicine, this site is a sure bet for authoritative information on food and nutrition topics, ongoing clinical trials, and other information. It also offers a medical dictionary and glossary.

www.quackwatch.com
Uncovers health frauds, including nutrition scams, dubious allergy treatments, and suspect alternative medicines.

Notes

1. Who's minding the lab? Tufts University Health and Nutrition Letter 1997, May:4.

2. Levine J, Gussow JD, et al. Authors' financial relationships with the food and beverage industry and their published positions on the fat substitute Olestra. Am J Pub Health 2003;93:664–9.

3. Steinbrook R. Searching for the right search—reaching the medical literature. N Engl J Med 2006;354:4–7

4. Lang IA, et al. Association of urinary bisphenol a concentration with medical disorders and laboratory abnormalities in adults. JAMA 2008; DOI:10.1001/jama.300.11.1303. Available at: http://jama.ama-assn.org.

5. Hill M. Bisphenol A: 9 questions and answers, available from: children, webmd.com/features/bisphenols-a-9questions-and-answers, accessed 1/20/09.

6. Al-Ubaydli M. Using search engines to find online medical information. PLoS Med 2(9):e228, 10/3/05.

7. Busey JC et al. Telehealth—opportunities and pitfalls. J Am Diet Assoc 2008;108: 1296–1302.

8. Pathways to excellence. Chicago: The American Dietetic Association; 1997.

9. Alger SA, Schwalberg MD, Bigaouette JM, Michalek AV, Howard LJ. Effect of a tricyclic antidepressant and opiate antagonist on binge-eating behavior in normal weight, bulimic, and obese, binge-eating subjects. Am J Clin Nutr 1991;53:865–71.

10. Shapiro J. Hair loss in women. N Engl J Med 2008;357:1620–30.

11. Wasko, CA et al. The sixty-second hair count for assessing hair loss in men. Arch Dermatol. 2008;144:759–762.

Answers to Review Questions

1. True, see page 3-2
2. True, see page 3-3
3. False, see page 3-3
4. True, see page 3-4
5. False, see page 3-5
6. False, see pages 3-6, 3-7
7. True, see page 3-10
8. False, see page 3-12, 3-17
9. True, see page 3-15
10. True, see pages 3-12, 3-17

NUTRITION | Up Close

Checking Out a Fat-Loss Product

Feedback for Unit 3

The product advertised earns five "yes" responses, making it highly unlikely that the product works. One or two "yes" responses provide a strong clue that advertised products may not work.

Understanding Food and Nutrition Labels

NUTRITION SCOREBOARD

	TRUE	FALSE
1 Nutrition labels are required on all foods and dietary supplements sold in the United States.		
2 Nutrition labeling rules allow health claims to be made on the packages of certain food products.		
3 Nutrition labels contain all of the information people need to make healthy decisions about what to eat.		

Key Concepts and Facts

- People have a right to know what is in the food they buy.

- The purpose of nutrition labeling is to give people information about the composition of food products so they can make informed food-purchasing decisions.

- Nutrition labeling regulations cover the type of foods that must be labeled and set the standards for the content and format of labels.

Nutrition Labeling

Illustration 4.1 Nutrition information for produce and meats can be presented on posters.

Seafood — Nutrition Facts

Cooked (by moist or dry heat with no added ingredients), edible weight portion. Percent Daily Values (%DV) are based on a 2,000 calorie diet.

Seafood Serving Size (84 g/3 oz)	Calories	Calories from Fat	Total Fat g / %DV	Saturated Fat g / %DV	Cholesterol mg / %DV	Sodium mg / %DV	Potassium mg / %DV	Total Carbohydrate g / %DV	Protein g	Vitamin A %DV	Vitamin C %DV	Calcium %DV	Iron %DV
Blue Crab	100	10	1 / 2	0 / 0	95 / 32	330 / 14	300 / 9	0 / 0	20g	0%	4%	10%	4%
Catfish	130	60	6 / 9	2 / 10	50 / 17	40 / 2	230 / 7	0 / 0	17g	0%	0%	0%	0%
Clams, about 12 small	110	15	1.5 / 2	0 / 0	80 / 27	95 / 4	470 / 13	6 / 2	17g	10%	0%	8%	30%
Cod	90	5	1 / 2	0 / 0	50 / 17	65 / 3	460 / 13	0 / 0	20g	0%	2%	2%	2%
Flounder/Sole	100	15	1.5 / 2	0 / 0	55 / 18	100 / 4	390 / 11	0 / 0	19g	0%	0%	2%	0%
Haddock	100	10	1 / 2	0 / 0	70 / 23	85 / 4	340 / 10	0 / 0	21g	2%	0%	2%	6%
Halibut	120	15	2 / 3	0 / 0	40 / 13	60 / 3	500 / 14	0 / 0	23g	4%	0%	2%	6%
Lobster	80	0	0.5 / 1	0 / 0	60 / 20	320 / 13	300 / 9	0 / 0	17g	2%	0%	6%	2%
Ocean Perch	110	20	2 / 3	0.5 / 3	45 / 15	95 / 4	290 / 8	0 / 0	21g	0%	2%	10%	4%
Orange Roughy	80	5	1 / 2	0 / 0	20 / 7	70 / 3	340 / 10	0 / 0	16g	2%	0%	4%	2%
Oysters, about 12 medium	100	35	4 / 6	1 / 5	80 / 27	300 / 13	220 / 6	6 / 2	10g	0%	6%	6%	45%
Pollock	90	10	1 / 2	0 / 0	80 / 27	110 / 5	370 / 11	0 / 0	20g	2%	0%	0%	2%
Rainbow Trout	140	50	6 / 9	2 / 10	55 / 18	35 / 1	370 / 11	0 / 0	20g	4%	4%	8%	2%
Rockfish	110	15	2 / 3	0 / 0	40 / 13	70 / 3	440 / 13	0 / 0	21g	4%	0%	2%	2%
Salmon, Atlantic/Coho/Sockeye/Chinook	200	90	10 / 15	2 / 10	70 / 23	55 / 2	430 / 12	0 / 0	24g	4%	4%	2%	2%
Salmon, Chum/Pink	130	40	4 / 6	1 / 5	70 / 23	65 / 3	420 / 12	0 / 0	22g	2%	0%	0%	4%
Scallops, about 6 large or 14 small	140	10	1 / 2	0 / 0	65 / 22	310 / 13	430 / 12	5 / 2	27g	2%	0%	4%	14%
Shrimp	100	10	1.5 / 2	0 / 0	170 / 57	240 / 10	220 / 6	0 / 0	21g	4%	4%	6%	10%
Swordfish	120	50	6 / 9	1.5 / 8	40 / 13	100 / 4	310 / 9	0 / 0	16g	2%	2%	0%	6%
Tilapia	110	20	2.5 / 4	1 / 5	75 / 25	30 / 1	360 / 10	0 / 0	22g	0%	2%	0%	2%
Tuna	130	15	1.5 / 2	0 / 0	50 / 17	40 / 2	480 / 14	0 / 0	26g	2%	2%	2%	4%

Seafood provides negligible amounts of trans fat, dietary fiber, and sugars.

U.S. Food and Drug Administration
(January 1, 2008)

Misleading messages, hazy health claims, and the slippery serving sizes that characterized food labels of the past have led to a revolution in nutrition labeling. Consumers—especially those responsible for buying food for the family; people with weight concerns, food allergies, or diabetes; and the health-conscious—made it clear they wanted to end the mystery about what's in many foods. Passage of the 1990 Nutrition Labeling and Education Act by Congress indicated that their concerns had been heard. In 1993 the Food and Drug Administration published rules for nutrition labeling,[1] and implementation and revisions of the new standards have been ongoing since then.

Key Elements of Nutrition Labeling Standards

Nutrition labeling regulations for foods cover five areas:

- the "Nutrition Facts Panel"
- nutrient claims
- health claims
- structure/function claims
- qualified health claims

The Nutrition Facts Panel is required on most foods sold in grocery stores. Rules established for the other four areas must be followed when a nutrition or health claim is made for the food on the packaging.

The Nutrition Facts Panel With the exception of foods sold in very small packages or in stores like local bakeries, foods containing more than one ingredient must display a Nutrition Facts panel. Single foods, like pears, a head of cabbage, or fresh shrimp, do not have to be labeled. It is encouraged, however, that nutrition information for these foods be presented on posters in grocery stores (Illustration 4.1).

Nutrition Facts panels provide specific information about a food's caloric value, nutrient content, and ingredients. Illustration 4.2 shows a Nutrition Facts panel and provides explanations of what various components of the panel mean. The panel highlights a product's content of fat, saturated fat, *trans* fat, cholesterol, sodium, dietary fiber, two vitamins (A and C), and two minerals (calcium and iron). The food's content of these nutrients must be based on a standard serving size as defined by the Food and Drug Administration (FDA).

All labeled foods must provide the information shown on the Nutrition Facts panel in Illustration 4.2. Additional information on specific nutrients can be added to the panel on a voluntary basis (Table 4.1). However, if the package makes a claim about the food's content of a particular nutrient that is not on the "mandatory" list, then information about that nutrient must be added to the

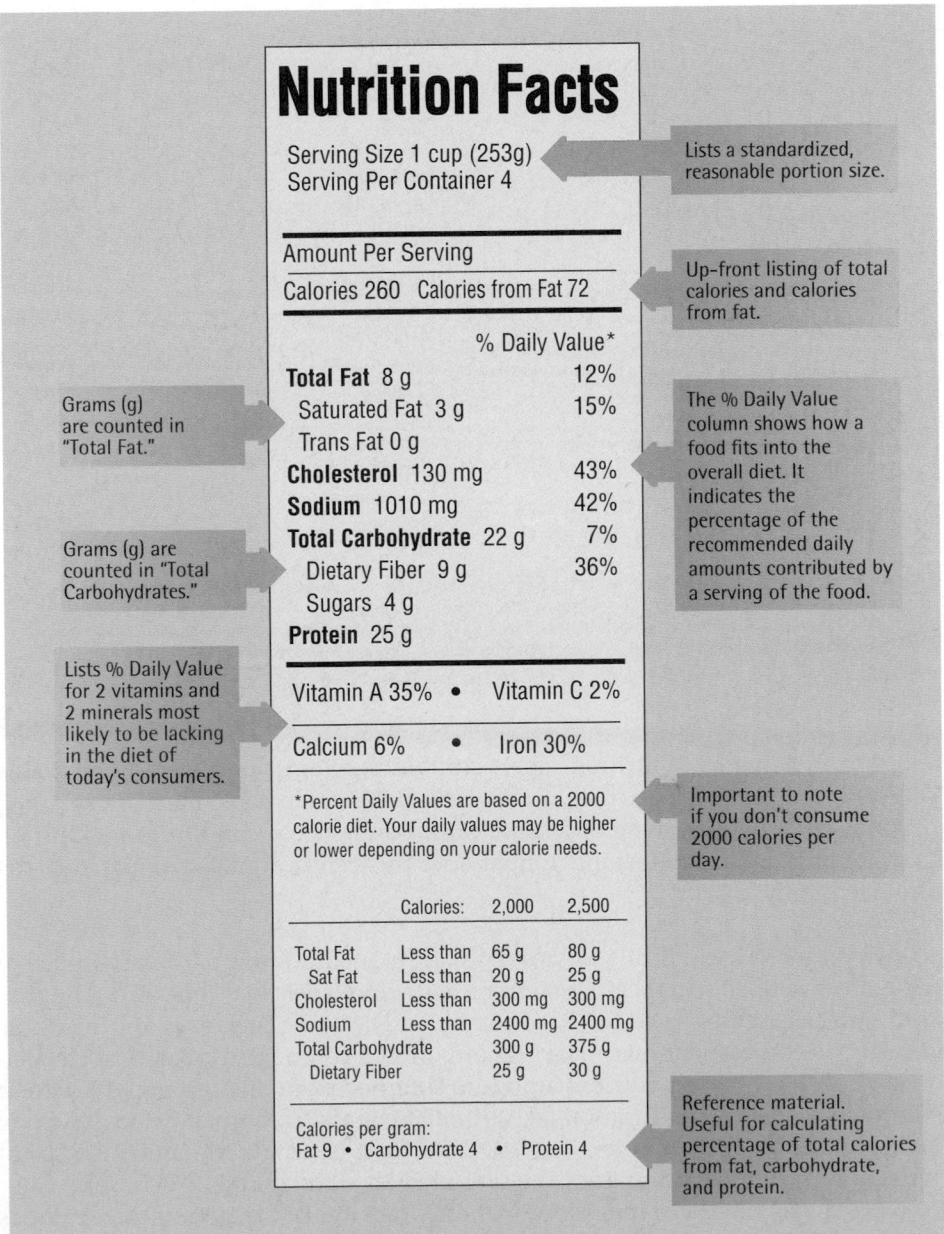

Illustration 4.2 What are you having for dinner? The Nutrition Facts panel lets you know.

Table 4.1

Mandatory and voluntary components of the Nutrition Facts panel and assigned Daily Values (DVs)

Components are listed in the order in which they must appear on the nutrition panel; unapproved components may not be listed.[a]

Mandatory	DV	Voluntary	DV
Total calories	—	Calories from saturated fat	
Calories from fat	30%	Polyunsaturated fat	
Total fat	65 g	Monounsaturated fat	
Saturated fat	20 g	Stearic acid	
Trans fat	—[b]		
Cholesterol	300 mg	Insoluble fiber	
Sodium	2400 mg		
Total carbohydrate	300 g	Other carbohydrates	
Dietary fiber	25 g	Soluble, Insoluble fiber	
Sugars	—[b]	Sugar alcohols (xylitol,	
Protein	—[b]	mannitol, sorbitol)	
Vitamin A	5000 IU	Vitamin D	400 IU
Vitamin C	60 mg	Vitamin E	30 IU
Calcium	1000 mg	Vitamin K	80 mcg
Iron	18 mg	Thiamin	1.5 mg
		Riboflavin	1.7 mg
		Niacin	20 mg
		Vitamin B6	2.0 mg
		Folate	400 mcg
		Vitamin B12	6 mcg
		Biotin	300 mcg
		Pantothenic acid	10 mg
		Phosphorus	1000 mg
		Iodine	150 mcg
		Magnesium	400 mg
		Zinc	15 mg
		Selenium	70 mcg
		Copper	2 mg
		Manganese	2 mg
		Chromium	120 mcg
		Molybdenum	75 mcg
		Chloride	3400 mg
		Potassium	3500 mg

[a]If the food package makes a claim about any of the voluntary components, or if the food is enriched or fortified with any of them, they become a mandatory part of the nutrition panel.
[b]DVs are not shown on Nutrition Facts panels for *trans* fats, sugars, or protein. *Trans* fats and sugars have not been assigned DVs, and, although there is a DV for protein, it is not listed because most people have adequate protein intakes.[2]

% Daily Value (%DV)

Daily Values are scientifically agreed-upon standards of daily intake of nutrients from the diet developed for use on nutrition labels. The "% Daily Values" listed in nutrition labels represent the percentages of the standards obtained from one serving of the food product.

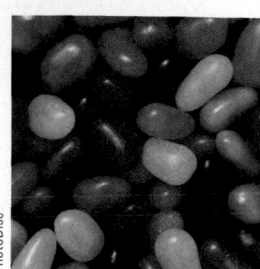
Nutrition Facts panel. Nutrition labels contain a column headed **% Daily Value (%DV)**. Figures given in this column are intended to help consumers answer such questions as "Does a serving of this macaroni and cheese contain more fat than the other brand?" and "How much fiber does this cereal provide compared to my daily need for it?"

■ **Daily Values (DVs)** Daily Values (DVs) are standard levels of dietary intake of nutrients developed specifically for use on nutrition labels (see Table 4.1). They are based on earlier editions of the Recommended Dietary Allowances and scientific consensus recommendations.[3] The %DV figures listed on labels represent the percentages of the standard nutrient amounts obtained from one serving of the food product. Standard values for total fat, saturated fat, and carbohydrate are based on a daily intake of 2000 calories. The %DV for total fat intake is based on 30% of total calories from fat (65 grams), the saturated fat standard on 10% of total calories (20 grams), and the standard for carbohydrate intake on 60% of total calories (300 grams). If, for example, a serving of a food product contains 10 grams of saturated fat, the %DV listed on the nutrition label would be 50% because the standard for saturated fat intake is 20 grams. In general, % Dietary Values of 5 or less are considered "low," and those listed as 20 or more "high."[2]

Nutrient Content   The packaging of approximately 40% of food products sold in grocery stores make one or more nutrition-related claims.[1] These claims are supposed to represent the scientific truth but they often do not (Illustration 4.3). Nutrition claims can only be trusted if they are approved by the FDA.

Statements used to label foods—such as "High fiber" and "Healthy"—must conform to standard definitions. For example, foods labeled "low fat"—a popular nutrient content claim made on food packages'—must contain 3 grams of fat or less per serving. Low-fat foods can be labeled with a "percent fat free" label, such as "98% fat free" turkey. This label means that the product contains approximately 2% fat on a weight basis. Meat products labeled "lean" must contain less than 10 grams of fat, 4.5 grams of saturated fat and *trans* fat combined, and 95 milligrams of cholesterol per serving. Some labels promote meat as lean based on the percentage of the meat's weight that consists of fat. So, a meatloaf mix that is 16% fat on a weight basis might be labeled as "lean." It would not be lean according to nutrition labeling standards. Illustration 4.4 provides an example of an appropriately and an inappropriately labeled meat.

Nutrient content claims cannot be used to make foods appear to be healthier than they really are. Foods often have to meet several standards for nutritional value before they qualify for a claim. You may see some food products inappropriately labeled with claims like "natural," "pure," or "0 calories" (see the Health Action Chart). The term "natural" is so often misused on food labels that it is now mistrusted.[5] Table 4.2 lists a variety of claims that are on food labels but that cannot be trusted because they have no standard definition or have been disapproved for use by the FDA.

Health Claims   On approval by the FDA, foods or food components with scientifically agreed-upon benefits to disease prevention can be labeled with a health claim (Table 4.3). However, the health claim must be based on the FDA's "model claim" statements. For instance, scientific consensus holds that diets high in fruits and vegetables may lower the risk of cancer, so a health claim to this effect is allowed. The FDA's model claim for labeling fruits and vegetables is "Low fat diets rich in fruits and vegetables may reduce the risk of some types of cancer, a disease associated with many factors." The FDA approves health claims only for

Illustration 4.3   This fuzzy photo (taken by the author) shows a "Ø₉ TRANS FAT" label on pineapples that don't contain trans fat in the first place.

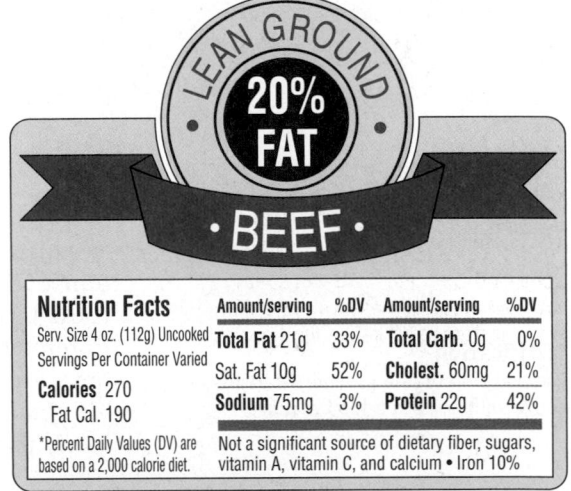

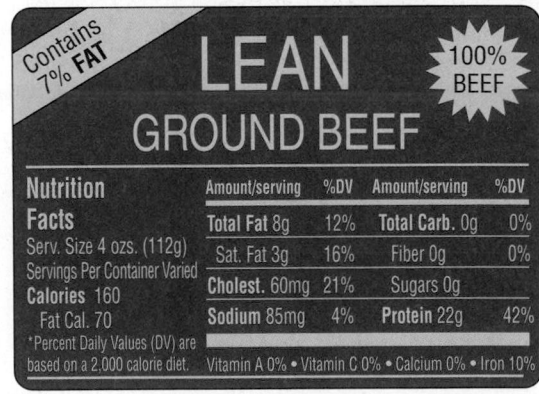

Illustration 4.4   Two examples of labels claiming "lean" ground beef. Only the 7% fat beef is actually lean. "20% fat" ground beef is far from lean, actually providing 21 grams of fat and 70% of total calories from fat per serving.

Table 4.2
Examples of claims not approved by the FDA for use on food labels
Natural
All natural
Pure
Antibiotic-free
Raised without antibiotics
Additive-free
Pesticide-free
Hormone-free
Nutritionally improved
No cholesterol (on plant foods)
Free-range
Eco-friendly
Pasture-fed

Table 4.3

FDA approved health claims[6]

1. Calcium, vitamin D, and osteoporosis

2. Dietary lipids (fats) and heart disease

3. Dietary saturated fat and cholesterol and risk of coronary heart disease

4. Dietary non-cariogenic carbohydrate sweeteners and dental caries

5. Fiber-containing grain products, fruits, vegetables and cancer

6. Folic acid and neural tube defects

7. Fruits, vegetables, and cancer

8. Fruits, vegetables and grain products that contain fiber, particularly soluble fiber, and risk of coronary hearth disease

9. Sodium and hypertension

10. Soluble fiber from certain foods and risk of coronary heart disease

11. Soy protein and risk of coronary heart disease

12. Stanols/sterols and risk of coronary heart disease

© Scott Goodwin Photography

enrichment
The replacement of thiamin, riboflavin, niacin, and iron lost when grains are refined.

fortification
The addition of one or more vitamins and/or minerals to a food product.

food products that are not high in fat, saturated fat, cholesterol, or sodium. Model health claims approved by the FDA are shown in Table 4.4.

■ **Labeling Foods as Enriched or Fortified** The vitamin and mineral content of foods can be increased by **enrichment** and **fortification**. Definitions for these terms were established more than 50 years ago. Enrichment pertains only to refined grain products, which lose vitamins and minerals when the germ and bran are removed during processing. Enrichment replaces the thiamin, riboflavin, niacin, and iron lost in the germ and bran. By law, producers of bread, cornmeal, pasta, crackers, white rice, and other products made from refined grains

Table 4.4

Examples of model health claims approved by the FDA for labels of foods that qualify based on nutrient content; model claims are often abbreviated on food labels

Food and Related Health Issues	Model Health Claim
Whole-grain foods and heart disease, certain cancers	Diets rich in whole-grain foods and other plant foods and low in fat, saturated fat, and cholesterol may reduce the risk of heart disease and certain cancers.
Sugar alcohols and tooth decay	Frequent between-meal consumption of food high in sugars and starches promotes tooth decay. The sugar alcohols in this food do not promote tooth decay.
Saturated fat, cholesterol, and heart disease	Development of heart disease depends on many factors. Eating a diet low in saturated fat and cholesterol and high in fruits, vegetables, and grain products that contain fiber may lower blood cholesterol levels and reduce your risk of heart disease.
Calcium, vitamin D and osteoporosis	Regular exercise and a healthy diet with enough calcium and vitamin D help maintain good bone health and may reduce the risk of osteoporosis later in life.
Fruits and vegetables and cancer	Low-fat diets rich in fruits and vegetables may reduce the risk of some types of cancer, a disease associated with many factors.
Folate and neural tube defects	Women who consume adequate amounts of folate daily throughout their childbearing years may reduce their risk of having a child with a brain or spinal cord defect.

Illustration 4.5 Examples of foods that are enriched (left) and fortified (right).

must use enriched flours. Beginning in 1998, federal regulations mandated that folate (a B vitamin) in the form of folic acid be added to refined grain products. This new regulation was put into effect to help reduce the risk of a particular type of inborn, structural problem in children (neural tube defects) related to low blood levels of folate early in pregnancy.

Any food product can be fortified with vitamins and minerals—and many are. One of the few regulations governing the fortification of foods is that the amount of vitamins and minerals added must be listed in the Nutrition Facts panel. Illustration 4.5 shows some examples of enriched and fortified foods.

Food enrichment and fortification began in the 1930s to help prevent deficiency diseases such as rickets (vitamin D deficiency), goiter (from iodine deficiency), pellagra (niacin deficiency), and iron-deficiency anemia. Today, foods are increasingly being fortified for the purpose of reducing the risk of chronic diseases such as osteoporosis, cancer, and heart disease. Other foods, such as energy bars, sports drinks, hydration fluids, and margarines are being fortified with an array of vitamins and minerals principally for sales appeal. Regular consumption of fortified foods increases the risk that people will exceed Tolerable Upper Intake Levels (ULs) of nutrients designated in the Recommended Dietary Allowances (RDAs). Regular use of multiple vitamin and mineral supplements, along with liberal intake of fortified foods, enhances the likelihood that excessive amounts of some nutrients will be consumed.

The Ingredient Label Still more useful information about the composition of food products is listed on ingredient labels. The label of any food that contains more than one ingredient must list the ingredients in order of their contribution to the weight of the food (Table 4.5).

On the Side

In 1984 the Kellogg Company launched an ad campaign for All-Bran cereal that announced, "eating the right foods may reduce your risk of some kinds of cancer." Sales of the high-fiber cereal increased 37% in one year, but then Kellogg had to withdraw the ads. The FDA ruled the All-Bran statement was equivalent to a claim for a drug. The campaign, however, started the nutrition and health claims revolution.[4]

On the Side

The Twinkie was introduced in 1930. The rumor is still spreading that they contain enough preservatives to keep them from spoiling for 30 years. The package of Twinkies shown here was purchased in 1991 and photographed when it was 15 years old.

Table 4.5

All foods with more than one ingredient must have an ingredient label. Here are the rules.

1. Ingredients must be listed in order of their contribution to the weight of the food, from highest to lowest.

2. Beverages that contain juice must list the percentage of juice on the ingredient label.

3. The terms "colors" and "color added" cannot be used. The name of the specific color ingredients must be given (for example, caramel color, turmeric).

4. Milk, eggs, fish, and five other foods to which some people are allergic must be listed on the label.

Table 4.6

Solving the mystery of ingredient label terms

Cake Mix Ingredient Label	Additive	Function
Ingredients: Sugar, enriched flour bleached (wheat flour, niacin), iron thiamin Mononitrait, (vitamin B₁), riboflavin (vitamin B₂), vegetable shortening (contains partially hydrogenated soybean cottonseed oil), **sodium aluminum phosphate**, dextrose, leavening, (baking soda, monocalcium phosphate, diacalcium phosphate, aluminum sulphate), wheat starch, **propylene glycol monoesters**, modified corn starch, salt, egg white, vanilla, dried corn syrup, polysorbate 60, nonfat milk, **mono and diglycerides**, sodium caseetrate, **xanthan gum**, soy lecithin.	Sodium aluminum phosphate	Gives baked products a light texture
	Propylene glycol monoesters	Helps blend ingredients uniformly, enhances moisture content and texture
	Mono- and diglycerides	Maintains product softness after baking
	Xanthan gum	Thickening agent, helps hold product together after baking

© Scott Goodwin Photography

The FDA now requires that ingredient labels include the presence of common food allergens in products. Potential food allergens consist of milk, eggs, fish, shellfish, tree nuts, wheat, peanuts, and soybeans. These foods, sometimes called the "Big Eight," account for 90% of food allergies. Precautionary statements, such as "may contain," or "processed in a facility with" are allowed.[8]

Food Additives on the Label Specific information about **food additives** must be listed on the ingredient label. Nearly 3000 chemical substances may be added to food to enhance its flavor, color, texture, cooking properties, shelf life, or nutrient content. Food additives on the FDA's GRAS (*Generally Recognized As Safe*) list can be used in food without preapproval. New additives must be approved by the FDA prior to use. Table 4.6 provides examples of functions of some additives used in food, and the nearby "On the Side" shows you an example of a highly preserved snack food. The most common food additives are sugar and salt, but trace amounts of polysorbate, potassium benzoate, and many other additives that are not so familiar are also included in foods. Although the new labeling regulations are more comprehensive than the old ones, they don't help consumers understand what some additives do. Appendix D lists many of the most common additives and indicates their function in foods.

Trace amounts of substances such as pesticides; hormones and antibiotics given to livestock; fragments of packaging materials such as plastic, wax, aluminum, or tin; very small fragments of bone; and insects may end up in foods. These are considered "unintentional additives" and do not have to be included on the label.

■ **Irradiated Foods** Food irradiation is an odd example of a food additive. It is actually a process that doesn't add anything to foods. Irradiation uses X-rays, gamma rays, or electron beams to kill insects, bacteria, molds, and other microorganisms in food. Food irradiation enhances the shelf life of food products and decreases the risk of foodborne illness.[9] Irradiation must be performed according to specific federal rules.

Irradiated foods retain no radioactive particles. The process is like having your luggage X-rayed at the airport. Your luggage doesn't become radioactive, nor has anything in it changed. The process leaves no evidence of having occurred. Actually, this lack of change creates a challenge because inspection agencies can't determine whether a food has been irradiated or not. All irradiated foods—except spices that are added to processed foods—are required to display the international "radura" symbol and to indicate that the food has been irradiated (Illustration 4.6). Irradiation

food additives
Any substances added to food that become part of the food or affect the characteristics of the food. The term applies to substances added both intentionally and unintentionally to food.

Illustration 4.6 The "radura," as the symbol is called, must be displayed on irradiated foods. The words "treated by irradiation, do not irradiate again" or "treated with radiation, do not irradiate again" must accompany the symbol.

Health Action

Some Examples of What "Front of the Package" Nutrient-Content Claims Must Mean

Term	Examples	Means That a Serving of the Product Contains:
Fresh	Fresh spinach	Foods are raw, not frozen or heated, and contain no preservatives.
Healthy	Healthy burritos, canned vegetables	No more than 60 milligrams of cholesterol, 3 grams of fat, and 1 gram of saturated fat; more than 10% of the Daily Value of vitamin A, vitamin C, iron, calcium, protein, or fiber. "Healthy" foods must also contain 360 milligrams or less of sodium.
Extra lean	Extra-lean pork, extra-lean hamburger	Fewer than 5 grams of fat, fewer than 2 grams of saturated fat and trans fat combined, *and* fewer than 95 milligrams of cholesterol (applies to meats only).
Lean	Lean beef, lean turkey	Fewer than 10 grams of fat, fewer than 4.5 grams of saturated fat and *trans* fat combined, *and* fewer than 95 milligrams of cholesterol (applies to meats only).
Free	Fat-free, trans fat-free, sugar-free, sodium-free foods	No—or negligible amounts of—fat, sugars, trans fat, or sodium.
Good source	Good source of fiber, good source of calcium or antioxidants	From 10 to 19% of the Daily Value for a particular nutrient or for vitamin A (beta-carotene), vitamin C, or vitamin E in the case of antioxidants.
High	High iron, high fiber foods	20% or more of the Daily Value for a particular nutrient.
Low cholesterol	Low-cholesterol egg product	20 milligrams or less cholesterol (applies to animal products only).
Low fat	Low-fat cheese, low-fat ice cream	3 grams or less of fat.
Low saturated fat	Low-saturated-fat pancake mix, low-saturated-fat eggnog	1 gram or less saturated fat and 0.5 grams or less *trans* fat.
Low sodium	Low-sodium soup, low-sodium hot dogs	140 milligrams or less sodium.

is approved for use on chicken, turkey, pork, beef, eggs, grains, fresh fruits and vegetables, and other foods in the United States.[9]

Irradiated foods have a long history but have not yet won the trust of some American consumers.[14] The desire to limit radioactive processes of all sorts, as well as radioactive waste products, is widespread; many people fear anything that involves radioactivity, especially when it comes to food.

Dietary Supplement Labeling

A **dietary supplement** is a product taken by mouth that contains a "dietary ingredient" intended to supplement the diet. The "dietary ingredients" in these products include vitamins, minerals, proteins, enzymes, herbs, hormones, and organ tissues. Nutrition labeling regulations place dietary supplements in a special category under the general umbrella of "foods," not drugs. Dietary supplements differ from drugs in that they do not have to undergo vigorous testing and obtain FDA approval before they are sold. In return, dietary supplement labels cannot claim that the products treat, cure, or prevent disease.[10]

dietary supplement
Any product intended to supplement the diet, including vitamins, minerals, proteins, enzymes, herbs, hormones, and organ tissues. Such products must be labeled "Dietary Supplement."

Nutrition Facts

Per 125mL (87g)

Amount	% Daily Value
Calories 80	
Fat 0.5 g	1%
Saturated 0 g + Trans 0 g	0%
Cholesterol 0 mg	
Sodium 0 mg	0%
Carbohydrate 18 g	6%
Fibre 2 g	8%
Sugars 2 g	
Protein 3 g	

Vitamin A 2%	Vitamin C 10%
Calcium 0%	Iron 2%

Illustration 4.7 Dietary supplement label.

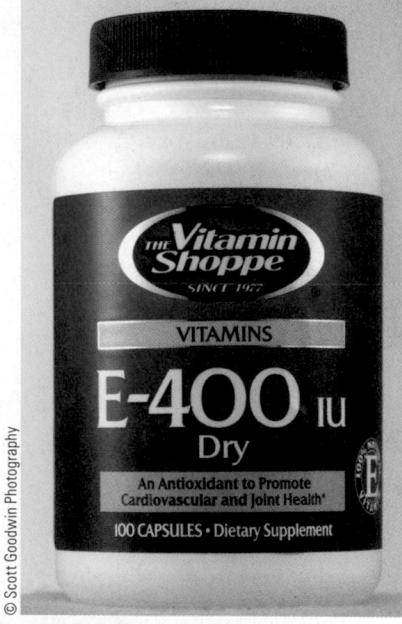

Illustration 4.8 Example of a structure, function claim.

Illustration 4.9 Foods certified as organic by the USDA can display this seal on food packages.

According to FDA regulations, dietary supplements must be labeled as such and include a "Supplement Facts" panel that lists serving size, ingredients, and percent Daily Value (%DV) of essential nutrient ingredients (Illustration 4.7). Like foods, qualifying dietary supplements can be labeled wih nutrient claims ("high in calcium," for example), and health claims (for instance, "diets low in saturated fat and cholesterol that include sufficient soluble fiber may reduce the risk of heart disease"). Dietary supplement labels can also include **structure/function claims.**

Structure/Function Claims Dietary supplements can be labeled with statements that describe effects the supplement may have on body structures or functions (shown in Illustration 4.8). Under this regulation, for example, certain supplements can be labeled with the following statements "Promotes healthy heart and circulatory function," "Helps maintain mental health," or "Supports the immune system." If a structure/function claim is made on the label or package inserts, the label or insert must acknowledge that the FDA does not support the claim: *"This statement has not been evaluated by the FDA. This product is not intended to diagnose, treat, cure, or prevent any disease."* Dietary supplements labeled with misleading or untruthful information, and those that are not safe, can be taken off the market and the manufacturers fined.

Structure and function claims do not have to be approved before they can appear on product labels. The FDA and FTC have taken action against companies responsible for fradulent claims.[11] The FDA has recently issued new guidelines that define the quality of scientific evidence needed to substantiate structure/function claims. [10]

The COOL Rule

USDA now requires retailers to display a country-of-origin label (COOL) on certain products (Illustration 4.11).[16] The rule is meant to expand informed consumer choices and to help track down food-borne illness outbreaks. The rule applies to meats, fish, seafoods, fruits, vegetables, many nuts, and some herbs.

Qualified health claims are being reviewed by the FDA and may be disallowed in the future.[10]

Organic Foods

For many years, consumers and producers of organic foods urged Congress to set criteria for the use of the term *organic* on food labels (Illustration 4.9). Consumers wanted to be assured that foods were really organic, and producers wanted to keep the business honest. Standards were needed because consumers cannot distinguish organically produced foods from others by looking at them, tasting them, or reading their nutrient values.[13] The U.S. Department of Agriculture (USDA) has developed and is implementing standards for organic foods.[14]

Labeling Organic Foods Rules that qualify foods as organic are shown in Illustration 4.10. If organic growers and processors qualify according to USDA approved certifying organizations, they can place the green-and-white USDA Organic seal on product labels. The USDA can impose financial penalties on companies that use the seal inappropriately. Organically grown and produced foods can be labeled in four other ways:

- "100% Organic" if they contain entirely organically produced ingredients

- "Organic" if they contain at least 95% organic ingredients

- "Made with organic ingredients" if they contain at least 70% organic ingredients

- "Some Organic Ingredients" if the product contains less than 70% organic ingredients

.eat right. **American Dietetic** **Association**

[hidden] The world's largest organization of food and nutrition professionals.

A Quick Guide to Nutrition Labeling

- ***Whole Grain Council Stamp***

  The Whole Grain Stamps feature a stylized sheaf of grain on a golden-yellow background with a bold black border. There are two different varieties of Stamps, the Basic Stamp and the 100% Stamp. The percentage of whole grains in the product determines whether the item is stamped with a 100% stamp (all grains are whole grains) or a basic stamp (contains at least eight grams — a half serving — of whole grains).

  Put on Pack by Whole Grains Council.

- ***Heart Check Symbol***

  This symbol is found on packaging that supports the American Heart Association's science and recommendations. Foods endorsed with this check have been screened and verified to meet the American Heart Association's certification criteria to be low in saturated fat and cholesterol for healthy people over age two.

  Put on pack with approval from the American Heart Association and displayed by food manufacturers listed at checkmark.heart.org/ProductsByManufacturer.

- ***Smart Spot Symbol***

  Developed by PepsiCo, this symbol of smart choices made easy was designed as a quick way for consumers to be sure that their choices in the grocery store are contributing to a healthier lifestyle. Every Smart Spot product meets nutrition criteria based on authoritative statements of the US Food and Drug Administration and the National Academy of Sciences.

- ***Sensible Solutions Symbol***

  Developed by Kraft Foods, the Sensible Solution flag was developed to assist consumers in choosing healthier choices among food and beverage products. To be labeled as a Sensible Solution, a food must meet criteria derived from the 2005 *Dietary Guidelines for Americans*, as well as authoritative statements from the Food and Drug Administration, National Academy of Sciences and other public health authorities. All Sensible Solution products contain limited amounts of calories, fat (including saturated and trans fat), sodium and sugar. Various Sensible Solution products meet specifications for "reduced," "low" or "free" in calories, fat, sodium or sugar and some products are fortified with micronutrients or deliver a functional benefit such as heart health.

- ***Eat Smart, Drink Smart***

  This <u>Unilever</u> logo is found on the labeling of foods and drinks that meet healthy eating criteria are based on US Dietary Guidelines. The Eat Smart, Drink Smart program was developed as part of the International Choices Foundation, a world-wide initiative with a goal of making the healthier choice the easy choice. This simple front-of-pack logo is placed on food products that have passed an evaluation against a set of qualifying criteria based on international dietary guidelines.

- ***NuVal System***

  The <u>NuVal System</u> is a food scoring system that helps consumers see at a glance the nutritional value of the food they buy. The NuVal System scores food on a scale of 1 to 100. The higher the score, the healthier the choice. This system summarizes the overall nutritional value of food and uses the Institute of Medicine's Dietary Reference Intakes and the *Dietary Guidelines for Americans* to quantify the presence of more than 30 nutrients including vitamins, minerals, fiber and antioxidants; sugar, salt, trans fat, saturated fat and cholesterol. The system also incorporates measures for the quality of protein, fat and carbohydrates, as well as calories and omega-3 fats.

- ***Smart Choices***

  Note: The Smart Choices program has <u>"voluntarily postpone[d] active operations."</u>

  The <u>Smart Choices</u> program was launched by a group of scientists, academicians, health and research organizations, food and beverage manufacturers and retailers to reduce the amount of independent, varying nutrition symbols currently seen on the packages of food. The program's goal was for Smart Choices to be the most widely used front-of-pack labeling program in the United States and assist people in making positive dietary changes. Products that qualified for the Smart Choices Program symbol would also display information on the front of the package, clearly stating calories per serving and number of servings per container. The goal was to help people stay within their daily calorie needs and make it easier for calorie comparisons within and across product categories.

Information provided by Dietitians in Business and Communications, a dietetic practice group of the American Dietetic Association.

1. Plants
   - Must be grown in soils not treated with synthetic fertilizers, pesticides, and herbicides for at least three years
   - Cannot be fertilized with sewer sludge
   - Cannot be treated by irradiation
   - Cannot be grown from genetically modified seeds or contain genetically modified ingredients
2. Animals
   - Cannot be raised in "factory-like" confinement conditions
   - Cannot be given antibiotics or hormones to prevent disease or promote growth
   - Must be given feed products that are 100% organic

Richard Anderson

Illustration 4.10 USDA rules for qualifying foods as organic.

Many people choose organic foods not so much for what is in them, but for what is not in them: hormones, antibiotics, and pesticide and herbicide residues. Most organic foods, however, are not totally free of some of these ingredients. Due to "pesticide drift" from sprayed crops, past use of pesticides on farmland, and the leaching of chemicals used on crops into groundwater, organically grown plants may contain traces of pesticides. However, conventionally grown crops are six times more likely to have traces of several pesticides than are organic crops.[15]

structure/function claim
Statement appearing primarily on dietary supplement labels that describes the effect a supplement may have on the structure or function of the body. Such statements cannot claim to diagnose, cure, mitigate, treat, or prevent disease.

The Nutrition Labeling Transition

Nutrition information labeling is in transition in the U.S. and other countries affected by the obesity and diabetes epidemics. The current U.S. nutrition information labeling system is viewed as being overly complex and sometimes confusing. Efforts are underway to simplify it.[17,18] The search is on for a labeling system that captures the public's attention while quickly and easily describing a product's caloric and nutrient value. It is reasoned that better-informed people can make better food choices than those left in the dark.

The New Wave of Nutrition Labels   If you are a nutrition label reader you have probably noticed that a wide variety of new labeling systems appear on the front of food packages. Three of these, "Nutrition at Glance," "Smart Choices Made Easy," and the "Smart Choices Program™," are shown in Illustration 4.12. The Smart Choices Program™ features a symbol that identifies more nutritious choices within specific food product categories, and lists the calories per serving and number of servings per package. This system has been selected for use by food industries in Europe as well as in the United States.[19] Many more nutrition labeling systems, including ones from the American Heart Association, various grocery store chains, and food companies are being used on food product packaging.

Industry initiated labeling systems are voluntary and generally employ science-based criteria for judging the nutrient qualities of food products.[20] The labels supplement the Nutrition Facts Panel for the food product. The primary weaknesses of the food industry-backed labeling systems are that they are not regulated publicly, criteria used to label foods vary, and the labels only appear on selected food products.

Calories on Display   New labeling systems and requirements are being instituted by local and state governments. Most of these efforts are aimed at providing point-of-purchase information about the calorie value to foods served in chain restaurants (Illustration 4.13). California was the first state to pass a law requiring chain restaurants with more than 20 locations to post the calorie content of menu items, and New York City adopted a similar measure earlier. The changes may dampen people's appetite for their favorite muffin that weighs in at 430 calories, or the 500-calorie cost of a large order of french fries.

© Scott Goodwin Photography

Illustration 4.11 Country-of-origin labels are now required on certain foods.

Illustration 4.12 A few examples of nutrition labeling systems now used on food packages.

Beyond Nutrition Labels: We Still Need to Think

Even with the new food labels, consumers can't be stupid.

—MAX BROWN, STUDENT

Understanding and applying the information on nutrition labels calls for more nutrition knowledge on the public's part. As highlighted in the nearby Reality Check, people need to know a good bit about nutrition before they can understand nutrition labels and incorporate labeled foods appropriately into an overall diet. Good diets include more than just foods with nutrition labels. The ice cream cone from the stand in the mall; the orange, potato, or fish we buy in the store; and the pizza delivered to the dorm are unlabeled parts of many diets. We need to know enough about the composition of unlabeled foods to fit them into a healthy diet. Nutrition labels often list only two vitamins and two minerals, but many more are required for health, so it's particularly important for people to know if their diet is varied enough to supply needed vitamins and minerals. Use of the MyPyramid Food Guide should go hand in hand with label reading and food choices.

Healthy diets consist of a wide variety of foods that may include high-fat or high-sodium foods on occasion. Not every food we eat has to have the "right" label profile. Serving mostly low-fat or low-calorie foods to children, for

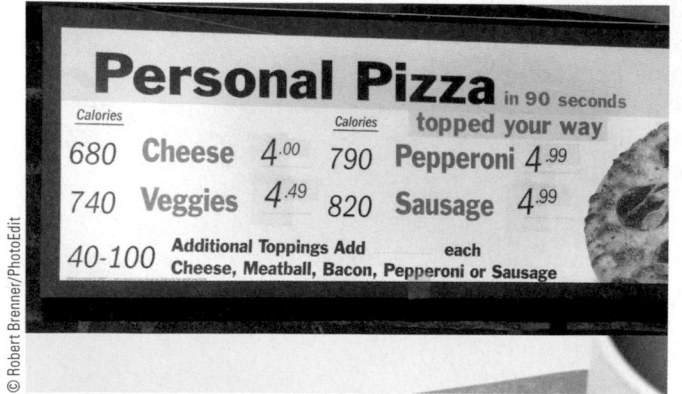

Illustration 4.13 An example of the new trend in calorie labeling of fast foods on menu boards.

example, might have unintended, unhealthy effects. Young children need calories and fat for growth and development. If diets are severely restricted, growth and development will be impaired.

Nutrition labels are an important tool for helping people make informed food-purchasing decisions. About half of adults use them to guide their food purchasing decisions.[21] However, labels do not now—nor will they ever—provide all the information needed to make wise decisions about food. Only people who are well informed about nutrition can do that.

NUTRITION Up Close

© Scott Goodwin Photography

Comparison Shopping

Focal Point: Comparing label information can help you choose the most nutritious product for your money.

Assume both cereals cost the same.
Rate the nutrition value of these two cereals by completing the second paragraph below using information from the Nutrition Facts panels. The first paragraph has been done for you.

Whole-Grain Cereal

Nutrition Facts
Serving Size 1 cup (50g)
Serving Per Container 10

Amount Per Serving

Calories 170   Calories from Fat 5

	% Daily Value*
Total Fat 0.5 g	1%
Saturated Fat 0 g	0%
Trans Fat 0 g	
Cholesterol 0 mg	0%
Sodium 0 mg	0%
Total Carbohydrate 41 g	14%
Dietary Fiber 5 g	21%
Sugars 0 g	
Protein 5 g	

Vitamin A 0%	Vitamin C 0%
Calcium 2%	Iron 8%

*Percent Daily Values are based on a 2000 calorie diet. Your daily values may be higher or lower depending on your calorie needs:
**Intake should be as low as possible.

	Calories:	2,000	2,500
Total Fat	Less than	65 g	80 g
Sat Fat	Less than	20 g	25 g
Cholesterol	Less than	300 mg	300 mg
Sodium	Less than	2400 mg	2400 mg
Total Carbohydrate		300 g	375 g
Dietary Fiber		25 g	30 g

Health Granola Cereal

Nutrition Facts
Serving Size 1 cup (50g)
Serving Per Container 10

Amount Per Serving

Calories 210   Calories from Fat 30

	% Daily Value*
Total Fat 3 g	5%
Saturated Fat 0 g	0%
Trans Fat 0 g	
Cholesterol 0 mg	0%
Sodium 120 mg	5%
Total Carbohydrate 43 g	14%
Dietary Fiber 3 g	12%
Sugars 16 g	
Protein 5 g	

Vitamin A 2%	Vitamin C 0%
Calcium 2%	Iron 10%

*Percent Daily Values are based on a 2000 calorie diet. Your daily values may be higher or lower depending on your calorie needs:
**Intake should be as low as possible.

	Calories:	2,000	2,500
Total Fat	Less than	65 g	80 g
Sat Fat	Less than	20 g	25 g
Cholesterol	Less than	300 mg	300 mg
Sodium	Less than	2400 mg	2400 mg
Total Carbohydrate		300 g	375 g
Dietary Fiber		25 g	30 g

The **Whole-Grain Cereal**, providing 170 **calories** per serving, contributes 5 **calories from fat** (or 1% of the % Daily Value). Each serving also provides 0 mg of **sodium** (or 0% of the % Daily Value), 5 g of **dietary fiber** (or 21% of the % Daily Value), and 0 g of **sugars**.

The **Health Granola Cereal**, providing _____ **calories** per serving, contributes _____ **calories from fat** (or _____ % of the % Daily Value). Each serving also provides _____ mg of **sodium** (or 5% of the % Daily Value), _____ g of **dietary fiber** (or _____ % of the % Daily Value), and _____ g of **sugars**.

Which cereal provides the better nutritional value?

FEEDBACK (answers to these questions) can be found at the end of Unit 4.

Nutrition labeling   Ronald got it. "Fat free" does not equal "calorie free."

Fat-free Caesar salad dressing, for example, provides 40 calories in a 2-tablespoon serving.

Jared:

Ronald:

[Key Terms

Daily Values (DVs), page 4-4
dietary supplement, page 4-9
enrichment, page 4-6

food additives, page 4-8
fortification, page 4-6

% Daily Value (%DV), page 4-4
structure/function claim, page 4-11

[Review Questions

TRUE FALSE

1. Almost all multiple-ingredient foods must be labeled with nutrition information.   ☐ ☐

2. Food manufacturers can list any serving size they want on Nutrition Facts panels.   ☐ ☐

3. In general, %DV listed for nutrients in Nutrition Facts panels of 10% or more are considered "low," and those listed as 50% or more are considered "high."   ☐ ☐

4. An overriding principle of nutrition labeling regulations is that nutrient content and health claims made about a food on the packaging must be truthful.   ☐ ☐

5. The term *enriched* on a food label means that extra vitamins and minerals have been added to the food to bolster its nutritional value.   ☐ ☐

6. The ingredient that makes up the greatest portion of a food product's weight must be listed first on ingredients labels.   ☐ ☐

TRUE FALSE

7. Food labeling regulations do *not* require food manufacturers to list the presence of major food allergens on ingredient labels.   ☐ ☐

8. Dietary supplements, such as those for herbs and vitamins, are considered drugs and are not regulated by FDA's Nutrition Labeling rules.   ☐ ☐

9. Foods, but *not* dietary supplements, can be labeled with nutrient content and health claims.   ☐ ☐

10. Animals providing meats labeled "organic" cannot be given antibiotics or hormones.   ☐ ☐

11. Foods bearing the USDA Organic seal are certified as organic by the U.S. Department of Agriculture.   ☐ ☐

12. Nutrition labels provide all the nutrition information we need to make healthful food choices.   ☐ ☐

Media Menu

www.cfsan.fda.gov/~dms/flquiz1.html
Test your food label knowledge and get answers to questions about identifying healthy food choices based on nutrition label information.

www.cfsan.fda.gov/~dms/foodlab.html
The FDA's main page is devoted to understanding and using food labels.

www.nutrition.gov
For quick access to high-quality information, search the terms "nutrition labels," "dietary supplement labels," "organic foods," and "food irradiation" from this Web site.

www.fda.gov/ora/inspect_ref/igs/nleatxt.html
Got a question about nutrition labeling, nutrient or health claims? You'll find the Nutrition Labeling and Education Act of 1990 at this address.

www.hc-sc.gc.ca/fn-an/label-etiquet/nutrition/index-eng.php
Health Canada's main page for nutrition labeling regulations.

http://www.hc-sc.gc.ca/fn-an/label-etiquet/nutrition/cons/interactive-eng.php
Health Canada provides interactive learning tools related to nutrition labels. One helps people get labeling facts right and another is a nutrition labeling quiz.

http://www.ams.usda.gov/nop
This address leads to the USDA's organic standards home page. Search "National Organic Standards" on Google if needed; Web address will change. You can call USDA's National Organic Program to get detailed information on organic food regulations: 202-720-3252.

Notes

1. Brecher SJ et al. Status of nutrition labeling, health claims, and nutrient content claims for processed foods. J Am Diet Assoc 2000;100:1057–62.

2. How to Understand Use the Nutrition Facts Label. www.cfsan.fda.gov/~dms/foodlab.html#nopercent, accessed 7/06.

3. Pennington JA, Hubbard VS. Derivation of daily values used for nutrition labeling. J Am Diet Assoc 1999;97:1407–12.

4. Marquart L et al. Solid science and effective marketing for health claims. Nutr Today 2001;36:107–14.

5. Manufacturers and consumers lose faith in natural label claims, www.nutraingredients-usa.com, 9/11/08, accessed 1/09.

6. Health claims that meet significant scientific agreement, 2008, www.cfsan.fda.dms/lab-ssa.html, accessed 1/09.

7. Position of the American Dietetic Association: food fortification and nutritional supplements. J Am Diet Assoc 2005;105:1300–11.

8. Food Allergen Labeling and Consumer Protection Act of 2004. www.cfsan.fda.gov/~dms/alrgact.html, accessed 7/06.

9. Shea KM. Technical report: irradiation of food. Pediatrics 2000;106:1505–9.

10. Guidance for Industry, Substantiation for Dietary Supplement Claims Made Under Section 403(r) (6) of the Federal Food, Drug, and Cosmetic Act, 12/08, available at www.cfsan.fda.gov/guidance.html, accessed 1/09.

11. Fraud supplement makers fined $16 m, 1/16/09, nutraingredients-usa.co/content/view/233187, accessed 1/09.

12. Heller L, When foods become drugs. Nutraingredients-use.com/content/view/229883, 12/11/08, accessed 12/08.

13. Kristensen M et al. Effect of plant cultivation methods on content of major and trace elements in foodstuffs and retention in rats, J Food Sci Agric 2008;88:2162–72.

14. The National Organic Program, www.ams.usda.gov/nop/indexNet.htm, accessed 7/06.

15. McCally M. et al. Irradiation of food. N Engl J Med 2004,351:402.

16. Hitti M. New food labels show country of origin, 1/10/08, www.medscape.com/viewarticle/581365, accessed 1/09.

17. Promoting brand simplicity in food and drinks: reducing product claims, brand dilution and private label threat, Business Insights, available from www.nutraingredient-usa.com, accessed 11/08.

18. Hasler CM. Health claims in the United States: an aid to the public or a source of confusion? J Nutr 2008;138:1216S-20S.

19. European backing for U.S. food label, NutraIngredientsUSA, www.nutraingredients-usa.com, 11/4/08.

20. Hitti M. "Smart Choices" food labels are coming, www.medscape.com/viewarticle/582810, accessed 10/08.

21. Golan E et al. Economics of Food Labeling, Agricultural Economic Report No. (AER793), January 2001, http://www.ers.usda.gov/Publications/AER793, accessed 1/09.

NUTRITION | Up Close

Comparison Shopping

Feedback for Unit 4

The **Health Granola Cereal**, providing 210 **calories** per serving, contributes 30 **calories from fat** (or 5% of the % **Daily Value**). Each serving also provides 120 mg of **sodium** (or 5% of the **% Daily Value**), 3 g of **dietary fiber** (or 12% of the **% Daily Value**), and 16 g of **sugars**. The **Whole-Grain Cereal** provides the better nutritional value because it is lower in calories, total fat, sodium, and sugar and higher in fiber than the **Health Granola Cereal**. Don't be fooled by an attractive-sounding product. The proof of nutritional value is in the label, not the name.

Nutrition, Attitudes, and Behavior

NUTRITION SCOREBOARD

	TRUE	FALSE
1 Food preferences are genetically determined.		
2 Food habits never change.		
3 Skipping breakfast does not affect school performance in well-nourished children.		

Key Concepts and Facts

- Most food preferences are learned.

- The value a person assigns to eating right has more effect on dietary behaviors than does knowledge about how to eat right.

- Food habits can and do change.

- The smaller and more acceptable the dietary change, the longer it lasts.

- Behavior and mental performance can be affected by diet.

Origins of Food Choices

Horse meat is a favorite food in a large area of north-central Asia. Pork, which is widely consumed in North and South America, Europe, and other areas, is rigidly avoided by many people in Islamic countries. Bone-marrow soup and sautéed snails are delicacies in France, while kidney pie is traditional in England. Dog is a popular food in Borneo, New Guinea, the Philippines and other countries, whereas snake is a delicacy in China. In some countries, people enjoy insects (illustration 5.1). Some people consider corn and soybeans fit only for animal feed. And then there are steamed clams and raw oysters—food passions for some, but absolutely disgusting to others.[3]

When did you first think "yecck!?" The food choices just described would elicit that response among people from a variety of cultures, but they would not necessarily be responding to the same foods.

Illustration 5.1
Grasshoppers are Mexican delicacies, served at Girasoles Restaurant in Mexico City.

© Gavriel Jecan/Getty Images/Photodisc

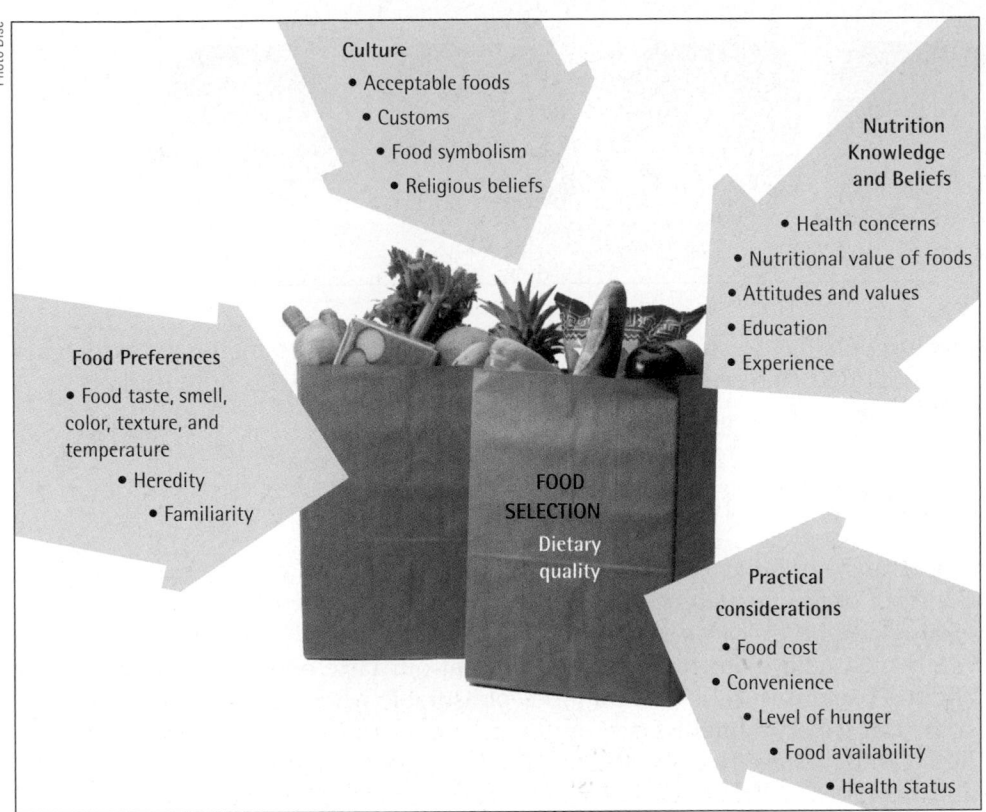

Illustration 5.2 Factors influencing food selection and dietary quality. Each of these sets of factors interacts with the others.[6]

Culture
- Acceptable foods
- Customs
- Food symbolism
- Religious beliefs

Nutrition Knowledge and Beliefs
- Health concerns
- Nutritional value of foods
- Attitudes and values
- Education
- Experience

Food Preferences
- Food taste, smell, color, texture, and temperature
- Heredity
- Familiarity

FOOD SELECTION
Dietary quality

Practical considerations
- Food cost
- Convenience
- Level of hunger
- Food availability
- Health status

Why do people eat what they do? People learn from their culture and the society in which they live what animals and plants are considered food and which are not.[4] Once items are identified as food, they develop a legacy of strong symbolic, emotional, and cultural meanings. Comfort foods, health foods, junk foods, fun foods, soul foods, fattening foods, mood foods, and pig-out foods, for example, have been identified in the United States. All cultures have their "super food": in Russia and Ireland, it's potatoes; in Central America, it's corn and yucca (a starchy root, also called manioc); in Somalia, it's rice. The designation refers to the cultural significance of the food and not to its nutritional value.[3]

In countries like the United States, where a wide variety of foods are available and people have the luxury of selecting which foods they will eat, food choices are influenced by a wide range of factors (Illustration 5.2). Of these factors, food preference has the largest impact.[5] Food preferences vary a good deal among individuals, and, as addressed in the next "On the Side" feature, lead to a wide array of specific food choices. Rather than being inborn, food preferences are primarily learned.[1]

We Don't Instinctively Know What to Eat

The food choices people make are not driven by a need for nutrients or guided by food selection genes. People deficient in iron, for example, do not seek out iron-rich foods. If we're overweight, no inner voice tells us to reject high-calorie foods. Women who are pregnant don't instinctively know what to eat to nourish their growing fetuses. No evidence indicates that young children, if offered a wide variety of foods, would select and ingest a well-balanced diet.[7]

Humans are born with mechanisms that help them decide when and how much to eat, however.[8] An inborn attraction to sweet-tasting foods, a dislike for bitter foods, and the response of thirst when water is needed all influence food and fluid intake to an extent (illustration 5.3).[9] There is evidence to suggest that people deficient in sodium experience an increased preference for salty foods.[10]

Why do we prefer the foods we do?

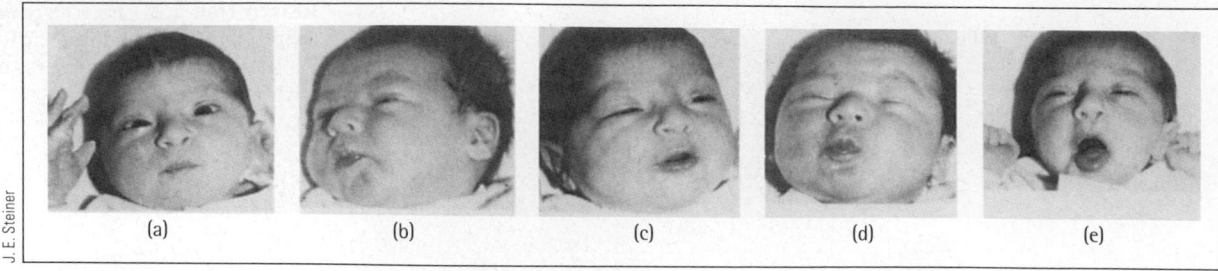

(a)   (b)   (c)   (d)   (e)

J. E. Steiner

Illustration 5.3 Newborn infants respond to different tastes: (a) Baby at rest, (b) tasting distilled water, (c) tasting sugar, (d) tasting something sour, and (e) tasting something bitter.

Source: Taste-induced facial expressions of neonate infants from the classic studies of J. E. Steiner, in *Taste and Development: The Genesis of Sweet Preference*, ed. J. M. Weiffenbach, HHS Publication no. NIH 77-1068 (Bethesda, Md.: U.S. Department of Health and Human Services, 1977), pp. 173–189, with permission of the author.

"You're going to eat that? Do you have any idea what's in that hot dog?"

"Yeah, I do. There's barbecue in the backyard, ball games with my mom and dad, and parties at my friend's house. The memories taste great!"

Due to individual differences in preferences for coffee, Starbucks offers more than 19,000 ways a cup of it can be served.[12]

© Michael Newman/PhotoEdit

On the Side

The first thing I remember liking that liked me back was food.
—RITA RUDNER, COMEDIAN

Most people have a strong aversion to foods they think "smell bad," which are likely spoiled. Whether this response is inborn or learned is not clear.

Food Choices and Preferences

The strong symbolic, emotional, and cultural meanings of food come to life in the form of food preferences. We choose foods that, based on our cultural background and other learning experiences, give us pleasure.[10] Foods give us pleasure when they relieve our hunger pains, delight our taste buds, or provide comfort and a sense of security. We find foods pleasurable when they outwardly demonstrate our superior intelligence, our commitment to total fitness, or our pride in our ethnic heritage. We reject foods that bring us discomfort, guilt, and unpleasant memories and those that run contrary to our values and beliefs.

The Symbolic Meaning of Food

Food symbolism, cultural influences, and emotional reasons for food choices are broad concepts that may become clearer with concrete examples. Here are a few examples to consider.

Status Foods   Vance Packard, in his book *The Status Seekers,* provided a memorable example of the symbolic value of food:

As a lad, this man had grown up in a poor family of Italian origin. He was raised on blood sausages, pizza, spaghetti, and red wine. After completing high school, he went to Minnesota and began working in logging camps, where—anxious to be accepted—he soon learned to prefer beef, beer, and beans, and he shunned "Italian" food. Later, he went to a Detroit industrial plant, and eventually became a promising young executive. . . . In his executive role he found himself cultivating the favorite foods and beverages of other executives: steak, whiskey, and seafood. Ultimately, he gained acceptance in the city's upper class. Now he began winning admiration from people in his elite social set by going back to his knowledge of Italian cooking, and serving them, with the aid of his manservant, authentic Italian treats such as blood sausage, spaghetti, and red wine![11]

Comfort Foods   Ice cream, apple pie, chicken noodle soup, boxed chocolates, meat loaf and mashed potatoes: these are the most popular comfort foods in the United States.[19] The feelings of security and love that came along with the tea and honey or chicken soup that your mother or father gave you when you had a cold, or with the ice cream and popsicles lovingly given to soothe a sore throat, are renewed with comfort foods. Some comfort foods can bring pleasure and reduce anxiety just by their image. (Illustration 5.4).

Once the symbolic value of a food is established, its nutritional value will remain secondary.[14] Food status is a strong determinant of food choices; and after all, as a noted nutritionist once said, "life needs a little bit of cheesecake."[15]

"Discomfort Foods" Memories of bad experiences with food, and expectations that certain foods will harm us in some way, each contribute to our learning about food and affect our food preferences. Eating a piece of blueberry pie right before an attack of the flu hits or overdosing on sweet pickles or olives, for example, may take these foods off your preferred list for a long time.

Cultural Values Surrounding Food

A team of scientists observed that the diet of certain groups in the Chin States of Upper Burma was seriously deficient in animal protein. After considerable study a way was found to improve the situation by cross-breeding the small, local black pigs raised by the farmers with an improved strain to obtain progeny, giving a greater yield of meat. The entire operation, however, completely failed to benefit the nutrition of the population because of one fact which had been viewed as irrelevant. The cross-bred pigs were spotted. And it was firmly believed—as firmly as we believe that to eat, say, mice would be disgusting—that spotted pigs were unfit to eat.[3]

© Anthony-Masterson/PictureArts/CORBIS

Illustration 5.4 Do you find comfort in this photo? Two restaurants in New York City offer only macaroni and cheese dishes as entrées.[15]

Dietary change introduced into a culture for the purpose of improving health can be successful only if it is accepted by the culture. Cultural norms are not easily modified.

Other Factors Influencing Food Choices and Preferences

Food preferences and selections are also affected by the desire to consume foods that are considered healthy. Reducing fat intake, eating more fruits and vegetables, and cutting down on sweets bring rewards and pleasures such as weight loss and maintenance, an end to constipation, lower blood cholesterol level, and a newly discovered preference for basic foods.

"When I was a kid, my parents made me eat the brussels sprouts they put on my plate. I finally ate them, but then I threw up. Nobody has ever made me eat brussels sprouts again."

Food Cost and Availability Food choices are also affected by the cost and availability of food. Researchers found that college students eating in dining halls have better diets when they prepay for their meals for the entire term rather than paying at each meal.[16] Grocery shoppers tend to select more low-fat and high-fiber foods when presented with a wide selection of those foods rather than just a few.[17]

Genetic influences Genetics plays an important role in food preference development. Some people are born with a strong sensitivity to bitterness and may reject foods such as brussels sprouts and broccoli (especially if they are overcooked!), bitter-tasting teas and wines, and tonic water. Some people with this genetic trait will continue to eat strong-flavored vegetables—particularly after they cover up the bitterness with seasonings or sauces.

Food Choices Do Change

Who says old dogs can't learn new tricks? Most Americans aren't eating the way they used to. Per person consumption of whole-grain products has increased 41% since 1970, while fresh egg intake has fallen from an average of 4½ eggs per week to 4. Low-fat milk sales have risen 112%, beef consumption dropped 22%, and broccoli consumption is skyrocketing. Per person consumption of broccoli has risen 1,007% since 1970—a bigger gain than for any other food. Americans are eating 168% more cheese and 20 to 25% more vegetables and

Photo Disc

Sugars +17%   Skim milk +25%   Chicken +33%   Margarine -42%   Oranges -18%

Illustration 5.5 Changes in Americans' food choices since 1990.[18]

fruits than in 1970.[18] Examples of changes in food choices since 1990 are shown in Illustration 5.5.

Food choices are largely learned and do change as we learn more about foods. Perhaps your food choices have changed over time. How do the food choices you make now compare with the choices you made five years ago?

How Do Food Choices Change?

What are the ingredients for change in food choices? Why do some people succeed in improving their food choices while other people find that very hard to do? Nutrition knowledge, attitudes, and values have a lot to do with changing food choices for the better.

Daily Tribune

Nutrition Course Improves Diets of College Students

The diets of college students improved after taking a basic nutrition course.

Results indicated that students decreased the amount of total fat from 82 to 68 grams and made other changes putting their diets more in line with the Dietary Guidelines.

Illustration 5.6 A true story.

Illustration 5.7 Why knowledge about a good diet may not be enough to improve food choices.

Nutrition Knowledge and Food Choices Sound knowledge about good nutrition necessarily precedes the selection of a healthful diet. But is knowledge enough to ensure that healthy changes in diet will be made? The answer is "yes" for some people and "no" for others. Higher levels of knowledge about diet and health have been related to healthier eating practices among students at James Madison University, the University of Texas at Austin, and the University of Vermont (Illustration 5.6).[19–21] For some Americans, news that animal fats raise blood cholesterol levels was enough to convince them to cut down on beef and whole milk. Knowledge that diets containing an adequate amount of fiber reduce the risk of certain types of cancer led many Americans to switch to high-fiber breakfast cereals. Information about good nutrition leads some people to modify their eating behavior some of the time and is more likely to change the food choices of women than men.[22] Knowledge alone is often not enough, however.

When Knowledge Isn't Enough Many people know far more about the components of a good diet than they put into practice. The problem is that between knowledge and practice lie multiple beliefs and experiences that act as barriers to change (Illustration 5.7). Change of any type is most likely to succeed when the

"I feel guilty about eating the foods I like."   "Eating right is too expensive."   "I tried eating better, but I didn't stick with it."   "I don't have the time to eat right."   "The vegetables I like aren't available."   "I'm healthy now... Why should I worry about my diet?"

benefits of making the change outweigh the disadvantages. This makes changes in food choices a very individual decision. Individuals decide whether a change is in their best interests. But what sorts of circumstances, in addition to increased knowledge, make changes in food choices worthwhile for individuals—and even highly desired?

Nutrition Attitudes, Beliefs, and Values   The value individuals place on diet and health is reflected in the food choices they make. A survey of restaurant patrons found that food choices varied according to the consumer's perceptions of the importance of diet to health:[22]

- "Unconcerned" consumers—people who are unconcerned about the connection between diet and health and who tend to describe themselves as "meat and potato eaters"—select foods for reasons other than health.

- "Committed" consumers believe that a good diet plays a role in the prevention of illness. They tend to consume a diet consistent with their commitment to good nutrition.

- "Vacillating" consumers—people who describe themselves as concerned about diet and health but who do not consistently base food choices on this concern—tend to vary their food choices depending on the occasion. These consumers are likely to abandon diet and health concerns when eating out or on special occasions, but they generally adhere to a healthy diet.

Avoiding illness and curing or diminishing current health problems are likewise strong incentives for changing food choices (Table 5.1).[24] One study found that nurses and dietitians based dietary changes primarily on the benefits of good diets to future health.[25] In almost all instances, the key to lasting improvements in diet is to make changes you can live with. Changes that bring you more pleasure than inconvenience or discomfort have staying power, while those that require a lot of willpower to maintain rarely last.[26]

Successful Changes in Food Choices

The primary reason efforts to improve food choices fail is that the changes attempted are too drastic. Improvements that last tend to be the smallest acceptable changes needed to do the job.[27,28]

The Process of Changing Food Choices   Assume you need to lose weight and want to keep it off by modifying your food choices. A promising plan to accomplish this goal would begin by identifying food choices you would like to change and lower-calorie food options you would be willing and able to eat (Table 5.2). Then you could make the plan more specific by identifying the changes that would be easiest to implement. For example, assume the low-calorie foods you like include frozen nonfat yogurt and oranges. You might decide to eat yogurt or an orange in place of your usual bedtime snack of ice cream. A specific dietary change such as this is much easier to implement than a broad notion, such as "eat less." Although weight loss will take a while, such a small acceptable change has a much better chance of working than a drastic change in diet.

Planning for Relapses   When making a change in your diet, be prepared for relapses. Relapses happen for a number of reasons, and they don't mean the attempt has failed. People often return to old habits because the change they attempted was too drastic or because they tried to make too many changes at once. If the change undertaken doesn't work out, rethink your options and make a midcourse correction.

Table 5.1

Factors that enhance food changes

- Attitude that nutrition is important
- Belief that diet affects health
- Perceived susceptibility to diet-related health problems
- Perception that benefits of change outweigh barriers to change

Perhaps the key in making dietary changes is to determine which changes are the easiest.

Table 5.2

Changing food choices for the better[27, 28]

The Process	An Example
1. Identify a healthful change in your diet you'd like to make.	1. I'd like to lower my fat intake.
2. Identify two food choices you make that should change because they contribute to the need for the healthful change you identified.	2. I eat at fast-food restaurants three times a week. I usually have a large order of fries, and once a week I eat fried chicken.
3. Identify two or more specific, acceptable options for more healthful food choices than the ones identified under number 2.	3. Options identified: • Order tossed salad with low-calorie dressing instead of fries. • Eat a grilled chicken sandwich every other week instead of fried chicken. • Eat Mexican fast-food more often.
4. Decide which option is easiest to accomplish and requires the smallest change to get the job done.	4. I love Mexican food. It would be easy to eat tocos instead of french fries or fried chicken.
5. Plan how to incorporate the change into your diet.	5. Mondays and Fridays, I'll eat tacos.
6. Implement the change. Be prepared for midcourse corrections.	6. Midcourse correction: On Fridays, when I'm with my friends, it's easier to eat at the restaurant they like. I'll order the grilled chicken sandwich and coleslaw.

Does Diet Affect Behavior?

Food affects behavior in some rather striking ways. Irritable, crying infants rapidly change into cooing, sleepy angels after they are fed. Low-on-sleep employees perk up after their morning coffee. A high-calorie lunch makes many people feel calm and sleepy.[29] Not only do our behaviors affect our diet, but our diets can affect our behaviors. Examples of associations between dietary characteristics and behavior are listed in Table 5.3. One common belief about food and behavior is the subject of this unit's "Reality Check."

Malnutrition and Mental Performance

Like growth and health, mental development and intellectual capacity can be affected by diet. The effects range from mild and short term to serious and lasting, depending on when the malnutrition occurs, how long it lasts, and how

Table 5.3

Examples of ways in which diet may affect behavior

Dietary characteristic	Behavioral Outcomes
Malnutrition, growth stunting in early childhood	lower intellectual functioning and school performance, increased antisocial behavior in childhood[30]
Nutritional supplementation of malnourished young children	improved growth and intellectual functioning in adulthood[31]
Very low carbohydrate intake (less than 20 grams per day)	reduced short-term memory, slower reaction times, increased attention span[32]
Ingestion of certain color additives and a preservative food additive	moderate increases in hyperactivity, and inattentive behaviors in children[33]
Lead consumption	higher risk of violent and aggressive behavior, hyperactivity, and mental and behavioral problems[34]
Iron deficiency in young children	long-term deficits in learning ability and social skills[35]

Can food be a love potion? Toward the end of a friend's birthday party, Glenda and Cassell got into a heated debate about the existence of food aphrodisiacs. Part of the conversation went like this:

Is it all in Glenda's head?
Answers on next page

PhotoDisc

Glenda:
Chocolate and vanilla are natural love potions, there's no doubt about it. If I eat a chocolate truffle and spritz on vanilla flavoring like perfume before a date, the guy always goes nuts for me!

PhotoDisc

Cassell:
Are you kidding? No way! I've heard about oysters and this tree bark that are supposed to work miracles, too. It's all in your head.

severe it is. The effects are most severe when malnutrition occurs while the brain is growing and developing.

Severe deficiency of protein, calories, or both early in life leads to growth retardation, low intelligence, poor memory, short attention span, and social passivity. When the nutritional insult is early and severe, some or all of these effects may be lasting (Illustration 5.8)[30,36]. In Barbados, for example, children who experienced protein-calorie malnutrition in the first year of life did not fully recover even with nutritional rehabilitation. Growth improved but academic performance did not. Compared to well-nourished children, those experiencing protein and calorie deficits during infancy were more likely to drop out of school and were four times more likely to have symptoms of ADHD (attention deficit hyperactivity disorder).[32]

Protein-calorie malnutrition that occurs later in childhood, after the brain has developed, produces behavioral effects that can be corrected with nutritional rehabilitation. Correction of other deficits that often accompany malnutrition, such as the lack of educational and emotional stimulation and harsh living conditions, hastens and enhances recovery.[30]

Protein-calorie malnutrition severe enough to cause permanent delays in mental development rarely occurs in the United States. When it does, malnutrition is usually due to neglect or inadequate caregiving. More common dietary events that impair learning in U.S. children are breakfast skipping, fetal exposure to alcohol, iron deficiency, and lead toxicity.

© Reuters./CORBIS

Illustration 5.8 Malnutrition in early childhood has long-lasting effects. Some children never fully recover.

Does Breakfast Help You Think Better? Short-term fasting, such as skipping breakfast, reduces the late-morning problem-solving performance of

Can food be a love potion? The idea that food can act as an aphrodisiac has been around since ancient times. Although many have looked and others have tried, no one has found a food that acts like a love potion.[37]

Glenda: 👎

Cassell: 👍

children. About one in seven young children in the United States doesn't eat breakfast regularly. (No one has yet studied the effects of breakfast skipping on college students' midterm exam scores. . . .) In general, children and adolescents who are breakfast eaters are less likely to be overweight than their peers who aren't.[2]

Early Exposure to Alcohol Affects Mental Performance Mental development can be permanently delayed by exposure to alcohol during fetal growth. Although growth is also retarded, the most serious effects of fetal exposure to alcohol are permanent delays in mental development and behavioral problems associated with them. Women are advised not to drink if they are pregnant or may become pregnant.[40]

On a global basis, an estimated 20% of men, 35% of women, and 40% of all children are anemic, primarily due to iron deficiency.[41]

Iron Deficiency Impairs Learning Most cases of iron-deficiency anemia in children result from inadequate intake of dietary iron. Iron-deficiency anemia in children is a widespread problem in developed and developing countries and likely is the most common single nutrient deficiency.[41] The potential impact of iron-deficiency anemia on the functional capacity of humans represents staggering possibilities.

Until recently, it was thought that the effects of iron-deficiency anemia on intellectual performance were short term and could be corrected by treating the anemia. It now appears that some of the effects may be lasting. Five-year-old children in Costa Rica treated for iron-deficiency anemia during infancy scored lower on hand–eye coordination and other motor skill tests than similar children without a history of anemia. Studies in the United States have detected shortened attention span and reduced problem-solving ability in iron-deficient children.[36]

Illustration 5.9 There are many opportunities for overexposure to lead. Young children are especially vulnerable.

Overexposure to Lead Our concern about exposure to lead has recently increased, due in large measure to information indicating that exposure to low levels of lead has long-term behavioral effects.[35] There are many opportunities for overexposure to lead. Approximately 84% of U.S. houses built before 1980 contain some lead-based paint.[42] Children living in or near these houses may eat the paint flakes (they taste sweet), or the old paint may contaminate the soil near the houses (Illustration 5.9). Lead also ends up in soil from industrial and agricultural chemicals, in water from lead-based pipes and solder, and in the air from the days when leaded gas was used.

Although the use of lead in cans, pipes, and gasoline has decreased dramatically, lead remains in the environment for long periods. Lead also stays in the body, stored principally in the bones, for a long time—20 years or more. It takes over a year of treatment to reduce blood lead levels. The effects of excessive exposure to lead include increased absenteeism from school, impaired reading skills, higher dropout rates, and increased aggressive behavior.[34] Blood lead levels in children have dropped substantially in recent decades. Despite this drop, however, half a million young children in the United States still have elevated blood lead levels.[44]

Food Additives, Sugar, and Hyperactivity The notion that certain food additives are related to hyperactivity in children has been popular since the related "Feingold Hypothesis" was announced in the mid-1970's. Research has failed to substantiate this claim...until recently. In a study involving a large group of healthy 3-, 8-, and 9-year-old children, intake of a beverage containing four food colorants (types of yellow, orange, and red color additives) and a preservative (sodium benzoate) was related to the development of hyperactivity. Signs of hyperactivity detected in the children consuming the additive-containing beverage included over-activity, impulsiveness, and short attention span. Not all children consuming the beverage demonstrated hyperactive behaviors, indicating that some children may be vulnerable to the effects of these food additives while others are not.[33]

Studies examining the effects of sugar intake on hyperactivity in children have not demonstrated that such a relationship exits.[45] The excitement that often accompanies high-sugar eating occasions such as Halloween and birthday parties—or the expectation that sugar causes hyperactivity—may be responsible for the reported effect. (See Illustration 5.10)

The Future of Diet and Behavior Research

Identifying the effects of nutrition on behavior is a tricky business. Many actors in addition to diet influence behavior, making it difficult to separate influences from social, economic, educational, and genetic influences. We still have much to learn, and many assumptions about diet and behavior must await confirmation through research.

© Ed Bock/Corbis

Illustration 5.10 Do food additives or lots of sugary food make children hyperactive?

Up Close

© Guy Cali/photolibrary

Improving Food Choices

Focal Point: Developing a plan for healthier eating.

Identify a change in your diet that you would like to make. Then develop a plan for making the change by thinking through and responding to each element of the dietary change process listed. (Refer to Table 5.2 for examples of responses.)

Dietary Change Process	**Your Response**
1. Identify a healthful change in your diet you'd like to make.	1. _____
2. Identify two food choices you make that should change because they contribute to the need for the healthful change you identified.	2. _____
3. Identify two specific, acceptable options for food choices more healthful than the ones identified under number 2.	3. _____
4. Decide which option is easiest to accomplish and requires the smallest change to get the job done.	4. _____
5. Plan how to incorporate the change into your diet.	5. _____

FEEDBACK (answers to these questions) can be found at the end of Unit 5.

Review Questions

TRUE FALSE

1. Food preferences are universal—everyone likes the same foods. ☐ ☐

2. Foods will be rejected by a population, no matter how nutritious they may be, if the foods don't fit into a culture's definition of what foods are appropriate to eat. ☐ ☐

3. Food choices are driven largely by a person's need for nutrients. ☐ ☐

4. Potatoes, rice, corn, and yucca are examples of "super foods" in specific countries due to their cultural significance. ☐ ☐

5. Nutrition knowledge is an important prerequisite for making healthful food choices. ☐ ☐

6. Broad dietary changes, such as a decision to simply "eat less," are more likely to produce lasting behavioral changes than are small changes in diets, such as snacking on favorite fruits rather than candy. ☐ ☐

TRUE FALSE

7. Changes in dietary intake that are acceptable to an individual and easy to implement are the types of changes that are most likely to last. ☐ ☐

8. Changing food choices for the better takes planning and includes individual decisions on specifically how the change in food choices will be implemented. ☐ ☐

9. Individuals need to plan for modifying their approach to improving food choices because even the best planned changes in food choices sometimes fail. ☐ ☐

10. Examples of ways in which dietary intake affects behavior include the relationship between sugar intake and hyperactivity in children. ☐ ☐

11. Results of recent research studies make it clear that the consumption of breakfast makes little difference to children's academic performance or weight status. ☐ ☐

Media Menu

www.nimh.nih.gov
The National Institute of Mental Health provides information on attention deficit hyperactivity and other disorders.

www.healthfinder.gov
Use this site to find reliable information about iron deficiency, fetal alcohol syndrome, and other disorders presented in this unit.

www.ncemch.org
This URL connects to the National Center for Education in Maternal and Child Health and provides information on lead poisoning and searchable databases.

www.faculty.de.gcsu.edu/~cbader/ ghprecwithinsects.html
Delicious grasshopper recipes are available at this site. It also provides a comprehensive resource list for insects as food.

search.ama-assn.org
The American Medical Association offers interactive tools for planning and implementing healthy diets and lifestyles. Included in the tool set are tips, action plans, and a healthy eating progress tracking calendar.

www.cfsan.fda.gov/~dms/col-toc.html
This is the FDA's site for information about color additives.

Notes

1. van den Bree MBM et al. Genetic and environmental influences on eating patterns of twins aged ≥50 y. Am J Clin Nutr 1999;70:456–65.

2. Rampersaud GC et al. Breakfast habits, nutritional status, body weight, and academic performance in children and adolescents. J Am Diet Assoc 2005:105:743–60.

3. Pyke M. Man and food. New York: McGraw-Hill; 1972.

4. Beauchamp GK, Mennella JA. Sensitive periods in the development of human flavor perception and preference. Ann Nestle 1998;56:19–31.

5. Rolls B. Obesity and Weight Control Symposium, Experimental Biology Annual Meeting, San Diego, CA, April 14, 2003.

6. Birch LL, et al. Effects of a nonenergy fat substitute on children's energy and macronutrient intake (Am J Clin Nutr 1993;58:326–33).

7. Story M, Brown JE. Do young children instinctively know what to eat? The studies of Clara Davis revisited. N Engl J Med 1987;316:103–6.

8. Birch LL et al. Effects of a nonenergy fat substitute on children's energy and macronutrient intake. Am J Clin Nutr 1993;58:326–33.

9. Steiner JE. The gustofacial response: observation on normal and anencephalic newborn infants. In: Bosma JF, ed. Fourth symposium on oral sensation and perception: development in the fetus and infant. Bethesda (MD): National Institute of Dental Research, DHEW Publication No. 73-546, US Dept HEW, NIH; 1973:254.

10. Factors influencing food choices in humans. Nutr Rev 1990;48:442.

11. V Packard, The status seekers (New York: Random House, 1959), 146.

12. Horovitz B. You want it your way. USA Today, Mar. 5, 2004, p. 1A.

13. Yankelovich Survey results reported in USA Today, 2000 Feb. 7:D1.

14. Parraga IM. Determinants of food consumption. J Am Diet Assoc 1990;90:661–3.

15. Stein K. Contemporary comfort foods: bringing back old favorites, J Am Diet Assoc 2008;108:412, 414.

16. Beerman KA. Variation in nutrient intake of college students: a comparison by students' residence. J Am Diet Assoc 1991;91:343–4.

17. Cheadle A et al. Community-level comparisons between the grocery store environment and individual dietary practices. Prev Med 1991;20:250–6.

18. Loss-adjusted food availability, www.ers.usda.gov/Data/Foodconsumption, accessed 7/06 and 1/09.

19. Brevard PB. Nutrition education improves the diet of college students. J Am Diet Assoc 1991, abs. A-47.

20. Davis JN et al. Is nutrition knowledge reflected in healthier dietary practices of college students? Experimental Biology 2003: Meeting Abstract 709.8, San Diego, April 2003.

21. Kolodinsky J et al. Knowledge of current dietary guidelines and food choices by college students: better eaters have higher knowledge of dietary guidance. J Am Diet Assoc 2007;107:1409–13.

22. Nutrition and you: trends 2008, American Dietetic Association, www.eatright.org, accessed 12/08.

23. Shepherd R, Stockley L. Nutrition knowledge, attitudes, and fat consumption. J Am Diet Assoc 1987;87:615–9.

24. Becker MH et al. The health belief model and prediction of dietary compliance: a field experiment. J Health Soc Behav 1977;18:348–66.

25. Holdt CS, et al. Knowledge and behaviors of allied health professionals regarding meat (abs.). J Am Diet Assoc 1991;91S, abs. A-13.

26. Handley B et al. Using action plans to help primary care patients adopt healthy behaviors: a descriptive study. J Am Board Fam Med 2006;19:224–31.

27. Hornick BA et al. Menu modeling with MyPyramid food patterns: incremental dietary changes lead to dramatic improvements in diet quality of menus. J Am Diet Assoc 208;108:2077–83.

28. Gould L. Consumers taking small steps toward big lifestyle changes. Results of Datmonitor's survey of diet trends in Europe and the U.S. Reported in nutraingredients.com/Europe, March 18, 2005.

29. Young SN. Nutrition 3. The fuzzy boundary between nutrition and psychopharmacology. Can Med Assoc J, Jan. 22, 2002, vol. 155.

30. Walker SP et al. Early childhood stunting is associated with poor psychological functioning in late adolescence and effects are reduced by psychological stimulation. J Nutr 2007;137:2464–9.

31. Stein AD et al. Nutritional supplementation in early childhood, schooling, and intellectual functioning in adulthood: a prospective study in Guatemala, Arch Pediatr Adolesc Med, 2008;162:612–8.

32. D'Anci KE et al. Low-carbohydrate weight-loss diets: effect on cognition and mood, Appetite 2009;52: 96–103.

33. McCann D et al. Food additives and hyperactive behaviour in

3-year-old and 8/9-year-old children in the community: a randomised, double-blinded, placebo-controlled trial, The Lancet 2007;370: 1560–7.

34. Wright JP et al. Association of prenatal and childhood blood lead concentrations with criminal arrests in early adulthood, PLoS Med 2008;5(5):e101, www.medscape.com/viewarticle/576717, accessed 1/09.

35. Bellinger DC, Very low lead exposures and children's neurodevelopment, Curr Opin Pediatr 2008;20:172–7.

36. Lozoff B. Iron deficiency and child development, Food Nutr Bull 2007; 28(4 Suppl):S560–71.

37. Parker G, et al. Mood state effects of chocolate. J Affect Disord 2006; 92:149–59.

38. Allen LH. Functional indicators of nutritional status of the whole individual or the community. Clin Nutr 1984;3:169–75.

39. Nicklas TA et al. Breakfast consumption affects adequacy of total daily intake in children. J Am Diet Assoc 1993;93:886–91.

40 Dietary Guidelines for Americans: Alcohol, 2005, available at www.nutrition.gov.

41. Lazoff B. Nutrition and behavior. Am Psychologist 1989;44:231–6.

42. Child Health USA, 1997. US Maternal and Child Health Bureau, PHS, Washington, DC, p. 29.

43 Dignam TA et al. Reduction in elevated blood lead levels in children in North Carolina and Vermont, 1996–1999, Envrion Health Perspect 2008;116:981–5.

44. Rogan WJ, Ware JH. Exposure to lead—how low is low enough? N Engl J Med 2003;348:1515–16.

45. Regalado M. Busting the sugar-hyperactivity myth. WebMD, www.webMD.com, 1999, accessed 1/01.

Answers to Review Questions

1. False, see page 5-2.
2. True, see pages 5-2, 5-4, 5-5.
3. False, see pages 5-3, 5-4.
4. True, see pages 5-3.
5. True, see pages 5-5, 5-6.
6. False, see page 5-7.
7. True, see page 5-7.
8. True, see page 5-7.
9. True, see page 5-7.
10. False, see page 5-11.
11. False, see page 5-10.

NUTRITION | # Up Close

Improving Food Choices

Feedback for Unit 5

There are no right or wrong answers to this unit's "Nutrition Up Close." Plans for dietary change that are specific, easy to accomplish, and highly acceptable are most likely to work in the long run.

Healthy Diets, Dietary Guidelines, MyPyramid, and More

NUTRITION SCOREBOARD

	TRUE	FALSE
1 Over half of U.S. adults fail to consume five servings of vegetables and fruits daily.		
2 The basic food groups include a "healthy snack" food group.		
3 "Supersized" fast-food meals often provide two to three times more calories than regular-sized fast food meals.		

Key Concepts and Facts

- Healthful diets are characterized by adequacy, variety, and balance.

- There are many types of healthful diets.

- The Dietary Guidelines for Americans and MyPyramid Food Guide provide foundation information for healthful diets and physical activity levels.

Answers to NUTRITION SCOREBOARD	TRUE	FALSE
1 Sixty percent of U.S. adults fail to meet this recommendation.[1]	✔	
2 The basic food groups do not include a "healthy snack" food group.		✔
3 Supersizing fast-food meals piles on calories.	✔	

This Is Just to Say I have eaten the plums that were in the icebox and which you were probably saving for breakfast
Forgive me they were delicious so sweet and so cold

—WILLIAM CARLOS WILLIAMS

Healthy Eating: Achieving the Balance between Good Taste and Good for You

May I have your attention please? For a moment, think about the foods in Illustration 6.1. If your mouth is watering and you're ready to go out and buy some ripe peaches, you have found the balance between good taste and good for you. Who said foods that taste good aren't good for you?

(a)

(b)

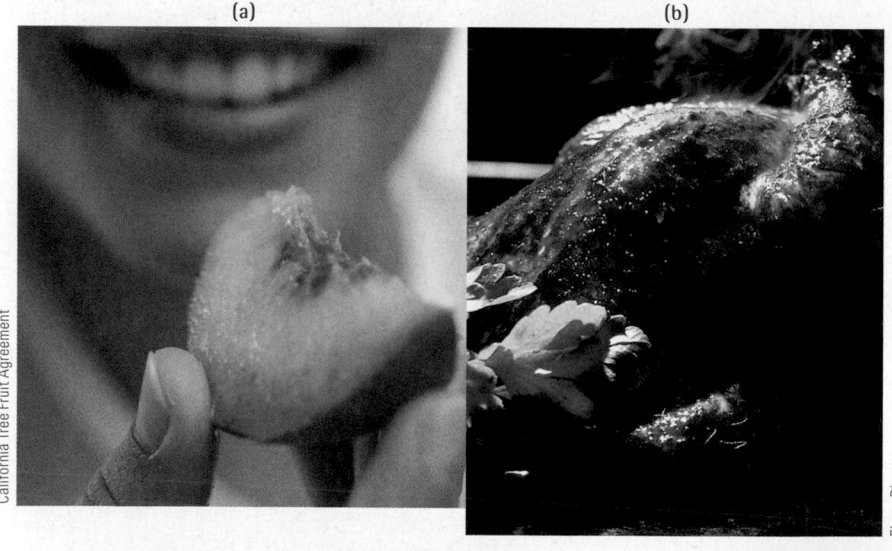

California Tree Fruit Agreement

Photo Disc

Illustration 6.1 Can you smell it? Can you taste it? A plump, golden peach. It's so ripe that juice spurts from it and drips down your chin when you take a bite.... A golden brown turkey just taken out of the oven. The wonderful smell fills the kitchen. A steaming loaf of homemade bread just set out to cool. A perfect ripe tomato just picked from the garden. It melts in your mouth.

Photo Disc

(c)

© Wally Eberhart/Getty Images/Visuals Unlimited

(d)

Characteristics of Healthful Diets

Healthful diets come in a variety of forms. They may be based on bread, olives, nuts, fruits, beans, vegetables, lamb, and chicken (as in Greece); rice, vegetables, and small amounts of fish and other meats (as in China); or black beans, rice, meat, and tropical fruits (as in Cuba and Costa Rica). Illustration 6.2 gives examples of the diverse foods that can be part of a healthy diet.

Although the types of foods that go into them can vary substantially, healthful diets all share three basic characteristics: Adequacy, variety, and balance.

Adequate diets include a wide variety of foods that together provide sufficient levels of calories and **essential nutrients**. What's sufficient? For calories, it's the number that maintains a healthy body weight. For essential nutrients, sufficiency corresponds to intakes that are in line with recommended intake

Illustration 6.2 Foods that contribute to healthy diets in different countries.   a. Dinner in Italy might be linguine primavera (pasta with vegetables). b. Pad Thai, rice noodles and vegetables, is a favorite dish in Thailand. c. Tamales are a celebration food in many Latin cultures. d. Dal, curry dishes, vegetables, and chicken are popular parts of the cuisine of India.

(a)

(b)

(c)

(d)

adequate diet
A diet consisting of foods that together supply sufficient protein, vitamins, and minerals and enough calories to meet a person's need for energy.

essential nutrients
Substances the body requires for normal growth and health but cannot manufacture in sufficient amounts; they must be obtained in the diet.

balanced diet
A diet that provides neither too much nor too little of nutrients and other components of food such as fat and fiber.

macronutrients
The group name for the energy-yielding nutrients of carbohydrate, protein, and fat. They are called macronutrients because we need relatively large amounts of them in our daily diet.

saturated fats
The type of fat that tends to raise blood cholesterol levels and the risk for heart disease. They are solid at room temperature and are found primarily in animal products such as meat, butter, and cheese.

trans fats
A type of unsaturated fat present in hydrogenated oils, margarine, shortening, pastries, and some cooking oils that increases the risk of heart disease.

levels represented by the *Recommended Dietary Allowances (RDAs)* and *Adequate Intakes (AIs)*. (These are shown in the Dietary Reference Intake tables inside the front cover of this book.) Recommended amounts of essential nutrients should be obtained from foods to reap the benefits offered by the variety of naturally occurring substances in foods that promote health.

Variety is a core characteristic of healthy diets because the essential nutrient and phytochemical content of foods differ. Consumption of an assortment of foods from each of the basic food groups increases the probability that the diet will provide enough nutrients. You could, for example, eat three servings of potatos a day to meet the recommended "3-a day" guideline. But, you would consume a much broader variety of vitamins, minerals, and plant antioxidants, for example, if you consumed spinach and tomatoes along with potatos.

A **balanced diet** provides calories, nutrients, and other components of food in the right proportion—neither too much nor too little. Diets that contain too much sodium or too little fiber, or are high in fat or sugar, for example, are out of balance. Diets that provide more calories than needed to maintain a healthy body weight are also out of balance.

Current dietary intake standards include guidelines to help individuals balance their intake of carbohydrate, protein, and fat—the **macronutrients**. Called the "Acceptable Macronutrient Distribution Ranges," or AMDRs, the guidelines indicate percentages of total caloric intake that should consist of carbohydrate, protein, and fat. AMDRs have also been set for added sugars, linoleic acid and alpha-linolenic acid (the two essential fatty acids). Illustration 6.3 shows the acceptable ranges of macronutrient, added sugars, and essential fatty acid intake, as well as the average intake levels of U.S. adults for each of them. It is recommended that diets be kept as low in **saturated fat** and **trans** fat as possible. The AMRDs apply to individuals over the age of four years.[2]

How Balanced Is the American Diet?

Comparisons of the recommended ranges of macronutrient intake with actual levels of intake by adults show that the U.S. diet is out of balance in several ways. The average intake of fat and added sugars by U.S. adults averages near the top of the acceptable intake ranges, indicating many adults consume more fat and added sugars than recommended. Foods high in fat or added sugars, such as desserts, sausages, potato chips, and regular soft drinks, tend to be high in calories and low in nutrients. Excess consumption of these foods can knock a diet out of balance. Although the average intake of fat is on the high side, intakes of the two essential fatty acids average near the bottom of the acceptable ranges. Adults consume a lot of fat, but too much of it is in the form of animal and milk fats that contain relatively low amounts of the essential fatty acids.[4]

Diets in the United States are out of balance in other respects. Only 40% of adults consume at least five servings of vegetables and fruits daily, and less

Illustration 6.3 The match between Acceptable Macronutrient Distribution Ranges (AMDRs) and average intake by adults in the United States.[2,3]

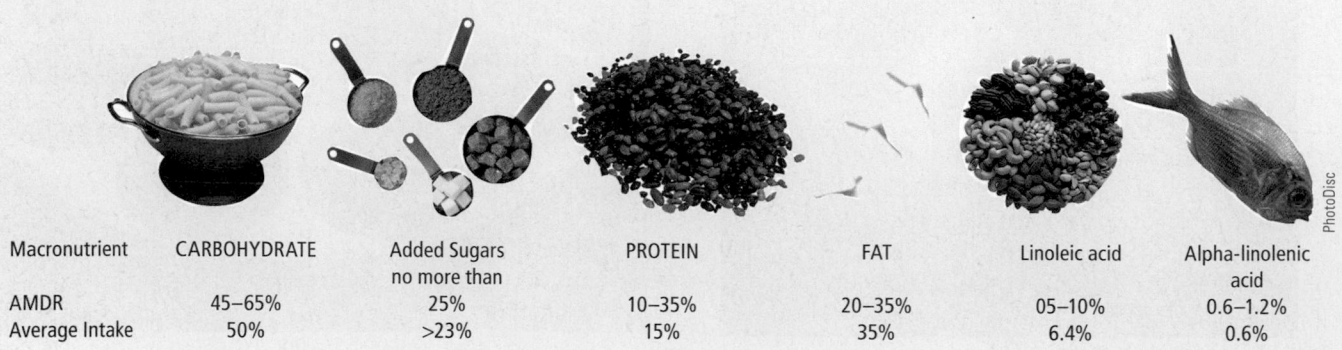

Macronutrient	CARBOHYDRATE	Added Sugars no more than	PROTEIN	FAT	Linoleic acid	Alpha-linolenic acid
AMDR	45–65%	25%	10–35%	20–35%	05–10%	0.6–1.2%
Average Intake	50%	>23%	15%	35%	6.4%	0.6%

than 15% include dark-green or colorful vegetables that are high in antioxidants. The vegetable most commonly consumed in the United States is french fries.[5] We tend to consume far less than the recommended amount of whole grain products, opting instead for breads and cereals made with refined flours. Caloric intake is becoming out of balance with body weight. Average calorie intake in the United States has increased steadily over the last few decades and the incidence of obesity is increasing along with it.[6] This dietary profile prompted one leading nutrition researcher to announce, "We're becoming an over-fed but undernourished population."[7]

Guides to Healthy Diets

Due to the impact of food choices on the health of individuals and population groups, many countries have established recommendations for dietary intake. Such recommendations are periodically updated as new discoveries about diet and health emerge. Many of the guidelines include recommendations for physical activity as well.

National recommendations for diet and physical activity usually apply to children over the age of two years and aren't appropriate for every individual in a population. General guidelines may not completely match the needs of individuals who, for example, are strict vegetarians, disabled, or have disorders such as hypertension or diabetes. A basic premise of national dietary guidelines is that nutrient needs should be met primarily through food consumption.

Benefits of population-based dietary and physical activity recommendations are multiple. The information is science-based, free, and made widely available to the public on the Web. Adherence to the information can help people stay healthy and lower their risk of developing disorders such as cancer, osteoporosis, heart disease, and obesity.

Some of the national guidelines also give credit to the cultural and social importance of food. "Enjoy your food!" is listed as Ireland's first dietary guideline, and the guidelines for Japanese people include "Happy eating makes for happy family life; sit down and eat together and talk; treasure family taste and home cooking."[8] Table 6.1 provides a summary of the dietary guidelines established for Canada and Japan. An Internet address that will link you to more examples of national dietary guidelines is given at the end of this unit. Additional information on dietary guidance for Canadians is given in Appendix F.

National diet and physical activity guidance in some countries is accompanied by information on how to select a healthful diet and achieve recommended levels of physical activity. In the United States, the national guidelines for diet and physical activity are called the "Dietary Guidelines for Americans," and the major how-to guide for consumers is "MyPyramid."[9]

Dietary Guidelines for Americans

The Dietary Guidelines for Americans provide science-based recommendations to promote health and to reduce the risk for major chronic diseases through diet and physical activity. Due to their credibility and focus on health promotion and disease prevention for the public, the Dietary Guidelines form the basis of federal food and nutrition education programs and policies.

The 2005 edition of the Dietary Guidelines stresses the importance of selecting nutrient-dense foods, balancing caloric intake with output, and increasing physical activity. This document includes the promise that the health of most individuals will be enhanced if the recommendations are followed. Each recommendation is part of an integrated whole—all the recommendations should be implemented for best results.

Table 6.1

Good advice from the dietary guidelines of Canada and Japan

Canadian Dietary Guidelines

A. **Summary of Nutrition Recommendations**
   The Canadian diet should:
   - provide energy consistent with the maintenance of body weight within the recommended range.
   - include essential nutrients in amounts specified in the Recommended Nutrient Intakes.
   - include no more than 30% of energy as fat (33 g/1000 kcal or 39 g/5000 kJ) and no more than 10% as saturated fat (11 g/1000 kcal or 13 g/5000 kJ).
   - provide 55% of energy as carbohydrate (138 g/1000 kcal or 165 g/5000 kJ) from a variety of sources.
   - be reduced in sodium content.
   - include no more than 5% of total energy as alcohol, or 2 drinks daily, whichever is less.
   - contain no more caffeine than the equivalent of 4 cups of regular coffee per day.
   - Community water supplies containing less than 1 mg/litre should be fluoridated to that level.

B. **Guidelines for Healthy Eating**
   - Enjoy a variety of foods.
   - Emphasize cereals, breads, other grain products, vegetables, and fruits.
   - Choose low-fat dairy products, lean meats, and foods prepared with little or no fat.
   - Achieve and maintain a healthy body weight by enjoying regular physical activity and healthy eating.
   - Limit salt, alcohol, and caffeine.

Japan's Dietary Guidelines for Health Promotion

---

1. Enjoy your meals
   - Have delicious and healthy meals that are good for your mind and body.
   - Enjoy communication at the table with your family or other people and participate in the preparation of meals.

2. Establish a healthy rhythm by keeping regular hours for meals.

3. Eat well-balanced meals with staple food, as well as main and side dishes.

4. Eat enough grains, such as rice and other cereals.

5. Combine vegetables, fruits, milk products, beans, and fish in your diet.

6. Avoid too much salt and fat.

7. Learn your healthy body weight and balance the calories you eat with physical activity.

8. Chew your food well and do not eat too quickly.

9. Enjoy nature's bounty and the changing seasons by using local food products and ingredients in seasons and by enjoying holiday and special-occasion dishes.

10. Reduce leftovers and waste through proper cooking and storage methods.

11. Assess your daily eating habits.
    - Learn and practice healthy eating habits at school and at home.
    - Promote appreciation of good eating habits from an early stage of life.

Focus Areas and Key Recommendations

The Dietary Guidelines for 2005 include nine "Focus Areas" and 23 "Key Recommendations." These are highlighted in the Health Action presented on page 6–7. "Key Recommendations for Specific Population Groups" are also provided by the Dietary Guidelines. Special population groups addressed are infants and young children, pregnant women, older adults, and people with weakened immune systems. These and other recommendations included in the Dietary Guidelines are available online as part of the full report (refer to the Media Menu section at the end of this unit).

Under legislative mandate, the Dietary Guideline for Americans must be updated every five years. This requirement was established over 25 years ago to ensure that up-to-date, scientifically based information on diet, physical activity, and health will be available to the U.S. public, health professionals, and policy makers.

Health Action

The 2005 Dietary Guidelines for Americans: Focus Areas and Examples of Key Recommendations

Find your way to a healthier you.

Adequate Nutrients within Calorie Needs

- Consume a variety of nutrient-dense foods and beverages within and among the basic food groups while choosing foods that limit the intake of saturated and *trans* fats, cholesterol, added sugars, salt, and alcohol.
- Meet recommended intakes within energy needs by adopting a balanced eating pattern, such as the USDA MyPyramid Food Guide or the DASH Eating Plan.

Weight Management

- To maintain body weight in a healthy range, balance calories from foods and beverages with calories expended.
- To prevent gradual weight gain over time, make small decreases in food and beverage calories and increase physical activity.

Physical Activity

- Engage in regular physical activiy and reduce sedentary activites to promote health, psychological well-being, and a healthy body weight.
- Engage in at least 30 minutes of moderate-intensity physical activity, above usual activity, at work or home on most days of the week.
- To help manage body weight engage in approximately 60 minutes of moderate- to vigorous-intensity activity on most days of the week.

Food Groups to Encourage

- Consume a sufficient amount of fruits and vegetables while staying within energy needs.
- Choose a variety of fruits and vegetables each day. In particular, select from all five vegetable subgroups (dark green, orange, legumes, starchy vegetables, and other vegetables) several times a week.
- Consume three or more ounce-equivalents of whole-grain products per day.
- Consume three cups per day of fat-free or low-fat milk or equivalent milk products.

Fats

- Consume less than 10% of calories from saturated fatty acids and less than 300 mg/day of cholesterol and keep trans fatty acid consumption as low as possible.
- Keep total fat intake between 20 to 35% of calories.

Carbohydrates

- Choose fiber-rich fruits, vegetables, and whole grains often.
- Choose and prepare foods and beverages with little added sugars or caloric sweeteners.

Sodium and Potassium

- Consume less than 2,300 mg (approximately 1 tsp of salt) of sodium per day.

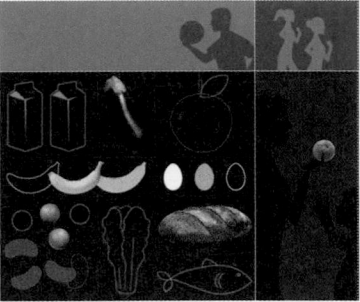

Dietary Guidelines for Americans 2005

U.S. Department of Health and Human Services
U.S. Department of Agriculture
www.healthierus.gov/dietaryguidelines

- Choose and prepare foods with little salt. At the same time, consume potassium-rich foods, such as fruits and vegetables.

Alcoholic Beverages

- Those who choose to drink alcoholic beverages should do so sensibly and in moderation—defined as the consumption of up to one drink per day for women and up to two drinks per day for men.

Food Safety

- To avoid microbial food-borne illness:
- Wash hands, food contact surfaces, and fruits and vegetables.
- Cook foods to a safe temperature to kill microorganisms.

Implementation of the Dietary Guidelines

The MyPyramid Food Guide is the major how-to tool intended to help the public implement the 2005 Dietary Guidelines. In addition, the Dietary Guidelines report identified the DASH (*D*ietary *A*pproaches to *S*top *H*ypertension) Eating Plan as being consistent with the dietary recommendations. MyPyramid covers both food selection and physical exercise, while the DASH Eating Plan addresses only dietary intake. Both are valuable tools that provide the framework for planning nutrient-dense, calorically appropriate diets that diminish the risk of chronic disease.[10]

MyPyramid Food Guide

Food group guides have been available in the United States since 1916. Known by names such as "Basic Four Food Groups" and "Food Guide Pyramid," the guides have been periodically updated since then. New releases of food group guides reflect existing scientific knowledge about nutrition and health and are modified to address emerging health problems.

The U.S. Department of Agriculture (USDA) released its newest version of the food guide in 2005. Called "MyPyramid," it is very different from previous guides (Illustration 6.4). This popular Web site is Internet-based and resource-filled, and some of the educational resources offered are interactive.

Illustration 6.4 MyPyramid's major graphic for the food guide is much different from those used in the past.

What's in MyPyramid?

MyPyramid presents recommendations for daily food choices and food amounts that are consistent with healthy diets. This guide recommends that people consume basic foods from five specific food groups in their most nutrient-dense form. Meat choices from the meat and bean group, for example, consist of lean meats, and meats and fish that are not prepared with fat. Vegetables are assumed to have no added butter or margarine, fruits no added sugar, dairy products are assumed to be low-fat and milk non-fat. The guide encourages consumption of dark-green- and orange-colored vegetables, whole grains and whole grain products, lean meats and fish, skim milk, dried beans, and fruit.[10]

Sweet desserts, pastries, soft drinks, fatty meats, processed meats, sugar, alcohol-containing beverages, and other energy-dense foods are not part of the basic food groups. But a limited amount of foods providing fat and sugar "discretionary calories" may be included in a diet if nutrient needs are met by basic foods and more calories are needed to maintain a healthy weight. Only a small proportion of Americans qualify for these extra calories.[4]

MyPyramid uses cups and ounces as the primary measures of how much food to consume daily and gives the recommended number of cups or ounces for food within each group. The unit of measure for fats, oils, and sugars is the teaspoon.

MyPyramid.gov: The Web Site

Log on to mypyramid.gov and be prepared to be amazed at what the site offers.

- Use "MyPyramid Menu Planner" to develop your own menus based on your food preferences and calorie need. After you answer questions about your age, sex, height, weight, and physical activity level, the planner calculates your calorie need and food group servings. You get to pick the foods that go into the menu and see how well those choices stack up against the recommendations.

- Click on "MyPyramid Plan" and enter your age, sex, and activity level. You will quickly be rewarded with food group recommendations based on your calorie need. The Web page that corresponds to the assessment displayed is shown Illustration 6.5.

- Go to "MyPyramid for Kids" and discover the MyPyramid Blast-Off game and other resources designed for elementary school-aged childern. A link titled "MyPyramid for Preschoolers" provides parents and caretakers information on good diets and eating behaviors in 1 to 4 year olds.

United States Department of Agriculture

MyPyramid.gov

| Home | About Us | News & Media | Site Help | Online Ordering | Contact Us | En Español |

Search MyPyramid.gov

[   ] Go

Subjects

- **MyPyramid Basics**
  - ○ *Inside the Pyramid*
  - ○ *Tips & Resources*
  - ○ *Print Materials*
  - ○ *Got a Question?*
- **Interactive Tools**
  - ○ *MyPyramid Plan*
  - ○ *Menu Planner*
  - ○ *MyPyramid Tracker*
  - ○ *Child Cost Calculator*
- **Multimedia**
  - ○ *Podcasts*
  - ○ *PSAs*
  - ○ *Animation*
- **Specific Audiences**
  - ○ *Preschoolers (2-5y)*
  - ○ *Kids (6-11y)*
  - ○ *Pregnancy & Breastfeeding*
  - ○ *General Population*
- **For Professional Use**
- **Steps to a Healthier Weight**
- **Dietary Guidelines**
- **Partnering with MyPyramid**
- **Related Links**

You are here: Home / MyPyramid Plan

MyPyramid Plan

Eat these amounts from each food group daily. This plan is a **2800** calorie food pattern. It is based on average needs for someone like you. (A **26** year old **male**, **5** feet **9** inches tall, **155** pounds, physically active **30 to 60 minutes** a day.) Your calorie needs may be more or less than the average, so check your weight regularly. If you see unwanted weight gain or loss, <u>adjust the amount you are eating</u>.

▶ **Grains** [1]	10 ounces	tips
▶ **Vegetables** [2]	3.5 cups	tips
▶ **Fruits**	2.5 cups	tips
▶ **Milk**	3 cups	tips
▶ **Meat & Beans**	7 ounces	tips

Click the food groups above to learn more.

[1] Make Half Your Grains Whole

Aim for at least 5 ounces of whole grains a day.

[2] Vary Your Veggies

Aim for this much every week:

Dark Green Vegetables = 3 cups weekly
Orange Vegetables = 2 1/2 cups weekly
Dry Beans & Peas = 3 1/2 cups weekly
Starchy Vegetables = 7 cups weekly
Other Vegetables = 8 1/2 cups weekly

Oils & Discretionary Calories

Aim for 8 teaspoons of oils a day.

Limit your extras (extra fats & sugars) to 425 Calories.

Physical Activity

Physical activity is also important for health. Adults should get at least 30 minutes of moderate level activity most days. Longer or more vigorous activity can provide greater health benefits. <u>Click here</u> to find out if you should talk with a health care provider before starting or increasing physical activity. <u>Click here</u> for more information about physical activity and health.

View, Print & Learn More:

- ▶ Click here to view and print a PDF version of **your results**.

- ▶ Click here to view and print a PDF of a helpful **Meal Tracking Worksheet**.

- ▶ For a more detailed assessment of your diet quality and physical activity go to the **MyPyramid Tracker**.

- ▶ You can view/print the **MyPyramid Calorie Results** and the **Food Tracking Worksheets** for any or all of the 12 calorie levels.

You will need the free Adobe Acrobat Reader plug-in to view and print the above PDF files.

Last Modified: April 15, 2009 04:55 PM

USDA.gov | FOIA | Accessibility Statement | Privacy Policy | Non-Discrimination Statement | Information Quality | USA.gov | White House

---

- Explore the "For Professionals" link. It will connect you to detailed information about using MyPyramid educational materials and seven days of sample menus that correspond to the MyPyramid recommendation for a 2000-calorie food pattern. Four of the seven days included in the menus are shown in Illustration 6.6.

Illustration 6.5 MyPyramid Plan results based on age, sex, weight, height, and physical activity level.

Illustration 6.5 (Continued)

MyPyramid
STEPS TO A HEALTHIER YOU

Based on the information you provided, this is your daily recommended amount from each food group.

GRAINS	VEGETABLES	FRUITS	MILK	MEAT & BEANS
10 ounces	3 1/2 cups	2 1/2 cups	3 cups	7 ounces

Make half your grains whole

Aim for at least **5 ounces** of whole grains a day

Vary your veggies

Aim for these amounts **each week:**

Dark green veggies = 3 cups

Orange veggies = 2 1/2 cups

Dry beans & peas = 3 1/2 cups

Starchy veggies = 7 cups

Other veggies = 8 1/2 cups

Focus on fruits

Eat a variety of fruit

Go easy on fruit juices

Get your calcium-rich foods

Go low-fat or fat-free when you choose milk, yogurt, or cheese

Go lean with protein

Choose low-fat or lean meats and poultry

Vary your protein routine—choose more fish, beans, peas, nuts, and seeds

Find your balance between food and physical activity

Be physically active for at least **30 minutes** most days of the week.

Know your limits on fats, sugars, and sodium

Your allowance for oils is **8 teaspoons a day.**

Limit extras—solid fats and sugars—to **425 calories a day.**

Your results are based on a 2800 calorie pattern.   Name: _____

This calorie level is only an estimate of your needs. Monitor your body weight to see if you need to adjust your calorie intake.

Illustration 6.6 Sample menus and nutrient analysis from the "For Professionals" link at MyPyramid.gov.

Sample Menus for a 2000 Calorie Food Pattern

Averaged over a week, this seven day menu provides all of the recommended amounts of nutrients and food from each food group. (Italicized foods are part of the dish or food that precedes it.)

Day 1

BREAKFAST

Breakfast burrito
  1 flour tortilla (7" diameter)
  1 scrambled egg (in 1 tsp soft margarine)
  1/3 cup black beans*
  2 tbsp salsa
1 cup orange juice
1 cup fat-free milk

LUNCH

Roast beef sandwich
  1 whole grain sandwich bun
  3 ounces lean roast beef
  2 slices tomato
  1/4 cup shredded romaine lettuce
  1/8 cup sauteed mushrooms (in 1 tsp oil)
  1 1/2 ounce part-skim mozzarella cheese
  1 tsp yellow mustard
3/4 cup baked potato wedges*
  1 tbsp ketchup
1 unsweetened beverage

DINNER

Stuffed broiled salmon
  5 ounce salmon filet
  1 ounce bread stuffing mix
  1 tbsp chopped onions
  1 tbsp diced celery
  2 tsp canola oil
1/2 cup saffron (white) rice
  1 ounce slivered almonds
1/2 cup steamed broccoli
  1 tsp soft margarine
1 cup fat-free milk

SNACKS

1 cup cantaloupe

Day 2

BREAKFAST

Hot cereal
  1/2 cup cooked oatmeal
  2 tbsp raisins
  1 tsp soft margarine
1/2 cup fat-free milk
1 cup orange juice

LUNCH

Taco salad
  2 ounces tortilla chips
  2 ounces ground turkey, sauteed in
  2 tsp sunflower oil
  1/2 cup black beans*
  1/2 cup iceberg lettuce
  2 slices tomato
  1 ounce low-fat cheddar cheese
  2 tbsp salsa
  1/2 cup avocado
  1 tsp lime juice
1 unsweetened beverage

DINNER

Spinach lasagna
  1 cup lasagna noodles, cooked (2 oz dry)
  2/3 cup cooked spinach
  1/2 cup ricotta cheese
  1/2 cup tomato sauce tomato bits*
  1 ounce part-skim mozzarella cheese
1 ounce whole wheat dinner roll
1 cup fat-free milk

SNACKS

1/2 ounce dry-roasted almonds*
1/4 cup pineapple
2 tbsp raisins

Day 3

BREAKFAST

Cold cereal
  1 cup bran flakes
  1 cup fat-free milk
  1 small banana
1 slice whole wheat toast
  1 tsp soft margarine
1 cup prune juice

LUNCH

Tuna fish sandwich
  2 slices rye bread
  3 ounces tuna (packed in water, drained)
  2 tsp mayonnaise
  1 tbsp diced celery
  1/4 cup shredded romaine lettuce
  2 slices tomato
1 medium pear
1 cup fat-free milk

DINNER

Roasted chicken breast
  3 ounces boneless skinless chicken breast*
1 large baked sweetpotato
1/2 cup peas and onions
  1 tsp soft margarine
1 ounce whole wheat dinner roll
  1 tsp soft margarine
1 cup leafy greens salad
  3 tsp sunflower oil and vinegar dressing

SNACKS

1/4 cup dried apricots
1 cup low-fat fruited yogurt

Day 4

BREAKFAST

1 whole wheat English muffin
  2 tsp soft margarine
  1 tbsp jam or preserves
1 medium grapefruit
1 hard-cooked egg
1 unsweetened beverage

LUNCH

White bean-vegetable soup
  1 1/4 cup chunky vegetable soup
  1/2 cup white beans*
2 ounce breadstick
8 baby carrots
1 cup fat-free milk

DINNER

Rigatoni with meat sauce
  1 cup rigatoni pasta (2 ounces dry)
  1/2 cup tomato sauce tomato bits*
  2 ounces extra lean cooked ground beef (sauteed in 2 tsp vegetable oil)
  3 tbsp grated Parmesan cheese
Spinach salad
  1 cup baby spinach leaves
  1/2 cup tangerine slices
  1/2 ounce chopped walnuts
  3 tsp sunflower oil and vinegar dressing
1 cup fat-free milk

SNACKS

1 cup low-fat fruited yogurt

- Enter the "Tips & Resources" page and head to a wealth of ideas that can help you get started toward eating a healthy diet. Included are tips for each food group, physical activity, and eating out.

- Visit "Inside the Pyramid" for explanations about each food group, discretionary calories, and physical activity recommendations. This site provides information on which foods are within the various groups and food measure equivalents so you can convert food amounts into cups and ounces, and oils into teaspoons. Table 6.2 lists food measure equivalents for common foods by food group shown in this section of Inside the Pyramid.

- Refer to the "MyPyramid for Pregnant and Breastfeeding Mothers" to get basic information on nutritious diets for pregnancy and breastfeeding. You can track, evaluate, and plan changes to your pregnancy or breastfeeding diet at this site.

- Use "MyPyramid Tracker" for dietary and physical activity assessments that provide information on your diet quality, physical activity status, related nutrition messages, and links to nutrient and physical activity information.

Table 6.2

MyPyramid food measure equivalents

Grains	bagel	1 mini bagel = 1 oz	fruits	apple	1 small = 1 cup
		1 large bagel = 4 oz			½ large = 1 cup
	biscuit	1–2" diameter = 1 oz		banana	1 large = 1 cup
		1–3" diameter = 2 oz		cantaloupe	$^{1}/_{8}$ = 1 cup
	bread	1 slice = 1 oz		grapes	12 = 1 cup
	cooked cereal	½ cup = 1 oz		grapefruit	1 = 1 cup
	crackers	5 whole wheat = 1 oz		orange	1 = 1 cup
		7 square/round = 1 oz		peach	1 = 1 cup
	English muffin	½ muffin = 1 oz		pear	1 = 1 cup
	muffin	1–2½" diameter = 1 oz		plums	3 = 1 cup
		1–3½" diameter = 3 oz		strawberries	8 large = 1 cup
	pancake	1–4½" diameter = 1 oz		watermelon	1" wedge = 1 cup
		2–3" diameter = 1 oz		dried fruit	½ cup = 1 cup
	popcorn	3 cups = 1 oz		fruit juice	1 cup = 1 cup
	breakfast cereal	1 cup flakes = 1 oz	Meats and Beans	steak	1–3½" × 2½" × ½" = 3 oz
		1¼ cups puffed = 1 oz		hamburger	1 small = 2 oz
	rice	½ cup = 1 oz			1 medium = 4 oz
	pasta	½ cup = 1 oz			1 large = 6 oz
	tortilla	1–6" diameter = 1 oz		chicken	½ breast = 3 oz
		1–12" diameter = 4 oz			1 thigh = 2 oz
Vegetables	cooked	1 cup = 1 cup			1 leg = 3½ oz
	carrots	2 medium = 1 cup		pork chops	1 medium = 3 oz
		12 baby carrots = 1 cup		fish	1 small can tuna = 3½ oz
	celery	1 large stalk = 1 cup			1 small fish = 3 oz
	corn on the cob	1–6" long = ½ cup			1 salmon steak = 5 oz
		1–9" long = 1 cup		seafood	5 large shrimp = 1 oz
	green/red peppers	1 large = 1 cup			10 medium clams = 3 oz
	potatoes	1 medium (3" diameter) = 1 cup			½ cup lobster = 2½ oz
	raw, leafy	2 cups = 1 cup		eggs	1 small = 1 oz
	tomato	1 large = 1 cup			1 large = 2 oz
Milk	milk	1 cup = 1 cup		nuts	12 almonds = 1 oz
	yogurt	1 cup = 1 cup			1 Tbsp peanut butter = 1 oz
	cheese	1½ oz hard = 1 cup		beans	¼ cup cooked, dry beans = 1 oz
		$^{1}/_{3}$ cup shredded = 1 cup			¼ cup tofu = 1 oz
		2 oz processed = 1 cup			2 Tbsp hummus = 1 oz
		½ cup ricotta = 1 cup			
		2 cups cottage cheese = 1 cup			
	pudding	1 cup = 1 cup			
	frozen yogurt	1 cup = 1 cup			
	ice cream	1½ cup = 1 cup			

Although the MyPyramid Food Guide is not designed for weight-loss diets, the MyPyramid Tracker tool can be used to estimate and monitor calorie intake and physical activity levels.

Food group recommendations generated by MyPyramid are based on an individual's calorie need and provide adequate amounts of essential nutrients. People who exceed the recommended intake levels of fats and sugars may consume too many calories, however, and gain weight as a result.

Limitations of MyPyramid Materials available for MyPyramid are almost entirely made available on the Web, making the information inaccessible to people who do not use the Internet. This situation leaves primarily low-income families without access to MyPyramid.[11] MyPyramid does not provide specific recommendations for infants, individuals on therapeutic diets, or strict vegetarians.

Menus offered as examples on MyPyramid may not correpond to individual food preferences and contain relatively few ethnic foods. As with past food guides, planning and evaluating how mixed dishes (such as stews, soups, salads, and various types of pizza) fit into the foods groups still can be perplexing.

In MyPyramid, total amounts of food recommended from each food group are given in cups or ounces, servings of oils are given in teaspoons, and a calorie number is given for the "Discretionary Calories" allotment. One problem with this system is that people are not used to thinking of serving sizes in ounces of bread or rice or as teaspoons of oils. Also, they may not know the calorie value of sweets and fats covered by the Discretionary Calorie allotment.

The DASH Diet

Originally published as a diet that helps control mild and moderate **hypertension** in experimental studies, the DASH Eating Plan also reduces the risk of cancer, osteoporosis, and heart disease. Improvements in blood pressure are generally seen within two weeks of starting this dietary pattern.[12,13]

The DASH dietary pattern emphasizes fruits, vegetables, low-fat dairy foods, whole-grain products, poultry, fish, and nuts. Only low amounts of fats, red meats, sweets, and sugar-containing beverages are included. This dietary pattern provides ample amounts of potassium, magnesium, calcium, fiber, and protein and limited amounts of saturated and trans fats.

The DASH Eating Plan is available as part of the 2005 Dietary Guidelines. Recommendations for types and amounts of foods included in this eating plan by calorie need are shown in Table 6.3 Although two calorie need levels are shown in the illustration, a DASH eating plan is available for 12 levels (1600 to 3200 calories) at the Web site listed in Table 6.3.

The Mediterranean Diet The Mediterranean diet ranks with the MyPyramid Food Guide and the DASH Eating Plan when it comes to health promotion and chronic disease prevention.[10,15] The Mediterranean diet was originally based on foods consumed by people in Greece, Crete, southern Italy, and other Mediterranean areas where rates of chronic disease were low and life expectancy long.[16]

The Mediterranean dietary pattern, shown in Illustration 6.7, emphasizes daily consumption of bread, pasta, fruits, vegetables, olive oil, cheese and yogurt, and dried beans and nuts. Daily physical activity, a traditional part of life in these areas, is included in the plan. Fish, poultry, eggs, and sweets are recommended weekly, and red meat monthly. Wine with meals is part of the Mediterranean diet. A number of studies have shown that this dietary pattern is associated with lower risks of heart disease, stroke, several forms of cancer, and overall mortality.[10,12,15]

© Foodcollection/Getty Images

Table 6.3

The DASH Eating Plan

Food Group	2000 Calories	2600 Calories	Serving Size
	NUMBER OF SERVINGS PER DAY		
Grains[a]	6–8	10–11	1 slice bread 1 oz dry cereal ½ cup cooked rice, pasta, cereal
Vegetables	4–5	5–6	1 cup raw leafy vegetable ½ cup raw or cooked vegetables ½ cup vegetable juice
Fruits	4–5	5–6	1 medium fruit ¼ cup dried fruit ½ cup fresh, frozen, or canned fruit ½ cup fruit juice
Fat-free or low-fat milk and milk products	2–3	3	1 cup milk 1 cup low-fat or fat-free yogurt 1½ oz low-fat or fat-free cheese
Lean meats, poultry, fish	2 or fewer	2	3 oz cooked meat, poultry, or fish
Nuts, seeds, legumes	4–5/week	1/day	1/3 cup or 1½ oz nuts 2 Tbsp peanut butter 2 Tbsp or ½ oz seeds ½ cup cooked dry beans or peas
Fats and oils	2–3	3	1 tsp soft margarine 1 Tbsp low-fat mayonnaise 2 Tbsp light salad dressing 1 tsp vegetable oil
Sweets	5/week	2 or fewer/day	1 Tbsp sugar, jelly, jam ½ cup sorbet and ices 1 cup lemonade

[a]whole grain products primarily

Source: Dietary Guidelines for Americans, Appendix A-1: The DASH Eating Plan at 1,600-, 2,000-, 2,600-, and 3,100- Caloric Levels. Available at www.health.gov/dietaryguidelines/dga2005/document/html/appendixA.htm.

Courtesy, Media Partners, Inc. www.mediapartnersinc.com

Portion Distortion

Although food measure equivalents needed to estimate how much food should be consumed daily from each food group are described in MyPyramid (see Table 6.2), the problem of estimating portion sizes remains. Individual ideas of normal serving amounts are based on past experiences at family meals, the size of portions provided by restaurants, and packaged food and beverage sizes. Supersized meals at fast-food restaurants, large portions served by other restaurants, large bakery products, and larger cups of soft drinks are contributing to the problem of portion distortion. Table 6.4 provides examples of how food portion sizes are expanding.

As portion sizes have grown, people don't really know what a cup of spaghetti looks like on a plate. We have no conception of what portions are anymore.

—ELLEN SCHUSTER, PH.D.[17]

Is Supersizing Leading to Supersized Americans? Supersizing fast foods can double or triple the caloric content of the foods compared to their regular-sized counterparts. A single, supersized meal of a cheeseburger, large fries and thick shake provides more calories (about 2,200) than many people need in a day. Larger portions don't cost restaurants much more than smaller portions, they increase sales volume, and they encourage people to eat more. Many Americans are already eating plenty, and it is suspected that rising rates of obesity may be partly related to increased portion sizes.[18] Some health activists and legislators think it's time for fast-food and other restaurants to let consumers know the caloric cost, as well as the price, of menu items (Illustration 6.8).[19]

Illustration 6.7 The Mediterranean
Diet Pyramid

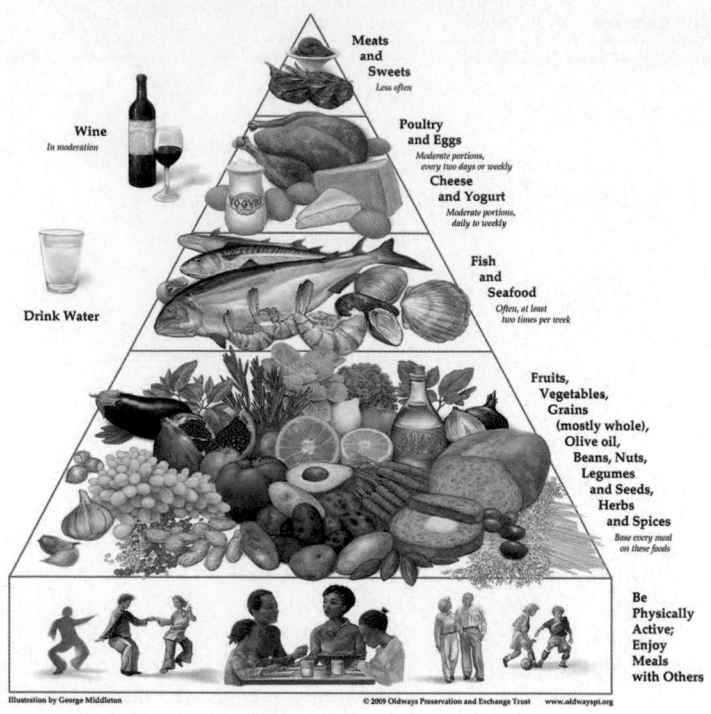

Mediterranean Diet Pyramid
A contemporary approach to delicious, healthy eating

Meats and Sweets
Less often

Wine
In moderation

Poultry and Eggs
Moderate portions, every two days or weekly

Cheese and Yogurt
Moderate portions, daily to weekly

Drink Water

Fish and Seafood
Often, at least two times per week

Fruits, Vegetables, Grains (mostly whole), Olive oil, Beans, Nuts, Legumes and Seeds, Herbs and Spices
Base every meal on these foods

Be Physically Active; Enjoy Meals with Others

Illustration by George Middleton   © 2009 Oldways Preservation and Exchange Trust   www.oldwayspt.org

A report commissioned by the Food and Drug Administration (FDA) concluded that restaurants are in a prime position to help improve the nation's diet and combat obesity. There are nearly a million restaurants and other eating establishments in the United States, but only a minority of them provide readily available nutrition information about the products they serve. Most people do not know how many calories are in the large portions of food they are eating in restaurants. Chances are good they are being served more calories than they realize. Simply letting consumers know how many calories are in the menu items offered may help them make better-informed decisions about what to eat.[20]

Table 6.4

Typical portion sizes in the marketplace

Marketplace Portion	Size
Bagel	5.8 oz
Muffin	6.5 oz
Hamburger bun	2.2 oz
Hamburger	3.9 oz
Steak	8.1 oz
French fries	5.3 oz
Pasta	2.9 C

The standard bakery size for a muffin used to be 1.5 ounces. It has grown to become a 6.5-ounce muffin.

Photo Disc

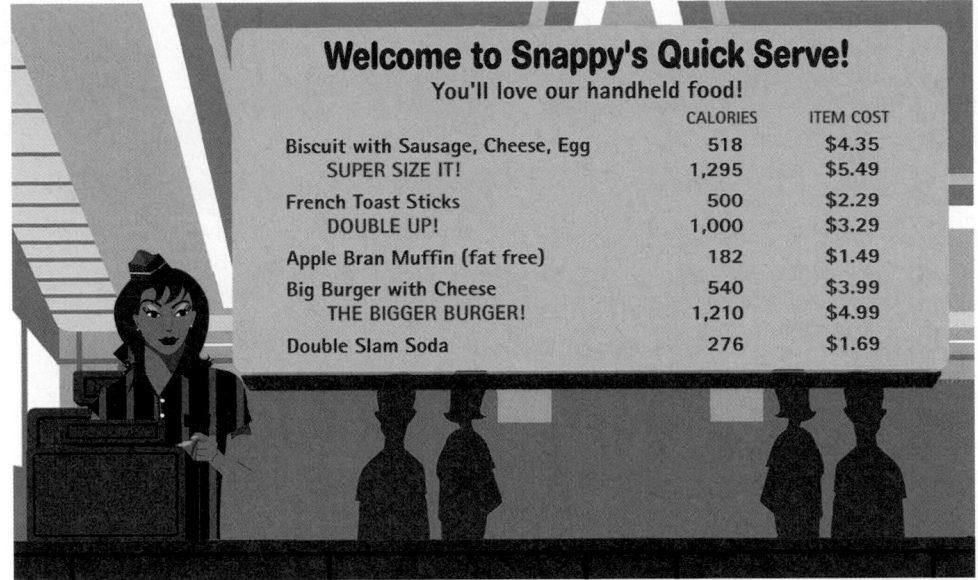

Welcome to Snappy's Quick Serve!
You'll love our handheld food!

	CALORIES	ITEM COST
Biscuit with Sausage, Cheese, Egg	518	$4.35
SUPER SIZE IT!	1,295	$5.49
French Toast Sticks	500	$2.29
DOUBLE UP!	1,000	$3.29
Apple Bran Muffin (fat free)	182	$1.49
Big Burger with Cheese	540	$3.99
THE BIGGER BURGER!	1,210	$4.99
Double Slam Soda	276	$1.69

Illustration 6.8 Should fast-food and chain restaurants be required to disclose the caloric content of menu items? Calorie labeling of fast food menu items is becoming a reality in some areas of the U.S.[19]

Can You Still Eat Right When Eating Out?

The question about what to eat often boils down to choosing the right restaurant. According to recent USDA data, 50% of Americans eat out every day. In general, foods eaten away from home have lower nutrient content and are higher in fat than foods eaten at home (see Illustration 6.9). In addition, children and teenagers who eat dinner with their families most days tend to have more healthful diets—including more vegetables, fruits, vitamins, minerals, and fiber, and less fat—than others who never or occasionally eat dinner with the family.[22]

Staying on Track while Eating Out   You'll find it easier to stick to a healthy diet if you decide what to eat before you enter a restaurant and look over the menu (Illustration 6.10). You could make the decision to order soup and a salad, broiled meat, a half-portion of the entrée, or no dessert *before* entering the restaurant. "Impulse ordering" is a hazard that can throw diets out of balance. If you're going to a party or business event where food will be served, decide before you go what types of food you will eat and what you will drink. If only high-calorie foods are offered, plan on taking a small portion and stopping there.

Illustration 6.9 Hamburgers, french fries, and pizza are the top-selling food items in U.S. restaurants.[20]

Portion Distortion   Mohammad and Kevin decided to eat at an Italian restaurant after soccer practice. Just for the fun of it, they agreed to guess how many ounces of spaghetti they would consume if they ate the portion served to them.

Answers on page 6-17

PhotoDisc

PhotoDisc

Mohammad: Let me see ... I'd guess it would be 6 ounces.

Kevin: I bet it's 16 ounces. It looks like it weighs a pound.

Photo Disc

Illustration 6.10   "No, no thank you. Extra cheese isn't part of what I planned to eat."

Can Fast Foods Be Part of a Healthy Diet?

As the information in Table 6.5 demonstrates, many of the foods served in fast-food restaurants deserve their reputation as being high in calories and fat. Most of the foods offered do not fit into a healthy diet if eaten on a regular basis. In recognition of this fact, and in response to legislative pressure to label the caloric value of fast-foods, lower calorie and more nutrient dense foods are being added to fast-food menus. Some of these foods are listed in Table 6.6 along with their content of calories and several key nutrients. Several higher calorie foods representing traditional offerings at fast-food restaurants are shown in the table for comparative purposes.

The addition of low-fat milk, apple slices, a variety of salads with low-fat dressings, and baked potatoes to fast-food menus makes it a bit easier to eat nutritiously when eating out. However, there is still room for improvement. Only 3% of kid's meals served at fast-food restaurants meet the National School Lunch Program's criteria for nutritional quality.[23]

The Slow Food Movement

An interesting trend in food preparation and consumption is making its way across the globe. The trend is away from fast and processed foods, and toward sustainable, eco-friendly agricultural practices and locally grown foods. The "Slow Food USA" movement represents some aspects of this trend. Part of an international group, Slow Food USA is an educational organization that supports ecologically sound food production; the revival of the kitchen and the table as centers of pleasure, culture, and community; and living a slower and more harmonious

Table 6.5

Calorie and fat content of some fast foods[a]

Food	Calories	% of Calories from Fat	Food	Calories	% of Calories from Fat
Sausage Breakfast Croissanwich	538	69	Hamburger	350	41
Bacon cheeseburger	610	58	Beef tostada	239	41
Chicken McNuggets® (6)	323	56	Beef n' Cheddar	490	39
Fried chicken breast	276	55	Beef burrito	357	38
Fried chicken drumstick	147	55	Roast beef sandwich	353	38
Quarter Pounder with cheese	525	55	Tostada	179	30
Big Mac	570	55	Grilled chicken breast on bun	320	28
Sausage McMuffin	427	55	Pizza with pepperoni (2 slices)	380	28
Whopper with cheese	711	54	Bean burrito	357	25
French fries, regular	227	52	Pizza with cheese (2 slices)	376	24
Cheeseburger	318	45	Vanilla shake	280	19
Burrito Supreme	457	43	Chocolate shake	447	16
Fillet-o-fish	373	43	Mashed potatoes, without gravy	62	15
Egg McMuffin	340	42	Baked potato, plain	250	7

[a]Calories and fat content may vary somewhat depending on the fast-food restaurant or chain.

Table 6.6

Improvements in the calorie and nutrient profiles of fast foods[a]

Newer items	Calories	Saturated fat	Fiber	Vegetable/Fruit
Side garden salad with low-fat dressing	135	1.5 g	1 g	1.0 c
Sliced apples	25	0 g	1 g	1.0 c
Raisins	130	0 g	2 g	0.5 c
Low-fat turkey sub, 6"	280	1.5 g	4 g	1.0 c
Fruit and walnut salad	210	1.5 g	2 g	1.0 c
Traditional items				
Quarter-pound hamburger on bun	410	7 g	2 g	0
Crispy chicken sandwich	530	3.5 g	3 g	0.25 c
Bacon ranch salad with crispy chicken and ranch dressing	540	8.5 g	4 g	2.5 c
Large french fries	550	3.5 g	6 g	2.0 c

[a]Calorie and nutrient content given per portion served by a fast food restaurant offering the item. Content may vary depending on the restaurant.

rhythm of life.[26,27] The trend is placing the topic of healthy eating in a new light for many individuals and communities (Illustration 6.11) and may help bring people closer to family, friends, and the environment.

What If You Don't Know How to Cook?

With so many convenience foods available and time at a premium, there is growing concern that we're becoming a nation of cooking illiterates. Cooking at home gives you control over what you eat and how it's prepared. Some people

Forsake fast food and convenience cuisine. Savor the enjoyment of ripe, locally grown produce and freshly prepared foods that have made a short, simple field-to-plate journey.
—SHARON PALMER, RD.[24]

I can be warmed by the memory of cooking something for somebody.
—MAYA ANGELOU, 2004

Portion Distortion The average portion size of pasta served by restaurants is nearly 3 cups—or 6 ounces.[25] For people who need 2000 calories, that's a day's allotment for the foods from the grain group.

Mohammad: 👍

Kevin: 👎

Illustration 6.11 The slow food movement is reaching urban areas.

© Jim West/PhotoEdit

immensely enjoy cooking and get a thrill out of making their specialties for friends and family. It's becoming a popular leisure-time activity: 43% of U.S. adults have taken it up for enjoyment.[28]

If you don't know how to cook, there are several ways to learn. You could start on your own by using the recipes on food packages like pasta, tomato sauce, or dried beans. You could search for recipes online or buy a basic cookbook and make simple dishes like salads, tacos, shish kebab, and lentil soup. You could even take a community education course. Illustration 6.12 shows some examples of good starter cookbooks. Read the sections on basic cooking skills and learn what types of equipment and utensils you need to prepare basic dishes. Get the foods and other supplies you need. Select the recipes that look doable and be sure they pass your taste and nutritional standards tests. Voilà! You're cooking.

Richard Anderson

Illustration 6.12 Some good starter cookbooks and recipe sources.

Bon Appétit!

Dietary guidelines from some other countries contain one other rule of healthy eating that would serve Americans well: "Enjoy your meals." Eating a healthy diet should be enjoyable. If it's too much of a struggle, the healthy diet won't last. The best diets are those that keep us healthy and enhance our sense of enjoyment and well-being. The trick is to remember the broad array of nutritious foods we like that give us good taste and enjoyment when we eat them. And remember not to feel guilty when you occasionally eat hamburgers or ice cream. Enjoy them to the utmost—as much as ripe oranges, papaya, homemade soups, roast turkey, hummus, and countless other nutritious delicacies.

Up Close

MyPyramid: A Balancing Act

Focal Point: Translating dietary intake into MyPyramid Food Group amounts

Planning a diet that supplies recommended amounts of foods from each food group requires that you be able to determine portion sizes and know how to use food measure equivalents. Here's an activity that will give you practice doing that.

Breakfast:	1 cup corn flakes ½ cup 2% milk
Lunch:	¼ pound cheeseburger on bun 1 cup potato chips 16 ounces iced tea (no sugar)
Dinner:	100 diameter white pizza with a thick crust 16 ounces cola
Snack:	2 cups ice cream

Using the information on Food Measure Equivalents in Table 6.2, and your best judgment, estimate how many cups or ounces of food are provided by the following one-day's diet for each MyPyramid food group.

Food Group

Grains	_____ ounces
Fruits	_____ cups
Vegetables	_____ cups
Meat and Beans	_____ ounces
Milk	_____ cups

FEEDBACK (answers to these questions) can be found at the end of Unit 6.

[Key Terms

adequate diet, page 6-4
balanced diet, page 6-4
essential nutrients, page 6-4

macronutrients, page 6-4
hypertension, page 6-12

saturated fats, page 6-4
trans fats, page 6-4

Review Questions

TRUE FALSE

1. Adequate diets are defined as those that provide sufficient calories to relieve hunger and maintain a person's body weight. ☐ ☐

2. The diets of Americans tend to be out of balance in a number of ways. ☐ ☐

3. The Mediterranean food guide is an example of a vegetarian diet plan that promotes health and decreases chronic disease risk. ☐ ☐

4. The 2005 Dietary Guidelines for Americans emphasize the importance of selecting nutrient-dense foods, balancing calorie intake with output, and increasing physical activity. ☐ ☐

5. "MyPyramid " is the major tool used in the United States to help people implement the Dietary Guidelines. ☐ ☐

6. The DASH Eating Plan represents a dietary pattern that is consistent with the 2005 Dietary Guidelines recommendations for dietary intake. ☐ ☐

TRUE FALSE

7. Reduction in salt (sodium) consumption is no longer recommended by the Dietary Guidelines. ☐ ☐

8. "Key Recommendations" are included for food safety in the Dietary Guidelines. ☐ ☐

9. To help stem the tide of increasing rates of obestiy in the United States, MyPyramid stresses the importance of lower caloric intake and increased physical activity. ☐ ☐

10. An advantage, as well as a limitation, of MyPyramid Food Guide is that it is principally available on the Internet. ☐ ☐

11. "Food Measure Equivalents" are used to identify standard serving sizes of foods with food groups. ☐ ☐

12. Fast-food resturants are now required to label food wrappers with the calorie value of the food in the wrapper. ☐ ☐

Media Menu

www.nutrition.gov
This site provides links to MyPyramid and the Dietary Guidelines.

www.dietitians.ca
Go from this Dietitians of Canada home page to find consumer information on advice for healthy eating and food preparation, visit a virtual grocery store, assess your nutrition profile, and find a dietitian.

www.mypyramid.gov
This is the U.S. Department of Agriculture's MyPyramid Food Guide main page, with links to dietary and physical activity assessments, menu planning information, interactive learning tools, and much more. Don't miss "MyPyramid Tracker"—it's fun.

www.slowfoodusa.org
Slow Food is an international association that promotes the pleasures of the table, locally produced food and sustainable agriculture, and defends food and agricultural biodiversity worldwide along with its parent, international group.

www.ars.usda.gov/foodsearch
This site presents USDA's What's In The Foods You Eat Search Tool. You can use it to view nutrient profiles of 13,000 foods.

www.healthydiningfinder.com/default.asp
The Healthy Dining Finder can help you locate healthy menu items at hundreds of fast food and other restaurants. Calorie and nutrient summaries are presented for the foods identified.

www.fao.org/ag/humannutrition/ nutritioneducation/fbdg/en
The Food and Agriculture Organization presents dietary guidelines and food guides they have received from around the world at this Web site.

www.ars.usda.gov/nutrientdata
USDA has the most expansive, highest quality, most frequently updated nutrient database on earth. Look up food sources of specific nutrients and find out the nutrient composition of foods at this interactive Web site.

www.dietaryguidelines.gov
The 2010 Dietary Guidelines for Americans will be released late in 2010 and can be located using this address. The address will also lead you to the 2010 Health Objectives for the Nation, also scheduled for release in 2010.

Notes

1. Food consumption and spending. Food Review, vol. 23, no. 3, www.ers.usda.gov/data/foodcomposition, accessed 8/03.

1. Guenther PM et al. Most Americans eat much less than the recommended amounts of fruits and vegetables, J Am Diet Assoc 2006;106:1371–9.

2. Dietary Reference Intakes. Energy, carbohydrate, fiber, fat, fatty acids, cholesterol, protein, and amino acids. Institute of Medicine, National Academy of Sciences. Washington, DC: National Academies Press, 2002.

3. What we eat in America, NHANES 2005–2006, available at www.ars.usda.gov/ba/bhnrc/fsrg, accessed 2/2/09.

4. Bachman JL et al. Sources of food group intake among the U.S. population. J Am Diet Assoc 2008;108:804–14.

5. Johnston CS et al. More Americans are eating "5-A-Day" but intake of dark green cruciferous vegetables remains low. J Nutr 2000;130:3063–67.

6. Kant AK et al. Secular trend in patterns of self-reported food consumption of adult Americans: NHANES 1971–1975 to NHANES 1999–2002, Am J Clin Nutr 206;84:1215–23.

7. Drewnowski A. The naturally nutrient-rich food index. Presented at the Experimental Biology annual conference, San Diego, Ca, April, 2008.

8. Asian Food Information Centre, Dietary Guidelines, www.afic.org, accessed 7/06.

9. Dietary Guidelines Executive Summary, www.health.gov/dietaryguide lines/dga2005/document/html/ executivesummary.htm, accessed 7/06.

10. Krebs-Smith SM et al. How does MyPyramid compare to other population-based recommendations for controlling chronic disease? J Am Diet Assoc 2007;107:830–7.

11. Johnston CS. Uncle Sam's diet sensation: My Pyramid—an overview and commentary. Medscape General Medicine 2005;7:10/21/05, accessed at www.medscape.com.

12. Dixon LB et al. Adherence to the USDA Food Guide, DASH Eating Plan, and Mediterranean Dietary Pattern reduces the risk of colorectal adenoma, J Nutr 2007;137:2443–50.

13. Nowson CA. et al. Blood pressure response to dietary modifications in free-living individuals J. Nutr. 134:2322–2329.

14. Tillotson JE. Fast-casual dining: our next eating passion? Nutr Today 2003; 38:91–4.

15. Dai J et al. Association between adherence to the Mediterranean diet and oxidative stress. Am J Clin Nutr 2008;88:1346–70.

16. Simopoulos AP. The Mediterranean diet: what is so special about the diet of Greece? The scientific evidence. J Nutr 2001;131 (Suppl):S3065–73.

17. Ellen Schuster is a professor of nutrition at Oregon State University. The quote was taken from a *Wall Street Journal* article published 5/20/03, p. D1. Fisher JO et al. Children's bite size and intake of an entrée are greater with large portions than with age-appropriate or self-selected portions. Am J Clin Nutr 2003;77:1164–70.

18. Wansink B et al. Portion size me: downsizing our consumption, J Am Diet Assoc 2007;107:1103–6.

19. Beck M. On the table: the calories lurking in restaurant food. Wall Street Journal 7/28/08;D1.

20. Bridges A. Restaurants should shrink portions. cnn.netscape.com/news/story, accessed 6/3/06.

21. Larson NI et al. Food preparation by young adults is associated with better diet quality, J Am Diet Assoc 2006;106:2001–7.

22. Gillman MW et al. Family dinner and diet quality among older children and adolescents. Arch Fam Med 2000;9:235–40.

23. O'Donnell SI et al. Nutrient quality of fast food kids meals. Am J Clin Nutr 2008;88:1388–95.

24. Palmer S. The slow food movement picks up speed. Today's Dietitian 2005;Nov.:36–9.

25. Young LR, Nestle M. Expanding portion sizes in the US marketplace: implications for nutrition counseling. J Am Diet Assoc 2003;103:231–4.

26. Slow food movement finally picking up speed, Aug. 24, 2008, www.msnbc.msn.com/id/26378691, accessed 2/2/09.

27. Slow Food USA, www.slowfoodusa.org, accessed 2/2/09.

28. Riley NS. The joys of cooking class. Wall Street Journal, March 9, 2006, page D9.

NUTRITION | # Up Close

MyPyramid: A Balancing Act

Feedback for Unit 6

If you came up with numbers close to the following you did a good job of breaking down a one-day diet into MyPyramid food group amounts:

Grains	20–21 ounces
Fruits	0 cups
Vegetables	½ cup
Meat and beans	3 ounces
Milk	4½ cups

Note: MyPyramid Tracker calculates that this day's diet provides 3125 calories.

How the Body Uses Food: Digestion and Absorption

	TRUE	FALSE
NUTRITION SCOREBOARD		
1 Almost all of the carbohydrate and fat you consume in foods is absorbed by the body, but only about half of the protein is absorbed.		
2 Disorders of the digestive system are a leading cause of hospitalizations and medical visits in the United States and Canada.		
3 Lactose maldigestion is a common digestive disorder.		

Key Concepts and Facts

- Our bodies are in a continuous state of renewal. Materials used to renew body tissues come from the food we eat in the form of nutrients.

- Digestion and absorption are processes that make nutrients in foods available for use by the body.

- Digestive disorders are common and often related to dietary intake.

La vie est une fonction chimique.
("Life is a chemical process.")
—Antoine Lavoisier, late 18th century

Photo Disc

The human body is an amazing machine. Mine is, anyway. For example, I regularly feed my body truly absurd foods, such as cheez doodles, and somehow it turns them into useful body parts, such as glands. At least I assume it turns them into useful body parts.
—Dave Barry, 1987

digestion
The mechanical and chemical processes whereby ingested food is converted into substances that can be absorbed by the intestinal tract and utilized by the body.

My Body, My Food

You are not the same person you were a month ago. Although your body looks the same and you don't notice the change, the substances that make up the organs and tissues of your body are constantly changing. Tissues we generally think of as solid and permanent, such as bones, the heart, blood vessels, and nerves, are continually renewing themselves. The raw materials used in the body's renewal processes are the nutrients you consume in foods.

Each day, about 5% of our body weight is replaced by new tissue. Existing components of cells are renewed, the substances in our blood are replaced, and body fluids are recycled. Taste cells, for example, are replaced about every seven days, and cells lining the intestinal tract are replaced every one to three days. All of the cells of the skin are replaced every month. Red blood cells turnover every 120 days. If you thought it was hard to maintain a car, an apartment, or a house, just imagine the maintenance of a body! Maintenance is just one of the body's ongoing functions that require nutrients as raw material.

How Do Nutrients in Food Become Available for the Body's Use?

The components of food that make "useful body parts" are nutrients. Through the processes of **digestion** and **absorption**, they are made available for use by every cell in the body.

The Internal Travels of Food: An Overview   The "food processor" of the body is the digestive system, shown in Illustration 7.1. It consists of a 25- to 30-foot-long muscular tube and organs such as the liver and pancreas that secrete digestive juices. The digestive juices break foods down into very small particles that can be absorbed and used by the body. The absorbable forms of carbohydrates are **monosaccharides**: glucose, fructose and galactose. Proteins are absorbed as amino acids and fats as fatty acids and glycerol. Vitamins and minerals are not broken down before they are absorbed; they are simply released from foods during digestion.

Much of the work of digestion is accomplished by **enzymes** manufactured by components of the digestive system such as the salivary glands, stomach,

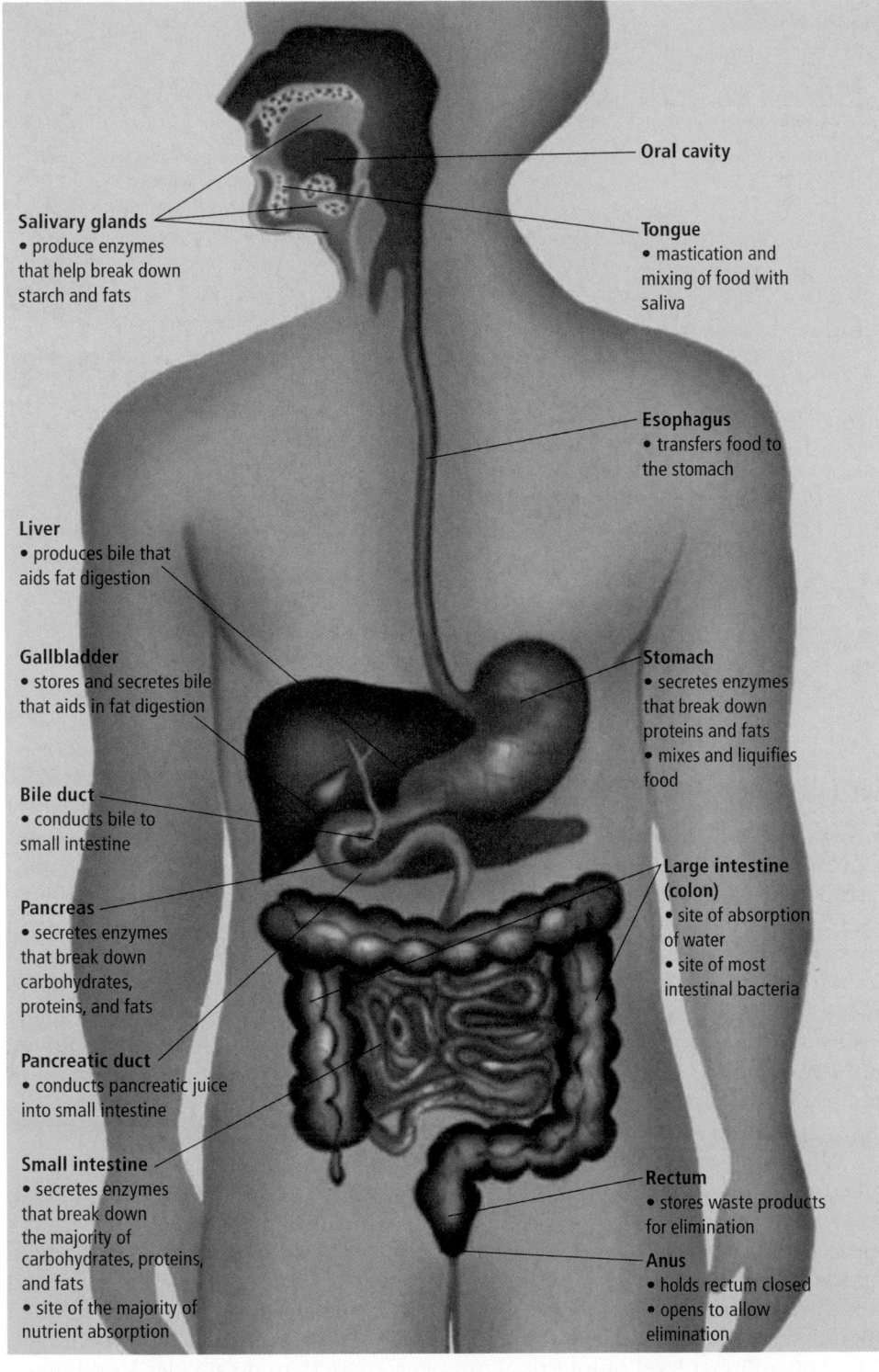

Illustration 7.1 The digestive system.

Oral cavity

Tongue
• mastication and mixing of food with saliva

Salivary glands
• produce enzymes that help break down starch and fats

Esophagus
• transfers food to the stomach

Liver
• produces bile that aids fat digestion

Stomach
• secretes enzymes that break down proteins and fats
• mixes and liquifies food

Gallbladder
• stores and secretes bile that aids in fat digestion

Bile duct
• conducts bile to small intestine

Large intestine (colon)
• site of absorption of water
• site of most intestinal bacteria

Pancreas
• secretes enzymes that break down carbohydrates, proteins, and fats

Pancreatic duct
• conducts pancreatic juice into small intestine

Small intestine
• secretes enzymes that break down the majority of carbohydrates, proteins, and fats
• site of the majority of nutrient absorption

Rectum
• stores waste products for elimination

Anus
• holds rectum closed
• opens to allow elimination

and pancreas. Enzymes are complex protein substances that speed up the reactions that break down food. A remarkable feature of enzymes is that they are not changed by the chemical reactions they affect. This makes them reusable.

Carbohydrates, proteins, and fat each have their own set of digestive enzymes. All together, over a hundred different enzymes participate in the digestion of carbohydrates, proteins, and fat. Table 7.1 presents information on some of the enzymes involved in digestion and highlights their specific roles. In Table 7.2 you will see how these enzymes are involved in the digestion of carbohydrate, fat, and protein.

absorption
The process by which nutrients and other substances are transferred from the digestive system into body fluids for transport throughout the body.

monosaccharides
(*mono* = one, *saccharide* = sugar) Simple sugars consisting of one sugar molecule. Glucose, fructose, and galactose are monosaccharides.

enzymes
Protein substances that speed up chemical reactions. Enzymes are found throughout the body but are present in particularly large amounts in the digestive system.

Table 7.1

Primary function of some digestive enzymes

Enzyme	Enzyme Function	Enzyme Source
A. Carbohydrate Digestion		
Amylase	Breaks down **starch** into smaller chains of glucose molecules	Produced in the salivary glands (salivary amylase) and the pancreas (pancreatic amylase)
Sucrase	Separates the **disaccharide** sucrose into glucose and fructose	Produced in the small intestine
Lactase	Splits the disaccharide lactose into glucose and galactose	Produced in the small intestine
Maltase	Separates maltose into two molecules of glucose	Produced in the small intestine
B. Fat Digestion		
Lipase	Breaks down fats into fragments of fatty acids and glycerol	Produced in salivary glands (lingual lipase) and the pancreas (pancreatic lipase). The action of lipase is enhanced by **bile**.
C. Protein Digestion		
Pepsin	Separates protein into shorter chains of amino acids	Produced by the stomach
Trypsin	Splits short chains of amino acids into molecules containing, one, two, or three amino acids	Produced by the pancreas

starch
Complex carbohydrates made up of complex chains of glucose molecules. Starch is the primary storage form of carbohydrate in plants. The vast majority of carbohydrate in our diet consists of starch, monosaccharides, and disaccharides.

disaccharide
Simple sugars consisting of two sugar molecules. Sucrose (table sugar) consists of a glucose and a fructose molecule, lactose (milk sugar) consists of glucose and galactose, and maltose (malt sugar) consists of two glucose molecules.

bile
A yellowish-brown or green fluid produced by the liver, stored in the gallbladder, and secreted into the small intestine. It acts like a detergent, breaking down globs of fat entering the small intestine to droplets, making the fats more accessible to the action of lipase.

A Closer Look As you chew, glands under the tongue release saliva that lubricates the food so that it can be swallowed and easily passed along the intestinal tract. Saliva also gets food digestion started. It contains salivary amylase and lipase that begin to break down carbohydrates and fats.

After food is chewed, it is swallowed and passed down the esophagus to the stomach. Muscles that act as valves at the entrance and exit of the stomach ensure that the food stays there until it's liquefied, mixed with digestive juices, and ready for the digestive processes of the small intestine. Solid foods tend to stay in the stomach for two to four hours, whereas most liquids pass through it in about 20 minutes.[4]

When the stomach has finished its work, it ejects 1 to 2 teaspoons of its liquefied contents into the small intestine through the muscular valve at its end. Stomach contents continue to be ejected in this fashion until they are totally released into the small intestine. These small pulses of liquefied food stimulate muscles in the intestinal walls to contract and relax; these movements churn and mix the food as it is digested by enzymes. When the diet contains sufficient fiber, the bulge of digesting food in the intestine tends to be larger. Larger food bulges stimulate a higher level of intestinal muscle activity than do smaller food bulges. Thus, high-fiber meals pass through the digestive system somewhat faster than low-fiber meals.

Digestion, as well as the absorption of nutrients, is greatly enhanced by the structure of the intestines (Illustration 7.2). Fingerlike projections called "villi" line the inside of the intestinal wall and increase its surface area tremendously. If laid flat, the surface area of the small intestine would be about the size of a baseball infield, or approximately 675 square feet. This large mass of tissue requires a high level of nutrients for maintenance. Much of this need (50% in the small intestine and 80% in the large intestine) is met by foods that are being digested.[6]

Table 7.2

Summary of the digestion of carbohydrates, fats, and proteins

	Mouth	Stomach	Small Intestine, Pancreas, Liver, and Gall Bladder	Large Intestine (Colon)
Carbohydrates (excluding fiber)	The salivary glands secrete saliva to moisten and lubricate food; chewing crushes and mixes it with salivary amylase that initiates starch digestion.	Digestion of starch continues while food remains in the stomach. Most alcohol is absorbed here. Acid produced in the stomach aids digestion and destroys bacteria in food.	Pancreatic amylase continues starch digestion. Sucrase, lactase, and maltase break down disaccharides into monosaccharides that are absorbed. Some alcohol is absorbed here.	Undigested carbohydrates reach the colon and can be partly broken down by intestinal bacteria.
Fiber	The teeth crush fiber and mix it with saliva to moisten it for swallowing.	No action.	Fiber binds cholesterol and some minerals.	Most fiber is excreted with feces; some fiber is digested by bacteria in the colon.
Fat	Fat-rich foods are mixed with saliva. Small amounts of lingual lipase accomplish some fat breakdown.	Fat tends to separate from the watery stomach fluid and foods and float on top of the mixture. Only a small amount of fat is digested. Fat is last to leave the stomach.	Bile readies fat for the action of lipase from the pancreas. Lipase splits fats into fatty acids and glycerol that are absorbed.	A small amount of fatty materials escapes absorption and is carried out of the body with other wastes.
Protein	In the mouth, chewing crushes and softens protein-rich foods and mixes them with saliva.	Stomach acid works to uncoil protein strands and to activate the stomach's protein-digesting enzyme. Pepsin breaks the protein strands into smaller fragments.	Trypsin splits protein into molecules containing one, two, or three amino acids. These amino acids are absorbed.	The large intestine carries undigested protein residue out of the body. Normally, almost all food protein is digested and absorbed.

Digestion is completed when carbohydrates, proteins, and fats are reduced to substances that can be absorbed and when vitamins and minerals are released from food. Between 40 to 70% of the alcohol consumed with a meal is absorbed from the stomach.[7] Water, sodium, and breakdown products of bacterial digestion of fiber (mostly fatty acids) are absorbed in the large intestine. Most nutrient absorption, however, takes place in the small intestine. The large intestine is home to many strains of bacteria that consume undigested fiber and other types of complex carbohydrates that are not broken down by human digestive enzymes. These bacteria excrete gas as well as fatty acids that are partly absorbed in the large intestine. Substances in food that cannot be digested or absorbed collect in the large intestine and are excreted in the stools.

On the Side

Get Your Juices Flowing
You don't have to actually eat food to start your digestive juices flowing. You just have to think about food or see it.[5] Put this information to the test. Clear your mind, and take a close look at Illustration 7.3.

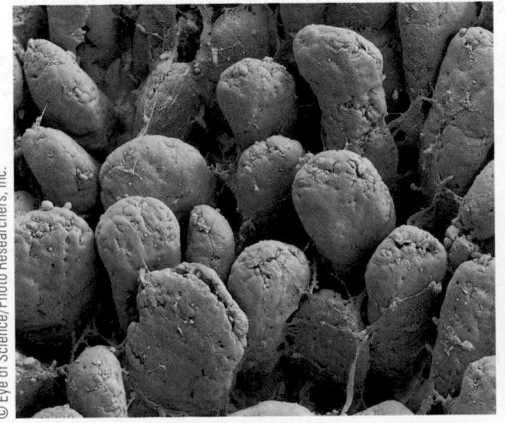 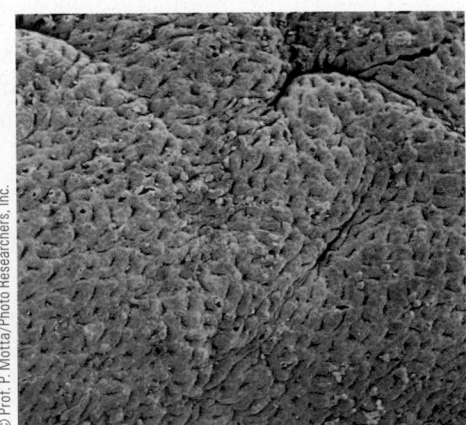

© Eye of Science/Photo Researchers, Inc.

© Prof. P. Motta/Photo Researchers, Inc.

Illustration 7.2 Scanning electron micrographs of cross sections of the small intestine (left) and the large intestine (right).
Note the high density of villi in the small intestine and the relative flatness of the lining of the large intestine.

Illustration 7.3 Testing, testing. This is a test of your salivary secretions. Did the lemon speak directly to your salivary glands? *If you want to turn the digestive processes off, quit thinking about food.*

© Michael Newman/PhotoEdit

lymphatic system
A network of vessels that absorb some of the products of digestion and transport them to the heart, where they are mixed with the substances contained in blood.

circulatory system
The heart, arteries, capillaries, and veins responsible for circulating blood throughout the body.

Absorption   Absorption is the process by which the end products of digestion are taken up by the **lymphatic system** (Illustration 7.4) and the **circulatory system** (Illustration 7.5) for eventual distribution to the cells of the body. Lymph vessels and blood vessels infiltrate the villi that line the inside of the intestines (Illustration 7.6) and transport absorbed nutrients toward the major branches of the lymphatic and circulatory systems. The breakdown products of fat digestion are largely absorbed into lymph vessels, whereas carbohydrate and protein breakdown products enter the blood vessels.

The nutrient-rich contents of the lymphatic system are transferred to the bloodstream at a site near the heart where vessels from both systems merge into one vessel. From there the lymph and blood mixture is sent to the heart and subsequently throughout the body by way of the circulatory system. The circulatory system reaches every organ and tissue in the body, thereby supplying cells with the nutrients obtained from food.

■ **Beyond Absorption.**   Cells can use nutrients directly for energy, body structures, or the regulation of body processes, they can also be converted into other usable substances. For example, glucose can be used "as is" for energy formation or converted to glycogen and stored for later use. Fatty acids, an end product of fat digestion, can be incorporated into cell membranes or used in the synthesis of certain hormones. Vitamins and minerals freed from food by digestion can be used by cells to control enzyme activity or can be stored for later use. The body has a limited storage capacity for some vitamins and minerals. Consequently, excessive amounts of certain vitamins and minerals such as vitamin C, thiamin, and sodium are largely excreted in urine.

■ **Digestion and Absorption Are Efficient.**   Approximately 99% of the carbohydrate, 92% of the protein, and 95% of the fat we consume in food are digested and absorbed. Dietary fiber, however, leaves the digestive system in much the same form as it entered. Humans don't have enzymes that break down fiber. It should be noted, however, that some fiber is digested in the large intestine by bacteria.

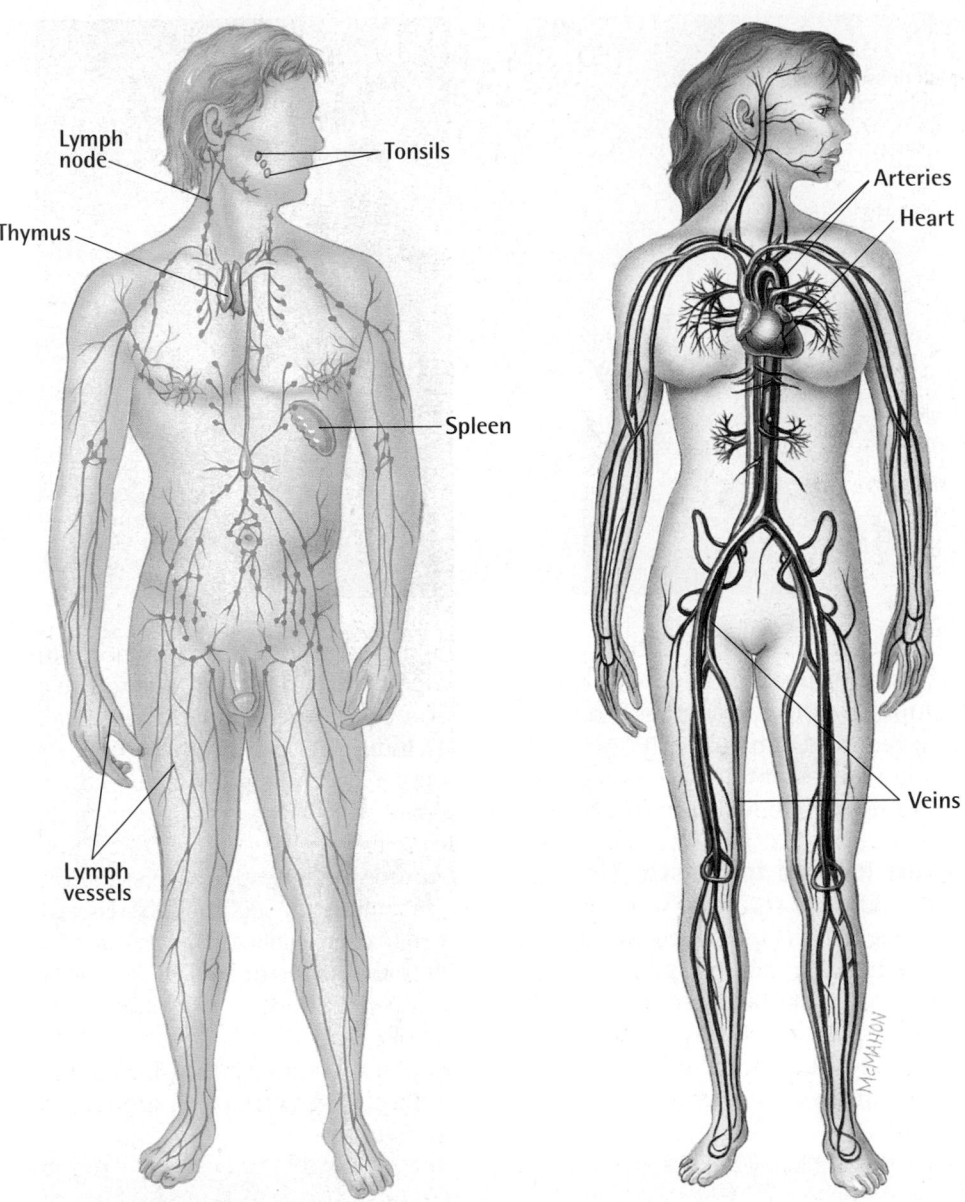

Illustration 7.4 (left) The lymphatic system.

Lymph node
Tonsils
Thymus
Spleen
Lymph vessels

Arteries
Heart
Veins

McMAHON

Illustration 7.5 (right) The circulatory system includes the heart and blood vessels. *This system serves as the nutrient transportation system of the body.*

Digestive Disorders Are Common

Excluding childbirth, digestive disorders—such as **heartburn**, **hemorrhoids**, **irritable bowel syndrome**, and **duodenal and stomach ulcers**—are the leading cause of hospitalization among U.S. and Canadian adults aged 20 to 44 years. Digestive disorders account for over 70 million medical visits each year in the United States alone.[12] Table 7.3 shows the percentages of U.S. adults who have common digestive disorders. Digestive disorders are common in children as well as adults. At least one-third of U.S. adults experience heartburn, and up to 28% of school children experience constipation.[8]

Constipation Almost everyone is constipated sometimes. Constipation is characterized by difficulty passing stools because they are hard and dry. People with constipation feel "blocked up" and lousy overall. They may see small amounts of bright red blood on the stool or toilet paper caused by bleeding **hemorrhoids** or a slight tearing of the anus. This generally disappears after constipation is controlled.[9]

There are a number of causes of constipation: the presence of a disorder or disease in the intestinal tract, immobility, medication use, the habitual use of

heartburn
A condition that results when acidic stomach contents are released into the esophagus, usually causing a burning sensation.

hemorrhoids (hem-or-oids)
Swelling of veins in the anus or rectum.

irritable bowel syndrome (IBS)
A disorder of bowel function characterized by chronic or episodic gas; abdominal pain; diarrhea, constipation, or both.

duodenal (do-odd-en-all) and stomach ulcers
Open sores in the lining of the duodenum (the uppermost part of the small intestine) or the stomach.

Illustration 7.6 Structure of villi, showing blood and lymph vessels.

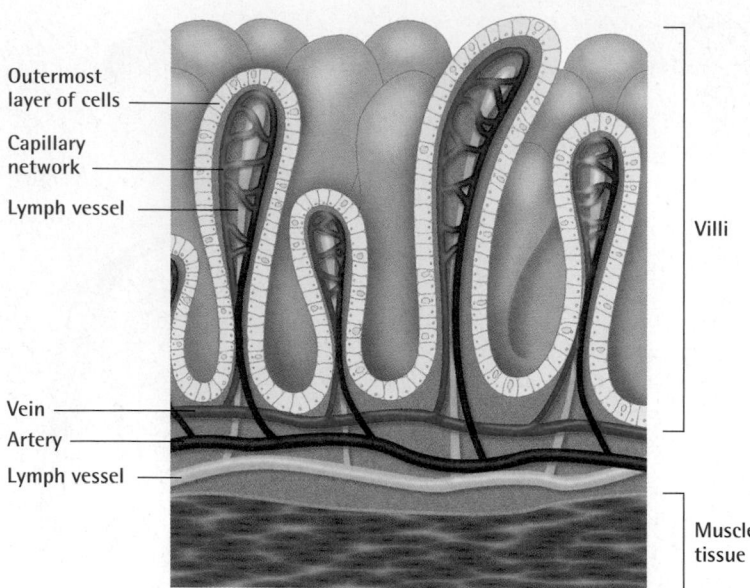

Outermost layer of cells

Capillary network

Lymph vessel

Vein

Artery

Lymph vessel

Villi

Muscle tissue

Table 7.3

Common digestive disorders[1]

U.S. Adults Affected[a]

Heartburn	33.0%
Hemorrhoids	12.8
Irritable bowel syndrome	10–15
Ulcers	3.5
Chronic constipation	3.0
Chronic diarrhea	1.2

[a]Noninstitutionalized adults experiencing the disorder in the past 12 months.

laxatives, or a slow transit time of food in the digestive tract. Constipation caused by a slow transit time of food through the digestive tract is often due to the consumption of too little fiber. This is a common cause of constipation and can be relieved, and subsequently prevented, by including 25–30 grams of fiber daily. Good sources of fiber are shown in Illustration 7.7. People with severe constipation may not benefit from high fiber intake, however.[9]

Myths Related to Constipation Some common beliefs about constipation do not hold up to scientific scrutiny (see Table 7.4 for a list of such beliefs). For example, there is no evidence that stools contain toxins that can be absorbed and cause harm to the body, or that you can improve your health by periodically "cleansing" or "detoxifying" the colon.[11] Unless dehydration is a problem, consuming plenty of fluids will not treat constipation.[9]

Some people are overly concerned about the regularity of bowel movements and constipation because they have been taught that you are supposed

Illustration 7.7 Food sources of dietary fiber.
Together, the foods shown provide 29 grams of dietary fiber, an amount that helps prevent constipation.

© Scott Goodwin Photography

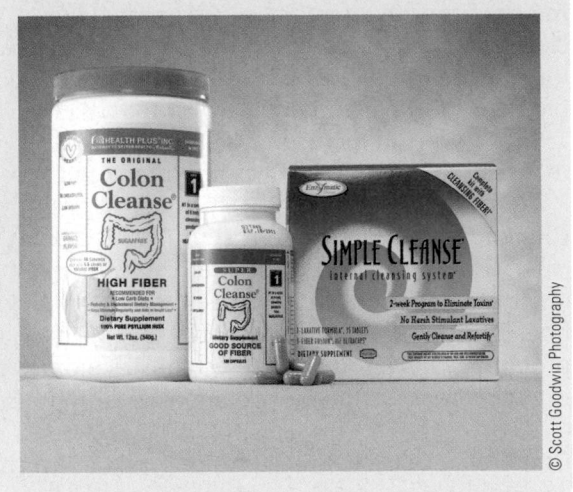
to empty your colon everyday. This isn't true. The range of normal bowel movements varies from three a day to three a week. A diagnosis of constipation is made when fewer than three bowel movements occur within a week. The presence of soft stools that are easily excreted is a strong sign that constipation does not exist.[9]

Ulcers and Heartburn   Ulcers develop when the protective barrier formed by cells lining the stomach and duodenum (the uppermost part of the small intestine) is damaged. This allows stomach acid and digestive enzymes to erode the lining of the stomach and duodenum and cause an "ulcer." Duodenal ulcers are ten times more common than stomach ulcers and are closely associated with the presence of *Helicobacter pylori (H. pylori)* bacteria.[9] *H. pylori* infects and irritates the lining of the stomach. The infection is acquired by the ingestion of foods and other substances contaminated with saliva, vomit, or feces from people harboring the bacteria in their stomach. Rates of *H. pylori* infection are highest in countries with poor sanitary conditions.[12]

A number of medications are available for the treatment of ulcers. Some work by reducing or eliminating *h. pylori* or controlling stomach acidity. The effectiveness of these medications in healing ulcers and preventing their re-development appears to be enhanced by the addition of **probiotics** to the diet. The appropriate types of probiotics for ulcer therapy can be consumed in specially formulated beverages, yogurt, or powders.[13]

probiotics
Nonharmful bacteria and some yeasts that help colonize the intestinal tract with beneficial microorganisms and that sometimes replace colonies of harmful microorganisms. The most common probiotic strains are *Lactobacilli* and *Bifidobacteria*.

Heartburn   Heartburn occurs daily in about 10% of American adults and 50% of pregnant women. It occurs occasionally in 30% of the population. Heartburn feels like a painful burning is the chest or throat near the area where the heart is located. The pain is due to stomach acid that has entered the esophagus. Stomach acid can be forced into the esophagus when the valve that closes the top of the stomach opens too often or does not close tightly. Occasional heartburn is not dangerous but chronic heartburn can develop into gastroesophageal reflux disease (GERD), which can cause serious damage to the esophagus.[9]

Many factors are related to the onset of heartburn (Table 7.5). Pregnancy, stress, smoking, certain medications, and overeating can bring on heartburn. Certain foods appear to relax the upper stomach valve and can trigger heartburn. These foods include tomatoes, citrus fruits, garlic, onions, alcohol, and caffeinated beverages. Foods high in fats and oils can also trigger heartburn.[9]

Avoiding stress, overeating, smoking, excess alcohol consumption, and offending foods can help prevent heartburn.[9–11] There are also medications available that can reduce symptoms.

Table 7.5

Foods and other factors that may be associated with the development of heartburn[9]

- Overeating
- High fat intake
- Consumption of "trigger" foods:
  - tomatoes
  - citrus fruit
  - garlic
  - onions
  - chocolate
  - coffee
  - alcohol
  - peppermint
- Stress
- Some medications
- Certain disorders
- Smoking

Irritable Bowel Syndrome   Irritable Bowel Syndrome, or IBS, has a name that matches its primary symptom: an irritated bowel. What causes the irritation is not yet clear, but the symptoms of abdominal pain, bloating, and diarrhea or constipation for three months or longer are common to the disorder.[14] Anxiety or depression is a problem for some individuals with IBS. IBS is a common disorder that affects 10 to 15% of people in the United State.[9] Although IBS is not associated with the development of serious disease, it has important, negative impacts on quality of life.[15]

There is no single, entirely effective therapy for IBS. It is usually treated with therapies that address each symptom.[16] Recently, a combination of soluble fiber (such as pysllium found in over the counter fiber supplement powders), probiotics, and peppermint oil has been found to alleviate the symptoms of IBS. Soluble fiber improves all IBS symptoms, peppermint oil relieves abdominal pain, and probiotics appear to help re-establish colonies of "healthy bacteria" in the large intestine that facilitate normal bowel functions.[14,17]

diarrhea
The presence of three or more liquid stools in a 24-hour period.

Diarrhea   **Diarrhea** is a common problem in the United States occuring an average of four times a year in adults.[9] It is a leading public health problem in developing countries. Most cases of diarrhea are due to bacterial- or viral-contaminated food or water, lack of immunizations against infectious diseases, and vitamin A, zinc, and other nutrient deficiencies. Children are particularly susceptible to diarrhea. Diarrhea can deplete the body of fluids and nutrients and produce malnutrition as well. If it lasts more than two weeks or is severe, diarrhea can lead to dehydration, heart and kidney malfunction, and death. An estimated 3.5 million deaths from diarrhea diseases occur each year to the world's population of children five years of age or under.[18]

The vast majority of cases of diarrhea can be prevented through food and water sanitation programs, immunizations, and adequate diets. The early use of oral rehydration fluids (for example, the formula provided by the World Health Organization and commercial formulas such as Pedialyte and Rehydralyte) shortens the duration of diarrhea. Rehydration generally takes four to six hours after the fluids are begun.[19]

Rather than "resting the gut" during diarrhea (as used to be recommended), children and adults, once rehydrated, should eat solid foods. Foods such as yogurt, lactose-free or regular milk, chicken, potatoes and other vegetables, dried beans, and rice and other cereals are generally well tolerated and provide nutrients needed for the repair of the intestinal tract. It is best to avoid sugary fluids such as soft drinks. High-sugar beverages tend to draw fluid into the intestinal tract rather than increase the absorption of fluid.[20]

Flatulence   Everyone experiences **flatulence**—it's normal. Gas can occur in the esophagus, stomach, small intestine, and large intestine due to swallowed air or bacterial breakdown of food in the large intestine. Air may be swallowed along with food and beverages or while chewing gum. Eating and drinking while in a rush generally increases air ingestion. Bacterial production of gas in the large intestine may be related to the ingestion of dried beans, broccoli, cauliflower, brussels sprouts, onions, corn, and other vegetables containing "resistant starch" that bacteria, but not humans, can break down. Fructose, which is used to sweeten a variety of food products and beverages, and sorbitol (used in some types of candy and gum) may lead to gas formation by bacteria that produce gas as a waste product of carbohydrate digestion. Heartburn and other gastrointestinal tract disorders and medications such as antibiotics are also associated with gas production.[1]

People often think they produce too much gas, even when they don't. The amount of gas swallowed and produced by gut bacteria varies a good deal among individuals and within the same individual. Gas production changes depending on what foods are eaten, the types of bacteria populating the large intestine, the medications used, and the presence of gastrointestinal tract disorders. Severe and painful symptoms related to gas production may signal the presence of a digestive disorder.[1]

Stomach Growling Gas in the stomach can make your stomach growl. When your stomach growls, you know that gas and food or fluids are mixing in your stomach. The growling tends to be louder when your stomach is empty, when there's no food to muffle the noise.

Lactose Maldigestion and Intolerance

A very common digestive disorder is **lactose maldigestion.** The lactose found in milk and milk products presents a problem for most of the world's adults, who cannot digest it, either partially or completely (Table 7.6).[23] The condition occurs more commonly in population groups that have no historical links to dairy farming and milk drinking.[22] Early humans in central and northwestern Europe and the regions of Africa and China highlighted in Illustration 7.8 tended to raise dairy animals and drink milk.

Lactose maldigestion is caused by a genetically determined low production of lactase, the enzyme that digests lactose. People who lack this enzyme end up with free lactose in their large intestine after they consume milk or milk products. The presence of lactose in the large intestine produces the symptoms of **lactose intolerance**: a bloated feeling, diarrhea, gas, and abdominal cramping.

Lactose maldigestion is rare in young children but affects adults to various degrees. Some adults produce little or no lactase and develop symptoms of lactose intolerance when they consume only small amounts of milk or milk products. Others produce some lactase and can tolerate limited amounts of lactose-containing milk and milk products, such as a cup of milk at a time or two cups of milk consumed with meals during the day. Regular consumption of milk may improve lactose digestion due to enhanced bacterial breakdown of lactose in the gut.[24] Lactase tablets can help improve lactose digestion, but the benefit is limited due to the inactivation of lactase by digestive processes in the stomach. Once lost, the body's ability to produce sufficient lactase cannot be restored.[25]

Many people who are lactose maldigesters have no trouble eating yogurt and other fermented milk products such as cultured buttermilk, kefir, and aged cheese. The bacteria used to culture yogurt can digest half or more of the lactose.

I can't drink milk. It tears me up inside.

lactose maldigestion
A disorder characterized by reduced digestion of lactose due to the low availability of the enzyme lactase.

lactose intolerance
The term for gastrointestinal symptoms (flatulence, bloating, abdominal pain, diarrhea, and "rumbling in the bowel") resulting from the consumption of more lactose than can be digested with available lactase.

Table 7.6

Estimated incidence of lactose maldigestion among older children and adults in different population groups[23]

Incidence of Lactose Maldigestion	
Asian Americans	90%
Africans	70
African Americans	70
Asians	65 or more
American Indians	62 or more
Mexican Americans	53 or more
U.S. adults (overall)	25
Northern Europeans	20
American Caucasians	15

Lactose digestion
Lactose maldigestion

Illustration 7.8 Lactose maldigestion is less common among descendants of people who consumed milk from domesticated animals during prehistoric times (light areas) than among people whose early ancestors did not drink milk (dark areas).[23]

Richard Anderson

Photo Disc

This reduction in lactose content is sufficient to prevent adverse effects in many people with lactose maldigestion.[25]

Milk solids, milk, and other lactose-containing components of milk may be added to foods you wouldn't expect. Consequently, it's best to examine food ingredient labels when in doubt. Milk, for instance, is a primary ingredient in some types of sherbet, and milk solids are added to many types of candy.

Do You Have Lactose Maldigestion? The single most reliable indicator of lactose maldigestion is the occurrence of lactose intolerance within hours after consuming lactose.[22] If you consistently experience symptoms of lactose maldigestion, visit your health care provider for a diagnosis. The symptoms could be due to lactose, other substances in milk, or another problem.

How Is Lactose Maldigestion Managed? Lactose maldigestion should *not* be managed by omitting milk and milk products from the diet! Doing so would exclude a food group that contributes a variety of nutrients that cannot easily be replaced by other foods. The omission of milk and milk products from the diet of people with lactose intolerance promotes the development of osteoporosis.[24] Rather, fortified soy milk, low-lactose cow's milk, milk pretreated with lactase drops, and yogurt and other fermented milk products (if tolerated) should be consumed. Illustration 7.9 shows a variety of dairy products that are generally well tolerated by people with lactose maldigestion.

Lactase tablets are also available and should be taken within 30 minutes of consuming lactose.[24] Lactase tablets can help improve lactose digestion, but the benefit is limited due to the inactivation of lactase by digestive processes in the stomach. Many people with lactose intolerance can drink a cup of regular milk with meals and feel fine. Repeated exposure to small quantities of milk and other lactose containing dairy products appears to enhance lactose digestion by fostering the growth of lactose consuming bacteria in the large intestine.[25]

Digestion is a remarkably complex and efficient process. Readers are encouraged to consult the Web sites listed at the end of the unit for additional information about digestion.

Up Close

© Digital Vision/Alamy

What are the end products of digestion?

Focal Point: Digestion makes carbohydrates, proteins, and fats available for absorption and utilization by the body.

You decide to develop a table that lists the primary end products of carbohydrate, protein, and fat digestion. You plan to use it as a study aid.

What would you list at the primary end products of carbohydrate, protein, and fat digestion?

	Primary end products of digestion
Carbohydrate	1.
	2.
	3.
Protein	1.
Fat	1.
	2.

FEEDBACK (Answers for the table are listed on the next page.)

[Key Terms

absorption, page 7-3
bile, page 7-4
circulatory system, page 7-6
diarrhea, page 7-10
digestion, page 7-2
disaccharide, page 7-4

duodenal and stomach ulcers, page 7-7
enzymes, page 7-3
flatulence, page 7-10
heartburn, page 7-7
hemorrhoids, page 7-7
irritable bowel syndrome (IBS), page 7-7

lactose intolerance, page 7-11
lactose maldigestion, page 7-11
lymphatic system, page 7-6
monosaccharides, page 7-3
probiotics, page 7-9
starch, page 7-4

Review Questions

TRUE FALSE

1. Cells lining the intestinal tract are replaced every 30 days. ☐ ☐

2. Enzymes are protein substances that speed up chemical reactions and are reusable. ☐ ☐

3. Digestion of proteins begins with the action of salivary amylase. ☐ ☐

4. The end products of digestion are taken up by the lymphatic and circulatory systems. ☐ ☐

5. The nutrient-rich contents of the lymphatic system do not mix with blood. These nutrients are distributed

TRUE FALSE

throughout the body by a separate system of lymph vessels. ☐ ☐

6. The body has a limited capacity to store some vitamins and minerals. ☐ ☐

7. Only about 75% of the protein, carbohydrate, and fat we consume in food is digested and absorbed. ☐ ☐

8. Tomatoes, high fat foods, and overeating may trigger heartburn. ☐ ☐

9. Most of the world's adult population has lactose maldigestion. ☐ ☐

Media Menu

www.healthfinder.gov
Search digestive diseases by name, or search "digestion."

www2.niddk.nih.gov
NIH's National Digestive Disease homepage provides links to information about specific digestive diseases, clinical trials, statistics, and

answers to questions such as "Why do I have gas?"

www.WebMD.com
Search "heartburn," "irritable bowel syndrome," "diarrhea" or other digestive disease topics for science-based information and practical advice on symptoms and management.

www.nlm.nih.gov/medlineplus/ lactoseintolerance.html
Get practical advice about lactose intolerance here.

www.nlm.nih.gov/medlineplus
This site highlights key aspects of digestion and absorption. Try using keyword searches.

Notes

1. Digestive disease statistics, National Digestive Disease Clearinghouse, http://digestive.niddk.nih.gov/statistics/statistics.htm, accessed 8/03 and 2/09.

2. Leading causes of hospitalization in Canada. Health Canada, Population and Public Health Branch, www.hc-sc.gc.ca, accessed 8/03.

3. Hertzler SR et al. Kefir improves lactose digestion and tolerance in adults with lactose maldigestion. J Am Diet Assoc 2003;103:582–74.

4. Camilleri M et al. Human gastric emptying and colonic filling of solids characterized by a new method, Am J Physiol Gastrointest Liver Physiol 1989;284–90.

5. Schneeman B. Nutrition and gastro-intestinal function. Nutr Today 1993; Jan/Feb:20–24.

6. Bengmark S. Econutrition and health maintenance: a new concept to prevent GI inflammation, ulceration, and sepsis. Clin Nutr 1996;15:1–10.

7. Cortot A et al. Gastric emptying and gastrointestinal absorption of alcohol ingested with a meal, Digest Disease Sci 1986;31343–8.

8. Borowitz SM et al. Precipitants of constipation during early childhood. J Am Fam Pract 2003;16:213–8.

9. National Institute of Digestive Diseases Health Information site, www2.niddk.nih.gov, accessed 2/09.

10. Kurata AN et al. Dyspepsia in primary care: perceived causes, reasons for improvement, and satisfaction with care. J Fam Practice 1997;44:281–8.

11. Muller-Lissner SA et al. Dispelling myths about constipation. Am J Gastroenterol 2005;100:232–42.

12. Suerbaum S et al. Helicobacter pylori infection. N Engl J Med 2002; 347:1175–86.

13. Lesbros-Pantoflickova et al. *Helicobacter pylori* and probiotics. J Nutr 2007;137:812S–18S.

14. American College of Gastroenterology IBS Task Force, Epidemiology, diagnosis, and treatments of IBS, Am J Gastroenterol 2009;104(Suppl):S1–S34.

15. Spiller R et al. Guidelines for the diagnosis and treatment of irritable bowel syndrome, Gut, May 8, 2007. Online First issue, available at www.medscape.com/viewarticle/556356, accessed 2/09.

16. Mayer EA, Irritable bowel syndrome, N Engl J Med 2008;358:1692–9.

17. Ford AC et al. Treatment of irritable bowel syndrome with fiber, peppermint oil, antispasmodics, and probiotics. BMJ, published online Nov. 14, 2008, accessed at www.medscape.com/viewarticle/583595, 2/09.

18. Kilgore PE et al. Trends in diarrheal disease associated mortality in U.S. children, 1968 through 1991. JAMA 1995; 274:1143–8.

19. Meyers A. Oral rehydration therapy: what are we waiting for? Am Fam Phys 1993;47:740–2.

20. Lima AAM et al. Persistent diarrhea in children: epidemiology, risk factors, pathophysiology, nutritional impact, and management. Epidemiol Rev 1992;34:222–42.

21. Stomach growling. Scientific American. Com ask the expert. www.scientific american.com, accessed 8/03.

22. Simmons FJ. Primary adult lactose intolerance and the milking habit: a problem in biological and cultural interrelationships. I. Review of the medical research. Am J Digestion 1981;14:819.

23. Inman-Felton AE. Overview of lactose maldigestion (lactose nonpersistence). J Am Diet Assoc 1999;99:481–9.

24. Lactose intolerance: a self-fulfilling prophecy leading to osteoporosis. Nutr Rev 2003;61:221–3.

25. Montalto M et al, Management and treatment of lactose malabsorption, Gastroenterol. 2006;12:187–91.

26. Ice cream headache. www.mayoclinic.com/health/ice-creamheadaches/DS00640, accessed 3/06.

Answers to Review Questions

1. False, see page 7-2.
2. True, see page 7-3.
3. False, see pages 7-4, 7-5.
4. True, see page 7-6.
5. False, see page 7-6.
6. True, see page 7-6.
7. False, see pages 7-6.
8. True, see page 7-9.
9. True, see page 7-11.

NUTRITION | Up Close

Primary end products of digestion

Feedback for Unit 7

Carbohydrate	1. glucose 2. fructose 3. galactose
Protein	1. amino acids
Fat	1. fatty acids 2. glycerol

Calories! Food, Energy, and Energy Balance

NUTRITION SCOREBOARD

		TRUE	FALSE
1	Carbohydrates provide more calories than do fats.		
2	A teaspoon of butter has a higher calorie value than a teaspoon of margarine.		
3	Energy can be neither created nor destroyed. It can, however, change from one form to another.		

Key Concepts and Facts

- A calorie is a unit of measure of energy.

- The body's sources of energy are carbohydrates, proteins, and fats (the "energy nutrients").

- Fats provide over twice as many calories per unit weight as carbohydrates and proteins do.

- Most foods contain a mixture of the energy nutrients as well as other substances.

- Weight is gained when caloric intake exceeds the body's need for energy. Weight is lost when caloric intake is less than the body's need for energy.

Energy!

It's high in energy, you say? Sure, I'll eat one, thanks. I thought those snack bars were loaded with calories.

When you think of calories, do you think of energy? That is the "scientifically correct" way to think about them. Energy is what calories are all about.

Calories Are a Unit of Measure

The **calorie** is like a centimeter or pound in that it is a unit of measure. Rather than serving as a measure of length or weight, the calorie is used as a measure of energy. Specifically, a calorie is the amount of energy needed to raise the temperature of 1 kilogram of water (about 4 cups) from 15°C to 16°C (59°F to 61°F)

Illustration 8.1

A calorie is the amount of energy needed to raise the temperature of 1 kilogram of water (about 4 cups) from 15°C to 16°C (59°F to 61°F).

(Illustration 8.1). This amount of energy is used as a standard for assigning caloric values to foods. Because calories are a unit of measure, they are not a component of food like vitamins or minerals. When we talk about the caloric content of a food, we're really talking about the caloric value of the food's energy content.

The caloric value of food is determined in a "bomb calorimeter" by burning it completely in a container surrounded by a specific amount of water (Illustration 8.2). The energy released by the food in the form of heat raises the temperature of the surrounding water. The rise in temperature indicates how many calories were

calorie (calor = heat)
A unit of measure used to express the amount of energy produced by foods in the form of heat. The calorie used in nutrition is the large "Calorie," or the "kilocalorie" (kcal). It equals the amount of energy needed to raise the temperature of 1 kilogram of water (about 4 cups) from 15 to 16°C (59 to 61°F). The term **kilocalorie**, or "calorie" as used in this text, is gradually being replaced by the "kilojoule" (kJ) in the United States; 1 kcal = 4.2 kJ.

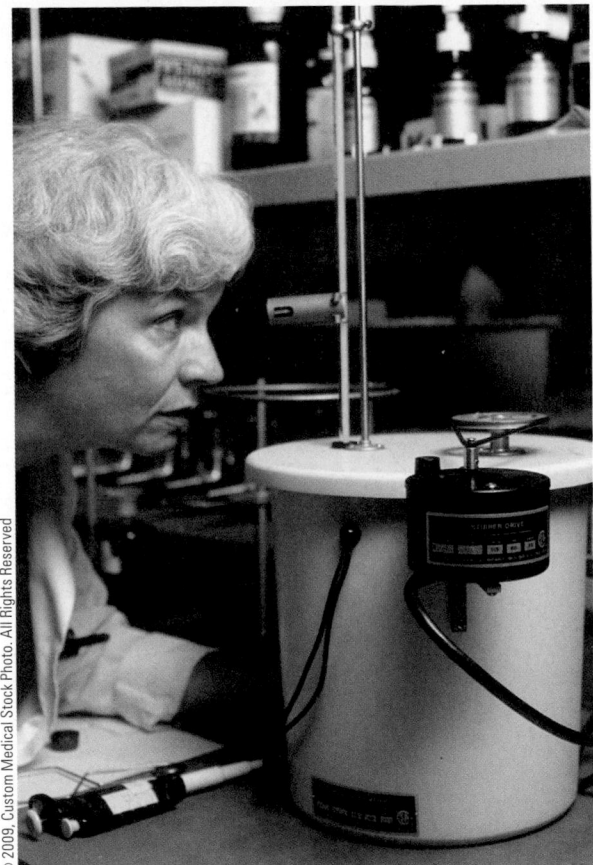

Illustration 8.2 A bomb calorimeter used to measure the calorie value of foods.
A food's calorie value is determined by the amount of heat released and transfered to water when the food is completely burned.

released from the portion of food. Although the body doesn't literally burn food, the amount of heat released by food while burning is approximately the same as the amount of energy it supplies to the body.

The Body's Need for Energy

The body uses energy from foods to fuel muscular activity, growth, and tissue repair and maintenance; to chemically process nutrients; and to maintain body temperature (to name a few examples). These needs for energy are subdivided into three categories: basal metabolism, physical activity, and dietary thermogenesis (Table 8.1 and Illustration 8.3).

The largest single contributor to energy need is basal metabolism or "resting metabolism," as it is also called. It accounts for 60 to 75% of the total need for calories in the vast majority of people.[1] Energy-requiring processes of **basal metabolism** include breathing, the beating of the heart, maintenance of body temperature, renewal of muscle and bone tissue, and other ongoing activities that sustain life and health. Growth is considered a component of basal metabolism. The

basal metabolism
Energy used to support body processes such as growth, health, tissue repair and maintenance, and other functions. Assessed while at rest, basal metabolism includes energy the body expends for breathing, the pumping of the heart, the maintenance of body temperature, and other life-sustaining, ongoing functions.

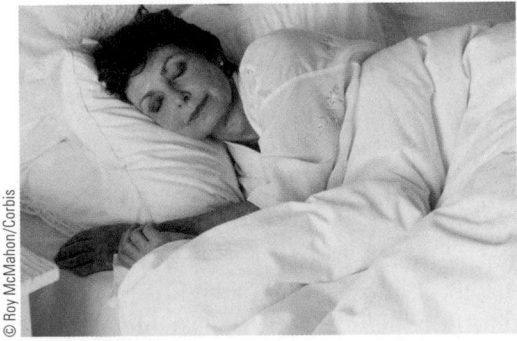

basal metabolism

(a)

physical activity

(b)

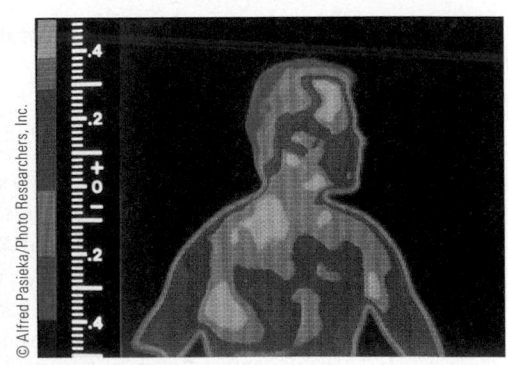

dietary thermogenesis

(c)

Illustration 8.3 Examples of the three types of energy-requiring processes in the body.
(a) basal metabolism (b) physical activity
(c) dietary thermogenesis

Is there any such thing as a slow metabolism?
Answers on next page

Mel:
I've got a slow metabolism. It doesn't matter how little I eat, I still can't lose weight.

Juanita:
Some of my friends who have trouble losing weight say that, too, but it can't be right. "Slow metabolism" is a myth.

proportion of total calories needed for basal metabolism is particularly high during the growing years.

Some organs and tissues in the body are more metabolically active, or require more energy to sustain their functions, than others. Body fat is less metabolically active than other tissues and accounts for under 20% of basal metabolic calorie expenditure in most people. The brain, liver, kidneys, and muscle are metabolically active; together, they account for 80% or more of the energy used for basal metabolism.[2]

Only a small percentage of people have unusually low or high **basal metabolic rates (BMR)**. In healthy individuals, slight differences in BMR rarely account for the ease with which weight is gained, maintained, or lost. Uncommon metabolic disorders, such as underactive thyroid and Cushing's syndrome, can modify basal metabolic processes and lead to changes in weight status.[3] The Reality Check for this unit addresses the topic of "slow" metabolism and weight loss.

Energy-using activities of basal metabolic processes require no conscious effort on our part; they are continuous activities that the body must perform to sustain life. The energy needed to carry out basal metabolic functions is assessed when the body is in a state of complete physical and emotional rest.

basal metabolic rate (BMR)
The rate at which energy is used by the body when it is at complete rest. BMR is expressed as calories used per unit of time, such as an hour, per unit of body weight in kilograms. Also commonly called *resting metabolic rate,* or *RMR.*

How Much Energy Do I Expend for Basal Metabolism?   You can quickly estimate the calories needed for basal metabolic processes:

- For men: Multiply body weight in pounds by 11.

- For women: Multiply body weight in pounds by 10.[4]

Thus, a man who weighs 170 pounds needs approximately 170 × 11, or 1870 calories per day for basal metabolic processes. A 135-pound woman needs 135 × 10, or 1350 calories.

This formula gives an estimate of calories used for basal metabolism in adults based on sex and weight. Physical activity level, muscle mass, height, health status, and genetic traits also influence calorie expenditure for basal metabolism to some extent. Consequently, results obtained using this quick formula may be 10 to 20% lower or higher than the true number of calories required for basal metabolism.[5–7]

How Much Energy Do I Expend in Physical Activity?   The caloric level needed for physical activity can vary a lot, depending on how active a person is. It usually accounts for the second highest amount of calories we expend. The energy cost of supporting a physically inactive lifestyle (Table 8.2) is about 30% of the number of

Table 8.2

Energy expenditure by usual level of activity[7,8]

Activity Level	Percentage of Basal Metabolism Calories
Inactive. Sitting most of the day; less than two hours of moving about slowly or standing	30%
Average. Sitting most of the day; walking or standing two to four hours, but no strenuous activity	50%
Active. Physically active four or more hours each day; little sitting or standing; some physically strenuous activities	75%

calories needed for basal metabolism. An "average" activity level requires roughly 50% of the calories needed for basal metabolism, and an "active" level requires approximately 75%.[5] A physically inactive person needing 1500 calories a day for basal metabolism, for example, would require about 450 calories (1500 calories × 0.30—or 30%) for physical activity.

■ **A Common Source of Error.** People have a tendency to overestimate time spent in physical activity. This in turn tends to overestimate calories needed for physical activity, and depending on the amount of overestimation, can lead to highly inaccurate estimates of total calorie need. Physical activity level should be based on time spent actually engaged in physical activity and not include time spent getting ready for the activity, on breaks, or between activities.

How Many Calories Does Dietary Thermogenesis Take? A portion of the body's energy expenditure is used for digesting foods, absorbing, utilizing, and storing nutrients, and transporting nutrients into cells. Some of the energy involved in such activities escapes as heat. These processes are referred to as **dietary thermogenesis.** Calories expended for dietary thermogenesis are estimated as 10% of the sum of basal metabolic and usual physical activity calories. For instance, say a person's basal metabolic need is 1500 calories and 450 calories are required for usual activity: 1500 calories + 450 calories = 1950 calories. Calories expended for dietary thermogenesis would equal approximately 10% of the 1950 calories, or 195 calories.

Adding It All Up Your estimated total daily need for calories is the sum of calories used for basal metabolism, physical activity, and dietary thermogenesis. In the preceding example, total calorie need would be 2145 calories (1500 + 450 + 195). A complete example of calculations involved in estimating a person's total calorie need is provided in Table 8.3. Although the caloric level calculated won't be exactly right, it should provide a reasonable estimate of your total caloric need.

dietary thermogenesis
Thermogenesis means "the production of heat." Dietary thermogenesis is the energy expended during the digestion of food and the absorption, utilization, storage, and transport of nutrients. Some of the energy escapes as heat. It accounts for approximately 10% of the body's total energy need. Also called *diet-induced thermogenesis* and *thermic effect of foods or feeding.*

Is there any such thing as a slow metabolism?
Mel's basal metabolism may be lower than average BMR for his weight, but his metabolism wouldn't be "slow." Calories needed for basal metabolism can be somewhat lower in people with relatively high amounts of body fat relative to muscle mass.[3]

Mel:

Juanita:

Table 8.3

Summary of calculations for estimating total calorie need of a 130-pound, inactive woman

	Calories
1. **Basal metabolism** Multiply body weight in pounds by 10. (For men the figure is 11.)	$130 \times 10 = 1300$
2. **Physical activity** Multiply basal metabolism calories by 0.30 (for 30%) based on the usual energy expenditure level of "inactive" (see Table 8.2).	$1300 \times 0.30 = 390$
3. **Dietary thermogenesis** Add calories needed for basal metabolism and physical activity together: 1300 + 390 = 1690 Multiply the result by 0.10 (for 10%).	$1690 = 0.10 = 169$
4. **Total calorie need** Add calories needed for basal metabolism, physical activity, and dietary thermogenesis together. 1300 + 390 + 169 =	**Total calorie need = 1859**

Where's the Energy in Foods?

Any food that contains carbohydrates, proteins, or fats (the "energy nutrients") supplies the body with energy. (That *includes* the foods listed in Table 8.4!) Carbohydrates and proteins supply the body with four calories per gram, and fat provides nine calories per gram. Alcohol also serves as a source of energy. There are seven calories in each gram of alcohol in the diet (Table 8.5).

If you enjoy grilling foods on an outdoor barbecue, you have probably observed firsthand the high level of stored energy in fats (Illustration 8.4). Unlike the drippings of low-fat foods such as shrimp or vegetables, drips from high-fat foods cause bursts of flames to shoot up from the grill. The high-energy content of alcohol can be seen in the alcohol-fueled flames that adorn cherries jubilee and bananas Foster.

If you know the carbohydrate, protein, fat, and (if present) alcohol content of a food or beverage, you can calculate how many calories it contains. For example, say a cup of soup contains 15 grams of carbohydrate, 10 grams of protein, and 5 grams of fat. To calculate the caloric value of the soup, multiply the number of grams of carbohydrate and protein by 4 and the number of grams of fat by 9. Then add the results together:

$$15 \text{ grams carbohydrate} \times 4 \text{ calories/gram} = 60 \text{ calories}$$
$$10 \text{ grams protein} \times 4 \text{ calories/gram} = 40 \text{ calories}$$
$$5 \text{ grams fat} \times 9 \text{ calories/gram} = \underline{45 \text{ calories}}$$
$$145 \text{ calories}$$

You can calculate the percentage of total calories from carbohydrate, protein, and fat in the soup by dividing the number of calories supplied by each nutrient by the total number of calories and then multiplying by 100:

- Carbohydrate: $\dfrac{60 \text{ calories}}{145 \text{ calories}} = 0.41 \times 100 = 41\%$

- Protein: $\dfrac{40 \text{ calories}}{145 \text{ calories}} = 0.28 \times 100 = 28\%$

- Fat: $\dfrac{45 \text{ calories}}{145 \text{ calories}} = 0.31 \times 100 = \underline{31\%}$

$$100\%$$

Table 8.4

Foods that *do* have calories

1. A candy bar eaten with a diet soda
2. Celery and grapefruit
3. Hot chocolate, cheesecake, or soft drinks consumed to make you feel better
4. Cookie pieces
5. Foods "taste-tested" during cooking
6. Foods you eat while on the run
7. Foods you eat straight from their original containers (like ice cream from the carton, milk from a jug, or chips from a large bag)

Table 8.5

Caloric values of the energy nutrients and alcohol

	cals/gm
Carbohydrate	4
Protein	4
Fat	9
Alcohol	7

Illustration 8.4 If you have observed the flames produced by fat dripping from a steak or hamburger on a grill, you have seen the powerhouse of energy stored in fat. *The carbohydrate and protein contents of grilled foods don't burn with nearly the same intensity. They have less energy to give.*

Given this information about the caloric value of carbohydrates, proteins, and fats, which of the items in Illustration 8.5 would you expect to be highest in calories?

College students have been found to miss this question 47% of the time.[9] It's the margarine! Margarine contains the most fat; it is made primarily from oil. *High-fat foods provide more calories ounce for ounce than foods that contain primarily carbohydrate or protein.* That means bread and potatoes (rich in carbohydrates), catsup (rich in water and a low-calorie vegetable), lean meats, and other low-fat or no-fat foods provide fewer calories than equal amounts of foods that contain primarily fat.

Most Foods Are a Mixture

Some foods (such as oil and table sugar) consist almost exclusively of one energy nutrient, but most foods contain carbohydrates, proteins, and fats in varying amounts.

Bread is high in complex carbohydrates ("starch"), but it also contains protein and a small amount of fat. Likewise, steak is not all protein. Although protein constitutes about 32% of the total weight of lean sirloin steak, fats make up about 8%, and the largest single ingredient is water—60% of the weight. Yet we don't think of steak as a "high-water food"; instead, it's often thought of as pure protein.

So, even though some foods provide relatively more carbohydrate, protein, or fat than other foods, most foods contain a mixture of energy nutrients.

Illustration 8.5 Which contains the most calories—a tablespoon of margarine, sugar, or pork? (Answers are on page 8-8.)

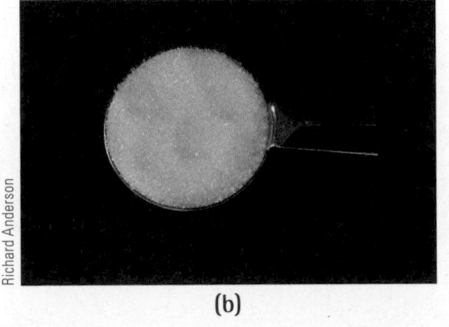

(a)

(c)

The Caloric Value of Foods: How Do You Know?

Some people say you can estimate the caloric value of food by how it tastes or by how appetizing it looks. Actually, it's not that simple. How many calories do you think are contained in a half-cup of peanuts; a boiled, medium-sized potato; and a cup of rice (Illustration 8.6)?

Taste, appearance, and reputation do not make good criteria for determining the caloric value of foods. Here's an example:

> I used to order the fish sandwich at fast-food restaurants because I was trying to avoid all the calories in hamburgers. Then I found out the fish sandwich had about the same number of calories as the quarter-pound hamburger! Where did I get the idea that fried fish loaded with tartar sauce has fewer calories than a hamburger?

This doesn't have to happen to you! By referring to Appendix A, you could determine that a typical fast-food fish sandwich and a quarter-pound hamburger weigh in at about 400 calories each. If you're interested in the caloric value of foods you frequently eat, look them up in this appendix.

Energy Density

The obesity epidemic in the United States is strongly related to increased calorie intake over recent decades. Levels of physical activity have declined, too, and have contributed to the increased rate of obesity.[10] Why are so many people consuming more calories than they need and gaining weight? Part of the answer appears to be related to the **energy density** (or calorie-density) of foods that have become a regular part of the U.S. diet.[11,12]

Energy density represents number of calories per gram of a food item. Twenty grams of potato chips (about 10 chips), for example, has 107 calories. Its energy density equals 107 calories divided by 20 grams, or 5.4. A 202 gram baked potato that provides 212 calories, on the other hand, has an energy density of 1.0 (212 calories divided by 202 grams).

Diets that regularly provide energy-dense foods are associated with overeating, excess calorie intake, weight gain, obesity, and type 2 diabetes.[11,12] Diets high in energy-dense foods may interfere with normal food intake regulation mechanisms by delaying the onset of satiety.[13] In addition, foods high in energy density tend to be nutrient-poor and those of low energy density nutrient-rich.[14] Because they are not energy-dense, nutrient-rich foods can be consumed in higher quantities while keeping calorie intake in check. Regular consumption of nutrient-rich foods that are low in energy density is associated with favorable nutrient intakes and reduced weight gain.[12] Vegetables, fruits, whole grain products, and

energy density
The number of calories in a gram of food. It is calculated by dividing the number of calories in a portion of food by the food's weight in grams.

Illustration 8.6 What's the caloric value of these foods?
One contains 420 calories, another 205, and the third 118 calories. Try matching the caloric values with the foods, and then check your answers on p. 8-10.

(a)

(b)

(c)

Health Action Moving toward foods lower in energy density

EAT MORE. CONSUME FEWER CALORIES!!!
No, this isn't an ad for a weight loss product. It's a public service announcement about the benefits of lower energy-dense foods. You *can* improve your nutrient intake while eating more food and reducing calorie intake. The key is to select foods that are relatively low in energy density often.

Higher energy-dense food	calories/g	Lower energy-dense food	calories/g
Taco shell	4.7	Corn tortilla	2.2
Bologna	3.1	Sliced turkey breast	0.9
Fried chicken	2.8	Grilled chicken	1.7
Fried pork chop	2.8	Broiled pork chop	2.0
Cheeseburger	2.7	Bean burrito	1.9
Hash brown potatoes	2.2	Boiled potato	0.9
Fried fish	2.2	Broiled fish	1.2
Fried rice	1.6	Rice	1.3
Potato salad	1.4	Tossed salad with salad dressing	1.1
Frozen, sweetened strawberries	1.1	Fresh strawberries	0.3

lean meats have lower energy densities than do foods like processed meats, fried foods, and high fat, sweet desserts.

Small differences in energy density make a big difference   Rather small differences in the energy density of foods can make a sizeable difference to calorie intake. On an equal weight basis, there is 38% more calories in macaroni and cheese (energy density = 1.2 calories/gram) than in Ramen noodles (energy density = 0.7 calories/gram). More examples of the energy density of foods are given in the Health Action feature.

How Is Caloric Intake Regulated by the Body?

Because energy is critical to survival, the body has a number of mechanisms that encourage regular caloric intake. It has less effective means of discouraging an excessive intake of calories.[13]

Mechanisms that encourage food intake don't depend on weight status. Rather, they are keyed to encouraging eating on a regular basis so that food, if available, will be consumed and carry the body through times when food isn't around. Humans developed on a schedule of "feast and famine." Those who could store enough fat to see them through the times when food was scarce had an advantage. So, no matter how thin or fat a person is, she or he experiences **hunger** if food is not consumed several times throughout the day. The "hungry" signal is thought to be sent by a series of complex mechanisms when cells run low on energy nutrients supplied by the last meal or snack.[15]

When we eat, we reach a point when we feel full and are no longer interested in eating. The signal is due to hormones and internal sensors in the brain, stomach, liver, and fat cells that indicate **satiety**—the feeling that we've had enough to eat.[16]

For some people, hunger and satiety mechanisms adjust energy intake to match the body's need for energy. However, for other people, internal signals that urge us to eat or to stop eating can be overridden. People can resist eating, no matter how strong the hunger pains. On the other hand, even after the "I'm full" siren has sounded, people can go on eating.[17] And sometimes people eat because they have an **appetite** for specific foods and the pleasure foods can bring.

hunger
Unpleasant physical and psychological sensations (weakness, stomach pains, irritability) that lead people to acquire and ingest food.

satiety
A feeling of fullness or of having had enough to eat.

appetite
The desire to eat; a pleasant sensation that is aroused by thoughts of the taste and enjoyment of food.

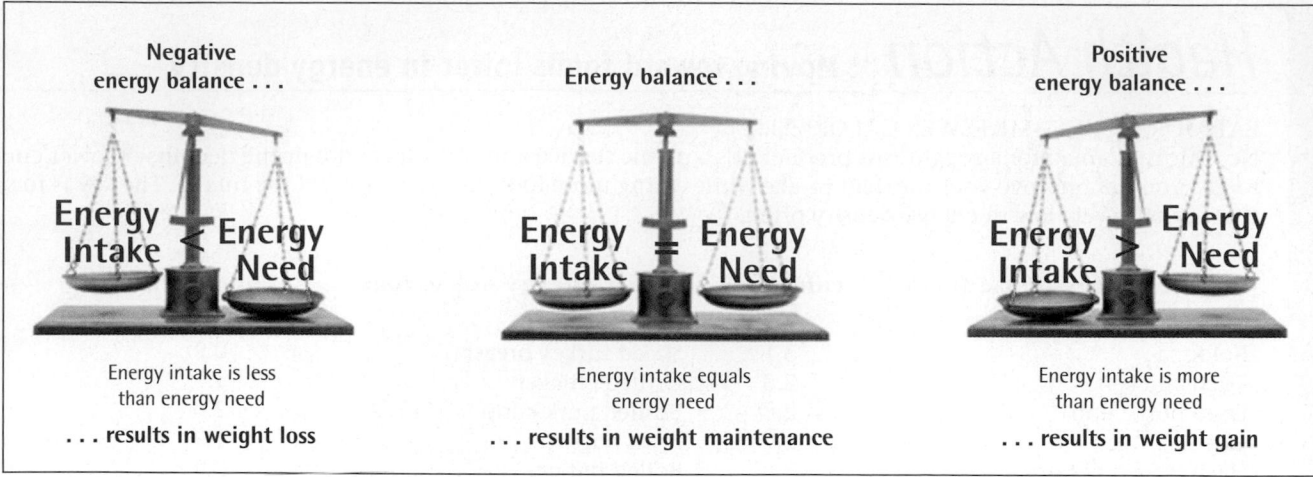

Illustration 8.7 The body's energy status.

People expend an average of 11 calories an hour chewing gum. If you chewed gum for an hour every day and didn't change any other component of energy balance, you'd lose about a pound a year.[18]

Photo Disc

Answer to Illustration 8.6:

	Calories
$\frac{1}{2}$ cup of peanuts	420
1 medium boiled potato	118
1 cup of white rice	205

Appetite may or may not be related to being hungry. It can be triggered when we smell or see a tasty food right after a meal or when we're really hungry. Have you ever seen those TV commercials for a juicy burger and fries that air around 11:00 P.M.? How many people who jump into their cars and head to the carry-out window are actually hungry? Or have you noticed how appealing the idea of eating is and how good food tastes when you're very hungry? That's appetite at work.

The Question of Energy Balance   Unless you are currently losing or gaining weight, the number of calories you need is the number you usually consume in your diet. Adults who maintain their weight are in a state of energy balance (Illustration 8.7). Because they are not losing weight (using fat and other energy stores) or gaining it (storing energy), their body's expenditure of energy and its intake of energy are balanced.

When energy intake is less than the amount of energy expended, people are in negative energy balance. In this case, energy stores are used and people lose weight. When a positive energy balance exists, weight and fat stores are gained because more energy is available from foods than is needed by the body. Slight changes in caloric need may not amount to noticeable fluctuations in body weight on a day-to-day basis. Over the course of a year, however, a *positive* or *negative* daily energy balance of 50 calories would result in a weight gain or loss of approximately five pounds.

Sometimes, a positive energy balance is a healthy and normal circumstance. For example, a positive energy balance is normal when growth is occurring, as in childhood or pregnancy, or when a person is regaining weight lost during an illness.

Keep Calories in Perspective

Calories is not a word that means "fattening" or "bad for you." Calories are a life and health-sustaining property of food. Diets compatible with health contain a mixture of foods providing various amounts of calories. It's not the caloric content of individual foods that makes them good or not. It's the sum of calories and nutrients in foods that make up our total diets.

Up Close

Food as a Source of Calories

Focal Point: Examine the distribution of calories in the foods you eat.

What are the sources of calories in food? Determine the caloric contribution of the fat, carbohydrate, and protein content of the following snack foods using the composition data and calorie conversion factors listed. Then answer the two questions that follow.

Which snack is lower in total calories?
_____

Which snack is lower in fat calories?
_____

PhotoDisc

PhotoDisc

Calorie conversion factors

1g fat   = 9 calories

1g carbohydrate = 4 calories

1g protein   = 4 calories

Potato Chips

Serving size: 1 oz (about 20 chips)

Fat: 10g × ____ = ____ calories from fat

Carbohydrate: 15g × ____ = ____ calories from carbohydrate

Protein: 2g × ____ = ____ calories from protein

_____ Total calories

% of calories from fat: _____ ÷ _____ = _____ × 100 = _____ %

Mini Pretzels

Serving size: 1 oz (about 17 pieces)

Fat: 0g × ____ = ____ calories from fat

Carbohydrate: 24g × ____ = ____ calories from carbohydrate

Protein: 3g × ____ = ____ calories from protein

_____ Total calories

% of calories from carbohydrate: _____ ÷ _____ = _____ × 100 = _____ %

FEEDBACK (answers to these questions) can be found at the end of Unit 8.

Key Terms

appetite, page 8–9
basal metabolic rate (BMR), page 8–4
basal metabolism, page 8–3

calorie, page 8–2
dietary thermogenesis, page 8–5
energy density, page 8–8

hunger, page 8–9
resting metabolic rate (RMR), page 8–4
satiety, page 8–9

Review Questions

	TRUE	FALSE
1. A calorie is also referred to as a kilocalorie (kcal).	☐	☐
2. Calories are a component of food that provide the body with energy.	☐	☐
3. A person's calorie need is based on his or her energy expenditure from basal metabolism, physical activity, and dietary thermogenesis.	☐	☐
4. Basal metabolic rate varies substantially among people due to differences in genetic traits and disease history.	☐	☐
5. George, a city bus driver, weighs 220 pounds. The number of calories he expends for basal metabolism daily would equal about 2420 calories.	☐	☐
6. A common source of error in estimations of physical activity level is underestimation of the amount of time spent in physical activities.	☐	☐

	TRUE	FALSE
7. Dietary thermogenesis accounts for approximately 30% of a person's total calorie need.	☐	☐
8. Say you ate a medium serving of french fries that contained 3 grams of protein, 38 grams of carbohydrate, and 14 grams of fat. The calorie value of the french fries would be 290.	☐	☐
9. The serving of french fries referred to in question 8 would provide 60% of calories from fat.	☐	☐
10. The composition of lean steak is nearly 100% protein.	☐	☐
11. *Appetite* is related to the desire for food, whereas *hunger* refers to a physiological need for food.	☐	☐
12. A person who is gaining weight is in positive energy balance.	☐	☐

Media Menu

www.ars.usda.gov/ba/bhnrc/ndl
At the homepage for USDA's food composition information, you can obtain nutrient data files and get answers to commonly asked questions about the calorie value of foods from this huge site.

http://exercise.about.com/od/ fitnesstoolscalculators/
This site offered by "About Health and Fitness," provides a BMI and calorie expenditure

calculator based on data you input. The site has a lot of ads, but the calculators are easy to use.

www.jacn.org/cgi/content/full/25/2/123
Have you gained weight during college? Want to know more about why? Go to this site and read the Journal of the American College of Nutrition's article titled "Changes in

Body Weight, Body Composition and Resting Metabolic Rate (RMR) in First-Year University Freshmen Students."

Notes

1. Poehlman ET et al. Energy needs: assessment and requirement in humans. In: Shils ME et al. Modern Nutrition in Health and Disease, 9e, Philadelphia: Lippincott, Williams, and Wilkins, 1998, p.96.

2. Hsu A et al. Larger mass of high-metabolic rate organs does not explain higher resting energy expenditure in children. Am J Clin Nutr 2003;77: 1506–11.

3. Slow metabolism: Is there any such thing? www.mayoclinic.com/health/slow-metabolism/AN00618, accessed 7/06.

4. Harris JA et al. A biometric study of basal metabolism in men. Publication no. 279 of the Carnegie Institute of Washington, 1919.

5. Sjodin AM, et al., The influence of physical activity on BMR, Med Sci Sports Exercise 1996;28:85–91.

6. Klausen B, et al., Age and sex effects on energy expenditure, Am J Clin Nutr 1997;65:895–907.

7. Frankenfield DC, et al., The Harris-Benedict studies of human basal metabolism: history and limitations, J Am Diet Assoc 1998;98:439–45.

8. Boothby WM et al. Studies of the energy of metabolism of normal individuals: a standard for basal metabolism, with a nomogram for clinical application. Am J Physiol 1936;116:468–84.

9. Melby CL et al. Reported dietary and exercise behaviors, beliefs and knowledge among university undergraduates. Nutr Res 1986;6:799–808.

10. Dietary Guidelines for Americans: Discretionary Calories, 2005, available from www.nutrition.gov.

11. Wang J et al. Dietary energy density predicts the risk of incident type 2 diabetes: the European Prospective Investigation of Cancer (EPIC)-Norfolk Study. Diabetes Care. 2008;31:2120–5.

12. Savage JS et al. Dietary energy density predicts women's weight change over 6 y, Am J Clin Nutr 2008;88:677–84.

13. Jacobs DR, Jr. Fast food and sedentary lifestyle: a combination that leads to obesity. Am J Clin Nutr 2006;83:189–90.

14. Schroder H et al. Low energy density diets are associated with favorable nutrient intake profile and adequacy in free-living elderly men and women. J Nutr 2008;138:1476–81.

15. Mattes RD et al. Nonnutritive sweetener consumption in humans: effects on appetite and food intake and their putative mechanisms. Am J Clin Nutr 2009;8:1–14.

16. McCrory MA et al. Dietary determinants of energy intake and weight regulation in healthy adults. J Nutr 2000; 130:276S–9S.

17. Marmonier C et al. Snacks consumed in a nonhungry state have poor satiating efficiency. Am J Clin Nutr 2002;76: 518–28.

18. Levine J et al. The energy expended in chewing gum. N Eng J Med 1999;341: 2100.

1. True, see page 8-2.
2. False, see page 8-2.
3. True, see page 8-3.
4. False, see page 8-4.
5. True, see page 8–4.
6. False, see page 8–5.
7. False, see page 8–5.
8. True, see page 8–6.
9. False, see page 8–6.
10. False, see page 8–7.
11. True, see page 8–9.
12. True. see page 8–10.

NUTRITION | Up Close

Food as a Source of Calories

Feedback for Unit 8

Potato Chips

Serving size: 1 oz (about 20 chips)

Fat: 10 g 3 __9__ = __90__ calories from fat

Carbohydrate: 15 g 3 __4__ = __60__ calories from carbohydrate

Protein: 2 g 3 __4__ = __8__ calories from protein

__158__ Total calories

% of calories from fat: $\dfrac{90}{158}$ = 0.57×100 = __57%__

Mini Pretzels

Serving size: 1 oz (about 17 pieces)

Fat: 0 g 3 __9__ = __0__ calories from fat

Carbohydrate: 24 g 3 __4__ = __96__ calories from carbohydrate

Protein: 3 g 3 __4__ = __12__ calories from protein

__108__ Total calories

% of calories from carbohydrate: $\dfrac{96}{108}$ = 0.89×100 = __89%__

The pretzels are lower in total calories and fat calories
than the potato chips.

Obesity to Underweight: The Highs and Lows of Weight Status

NUTRITION SCOREBOARD

	TRUE	FALSE
1 There is usually little difference between ideal body weights defined by cultural norms and those defined by science.		
2 Fat stores located in the stomach area present a greater hazard to health than do fat stores located around the hips and thighs.		
3 The basic cause of obesity is calorie intake that exceeds calorie needs.		
4 Healthy underweight people often find it nearly impossible to gain weight.		

Key Concepts and Facts

- Ideal body weight and shape are defined by culture and by health measures. The operating definition should be based on health.

- Rates of overweight and obesity are increasing worldwide. Diseases and disorders related to excess body fat are also increasing.

- The location of body fat stores, as well as the amount of body fat, are important to health.

- The causes of obesity are complex and not completely understood. Diet, physical activity, environmental exposures, and genetic factors influence the development of obesity.

- Underweight in the United States usually results from a genetic tendency to be thin or from poverty, illness, or the voluntary restriction of food intake.

Answers to **NUTRITION SCOREBOARD**	TRUE	FALSE
1 Cultural norms of ideal body weights are often at odds with the "healthy" weights defined by science.		✔
2 Health is affected by the location of body fat stores as well as by obesity.[1] Adults who store body fat in the stomach, or central area of the body, are at higher risk for a number of health problems than are adults who store fat primarily in their hips and thighs.	✔	
3 Many factors contribute to the development of obesity. However, the basic cause of obesity is calorie intakes that exceed calorie needs.	✔	
4 Gaining weight is as difficult for many underweight people as losing weight is for many overweight people.	✔	

Variations in Body Weight

Wouldn't it be terrific if your body adjusted your food intake based on a healthy level of body fat stores? Then you wouldn't have to worry about being too thin or getting too fat. Why doesn't that happen?

For humans, one size does not fit all. Human bodies come in a range of sizes that vary from heavy and tall to thin and short. Over recent decades, however, the distribution of body sizes has become lopsided. The proportion of people in the United States and other countries classified as overweight or obese is increasing, and rates of underweight are dropping. What lies underneath these trends? Why are humans so susceptible to gaining weight?

Obesity appears to represent a weak link in the biological evolution of humans. Body processes that regulate food intake developed over 40,000 years ago when "feast and famine" cycles were common. Being underweight was a distinct disadvantage. The more body fat people could store after a feast, the better their chances of surviving the subsequent famines. Consequently, multiple mechanisms that favored food intake and body fat storage developed.[2] These body mechanisms continue to encourage food intake even when food is constantly available and obesity poses a greater threat to survival than does famine. In the words of Theodore Van Itallie, a noted obesity researcher, in environments with an abundant supply of food and no requirement for vigorous physical activity, "perhaps thin people are the ones who are abnormal."[3]

Circumstances today are very different than they were 40,000 years ago. These differences may go a long way to explaining why obesity is a major problem in many countries.

© Neville Lockhart/Getty Images/Gallo Images

How Is Weight Status Defined?

Culture and science define the appropriateness of body size. In the 15th century, moon-faced and pear-shaped women were considered beautiful.[4] In the 1930s, full-figured women such as Jean Harlow and Mae West were box-office stars, and Marilyn Monroe—also a full-figured woman—was an icon of the 1950s. Then came the ulta-thin Twiggy and Keira Knightley. Thinness beyond health boundaries has become the standard of beauty. With "too thin" as the cultural ideal, many people regard weights within the normal range as being too high. Men feel the pressure, too. For them, visible body fat and weak-looking bodies are culturally taboo.[5]

Science defines standards for body weight for adults primarily based on the risk of death from all causes. Death rates are highest among adults who have very high body weights for height, the next highest rates are for underweight adults, and the lowest rates are among adults who are normal weight for height (Illustration 9.1).[1] You can get an idea of what a normal weight for height is by following the steps detailed in the "Health Action."

In the past, standards used to identify healthy body weights for adults were developed by the insurance industry. Height and weight tables were created to estimate life and death expectancy and the risk of death for adults. These tables have been replaced by standards that employ **body mass index**.

Body Mass Index

Most commonly referred to as BMI, body mass index is a measure of weight for height that provides a fairly good estimate of body fat content.[8] Ranges of BMI are used to define weights for height that correspond to underweight, normal weight, overweight, and obesity in adults (Table 9.1). BMI has the advantage of being calculated the same way for adult males and females. Calculating BMI involves dividing weight in pounds by height in inches, then dividing this result by height in inches again, and then multiplying that result by 703. An example of BMI calculation is given in Table 9.2. You can use the chart shown in Illustration 9.2 to identify weight status based on BMI for information on weight and height.

Assessing Weight Status in Children and Adolescents Standards used to assess weight status in children and adolescents employ BMI percentile ranges for girls and boys (Illustration 9.3). Percentiles of BMI are based on the proportion of children and adolescents who have different levels of BMI at given ages. For example, if a child's BMI is at the 50th percentile for his or her age, then half of children will have BMIs that are below and half will have values that are above.

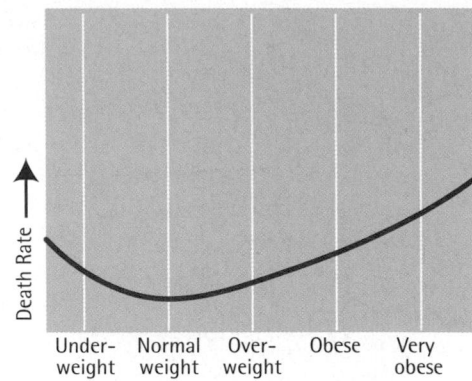

Illustration 9.1 The relationship between body weight status and deaths from all causes for adults. (Graph was drawn by author from data presented by Pischon et al).[1]

body mass index (BMI)
An indicator of the status of a person's weight for their height. It is calculated by dividing weight in kilograms by height in meters squared. It can also be calculated using inches and pounds as shown in Table 9.2.

Table 9.1

Classifying weight status by body mass index.[9]

	Body Mass Index
Underweight	under 18.5 kg/m²
Normal weight	18.5–24.9 kg/m²
Overweight	25–29.9 kg/m²
Obese	30 kg/m² or higher[a]

[a]Obesity is subdivided for some purposes into the BMI groups of moderate obesity (30–34.9 kg/m²), severe obesity (35–39.9 kg/m²), and very severe obesity (40+ kg/m²).

Health Action   Estimating Normal Weight for Height

There is a quick way to estimate within ±10% what is a normal, or healthy, weight for height. It is called the Hamwi[7] method.

Women
Begin with 5 feet equals 100 pounds, and then add 5 pounds for each additional inch of height. Here's an example of how you would estimate a healthy weight for a woman who is 5 feet, 7 inches tall:

5 feet = 100 pounds
7 inches × 5 pounds = 35 pounds
100 pounds + 35 pounds = 135 pounds

Men
For men, five feet equals 106 pounds, and each additional inch of height adds on 6 pounds. So, for example, we would estimate a healthy weight for a man who is 5 feet 10 inches tall the following way:

5 feet = 106 pounds
10 inches × 6 pounds = 60 pounds
100 pounds + 66 pounds = 166 pounds

Table 9.2

Calculating BMI: An example

Say Chris weighs 140 pounds and is 5 feet, 3 inches tall. To calculate BMI:

Figure out how many inches tall Chris is:

5 feet × 12 inches per foot = 60 inches
60 inches + 3 inches = 63 inches

Divide Chris's weight by his height in inches:

$$\frac{140 \text{ pounds}}{63 \text{ inches}} = 2.22$$

Divide the result (2.22) by Chris's height in inches again:

$$\frac{2.22}{63 \text{ inches}} = 0.035$$

Multiply this result by 703:

0.035 × 703 = 24.8 This is Chris's BMI.

Table 9.3 shows BMI percentile ranges that correspond to underweight, at risk of underweight, normal weight, at risk of overweight, and overweight in children and adolescents. Obesity is not included in this table. Body fat content, rather than BMI percentile ranges, should be used to diagnose obesity in children and adolescents.[10]

BMI percentile ranges for children and adolescents do not provide information on growth progress in terms of height. Consequently, growth in height cannot be assessed using BMI. Other standards for assessment of height in children and adolescents are available.

Most Adults in the United States Weigh Too Much

Being **overweight** or **obese** is the norm in the United States. The combined incidence of overweight and obesity among adults is 66%. (Illustration 9.4).

Over one in six children and adolescents in the United States are overweight, and this percentage may be rising as you read this. Overweight and obesity are becoming America's number one health problems. Obesity, which represents the largest risks to health, varies considerably among U.S. states (Illustration 9.5). In 2007, for example, rates of obesity were highest in Mississippi, Tennessee, and Alabama and lowest in Colorado.[11]

Illustration 9.2 BMI Chart

Body Mass Index (BMI)

Height	18	19	20	21	22	23	24	25	26	27	28	29	30	31	32	33	34	35	36	37	38	39	40
	Body Weight (pounds)																						
4′10″	86	91	96	100	105	110	115	119	124	129	134	138	143	148	153	158	162	167	172	177	181	186	191
4′11″	89	94	99	104	109	114	119	124	128	133	138	143	148	153	158	163	168	173	178	183	188	193	198
5′0″	92	97	102	107	112	118	123	128	133	138	143	148	153	158	163	168	174	179	184	189	194	199	204
5′1″	95	100	106	111	116	122	127	132	137	143	148	153	158	164	169	174	180	185	190	195	201	206	211
5′2″	98	104	109	115	120	126	131	136	142	147	153	158	164	169	175	180	186	191	196	202	207	213	218
5′3″	102	107	113	118	124	130	135	141	146	152	158	163	169	175	180	186	191	197	203	208	214	220	225
5′4″	105	110	116	122	128	134	140	145	151	157	163	169	174	180	186	192	197	204	209	215	221	227	232
5′5″	108	114	120	126	132	138	144	150	156	162	168	174	180	186	192	198	204	210	216	222	228	234	240
5′6″	112	118	124	130	136	142	148	155	161	167	173	179	186	192	198	204	210	216	223	229	235	241	247
5′7″	115	121	127	134	140	146	153	159	166	172	178	185	191	198	204	211	217	223	230	236	242	249	255
5′8″	118	125	131	138	144	151	158	164	171	177	184	190	197	203	210	216	223	230	236	243	249	256	262
5′9″	122	128	135	142	149	155	162	169	176	182	189	196	203	209	216	223	230	236	243	250	257	263	270
5′10″	126	132	139	146	153	160	167	174	181	188	195	202	209	216	222	229	236	243	250	257	264	271	278
5′11″	129	136	143	150	157	165	172	179	186	193	200	208	215	222	229	236	243	250	257	265	272	279	286
6′0″	132	140	147	154	162	169	177	184	191	199	206	213	221	228	235	242	250	258	265	272	279	287	294
6′1″	136	144	151	159	166	174	182	189	197	204	212	219	227	235	242	250	257	265	272	280	288	295	302
6′2″	141	148	155	163	171	179	186	194	202	210	218	225	233	241	249	256	264	272	280	287	295	303	311
6′3″	144	152	160	168	176	184	192	200	208	216	224	232	240	248	256	264	272	279	287	295	303	311	319
6′4″	148	156	164	172	180	189	197	205	213	221	230	238	246	254	263	271	279	287	295	304	312	320	328
6′5″	151	160	168	176	185	193	202	210	218	227	235	244	252	261	269	277	286	294	303	311	319	328	336
6′6″	155	164	172	181	190	198	207	216	224	233	241	250	259	267	276	284	293	302	310	319	328	336	345
	Underweight (<18.5)	Healthy Weight (18.5-24.9)						Overweight (25-29.9)					Obese (30)										

Find your height along the left-hand column and look across the row until you find the number that is closest to your weight. The number at the top of that column identifies your BMI. The area shaded in green represents healthy weight ranges.

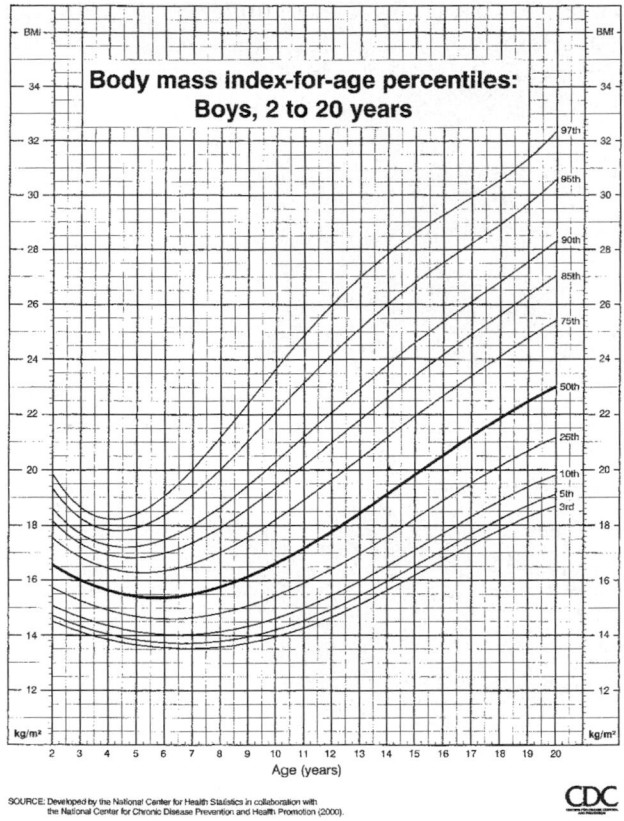

Illustration 9.3 An example of the CDC's 2000 Growth Charts: BMI-for-age.
Copies of all growth charts can be obtained from www.cdc.gov/growthcharts

Table 9.3

Weight status standards for 2- to 20-year-olds based on BMI-for-age.[10]

	Percentile Range(s)
Underweight	5th or less
At risk of underweight	5th-15th
Normal weight	15th-85th
At risk of overweight	85th-95th
Overweight	95th or higher

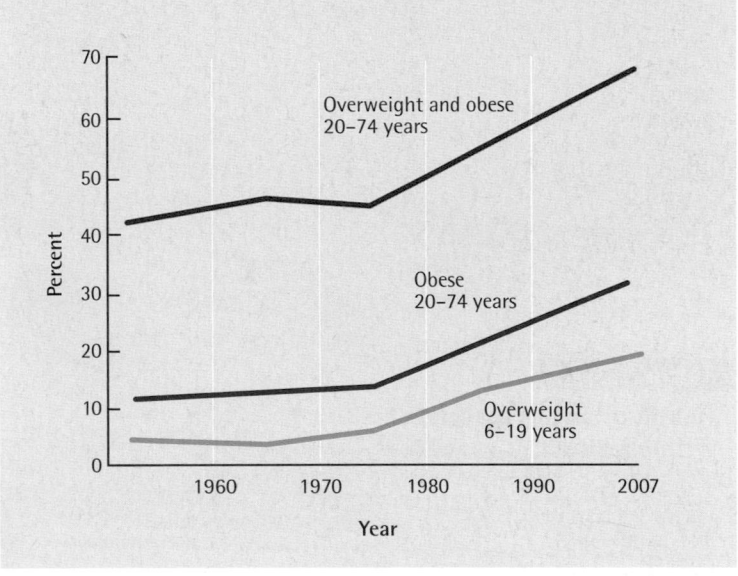

Illustration 9.4 Overweight and obesity by age in the United States, 1960-2007.[11]

The trend toward higher rates of overweight and obesity in the United States is now shared by most other countries of the world.[14] A sampling of overweight plus obesity rates by country is shown in Table 9.4. The high and growing incidence of overweight and obesity in the United States and around the world represents a time bomb for a future explosion in obesity-related disease rates.

The Influence of Obesity on Health People who are obese are much more likely to experience diabetes, heart disease, certain types of cancer, hypertension, and other disorders (listed in Table 9.5) than are people of normal weight.[15,16] The increased risk of disease appears to be primarily due to a higher prevalence of **metabolic** abnormalities in obese people. Approximately 70% of obese persons have two or more metabolic abnormalities such as:

- hypertension
- elevated triglycerides, glucose, and/or insulin
- low HDL-cholesterol (the "good cholesterol")
- high **C-reactive protein** (a key marker of inflammation)

About 23% of normal weight adults, and 50% of overweight individuals have two or more metabolic abnormalities that increase disease risk.[17]

Weight loss of 10–15% of initial body weight, paired with exercise that improves physical fitness level, reduces metabolic abnormalities and the risk of disease.[15]

overweight A high weight-for-height.

obese A condition characterized by excess body fat.

metabolism
The chemical changes that take place in the body. The conversion of glucose to energy or to body fat is an example of a metabolic process.

C-reactive protein (CRP)
A key inflammatory factor produced in the liver in response to infection or inflammation. Elevated concentrations of CRP are associated with heart disease, obesity, diabetes, inactivity, infection, smoking, and inadequate antioxidant intake.

Table 9.4

Prevalence of adult overweight plus obesity in a sampling of countries[11-13]

Country	Percent Overweight and Obese
Albania	79
Czech Republic	53
United States	66
Iran	48
England	61
Costa Rica	59
Germany	60
Canada	59
Netherlands	46
Australia	60
France	49
Israel	62
China	19
Japan	24

Table 9.5

Health problems associated with obesity[15,16]

Type 2 diabetes
Chronic inflammation
Metabolic syndrome
Hypertension
Stroke
Elevated cholesterol level
Low HDL-cholesterol level
Heart disease
Certain types of cancer
Gallbladder disease
Fatty liver disease
Shortened life expectancy
Discrimination
Depression
Infertility
Accidents
Skin disorders
Sleep disorders

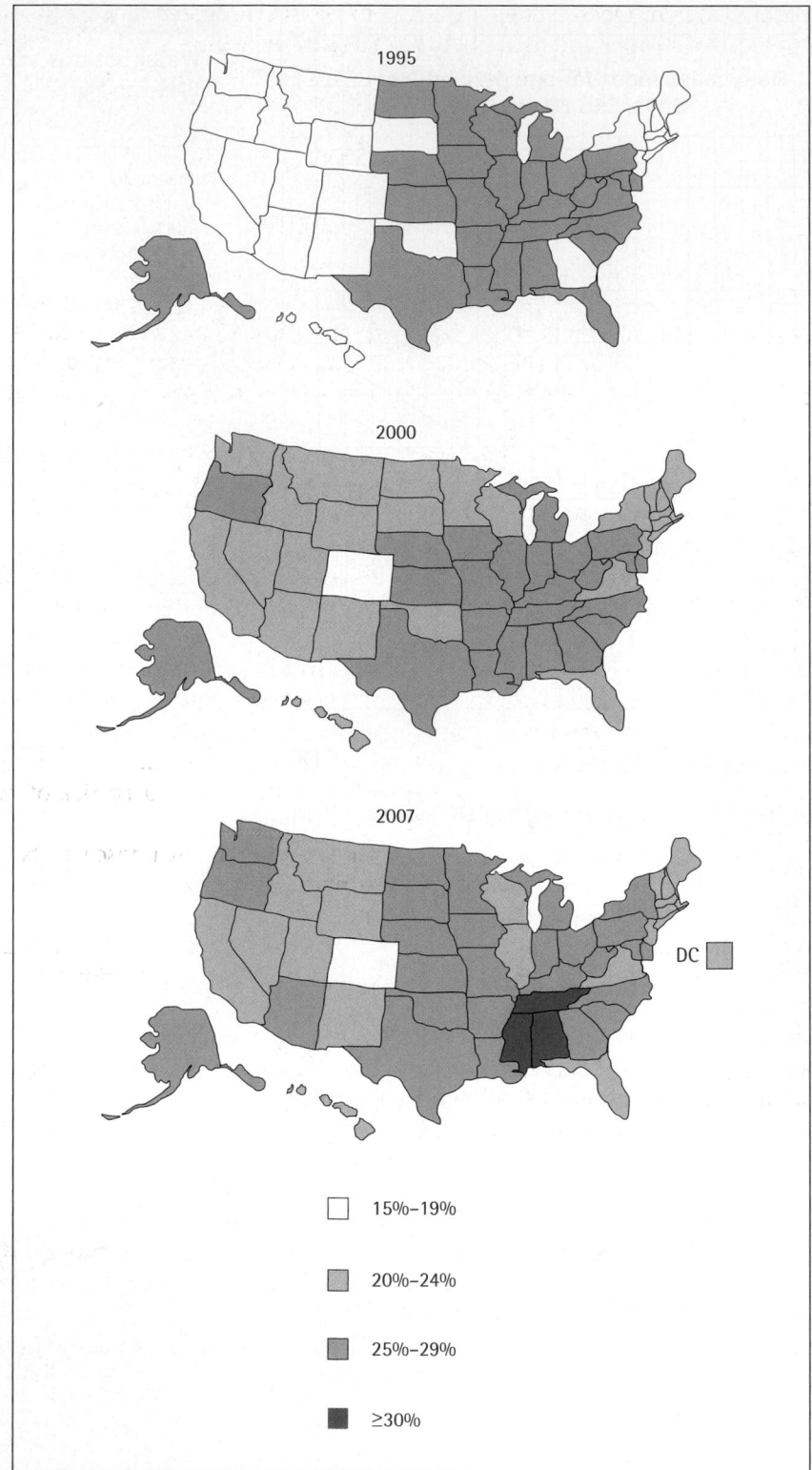

Illustration 9.5 Percent of obese (BMI ≥ 30kg m^2) adults by state in 1995, 2000, and 2007.[11]

Obesity and Psychological Well-Being Another consequence of obesity is the ingrained cultural prejudice to which obese people are subjected. Children who are obese are more likely to suffer unfair or indifferent treatment from teachers than other children. They experience more isolation, rejection, and feelings of inferiority than other children. Obese adults are likely to be discriminated against in hiring and promotion decisions and to be thought of as lazy or lacking in self-control, even by the health professionals who care for them.[18] Society's prejudice against people who don't conform to the cultural ideal of body size may be the most injurious consequence of obesity.

Body Fat and Health: Location, Location, Location

It is becoming increasingly clear that many of the health problems associated with obesity are directly related to where excess fat is stored. Humans store fat in two major locations: under the skin over the hips, upper arms, and thighs, and in the abdomen. Fat stored under the skin is called **subcutaneous fat** and that stored in the abdomen under the skin and a layer of muscle **visceral fat** (see Illustration 9.6). People who store fat primarily in their hips, upper arms, and thighs are said to have a "pear shape" and those who store fat principally in the abdomen an "apple shape." These body shapes are shown in Illustrations 9.7. You may be better off healthwise if you're a "pear" rather than an "apple."

Visceral fat is much more metabolically active, and more strongly related to disease risk than is subcutaneous fat.[19] Metabolic processes initiated by visceral fat produce **chronic inflammation** and oxidation reactions that disrupt normal body functions. These disruptions promote the development of **insulin resistance**, **metabolic syndrome**, elevated blood glucose and triglyceride concentrations, high blood pressure, and hardening of the arteries. These changes, in turn, can lead to the development of heart disease, some types of cancer, **type 2 diabetes**, hypertension, **fatty liver disease**, and other disorders.[20] Normal weight and overweight individuals with excessive visceral fat deposits are also at increased risk of metabolic abnormalities and diseases associated with them.[1]

Metabolic abnormalities and disease risk associated with visceral fat can be reduced by regular exercise and improved physical fitness level. Higher levels of health benefits are achieved if both aerobic (walking, jogging, swimming, gardening) and resistance (strength training) exercises are part of the program.[22] Weight loss combined with exercise is the bonus pack for large reductions in risks.[1]

Visceral Fat and Waist Circumference

The size of visceral fat deposits can be closely estimated by measuring waist circumference (Illustration 9.8).[23] In men, waist circumferences over 40" (102 cm), and in women, over 35" (88 cm) are related to excess visceral fat.[24]

subcutaneous fat
(Pronounced sub-q-tain-e-ous) Fat located under the skin.

visceral fat
(Pronounced vis-sir-el) Fat located under the skin and muscle of the abdomen.

chronic inflammation
Low-grade inflammation that lasts weeks, months, or years. Inflammation is the first response of the body's immune system to infection or irritation. Inflammation triggers the release of biologically active substances that promote oxidation and other potentially harmful reactions in the body.

insulin resistance
A condition in which cells "resist" the action of insulin in facilitating the passage of glucose into cells.

metabolic syndrome
A constellation of metabolic abnormalities generally characterized by insulin resistance, abdominal obesity, high blood pressure and triglyceride levels, low levels of HDL cholesterol, and impaired glucose tolerance. Metabolic syndrome predisposes people to the development of type 2 diabetes, heart disease, hypertension, and other disorders. It is common—one in five U.S. adults has metabolic syndrome.[26]

fatty liver disease
A reversible condition characterized by fat infiltration of the liver (10% or more by weight). If not corrected, fatty liver disease can produce liver damage and other disorders. The condition is primarily associated with obesity, diabetes, and excess alcohol consumption. The disease is called "steatohepatitis" when accompanied by inflammation.

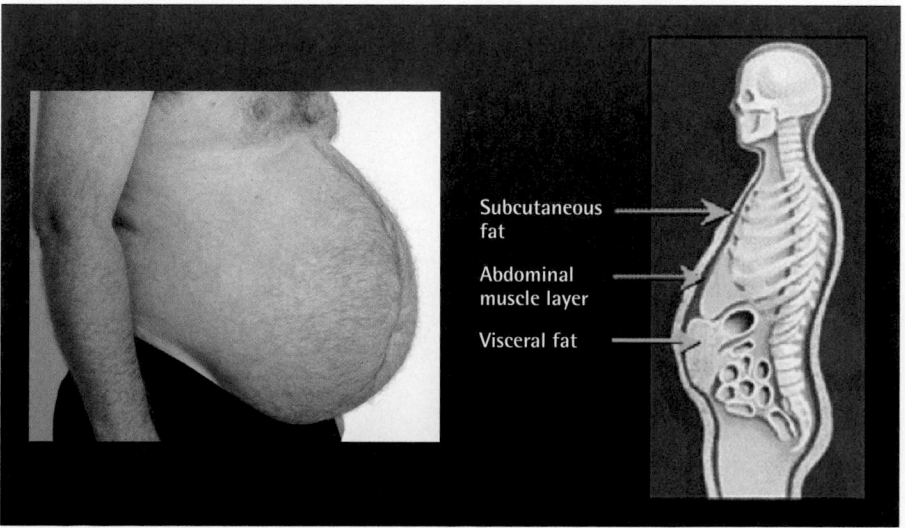

Subcutaneous fat

Abdominal muscle layer

Visceral fat

Illustration 9.6 A diagram of the location of subcutaneous and visceral fat.

Source: Caballero E. Dyslipidemia and vascular function: The rationale for early and aggressive intervention. www.medscape.com, 4/18/06

Illustration 9.7 Basic body shapes.
The pear normally has narrow shoulders, a small chest, and an average-size waist. Fat is concentrated in the hips, upper arms, and thighs.

Apples are round in the middle. The apple shape is riskier than the pear shape due to the presence of visceral fat.

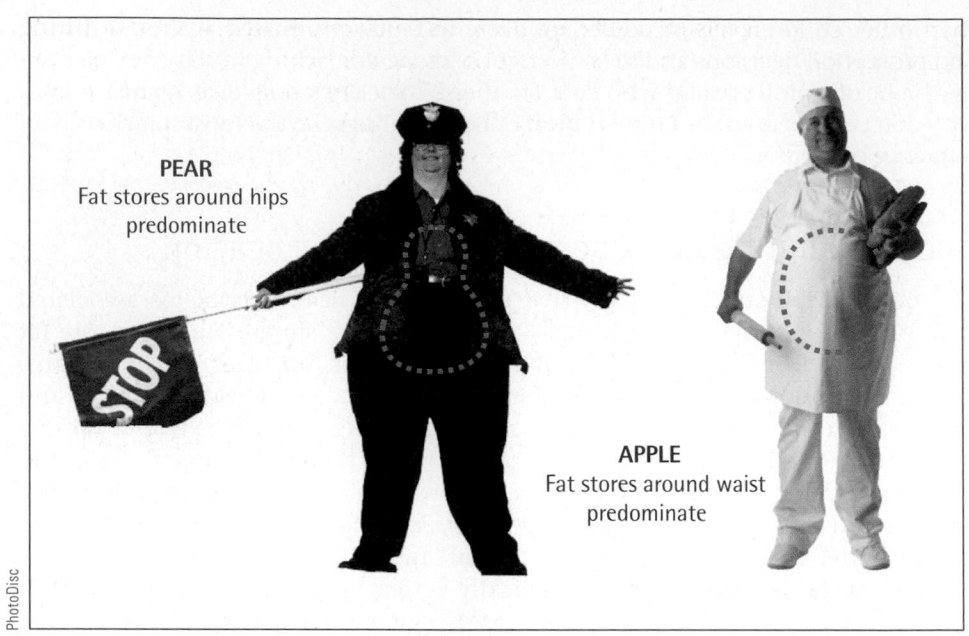

PEAR
Fat stores around hips predominate

APPLE
Fat stores around waist predominate

PhotoDisc

type 2 diabetes
A disease characterized by high blood glucose levels due to the body's inability to use insulin normally or to produce enough insulin (previously called adult-onset diabetes).

On the Side

Waist circumference may not accurately estimate visceral fat content in large, muscular individuals.[23]

Because waist circumference is such a strong indicator of disease risk, its measurement in clinical practice is being encouraged.[23] Japan is attempting to prevent and control the country's quickly growing rates of obesity and health care expenditures by population-wide screening of central body fat. The country has instituted a policy that requires all citizens aged 50 to 74 years (that's 56 million people) have their waist circumference measured yearly (Illustration 9.9). Individuals with high circumferences and weight-related health problems are given dieting guidance.[25]

Assessment of Body Fat Content

Body mass index is commonly used to approximate body fat content because the two measures correspond closely in groups of people. This is not always the case for individuals. Take a 165-pound, 30-year-old woman who is 5 feet 6 inches tall and weight trains heavily. Her body weight for height would indicate obesity, but she may have a low body fat content and be healthy. If BMI were used

Illustration 9.8 Determining your waist-circumference.

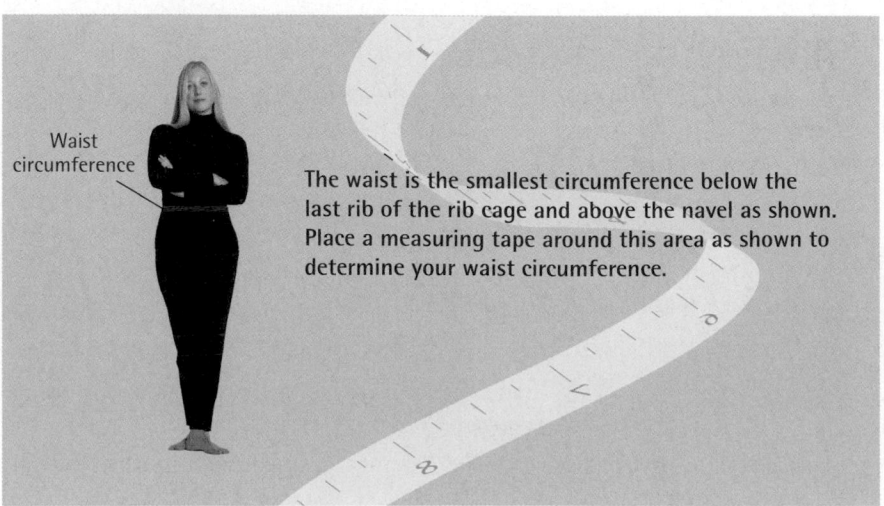

Waist circumference

The waist is the smallest circumference below the last rib of the rib cage and above the navel as shown. Place a measuring tape around this area as shown to determine your waist circumference.

to assess body fat levels in professional football players, most all of them would be wrongly categorized as obese.[23] Sometimes people who are classified as normal weight or underweight by BMI standards have too much body fat because they are physically inactive. Certain medications make people retain fluid. Their weight-for-height may qualify them as overweight, but their body fat content may actually be very low. Obviously, measures of body fat content are better estimators of health status than are measures of weight-for-height.

Methods for Assessing Body Fat Content   Easy to use, accurate, and inexpensive tools for assessing body fat content are available, and their use is spreading. Standards for classifying percent body fat, or the percent of weight that consists of fat, have been developed (Table 9.6) and will be refined as additional studies on the relationships between body fat and health risk are conducted.

Here are the most common methods for determining body fat content:

- skinfold thickness measures

- bioelectrical impedance analysis (BIA)

- underwater weighing

- magnetic resonance imaging (MRI)

- computerized axial tomography (CT or CAT scans)

- dual-energy X-ray absorptiometry (DEXA)

- whole body air displacement (BOD POD)

The theories underlying each of these methods and its advantages and limitations are presented in Table 9.7. The tests are most likely to provide accurate results when they are performed by skilled, experienced technicians using proper, well-maintained equipment and when the measurements are converted into percent body fat by the appropriate formulas.

Everybody Needs Some Body Fat   A certain amount of body fat—3 to 5% for men and 10 to 12% for women—is needed for survival. Body fat serves essential roles in the manufacture of hormones; it's a required component of every cell in the body; and it provides a cushion for internal organs. Fat that serves these purposes is not available for energy formation no matter how low energy reserves become. Low body fat levels are associated with delayed physical maturation during adolescence, infertility, accelerated bone loss, and problems that accompany starvation.[29]

What Causes Obesity?

Simply stated, obesity results when the intake of calories exceeds caloric expenditure. But the cause of obesity is not that simple. Whether people accumulate excess body fat or not is due to complex and interacting factors that include

- diet

- physical activity

- environmental exposures

- genetic background[30]

Some medications such as antipsychotics, antidepressants, insulin, and beta-blockers used for hypertension are also associated with the development of obesity.[31] Their contribution to the overall incidence of obesity is small (although meaningful to the affected people).

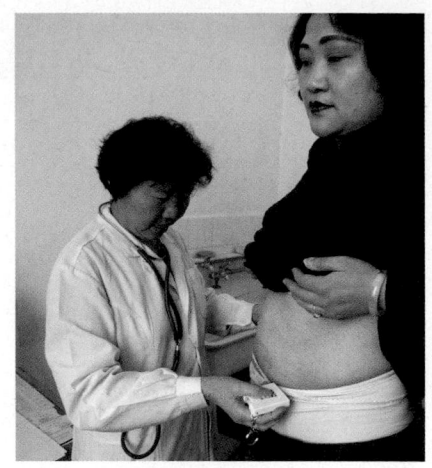

AP Images/Greg Baker

Illustration 9.9

Table 9.6

Percentages of body weight as fat considered low, average, and high[28]

	% Body Fat		
	Low	Average	High
Women	less than 12	32	35 or more
Men	less than 5	22	25 or more

On the Side

We're getting bigger. According to the CDC, men now weigh 191 pounds and are 5'9½" tall on average. In 1960, the average weight of men was 166 pounds and height was 5'8". Women are now 5'4" tall and weigh 164 pounds on average. In 1960, those figures were 5'3" and 140 pounds.

Table 9.7

Commonly used methods for assessing body fat content

Tom Pantages

Skinfold Measurement

Body fat content can be estimated by measuring the thickness of fat folds that lie underneath the skin. Calipers are used to measure the thickness of fat folds, preferably over several sites on the body. Body fat content is estimated by "plugging" the thicknesses into the appropriate formula.

Advantages: Calipers are relatively inexpensive (they cost about $250–$350). The procedure is painless if done correctly and can yield a good estimate of percent body fat.

Limitations: This technique is often performed by untrained people. Skinfolds may be difficult to isolate in some individuals.

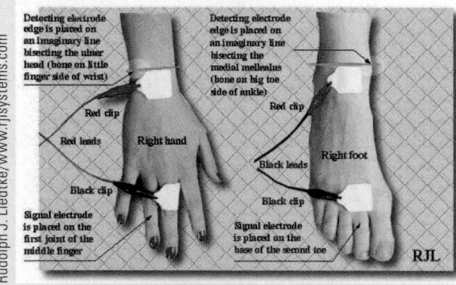

Rudolph J. Liedtke/www.rjlsystems.com

Bioelectrical Impedance Analysis (BIA)

Because fat is a poor conductor of electricity and water and muscles are good conductors, body fat content can be estimated by determining how quickly electrical current passes from the ankle to the wrist.

Advantages: The equipment required is portable, and the test is easy to do and painless. The results are fairly accurate for people who are not at the extremes of weight-for-height.

Limitations: Equipment may be expensive; inferior equipment produces poor results. Hydration status and meal ingestion may affect electrical conductivity and produce inaccurate results, as may inaccurate formulas used to calculate body fat content from test results.

© Yoav Levy/Phototake

Underwater Weighing

The subject is first weighed on dry land; next he or she is submerged in water and exhales completely; then his or her weight is measured. The less the person weighs under water compared to the weight on dry land, the higher the percent body fat. (Fat, but not muscle or bone, floats in water.)

Advantages: If undertaken correctly and if appropriate formulas are used in calculations, this technique gives an accurate value of percent body fat.

Limitations: The equipment required is expensive and not easily moved. The test doesn't work well for people who don't swim or who are ill or disabled in some way.

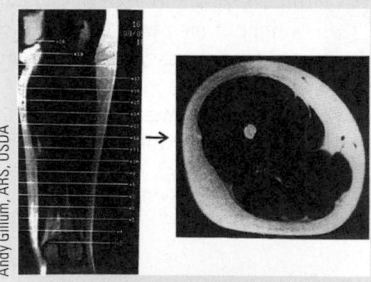
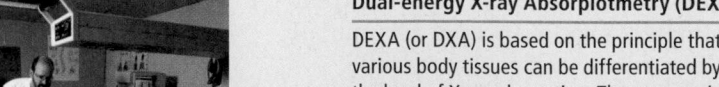

Andy Gillum, ARS, USDA

Magnetic Resonance Imaging (MRI)

MRI and CT or CAT scans provide similar results. Using this technology, a person's body fat and muscle mass can be photographed from cross-sectional images obtained when the body, or parts of it, is exposed to a magnetic field (or, in the case of CAT scans, to radiation). Based on the volume of fat and muscle observed, total fat and muscle content can be determined.

Advantages: Provides highly accurate assessment of fat and muscle mass.

Limitations: The test is expensive (around $1,000–$1,500 per assessment) and largely used for research purposes.

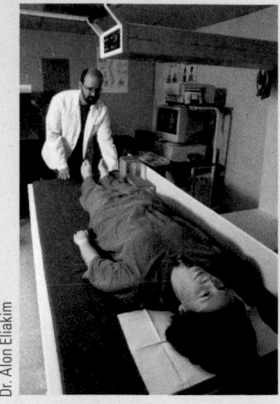

Dr. Alon Eliakim

Dual-energy X-ray Absorpiotmetry (DEXA)

DEXA (or DXA) is based on the principle that various body tissues can be differentiated by the level of X-ray absorption. The measure is made by scanning the body with a small dose of X-rays (similar to the level of exposure from a transcontinental flight) and then calculating body fat content based on the level of X-ray absorption.

Advantages: Provides highly accurate results when measurements are undertaken correctly. DEXA is safe and "user friendly" for people being measured and can also be used to assess bone mineral content and lean tissue mass.

Limitations: DEXA machine is expensive, and so is the cost of individual assessments ($200 per person). Machine must be operated by a trained and certified radiation technologist in many states.

Table 9.7

Commonly used methods for assessing body fat content *(continued)*

Courtesy, Life Measurement, Inc., www.bodpod.com

Whole Body Air Displacement

This is an established method that has become practical for broader use due to development of the BOD POD. The method is similar to that of underwater weighing but uses air displacement for determining percent body fat. Individuals sit in an enclosed "cabin" for about five minutes while wearing a tight-fitting swim suit and cap. Computerized sensors determine body weight and the amount of air that is displaced by the body. The PEA POD is used for assessing body fat in infants.

Advantages: This method provides a quick, comfortable, automated, and reasonably accurate way to assess body fat content. It is suitable for disabled individuals, the elderly, and children.

Limitations: Results can be modified by drinking, eating, or exercising before testing; a full bladder or failure to adhere to test procedures may lead to error in results. The test is costly but less expensive than DEXA. A BOD POD machine costs around $40,000.

Are Some People Born to Be Obese?

With very rare exception, people are not genetically destined to become obese. The current epidemic of obesity is primarily driven by environmental factors and not by our genes.[32] However, genetic traits inherited at birth can influence a person's susceptibility to becoming obese.

Genetic influences on obesity take several paths. Some people are born with errors in metabolism that produce obesity (this appears to be rare). Others are born with multiple genetic traits that predispose them to becoming obese (probably common). A predisposing genetic trait is expressed when the right **environmental trigger** exists. For example, a person with a genetic predisposition to becoming obese may maintain normal weight as long as she or he is physically active or consumes a low-fat diet. If activity level becomes low or the diet changes to one high in fat (two types of environmental triggers), the genetically susceptible person then gains weight. Genetically based differences in the affects of environmental triggers on food intake and energy utilization appear to influence a person's susceptibility to obesity.[32,33]

We are just beginning to understand the nature of the interactions between environmental factors and genes. The day will come, however, when people with inborn tendencies toward environmental triggers and obesity can be identified by an examination of their genes.

environmental trigger
An environmental factor, such as inactivity, a high-fat diet, or a high sodium intake, that causes a genetic tendency toward a disorder to be expressed.

Do Obese Children Become Obese Adults?

The link between early and later obesity is weak for young children. It becomes stronger among older children and adolescents, especially if one or both parents are obese.[34,35] Only 8% of obese children who are heavy at one to two years of age and who do not have an obese parent are obese as adults. However, nearly 80% of children who are obese between the ages of 10 and 14 and have at least one obese parent are obese as adults.[36]

The Role of Diet in the Development of Obesity

Regardless of the cause of obesity, weight gain results when more energy is consumed than expended. Americans are consuming more energy than in previous years: Caloric intake is up by an average of 340 per day while physical activity level has remained about the same. Fruits, vegetables, and fiber are often missing from diets that provide too many high-calorie, energy-dense foods.[37]

The generous availability of inexpensive, energy-dense foods; eating out regularly at fast-food restaurants and all-you-can-eat buffets, and large portions of foods tend to increase calorie intake.[38] Adults served large portions of food consume an average of 30 to 50% more food than when presented with small food portions.[39] When offered a lot of tasty food, people tend to eat beyond the feeling of satiety and past the point where food continues to taste good.[40]

Low Levels of Physical Activity Promote Obesity   Low levels of physical activity are related to the high and increasing incidence of obesity among Americans.[41] To a large extent, physical activity has become voluntary. Many farmers now plow, sow, and reap in air-conditioned tractors with power steering. Lawn mowers propel themselves and sometimes their operators, too. Instead of walking or biking two and a half miles to the store and back, we drive. An average-size adult driving 2.5 miles burns about 17 calories. Biking that distance would use seven times more calories (122). But walking 2.5 miles would burn around 210 calories, more than 12 times the calories it takes to drive! We have traded physical activity for convenience, time, and, perhaps, personal fat stores. Too much television watching, in particular, has been blamed for obesity in children.[42]

Obesity: The Future Lies in Its Prevention

Whether obesity is related to genetic predisposition, environmental factors, or a combination of both, certain steps can be taken to help prevent it.

Preventing Obesity in Children

For children, the prevention of obesity includes the early development of healthy eating and activity habits. Parents should offer a nutritious selection of food, but children themselves should be allowed to decide how much they eat.[43] Physical activities that are fun for every child—not just those who show athletic promise—should be routine in schools and summer programs.

Interactions between parents and children around eating and body weight can set the stage for the prevention or the promotion of overweight in children. Parents who overreact to a child's weight by focusing on it, restricting food access, and making negative comments to the child may increase the likelihood that eating and weight problems will develop or endure. Lifestyle changes for the whole family—such as incorporating fun physical activities into daily schedules; making a wide assortment of nutritious foods available in the home; and decreasing a focus on eating, foods, and weight—are some of the positive changes families can make to promote healthy eating and exercise habits, and normal weight in children.[43]

Preventing Obesity in Adults

Action needs to be taken to prevent weight gain during the adult years. Many adults gain weight at a slow pace (about a pound per year) as they age, whereas others gain substantial amounts of weight over short periods of time.[44] Data from a national nutrition and health survey indicate that major gains in weight are most likely to occur in adults between the ages of 25 and 34 years.[45] Regular, vigorous exercise may prevent or lessen the amount of weight gain that occurs with age, as may decreased portion sizes at home and in restaurants.[46] Paying attention to the "I'm hungry" and "I'm full" signals can help moderate food intake.[40] For some people, regularly getting 8 hours of sleep at night appears to reduce weight gain.[48] For more on this topic, visit the "Reality Check" on page 9-14.

Changing the Environment

Environmental changes in food portion sizes, the increased availability of energy-dense, inexpensive foods, and sedentary lifestyles are among the key factors that underlie the current obesity epidemic.[32] From a public health point of view, it is being reasoned that if environmental changes got us into the obesity epidemic, then environmental changes will help get us out of it.[49]

Many communities are taking action to promote healthy environments. These actions include limiting access to "junk" foods in schools, requiring calorie labeling on foods sold in fast food and other chain restaurants, and the development of community gardens in urban areas. Sidewalks, bicycle and walking paths, and nature trails are being planned or added in urban residential areas.[32] Fast food and other restaurants are gradually making changes toward selling smaller portions of energy-dense foods and offering more nutrient-dense items. Some restaurants are adding a wider selection of not-fried entrees and half portions to their menus. The general trend in weight- and health-friendly environmental change is toward providing individuals and families with widespread opportunities for making healthful choices.[50]

Some People Are Underweight

In contrast to the desperate situations faced by many people in economically emerging countries, underweight in developed nations largely results from illnesses such as HIV/AIDS, pneumonia, and cancer; an eating disorder (anorexia nervosa); or the voluntary restriction of food. An important and preventable cause of underweight in the United States and some economically developed nations is poverty.

underweight
Usually defined as a low weight-for-height. May also represent a deficit of body fat.

Illustration 9.10 Some underweight people are genetically thin and are healthy.

Underweight Defined

People who are **underweight** have too little body fat, or less than 12% body fat in adult females and 5% body fat in males.[28] They have BMIs below 18.5 kg/m^2 as indicated in Table 9.1. A portion of the 2% of U.S. adults classified as underweight by BMI will not actually be underweight, just as a subset of people categorized as obese by BMI are not really obese.[51]

Some people assessed as underweight for height are healthy and have a normal body composition. Like the person in Illustration 9.10, they are probably genetically thin. People who are naturally thin often have as much difficulty gaining weight as obese people have losing it.[37] People who are thin and unhealthy are more likely than others to experience apathy, fatigue, and illnesses frequently, and they take longer to recover from illness. They tend to have reduced bone mineral density and more bone fractures, be intolerant of cold temperatures, and have impaired concentration.[33,52]

© Richard Hutchings/Photo Researchers,

Underweight and Longevity in Adults

Longevity can be extended in adult mice and monkeys by feeding them a nutritious, calorie-restricted diet that produces underweight.[53] Although it is not known whether caloric restriction and underweight would serve as a

The "I'm full" feeling

How do you know when you've had enough to eat?
Answers on next page

Carlos
I'm on automatic pilot for eating until my plate is clean.

Sandra
I stop eating when, like boom!—the "I'm full" signal hits.

fountain of youth for humans, there are groups of adults who believe that it will. Devotees of calorie restriction for longer life tightly control their food intake and activity level to maintain a lean body. Calorie restrictors tend to consume 1,600 to 1,700 calories per day, emphasize nutrient-rich foods, and exercise regularly.[54] They in general appear thin but healthy (see Illustration 9.11).

Adults following calorie-restricted diets tend not to have chronic inflammation nor the health problems associated with it. They are at risk of iron deficiency, osteoporosis, infertility, infections, and of becoming irritable. The theory that life expectancy would be increased by calorie restriction is challenged by data on humans that show increased disease and death rates among underweight compared to normal weight people.[53] Several long-term studies of health outcomes related to calorie restriction in men are under way. Relatively few women follow this diet and lifestyle; little is known about its effects on their health.

"The war is on obesity, not the obese."[56]

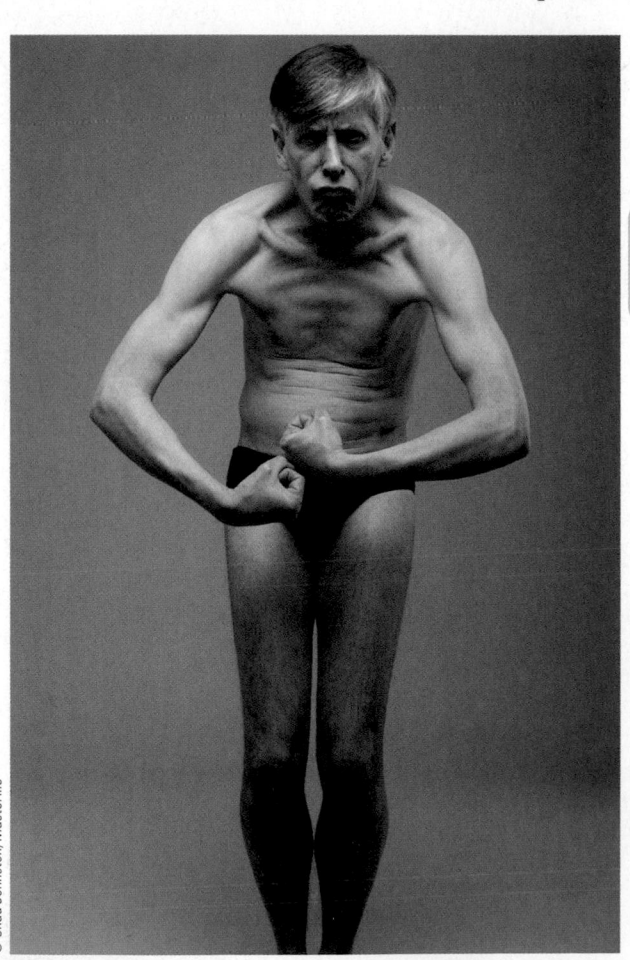

Illustration 9.11 The appearance of one man who follows a calorie-restricted diet.

Toward a Realistic View of Body Weight

A widespread belief among Americans is that individuals can achieve any body weight or shape they desire if they just diet and exercise enough. It's a myth. People naturally come in different weights and shapes, and these can only be modified so much (Illustraction 9.12).[55] Half of all women in the United States wear sizes 14 to 26, yet many clothing models are very underweight. Many men, no matter how hard they work out, will never have a washboard stomach or fit into "slim jeans."

Size Acceptance

The U.S. obsession with body weight and shape is spreading to other industrialized countries as part of popular culture. Ironically, strong societal bias against certain body sizes may contribute broadly to weight and health problems. Intolerance of overweight and obese children and adults tends to increase discrimination against them. This type of societal bias lowers the individual's feeling of self-worth and may promote eating disorders, including the consumption of too much food. Females are hardest hit by negative attitudes about body size. Although the incidence of overweight and obesity tends

Illustration 9.12 People come in many different sizes and shapes.

to be higher in males, obesity in females carries with it many more negative stereotypes.[12,57] Overreactive parents make things worse.

Acceptance of people of different sizes and a more realistic view of obtainable body weights and shape may be two of the most important things society can do to prevent obesity.

The Health at Every Size Program A kinder and more effective approach to health improvement in obese people has been developed. Called "Health at Every Size," the program is gaining acceptance among consumers and health professionals in the United States and Canada.[58] The program emphasizes eating when hungry and stopping when full; enjoyable, life-enhancing physical activities; body size acceptance; healthy eating behaviors; a peaceful relationship with food; self-esteem; and social support networks. The program reduces a number of health problems related to obesity even though it does not lead to weight loss in the short term. Health at Every Size program participation is associated with reduced blood pressure and LDL-cholesterol, increased HDL-cholesterol levels, improved self-esteem, decreased disordered eating behaviors, and improved body image.[59] The Health at Every Size philosophy and program components could help future generations of children, men, and women achieve healthful eating patterns and high levels of well-being and quality of life, as well as improved long-term health.[60]

The "I'm full" feeling

Even favorite foods lose some or all of their appeal after we're full. The "I'm full" feeling will let you know when that happens—if you pay attention.[40]

Carlos:

Sandra:

© Ilene MacDonald/ Alamy

Up Close

Are You an Apple?

Focal Point: Determining your waist circumference.

Follow the directions given in Illustration 9.9 for determining your waist circumference. If you don't have a measuring tape, use a string. Mark the string where the two ends intersect when placed around your waist. Then use a ruler to measure the distance between the end and the mark on the string. Are you an apple? See the feedback section.

FEEDBACK can be found at the end of Unit 9.

[Key Terms

body mass index (BMI), page 9-3
chronic inflammation, page 9-7
c-reactive protein (CRP), page 9-5
environmental trigger, page 9-11
fatty liver disease, page 9-7

insulin resistance, page 9-7
metabolic syndrome, page 9-7
metabolism, page 9-5
obese, page 9-5
obesity, page 9-2

overweight, page 9-5
subcutaneous fat, page 9-7
type 2 diabetes, page 9-7
underweight, page 9-13
visceral fat, page 9-7

[Review Questions

	TRUE	FALSE
1. Death rates are lowest among adults who remain underweight as they age.	☐	☐
2. According to the Hamwi method for estimating normal weight in adults, a 6'1" male should weigh approximately 204 pounds.	☐	☐
3. A child with a body mass index higher than the 95th percentile on the CDC growth charts is considered obese.	☐	☐
4. The rate of overweight and obesity in the United States is higher than that reported for any other country.	☐	☐
5. Excess visceral fat poses higher risks to health than excess subcutaneous fat.	☐	☐
6. Features of metabolic syndrome can include excess abdominal fat and insulin resistance.	☐	☐
7. Women but not men require a minimal amount of body fat stores for health.	☐	☐

	TRUE	FALSE
8. If undertaken correctly, and if appropriate formulas are used, underwater weighing produces accurate values of a person's percent body fat.	☐	☐
9. The use of medications that cause weight gain is a leading reason for the obesity epidemic in the United States.	☐	☐
10. Vegetables, fruits, and fiber are often consumed in low amounts by people who regularly eat energy-dense, high calorie foods.	☐	☐
11. Adults tends to eat 30 to 50% more food when given large portions of food than when given smaller portions.	☐	☐
12. Calorie-restricted diets have been shown to prolong life in mice, monkeys, and humans.	☐	☐

Media Menu

www.naafa.org
The National Association to Advance Fat Acceptance is dedicated to improving the quality of life for fat people. The organization takes on issues of size discrimination and policies against the interest of fat people, and it promotes fat acceptance through media and legislative routes.

www.sizewise.com
This site provides information, inspirational stories, updates, and resources related to size acceptance.

www.healthfinder.gov
Search "obesity" to find Quick Guides to Healthy Living, clinical guidelines for diagnosis and treatment of obesity, the role of genes in

the development of obesity, national efforts aimed to reduce obesity, and more.

www.shapeup.org
The "Shape Up America" site offers information, advice, and products for weight management. Provides a BMI calculator for kids and information on "obesity and adult health."

www.cdc.gov/growthcharts
CDC's growth charts for children and adolescents from two to 20 years old, including BMI-for-age graphs, are available at this site.

www.nhlbi.nih.gov/health/public/heart/ obesity/lose_wt
Components of the "Aim for a Healthy Weight" Obesity Education Initiative developed by the National Institutes of health are described on

this site. This initiative aims to help reduce the prevalence of overweight and physical inactivity.

www.nhlbi.nih.gov/guidelines/obesity/ prctgd_b.pdf
The Practical Guide to the Identification, Evaluation, and Treatment of Obesity is offered at this site. The Guide, developed by the National Institute of Health, provides a breadth of information on obesity, weight loss, and physical activity.

www.cdc.gov/nchs/fastats/overwt.htm
Get data on rates of obesity, overweight, and the health status of Americans quickly at this site.

Notes

1. Pischon T et al. General and abdominal adiposity and risk of death. N Engl J Med 2008;359:2105–20.

2. Brown PJ, Konner M. An anthropological perspective on obesity. Annals of NY Acad Sci 1987;499:29–46.

3. Van Itallie TB. Obesity: the American disease. Food Technol 1979;(Dec):43–47.

4. Rossner S. Ideal body weight—for whom? (editorial). Acta Medica Scand 1984;216:241–2.

5. Fallon A, Rozin P. Sex differences in perception of desirable body shape. J Abnormal Psychology 1985;94:102–5.

6. Stevens J et al. Evaluation of WHO and NHANES II standards for overweight using mortality rates. J Am Diet Assoc 2000;100:825–7.

7. Defining overweight and obesity, www.cdc.gov/nccdphp/dnpa/obesity/defining.htm, accessed 7/03.

8. Fernández JR et al. Is percentage body fat differentially related to body mass index? Am J Clin Nutr 2003;77:71–5.

9. Defining overweight and obesity (www.cdc.gov/nccdphp/dnpa/obesity/defining.htm).

10. CDC growth charts: United States. Advance Data, No. 314, 12/4/00 (Revised), and 9/09, www.cdc.gov/nchs. and www.cdc.gov/obesity/childhood/defining.html.

11. CDC Fastats, overweight, www.cdc.gov/nchs/fastats/overwt.htm, accessed 2/09.

12. International obesity task force data, Global prevalence of obesity, www.iotf.org/database/index.asp, accessed 2/12/09.

13. Janghorbani M et al. First nationwide survey of prevalence of overweight, underweight, and abdominal obesity in Iranian adults, Obesity (Silver Spring). 2007;15:2797–808.

14. Hossain P et al. Obesity and diabetes in the developing world—a growing challenge, N Engl J Med 2007;365:213-5.

15. Pi-Sunyer X et al. Obesity associated inflammation, presented at the Experimental Biology annual meetings, Washington, DC, 4/03/07.

16. Bray GA. Physiology and consequences of obesity. Medical Education Collaborative, Diabetes and endocrinology clinical management, www.medscape.com/Medscape/endocrinology/Clinical Mgmt/Cm.v03/public/index.CM.v03.html, accessed 1/8/01.

17. Wildman RP et al. The obese without cardiometabolic risk factor clustering and the normal weight with cardiometabolic risk factor clustering: prevalence and correlates of 2 phenotypes among the U.S. population: (NHANES 1999–2004), Arch Intern Med. 2008;168:1617–1624.

18. Price J et al. Family practice physicians' beliefs, attitudes and practices regarding obesity, Am J Prev Med 1987;3: 339–45.

19. Demerath EW et al. Visceral adiposity and its anatomical distribution as predictors of the metabolic syndrome and cardiometabolic risk factor levels, Am J Clin Nutr 2008;88:1263–71.

20. Forgarty AW et al. A prospective study of weight change and systemic inflammation over 9 y. Am J Clin 2008;87:30–5.

21. Stefan N et al. Identification and characterization of metabolically benign obesity in humans. Arch Intern Med. 2008;168:1609–1616.

22. Davidson LE et al. Resistance Plus Aerobic Exercise May Be Best for Sedentary, Abdominally Obese Older Adults, Arch Intern Med. 2009;169:122–131.

23. Steinberg BA et al. Measuring waist circumference, Medscape Cardiology 2006;10(2), www.medscape.com/viewarticle/542635, accessed 10/06.

24. Farin HMK et al. Body mass index and waist circumference both contribute to differences in insulin-mediated glucose disposal in nondiabetic adults. Am J Clin Nutr 2006;83:47–51.

25. Japan legislates mandatory national waist circumference measurement, 6/13/08, amicor.blogspot.com/2008/06/japan-legislates-mandatory-national.html, accessed 2/09.

26. Broom I. Thinking about abdominal obesity and cardiovascular risk. Br J Vasc Dis 2006;6:58–61.

27. German AJ. The growing problem of obesity in dogs and cats. J Nutr 2006;136:1940S–6S

28. Appendix H, Table H-1. Body measurement summary statistics. In: Dietary Reference Intakes for Energy through Amino Acids, Washington,

DC: The National Academies Press, 1999.

29. Gibson RS, Principles of nutritional assessment, New York: Oxford University Press; 1990.

30. Romao I et al. Genetic and environmental interactions in obesity and type 2 diabetes, J Am Diet Assoc 2008;108:S24–S28.

31. Malone M et al. Medication associated with weight gain may influence outcome in weight management program. Ann pharmacotherapy 2005;39:1204–13.

32. Leibel RL. Energy in, energy out, and the effects of obesity-related genes. N Engl J Med 2008;359:2603–4.

33. Faroogi S. Genetic basis of human obesity, presented at the Experimental Biology annual meeting, San Diego, CA, 4/6/08.

34. Nooyens A et al. Adolescent skinfold thickness is a better predictor of body fatness in adults than is BMI: the Amsterdam Growth and Health Longitudinal Study. Am J Clin Nutr 2007;85:1533–9.

35. Li L et al. Intergenerational influences on childhood body mass index: the effect of parental BMI trajectories, Am J Clin Nutr 2009;89:551–7.

36. Whitaker JA et al. Predicting obesity in young adulthood from childhood parental obesity. N Engl J Med 1997; 337:869–73.

37. Ledikwe JH et al. Dietary energy density is associated with energy intake and weight status in U.S. adults. Am J Clin Nutr 2006;83:1362–8.

38. Duerksen SC et al. Family restaurant choices are associated with child and adult overweight status in Mexican-American families, J Am Diet Assoc 2007;107:849–53.

39. Rolls BJ et al. Larger portion sizes lead to sustained increases in energy intake over 2 days. J Am Diet Assoc 2006;106:543–9.

40. Yanover T et al. Eating beyond satiety and body mass index, Eat Weight Disord. 2008;13:119–28.

41. Levine JA et al. The role of free-living daily walking in human weight gain and obesity, Diabetes 2008;57:548–54.

42. Danner FW. A national longitudinal study of the association between hours of TV viewing and the trajectory of BMI growth among U.S. children. J Pedaitr Psychol 2008;33:1100–7.

43. Position of the American Dietetic Association: Individual-, family-, school-, and community-based interventions for pediatric overweight. J Am Diet Assoc 2006;106:925–45.

44. Jeffery RW et al, Prevalence and correlates of large weight gains and losses in adults, Int J Obes 2002;26:969–72.

45. Costanzo PR, Schiffman SS. Thinness— not obesity—has a genetic component. Neuroscience and Biobehavioral Reviews 1989;13:55–58.

46. Rolls BJ. W. O. Atwater Memorial Lecture, presented at the Experimental Biology annual meetings, Washington, DC, 5/1/07.

47. Sui X et al. Cardiorespiratory fitness and adiposity as mortality predictors in older adults. JAMA 2007;298:2507–16.

48. Nedeltcheva AV et al. Sleep curtailment is accompanied by increased intake of calories from snacks. Am J Clin Nutr 2009;89:126–33.

49. Byers T et al. Public Health response to the obesity epidemic too soon or too late? J Nutr 2007;137:488–92.

50. Kumanyika SK et al. American Heart Association's Obesity Statement: Population-based prevention of obesity, Circulation 2008, available at http://circ.ahajournals.org.

51. State-specific prevalence of obesity among adults—United States, 2005. Centers for Disease Control and Prevention, MMWR. 2006;55:985–988.

52. Sabia S et al. BMI over the adult life course and cognition in late midlife: the Whitehall II Cohort Study, Am J Clin Nutr 2009;89:601–7.

53. Shapses SA et al. Bone, body weight, and weight reduction: What are the concerns? J Nutr 2006;136:1453–6.

54. Meyer TE et al. Long-term calorie restriction ameliorates the decline in diastolic functions in humans. J Am Coll Cardio 2006;47:398–402.

55. Satter EM. Internal regulation and the evaluation of normal growth as the basis for prevention of obesity in children. J Am Diet Assoc 1996;96:860–4.

56. Friedman J. Leptin, the most recent advances, presented at the Experimental Biology annual meetings, San Diego, CA, 4/6/08. Nutr 2008;87:30–5.

57. UK panel calls for media to end pressure on girls to be thin, Report of the UK's Cabinet's Body Image Summit, Reuter Health 2000; June 22.

58. Chapman GE et al. Canadian dietitians' approaches to counseling adult clients seeking weight-management advice. J Am Diet Assoc, 2005;105:1275–9.

59. Bacon L et al. Size acceptance and intuitive eating improve health for obese, female chronic dieters. J Am Diet Assoc 2005;105:929–36.

60. Robinson J. Health at every size: toward a new paradigm of weight and health. www.medscape.com-viewarticle/506299, accessed 9/2/05.

Answers to Review Questions

1. False, see page 9-3.
2. False (it's 184 pounds), see page 9-3.
3. True, see page 9-5.
4. False, see Table 9.4, page 9-6.
5. True, see page 9-7.
6. True, see page 9-7.
7. False, see page 9-9.
8. True, see page 9-10.
9. False, see page 9-9.
10. True, see page 9-12.
11. True, see page 9-12.
12. False, see pages 9-13 and 9-14.

NUTRITION | Up Close

Are You an Apple?

Feedback for Unit 9

If your waist circumference is over 35" (88 cm) and you're a female, and over 40" (120 cm) if male, you're an apple. (You are all a peach for doing this exercise.)

Weight Control: The Myths and Realities

NUTRITION SCOREBOARD

	TRUE	FALSE
1 Anybody who really wants to can lose weight and keep it off.		
2 Weight loss is the cure for obesity.		
3 Weight loss can be accomplished using many types of popular diets. Such diets rarely help people maintain the weight loss in the long run.		
4 Weight-loss products and services must be shown to be safe and effective before they can be marketed.		
5 Small and acceptable improvements in eating and exercise behaviors are more likely to produce weight loss and weight loss maintenance than are large and unpleasant changes in behaviors.		

Key Concepts and Facts

- The effectiveness of weight-control methods should be gauged by their ability to prevent weight regain.

- That a weight-loss product or service is widely publicized and utilized doesn't mean it works.

- Successful weight control is characterized by gradual weight loss from small, acceptable, and individualized changes in eating and activity.

Answers to NUTRITION SCOREBOARD	TRUE	FALSE
1 If this claim were true, hardly anybody would be obese.		✔
2 Maintenance of weight loss is the cure for obesity.		✔
3 Many popular diets can lead to weight loss, but none successfully help people prevent weight regain.[1]	✔	
4 Unfortunately, many weight-loss products and services on the market have not been shown to be safe or effective. Laws and regulations do not fully protect the consumer from the introduction of bogus products and services.		✔
5 Small and acceptable behavioral changes are easier to live with over time than are drastic and disliked changes.[2]	✔	

Photo Disc

Baseball, Hot Dogs, Apple Pie, and Weight Control

Americans are preoccupied with their weight. On any given day, 40% of adults, and a majority of individuals who are overweight or obese, are trying to lose weight. To help get the weight off, Americans spend over *$60 billion* annually (an average of approximately $220 per person each year) on weight-loss products and services.[3] Yet Americans are gaining weight faster than they are losing it, and the incidence of obesity in the United States is on the rise.[4] Roughly 5 to 10% of people who lose weight keep it off.[5] Consumers are paying handsomely for weight loss without experiencing the desired cosmetic changes or health benefits that come when weight loss is maintained.

Why do so many Americans fail at weight control? For some people, achieving permanent weight reduction on their own may be truly impossible. For others, the problem is the ineffective methods employed, not the people who use them.

"Jess—did I hear you say you wanted to lose some weight? It happens I know about a terrific diet! A couple of months ago I went on Dr. Quick's Amino Acid Diet and lost 15 pounds! You eat nothing but fish, papaya, and broccoli, and the pounds just melt away!"

(Continued)

Weight Loss versus Weight Control

Dietary torture and weight loss are not the cure for overweight. If losing weight was all it took to achieve the cultural ideal of thinness, nobody would be overweight. Any popularized approach to weight loss (and some get pretty spectacular) that calls for a reduction in caloric intake can produce weight loss in the short run. These methods fail in the long run, however, because they become too unpleasant. Feelings of hunger, deprivation, and depression that often occur while on a weight-loss diet eventually lead to a breakdown of control, eating binges, a return to previous habits, and weight regain.[6] Humans are creatures of pleasure and not pain. Any painful approach to weight control is bound to fail. Improved and enjoyable eating and exercise habits are needed to keep excess weight off. Quick weight-loss approaches don't change habits.

Unfortunately, many dieters think weight-control methods are successful if they lead to a rapid loss of weight.[7] When the lost weight is regained, dieters tend to blame themselves and not the faulty method. Blaming themselves for the failure, many people are ready to try other quick weight-loss methods. But they usually fail, too.[8]

The Business of Weight Loss   More than 29,000 weight-loss products and services are available. Some of these are shown in Illustration 10.1. Most of them either don't work at all or don't prevent weight regain. "Quick-fix" weight-loss approaches that don't lead to long-term changes in behavior are the primary reason approximately 90 to 95% of people who lose weight gain some or all of it back. The demand for such products and services is so great, however, that many are successful—financially.

One reason so many weight-loss products and services are available is that almost none of them work. If any widely advertised approach helped people lose weight and keep it off, manufacturers of bogus methods would go out of business. The weight-loss industry also thrives because of the social pressure to be thin. Many people try new weight-loss methods even though they sound strange or too good to be true. Often people believe that a product or service must be effective, or it wouldn't be allowed on the market. Although reasonable, this belief is incorrect.

The Lack of Consumer Protection   The truth is that general societal standards for consumer protection do not apply to the weight-loss industry.[8] No laws

Jess was shocked. Selma looked heavier than she did last year when they were in class together.

"But Selma," Jess inquired, "what happened to the weight you lost?"

"Oh, I gained it back. I ran out of willpower and started eating everything in sight. I'm going to get back on track, though. Tomorrow I start the Slim Chance Diet."

Illustration 10.1   There is no lack of weight-loss books and products. There is a lack of popular approaches that help people keep weight off.

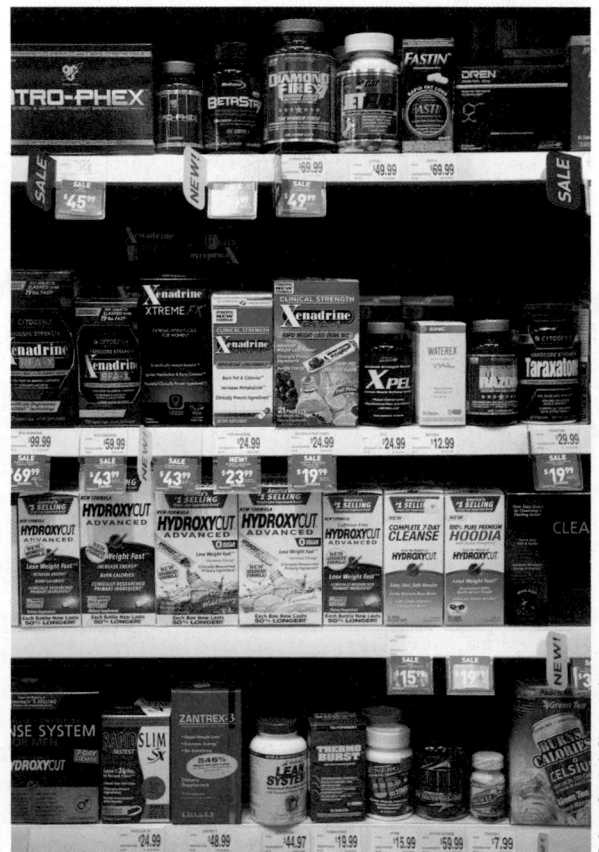

Do some foods have negative calories? You're looking for a way to shed the 10 pounds you gained during your freshman year and are seriously thinking about adding grapefruit and vinegar to your diet. You have heard these foods have "negative calories" because they make the body burn fat.

Answers on next page

Natalie
I know vinegar cleans the grease off windows. Maybe it will melt away my fat, too.

Chuck:
My grandfather lives in Florida and eats a lot of grapefruit. He's as thin as a rail.

Table 10.1

A brief history of discontinued weight-loss methods

Year	Method	Reason for Discontinuation
1940s–1960s	Amphetamines	Highly addictive, heart and blood pressure problems
	Vibrating machines	Did not work
1950s	Jejunoileal; bypass surgery	Often caused chronic diarrhea, vitamin and mineral deficiencies, kidney stones, liver failure, arthritis
1960s	Liquid protein diet	Poor-quality protein caused heart failure, deaths
1980s	Intestinal bypass surgery	Excessive risk of serious health problems
1990	Oprah Winfrey liquid diet success (she lost 67 pounds)	
1991	Oprah Winfrey gains 67 pounds and declares "No more diets!"	
1997	Phen-Fen (Redux)	Heart valve defects, hypertension in lung vessels
2004	Ephedra	Excessive risk of stroke, heart attack, and psychiatric illnesses

Table 10.2

Loads of false and misleading weight-loss advertisements appear in the media. Here is a list of the top six features of weight-loss ads that make false or misleading claims.[11]

1. Use testimonials, before-and-after photos.

2. Promise rapid weight loss.

3. Require no special diet or exercise.

4. Guarantee long-term weight loss.

5. Include a "clinically proven" or "doctor approved" statement.

6. Make a "safe," "natural," or "easy" claim.

require a product to be effective; in most cases, companies do not have to show that weight-loss products or services actually work before they can be sold. That's why products like herbal remedies, forks with stop and go lights, weight-loss skin patches, colored "weight-loss" glasses, electric cellulite dissolvers, mud and plastic wraps, and inflatable pressure pants are available for sale. These, like many other weight-loss products, do not work, and there is no reason why they should. Promotions for the products just have to make it *seem* as though they might work wonders on appetite or body fat.

Furthermore, weight-loss products and services are usually not tested for safety before they reach the market. The history of the weight-loss industry is littered with abject failures: fiber pills that can cause obstructions in the digestive tract, very-low-calorie diets and intestinal bypass surgeries that lead to nutrient-deficiency diseases and serious health problems, amphetamines that can produce physical addiction, diet pills that cause heart valve problems, and liquid protein diets that have led to heart problems and death from heart failure.[10] A brief history of weight-loss method failures is chronicled in Table 10.1.

Pulling the Rug Out Fraudulent weight-loss products and services may be investigated and taken off the market. Currently, the Federal Trade Commission (FTC) monitors deceptive practices and weight-loss claims on a case-by-case basis. The FTC has filed suits against companies that make exaggerated claims, and has identified the most common dubious claims made by the industry (Table 10.2). The FTC performs most investigations in response to consumer complaints. Illustration 10.2 shows three examples of products that were taken off the market and explains why.

Fat magnet pills were purported to break into thousands of magnetic particles once swallowed. When loaded with fat, the particles simply flushed themselves out of the body. The FTC found the product too hard to swallow. The company took the product off the market and made $750,000 available for customer refunds.

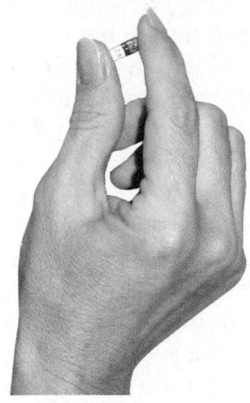

"Blast" away 49 pounds in less than a month with Slim Again, Absorbit-All, and Absorbit-AllPlus pills, claimed ads in magazines, newspapers, and on the Internet. The company got blasted by the FTC to the tune of $8 million for making fraudulent claims.

Take a few drops of herbal liquid, put them on a bandage patch, and voilà—a new weight-loss product! The herbal liquid was supposed to reach the appetite center of the brain and turn the appetite off. Federal marshals weren't impressed. They seized $22 million worth of patch kits and banned the sale of others.

Illustration 10.2   Examples of bogus weight-loss products.

Requirements for truth in labeling keep many bogus products from including false or misleading information on their labels. These laws, however, do not keep outrageous claims from being made on television or printed in pamphlets, books, magazines, and advertisements.

What if the Truth Had to Be Told?   Suppose the weight-loss industry had to inform consumers about the results of scientific tests of the effectiveness and safety of weight-loss products and services. What if this information had to be routinely included on weight-loss product labels (Illustration 10.3)? What do you think would happen to consumer choices and the weight-loss industry? Increased federal enforcement of truth-in-advertising laws may help put an element of honesty into promotions for many weight-loss products. In addition, the FTC has proposed that claims about long-term weight loss "must be based on the experience of patients followed for at least two years after they complete the [weight-loss] program."[12] If this proposal ever becomes law, it will change the weight-loss industry in the United States.

Diet Pills   Two types of prescription diet pills are currently approved for long-term use in obese people in the United States. One is Meridia (sibutramine), and the other is Xenical (orlistat).[13] Meridia works by enhancing satiety (which

Do some foods have negative calories?   The vinegar, grapefruit, negative calorie myth lives on because it strikes many people as reasonable. Neither vinegar nor grapefruit, nor any other food has negative calories or causes the body to gear up metabolism and burn fat.[9]

Juanita

Chuck:

Illustration 10.3 Just imagine what would happen if weight-loss approaches were required to divulge their effectiveness and risks.

reduces food intake), and Xenical works by partially blocking fat absorption in the intestines. Both have side effects. Meridia may increase blood pressure and heart rate, cause headaches, and lead to dry mouth. Xenical's primary side effect is oily stools, which are particularly bothersome if high-fat meals are consumed. Also, the malabsorption of fat caused by Xenical reduces absorption of a number of fat-soluble nutrients such as vitamin D, E, and beta-carotene.[14]

It is recommended that Meridia and Xenical be used in conjunction with reduced-calorie diets and exercise. Both reduce body weight to some extent. Merida use is associated with a 10 pound weight loss per year, while Xenical is related to a loss of five pounds up to a total of about 10% of initial body weight. The over-the counter version of Xenical can produce weight loss of up to 5% of initial body weight.[13–15] Weight regain after pill use stops is common.[16]

Many other diet drugs are under development or being tested. Much activity currently centers around medications that decrease appetite and increase satiety. Even though people using prescribed diet drugs are carefully screened and monitored by medical professionals, serious side effects can develop. No diet drug is absolutely safe, and none are known to cure obesity forever.[16]

Popular Diets

There is a never-ending supply of diet books on the market. More than 1500 of them are available at any one time.[17] Although the safety and effectiveness of almost all of the diets described in popular weight loss books have not been evaluated, a few have been selected for examination due to their popularity. Five popular diets that have undergone scrutiny by researchers are Dr. Atkins's New Diet Revolution, Weight Watchers Pure Points Program, the Slim-Fast Plan, the Zone Diet, and Rosemary Conley's Eat-Yourself-Slim Diet (Illustration 10.4). Studies have compared weight loss, adherence to the diet, and health effects of these popular diets.[18–20]

Illustration 10.4 Some of these have been popular weight loss diets evaluated by research studies.

Approaches to weight loss described in these books represent a mix of balanced, low-carbohydrate, high-protein diets (see Illustration 10.5); and low-fat diets. All of the diets were found to produce a reduction in calorie intake if followed and weight loss. Weight loss varied by how long a person stayed on the diet and averaged about 7 pounds. Most of the people in the studies were unable to stick to the diet for a year, primarily because it was too hard to follow.[18,19]

Organized weight loss programs such as Weight Watchers and Jenny Craig offer foods and food plans that are reasonably well balanced and low in calories. The programs also provide support services for users. Whether they work for weight loss maintenance is not clearly known because study results are not available.

Fad Diets

Many other weight loss books are offered to the public and almost always contain made-up approaches to weight loss. Some examples of outrageous approaches offered by these books are given in Table 10.3. Some fad diets become popular because movie stars have used them or because they promise rapid weight loss. None have been shown to lead to sustained weight loss or weight loss maintenance.

Illustration 10.5   An example meal from a low-carbohydrate, high-protein diet.

Richard Anderson

Internet Weight Loss Frauds

Other than a lipase inhibitor (alli) that can be obtained without a prescription, there is no dietary supplement or medication sold over the Internet that is approved by the FDA as being a safe and effective for weight loss (Illustration 10.6).[21] Internet weight loss aides are intended to fatten wallets and not to reduce people's waistlines.[3]

Weight Regain: A Shared Characteristic of Popular Diets   The behavioral changes required to stick to popular dietary and exercise programs turn out to be too difficult and unpleasant for most people to follow in the long run. Differences in the amounts of weight lost for this or that diet and the health benefits achieved disappear over time as weight is regained. The situation has led James Hill, one of the directors of the National Weight Control Registry, to conclude, "Our real challenge is not helping people lose weight but is helping them keep it off."[14,26]

Table 10.3

Examples of fad diets

1. The Popcorn Diet: Like popcorn? You'll get to eat all the unbuttered, unsalted popcorn on this diet you want. The diet also emphasizes fruits, vegetables, smaller portions, and exercise.

2. The Grapefruit Diet: This diet is based on the myth that, when combined with protein, grapefruit triggers fat burning and weight loss. The high grapefruit-high protein diet has been around since the 1930s.

3. The Chocolate Diet: A weight loss diet for chocolate addicts, this nonsense diet categorizes chocolate lovers into types and provides a diet plan for each type. Includes a liquid chocolate diet shake.

4. The Metabolism Diet: As the name implies, this diet purports to produce weight loss by speeding up metabolism. The low carbohydrate, low calorie diet recommended can only be used for 7 days.

5. The 3 Day Diet: If you follow the food prescriptions for what to eat, how much to eat, and when to eat, you'll supposedly be rewarded with ramped-up metabolism and a 10 pound weight loss in 3 days. But wait, there's more. The diet also provides internal cleansing, cholesterol reduction, and more energy.

6. The Cabbage Soup Diet: It's a 7-day weight loss plan based on cabbage soup with other vegetables.

7. Japanese Morning Banana Diet: You get to eat all the bananas you want on this diet—and nothing else.

There are indications that consumers are growing tired of popular, money-making approaches to weight loss. A survey undertaken by the food industry in the United Kingdom and the United States found that 29% and 44% of the adults studied were aware that extreme diets *cannot* produce sustained weight loss. Instead of popular weight-loss diets, these consumers are deciding that the path to weight control is paved by small, easy changes to their diets and physical activities.[26]

Physical Activity and Weight Control

Physical activity can play an important role in the prevention of weight gain, weight loss, and weight maintenance after weight loss. Even without weight loss, regular physical activity promotes health by decreasing abdominal fat stores, improving blood cholesterol concentrations, insulin sensitivity, and blood pressure. A regular program of resistance (strength building) exercise alone increases lean body mass and reduces fat mass even without weight loss.[27]

Recommendations for physical activity in overweight and obese adults for the prevention of additional weight gain, weight loss, and maintenance of weight loss have been developed by American College for Sports Medicine.[27] These recommendations are summarized in Table 10.4. The minutes of physical activity listed in the table refer to moderate intensity activities such as brisk walking, soccer, jogging, tennis, calisthenics, rope skipping, and aerobic dancing. Low intensity physical activities are also related to improvements in health risks and reductions in weight gain, but to a lesser extent than are moderate intensity activities.[28]

Calorie intake reductions of 100 or more daily combined with regular physical activity are related to higher levels of weight loss than exercise alone because they create a larger calorie deficit and a greater use of fat stores for energy.[2] The increased physical activity approach to weight maintenance may be an effective method for weight management among people who prefer exercise to cutting back on calories. It will work as long as increases in energy expended in physical activity are not exceeded by increases in calorie intake.

Although uncommon, some people have health problems requiring medical supervision of exercise programs. Individuals with health concerns should get an "all clear" from their health care provider before undertaking a higher level of physical activity than usual.

www.ftc.gov/bcp/edu/pubs/business/adv/bus60.pdf

Illustration 10.6 The Internet is home to hundreds of bogus weight loss sites.

Table 10.4	
American College of Sports Medicine's recommendation for moderate physical activity and weight management in overweight and obese adults[27]	
Weight outcome for most people	**Average number of minutes per day of moderate physical activity**
Prevention of weight gain	21 minutes or more
Weight loss	32 to 60 minutes
Weight loss maintenance	29 to 43 minutes

Weight Loss Benefits

Even modest amounts of weight loss offer substantial health benefits to people who are overweight and obese. Loses of 5 to 10% of initial body weight are considered "clinically significant." These loses are related to lowered blood concentrations of insulin, triglycerides, and glucose. Other benefits include increased insulin sensitivity and HDL cholesterol levels, and reduced concentrations of C-reactive protein (a marker of inflammation).[18,19] These changes decrease the likelihood that diseases such as heart disease and type 2 diabetes will develop. The benefits strengthen as a person gets closer to a normal weight.[22]

Weight loss reduces calorie need somewhat because of metabolic adaptations the body makes to conserve energy and the loss of muscle mass. About 75 to 80% of weight loss consists of fat and the rest is from lean body tissues such as muscle. Lean body mass, including muscle mass, can be restored with physical activity, especially strength-building activities. Increased muscle mass drives calorie need up because it takes more energy to maintain muscle than fat tissue.[8,19,28]

Obesity Surgery

Weight-loss surgery, referred to as **bariatric** surgery in the medical field, is a weight control method of last resort. Weight-control surgery is reserved for people with BMIs over 40 or for people with BMIs of 35 to 40 who have serious health problems related to their weight. Although several methods of weight-loss surgery exist, gastric bypass surgery and adjustable gastric banding (Lap Band) are the most frequently performed worldwide.[29] Bariatic surgery is becoming increasingly popular. In 2007, 205,000 operations were performed in the United States.[30] Some people who don't qualify are gaining weight to get past the BMI thresholds.[25]

bariatrics
The field of medicine concerned with weight loss.

Gastric Bypass Surgery   Gastric bypass surgery is the most effective method for weight loss and weight maintenance available. On average, individuals undergoing this surgery lose 50–60% of excess body weight, and they often maintain much of the loss over the long term.[23] Health status generally improves dramatically as a result of the weight loss. Resolution of type 2 diabetes, hypertension, sleep disorders, and elevated LDL cholesterol blood levels often follow gastric bypass surgery.[23]

In gastric bypass surgery (Illustration 10.7), most of the stomach is stapled shut, leaving a pouch at the top that can hold about two tablespoons of food. A section of the small intestine that connects to the bottom of the stomach is cut

Illustration 10.7   Gastric bypass surgery (left), Lap Band surgery (right).

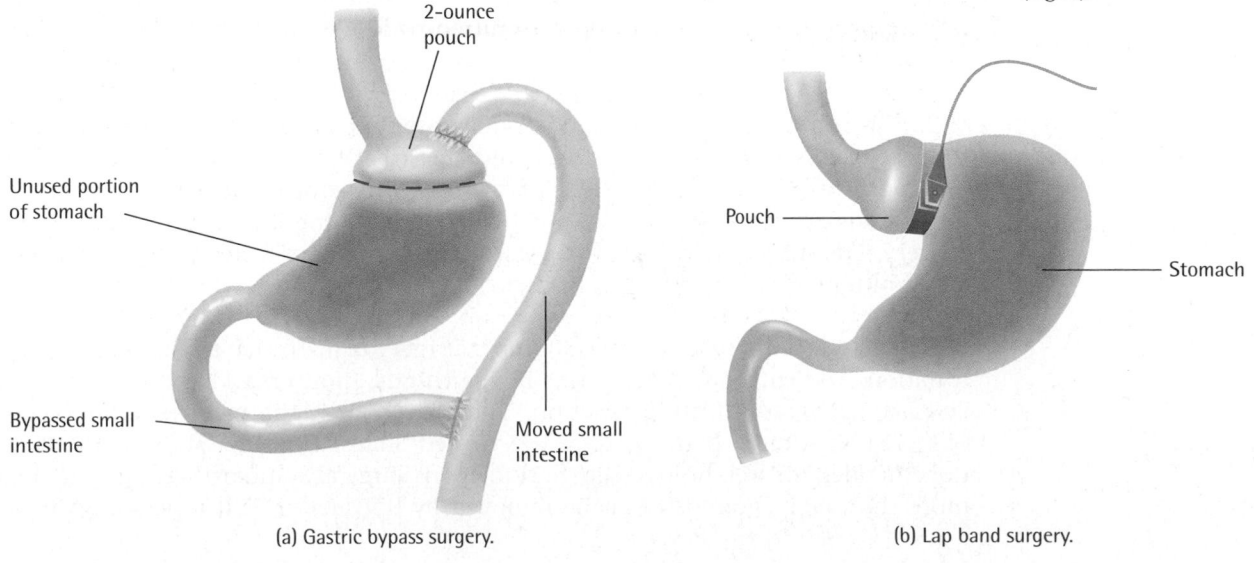

2-ounce pouch

Unused portion of stomach

Bypassed small intestine

Moved small intestine

Pouch

Stomach

(a) Gastric bypass surgery.

(b) Lap band surgery.

off and attached as a drain for the stomach pouch. The loose end of the small intestine is then reattached to the section of the small intestine that leads from the pouch. This way, digestive juices from the stomach and the upper part of the small intestine are made available for digesting food. People who have their stomachs stapled can eat only a small amount of food. If they eat too much, they feel nauseous and dizzy and vomit.

Lap Band Surgery

Lap band surgery produces a small stomach pouch by constricting the upper part of the stomach with a band (Illustration 10.7). The band can be inflated by the injection of saline water and the amount of food allowed to enter the stomach limited further.[22] Lap band surgery is performed "laparoscopically," or by inserting a tube through small incisions made in the abdomen. Individuals receiving this surgery tend to lose less weight (48% of excess body weight on average) than do people having gastric bypass surgery.[23]

■ **Concerns Related to Bariatric Surgery.** Gastric bypass surgery is not a panacea for weight control. It comes at a high cost ($30,000 or more) and is accompanied by high rates of postsurgery complications. There is a 0.5–1.1% chance of dying from this operation, but that risk is considered worth it because obese people who receive the surgery live significantly longer than do those who stay obese.[23] Rates of complications from the surgery are 22% during the postsurgery hospital stay and 40% within the six months that follow surgery. Readmissions for complications can increase costs associated with gastric bypass surgery to over $60,000.[29]

Some of the complications arising from bariatric surgery are related to nutrient deficiencies. The small stomach and changes in absorption that results from the surgey leads to malabsorption of a number of vitamins and minerals, most notably vitamin D, vitamin B_{12}, folate, calcium, and iron. Multivitamin and mineral supplements are a routine component of care after bypass surgery.[31] Other complications related to adverse consequences of anesthesia—such as infection, nauses, vomiting, dehydration, and gallstones—may also develop.[32]

Some amount of lost weight is usually regained in the years following bariatric surgery. The reason for the gain is increased volume of the stomach pouch. Increasing food intake can stretch people's stomachs. The 2-tablespoon pouch left after the surgery can be enlarged to hold 1/2 to 2/3 cup of food.[33] (A regular-sized stomach has a capacity of about four cups.)

About half of the individuals who undergo bypass surgery struggle with emotional issues around food when old habits begin to come back. People who undergo this surgery have to be committed to long-term lifestyle changes and follow-up care.[29]

■ **Body Contouring Surgery.** Gastric bypass surgery leaves some people with folds of excess skin where fat stores were lost (see Illustration 10.8). Skin tissue previously stretched by high levels of fat stores does not retract on its own after fat stores are lowered. The only known way of removing it is "body contouring" surgery. The surgery may have to be repeated a number of times and can come at a cost of tens of thousands of dollars.

Liposuction At a cost of over $3000 per surgery (prices vary by fat deposit site, inflation, and surgeon), fat deposits in the thighs, hips, arms, back, or chin can be partially removed by liposuction. The procedure is the most common type of cosmetic surgery performed in the United States. Considered cosmetic, it is not intended for weight loss (Illustration 10.9). Surgical standards require that no more than eight pounds of fat be removed by liposuction.[34] If a person gains a

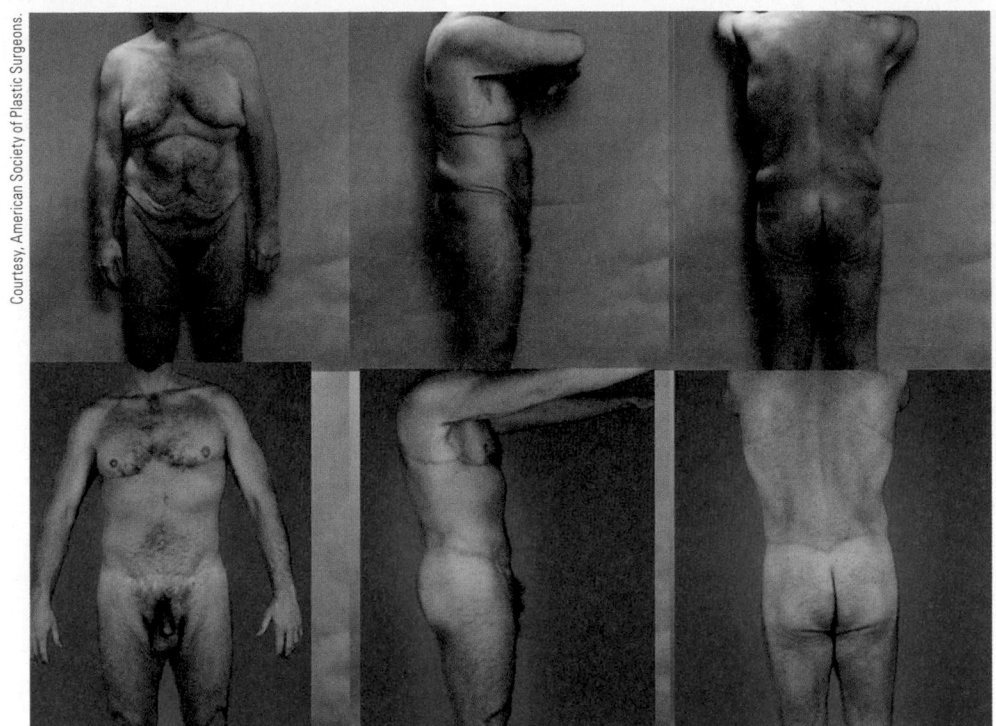

Courtesy, American Society of Plastic Surgeons.

Illustration 10.8 The photos in the top row show excess skin folds that developed after weight loss surgery. The photos on the bottom show the results of body contouring surgery.

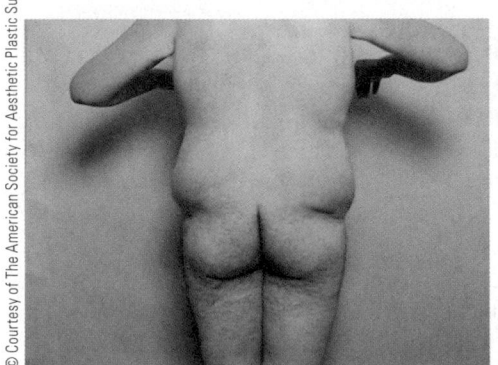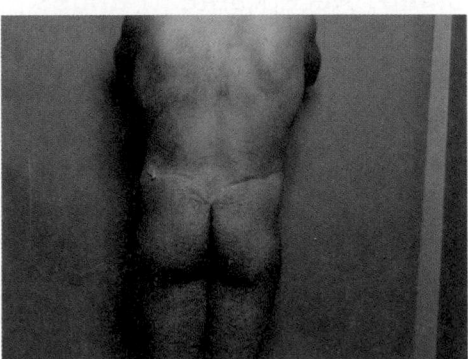

© Courtesy of The American Society for Aesthetic Plastic Surgerry

Illustration 10.9 The effects of liposuction.
The photo on the left shows the "spare tire" of a man before liposuction, and the photo on the right shows the same man's waist after liposuction.

good deal of weight after liposuction, fat will be deposited to some extent in the breasts and other areas not operated on. These deposits can lead to a return of an undesired body shape.[35] In addition, surgery always carries a risk of infection and other complications, so it cannot be taken lightly.

Weight Loss: Making It Last

An approach to weight loss can be considered successful only if it is safe, healthful, and prevents weight regain. Successful approaches focus on healthy eating and exercise for a lifetime, rather than on "dieting." People who lose weight and maintain their weight afterward have found physical activities they enjoy and they exercise regularly (Illustration 10.10). That is the major commonality among people who maintain their weight after weight loss.[24] Other characteristics of

(a)

(b)

(c)

Illustration 10.10 Small, acceptable changes are the key to a successful weight-loss/weight-maintenance program. Try to find activities that you enjoy; get out there and play!

people who lose weight and keep it off and characteristics of people who regain lost weight are listed in the nearby Health Action.

Safe, healthful, and effective methods of weight loss and maintenance described here are not the subject of most popular diet books and weight loss programs. The approach that works in the long run depends on individual decisions about changes in diet and physical activiy that can be made and kept for the rest of one's life. Changes in diet and physical activity most likely to be maintained are small, easy to implement, and acceptable—even preferable—to existing behaviors.[24,36]

Small, Acceptable Changes

For most people, excess body fat accumulates slowly over time. Consuming just an extra 50 calories a day, for example, will lead to a gain of approximately five pounds in one year. (Approximately 3500 excess calories will produce a weight gain of one pound, and a 3500-calorie deficit will produce a loss of one pound in body weight in many people.) Excess weight is rarely all gained over the course of a few weeks or months. Fat is put on slowly, and that's the best way to take it off.[2]

Gradual losses in body fat do not require dramatic changes in diet or activity level. Only small changes in diet and activity are needed. By cutting back food intake by 100 calories per day, a person could lose 10 pounds in a year. Table 10.5 provides examples of small changes in diet and physical actvity level that, if all else about a person's diet and physical activity remain the same, would produce about a 100-calorie deficit a day.

Health Action   Weight-Loss Maintainers versus Weight Regainers[24,37]

Weight-Loss Maintainers

- Exercise regularly.
- Make small and comfortable changes in diet and physical activity.
- Eat breakfast.
- Choose low-fat foods.
- Keep track of their weight, dietary intake, and physical activity level.

Weight Regainers

- Exercise little.
- Use popular diets.
- Make drastic and unpleasant changes in their diets and physical activity levels.
- Take diet pills.
- Cope with problems and stress by eating.

Table 10.5

Small changes in diet and physical activity worth about 100 calories[a]

Diet	Physical Activity
1. Consume 1 cup of fat-free yogurt with fruit instead of a cup of low-fat yogurt with fruit	1. Walk an additional 20 minutes.
2. Drink 1½ cups of skim milk rather than 1½ cups of whole milk.	2. Lift arm weights for 15 minutes.
3. Eat a roasted or grilled chicken sandwich rather than a fried chicken sandwich.	3. Garden for 15 minutes.
4. Eat ½ cup rather than a cup of rice.	4. Play Frisbee for 30 minutes.
5. Order a regular rather than a large fish sandwich.	5. Clean house for 30 minutes.

[a]Physical activity calories are based on a 150-pound person.

Identifying Small, Acceptable Changes To identify changes in eating and activity that have staying power, first list the weak points in your diet and activity. Dietary weak points might include the consumption of high-fat foods due to eating out often, or relying on high-fat convenience foods that can be heated up in seconds. Another weak point might be skipping breakfast and overeating later in the day because of extreme hunger. Weak points in physical activity might include driving instead of walking, not engaging in sports, or spending too little time playing outside.

For each weak point, identify options that seem acceptable and enjoyable. A person who enjoys broiled chicken with barbecue sauce might not mind eating that at restaurants instead of fried chicken. That's a change people can make if they plan ahead. A person who gets too full from a large serving of fries might be happier ordering a small serving and not eating so many. A breakfast skipper might find grabbing a piece of fruit and a slice of cheese for breakfast acceptable and doable. People who enjoy walking may not mind leaving the car or bus behind and letting their feet carry them to class, the grocery store, or a friend's house.

Many acceptable options for making small improvements in diet and activity may be available (see the "Take Action—Small Steps Can Make a Big Difference"). The easiest changes to accomplish are the ones that should be incorporated into the overall lifestyle improvement plan. Some people, for example, lose weight and keep it off simply by consciously cutting down on portion sizes. Others avoid eating too much at any meal and walk more. Simply adding breakfast helps some people lose weight and maintain the loss.[38] Increasingly, people are losing weight and keeping it off by eating more nutrient-dense foods like vegetables and fruits, and fewer high fat, high sugar, energy-dense foods.[39] This change doesn't require that people eat less food but rather select and prepare foods that are nutrient-dense rather than calorie-dense. The easier the changes are to follow, the more likely they are to succeed.

Individualized plans that don't work out often include unacceptable or unenjoyable changes. The changes may be too large or too different from the usual. In that case, go back to the drawing board and modify the plan to include small changes that are acceptable in the long run. Perhaps the original plan included jogging, but it turns out that you don't enjoy jogging. In that case, take jogging out of the plan! Replace it by any other physical activity that would be enjoyed. Midcourse corrections should be expected. Some experimentation may be required to identify the small changes that will last.

What to Expect for Weight Loss If you're happy with the changes and have found the right levels of calorie intake and physical activity, weight loss will be gradual but lasting. The pattern of loss should be somewhat like that graphed in Illustration 10.11, where the person lost 18 pounds over 30 weeks. The pattern will

Take Action

Small Steps Can Make a Big Difference

Trying to lose weight or keep it off? Pick one eating and one physical activity option from the lists below that you find attractive. Give it a try for a day or two and see how it works out for you. Write down the changes you made and how you felt about making them. Could you see yourself making the changes for a week? The rewards offered by small steps may be greater than you expect.

Small Steps for Healthy Eating

1. Eat cereal for breakfast
2. Eat a piece of fruit before dinner
3. Eat larger portions of vegetables than meat
4. Drink skim milk
5. Switch from soft drink to flavored water. You can make your own by diluting fruit juice with water.
6. Stop eating as soon as you start to feel full
7. Eat only when you feel hungry
8. Dish up a smaller than usual portion of dessert

9. If you're right-handed, eat using your spoon or fork in your left hand.

Do it the other way if you're left-handed.

Small Steps for Physical Activity

1. Walk an additional 2000 steps (20 minutes)
2. Park farther from the store than usual
3. Take a 10-minute walk in a park and look for flowers
4. Lift a weight (or bottle of water, your textbook, a phone book) for 10 minutes while watching TV
5. Do 10 sit-ups in the morning
6. Use the stairs
7. Take your neighbor's dog for a walk
8. Jump rope for two minutes
9. Dance for five minutes while no one is watching

include peaks, valleys, and plateaus—not a straight downward curve. Sometimes a bit of weight will be gained, and other times more weight than expected will be lost. It's more important to enjoy and continue improved eating and activity patterns than to concentrate on the number of pounds lost.

If diet and exercise behaviors are improved in acceptable ways, there may be little need to become preoccupied with the number of calories consumed, the number of calories burned off in a bout of exercise, or the number of pounds lost last week. The goal is reached when the small changes become an enjoyable part of life on a day-to-day basis. Improved eating and activity patterns offer many benefits. Weight loss is only one of them.

Illustration 10.11 People who lose weight gradually are more likely to keep it off than those who lose weight rapidly. The weight loss graphed here averages half a pound per week.

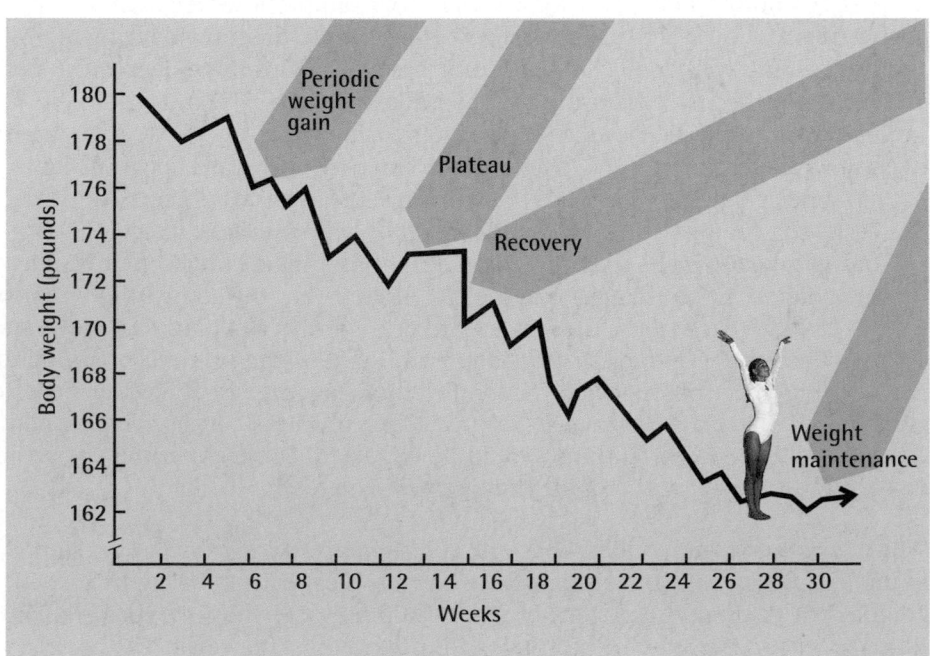

Up Close

©Jose Luis Pelaez Inc / Getty Images/
Blend Images

Setting Small Behavior Change Goals

Small changes in diet and exercise can return large benefits to health over time. What's the best way to get started on small and important behavioral changes? For many people, it begins with setting small but concrete goals.

This activity asks you to practice developing two small behavioral change goals, one related to diet and the other physical activity. Each small change goal should:

   a. state an activity

   b. state when the activity will be performed. For a physical activity goal, it should also:
   c. state how long the activity will be performed.

Here are two examples of small and concrete behavioral change goals:

"I will eat my vegetables first at dinner three times a week."
"I will work in the garden twice a week for 20 minutes."

Your Small Change Goals

Goal 1 - Diet change:

_____

Goal 2 - Physical activity change:

_____

Check list: Do your goals meet the criteria listed above? Do they represent small rather than large changes? If not, re-work the goals so they do.

FEEDBACK can be found at the end of this Unit.

[Key Terms

bariatrics, page 10-9.

Review Questions

1. Claims made for the effectiveness of weight loss products in advertisements must, by federal regulation, be true. ☐ ☐

2. Commerical weight-loss services and products must be tested for safety before they are made available to the public. ☐ ☐

3. Some foods, including celery and grapefruit, have negative calories and help overweight people burn fat. ☐ ☐

4. It is recommended that people using prescription diet pills for weight loss should also follow a reduced-calorie diet and exercise routine. ☐ ☐

5. Many different types of weight loss diets lead to weight loss if followed. ☐ ☐

6. Liposuction is a recommended method for weight loss and weight loss maintenance. ☐ ☐

7. The major problem shared by all popular weight-loss approaches is that they do not help people maintain weight in the long run. ☐ ☐

8. A person who weighs 250 pounds and has 95 pounds of excess body weight before gastric bypass sugery and loses the average amount of excess body weight years later would be expected to weigh 170 pounds. ☐ ☐

9. The best way to lose weight and keep it off is by making small and acceptable changes in diet and physical activity. ☐ ☐

Media Menu

www.thedietchannel.com/faddiets.htm
Read up on how popular diets compare to the Dietary Guidelines, specific strengths and weaknesses of selected fad diets, and guidance on how to lose weight and keep it off here.

http://nutrition.about.com/library/bl_fad_diets.htm
Take the nutrition quiz on fad diets at this site developed by About Fitness & Health.

www.nhlbi.nih.gov/health/public/heart/obesity/lose_wt/behavior.htm
A Guide to Behavior Change for weight loss from the National Heart, Blood, and Lung Institute is available at this site. It also provides helpful information on weight loss and weight maintenance.

www.healthfinder.gov/prevention/ViewTopic.aspx?topicID=25
This site offers descriptions of small changes you can make to lose weight and keep it off.

www.nlm.nih.gov/medlineplus/weightcontrol.html
This page will lead you to scientifically credible information on weight loss and weight maintenance offered by the National Institute of Diabetes and Digestive and Kidney Diseases and other organizations.

www.win.niddk.nih.gov/publications/myths.htm
NIH's Weight Control Information network presents leading myths about weight control, weight loss diets, food, and physical activity. It's fun and informative.

www.ftc.gov.redflag
The FTC updates this site about bogus weight loss advertisements and guides for industries that may produce them.

www.ars.usda.gov/ba/bhnrc/ndl
At the homepage for USDA's food composition information, you can obtain nutrient data files

and get answers to commonly asked questions about the calorie value of foods.

www.nwcr.ws
The National Weight Control Registry was developed to identify and investigate the characteristics of individuals who succeed at long-term weight loss. It is tracking over 5000 individuals who have lost significant amounts of weight and kept it off for long periods of time. Learn more about characteristics of long-term weight losers or share your story.

www.ediets.com
This commercial site offers personalized diet plans and other resources developed largely by dietitians.

www.shapedown.com
This commercial site provides background information and methods for the tested Shapedown Program for weight management in children and adolescents.

Notes

1. Franz MJ et al. Weight-loss outcomes: a systemic review and meta-analysis of weight-loss clinical trials with a minimum of 1-year follow-up. J Am Diet Assoc 2007; 107:1755–67.

2. Hill JO. Can a small-changes approach help address the obesity epidemic? A report of the joint task force of the ASN, IFT, and the IFIC, Am J Clin Nutr 2009;89:477–84.

3. Ethics opinion: weight loss products and medications. J Am Diet Assoc 2008;108:2109–2113.

4. CDC Fastats, overweight, www.cdc.gov/nchs/fastats/overwt.htm, accessed 2/09.

5. Sarlio-Lahteenkorva SS et al. A descriptive study of weight loss maintenance: 6 and 15 year follow-up of initially overweight adults. Int J Obesity 2000;24:116–25.

6. Goodrick GK et al. Why treatments for obesity don't last. J Am Diet Assoc 1991;91:234–47.

7. Womble L et al. Unrealistic weight loss goals and feelings of failure. Presentation at the North American Association for the Study of Obesity, Long Beach, CA, 2000 Nov.

8. Wooley SC et al. Obesity treatment: the high cost of false hope. J Am Diet Assoc 1991;91:1248–51.

9. Cunningham A et al. Is it possible to burn calories by eating grapefruit or vinegar?

10. Wadden TA, Stunkard AJ, Brownell KD. Very low calorie diets: their efficacy, safety, and future. Ann Intern Med 1983;99:675–84.

11. Current trends in weight-loss advertising. J Am Diet Assoc 2003;103:150.

12. Taubes G. Dietary approaches to weight control. What works? Experimental Biology Annual Meeting, San Diego, April 14, 2003.

13. Hulisz DT et al. The Skinny on Weight Loss Supplements: Fact or Fantasy? www.medscape.com/viewprogram/12613, accessed 2/09.

14. Pray WS, New nonprescription weight loss product, U.S. Pharmacist 2007;32:10–5.

15. Hawthrone F. Diverting a diet drug, The Scientist 2008; Jun:41–7.

16. Yanovski SZ, Yanovski JA. Obesity. N Engl J Med 2002;346:591–602.

17. Van Horn L. Nutritional research: the power behind the fad-free diet. J Am Diet Assoc 2007;107:371.

18. Dansinger MG et al. Comparisons of the Atkins, Ornish, Weight Watchers, and Zone diets for weight loss and heart disease risk reduction: a randomized trial. JAMA 2005;293:43–50.

19. Arterburn D et al. Comparison of commercial weight loss diets. JAMA 2006;292:101–3,

20. Buchholz AC, Schoeller DA. Is a calorie a calorie? Am J Clin Nutr 2004;79(suppl):899S–906S.

21. Food and Drug Administration. FDA expands warnings to consumers about tainted weight loss pills, 1/8/08, www.fad.gov/bbs/topics/NEWS/2008/NEW01933.html, accessed 2/09.

22. Position of the American Dietetic Association: weight management, J Am Diet Assoc 2009;109:330–46.

23. Buchwald H. Surgical intervention of the treatment of morbid obesity, Future Lipidol 2007;2:513–25.

24. Hill JO et al. Weight maintenance: what's missing? J Am Diet Assoc 2005; 105:S63–S66.

25. Parker-Pope T. A desperate diet: putting on pounds to qualify for weight loss surgery. Wall Street Journal, 7/8/03, p. D1.

26. Gould L. Consumers taking small steps toward big lifestyle changes. Datmonitor survey of diet trends and advertising in the United Kingdom and the United States. NUTRAingredients.com/europe, 3/18/05.

27. American College of Sports Medicine, Updated guidelines for appropriate physical activity interventions for weight loss and prevention of weight regain, Med Sci Sports Exer 2009;41:459–71.

28. Rosenbaum M et al. Long-term persistence of adaptive thermogenesis in subjects who have maintained a reduced body weight. Am J Clin Nutr 2008;88:906–12.

29. DeMaria EJ. Bariatric surgery for morbid obesity, N Engl J Med 2007;356:2176–83.

30. Aasheim ET. Wernicke Encephalopathy After Bariatric Surgery: A Systematic Review Ann Surg 248(5):714–720, 2008.

31. Malinowski SS. Nutritional and metabolic complications of bariatric surgery, Am J Med Sci. 2006 Apr;331(4):219–25.

32. Marcason W. What are the dietary guidelines following bariatric surgery? J Am Diet Assoc 2004;104:487–8.

33. Eisenberg D et al. Update on obesity surgery, World J Gastroenterol 2006;12: 3196–3203.

34. Liposuction comes of age, and New liposuction technique faster, less invasive. www.healthfinder.gov, accessed 9/03.

35. Cohen S. Lipolysis: pitfalls and problems in a series of 1246 procedures. Aesthetic Plast Surg 1985;9:207.

36. Wing RR, Phelan S. Long-term weight maintenance. Am J Clin Nutr 2005;82(suppl):222S–5S.

37. Weinsier RL et al. Free-living energy expenditure in women successful and unsuccessful at maintaining a normal body weight, Am J Clin Nutr 2002.

38. Schlundt DG et al. The role of breakfast in the treatment of obesity: a randomized trial. Am J Clin Nutr 1992;55:645–51.

39. Ello-Martin JA et al. Dietary energy density in the treatment of obesity: a year-long trial comparing 2 weight-loss diets. Am J Clin Nutr 2007;85: 1465–77.

NUTRITION | # Up Close

Setting Small Behavioral Change Goals

Feedback for Unit 10

FEEDBACK

The specific, small steps in behavioral change that work will be different for different individuals. They can be changed over time, and may end up being a rewarding part of an enhanced lifestype.

If you need more practice writing small behavioral change goals, take a look at a few more examples:

I'll use a low-calorie oil and vinegar dressing on my sandwich rather than mayonnaise at lunch three times a week.

I'll brighten up my plate and my diet with a serving of one of my favorite brightly colored vegetables three times a week.

I'll walk up the stairs to get to my office rather than take the elevator daily.

I'll lift a 10-pound weight for 15 minutes five times a week while I watch the nightly news.

Disordered Eating: Anorexia Nervosa, Bulimia, and Pica

NUTRITION SCOREBOARD

		TRUE	FALSE
1	The United States has one of the world's highest rates of anorexia nervosa.		
2	Eating disorders result from psychological, and not biological, causes.		
3	People in many different cultures may consume clay, dirt, and other nonfood substances.		

Key Concepts and Facts

- Anorexia nervosa, bulimia nervosa (bulimia), binge-eating disorder, and pica are four specific eating disorders. They may seriously threaten health.

- Eating disorders are much more common in females than males.

- The incidence of eating disorders in a society is related to the value placed on thinness by that society.

- An important route to the prevention of anorexia nervosa and bulimia is to change a society's cultural ideal of thinness and to eliminate biases against people (especially women) who are not thin.

The Eating Disorders

Three square meals a day, an occasional snack or missed meal, and caloric intakes that average out to match the body's need for calories—this set of practices is considered "orderly" eating. Self-imposed semistarvation, feast and famine cycles, binge eating, **purging**, and the regular consumption of nonfood substances such as paint chips and clay—these behaviors are symptoms of disordered eating.

Four specific types of disordered eating patterns are officially recognized as eating disorders and have been assigned diagnostic criteria. They are (1) anorexia nervosa, (2) bulimia nervosa, (3) binge-eating disorder, and (4) pica.[2] Other forms of disordered eating such as compulsive overeating and nighttime-eating syndrome have been observed, but too little research exists to establish criteria for diagnosis.

purging
The use of self-induced vomiting, laxatives, or diuretics (water pills) to prevent weight gain.

Anorexia Nervosa

It's about 9:30 on a Tuesday night. You're at the grocery store picking up sandwich fixings and some milk. Although your grocery list contains only four items, you arrive at the checkout line with a half-filled cart. The woman in front of you has only five items: a bag with about 10 green beans, an apple, a bagel, a green pepper, and a 4-ounce carton of nonfat yogurt. As she carefully places each item into her shopping bag, you notice that she is dreadfully thin.

The woman is Alison. She has just spent half an hour selecting the food she will eat tomorrow. Alison knows a lot about the caloric value of foods and makes only low-calorie choices. Otherwise, she will never get rid of her excess fat. To Alison, weight is everything—she cannot see the skeleton-like appearance others see when they look at her.

Alison has an intense fear of gaining weight and of being considered fat by others. She is annoyed when her parents and friends express their concern about her weight.

You didn't know this about Alison when you saw her. There is much more to anorexia nervosa than meets the eye.

Illustration 11.1 A day's diet? For a person with anorexia nervosa, it was. The foods shown provide approximately 562 calories.

Individuals with **anorexia nervosa** starve themselves (Illustration 11.1). They can never be too thin—no matter how emaciated they may be. As shown in Illustration 11.2, people with anorexia nervosa look extraordinarily thin from the neck down. The face and the rest of the head usually look normal because the head is the last part of the body to be affected by starvation.

People with anorexia nervosa have little fat (7 to 18% of body weight). They keep their body fat content low by consuming between 1000 and 1200 calories a day.[4] They become cold easily and have unusually low heart rates and sometimes an irregular heartbeat, dry skin, and low blood pressure. Women with anorexia nervosa experience absent or irregular menstrual cycles, infertility, and poor pregnancy outcomes (Table 11.1). Low testosterone levels in males with this eating disorder produce diminished sexual drive and impaired fertility.[3] Approximately 9 in 10 women with anorexia nervosa have significant bone loss, and 38% have osteoporosis. The extent of bone loss correlates strongly with undernutrition: the lower the body weight, the lower the bone density. Males with anorexia nervosa lose bone mass, too. Improving calcium and vitamin D intake is recommended along with nutritional rehabilitation experience.[4]

The Female Athlete Triad Pediatricians, nutritionists, and coaches are beginning to be on the lookout for eating disorders, menstrual cycle dysfunction, and decreased bone mineral density in young female athletes. Low caloric intakes and underweight related to eating disorders can lower estrogen levels and disrupt menstrual cycles. The lack of estrogen decreases calcium deposition in bones and reduces bone density at a time when peak bone mass is accumulating.[5]

Irregular or absent menstrual cycles used to be thought of as "no big deal." That attitude has changed, however, due to research results indicating that abnormal cycles in young females are related to delayed healing of bone and connective tissue injuries, and to bone fractures and osteoporosis later in life.[6]

Motivations Underlying Anorexia Nervosa The overwhelming desire to become and remain thin drives people with anorexia nervosa to refuse to eat, even when ravenously hungry, and to exercise

anorexia nervosa
An eating disorder characterized by extreme weight loss, poor body image, and irrational fears of weight gain and obesity.

Illustration 11.2 Eating disorders occur in males as well as females, but females make up approximately 90% of all cases.

Photo Disc

The average size of female gymnasts on the U.S. Olympic team shrank from 5 feet 3 inches tall and 105 pounds in 1976 to 4 feet 9 inches tall and 88 pounds in 1992.

On the Side

Table 11.1

Features of anorexia nervosa

A. Essential Features

1. Refusal to maintain body weight at or above 85% of normal weight for age and height

2. Intense fear of gaining weight or becoming fat, despite being underweight

3. Disturbance in the way in which body weight or shape is experienced, undue influence of body weight or shape on self-evaluation, or denial of the seriousness of current low body weight

4. Lack of menstrual periods in teenage females and women (missing at least three consecutive periods)

Restricting type: Person does not regularly engage in binge-eating or purging behavior.
Binge-eating type: Person regularly engages in binge-eating or purging behavior (self-induced vomiting; laxative, diuretic, or enema use).

B. Common Features in Females

1. Low-calorie diet, extensive exercise, low body fat

2. Soft, thick facial hair, thinning scalp hair

3. Loss of heart muscle; irregular, slow heartbeat

4. Low blood pressure

5. Increased susceptibility to infection

6. Anemia

7. Constipation

8. Low body temperature (hypothermia)

9. Dry skin

10. Depression

11. History of physical or sexual abuse

12. Low estrogen levels

13. Low bone density

14. Infertility, poor pregnancy outcome

C. Common Features in Males

1. Most of the common features in females

2. Substance abuse

3. Mood and other mental disorders, self-loathing

4. Decreased testosterone level, sex drive, and fertility

Source: Reprinted with permission from the Diagnostic and Statistical Manual of Mental Disorders, Fourth Edition, Text Revision. Copyright 2000 American Psychiatric Association.

binge eating
The consumption of a large amount of food in a small amount of time.

intensely. Half of the people with anorexia turn to **binge eating** and purging—features of bulimia nervosa—in their efforts to lose weight.[2] Preoccupied with food, people with anorexia may prepare wonderful meals for others but eat very little of the food themselves. Family members and friends, distressed by their failure to persuade the person with anorexia to eat, report high levels of anxiety. Although adults often describe people with anorexia as "model students" or "ideal children," their personal lives are usually marred by low self-esteem, social isolation, and unhappiness.[7]

What Causes Anorexia Nervosa?   The cause of anorexia nervosa isn't yet clear. It is likely that many different conditions, both psychological and biological, predispose an individual to become totally dedicated to extreme thinness. The

Richard Anderson

Illustration 11.3 There is a need for a more realistic body shapes on television and in fashion magazines.

—Vivienne Nathanson, British physician, 2000

value that Western societies place on thinness, the need to conform to society's expectations of acceptable body weight and shape, low self-esteem, and a need to control some aspect of one's life completely are commonly offered as potential causes for this disorder (Illustration 11.3).[4,8]

How Common Is Anorexia Nervosa? It is estimated that 1% of adolescent and young women in the Western world and less than 0.1% of young males have anorexia nervosa. The disorder has been reported in girls as young as five and in women through their forties;[10] however, it usually begins during adolescence. It is estimated that one in ten females between the ages of 16 and 25 has "subclinical" anorexia nervosa, or exhibits some of the symptoms of the disorder.[9]

Certain groups of people are at higher risk of developing anorexia nervosa than others (Table 11.2). People at risk come from all segments of society but tend to be overly concerned about their weight and food and have attempted weight loss from an early age.[9]

Treatment There is no "magic bullet" treatment that cures anorexia nervosa quickly and completely. In all but the least severe cases, the disorder generally takes 5 to 7 years and professional help to correct. Treating the disorder is often difficult because few people with anorexia believe their weight needs to be increased.[2]

Treatment programs for anorexia nervosa generally focus on the prompt restoration of nutritional health and body weight, psychological counseling to improve self-esteem and attitudes about body weight and shape, antidepressant or other medications, family therapy, and normalizing eating and exercise behaviors. These programs are successful in 50% of people and partially successful in most other cases.[4] One-third of people who recover fully from anorexia nervosa will relapse within 7 years or less. Eight years after diagnosis, 3% of people with anorexia nervosa will have died from the disorder, and it claims the lives of 18% 33 years later. Results of treatment are often excellent when the disorder is treated early.[11] Unfortunately, many people with the condition deny that problems exist and postpone treatment for years. Initiation of treatment is often prompted by a relative, coach, or friend.[12]

Table 11.2

Risk groups for anorexia nervosa[9]

- Dieters
- Ballet dancers
- Competitive athletes (gymnasts, figure skaters)
- Fitness instructors
- Dietetics majors
- People with type 1 (insulin dependent) diabetes

Bulimia Nervosa

Finally home alone, Lisa heads to the pantry and then to the freezer. She has carefully controlled her eating for the last day and a half and is ready to eat everything in sight.

It's a bittersweet time for her. Lisa knows the eating binge she is preparing will be pleasurable, but that she'll hate herself afterward. Her stomach will ache from the volume of food she'll consume, she'll feel enormous guilt from losing control, and she'll be horrified that she may gain weight and will have to starve herself all over again. Lisa is so preoccupied with her weight and body shape that she doesn't see the connection between her severe dieting and her bouts of uncontrolled eating. To get rid of all the food she is about to eat, she will do what she has done several times a week for the last year. Lisa avoids the horrible feelings that come after a binge by "tossing" everything she ate as soon as she can.

In just 10 minutes, Lisa devours 10 peanut butter cups (the regular size), a 12-ounce bag of chocolate chip cookies, and a quart of ice cream. Before 5 more minutes have passed Lisa will have emptied her stomach, taken a few deep breaths, thrown on her shorts, and started the 5-mile route she jogs most days. As she jogs, she obsesses about getting her 138-pound, 5-foot 5-inch frame down to 115 pounds. She will fast tomorrow and see what news the bathroom scale brings.

Lisa is not alone. **Bulimia nervosa** occurs in 1 to 3% of young women and in about 0.5% of young males in the United States.[4] The disorder is characterized by regular episodes of dieting, binge eating (see Illustration 11.4) and attempts to prevent weight gain by purging; use of laxatives, diuretics, or enemas; dieting; and sometimes exercise. In most cases, bulimia nervosa starts with voluntary dieting to lose weight. At some point, voluntary control over dieting is lost, and people feel compelled to engage in binge eating and vomiting.[4] The behaviors become cyclic: Food binges are followed by guilt, purging, and dieting. Dieting leads to a feeling of deprivation and intense hunger, which leads to binge eating, and so on. Once a food binge starts, it is hard to stop.

Table 11.3 lists the features of bulimia nervosa. Approximately 86% of people with this condition vomit to prevent weight gain and avoid postbinge anguish. A smaller proportion of people use laxatives, ipecac (a vomiting-inducing

Illustration 11.4 Bulimia nervosa is characterized by the consumption of a large amount of food (such as shown here) followed by purging and dieting.

© Scott Goodwin Photography

Table 11.3

Features of bulimia nervosa[2]

A. Essential Features

1. Recurrent episodes of binge eating. An episode of binge eating is characterized by both of the following:

   a. Eating an amount of food within a two-hour period of time that is definitely larger than most people would eat in a similar amount of time and under similar circumstances.

   b. A sense of lack of control over eating during the episode; a feeling that one cannot stop eating or control what or how much one is eating.

2. Recurrent inappropriate compensatory behavior in order to prevent weight gain, such as self-induced vomiting; misuse of laxatives, diuretics, enemas, or other medications; fasting; or excessive exercise.

3. The binge eating and inappropriate compensatory behaviors both occur, on average, at least twice a week for three months.

4. Self-evaluation is unduly influenced by body weight and shape.

5. The disturbance does not occur exclusively during episodes of anorexia nervosa.

Purging type: The person regularly engages in self-induced vomiting or the misuse of laxatives, diuretics, or enemas.

Nonpurging type: The person regularly engages in fasting or excessive exercise but does not regularly engage in self-induced vomiting or the misuse of laxatives, diuretics, or enemas.

B. Common Features

1. Weakness, irritability

2. Abdominal pain, constipation, bloating

3. Dental decay, tooth erosion

4. Swollen cheeks and neck

5. Binging on high-calorie foods

6. Eating in secret

7. Normal weight or overweight

8. Guilt and depression

9. Substance abuse

10. Dehydration

11. Impaired fertility

12. History of sexual abuse

Source: Reprinted with permission from the Diagnostic and Statistical Manual of Mental Disorders, Fourth Edition, Text Revision. Copyright 2000 American Psychiatric Association.

medication), diuretics (water pills), or enemas alone or in combination with vomiting.[2] These approaches do not prevent weight gain, however, and their regular use can be harmful. The habitual use of laxatives and enemas causes "laxative dependency"—these products become necessary for bowel movements. Long-term use of ipecac may damage heart muscle, and diuretics can cause illnesses by disturbing the body's fluid balance.[4]

The lives of people with bulimia nervosa are usually dominated by conflicts about eating and weight. Some affected individuals are so preoccupied with food that they spend days securing food, bingeing, and purging. Others experience only occasional episodes of binge eating, purging, and fasting.[2]

Unlike those with anorexia nervosa, people with bulimia usually are not underweight or emaciated. They tend to be normal weight or overweight. Like anorexia nervosa, bulimia nervosa is more common among athletes (including gymnasts, weight lifters, wrestlers, jockeys, figure skaters, physical trainers, and distance runners) and ballet dancers than in other groups.[4]

"Do you follow a special diet?" asks the dietitian at the eating disorder clinic.

"Yes," answers the client with bulimia. "Feast or famine."

restrained eating
The purposeful restriction of food intake below desired amounts in order to control body weight.

binge-eating disorder
An eating disorder characterized by periodic binge eating, which normally is not followed by vomiting or the use of laxatives. People must experience eating binges twice a week on average over a period of six months to qualify for the diagnosis.

Table 11.4
Features of binge-eating disorder[2]
1. Rapid consumption of extremely large amounts of food (several thousand calories) in a short period of time
2. Two or more such episodes of binge eating per week over a period of six months
3. Binge eating by oneself
4. Lack of control over eating or an inability to stop eating during a binge
5. Post binge-eating feelings of self-hatred, guilt, and depression or disgust
6. Purging, fasting, excessive exercise, or other compensation for high-calorie intakes not present

Bulimia nervosa leads to major changes in metabolism. The body must constantly adjust to feast and famine cycles and mineral and fluid losses. Salivary glands become enlarged, and teeth may erode due to frequent vomiting of highly acidic foods from the stomach.[2]

Is the Cause of Bulimia Nervosa Known? The cause of bulimia nervosa is not known with certainty, but the scientific finger is pointing at depression, abnormal mechanisms for regulating food intake, and feast-and-famine cycles as possible causes. Fasts and **restrained eating** may prompt feelings of deprivation and hunger that may trigger binge eating.[4] The ideal thinness may become more and more difficult to achieve as the feast-and-famine cycles continue.

Treatment The goal of bulimia treatment is to break the feast-and-famine cycles via nutrition and psychological counseling. Replacing the disordered pattern of eating with regular meals and snacks often reduces the urge to binge and the need to purge. Psychological counseling aimed at improving self-esteem and attitudes toward body weight and shape goes hand in hand with nutrition counseling. In many cases, antidepressants are a useful component of treatment.[8] The full recovery of women with bulimia nervosa is higher than that of women with anorexia nervosa. Nearly all women with bulimia achieve partial recovery, but one-third will relapse into bingeing and purging within 7 years.[13] Bulimia nervosa usually improves substantially during pregnancy; about 70% of women with the condition will improve their eating habits for the sake of their unborn baby.[14]

Binge-Eating Disorder

Psychiatrists now recognize an eating disorder called **binge-eating disorder** (Table 11.4). People with this condition tend to be overweight or obese, and it affects an equal number of males and females.[4] Like individuals with bulimia nervosa, people with binge-eating disorder eat several thousand calories' worth of food within a short period of time during a solitary binge, feel a lack of control over the binges, and experience distress or depression after the binges occur. People must experience eating binges twice a week on average over a period of 6 months to qualify for the diagnosis. Unlike individuals with bulimia nervosa, however, people with binge-eating disorder don't vomit, use laxatives, fast, or exercise excessively in an attempt to control weight gain.[4]

It is estimated that 9 to 30% of people in weight-control programs and 30 to 90% of obese people have binge-eating disorder.[15] The condition is far less common (2 to 5%) in the general population.[14] Stress, depression, anger, anxiety, and other negative emotions appear to prompt binge-eating episodes. Preliminary evidence indicates that binge-eating disorder aggregates in families and has both genetic and environmental orgins.[4]

Reality Check

Close to home Although she hides it, you are sure your sister has bulimia nervosa and that she is not getting help. You are deeply concerned for her health and well-being but don't know what to do about it.

Here's what Heather and Crystal say they would do:

Who do you think has the better idea?
Answers on next page

Photo Disc

Heather
I'd talk with her about getting help.

Photo Disc

Crystal
I'd spend more time with her to let her know I love her.

The Treatment Approach to Binge-Eating Disorder

The treatment of binge-eating disorder focuses on both the disordered eating and the underlying psychological issues. Persons with this condition will often be asked to record their food intake, indicate bingeing episodes, and note feelings, circumstances, and thoughts related to each eating event (Illustration 11.5). This information is used to identify circumstances that prompt binge eating and practical alternative behaviors that may prevent it. Individuals being treated for binge-eating disorder are usually given information about it, attend individual and group therapy sessions, and receive nutrition counseling on normal eating, hunger cues, and meal planning. Antidepressants may be part of the treatment. Treatment is successful in 85% of women treated for binge-eating disorder.[4,16]

Resources for Eating Disorders

Information and services related to eating disorders are available from a variety of sources. Services are best delivered by health care teams that specialize in the treatment of eating disorders. Contact with a primary care physician, dietitian, or nurse practitioner is often a good start to the process of identifying qualified health care teams. Reliable sources of information about eating disorders, support groups, nearby treatment centers, and hotlines can be found on the Internet. (See "Media Menu" at the end of the unit.)

One of the most important resources for people with an eating disorder may be a trusted friend or relative. This unit's "Reality Check" explores this resource in a very personal way—by putting you in the shoes of a person whose sister has bulimia.

Daily Food Record

Date_____

Time	Type and amount of food and beverage	Meal, Snack, Binge?	Eating triggers (feelings, situation)
7:30 am	coffee, 2 cups sugar 2 tsp cornflakes, 2 cups skim milk, 1 cup	M	Hunger!
11:30 am	tuna sandwich ice tea, 2 cups	M	Bored, hungry
7:30 pm	3 hamburgers 2 large fries 24 oreos 1/2 gallon ice cream	B	Stressed out, angry at my coach

Illustration 11.5 Example of a food dairy of a person with binge-eating disorder.

Undieting: The Clash between Culture and Biology

The pressure to conform to society's standard of beauty and acceptability is thought to be a primary force underlying the development of eating disorders.[4] Children acquire prevailing cultural values of beauty before adolescence. As early as age 5, American children learn to associate negative characteristics with people who are overweight and positive characteristics with those who are thin.[10] Standards of beauty defined by models and movie and television stars

There is a saying that women underreport their weight and men overreport their height. Clearly there are cultural norms at work here.

—L. Cohen, 2001[17]

often include thinness, but the body shape portrayed as best is often unhealthfully thin and unattainable by many.[8] The disparity between this ideal and what people normally weigh has led to widespread discontent. Approximately 50% of normal-weight adult women are dissatisfied with their weight; many diet, binge, purge or fast occasionally in an attempt to reach the standard of beauty set for them.[18] Less affected by thinness, men are more likely to express discontent with their body shape. They may strive obsessively for larger, more muscular bodies.[19]

A movement toward acceptance of body size, fashionable attire for larger people, full-size models, and a more realistic view of individual differences in body shapes is emerging in the United States and Europe[20] (Illustration 11.6). Acceptance of a realistic standard of body weight and shape—one that corresponds to health and physical fitness—and respect for people of all body sizes may be the most effective measures that can be taken to prevent anorexia nervosa, bulimia nervosa, and binge-eating disorder.

Illustration 11.6 The trend toward size acceptance. Acceptance of a realistic standard of body weight and shape—one that corresponds to health and physical fitness—and respect for people of all body sizes may be the most effective measures that can be taken to prevent anorexia nervosa, bulimia nervosa, and binge-eating disorder.

Pica

When did I start eating clay? I know it might sound strange to you, but I started craving clay in the summer of '58. It was a beautiful spring morning—it had just rained. I smelled something really sweet in the breeze coming in my bedroom window. I went outside and knew instantly where the sweet smell was coming from. It was the wet clay that lies all around my house. I scooped some up and tasted it. That's when and how I started my craving for that sweet-smelling clay. I keep some in the fridge now because it tastes even better cold.

A most intriguing type of eating disorder, **pica** has been observed in chimpanzees and in humans in many different cultures since ancient times.[21] The history and persistence of pica might suggest that the practice has its rewards. Nevertheless, important health risks are associated with eating many types of nonfood substances.

The characteristics of pica are summarized in Table 11.6. Young children and pregnant women are most likely to engage in the practice; for unknown reasons, it rarely occurs in men.[22] It most commonly takes the form of **geophagia** (clay or dirt eating), **pagophagia** (ice eating), **amylophagia** (laundry starch and cornstarch eating), or **plumbism** (lead eating). A potpourri of nonfood substances, listed in Table 11.7, may be consumed. It is not clear why pica exists, although several theories have been proposed.

Geophagia

Some people very much like to eat certain types of clay or dirt. Those who do often report that the clay or dirt tastes or smells good, quells a craving, or helps relieve nausea or an upset stomach. The belief that certain types of clay provide relief from stomach upsets may have some validity: A component of some clays is used in nausea and diarrhea medicines. There is no evidence that geophagia is motivated by a need for minerals found in clay or dirt, however.[29]

pica (pike-eh)
The regular consumption of nonfood substances such as clay or laundry starch.

geophagia (ge-oh-phag-ah)
Clay or dirt eating.

pagophagia (pa-go-phag-ah)
Ice eating.

amylophagia (am-e-low-phag-ah)
Laundry starch or cornstarch eating.

plumbism
Lead (primarily from old paint flakes) eating.

Pica permits the mind no rest until it is satisfied.[23]

Table 11.6

Characteristics of pica

A. **Essential features:** Regular ingestion of nonfood substances such as clay, paint chips, laundry starch, paste, plaster, dirt, or hair.

B. **Other common features:** Occurs primarily in young children and pregnant women in the southern United States.

Table 11.7

A partial list of nonfood substances reported to be consumed by individuals with pica

Animal droppings	Coffee grounds	Leaves	Plaster
Baking soda	Cornstarch	Mothballs	Sand
Burnt matches	Crayons	Nylon stockings	String
Cigarette butts	Dirt	Paint chips	Wool
Clay	Foam rubber	Paper	
Cloth	Hair	Paste	
Coal	Laundry starch	Pebbles	

Photo Disc

GLUE

CRAFT PASTE

Although the reasons given for clay and dirt ingestion make the practice understandable and helps explain its acceptance in some cultures, the consequences to health outweigh the benefits. Clay and dirt consumption can block the intestinal tract and cause parasitic and bacterial infections.[2] The practice is also associated with iron-deficiency and sickle-cell anemia in some individuals.[25]

Pagophagia

Have you ever known somebody who constantly crunches on ice? That person may have a 9-in-10 chance of being iron deficient. Regular ice eating, to the extent of one or more trays of ice cubes a day, is closely associated with an iron-deficient state. Ice eating usually stops completely when the iron deficiency is treated.[25]

Ice eating may be common during pregnancy. In one study of women from low-income households in Texas, 54% of pregnant women reported eating large amounts of ice regularly. Ice eaters had poorer iron status than other pregnant women who did not eat ice.[26]

Amylophagia

The sweet taste and crunchy texture of flaked laundry starch are attractive to a small number of women, especially during pregnancy. If the laundry starch preferred is not available, cornstarch may be used in its place. Laundry starch is made from unrefined cornstarch. The taste for starch almost always disappears after pregnancy.[22]

Laundry starch and cornstarch have the same number of calories per gram as do other carbohydrates (4 calories per gram). Consequently, starch eating provides calories and may reduce the intake of nutrient-dense foods. In addition, starch may contain contaminants because it is not intended for consumption. Starch eaters' diets are generally inferior to the diets of pregnant women who don't consume starch, and their infants are more likely to be born in poor health.[22]

Plumbism

The consumption of lead-containing paint chips poses a major threat to the health of children in the United States and many other countries (Illustration 11.7). Many older homes and buildings, especially those found in substandard housing areas, are covered with lead-based paint and its dried-up flakes. Children may develop lead poisoning if they eat the sweet-tasting paint flakes or inhale lead from contaminated dust and soil near the buildings. Approximately 1.4% of young children in the United States have elevated blood lead levels.[27]

High levels of exposure to lead can cause profound mental retardation and death in young children. Low levels of exposure can lead to hearing problems, growth retardation, reduced intelligence, and poor classroom performance. Children with lead poisoning are more likely to fail or drop out of school than children not exposed to lead in their environment.[27]

Eating disorders affect the health and well-being of over a million people in the United States. Although there are treatment strategies, such as counseling and the removal of lead-based paints from old houses and apartments, the solution to eating disorders lies in their prevention. With the exception of certain types of pica discussed here, the most effective way to prevent eating disorders may be to adjust our expectations and cultural norms to reflect reality.

Illustration 11.7 The regular consumption of lead-based paint chips from old houses is a major cause of lead poisoning in young children.

Proposed Eating Disorders   Several other patterns of abnormal eating behaviors have been described and may become official diagnoses in the future.[28,30] These tentative eating disorders include:

- Nighttime eating syndrome—high food consumption during the night accompanied by sleep disturbances and psychological distress

- Compulsive overeating—the uncontrolled ingestion of large amounts of food, as found in binge-eating disorder

- Purging disorder—frequent purging without binge eating

- Restrained eating—the consistent limitation of food intake to avoid weight gain

- Orthorexia nervosa (pronounced ortho-rex-e-ah)—an unhealthy fixation with the health value and purity of food

These disorders exist, and experimental prevention and treatment services are available. Web sites listed under "Media Menu" will lead you to more information on these lesser-known, proposed eating disorders.

Up Close

Eating Attitudes Test

Focal Point: Discover if your eating attitudes and behaviors are within a normal range.

Date _____ Age _____ Gender _____
Height _____ Present weight _____ How long at present weight? _____
Highest past weight _____ How long ago? _____
Lowest past weight _____ How long ago? _____

Answer the following questions using these responses:

A = always   S = sometimes   U = usually   R = rarely   O = often   N = never

_____ 1. I am terrified of being overweight.
_____ 2. I avoid eating when I am hungry.
_____ 3. I find myself preoccupied with food.
_____ 4. I have gone on eating binges where I feel that I may not be able to stop.
_____ 5. I cut my food into very small pieces.
_____ 6. I am aware of the calorie content of the foods I eat.
_____ 7. I particularly avoid foods with a high-carbohydrate content.
_____ 8. I feel that others would prefer that I ate more.
_____ 9. I vomit after I have eaten.
_____ 10. I feel extremely guilty after eating.
_____ 11. I am preoccupied with a desire to be thinner.
_____ 12. I think about burning up calories when I exercise.
_____ 13. Other people think I am too thin.
_____ 14. I am preoccupied with the thought of having fat on my body.
_____ 15. I take longer than other people to eat my meals.
_____ 16. I avoid foods with sugar in them.
_____ 17. I eat diet foods.
_____ 18. I feel that food controls my life.
_____ 19. I display self-control around food.
_____ 20. I feel that others pressure me to eat.
_____ 21. I give too much time and thought to food.
_____ 22. I feel uncomfortable after eating sweets.
_____ 23. I engage in dieting behavior.
_____ 24. I like my stomach to be empty.
_____ 25. I enjoy trying new rich foods.
_____ 26. I have the impulse to vomit after meals.

Feedback (including scoring) can be found at the end of Unit 11.

Source: The EAT-26 has been reproduced with permission. Garner, D. M., Olmsted, M. P., Bohr, Y., and Garfinkel, P. E. (1982). The Eating Attitudes Test: Psychometric features and clinical correlates. Psychological Medicine, 12, 871–878.

Key Terms

amylophagia, page 11-11
anorexia nervosa, page 11-3
binge eating, page 11-4
binge-eating disorder, page 11-8

bulimia nervosa, page 11-6
geophagia, page 11-11
pagophagia, page 11-11
pica, page 11-11

plumbism, page 11-11
purging, page 11-2
restrained eating, page 11-8

Review Questions

TRUE   FALSE

1. A critical aspect of anorexia nervosa is an intense fear of gaining weight or becoming fat. ☐ ☐

2. One of the most common health consequences of anorexia nervosa is blurred vision. ☐ ☐

3. The "female athlete triad" refers to female athletes who have an eating disorder, abnormal menstrual function, and decreased bone mineral density. ☐ ☐

4. More individuals have anorexia nervosa than any other recognized eating disorder. ☐ ☐

5. Essential features of bulimia nervosa are binge eating and excessive sleeping. ☐ ☐

TRUE   FALSE

6. Obese people are more likely to experience binge-eating disorder than are people who are not obese. ☐ ☐

7. Treatment is effective for 85% of women treated for binge-eating disorder. ☐ ☐

8. *Pica* refers to the regular consumption of raw whole grains, grasses, and other foods generally used to feed livestock. ☐ ☐

9. People who consume lead, such as from lead-based paints, have a condition called "geophagia." ☐ ☐

Media Menu

www.edap.org
The National Eating Disorders Association is the largest not-for-profit organization in the United States. The organization works to prevent eating disorders and provide treatment referrals to those suffering from them.

www.anad.org
The National Association of Anorexia Nervosa and Associated Disorders (ANAD) provides help for those struggling with eating disorders. Resources include a hotline (847–831–3438, Monday–Friday, 9A.M.–5P.M, Central Time), support groups, referrals to health care professionals, and education and prevention

programs to promote self-acceptance and healthy lifestyles. All services are free.

www.bulimiaguide.org
A resource guide for families and friends of individuals with an eating disorder is presented on the Web site.

www.nedic.ca
The National Eating Disorder Info Center in Toronto provides information and resources on eating disorders and weight preoccupation, a telephone support line, information on support groups, and listings of Canada-wide treatment resources.

www.edreferral.com
This site includes basic information on eating disorders along with specific information on treatment and recovery for men, pregnant women, and others with eating disorders.

www.naafa.org
From the National Association to Advance Fat Acceptance, Inc., this site is dedicated to improving the quality of life for fat people. It takes on policies and practices related to size discrimination and advocates size acceptance by individuals and society.

Notes

1. Lake AJ et al. Effect of Western culture on women's attitudes to eating and perceptions of body shape. Int J Eat Disord 2000;27:83–9.

2. American Psychiatric Association. Diagnostic and statistical manual of mental disorders: DMS-IV, 4th ed., Text Revision. Washington, DC: 2000.

3. Misra M et al. Percentage extremity fat, but not percentage trunk fat, is lower in adolescent boys with anorexia nervosa than in healthy adolescents, Am J Clin Nutr 2008;88:1478–84.

4. Position of the American Dietetic Association: nutrition intervention in the treatment of anorexia

nervosa bulimia nervosa, and other eating disorders, J Am Diet Assoc 2006;10:2073–82.

5. Barrack MT et al. Dietary restraint and low bone mass in female adolescent endurance runners, Am J Clin Nutr 2008;87:36–43.

6. Beals KA, Manore MM. Disorders of the female athlete triad among collegiate athletes. Int J Sports Nutr Exer Metab 2002;12:281–93.

7. Omizo SA, Oda EA. Anorexia nervosa: psychological considerations for nutrition counseling. J Am Diet Assoc 1988; 88:49–51.

8. Field AE et al. Factors related to bingeing and purging in girls and boys: the Growing Up Today Study, Arch Pediatr Adolesc Med 2008;162:574–9.

9. Grinspoon S et al. Prevalence and predictive factors for regional osteopenia in women with anorexia nervosa. Ann Intern Med 2000;133:790–4.

10. Feldman W et al. Culture versus biology: children's attitudes toward thinness and fatness, Pediatrics 1988;81:190–4.

11. Yager J et al. Anorexia nervosa. N Engl J Med 2005;353:1481–8.

12. Treating eating disorders, Harvard Women's Health Watch 1996 May:4–5; and When eating goes awry: an update on eating disorders, Food Insight 1997 Jan/Feb:35.

13. Herzog DB, Copeland PM. Bulimia nervosa—psyche and satiety (editorial). N Engl J Med 1988;319:716–8.

14. Hohlstein LA. Eating disorders program. American Dietetic Association annual meeting, Boston, 1997 Oct 27.

15. Branson R et al. Binge eating as a major phenotype of melanocortin 4 receptor gene mutations. N Engl J Med 2003;348:1096–103.

16. Fairburn CG et al. Distinctions between binge eating disorder and bulimia nervosa. Arch Gen Psychiatry 2000;57:659–65.

17. Cohen, LA. Nutrition and cancer prevention. Nutr Today 2001;36:78–9.

18. Zuckerman DM et al. The prevalence of bulimia among college students. Am J Public Health 1986;76:1135–7.

19. Pickett TC et al. Men, muscles, and body image: comparisons of competitive bodybuilders, weight trainers, and athletically active controls, Br J Sports Med 2005;39:217–22.

20. Passareillo C et al. French law targets extreme thinness, Wall Street Journal, p. A11, 4/16/08.

21. Cooper M. Pica: a survey of the historical literature as well as reports from the fields of veterinary medicine and anthropology. Springfield (IL): Charles C. Thomas; 1957.

22. Edwards CH et al. Clay- and cornstarch-eating women. J Am Diet Assoc 1959;35:810–15.

23. Craign FW. Observations on cachexia Africana or dirt-eating. Am J Med Sci 1935;17:365.

24. Johns T. Duquette M. Detoxification and mineral supplementation as functions of geophagy. Am J Clin Nutr 1991;53: 448–56.

25. Coltman CA Jr. Pagophagia and iron lack, JAMA 1969;207:513–16.

26. Rainville AJ. Pica practices of pregnant women are associated with lower maternal hemoglobin level at delivery. J Am Diet Assoc 1998;98:293–6.

27. Jones RL et al. Blood lead levels of children in the United States, Pediatrics 2009;123:e376–85.

28. Gluck ME at al. Nighttime eating: commonly observed and related to weight gain in an inpatient food intake study, Am J Clin Nutr 2008;88:900–5.

29. Mathieu J. What is orthorexia? J Am Diet Assoc 2005;105:1510–2.

30. Keel PK et al. Purging disorder as a distinct eating disorder. Int J Eat Disord 2005;38:191–9.

Answers to Review Questions

1. True, see page 11-4.
2. False, see pages 3, 11-4.
3. True, see page 11-3.
4. False, see pages 11-5, 11-7, 11-8.
5. False, see page 117.
6. True, see page 11-8.
7. True, see page 11-9.
8. False, see page 11-11.
9. False, see page 11-12.

NUTRITION | Up Close

Eating Attitudes Test

Feedback for Unit 11

Never	= 3
Rarely	= 2
Sometimes	= 1
Always, usually, and often	= 0

A total score under 20 points may indicate abnormal eating behavior. If you think you have an eating disorder, it is best to find out for sure. Careful evaluation by a qualified health professional is necessary to exclude any possible underlying medical reasons for your symptoms. Contacting a physician, nurse practitioner, dietitian, or the student health center is an important first step. You may wish to show your Eating Attitudes Test to the health professional.

NUTRITION SCOREBOARD

	TRUE	FALSE
1 Pasta, bread, and potatoes are good sources of complex carbohydrates.		
2 Ounce for ounce, presweetened breakfast cereals and unsweetened cereals provide about the same number of calories.		
3 A 12-ounce can of soft drink contains about 3 tablespoons (9 teaspoons) of sugar.		
4 Lettuce, onions, and celery are high in dietary fiber.		
5 Cooking vegetables destroys their fiber content.		
6 Sugar consumption is the leading cause of tooth decay.		

Key Concepts and Facts

- Simple sugars, the "starchy" complex carbohydrates, and dietary fiber are members of the carbohydrate family.

- Ounce for ounce, sugars and the starchy complex carbohydrates supply fewer than half the calories of fat.

- Tooth decay and poor-quality diets are related to high sugar intake.

- Fiber benefits health in a number of ways.

Answers to

NUTRITION SCOREBOARD

		TRUE	FALSE
1	If you get this right, you may be in the minority. In one study of college students, only 38% could identify good sources of complex carbohydrates. Rice, crackers, grits, dried beans, corn, peas, tortillas, biscuits, and oatmeal are also good sources of complex carbohydrates.	✔	
2	That's true! Both sweetened and unsweetened cereals consist primarily of carbohydrate. A gram of carbohydrate provides 4 calories whether the source is sugar or flakes of corn.	✔	
3	That's a lot of sugar!	✔	
4	Although these are healthy food choices, they do not contain very much dietary fiber. (Not all that goes "crunch" is a good source of fiber.)		✔
5	Cooking doesn't destroy dietary fiber.		✔
6	Rates of tooth decay increase in populations as sugar intake increases.[1]	✔	

The Carbohydrates

carbohydrates
Chemical substances in foods that consist of a simple sugar molecule or multiples of them in various forms.

Carbohydrates are the major source of energy for people throughout the world. They are the primary ingredient of staple foods such as pasta, rice, cassava, beans, and bread. On average, Americans consume less carbohydrate than people in much of the world: approximately 50% of total calories.[2] This level of intake is on the low end of the recommended range of carbohydrate intake of 45 to 65% of total calories.[3]

The carbohydrate family consists of three types of chemical substances:

1. Simple sugars

2. Complex carbohydrates ("starch")

3. Total fiber

Some food sources of these different types of carbohydrate are shown in Illustration 12.1. Carbohydrates consist of carbon, hydrogen, and oxygen. They perform a number of functions in the body, but their primary function is to serve as an energy source. Simple sugars and complex carbohydrates supply the body with four calories per gram. Dietary fiber, on average, supplies two calories per gram.[4] Although humans cannot digest any of them, bacteria in the colon can digest some types of dietary fiber. These bacteria excrete fatty acids as a waste product from fiber digestion. The fatty acids are absorbed and used as a source of energy.[4] The total contribution of fiber to our energy intake is modest (around 50 calories) and supplying energy is not a major function of fiber. Certain

(a)

(b)

carbohydrates perform roles in the functioning of the immune system, reproductive system, and blood clotting. One simple sugar (ribose) is a key a component of the genetic material DNA.

Alcohol sugars and alcohol (ethanol) are similar in chemical structure to carbohydrates (Illustration 12.2) and are presented in this text. The alcohol sugars are presented in this Unit and alcohol in Unit 14. Increasingly, carbohydrates and carbohydrate-containing foods are being classified by their "glycemic index," or the extent to which they increase blood glucose levels. This topic is introduced in this unit and explored further in Unit 13 on diabetes.

Simple Sugar Facts

Simple sugars are considered "simple" because they are small molecules that require little or no digestion before they can be used by the body. They come in two types: **monosaccharides** and **disaccharides**. The monosaccharides consist of one molecule and include glucose ("blood sugar" or "dextrose"), fructose ("fruit sugar"), and galactose. Disaccharides consist of two monosaccharide molecules (see Table 12.1). The combination of a glucose molecule and a fructose molecule makes sucrose (or "table sugar"); maltose ("malt sugar") is made from two glucose molecules; and lactose ("milk sugar") consists of a glucose molecule plus a galactose molecule. Honey, by the way, is a disaccharide. It is composed of glucose and fructose just as sucrose is, but it's a liquid rather than a solid because of the way the two molecules of sugar are chemically linked together. Disaccharides are broken down into their monosaccharide components during digestion; only glucose, fructose, and galactose are absorbed into the bloodstream.

Illustration 12.1 The carbohydrate family. Some food sources of simple sugars (left), and some food sources of starch and dietary fiber (right) are shown.

simple sugars
Carbohydrates that consist of a glucose, fructose, or galactose molecule, or a combination of glucose and either fructose or galactose. High-fructose corn syrup and alcohol sugars are also considered simple sugars. Simple sugars are often referred to as "sugars."

monosaccharides
(*mono* = one, *saccharide* = sugar): Simple sugars consisting of one sugar molecule. Glucose, fructose, and galactose are common examples of monosaccharides.

disaccharides
(*di* = two, *saccharide* = sugar): Simple sugars consisting of two molecules of monosaccharides linked together. Sucrose, maltose, and lactose are disaccharides.

Illustration 12.2 The similar chemical structures of glucose, fructose, xylitol (an alcohol sugar), and ethanol (an alcohol).

Table 12.1

The monosaccharides and the disaccharides they form

Monosaccharides	Disaccharide formed
glucose + glucose	maltose
glucose + fructose	sucrose
glucose + galactose	lactose

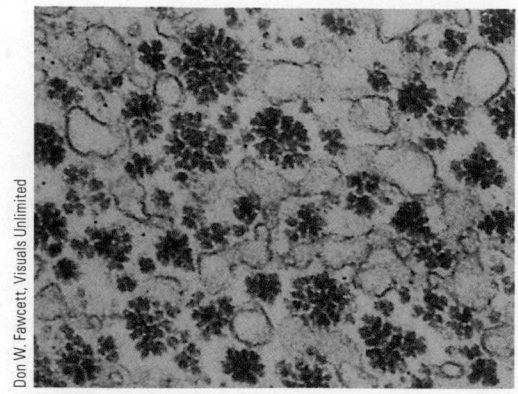

Illustration 12.3 Glycogen in a liver cell. The black "rosettes" are aggregates of glycogen molecules. This cell was photographed under an electron microscope at a magnification of 65,000X.

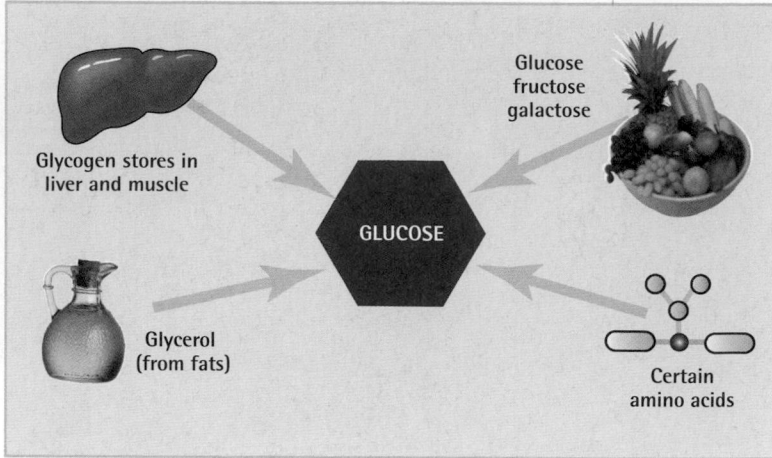

Illustration 12.4 The body's sources of glucose.

glycogen
The body's storage form of glucose. Glycogen is stored in the liver and muscles.

High-fructose corn syrup—a liquid sweetener used in many soft drinks, fruit drinks, breakfast cereals, and other products is also considered a simple sugar. It generally consists of 55% fructose and 45% glucose.[5] Most of the simple sugars have a distinctively sweet taste.

The simple sugars the body uses directly to form energy are glucose and fructose. Galactose is readily converted by the body to glucose. When the body has more glucose than it needs for energy formation, it converts the excess to fat and to **glycogen**, the body's storage form of glucose. Glycogen is a type of complex carbohydrate. It consists of chains of glucose units linked together in long strands. Glycogen is produced only by animals and is stored in the liver (Illustration 12.3) and muscles. When the body needs additional glucose, glycogen is broken down, making glucose available for energy formation. Glucose can also be derived from certain amino acids and the glycerol component of fats. Illustration 12.4 shows the various ways glucose becomes available to the body. A constant supply of glucose is needed because the brain, red blood cells, white blood cells, and specific cells in the kidneys require glucose as an energy source.[2]

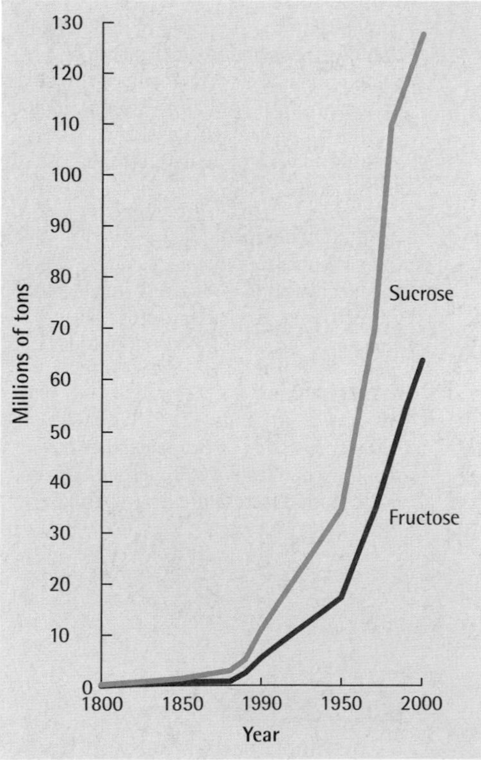

Simple Sugar Intake Most of the simple sugar in our diet comes from foods and beverages sweetened with sucrose and high fructose corn syrup. As a matter of fact, simple sugars are the most commonly used food additive.[6] Per person consumption of added sucrose and fructose has increased dramatically across the globe over time (Illustration 12.5). Added sugars now make up 17% of the total calorie intake of Americans.[8] Combined with naturally occurring simple sugars in fruits and some vegetables, total simple sugar consumption in the United States averages 23% of total calories.[2] That's a lot of sugar and far more than is good for health.[7,9]

The largest source of simple sugar in the diet is beverages, particularly soft drinks.[9] One 12-ounce serving of soft drink contains about 9 teaspoons of simple sugar. Sucrose and high fructose corn syrup are commonly added to fruit drinks, breakfast cereals, and candy. Simple sugars are also present in fruits, and small amounts are found in some vegetables (Table 12.2). Milk is the only animal product that contains a simple sugar (lactose).

Nutrition Labeling of Sugars Nutrition labels must list the total amount of mono- and diglycerides per serving of food under the heading "sugars" (Illustration 12.6). In addition, in the ingredient list, all simple sugars contained in the product must be listed in order of weight.[10] Labels contain information on total sugars per serving and do not distinguish between sugars naturally present in foods and added sugars.

Illustration 12.5 Worldwide trends in sucrose and fructose consumption between 1800 and 2000.[a]

[a] Source: Figure drawn from data presented in Bray GA.[7]

Table 12.2

The simple sugar content of some common foods

	Amount	Simple Sugars (grams)[a]	% Total Calories from Simple Sugars
Sweeteners:			
Corn syrup	1 tsp	5	100%
Honey	1 tsp	6	100
Maple syrup	1 tsp	4	100
Table sugar	1 tsp	4	100
Fruits:			
Apple	1 medium	16	91
Peach	1 medium	8	91
Watermelon	1 wedge (4″ × 8″)	25	87
Orange	1 medium	14	86
Banana	1 medium	21	85
Vegetables:			
Broccoli	½ cup	2	40
Corn	½ cup	3	30
Potato	1 medium	1	4
Beverages:			
Fruit drinks	1 cup	29	100
Soft drinks	12 oz	38	100
Skim milk	1 cup	12	53
Whole milk	1 cup	11	28
Candy:			
Gumdrops	1 oz	25	100
Hard candy	1 oz	28	100
Caramels	1 oz	21	73
Fudge	1 oz	21	73
Milk chocolate	1 oz	16	44
Breakfast cereals			
Apple Jacks	1 oz	13	52
Raisin Bran	1 oz	19	40
Cheerios	1 oz	14	4

[a]4 grams sucrose 5 1 teaspoon.

What's So Bad about Sugar?   Foods to which simple sugars have been added are often not among the top sources of nutrients. Simple sugars are among the few foods that provide only calories. Many foods high in simple sugars, such as cake, sweet rolls, cookies, pie, candy bars, and ice cream, are also high in fat. The likelihood that diets will provide insufficient amounts of vitamins and minerals increases along with sugar intake.[2] High levels of consumption of sucrose and fructose are related to increased blood levels of triglycerides.[5] Neither sucrose or high fructose corn syrup has been found to cause obesity or type 2 diabetes. However, high intakes of these sweeteners may contribute to the risk of both disorders due to excessive calorie intake.[9,11] Furthermore, it is perfectly clear that the frequent consumption of sticky sweets causes tooth decay.[6]

Advice on Sugar Intake   What's the bottom line on eating sugary foods? Enjoy them in limited amounts as part of a healthful diet. The World Health Organization and other health authorities in 26 countries recommend limiting added sugars to 10% of calories or less.[13] If you are concerned about the amount of sugar in your diet and your teeth, don't wait until it's time for a New Year's resolution. Get some suggestions for small changes that will have a positive impact in this Unit's Take Action feature.

Illustration 12.6 Labeling the sugar content of breakfast cereal

	% Daily Value**	
Total Fat 1 g*	2%	2%
Saturated Fat 0 g	0%	0%
Monounsaturated Fat 0 g		
Polyunsaturated Fat 0.5 g		
Trans Fat 0 g		
Cholesterol 0 mg	0%	0%
Sodium 5 mg	0%	3%
Potassium 200 mg	6%	12%
Total Carbohydrate 48 g	16%	18%
Dietary Fiber 6 g	24%	24%
Sugars 12 g		
Other Carbohydrate 30 g		
Protein 6 g		
Vitamin A	0%	4%
Vitamin C	0%	0%
Calcium	0%	15%
Iron	90%	90%
Thiamin	25%	30%
Riboflavin	25%	35%
Niacin	25%	25%
Vitamin B$_6$	25%	25%
Folic Acid	25%	25%
Vitamin B$_{12}$	25%	35%
Phosphorus	15%	25%
Magnesium	15%	20%
Zinc	10%	15%
Copper	10%	10%

		Calories	2,000	2,500
Total Fat	Less than		65 g	80 g
Saturated Fat	Less than		20 g	25 g
Cholesterol	Less than		300 mg	300 mg
Sodium	Less than		2,400 mg	2,400 mg
Potassium			3,500 mg	3,500 mg
Total Carbohydrate			300 g	375 g
Dietary Fiber			25 g	30 g
Calories Per gram: Fat 9 • Carbohydrate 4 • Protein 4				

INGREDIENTS: WHOLE GRAIN WHEAT, SUGAR, HIGH FRUCTOSE CORN SYRUP, GELATIN, **VITAMINS AND MINERALS:** REDUCED IRON, NIACINAMIDE, ZINC OXIDE, PYRIDOXINE HYDRO-CHLORIDE (VITAMIN B$_6$), RIBOFLAVIN (VITAMIN B$_2$), THIAMIN HYDROCHLORIDE (VITAMIN B$_1$), FOLIC ACID AND VITAMIN B$_{12}$, TO MAINTAIN QUALITY, BHT HAS BEEN ADDED TO THE PACKAGING.

The Alcohol Sugars—What Are They?

alcohol sugars
Simple sugars containing an alcohol group in their molecular structure. The most common are xylitol, mannitol, and sorbitol. They are a subgroup of chemical substances called "polyols."

Nonalcoholic in the beverage sense, the **alcohol sugars** (or "polyols") are like simple sugars except that they include a chemical component of alcohol. Like simple sugars, the alcohol sugars have a sweet taste. Xylitol is by far the sweetest alcohol sugar—it's much sweeter than the other two common alcohol sugars, mannitol and sorbitol.

Alcohol sugars are found naturally in very small amounts in some fruits. They are mostly used as sweetening agents in gums and candy (Illustration 12.7). Unlike the simple sugars, xylitol, mannitol, and sorbitol do not promote tooth

Take Action   Lower Your Sugar Intake

Concerned about your sugar intake? Want to consider some ways to lower it and to protect your teeth?

Consider these actions and check the options you'd be willing to try.

_____ 1. When you want something sweet to eat, try:

_____ raisins   _____ a mango
_____ sweet cherries   _____ unsweetened
_____ melon   applesauce
_____ a banana
_____ other fruit: _____

_____ 2. Replace a serving of soft drink or fruit drink with a no-added sugar, 100% juice serving such as:

_____ tomato juice   _____ pineapple juice
_____ vegetable juice   _____ cranberry juice
_____ dark grape juice _____ grapefruit-
_____ apple cider/juice   tangerine juice
_____ other juice: _____

_____ 3. Taste-test beverages or gum sweetened with alcohol sugars or artificial sweeteners for acceptability. Try:

_____ ice tea sweetened with aspartame or sucrolose
_____ soft drinks sweetened with Reb A or aspartame
_____ gum sweetened with xylitol or other alcohol sugar

_____ 4. Keep sugar off your teeth. After you eat a food with added sugar:

_____ rinse your mouth with water
_____ brush your teeth
_____ floss in between your teeth

decay because bacteria in the mouth that cause tooth decay cannot digest them.[6] Foods sweetened with alcohol sugars can use the health claim "Does not promote tooth decay" on labels.

Like dietary fiber, the alcohol sugars are slowly and incompletely broken down in the gastrointestinal tract, and provide fewer calories per gram than other carbohydrates. On average, alcohol sugars provide two calories per gram, so foods labeled "sugar free" will not be calorie-free. High intake of alcohol sugars can, like fiber, cause diarrhea. This characteristic limits their use in foods. The "diarrhea dose" for alcohol sugars defined by the FDA equals 50 grams of sorbitol, and 20 grams of mannitol. A food product's content of alcohol sugars per serving must be listed on the nutrition label as "Sugar Alcohols" or by the name of the sugar alcohol.[10]

Artificial Sweetener Facts

Unwanted calories in simple sugars, the connection of sucrose with tooth decay, the need for a sugar substitute for people with diabetes, and sugar shortages such as occurred during the two world wars have all provided incentives for developing sugar substitutes. Six artificial sweeteners are currently on the market in the United States, (Table 12.3) more are being developed. None of the artificial sweeteners that are currently approved for use exactly mimic the taste and properties of sugar.[14,15]

Artificial sweeteners are also known as non-nutritive sweeteners because they are not a significant source of energy or nutrients.[14] Although they are not carbohydrates, they have chemical properties that invoke an intensively sweet taste on the tongue. Gram for gram, artificial sweeteners are 160 to 13,000 times sweeter than sucrose and only small amounts are needed to sweeten food products. Of the artificial sweeteners, only aspartame provides calories (4 calories/gram).[6]

Table 12.3		
Artificial sweeteners currently approved for use in the United States		
Trade name	**Product name**	**Calories/gram**
Saccharin	Sweet and Low	0
Aspartame	NutraSweet, Equal Sugar Twin	4
Sucralose	Splenda	0
Acesulfame potassium	Acesulfame K, Sunnette, Sweet One	0
Neotame	–	0
Rebiana	Reb-A, Truvia, PureVia	0

Illustration 12.8 Some of the thousands of foods that contain artificial sweeteners.

©Scott Goodwin Photography

A few drops of sucrose solution on an infant's tongue have a pain-relieving effect that persists for several minutes. Sugar solution is being increasingly used in hospitals to control pain in infants undergoing heel pricks for blood collection and other minor procedures.[17]

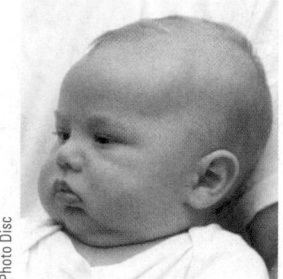

Photo Disc

phenylketonuria (feen-ol-key-tone-u-re-ah), PKU
A rare genetic disorder related to the lack of the enzyme phenylalanine hydroxylase. Lack of this enzyme causes the essential amino acid phenylalanine to build up in blood.

The artificial sweeteners currently on the market do not promote tooth decay because they are not utilized by bacteria in the mouth that cause decay.[6] They do not appear to promote weight loss without calorie restriction.[14] Artificial sweeteners are used to sweeten thousands of products, a few of which are shown in Illustration 12.8.

Saccharin Saccharin was the first artificial sweetener developed. Did you know that it was discovered in a laboratory in the late 1800s? That's right—saccharin is over 100 years old. The availability of this artificial sweetener, which is 300 times as sweet as sucrose, helped relieve the sugar shortages that occurred during World Wars I and II.

In 1977 saccharin was taken off the market after very high doses were found to cause cancer in laboratory animals. At that time, however, saccharin was the only no-calorie, artificial sweetener available, and its removal sparked a public outcry. After many people complained to Congress, saccharin was returned to the market by congressional mandate. Saccharin was deemed safe in 2000 after scientists concluded there was no clear evidence that it causes cancer in humans.[16]

Aspartame Early in the 1980s, the artificial sweetener aspartame was approved for use in the United States and more than 90 other countries. Primarily known as Nutrasweet, this artificial sweetener is about 200 times sweeter than sucrose.

Aspartame is made from two amino acids (phenylalanine and aspartame). Both are found in nature, but it took chemists to arrange their chemical partnership. Because aspartame is made from amino acids (the building blocks of protein), it supplies four calories per gram. Aspartame is so sweet, however, that very little is needed to sweeten products. Illustration 12.9 shows the relative sweetening power of various artificial sweeteners and naturally occurring sugars.

Aspartame is used in more than 6000 products worldwide, including soft drinks, whipped toppings, jellies, cereals, puddings, and some medicines. Products containing aspartame must carry a label warning people with **phenylketonuria** (an inherited disease) and others with certain liver conditions about the presence of phenylalanine. People with these disorders are unable to utilize the amino acid phenylalanine, causing it to build up in the blood. Because high temperatures tend to break down aspartame, it is not used in baked or heated products.[6]

■ **Is Aspartame Safe?** A safe level of aspartame intake is defined as 50 milligrams per kilogram of body weight per day in the United States and as

40 milligrams per kilogram of body weight in Canada.[18] In food terms, the limit in the United States is equivalent to approximately 20 aspartame-sweetened soft drinks or 55 desserts per day (Illustration 12.10). The average intake of aspartame in the United States, Canada, Germany, and Finland, for example, ranges from 2 to 10 milligrams per day, well below the level of intake considered safe.[21] A small proportion of individuals, however, report that they are sensitive to aspartame and develop headaches, dizziness, or anxiety when they consume small amounts. Studies have failed to confirm these effects.[19] Aspartame has not been found to promote cancer, nerve disorders, or other health problems in humans.[20]

Sucralose This noncaloric, intense sweetener is made from sucrose, is safe, is very sweet (600 times sweeter than sucrose), and does not leave a bitter aftertaste. Known primarily as "Splenda" on product labels, it is used in both hot and cold food products, including soft drinks, baked goods, frosting, pudding, and chewing gum.

Acesulfame Potassium Acesulfame potassium, also known as acesulfame K, "Sunette" and "Sweet One," was approved by the FDA in 1988. It is added to at least 4000 foods and is used in food production in about 90 countries. It is 200 times as sweet as sucrose, provides zero calories, and does not break down when heated.

Neotame Neotame is derived from the same amino acids as aspartame and is extraordinarily sweet. (Its sweetness potency is 7000 to 13,000 times that of sucrose.) Only minute amounts of neotame are absorbed and it is not considered harmful to individuals with PKU. It is marketed as having a clean sweet taste with little bitter or metallic aftertaste.[6]

Rebiana In December 2008 the FDA approved Rebiana for use as an artificial sweetener. Called Reb-A, Truvia, and Purevia, this sweetener is derived from the herb stevia (Illustration 12.11). Stevia grows in sub-tropical and tropical areas and has been known in these areas as "sugar leaf" for centuries.

Stevia leaves contain the chemical rebaudioside that imparts a sweet taste and it is used in purified form in Reb-A. It is currently used to sweeten beverages (Illustration 12.12) and reportedly tastes best with citrus flavors.[15]

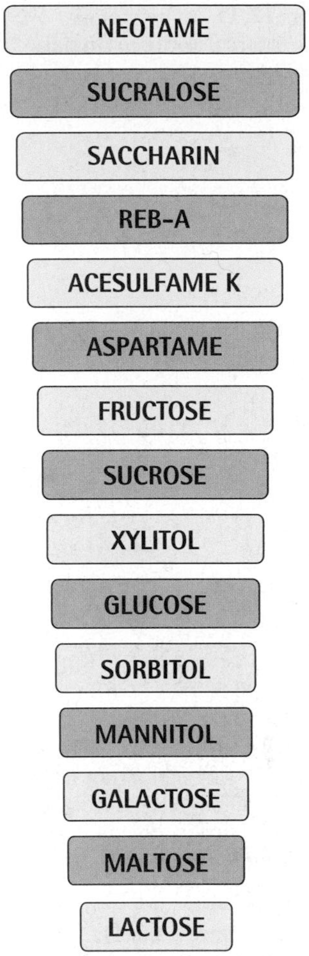

NEOTAME

SUCRALOSE

SACCHARIN

REB–A

ACESULFAME K

ASPARTAME

FRUCTOSE

SUCROSE

XYLITOL

GLUCOSE

SORBITOL

MANNITOL

GALACTOSE

MALTOSE

LACTOSE

Illustration 12.9 A ranking of various types of artificial sweeteners and naturally occurring sugars in order of sweetness.

Illustration 12.10 You would have to consume more than 20 cans of soft drinks sweetened with aspartame (Nutrasweet) a day to exceed the safe limit set for this artificial sweetener.

Richard Anderson

Illustration 12.11 The stevia plant is the source of the artificial sweetener rebiana.

© Steffen Hauser / botanikfoto / Alamy

complex carbohydrates
The form of carbohydrate found in starchy vegetables, grains, and dried beans and in many types of dietary fiber. The most common form of starch is made of long chains of interconnected glucose units.

polysaccharides
(*poly* = many, *saccharide* = sugar): Carbohydrates containing many molecules of monosaccharides linked together. Starch, glycogen, and dietary fiber are the three major types of polysaccharides. Polysaccharides consisting of 3 to 10 monosaccharides may be referred to as oligosaccharides.

Illustration 12.12 Example of beverages sweetened with rebiana (Reb-A).

Complex Carbohydrate Facts

Starches, glycogen, and dietary fiber constitute the **complex carbohydrates** known as **polysaccharides.** Only plant foods such as grains, potatoes, dried beans, and corn that contain starch and dietary fiber are considered dietary sources of complex carbohydrates (Table 12.4). Very little glycogen is available from animal products.

Which Foods Have Carbohydrates?   Food sources of complex carbohydrates include whole-grain breads, cereals, pastas, and crackers, as well as these same foods produced from refined grains. Whole grain products provide more fiber and beneficial substances naturally present in grains than do refined grain products. Whole grain foods reduce the risk of heart disease and some types of cancer.[21]

©Scott Goodwin Photography

Table 12.4

The complex carbohydrate content of some common foods

	Amount	Complex Carbohydrate (grams)	% Total Calories from Complex Carbohydrates
Grain and grain products:			
Rice (white), cooked	$^1/_2$ cup	21	83%
Pasta, cooked	$^1/_2$ cup	15	81
Cornflakes	1 cup	11	76
Oatmeal, cooked	$^1/_2$ cup	12	74
Cheerios	1 cup	11	68
Whole wheat bread	1 slice	7	60
Dried beans (cooked):			
Lima beans	$^1/_2$ cup	11	64
White beans	$^1/_2$ cup	13	63
Kidney beans	$^1/_2$ cup	12	59
Vegetables:			
Potato	1 medium	30	85
Corn	$^1/_2$ cup	10	67
Broccoli	$^1/_2$ cup	2	40

Are They Fattening?   Starchy foods are caloric bargains (Illustration 12.13, next page). A medium baked potato weighs in at only 122 calories, a half-cup of corn at 85 calories, and a slice of bread at 70 calories. You can expand the caloric value of complex carbohydrates quite easily by adding fat, sauces, and cheese. One cup of macaroni (about 200 calories) gains around 180 calories when it comes as macaroni and cheese. Adding a quarter-cup of gravy to potatoes elevates calories by 150.

United States Fiber Facts   What is low in calories, prevents constipation, may lower the risk of heart disease, obesity, and diabetes; and is generally undercon-sumed by people in the United States? The answer is fiber.

Total fiber intake by U.S. children and adults (15 grams per day) is well below the amount recommended (28 grams for women and 35 grams for men).[41] People who consume the recommended amount of fiber tend to select whole-grain breads, high-fiber cereal, and dried beans most days and eat at least five serv-ings of vegetables and fruits daily.[22] Food sources of fiber are listed in Table 12.5. It doesn't matter whether the fiber foods are mashed, chopped, cooked, or raw. They retain their fiber value through it all.

In the past it was assumed that fiber had no calorie value because it is not broken down by human digestive enzymes. Recent studies suggest that the calo-rie contribution of dietary fiber be re-considered. Bacteria in the colon are able to break down many types of fiber to some extent. The bacteria excrete fatty acids as a waste product and they are used as an energy source by the colon and the rest of the body. On average, fiber provides two calorie per gram.[4] Although not yet required on U.S. or Canadian nutrition information labels, those in the European Union must include the calorie contribution of fiber in food products.[23]

Types of Fiber   A new classification system for defining edible fibers based on source of fiber and effects on body processes has been developed.[3] Fibers are clas-sified as **functional fiber**, **dietary fiber**, and **total fiber**. All fiber shares the property of not being digested by human digestive enzymes.

Functional fibers perform specific, beneficial functions in the body, includ-ing decreasing food intake by providing a feeling of fullness, reducing postmeal rises in blood glucose levels, preventing constipation, and decreasing fat and

functional fiber
Specific types of nondigestible carbohydrates that have beneficial effects on health. Two examples of functional fibers are psyllium and pectin.

dietary fiber
Naturally occurring, intact forms of nondigestible carbohydrates in plants and "woody" plant cell walls. Oat and wheat bran, and raffinose in dried beans, are examples of this type of fiber.

total fiber
The sum of functional and dietary fiber.

Illustration 12.13 Which has more calories?
Check your answers below.[a]

One medium baked potato (four ounces) OR three ounces of lean hamburger?

One slice of bread OR a half-cup of low-fat cottage cheese?

One cup of spaghetti noodles OR 17 french fries (three ounces)?

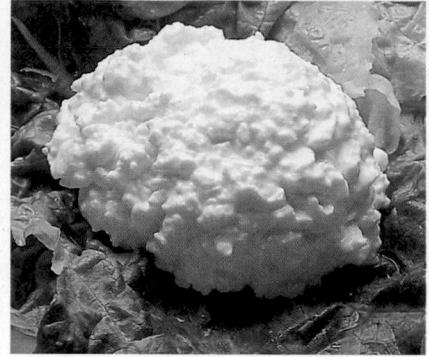

[a]*Answers*
Potato = 122 calories;
Lean hamburger = 239 calories.
Bread = 70 calories;
Cottage cheese = 102 calories.
Spaghetti = 197 calories;
French fries = 265 calories.

Carbohydrate craving   Have you ever pined or whined for something sweet? Is a meal not complete unless there's a "starch?" Do you think you can get addicted to carbohydrates?

Who gets thumbs up?
Answers on page 12-14

Terry:
Maybe you could really love to eat them, but addicted? I don't think so.

Wolfgang:
The more I don't eat candy, the more I want to eat some. I swear, I'm addicted.

Table 12.5

Examples of good sources of fiber

The recommended intake of total fiber for men is 38 grams per day and 25 grams per day for women.

	Amount	Fiber (grams)
Grain and grain products:		
Bran Buds	$^1/_2$ cup	12.0
All Bran	$^1/_2$ cup	11.0
Raisin Bran	1 cup	7.0
Granola (homemade)	$^1/_2$ cup	6.0
Bran Flakes	$^3/_4$ cup	5.0
Oatmeal	1 cup	4.0
Spaghetti noodles	1 cup	4.0
Shredded Wheat	1 biscuit	2.7
Whole wheat bread	1 slice	2.0
Bran (dry; wheat, oat)	2 tbs	2.0
Fruits:		
Raspberries	1 cup	8.0
Avocado	$^1/_2$ medium	7.0
Mango	1 medium	4.0
Pear (with skin)	1 medium	4.0
Apple (with skin)	1 medium	3.3
Banana	6" long	3.1
Orange (no peel)	1 medium	3.0
Peach (with skin)	1 medium	2.3
Strawberries	10 medium	2.1
Vegetables:		
Lima beans	$^1/_2$ cup	6.6
Green peas	$^1/_2$ cup	4.4
Potato (with skin)	1 medium	3.5
Brussels sprouts	$^1/_2$ cup	3.0
Broccoli	$^1/_2$ cup	2.8
Carrots	$^1/_2$ cup	2.8
Green beans	$^1/_2$ cup	2.7
Collard greens	$^1/_2$ cup	2.7
Cauliflower	$^1/_2$ cup	2.5
Corn	$^1/_2$ cup	2.0
Nuts:		
Almonds	$^1/_4$ cup	4.5
Peanuts	$^1/_4$ cup	3.3
Peanut butter	2 tbs	2.3
Dried beans (cooked):		
Pinto beans	$^1/_2$ cup	10.0
Peas, split	$^1/_2$ cup	8.2
Black beans (turtle beans)	$^1/_2$ cup	8.0
Lentils	$^1/_2$ cup	7.8
Kidney or navy beans	$^1/_2$ cup	6.9
Black-eyed peas	$^1/_2$ cup	5.3
Fast foods:		
Big Mac	1	3
French fries	1 regular serving	3
Whopper	1	3
Cheeseburger	1	2
Taco	1	2
Chicken sandwich	2	1
Egg McMuffin	1	1
Fried chicken, drumstick	1	1

Photo Disc

Have you ever had buckwheat pancakes (three grams of fiber per serving)? Did you know that buckwheat is a member of the rhubarb family—not a grain?

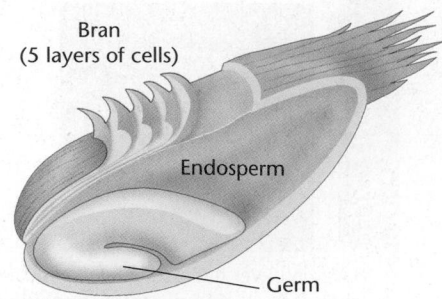

Illustration 12.14 Diagram of a grain of wheat showing the bran that is a rich source of dietary fiber. The germ contains protein, unsaturated fats, thiamin, niacin, riboflavin, iron, and other nutrients. (The bran and germ are removed in the refining process.) The endosperm primarily contains starch, the storage form of glucose in plants.

cholesterol absorption.[3] This type of fiber can be extracted from foods or produced commercially for use in fortifying foods with fiber, as well as in fiber supplements.[4] Psyllium, pectin, gels, and seed and plant gums are classified as functional fibers.

Dietary fiber consists of nondigestible carbohydrates found in plant foods. Because functional fibers are components of plant foods, this type of fiber comes with all of the other beneficial nutrients and other substances found in plants. Dietary fibers are found in the bran component of oats and wheat (Illustration 12.14), in cellulose (a rigid component of plant cell walls), in vegetables and fruits, and in the nondigestible starch components of dried beans. The recommended daily intake of fiber is based on total fiber, which is the sum of functional plus dietary fiber intake.[3]

■ **Soluble and Insoluble Fiber.** Soluble fiber is a functional fiber because it benefits health in several ways. Soluble fiber slows glucose absorption, thereby lowering peak blood levels of glucose, and reduces fat and cholesterol absorption. Soluble fibers are found in oats, barley, fruit pulp, and psyllium. They are called "soluble" because of their ability to combine chemically with water.[41]

Insoluble fibers, such as wheat bran and other brans, and the fiber in legumes, do not combine chemically with water. They offer some of the same health benefits as do soluble fiber and are particularly beneficial in preventing constipation.[41]

Be Cautious When Adding More Fiber to Your Diet Newcomers to adequate fiber diets often experience diarrhea, bloating, and gas for the first week or so of increased fiber intake. These side effects can be avoided. They occur when too much fiber is added to the diet too quickly. Adding sources of dietary fiber to the diet gradually can prevent these side effects. In addition, dietary fiber can be constipating if consumed with too little fluid. Your fluid intake should increase along with your intake of dietary fiber. You know you've got the right amount of fiber in your diet when stools float and are soft and well formed.

The "Health Action" on the next page suggests some food choices that will put fiber into your meals and snacks. Fiber should be added carefully to children's diets, because the large volume of some high-fiber diets can fill the children up before they consume enough calories.[24]

Glycemic Index of Carbohydrates

In the not-too-distant past it was assumed that "a carbohydrate is a carbohydrate is a carbohydrate." It was thought that all types of carbohydrates had the same effect on blood glucose levels and health, so it didn't matter what type was consumed. As is the case with many untested assumptions, this one fell by the wayside. It is now known that some types of simple and complex carbohydrates in

Carbohydrate craving The popular theory about carbohydrate craving and addiction goes like this: Carbohydrates increase blood levels of glucose, then insulin level surges to drop blood glucose levels. Resulting low blood glucose levels send a message to your brain telling you that carbohydrates are needed to bring your blood glucose levels back up. The truth is there are no biological processes that make healthy people crave carbohydrate. (There may be psychological cravings, however.) Any source of calories will relieve hunger.[21]

Terry 👍

Wolfgang 👎

Health Action   Putting the Fiber into Meals and Snacks

HIGH-FIBER OPTIONS FOR BREAKFAST

Whole grain toast		2 g per slice
Bran cereal:		
Bran flakes	1 cup	7 g
All Bran	½ cup	11 g
Raisin bran	1 cup	7 g
Oat bran	⅓ cup	5 g
Bran muffin, with fruit	1 small	3 g
Strawberries	10	2 g
Raspberries	½ cup	3 g
Bananas	1 medium	2 g

LUNCHES THAT INCLUDE FIBER

Whole grain bread		2 g per slice
Baked beans	½ cup	10 g
Carrot	1 medium	2 g
Raisins	¼ cup	2 g
Peas	½ cup	4 g
Peanut butter	2 tablespoons	2 g

FIBER ON THE MENU FOR SUPPER

Brown rice	½ cup	2 g
Potato	1 medium	3 g
Dried cooked beans	½ cup	8 g
Broccoli	½ cup	3 g
Corn	½ cup	3 g
Tomato	1 medium	2 g
Green beans	½ cup	3 g

Photo Disc

FIBER-FILLED SNACKS

Peanuts	¼ cup	3 g
Apple	1 medium	2 g
Pear	1 medium	4 g
Orange	1 medium	3 g
Prunes	3	2 g[a]
Sunflower seeds	¼ cup	2 g
Popcorn	2 cups	2 g

[a]Prunes contain fiber, but their laxative effect is primarily due to a naturally occurring chemical substance that causes an uptake of fluid into the intestines and the contraction of muscles that line the intestines.

foods elevate blood glucose levels more than do others. Such differences are particularly important to people with disorders such as **insulin resistance** and **type 2 diabetes.**[25]

Carbohydrates and cabohydrate-containing foods are now being classified by the extent to which they increase blood glucose levels. This classification system is called the **glycemic index.** Carbohydrates that are digested and absorbed quickly have a high glycemic index and raise blood glucose levels to a higher extent than do those with lower glycemic index values. Carbohydrates and carbohydrate-containing foods with high glycemic index values include glucose, white bread, baked potatoes, and jelly beans. Fructose, xylitol, hummus, apples, and all-bran cereal are examples of carbohydrates and carbohydrate-containing

insulin resistance
A condition is which cell membranes have a reduced sensitivity to insulin so that more insulin than normal is required to transport a given amount of glucose into cells.

type 2 diabetes
A disease characterized by high blood glucose levels due to the body's inability to use insulin normally or to produce enough insulin.

glycemic index
A measure of the extent to which blood glucose is raised by a 50-gram portion of a carbohydrate-containing food compared to 50 grams of glucose or white bread.

foods with low glycemic indexes. Fructose has a low glycemic index because it does not raise blood glucose levels. It is absorbed as fructose and does not require insulin for passage into cells.[26]

Diets providing low glycemic index carbohydrates have been found to improve blood glucose control in people with diabetes, to reduce elevated levels of blood cholesterol and triglycerides, increase levels of beneficial HDL cholesterol, and decrease the risk of developing type 2 diabetes, some types of cancer, and heart disease.[27]

Carbohydrates and Your Teeth

We tried the white man's food, and we liked it. Now we have toothaches.

—Chief Senebura, Xavante people in Pimentel Barbosa, Brazil[20]

tooth decay
The disintegration of teeth due to acids produced by bacteria in the mouth that feed on sugar. Also called dental caries or cavities.

The relationship between sugar and **tooth decay** is very close, and the history of tooth decay closely parallels the availability of sugar. The incidence of tooth decay is estimated to have been very low (less than 5%) among hunter-gatherers who had minimal access to sugars.[29] Tooth decay did not become a widespread problem until the late 17th century, when great quantities of sucrose were exported from the New World to Europe and other parts of the world. When sugar shortages occurred in the United States and Europe during World War I and World War II, rates of tooth decay declined; they rebounded when sugar became available again.[29] Rates of tooth decay in children vary substantially among countries, but the highest rates are in countries where sugar is widely available in processed foods and beverages. Tooth decay is spreading rapidly in developing countries where sugar, candy, soft drinks, and fruit drinks are becoming widely available.[30]

Sweets are not the only culprit. Simple sugars that promote tooth decay can also come from starchy foods, especially pretzels, crackers, and breads that stick to your gums and teeth. Some of the starch is broken down to simple sugars by enzymes in the mouth.

To reduce the incidence of tooth decay, a number of countries have developed campaigns to help inform consumers about cavity-promoting foods. Switzerland and other countries label foods that are safe for the teeth with a "happy tooth" symbol (Illustration 12.15) and encourages the use of alcohol sugars (which don't promote tooth decay) in sweets.[23] Other countries recommend that sweets be consumed with meals or that teeth be brushed after sweets are eaten.

There's More to Tooth Decay Than Sugar Per Se How frequently sugary and starchy foods are consumed and how long they stick to gums and teeth make a difference in their tooth-decay-promoting effects. Marshmallows, caramels, and taffy, for example, are much more likely to promote tooth decay than are apples and milk chocolate. Nevertheless, all of the foods listed in Table 12.6 can promote

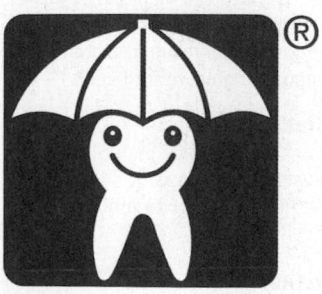

Illustration 12.15 Switzerland's "happy tooth" symbol. It has became an internationally used symbol.

Table 12.6

The "stickiness" value of some foods

The stickier the food, the worse it is for your teeth.

Very Sticky	Sticky	Somewhat Sticky	Barely Sticky
Caramels	Doughnuts	Bagels	Apples
Chewy cookies	Figs	Cake	Bananas
Crackers	Frosting	Cereal	Fruit drinks
Cream-filled cookies	Fudge	Dry cookies	Fruit juices
Granola bars	Hard candy	Milk chocolate	Ice cream
Marshmallows	Honey	Rolls	Oranges
Pretzels	Jelly beans	White bread	Peaches
Taffy	Pastries		Pears
	Raisins		
	Syrup		

Table 12.7

Foods that don't promote tooth decay.[3,39,40]

Artificial sweeteners	Gum and candy sweetened with	Peanut butter
Cheese	alcohol sugars	Tea
Coffee (no sugar)	Meats	Water
Eggs	Milk	Yogurt (plain)
Fats and oils	Nuts	Vegetables
		Fresh fruit

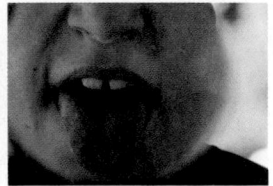
tooth decay if allowed to remain in contact with the gums and teeth. Drinking coffee or tea with sugar throughout the day or consuming three or more regular soft drinks between meals hastens tooth decay (more than if these beverages are consumed with meals) Candy, cookies, and crackers eaten between meals are much more likely to promote tooth decay than are the same foods consumed as part of a meal. Chewing as few as two sticks of sugar-containing gum a day also significantly increases tooth decay.[31,32]

Why Does Sugar Promote Tooth Decay?

Sugar promotes tooth decay because it is the sole food for certain bacteria that live in the mouth and excrete acid that dissolves teeth. In the presence of sugar, bacteria in the mouth multiply rapidly and form a sticky, white material called **plaque.** Tooth areas covered by plaque are prime locations for tooth decay because they are dense in acid-producing bacteria. Acid production by bacteria increases within 5 minutes of exposure to sugar. It continues for 20 to 30 minutes after the bacteria ingest the sugar.[34] If teeth are frequently exposed to sugar, the acid produced by bacteria may erode the enamel, producing a cavity. If the erosion continues, the cavity can extend into the tooth and allow bacteria to enter the inside of the tooth. That can cause an infection and the loss of the tooth. It can be prevented if the plaque is removed before the acid erodes much of the enamel. Teeth are capable of replacing small amounts of minerals lost from enamel.[35] Table 12.7 lists foods that do not promote tooth decay.

plaque
A soft, sticky, white material on teeth; formed by bacteria.

Water Fluoridation

In the early 1930s, lower rates of tooth decay were observed among children living in areas where water naturally contained fluoride. This provided the initial evidence that led to the fluoridation of many community water supplies. Fluoridated water reduces the incidences of tooth decay by 50% or more and is primarily responsible for declining rates of tooth decay and loss.[36]

"Water fluoridation is one of the ten great public health achievements of the twentieth century."

—W. Bailey[36]

Credit for declines in tooth decay and tooth loss in the United States is also shared by fluoride supplements, toothpastes, rinses and gels, protective sealants, and improved dental hygiene and care.[1] Further improvements in rates of dental caries will occur with reduced intake of sugars and sticky carbohydrates. Fluoridation is a safe, effective, and cheap method of controlling dental disease; providing fluoridated community water supplies costs about 50 cents per person per year.[38] Despite the advantages of fluoride, 31% of Americans consume water from a less than optimally fluoridated water supply.[36]

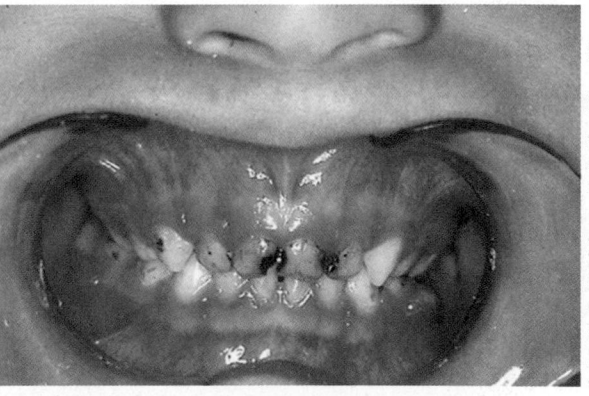

© 2009 K.L. Boyd DDS/Custom Medical Stock Photo. All Rights Reserved

Illustration 12.16 "Baby bottle caries" (also called "nursing bottle syndrome") occurs in infants who habitually receive sweet fluids or milk in bottles when they go to sleep. Cavities occur first in the upper front teeth because that's where fluid pools when babies sleep.

Baby Bottle Caries

A startling example of the effect that frequent and prolonged exposure to sugary foods can have is "baby bottle caries" (Illustration 12.16). Infants and young children who routinely fall asleep while sucking a bottle of sugar water, fruit drink, milk, or formula—or while breastfeeding—may develop severe decay. After the child falls asleep, the fluid may continue to drip into the mouth. A pool of the fluid collects between the tongue and the front teeth, bathing the teeth in the sweet fluid for as long as the child sleeps. The upper front teeth become decayed first because the tongue protects the lower teeth. Baby bottle caries occur in 5 to 10% of infants and young children and can lead to the destruction of all baby teeth.[38]

A number of other health problems related to carbohydrates are presented in other Units within this text. Diabetes, insulin resistance, and hypoglycemia are covered in Unit 13, and lactose intolerance in Unit 7.

© Tony Freeman/PhotoEdit

Up Close

Does Your Fiber Intake Measure Up?

Focal Point: Approximate the amount of fiber your diet contains.

Are you meeting your fiber quota, or do you consume the typical low-fiber American diet? To determine if your fiber intake is adequate, award yourself the allotted number of points for each serving of the following foods that you eat in a typical day. For example, if you normally eat one slice of whole grain bread each day, give yourself two points. If you eat two slices daily, give yourself four points. After tallying your score, refer to the Feedback section for the results.

High-fiber food choices

Fruits:

   1 whole fruit (e.g., apple, banana)

   $1/2$ cup cooked fruit

   $1/4$ cup dried fruit

Grains:

   $1/2$ cup cooked brown rice

   1 whole grain slice of bread, roll, muffin, or tortilla

   $1/2$ cup hot whole-grain cereal (e.g., oatmeal)

   $3/4$ cup cold whole-grain cereal (e.g., Cheerios)

   2 cups popcorn

Nuts and seeds:

   $1/4$ cup seeds (e.g., sunflower)

   2 tablespoons peanut butter

Vegetables: **3 points for each serving**

   1 whole vegetable (e.g., potato)

   $1/2$ cup cooked vegetable (e.g., green beans)

Bran cereals: **7 points for each serving**

   $1/2$ cup cooked oat bran cereal

   1 cup cold bran cereal

Legumes: **8 points for each serving**

   $1/2$ cup cooked beans (e.g., baked beans, pinto beans)

                       Total score

2 points for each serving

_____

_____

_____

_____

_____

_____

_____

_____

_____

_____

_____

_____

_____

_____

_____

Special note: You can also calculate your fiber intake using the Diet Analysis Plus software. Input your food intake for one day. Then go to the Analyses/Reports section to view the total number of grams of fiber in your diet on that day.

FEEDBACK (including scoring) can be found at the end of Unit 12.

[Key Terms

alcohol sugars, page 12-6
carbohydrates, page 12-2
complex carbohydrates, page 12-10
dietary fiber, page 12-11
disaccharides, page 12-3
functional fiber, page 12-11

glycemic index, page 12-15
glycogen, page 12-4
insulin resistance, page 12-15
monosaccharides, page 12-3
phenylketonuria (PKU), page 12-8
plaque, page 12-17

polysaccharides, page 12-10
simple sugars, page 12-3
tooth decay, page 12-16
total fiber, page 12-11
type 2 diabetes, page 12-15

[Review Questions

TRUE FALSE

1. Excluding fiber, one gram of carbohydrate provides four calories. ☐ ☐

2. Fructose is a monosaccharide, maltose is a diasaccharide, and glycogen is a trisaccharide. ☐ ☐

3. When the body has more glucose than it needs for energy formation, the excess glucose is converted to glycogen and fat. ☐ ☐

4. Sugars are the most commonly used food additive. ☐ ☐

5. Excess sugar intake causes obesity. ☐ ☐

6. Xylitol and sorbitol are examples of alcohol sugars. ☐ ☐

TRUE FALSE

7. Excess intake of aspartame (also called Nutrasweet) causes headaches in a majority of children and adults. ☐ ☐

8. Complex carbohydrates are also known as polysaccharides. ☐ ☐

9. All types of fiber share the characteristic of providing four calories per gram. ☐ ☐

10. Carbohydrates and carbohydrate-containing foods are classified by their glycemic index, or the extent to which they increase blood glucose levels. ☐ ☐

11. Declining rates of dental caries in the United States are primarily related to increased access to fluoridated water supplies. ☐ ☐

[Media Menu

www.cancer.gov/cancertopics/factsheet/ Risk/artificial-sweeteners
The National Cancer Institute provides information on artificial sweeteners and cancer from this site.

www.nlm.nih.gov/medlineplus/dietaryfiber. html
This site offers specific suggestions on how to fit more fiber into your diet, fiber and health information, fiber content of foods, and the latest research on fiber.

www.mendosa.com/gilists.htm
This site provides a table of the glycemic index of foods.

www.ada.org
It's not the American Dietetic Association, it's the American Dental Association; they have a column called "The Public" that provides information on oral health, finding a dentist, videos, tooth whitening, and tips for teachers.

www.nlm.nih.gov/medlineplus/dentalhealth. html
The long menu of topics on this site includes answers to FAQs about caring for your teeth and gums; information on diet and dental health, gum chewing and caries prevention, causes of periodontal disease, facts about fluoride, and an atlas of teeth.

www.adha.org
The American Dental Hygienists Association Web site provides dental health education tools and information on dental health and care here.

www.ific.org
This is an industry-sponsored site offering information on carbohydrates and sugars, artificial sweeteners, and oral health.

www.mchoralhealth.org
The Bureau of Maternal and Child Health offers webcasts and information on practices and programs aimed at improving oral health in children.

Notes

1. Touger-Decker R et al. Sugars and dental caries. Am J Clin Nutr 2003;78(suppl):881S–92S.

2. What we eat in America, NHANES, 2005–2006, USDA, 2008, available at www.are.usda.gov/ba/bhnrc/fsrg.

3. Dietary reference intakes: energy, carbohydrate, fiber, fat, fatty acids, cholesterol, protein, and amino acids. Institute of Medicine, National Academies of Sciences. Washington, DC: National Academies Press, chapter 11, 2002.

4. Grabitske HA et al. Low-digestible carbohydrates in practice. J Am Diet Assoc 2008;108:1677–81.

5. Fulgoni III, V. High-fructose corn syrup: everything you wanted to know, but were afraid to ask, Am J Clin Nutr 2008;88(suppl):1715S.

6. Position of the American Dietetic Association: Use of nutritive and nonnutritive sweeteners. J Am Diet Assoc 2004;104:255–58.

7. Bray GA. Fructose—how worried should we be? Medscape J Med 2008;10:159, www.medscape.com/viewarticle/575891, accessed 2/09.

8. Duffey KJ et al. High-fructose corn syrup: is this what's for dinner? Am J Clin Nutr 2008;88(suppl):1722S–32S.

9. Vos MB et al. Dietary fructose consumption among U.S. children and adults: the Third National Health and Nutrition Examination Survey, Medscape J Med 2008, www.medscape.com/viewarticle/575891, accessed 3/09.

10. Food and Drug Administration (Department of Health and Human Services), Nutrition labeling, Federal Register 1991 Nov 27.

11. American Medical Association, AMA finds high fructose corn syrup unlikely to be more harmful to health than other caloric sweeteners, AMA Press Release, 12/08, www.ama.org, accessed 12/08.

12. Dietary reference intakes, NAS, chap. 11; and National Institutes of Health Consensus Development Conference Statement: Diagnosis and management of dental caries throughout life. March 28, 2001; 18:1–24.

13. Dietary reference intakes, NAS, chap. 11; and Joint WHO/FAO Expert Consultation on Diet, Nutrition, and the Prevention of Chronic Diseases, Geneva, Switzerland; 2002.

14. Mattes RD et al. Nonnutritive sweetener consumption in humans: effects on appetite and food intake and their putative mechanisms. Am J Clin Nutr 2009;8:1–14.

15. McKay B. Beverage wars take on new flavor, Wall Street Journal 8/31/08, p. B12.

16. Saccharin deemed safe. Community Nutrition Institute, 2000; June 2:8.

17. Taddio A et al. Sucrose as a pediatric analgesia, Pediatrics 2009;123:e425–e429.

18. Butchko HH et al. Acceptable daily intake vs. actual intake: the aspartame example. J Am Coll Nutr 1991;10:258–66.

19. Spiers PA et al. Aspartame: neuropsychologic and neurophysiologic evaluation and chronic effects. Am J Clin Nutr 1998;68:531–7.

20. Magnuson B et al. Safety of Aspartame, Crit Rev Toxicol 10/07, accessed at www.medscape.com/viewarticle/564923, 3/09.

21. American Association of Cereal Chemists Report: All fibers are essentially functional, Cereal Foods World 2003;48:128.

22. Marlett JA et al. Position of the American Dietetic Association: health implications of dietary fiber. J Am Diet Assoc 1997;97:1157–9.

23. Scott-Thomas C. Prepare for higher calorie count for fibre, say scientists, 19-Dec-2008, www.foodnavigator/europe.com, accessed 12/08.

24. Williams CL. A summary of conference recommendations on dietary fiber in childhood. Pediatrics 1995; 96:1023–8.

25. Rizkella SW et al. Low glycemic index diet and glycemic control in type 2 diabetes. Diabetes Care 2004;27:1866–72.

26. Stanhope KL et al. Endocrine and metabolic effects of consuming beverages sweetened with fructose, glucose, sucrose, or high-fructose corn syrup, Am J Clin Nutr 2008;88(suppl):1733S–7S.

27. Riccardi G et al. Role of glycemic index and glycemic load in the healthy state, in prediabetes, and in diabetes. Am J Clin Nutr 2008;87(Suppl):269S–74S.

28. Jordan M. Colgate brings dental care to Brazilian Indian tribes; ravages of tobacco and rice. Wall Street Journal, 7/23/02, page B1.

29. Cornero S et al. Diet and nutrition of prehistoric populations at the alluvial banks of the Parana River. Medicina 2000;60:109–14.

30. Parajas IL. Sugar content of commonly eaten snack foods of school children in relation to their dental health status. J Phillipine Dent Assoc 1999;51:4–21.

31. Ismail A. Food cariogenicity in Americans aged from 9 to 29 years accessed in a national cross-sectional study, 1971–1974, J Dental Res 1986; 65:1435–40.

32. Edgar WM. Sugar substitutes, chewing gum, and dental caries—a review, British Dental J 1998;184:29–32.

33. Bad breath. Nutr Today 2000;35:6.

34. Schachtele CF et al. Will the diets of the future be less cariogenic? J Canadian Dental Assoc 1984;3:213–9.

35. Palmer CA et al. Position of the American Dietetic Association: the impact of fluoride on dental health. J Am Diet Assoc 2005;105: 1620–28.

36. Bailey W et al. Populations receiving optimally fluoridated public drinking water—United States, 1992–2006. MMWR 2008;57(27):737–41.

37. Position of the American Dietetic Association: oral health and nutrition. J Am Diet Assoc 2003;103:615–25.

38. Trends in children's oral health. National Maternal and Child Oral Health Resource Center, www.ncemch.org, 1999.

39. Moynihan P et al. Diet, nutrition and the prevention of dental diseases, Public Health Nutr 2004;7:201–26.

40. Sheihan A. Dietary effects on dental diseases, Public Health Nutr 2001;4:569–91.

41. Position of the American Dietetic Association: health implications of dietary fiber, J Am Diet Assoc 2008;108:1716–31.

NUTRITION | # Up Close

Does Your Fiber Intake Measure Up?

Feedback for Unit 12

The total number of points you scored approximates the **grams** of total fiber you typically consume daily.[a] Use this scale to find out if your fiber intake meets the recommended goal:

- **0–10 grams:** You consume less than the average American. Increase your fiber intake by including more fruits, vegetables, whole grains, and legumes in your diet overall.
- **11–15 grams:** Like other Americans, you consume too little fiber. Increase the number of servings of high-fiber foods you already enjoy, while substituting more high-fiber foods for refined food products. A quick way to add fiber to your diet is to consume more of the two fiber powerhouses: legumes and bran cereal.
- **15–20 grams:** You currently consume more fiber than the average American. Make sure you're including 5 or more servings of fruits and vegetables. Eat 6 to 11 servings of bread, cereal, rice, and pasta daily; choose whole-grain versions of these foods often.
- **20–40 grams:** Congratulations! Your dietary fiber intake is in the vicinity of that recommended. Keep up the good work.

[a]Because different foods within a food group contribute varying amounts of dietary fiber, the point values have been average. Make sure to check the Nutrition Facts panel on bran cereals because these cereals vary in the amount of dietary fiber they contain.

Diabetes Now

NUTRITION SCOREBOARD

	TRUE	FALSE
1 Excess sugar consumption causes diabetes.		
2 Diabetes generally develops over the course of many years.		
3 Glycemic index is a measure of the simple sugar content of foods.		

Key Concepts and Facts

- Diabetes is related to abnormal utilization of glucose by the body.

- The three main forms of diabetes are type 1, type 2, and gestational diabetes.

- Rates of type 2 diabetes increase as obesity does.

- Diet is always a part of the treatment of diabetes.

- Weight loss and physical activity can prevent or delay the onset of type 2 diabetes in many people.

The Diabetes Epidemic

What AIDS was in the last 20 years of the 20th century, diabetes is to be in the first 20 years of this century.

—PAUL ZIMMET, INTERNATIONAL DIABETES INSTITUTE[4]

Illustration 13.1 Diabetes headlines say it all.

It's not the plague, yellow fever, or heart disease. The latest worldwide disease epidemic is **diabetes,** and the rising rates are directly related to the global increase in obesity (Illustration 13.1). Diabetes affects 6% of adults worldwide,[5] and 10% of U.S. adults.[6] Less than 1% of U.S. adults were diagnosed with type 2 diabetes in 1960. One third of the 10% of U.S. adults with diabetes have yet to be diagnosed and are not receiving health care services for the disorder.[6]

There are three major forms of diabetes: **type 1**, **type 2**, and **gestational diabetes**. Type 2 diabetes is the most common by far and is fueling the diabetes epidemic. Table 13.1 summarizes key features of type 1 and type 2 diabetes. Both types are diagnosed when fasting levels of blood glucose are 126 milligrams/deciliter (mg/dl) and higher; these types generally take years to develop.[2] In all cases of diabetes, the central defect is elevated blood glucose level caused by an inadequate supply of insulin, an ineffective utilization of insulin, or both.[7]

Insulin is a hormone produced by the pancreas that performs many functions, one of which is to reduce blood glucose levels after meals. By facilitating the passage of glucose into cells, insulin keeps a steady supply of glucose going into cells. Glucose is needed by cells as a source of energy for thousands of chemical reactions that participate in the maintenance of ongoing body functions and health. If insulin is produced in insufficient amounts, or if cell membranes are not sensitive to the action of insulin, cells become starved for glucose. Functional levels of multiple tissues and organs in the body degrade as a result. High levels of blood glucose are related to adverse side effects in the body, too, such as elevated blood levels of triglycerides, **chronic inflammation**, increased blood pressure, and hardening of the arteries.[9]

Table 13.1

Key characteristics of type 1 and type 2 diabetes[8]

Characteristic	Type 1 Diabetes	Type 2 Diabetes
Insulin deficiency?	Yes	Occurs in advanced stages of the disease
Proportion of cases	5–10%	90–95%
Risk factors	Viral infection early in life (or other triggers in genetically sensitive individuals) that destroys the part of the pancreas that produces insulin, certain medications, poor vitamin D status.	Obesity (especially abdominal fatness), sedentary lifestyle, insulin resistance, low weight at birth, certain ethnicities, family history, older age
Treatment	Insulin, individualized diet and exercise	Weight loss (in most cases), increased physical activity, individualized diet, sometimes oral medications, and/or insulin

Health Consequences of Diabetes

Health effects of diabetes vary depending on how well blood glucose levels are controlled and on the presence of other health problems such as hypertension or heart disease. In the short run, poorly controlled and untreated diabetes produces blurred vision, frequent urination, weight loss, increased susceptibility to infection, delayed wound healing, and extreme hunger and thirst. In the long run, diabetes contributes to heart disease, hypertension, nerve damage, blindness, kidney failure, stroke, and the loss of limbs due to poor circulation. The number one cause of death among people with diabetes is heart disease. Many of the side effects of diabetes can be prevented or delayed if blood glucose levels are maintained within the normal range.[2]

Type 2 Diabetes

The development of this common type of diabetes is most likely to occur in overweight and obese, inactive people (Illustration 13.2). The term *diabesity* is being used to describe the close relationship between obesity and type 2 diabetes, and the current surge in worldwide rates of both is being called the "diabesity epidemic."[10] Approximately 60% of people who develop type 2 diabetes are obese, and 30% are overweight.[11,12] Illustration 13.3 shows the close relationship between rising rates of obesity and type 2 diabetes in the United States.

Although most often diagnosed in people over the age of 40, type 2 diabetes is becoming increasingly common in children and adolescents.[13] There are genetic components to this disease, as evidenced by the fact that it tracks in families and is more likely to occur in certain groups (Hispanic American, African American, Asian and Pacific Islanders, and Native Americans) than others.[14] Rather than inheriting genes that cause diabetes, certain individuals inherit or acquire very early in life multiple genetic traits that increase the likelihood that diabetes will develop. These genetic traits lead to metabolic changes that promote the development of diabetes when a person is exposed to certain environmental triggers, such as a diet high in saturated fats or excess calories.[15]

Type 2 diabetes takes years to develop. Its development is characterized by a number of metabolic changes that can be identified before blood levels of glucose become high enough to qualify for the diagnosis. Individuals on their way

diabetes
A disease characterized by abnormal utilization of carbohydrates by the body and elevated blood glucose levels. There are three main types of diabetes: type 1, type 2, and gestational diabetes. The word *diabetes* in this unit refers to type 2, which is by far the most common form of diabetes.

type 1 diabetes
A disease characterized by high blood glucose levels resulting from destruction of the insulin-producing cells of the pancreas. This type of diabetes was called juvenile-onset diabetes and insulin-dependent diabetes in the past, and its official medical name is type 1 diabetes mellitus.

type 2 diabetes
A disease characterized by high blood glucose levels due to the body's inability to use insulin normally, or to produce enough insulin. This type of diabetes was called adult-onset diabetes and non-insulin-dependent diabetes in the past, and its official medical name is type 2 diabetes mellitus.

gestational diabetes
Diabetes first discovered during pregnancy.

chronic inflammation
Low-grade inflammation that lasts weeks, months, or years. Inflammation is the first response of the body's immune system to infection or irritation. Inflammation triggers the release of biologically active substances that promote oxidation and other potentially harmful reactions in the body.

to developing type 2 diabetes tend to have somewhat higher that normal blood glucose levels and a disorder called **insulin resistance**.

Prediabetes

Elevated fasting blood glucose levels that are somewhat below the cut point used to diagnose type 2 diabetes characterize **prediabetes**. Approximately 26% of U.S. adults, and 314 million people worldwide, are at risk of type 2 diabetes due to this condition.[16] The presence of prediabetes increases a person's odds of developing type 2 diabetes by 10% per year.[17] Abdominal obesity, physical inactivity, and genetic predispositions are common risk factors for the development of prediabetes.[20]

The American Diabetes Association recommends that individuals with prediabetes receive individualized medical nutrition therapy aimed at reducing body weight about 7% (if needed), increasing physical activity level to 150 minutes per week, and instituting healthy eating practices. If accomplished, these measures help prevent the progression of prediabetes into type 2 diabetes.[18] Type 2 diabetes is diagnosed when fasting blood glucose values consistently reach 126 mg/dL or higher.[8]

Insulin Resistance   Most people with prediabetes or type 2 diabetes have insulin resistance. Insulin resistance is due to abnormalities in the way the body uses insulin. Normally, insulin is able to lower blood glucose levels after meals by binding to receptors on cell membranes. These receptors are activated by insulin and allow glucose to pass into cells. With insulin resistance, cell membranes "resist" the effects of insulin, and that lowers the amount of glucose transported into cells. The pancreas responds to the low availability of glucose within cells by producing additional insulin. Higher than normal levels of insulin are generally sufficient to keep blood glucose levels under control for a number of years. Cells in the pancreas, however, may become exhausted from years of overwork. In such cases, the production of insulin slows and may eventually stop, leading to elevated blood glucose levels.[8]

Many other adverse health affects of insulin resistance have been identified. The reduction in glucose availability to cells caused by insulin resistance forces the body to mobilize fat from fat stores and use it as the primary source of energy. High levels of fat mobilization from fat stores in people with insulin resistance contribute to the development of elevated blood levels of free fatty acids and triglycerides. These changes increase insulin resistance further and promote the development of **fatty liver disease**, chronic inflammation, hypertension, and plaque formation in arteries.[19]

Insulin resistance is also related to the development of a spectrum of metabolic abnormalities that have far-reaching effects. Collectively, the adverse effects of insulin resistance are included in a disorder called **metabolic syndrome**.

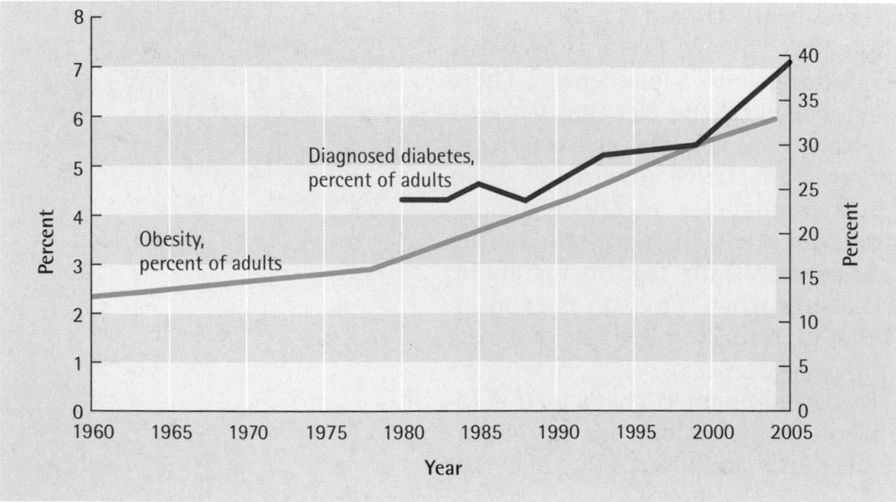

Metabolic Syndrome   Physicians have known for decades that obese people with hypertension and type 2 diabetes are at high risk of heart disease. What they didn't know is why. Over time, research studies discovered that part of the answer to the "why" question was insulin resistance.[21] There is currently no simple and inexpensive test for insulin resistance. Clinically, it is assumed to be present in people with high waist circumferences.[22] Insulin resistance is

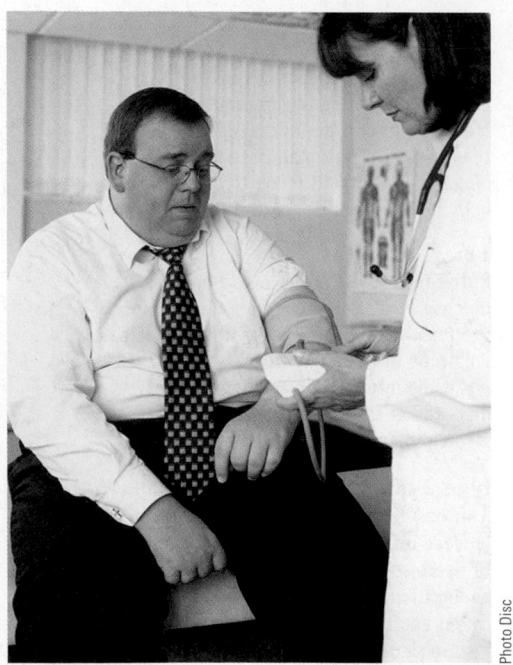

Photo Disc

Illustration 13.2   Obesity characterized by central-body fat stores and physical inactivity is a strong risk factor for type 2 diabetes.

insulin resistance
A condition in which cell membranes have reduced sensitivity to insulin so that more insulin than normal is required to transport a given amount of glucose into cells. It is characterized by elevated levels of serum insulin, glucose, and triglycerides, and increased blood pressure.

Illustration 13.3   The incidence of type 2 diabetes increases as rates of obesity increase.[6]

Diagnosed diabetes, percent of adults

Obesity, percent of adults

Percent

Percent

Year

1960 1965 1970 1975 1980 1985 1990 1995 2000 2005

related to a cluster of metabolic abnormalities included in the metabolic syndrome that increase the risk of type 2 diabetes as well as heart disease. Symptoms related to the metabolic syndrome include:

- Waist circumference 40" or over in males, 35" or over in females

- High blood pressure (130/85 mm Hg or higher)

- Elevated blood triglycerides levels (150 mg/dL or higher)

- Low levels of protective HDL cholesterol (less than 50 mg/dL in women and 40 mg/dL in men)

- Elevated fasting blood glucose levels (110 mg/dL or higher)

The diagnosis of metabolic syndrome is made when three or more abnormalities are identified. Individuals with four or five metabolic abnormalities have a 3.7 times greater risk of heart disease, and a 25 times higher risk for diabetes than do people with no abnormalities.[9] It is estimated that 25% of men and women in the United States have metabolic syndrome, with the bulk of cases being made up of overweight and obese inactive adults.[23] Weight loss, exercise, adequate vitamin D status, adequate magnesium intake, and the other factors listed in Illustration 13.4 are the key components of preventing and managing metabolic syndrome.[14,24]

Managing Type 2 Diabetes

As is the case for all forms of diabetes, diet is a cornerstone of the treatment of type 2 diabetes.[25] Modest weight loss alone (5 to 10% of body weight) has been repeatedly shown to significantly improve blood glucose control in overweight and obese people with type 2 diabetes.[26]

In general, diets developed for diabetes emphasize

- Calorie reduction if overweight or obese

- Complex carbohydrates, including whole-grain breads and cereals, and other high-fiber foods, vegetables, fruits, low-fat milk and meats, and fish

- Unsaturated fats, especially monounsaturated fats

- Regular meals and snacks[27]

Dietary management of diabetes should focus on heart disease risk reduction as well as blood glucose control. Food sources of monounsaturated fats, such as olive oil, nuts, and seeds, are recommended over foods high in saturated or *trans* fats. Monounsaturated fats raise blood levels of HDL cholesterol while reducing

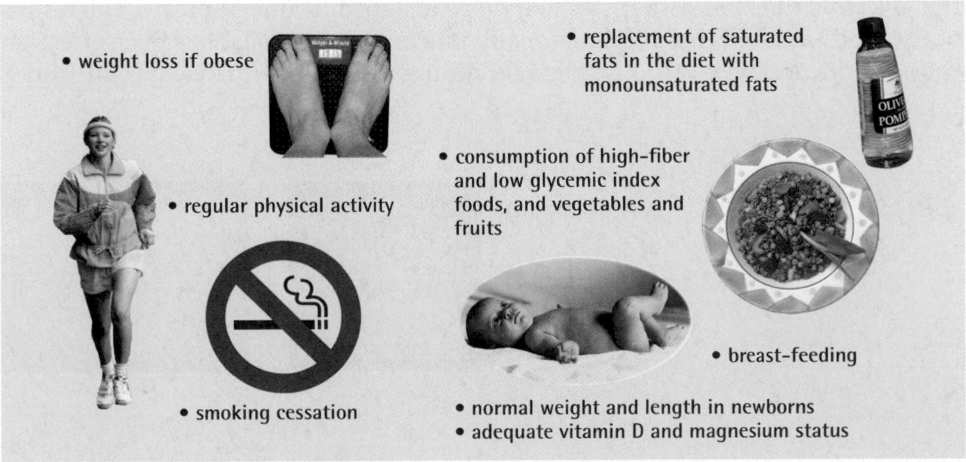

prediabetes
A condition in which blood glucose levels are higher than normal but not high enough for the diagnosis of diabetes. It is characterized by impaired glucose tolerance, or fasting blood glucose levels between 110 and 126 mg/dL.

fatty liver disease
A reversible condition characterized by fat infiltration of the liver (10% or more by weight). If not corrected, fatty liver disease can produce liver damage and other disorders. The condition is primarily associated with obesity, diabetes, and excess alcohol consumption.

metabolic syndrome
A constellation of metabolic abnormalities that increase the risk of heart disease and type 2 diabetes. Metabolic syndrome is characterized by insulin resistance, abdominal obesity, high blood pressure and triglycerides levels, low levels of HDL cholesterol, and impaired glucose tolerance. It is also called *Syndrome X* and *insulin resistance syndrome*.

- weight loss if obese
- regular physical activity
- smoking cessation
- replacement of saturated fats in the diet with monounsaturated fats
- consumption of high–fiber and low glycemic index foods, and vegetables and fruits
- breast-feeding
- normal weight and length in newborns
- adequate vitamin D and magnesium status

PhotoDisc

Illustration 13.4 Key components of the prevention and management of metabolic syndrome.

LDL cholesterol level. They also improve insulin resistance and lower blood glucose levels somewhat.[28] Diet and weight-loss interventions may be supplemented by oral medications that decrease insulin resistance and blood lipids and by insulin if needed.[26]

Regular physical activity is an important component of the management of type 2 diabetes. In addition to facilitating weight loss, physical activity reduces insulin resistance, decreases blood pressure and body fat content, and improves blood lipid and glucose levels.[29] Moderate intensity aerobic and strength-building activities, such as weight lifting, jogging, fast walking, aerobic dancing, and swimming are recommended. The target generally set for the duration of physical activity is 150 minutes per week, or an average of 21 minutes per day.[29]

Knowledge of the role of diet, exercise, insulin, and other factors in the management of diabetes is expanding so rapidly that the American Diabetes Association updates management recommendations yearly. Dietary recommendations are currently not consistent across developed countries, indicating that scientific consensus is yet to be reached on a number of important issues related to diet and diabetes.[25]

Glycemic Index and Glycemic Load   Carbohydrate-containing foods have a range of effects on blood glucose levels—some cause a rapid rise, and others do not. Foods that increase blood glucose to relatively high levels require more insulin to move glucose into cells than do foods that produce lower levels of glucose. Over the past 25 years, many carbohydrate-containing foods have been tested for their effect on blood glucose level and assigned a **glycemic index** value. Compared to high glycemic index (GI) carbohydrate sources, foods with low GI values increase blood glucose levels to a lower extent, and decrease insulin need.[32]

The glycemic index of carbohydrate-containing foods is determined by assessing the elevation in blood glucose level caused by ingestion of 50 grams of a carbohydrate-containing food compared to the rise in blood glucose level that results from consuming 50 grams of glucose. (Sometimes the standard for comparison is white bread.) Table 13.2 shows the glycemic index of a number of foods, using 50 grams of glucose as the standard for comparison. As you read over the list of foods and their glycemic index values, you may find some surprises. Sucrose, honey, fructose, and other simple sugars are not high glycemic index foods, and many fruits we think of as sweet do not cause a relatively high rise in blood glucose level. Coarse-ground whole-wheat breads and dried beans have medium-to-low glycemic index values. It probably comes as no surprise that glucose has a glycemic index of 100, but it is infrequently found by itself in foods.

Foods providing carbohydrates are usually consumed as part of a meal, but glycemic index is determined for individual foods. It is possible that the protein and fat content of other foods in a meal affect glycemic index values of the carbohydrate-containing foods. To determine if this is the case, scientists assessed the glycemic index of carbohydrate foods consumed in mixed meals. The results of the study showed that the protein and fat content of a meal has little effect on the glycemic index of carbohydrate-containing foods consumed. Glycemic index

glycemic index (GI)
A measure of the extent to which blood glucose level is raised by a 50-gram portion of a carbohydrate-containing food compared to 50 grams of glucose or white bread.

Reality Check

Will the real whole grain please stand up?   Which bread is made from whole grains?

Who gets thumbs up?
Answers on page 13-8

Richard Anderson   Richard Anderson

Table 13.2

Glycemic index (GI) of selected foods[3,33]

High GI	(70–100)	Medium GI	(56–69)	Low GI	(55 or lower)
glucose	100	orange soda	68	honey	55
French bread	95	sucrose	68	oatmeal	54
scone	92	croissant	67	corn	53
potato, baked	85	Cream of Wheat	66	cracked wheat bread	53
potato, instant mashed	85	couscous	65	orange juice	52
Corn Chex	83	chapati	62	banana	52
pretzel	83	sweet potato	61	mango	51
Rice Krispies	82	muffin, blueberry	59	potato, boiled	50
cornflakes	81	Coca-Cola	58	green peas	48
Corn Pops	80	rice, white or brown	60	pasta	48
Gatorade	78	breadfruit	69	carrots, raw	47
jelly beans	78	taco shells	68	lactose	46
doughnut, cake	76	angel food cake	67	milk chocolate	43
waffle, frozen	76	Quaker Quick Oats	65	All-Bran	42
french fries	75	French bread with butter and jam	62	orange	42
Shredded Wheat	75	couscous	61	peach	42
Cheerios	74	Raisin bran	61	apple juice	40
popcorn	72	bran muffin	60	apple	38
watermelon	72	Just Right cereal	60	pear	38
Grape Nuts	71	power bar	56	tomato juice	38
wheat bread	70			yam	37
white bread	70			yogurt	31
				dried beans	25
				grapefruit	25
				milk	25
				fructose	19
				xylitol	8
				hummus	6

values for the carbohydrate-containing foods in the meals were within 90% of the values reported for the individual foods.[34]

Some high-GI foods such as baked potatoes, French bread, and cornflakes are good sources of a number of nutrients. Just because a food has a high glycemic index doesn't mean it should not be consumed as part of a balanced diet. Adjusting food choices toward selection of mainly low-GI foods is most helpful for people attempting to prevent or control type 2 diabetes or to diminish the effects of insulin resistance.[25,35]

■ **Glycemic Load** Because of the way glycemic index is calculated, it is difficult to know the extent to which blood glucose will be raised by consumption of a particular amount of a food. Another index of the blood-glucose-raising potential of carbohydrate-containing foods has been developed to help straighten out this confusion. It's called **glycemic load**, and it represents the blood-glucose-raising potential of the specific amount of food consumed. Glycemic load is calculated by multiplying the grams of carbohydrate in a specific amount of food times the food's glycemic index. This result is then divided by 100 to calculate glycemic load. A carrot provides approximately seven grams of carbohydrate and has a glycemic index of 47. Its glycemic load would be calculated as:

$$7 \times 47 = 329$$
$$329/100 = 3.29$$
$$\text{Glycemic load} = 3.29$$

glycemic load (GL)
A measure of the extent to which blood glucose level is raised by a given amount of a carbohydrate-containing food. GL is calculated by multiplying a food's GI by its carbohydrate content.

Will the real whole grain please stand up?
You can't tell by looking, but the slice of bread on the right contains more fiber (four vs. two grams). Want to find whole-grain breads? Look for the term "whole grain" or "whole wheat" on the label. Bread products carrying that label contain 51% or more whole grains. The ingredient label on whole-grain products will list one or more whole grains before listing enriched flour.

Wheat Bread

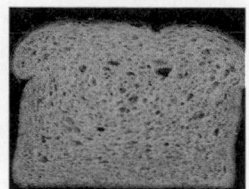

Whole Wheat Bread

The blood-glucose-raising effect of one carrot doesn't amount to much. If you consumed 4 slices (4 ounces) of French bread, the result would tell a different story. The glycemic load supplied by this amount of French bread is 49.4, a level that raises blood glucose and insulin levels far more than a carrot or 1 slice of French bread.

Consumption of low-GI foods is a recommended component of the dietary management of type 2 and gestational diabetes in a number of countries, and is becoming increasingly recommended in the United States.[36] Consumption of low-GI foods is viewed as a useful part of the management of insulin for resistance and metabolic syndrome, and as a secondary aid to blood glucose control for people with all forms of diabetes.[36,38]

Sugar Intake and Diabetes

Does sugar intake cause diabetes? The answer to this reasonable and often asked question is no. High sugar intakes have not been found to directly cause type 2 diabetes.[1] However, high sugar diets that provide excessive levels of calories could contribute to diabetes by promoting weight gain.[30] Regular consumption of high glycemic index foods has been related to the development of type 2 diabetes in genetically susceptible people.[30] Should people with diabetes exclude sugars from their diet? Sugar does not have to be eliminated from the diet of people with diabetes but the amount consumed should be limited due to its effect on blood glucose level and insulin need.[27,36] Intake of total carbohydrates and the glycemic index of the carbohydrate-containing foods, rather than sugar intake specifically, are most strongly related to blood glucose levels.[39] High-sugar diets increase blood triglyceride levels in people with metabolic syndrome, and that may increase the risk of heart disease.[40]

Prevention of Type 2 Diabetes

The effects of weight loss and exercise in preventing type 2 diabetes can be quite dramatic. In one large study that took place over a 3-year period, people with pre-diabetes reduced their risk of developing type 2 diabetes by over 50% by losing around 7% of body weight and exercising for 150 minutes a week.[18]

Diets rich in whole-grain and high-fiber foods are protective against the development of type 2 diabetes and appear to aid weight loss (Illustration 13.5). Components of high-fiber, whole grain foods raise blood glucose levels marginally and appear to provide nutrients and other biologically active substances that lessen the risk of this disease.[41] Consumption of regular or decaffeinated coffee (1 to 4 cups daily) and moderate alcohol intake (1 to 2 drinks per day

PhotoDisc

Illustration 13.5 Whole-grain products, other high-fiber foods, and ample servings of vegetables and fruits can help prevent type 2 diabetes.

Health *Action*   **Preventing Type 2 Diabetes**

Are You at Risk?

Check each category that applies to you:

___ I have a brother, sister, mother, or father with type 2 diabetes.

___ My BMI is 25 or higher.

___ My waist cicrumfrence is over 35 inches (if female) or 40 inches (if male).

___ I have been diagnosed with prediabetes or insulin resistance.

___ I am habitually inactive.

___ I have been diagnosed with high blood pressure.

___ I have been diagnosed as having high bood triglyceride levels.

If you checked one or more of the categories above, you are at increased risk of developing type 2 diabetes.

What to Do Now?

1. Recognize that there are no quick fixes for the prevention of type 2 diabetes.
2. If you are overweight or obese, gradually cut back on your calorie intake by making small and acceptable changes in your diet. Aim for a reduction in caloric intake of 100 to 200 calories a day.
3. Gradually increase your physical activity level until you are moderately to vigorously active for at least 150 minutes per week (21 minutes per day). Select activities you enjoy and will keep doing in the long run.
4. Choose low glycemic index foods for your carbohydrate-containing food choices as often as possible. (You can use Table 13.2 to check out your options for low glycemic index foods).
5. Consume enough fiber (26 grams daily).
6. Eat your favorite vegetables and fruits often.
7. Choose low-fat dairy products.
8. Select and consume whole-grain products rather than refined grain products.
9. See your health care provider if you think you have a problem with your blood glucose levels.

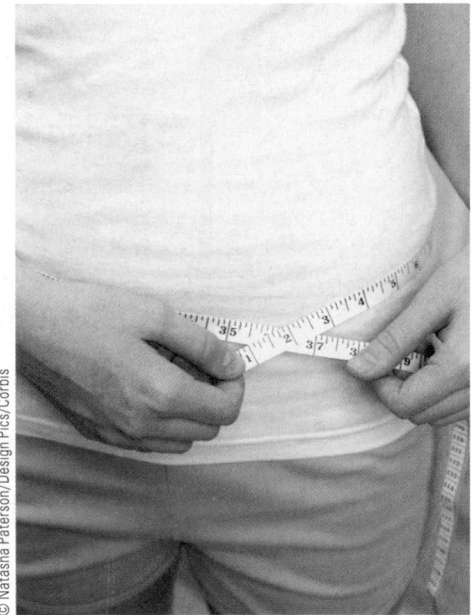

© Natasha Paterson/Design Pics/Corbis

if appropriate) also appear to decrease the risk of developing type 2 diabetes.[42] Finally, type 2 diabetes may be prevented or postponed by not gaining weight during the young adult years. Weight gains of 10 to 15 pounds between the ages of 25 and 40 years have been found to increase the risk of type 2 diabetes sevenfold compared to not gaining weight or gaining it after the age of 40.[43]

It is anticipated that 800,000 new cases of type 2 diabetes may develop each year in the United States due to rising rates of obesity.[14] The additional burdens such an increase would place on individuals and health care costs clearly convey the message that prevention is urgently needed. Public health campaigns are now under way to encourage people to lose weight if overweight, to exercise regularly, and to select whole-grain products and other high-fiber foods along with ample intake of vegetables and fruits.

Are you at risk of developing type 2 diabetes? Find out if you are, and what people at risk can do to help prevent it in the Health Action.

1500 B.C Early healers noticed that ants and files were attracted to the urine of people with a mysterious, emaciating disease. The attraction was later found to be glucose and the emaciating disease, diabetes.[46]

On the Side

©Jan Bengtsson/Corbis

Illustration 13.6 An insulin pump with a built-in glucose sensor can warn individuals about potentially harmful changes in blood glucose levels.

AP Images/Tom Strattman

Type 1 Diabetes

Type 1 diabetes is an **autoimmune disease** that produces a deficiency of insulin.[31] It accounts for 5–10% of cases of diabetes. The onset of type 1 peaks around the ages of 11 to 12 years and usually occurs before the age of 40.[8] Unlike type 2 diabetes, worldwide rates of type 1 diabetes are stable.[44]

The insulin deficiency that marks type 1 diabetes appears to develop when a person's own **immune system** destroys beta cells in the pancreas that produce insulin. The destruction is likely triggered by a viral infection (mumps, rubella, measles, the flu) in genetically susceptible people.[31] Medication used to treat high blood pressure, arthritis, and other conditions may also contribute to the development of type 1 diabetes.

Breast-feeding infants for the first four months or more of life, adequate vitamin D status, and ample intake of omega-3 fatty acids confer a level of protection against the development of type 1 diabetes.[45,46]

Managing Type 1 Diabetes

The main goals of the management of type I diabetes are blood glucose control and health maintenance.[26] Blood glucose levels are primarily managed by regular, healthful meals controlled in carbohydrate content that are consumed at planned times and in specific amounts. Diets are designed to match insulin dose so that blood glucose levels remain within normal ranges. They are controlled in carbohydrate content because carbohydrates raise blood glucose levels and increase insulin need to a greater extent than do protein or fats.[27]

People with type 1 diabetes are urged to replace simple sugars with reasonable amounts of artificial sweeteners. Foods low in glycemic index and high in fiber (especially soluble fiber such as oatmeal) are encouraged, as are brightly colored fruits and vegetables, low fat meat and dairy products, fish, dried beans, and nuts and seeds.[27,36] Reduced calorie diet plans should be included as part of the care for individuals with type 1 diabetes who would benefit from weight loss.

Physical activity is generally part of a diabetes care plan because it improves blood glucose levels, physical fitness, and insulin utilization. Both strength and aerobic exercises are recommended.[27] Individualized meal and physical activity plans, and follow-up care for individuals with type I diabetes should be provided by an experienced health care team that includes a registered dietitian.[36]

Insulin and New Technologies in the Management of Type 1 Diabetes People with type 1 diabetes require insulin to control blood glucose levels. The amount and type of insulin required depends on diet and physical activity level and the presence of conditions such as pregnancy, stress, illness, and physical activity. To get the insulin dose right, individuals with type 1 diabetes measure their blood glucose level a number of times daily. Blood glucose levels have been traditionally tested using a pinprick blood sample and a device that measures blood glucose level in a droplet of blood. The glucose measurement would be followed by the appropriate amount of insulin delivered in an injection. Although painless if very small needles are used, many people hate the idea of getting pricked by a needle.[47]

Technological advances are taking some of the perceived sting out of type 1 diabetes management. Insulin pumps that deliver programmed doses of insulin are now combined with continuous glucose monitors that assess the body's glucose level (Illustration 13.6).

Continuous glucose monitors issue a warning signal if glucose levels are rising or dropping too much. The signal gives the individual enough advance warning to take appropriate action, like eating or using insulin, before a problem develops. It must be used with self-monitoring of blood glucose levels.[29,48]

Many new types of insulin that better manage blood glucose levels are now available, and it is hoped that in the future a cure for type 1 diabetes will be identified. The cure may take the form of pancreatic and pancreatic islet cell transplants. These techniques are being intensively studied now and preliminary results are promising.[49,50]

Gestational Diabetes

Approximately 5 to 6% of women develop gestational diabetes during pregnancy, but the incidence varies a good deal based on age, body weight, and ethnicity. Native Americans, African Americans, women over the age of 35 years, obese women, and those with habitually low levels of physical activity are at higher risk than other women.[51] Gestational diabetes is likely caused by the same basic environmental and genetic predisposition factors as is type 2 diabetes.[48] Infants born to women with poorly controlled diabetes may be excessively fat at birth, require a cesarean delivery, and have blood glucose control problems after delivery. They are at greater risk for developing diabetes later in life.[51]

As is the case for type 2 diabetes, women with gestational diabetes are insulin resistant and can often control their blood glucose levels with an individualized diet and exercise plan. Some women require daily oral medication or insulin injections for blood glucose control.[52]

Gestational diabetes often disappears after delivery, but type 2 diabetes may appear later in life. Exercise, maintenance of normal weight, and consumption of a healthy diet reduce the risk that diabetes will return.[53]

Hypoglycemia: The Low Blood Sugar Blues?

Hypoglycemia is due to abnormally low blood glucose levels. It is thought to be a rare disorder because it is not often diagnosed. The diagnosis is tricky—blood tests for glucose should be conducted when the symptoms of hypoglycemia are present rather than during a scheduled appointment.[54] The true incidence of hypoglycemia is not known.

Hypoglycemia is most often caused by an excessive availability of insulin in the blood. The oversupply of insulin may be caused by certain tumors that secrete insulin, by other health problems, by high alcohol intake on an empty stomach, or, in people with diabetes, by an insulin dose that is too high. Hypoglycemia also occurs during prolonged starvation or fasting.[55]

Symptoms of hypoglycemia include weakness, sweating, nervousness, confusion, and irritability. Among individuals using insulin, these symptoms are often related to the presence of abnormally low blood glucose levels, or true hypoglycemia. Ingestion of small amounts (about 1–2 Tablespoon) of glucose or other sugar can effectively increase blood glucose levels in about 20 minutes. Care should be taken to assure that blood glucose levels remain in the normal range after an episode of hypoglycemia.[56]

In people not using insulin, the symptoms appear to be related to declining, rather than to abnormally low blood glucose values. Apparently people vary considerably in their response to low-normal blood glucose levels. Symptoms tend to occur 1 to 3 hours after a meal, and one study suggests that they are more likely to occur if the meal was rich in sugar.[57] Symptoms disappear when blood glucose levels rebound.[57]

hypoglycemia
A disorder resulting from abnormally low blood glucose levels. Symptoms of hypoglycemia include irritability, nervousness, weakness, sweating, and hunger.

Diabetes in the Future

The anticipated surge in the worldwide incidence of type 2 diabetes shown in Illustration 13.7 is not inevitable. It could be lowered substantially by environmental and lifestyle changes that reduce the risk for, and incidence of, overweight and obesity. Increased awareness of the connection between diabetes and body weight may help. Only a small proportion of people are aware of the connection now.[59] The hoped-for future of diabetes would be the one that negates the dire forecasts of the experts.

Illustration 13.7 The red bars indicate millions of cases of type 2 diabetes in 2000 and the dark blue bars the projections for 2030. The projected percentage increases in cases of type 2 diabetes are shown below the bars.[58]

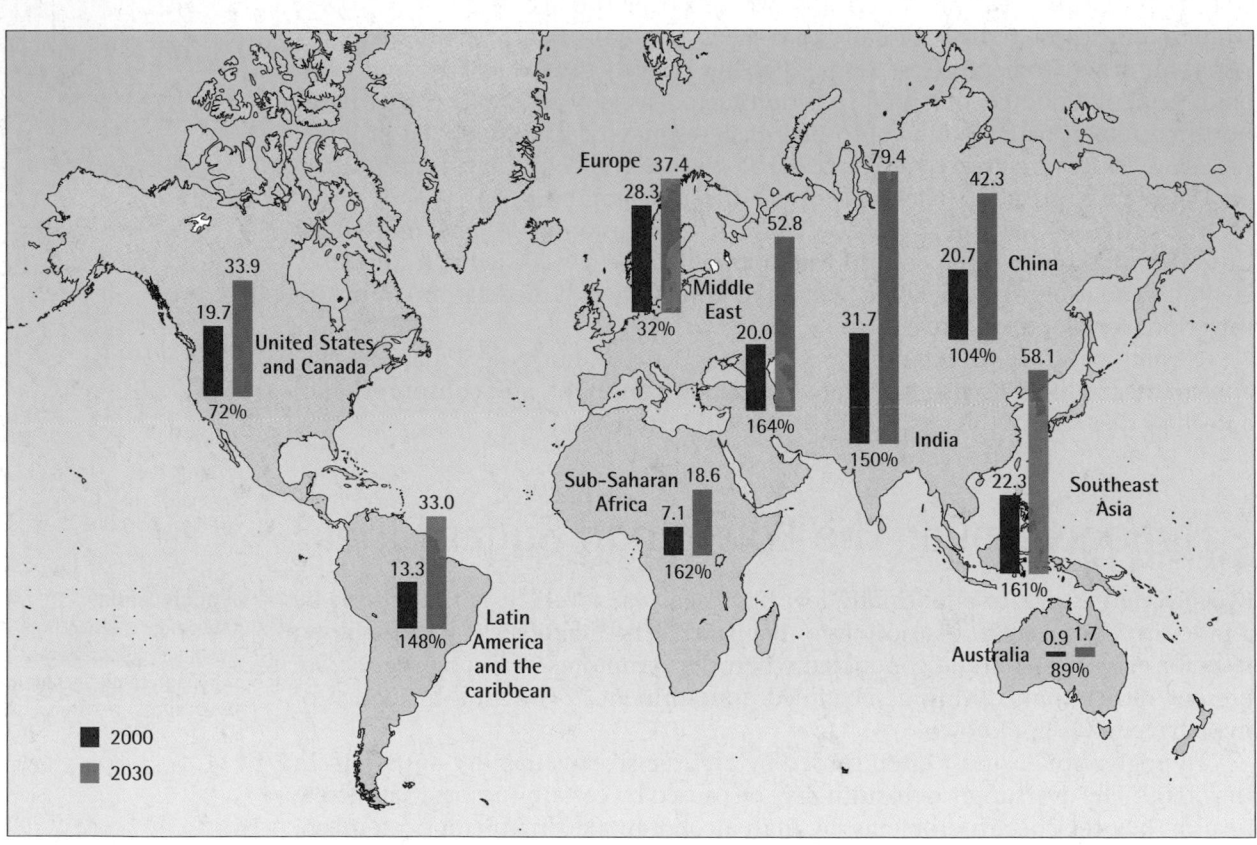

© Scott Goodwin Photography

NUTRITION Up Close

Calculating Glycemic Load

Focal Point: To gain an appreciation of the effect of source and amount of carbohydrate consumed on blood glucose levels.

Glycemic index provides an estimate of the rise in blood glucose level expected from consuming 50 grams of a carbohydrate containing food. Glycemic load, on the other hand, is a measure of the expected rise in blood glucose related to the ingestion of other amounts of this food. Consequently, glycemic load estimates the blood-glucose-raising potential of the amount of carbohydrate-containing food actually consumed.

1. Use Appendix A to look up the carbohydrate content of each food for the serving size listed.
2. Use Table 13.2 to identify the glycemic index for each food.
3. Calculate glycemic load based on this formula:

Glycemic load = grams carbohydrate × glycemic index of the food, divided by 100

Here's an example for 2 tsp sucrose:

grams carbohydrate in 2 tsp sucrose = 8
glycemic index of sucrose = 68
8 × 68 = 544
glycemic load = 544 ÷ 100 = 5.44

Food	Serving Size	grams carbohydrate ×	Glycemic Index ÷	Glycemic Load 100
cola beverage	36 oz (3 cans)	_____	_____	_____
potato, baked, no skin	1	_____	_____	_____
apple juice, bottled	1 cup	_____	_____	_____
milk chocolate, plain	1 oz	_____	_____	_____
hummus	1/2 cup	_____	_____	_____

FEEDBACK Glycemic load answers can be found at the end of Unit 13.

[Key Terms

autoimmune disease, page 13-10
chronic inflammation, page 13-3
diabetes, page 13-3
fatty liver disease, 13-5
gestational diabetes, page 13-3

glycemic index, page 13-6
glycemic load, page 13-7
hypoglycemia, page 13-11
immune system, page 13-10
insulin resistance, page 13-4

metabolic syndrome, page 13-5
prediabetes, page 13-5
type I diabetes, page 13-3
type 2 diabetes, page 13-3

Review Questions

1. The incidences of type 2 diabetes and of gestational diabetes increase as rates of overweight and obesity increase. ☐ ☐

2. Insulin facilitates the passage of vitamins into cells. ☐ ☐

3. Individuals with type 2 diabetes and gestational diabetes generally have the condition known as insulin resistance. ☐ ☐

4. The most effective strategies for the prevention and control of type 2 diabetes are weight loss and exercise. ☐ ☐

5. Prediabetes is defined as a condition in which fasting blood glucose levels are between 110 and 126 mg/dL. ☐ ☐

6. A female with a waist circumference of 29 inches, high blood pressure, and low levels of LDL cholesterol would be diagnosed as having "metabolic syndrome." ☐ ☐

7. Gestational diabetes appears to be caused by the same environmental and genetic predisposition factors that apply to type 2 diabetes. ☐ ☐

8. All foods high in complex carbohydrates have low glycemic indices. ☐ ☐

9. Excess sugar intake from soft drinks is a direct cause of type 2 diabetes. ☐ ☐

10. Hypoglycemia is most often caused by an excess availability of insulin. ☐ ☐

Media Menu

www.diabetes.org/risk-test.jsp
The American Diabetes Association has developed an interactive questionnaire for estimating your risk of having prediabetes or diabetes. It's simple and easy to use.

www2.niddk.nih.gov
The National Institute of Diabetes, Digestive, and Kidney Disease provide a broad array of information about diabetes and a number of down-loadable MP3 files on diabetes prevention and management in children and adults.

www.eatright.org
www.diabetes.org
These organizations have produced and sell a well-respected guide for meal planning for diabetes. It's called "Choose Your Foods: Exchange Lists for Diabetes." Recently updated,

it is often used in conjunction with meal planning by dietitians and people with diabetes. These organizations provide other resources related to diabetes.

www.dce.org/links/jada/00449.htm
Exercises appropriate for individuals with diabetes are listed at this site.

www.webmd.com
Go to "Diabetes," or to "Conditions A–Z," select diabetes, and hit "go." View a video, get the latest information on the treatment or a risk assessment for diabetes, and test your diabetes IQ.

www.diabetes.com
Part of WebMD, provides videos, audios, and print material on diabetes facts, tips for eating

well, information on insulin resistance, and more.

www.healthfinder.gov
Search insulin resistance, diabetes, metabolic syndrome, and hypoglycemia to find reliable reports.

www.ndep.nih.gov
The home page of the National Diabetes Education Program leads you to information and resources on diabetes, prediabetes, diabetes in children and youth, prevention strategies, programs, and more.

Notes

1. Liu S et al. Sugar intake and diabetes risk. Diabetes Care 2003;26:1008–15.

2. Nathan DM. Initial management of glycemia in type 2 diabetes mellitus. N Engl J Med 2002;347:1242–50.

3. Foster-Powell K et al. International table of glycemic index and glycemic load values: 2002. Am J Clin Nutr 2002; 76:5–56.

4. World seen facing diabetes catastrophe, impact may outpace AIDS. International Diabetes Federation Conference, Paris 2003. Reported in www.medscape.com, 9/6/03.

5. www.nlm.nih.gov/medlineplus/news/ fullstory_34711.html, 7/06. Type 2 affects 6% of adults worldwide.

6. CDC Fastats A to Z, www.cdc.gov/nchs/ fastats.htm, accessed 3/09.

7. Rother KI. Diabetes treatment—bridging the divide, N Engl J Med 2007; 356: 1499–1501.

8. Diabetes. http://diabetes.niddk.nih. gov/dm/pubs/statistics/index.htm#7, accessed 8/06.

9. Alexander CM et al. NCEP-defined metabolic syndrome, diabetes, and prevalence of coronary heart disease among NHANES III participants age 50 years and older. Diabetes 2003;52: 1210–14.

10. Jones V. The "Diabesity" epidemic: Let's rehabilitate America. Medscape Gen Med 2006; 8:34.

11. Incidence of diagnosed diabetes rising in the US. www.cdc.gov, 10/05.

12. Graph produced based on information in: The CDC's Database Modeling Project: developing a new tool for chronic disease prevention and control, www.cdc.gov, accessed 8/06

13. Song SH et al. Early-onset diabetes mellitus: an increasing phenomenon of elevated cardiovascular risk, Expert Rev Cardiovasc Ther 2008;6:315–22.

14. Health, United States, 2005. www.cdc. gov, accessed 8/06.

15. Marin, C et al. The Ala54Thr polymorphism of the fatty-acid binding protein 2 gene is associated with a change in insulin sensitivity after a change in the type of dietary fat. Am J Clin Nutr 2005; 82:196–200.

16. Rolka DR et al. Self-reported predicated and risk reduction activities—United States, 2006. MMWR 2008;57(44):1203–5.

17. Cowie CC, MMWR, CDC Surveillance System, 2003:52:833–7.

18. Diabetes Prevention Program Research Group, Reduction in the incidence of type 2 diabetes with lifestyle intervention or metformin, N Engl J Med 2002.

19. Ginsberg HN. Comprehensive insights into the pathophysiology of mixed dyslipidemia: The role of obesity and insulin resistance, mp.medscape.com/ cgi-bin1/DM/y/eBrH30U1roMODy0Jalu 0ET&uac=61213SX, released 10/08.

20. Ronti T et al. The endocrine function of adipose tissue. Clin Endocrinol 2006;64:355–65.

21. Ezmaillzadeh A et al. Dietary patterns, insulin resistance, and prevalence of metabolic syndrome in women, Am J Clin Nutr 2007;85: 910–8.

22. Ehrmann DA. Polycystic ovary syndrome. N Engl J Med 2005;352:1223–36.

23. Janiszewski PM et al. Themed review: lifestyle treatment of the metabolic syndrome, Am J Lifestyle Med 2008;2:99–108.

24. Pittas AG et al. High levels of vitamin D, calcium may lower type 2 diabetes risk. Diabetes Care 2006;29:650–56.

25. Wolever TMS et al. The Canadian Trial of Carbohydrates in Diabetes, a 1-y controlled trial of low-glycemic index dietary carbohydrate in type 2 diabetes, Am J Clin Nutr 2008;87:114–25.

26. Ripsin CM et al. Review of blood glucose management for type 2 diabetes, Am Fam Physician 2009;79:29–36.

27. American Dietetic Association. Evidence-based Nutrition Practice Guidelines, www.adaevidencelibrary. com/topic.cfmf?cat=3731, accessed 2/09.

28. Lopez S et al. Distinctive postprandial modulation of beta cell function and insulin sensitivity by dietary fats: monounsaturated compared with saturated fatty acids. Am J Clin Nutr 2008;88:638–44.

29. Hayes C et al. Role of physical activity in diabetes management, J Am Diet Assoc 2008;108:S19–S23.

30. Villegas AR et al. High intake of high glycemic index food and the development of type 2 diabetes in women of Chinese descent, Arch Intern Med 2007;167: 2310–6.

31. Fillippi CM et al. Viral triggers for type 1 diabetes: pros and cons, Diabetes 2008;57:2863–71.

32. Jenkins DJA et al. Almonds decrease postprandial glycemia, insulinemia, and oxidative damage in healthy individuals, J Nutr 2006;136:2987–92.

33. Brand-Miller JC et al. Glycemic index, postprandial glycemia, and the shape of the curve in healthy subjects: analysis of a database or more than 1000 foods, AM J Clin Nutr 2009;89:97–105, 49.

34. Wolever TMS et al. Food glycemic index, as given in Glycemic Index tables, is a significant determinant of glycemic response elicited by a composite breakfast meal. Am J Clin Nutr 2006;83:1306–12.

35. Riccardi G et al. Role of glycemic index and glycemic load in the healthy state, in prediabetes, and in diabetes. Am J Clin Nutr 2008;87(Suppl):269S–74S.

36. Bantle JP et al. American Diabetes Association updated guidelines for Medical Nutrition Therapy to prevent diabetes , manage existing diabetes, and prevent or slow the rate of development of diabetes complications, Diabetes Care 2008;31(Suppl):S61–78.

37. Nansel TR et al. Effect of low glycemic index diet on glucose excursions in children and adolescents with type 1 diabetes, Diabetes Care 2008;31:695–7.

38. Livesey G et al. Glycemic response and health—a systematic review and meta-analysis: relations between dietary glycemic properties and health outcomes, Am J Clin Nutr 2008;87(Suppl):258S–68S.

39. Ludwig DS, Glycemic load comes of age. J Nutr 2003;133:2728–32.

40. Franz MJ et al. Evidence-based nutrition principles and recommendations for the treatment and prevention of diabetes and related complications. Diabetes Care 2002;25:148–66.

41. McKeown N et al., Whole-grain intake is favorably associated with metabolic risk factors for type 2 diabetes and cardiovascular disease, Am J Clin Nutr 2002;76:390–8.

42. Van Dam RM et al. Coffee and the risk of type 2 diabetes in women. Diabetes Care 2006;29:398:403.

43. Schienkiewitz A et al. Body mass index history and risk of type 2 diabetes: results from the European Prospective Investigation into Cancer and Nutrition

(EPIC)-Potsdam Study. Am J Clin Nutr 2006;84:427–33.

44. Gonzalez ELM et al. Incidence of type 2 diabetes in the United Kingdom, J Epidemiol Community Health 2009, www.medscape.com/viewarticle/588613, accessed 3/09.

45. Virtanen SM et al. Infant feeding in Finnish children over the age of 4 years with newly diagnosed IDDM. Diabetes Care 1991;14:415–7.

46. Norris JM et al. Omega-3 polyunsaturated fatty acid intake and islet autoimmunity in children at increased risk for type 1 diabetes. JAMA 2007; 298: 1420–1428.

47. Armstrong D, Dagogo-Jack S et al. Therapeutic options to improve diabetes in African Americans: clinical case discussion, A continuing medical education program offered by Medscape, www.medscape.com, accessed 3/09.

48. Menato G et al. Current management of gestational diabetes mellitus, Expert Rev of Obstet Gynecol 2008;3:73–91.

49. Bromberg JS et al. Diabetes cure—is the glass half full? N Engl J Med 2006;355:1372–4.

50. Heimberg H. Boosting beta-cell numbers, N Engl J Med 2008;359:2723–4.

51. What is gestational diabetes? www.nichd.nih.gov/publications/pubs/gdm/sub1.htm, accessed 8/06.

52. Reader D et al. Impact of gestational mellitus nutrition practice guidelines implemented by registered dietitians on pregnancy outcomes. J Am Diet Assoc. 2006;106:1426–33.

53. Buchanan TA et al. Gestational diabetes mellitus. J Clin Invest 2005;115: 485–91.

54. Amiel S. Reversal of unawareness of hypoglycemia. N Engl J Med 1993; 329:876–7.

55. Blumberg S. Should hypoglycemia patients be prescribed a high-protein diet? J Am Diet Assoc 2005;105:196–7.

56. Albright A et al. Revised nutrition guidelines for diabetes prevention, Diabetes Care 2006;29:2140–57.

57. Simpson EJ et al. Interstitial glucose profiles associated with symptoms attributed to hypoglycemia by otherwise healthy women, Am J Clin Nutr 2008;87:354–61.

58. Wild S et al. Global prevalence of diabetes: estimates for the year 2000 and projections for 2030, Diabetes Care 2004;27:1047–53.

59. Few overweight and obese people aware of diabetes risk. www.medscape.com, accessed 9/03.

Answers to Review Questions

1. True, see pages 13-3, 13-11.
2. False, see page 13-2
3. True, see pages 13-4, 13-11.
4. True, see pages 13-8, 13-11.
5. True, see page 13-5.
6. False, see page 13-5.
7. True, see page 13-11.
8. False, see page 13-6.
9. False, see pages 13-8
10. True, see page 13-11.

NUTRITION | Up Close

Calculating Glycemic Load

Feedback for Unit 13

Food	Glycemic Load
cola beverage	69.6
baked potato	28.9
apple juice	11.6
milk chocolate	7.3
hummus	1.5

Effects of carbohydrate-containing foods on blood glucose levels vary depending on the glycemic index of these foods and the amount of them we consume.

UNIT 14

Alcohol: The Positives and Negatives

NUTRITION SCOREBOARD

	TRUE	FALSE
1 Alcohol is produced by the fermentation of carbohydrates.		
2 You can protect your body from the harmful effects of consuming excessive amounts of alcohol by eating a nutritious diet.		
3 Alcohol abuse plays a major role in injuries and deaths in the United States.		
4 Dark beer provides more calories than the same amount of regular beer.		

Key Concepts and Facts

- Alcohol is both a food and a drug and can have positive or negative effects on health.

- Alcohol is produced from carbohydrates.

- Alcohol abuse is harmful to the body and is associated with a high proportion of acts of violence and accidents.

- Both genetic and environmental factors are associated with the development of alcoholism.

Answers to
NUTRITION SCOREBOARD

		TRUE	FALSE
1	Alcohol (actually ethanol) is produced by the fermentation of carbohydrates in grains, fruits, and other plant foods.	✔	
2	High intakes of alcohol are harmful to the body, regardless of the quality of the diet.		✔
3	The statistics on alcohol abuse, injury, and death are startling. Alcohol abuse is a major personal, social, and public health problem in the United States.	✔	
4	It's true. Dark beers provide more calories than regular beer. See Table 14.1 on page 14–5.	✔	

Alcohol Facts

Alcohol can be considered either a tonic or a toxin in a dose-dependent fashion.[1]

fermentation
The process by which carbohydrates are converted to ethanol by the action of the enzymes in yeast.

Alcohol is both a food and a drug. It's a food because alcohol is made from carbohydrates, and the body uses it as an energy source. Alcohol is a drug because it modifies various body functions.

The type of alcohol people consume in beverages is ethanol. (We refer to ethanol by the broader term *alcohol* in this unit.) Alcohol is produced from carbohydrates in grains, fruits, and other foods by the process of **fermentation**. Wines, brews made from grains, and other alcohol-containing beverages are a traditional part of the food supply of many cultural groups.[2] In high doses, however, alcohol is harmful to the body and can cause a wide variety of nutritional, social, and physical health problems.

The Positive

chronic inflammation
Inflammation that is low-grade and lasts weeks, months, or years. Inflammation is the first response of the body's immune system to infection or irritation. It triggers the release of biologically active substances that promote oxidation and other potentially harmful reactions in the body.

Whether alcohol has harmful effects on health depends on how much is consumed. The consumption of moderate amounts of alcohol by healthy adults who are not pregnant appears to cause no harm. In fact, moderate alcohol consumption is associated with a significant level of protection against heart disease, type 2 diabetes, hypertension, stroke, dementia (cognitive decline), and all cause mortality.[3] A moderate level of alcohol consumption is considered to be one standard-sized drink per day for women and two drinks for men (Illustration 14.1).[1] The beneficial effects of moderate alcohol intake on health are related to the biological effects of alcohol and of phytochemicals present in many alcohol-containing beverages. Alcohol at moderate doses increases HDL-cholesterol levels (the "good" cholesterol) and decreases **chronic inflammation**. Decreased inflammation helps prevent the formation of plaque in arteries and improves circulatory function and maintenance of normal cell health. Alcohol improves the body's utilization of insulin and glucose, lowers post-meal blood glucose levels somewhat, and improves cognitive function.[1,4]

Some of the beneficial effects of alcohol on health are related to the phytochemical content of the fruit, vegetable, or grain fermented to produce it. Red wine, and to a lesser extent beer and white wine, contain pigments and other phytochemicals that act as powerful antioxidants and decrease inflammation and artery plaque formation.[5] Purple grape juice and other purple and blue colored fruit juices also provide antioxidants and have anti-inflammatory effects that benefit health, although to a lesser degree than does red wine.[1,6]

People don't have to consume alcohol to reduce their risk of heart disease. Diets low in saturated and *trans* fat, liberal intakes of vegetables and fruits, ample physical activity, and not smoking also reduce the risk of heart disease.[7]

The Negative

Heavy drinking, often defined as the consumption of five or more drinks per day, poses a number of threats to the health of individual drinkers and often to other people as well. Although health can be damaged by the regular consumption of high amounts of alcohol, the ill effects of alcohol are most obvious in people with **alcoholism.**

Habitually high alcohol intakes and alcoholism increase the risk of developing high blood pressure, stroke, and dementia; throat, stomach, and bladder cancer; central nervous system disorders; and vitamin and mineral deficiency diseases.[1] Alcohol abuse is associated with a high proportion of deaths from homicide, drowning, fires, traffic accidents, and suicide (Illustration 14.2). It is also involved in a large proportion of rapes and assaults and can devastate families.[8] **Alcohol poisoning** from the consumption of a large amount of alcohol in a short period of time can cause death—and does to about 1,700 college students each year.[5] You can read more about the important topic of alcohol poisoning in the Health Action.

Long-term, excessive alcohol intake is related to the development of **steatohepatitis.** This condition is marked by inflammation, the build-up of fat in the liver (Illustration 14.3), and the eventual development of **cirrhosis.**

Illustration 14.1 Standard serving sizes of alcohol-containing beverages.
Standard servings shown each contain 0.6 oz (17g) of alcohol.

alcoholism
An illness characterized by a dependence on alcohol and by a level of alcohol intake that interferes with health, family and social relations, and job performance.

alcohol poisoning
A condition characterized by mental confusion, vomiting, seizures, slow or irregular breathing, and low body temperature due to the effects of excess alcohol consumption. It is life-threatening and requires emergency medical help.

steatohepatitis
Steatohepatitis (pronounced ste-at-oh-hep-ah-tie-tis) is a disease characterized by inflammation of, and fat accumulation in the liver. It is associated with alcoholism and may occur in obesity and diabetes. Steatohepatitis may progress to cirrhosis.

cirrhosis
Cirrhosis (pronounced sear-row-sis) is a disease of the liver characterized by widespread fibrous tissue buildup and disruption of normal liver structure and function. It can be caused by a number of chronic conditions that affect the liver.

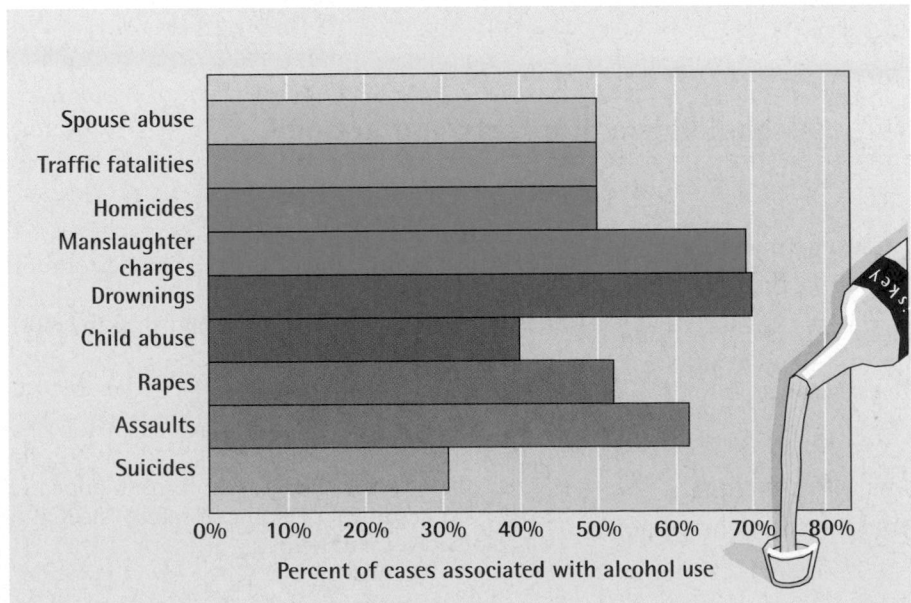

Illustration 14.2 Violence and injuries associated with alcohol.

Source: National Institute on Alcohol Abuse and Alcoholism, 2001, 2006.

Illustration 14.3. Normal liver tissue is shown in the left-hand photo. The photo on the right shows alcohol-related fatty liver tissue, or steatohepatitis.

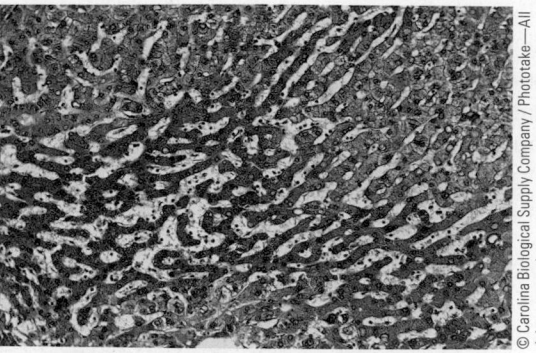

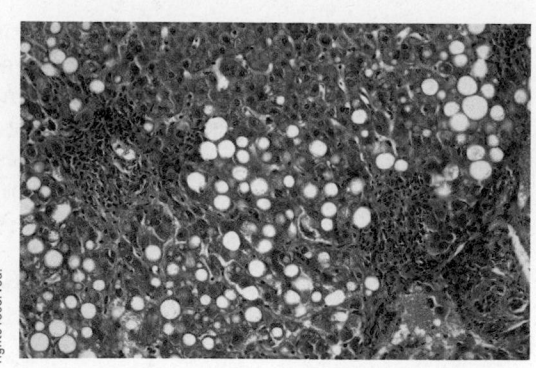

Illustration 14.4 Children with fetal alcohol syndrome experience growth and mental retardation, in addition to specific facial characteristics. Alcohol-containing beverages must show a warning statement on labels.

Drinking during pregnancy may harm the fetus. Women who binge drink or drink regularly during pregnancy are at risk of delivering an infant with signs of fetal alcohol syndrome (Illustration 14.4). Children with fetal alcohol syndrome experience long-term growth and mental retardation. The severity of the condition depends on how much alcohol was consumed during pregnancy, whether the mother is genetically susceptible to adverse effects of alcohol, and if excessive intake occurred early or late in pregnancy. Because there is no known safe level of alcohol intake, it is recommended that women who are or may become pregnant not drink.[10]

Alcohol Intake, Diet Quality, and Nutrient Status

Alcohol provides seven calories per gram, making alcohol-containing beverages rather high in caloric content (Table 14.1). Because many alcohol-containing beverages provide calories and few or no nutrients, they are considered energy-dense, empty-calorie foods. On average, alcohol accounts for 3 to 9% of the caloric intake of U.S. adults who drink. The average goes up to around 50% among heavy drinkers.[11] Although beer, wine, and mixed drinks are known to contain alcohol and to provide calories, there exists some confusion about whether calories from alcohol contribute to weight gain. This issue is addressed in the "Reality Check."

Although diet quality tends to be better than average in moderate drinkers, as caloric consumption from alcohol-containing beverages increases, the quality of the diet generally decreases. Diets of heavy drinkers frequently provide too

---

Health Action   Alcohol Poisoning: Facts and Action[18]

Facts

- Alcohol poisoning is a serious and sometimes deadly result of drinking excessive amounts of alcohol, usually in an episode of binge drinking.
- A high level of alcohol in the blood depresses breathing, promotes choking and vomiting, and slows heart rate.
- Blood alcohol levels continue to rise after a person passes out.

CRITICAL SIGNS of alcohol poisoning include:

1. Mental confusion and loss of consciousness; sometimes the person cannot be roused.
2. Vomiting
3. Seizures
4. Slow or irregular breathing (10 seconds or more between breaths)
5. Off-colored skin (paleness or bluish skin color)

ACTION

1. CALL 911 for help if there is any suspicion of an alcohol overdose.
2. DO NOT wait for all the symptoms to appear.
3. DO NOT assume a person can safely "sleep it off" or that a cold shower will help.

Do alcohol calories count? Perhaps you've heard the popular opinion that alcohol intake does not increase the risk of obesity. Is that the same as the not-so-popular opinion of scientists?

Do your thoughts side with Pedro or Erik?
Answers on next page

Pedro:
I started drinking a beer at night over the summer, and my weight never changed.

Erik:
The six-pack around my abdomen is really a six-pack.

---

little thiamin, niacin, vitamins B_{12}, A and C, and folate.[1, 12] Deficiencies of nutrients, as well as direct, toxic effects of high levels of alcohol ingestion, produce most of the physical health problems associated with alcoholism. The lack of thiamin, for example, impairs the brain's utilization of glucose. When people with alcoholism initially withdraw from alcohol, the thiamin deficiency may result in "delirium tremens," a condition called the "DTs" by people who staff detoxification centers. People with delirium tremens experience convulsions and hallucinations and are severely confused. Thiamin injections are a key component of treatment for delirium tremens.[13] Because alcohol in excess is directly toxic to body tissues, consuming an adequate diet protects heavy drinkers from only some of the harmful effects of alcoholism.[14]

How the Body Handles Alcohol

Alcohol is easily and rapidly absorbed in the stomach and small intestine. Within minutes after it is consumed, alcohol enters the circulatory system and is on its way to the liver, brain, and other tissues throughout the body. Alcohol remains in blood and body tissues until it is broken down and used for energy or is converted into fat and stored. The process of converting alcohol into a source of energy takes several hours or more to complete, depending on the amount of alcohol consumed. Because of the lag time between alcohol intake and utilization, blood levels of alcohol build up as drinking continues (Table 14.2).

Table 14.1

Caloric value of common alcohol-containing beverages

	Serving	Calories
Beer, regular	12 oz	150
Beer, light	12 oz	110
Beer, dark	12 oz	168
Malt beverage	12 oz	225
80-proof liquor	1.5 oz	100
Wine, red	5 oz	105
Wine, white	5 oz	100
Wine cooler	12 oz	215

Table 14.2

Alcohol doses and estimated percent blood alcohol levels

Number of Drinks[a]	Percent Blood Alcohol by Body Weight					
	100 LB	120 LB	140 LB	160 LB	180 LB	200 LB
1	0.04	0.03	0.03	0.02	0.02	0.02
2	0.04	0.03	0.03	0.03	0.02	0.02
3	0.07	0.06	0.05	0.05	0.04	0.04
4	0.11	0.09	0.08	0.07	0.06	0.06
5	0.14	0.12	0.10	0.09	0.08	0.07
6	0.18	0.15	0.13	0.11	0.10	0.09

[a]Taken within an hour or so; each drink equal to $1/2$ ounce pure ethanol. Affects may vary based on food intake, sex, and other factors.
Source: University of Oklahoma Police Dept., Blood Alcohol Calculator, www.ou.edu/oupd/bac.htm, accessed 8/06.

Do alcohol calories count? Maybe you've never seen an obese person with alcoholism, so you're tempted to think alcohol calories don't count. Chronic alcohol abuse is associated with weight loss and muscle wasting, even though the calorie intake of heavy drinkers is high. The effect appears to be due to an inhibition of fat tissue accumulation. The calories do count for light and moderate drinkers, however.[1]

Pedro:

Erik:

The intoxicating effects of alcohol correspond to blood alcohol levels. A drink or two in an hour raises blood levels of alcohol to approximately 0.03% in most people who weigh about 140 pounds. Blood alcohol levels of 0.03% correspond to mild intoxication. At this level, people lose some control over muscle movement and have slowed reaction times and impaired thought processes. A person's ability to drive or operate equipment in a safe manner is decreased at this level of blood alcohol content (Illustration 14.5). Blood alcohol levels of around 0.06% are associated with an increased involvement in traffic accidents. The legal limit for intoxication according to all states' highway safety ordinances is 0.08%—beyond the point where driving is impaired. When blood alcohol content increases to 0.13%, speech becomes slurred, "double vision" occurs, reflexes are dulled, and body movements become unsteady. If blood alcohol level continues to increase, drowsiness occurs and people may lose consciousness. Levels of blood alcohol above 0.6% can cause death.[5]

A given amount of alcohol intake among women produces higher blood levels of alcohol than for men of the same body weight. Pound for pound, women's bodies contain less water than men's bodies, so blood alcohol levels in women increase faster than in men.[1]

Over 150 medications, including sleeping pills, antidepressants, and painkillers, interact harmfully with alcohol. Combining three or more drinks per day with aspirin or nonaspirin pain relievers (acetaminophen, ibuprofen) may cause stomach ulcers or liver damage.[16]

How to Drink Safely if You Drink Many of the problems related to alcohol intake can be prevented by not drinking or by drinking responsibly. That means:

- Not drinking if you are or could become pregnant.

- Not drinking on an empty stomach (which can make you intoxicated surprisingly fast).

- Slowly sipping rather than gulping drinks.

- Limiting alcohol to an amount that doesn't make you lose control over your mind and body.

- Never driving a car or boat, hunting, or operating heavy equipment while under the influence of alcohol.

What Causes Alcoholism?

One in 13 adults in the United States abuse alcohol or has alcoholism.[18] Alcoholism tends to run in families (about half of alcohol-dependent people have a family history of the disease), and there is a documented genetic component to alcoholism.[19] Its development is also influenced by environmental factors. In general, the younger individuals are when they begin to drink, the greater likelihood that

Illustration 14.5   The legal limit for intoxication is 0.08%—beyond the point where driving is impaired.

Illustration 14.6   The younger a person is when drinking begins, the higher the probability that a drinking problem will develop.

they will develop a drinking problem at some point in life.[9] Individuals who begin drinking before the age of 15, for example, are four times more likely to become alcohol dependent than are people who do not drink before age 21 (Illustration 14.6). Close association with friends or peers who drink, high levels of stress, and availability of alcohol may also increase the risk of alcoholism. Television ads depicting youth-oriented parties, fun, and beer may increase underage drinking.[20]

Alcohol Use Among Adolescents   Alcohol use among adolescents is increasing, and the age when teens begin drinking is going down. Underage drinking accounts for 20% of all the alcohol consumed in the United States. The average age when teens begin drinking is now 14 years. These trends are particularly disturbing because they may lead to higher rates of alcoholism and alcohol-related problems in the near future.[21] Reduction in alcohol intake by adolescents is a major public health initiative of the Health Objectives for the Nation.

Help for Alcohol Dependence   Alcoholism is a chronic disease that can be successfully managed but not always cured.[2] Many options for treatment are available and a number of them can be accessed through the Web sites listed in the Media Menu for this unit. Treatments generally involve behavioral therapy, medications, or both. Behavioral therapy is successful in about one-third of people with alcoholism. Medications now available successfully treat alcohol dependency in certain individuals with a genetic predisposition toward developing the disease. Other medications that are under development would act by reducing the craving for, and intake of, alcohol among individuals without a genetic predisposition for the disease.[2]

NUTRITION

© Stewart Cohen / Getty Images/ Riser

Up Close

Effects of Alcohol Intake

Focal Point: Estimating blood alcohol levels and side effects.

Scenario: Ligia and Mark attend a wedding reception. Prior to the meal, they both drink a glass of champagne to toast the bride. Fifteen minutes later, they drink another glass to toast the groom. Ligia weighs 140 pounds and Mark 180.

Feedback (answers to these questions) can be found at the end of Unit 14.

Questions: Using the information in Table 14.2 and the information on "How the Body Handles Alcohol" (p. 14–5), answer the following questions:

A. After two glasses of champagne, what would be Ligia's estimated blood alcohol level? _____% blood alcohol. What would be Mark's? _____% blood alcohol

B. List three side effects of 0.03% blood alcohol content:
   1)   2)   3)

[Key Terms

alcoholism, page 14-3
alcohol poisoning, page 14-3

cirrhosis, page 14-3
chronic inflammation, page 14-2

fermentation, page 14-2
steatohepatitis, page 14-3

[Review Questions

TRUE FALSE

1. Moderate consumption of alcohol-containing beverages decreases the risk of heart disease. ☐ ☐

2. A "moderate" intake of alcohol-containing beverages is considered to be two standard servings daily for men and one for women. ☐ ☐

3. Alcohol is considered a food because it is a good source of a variety of vitamins and minerals. ☐ ☐

TRUE FALSE

4. Alcohol is used by the body for energy or is converted to glycogen. ☐ ☐

5. Alcohol poisoning represents an emergency situation requiring medical care. ☐ ☐

6. The idea that alcohol is absorbed more slowly if you drink after you eat or while eating is a myth. ☐ ☐

7. Some people are genetically susceptible to developing alcoholism. ☐ ☐

Media Menu

www.niaaa.nih.gov
The National Institute on Alcohol Abuse and Alcoholism offers this site for exploration of alcohol and health issues, quick facts, college drinking prevention programs, current research projects, and answers to common questions on alcohol abuse.

www.MayoClinic.com
Search "alcohol" and get the latest information on alcoholism, treatment programs, taking control, pros and cons of alcohol use, and an alcohol quiz.

www.samhsa.gov
This site from the national Substance Abuse and Mental Health Services Administration offers a Quick Guide to Finding Effective Alcohol and Drug Addiction Treatment resources. You may find this site useful if you or someone you care about is dependent on alcohol or drugs and needs treatment. You can also gain access to a four-page brochure titled Alcohol and Drug Treatment: How It Works, and How It Can Help You from this site.

www.aa.org
Information about services for alcoholism available from Alcoholics Anonymous World Service is available at this address.

www.al-anon.org
Al-Anon/Alateen alcohol treatment services can be found at this address.

www.ou.edu/oupd/bac.htm
The University of Oklahoma Police Department has developed an online tool for estimating blood alcohol level based on alcohol consumption and other factors.

www.e-chug.com
The e-CHUG is an evidence-based, online alcohol intervention and personalized feedback tool developed by counselors and psychologists at San Diego State University. It provides personalized information about an individual's drinking level and risk factors.

www.nlm.nih.gov/medlineplus
Search the word *alcohol* and be greeted by a large selection of topics related to alcohol such as women and alcohol, fetal alcohol syndrome, and alcohol and youth.

Notes

1. Ferreira MP, Weems MKS. Alcohol consumption by aging adults in the United States: health benefits and detriments. J Am Diet Assoc 2008;108:1668–76.

2. Heilig M. Triggering addiction. The Scientist 2008;Dec:28–35.

3. O'Keefe JH et al. Advantages and disadvantages of alcohol intake on cardiovascular health, J Am Coll Cardiol. Published online August 23, 2007, accessed from www.medscape.com/viewarticle/562354.

4. McClelland RL et al. Alcohol and coronary artery calcium prevalence, incidence, and progression: results from the Multi-Ethnic Study of Atherosclerosis (MESA). Am J Clin Nutr 2008;88:1593–601.

5. Sacanella E at al. Down-regulation of adhesion molecules and other inflammatory biomarkers after moderate wine consumption in healthy women: a randomized trial. Am J Clin Nutr 200;86:1463–9.

6. Ovaskainen M-L et al. Dietary intake and major food sources of polyphenols in Finnish adults. J Nutr 2008;138:562–9.

7. Goldberg IJ. To drink or not to drink? N Engl J Med 2003;348:163–64.

8. Markel H. Dying for a drink: alcohol on campus. Medscape Public Health and Prevention 2006;4(1), 6/07/2006.

9. Hingson RW. College-age and under-age drinking, American Society of Addiction Medicine 39th Annual Medical-Scientific Conference. April 10–13, 2008, accessed from www.medscape.com/viewarticle/573540.

10. Sood B et al. Prenatal alcohol exposure and childhood behavior at age 6 to 7 years: I. Dose-response effect. Pediatrics 2001;108(2), available at www.pediatrics.org/cgi/content/full/108/2/e34, accessed 12/08.

11. Kesse E et al., Do eating habits differ according to alcohol consumption? Am J Clin Nutr 2001;74:322–7.

12. Gibson A et al. Alcohol increases homocysteine and reduces B vitamin concentration in healthy male volunteers—a randomized, crossover intervention study. QJM 2008;101:881–7.

13. Lieber CS. The influence of alcohol on nutritional status. Nutr Rev 1988;46:241–54.

14. Grant LP et al. Nutrition education is positively associated with substance abuse treatment program outcomes. J Am Diet Assoc 2004;104:604–10.

15. Charness ME, Simon RP, Greenberg DA. Ethanol and the nervous system. N Engl J Med 1989;321:442–53.

16. Alcohol and your health. Weighing the pros and cons. www.mayoclinic.com/health/alcohol/SC00024, accessed 8/06.

17. Franzen J. How do you determine the alcohol proof of a given liquid? MadSci Network: Chemistry, 2/98.

18. Alcoholism: getting the facts. National Institute on Alcohol Abuse and Alcoholism. www.niaa.nih.gov, accessed 6/03.

19. Moss HB. NESARC findings changing understanding of alcoholism, www.medscape.com/viewarticle584983, accessed 12/08.

20. Alcohol Alert. Underage drinking. www.niaaa.nih.gov, accessed 1/06.

21. Alcohol consumption and expenditures for underage drinking and adult excessive drinking. Alcoholism & Drug Abuse Weekly 2003;15(9):3–4.

NUTRITION | # Up Close

Effects of Alcohol Intake

Feedback for Unit 14

A. Ligia's estimated % blood alcohol = 0.03%.
   Mark's estimated % blood alcohol = 0.02%.
B. Three side effects of these blood alcohol levels:
   1. Loss of some control over muscle movements
   2. Slowed reaction time
   3. Impaired thought processes

Proteins and Amino Acids

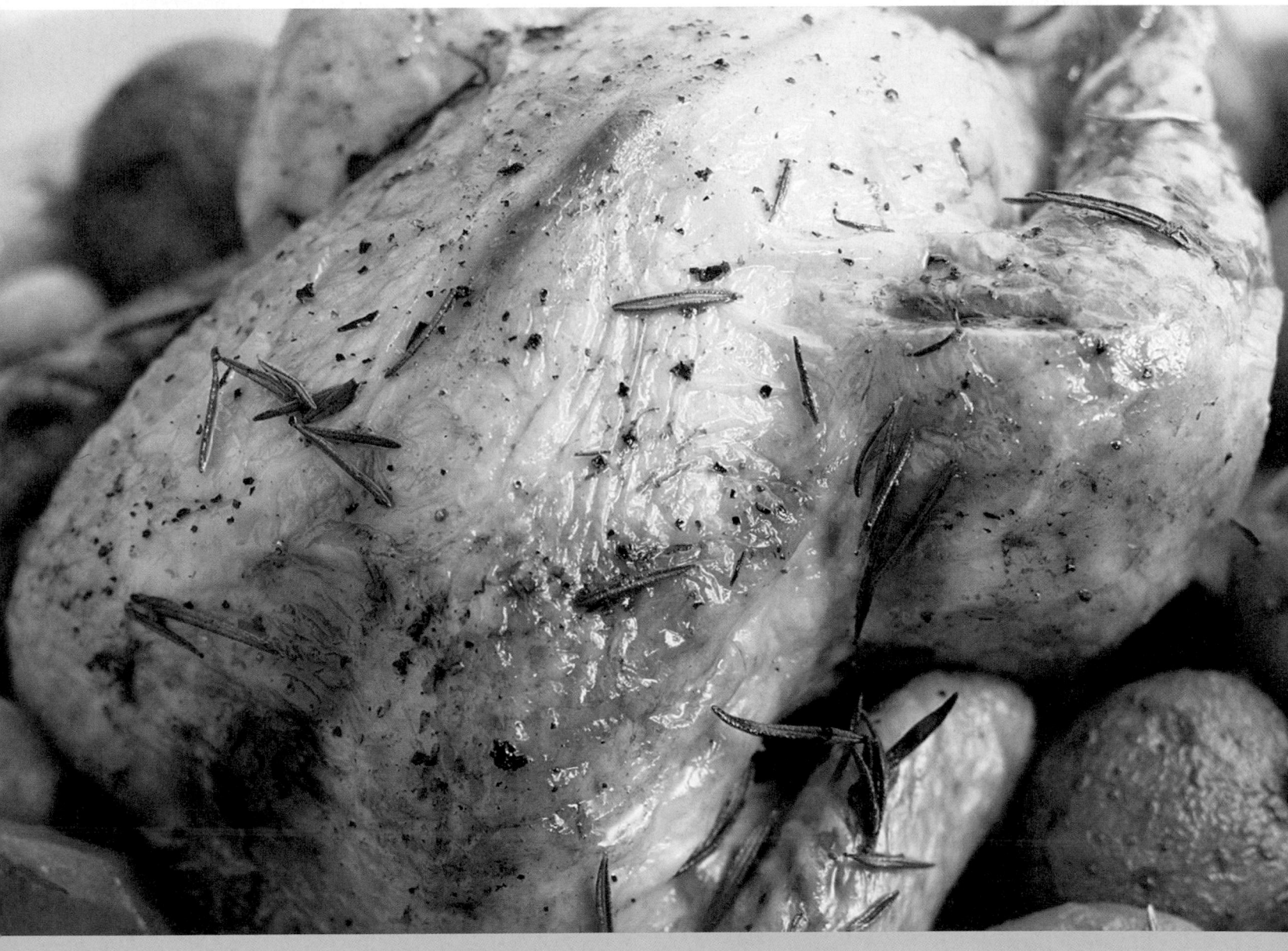

NUTRITION SCOREBOARD

	TRUE	FALSE
1 The primary function of protein is to provide energy.		
2 "Nonessential amino acids" are not required for normal body processes. Only "essential amino acids" are.		
3 High-protein diets and amino acid supplements by themselves increase muscle mass and strength.		

Key Concepts and Facts

- Proteins are made of amino acids. Some amino acids are "essential" (required in the diet), and some are "nonessential" (not a required part of diets).

- Although protein can be used for energy, its major functions in the body involve the construction, maintenance, and repair of protein tissues.

- Protein tissue construction in the body proceeds only when all nine essential amino acids are available.

- Appropriate combinations of plant foods can supply sufficient quantities of all the essential amino acids.

---

protein
Chemical substance in foods made up of chains of amino acids.

hormone
A substance, usually a protein or steroid (a cholesterol-derived chemical), produced by one tissue and conveyed by the bloodstream to another. Hormones affect the body's metabolic processes such as glucose utilization and fat deposition.

immunoproteins
Blood proteins such as antibodies that play a role in the functioning of the immune system (the body's disease defense system). Antibodies attack foreign proteins.

Illustration 15.1 The protein perception.

Protein

Other nutrients

Photo Disc

Protein's Image versus Reality

The term **protein** is derived from the Greek word *protos*, meaning "first." The derivation indicates the importance ascribed to this substance when it was first recognized. Protein is an essential structural component of all living matter and is involved in almost every biological process in the human body. Protein has a very positive image (Illustration 15.1). It's so positive that you don't have to talk about the importance of protein—people are already convinced of it.

Rich or poor, nearly all people in the United States get enough protein in their diets. Actually, most people consume more protein than they need. Average intakes of protein exceed the Recommended Dietary Allowance (RDA) level for all age and sex groups. Approximately 15% of total calories in the average U.S. adult diet are supplied by protein.[2]

High-protein intakes are generally accompanied by high-fat and low-fiber intakes. That's because foods high in protein such as hamburger, cheese, nuts, and eggs are high in fat and contain little or no fiber. Even lean meats provide a considerable proportion of their total calories as fat (Illustration 15.2).

Functions of Protein

Proteins perform four major functions in the body (Table 15.1). They are an integral structural component of skeletal muscle, bone, connective tissues (skin, collagen, and cartilage), organs (such as the heart, liver, and kidneys), red blood cells and hemoglobin, hair, and fingernails. Proteins are the basic substance that make up thousands of enzymes in the human body; they are a major component of **hormones** such as insulin and growth hormone, and serve as other substances that perform important biological functions. Tissue maintenance and the repair of organs and tissues damaged due to illness or injury are functions of different types of protein. Albumin, a protein made by the liver, is the blood's "tramp steamer." It attaches to and transports fatty acids, calcium, and other substances through the circulatory system to cells throughout the body.[3] Finally, protein serves as an energy source at the level of four calories per/gram.

The body of a 154-pound man contains approximately 24 pounds of protein. Nearly half of the protein is found in muscle, while the rest is present in the skin, collagen, blood, enzymes, and **immunoproteins**; organs such as the heart, liver,

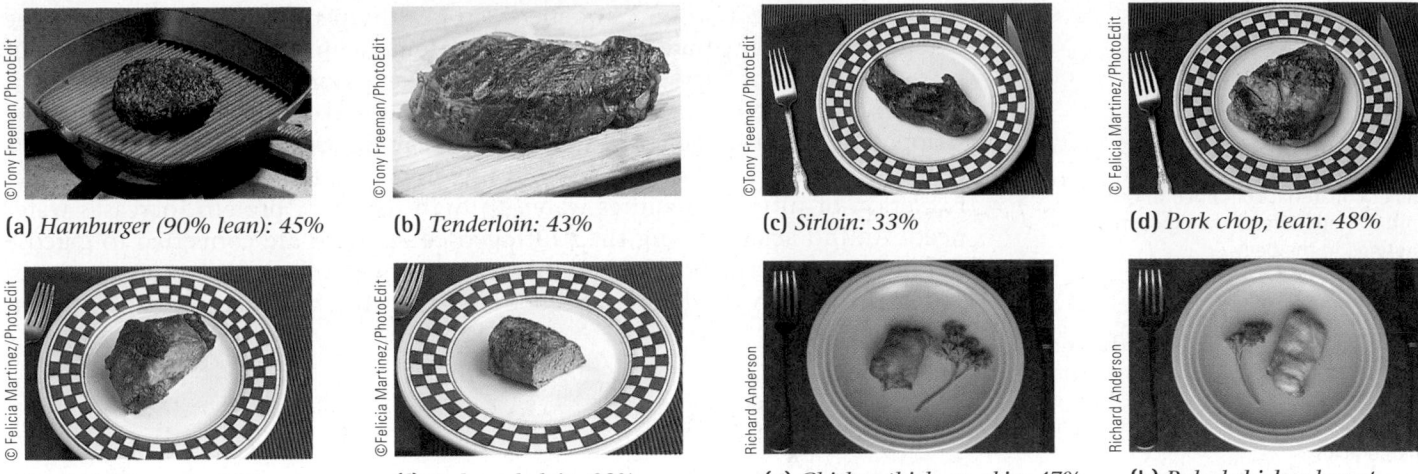

(a) *Hamburger (90% lean): 45%* **(b)** *Tenderloin: 43%* **(c)** *Sirloin: 33%* **(d)** *Pork chop, lean: 48%*

(e) *Pork loin roast: 36%* **(f)** *Pork tenderloin: 28%* **(g)** *Chicken thigh, no skin: 47%* **(h)** *Baked chicken breast, no skin: 19%*

Illustration 15.2   The fat content of three-ounce portions of "lean" meats. The percentage of calories from fat is indicated for each portion. (A three-ounce portion of meat is about the size of a deck of cards.) Each portion of meat provides approximately 21 grams of protein.

and intestines; and other body parts. All protein in the body is continually being turned over, or broken down and rebuilt. This process helps maintain protein tissues in optimal condition so they continue to function normally. The process of protein turnover utilizes roughly nine ounces of protein each day. Yet, we consume only two to three ounces of protein daily. Most of the protein used for maintenance is recycled from muscle and other protein tissues being turned over. Proteins play key roles in the repair of body tissues by serving as substances—such as fibrin—that help blood clot (Illustration 15.3) and by replacing tissue proteins damaged by illness or injury.[4]

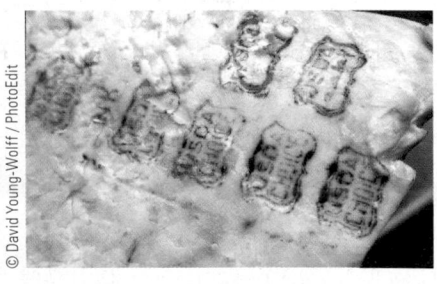

Table 15.1

Functions of protein

1. Serves as a structural material in muscles, connective tissue, organs, and hemoglobin
2. Serves as the basic component of enzymes, hormones, and other biologically important chemicals
3. Maintains and repairs protein-containing tissues
4. Serves as an energy source

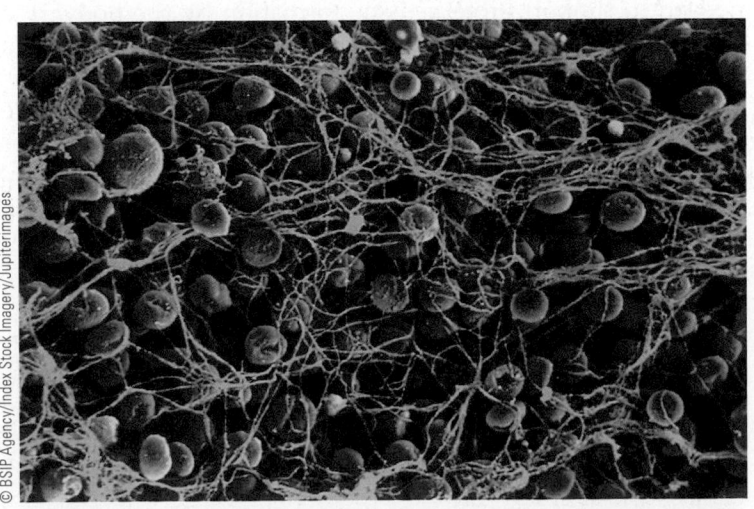

Illustration 15.3   Red blood cells enmeshed in fibrin in a color-enhanced microphotograph. Red blood cells and fibrin (which helps stop bleeding by causing blood to clot) are made primarily from protein.

Protein serves as a source of energy in healthy people, but not nearly to the extent that carbohydrates and fats do. Protein is unlike carbohydrate and fat in that it contains nitrogen and does not have a storage form in the body. In order to use protein for energy, amino acids that make up proteins must first be stripped of their nitrogen. The free nitrogen can be used as a component of protein formation within the body; or, if present in excess, it is excreted in urine. Excretion of nitrogen requires water, so high intake of protein increases water need. Amino acids missing their nitrogen component are converted to glucose or fat that then can be used to form energy. A small amount of protein (1%) can be obtained from the liver and blood and used to cover occasional deficits in protein intake.[5]

Amino Acids

The "building blocks" of protein are amino acids. Protein consumed in food is broken down by digestive enzymes and absorbed into the bloodstream as amino acids. There are 20 common amino acids (Table 15.2) that form proteins when linked together. Every protein in the body is composed of a unique combination of amino acids linked together in chains (Illustration 15.4). The organization of amino acids into the chains is orchestrated by **DNA**, the genetic material within each cell that directs protein synthesis. Once formed, the chains of amino acids may fold up into a complex shape. Some proteins are made of only a few amino acids, while other proteins contain over 2,500. Whatever the number of amino acids, the specific amino acids involved and their arrangement determine whether the protein is an enzyme, a component of red blood cells, a muscle fiber, or another tissue made from protein.

Nine of the 20 common amino acids are considered **essential**, and 11 are **nonessential**. Despite the labels, all 20 amino acids are required to build and maintain protein tissues. The essential amino acids are called "essential" because the body cannot produce them, or produce enough of them, so they must be provided by the diet. Our bodies can produce nonessential amino acids, so we don't require a dietary source of them. Proteins in foods contain both essential and nonessential amino acids.

Proteins Differ in Quality

The ability of proteins to support protein tissue construction in the body varies depending on their content of essential amino acids. How well dietary proteins support protein tissue construction is captured by tests of the protein's "quality."

Proteins of high quality contain all the essential amino acids in the amounts needed to support protein tissue formation by the body. If any of the essential amino acids are missing in the diet, proteins are not formed—even those proteins that could be produced from available amino acids. Shutting off all protein formation for want of an amino acid or two may appear inefficient; but if the

DNA (deoxyribonucleic acid)
Genetic material contained in cells that initiates and directs the production of proteins in the body.

essential amino acids
Amino acids that cannot be synthesized in adequate amounts by humans and therefore must be obtained from the diet. They are sometimes referred to as "indispensable amino acids."

nonessential amino acids
Amino acids that can be readily produced by humans from components of the diet. Also referred to as "dispensable amino acids."

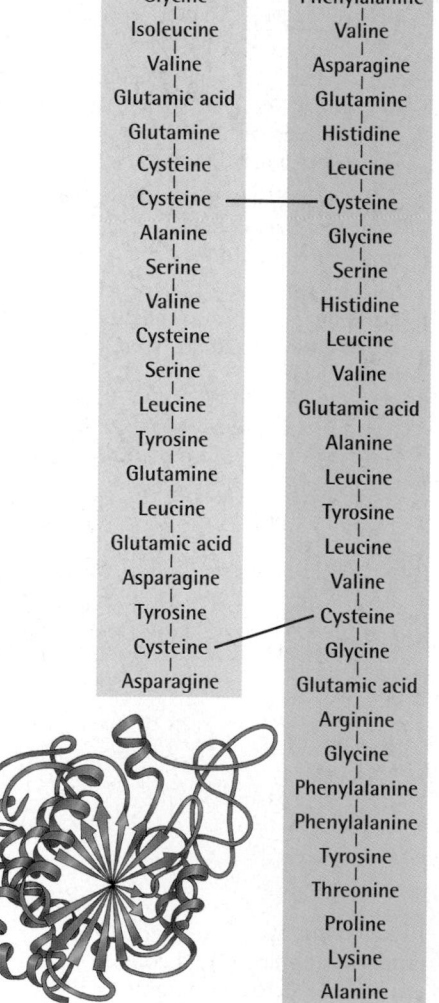

Illustration 15.4 Amino acid chains in the protein insulin and diagram of the structure of insulin.

Table 15.2

Essential and nonessential amino acids

Essential		Nonessential	
Histidine	Threonine	Alanine	Glutamine
Isoleucine	Tryptophan	Arginine	Glycine
Leucine	Valine	Asparagine	Proline
Lysine		Aspartic acid	Serine
Methionine		Cysteine	Tyrosine
Phenylalanine		Glutamic acid	

body did not cease all protein formation, cells would end up with an imbalanced assortment of proteins. This would seriously affect cell functions. When the required level of an essential amino acid is lacking, the remaining amino acids are primarily used for energy.

Amino acids cannot be stored very long in the body, so we need a fresh supply of essential amino acids daily. This means we need to consume foods that provide a sufficient amount of all essential amino acids every day.

Complete Proteins   Food sources of high-quality protein (meaning they contain all the essential amino acids in the amount needed to support protein formation) are called **complete proteins**. Proteins in this category include those found in animal products such as meat, milk, and eggs (Illustration 15.5). **Incomplete proteins** are deficient in one or more essential amino acids. Proteins in plants are "incomplete," although soybeans are considered a complete source of protein for adults.[1] (Soybeans may not meet the essential amino acid requirements of young infants.) You can "complement" the essential amino acid composition of plant sources of protein by combining them to form a "complete" source of protein. Illustration 15.6 shows a few complementary plant food combinations that produce complete proteins.

Vegetarian Diets   Diets consisting only of plant foods can provide an adequate amount of complete proteins. The key to success is eating a variety of complementary sources of protein each day. Vegetarian diets have been practiced for centuries by some religious and cultural groups, bearing testimony to their general adequacy and safety. (Unit 16 on vegetarian diets expands on this topic.)

Amino Acid Supplements

Because amino acids occur naturally in foods, people often assume they are harmless, no matter how much is taken (Illustration 15.7). Researchers have known for decades, however, that high intakes of individual amino acids can harm health. High amounts may disrupt normal protein production by overwhelming cells with a surplus of some amino acids and a relative deficit of

Illustration 15.5   Animal sources of protein supply "complete proteins."
Each food shown is a source of complete protein.

complete proteins
Proteins that contain all of the essential amino acids in amounts needed to support growth and tissue maintenance.

incomplete proteins
Proteins that are deficient in one or more essential amino acids.

Illustration 15.6   Each of these combinations of plant foods provides complementary protein sources.

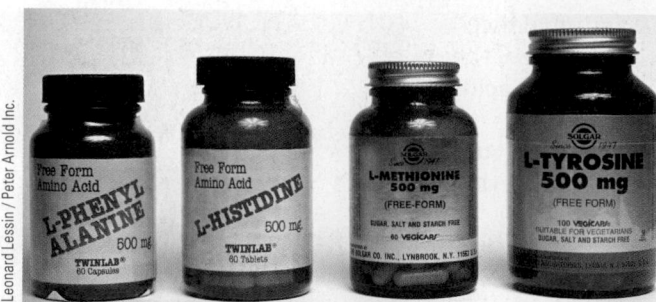

Illustration 15.7 A wide array of amino acid supplements are available over the counter, but the safety of these supplements is unknown.

others.[2] Excess consumption of the amino acid methionine, in particular, causes a host of problems. Intake levels two to five times above the amount normally consumed from foods worsen the symptoms of schizophrenia, promote hardening of the arteries, impair fetal and infant development, and lead to nausea, vomiting, bad breath, and constipation.[8, 9] Adverse health effects associated with cysteine and phenylalanine supplements have also been reported.[11,12] Use of amino acid supplements should be supervised by a physician.

Safe doses of certain amino acids or their derivatives are being used to manage certain conditions. For example, melatonin (a derivative of the essential amino acid tryptophan), is used to promote sleep.[13,14] Melatonin supplements have been found to reset the sleep clock in shift-workers, pilots, jet-lagged travelers, and people with sleep disorders.

The biggest users of amino acids supplements are athletes who believe they will increase muscle mass and strength.[12]

Amino Acid Supplements and Muscle Mass You can't just consume amino acids or protein powders and watch your muscles grow (no matter how convincing the ads that sell such products are). If that happened, everyone who wanted a rippled stomach and bulging triceps could have them. Neither essential amino acids nor protein supplements by themselves increase muscle size and strength. Muscle size and strength are built slowly from the raw ingredients of a healthy diet and resistance training (Illustration 15.8).[15] Current research indicates, however, that you may be able to enhance the build-up in muscle mass by consuming 20 grams of high quality protein immediately following resistance exercise workouts.[16,17] Skim milk, lean meat, fish, eggs, and beans and rice are examples of high quality protein foods that could be consumed. Protein intakes over 20 grams following exercise may not offer additional benefit.[16]

As is the case for other dietary supplements, the purity, dose, and safety of supplements available on the Web and in stores is not guaranteed (Illustration 15.9). Long-term effectiveness and safety studies are needed to take all of the worry out of using these supplements.

Illustration 15.8 Building muscles like these takes time, a good diet, and lots of resistance exercise.

Food as a Source of Protein

The average intake of protein of adults in the United States is 82 grams per day, exceeding the RDA for men of 56 grams and that for women of 46 grams.[2] Approximately 70% of the protein consumed by Americans comes from meats, milk, and other animal products.[4] Dried beans and grains are not as well known for their protein, but are nevertheless good sources (Table 15.3). Plant sources of protein are generally low in fat, making them a wise choice for consumers who are trying to limit their intake of fat. Nearly all food sources of protein provide an assortment of vitamins and minerals as well. Beef and pork are particularly good sources of iron, a mineral often lacking in the diets of women (Table 15.4).

What Happens When a Diet Contains Too Little Protein?

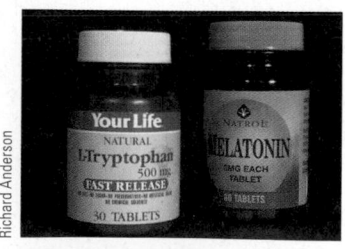

Illustration 15.9 Tryptophan supplements, although banned in the United States and other countries, are available on the Web and by prescription. Melatonin is a derivative of tryptophan and can be sold.

Protein deficiency can occur by itself or in combination with a deficiency of calories and nutrients. Because food sources of protein generally contain essential nutrients such as iron, zinc, vitamin B$_{12}$, and niacin, diets that produce protein deficiency usually cause a variety of other deficiencies, too. Protein does not generally serve as an important source of energy, but body protein will be used as a major energy source during starvation. To meet the need for energy, the body

Pure protein   You've got a nutrition exam coming up, so you and your classmate Carole have gotten together to study. You get into a discussion about food sources of protein that goes like this:

Who gets the "thumbs up"?
Answers on next page

Carole:
Lean meats are the best source. They're pure protein!

Lauren:
Pure protein? Even the driest, toughest meats contain more than protein.

Table 15.3

Food sources of protein

The adult RDA is 46 grams for women aged 19 to 30 years and 56 grams for men aged 19 to 30 years.

Food	Amount	Grams	Protein Content Percentage of Total Calories
Animal products			
Tuna (water packed)	3 oz	24	89%
Shrimp	3 oz	11	84
Cottage cheese (low-fat)	1/2 cup	14	69
Beef steak (lean)	3 oz	26	60
Chicken (no skin)	3 oz	24	60
Pork chop (lean)	3 oz	20	59
Beef roast (lean)	3 oz	23	45
Skim milk	1 cup	9	40
Fish (haddock)	3 oz	19	38
Leg of lamb	3 oz	22	37
Yogurt (low-fat)	1 cup	13	34
Hamburger (lean)	3 oz	24	34
Egg	1 medium	6	32
Swiss cheese	1 oz	8	30
Sausage (pork links)	3 oz	17	28
2% milk	1 cup	8	26
Cheddar cheese	1 oz	7	25
Whole milk	1 cup	8	23
Dried beans and nuts			
Tofu	1/2 cup	14	38
Soybeans (cooked)	1/2 cup	10	33
Split peas (cooked)	1/2 cup	5	31
Lima beans (cooked)	1/2 cup	6	27
Dried beans (cooked)	1/2 cup	8	26
Peanuts	1/2 cup	9	17
Peanut butter	1 tbs	4	17
Grains			
Corn	1/2 cup	3	29
Egg noodles (cooked)	1/2 cup	4	25
Oatmeal (cooked)	1/2 cup	3	15
Whole-wheat bread	1 slice	2	15
Macaroni (cooked)	1/2 cup	3	13
White bread	1 slice	2	13
White rice (cooked)	1/2 cup	2	11
Brown rice (cooked)	1/2 cup	2	10

Table 15.4

Iron content in a 3-ounce serving of various meats

The RDA for women aged 19 to 30 years is 15 milligrams. The RDA for men aged 19 to 30 years is 10 milligrams.

Meat	Iron Content (mg)
Pork chop (lean)	3.4
Round steak (lean)	3.1
Hamburger (lean)	3.0
Shrimp	2.6
Tuna	1.6
Baked chicken (no skin)	1.4
Lamb (lean)	1.3

Pure protein   Some people think muscle and lean meat consist only of protein. They don't. By weight, lean cooked sirloin steak is 29% protein, 8% fat, and 62% water. Lean pork is 29% protein, 9% fat, and 61% water.

Carole:

Lauren:

will extract protein from the liver, intestines, heart, muscles, and other organs and tissues. Loss of more than about 30% of body protein results in reduced body strength for breathing, susceptibility to infection, abnormal organ functions, and death.[16] Inadequate protein intake is related to decreased growth in children and loss of muscle mass and strength in adults.[18,19]

In the past it was thought that **kwashiorkor**, a devastating disease that can affect severely undernourished children, was primarily due to a protein deficiency. Although kwashiorkor is related to protein-calorie malnutrition, the disease does not appear to be due to a lack of protein. This conclusion is supported by studies that show that improving protein intake in children with kwashiorkor does not correct the disease. It appears that some children with protein-calorie malnutrition develop kwashiorkor due to an inability to utilize protein and fat normally during starvation.[20,21]

kwashiorkor
A form of severe protein-energy malnutrition in young children. It is characterized by swelling, fatty liver, susceptibility to infection, profound apathy, and poor appetite. The cause of kwashiorkor is unclear.

How Much Protein Is Too Much?

Adults can consume a substantial amount of protein—approximately 35% of total calories—for months at a time without ill effects. This observation is based on studies of the diets of Eskimos, explorers, trappers, and hunters in northern America. The very high-protein diets would generally contain a good deal of fat in the form of whale blubber, lard, or fat added to dried meat. Consumption of 45% of total calories from protein is considered too high and is related to the development of symptoms such as nausea, weakness, and diarrhea. Diets very high in protein result in death after several weeks. A complex disease resulting from excess protein intake was termed "rabbit fever" after it occurred in trappers attempting to exist on wild rabbit only.[4]

High-protein diets have been implicated in the development of weak bones, kidney stones, cancer, heart disease, and obesity. The National Academy of Sciences has concluded that the risk of such disorders does not appear to be increased among individuals consuming 10 to 35% of total calories from protein, and on average adults consume 15%.[2]

© Envision/Corbis

NUTRITION | Up Close

My Protein Intake

Focal Point: Determine the amount of protein in your diet yesterday.

For *each serving* of a food item you ate yesterday, write the grams of protein the food contains in the corresponding blank. For example, a standard serving of meat is three ounces (about the size of the palm of your hand or a deck of cards). If you had *one* three-ounce pork chop yesterday, write *20 grams* in the corresponding blank. If you had *two* three-ounce pork chops, write *40 grams*. If a protein food you ate yesterday is not included, choose the item on the list closest to it. Then, total the grams of protein you ate yesterday from both plant and animal sources. Finally, compare your protein intake with the RDA of 46 grams for women or 56 grams for men.

Food	One Serving	Protein in One Serving (grams)	Protein You Ate (grams)
Animal products			
Milk (whole)	1 c (8 oz)	8	_____
Yogurt	1 c (8 oz)	13	_____
Cottage cheese	1/2 c (4 oz)	14	_____
Hard cheese	1 oz	7	_____
Hamburger (lean)	3 oz	24	_____
Beef steak (lean)	3 oz	26	_____
Chicken (no skin)	3 oz	24	_____
Pork chop (lean)	3 oz	20	_____
Fish	3 oz	19	_____
Hot dog	1	6	_____
Sausage	3 oz	17	_____
		Subtotal from animal foods:	_____
Plant products			
Bread	1 slice	2	_____
Rice	1/2 c (4 oz)	2	_____
Pasta	1/2 c (4 oz)	3	_____
Cereals	1/2 c (4 oz)	3	_____
Vegetables	1/2 c (4 oz)	2	_____
Peanut butter	1 tbs	4	_____
Nuts	1/4 c (2 oz)	7	_____
Cooked beans (legumes)	1/2 c (4 oz)	8	_____
		Subtotal from plant foods:	_____
		Total grams of protein from plant and animal foods:	_____
		Amount above/below RDA:	_____

Special note: You can also calculate your protein intake using Cengage's Diet Analysis Plus software. Input your food intake for one day. Then go to the Analyses/Reports section to view the total number of grams of protein in your diet.

FEEDBACK can be found at the end of Unit 15.

Key Terms

complete proteins, page 15-5
DNA (deoxyribonucleic acid), page 15-4
essential amino acids, page 15-4
hormone, page 15-2

immunoprotein, page 15-2
incomplete proteins, page 15-5
kwashiorkor (kwa-she-or-kor), page 15-8
marasmus, page 15-8

nonessential amino acids, page 15-4
protein, page 15-2

Review Questions

TRUE FALSE

1. The only known function of protein is to serve as a structural component of muscle, bones, and other solid tissues. ☐ ☐

2. Low protein intake is a fairly common problem among adults and children in the United States. ☐ ☐

3. Immunoproteins are a part of the body immune system. ☐ ☐

4. Fibrin is a protein that helps blood clot. ☐ ☐

5. The nitrogen component of amino acids must be removed before they can be used for energy. ☐ ☐

6. There are 22 amino acids, and half of them are "essential." ☐ ☐

TRUE FALSE

7. DNA provides the blueprint for the production of specific proteins by the body. ☐ ☐

8. People need to eat animal sources of protein in order to consume enough high-quality "complete" proteins. ☐ ☐

9. Individual amino acids supplements are uniformly hazardous to health. ☐ ☐

10. Amino acid supplements plus resistance exercise are the keys to building muscle mass and strength. ☐ ☐

11. Kwashiorkor is the name of the protein deficiency disease. ☐ ☐

12. Protein intake should constitute 10 to 35% of total calorie intake. ☐ ☐

Media Menu

www.healthfinder.gov
Search topics related to protein on this site.

**www.fsis.usda.gov/Factsheets/
Inspection_&_Grading/index.asp**
accessed 3/09.
Interested in USDA's meat inspection and grading systems? Find out more about these topics at this site.

www.iom.edu
Select "Food & Nutrition" to gain access to the 2002 Dietary Reference Intake report on macronutrients and the chapter on Protein.

Notes

1. Phillips SM et al. Mixed muscle protein synthesis and breakdown after resistance exercise in humans. Am J Physiol 1997;273:E99–107.

2. What we eat in America, NHANES, 2005–2006, USDA, 2008, available at www.are.usda.gov/ba/bhnrc/fsrg.

3. Van den Akker CHP et al. Human fetal albumin synthesis rates during different periods of gestation, 2008;88:997–1003.

4. Dietary Reference Intakes. Energy, carbohydrate, fiber, fat, fatty acids, cholesterol, protein, and amino acids. Institute of Medicine, National Academy of Sciences, Washington, DC: National Academies Press; 2002.

5. Matthews DE. Proteins and amino acids. In: Modern nutrition in health and disease, 9th edition, Shils ME et al., eds. Philadelphia: Lippincott, Williams & Wilkins, 1998, pp. 11–48.

6. Barnes S. Nutritional genomics, polyphenols, diets, and their impact on dietetics, J Am Diet Assoc 2008; 108:1888–95.

7. High protein intake harms the kidneys. www.healthfinder.gov/high protein diet, accessed 9/03.

8. Garlick PJ. Toxicity of methionine in humans. J Nutr 2006;136:1722S–25S.

9. Jakubowski H. Pathophysiological consequences of homocysteine excess, J Nutr 2006;136:1741–9S.

10. USDA's meat inspection and grading systems, www.fsis.usda.gov/Factsheets/ Inspection_&_Grading/index.asp, accessed 3/09.

11. Digler RN et al. Excess dietary L-cysteine, but not L-cystine, is lethal for chicks but not for rats or pigs, J Nutr 2007;137:331–8.

12. Pencharz PB et al. An approach to defining the upper safe limits of amino acid intake, J Nutr 2009;138:1995S–2002S.

13. Kay LK. Melatonin. Today's Dietitian 2004: Nov., p. 54–6.

14. Circardian rhythms: going back in time, The Economist Dec. 16, 2008, p. 99.

15. Iglay HB et al. Resistance training and dietary protein: effects on glucose tolerance and contents of skeletal insulin signaling proteins in older persons, Am J Clin Nutr 2007; 85:1005–13.

16. Moore DR et al. Ingested protein dose response of muscle and albumin protein synthesis after resistance exercise in young men, Am J Clin Nutr 2009; 8:161–8.

17. Verdijk LB et al. Protein supplementation before and after exercise does not further augment skeletal muscle hypertrophy after resistance exercise training in elderly men, Am J Clin Nutr 2009;89:608–16.

18. Thalacker-Mercer AE et al. Inadequate protein intake affects skeletal muscle transcript profiles in older humans, Am J Clin Nutr 2007;85:1344–52.

19. Hass VK et al. Total body protein in healthy adolescent girls: validation of estimates derived from simpler measures with neutron activation analysis, Am J Clin Nutr 2007;85:66–72.

20. Amadi B et al. Reduced production of sulfated glycosaminoglycans occurs in Zambian children with kwashiorkor but not marasmus, Am J Clin Nutr 2009; 89:592–600.

21. Badaloo AV et al. Lipid kinetic differences between children with kwashiorkor and those with marasmus, Am J Clin Nutr 2006;83:1283–8.

NUTRITION | # Up Close

My Protein Intake

Feedback for Unit 15

Compare your subtotals to find out which protein source you prefer, plant or animal. Protein from animal products is often accompanied by fat. If you are concerned about calories and fat in your diet, choose plant protein sources more often. And, if you are similar to many Americans, your intake of protein will exceed the RDA by quite a bit.

UNIT 16

Vegetarian Diets

NUTRITION SCOREBOARD

	TRUE	FALSE
1 The human body developed to function best on a vegetarian diet.		
2 People who don't eat meat have more health problems than people who do.		
3 Macrobiotic diets cure some types of cancer.		
4 In order to consume enough high-quality protein, vegetarians need to consume combinations of plant foods that provide a complete source of protein at every meal.		

Key Concepts and Facts

- Vegetarianism is more than a diet. It is often part of a value system that influences a variety of attitudes and behaviors.

- Appropriately planned vegetarian diets are health-promoting.

- Vegetarian diets that lead to caloric and nutrient deficiencies generally include too narrow a range of foods.

Answers to NUTRITION SCOREBOARD	TRUE	FALSE
1 Early humans developed on an omnivorous diet (meat and plant diet). However, health can be fostered on an omnivorous or a vegetarian diet.		✔
2 Persons who consume appropriately planned vegetarian diets do not have more health problems than other people. Furthermore, such diets may reduce the risk of developing several chronic diseases.[1]		✔
3 No type of vegetarian or quasi-vegetarian diet, including macrobiotic and vegan diets, has been found to cure cancer.		✔
4 Vegetarians do need to plan their diets to ensure that they obtain enough high-quality protein, but they need to consume complete sources of protein daily, not at every meal. Appropriately planned vegetarian diets provide sufficient amounts of high-quality protein.		✔

Perspectives on Vegetarianism

... unless vegetarianism comes from the soul, it will just be a passing fad.
—Janis Barkas, 1978

It wouldn't be any fun to live without meat.... I do not believe in this meatless, dairyless, butterless society.
—Julia Child, 1991

Table 16.1

Reasons for vegetarianism[3,5]

- Lack of availability or affordability of animal products
- Desire not to cause harm to animals
- Religious beliefs
- Desire to "eat low on the food chain"
- Desire to preserve the world's food supply
- Health
- Desire to omit hormones, antibiotics, and possible contaminants in meats

Vegetarianism in the United States, Canada, and other economically developed countries is moving from the realm of counterculture to the mainstream.[1] Yet even with the increased acceptance of vegetarian diets, people tend to be for vegetarianism or against it, often without knowing much about it. Few of those opposed to vegetarianism have tried to learn about the vegetarian way of life or understand that vegetarian diets can be healthful. A small percentage of health professionals are vegetarians, and those who are not are often skeptical about how healthy a vegetarian diet can be. The possibility that something so different from the customary diet can be nourishing to the body may be rejected out of hand.

An objective look at vegetarianism reveals that both appropriately planned vegetarian diets and the lifestyle often followed by vegetarians can be very good for health.[2] Vegetarians tend to be health conscious; often avoid using alcohol, tobacco, and illicit drugs; and engage in regular physical activity.[3] Diet is usually one of several characteristics shared by people practicing particular types of vegetarianism.

Reasons for Vegetarianism

Worldwide, vegetarians number in the hundreds of millions.[4] Much of the world's population subsists on vegetarian diets because meat and other animal products are scarce or too expensive (Table 16.1). In other societies, people have the luxury of choosing a healthy assortment of food from an abundant and affordable food supply that includes a wide variety of items acceptable to vegetarians (Illustration 16.1). When food availability is not an issue, people tend to adopt vegetarian diets because of a desire to cause no harm to animals, a personal commitment to preserve the environment and the world's food supply by "eating low on the food chain," or a belief that animal products are unhealthful or unsafe. They may avoid animal products as part of a value or religious belief system. Others follow vegetarian diets to keep their weight down or to lower the risk of developing specific diseases such as diabetes or heart disease.[5]

Photo Disc

Vegetarian Diets Come in Many Types   There is no one vegetarian diet. People who consider themselves to be vegetarians range from those who eat all foods except red meat (mainly beef) to those who exclude all foods from animal sources, including honey. According to the American Vegetarian Society, a person is a vegetarian only if she or he eats *no* meat.

Vegetarian Diet Options   The types of foods included in common vegetarian diets are summarized in Table 16.2. The least restrictive form of vegetarian diet has been unofficially labeled "far vegetarian" because it excludes only red meats. This diet is very much like that consumed by omnivores, or meat and plant eaters. Quasi-vegetarian diets (also called semivegetarian and flexitarian diets) vary somewhat but generally exclude beef, pork, and poultry while including fish, eggs, dairy products, and plant foods. The lacto-ovo vegetarian diet includes only dairy foods, eggs, and plant foods. Individuals practicing this type of diet exclude all meats. The lacto-vegetarian diet, as the name implies, includes only milk and milk products and plant foods. Vegan (pronounced *vee-gun*) and raw food diets are a more restrictive type of vegetarian diet. Vegans and raw food dieters eat only plant foods; in addition, vegans may avoid honey and clothes made from wool, leather, or silk.

Yeah, I guess I'm kind of a vegetarian, too. I try not to eat meat—well, I love chicken but I don't eat red meat. I mean, I eat hamburgers. I can't remember the last time I had a steak.

—A 17-YEAR-OLD "VEGETARIAN"

Illustration 16.1   The growing selection of vegetarian foods in supermarkets.

Richard Anderson

Table 16.2

Vegetarian diets come in many types

Type of Diet	Beef Lamb, Pork ("Red Meat")	Poultry	Fish	Eggs	Milk and Milk Products	Plant Foods
"Far" vegetarian		x	x	x	x	x
Quasi-vegetarian[a]		x	x	x	x	x
Lacto-ovo vegetarian[b]				x	x	x
Lacto-vegetarian					x	x
Macrobiotic						x
Vegan						x
Raw food						x

[a]Quasi-vegetarian diets (also called semivegetarian) may include poultry; they tend to vary in the type of animal products consumed.
[b]Macrobiotic diets may include fish and other animal products.

What's in the bowl? It depends on what type of vegetarian diet you are following.

Photo Disc

■ **Macrobiotic Diets.** Macrobiotic diets fall somewhere between quasi-vegetarian and vegan diets. The formulation of macrobiotic diets has changed dramatically over the past few decades. In addition to brown rice, other grains, and vegetables, these diets now include fish, dried beans, spices, fruits, and many other types of foods. No specific foods are prohibited, and locally grown and whole foods are emphasized.

Persons adhering to the macrobiotic philosophy place value on consuming organic foods and balancing the intake of "yin" and "yang" foods. Foods are classed as yin or yang based on beliefs about the food's relationship to the emotions and the physical condition of the body. Yin foods such as corn, seeds, nuts, fruits, and leafy vegetables are considered negative, dark, cold, and feminine. Yang foods represent opposing positive forces of light, warmth, and masculinity. Poultry, fish, eggs, and cereal grains such as buckwheat are yang foods.[6]

■ **Raw Food Diets.** Diets consisting primarily of raw foods have come in and out of fashion for hundreds of years. Raw food diets currently in vogue center on the health benefits of plant-based diets and the belief that raw foods provide enzymes helpful for digestion. Enzymes in raw foods do not promote normal digestion, however. Enzymes in food are inactivated by stomach acid and by digestive system enzymes that break down proteins.[7] Although there is no formal definition, raw food diets are usually described as an uncooked vegan diet. The diet consists of vegetables, fruits, nuts, seeds, sprouted grains, and beans. Uncooked foods usually make up between 50 and 100% of the diet.[7]

Extremely restrictive raw food diets are associated with impaired growth in children. Raw food diets, in general, lower beneficial HDL cholesterol levels and may lead to vitamin B_{12} deficiency and loss of bone mineral density. On the positive side, they are related to low body weights for height and healthy blood levels of triglycerides and total cholesterol.[8, 9]

■ **Other Vegetarian Diets.** For a number of other vegetarian regimes, the spiritual or emotional importance assigned to certain foods supersedes consideration of their contribution to an adequate diet.[10] Vegetarians adhering to the "living foods diet" consume uncooked and fermented plant foods only. This

Reintroducing meat Larry has been a vegetarian for the last seven years and wants to eat a hamburger again to see if it tastes as good as he remembers. He's a bit nervous about doing it because he thinks meat might make him feel sick.

Who gets thumbs up?
Answers on next page

Susan:
Larry should eat the hamburger if he wants to. It's a food, not an indigestion time bomb.

Doug:
Larry's stomach isn't used to the heaviness of meat. It will make him nauseated.

---

diet is inadequate in a number of nutrients, including vitamin B_{12}.[11] **Fruitarians** consume only fruit and olive oil. People adopting this type of diet rarely stick with it for long—it does not sustain health.

Vegetarian Diets and Health

Long-standing vegetarian dietary regimes that promote health in India and China have been tested by time (if not by science) and found to be adequate in essential nutrients.[1] In fact, vegetarian diets are generally healthier than those of nonvegetarians.[13] On the other hand, highly restrictive vegetarian diets—such as fruitarian regimes, the raw food diet, and other popular vegetarian diets that stray from established regimes—can be harmful to health.[1] In general, the more restrictive the diet, the more likely it is to be inadequate and lead to health problems. This is especially true for pregnant women, children, and persons who are ill, all of whom have relatively high needs for nutrients. Reports of caloric and nutrient deficiencies in young children of vegan parents are more common than is the case with children of parents with less restrictive vegetarian diets.[1]

Availability of a large assortment of vegan foods in the United States and other developed countries is making it less likely that vegetarian diets of children and adults will be inadequate.[12] Vegetarians in economically developed countries are not at greater risk for iron deficiency than nonvegetarians. They generally obtain sufficient iron from plants and have ample intake of vitamin C, which enhances absorption of iron from plants. In contrast, the diets of vegetarians in developing countries often include too few iron-rich plants, making iron deficiency common. Both the quality and the quantity of protein in vegetarian diets can also be a source of concern. However, vegetarians in developed countries generally have adequate protein intakes.[12] Vitamin B_{12} has consistently been shown to be the most likely nutrient to be lacking in the diets of individuals who do not consume animal products.[12] Vegans (who exclude milk and milk products) appear to be at greater risk for inadequate calcium intake and bone fractures than do vegetarians who consume these foods.[14] Increasingly, vegetarian diets are being recognized for their beneficial effects on health and disease prevention.

What Are the Health Benefits of Vegetarian Diets?

Compared to the usual American diet and lifestyle, vegetarianism is associated with a lower risk of developing heart disease, stroke, hypertension, type 2 diabetes, chronic bronchitis, and gallstones and kidney stones. People who follow a vegetarian diet rarely become obese or develop high blood cholesterol levels.[15,16]

fruitarian
A form of vegetarian diet in which fruits are the major ingredient. Such diets provide inadequate amounts of a variety of nutrients.

Appropriately planned vegetarian diets have been shown to be healthful, nutritionally adequate, and beneficial in the prevention and treatment of certain diseases.[12]

I am a great eater of beef, and I believe that does great harm to my wit.

—William Shakespeare, 1589

Reintroducing meat   The thought that meat could make you sick may be enough to trigger indigestion. Many self-defined vegetarians consume meat on rare occasion without reported ill effects.[12]

Susan: 👍

Doug: 👎

For children and adults at risk of early heart disease, the low-fat and high-fiber vegetable and fruit content of vegetarian diets may help reduce blood cholesterol levels and the risk of heart disease.[17] Benefits of vegetarian diets, macrobiotic diets, and other plant-based diets in the prevention and treatment of cancer have not been demonstrated.[18]

Dietary Recommendations for Vegetarians

Because vegetarian diets exclude one or more types of foods, it's important that the foods included provide sufficient calories and the assortment and quantity of nutrients needed for health. No matter what the motivation underlying the assortment of foods included, vegetarian diets that fail to provide all the nutrients humans need in the required amounts will not sustain health.

Well-Planned Vegetarian Diets

There are a number of types of vegetarian diets, so there is no one set of dietary recommendations that is appropriate for all of them. Dietary guidelines for vegan diets generally recommend a variety of foods that includes grains, legumes, vegetables, fruits, fats and oils, and sweets (Illustration 16.2). Additional nutrition guidance, such as that given by the American and Canadian Dietetic Associations (Table 16.3) is offered to help vegetarians make decisions about specific types of foods to consume within each food group and good ways to meet key nutrient needs.

Table 16.3

The American and Canadian dietetic associations' guidelines for vegetarian meal planning[12]

1. Choose a variety of foods, including whole grains, vegetables, fruits, legumes, nuts, seeds, and, if desired, dairy products and eggs.

2. Choose whole, unrefined foods often and minimize intake of highly sweetened, fatty, and highly refined foods.

3. Choose a variety of fruits and vegetables.

4. If animal foods such as dairy products and eggs are used, choose lower-fat versions of these foods. Cheeses and other high-fat dairy foods and eggs should be limited in the diet because of their saturated fat content and because their frequent use displaces plant foods in some vegetarian diets.

5. Vegans should include a regular source of vitamin B_{12} in their diets along with a source of vitamin D if sun exposure is limited.

6. Do not restrict dietary fat in children younger than two years. For older children, include some foods higher in unsaturated fats (for example, nuts, seeds, nut and seed butters, avocado, and vegetable oils) to help meet nutrient and energy needs.

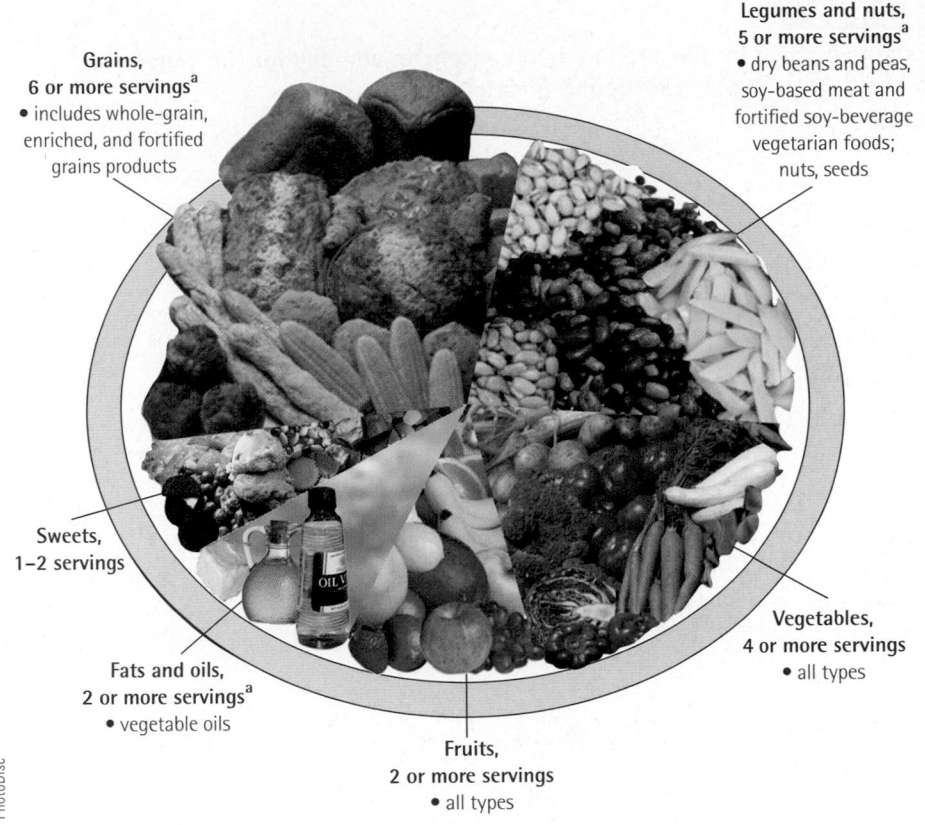

Grains,
6 or more servings[a]
• includes whole-grain,
enriched, and fortified
grains products

Legumes and nuts,
5 or more servings[a]
• dry beans and peas,
soy-based meat and
fortified soy-beverage
vegetarian foods;
nuts, seeds

Sweets,
1–2 servings

Fats and oils,
2 or more servings[a]
• vegetable oils

Fruits,
2 or more servings
• all types

Vegetables,
4 or more servings
• all types

PhotoDisc

Illustration 16.2 Foundation foods
for well-planned vegetarian diets.[19–22]
VEGAN DIETS
GRAINS, 6 or more servings[a]
• includes whole-grain, enriched, and
fortified grain products
LEGUMES AND NUTS, 5 or more
servings[a]
• dry beans and peas, soy-based meat
and fortified soy-based beverage
vegetarian foods; nuts, seeds
VEGETABLES, 4 or more servings
• all types
FRUITS, 2 or more servings
• all types
FATS AND OILS, 2 or more servings[a]
• vegetable oils
SWEETS, 1–2 servings

[a]Number of serving depends on calorie need.

Vegetarians who consume fish, dairy products, or eggs can reduce their reliance on legumes, nuts, and grains as protein and nutrient sources. Animal products are a major source of **complete proteins**, vitamin B_{12}, vitamin D, calcium, and EPA (eicosapentaenoic acid) and DHA (docosahexaenoic acid). EPA and DHA are two important fatty acids primarily found in fish and seafood. Vitamin B_{12}, vitamin D, calcium, and EPA and DHA are the nutrients most likely to be lacking in the diets of vegetarians.[18, 22]

A specific example of the food and nutrient composition of one day of a vegetarian's diet are shown in Table 16.4 (p.16-8). The diet provides low amounts of cholesterol and saturated fat, is high in fiber, and provides adequate amounts of protein and most vitamins and minerals. These are fairly typical results for a vegetarian diet.[15] Of the four key nutrients most likely to be missing in the diet of vegetarians, this day's diet provides the recommended amount of vitamin D, but not of vitamin B_{12}, calcium, or EPA/DHA. Vitamin D intake reaches 6 mcg (240 IU) and B_{12} levels 1.9 mcg in this diet due to the inclusion of fortified soy milk.

Vegetarians can meet their needs for vitamin B_{12}, vitamin D, calcium, and EPA/DHA by consuming fortified foods (such as those shown in Table 16.5) or by use of B_{12}-fortified yeast, DHA algae supplements, or fortified plant foods (such as DHA-fortified peanut butter or margarine). EPA can be obtained from sources of DHA. EPA and DHA are closely related chemicals, and the body can convert one to the other to an extent.[23] Vitamin D is becoming easier to come by for vegetarians. The FDA recently approved the addition of vitamin D to soy-based foods such as soy beverages, tofu, burgers, and desserts. The change will be reflected on food labels of soy vegetarian products.[24] Direct exposure of the skin to sunlight for about 15 minutes a day is one of the best ways to get vitamin D.

Vitamin and mineral supplements can be, and often are, used by vegetarians to meet specific nutrient needs. If needed, supplementary intake levels of vitamins B_{12} and D, calcium, and EPA/DHA should approximate those shown in Table 16.6. The best way to know if supplemental nutrients are needed is to

complete proteins
Proteins that contain all of the nine essential amino acids in amounts sufficient to support protein tissue construction by the body.

Table 16.4

Example of food intake and nutrient content in one day for the vegetarian diet of a 31-year-old, 130-pound female

One day's diet:

Oatmeal, 1 cup	Salad dressing, 3 tablespoons
Banana, 1 medium	Soy milk, 1 cup
Soy milk, 1 cup	Veggie burger, 1 patty
Brown sugar, 1 Tablespoon	Bun, 1
Almonds, 1 ounce (22 almonds)	French fries, 1 cup
Black beans, 2 cups	Herbal tea, 1 cup
Brown rice, 1 cup	Hummus, 1/2 cup
Lettuce salad with mixed vegetables, 1 cup	Baby carrots, 10

Selected nutrient analysis results

NUTRIENT	AMOUNT CONSUMED	RECOMMENDED INTAKE
Protein, g	87	46
Total fiber, g	53	25
Total fat, g	100	53–92
Saturated fat, g	20	<26
Cholesterol, mg	27	<300
EPA + DHA, mg[a]	50	300–600
Vitamin A, mcg	723	700
Vitamin C, mg	54	75
Vitamin E, mg	14	15
Vitamin B_6, mg	3.2	1.3
Vitamin B_{12}, mcg	1.9	2.4
Thiamin, mg	3.1	1.1
Riboflavin, mg	1.6	1.1
Niacin, mg	19	14
Folate, mcg	561	400
Vitamin D, mcg[a]	6	5
Calcium, mg	491	1000
Magnesium, mg	614	320
Iron, mg	18.5	18
Zinc, mg	12.2	8
Selenium, mg	37	55

[a]EPA and DHA availability are estimated based on alpha-linolenic acid intake of 2.8 g; vitamin D intake is estimated based on fortified soymilk intake. Analysis is from mypyramidtracker.gov, 9/06.

Table 16.5

Key nutrient fortification levels of some vegetarian foods[a]

	Key Nutrients		
	VITAMIN B_{12}	VITAMIN D	CALCIUM
Recommended intake for adults:	2.4 mcg	5 mcg	1000 mg
Food			
Rice Dream, 1 cup	1.5 mcg	2.5 mcg	300 mg
Soy milk, 1 cup	3 mcg	3 mcg	300 mg
Meat analogs[b]	0–1.4	0	0
Fully fortified breakfast cereals	6 mcg	1 mcg	1000 mg
Other breakfast cereals	1.5 mcg	1 mcg	0–100 mg

[a]Check product nutrition information lablels for product-specific information. None of the foods listed are yet fortified with EPA and DHA.
[b]Includes burgers, hot dogs, mixed dishes, and other foods made with plant foods.

Table 16.6

Amounts of key nutrients that may be needed by vegetarians daily[a]

Vitamin B_{12}	2–6 mcg
Vitamin D	200–400 IU (10 mcg)
Calcium	500 mg
EPA and DHA	300 mg

[a]Fortified foods may also supply these nutrients.

carefully perform a dietary assessment that covers several typical days of food intake. (The Cengage/Wadsworth Diet Analysis Plus Program or USDA's dietary analysis program available at mypyramidtracker.gov can be used for the dietary assessment.)

Complementary Protein Foods   Most well-planned vegetarian diets provide adequate amounts of protein, but the quality of the protein varies depending on which plant foods are consumed.

Animal products such as meat, eggs, and milk provide all of the nine **essential amino acids** in sufficient quantity to qualify as complete sources of protein. In addition, tests have shown soy proteins to be complete protein sources for children and adults.[25] The diet must include complete proteins because the body needs a sufficient supply of each essential amino acid if it is to build and replace protein substances such as red blood cells and enzymes. If any of the essential amino acids are missing in the diet, protein tissue construction stops, and the available amino acids will be used for energy instead. Essential amino acids consumed in foods are not stored for long, so the body needs a fresh supply each day or so.

Vegetarians who don't consume animal products can meet their need for essential amino acids by combining plant foods to yield complete proteins. This is done by consuming plant foods that *together* provide all the essential amino acids, although each individual food is missing some of these essential nutrients. The goal is to "complement" plant sources of essential amino acids, or to consume **complementary protein sources** from plant foods regularly.[5]

Many different combinations of plant foods yield complete proteins. Basically, complete sources of protein can be obtained by combining grains such as rice, bulgur (whole wheat), millet, or barley with dried beans, tofu, or green peas; or corn with lima beans or dried beans; or seeds with dried beans. Some examples of complementary sources of plant proteins are shown in Illustration 16.3. Milk, meat, and eggs contain complete proteins and will complement the essential amino acids profile of any plant source of protein.

Where to Go for More Information on Vegetarian Diets

You can find information about vegetarian diets in a variety of sources. World Wide Web resources appear at the end of this unit, and several cookbooks that offer information about vegetarian nutrition and delicious recipes are pictured in Illustration 16.4. Unfortunately, some of the information available on

On the Side

There are more than 70 varieties of dried beans. They are an important part of the cuisine of almost all cultures. Bean cuisine includes chili, baked beans, beans and rice, hummus, and refried beans.

Photo Disc

essential amino acids
Amino acids that cannot be synthesized in adequate amounts by the human body and must therefore be obtained from the diet.

complementary protein sources
Plant sources of protein that together provide sufficient quantities of the nine essential amino acids.

Illustration 16.3   Some combinations of plant foods that provide complete protein:

- *Rice and dried beans*
- *Rice and green peas*
- *Bulgur (wheat) and dried beans*
- *Barley and dried beans*
- *Corn and dried beans*
- *Corn and green peas*
- *Corn and lima beans (succotash)*
- *Soybeans and seeds*
- *Peanuts, rice, and dried beans*
- *Seeds and green peas*

Rice and black beans.

Photo Disc

Hummus and bread.

© Amy C. Erra/PhotoEdit

Corn and black-eyed peas.

© Michael Newman/PhotoEdit

Bulgur (whole wheat) and lentils.

PhotoDisc

Tofu and rice.

Photo Disc

Corn and lima beans (succotash).

© David Young-Wolff/PhotoEdit

Tortilla with refried beans (e.g., a bean burrito).

© Michael Newman/PhotoEdit

Pea soup and bread.

Photo Disc

Illustration 16.4 Many excellent vegetarian cookbooks are available; most provide information on diet as well as recipes. The books pictured here are *Moosewood Restaurant Low-Fat Favorites: Flavorful Recipes for Healthful Meals,* edited by Pam Krauss; *Vegetarian Times Complete Cookbook,* by the editors of *Vegetarian Times; Vegetarian Cooking for Everyone,* by Deborah Madison; *The Whole Soy Cookbook,* by Patricia Greenberg; and *Vegetable Heaven,* by Mollie Katzen.

vegetarianism is wrong or misconstrued. Some vegetarian organizations are more committed to selling particular beliefs, memberships, or books and magazines than to promoting healthy vegetarian diets. Beware of vegetarian diets that claim to cure cancer, AIDS, or other serious illnesses or promise you'll experience inner peace or spiritual renewal. Appropriately planned vegetarian diets are health promoting, but they are not magic bullets that will cure all the ills of body and soul.

NUTRITION | Up Close

Vegetarian Main Dish Options

Focal Point: Serving up vegetarian alternatives to meat dishes.

Assume you are a member of the "vegetarian option" planning committee for your college. You are asked to identify three meatless main dishes that could be served in the dining halls for lunch and another three that could be served for dinner. The one stipulation is that they should be main dishes you would enjoy eating.

What meatless dishes would you identify as options for lunch and for dinner?

Lunch Dishes	**Dinner Dishes**
1.	1.
2.	2.
3.	3.

Feedback (answers to these questions) can be found at the end of Unit 16.

Key Terms

complementary protein sources, page 16–9

complete proteins, page 16–7
essential amino acids, page 16–9

fruitarian, page 16–5

Review Questions

TRUE FALSE

1. The emotional and spiritual importance of certain foods in some vegetarian regimes supersedes any contribution the foods may make to an adequate diet. ☐ ☐

2. Raw food diets supply ample amounts of all vitamins and required minerals and improve indigestion in people with digestive problems. ☐ ☐

3. Compared to omnivore diets, vegetarian diets are associated with lower rates of obesity, heart disease, hypertension, and type 2 diabetes. ☐ ☐

4. Vitamin B_{12} and EPA/DHA are largely found in animal foods. ☐ ☐

TRUE FALSE

5. The four nutrients most likely to be lacking in the diets of vegetarians are vitamin B_{12}, EPA/DHA, magnesium, and vitamin E. ☐ ☐

6. Well-planned vegetarian diets provide adequate amounts of iron and protein. ☐ ☐

7. When forming a new protein tissue, the body will respond to a deficiency of an essential amino acid by replacing it with another essential amino acid that is available. ☐ ☐

8. Corn combined with cooked, dried beans is an example of a complementary protein source. ☐ ☐

Media Menu

www.nal.usda.gov/fnic/pubs/bibs/gen/vegetarian.htm
An extensive listing of, and links to, information on vegetarian diets and health and weight status, education materials, scientific reports, and other resources are available at this site.

www.mypyramid.gov/tips_resources/vegetarian_diets.html
USDA's site for vegetarian diet information is based on MyPyramid food group recommendations.

www.vrg.org
Contact the Vegetarian Resource Group for dietary advice, healthy cooking and shopping tips, and other information for the vegetarian community.

www.vegweb.com
This site provides a guide to vegetarianism, recipes and books, and much more (including ads).

www.mayoclinic.com/health/vegetarian-diet/HQ01596
This is the Mayo Clinic's site on vegetarian diets. It includes information on meal planning and a colorful vegetarian food guide pyramid.

www.nlm.nih.gov/medlineplus/vegetariandiet.html
Medline presents background information on vegetarian diets, nutrient needs, and the health-promoting properties of well-planned vegetarian diets.

Notes

1. Sabate J. The contribution of vegetarian diets to health and disease: a paradigm shift? Am J Clin Nutr 2003;78(suppl): 501S–7S.

2. Cade JE et al. The UK Women's Cohort Study: comparison of vegetarians, fish-eaters and meat-eaters. Public Health Nutr. 2004;7:871–8.

3. Buffonge I. The vegetarian/vegan lifestyle. West Indian Med J 2000; 49:17–19.

4. Antony AC. Vegetarianism and vitamin-12 (cobalamin) deficiency. Am J Clin Nutr 2003;78:3–6.

5. Barr SL et al. Perceptions and practices of self-defined current vegetarian, former vegetarian, and nonvegetarian women. J Am Diet Assoc 2002;102: 354–6.

6. Kushi LH et al. The macrobiotic diet and cancer. J Nutr 2001. 131:3056S–64S.

7. Cunningham E. What is a raw food diet and are there any risks or benefits associated with it? J Am Diet Assoc 2004; 104:1623.

8. Koebnick C. Long-term consumption of a raw food diet is associated with favorable serum LDL cholesterol and triglycerides but also with elevated plasma homocysteine and low serum HDL cholesterol in humans. J Nutr. 2005; 135:2372–8.

9. Fontana L et al. Low bone mass in subjects with long-term raw vegetarian diet. Arch Int Med 2005;165:684–9.

10. Mutch PB. Food guides for the vegetarian. Am J Clin Nutr 1988;48:913–19.

11. Rauma A-L et al. Vitamin B-12 status of long-term adherents of a strict, uncooked vegan diet ("living foods diet") is compromised. J Nutr 1995;125:2511–15.

12. Position of the American Dietetic Association and Dietitians of Canada: vegetarian diets. J Am Diet Assoc 2003;103:748–65.

13. Haddad EH, Tanzman JS. What do vegetarians in the United States eat? Am J Clin Nutr 2003;78(suppl):626S–32S.

14. Appleby P et al. Comparative fracture risk in vegetarians and nonvegetarians in EPIC-Oxford, Eur J Clin Nutr 2007;61:1400–6.

15. Aronson D. Vegetarian nutrition: What every dietitian should know. Today's Dietitian 2005;Mar:31–7.

16. Berkow SE et al. Vegetarian diets and weight status. Nutr Rev. 2006;64:175–88.

17. Dunham L et al. Vegetarian eating for children and adolescents. J Pediatr Health Care 2006;20:27–34.

18. Key TJ et al. Health effects of vegetarian and vegan diets. Proc Nutr Soc. 2006;65:35–41.

19. Messina V et al. A new food guide for North American vegetarians. J Am Diet Assoc 2003;103:771–5.

20. Haddad EH et al. Vegetarian food guide pyramid: a conceptual framework. Am J Clin Nutr 1999;70(suppl): 615S–9S.

21. Vegetarian food guide pyramid, http://fnic.nal.usda.gov, accessed 9/06.

22. Leitzmann C. Vegetarian diets: what are the advantages? Forum Nutr. 2005;57:147–56.

23. Arterburn LM et al. Distribution, interconversion, and dose response of n-3 fatty acids in humans. Am J Clin Nutr 2006;83:(suppl):1467S–76S.

24. Heller L. FDA approves vitamin D fortification of soy foods, nutraingredientsusa.com, accessed 3/09.

25. Dietary reference intakes for carbohydrates, fiber, fat, fatty acids, cholesterol, protein, and amino acids. Washington, DC: National Academies Press; 1999.

Answers to Review Questions

1. True, see page 16–4.
2. False, see page 16–4.
3. True, see page 16–5.
4. True, see page 16–7.
5. False, see page 16–7.
6. True, see pages 16–5.
7. False, see page 16–9.
8. True, see page 16–9.

NUTRITION | Up Close

Vegetarian Main Dish Options

Feedback for Unit 16

Vegetarians definitely had the advantage in completing this exercise! For the meat eaters, were the main dishes you identified meat-free? Here are six vegetarian options for lunch and dinner main dishes:

- Eggplant parmesan
- Falafel
- Red beans and rice
- Pasta with broccoli
- Melted cheese, tomato, and sprout sandwiches
- Vegetable lasagna
- Veggie burgers
- Vegetable pizza

Food Allergies and Intolerances

NUTRITION SCOREBOARD

	TRUE	FALSE
1 About one in every three Americans is allergic to at least one food.		
2 Food ingredient labels must indicate the presence of major food allergens in food products.		
3 Skin prick tests are an accurate way to diagnose specific food allergies.		
4 Food intolerances cause less severe reactions than food allergies do.		

Key Concepts and Facts

- True food allergies involve a response by the body's immune system to a particular substance in food.

- Food allergies may be caused by hundreds of different foods. However, 90% of food allergies are due to eight foods: nuts, eggs, wheat, milk, peanuts, soy, shellfish, and fish.

- Food intolerances cover an array of adverse reactions that do *not* involve the body's immune system.

- The most accurate method of identifying food allergies and intolerances in most cases is the double-blind, placebo-controlled food challenge.

Answers to

NUTRITION SCOREBOARD

		TRUE	FALSE
1	About one in three Americans *believes* he or she is allergic to at least one food. Probably 2 or 3 in 100 adults and 5 in 100 infants and young children actually are.[1]		✔
2	Listing potential, relatively common food allergens on food ingredient labels is required for packaged foods.	✔	
3	Skin prick test results assist in the diagnosis of food allergies but generally do not identify the specific food allergen that causes the allergic reaction.[2]		✔
4	Food allergies and food intolerances have different causes, but both can produce life-threatening reactions.	✔	

Food Allergy Mania

Talk about your controversial subjects. This one is very hot. People use the term *food allergy* to refer to virtually any type of problem they have with food. At one extreme are people who believe allergies to milk, wheat, and sugar are to blame for hyperactivity and a host of other behavioral problems in children. At the other extreme are some health professionals who think people who complain of food allergies need to have their heads examined. Despite the controversies, food allergies have become a major health concern in the United States and many other Westernized countries.[3]

Food allergies are real and can be very serious. At the minimum, true food allergies can cause a rash or an upset stomach. At the maximum, they can result in death. Unreal food allergies can cause problems, too. They can lead people to eliminate nutritious foods from their diet unnecessarily, resulting in inadequate diets and eventually in health problems. One of the most intriguing aspects of food allergies is the frequency and ease with which foods are falsely blamed for a variety of mental and physical health problems.

Adverse Reactions to Foods

People experience adverse reactions to food for three primary reasons. One is food poisoning. The other two involve the body's reaction to substances in food that are normally harmless. With **food allergies**, the body's **immune system** reacts to a substance in food (almost always a protein) that it identifies as harmful (Illustration 17.1). The immune system is not involved in **food intolerances**, which encompass other adverse reactions to normally harmless substances in food.

Why Do Food Allergies Involve the Immune System? It may seem odd that a system that helps the body conquer bacteria and viruses is involved in

food allergy
Adverse reaction to a normally harmless substance in food that involves the body's immune system. (Also called *food hypersensitivity*.)

immune system
Body tissues that provide protection against bacteria, viruses, and other substances identified by cells as harmful.

food intolerance
Adverse reaction to a normally harmless substance in food that does *not* involve the body's immune system.

Illustration 17.1 What do these foods have in common? Each food shown, plus many more not shown, may cause adverse reactions in some individuals.

"protecting" us from normal constituents of food. In people with allergies, however, the cells recognize some food components as harmful—just as they recognize bacteria and viruses. Components of food that trigger the immune system are called **food allergens**.

For some susceptible people, exposure to trace amounts of an allergen triggers an allergic reaction. Highly sensitive people can develop an allergic reaction simply by being in the room when the food to which they are allergic is being prepared, or by touching the food. Allergic reactions to peanuts have been triggered by kissing someone who has recently consumed them and by breathing peanut vapors on airplanes.[4]

In response to the allergen, the immune system forms **antibodies** (Illustration 17.2). The antibodies attach to cells located in the nose, throat, lungs, skin, eyes, and other areas of the body. When the allergen appears again, the body is ready. The previously formed antibodies recognize the allergen, attach to it, and signal the body to secrete **histamine** and other substances that cause the physical signs of allergic reactions. Most commonly, allergic reactions cause a rash, diarrhea, congestion, or wheezing (Table 17.1), but the symptoms may be much more serious.

Exposure to trace amounts of an allergen in peanuts, nuts (peanuts are actually a legume, not a nut), fish, and shellfish can cause **anaphylactic shock**

Illustration 17.2 The development of allergic reactions to foods.

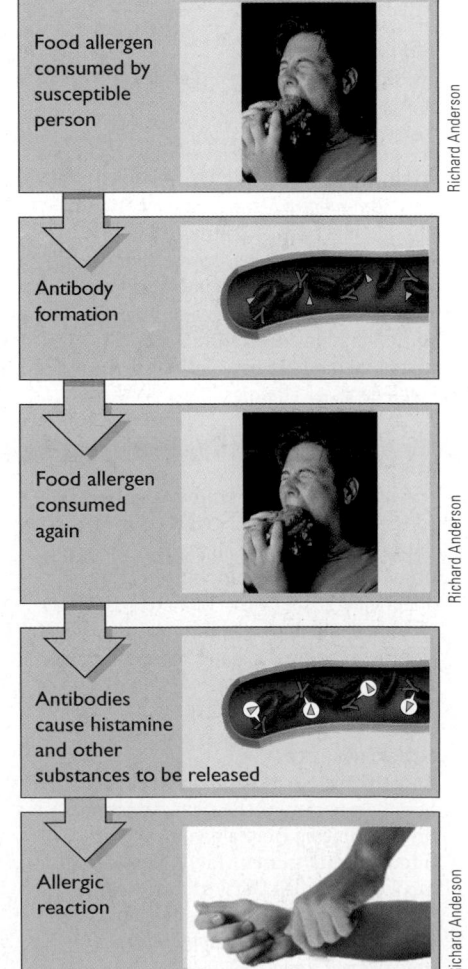

Food allergen consumed by susceptible person

Antibody formation

Food allergen consumed again

Antibodies cause histamine and other substances to be released

Allergic reaction

Table 17.1

The three most common types of symptoms caused by food allergies

Some people experience more than one type of reaction.[4]

Symptom	Percentage of People with Food Allergies Who Develop the Symptom
• Skin eruptions: rash, hives	84%
• Gastrointestinal upsets: diarrhea, vomiting, cramps, nausea	52
• Respiratory and other problems: congestion, swelling of the tongue and throat, runny nose, cough, wheezing, asthma	32

Illustration 17.3 These foods can cause anaphylactic shock—a severe allergic reaction—in highly sensitive people.
(Most food allergies cause milder symptoms.)

food allergen
A substance in food (almost always a protein) that is identified as harmful by the body and elicits an allergic reaction from the immune system.

antibodies
In the case of allergies, proteins the body makes to combat allergens.

histamine
(hiss-tah-mean) A substance released in allergic reactions. It causes dilation of blood vessels, itching, hives, and a drop in blood pressure and stimulates the release of stomach acids and other fluids. Antihistamines neutralize the effects of histamine and are used in the treatment of some cases of allergies.

anaphylactic shock
(an-ah-fa-lac-tic) Reduced oxygen supply to the heart and other tissues due to the body's reaction to an allergen in food or other "foreign" substance. Symptoms of anaphylactic shock (or "anaphylaxis") may include abdominal cramps, vomiting, chest tightness, paleness, weak and rapid pulse, and difficulty breathing.

celiac disease
An autoimmune disease characterized by inflammation of the small intestine lining resulting from a genetically based intolerance to gluten. The inflammation produces diarrhea, fatty stools, weight loss, and vitamin and mineral deficiencies. (Also called *celiac sprue and gluten-sensitive enteropathy.*)

(Illustration 17.3). This massive reaction of the immune system can result in death, caused by the cutoff of the blood supply to tissues throughout the body. People experiencing anaphylactic shock can be revived with an injection of epinephrine, and may also require antihistamines and other medications. Follow-up care should be provided because symptoms of anaphylactic shock may reoccur up to 24 hours later.[5]

The incidence of peanut allergy in the United States is increasing—about 1 in 150 people are affected. Despite ongoing studies, it is not yet clear why the rate of peanut allergy is increasing.

Symptoms of food allergies may occur within seconds after the body is exposed to the food allergen, or they may take several hours to occur. The symptoms often disappear within two hours after they begin.[1]

What Foods Are Most Likely to Cause Allergic Reactions? Many foods may cause allergic reactions in genetically susceptible individuals (Illustration 17.4). Nevertheless, approximately 90% of all food allergies are caused by eight foods: nuts, eggs, wheat, milk, peanuts, soy, shellfish, and fish. The nearby "On the Side" feature gives you an idea for how to remember these foods. Together, these are considered "the big eight" food allergies.[6]

Passage of the Food Allergen Labeling and Consumer Protection Act of 2004 is helping consumers with food allergies find offending foods and ingredients. Food ingredient labels are now required to state clearly whether the food contains one or more of the big eight allergenic foods.[6]

■ A Closer Look at Wheat Allergy (Celiac Disease).

Minnie is 46 years old and has three sisters and two brothers, all of whom are taller than she is. (She considers herself the "runt of the litter.") Since childhood, Minnie has had problems with diarrhea and cramps. Over the past two years, these problems have become worse, and she has been chronically tired. Assuming that she was lactose intolerant, Minnie stopped eating all dairy products. Her problems persisted, however. When she almost lost her job as a receptionist due to her frequent bathroom breaks and her obvious fatigue, she decided to see a doctor again.

Diagnosing anemia, the doctor sent Minnie to a registered dietitian. Hearing Minnie's story, the dietitian suspected celiac disease caused by an allergy to a component of wheat and other grains. After two weeks on an allergen-free diet, Minnie started to regain her strength and spent much less time in the bathroom. Diagnostic tests subsequently confirmed that Minnie had celiac disease, a condition that likely had existed since childhood.

An allergy to wheat, or more specifically to a component of gluten found in wheat and other grains such as barley, rye, and triticale has been underdiagnosed in the past. Called **celiac disease**, it affects approximately 1% of people worldwide.[7] It differs somewhat from other allergies in that the immune system reaction to gluten is localized in the lining of the small intestine.[7] Celiac disease can be difficult to diagnose because the symptoms (diarrhea, weight loss, cramps, anemia) are similar to those of other diseases and because symptoms are silent in some individuals. With new awareness about the disorder, doctors are ordering diagnostic tests more frequently, and the rate of confirmed cases of celiac disease is increasing.[8] The gold standard, diagnostic test for celiac disease is small bowel biopsy and examination of cells for signs of damage due to the disease. The test has to be undertaken while individuals are consuming their normal diets, and not a gluten-free one. Removal of gluten from the diet corrects intestinal cell damage and may lead to an incorrect diagnosis.[7]

NUTS

EGGS

PEANUTS

SOY

WHEAT

MILK

MILK

SEAFOOD

FISH

Photo Disc

Illustration 17.4 The "big eight" foods that can cause allergies.

Celiac disease is treated with a lifelong, but rewarding, gluten-free diet. Because gluten may be included in food additives, and processed and prepared foods as well as grain products, it generally takes the help of an experienced, registered dietitian to design and monitor gluten-free diets.[2] Table 17.2 provides examples of foods that contain gluten and some that don't.

The term *gluten-free* is standardized and allowed on food labels for qualifying foods. This labeling regulation is voluntary but is expected to be widely employed because the demand for gluten-free foods is increasing. Since the regulation is voluntary, checking for a "gluten-free" label to identify foods that can be safely consumed may not always work. Ingredient labels listing wheat or products made from wheat, rye, barley, or triticale should not be consumed, nor should foods labeled as containing, or potentially containing, these ingredients.

Table 17.2

Which foods contain gluten? Listed here are examples of foods that do or do not contain gluten.[7,9]

Gluten-containing foods	Gluten-free foods[a]
Beer, ale	Fruit
Barley	Vegetables
Broth, bouillon powder/cubes	Dried beans
Brown rice syrup	Buckwheat
Bulgur	Cassava
Commercial soups	Grits
Bread, other wheat/rye flour products	Cornmeal
Imitation seafood	Nuts
Cakes, pies, cookies	Quinoa
Processed meats	Fresh meats, fish
Soy sauce	Soy flour, cereals
Wheat starch	Wild rice
Pizza	Eggs
Macaroni and cheese	Nuts, seeds
Seasonings	Cheese (not processed)
Marinades	Popcorn
Rye-containing products	Milk
Vegetarian meat substitutes	Chips (100% corn, potato)

[a] Assumes foods have not been contaminated with gluten during processing and are free of gluten-containing ingredients.

How Common Are Food Allergies?

Infants and young children experience food allergies more often than adults do. Approximately 5% of infants and young children develop a food allergy, which most often is due to a protein in cow's milk or egg white.[2] Some infants and young children are at higher risk for developing allergies to foods due to the history of food allergy in one or both parents.[10] To prevent allergic reactions in infants and young children at risk of allergies, it is recommended that cow's milk and dairy products be excluded from the infant's diet during the first year of life. Eggs should be introduced after two years, and peanuts, nuts, fish, and shellfish should not be given for the first three years.[10] Most infants outgrow their allergy to cow's milk, eggs, soy, and wheat later in childhood. They often do not outgrow allergies to peanuts, nuts, fish, and shellfish.[1]

New recommendations related to food allergies in infants who are not at increased risk have been released by the American Academy of Pediatrics.[11] The new recommendations relax previous restrictions on the introduction of cow's milk, fish, nuts, and peanuts in the first year. These foods can be introduced along with other solid foods after the age of 4 to 6 months. The recommendations are based on research results that show that early introduction of potential allergens may help "program" the developing immune system to accept them rather than reject them as foreign.[3,11] It appears that earlier introduction of potentially allergenic foods may help prevent food allergies in many infants. Exclusive breast-feeding for at least the first 4 to 6 months of life is recommended, in part because it effectively reduces the risk of allergies in infants.[11]

The incidence of food allergies in adults is estimated to be 2 to 3%, and this number is increasing.[3] Nevertheless, around one-third of Americans believe they are allergic to one or more foods. About 60% of complaints of food allergy fail to be confirmed by testing.[12] Here are two, real-life examples of self-diagnosed food allergies that went awry:

1. Isaiah, a lover of blueberries and blueberry pie, hasn't touched a blueberry since 1995. That year, Isaiah had a piece of blueberry pie in a restaurant and later became violently sick to his stomach. Bingo! Isaiah thought he must be allergic to blueberries.

   Actually, Isaiah gave up blueberries for no good reason. He was coming down with the flu when he ate the pie. Nevertheless, to this day when he thinks of blueberries, he gets a bit queasy.

2. For 11 years, Emilia rigidly avoided even small amounts of cow's milk. She was convinced that just a few drops of cow's milk would cause pressure in her head, blurred vision, dizziness, cramps, and nausea.

   Finally, Emilia's presumed allergy to cow's milk was put to the test. She was given liquid through a dark tube inserted into her stomach from her mouth. When she was told the fluid was cow's milk—even though it was actually water—the familiar symptoms appeared within 10 minutes. When she was given milk but told it was water, no symptoms appeared.[13]

   Emilia was shocked. For 11 years she had scrutinized nearly everything she ate and had wasted hundreds of dollars on special food products and supplements. Finishing a frosted brownie and a glass of milk, Emilia contemplated the practical realities of the power of suggestion.

It is essential to realize that the [allergic] reactions occur in human beings and are therefore enmeshed in the interplay between body, mind and soul.... Only through experience with blind food challenges can the amazing power of self-deception be appreciated.[13]

double-blind, placebo-controlled food challenge
A test used to determine the presence of a food allergy or other adverse reaction to a food. In this test, neither the patient nor the care provider knows whether a suspected offending food or a placebo is being tested.

Diagnosis: Is It a Food Allergy?

A variety of tests are used to diagnose food allergies, but there is only one "gold standard." That test is the **double-blind, placebo-controlled food challenge.**[1]

Suppose you suspect you are allergic to grapes because you got an annoying rash twice after eating them. Your doctor arranges an appointment for you at an

allergy clinic. At the clinic, you will be given grapes, either in a concentrated pill form or blended into a liquid, and placebo pills (or a liquid) without being told which is the grapes and which the placebo. After an appropriate amount of time, your reaction to the grapes and to the placebo will be noted. If the rash appears after the grapes but not after the placebo, you have an adverse reaction to grapes that may be an allergy.

Food challenges are best undertaken under medical supervision. True food allergies can cause serious reactions, and immediate help may be needed. Foods suspected of causing anaphylactic shock in the past should not be given in food challenge tests. They should be totally omitted from the diet.[1]

The Other Tests for Food Allergy

Other tests for food allergy are reliable to varying degrees. None of them is always accurate, and the results of bogus tests are not to be trusted at all.

Immunoglobulin E Tests Immunoglobulin E (IgE) is a protein produced by the immune system in response to an allergen. Identification of specific types of IgE in a person's blood can help identify the presence of a food allergy, but often cannot identify the food allergen.[2]

Skin Prick Tests Skin prick tests (also called prick skin tests) for food allergy are commonly employed and are useful for identifying *the absence* of a food allergy. A positive test result isn't proof positive that an allergy exists, however, because positive test results are inaccurate about 50% of the time.[1]

For this test, a few drops of food extract are placed on the skin, and the skin is then pricked with a needle. At approximately the same time, another area of skin is pricked using only water. If the area around the food skin prick becomes redder and more swollen than the area pricked with water (Illustration 17.5), it is concluded that the person *might* be allergic to the food tested. Skin prick tests are useful for ruling out allergies and for reducing the number of foods that may have to be tested by double-blind, placebo-controlled food challenges.

Illustration 17.5 A positive reaction to a skin prick test for a food allergy usually looks like this. The result indicates that an allergy to the food tested may be present but doesn't guarantee it.

Bogus Tests A number of companies offer food allergy tests through the mail. In one study, researchers sent five of these companies the required samples, nine from adults who were allergic to fish and nine from adults who had no allergies. All five companies missed the fish allergy. When duplicate samples from the same adults were sent separately to each company, reports came back with different results for each of the pairs of samples.

ANSWERS TO
Reality Check

I think I'm allergic to... Some people, like Trevor may be sensitive to MSG. It doesn't elicit an immune system reaction, so it's not a food allergy.

Haley: 👎

Darrel: 👍

In addition, the companies reported their laboratories had identified various food allergies that none of the adults actually had.[13] Mail-order allergy tests are a waste of money; they do not use reliable assessment techniques. Other unreliable methods for identifying food allergies include hair analysis, cytotoxic blood tests, iridology, and sublingual (under the tongue) and food injection provocation tests.[23]

What's the Best Way to Treat Food Allergies?

After a food allergy is confirmed, the food is eliminated from the diet. This is the only treatment available for food allergies. "Allergy" shots for food allergens are not yet available but may be in the future.[3] If the eliminated food is an important source of nutrients or is found in many food products (such as milk or wheat), consultation with a registered dietitian is recommended.

Some food allergies don't last forever. Many infants and young children outgrow allergies to cow's milk, wheat, and eggs by late childhood. Children and adults with a severe allergy to peanuts, nuts, fish, or shellfish may have to eliminate the food from their diet for a lifetime.[1]

Food Intolerances

Food intolerances produce some of the same reactions as food allergies, but the reactions develop by different mechanisms. Food intolerance reactions do not involve the immune system; they are due to a missing enzyme or other cause.[23]

It is clear that some people are intolerant of lactose, sulfite, histamine (a component of red wine and aged cheese), and other foods (Table 17.3). It is also clear that not all food intolerances are real. Just as with food allergies, the best way to separate the real from the unreal is the double-blind, placebo-controlled food challenge. True food intolerances produce predictable reactions. Problems such as headache, diarrhea, swelling, or stomach pain will occur every time a person consumes a sufficient amount of the suspected food.

Lactose Maldigestion and Intolerance

Lactose maldigestion, or the inability to break down lactose in dairy products due to the lack of the enzyme lactase, results in the condition known as **lactose intolerance.** Symptoms of lactose intolerance, such as flatulence, bloating, abdominal pain, diarrhea, and "rumbling in the bowel," occur in lactose maldigesters within several hours of consuming more lactose than can be broken down by the available lactase. These symptoms are due to the breakdown of undigested lactose by bacteria in the lower intestines (which produce gas as a by-product of lactose ingestion) and by fluid accumulation.[16]

Table 17.3

Foods and substances in them linked to food intolerance reactions

- Aged cheese
- Anchovies
- Beer
- Catsup
- Chocolate
- Dried beans
- Food coloring
- Lactose
- Mushrooms
- Pineapple
- Red wine
- Sausage (hard, cured)
- Soy sauce
- Spinach
- Tomatoes
- Yeast

lactose maldigestion
A disorder characterized by reduced digestion of lactose due to the low availability of the enzyme lactase.

lactose intolerance
The term for gastrointestinal symptoms (flatulence, bloating, abdominal pain, diarrhea, and "rumbling in the bowel") resulting from the consumption of more lactose than can be digested with available lactase.

Reduced intake of lactose-containing dairy products (milk, ice cream, and cottage cheese, for example) may prevent lactose intolerance symptoms. Foods such as hard cheese, low- or no-lactose milk, buttermilk, and yogurt without added milk solids contain low amounts of lactose and can generally be consumed. Small amounts of milk or other lactose-containing dairy products are generally well tolerated by people with lactose maldigestion.[16] Care should be taken to ensure adequate intake of calcium and vitamin D if milk and dairy product intake is restricted.

Lactose maldigestion is a very common disorder. It occurs in about 25% of U.S. adults, but the incidence varies dramatically by population group. By one set of estimates, 62 to 100% of Native American adults, 90% of Asian Americans, 80% of African Americans, 53% of Mexican Americans, and 15% of Caucasian Americans are lactose maldigesters.[17]

Sulfite Sensitivity

Sulfite is a food additive used by food manufacturers in the past to keep vegetables and fruits looking fresh and to prevent mold growth. It is also added to some beers, wines, processed foods, and medications as a preservative (see Table 17.4). Very small amounts of sulfite can cause anaphylactic shock and bring on an asthma attack in sensitive people.[18] Because many foods have added sulfite, people sensitive to sulfite should read food ingredient labels carefully. The FDA requires that processed foods containing sulfites list sulfites on the ingredient label. It also prohibits the use of sulfite on fresh vegetables and fruits.

Red Wine, Aged Cheese, and Migraines

Some people develop migraine headaches when they consume red wine. The presence of histamine in wine has been blamed for the headaches. People with this intolerance are unable to break down histamine during digestion, so it accumulates in the blood and causes headaches.[19] Histamine is also found in beer, sardines, anchovies, hard cured sausage, pickled cabbage, spinach, and catsup. Tyramine, a compound closely related to histamine and found in aged cheese, soy sauce, and other fermented products, causes migraine headaches in sensitive people. Chocolate, too, may bring on migraine headaches in sensitive individuals, although this finding is controversial.[20]

MSG and the "Chinese Restaurant Syndrome"

Some people appear to be sensitive to MSG (monosodium glutamate), a flavor enhancer used on meats and in soups, stews, and many Chinese food items. Because symptoms are reported to occur shortly after sensitive individuals eat Chinese food, sensitivity to MSG has been dubbed the "Chinese restaurant syndrome." Symptoms associated with MSG sensitivity include dizziness, sweating, flushing, a rapid heartbeat, and a ringing sound in the ears.[21]

Precautions

People with food allergies or intolerances have to be very careful about what they eat and should have a plan of action ready in case they develop a serious reaction. Becoming highly knowledgeable about which foods and food products contain ingredients that cause an adverse reaction is essential. When eating out, people with allergies and intolerances should ask a lot of questions to make sure they know what is being served. They have to become students of food ingredient labels.

Table 17.4

Some foods that may contain sulfite

- Wine
- Beer
- Hard cider
- Tea
- Fruit juices
- Vegetable juices
- Guacamole
- Dried fruit
- Potato products
- Canned vegetables
- Baked goods
- Spices
- Gravy
- Soup mixes
- Jam
- Trail mix
- Fish and seafood

Photo Disc

On the Side

Foods may be cross contaminated with potential allergens during production or preparation. Yogurt-covered peanuts have been accidentally mixed with yogurt-covered raisins, and milk chocolate has been found to contain peanut fragments.

Photo Disc

Take Action   To Prevent Anaphylaxis

Are you at risk for anaphylactic shock due to a food allergy and unsure how to manage it? If yes, medical experts have advice to share with you about how to be prepared.[1,2,22] The advice can be translated into action by completing the following activities.

1. Scrupulously learn about and avoid eating foods that may trigger anaphylaxis. Get an appointment with a dietitian if needed.

2. Talk with your doctor about the symptoms of anaphylaxis and what you should do if it happens.

3. Get a prescription for two EpiPens and any other medications such as antihistamines and inhalers that may be needed from your doctor.

4. Ask your doctor to teach you how to use injectable epinephrine and any other medications needed when anaphylaxis develops.

5. Obtain these medications and carry the EpiPen and other prescribed, emergency medications with you at all times.

6. Get and wear a medical ID bracelet that contains information on the allergy.

6. Get and wear a medical ID bracelet that contains information on the allergy.

AP Images/Stephen Morton

Planning ahead is the order of the day for individuals with a history of anaphylactic shock. Although preparedness is the key factor related to quick treatment of anaphylaxis, individuals at risk are often not adequately prepared. Physicians frequently do not prescribe EpiPens (injectable epinephrine) to their patients with a history of anaphylactic shock and may not instruct patients on its use.[5] To overcome these and other potential obstacles it is recommended that individuals with a history of anaphylactic shock have a plan for handling an emergency situation.[1,2,22] This topic is the subject of this Unit's Take Action feature.

One final thought about food allergies and intolerances—they are never caused by studying about them.

© Dynamic Graphics/jupiterimages

Up Close

Gluten-Free Cuisine

Focal Point: How to serve gluten-free meals and snacks

You're having a ball game viewing party and want to offer snacks and a meal that can also be enjoyed by your best friend who happens to have celiac disease. What should you serve?

Place a check mark in front of the foods that are naturally gluten-free and could be served. Assume you prepare the foods without adding gluten-containing ingredients. (Hint: refer to Table 17.2.)

____ fresh meats, fish
____ tomatoes
____ pizza
____ macaroni and cheese
____ eggs
____ corn grits
____ corn chips
____ sourdough rolls
____ baked apple with sugar
____ hamburger
____ grapes

____ tossed salad with oil and vinegar dressing
____ apple pie
____ black beans and rice
____ canned soup
____ sausage
____ popcorn
____ cheddar cheese
____ veggie burgers
____ lemonade
____ broccoli

FEEDBACK Answers can be found at the end of this Unit.

[Key Terms

anaphylactic shock, page 17-4
antibodies, page 17-4
celiac disease, page 17-4
double-blind, placebo-controlled food
    challenge, page 17-6

food allergen, page 17-4
food allergy, page 17-2
food intolerance, page 17-2
histamine, page 17-4
immune system, page 17-2

lactose intolerance, page 17-8
lactose maldigestion, page 17-8

Review Questions

TRUE FALSE

1. Peanuts, shellfish, cow's milk, and eggs are among the "big eight" foods that can cause food allergies. □ □

2. Adverse reactions to sulfite in foods are prompted by immune system-mediated processes. □ □

3. The prevalence of food allergies appears to be increasing in the United States. □ □

4. The Food Allergen Labeling and Consumer Protection Act requires that manufacturers remove all potentially allergic ingredients from their foods. □ □

5. Skin prick tests definitively identify allergies to specific foods. □ □

TRUE FALSE

6. Symptoms of food allergy can be triggered by exposure to a very small amount of a food allergen. □ □

7. Food allergies can be treated with "allergy shots." □ □

8. Use of the term *gluten-free* on labels of qualifying food is allowed. □ □

9. Allergies to peanuts, nuts, fish, and shellfish are likely to last a lifetime. □ □

10. Recent research proves that the "Chinese Restaurant Syndrome does *not* actually exist. □ □

Media Menu

www.aaaai.org
This is the home page for the American Academy of Allergy, Asthma, and Immunology. Go from there to tips to remember, information on diagnosis, treatment, and prevention.

www.celiac.com
Get updated news about celiac disease and a list of gluten-free foods at this site.

www.celiac.org
The Celiac Disease Foundation offers information about celiac disease and gluten-free foods.

www.cfsan.fda.gov/~dms/wh-alrgy.html
This address leads you to the FDA Center for Food Safety and Applied Nutrition and information about food allergies.

www.foodallergy.org
The Food Allergy and Anaphylaxis Network provides lots of information on food allergies and intolerances, how to avoid food allergens, advice on how to deal with school food service, support services, and recipes.

www.healthfinder.gov
Search food allergy and food intolerance to find facts, specific information on peanut allergy, avoiding allergies, and treatment.

www.mayoclinic.com
This site is a great resource for scientific information and practical tips on food allergies.

www.nlm.nih.gov/medlineplus/foodallergy.html
Medline's site is filled with useful and interesting information about food allergies.

www.niaid.nih.gov
The National Institute of Allergy and Infectious Diseases offers information about specific types of food allergies at this site.

Notes

1. Kurowski K et al. Food allergies, Am Fam Physician 2008;77:1678–86,1687–88.

2. Lack G. Food allergy, N Engl J Med 2008;359:1252–60.

3. Wahn HU. Strategies for atopy prevention, J Nutr 2008;138:1770S–72S.

4. Formanek R. Food allergies: when food becomes the enemy. FDA Consumer Magazine, April 2004 update, www.fda.gov/FDAC/features/2001/401_food.html, accessed 3/09.

5. Peak I et al. Physician prescriptions for EpiPen among patients with a history of anaphylaxis, Paper presented at the American Academy of Asthma, Allergy, and Immunology Annual Meeting, 2009, 3/15/09.

6. Bren L. Food labels identify allergens more clearly. FDA Consumer Magazine, 2006;Mar–Apr.

7. Catassi C et al. Celiac disease, Curr Opin Gastroenterol 2008;24:687–91.

8. Telega G et al. Asymptomatic patients with celiac disease. Arch Pediatr Adolesc Med 2008;162:164–8.

9. Burrowes JD, Helping adults with celiac disease to eat well, Nutr Today 2008;43:250–6.

10. Fiocchi A et al. Consensus document for introducing solid foods into an infant's diet to avoid food allergies. Ann Allergy Asthma Immunol 2006;97:10–21.

11. Greer F et al. American Academy of Pediatrics: Policy Statement on nutritional options during pregnancy, lactation, and the first year of life related to atopic disease, Pediatrics, 2008;121:183–91.

12. Allergy statistics (www.nal.usda.gov/fnic); and Food allergies at a glance, p. 47, accessed 8/03.

13. May CD. Food sensitivity: facts and fancies. Nutr Rev 1984;42:72–8.

14. Murray JA. The widening spectrum of celiac disease. Am J Clin Nutr 1999;69:354–65. Ferguson A. Food sensitivity or self-deception? N Engl J Med 1990;323:476–8.

15. Chapman JA et al. New guidelines for food allergy. Ann Allergy Asthma Immunol 2006;96:S1–S68.

15. Tolstoi LG. Adult-type lactase deficiency. Nutr Today 2000;35:134–42.

17. Inman-Felton AE. Overview of lactose maldigestion (lactose nonpersistence). J Am Diet Assoc 1999;99:481–89.

18. Yang WH et al. Adverse reactions to sulfites. Can Med Assoc J 1985;133:865–7.

19. Jarisch R et al. Wine and headache. Int Arch Allergy Immunol 1996;110:7–12.

20. Hannington E. Preliminary report on tyramine headache, Br Med J 1967;2:550–1; and Chandra, Food hypersensitivity and allergic disease.

21. Definition of Chinese restaurant syndrome, MedicineNet.com, http://www.medterms.com/script/main/art.asp?articlekey=15584, accessed 9/06.

22. Tang MLK et al. A review of anaphylaxis episodes in children, Allergy 2008;63:1071–6.

22. Ferguson A. Food sensitivity or self-deception? N Engl J Med 1990; 323:476–8.

NUTRITION | # Up Close

Gluten-Free Cuisine

Feedback for Unit 17

Checkmarks should be placed in front of these gluten-free foods:

✔ fresh meats, fish
✔ tomatoes
___ pizza
___ macaroni and cheese
✔ eggs
✔ corn grits
✔ corn chips
___ sourdough rolls
✔ baked apple with sugar
✔ hamburger
✔ grapes

✔ tossed salad with oil and
    vinegar dressing
___ apple pie
✔ black beans and rice
___ canned soup
___ sausage
✔ popcorn
✔ cheddar cheese
___ veggie burgers
✔ lemonade
✔ broccoli

Additional information about gluten-free foods can be found on food package labels and from the Internet at the Web addresses listed in the Media Menu.

Fats and Cholesterol in Health

NUTRITION SCOREBOARD

	TRUE	FALSE
1 It is currently recommended that adults consume diets providing less than 30% of calories from fat.		
2 The types of fat consumed are more important to health than is total fat intake.		
3 Most all of the cholesterol in our diets comes from animal products.		
4 Saturated fat intake has a much stronger influence on blood cholesterol levels than does cholesterol intake.		

Key Concepts and Facts

- Fats are our most concentrated source of food energy. They supply nine calories per gram.

- Dietary fats "carry" the essential fatty acids, fat-soluble vitamins, and healthful phytochemicals along with them in foods.

- Fats are not created equal. Some types of fat have positive effects, and some have negative effects on health.

- Saturated fats and trans fats raise blood cholesterol levels more than does dietary cholesterol or any other type of fat.

Answers to **NUTRITION SCOREBOARD**	TRUE	FALSE
1 The acceptable range of fat intake is 20 to 35% of total calories.		✔
2 The types of fat consumed are more important to health than the total amount of fat.[1]	✔	
3 See Table 18.6.	✔	
4 Saturated fat, which is found primarily in animal products, raises blood cholesterol levels a good deal more than does dietary cholesterol.	✔	

Changing Views about Fat Intake and Health

Scientific evidence and opinions related to the effects of fat on health have changed substantially in recent years—and so have recommendations about fat intake. In the past, it was recommended that Americans aim for diets providing less than 30% of total calories from fat. However, recent evidence indicates that the type of fat consumed is more important to health than total fat intake. The watchwords for thinking about fat have become "Not all fats are created equal: Some are better for you than others." American adults are being urged to select food sources of "healthy" fats while keeping fat intake within the range of 20 to 35% of total caloric intake. Concerns that high-fat diets encourage the development of obesity have been eased by studies demonstrating that excessive caloric intakes—and not just diets high in fat—are related to weight gain.[2]

New recommendations regarding fat intake do not encourage increased fat consumption. Rather, they emphasize that healthy diets include certain types of fat and that total caloric intake and physical activity are the most important components of weight management. Diets providing as low as 20% of calories from fat, and those providing 30 to 35%, can be healthy—depending on the types of fat consumed and the quality of the rest of the diet.[1] This unit provides facts about fats, explains the reasons behind recent changes in recommendations for fat intake, and addresses the practical meaning of it all.

Facts about Fats

Fats are a group of substances found in food. They have one major property in common: They are not soluble (or, in other words, will not dissolve) in water. If you have ever tried to mix vinegar and oil when making salad dressing, you have observed the principle of water and fat solubility firsthand.

Fats are actually a subcategory of the fat-soluble substances known as **lipids**. Lipids include fats, oils, and cholesterol. Dietary fats such as butter, margarine, and shortening are often distinguished from oils by their property of being solid at room temperature. This physical difference between fats and oils is due to their chemical structures.

lipids
Compounds that are insoluble in water and soluble in fat. Triglycerides, saturated and unsaturated fats, and essential fatty acids are examples of lipids, or "fats."

Functions of Dietary Fats

Fats in Foods Supply Energy and Fat-Soluble Nutrients Dietary fats are a concentrated source of energy. Each gram of fat consumed supplies the body with 9 calories worth of energy. That's enough energy for a 160-pound person to walk casually for a little over two minutes or to jog at a slow pace for about a minute. Fats in food supply the **essential fatty acids** (linoleic acid and alpha-linolenic acid) and provide the fat-soluble vitamins D, E, K, and A (the "deka" vitamins). So, part of the reason we need fats in our diet is to get a supply of the essential nutrients they contain (Table 18.1). Diets containing little fat (less than 20% of total calories) often fall short on delivering adequate amounts of essential fatty acids and fat-soluble vitamins.[1]

Fat Contributes to the Body's Energy Stores Fat consumed as part of a dietary intake that exceeds calorie need is converted to triglycerides and stored in fat cells (Illustration 18.1). A pound of body fat can provide approximately 3,500 calories of energy to the body when needed.

Body fat is not just skin deep. Fat is also located around organs such as the kidneys and heart. It's there to cushion and protect the organs and keep them insulated. Cold-water swimmers can attest to the effectiveness of fat as an insulation material. They purposefully build up body fat stores because they need the extra layer of insulation (Illustration 18.2).

Fats Increase the Flavor and Palatability of Foods Although "pure" fats by themselves tend to be tasteless, they absorb and retain the flavor of substances that surround them. Thus, fats in meats and other foods pick up flavors from their environment and give those flavors to the food. This characteristic of fat is why butter, if placed next to the garlic in the refrigerator, tastes like garlic.

AP Images

Illustration 18.2 Although their body fat stores don't fit the image of the superathlete, cold-water swimmers need the fat to help stay warm.
Pictured here is the English swimmer Mike Read, who had swum the English Channel 20 times by age 39. The narrowest width of the English Channel is 22 miles, or 35 km.

Fats Contribute to the Sensation of Feeling Full As they should, at 9 calories per gram! Fats tend to stay in the stomach longer than carbohydrates or proteins and are absorbed over a longer period of time. Their presence in the stomach and small intestine triggers a feeling of fullness. That's why foods with fat "stick to your ribs."[3]

Fats are a Component of Cell Membranes, Vitamin D, and Sex Hormones Some types of fats give cell membranes flexibility and help regulate the transfer of nutrients into and out of cells.[4] Others serve as precursors to vitamin D and sex hormones, such as estrogen and testosterone.

Fats Come in Many Varieties

There are many types of fat in food and our bodies (Table 18.2). Of primary importance are *triglycerides* (or "triacylglycerols"), *saturated and unsaturated fats,*

essential fatty acids
Components of fats (linoleic acid, pronounced *lynn-oh-lay-ick,* and alpha-linolenic acid, pronounced *lynn-oh-len-ick*) required in the diet.

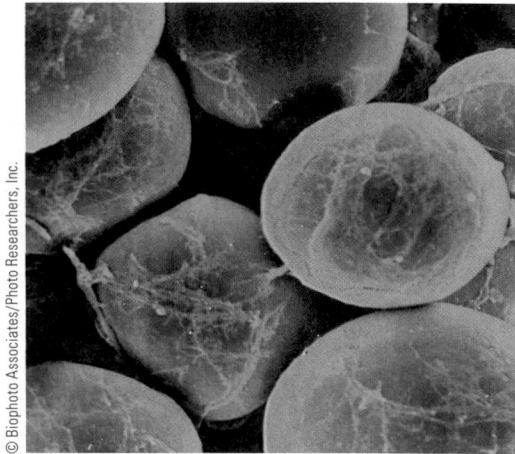

© Biophoto Associates/Photo Researchers, Inc.

Illustration 18.1 A close look at fat cells (color-enhanced microphotograph).

Table 18.2

Basic facts about the types of fat

Fats can be:
- Monoglycerides
- Diglycerides
- Triglycerides

Fats can be:
- Saturated
- Monounsaturated
- Polyunsaturated

Unsaturated fats come in:
- "Cis" forms
- "Trans" forms

Table 18.3

A glossary of fats

Triglycerides: Fats in which the glycerol molecule has three fatty acids attached to it; also called triacylglycerol. Triglycerides are the most common type of fat in foods and in body fat stores.

Saturated fats: Molecules of fat in which adjacent carbons within fatty acids are linked only by single bonds. The carbons are "saturated" with hydrogens; that is, they are attached to the maximum possible number of hydrogens. Saturated fats tend to be solid at room temperature. Animal products and palm and coconut oil are sources of saturated fats.

Unsaturated fats: Molecules of fat in which adjacent carbons are linked by one or more double bonds. The carbons are not saturated with hydrogens; that is, they are attached to fewer than the maximum possible number of hydrogens. Unsaturated fats tend to be liquid at room temperature and are found in plants, vegetable oils, meats, and dairy products.

Glycerol: A syrupy, colorless liquid component of fats that is soluble in water. It is similar to glucose in chemical structure.

Cholesterol: A fat soluble, calories liquid primarily found in animals. Cholesterol is used by the body to form hormones such as testosterone and estrogen and is a component of cell membranes. Cholesterol is present in plant cell membranes but the quantity is small and plants are not considered to be a significant dietary source of cholesterol.

Diglyceride: A fat in which the glycerol molecule has two fatty acids attached to it; also called diacylglycerol.

Monoglyceride: A fat in which the glycerol molecule has one fatty acid attached to it; also called monoacylglycerol.

Monounsaturated fats: Fats that contain a fatty acid in which one carbon-carbon bond is not saturated with hydrogen.

Polyunsaturated fats: Fats that contain a fatty acid in which two or more carbon–carbon bonds are not saturated with hydrogen.

Trans **fats:** Fats containing fatty acids in the trans form. Also called trans fatty acids.

Glycerol + 3 fatty acids = Triglyceride

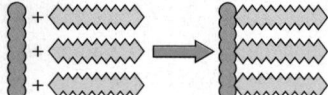

Illustration 18.3 A triglyceride.

cholesterol, and *trans* fats (for definitions, see Table 18.3). The different types of fats have different effects on health.

Triglycerides, which consist of one *glycerol* unit (a glucose-like substance) and three fatty acids (Illustration 18.3), make up 98% of our dietary fat intake and the vast majority of our body's fat stores. Triglycerides are transported in blood attached to protein carriers and are used by cells for energy formation and tissue maintenance. A minority of fats take the form of *diglycerides* (glycerol plus two fatty acids) and *monoglycerides* (glycerol and one fatty acid). Diglycerides are present in some oils and small amounts are used in food products as emulsifiers—or to increase the blending of fat- and water-soluble substances. Monoglycerides are present in small amounts in some oils; we don't consume very much of them in foods.

As far as health is concerned, the glycerol component of fat is relatively unimportant. It's the fatty acids that influence what the body does with the fat we eat; and they are responsible, in part, for how fat affects health. Many different types of fatty acids are found in triglycerides. You've heard of the major ones: those that make fat "saturated" or "unsaturated."

Saturated and Unsaturated Fats   Fatty acids found in fats consist primarily of hydrogen atoms attached to carbon atoms (Illustration 18.4). When the carbons are attached to as many hydrogens as possible, the fatty acid is "saturated"—that is, saturated with hydrogen. Saturated fats tend to be solid at room temperature. Except for palm and coconut oil, only animal products are rich in saturated fats (Illustration 18.5, p. 18-6). Fatty acids that contain fewer hydrogens than the maximum are "unsaturated." They tend to be liquid at room temperature. By and large, plant foods are the best sources of unsaturated fats.

Unsaturated fats are classified by their degree of unsaturation. If only one carbon–carbon bond in the fatty acid is unsaturated, the fat is called

Illustration 18.4 A look at the difference between a saturated and an unsaturated fatty acid.

Hydrogen
Oxygen
Oxygen Carbon **Saturated fatty acid** CH₃

CH CH
Monounsaturated fatty acid

CH CH   CH CH
Polyunsaturated fatty acid

Two hydrogens are missing from each of these carbon–carbon links, making the fatty acid polyunsaturated. With fewer hydrogens to attach to, these carbons are doubly bonded to each other. Monounsaturated fatty acids have only one carbon–carbon bond that is "unsaturated" with hydrogen atoms.

"monounsaturated." If two or more carbon–carbon bonds are unsaturated with hydrogen, the fat qualifies as "polyunsaturated."

The Omega-6 and Omega-3 Fatty Acids

The essential fatty acids linoleic acid and alpha-linolenic acid are members of the fatty acid families of omega-6 (also called n-6 fatty acids) and omega-3 fatty acids (also known as n-3 fatty acids), respectively. Both are polyunsaturated, can be used as a source of energy, and are stored in fat tissue. Because they are essential, both linoleic and alpha-linolenic acid are required in the diet.

Linoleic acid is required for growth, maintenance of healthy skin, and normal functioning of the reproductive system. It is a component of all cell membranes and is found in particularly high amounts in nerves and the brain. A number of biologically active compounds produced in the body that participate in regulation of blood pressure and blood clotting are derived from linoleic acid. The major food sources of linoleic acid are sunflower, safflower, corn, and soybean oils.

Alpha-linolenic acid is a structural component of all cell membranes and is found in high amounts in the brain and other nervous system tissues. It also forms biologically active compounds used in the regulation of blood pressure and blood clotting; but these compounds have the opposite effects on blood pressure and blood clotting as do some of the derivatives of linoleic acid.[5] Omega-3 fatty acids are found in walnuts, dark, leafy green vegetables; and flaxseed, canola, and soybean oils in the form of alpha-linolenic acid.

Other, biologically important omega-3 fatty acids exist, and the two primary ones are EPA (eicosapentaenoic acid, pronounced *e-co-sah-pent-tah-no-ick*) and DHA (docosahexaenoic acid, pronounced *dough-cos-ah-hex-ah-no-ick*). These omega-3 fatty acids can be produced from alpha-linolenic acid, but the conversion process is slow and results in the availability of relatively small amounts of EPA and DHA.[6]

EPA and DHA DHA and EPA are primarily ingredients of fish oils and perform a number of important functions in the body. DHA is a structural component of the brain and is found in high amounts in the retina of the eye. During the

On the Side

Fried fish is generally not a good source of EPA and DHA. The fish used is typically low-fat and a poor source of these omega-3 fatty acids.[7]

© Brand X / jupiterimages

Omega-3 fatty acids are not just good fats, they affect heart health in positive ways.

—PENNY KRIS-ETHERTON, DISTINGUISHED PROFESSOR OF NUTRITION

Illustration 18.5 Fat profiles of selected foods.

Source: Data compiled by author from USDA nutrient database.[11]

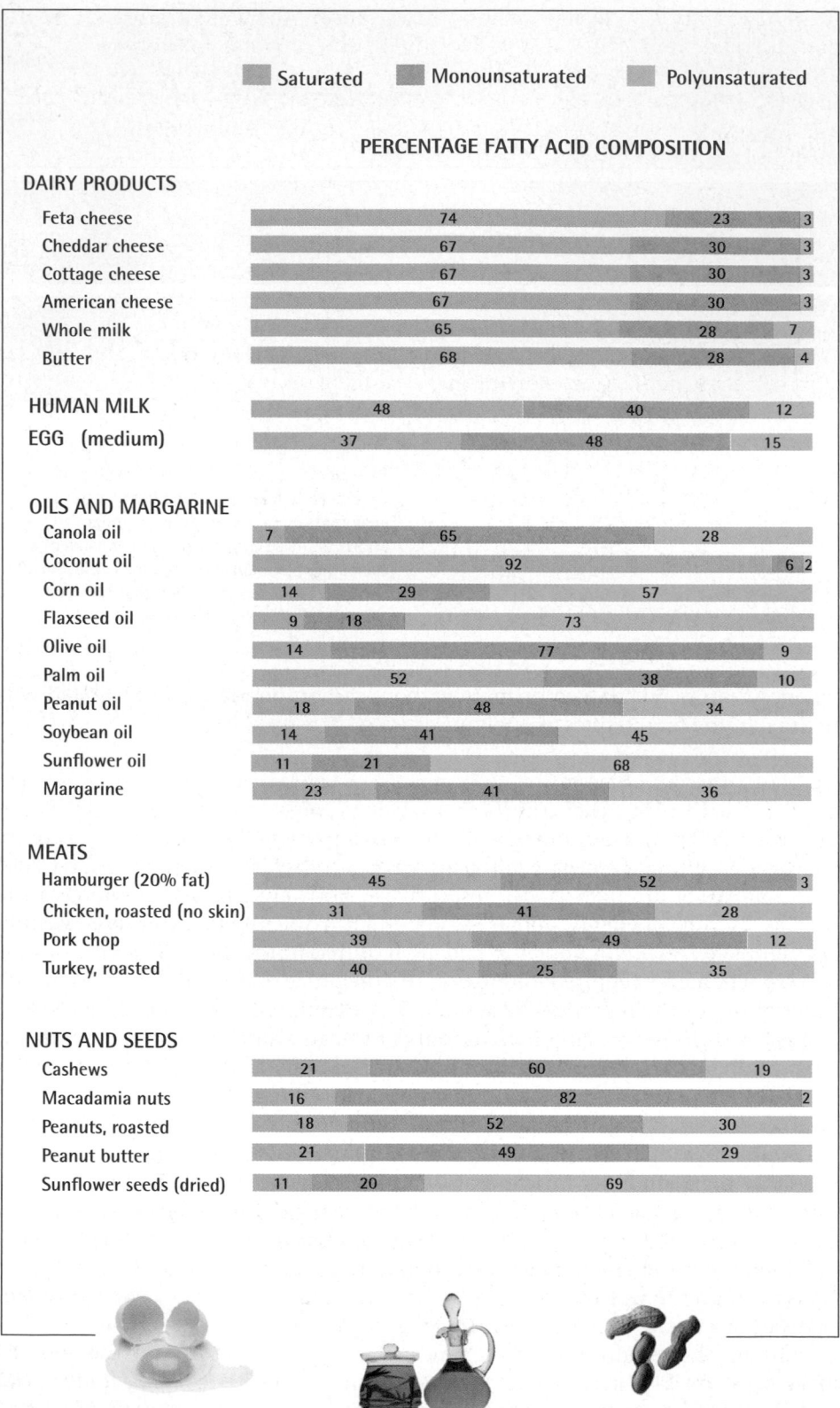

PERCENTAGE FATTY ACID COMPOSITION

Legend: ■ Saturated ■ Monounsaturated ■ Polyunsaturated

DAIRY PRODUCTS

Food	Saturated	Monounsaturated	Polyunsaturated
Feta cheese	74	23	3
Cheddar cheese	67	30	3
Cottage cheese	67	30	3
American cheese	67	30	3
Whole milk	65	28	7
Butter	68	28	4

	Saturated	Monounsaturated	Polyunsaturated
HUMAN MILK	48	40	12
EGG (medium)	37	48	15

OILS AND MARGARINE

Food	Saturated	Monounsaturated	Polyunsaturated
Canola oil	7	65	28
Coconut oil	92	6	2
Corn oil	14	29	57
Flaxseed oil	9	18	73
Olive oil	14	77	9
Palm oil	52	38	10
Peanut oil	18	48	34
Soybean oil	14	41	45
Sunflower oil	11	21	68
Margarine	23	41	36

MEATS

Food	Saturated	Monounsaturated	Polyunsaturated
Hamburger (20% fat)	45	52	3
Chicken, roasted (no skin)	31	41	28
Pork chop	39	49	12
Turkey, roasted	40	25	35

NUTS AND SEEDS

Food	Saturated	Monounsaturated	Polyunsaturated
Cashews	21	60	19
Macadamia nuts	16	82	2
Peanuts, roasted	18	52	30
Peanut butter	21	49	29
Sunflower seeds (dried)	11	20	69

last three months of pregnancy and during infancy, DHA accumulates in these tissues and promotes optimal intellectual and visual development.[8]

EPA serves as a precursor of a number of biologically active compounds involved in blood pressure regulation, blood clotting, and anti-inflammatory reactions. Inflammation is a central component of many chronic diseases, including heart disease, type 2 diabetes, osteoporosis, cancer, Alzheimer's disease, and rheumatoid arthritis. It is part of the body's response to the presence of infectious agents or irritants. A by-product of the body's inflammatory processes, however, are oxidation reactions that can harm cells and tissues. Derivatives of EPA limit the harmful effects of inflammatory and oxidation reactions.[9]

Adequate intake of EPA and DHA for adults is considered to be 500 mg per day.[8] This level of intake can be achieved by consuming 8 ounces of fatty fish weekly. Fish and shellfish content of EPA and DHA is listed in Table 18.4. The table only includes fish and shellfish that contain relatively low amounts of mercury. Mercury can cause nerve tissue damage and learning problems if consumed in excessive amounts. Pregnant and breast-feeding women are advised to limit their intake to 12 ounces per week of low-mercury fish and shellfish.[10]

Consumption of two fish meals a week reduces the risk of heart disease, heart attack, sudden death, and stoke and improves fetal and infant development.[8] Higher amounts of EPA and DHA (1 to 3 grams), generally provided by fish oils, are being used clinically to lower blood triglyceride levels, reduce the incidence of heart disease in people at risk, and reduce the need for anti-inflammatory drugs in people with rheumatoid arthritis.[12] The Food and Drug Administration recommends that consumers ingest no more than a total of 3 grams of EPA and DHA daily and limit supplementary intakes to 2 grams a day. Fish liver oils should be used with caution because they contain relatively high amounts of vitamins A and D. Fish oils, made from the body of the fish, do not.[4]

Table 18.4	
EPA and DHA content of fish and seafoods containing, on average, less than 0.2 ppm mercury in a 3-ounce serving.[11]	
	EPA + DHA, mg
Shad	2,046
Salmon, farmed	1,825
Anchovies	1,747
Herring	1,712
Salmon, wild	1,564
Whitefish	1,370
Mackerel	1,023
Sardines	840
Tilefish	796
Whiting	440
Flounder	426
Trout, fresh water	420
Oysters	375
Snapper	273
Shrimp	268
Clams	241
Haddock	202
Catfish, wild	201
Crawfish	187
Sheepshead	162
Tuna, light, canned in oil	109
Lobster	71

Increasing Omega-3 Fatty Acid Intake   In the past it was thought that consuming high amounts of omega-6 fatty acids compared to omega-3 fatty acids could interfere with the availability of omega-3 fatty acids, particularly EPA and DHA. Although still somewhat controversial, it appears that rather high intakes of omega-6 fatty acids do not interfere with the availability of EPA and DHA.[5] What interferes most with the availability of EPA and DHA for body functions is inadequate intake. On average, adults in the United States and Canada consume 100 mg EPA plus DHA daily, far short of the recommended intake of 500 mg daily.[8]

■ **EPA and DHA Fortified Foods.**   Fatty fish are clearly the richest sources of EPA and DHA. But what if you don't like fish? In that case you can turn to EPA and DHA fortified foods.[13] (For more information on specific EPA and DHA fortified foods see this Unit's Take Action feature.) Purified fish oils with no fishy taste are increasingly being added to products from fruit juice to yogurt, and to animal feeds. The EPA and DHA in feed are incorporated into the animal's tissues. Consequently, beef, pork, eggs, milk and milk products consumed from animals "fortified" with EPA and DHA are fortified, too. Some animal feeds contain DHA from algae and only provide DHA in their food products.

To make sure you're choosing foods with EPA and DHA, or DHA, confirm that the label specifies that these fatty acids are contained in the product. Just because a product announces "Omega-3" on the label doesn't mean it contains EPA and DHA. It may contain flax oil, walnut oil, alpha-linolenic acid, or other sources of omega-3 fatty acids that are not equivalent to EPA or DHA.

Hydrogenated Fats

Unsaturated fats aren't as stable as saturated fats. They are more likely to turn rancid with time and exposure to air and heat than are saturated fats. Additionally, solid fats are preferable to oils for some cooking applications. These problems with unsaturated fats have a solution. It's called hydrogenation.

Take Action To Consume Enough EPA and DHA from Foods Other than Fish

If you would like to increase your EPA and DHA intake, here are some of your non-fish food choices. Indicate with a check mark foods you would likely consume. At the bottom of the list, answer this question: Based on your choices, about how much EPA and DHA would you be adding to your daily diet by consuming the foods you checked?

	EPA and DHA/DHA content, mg
___ DHA fortified eggs, 1	150
___ Healthy Heart Omega-3 orange juice, 8 oz	50
___ Omega Farm nonfat yogurt, 8 oz	75
___ Egg Creations Liquid, $\frac{1}{4}$ cup	260
___ Smart Balance Omega Buttery Spread, 1 Tbsp	32
___ Italica Omega-3 Olive Oil, 1 Tbsp	120
___ Omega Farms Low-Fat Milk, 1 cup	75
___ Omega Farms Mild Cheddar, 1 oz	75
___ Smart Balance Lactose-Free Milk with Omega-3, 1 cup	32
___ Minute Maid 100% Fruit Juice Blend with Omega 3/DHA, 1 cup	50
___ Shrimp, 3 oz	268
___ Clams, 3 oz	241
___ Crab, 3 oz	375
___ Scallops, 3 oz	161

The approximate amount of EPA and DHA I would be adding to my diet daily based on my food choices is: _____ mg. If not at 500 mg, you would be getting closer.

---

hydrogenation
The addition of hydrogen to unsaturated fatty acids.

trans fats
Unsaturated fatty acids in fats that contain atoms of hydrogen attached to opposite sides of carbons joined by a double bond:

```
      H
 —C=C—          H H
      H         —C=C—
Trans fatty     Cis fatty
   acid            acid
```

Fats containing fatty acids in the *trans* form are generally referred to as *trans* fats. *Cis* fatty acids are the most common, naturally occurring form of unsaturated fatty acids. They contain hydrogens located on the same side of doubly bonded carbons.

What's Hydrogenation? **Hydrogenation** is a process that adds hydrogen to liquid unsaturated fats, thereby making them more saturated and solid. The shelf life, cooking properties, and taste of vegetable oils are improved in the process. Hydrogenation has two drawbacks, however. Hydrogenated vegetable oils contain more saturated fat than the original oil. Corn oil, for example, contains only 6% saturated fats; but corn oil margarine has 17%. The other negative is that hydrogenation causes a change in the structure of the unsaturated fatty acids. Specifically, hydrogenation converts some unsaturated fats into **trans fats.**

Trans Fatty Acids The bulk of *trans* fats in our diets comes from hydrogenated vegetable oils. Hydrogenation causes some of the unsaturated fatty acids to be converted from their naturally occurring *cis* to the *trans* form. Ruminant animals like cows, goats, and sheep form a small amount of *trans* fats in their stomachs. Consequently, milk and milk products from these animals will contain *trans* fats.[14]

Repositioned hydrogen molecules in *trans* fats change the way the body uses the fat. *Trans* fats raise blood cholesterol levels more than any other type of fat. They increase the risk of heart disease, stroke, sudden death from heart disease, and type 2 diabetes, and they promote inflammation. Intake of *trans* fats as low as 1% of total calories per day strongly increases the risk of heart disease. An intake of 2.2 grams of *trans* fat daily would place a person consuming 2000 calories a day at increased risk of heart disease.[15]

It is recommended that Americans consume as little *trans* fats as possible,[1] and nutrition information labeling requirements that began in 2006 are making that easier to accomplish. Nutrition Facts panels must include the *trans* fat content of food products (Illustration 18.6). The %DV column (for percent of Daily Value) is not used for *trans* fats because there is no recommended level of intake. Products labeled "*trans* fat-free" (Illustration 18.7) must contain less than 0.5 gram of both *trans* and saturated fats. The requirement to label the *trans* fat content of food products, regulations by various states and cities that ban their use in restaurant foods, and increased consumer awareness of the adverse effects of trans fat have forced food producers to take trans fats out of many prepared and processed foods.[16,17] Prepared foods contain less trans fats than they did just a few years ago. A sampling of some of the remaining food products that contain trans fats is given in Table 18. 5.

Checking Out Cholesterol

Cholesterol is a lipid found *primarily* in animal products. It is tasteless and odorless and contained in both the lean and fat parts of animal products. Table 18.6 lists some sources of cholesterol. Cholesterol is present in plant cell membranes, but the quantity is small and plants are not considered to be a significant dietary source of cholesterol.

Sources of Cholesterol

The cholesterol used by the body comes from two sources. Most (about two-thirds) of the cholesterol available to the body is produced by the liver. The rest comes from the diet (Illustration 18.8). Because the liver produces cholesterol from other substances in our diet, it does not qualify as an essential nutrient.

The Contributions of Cholesterol

Would you be surprised to learn that cholesterol:

- is found in every cell in your body?

- serves as the building block for estrogen, testosterone, and the vitamin D that is produced in your skin on exposure to sunlight?

- is a major component of nerves and the brain?

- cannot be used for energy (so it provides no calories)?

The body has many uses for cholesterol (Table 18.7). It doesn't just accumulate in arteries!

Nutrition Facts		
Serving Size 1 Entree		
Serving Per Container 1		
Amount Per Serving		
Calories 380 Calories from Fat 170		
		%Daily Value
Total Fat 19g		**29%**
Saturated Fat 10g		**50%**
Trans Fat 2g		
Cholesterol 85mg		**28%**
Sodium 810mg		**34%**
Total Carbohydrate 33g		**11%**
Dietary Fiber 3g		**12%**
Sugars 5g		
Protein 20g		
Vitamin A 10%		Vitamin C 0%
Calcium 10%		Iron 15%

Percent Daily Values are based on a 2000 calorie diet. Your daily values may be higher or lower depending on your calorie needs:

	Calories	2000	2500
Total Fat	Less Than	65g	80g
Sat Fat	Less Than	20g	25g
Cholesterol	Less Than	300mg	300mg
Sodium	Less Than	2400mg	2400mg
Total Carbohydrate		300g	375g
Dietary Fiber		25g	30g

Illustration 18.6 *Trans* fat: the newest addition to nutrition facts panels.

Richard Anderson

Illustration 18.7 Products that feature "no *trans* fats" and "*trans* fat-free" labels.

Table 18.5

Where are the trans fats now?

Values may change as companies lower the trans fat content of foods.

Food	Trans Fatty Acids (Grams)
Frosting, canned, 2 Tbsp	2
Pound cake, 2.5 oz	1
Lemon Bar mix, 1–2″ bar	1
Chocolate syrup Brownie mix, 1–2″ bar	1
Sugar cookies mix, 2 cookies	1

Source: Trans fat content of foods is based on a "trans fat search" conducted by the author in supermarkets, 4/09.

Table 18.6

Food sources of cholesterol

Note that all the leading food sources are animal products. Cholesterol in foods is a clear, oily liquid found in the fat and lean portions of many animal products.

Animal product	Amount	Cholesterol (Milligrams)
Brain	3 oz	1746
Liver	3 oz	470
Egg	1	212
Veal	3 oz	128
Shrimp	3 oz	107
Prime rib	3 oz	80
Chicken (no skin)	3 oz	75
Turkey (no skin)	3 oz	65
Hamburger, regular	3 oz	64
Pork chop, lean	3 oz	60
Fish, baked (haddock, flounder)	3 oz	58
Ice cream	1 cup	56
Sausage	3 oz	55
Hamburger, lean	3 oz	50
Milk, whole	1 cup	34
Crab, boiled	3 oz	33
Lobster	3 oz	29
Cheese (cheddar)	1 oz	26
Milk, 2%	1 cup	22
Yogurt, low-fat	1 cup	17
Milk, 1%	1 cup	14
Butter	1 tsp	10
Milk, skim	1 cup	7

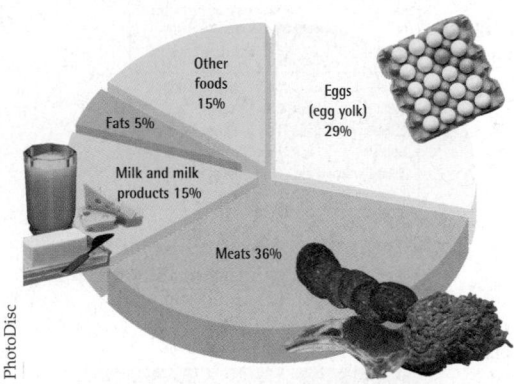

PhotoDisc

Illustration 18.8 Food sources of cholesterol in the U.S. diet.[11] Percentages indicate the proportion of cholesterol each type of food contributes to the diet.

Finding Out about the Fat Content of Food

Not all of the fat in food is visible. To avoid being fooled, it helps to use a reference on the fat composition of foods. Table 18.8 lists the fat content of common food sources of fat, including candy. Vegetables and fruits (except avocado and coconut) and grains are not listed because they contain relatively little fat. Other references can also be used, such as the food composition tables in Appendix A, the Diet Analysis Plus Program software, and the nutrition labels on food products.

Knowledge of the caloric and fat content of a food can be used to calculate the percentage of calories provided by fat. For example, suppose that a slice of cherry pie provides 350 calories and 15 grams of fat. To calculate the percentage of fat calories, multiply 15 grams by 9 (the number of calories in each gram of fat), divide the result by 350 calories, and then multiply this result by 100:

$$15 \text{ grams fat} \times 9 \text{ calories/gram} = 135 \text{ calories}$$
$$135 \text{ calories}/350 = 0.39$$
$$0.39 \times 100 = 39\% \text{ of total calories from fat}$$

Fat Labeling

Nutrition labeling regulations for fat require that food manufacturers adhere to standard definitions of "low fat," "fat-free," and related terms used on food labels. Similarly, claims made about the cholesterol content of food products must comply with standard definitions (Table 18.9, p. 18-12). If a claim is made about the fat content of a food, the nutrition facts panel must specify the food's fat, saturated fat, *trans* fat, and cholesterol content. If a claim is made about cholesterol content

Table 18.7

How the body uses cholesterol

- Cholesterol is a component of all cell membranes, the brain, and nerves.
- Cholesterol is needed to produce estrogen, testosterone, and vitamin D.

Table 18.8

The fat content of some foods

Food	Amount	Grams	Percentage of Total Calories From Fat
Fats and oils			
Butter	1 tsp	4.0	100%
Margarine	1 tsp	4.0	100
Oil	1 tsp	4.7	100
Mayonnaise	1 tbs	11.0	99
Heavy cream	1 tbs	5.5	93
Salad dressing	1 tbs	6.0	83
Meats and fast foods			
Hot dog	1 (2 oz)	17.0	83
Bologna	1 oz	8.0	80
Sausage	4 links	18.0	77
Bacon	3 pieces	9.0	74
Salami	2 oz	11.0	68
Hamburger, regular (20% fat)	3 oz	16.5	62
Chicken, fried with skin	3 oz	14.0	53
Big Mac	6.6 oz	31.4	52
Quarter Pounder with cheese	6.8 oz	28.6	50
Whopper	8.9 oz	32.0	48
Steak (rib eye)	3 oz	9.9	47
Veggie pita	1	17.0	38
Chicken, baked without skin	3 oz	4.0	25
Flounder, baked	3 oz	1.0	13
Shrimp, boiled	3 oz	1.0	10
Milk and milk products			
Cheddar cheese	1 oz	9.5	74
American cheese	1 oz	6.0	66
Milk, whole	1 cup	8.5	49
Cottage cheese, regular	1/2 cup	5.1	39
Milk, 2%	1 cup	5.0	32
Milk, 1%	1 cup	2.7	24
Cottage cheese, 1% fat	1/2 cup	1.2	13
Milk, skim	1 cup	0.4	4
Yogurt, frozen	3/4 cup	0.0–6.6	0–3
Other			
Olives	4 medium	1.5	90
Avocado	1/2	15.0	84
Almonds	1 oz	15.0	80
Sunflower seeds	1/4 cup	17.0	77
Peanuts	1/4 cup	17.5	75
Cashews	1 oz	13.2	73
Egg	1	6.0	61
Potato chips	1 oz (13 chips)	11.0	61
French fries	20 fries	20.0	49
Taco chips	1 oz (10 chips)	6.2	41
Candy			
Peanut butter cups, 2 regular	1.6 oz	15.0	54
Milk chocolate	1.6 oz	14.0	53
Almond Joy	1.8 oz	14.0	50
Kit Kat	1.5 oz	12.0	47
M & M's, peanut	1.7 oz	13.0	47

(and claims can be made only for products that normally contain a meaningful amount of cholesterol), the nutrition panel must also reveal the product's fat and saturated fat content. To prevent the use of unrealistically small serving sizes as a way to appear to cut down on a product's fat content, standard serving sizes must be used on food labels.

Recent Changes in Recommendations for Fat and Cholesterol Intake

Illustration 18.9 A look at the cuisine of the Mediterranean diet.

PhotoDisc

Adherence to certain types of diets that are relatively high in total calories from fats, such as the Mediterranean diet highlighted in Illustration 18.9, reduce the risk of heart disease, stroke, obesity, or a number of other diseases.[18] Evidence established on the healthful effects of the Mediterranean diet and other examples of dietary fat intake and health in different populations prompted the development of new recommendations based primarily on the healthfulness of various types of fats.[1]

"Good" Fats, "Bad" Fats Fats come in many types in foods, and with few exceptions, they serve as a source of energy and provide a number of essential functions in the body. With regard to raising or lowering the risk of heart disease and stroke, however, fats differ. Those that elevate total cholesterol and LDL-cholesterol levels are regarded as "bad" or unhealthy fats. Those that lower total cholesterol and LDL-cholesterol and raise blood levels of HDL-cholesterol (the one that helps the body get rid of cholesterol in the blood) are considered "good" or healthy."[19]

The list of unhealthy fats includes *trans* fats, saturated fats, and cholesterol. Fats labeled "bad" are generally solid at room temperature and are included in foods such as high-fat meats and dairy products, hard margarines, shortening, and crispy snack foods.[19] Monounsaturated fats, polyunsaturated fats, alpha-linolenic acid, DHA, and EPA are considered healthy fats and are present in food in the form of oils (Table 18.10).

Recommendations for Fat and Cholesterol Intake

Current recommendations for adults call for consumption of 20 to 35% of total calories from fat. Average fat consumption in the United States is at the upper end of this range (35% of total calories).[20] The AIs (Adequate Intakes) for the essential

Healthy and unhealthy fats and examples of food sources

Healthy Fats	Unhealthy Fats
DHA, EPA (omega-3 fatty acids) fish and seafood	***Trans* fats** Snack and fried foods, bakery goods
Monounsaturated fats Olive and peanut oil, nuts, avocados	**Saturated fats** Animal fats
Polyunsaturated fats Vegetable oils	**Cholesterol** eggs, seafood, and meat
Alpha-linolenic acid Soybeans, walnuts, flaxseed	

fatty acid linoleic acid is set at 17 grams a day for men and 12 grams for women. AIs for the other essential fatty acid, alpha-linolenic acid, are 1.6 grams per day for men and 1.1 grams for women. It is recommended that intake of *trans* fats and saturated fats be as low as possible while consuming a nutritionally adequate diet. Only a small proportion of Americans consume too little linoleic acid, but intakes of alpha-linolenic acid tend to be low. Americans are being encouraged to increase consumption of EPA and DHA by eating fish more often. In addition, saturated fat intake averages 11 to 12% of calories, an amount that increases the risk of heart disease.[8, 20]

There is no recommended level of cholesterol intake, because there is no evidence to indicate that cholesterol is required in the diet. The body is able to produce enough cholesterol, and people do not develop a cholesterol deficiency disease if it is not consumed. Because blood cholesterol levels tend to increase somewhat as consumption of cholesterol increases, it is recommended that intake should be minimal. Although cholesterol intake averages 237 mg per day in the United States,[20] a more health-promoting level of intake would be less than 200 mg a day.[21]

Recent recommendations for fat intake represent an unusually large but necessary change in dietary intake guidance. Much remains to be understood about the effects of dietary fats on health, and how other components of diet, lifestyle, and genetic traits modify relationships between fat intake and health.

Reality Check

Good fats, bad fats   What foods provide "healthy" fats?

Who gets thumbs up?
Answers on page 18–14

Photo Disc

Kristen:
How can I be wrong? Low-fat food products are best for healthy fat because they contain almost no fat!

Butch:
I'm thinking foods like fish, peanut butter, and *trans* fat–free margarine contain healthy fats.

"Good" fats, "bad" fats Low-fat foods contain less fat than the regular version of the foods. But that doesn't mean the products contain no fat, or only good fats. Food sources of fish oils, unsaturated fat, and *trans* fat–free products provide the healthy fats. As always, healthy diets aren't based on individual foods, they are based on overall diets. You can emphasize foods providing healthy fats without feeling bad about occasionally eating foods branded with the "bad fat" label.

Kristen:

Butch:

Photo Disc

NUTRITION

© James And James/Getty Images/FoodPix

Up Close

The Healthy Fats in Your Diet

FOCAL POINT: Identify your healthy fat food choices.

Are the fats in your diet the healthy type? Check it out by answering these questions:

How Often Do You Eat:	Seldom or Never	1–2 Times per Week	3–5 Times per Week	Almost Daily
1. Sausage, hot dogs, ribs, and luncheon meats?	☐	☐	☐	☐
2. Heavily marbled steaks or roasts and chicken with the skin?	☐	☐	☐	☐
3. Soybean products such as tofu or soynuts?	☐	☐	☐	☐
4. Nuts or seeds?	☐	☐	☐	☐
5. Whole milk, cheese, or ice cream?	☐	☐	☐	☐
6. Soft margarine or olive oil?	☐	☐	☐	☐
7. French fries, snack crackers, commercial bakery products?	☐	☐	☐	☐
8. Rich sauces and gravies?	☐	☐	☐	☐
9. Fish or seafood?	☐	☐	☐	☐
10. Peanut butter?	☐	☐	☐	☐

FEEDBACK (including scoring) can be found at the end of Unit 18.

Key Terms

essential fatty acids, page 18–3
hydrogenation, page 18–8

lipids, page 18–2

trans fats, page 18–8

Review Questions

TRUE   FALSE

1. Animal products are by far the leading source of saturated fats in the American diet. ☐ ☐

2. Flaxseed oil and dark, leafy green vegetables are excellent sources of the essential fatty acid alpha-linolenic acid. ☐ ☐

3. The best food sources of EPA and DHA are fish and shellfish. ☐ ☐

4. Cholesterol is not a required nutrient because it serves no essential, or life-sustaining, function in the body. ☐ ☐

5. Thanks in part to nutrition labeling requirements, there are lower amounts of *trans* fats in food products now than in the recent past. ☐ ☐

TRUE   FALSE

6. *Trans* fats raise blood cholesterol levels more than any other type of fat. ☐ ☐

7. If a claim is made on a food package about the product's fat content, the nutrition facts panel must list the food's fat, saturated fat, *trans* fat, and cholesterol content. ☐ ☐

8. The type of fat consumed is more important to health than is total fat consumption. ☐ ☐

9. Unhealthful, or "bad," fats include *trans* fats and saturated fats. ☐ ☐

10. It is recommended that adults consume 20 to 25% of total calories from fat. ☐ ☐

Media Menu

www.nlm.nih.gov/medlineplus/dietaryfat.html
This site provides a menu that connects you to other sites such as good and bad fats, interpreting blood lipid profiles, benefits of omega-3 fatty acids, and fat substitutes.

www.nal.usda.gov/fnic
Find out more about fats and fat replacers from the extensive list of topics covered under the search term "fats."

www.healthfinder.gov
Here is a good source of information on fats, *trans* fat, and cholesterol through search terms such as dietary fat and healthy fats.

www.mayoclinic.com
Healthy fats, bad fats, know your fats, fats and heart disease, and other topics are intelligently covered in sites available through the Mayo Clinic's home page.

www.epa.gov/waterscience/fish
The Environmental Protection Agency's site is good for looking up national and local advisories on fish contamination and consumption.

1. Dietary Reference Intakes, Energy, carbohydrate, fiber, fat, fatty acids, cholesterol, protein, and amino acids. Institute of Medicine, National Academy of Sciences, Washington, DC: National Academies Press; 2002

2. Shai I et al. Weight loss with a low-carbohydrate, Mediterranean, or low-fat diet. N Engl J Med. 2008;359:229–241.

3. Schwartz GJ et al. The lipid messenger OEA links dietary fat intake to satiety, Cell Metab, 2008;8:281–8.

4. Oh R. Practical applications of fish oil (omega-3 fatty acids) in primary care. J Am Board Fam Pract 2005;18:28–36.

5. Harris WS et al. Omega-6 fatty acids and risk for cardiovascular disease, Circulation 2009; available at http://circ.ahajournals.org, accessed 1/09.

6. Stark AH et al. Update on alpha-linolenic acid, Nutr Rev 2008;66:326–32.

7. Chung H et al. Frequency and type of seafood consumed influence plasma (n-3) fatty acid concentrations J Nutr 2008;138:2422–7.

8. Position of the American Dietetic Association and Dietitians of Canada:
dietary fatty acids, J Am Diet Assoc 2007;107:1599–1611.

9. Kornman KS. Interleukin 1 genetics, inflammatory mechanisms, and nutrigenetic opportunities to modulate diseases of aging. Am J Clin Nutr 2006;83(suppl):475S–83S.

10. FDA affirms position on mercury in fish. www.fda.gov, issued 6/7/06.

11. Nutrient Composition of foods, USDA Nutrient Database, www.ars.usda.gov/nutrientdata and www.ars.usda.gov/ba/bhnrc/ndl, accessed 6/10/06, 9/08.

12. American Heart Association. Guidelines for the secondary prevention of cardiovascular disease. Circulation, posted online May 15, 2006, www.medscape.com/viewarticle/532327.

13. Harris WS. International recommendations for consumption of long-chain omega-3 fatty acids, J Cardiovasc Med 2007;8 Suppl:1:S50–2.

14. Stuppy P. Transitioning away from trans fatty acids, Today's Dietitian 2003;Jan.:12–14; and Dietary Reference Intakes.

15. Mozaffarian D et al. Trans fatty acids and cardiovascular disease. N Engl J Med 2006;345:1602–12.

16. Eckel RH et al. Americans' awareness, knowledge, and behaviors regarding fats 2006–2007, J Am Diet Assoc 2009;109:288–96.

17. Okie S. New York to trans fats: You're out! N Engl J Med 2008;356:2017–12.

18. Trichopoulou A et al. Adherence to a Mediterranean diet and survival in a Greek population. N Engl J Med 2003;348:2599–608.

19. Fats: the good and the bad (www.mayoclinic.com/invoke.cfm?id=NU00262); and Kris-Etherton et al., New guidelines.

20. What we eat in America, NHANES, 2005–2006, USDA, 2008, available at www.are.usda.gov/ba/bhnrc/fsrg.

21. Krauss RM et al., Revision 2000: a statement for healthcare providers from the Nutrition Committee of the American Heart Association, J Nutr 2001;131:132–46.

Answers to Review Questions

1. True, see page 18-4.
2. True, see page 18-5.
3. True, see page 18-7.
4. False, see page 18-13.
5. True, see page 18-9.
6. True, see page 18-8.
7. True, see page 18-10.
8. True, see pages 18-2. 18-12.
9. True, see page 18–13.
10. False, see page 18–12.

NUTRITION | Up Close

The Healthy Fats in Your Diet

Feedback for Unit 18

Give yourself a point for each time you checked the "3–5 Times per Week" or "Almost Daily" columns for numbers 3, 4, 6, 9, and 10. These foods are sources of healthy unsaturated fats or DHA and EPA. Take a point away for each time your answer ended up in the same columns for foods listed in numbers 1, 2, 5, 7, and 8. These foods provide saturated or *trans* fats. If you have any points left, your selection of food sources of fat regularly include healthy fats.

Nutrition and Heart Disease

NUTRITION SCOREBOARD

	TRUE	FALSE
1. Heart disease is the leading cause of death among men and women in the United States.		
2. The most effective way for most people to lower blood cholesterol is to exercise regularly.		
3. Plaque that causes hardening of the arteries develops on the inside surface of artery walls.		
4. Chronic inflammation is a major contibutor to the development and progression of heart disease.		
5. The risk factors for heart disease are the same for women as for men.		

Key Concepts and Facts

- Heart disease is the leading cause of death in men and women in the United States.

- Dietary and lifestyle factors are among the most important contributors to heart disease.

- Moderate-fat diets that provide primarily "healthy fats" decrease heart disease risk to a greater extent than do low-fat, high-carbohydrate diets.

- Lowering high blood LDL-cholesterol and raising HDL-cholesterol levels reduces the risk of heart disease.

		TRUE	FALSE
Answers to **NUTRITION SCOREBOARD**			
1	About 27% of the deaths of men and women in the United States each year are related to heart disease.[1]	✔	
2	The most effective way for most people to lower blood cholesterol level is to reduce saturated and *trans* fat intake.[2]		✔
3	Plaque develops underneath the inside surface of artery walls.		✔
4	Chronic inflammation in artery walls plays a major role in the development and progression of heart disease. Diet is a major factor influencing inflammation.[3]	✔	
5	Risk factors for heart disease differ between women and men.[4]		✔

The Diet–Heart Disease Connection

Suspicions that dietary fat may be related to heart disease were first raised over 200 years ago. During the late 18th century, physicians noted that people who died of heart attacks had fatty streaks and deposits in the arteries that led to the heart. Unlike many early theories about the causes of diseases, this one has basically held up over time. It is clearly established that diets high in saturated fat and *trans* fat (rather than total fat as originally suspected) are a major risk factor for heart disease in the United States.[2] Other risk factors for heart disease have been identified and—except for genetic tendencies, sex, and age—can be modified. That makes heart disease a highly preventable cause of death and disability.

Declining Rates of Heart Disease

As Americans have become more aware of the relationship between diet and heart disease, food choices have changed, and the average blood cholesterol level in adults has dropped from 240 milligrams in the early 1970s to just under 200 in recent years. As cholesterol levels have declined, so has the rate of heart disease deaths in the United States (Illustration 19.1). The 60% drop in deaths from heart disease since 1950 is primarily related to declines in blood cholesterol levels, reduced rates of smoking, improved blood pressure control, and advances in medications and medical care.[6]

Despite the general decline in blood cholesterol levels and rate of heart disease deaths, cholesterol levels of U.S. adults is still too high. Every 1% drop in average cholesterol level in countries with high rates of heart disease produces a 2% decline in heart disease rate. Raising levels of beneficial HDL cholesterol by 1 milligram per deciliter reduces rates by 2 to 3%.[7]

No disease that can be treated by diet should be treated with any other means.

—MAIMONIDES, 1135–1204

Photo Disc

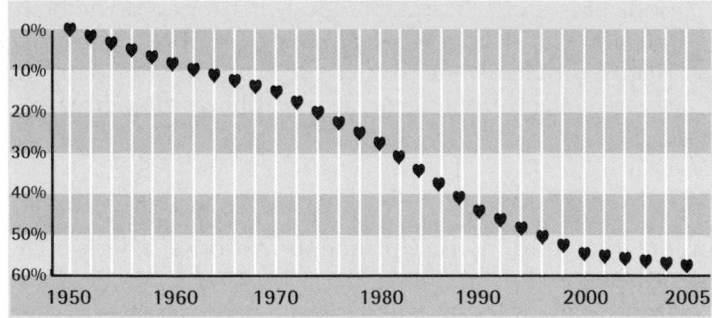

Illustration 19.1 The percentage decline in death rates for heart disease since 1950.[1]

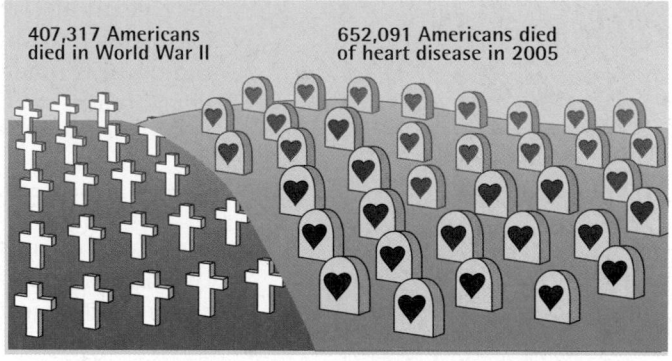

407,317 Americans died in World War II

652,091 Americans died of heart disease in 2005

Illustration 19.2 The impact of heart disease.
If deaths from stroke and other diseases related to atherosclerosis are included, the annual death toll reaches 795,670.[1]

A Primer on Heart Disease

There is no bigger health problem in the United States and other developed countries than heart disease (Illustration 19.2). Worldwide, heart disease accounts for one out of every four deaths.[8] It is an "equal opportunity" disease, striking as many women as men, although women on average die 10 years later from heart disease than do men. It is an *un*equal opportunity disease in that African Americans and Hispanic Americans are more likely to die from heart disease than are Caucasian Americans.[9]

What Is Heart Disease?

Heart disease, or more correctly "coronary" heart disease, is a term referring to several disorders that result from inadequate blood circulation to parts of the heart. Heart disease is almost always due to a narrowing of the arteries leading to the heart. Arteries become narrow due to a buildup of **plaque** (Illustration 19.3). People with narrowed arteries have **atherosclerosis**, or "hardening of the arteries" as it is often called. Heart disease develops silently over time (usually decades).

When arteries are narrowed by 50% or more, the shortage of blood to the heart can produce chest pain (called "angina"). A heart attack occurs when an artery leading to the heart becomes clogged by a piece of plaque released by a ruptured portion of the artery wall or by a blood clot (Illustration 19.4). Although heart disease primarily affects individuals over the age of 55, it's a progressive disease that may begin in childhood.[10]

Arteries leading to the heart aren't the only ones affected by atherosclerosis. Plaque can also build up in the arteries in the legs, neck, brain, and other body parts. If the blood supply to the legs is reduced, pain and muscle cramps

heart disease
One of a number of disorders that result when circulation of blood to parts of the heart is inadequate. Also called *coronary heart disease*. ("Coronary" refers to the blood vessels at the top of the heart. They look somewhat like a crown.)

plaque
Deposits of cholesterol, other fats, calcium, and cell materials in the lining of the inner wall of arteries.

atherosclerosis
"Hardening of the arteries" due to a build-up of plaque.

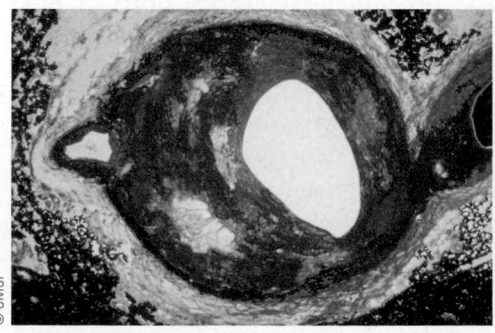

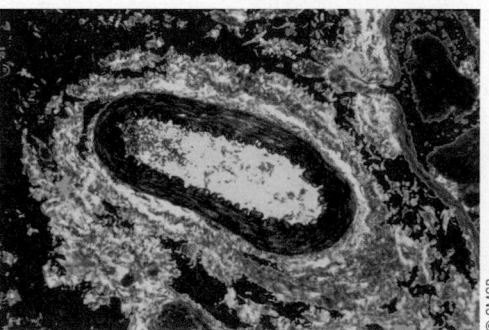

Illustration 19.3 The progression of atherosclerosis.
As plaque builds up, arteries narrow, reducing or stopping the supply of blood to the heart, brain, muscle, or other affected parts of the body.

may result after brief periods of exercise. Plaque buildup in arteries of the brain contributes to stroke—an event that occurs when the blood supply to a part of the brain is inadequate. Health problems due to atherosclerosis in arteries of the heart, brain, neck, and legs are collectively referred to as **cardiovascular disease**.

What Causes Atherosclerosis?

A number of conditions are known to increase plaque formation in arteries. The two most influential identified so far are elevated blood levels of cholesterol and **chronic inflammation** in the inner walls of arteries leading to the heart. High blood cholesterol levels and inflammation are interrelated and act together to increase atherosclerosis.[3]

Blood Cholesterol Levels and Heart Disease   In general, the higher the blood cholesterol level, the more likely it is that plaque will build up in the arteries and heart disease will occur (Illustration 19.5). (Some people with atherosclerosis don't have high blood levels of cholesterol, however.)[10] A person's blood cholesterol level is determined by a number of factors, including dietary intake, smoking, exercise, and genetic traits. Diets high in saturated fat elevate cholesterol levels in most people. Such diets are characterized by the regular consumption of high-fat milk and cheese, eggs, and beef. One type of unsaturated fat raises blood cholesterol levels more than saturated fat do, and that is *trans* fat. This type of fat is produced when vegetable oils are hydrogenated—made solid by the addition of hydrogen. Blood cholesterol levels can also be raised by high cholesterol intakes. But blood cholesterol responds less to dietary cholesterol intake than to *trans* or saturated fat intake.[2]

All Blood Cholesterol Is Not Equal   Cholesterol is soluble in fat, but blood is mostly water. Therefore, cholesterol must be bound to compounds that mix with water, or it would float in blood. For this reason, cholesterol present in blood is bound to protein, which is soluble in water. The resulting combination is called a "lipoprotein," and there are a number of different types. Two lipoproteins have gained notoriety by virtue of their coverage in the popular press and their role in the development of heart disease. One is HDL cholesterol (for "high-density lipoprotein" cholesterol—it could be nicknamed "Heart-Disease-Lowering" cholesterol). It is the "good" type of cholesterol, and you want high levels of it in your blood. LDL cholesterol (low-density lipoprotein cholesterol) is the villain. The composition and roles of both types of lipoproteins are summarized in Illustration 19.6.

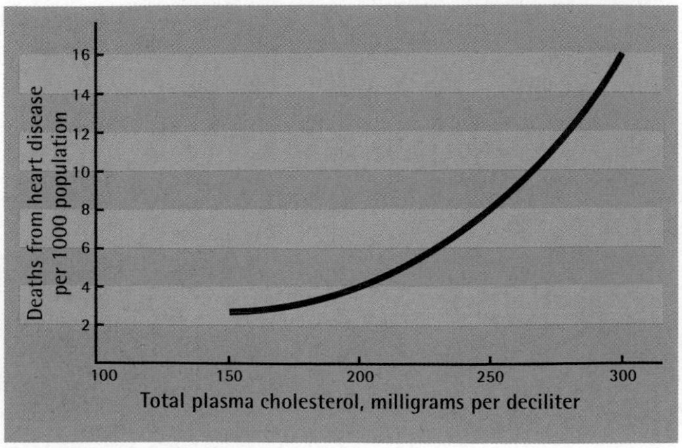

■ **Understanding HDL and LDL.**   HDL gets its reputation as the good cholesterol because it helps remove cholesterol from the blood. An avid cholesterol acceptor, HDL escorts cholesterol to the liver for its eventual excretion from the body. High HDL-cholesterol levels (over 40 milligrams per deciliter in men and over 50 in women) are protective against heart disease. LDL cholesterol carries more cholesterol than does HDL, and its cholesterol can be incorporated into plaque. The higher the LDL-cholesterol level, the more likely it is that atherosclerosis will develop and progress into heart disease.[10]

LDL- and HDL-cholesterol are components of total blood cholesterol measures. High levels of beneficial HDL-cholesterol may elevate total cholesterol values and make it appear that a person is at risk of heart disease. High levels of HDL-cholesterol (particularly those over 60 mg/dL) are strongly protective against heart disease.[11] Risk of heart

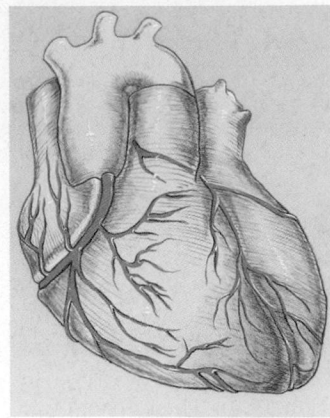

Illustration 19.4   The heart after a heart attack.
The dark portion at the base of the heart is affected by the blockage in blood flow.

cardiovascular disease
Disorders related to plaque buildup in arteries of the heart, brain, and other organs and tissues.

chronic inflammation
Inflammation that lasts weeks, months, or years. Inflammation is the first respose of the body's immune system to infection or irritation. It triggers the release of biologically active substances that promote oxidation and other potentially harmful reactions in the body.

Illustration 19.5   The relationship between blood cholesterol level and death from heart disease.

Source: Grundy SM. Cholesterol and coronary heart disease: a new era. JAMA 1986;256:2849–58. Copyright 1986, American Medical Association.

(graph: x-axis "Total plasma cholesterol, milligrams per deciliter" ranging 100 to 300; y-axis "Deaths from heart disease per 1000 population" ranging 2 to 16)

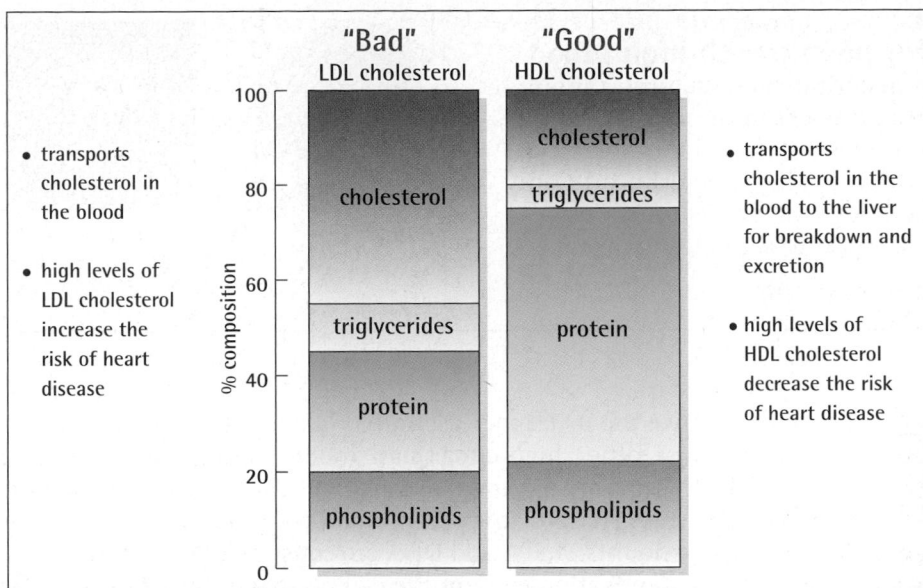

"Bad"
LDL cholesterol

"Good"
HDL cholesterol

- transports cholesterol in the blood

- high levels of LDL cholesterol increase the risk of heart disease

cholesterol

triglycerides

protein

phospholipids

cholesterol

triglycerides

protein

phospholipids

% composition

- transports cholesterol in the blood to the liver for breakdown and excretion

- high levels of HDL cholesterol decrease the risk of heart disease

Illustration 19.6 The "bad" cholesterol (LDL) and the "good" cholesterol (HDL) found in blood.

disease is more accurately predicted based on individual measures of LDL- and HDL-cholesterol than by total blood cholesterol results.

Triglycerides and Heart Disease Risk. Triglycerides are transported in blood attached to VLDL (very-low-density lipoprotein) cholesterol. Until recently, the risk of heart disease posed by elevated blood levels of triglycerides had taken a back seat to blood cholesterol. Newer study results confirm that high blood levels of triglycerides increase heart disease risk and that efforts to prevent and treat heart disease should include a focus on blood triglyceride levels.[12] In addition to increasing the risk of heart disease, elevated triglyceride levels may signal the development or presence of metabolic syndrome or type 2 diabetes. These conditions are characterized by elevated blood levels of triglycerides and glucose, insulin resistance, abdominal obesity, and low levels of HDL cholesterol. People with metabolic syndrome are at particularly high risk for heart disease.[13]

Genetic Effects on Blood Cholesterol Levels. Genetic traits play important roles in how diet and exercise affect blood lipid levels. For example, some people are born with one or more genetic traits that reduces cholesterol absorption and therefore blood cholesterol levels. People who absorb cholesterol poorly

Is shrimp off the menu for people with high blood cholesterol levels? Shrimp contains cholesterol—about 100 mg in three ounces. But that doesn't mean you should not eat it if you're watching your cholesterol intake. Shrimp also contains DHA and EPA, which are heart healthy, and only a trace of saturated fat. Blood cholesterol level is much less responsive to dietary cholesterol than to saturated fat intake.[2]

Lupe: 👍

Sharon: 👎

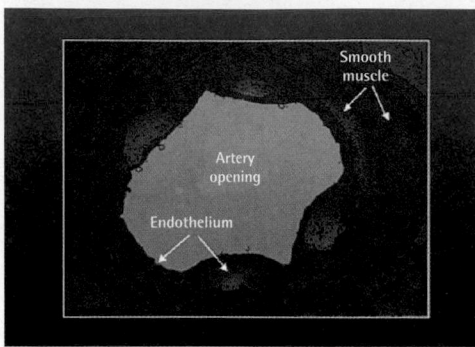

Illustration 19.7 The endothelium consists of cells that form the inside surface of arteries.

Illustration 19.8 Plaque (shown in yellow) forms under the endothelium and thickens, hardens, and narrows the artery.

endothelium
(Pronounced *n-dough-theil-e-um*) The layer of cells lining the inside of blood vessels.

do not experience the usual increase in blood cholesterol levels when they consume eggs or other high-cholesterol foods.[14] Other individuals have genetic traits that lower improvements in HDL-cholesterol levels that generally come with exercise or that keep blood triglyceride levels high even though ample amounts of EPA and DHA are consumed.[15]

Nutritional deprivation early in life can permanently modify the function of certain genes. Men born with low birth weight, for instance, tend to develop lower levels of HDL-cholesterol when they consume saturated fats than do men born with normal birth weights.[16] Most of the differences in genetic traits and functions identified so far are represented in a minority of the American population. Knowledge of individual genetic traits are, however, being increasingly used to guide decisions about dietary and physical activity approaches to the prevention and treatment of heart disease.

Chronic Inflammation and Heart Disease The link between chronic inflammation and heart disease appears to be related to oxidation reactions that occur as a result of prolonged inflammatory processes.[3] Inflammatory reactions take place within the **endothelium**—the layer of cells that line the inside of artery walls (Illustration 19.7). Some of the LDL-cholesterol in people with high blood levels of this lipoprotein enters the endothelium and triggers inflammation. Inflammatory reactions lead to oxidation of LDL-cholesterol. Oxidized LDL-cholesterol is reactive; it damages cells within artery walls and is far more likely to become part of plaque than is unoxidized LDL-cholesterol. As shown in Illustration 19.8, plaque accumulates within artery walls and causes a thickening, hardening, and narrowing of the arteries.[17] Antioxidants consumed from plant foods and those produced in the body help lower inflammation by quenching oxidation reactions or by repairing their harmful effects.

Nutrients and other components of foods that decrease inflammation are summarized in Table 19.1. Antioxidants are principally found in plants foods such as fruits, vegetables, and whole grains. Some of the strongest natural antioxidants are the phytochemicals that add color to fruits and vegetables. Nutrients that function as antioxidants include vitamins C and E, selenium, and beta-carotene.[3]

Who's at Risk for Heart Disease?

Frankly, you may be. Take a look at the Health Action feature and see. Most Americans are at risk for heart disease because they have high LDL cholesterol levels, hypertension, or a family history of heart attack before the age of 55; or because they smoke, are physically inactive, are obese, or have diabetes or high blood pressure (Illustration 19.9).

Some of the risk for heart disease may arise very early in life. Less-than-optimal availability of energy and nutrients during critical periods of fetal growth

Table 19.1

Nutrients and other substances that reduce chronic inflammation[2, 19, 20]

- EPA and DHA
- Vitamin D
- Nuts, tea, most fruits and vegetables, wine, alcohol, whole grains, fiber
- Statins (the primary type of drug used to treat heart disease)
- Antioxidant nutrients

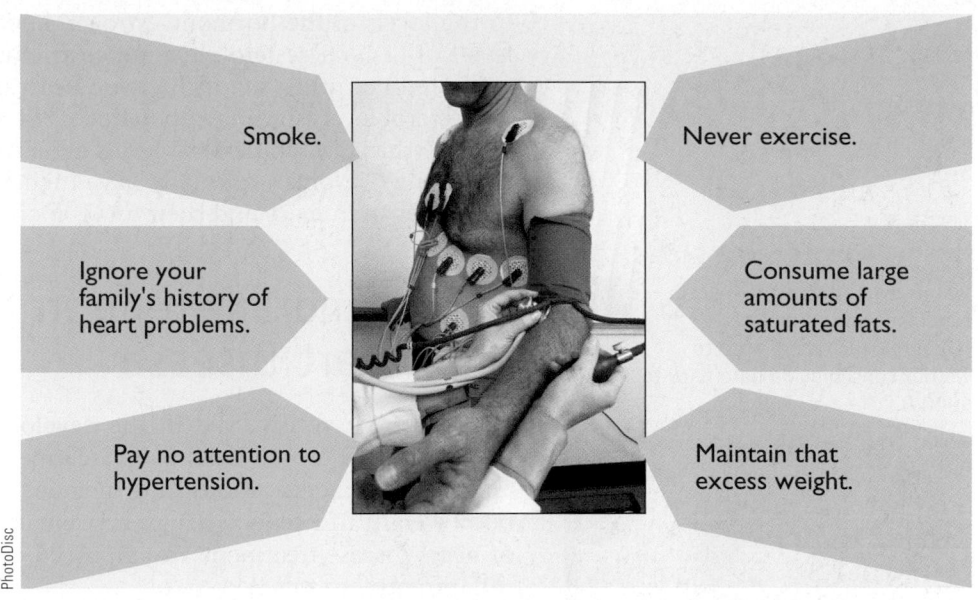

Smoke.

Never exercise.

Ignore your family's history of heart problems.

Consume large amounts of saturated fats.

Pay no attention to hypertension.

Maintain that excess weight.

PhotoDisc

Illustration 19.9 How to have a heart attack.

and development and early in childhood can modify gene functions in ways that increase the risk of heart disease later in life.[18] Infants who weigh less than 5.5 pounds at birth are at increased risk for heart disease. Although researchers have long believed that heart disease begins in childhood, the risk for heart problems may begin before birth.[16]

Are the Risks the Same for Women as for Men?

It is generally assumed that risk factors for heart disease identified in men apply to women as well. This assumption, along with the notion that heart disease represents a major health problem only for men, led to the exclusion of women from research studies. Both of these assumptions have been shown to be incorrect. Heart disease is a major health problem of women. Indeed, it is the leading cause of death among women in many countries.[21] Furthermore, risk factors for heart disease identified in studies of men do not always apply to women.

Results of recent studies involving women reveal that high blood cholesterol levels are a weak predictor of heart disease in older women and that low HDL-cholesterol levels are a stronger predictor than in men. Diabetes, obesity, high blood triglyceride levels, and increasing age are particularly strong risk factors for heart disease in women.[22]

The risk for heart disease in women increases substantially after menopause, which occurs around the age of 50 years. Menopause is characterized by declines

Health Action   Leading Risk Factors for Heart Disease[2, 12]

- High blood cholesterol level (particularly LDL-cholesterol level)
- Low HDL-cholesterol level
- Diet high in saturated fat, *trans* fat and cholesterol
- Family history of early heart attack
- Diet low in fruits, vegetables, and whole grains
- Elevated levels of inflammation markers

- High blood triglyceride level
- Hypertension
- Smoking
- Physical inactivity
- Obesity (especially central fat)
- Diabetes
- Age over 45 for men, over 55 for women

in the levels of the hormone estrogen and HDL cholesterol and increases in the levels of LDL-cholesterol. It is not clear, however, whether these increases in LDL cholesterol or other factors increase heart-disease risk, or whether lowering LDL-cholesterol levels significantly reduces risk in older women. Compared to men, it appears that LDL-cholesterol levels are a much weaker predictor of heart disease in women.[23] Women normally have higher levels of HDL cholesterol throughout their lives than men, and their total blood cholesterol level is a bit higher, too.[24]

Walnuts are one of the oldest foods known to humans. Historical records related to walnut consumption date back to 7000 B.C.[25]

On the Side

Diet and Lifestyle in the Management of Heart Disease

Many factors are involved in the development of heart disease, so approaches to treatment need to be broad. Treatment begins and continues with dietary and lifestyle modifications and includes reduction of high blood pressure and body weight (if needed), drugs in some cases, and smoking cessation. The goals of heart-disease treatment are improved overall health, blood lipid profiles, and inflammation status.[11, 26]

Diets and lifestyles that reduce the risk of heart disease benefit overall health in many ways. For most people, heart-healthy diets and lifestyles reduce the risk of diabetes, stroke, some types of cancer, osteoporosis, dementia, and obesity. When implemented, dietary recommendations for the treatment of heart disease lower plaque formation and improve health and physical fitness levels in a large majority of people.[27]

Modification of Blood Lipid Levels

Blood levels of total, LDL, and HDL cholesterol, and triglycerides considered desirable and abnormal have been defined (Table 19.2) and are used to assess heart disease risk and treatment progress. Total cholesterol represents the sum of the levels of LDL, HDL, and VLDL cholesterol (the lipoprotein that transports triglycerides in blood).

Table 19.3 summarizes the effects of dietary components and lifestyle factors on blood levels of LDL cholesterol, HDL cholesterol, and triglycerides. Some types of fat have both good and bad effects on blood lipid levels. Dietary recommendations for heart disease focus on consumption of specific types of fat and foods that lower levels of LDL cholesterol and triglycerides and raise levels of HDL cholesterol (Table 19.4).

High levels of LDL cholesterol are reduced by limiting saturated fat intake to less than 7% of total calories and by excluding processed foods that contain trans

---

Table 19.2

Risk categories for total, LDL- and HDL-cholesterol levels and triglyceride levels in adults.[28]

	Total Cholesterol (mg/dL)	LDL Cholesterol (mg/dL)	HDL Cholesterol (mg/dL)	Triglyceride (mg/dL)
Desirable/optimal	<200	<100	>50 in women, >40 in men	<150
Near optimal	—	100–129	<40	—
Borderline high	200–239	130–159	(low) <40	150–200
High[a]	240+	160–189	—	200–499
Very high	—	190+	—	500+

[a]High cholesterol levels among women and people over the age of 70 may be a poor predictor of heart disease risk.

Table 19.3

Effects of dietary components and lifestyle factors on blood lipid level[2]

Blood Lipid	Dietary Components	Lifestyle Factors
LDL cholesterol (low levels are heart healthy)	**Increases levels** • *trans* fats • saturated fats • dietary cholesterol	
	Decreases levels • mono- and polyunsaturated fats • whole-grain products • fiber • vegetables and fruits • plant stanols/sterols • nuts	**Decreases levels** • weight loss (if needed) • physical activity
HDL cholesterol (high levels are heart healthy)	**Increases levels** • moderate fat intake • saturated fats • alcohol (one–two drinks per day if appropriate) • moderate carbohydrate intake	**Increases levels** • physical activity • weight loss (if needed)
	Decreases levels • polyunsaturated fats • high-carbohydrate diets • *trans* fats	
Triglycerides (low levels are heart healthy)	**Increases levels** • high-carbohydrate diets • high alcohol intake	**Decreases levels** • weight loss (if needed) • physical activity
	Decreases levels • moderate fat diets • DHA and EPA	

Table 19.4

Characteristics of diets that lower heart-disease risk[2, 28]

- Provide 20–35% of total calories from fat, 10–25% from protein, and 45–65% from carbohydrates
- Emphasize healthy fats
  > plant sources of mono- and polyunsaturated fats
  > DHA and EPA from fish and shellfish
- Limit intake of unhealthy fats
  > *trans* fat in processed foods (< 1% of calories)
  > saturated fat from animal products, palm and coconut oil (< 7% of calories)
  > dietary cholesterol (< 200 mg per day)
- Are high in fiber; 35 grams per day for men and 25 grams per day for women
- Contain over three servings of vegetables and two servings of fruit daily
- Include whole-grain products and nuts
- Are moderate in carbohydrates and contain limited amounts of added sugars
- Include spreads with plant stanols or sterols (see Illustration 19.10)
- Contain alcohol in moderation (one–two drinks per day if appropriate)

fats. These fats are replaced in the diet by unsaturated fats and in particular by good food sources of monounsaturated fats like canola oil, olive oil, safflower oil, nuts, and avocados. Monounsaturated fats are preferred for lowering LDL cholesterol because they do not decrease HDL-cholesterol levels the way polyunsaturated fats do.

Whole-grain products, fiber, vegetables, fruits, and **plant stanols** and **sterols** also lower LDL cholesterol without decreasing HDL levels. Plant stanols and sterols lower LDL-cholesterol levels by reducing cholesterol absorption.[2] Spreads containing stanols and sterols, used to replace margarine and butter, are widely available in supermarkets (Illustration 19.10). Dietary recommendations for

plant stanols and sterols
Substances in corn, wheat, oats, rye, olives, wood, and some other plants that are similar in structure to cholesterol but that are not absorbed by the body. They decrease cholesterol absorption.

Illustration 19.10 Examples of spreads containing LDL-cholesterol-lowering plant stanols and sterols.

LDL-cholesterol reduction include limiting cholesterol intake to less than 200 mg daily. Improvements in weight status and increased physical activity result in lower levels of this risky lipoprotein.[11]

HDL-cholesterol levels can be increased by exercise, weight loss, and moderate alcohol consumption (one to two drinks per day if appropriate) and by including nuts in the diet. It's better to increase HDL cholesterol through these means than by consuming foods with saturated fats because saturated fats also increase LDL cholesterol.[11]

Low-fat, high-carbohydrate diets were a mainstay of heart-disease prevention and treatment approaches in the past, principally because this type of diet lowers LDL-cholesterol levels. However, because low-fat, high-carbohydrate diets also raise triglyceride levels and lower HDL-cholesterol levels, the recommendation has changed. Advice to consume low-fat, high-carbohydrate diets is being replaced by recommendations to consume healthy fats and foods that lower LDL cholesterol while maintaining triglyceride and HDL cholesterol at acceptable levels.[5] Food choices that are compatible with current recommendations for the dietary treatment of heart disease are shown in Illustration 19.11 and listed in Table 19.5.

Cholesterol-lowering drugs are indicated for the treatment of heart disease if blood lipid changes achieved by diet and lifestyle improvements are insufficient.[5]

Modification of Chronic Inflammation and Oxidation

Table 19.1, presented earlier in this unit, summarizes nutrients, foods, and other substances that reduce chronic inflammation. The omega-3 fatty acids EPA and DHA, found in fish and shellfish, reduce inflammation because they lead to the formation of anti-inflammatory chemicals. Lower levels of inflammation that result from the presence of these substances reduce plaque formation and the progression of heart disease. EPA and DHA also lower the risk of heart disease by decreasing blood triglyceride levels.[29] Recent evidence indicates that vitamin D has anti-inflammatory properties[30] Chronic inflammation is also reduced by the loss of excess body fat, by regular exercise, and by foods such as fruits, vegetables, nuts, teas, and whole-grain products.[19, 20] Antioxidants present in these

Illustration 19.11
Heart-healthy food

plant foods appear to contribute to the prevention of artery wall damage, and that reduces the accumulation of plaque.[3]

In the recent past it was thought that supplemental doses of folic acid and vitamins B_{12}, C, and E reduce plaque buildup and decrease inflammation and oxidation. Their roles in prevention of heart disease are in doubt.[2, 30] It is not recommended that supplemental doses of these vitamins be used to specifically help prevent or treat heart disease.[2]

The Statins

Statins have been hailed as wonder drugs for treating heart disease. These drugs, known by the names of Lipitor, Vytorin, Zetia, and Crestor, markedly reduce cholesterol production in the liver; they also combat the ill effects of heart disease in other ways. Their use is related to a 30% drop in LDL-cholesterol levels and a 30 to 40% reduction in heart attack and stroke in both women and men.[31] Statins are not a substitute for diet and lifestyle changes. They improve blood lipid levels more when combined with dietary and lifestyle changes than when used alone.[10]

Statins are used widely but are expensive and have side effects such as muscle pain and weakness, liver disease, and kidney failure.[32] The cost and side effects of statins have prompted researchers to take a close look at alternatives—like extreme cholesterol-lowering diets. Table 19.6 shows an example of the "portfolio diet," used to decrease LDL-cholesterol levels. A menu for the type of diet more typically used to lower LDL cholesterol is provided in the table for comparison. The portfolio diet is vegetarian and based on soy milk and soy-based

Table 19.5
Food choices that promote health and lower heart-disease risk
• Oils: canola, peanut, olive,
• safflower, flaxseed
• Fish and shellfish
• Nuts
• Vegetables
• Fruits
• Whole-grain products and other high fiber foods
• Lean, unfried, and unprocessed meats
• Low-fat dairy products
• Dried beans
• Spreads containing plant stanols or sterols

TABLE 19.6

Type of food included in the portfolio diet for substantial reduction in LDL cholesterol[33]

Preventing Heart Disease	LDL Cholesterol Lowering
Breakfast	
breakfast cereal (fortified)	oat bran cereal with added psyllium fiber
skim milk	soy milk
fruit	oat bran toast
whole-grain toast	plant-sterol enriched margarine
soft margarine	jam, strawberries
coffee/tea	
Lunch	
turkey chili and onions	bean soup with rice
whole-grain roll	oat bran bread
soft margarine	plant-sterol enriched margarine
cole slaw	soy milk
skim milk	
Dinner	
salmon steak	spicy sautéed tofu
tartar sauce	ratatouille
green peas and corn	cooked barley
whole-grain pasta	broccoli, cauliflower
tossed salad with avocado and Italian dressing	plant-sterol enriched margarine
angel food cake with fruit	soy milk
wine/coffee/tea	almonds
Snacks	
nuts, popcorn,	nuts, soy bar,
skim milk, soy milk,	psyllium in juice,
fruit, raw vegetables,	fruit, raw vegetables
peanut butter on whole-grain bread	

foods, fiber, oat bran, nuts, plant sterols and stanols, and vegetables and fruits. Far from representing usual food preferences, the diet may be a challenge to follow in the long run. The diet does, however, reduce LDL cholesterol to levels that are achieved by statins[2]. The results of one study showed this diet's dramatic effects of on blood lipids. The diet was then widely covered as a "wonder diet" in national newspapers. The end result is that therapeutic diets are increasingly viewed as one way to dramatically lower LDL-cholesterol levels while keeping the required dose of statin as low as possible.[33]

Looking toward the Future

Approaches to the prevention and treatment of heart disease have changed rather dramatically over recent years and will continue to evolve. Concerns about the cost of cholesterol-lowering drugs and their side effects, and the availability of low-cost preventive and treatment approaches, will factor into these changes. Approaches that use diet and lifestyle modification, changes in the quality of the food supply in stores and restaurants, and increased consumer involvement in risk reduction may lead the way to higher rates of decline in heart disease.

An end to escalating rates of obesity and physical inactivity would serve our collective heart especially well. It is anticipated that escalating rates of child and adolescent obesity in the United States will lead to a 5 to 16% increase in heart disease prevalence by 2035.[34]

NUTRITION

PhotoDisc

Up Close

Score Your Diet for Fat

Focal Point: Assess your chances for developing heart disease.

To estimate your risk for developing heart disease, circle the one number in each column that best describes you. Then, total the numbers from all categories to determine if your lifestyle encourages heart health or heart disease.

Heredity	Exercise	Age	Weight	Habits of Tobacco	Eating Fat
1	1	1	0	0	1
No known familial history of heart disease	Intensive exercise at work and during recreation	15–23	More than 5 lb below standard weight	Nonuser	No animal fat
2	2	2	1	1	2
One immediate family member who developed heart disease over age 55	Moderate exercise at work and during recreation	24–34	standard weight	Cigar or pipe	Very little animal fat (<10% of total calories)
3	3	3	2	2	3
Two immediate family members who developed heart disease over age 55	Sedentary work with intensive exercise during recreation	35–44	6–20 lb overweight	10 cigarettes or fewer per day	Some animal fat (11–20% of total calories)
4	5	4	4	4	4
One immediate family member who developed heart disease under age 55	Sedentary work with moderate exercise during recreation	45–54	21–35 lb overweight	20 cigarettes or more per day	Moderate animal fat (21–30% of total calories)
6	6	6	6	6	6
Two immediate family members who developed heart disease under age 55	Sedentary work with light exercise during recreation	55+ years	36–50 lb overweight	30 cigarettes or more per day	Excessive animal fat (>30% of total calories)

Total of all categories _____

FEEDBACK (including scoring) can be found at the end of Unit 19.

Source: Adapted from Heart Health Quiz (Loma Linda University).

Key Terms

atherosclerosis, page 19-3
cardiovascular disease, page 19-4
chronic inflammation, page 19-4

endothelium, page 19-6
heart disease, page 19-3

plant stanols or sterols, page 19-9
plaque, page 19-3

Review Questions

TRUE FALSE

1. Heart disease is the leading cause of death in women and men in developed countries. ☐ ☐

2. The two most important factors known to increase the risk of atherosclerosis are low carbohydrate diets and inadeguate intake of vitamin C. ☐ ☐

3. The term *cardiovascular disease* refers to disorders related to plaque buildup in arteries of the heart, brain, and other organs and tissues. ☐ ☐

4. High levels of LDL cholesterol increase the risk of heart disease just as low levels of HDL cholesterol do. ☐ ☐

5. Vitamin E and C supplements do *not* appear to decrease the risk of heart disease. ☐ ☐

TRUE FALSE

6. High levels of total cholesterol in the blood always represent a major risk factor for heart disease. ☐ ☐

7. Diabetes, obesity, low HDL-cholesterol levels, and high blood triglyceride levels are stronger risk factors for heart disease in women than in men. ☐ ☐

8. The major drawback to heart healthy diets is that they help prevent heart disease only. ☐ ☐

9. Loss of excess body weight and physical activity decrease blood levels of HDL-cholesterol and triglycerides. ☐ ☐

10. Heart healthy diets include ample fiber, whole-grain products, fish, and lean, unprocessed meats. ☐ ☐

Media Menu

hp2010.nhlbihin.net/atpiii/calculator.asp?usertype=prof
The Framingham Risk Calculator that pops up at this site asks you seven questions and returns an estimate for your 10-year risk of heart disease. You need to know your total cholesterol, HDL cholesterol, and blood pressure to answer the questions.

www.americanheart.org
The Web site for the American Heart Association serves as a gateway to dozens of interactive tools, videos, and education materials focusing on heart health.

www.hearthub.org
CVD risk assessment tools, brief educational videos, and an "Ask the Expert" Q&A feature are available on this easy-to-navigate, award-winning Web site sponsored by the American Heart Association.

www.nlm.nih.gov/medlineplus/coronarydisease.html
Links to information on coronary heart disease, coronary artery disease, risk factor reduction,

current news, and latest study results can be found on this site.

www.nlm.nih.gov/medlineplus/cholesterol.html
Get the latest news about good and bad cholesterol, *trans* fats, statins, dietary and blood cholesterol, and heart disease at this site—or visit the Virtual Fitness Room.

www.nlm.nih.gov/medlineplus/heartdiseasesprevention.html
This site from the National Institute of Health is dedicated to presenting information on heart-disease prevention.

www.deliciousdecisions.org
American Heart Association's nutrition Web site includes links to sites on nutrition basics, a cookbook tailored to people with heart disease or those who want to prevent it, and tips for grocery shopping and eating out.

www.medscape.com
This is an excellent site for health, medical, and nutrition information searches related to heart disease, stroke, and arteriosclerosis.

www.womenheart.org
The National Coalition for women with heart disease provides news, fact sheets, heart hospital ratings, heart healthy recipes, and more at this site.

www.nhlbi.nih.gov
The National Heart, Blood, and Lung Institute offers state-of-the-science information and papers on heart disease and stroke.

[Notes

1. CDC Fastats, www.cdc.gov/nchs/fastats, accessed 3/09.

2. Van Horn L et al. The evidence for dietary prevention and treatment of cardiovascular disease, J Am Diet Assoc 2008;108:287–331.

3. Kornman KS. Interleukin 1 genetics, inflammatory mechanisms, and nutrigenetic opportunities to modulate diseases of aging. Am J Clin Nutr 2006;83(suppl):475S–83S.

4. Detection, evaluation, and treatment of high blood cholesterol in adults. Adult Treatment Panel III. National Heart, Lung, and Blood Institute of the National Institutes of Health, 2001.

5. Dietary Reference Intakes. Energy, carbohydrate, fiber, fat, fatty acids, cholesterol, protein, and amino acids. Institute of Medicine, National Academy of Sciences, Washington, DC: National Academies Press, 2002. Schober SE et al. High serum cholesterol—an indicator for monitoring cholesterol-lowering efforts; US adults, 2005–2006, 2007, available at www.cdc.goc/nchs/data/databriefs/db02.pdf, accessed 3/09.

6. Ford ES et al. Explaining the decrease in U.S. deaths from coronary disease, 1980–2000, N Engl J Med 2007;356:2388–98.

7. Jellinger PS et al. The American Association of Clinical Endocrinologists medical guidelines for the diagnosis and treatment of dyslipidemia and prevention of atherogenesis. Endocrin Prac 2000;6:1–213.

8. Dwyer T et al. Differences in HDL cholesterol concentrations in Japanese, American, and Australian children. Circ 1997;76:2830–6.

9. Heart disease statistics, www.cdc.gov/nchs/fastats/heart.htm, accessed 9/08.

10. Caballero E. Dyslipidemia and vascular function: The rationale for early and aggressive intervention. www.medscape.com, 4/18/06.

11. Davidson MH. New tactics, new targets: the changing landscape of dyslipidemia management in coronary prevention, Medscape CME Activity, www.medscape.com, accessed 5/03.

12. Jacobson TA. Role of n-3 fatty acids in the treatment of hypertriglyceridemia and cardiovascular disease, Am J Cin Nutr 2008;87(suppl):1981S–90S.

13. Pande R et al. Association of insulin resistance and inflammation with peripheral artery disease, Circulation 2008, available at circ.ahajournals.org, accessed 3/09.

14. Herron KL et al. The ABCG5 polymorphism contributes to individual responses to dietary cholesterol and carotenoids in eggs. J Nutr 2006;136:1161–5.

15. Ashen MD et al. Low HDL cholesterol levels. N Engl J Med 2005;353:1125260.

16. Robinson SM et al. Combined effects of dietary fat and birth weight on serum cholesterol concentrations: the Hertfordshire Cohort Study. Am J Clin Nutr 2006;84:237–44.

17. Libby P. Inflammation and cardiovascular disease mechanisms. Am J Clin Nutr 2006;83:456S–60S.

18. Yajnik C. Interactions of perturbations in intrauterine growth and growth during childhood on the risk of adult-onset disease, Proc Nutr Soc 2000;59:257–65.

19. Mena M-P et al. Inhibition of circulating immune cell activation: a molecular anti-inflammatory effect of the Mediterranean diet, Am J Clin Nutr 2009;I89:248–56.

20. Nettleton JA et al. Dietary patterns are associated with biochemical markers of inflammation and endothelial activation in the Multi-Ethnic Study of Atherosclerosis. Am J Clin Nutr 2006;83:1369–79.

21. Kereiakes DJ et al. Managing cardiovascular disease in women. Medscape Women's Health, http://womenshealth.medscape.com/CMECircleWomensHealth/2000/CME02/pnt-CME02.html, accessed 1/01.

22. Sprecher DL et al. Metabolic coronary risk factors and mortality after bypass surgery. J Am Coll Cardiol 2000;36:1159–65.

23. Knopp RH et al. Saturated fat prevents coronary heart disease? An American paradox. Am J Clin Nutr 2004;80:1102–3.

24. Ernst ND et al. Consistency between US dietary fat intake and serum total cholesterol concentrations: the National Health and Nutrition Examination Surveys. Am J Clin Nutr 1997;66(suppl):965S–72S.

25. Dreher ML et al. The traditional and emerging role of nuts in healthful diets. Nutr Rev 1996;54:241–5.

26. Hilbert KF et al. Lipid response to a low-fat diet with or without soy is modified by C-reactive protein status in moderately hypercholesterolemic adults, J Nutr 2005;135:1075–9.

27. Varady KA et al. Combination diet and exercise interventions for the treatment of dyslipidemia: an effective preliminary strategy to lower cholesterol levels? J Nutr 2005;135:1829–35.

28. American Heart Association. Guidelines for the secondary prevention of cardiovascular disease. Circulation, posted online May 15, 2006, www.medscape.com/viewarticle/532327.

29. Hay JW. Pharmacoeconomics of elevated triglycerides and their management, Am Family Med 2008, posted 6/08, available at www.medscape.com/viewarticle/575862, accessed 9/08.

30. Holick MF. Vitamin D: Important for prevention of osteoporosis, cardiovascular disease, type 1 diabetes, autoimmune disease, and some cancers. www.medscape.com/ viewarticle/516238, accessed 9/06. Neuhouser ML et al. influence of multivitamin use on the risk of common cancers, Arch Intern Med 2/9/09, available at www.medscape.com/viewarticle/588127, accessed 2/09.

31. Steinberg D. The statins in preventive cardiology, N Engl J Med 2008;359:1426–8.

32. Vega C. Diet may lower cholesterol as much as statins, Medscape CME Activity, www.medscape.com, accessed 8/03.

33. Jenkins DJA et al. Assessment of the longer-term effects of a dietary portfolio of cholesterol-lowering foods in hypercholesterolemia. Am J Clin Nutr 2006:83:582–91.

34. Adolescent overweight and future adult coronary heart disease, N Engl J Med 2007;357:2371–80.

NUTRITION | # Up Close

Heart Health Quiz

Feedback for Unit 19

Find your score below to learn your estimated risk of developing heart disease:

4–9 very remote
10–15 less than average
16–20 average
21–25 moderately increased
26–30 excessive
31–35 too high—reduce score!

You have no control over certain risk factors for heart disease, such as heredity and age. Other conditions that detract from heart health, such as high blood pressure, diabetes, and high blood cholesterol, should be evaluated by your physician. Some risk factors are within your control, however. These lifestyle changes can help reduce your risk of developing heart disease:

- Keep your weight within a healthy range.
- Exercise regularly.
- Don't smoke.
- Reduce the amount of saturated fat in your diet if it is high.
- Eat at least five servings of vegetables and fruits each day.

Source: Adapted from Heart Health Quiz (Loma Linda University).

Vitamins and Your Health

NUTRITION SCOREBOARD	TRUE	FALSE
1 The only documented benefit of consuming sufficient amounts of vitamins is protection against deficiency diseases.		
2 Vitamins provide energy.		
3 Vitamin C is found only in citrus fruits.		
4 Nearly all cases of illness due to excessive intake of vitamins result from the overuse of vitamin supplements.		

Key Concepts and Facts

- Vitamins are chemical substances found in food that are required for normal growth and health.

- Adequate intakes of vitamins protect people against deficiency diseases and help prevent a number of chronic diseases and disorders.

- Each vitamin has a range of intake in which it functions optimally. Intakes below and above the range impair health.

- Eating five or more servings of fruits and vegetables each day is a good way to get enough vitamins in your diet.

Vitamins: They're on Center Stage

These are exciting times for people interested in vitamins. Long relegated to the role of preventing deficiency diseases, vitamins are now being viewed from a very different vantage point. These essential nutrients clearly do more than the important job of protecting us from vitamin deficiency diseases. Vitamins are taking a preeminent position as protectors against a host of ills, ranging from certain birth defects and cataracts to osteoporosis and cancer.[1]

But first, what are vitamins? How much of them do we need? Where do we get them? And what's behind this expanding interest in them? Background information essential to understanding the current interest in the benefits of vitamins is next.

Vitamin Facts

vitamins
Chemical substances that perform specific functions in the body.

Vitamins are chemical substances that perform specific functions in the body. They are essential nutrients because, in general, the body cannot produce them or produce sufficient amounts of them. If we fail to consume enough of any of the vitamins, specific deficiency diseases develop.

Fourteen vitamins have been discovered so far, and they are listed in Table 20.1. It is possible that a few more substances will be added to the list of vitamins in years to come.

Water- and Fat-Soluble Vitamins Vitamins come in two basic types—those soluble in water (the B-complex vitamins and vitamin C) and those that dissolve in fat (vitamins D, E, K, and A, or the "deka" vitamins). Their key features are summarized in Table 20.2, starting on page 20-4, which provides an "intensive

Table 20.1

Fourteen vitamins are known to be essential for health. Ten are water soluble—that is, they dissolve completely in water—and four dissolve only in fats.

The Water-Soluble Vitamins	The Fat-Soluble Vitamins
B-complex vitamins	Vitamin A (retinol) (provitamin is beta-carotene)
Thiamin (B_1)	Vitamin D (1,25 dihydroxy-cholecalciferol)
Riboflavin (B_2)	Vitamin E (tocopherol)
Niacin (B_3)	Vitamin K (phylloquinone, menaquinone)
Vitamin B_6 (pyridoxine)	
Folate (folacin, folic acid)	
Vitamin B_{12} (cyanocobalamin)	
Biotin	
Pantothenic acid (pantothenate)	
Choline	
Vitamin C (ascorbic acid)	

course" on vitamins. With the exception of vitamin B_{12}, the water-soluble vitamins can be stored in the body only in small amounts. Consequently, deficiency symptoms generally develop within a few weeks to several months after the diet becomes deficient in water-soluble vitamins. Vitamin B_{12} is unique in that the body can build up stores that may last for a year or more after intake of the vitamin stops. Of the water-soluble vitamins, niacin, vitamin B_6, choline, and vitamin C are known to produce ill effects if consumed in excessive amounts.

The fat-soluble vitamins are stored in body fat, the liver, and other parts of the body. Because the body is better able to store these vitamins, deficiencies of fat-soluble vitamins generally take longer to develop than deficiencies of water-soluble vitamins when intake from food is too low.

Bogus Vitamins Some substances that don't belong on the list of vitamins and won't end up there in years to come are listed in Table 20.3 on page 20-11. This list includes some of the most common substances claimed to be vitamins by enterprising quacks, misdirected manufacturers of supplements, and some weight-loss and cosmetic product producers. Although these substances may help sales, they aren't essential and therefore cannot be considered vitamins. People do not develop deficiency diseases when they consume too little of the bogus "vitamins."

What Do Vitamins Do?

For starters, vitamins *don't* provide energy or, with the exception of choline, serve as components of body tissues such as muscle and bone. A number of vitamins do play critical roles as **coenzymes** in the conversion of proteins, carbohydrates, and fats into energy. Coenzymes are also involved in reactions that build and maintain body tissues such as bone, muscle, and red blood cells. Thiamin, for example, is needed for reactions that convert glucose into energy. People who are thiamin deficient tire easily and feel weak (among other things). Folate, another B-complex vitamin, is required for reactions that build body proteins. Without enough folate, proteins such as those found in red blood cells form abnormally and function poorly. Vitamin A is needed for reactions that generate new cells to replace worn-out cells lining the mouth, esophagus, intestines, and eyes. Without enough vitamin A, old cells aren't replaced, and the affected tissues are damaged. And vitamin C is required for reactions that build and maintain collagen, a protein found in skin, bones, blood vessels, gums, ligaments, and cartilage.

coenzymes
Chemical substances, including many vitamins, that activate specific enzymes. Activated enzymes increase the rate at which reactions take place in the body, such as the breakdown of fats or carbohydrates in the small intestine and the conversion of glucose and fatty acids into energy within cells.

Table 20.2

An intensive course on vitamins

The Water-Soluble Vitamins

	PRIMARY FUNCTIONS		CONSEQUENCES OF DEFFICIENCY
Thiamin (vitamin B$_1$) AI[a] women: 1.1 mg men: 1.2 mg	• Helps body release energy from carbohydrates ingested • Facilitates growth and maintenance of nerve and muscle tissues • Promotes normal appetite	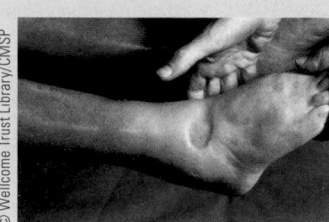	• Fatigue, weakness • Nerve disorders, mental confusion, apathy • Impaired growth • Swelling • Heart irregularity and failure A look at beriberi, a thiamin-deficiency disease.
Riboflavin (vitamin B$_2$) AI women: 1.1 mg men: 1.3 mg	• Helps body capture and use energy released from carbohydrates, proteins, and fats • Aids in cell division • Promotes growth and tissue repair • Promotes normal vision	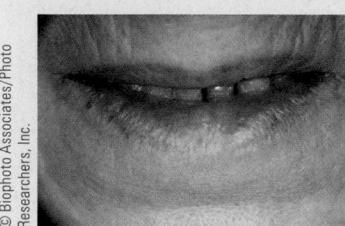	• Reddened Lips, cracks at both corners of the mouth • Fatigue
Niacin (vitamin B$_3$) RDA women: 14 mg men: 16 mg UL: 35 mg (from supplements and fortfied foods)	• Helps body capture and use energy released from carbohydrates, proteins, and fats • Assists in the manufacture of body fats • Helps maintain normal nervous system functions	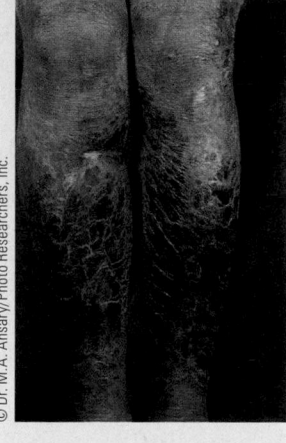	• Skin disorders • Nervous and mental disorders • Diarrhea, indigestion • Fatigue Pellagra: the niacin-deficiency disease.
Vitamin B$_6$ (pyridoxine) AI women: 1.3 mg men: 1.3 mg UL: 100 mg	• Needed for reactions that build proteins and protein tissues • Assists in the conversion of tryptophan to niacin • Needed for normal red blood cell formation • Promotes normal functioning of the nervous system		• Irritability, depression • Convulsions, twitching • Muscular weakness • Dermatitis near the eyes • Anemia • Kidney stones

[a](Adequate Intakes) and RDAs (Recommended Dietary Allowances) are for 19–30 year olds; UL (Upper Limits) are for 19–70 year olds,

Table 20.2

An intensive course on vitamins (*continued*)

The Water-Soluble Vitamins

CONSEQUENCES OF OVERDOSE	PRIMARY FOOD SOURCES	HIGHLIGHTS AND COMMENTS
• High intakes of thiamin are rapidly excreted by the kidneys. Oral doses of 500 mg/day or less are considered safe.	• Grains and grain products (cereals, rice, pasta, bread) • Ready-to-eat cereals • Pork and ham, liver • Milk, cheese, yogurt • Dried beans and nuts	• Need increases with carbohydrate intake • There is no "e" on the end of thiamin! • Deficiency rare in the United States; may occur in people with alcoholism • Enriched grains and cereals prevent thiamin deficiency
• None known. High doses are rapidly excreted by the kidneys.	• Milk, yogurt, cheese • Grains and grain products (cereals, rice, pasta, bread) • Liver, poultry, fish, beef • Eggs	• Destroyed by exposure to light
• Flushing, headache, cramps, rapid heart-beat, nausea, diarrhea, decreased liver function with doses above 0.5 g per day	• Meats (all types) • Grains and grain products (cereals, rice, pasta, bread) • Dried beans and nuts • Milk, cheese, yogurt • Ready-to-eat cereals • Coffee • Potatoes	• Niacin has a precursor—tryptophan. Tryptophan, an amino acid, is converted to niacin by the body. Much of our niacin intake comes from tryptophan. • High doses raise HDL-cholesterol levels, lower LD L-cholesterol and triglycerides.
• Bone pain, loss of feeling in fingers and toes, muscular weakness, numbness, loss of balance (mimicking multiple sclerosis)	• Oatmeal, bread, breakfast cereals • Bananas, avocados, prunes, tomatoes, potatoes • Chicken, liver • Dried beans • Meats (all types), milk • Green and leafy vegetables	• Vitamins go from B_3 to B_6 because B_4 and B_5 were found to be duplicates of vitamins already identified.

(continued)

Table 20.2

An intensive course on vitamins (*continued*)

The Water-Soluble Vitamins

	PRIMARY FUNCTIONS	CONSEQUENCES OF DEFFICIENCY	

Folate (folacin, folic acid) RDA:
    women: 400 mcg
    men: 400 mcg
    UL: 1000 mcg (from supplements and fortified foods)
	• Needed for reactions that utilize amino acids (the building blocks of protein) for protein tissue formation
	• Promotes the normal formation of red blood cells
		• Megaloblastic anemia
		• Diarrhea
		• Red, sore tongue
		
		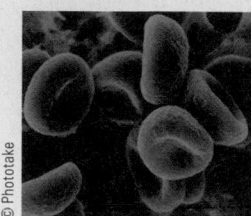
		Normal red blood cells.
			• Neural tube defects, low birth weight (in pregnancy), preterm delivery
			• Elevated blood levels of homocysteine
			
			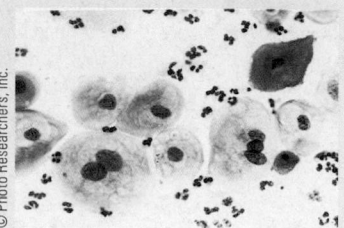
			Red blood cells in megaloblastic anemia

Vitamin B₁₂ (cyanocobalamin) AI women: 2.4 mcg men: 2.4 mcg	• Helps maintain nerve tissues • Aids in reactions that build up protein tissues • Needed for normal red blood cell development	• Neurological disorders (nervousness, tingling sensations, brain degeneration) • Pernicious anemia • Fatigue • Elevated blood level of homocysteine	
Biotin AI women: 30 mcg men: 30 mcg	• Needed for the body's manufacture of fats, proteins, and glycogen	• Seizures • Vision problems • Muscular weakness • Hearing loss	
Pantothenic acid (pantothenate) AI women: 5 mg men: 5 mg	• Needed for the release of energy from fat and carbohydrates	• Fatigue, sleep disturbances, impaired coordination • Vomiting, nausea	

Vitamin C (ascorbic acid)
RDA women: 75 mg
      men: 90 mg
      UL: 2000 mg
	• Needed for the manufacture of collagen
	• Helps the body fight infections, repair wounds
	• Acts as an antioxidant
	• Enhances iron absorption
		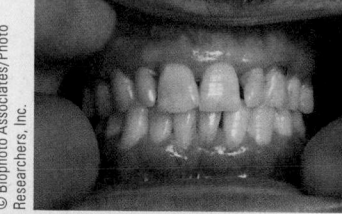
		Gums that are swollen and bleed easily are signs of scurvy, the vitamin C-deficiency disease.
			• Bleeding and bruising easily due to weakened blood vessels, cartilage, and other tissues containing collagen
			• Slow recovery from infections and poor wound healing
			• Fatigue, depression

Choline
AI   women: 425 mg
      men: 550 mg
      UL: 3.5 g
	• serves as a structural and signaling component of cell membranes
	• required for the normal development of memory and attention processes during early life
	• required for the transport and metabolism of fat and cholesterol
		• fatty liver
		• infertility
		• hypertension	

Table 20.2

An intensive course on vitamins (*continued*)

The Water-Soluble Vitamins

CONSEQUENCES OF OVERDOSE	PRIMARY FOOD SOURCES	HIGHLIGHTS AND COMMENTS
• May cover up signs of vitamin B_{12} deficiency (pernicious anemia)	• Fortified, refined grain products (bread, flour, pasta) • Ready-to-eat cereals • Dark green, leafy vegetables (spinach, collards, romaine) • Broccoli, brussels sprouts • Oranges, bananas, grapefruit • Milk, cheese, yogurt • Dried beans	• *Folate* means "foliage." It was first discovered in leafy green vegetables. • This vitamin is easily destroyed by heat. • Synthetic form (folic acid) added to fortified grain products is better absorbed than naturally occurring folates. • Very low and very high intakes of folic acid may be related to the risk of some forms of cancer.
• None known. Excess vitamin B_{12} is rapidly excreted by the kidneys or is not absorbed into the bloodstream. • Vitamin B_{12} injections may cause a temporary feeling of heightened energy.	• Animal products: beef, lamb, liver, clams, crab, fish, poultry, eggs • Milk and milk products • Ready-to-eat cereals	• Older people and vegans are at risk for vitamin B_{12} deficiency. • Some people become vitamin B_{12} deficient because they are genetically unable to absorb it. • Vitamin B_{12} is found in animal products and microorganisms only.
• None known. Excesses are rapidly excreted.	• Grain and cereal products • Meats, dried beans, cooked eggs • Vegetables	• Deficiency is extremely rare. May be induced by the overconsumption of raw eggs.
• None known. Excesses are rapidly excreted.	• Many foods, including meats, grains, vegetables, fruits, and milk • Required for the conversion of homocysteine to methionine	• Deficiency is very rare.
• Intakes of 1 g or more per day can cause nausea, cramps, and diarrhea and may increase the risk of kidney stones.	• Fruits: oranges, lemons, limes, strawberries, cantaloupe, honeydew melon, grapefruit, kiwi fruit, mango, papaya • Vegetables: broccoli, green and red peppers, collards, cabbage, tomato, asparagus, potatoes • Ready-to-eat cereals	• Need increases among smokers (to 110–125 mg per day). • Is fragile; easily destroyed by heat and exposure to air. • Supplements may decrease duration and symptoms of colds in some people. • Deficiency may develop within three weeks of very low intake.
• low blood pressure • sweating, diarrhea • fishy body odor • liver damage	• beef • eggs • pork • dried beans • fish • milk	• Most of the choline we consume from foods comes from its location in cell membranes. • Lecithin, an additive commonly found in processed foods, is a rich source of choline. • Choline is primarily found in animal products. • It is considered a B-complex vitamin.

(*continued*)

Table 20.2

An intensive course on vitamins (*continued*)

The Fat-Soluble Vitamins

	PRIMARY FUNCTIONS	CONSEQUENCES OF DEFICIENCY	
Vitamin A **I. Retinol** RDA women: 700 mcg men: 900 mcg UL: 3000 mcg	• Needed for the formation and maintenance of mucous membranes, skin, bone • Needed for vision in dim light	• Increased susceptibility to infection, increased incidence and severity of infection (including measles) • Impaired vision, blindness • Inability to see in dim light	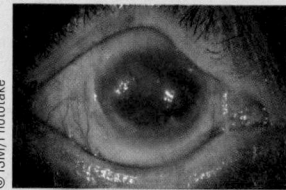 Xerophthalmia. *Vitamin A deficiency is the leading cause of blindness in developing countries.*
2. Beta-carotene (a vitamin A precursor or "provitamin") No RDA; suggested intake: 6 mg	• Acts as an antioxidant; prevents damage to cell membranes and the contents of cells by repairing damage caused by free radicals	• Deficiency disease related only to lack of vitamin A	
Vitamin E **(alpha-tocopherol)** RDA women: 15 mg men: 15 mg UL: 1000 mg	• Acts as an antioxidant, prevents damage to cell membranes in blood cells, lungs, and other tissues by repairing damage caused by free radicals • Reduces the ability of LDL cholesterol (the "bad" cholesterol) to form plaque in arteries	• Muscle loss, nerve damage • Anemia • Weakness • Many adults may have non-optimal blood levels.	
Vitamin D **(vitamin D₂ = ergocalciferol, vitamin D₃ = cholecalciferol)** AI women: 5 mcg (200 IU) men: 5 mcg (200 IU) UL: 50 mcg (2000 IU)	• Needed for the absorption of calcium and phosphorus, and for their utilization in bone formation, nerve and muscle activity. • inhibits inflammation	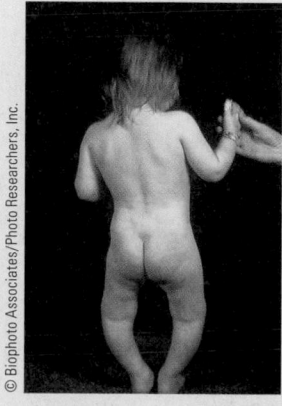	• Weak, deformed bones (children) • Loss of calcium from bones (adults), osteoporosis • Increased risk of cardiovascular disease, type 1- and type 2 diabetes, some cancers, and other inflammation-related disorders. The vitamin D-deficiency disease: rickets

Table 20.2

An intensive course on vitamins (*continued*)

The Fat-Soluble Vitamins

CONSEQUENCES OF OVERDOSE	PRIMARY FOOD SOURCES	HIGHLIGHTS AND COMMENTS	
• Vitamin A toxicity (hypervitaminosis A) with acute doses of 500,000 IU, or long-term intake of 50,000 IU per day. Limit retinol use in pregnancy to 5000 IU daily. • Nausea, irritability, blurred vision, weakness, headache • Increased pressure in the skull, hip fracture • Liver damage • Hair loss, dry skin • Birth defects	• Vitamin A is found in animal products only. • Liver, butter, margarine, milk, cheese, eggs • Ready-to-eat cereals	• Symptoms of vitamin A toxicity may mimic those of brain tumors and liver disease. Vitamin A toxicity is sometimes misdiagnosed because of the similarities in symptoms. • 1 IU vitamin A = 0.3 mcg retinol or 3.6 mcg beta-carotene	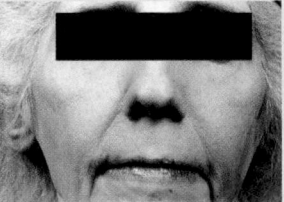 Brittle hair and dry, rough, scaly, and cracked skin from vitamin A overdose. From: American Journal of Clinical Nutrition, Vol 71, No 4, 878–884. April 2000, Robert M. Russell.
• High intakes from supplements may increase lung damage in smokers. • With high intakes and supplemental doses (over 12 mg/day for months), skin may turn yellow-orange.	• Deep orange, yellow, and green vegetables and fruits are often good sources. • Carrots, sweet potatoes, pumpkin, spinach, collards, red peppers, broccoli, cantaloupe, apricots, tomatoes, vegetable juice	• The body converts beta-carotene to vitamin A. Other carotenes are also present in food, and some are converted to vitamin A. Beta-carotene and vitamin A perform different roles in the body, however. • May decrease sunburn in certain individuals.	
• Intakes of up to 800 IU per day are unrelated to toxic side effects; over 800 IU per day may increase bleeding (blood-clotting time). • Avoid supplement use if aspirin, anticoagulants, or fish oil supplements are taken regularly.	• Oils and fats • Salad dressings, mayonnaise, margarine, shortening, butter • Whole grains, wheat germ • Leafy, green vegetables, tomatoes • Nuts and seeds • Eggs	• Vitamin E is destroyed by exposure to oxygen and heat. • Oils naturally contain vitamin E. It's there to protect the fat from breakdown due to free radicals. • Eight forms of vitamin E exist, and each has different antioxidant strength. • Natural form is better absorbed than synthetic form: 15 IU alpha-tocopherol = 22 IU d-alpha tocopherol (natural form) and 33 IU synthetic vitamin E.	
• Mental retardation in young children • Abnormal bone growth and formation • Nausea, diarrhea, irritability, weight loss • Deposition of calcium in organs such as the kidneys, liver, and heart • Toxicity possible with doses 10,000 IU daily	• Vitamin D-fortified milk and margarine • Butter • Fish • Eggs • Mushrooms • Milk products such as cheese, yogurt, ice cream; breakfast cereals, bread, and other products fortified with vitamin D	• Vitamin D_3, the most active form of the vitamin, is manufactured from cholesterol in cells beneath the surface of the skin upon exposure of the skin to sunlight. • Poor vitamin D status appears to be common in all age groups. • Breast-fed infants with little sun exposure benefit from vitamin D supplements.	

(continued)

Table 20.2

An intensive course on vitamins (*continued*)

The Fat-Soluble Vitamins		
	PRIMARY FUNCTIONS	**CONSEQUENCES OF DEFICIENCY**
Vitamin K **(phylloquinone, menaquinone)** AI women: 90 mcg men: 120 mcg	• Is an essential component of mechanisms that cause blood to clot when bleeding occurs • Aids in the incorporation of calcium into bones	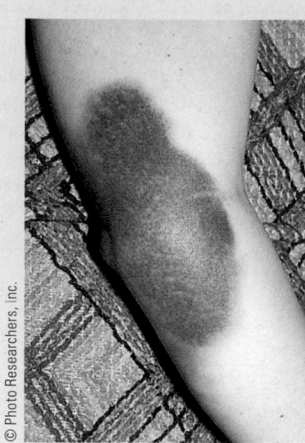 • Bleeding, bruises • Decreased calcium in bones • Deficiency is rare. May be induced by the long-term use (months or more) of antibiotics.

The long-term use of antibiotics can cause vitamin K deficiency. People with vitamin K deficiency bruise easily.

Table 20.3

Nonvitamins

The "real" vitamins are listed in Table 20.1. But a number of bogus vitamins are also on the market. They turn up in supplements, skin care creams, weight-loss and hair care products, and other items. These are some of the more popular nonvitamins.

Bioflavonoids
  (vitamin P)
Coenzyme Q_{10}
Gerovital H-3
Hesperidin
Inositol
Laetrile (vitamin B_{17})
Lecithin
Lipoic acid
Nucleic acids
Pangamic acid
  (vitamin B_{15})
Para-amino benzoic
  acid (PABA)
Provitamin B_5
  complex
Rutin

dementia
(Pronounced *di-men-cha*.) A usually progressive condition (such as Alzheimer's disease) marked by the development of memory impairment and an inability to use or comprehend words or to plan and initiate complex behaviors.

(Approximately 30% of the total amount of protein in the body is collagen.) With vitamin C deficiency, collagen becomes weak, causing tissues that contain collagen to weaken and bleed easily.

The examples just given all relate to the physical effects of vitamins. Vitamins participate in reactions that affect behaviors, too. Alterations in behaviors such as reduced attention span, poor appetite, irritability, depression, or paranoia often precede the physical signs of vitamin deficiency.[2] Vitamins are truly "vital" for health.

Protection from Vitamin Deficiencies and More

Current research on vitamins centers around their effects on disease prevention and treatment. It is a very active area of research and, no doubt, new and important results will be announced after this text goes to press. (Nutrition textbooks should really be updated every few weeks. Stay tuned to your instructor for the latest developments.) Here are a few examples of recent developments in research on vitamins and disease prevention and treatment:

Folate, Neural Tube Defects, Dementia, and Cancer   Daily consumption of 400 micrograms of folic acid (the synthetic form of folate added to refined grain products) before and early in pregnancy significantly reduces the incidence of neural tube defects in newborns.[3] Neural tube defects are abnormalities of the spinal cord and brain (Illustration 20.1). They are one of the most common types of malformation of newborns in the United States. Adequate intake of folate also reduces the risk of developing **dementia** and certain types of cancer such as childhood brain tumors and breast, ovary, and stomach cancer.[4,5] Very high intakes of folic acid, on the other hand, may increase the development of certain types of cancer.[5]

In 1998 manufacturers started fortifying refined grain products such as bread, pasta, and rice with folic acid. The addition of folic acid to these foods has produced substantial gains in people's folate status. Blood levels of folate have nearly doubled, and the prevalence of poor folate status has declined by 84%.[6]

Table 20.2

An intensive course on vitamins (continued)

The Fat-Soluble Vitamins

CONSEQUENCES OF OVERDOSE	PRIMARY FOOD SOURCES	HIGHLIGHTS AND COMMENTS
• Toxicity is only a problem when synthetic forms of vitamin K are taken in excessive amounts. That may cause liver disease.	• Leafy, green vegetables • Grain products	• Vitamin K is produced by bacteria in the gut. Part of our vitamin K supply comes from these bacteria. • Newborns are given a vitamin K injection because they have "sterile" guts and consequently no vitamin K–producing bacteria.

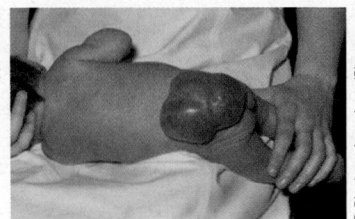

Illustration 20.1 A baby with spina bifida, a form of neural tube defect associated with poor folate status early in pregnancy.

Vitamin A—From Measles to "Liver Spots" Studies undertaken in both developing countries and the United States indicate that adequate vitamin A status decreases the severity of measles and other infectious diseases.[7] (It has been known for decades that adequate vitamin A intake also prevents blindness, an all-too-common consequence of vitamin A deficiency in developing nations.) This vitamin is being used successfully in the treatment of serious cases of acne, skin wrinkles, and "liver" or "aging" spots on the skin due to overexposure to the sun.[8, 9]

Vitamin D: From Osteoporosis to Chronic Inflammation Vitamin D is best known as the sunshine vitamin that helps build strong bones by facilitating the absorption and utilization of calcium. Vitamin D does much more than that, however. It plays key roles as a hormone in combating **chronic inflammation**. Low grade, chronic inflammation is at the core of the development of disorders such as type 2 and type 1 diabetes, cardiovascular disease, multiple sclerosis, certain cancers, and rheumatoid arthritis.[10-12] Vitamin D reduces inflammation by entering cells and turning genes that produce inflammatory substances "off," and those that produce substances that reduce inflammation "on."[13]

■ **Recommended Intake of Vitamin D.** Recommendations for vitamin D intake are changing. Currently, it is recommended that adult women and men consume 5 mcg (200 IU) of vitamin D daily. For people who do not meet their vitamin D needs via sun exposure, this level of Vitamin D intake is not high enough to lower the risk of osteoporosis, heart disease, type 1 and type 2 diabetes, and inflammation disorders related to vitamin D.[14,15] A number of studies have demonstrated that, in the absence of direct sun exposure, vitamin D intakes around 800 IU per day in adults are associated with lowered disease risk.[15-17] Optimal levels of vitamin D intake in infants, young children, and older adults also appear to be higher than currently recommended.[16,18]

chronic inflammation
Low grade inflammation that lasts weeks, months, or years. Inflammation is the first response of the body's immune system to infection or irritation. Inflammation triggers the release of biologically active substances that promote oxidation and other potentially harmful reactions in the body.

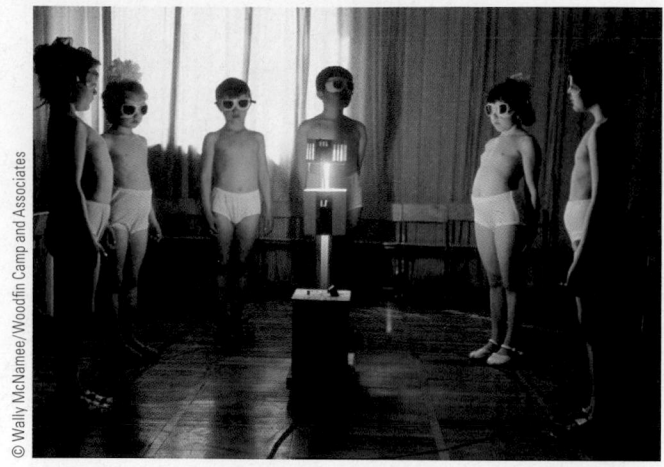

Illustration 20.2 Russian children are exposed to a quartz UV lamp to prevent vitamin D deficiency during the long winter.

Vitamin D status is more likely to be inadequate in northern areas of the globe where sunlight is indirect and weak for much of the year.[19] The ultraviolet (UV) rays emitted by the sun that initiate the formation of vitamin D in the skin (Illustration 20.2) do not penetrate glass, clear plastic or other transparent materials.

The vitamin D status of most people in the United States and Canada is below the level associated with reduced disease risk.[20] You may be among that group if you do not consume vitamin D fortified foods and other good sources of vitamin D (see Table 20.4, which starts on p. 20-15), take a vitamin D supplement, or get some direct sun exposure on your skin regularly. If you're thinking "that could be me" divert your attention to this Unit's Take Action insert.

It is expected that the recommended intake levels for vitamin D will be increased by the Institute of Medicine. An increased assortment of vitamin D fortified foods is becoming available on grocery store shelves in preparation for the anticipated increase.[21] Illustration 20.3 provides examples of newly vitamin D fortified foods that are available.

■ **The Sun as a Source Vitamin D.** It is estimated that exposing the whole body to direct sunlight for 10 to 15 minutes generates around 500 mcg (20,000 IU) of vitamin D.[22] Maximum production of vitamin D in the skin is achieved before changes in skin color occur. You can get a healthy dose of vitamin D without risking a sun burn and skin damage.[23] And, you cannot get too much vitamin D from the sun. Production of the vitamin stops when adequate amounts have been produced.[22]

Vitamin D production in light-skinned people is higher than in those with darker skin because more of the sun's UV light can penetrate the top layers of light skin. Consequently, it takes at least twice as much sun exposure to produce high levels of vitamin D in people with dark versus light skin.[24] On the plus side is the fact that people with darker skin are less susceptible to harmful effects of UV light than are their light-skin counterparts. Vitamin D rich food and supplements, if needed, are recommended for individuals whose skin is sensitive to even short durations of sun exposure.[23]

Illustration 20.3 More foods are now fortified with Vitamin D than in the past. Check the labels on these foods for vitamin D fortification levels.

How do you become a person that stands out from the average in terms of vitamin D adequacy? If you're not that person already, here are some tips that will get you on your way.

Choose and check two options for getting more vitamin D that appeal to you most.

I would:

_____ substitute a cup of skim milk for a sweetened beverage at one meal or snack a day.

_____ eat salmon once a week at dinner.

_____ have a vitamin D-fortified yogurt with lunch every other day.

_____ eat a vitamin D-fortified breakfast cereal for breakfast on weekdays.

_____ take a vitamin D supplement (400-600 IU) daily until I am able to get enough vitamin D in my diet or by brief exposure to direct sunshine.

_____ exercise in sunshine for 10 minutes four times a week when the weather is warm while wearing only shorts and a top.

_____ take a walk in the sunshine with bare arms and legs for 10 minutes three times a week.

Vitamin C and the Common Cold Revisited   Recent research confirms that vitamin C, the popular cold remedy, reduces the symptoms and duration of the common cold modestly but does *not* affect how often colds occur.[25]

The Antioxidant Vitamins

Beta-carotene (a **precursor** to vitamin A), vitamin E, and vitamin C function as **antioxidants.** This means they prevent or repair damage to components of cells caused by exposure to **free radicals.** A free radical is formed when an atom of hydrogen or oxygen is missing an electron. Without the electron, there is an imbalance between the atom's positive and negative charges. This makes the atom reactive—it needs to steal an electron from a nearby atom or molecule to reestablish a balance between its positive and negative charges. Atoms and molecules that have lost electrons to free radicals are said to be "oxidized." These oxidized substances are reactive and can damage lipids, cell membranes, DNA, and other cell components. Antioxidants such as beta-carotene, vitamin E, and vitamin C donate electrons to stabilize oxidized molecules or repair them in other ways. The effectiveness of antioxidant vitamins appears to be increased when they are consumed in plant foods rather than supplements.

A number of studies have shown that beta-carotene, vitamin C, and vitamin E supplements do not decrease the risk of cardiovascular disease, Alzheimer's disease, or cancer. Consumption of diets rich in these nutrients, however, is related to lower risk of these diseases.[26,28,29] Fruits, vegetables, whole grains, and other plant foods contain thousands of naturally occuring antioxidants that work together with the antioxidant vitamins in disease prevention.[26]

Why do free radicals exist in the body? Free radicals play a number of roles in the body, so they are always present. They are produced during energy formation, by breathing, and by the immune system to help destroy bacteria and viruses that enter the body. They can also be formed when the body is exposed to alcohol, radiation emitted by the sun, smoke, ozone, smog, and other environmental pollutants.

precursor
In nutrition, a nutrient that can be converted into another nutrient (also called provitamin). Beta-carotene is a precursor of vitamin A.

antioxidants
Chemical substances that prevent or repair damage to cells caused by exposure to free radicals. Beta-carotene, vitamin E, and vitamin C function as antioxidants.

free radicals
Chemical substances (usually oxygen) that are missing an electron. The absence of the electron makes the chemical substances reactive and prone to oxidizing nearby atoms or molecules by stealing an electron from them.

On the Side

Apple slices exposed to air turn brown due to oxidation. Coating the slice with a vitamin C solution or lemon juice prevents the oxidation.

© Scott Goodwin Photography

Vitamins: Getting Enough without Getting Too Much

Vitamins are widely present in basic foods (Table 20.4). Adequate amounts of them can be obtained from diets that include the variety of foods recommended in the MyPyramid. Most fruits and vegetables are good sources of vitamins, and eating five or more servings a day is one way for Americans to get their vitamins. In the United States, 80% of adults consume two or fewer servings of vegetables and fruits daily, and 10% consume none.[27]

Fortified foods such as ready-to-eat cereals, fruit juices and drinks, and snack bars are not listed in Table 20.4. They can increase vitamin intake substantially, and concern has been expressed that fortified food consumption may increase intake too much. Although an important concern, overdose reactions from the new generation of vitamin-fortified foods have not yet been reported.

Preserving the Vitamin Content of Foods

The vitamin content of foods can be affected by food preparation and storage methods (Table 20.5 on p. 20-20).

Food storage and preparation methods that involve heat lead to higher losses of heat-sensitive vitamins such as vitamin C and folate, and low or no loss of vitamin B_{12} and choline which are much less sensitive to heat. Vitamins in foods boiled in water can be lost down the drain if the water is thrown out. Vitamins in foods dissolve in cooking water to some extent and escape into the cooking fluid. In general, boiling or steaming foods using a small amount of water, using the cooking water in soups, stews, or sauces; and stir-frying lead to superior vitamin retention.[30]

Recommend Intake Levels of Vitamins Updated recommendations for vitamin intakes associated with the prevention of deficiency and chronic diseases are represented by standards called Dietary Reference Intakes, or DRIs. DRIs include Recommended Dietary Allowances (RDAs) for vitamins for which convincing scientific data exist for establishing intake standards. Adequate Intakes, or AIs, are assigned to vitamins for which scientific information about levels of intake associated with chronic disease prevention is less convincing. Tolerable Upper Levels of Intake, abbreviated ULs, are also assigned to vitamins and indicate levels of vitamin intake from foods, fortified foods, and supplements that should not be exceeded. The RDAs or AIs and the ULs for the vitamins are given in Table 20.2.

Although people can get all the vitamins they need from supplements, it makes more sense to get them from basic foods. Foods offer fiber, minerals, and other healthful ingredients that don't come in supplements, and they certainly taste better on the way down!

Table 20.4

Food sources of vitamins

Thiamin

FOOD	SERVING SIZE	THIAMIN (MILLIGRAMS)
Meats:		
Pork roast	3 oz	0.8
Beef	3 oz	0.4
Ham	3 oz	0.4
Liver	3 oz	0.2
Nuts and seeds:		
Sunflower seeds	¼ cup	0.7
Peanuts	¼ cup	0.1
Almonds	¼ cup	0.1
Grains:		
Bran flakes	1 cup (1 oz)	0.6
Macaroni	½ cup	0.1
Rice	½ cup	0.1
Bread	1 slice	0.1
Vegetables:		
Peas	½ cup	0.3
Lima beans	½ cup	0.2
Corn	½ cup	0.1
Broccoli	½ cup	0.1
Potato	1 medium	0.1
Fruits:		
Orange juice	1 cup	0.2
Orange	1	0.1
Avocado	½	0.1

Beef and broccoli are good sources of thiamin.

Riboflavin

FOOD	SERVING SIZE	RIBOFLAVIN (MILLIGRAMS)
Milk and milk products:		
Milk	1 cup	0.5
2% milk	1 cup	0.5
Yogurt, low-fat	1 cup	0.5
Skim milk	1 cup	0.4
Yogurt	1 cup	0.1
American cheese	1 oz	0.1
Cheddar cheese	1 oz	0.1
Meats:		
Liver	3 oz	3.6
Pork chop	3 oz	0.3
Beef	3 oz	0.2
Tuna	3 oz	0.1
Vegetables:		
Collard greens	½ cup	0.3
Broccoli	½ cup	0.2
Spinach, cooked	½ cup	0.1
Eggs:		
Egg	1	0.2
Grains:		
Macaroni	½ cup	0.1
Bread	1 slice	0.1

Milk and milk products are good sources of riboflavin.

(continued)

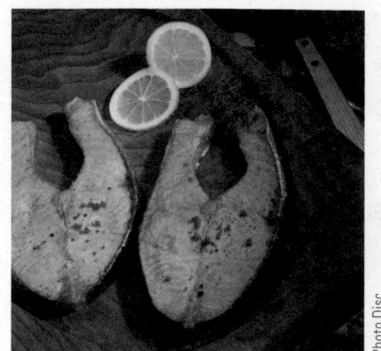

Salmon is a good source of niacin.

Table 20.4

Food sources of vitamins (continued)

Niacin

FOOD	SERVING SIZE	NIACIN (MILLIGRAMS)
Meats:		
Liver	3 oz	14.0
Tuna	3 oz	10.3
Turkey	3 oz	9.5
Chicken	3 oz	7.9
Salmon	3 oz	6.9
Veal	3 oz	5.2
Beef (round steak)	3 oz	5.1
Pork	3 oz	4.5
Haddock	3 oz	2.7
Scallops	3 oz	1.1
Nuts and seeds:		
Peanuts	1 oz	4.9
Vegetables:		
Asparagus	½ cup	1.5
Grains:		
Wheat germ	1 oz	1.5
Brown rice	½ cup	1.2
Noodles, enriched	½ cup	1.0
Rice, white, enriched	½ cup	1.0
Bread, enriched	1 slice	0.7
Milk and milk products:		
Cottage cheese	½ cup	2.6
Milk	1 cup	1.9

Vitamin B$_6$

FOOD	SERVING SIZE	VITAMIN B$_6$ (MILLIGRAMS)
Meats:		
Liver	3 oz	0.8
Salmon	3 oz	0.7
Other fish	3 oz	0.6
Chicken	3 oz	0.4
Ham	3 oz	0.4
Hamburger	3 oz	0.4
Veal	3 oz	0.4
Pork	3 oz	0.3
Beef	3 oz	0.2
Eggs:		
Egg	1	0.3
Legumes:		
Split peas	½ cup	0.6
Dried beans, cooked	½ cup	0.4
Fruits:		
Banana	1	0.6
Avocado	½	0.4
Watermelon	1 cup	0.3
Vegetables:		
Turnip greens	½ cup	0.7
Brussels sprouts	½ cup	0.4
Potato	1	0.2
Sweet potato	½ cup	0.2
Carrots	½ cup	0.2

Bananas are a good source of vitamin B$_6$.

Table 20.4

Food sources of vitamins (*continued*)

Folate

FOOD	SERVING SIZE	FOLATE (MICROGRAMS)
Vegetables:		
Garbanzo beans	½ cup	141
Spinach, cooked	½ cup	131
Navy beans	½ cup	128
Asparagus	½ cup	120
Brussels sprouts	½ cup	116
Black-eyed peas	½ cup	102
Collard greens, cooked	½ cup	89
Romaine lettuce	1 cup	86
Lima beans	½ cup	71
Peas	½ cup	70
Peas	½ cup	47
Sweet potato	½ cup	43
Broccoli	½ cup	43
Fruits:		
Cantaloupe	¼ whole	100
Orange juice	1 cup	87
Orange	1	59
Grains:[a]	1 cup/1 oz	100–400
Ready-to-eat cereals	½ cup	97
Oatmeal	½ cup	77
Rice	½ cup	45
Noodles	2 tbs	40
Wheat germ		

Beans are an especially rich source of folate.

[a]Fortified, refined grain products such as bread, rice, pasta, and crackers provide approximately 60 micrograms of folic acid per standard serving.

Vitamin B$_{12}$

FOOD	SERVING SIZE	VITAMIN B$_{12}$ (MICROGRAMS)
Meats:		
Liver	3 oz	6.8
Trout	3 oz	3.6
Beef	3 oz	2.2
Clams	3 oz	2.0
Crab	3 oz	1.8
Lamb	3 oz	1.8
Tuna	3 oz	1.8
Veal	3 oz	1.7
Hamburger, regular	3 oz	1.5
Milk and milk products:		
Skim milk	1 cup	1.0
Milk	1 cup	0.9
Yogurt	1 cup	0.8
Cottage cheese	½ cup	0.7
American cheese	1 oz	0.2
Cheddar cheese	1 oz	0.2
Eggs:		
Egg	1	0.6

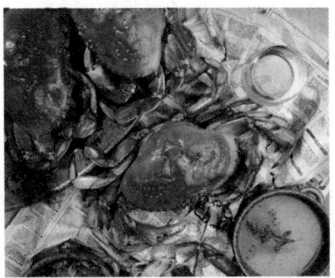

Crab is one food source of vitamin B$_{12}$.

(*continued*)

Citrus fruits are an excellent source of vitamin C.

Table 20.4

Food sources of vitamins (continued)

Vitamin C

FOOD	SERVING SIZE	VITAMIN C (MILLIGRAMS)
Fruits:		
Orange juice, vitamin C-fortified	1 cup	108
Kiwi fruit	1 or ½ cup	108
Grapefruit juice, fresh	1 cup	94
Cranberry juice cocktail	1 cup	90
Orange	1	85
Strawberries, fresh	1 cup	84
Orange juice, fresh	1 cup	82
Cantaloupe	¼ whole	63
Grapefruit	1 medium	51
Raspberries, fresh	1 cup	31
Watermelon	1 cup	15
Vegetables:		
Green peppers	½ cup	95
Cauliflower, raw	½ cup	75
Broccoli	½ cup	70
Brussels sprouts	½ cup	65
Collard greens	½ cup	48
Vegetable (V-8) juice	¾ cup	45
Tomato juice	¾ cup	33
Cauliflower, cooked	½ cup	30
Potato	1 medium	29
Tomato	1 medium	23

Choline

FOOD	SERVING SIZE	CHOLINE MILLIGRAMS
Meats:		
Beef	3 oz	111
Pork chop	3 oz	94
Lamb	3 oz	89
Ham	3 oz	87
Beef	3 oz	85
Turkey	3 oz	70
Salmon	3 oz	56
Eggs:		
Egg	1 large	126
Vegetables:		
Baked beans	½ cup	50
Navy beans, boiled	½ cup	41
Collards, cooked	½ cup	39
Black-eyed-peas (Cowpeas)	½ cup	39
Chickpeas (garbanzo beans)	½ cup	35
Brussels sprouts	½ cup	32
Broccoli	½ cup	32
Collard greens	½ cup	30
Refried beans	½ cup	29
Milk and milk products		
Milk, 2%	1 cup	40
Cottage cheese, low-fat	½ cup	37
Yogurt, low fat	1 cup	35

Table 20.4

Food sources of vitamins (continued)

Vitamin A (Retinol)

FOOD	SERVING SIZE	VITAMIN A (MICROGRAMS RE)[b]
Meats:		
Liver	3 oz	9124
Salmon	3 oz	53
Tuna	3 oz	14
Eggs:		
Egg	1 medium	84
Milk and milk products:		
Skim milk, fortified	1 cup	149
2% milk	1 cup	139
American cheese	1 oz	82
Whole milk	1 cup	76
Swiss cheese	1 oz	65
Fats:		
Margarine, fortified	1 tsp	46
Butter	1 tsp	38

Vitamin A and beta-carotene

Eggs are one food source of vitamin A (retinol).

Beta-Carotene

FOOD	SERVING SIZE	VITAMIN A VALUE (MICROGRAMS RE)[b]
Vegetables		
Pumpkin, canned	½ cup	2712
Sweet potato, canned	½ cup	1935
Carrots, raw	½ cup	1913
Spinach, cooked	½ cup	739
Collard greens, cooked	½ cup	175
Broccoli, cooked	½ cup	109
Winter squash	½ cup	53
Green peppers	½ cup	40
Fruits:		
Cantaloupe	¼ whole	430
Apricots, canned	½ cup	210
Nectarine	1 medium	101
Watermelon	1 cup	59
Peaches, canned	½ cup	47
Papaya	½ cup	20

Spinach and winter squash are two sources of beta-carotene.

Vitamin E

FOOD	SERVING SIZE	VITAMIN E (IU)[c]
Oils:		
Oil	1 tbs	6.7
Mayonnaise	1 tbs	3.4
Margarine	1 tbs	2.7
Salad dressing	1 tbs	2.2
Nuts and seeds:		
Sunflower seeds	¼ cup	27.1
Almonds	¼ cup	12.7
Peanuts	¼ cup	4.9
Cashews	¼ cup	0.7

[b]RE (retinol equivalent) = 3.33 IU.
[c]15 milligrams alpha-tocopherol = 22 IU d-alpha tocopherol (natural form) and 33 IU synthetic vitamin E.

Seafood and asparagus both provide vitamin E.

(continued)

Shrimp and other shellfish are good sources of vitamin D.

Photo Disc

Table 20.4

Food sources of vitamins (*continued*)

Vegetables:

Sweet potato	½ cup	6.9
Collard greens	½ cup	3.1
Asparagus	½ cup	2.1
Spinach, raw	1 cup	1.5

Grains:

Wheat germ	2 tbs	4.2
Bread, whole wheat	1 slice	2.5
Bread, white	1 slice	1.2

Seafood:

Crab	3 oz	4.5
Shrimp	3 oz	3.7
Fish	3 oz	2.4

Vitamin D

FOOD	SERVING SIZE	VITAMIN D (IU)[d]
Milk:		
Milk, whole, low-fat, or skim	1 cup	100
Fish and seafoods:		
Salmon	3 oz	340
Tuna	3 oz	150
Shrimp	3 oz	127
Organ meats:		
Beef liver	3 oz	42
Chicken liver	3 oz	40
Eggs:		
Egg yolk	1	27

[d]40 IU = 1 microgram.

Table 20.5

Percent of original vitamin content lost in fruits by food storage method and in dried beans by cooking method.

	Fruits			Dried Beans Boiled 2–2.5 hours	
	Canned	Frozen	Dried	Water drained	Water used
Vitamin C	50%	30%	80%	35%	30%
Thiamin	15%	10%	10%	60%	55%
Riboflavin	5%	5%	5%	25%	20%
Niacin	10%	5%	5%	45%	40%
Vitamin B_6	10%	10%	5%	50%	45%
Folate	35%	25%	15%	70%	65%
Choline	0%	0%	0%	0%	0%
Vitamin B_{12}	0%	0%	0%	0%	0%
Vitamin A	5%	5%	10%	15%	10%

Source: Table prepared by author from data presented in USDA's Nutrient retention in foods tables.[30]

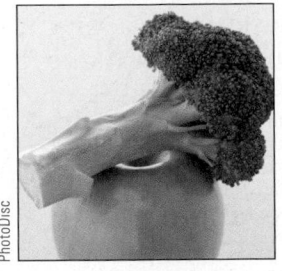
PhotoDisc

NUTRITION

Up Close

Antioxidant Vitamins–How Adequate Is Your Diet?

Focal Point: Determine if you eat enough antioxidant-rich foods.

Vitamin C, beta-carotene, and vitamin E, the antioxidant vitamins, help to maintain cellular integrity in the body. Good food sources of these antioxidants reduce the risk of heart disease, certain cancers, and other ailments. Check below to find out how frequently you consume foods containing these important, health-promoting nutrients.

How Often Do You Eat:	Seldom or Never	1–2 Times per Week	3–5 Times per Week	Almost Daily
Vitamin C–rich foods:				
1. Grapefruit, lemons, oranges, or pineapple?	☐	☐	☐	☐
2. Strawberries, kiwi, or honeydew melon?	☐	☐	☐	☐
3. Orange juice, cranberry juice cocktail, or tomato juice?	☐	☐	☐	☐
4. Green, red, or chili peppers?	☐	☐	☐	☐
5. Broccoli, Chinese cabbage, or cauliflower?	☐	☐	☐	☐
6. Asparagus, tomatoes, or potatoes?	☐	☐	☐	☐
Beta-carotene–rich foods:				
7. Carrots, sweet potatoes, pumpkin, or winter squash?	☐	☐	☐	☐
8. Spinach, collard greens, or chard?	☐	☐	☐	☐
9. Cantaloupe, papayas or mangoes?	☐	☐	☐	☐
10. Nectarines, peaches, or apricots?	☐	☐	☐	☐
Vitamin E–rich foods:				
11. Whole-grain breads, whole-grain cereals, or wheat germ?	☐	☐	☐	☐
12. Crab, shrimp, or fish?	☐	☐	☐	☐
13. Peanuts, almonds, or sunflower seeds?	☐	☐	☐	☐
14. Oils, margarine, butter, mayonnaise, or salad dressing?	☐	☐	☐	☐

FEEDBACK (including scoring) can be found at the end of Unit 20.

[Key Terms

antioxidants, page 20–13
chronic inflammation, page 20–11
coenzymes, page 20–3

dementia, page 20–10
free radicals, page 20–13

precursor, page 20–13
vitamins, page 20–2

Review Questions

1. Vitamins are essential. Specific deficiency diseases develop if we fail to consume enough of them. ☐ ☐

2. Vitamin D, E, C, and B$_6$ are fat soluble. ☐ ☐

3. The niacin deficiency disease is called pellagra. ☐ ☐

4. Some people become deficient in vitamin B$_{12}$ because they are genetically unable to absorb it. ☐ ☐

5. Vitamin A toxicity causes brain tumors. ☐ ☐

6. Three good sources of vitamin D are sunshine, milk, and seafood. ☐ ☐

7. The effectiveness of antioxidant vitamins appears to be increased when they are consumed in plant foods rather than supplements. ☐ ☐

8. Vitamin D acts as a hormone. ☐ ☐

9. Adequate vitamin D status reduces chronic inflammation. ☐ ☐

10. Excessively high intakes of every vitamin have been found to cause toxicity disease. ☐ ☐

Media Menu

www.ods.nih.gov/factsheets/vitamind.asp
The Office of Dietary Supplements presents summary information about the different vitamins at this site. Get reliable information on vitamin D, C, E, and others here.

www.nal.usda.gov/fnic/foodcomp
Looking for a comprehensive source on the vitamin content of foods? You can get it from this USDA site.

www.ars.usda.gov/Services/docs. htm?docid=9448.
Find out which vitamins are lost to some extent or spared by different cooking and preparation methods. USDA Tables of Nutrient Retention in Foods is located at this address.

www.healthfinder.gov
Search vitamins.

www.ars.usda.gov/ba/bhnrc/ndl
Look up the vitamin and other nutrient content of thousands of foods on this site.

www.iom.edu/fnb
This site provides updated information on the DRIs.

www.merckhomeedition.com
The Merck Manual of Medical Information is free and searchable. It can be used to find out more about health problems related to vitamins.

Notes

1. Hathcock JN. Vitamins and minerals: efficacy and safety. Am J Clin Nutr 1997;66:427–37.

2. Buzina R et al. Workshop on functional significance of mild-to-moderate malnutrition. Am J Clin Nutr 1989;50:172–6.

3. American Academy of Pediatrics. Folic acid and the prevention of neural tube defects. Pediatrics 1999;104:325–27.

4. Yoon J-S et al. Associations among folate, homocysteine, and dementia in elderly patients, J Neurol Neurosurg Psychiatry 2008;79:864–8.

5. Ulrich CM. Folate and cancer prevention: a closer look at a complex picture, Am J Clin Nutr 2007;86:271–3.

6. Ganji V et al. Trends in serum folate, RBC folate, and circulating total homocysteine concentrations in the US. J Nutr 2006;136:153–8.

7. Frieden TR et al. Vitamin A levels and severity of measles. Am J Dis Child 1992;146:182–6.

8. Topical retinoids in primary care practice. www.medscape.com, 8/03.

9. Bischoff-Ferrari HA et al. Estimation of optimal serum concentrations of 25-hydroxyvitamin D for multiple health outcomes. Am J Clin Nutr 2006;84:18–28.

9. Kafi R et al. Topical vitamin A (retinol) for the treatment of wrinkles associated with natural aging, Arch Dermatol 2007;143:606–12.

10. Holick MF. High prevalence of vitamin D inadequacy and implications for health, Mayo Clin Proc 2006;81:353–73.

11. Knekt P et al. Serum 25-hydroxyvitamin D levels and the risk of type 2 diabetes, Diabetes Care 2007;30:2569–70.

12. John EM et al. Exposure to sunlight and the risk of advanced breast cancer, Am J Epidemiol 2007, Oct. 29, 2007, available at www.medscape.com/viewarticle/565041, accessed 3/09

13. Arson Y et al. Vitamin D and autoimmunity: new etiological and therapeutic considerations, Ann Rheum Dis 2007;June 8, Epub ahead of print.

14. Holick MF. Vitamin D deficiency, N Engl J Med 2007;357:266–80.

15. National Report on Biochemical Indicators of Diet and Nutrition in the U.S. Population, www.cdc.gov/nutritionreport, accessed 9/08.

16. Nelson ML et al. Supplements of 20 ug/d cholecalciferol optimized serum 25-hydroxyvitamin D concentrations in 80% of premenopausal women in winter, J Nutr 2009;139:540–6.

18. Gordon CM et al. Prevalence of vitamin D deficiency in infants and toddlers, Arch Pediatr Adol Med 2008;162:505–12, 583-4.

19. Cashman KD et al. Estimation of the dietary requirement for vitamin D in healthy adults, Am J Clin Nutr 2008;88:1535–42.

20. Gozdzik A et al. Low wintertime vitamin D levels in a sample of healthy young adults of diverse ancestry living in the Toronto area: associations with vitamin D intake and skin pigmentation, BMC Public Health, 2008, www.medscape.com/viewarticle/584504, accessed 1/09.

21. Holick MF et al. Vitamin D deficiency: a worldwide problem with health consequences, Am J Clin Nutr 2008;87(suppl):1080S–6S.

22. Hollis BW. Circulating 25-hydroxyvitamin D levels indicative of vitamin D sufficiency: implications for establishing a new effective dietary intake recommendations for vitamin D. J Nutr 2005;135:31–22.

23. Gilchrest BA. Sun exposure and vitamin D deficiency, Am J Clin Nutr 2008;88(suppl):570S–7S.

24. Talwar SA et al. Dose response to vitamin D supplementation among postmenopausal African American women, Am J Clin Nutr 2007;86:1657-62.

25. Douglas RM et al. Vitamin C for preventing and treating the common cold. Cochrane Database Syst Rev. 2004 Oct 18;(4):CD000980. Halvorsen BL et al. Content of redox-active compounds (ie, antioxidants) in foods consumed in the United States. Am J Clin Nutr 2006;84:95–135.

26. Nutrient Data Laboratory, Oxygen radical absorbance capacity (ORAC) of selected foods—2007, www.ars.usda.gov/nutrientdata, accessed 8/08.

27. Hercberg S. The history of beta-carotene and cancers: from observational to intervention studies. What lessons can be drawn for future research on polyphenols? Am J Clin Nutr 2005;81 (suppl):218S–22S.

28. Larrson SC et al. Vitamin A, retinol, carotenoids and the risk of gastric cancer: a prospective cohort study, Am J Clin Nutr 2007;85:497–503.

29. Mahabir S et al. Comparison of risk of various forms of dietary vitamin E on lung cancer risk, Int J Cancer 2008;123:1173–80.

30. USDA Tables of Nutrient Retention in Foods, Release 6 (2007), www.ars.usda.gov/Services/docs.htm?docid=9448.

NUTRITION | # Up Close

Antioxidant Vitamins—How Adequate Is Your Diet?

Feedback for Unit 20

Several responses in the last two columns indicate adequate antioxidant vitamin consumption. If you need to boost your intake, increase the overall amount of fruits, vegetables, nuts, oils, and whole grains in your diet.

UNIT 21 | Phytochemicals and Genetically Modified Food

NUTRITION SCOREBOARD

	TRUE	FALSE
1 Phytochemicals are found only in plants.		
2 Phytochemicals taken as supplements provide the same health benefits as do phytochemicals consumed in plant foods.		
3 Humans eat lots of DNA every day.		
4 Some chemical substances that occur naturally in food may be harmful to health.		

Key Concepts and Facts

- Plants contain thousands of substances in addition to essential nutrients that affect body processes and health.

- Diets containing lots of vegetables, fruits, whole grains, and other plant foods are strongly associated with the prevention of chronic diseases such as heart disease and cancer.

- Biotechnology is rapidly changing the characteristics and types of foods available to consumers.

- Not all substances that occur naturally in foods are safe to eat.

Answers to **NUTRITION SCOREBOARD**	TRUE	FALSE
1 *Phytochemicals* means "plant chemicals."	✔	
2 Beneficial effects of phytochemicals on health appear to result primarily from complex interactions among them in foods. Supplements do not appear to provide these benefits.[1]		✔
3 The average meal contains approximately 250,000 miles (150,000 kilometers) of uncoiled DNA when it's digested.	✔	
4 Some foods contain "naturally occurring toxins" that can be harmful if consumed in excess.		✔

Things don't happen by accident in nature. If you observe it, it has a reason for being there.
—NORMAN KRINSKY, TUFTS UNIVERSITY MEDICAL CENTER

Photo Disc

phytochemicals
(phyto = *plant*) Chemical substances in plants, some of which perform important functions in the body.

zoochemicals
Chemical substances in animal foods, some of which likely perform important functions in the body.

Phytochemicals: Nutrition Superstars

As recently as 25 years ago, the science of nutrition focused on the study of the actions and health effects of protein, vitamins, minerals, and the other essential nutrients. Those days are gone. Now nutrition scientists are investigating the effects of thousands of other substances in food. Although we have barely scratched the surface of knowledge about these substances, current research shows that impressive benefits can be gained by diets high in plant foods. People who habitually consume lots of vegetables, fruits, whole grains, and other plant foods are less likely to develop heart disease, cancer, type 2 diabetes, infections, eye disease, premature aging, and a number of other health problems than are people who do not.[2, 3] Research results are sufficiently positive to change the way nutritionists and other scientists are thinking about food and health relationships.

At center stage in this new era in nutrition are the **phytochemicals**, or chemical substances found in plants (Illustration 21.1). The number of potentially beneficial substances in plants is mind boggling. There are thousands of individual phytochemicals in plants, and some foods contain hundreds of them. A sampling of the phytochemicals and other substances in two foods is shown in Illustration 21.2. Meats, eggs, dairy products, and other foods of animal origin also contain chemical substances that are not considered essential nutrients but nevertheless affect body processes. Much less is known about these **zoochemicals** because their effects on health are not yet as clear as those of phytochemicals.

Phytochemicals are not considered to be essential nutrients for humans, because we do not develop a deficiency disease if we consume too little of them.

Illustration 21.1 The foods shown here have star qualities. They have been featured in recent research reports on the benefits of phytochemicals in food. Foods shown are grape juice, green tea, tomatoes, cabbage, carrots, tofu, crushed garlic; green, red, and yellow peppers; brown rice, whole grain pita bread, and peanut butter. (Food preparation courtesy of The Blue Moon Cafe, Menomonie, WI.)

© Scott Goodwin Photography

They are similar to essential nutrients in that the body cannot make them—or enough of them. Consequently, they must be obtained from the diet. The chemical properties of most phytochemicals tend to make them heat and light stable, so they are not easily destroyed by cooking or storage.[4]

Many phytochemicals are excreted within a day or two after ingestion, so intake of vegetables, legumes, nuts, fruits, and other food sources should be maintained.[5] Cooking vegetables, or consuming them with a small amount of fat, increases the body's absorption of some phytochemicals.

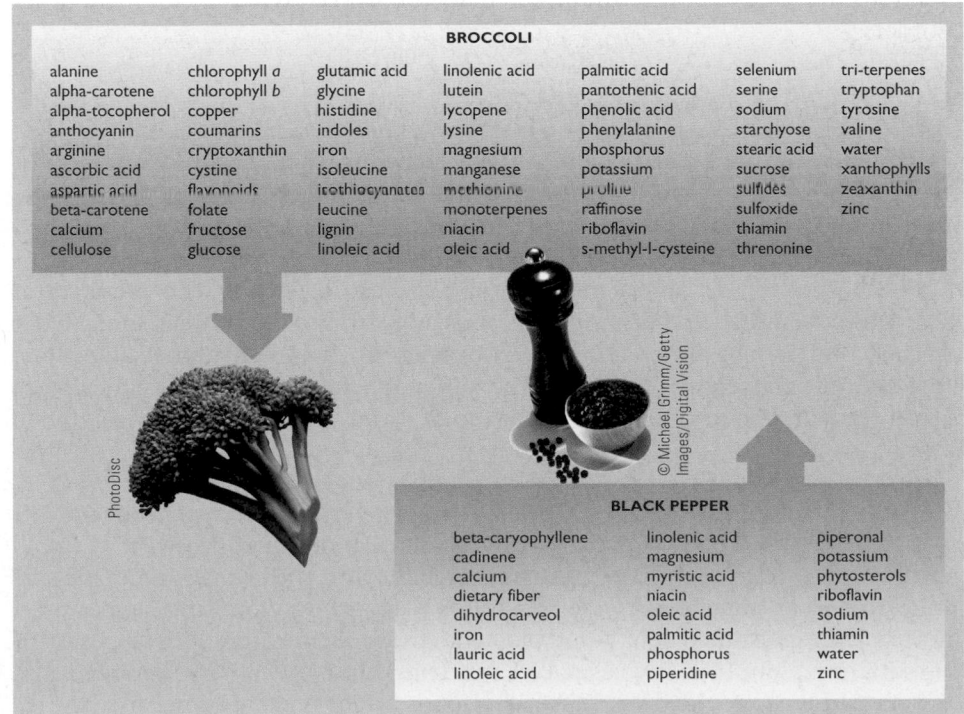

BROCCOLI

alanine	chlorophyll *a*	glutamic acid	linolenic acid	palmitic acid	selenium	tri-terpenes
alpha-carotene	chlorophyll *b*	glycine	lutein	pantothenic acid	serine	tryptophan
alpha-tocopherol	copper	histidine	lycopene	phenolic acid	sodium	tyrosine
anthocyanin	coumarins	indoles	lysine	phenylalanine	starchyose	valine
arginine	cryptoxanthin	iron	magnesium	phosphorus	stearic acid	water
ascorbic acid	cystine	isoleucine	manganese	potassium	sucrose	xanthophylls
aspartic acid	flavonoids	isothiocyanates	methionine	proline	sulfides	zeaxanthin
beta-carotene	folate	leucine	monoterpenes	raffinose	sulfoxide	zinc
calcium	fructose	lignin	niacin	riboflavin	thiamin	
cellulose	glucose	linoleic acid	oleic acid	s-methyl-l-cysteine	threonine	

PhotoDisc

© Michael Grimm/Getty Images/Digital Vision

BLACK PEPPER

beta-caryophyllene	linolenic acid	piperonal
cadinene	magnesium	potassium
calcium	myristic acid	phytosterols
dietary fiber	niacin	riboflavin
dihydrocarveol	oleic acid	sodium
iron	palmitic acid	thiamin
lauric acid	phosphorus	water
linoleic acid	piperidine	zinc

Illustration 21.2 A sampling of the chemical substances in two foods.

Characteristics of Phytochemicals

Phytochemicals serve a wide variety of functions in plants. They provide color and flavor and protect plants from insects, microbes, and oxidation due to exposure to sunlight and oxygen. Some phytochemicals are components of a plant's energy-making processes, and others act as plant hormones. More than 2000 types of phytochemicals that act as pigments have been identified. There are over 700 types of carotenoids, for example. These pigments give plants yellow, orange, and red color; they primarily function as antioxidants.

The amount and type of phytochemicals present in plants vary a good deal, depending on the plant. Some plant foods, such as citrus fruits and dried beans, which are considered rich sources of vitamins and minerals, contain an abundance of phytochemicals. Apples, celery, green tea, and potatoes, foods considered by nutrient composition tables to be relative "nutrient weaklings," also provide ample amounts of phytochemicals.

Not all of the phytochemicals in plants are beneficial to health, however. Some are "naturally occurring toxins" and can be harmful. This type of phytochemical is discussed later in the unit.

Phytochemicals and Health

It has been known for more than 30 years that diets rich in vegetables and fruits are protective against heart disease and certain types of cancer.[5] For most of this time, the benefits of such diets were attributed to the vitamin, mineral, or dietary fiber content of vegetables and fruits. More recently, scientists discovered that the essential nutrient content of the diet didn't explain all of the differences in disease incidence between people consuming diets low in plants and those eating diets rich in plants. As researchers looked for the reason for the difference, they discovered that other chemical substances in plants might be responsible. Many chemically active substances that could contribute to disease prevention were identified and their actions in the body described. Phytochemicals are associated with a reduced risk of developing heart disease, certain types of cancer (lung, breast, cervical, esophageal, stomach, and colon cancer, for example), **age-related macular degeneration**, (Illustration 21.3) **cataracts**, (Illustration 21.4) infectious diseases, osteoporosis, type 2 diabetes, stroke, hypertension, and other disorders.[6, 7]

Phytochemicals Work in Groups

News about the potential benefits of phytochemicals has sent consumers in search of pills that contain them. The market is replete with such products (Illustration 21.5), even though there is little solid evidence that individual phytochemicals extracted from foods benefit health.[1]

The absorption of phytochemicals appears to depend on the presence of other phytochemicals and nutrients in foods. Most (if not all) phytochemicals act together, producing a desired effect in the body if consumed at the same time. Since the optimal combinations of the different types of phytochemicals are not yet known, it is recommended that foods, rather than supplements, provide them.[1]

The Case of Beta-Carotene Supplements The lesson that phytochemicals act together was learned the hard way. Numerous studies had shown that smokers consuming diets high in vegetables and fruits and those with high levels of beta-carotene in their blood were much less likely to develop lung cancer than smokers who consumed few vegetables and fruits. Beta-carotene is a powerful antioxidant present in many vegetables and fruits, and researchers theorized that the beta-carotene provided by these plant foods might prevent lung cancer.

---

age-related macular degeneration
Eye damage caused by oxidation of the macula, the central portion of the eye that allows you to see details clearly. It is the leading cause of blindness in U.S. adults over the age of 65. Antioxidants provided by the carotenoids may help prevent and treat macular degeneration.[6]

cataracts
Complete or partial clouding over the lens of the eye (shown in illustration).

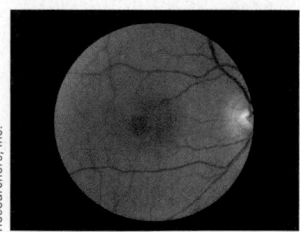

Illustration 21.3 Photograph of damage to the inside wall of an eyeball due to macular degeneration.

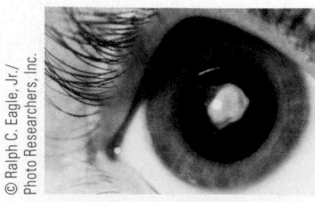

Illustration 21.4 Note the cataract forming in the center of the lens.

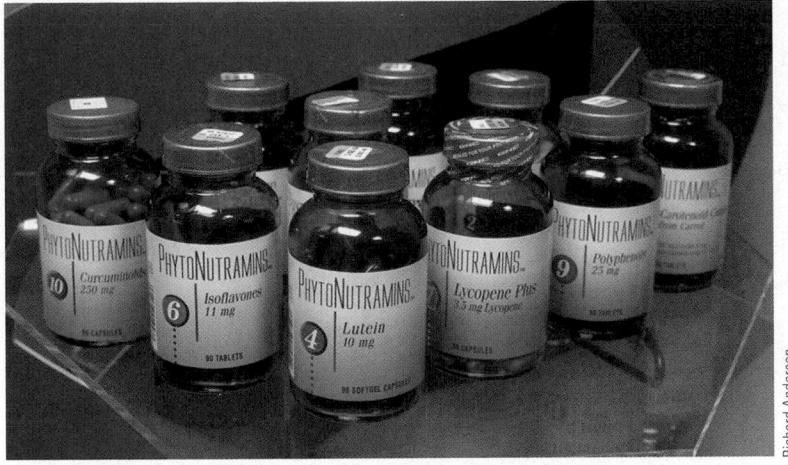

Richard Anderson

Keith Weller/ARS/USDA

Illustration 21.5 Consumers have their choice of phytochemicals in pills. Whether they benefit health is unclear.

Three large and expensive clinical trials of beta-carotene supplementation among male smokers were conducted to test this theory. The results? None of the trials showed a benefit from the beta-carotene supplements, and two studies found a higher rate of lung cancer among the groups using the supplements.[11]

Scientists drew two major conclusions from these studies:

1. High blood levels of beta-carotene found in the group of smokers who did not develop cancer were likely a marker for a high intake of vegetables and fruits and not the cause of the reduced risk.

2. Other phytochemicals in plants, or in a combination of plant foods, are likely responsible for the reduced risk of cancer.[1]

Vegetable Extracts and Essences Dehydrated, powdered extracts of vegetables high in certain phytochemicals and of the parts of vegetables richest in phytochemicals are also widely available. As is the case for supplements of individual phytochemicals, there is no solid evidence that these extracts benefit health. One such product called "Vegetable Essence" claimed to have "200 pounds of vegetables in a bottle." That is, of course, impossible. "Broccoli Concentrate," another vegetable extract, was found to contain primarily sulforaphane, one of the many phytochemicals in vegetables of the **cruciferous family** (Illustration 21.6). The problem with this and other vegetable concentrates is that you can fit only so much of a vegetable into a capsule. You would have to consume approximately 100 of the pills to get the amount of sulforaphane present in one serving of broccoli.[12] Even if you did that, you would still be missing out on the other phytochemicals and essential nutrients that are part of those little green trees.

How Do Phytochemicals Work?

A variety of body processes involved in disease development are affected by phytochemicals. Some of these processes, and the phytochemicals responsible, are listed in Table 21.1. In general, phytochemicals can

1. Act as hormone-inhibiting substances that prevent the initiation of cancer

2. Serve as antioxidants that prevent and repair damage to cells due to oxidation

cruciferous vegetables
Sulfur-containing vegetables whose outer leaves form a cross (or crucifix). Vegetables in this family include broccoli, cabbage, cauliflower, brussels sprouts, mustard and collard greens, kale, bok choy, kohlrabi, rutabaga, turnips, broccoflower, and watercress.

Richard Anderson

Illustration 21.6 Some examples of cruciferous vegetables.

Table 21.1

Examples of phytochemicals, their food sources, and potential mechanisms of action in disease prevention

Phytochemical	Comments	Food Sources	Proposed Action
Indoles, isothiocynates (sulforaphanes)	These sulfur-containing compounds may be particularly protective against breast cancer.	Cruciferous vegetables (broccoli, brussels sprouts, cabbage, cauliflower)	Interfere with the cancer growth-promoting effects of genes on cells; increase the body's ability to neutralize cancer-causing substances.
Allicin	Health benefits are associated with as little as a half clove of garlic per day on average.	Garlic, onions, leeks, shallots, chives, scallions	Interferes with the replication of cancer cells; decreases the production of cholesterol by the liver, reduces blood clotting.
Terpenes (monoterpenes, limonene)	These substances give citrus fruits a slightly bitter taste. Limonene is from the same family of compounds as tamoxifen, a drug used to treat breast cancer.	Oranges, lemons, grapefruit, and their juices	Facilitate the excretion of cancer-causing substances; decrease tumor growth.
Phytoestrogens (plant estrogens; isoflavones, genistein, daidzein, lignans)	May decrease risk of some cancers, heart disease, and osteoporosis. Estrogen effects, if any, are weak. High doses of some individual phytoestrogens may have adverse effects on health.	Soybeans, soy food, chickpeas, other dried beans, peas, peanuts	Interfere with cancer growth-promoting effects of estrogen; block the action of cancer-causing substances; lower blood cholesterol; decrease menopausal symptoms and bone loss, act as antioxidants.
Lignans (phytoestrogens)	Flaxseed is an extremely rich source of lignans. In the gut, lignans are converted to substances that may help prevent breast cancer.	Flaxseed, flaxseed oil, seaweed, soybeans and other dried beans, bran	Interfere with the action of estrogen.
Saponins	Nearly every type of dried beans is an excellent source of saponins.	Dried beans, whole grains, apples, celery, strawberries, grapes, onions, green and black tea, red wine	Neutralize certain potentially cancer-causing enzymes in the gut.
Flavonoids (tannins, phenols)	There are over 4000 flavonoids, some of which are plant pigments. Originally called "vitamin P." Gives red wines and dark teas astringent or bitter taste.	Apples, celery, strawberries, grapes, onions, green and black tea, red wine, soy, purple grape juice, broccoli, dark chocolate	Protect cells from inflammation and oxidation; decrease plaque formation and blood clotting; increase HDL cholesterol; decrease DNA damage related to cancer development.
Carotenoids (alpha-carotene, beta-carotene, lutein, zeaxanthin, beta-cryptoxanthin, lycopene)	There are more than 700 types of colorful carotenoids in plants. An orange contains at least 20 types. Dark green vegetables are often good sources; the green chlorophyll obscures the colors of carotenoids in these plants. Fat intake increases absorption.	Dark green vegetables, orange, yellow, and red vegetables and fruits	Neutralize oxidation reactions that can damage eyes and promote macular degeneration and cataracts; increase LDL cholesterol and cancer risk.
Plant stanols and sterols	Structurally similar to cholesterol; they block cholesterol absorption.	Edible and nonedible oils	Decrease blood LDL-cholesterol level.

3. Block or neutralize enzymes that promote the development of cancer and other diseases

4. Modify the absorption, production, or utilization of cholesterol

5. Decrease formation of blood clots

Some plant pigments are powerful antioxidants (see Table 21.2 for the top food sources of antioxidants). Zeaxanthin (pronounced *ze-ah-zan-thun*), which gives plants a corn yellow color, anthocyanin (*an-tho-sigh-ah-nin*, the blue in blueberries and grapes), and lycopene (*lie-co-peen*), which helps make tomatoes, strawberries, and guava red, are all strong antioxidants. Dark chocolate contains flavonoids, which decrease inflammation. Significant reductions in creative protein (CRP), a marker of inflammation, have been observed among individuals consuming a two-third ounce of dark chocolate three days a week.[12] Some of the phytochemicals that act as antioxidants help reduce plaque formation in arteries and the risk of cardiovascular disease, type 2 diabetes, osteoporosis, and some cancers.[3, 14] Because of their importance to health, this Unit's Take Action asks students to think about their own favorite anti-oxidant-rich foods.

Diets High in Plant Foods

The discovery of a wide array of health-promoting phytochemicals in plants is one more very important reason why vegetables, fruits, whole grains, and other plant foods should be a major part of the diet. They are a minor part now for many Americans. On average, Americans consume 1.7 cups of vegetables and 1 cup of fruit daily.[3] In reality, people would probably be healthier if they consumed more than the five servings of vegetables and fruits each day.

Naturally Occurring Toxins in Food

All that occurs naturally in foods is not necessarily good. There are many examples of foods containing phytochemicals that can have negative side effects. Spinach, collard greens, rhubarb, and other dark green, leafy vegetables contain oxalic acid. Eating too much of these foods can make your teeth feel as though they are covered with sand and give you a stomachache. Have you ever seen a

Table 21.2

Top food sources of antioxidants in a typical serving size[1, 3]

Pomegranate	Red cabbage
Blackberries	Pecans
Walnuts	Cloves, ground
Blueberries, wild	Grape juice
Strawberries	Chocolate, dark
Raspberries	Cranberry juice
Artichokes	Wine, red
Cranberries	Pineapple juice
Blueberries, domestic	Guava nectar
Coffee, brewed	Mango nectar

Take Action   To select and consume antioxidant-rich vegetables and fruits

Sometimes we simply forget about some vegetables, fruits, and nuts we love to eat. Read through this list of the "best" sources of antioxidants and place a check in front of the ones you forgot you love, or you would like to try.

_____	Pomegranate	_____	Red cabbage
_____	Blackberries	_____	Pecans
_____	Walnuts	_____	Cloves, ground
_____	Blueberries	_____	Grape juice
_____	Strawberries	_____	Chocolate, dark
_____	Raspberries	_____	Cranberry juice
_____	Artichokes	_____	Cherries, sour
_____	Cranberries	_____	Pineapple juice
_____	Spinach	_____	Guava nectar
_____	Oranges	_____	Mango nectar

© Scott Goodwin Photography

Illustration 21.7 Potatoes grown partly above ground develop a green color in the part exposed to the sun. The green section contains solanine, a naturally occurring, toxic phytochemical.

© Conde Nast Archive/Corbis

Illustration 21.8 Ackee fruit and seeds.

potato that was partly colored green? (If not, take a look at Illustration 21.7) The green area contains solanine, a bitter-tasting, insect-repelling phytochemical that is normally found only in the leaves and stalks of potato plants. Small amounts of solanine are harmless, but large quantities (an ounce or so) can interfere with the transmission of nerve impulses.

Phytates are an example of a naturally occurring substance that can have harmful effects. Phytates are found in whole grains; they tightly bind zinc, iron, and other minerals, making them unavailable for absorption. Although not toxic, phytates do reduce the availability of some minerals in whole grain products. Cassava, a root consumed daily in many parts of tropical Africa, can be very toxic if not prepared properly because it contains cyanide. Soaking cassava roots in water for three nights gets rid of the cyanide, but soaking for shorter periods does not. When the soaking time is cut to one or two nights, as sometimes happens during periods of food shortage, enough of the toxin remains in the root to cause konzo, a disease caused by the cyanide overdose.[15] Konzo is characterized by permanent, spastic paralysis.

Beware of Ackee Fruit Ackee fruit is another potential hazard to health. If you're from Jamaica or Africa, chances are excellent that you love the taste of the core of ackee—and know the fruit can be deadly. The national fruit of Jamaica, the yellow fleshy part around the seeds tastes like butter and looks like scrambled eggs (Illustration 21.8). The rest of the fruit is not edible. The ackee fruit—and the fruit of unopened, unripe ackee in particular—contains high concentrations of phytochemicals that cause severe vomiting and a drastic drop in blood glucose levels. Ingestion of the fruit has caused hundreds of deaths in Jamaica. Its sale was banned in the United States until 2000.[16]

What's the Scoop on Caffeine? Caffeine is another phytochemical in plants (primarily, coffee beans) that some people consider to be toxic. To others, it's a gift from the gods every morning. Does caffeine belong in the list of naturally occurring, harmful substances in food?

Caffeine is one of many phytochemicals in coffee beans, cocoa beans, cola nuts, and tea leaves. Coffee, however, is far and away the leading source of caffeine in the diet. Table 21.3 displays the caffeine content of various beverages, chocolate, and nonprescription drugs. Although the effects of caffeine on health are widely debated, it has received a clean bill of health with a couple of exceptions. Too much caffeine causes sleeplessness and "caffeine jitters" in many people.[17] Large amounts of coffee (more than four cups a day) may somewhat delay the time it takes women to become pregnant and may slightly increase the risk of miscarriage.[18] Suspicions that coffee or caffeine is related to heart disease, cancer, or osteoporosis

Table 21.3

Caffeine content of foods, beverages, and some drugs

Source	Caffeine (mg)
Coffee, one cup	
Drip	115–175
Decaffeinated (ground or instant)	0.5–4.0
Instant	61–70
Percolated	97–140
Espresso (2 ounces)	100
Tea, one cup	
Black, brewed 5 minutes, U.S. brands	32–144
Black, brewed 5 minutes, imported brands	40–176
Green, brewed 5 minutes	25
Instant	40–80
Soft drinks	
Coca-Cola (12 ounces)	47
Cherry Coke (12 ounces)	47
Diet Coca-Cola (12 ounces)	47
Dr. Pepper (12 ounces)	40
Ginger ale (12 ounces)	0
Mountain Dew (12 ounces)	54
Pepsi-Cola (12 ounces)	38
Diet Pepsi-Cola (12 ounces)	37
7-Up (12 ounces)	0
Energy drinks	
Ripped force (8 ounces)	120
Power shot (1 ounce)	100
Red Bull (8 ounces)	80
Full throttle (8 ounce)	72
Jolt (12 ounces)	72
Kick (12 ounces)	56
Surge (8 ounces)	35
Chocolate	
Cocoa, chocolate milk, one cup	10–17
Milk chocolate candy, one ounce	1–15
Chocolate syrup, one ounce (2 tablespoons)	4
Nonprescription drugs, two tablets	
Nodoz	200
Vivarin	200
Excedrin	130
Weight-control pills	150

have not been confirmed. Habitual coffee consumption (one to three cups a day) appears to decrease the risk of cardiovascular disease, type 2 diabetes, and declines in mental functioning with age.[19–22] The beneficial effects of coffee may be related to its content of phytochemicals that have antioxidant and anti-inflammatory properties and not to caffeine.[23]

Regular coffee drinking can be habit forming. Cessation of coffee intakes that average two-and-a-half cups a day leads to headaches in about 50% of people.[23]

© Peter Donaldson/Alamy

Genetically Modified Foods

Through **biotechnology**, researchers can modify the phytochemical and nutrient makeup of foods by altering the genetic makeup of plants. Resulting plants are called "genetically modified" or "GM" plants (Illustration 21.9). The process of biotechnology usually entails identifying a favorable genetic trait in one plant and transplanting that gene into another plant that lacks the characteristic.

The genetic makeup of plants has actually been modified for centuries. Some 8000 years ago, Native Americans increased corn yields by cross-fertilizing corn plants. Over the centuries, thousands of plant hybrids have been developed. Hybrids are achieved by combining the genes of different plant species through cross-pollination. Advances related to genetic engineering in recent years have refined techniques so that single genes, rather than the full complements of a plant's genetic makeup, can be transferred to another plant.[24]

Genetic engineering is now used to transfer disease-resistant genes from one plant to another plant, conferring upon it an improved ability to resist disease. Flavor genes can be transported from one plant to another, creating new taste sensations like garbanzo beans with a nutty, peppery taste and basil preseasoned with cinnamon. Watermelon and oranges have had their seeds removed through genetic engineering. Colors of vegetables and fruits can be modified by transferring the appropriate genes from one plant to another. Thanks to two genes from the snap dragon, you can buy purple tomatoes rich in anthocyanin, a dark–blue antioxidant pigment.[25] Carrots have been modified to be dark red so they would match Texas A&M's school color. Tomatoes have been genetically altered to stay firm during shipment and to produce 10 times the normal amount of lycopene; and rice has been altered to be rich in beta-carotene, the precursor of vitamin A. High beta-carotene rice was produced to make a good source of vitamin A available to people in countries where vitamin A deficiency is widespread. Called "golden rice," it contains two genes extracted from daffodils and one bacterial gene that together lead to the production of beta-carotene within rice seeds.[26]

biotechnology
As applied to food products, the process of modifying the composition of foods by biologically altering their genetic makeup. Also called *genetic engineering* of foods. The food products produced are sometimes referred to as "GM" and GMOs (genetically modified organisms).

Illustration 21.9 Genetic modification of plants.

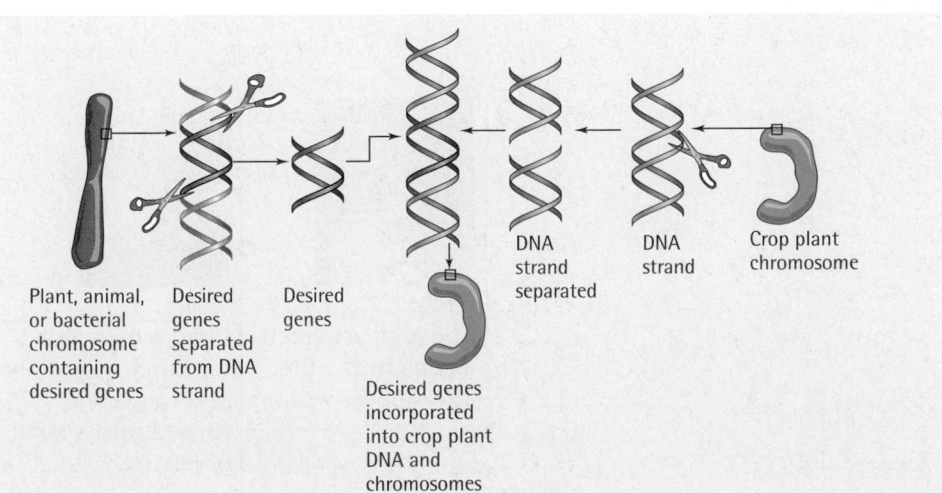

Plant, animal, or bacterial chromosome containing desired genes

Desired genes separated from DNA strand

Desired genes

Desired genes incorporated into crop plant DNA and chromosomes

DNA strand separated

DNA strand

Crop plant chromosome

Genetic Modification of Animals

Food biotechnology applies to foods derived from animals as well as plants. Scientists have engineered the DNA of Atlantic salmon so that they grow to market weight in 18 rather than the usual 24 to 30 months. Pigs that produce less smelly stools and gas, cattle with leaner muscles, and hens that lay more eggs are also products of animal genetic engineering.[27]

Animal Clones   Some of the new, genetically modified animal foods are the result of cloning. Cloning is accomplished by removing the nucleus of a donor egg obtained from a cow or pig, for example, and replacing it with the nucleus of an animal with desired characteristics, such as tender meat or high milk yield. The new nucleus is fused to the donor egg by electricity, is allowed to divide several times, and is then transplanted into a surrogate mother. The offspring produced is an identical copy of the animal whose nucleus was transplanted and is able to reproduce.[28]

Cloned animals are expensive ($20,000 to $80,000 each) and are not being used to produce food products. Their offspring are used to a limited extent. Currently, animal products obtained from cloned animals make up only a small fraction of meat sales.[28] Production of cloned animals, and sale of meat products from these animals, are regulated by the United States Department of Agriculture (USDA) and the Food and Drug Administration (FDA).[31]

GM Foods: Are They Safe and Acceptable?

People have a fear of eating DNA. We eat 150,000 kilometers [250,000 miles] of DNA in an average meal.

—David Cove, PhD, Leeds University's Genetics Department

People generally don't think of DNA as a normal component of food, but since we eat cells when we consume foods, we get hefty helpings of DNA and chromosomes in every meal. People have been consuming foods with novel DNA in hybrid plants for many years without ill effects. Obviously, consuming DNA is not harmful by itself. Concerns about the safety of GM foods have less to do with consuming them and more to do with their potential impact on people's lives and the environment.

Most Americans are unaware of the broad presence of GM foods in the marketplace. Over 60% of processed foods contain GM ingredients, but only 19% of adults believe they have eaten a GM food.[29,30] Over 80% of all soybeans and a third of corn grown in the United States are from genetically modified seeds. U.S. consumers regularly purchase and eat tomatoes, squash, cantaloupe, and potatoes that have been genetically modified.[29] Although a relatively small percentage of U.S. consumers (27%) are concerned about the safety of GM foods, several baby and snack food producers have excluded them from their products.

Reality Check

Eating Genes   Is it safe to eat genes?

Who gets thumbs up?
Answers on page 21–11

Photo Disc

Hyde:
I have never eaten a gene in my life, and I never will.

Ejay:
I had a huge salad for lunch yesterday and I haven't turned into a head of lettuce yet.

Individuals preferring GM-free foods in the United States and abroad are opting for organic products, which are GM-free, (Illustration 21.10) and locally grown produce.[29]

Foods derived from biotechnology are more heavily regulated than any other new foods. They are considered safe to eat, pose little threat of producing allergic reactions, and have real and potential benefits (which are listed in Table 21.4).[29,30] Important concerns about GM crops and foods exist, and they include the inability of small farmers to afford seed license fees and the fate of locally grown crops (these concerns are also listed in Table 21.4).

In 2008 the FDA ruled that GM animal foods are safe to eat and, against consumer sentiment, do not have to be labeled as genetically modified.[31] Concerns that reproduction by GM animals could introduce novel and potentially damaging traits in offspring, as well as ethical and animal welfare considerations, are making GM animal foods a hotly debated topic.[32, 33]

AP Images/Paul Sakuma

Illustration 21.10 Some organic products in England sport a GM-free label.

Table 21.4

Benefits and concerns related to GM crops and foods[29, 30]

Real and Potential Benefits	Real and Potential Concerns
• Reduced herbicide and pesticide use • Increased availability of crops that can grow in dry or salty soils and in hot or cold climates • Increased crop yield • Decreased waste due to spoilage • Improved nutritional content of foods • Improved food flavors • Source of antibodies that could help people become immune to food allergens	• GM crop traits may cross-pollinate with local crops and reduce the market for locally grown foods. • GM crop seeds are relatively expensive and perhaps too expensive for small farmers. • Nutritionally superior GM foods have yet to reach consumers. • Insects may become resistant to genetic modifications of plants intended to keep them away. • Foods produced may be unacceptable to the consumers they are intended to benefit. • Long-term effects of GM organisms on local plants and crops, insects, and animals are unknown. • A belief by some that humans should not mess with nature by changing the genetic makeup of life-forms.

ANSWERS TO **Reality Check**

Eating Genes Genes in food are broken down in the digestive tract and rendered inactive. Characteristics transferred to foods by genes are not acquired by humans.

Hyde:
👎

Ejay:
👍

© Scott Goodwin Photography

Up Close

Have You Had Your Phytochemicals Today?

FOCAL POINT: Consuming good sources of the "other" beneficial components of food.

Good food sources of beneficial phytochemicals are listed below. Indicate foods you consumed at least twice last week, foods you like, and those you have never tried.

	Foods Eaten at Least Twice Last Week	Foods You Like	Foods You've Never Tried
Broccoli			
Cabbage			
Carrots			
Cauliflower			
Celery			
Collard greens			
Garlic			
Onions			
Tomatoes			
Turnips			
Apple/juice			
Orange/juice			
Grapefruit/juice			
Grapes/juice			
Strawberries			
Brown rice			
Dried beans			
Tofu			
Whole-grain bread, cereal			

FEEDBACK (answers to these questions) can be found at the end of Unit 21.

[Key Terms

Review Questions

TRUE FALSE

1. Phytochemicals that benefit health are found in high amounts in fish and organ meats. ☐ ☐

2. Some plant foods that contain relatively low amounts of vitamins and minerals are rich sources of beneficial phytochemicals. ☐ ☐

3. Individual supplements of lutein, lycopene, and zeaxanthin more effectively prevent disorders such as macular degeneration and prostate cancer than do foods that naturally contain these phytochemicals. ☐ ☐

4. The top two food sources of antioxidants are dark chocolate and brewed coffee. ☐ ☐

5. The best way to achieve the health benefits of phytochemicals is through the consumption of five or more servings of vegetables and fruits daily. ☐ ☐

TRUE FALSE

6. If a chemical substance occurs naturally in a plant food, it can be considered harmless to health. ☐ ☐

7. Most of the health effects related to coffee consumption are due to caffeine. ☐ ☐

8. Genetically modified foods have been found to harm human health. ☐ ☐

9. Disease resistance, improved flavor and nutrient content, and reduced pesticide use are benefits of genetically modified plant foods. ☐ ☐

10. Three potential problems related to genetically modified (GM) foods include the loss of small farms, cross-pollination with non-GM plants, and lack of consumer acceptance of these foods. ☐ ☐

Media Menu

fnicsearch.nal.usda.gov
For current information about topics covered in this Unit, enter search terms such as genetically modified foods, or names of specific phytochemicals or antioxidants in the search feature at this site.

www.ers.usda.gov/briefing/biotechnology
Use this address to enter USDA's Briefing Rooms on Agricultural Biotechnology. You can get a briefing on adoption and impacts of biotechnology and food production, marketing, labeling, and trade, and research.

www.ars.usda.gov/Main/docs.htm?docid=15869
USDA's expanded tables of nutrient composition of foods includes food sources of a number of phytochemicals, including lycopene, leutein, and zeaxanthin.

http://5aday.gov
The National Cancer Institute's 5-a-day program site offers encouragement and tips on selecting colorful vegetables and fruit daily for better health.

www.nas.edu
Search "food biotechnology" or "genetically modified foods" to get reports on results from the National Academy of Science's subcommittee on genetically modified plants and animals.

Notes

1. Halvorsen BL et al. Content of redox-active compounds (ie, antioxidants) in foods consumed in the United States. Am J Clin Nutr 2006;84:95–135.

2. Thompson HJ et al. Dietary botanical diversity affects the reduction of oxidative biomarkers in women due to high vegetable and fruit intake. J Nutr 2006;136:2207–12.

3. Wolfe KL et al. Cellular antioxidant activity of common fruits, J Agric Food Chem 2008;56:8418–26.

4. Gross M. Flavonoids, platelets, and the risk of cardiovascular disease. Epidemiology Seminar, Minneapolis, 1999 Nov. 10.

5. Craig WT. Phytochemicals: guardians of our health. J Am Diet Assoc 1997;97 (suppl 2):S199–S204.

6. Jager RD et al. Age-related macular degeneration N Engl J Med 2008;358:2606–17.

7. Tan AG et al. Antioxidant nutrient intake and long-term incidence of age-related cataract: the Blue Mountain Eye Study, Am J Clin Nutr 2008;87:1899–905.

8. Hercberg S. The history of beta-carotene and cancers: from observational to intervention studies. What lessons can be drawn for future research on polyphenols? Am J Clin Nutr 2005;81(suppl):218S–22S.

9. Johnston CS et al. People with marginal vitamin C status are at high risk of developing vitamin C deficiency. J Am Diet Assoc 1999;99:854–6.

10. Hayes JD et al. The cancer chemopreventive actions of phytochemicals derived from glucoosinolates, Eur J Nutr 2008;47 Suppl 2:73–88.

11. Knekt P et al. Dietary flavonoids and the risk of lung cancer and other malignant neoplasms. Am J Epidemiol 1997;146:223–30.

12. Herber D. The stinking rose: organo-sulfur compounds and cancer. M J Clin Nutr 1997;66:425–6.

13. Di Giuseppe R et al. Regular consumption of dark chocolate is associated with low serum concentrations of C-reactive protein in a healthy Italian population. J Nutr 2008;138:1939–1945.

14. Valtuena S et al. Food selection based on total antioxidant capacity can modify antioxidant intake, systemic inflammation, and liver function without altering markers of oxidative stress. Am J Clin Nutr 2008;87:1290–7.

15. Tylleskar T et al. Dietary determinants of a non-progressive spastic paparesis (konzo). Int'l J Epidemiol 1995;24:949–56.

16. Holson D. Toxicity, plants: ackee fruit. www.emedicine.com, accessed 9/02.

17. Shirlow MJ et al. A study of caffeine consumption and symptoms: indigestion, palpitations, tremor, headache, and insomnia. Int'l J Epidemiol 1985;14:239–48.

18. Grunebaum A. Coffee and pregnancy: a bad mix? www.webmd.com/content/article/41/3606_696.htm, accessed 9/06.

19. Katan MB et al. Caffeine and arrhythmia. Am J Clin Nutr 2005;81:539–40.

20. Lopez-Garcia E et al. Coffee consumption and markers of inflammation and endothelial dysfunction in healthy diabetic women, Am J Clin Nutr 2006;84:888–93.

21. Eskelinen MH et al. Midlife coffee drinking and the risk of later life dementia: A population-based CAIDE Study, Alzheimer's Disease, 2009;16(1), released 1/16/09.

22. Hu F et al. Long-term coffee consumption related to reduced risk for type 2 diabetes. Harvard School of Public Health Press release, Jan 5, 2004, www.hsph.harvard.edu/press/releases/press01052004.html. accessed 9/06.

23. Silverman K et al. Withdrawal syndrome after the double-blind cessation of caffeine consumption. N Engl J Med 1992;327:1109–14.

24. Babcock BC et al. Solving global nutrition challenges requires more than new biotechnologies. J Am Diet Assoc 2000;100:1308–11.

25. Martin C et al. GM tomatoes boost life of cancer-prone rats, Nature Biotech 2008, Oct. 26 issue.

26. Falk MC et al. Food biotechnology: benefits and concerns. J Nutr 2002;132:1384–90.

27. Lewis C. A new kind of fish story: the coming of biotech animals. FDA Consumer, Jan.–Feb. 2001.

28. Zhang J et al. Animal clones are in food supply, Wall Street Journal, 9/2/08, p. A 3.

29. Position of the American Dietetic Association: agricultural and food biotechnology. J Am Diet Assoc 2006;106:285–93.

30. Palmer S. GE foods under the microscope. Today's Dietitian 2005; May:34–9.

31. Byrne J. Labeling of GM animals sought, 1/16/09, www.foodnavigator.com/content/view233126, accessed 1/21/09.

32. Caplan A. Genetically engineered animals as food, msnbc.com, 11/12/08, accessed 12/1/08.

33. Organizations clamor for more thorough GE controls, www.foodnavigator.com, released 3/23/09, accessed 3/24/09.

Answers to Review Questions

1. False, see page 21-2.
2. True, see page 21-4.
3. False, see page 21-4.
4. False, see Table 21.4, page 21-7.
5. True, see page 21-7.
6. False, see page 21-7.
7. False, see page 21-9.
8. False, see page 21-10.
9. True, see page 21-11 and Table 21.4.
10. True, see Table 21–4, pages 21–10, 21–11.

NUTRITION | Up Close

Have You Had Your Phytochemicals Today?

Feedback for Unit 21

Did you eat four or more of the foods listed at least twice last week? If yes, listen carefully and you'll hear your cells say "thank you."

If you didn't eat four or more twice last week, go for the foods you didn't eat but like. Be adventurous! Try some of the phytochemical-rich foods you have never eaten before.

Diet and Cancer

NUTRITION SCOREBOARD

	TRUE	FALSE
1 Some types of cancer are contagious.		
2 Cancer is primarily an inherited disease.		
3 People who eat plenty of fruits and vegetables are less likely to get cancer than people who don't.		
4 High levels of body fat contribute to the development of many types of cancer.		
5 Two out of three Americans never get cancer.		

Key Concepts and Facts

- Cancer has many different causes. Diet is a major factor that influences the development of most types of cancer.

- Diets primarily based on plant foods that include lean meats, fish, and low-fat dairy products; regular physical activity; and normal levels of body fat reduce cancer risk.

- Cancer is largely preventable, but there are no absolute guarantees that an individual will not develop cancer.

Answers to NUTRITION SCOREBOARD	TRUE	FALSE
1 Cancer doesn't spread from person to person.		✔
2 Cancer is primarily related to environmental factors such as diet, smoking, and physical activity. Genetic background plays a role by placing some individuals at increased risk for developing cancer.[1]		✔
3 Among the arsenal of cancer-protective measures, the consumption of five or more servings of fruits and vegetables a day stands out.[1]	✔	
4 High levels of body fat are related to the development of many types of cancer.[1]	✔	
5 Statistics from the American Cancer Society indicate that most people don't get cancer.[2]	✔	

What Is Cancer?

cancer
A group of diseases in which abnormal cells grow out of control and can spread throughout the body. Cancer is not contagious and has many causes.

prostate
A gland located above the testicles in males. The prostate secretes a fluid that surrounds sperm.

Cancer, the second leading cause of death in the United States, is a group of conditions that result from the uncontrolled growth of abnormal cells. Although these cells can begin to grow in any tissue in the body, the lungs, colon, **prostate**, and breasts are the most common sites for cancer development (Illustration 22.1). Some forms of cancer are highly curable.

How Does Cancer Develop?

Cancer develops by complex processes that are not yet fully understood.[3] Adding to the complexity is the fact that cancer development often does not proceed in a straight line—cancer can progress two steps forward and then take a step or two back.

Illustration 22.2 summarizes the processes involved in the development of cancer, as we currently understand them. Cancer begins when something goes wrong within cells that modifies cell division. Every minute, 10 million cells in the body divide. Usually, the cells divide the right way and on schedule due to a set of regulatory mechanisms that control the replication (duplication) of DNA, the genetic material that becomes part of new cells.

initiation phase
The start of the cancer process; it begins with the alteration of DNA within cells.

In the **initiation phase** of cancer development, DNA becomes damaged due to the presence in cells of toxic substances. These toxic substances can be generated by exposure to tobacco smoke, obesity, dietary components, certain viruses, or iodizing radiation; or by other environmental factors.[1] Damaged DNA can be repaired during the initiation phase, and that stops cancer from developing. If DNA is not repaired, the development of cancer continues to the **promotion phase**. During the *promotion* phase of cancer, cells with altered DNA divide, eventually producing large numbers of abnormal cells. This phase of cancer development commonly takes place over a span of 10 to 30 years. Unless they are hindered by the body or corrected by some other means, abnormal cells continue to divide, leading to the *progression* phase of cancer development.

The progression phase of cancer development is marked by loss of control over the abnormal cells, and their numbers increase rapidly. Eventually, the cells become so numerous that they erode the normal functions of the tissue where they are growing. During this phase, the abnormal cells may migrate to other tissues and cause DNA damage and abnormal cell development in these tissues, too.[4]

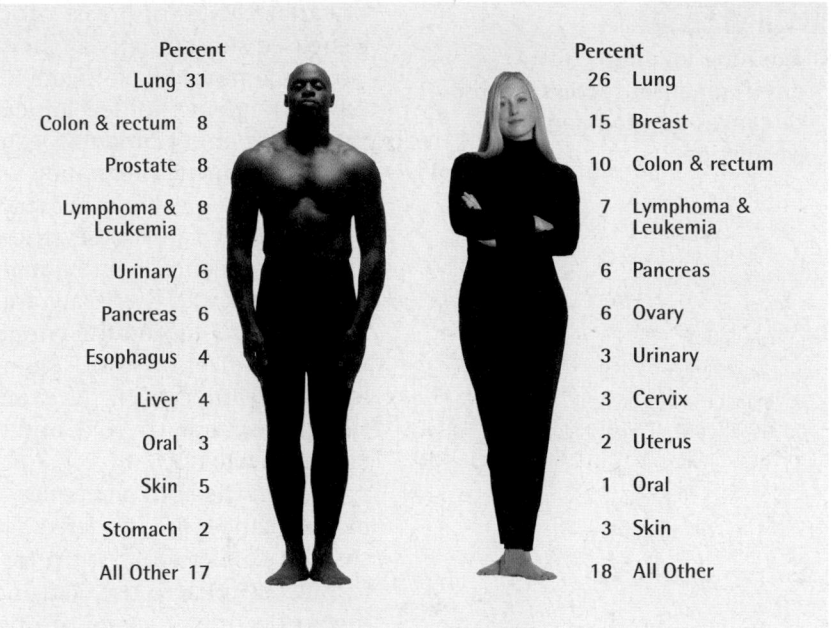

Percent			Percent	
Lung	31		26	Lung
Colon & rectum	8		15	Breast
Prostate	8		10	Colon & rectum
Lymphoma & Leukemia	8		7	Lymphoma & Leukemia
Urinary	6		6	Pancreas
Pancreas	6		6	Ovary
Esophagus	4		3	Urinary
Liver	4		3	Cervix
Oral	3		2	Uterus
Skin	5		1	Oral
Stomach	2		3	Skin
All Other	17		18	All Other

Photo Disc

Illustration 22.1 Percentage of cancer deaths by selected sites and sex, 2008.
In 2005, 23% of deaths in the United States were due to cancer.[7]

What Causes Cancer?

About 80 to 90% of all cancers are related to environmental factors that modify the structure and function of DNA.[4] Many environmental factors appear to play a role in cancer development, and most are modifiable. The top nine modifiable risk factors for cancer development worldwide are listed in Table 22.1.

Diet is one of the major environmental factors, and it accounts for approximately 40% of cancer risk.[5] Some of the most convincing evidence of the robust relationship between environmental factors and cancer come from studies of cancer rates in people migrating to another country.[6] Rates of breast cancer, for example, are low in rural Asia. When individuals from rural parts of Asia immigrate to the United States, rates of breast cancer become the same or higher than the U.S. rate by the third generation. Rates of prostate cancer similarly increase as people move from countries with low rates to countries with high rates. Westernization of dietary intake and lifestyles increases the risk of many types

promotion phase
The period in cancer development when the number of cells with altered DNA increases.

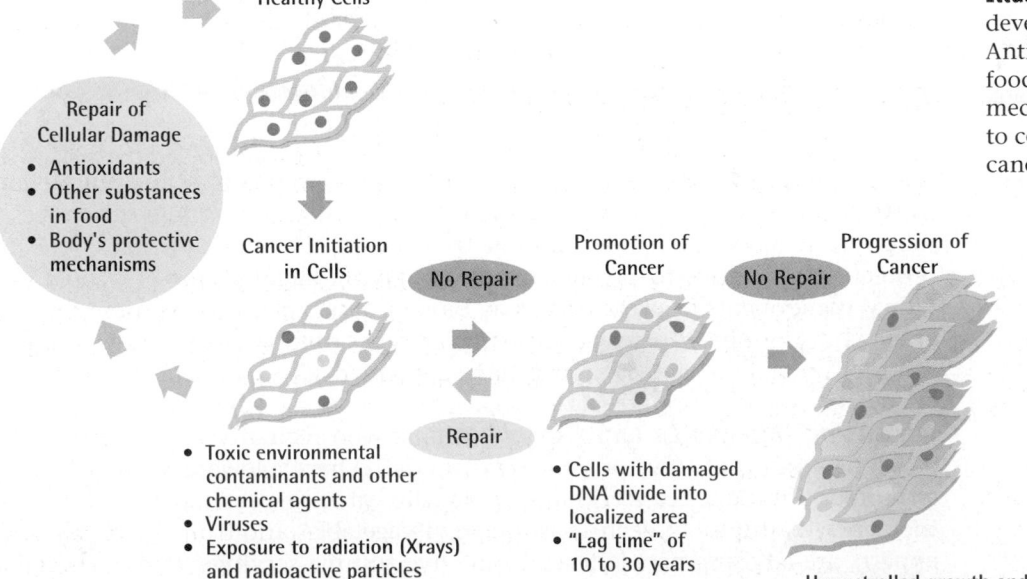

Illustration 22.2 Steps in the development of cancer.[2] Antioxidants, other substances in food, and the body's protective mechanisms may repair damage to cells and halt the progression of cancer.

Healthy Cells

Repair of Cellular Damage
- Antioxidants
- Other substances in food
- Body's protective mechanisms

Cancer Initiation in Cells

No Repair

Promotion of Cancer

No Repair

Progression of Cancer

Repair

- Toxic environmental contaminants and other chemical agents
- Viruses
- Exposure to radiation (Xrays) and radioactive particles
- Abnormal hormonal changes

- Cells with damaged DNA divide into localized area
- "Lag time" of 10 to 30 years

Uncontrolled growth and spread of abnormal cells

of cancer. Rates of breast cancer in Japanese and Alaskan Eskimo women have increased substantially as they have adopted Westernized diets and lifestyles.[5]

Some people have a genetic predisposition toward cancer, which means they have a tendency to develop cancer if regularly exposed to certain substances in the diet or environment. A genetically based susceptibility to cancer can develop in a fetus during pregnancy and in infancy. Exposure to calorie deficits, certain viruses, and specific other substances during these periods of rapid growth and development can modify the function of genes that help protect people from developing cancer.[9] Individuals with one or more gene types that decrease the body's ability to use folate are at increased risk of developing colorectal cancer if folate intake is low.[10] Genetic factors appear to account for 42% of the risk for prostate cancer, five to 27% of the risk for breast cancer, and 36% of the risk for pancreatic cancer, for example. Endometrial cancer (cancer of the lining of the uterus), oral, thyroid, and bone cancer do not appear to be related to genetic factors.[4]

Given the high percentage of cancers related to diet and other environmental factors, cancer is considered a largely preventable disease. Increasing rates of new cases of cancer took a turn for the better after 1992 and correspond to declines in rates of tobacco use. The incidence of cancer in the United States is continuing to decline at a rate of one to two percent per year.[11] It is anticipated that other improvements in lifestyles and diets will lead to further declines in cancer rates.[12]

Fighting Cancer with a Fork

Leading the list of cancer-promoting diets are those low in vegetables and fruits.[1] High intakes of beef, processed meat, charred meat, saturated fat, and alcohol are also related to cancer development.[13] Other, major risk factors for many types of cancer include smoking, physical inactivity, and excess body fat.[1, 13] Table 22.2 summarizes characteristics of diets and lifestyle that are related to a reduced risk of cancer.

Frequent consumption of certain types of food is sometimes more strongly related to particular cancers than to other types. For example, regular consumption of tomato products is related in particular to decreased risk of prostate cancer, and regular intake of black and green tea appears to contribute to breast and ovarian cancer reduction.[14] Diets and lifestyles that best prevent cancer are represented by a set of characteristics and not by hard rules about specific foods, dietary restrictions, or types of physical activities.

Dietary Risk Factors for Cancer: A Closer Look

Foods contain a variety of vitamins and minerals, as well as fiber and phytochemicals that participate in protecting the body against cancer. These substances in food, particularly plant foods, appear to work together in ways that confer the protection. Attempts to prevent cancer by giving large groups of people individual components of plants that may biologically explain the plant's beneficial effects have not been successful. Particular types of food clearly provide greater levels of protection against cancer than supplements.[15]

Fruits and Vegetables and Cancer   People who regularly consume plenty of vegetables and fruits (five or more servings daily) have a lower risk of developing a number of types of cancer than people who eat fewer servings. Because cancer incidence continues to decline as intake of vegetables and fruits increases, some experts are advising people to consume five to nine servings daily.[16] Vegetables and fruits contain antioxidant vitamins (vitamins C and E, for example) and phytochemicals that act as antioxidants and reduce inflammation. These substances

Kumquats and Cancer "Kumquats Prevent Cancer" reads the headline in the paper. Researchers speculate that "tartaphil," the substance that makes kumquats tangy, may be the reason. Should you eat kumquats?

Who gets thumbs up?
Answers on page 22–7

Sebastian:
Kumquats are too sour. I'd take a tartaphil supplement though.

Flora:
I already eat about three fruits every day. I'll include kumquats.

---

may reduce cancers related to inflammation and oxidative damage to DNA.[17] Damaged DNA may program the abnormal production of cell components and cells. Cancer can develop if abnormal cell multiplication directed by damaged DNA is allowed to continue. Antioxidants and anti-inflammatory compounds in vegetables and fruits assist in preventing oxidation of DNA and participate in suppression of abnormal cell multiplication.[13]

Consuming three servings per week of vegetables from the cruciferous family (broccoli, cabbage, cauliflower, and brussels sprouts) substantially reduces the risk of lung, bladder, stomach, and prostate cancer (Illustration 22.3).[18] If you enjoy eating these vegetables, consider yourself lucky.

Color-Coding Vegetable and Fruit Choices Many of the phytochemicals participating in mechanisms that likely help to prevent cancer can be identified by their color. This feature of some phytochemicals, and the advantages of eating a variety of vegetables and fruits, has led to the advice to select and consume colorful vegetables and fruits daily.[19] Table 22.3 lists phytochemicals, the color they impart to mature vegetables and fruits, and their sources. The phytochemicals listed all act as antioxidants.

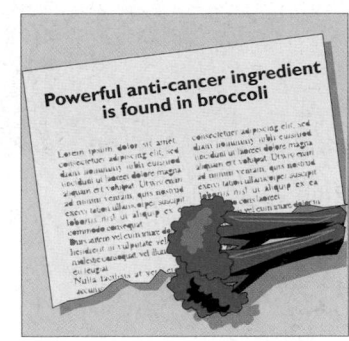

Illustration 22.3 An example of headline coverage of the cancer-protective effects of vegetables.

Table 22.3

Color-Coding Your Vegetables and Fruits

Color	Phytochemical Antioxidant	Vegetable and Fruit Sources
Red	lycopene (*lie-co-peen*)	tomatoes, red raspberries, watermelon, strawberries, red peppers
Yellow-green	lutein zeaxanthin (*lou-te-in, ze-ah-zan-thun*)	leafy greens, avocado, honeydew melon, kiwi fruit
Red-purple	anthocyanins (*antho-sigh-ah-nin*)	grapes, berries, wine, red apples, plums, prunes
Orange	beta-carotene	carrots, mangos, papayas, apricots, pumpkins, yams
Orange-yellow	flavonoids (*fla-von-oids*)	oranges, tangerines, lemons, plums, peaches, cantaloupe
Green	glucosinolates (*glu-co-sin-oh-lates*)	broccoli, brussels sprouts, kale, cabbage
White-green	allyl sulfides (*al-lill sulf-ides*)	onions, leeks, garlic

Illustration 22.4 The charred and black, oily coating on grilled or broiled meats is the part you shouldn't eat.

Table 22.4

Types of foods included in dietary patterns that lower the risk of cancer and heart disease

Breakfast

Oat flakes with banana and skim/low-fat milk
Cracked wheat toast with soft margarine
    and grape jam
Orange juice
Tea/coffee

Lunch

Skinless chicken sandwich on whole-grain
    bread with mozzarella cheese
Broccoli, cauliflower, raisin salad
Carrot sticks
Skim or low-fat milk

Dinner

Tomato soup
Broiled tuna with olive oil–lemon sauce
Baked beans
Coleslaw
Ice milk with strawberries
Skim or low-fat milk, coffee, tea,

Snacks

Nuts, seeds
Yogurt
Raw vegetables
Fruits
Skim or low-fat milk

Whole Grains and Cancer Whole grains contain vitamins, minerals, unsaturated fatty acids, fiber, and phytochemicals that play roles in cancer prevention.[25] Effects of whole grains and whole-grain products on cancer risk are related to the combined action of these substances. If you isolate a single substance, you in effect are destroying its ability to function in cancer prevention. Americans are advised to include three or more whole-grain products in their daily diet.[20]

Saturated Fat and Cancer Promotion High intake of saturated fats from meats and dairy products increases the risk of cancer.[21] Consequently, recommendations for diet and cancer prevention include the use of plant sources of protein because they provide unsaturated fats. There are additional advantages to increased reliance on plant foods in the diet. Plant foods such as dried beans, soy products, nuts, and seeds also provide vitamins, minerals, fiber, and phytochemicals that help ward off cancer progression.[22]

Nitrate-Preserved and Grilled Meats and Cancer Cancer of the stomach and liver appears to be related to the regular consumption of foods such as hot dogs, luncheon meats, bacon, and pickled eggs and vegetables preserved with nitrates.[13] (You can identify foods that contain nitrates by examining food ingredient labels.) Most cases of cancer associated with nitrate use in smoked, salted, and pickled foods now occur in China, parts of the former Soviet Union, and Central and South America, where such foods are commonly consumed.[23]

Certain nitrogen-containing and other substances in beef, chicken, fish, and other meats may become cancer promoting if heated to a high temperature. Such high temperatures can be reached by broiling and grilling food. The potentially cancer-promoting substances formed by meats are found in the charred portions of the meat and the fatty coating that accumulates on meat when fat drips into the heat source and smokes (Illustration 22.4).[13] People should be careful not to eat the charred portions of meat and to wipe or rinse off any coating on meat that results from fatty flare-ups on the grill.

Diet and Cancer Guidelines

Dietary patterns and lifestyle changes that reduce the risk of cancer (Table 22.1) are similar to dietary recommendations that reduce the risk of heart disease.[24] Considered together, dietary recommendations for cancer prevention can be transferred to dietary intake by selection of the types of foods shown in the example menu in Table 22.4.

Alcohol Intake and Cancer Consumption of alcoholic beverages has been linked to cancers of the breast, mouth, throat, and liver.[25] The risk of developing cancers of the mouth and throat increases for people who smoke cigarettes or chew tobacco as well as drink.[4]

Excess Body Fat and Cancer Obesity, and in particular obesity characterized by excess stores of central body fat, increases the risk of cancer at several sites. High levels of central fat appear to alter metabolism of hormones such as estrogen, testosterone, and insulin in ways that promote the growth of abnormal cells. Calorie-reduced diets combined with at least 30 minutes of moderate or vigorous activity five or more days a week are recommended for cancer prevention.[26]

Kumquats and Cancer Phytochemical supplements do not appear to prevent cancer. Diets that help prevent cancer are not based on one or a few foods. They are based on day-to-day intake of a healthy array of foods. P.S. The story about kumquats is bogus.

Sebastian:

Flora

Bogus Cancer Treatments

Unorthodox, purported cancer cures such as macrobiotic diets; hydrogen peroxide ingestion; laetrile tablets; vitamin, mineral, and herbal supplements; and animal gland therapy have not been shown to be effective treatments for cancer. Such remedies have been promoted since the early 1900s. They still exist because, although not proven to work, they offer some cancer patients their last ray of hope. They should not be used as a substitute for conventional cancer treatments.[27]

Eating to Beat the Odds

Two out of every three people in the United States do not develop cancer. You can likely improve your odds of being among the two by not smoking or drinking excessively, by regularly consuming five or more servings of fruits and vegetables each day, by sticking to a low-saturated-fat diet, and by being physically active and achieving or maintaining a normal level of body fat. Although there are no guarantees, people can help themselves prevent cancer by adhering to a good diet and a healthy lifestyle.

© Digital Vision/Getty Images

NUTRITION | Up Close

A Cancer Risk Checkup

Focal Point: Reducing cancer risk.

A number of behaviors that help protect people from developing cancer are listed below. Check those that apply to you.

	Yes	No	Don't Know
1. I eat five or more servings of vegetables and fruits daily.	_____	_____	_____
2. I consume whole-grain products.	_____	_____	_____
3. I eat a high-fat diet.	_____	_____	_____
4. I smoke or chew tobacco.	_____	_____	_____
5. I exercise regularly.	_____	_____	_____
6. I have too much body fat.	_____	_____	_____

FEEDBACK (answers to these questions) can be found at the end of Unit 22.

[Key Terms

cancer, page 22–2
initiation phase, page 22–2

promotion phase, page 22–3

prostate, page 22–2

[Review Questions

	TRUE	FALSE
1. Cancer development is basically related to changes in the structure function of DNA.	☐	☐
2. Little can be done to prevent cancer.	☐	☐
3. Most cases of cancer are caused by inherited genetic traits.	☐	☐
4. Diets providing five or more servings per day of vegetables and fruits reduce the risk of cancer.	☐	☐
5. Regular consumption of charred meat and high-saturated-fat foods increases the risk of cancer.	☐	☐

	TRUE	FALSE
6. Tomato sauce and red berries are a good source of lutein, a phytochemical that gives plant foods a red color.	☐	☐
7. Regular consumption of vegetables in the crucifer family reduces the risk of cancer of the lung and stomach.	☐	☐
8. Obesity has *not* been found to be related to the development of cancer.	☐	☐

Media Menu

www.nlm.nih.gov/medlineplus/cancergeneral.html
This address sends you to summarized, scientific information about cancer. Changing the term *cancergeneral* to *breastcancer, coloncancer, alternativemedicine,* and so on takes you to related topics.

www.cancer.gov
The National Cancer Institute's gateway to information about cancer can be found here.

www.nci.nih.gov
The Web site for the National Cancer Institute provides evidence-based information about cancer prevention and treatment, and research updates.

www.cancer.org
The American Cancer Society's site provides a wealth of information on cancer clinics, treatments, alternative therapies, prevention, and statistics.

www.healthfinder.gov
This is the search engine for reliable sources of information on cancer and other health concerns.

www.fnicsearch.nal.usda.gov
Search "cancer" and find recent results related to diet, nutrition, and cancer development.

www.cdc.gov/cancer/az
CDC's cancer A to Z

This site provides information on screening, prevention, and trends in cancer incidence.

www.aicr.org
The American Institute of Cancer Research offers a comprehensive diet and cancer library, dietary recommendations for cancer prevention, recipes, and free consultation with a Registered Dietitian by phone or email.

Notes

1. World Cancer Research Fund/American Institute for Cancer Research. Food, nutrition, physical activity, and the prevention of cancer: a global perspective, Washington, D.C.:AIRC, 2007, available at www.medscape.com/viewarticle/565197, accessed 4/09.

2. American Cancer Society. Cancer facts and figures 2008, www.cancer.org, accessed 4/09.3. Finn OJ. Cancer immunology, N Engl J Med 2008;358:2704–15

3. Finn OJ. Cancer immunology, N Engl J Med 2008;358:2704–15.

4. Lichtenstein P et al. Environmental and heritable factors in the causation of cancer. N Engl J Med 2000;343:78–85.

5. Hoover RN. Cancer-nature, nurture, or both, N Eng J Med 2000;343:135–6.

6. George SM et al. Fruit and vegetable intake and risk of cancer: a prospective cohort study, Am J Clin Nutr 2009;89:347–53.

7. Deaths: final data for 2005, www.cdc.gov/nchs, accessed 4/09.

8. Danaei G et al. One-third of cancer deaths may be attributable to nine modifiable risk factors. Lancet 2005;366:1784–93.

9. Jackson AA. Integrating the ideas of life across cellular, individual, and population levels in cancer causation. J Nutr 2006;135:2927S–33S.

10. Guerreriro et al. Risk of colorectal cancer associated with the C677T polymorphism in 5,10-methylenetetrahyrdofolate reductase in Portuguese patients depends on the intake of methyl-donor nutrients. Am J Clin Nutr 2008;88:1413–8.

11. American Cancer Society, the Centers for Disease Control and Prevention, the National Cancer Institute, the North American Association of Central Cancer Registries. Annual report to the Nation on the status of cancer, J Nat Cancer Inst 2008;100:1672–94.

12. World Cancer Research Fund/American Institute for Cancer Research. Policy and action for cancer prevention, www.medscape.com/viewarticle/588779, accessed 4/09.

13. Go ELW et al. Review of the International Research Conference on Food, Nutrition, and Health, 2006. J Nutr 2007;137:159S–160S.

14. Larsson SC et al. Tea consumption and the risk for ovarian cancer. Arch Intern Med 2005;165:2683–6.

15. Neuhouser ML et al. Multivitamin use and risk of cancer and cardiovascular disease in the Women's Health Initiative Cohorts. Arch Intern Med 2009;169:294–304.

16. Herber D, Bowerman S. Applying science to changing dietary patterns. J Nutr 2001;131:3078S–81S.

17. Wolfe KL et al. Cellular antioxidant activity of common fruits, J Agric Food Chem 2008;56:8418–26.

18. Lin J et al. Supplementation with vitamin C, vitamin E, or beta-carotene and the primary prevention of total cancer incidence or mortality, J Natl Cancer Inst 2009;101:2–4, 14–23.

19. Gerber M. The comprehensive approach to diet: a critical review. J Nutr 2001;131:3051S–5S.

20. Adams SM et al. What should we eat? Evidence from observational studies. South Med J 2006;99:744–8.

21. Stoeckli R et al. Nutritional fats and the risk of type 2 diabetes and cancer. Physiol Behav 2004;83:611–5.

22. Frankenfeld CA et al. Dietary flavonoid intake and non-Hodgkin lymphoma risk, Am J Clin Nutr 2008;87:1439–45.

23. Food, nutrition and the prevention of cancer: a global perspective. Washington, D.C. World Cancer Research Fund, American Institute for Cancer Research; 1997.

24. Byers T et al. Guidelines on nutrition and physical activity for cancer prevention: reducing the risk of cancer with healthy food choices and physical activity CA Cancer J 2002;52:92–119.

25. Allen N et al. Alcohol consumption and the risk of cancer in women, J Natl Cancer Inst 2009, Feb. 24, available at www.medscape.com/viewarticle/588649, accessed 4/09.

26. Friendenreich CM et al., Physical activity and cancer prevention: etiologic evidence and biological mechanisms, J Nutr 2002;132:3456S–64S.

27. Straus SE. Complementary and alternative medicine and cancer: what you should know, www.peoplelivingwithcancer.org, accessed 10/03.

NUTRITION | # Up Close

A Cancer Risk Checkup

Feedback for Unit 22

Best answers:

1. Yes		4. No	
2. Yes		5. Yes	
3. No		6. No	

If you responded "Don't Know" to statements 1, 2, or 3, analyze your diet using the Diet Analysis Plus program. "Regular exercise" in statement 5 means exercising 20 to 30 minutes three times per week. If you didn't know the answer to statement 6, can you pinch an inch of fat above your ribs while you are standing? If yes, you probably have too much body fat.

Good Things to Know about Minerals

NUTRITION SCOREBOARD

	TRUE	FALSE
1 The sole function of minerals is to serve as a component of body structures such as bones, teeth, and hair.		
2 Bones continue to grow and mineralize through the first 30 years of life.		
3 Ounce for ounce, spinach provides more iron than beef does.		
4 Worldwide, the most common nutritional deficiency is iron deficiency.		
5 More than one in four Americans have hypertension.		

Key Concepts and Facts

- Minerals are single atoms that cannot be created or destroyed by the human body or by any other ordinary means.

- Minerals serve as components of body structures and play key roles in the regulation of body processes.

- Deficiency diseases occur when too little of any of the 15 essential minerals are provided to the body, and overdose reactions occur when too much is provided.

- Inadequate intakes of certain minerals are associated with the development of chronic disorders, including osteoporosis, iron deficiency, and hypertension.

Mineral Facts

What substances are neither animal nor vegetable in origin, cannot be created or destroyed by living organisms (or by any other ordinary means), and provide the raw materials from which all things on Earth are made? The answer is the **mineral** elements, and they are displayed in full in the periodic table presented in Illustration 23.1. Minerals considered "essential," or required in the diet, are highlighted.

The body contains 40 or more minerals. Only 15 are an essential part of our diets; we obtain the others through the air we breathe or from other essential nutrients in the diet such as protein and vitamins.

Minerals are unlike the other essential nutrients in that they consist of single atoms. A single atom of a mineral typically does not have an equal number of protons (particles that carry a positive charge) and electrons (particles that carry a negative charge), and it therefore carries a charge. The charge makes minerals reactive. Many of the functions of minerals in the body are related to this property.

Getting a Charge out of Minerals

The charge carried by minerals allows them to combine with other minerals of the opposite charge and form fairly stable compounds that become part of bones, teeth, cartilage, and other tissues. In body fluids, charged minerals serve

minerals
In the context of nutrition, minerals are specific, single atoms that perform particular functions in the body. There are 15 essential minerals—or minerals required in the diet.

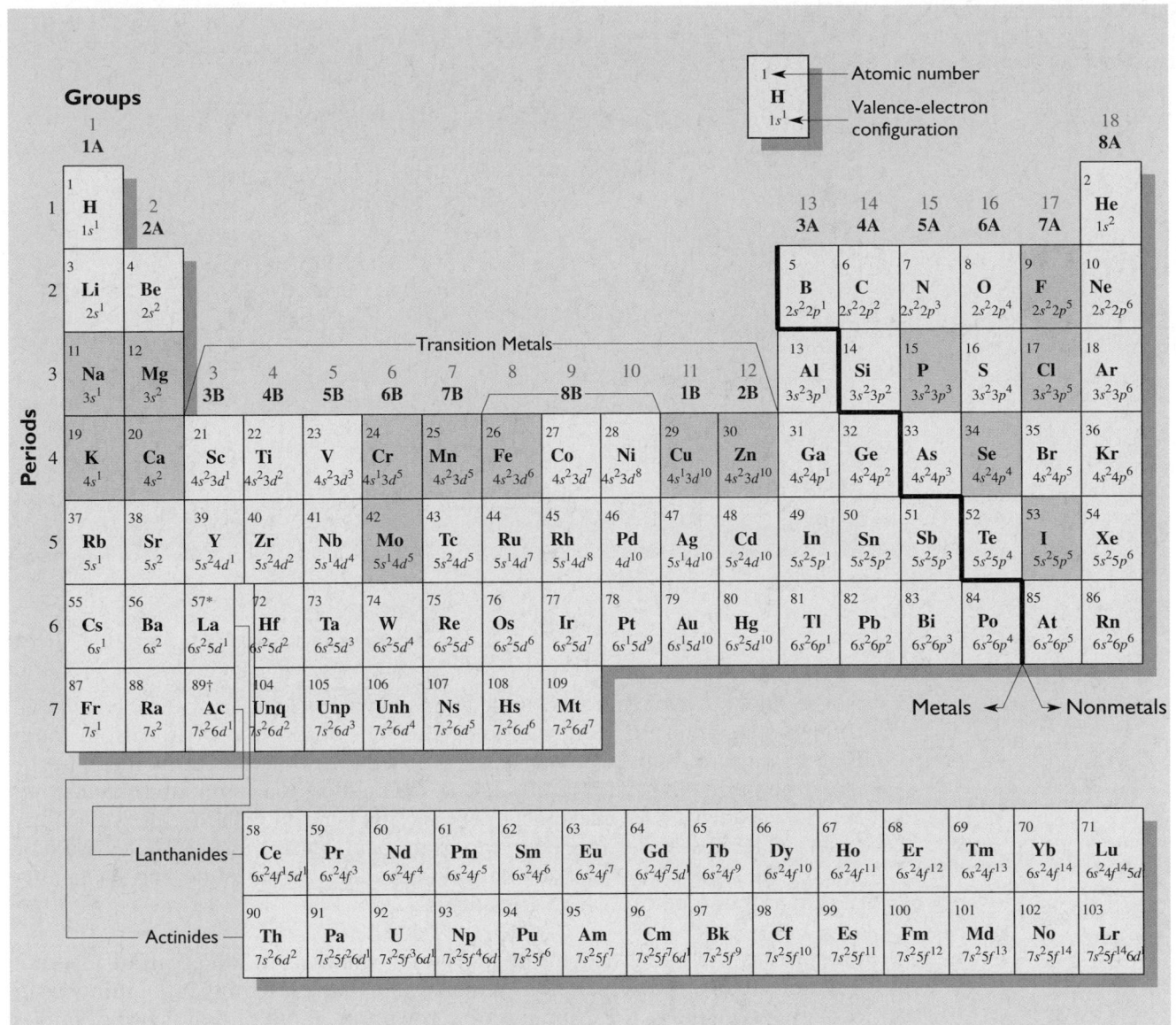

Groups

	1 **1A**																		18 **8A**
1	1 **H** $1s^1$	2 **2A**											13 **3A**	14 **4A**	15 **5A**	16 **6A**	17 **7A**	2 **He** $1s^2$	

Atomic number — 1 **H** $1s^1$ — Valence-electron configuration

Transition Metals

Metals ← → Nonmetals

Lanthanides

Actinides

Illustration 23.1 The periodic table lists all known minerals. The highlighted minerals are required in the human diet.

---

as a source of electrical power that stimulates muscles to contract and nerves to react. The electrical current generated by charged minerals when performing these functions can be recorded by an electrocardiogram (abbreviated EKG or ECG) or an electroencephalogram (EEG). Abnormalities in the pattern of electrical activity in EKGs signal pending or past problems in the heart muscle. An EKG recording is shown in Illustration 23.2. Electroencephalograms similarly record electrical activity in the brain.

The charge minerals carry is related to many other functions. It helps maintain an adequate amount of water in the body and assists in neutralizing body fluids when they become too acidic or basic. Minerals that perform the roles of **cofactors** are components of proteins and enzymes, and they provide the "spark" that initiates enzyme activity.

Charge Problems Because minerals tend to be reactive, they may combine with other substances in food and form highly stable compounds that are not easily absorbed. Absorption of zinc from foods, for example, can vary from zero to 100%, depending on what is attached to it. Zinc in whole-grain products is very poorly absorbed because it is bound tightly to a substance called phytate. In contrast, zinc in meats is readily available because it is bound to protein. People

cofactors
Individual minerals required for the activity of certain proteins. For example:

- Iron is needed for hemoglobin's function in oxygen and carbon dioxide transport.
- Zinc is needed to activate or is a structural component of over 200 enzymes.
- Magnesium activates over 300 enzymes involved in the formation of energy and proteins.

Illustration 23.2 The electrical current measured by an EKG results from the movement of charged minerals across membranes of the muscle cells in the heart.

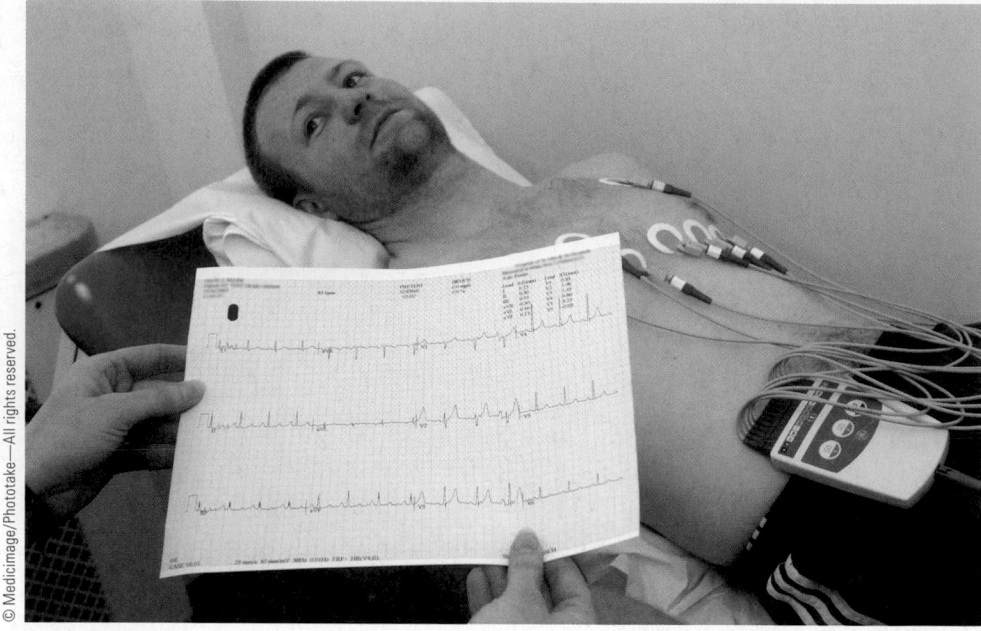

whose sole source of zinc was whole grains have developed zinc deficiency, even though their intake of zinc was adequate.[4] The absorption of iron from foods in a meal decreases by as much as 50% if tea is consumed with the meal. In the intestines, iron binds with tannic acid in tea and forms a compound that cannot be broken down.[5] The calcium present in spinach and collard greens is poorly absorbed because it is firmly bound to oxalic acid. Many more examples could be given. The point is that you don't always get what you consume; the availability of minerals in food can vary a great deal.

Preserving the Mineral Content of Food   Minerals in foods can be lost during food storage and preparation primarily due to the leaching out of minerals in cooking water and in the drippings from the meats. Table 23.1 gives two examples of the percentage of minerals lost by fruit storage method and the method used to prepare dried beans. Dried foods retain minerals well, and those lost in cooking fluids can be recovered if the cooking water is minimized and consumed.

The Boundaries of This Unit   All of the minerals could be the subject of fascinating stories, but in this unit we will concentrate on just three. In addition, a summary of the main features of all the essential minerals is provided in Table 23.2 on page 23–6. This table lists the recommended intake level, functions, consequences of deficiency and overdose, and food sources of each mineral. Table 23.12 at the end of the unit lists food sources for many of the essential minerals.

Minerals highlighted are calcium, iron, and sodium. They have been selected primarily because they play important roles in the development of osteoporosis, iron-deficiency anemia, and hypertension, respectively. These disorders are widespread in the United States and many other countries, and improved diets offer a key to their prevention and treatment.

Table 23.1

Percent of original mineral content lost in fruits by food storage method and in dried beans by cooking method.[6]

	FRUITS			DRIED BEANS	
				Boiled 2–2.5 hours	
	Canned	Frozen	Dried	Water drained	Water used
Calcium	5%	5%	0%	35%	30%
Iron	0%	0%	0%	25%	20%
Magnesium	0%	0%	0%	30%	25%
Potassium	10%	10%	0%	35%	30%
Zinc	0%	0%	0%	15%	10%
Copper	10%	10%	0%	45%	40%

Selected Minerals: Calcium

What you've heard about calcium is true: It's good for bones and teeth. About 99% of the three pounds of calcium in the body is located in bones and teeth. The remaining 1% is found in blood and other body fluids. We don't hear so much about this 1%, but it's very active. Every time a muscle contracts, a nerve sends out a signal, or blood clots to stop a bleeding wound, calcium in body fluids is involved. Calcium's most publicized function, however, is its role in bone formation and the prevention of osteoporosis.

A Short Primer on Bones

Most of the bones we see or study are hard and dead. As a result, people often have the impression that bones in living bodies are that way. Nothing could be further from the truth. The 206 bones in our bodies are slightly flexible living tissues infiltrated by blood vessels, nerves, and cells.

The solid parts of bones consist of networks of strong protein fibers (called the "protein matrix") embedded with mineral crystals (Illustration 23.3). Calcium is the most abundant mineral found in bone, but many other minerals, such as phosphorus, magnesium, and carbon, are also incorporated into the protein matrix. The combination of water, the tough protein matrix, and the mineral crystals makes bone very strong yet slightly flexible and capable of absorbing shocks.

Teeth Are a Type of Bone   Teeth have the same properties as other bone plus a hard outer covering called enamel, which is not infiltrated by blood vessels or nerves. Enamel serves to protect the teeth from destruction by bacteria and mechanical wear and tear.

Remodeling Your Bones   Bones slowly and continually go through a repair and replacement process known, appropriately enough, as **remodeling**. During remodeling, the old protein matrix is replaced and remineralized. If insufficient calcium is available to complete the remineralization, or if other conditions prevent calcium from being incorporated into the protein matrix, **osteoporosis** results.

Osteoporosis

If you are female, you have a one in four chance of developing osteoporosis in your lifetime. If you are a Caucasian, Hispanic, or Asian female, you have a higher risk of developing osteoporosis than if you are an African-American woman or a male. If you are male, your chance of developing osteoporosis is one in eight.[7] Approximately 8 million adults in the United States have osteoporosis,[9] and 1.5 million suffer a broken wrist or hip or crushed spinal vertebrae each year because of the disease.[9] Far more adults, 34 million in all, are at risk of fractures due to osteopenia, a condition characterized by borderline low bone mineral density.[10]

Osteoporosis is a disabling disease that reduces quality of life and dramatically increases the need for health care.[9] Because the incidence of osteoporosis increases with age, its importance as a personal and public health problem is

remodeling
The breakdown and buildup of bone tissue.

osteoporosis
(*osteo* = bones; *poro* = porous, *osis* = abnormal condition) A condition characterized by porous bones; it is due to the loss of minerals from the bones.

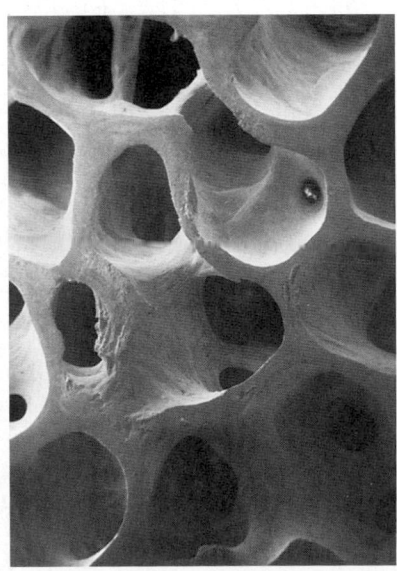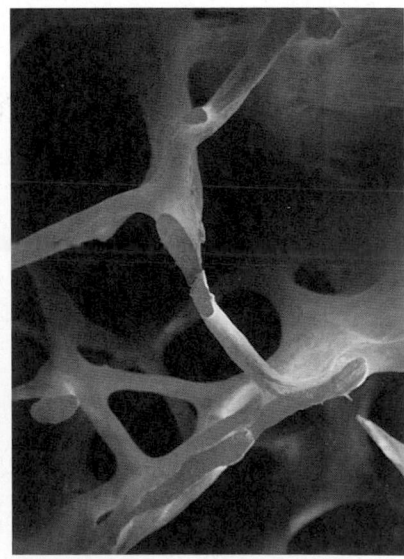

Illustration 23.3   (left) Electron micrograph of healthy bone. (right) Electron micrograph of bone affected by osteoporosis.

Reproduced with permission from Dempster et al, J Bone Min. Res 1, 15–21, 1986.

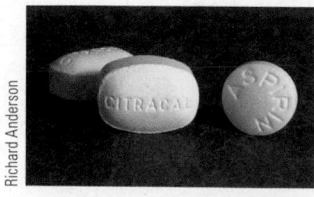

Calcium supplements are large because our daily need for calcium is high (1000 milligrams/day). Four pills provide 800 milligrams of calcium, the amount in $2\frac{2}{3}$ cups of milk. An aspirin is shown for comparison.

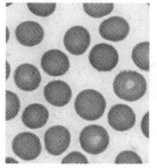 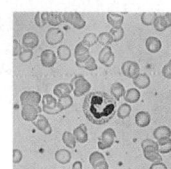

Iron-deficiency anemia is characterized by microcytic anemia and small, pale red blood cells (right photo). Normal red blood cells are shown on the left.

Table 23.2

An intensive course on essential minerals

			PRIMARY FUNCTIONS	CONSEQUENCES OF DEFICIENCY
Calcium			• Component of bones and teeth • Needed for muscle and nerve activity, blood clotting	• Poorly mineralized, weak bones (osteoporosis) • Rickets in children • Osteomalacia (rickets in adults) • Stunted growth in children • Convulsions, muscle spasms
	AI[a] women:	1000 mg		
	men:	1000 mg		
	UL:	2500 mg		
Phosphorus			• Component of bones and teeth • Component of certain enzymes and other substances involved in energy formation • Needed to maintain the right acid–base balance of body fluids	• Loss of appetite • Nausea, vomiting • Weakness • Confusion • Loss of calcium from bones
	RDA women:	700 mg		
	men:	700 mg		
	UL:	4000 mg		
Magnesium			• Component of bones and teeth • Needed for nerve activity • Activates enzymes involved in energy and protein formation	• Stunted growth in children • Weakness • Muscle spasms • Personality changes
	RDA women:	310 mg		
	men:	400 mg		
	UL:	350 mg (from supplements only)		
Iron			• Transports oxygen as a component of hemoglobin in red blood cells • Component of myoglobin (a muscle protein) • Needed for certain reactions involving energy formation	• Iron deficiency • Iron-deficiency anemia • Weakness, fatigue • Pale appearance • Reduced attention span and resistance to infection • Mental retardation, developmental delay in children • Hair loss
	RDA women:	18 mg		
	men:	8 mg		
	UL:	45 mg		
Zinc			• Required for the activation of many enzymes involved in the reproduction of proteins • Component of insulin, many enzymes	• Growth failure • Delayed sexual maturation • Slow wound healing • Loss of taste and appetite • In pregnancy, low-birth-weight infants and preterm delivery
	RDA women:	8 mg		
	men:	11 mg		
	UL:	40 mg		

[a]AI (Adequate Intakes) ad RDAs (Recommended Dietary Allowances are for 19–30-year-olds: UL (Upper Limits) are for 19–70-year-olds, 1997–2004.

Table 23.2

An intensive course on essential minerals (*continued*)

CONSEQUENCES OF OVERDOSE	PRIMARY FOOD SOURCES	HIGHLIGHTS AND COMMENTS
• Drowsiness • Calcium deposits in kidneys, liver, and other tissues • Suppression of bone remodeling • Decreased zinc absorption	• Milk and milk products (cheese, yogurt) • Broccoli • Dried beans • Calcium-fortified foods (some juices, breakfast cereals, bread, for example)	• The average intake of calcium among U.S. women is approximately 60% of the DRI. • One in four women and one in eight men in the U.S. develop osteoporosis. • Adequate calcium and vitamin D status must be maintained to prevent bone loss.
• Muscle spasms	• Milk and milk products (cheese, yogurt) • Meats • Seeds, nuts • Phosphates added to foods	• Deficiency is generally related to disease processes.
• Diarrhea • Dehydration • Impaired nerve activity due to disrupted utilization of calcium	• Plant foods (dried beans, tofu, peanuts, potatoes, green vegetables) • Milk • Bread • Ready-to-eat cereals • Coffee	• Magnesium is primarily found in plant foods where it is attached to chlorophyll. • Average intake among U.S. adults is below the RDA.
• Hemochromatosis ("iron poisoning") • Vomiting, abdominal pain • Blue coloration of skin • Liver and heart damage, diabetes • Decreased zinc absorption • Atherosclerosis (plaque buildup) in older adults	• Liver, beef, pork • Dried beans • Iron-fortified cereals • Prunes, apricots, raisins • Spinach • Bread • Pasta	• Cooking foods in iron and stainless steel pans increases the iron content of the foods. • Vitamin C, meat, and alcohol increase iron absorption. • Iron deficiency is the most common nutritional deficiency in the world. • Average iron intake of young children and women in the United States is low.
• Over 25 mg/day is associated with nausea, vomiting, weakness, fatigue, susceptibility to infection, copper deficiency, and metallic taste in mouth. • Increased blood lipids	• Meats (all kinds) • Grains • Nuts • Milk and milk products (cheese, yogurt) • Ready-to-eat cereals • Bread	• Like iron, zinc is better absorbed from meats than from plants. • Marginal zinc deficiency may be common, especially in children. • Zinc supplements may decrease duration and severity of the common cold.

(continued)

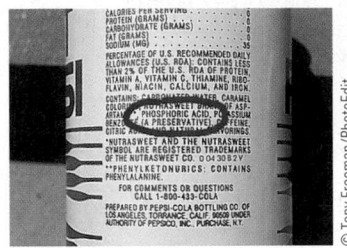

Phosphates are a common, multipurpose food additive.

Magnesium is to chlorophyll in plants as iron is to hemoglobin in humans.

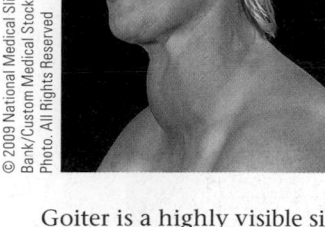

Goiter is a highly visible sign of iodine deficiency

Table 23.2

An intensive course on essential minerals (*continued*)

			PRIMARY FUNCTIONS	CONSEQUENCES OF DEFICIENCY
Fluoride			• Component of bones and teeth (enamel) • Helps rebuild enamel that is beginning to decay	• Tooth decay and other dental diseases
	AI women:	3 mg		
	men:	4 mg		
	UL:	10 mg		
Iodine			• Component of thyroid hormones that help regulate energy production and growth • Required for normal brain development	• Goiter, thyroid disease • Cretinism (mental retardation, hearing loss, growth failure)
	RDA women:	150 mcg		
	men:	150 mcg		
	UL:	1100 mcg		
Selenium			• Acts as an antioxidant in conjunction with vitamin E (protects cells from damage due to exposure to oxygen) • Needed for thyroid hormone production	• Anemia • Muscle pain and tenderness • Keshan disease (heart failure), Kashin-Beck disease (joint disease)
	RDA women:	55 mcg		
	men:	55 mcg		
	UL:	400 mcg		
Copper			• Component of enzymes involved in the body's utilization of iron and oxygen • Functions in growth, immunity, cholesterol and glucose utilization, brain development	• Anemia • Seizures • Nerve and bone abnormalities in children • Growth retardation
	RDA women:	900 mcg		
	men:	900 mcg		
	UL:	10,000 mcg		
Manganese			• Needed for the formation of body fat and bone	• Weight loss • Rash • Nausea and vomiting
	AI women:	2.3 mg		
	men:	1.8 mg		
Chromium			• Required for the normal utilization of glucose and fat	• Elevated blood glucose and triglyceride levels • Weight loss
	AI women:	35 mcg		
	men:	25 mcg		
Molybdenum			• Component of enzymes involved in the transfer of oxygen from one molecule to another	• Rapid heartbeat and breathing • Nausea, vomiting • Coma
	RDA women:	45 mcg		
	men:	45 mcg		
	UL:	2000 mcg		

Table 23.2

An intensive course on essential minerals (continued)

CONSEQUENCES OF OVERDOSE	PRIMARY FOOD SOURCES	HIGHLIGHTS AND COMMENTS
• Fluorosis • Brittle bones • Mottled teeth • Nerve abnormalities	• Fluoridated water and foods and beverages made with it • Tea • Shrimp, crab, fish • Bread	• Toothpastes, mouth rinses, and other dental care products may provide fluoride. • Fluoride overdose has been caused by ingestion of fluoridated toothpaste. • Fluoridated water is not related to cancer.
• Over 1 mg/day may produce pimples, goiter, decreased thyroid function, and thyroid disease.	• Iodized salt • Milk and milk products • Seaweed, seafoods • Bread from commercial bakeries	• Iodine deficiency was a major problem in the United States in the 1920s and 1930s. Deficiency remains a major health problem in some developing countries. • Amount of iodine in plants depends on iodine content of soil. • Most of the iodine in our diet comes from the incidental addition of iodine to foods from cleaning compounds used by food manufacturers.
• "Selenosis." Symptoms of selenosis are hair and fingernail loss, weakness, liver damage, irritability, and "garlic" or "metallic" breath.	• Meats and seafoods • Eggs • Whole grains	• Content of foods depends on amount of selenium in soil, water, and animal feeds. • May play a role in the prevention of some types of cancer.
• Wilson's disease (excessive accumulation of copper in the liver and kidneys) • Vomiting, diarrhea • Tremors • Liver disease	• Bread • Potatoes • Grains • Dried beans • Nuts and seeds • Seafood • Ready-to-eat cereals	• Toxicity can result from copper pipes and cooking pans. • Average intake in the United States is below the RDA.
• Infertility in men • Disruptions in the nervous system (psychotic symptoms) • Muscle spasms	• Whole grains • Coffee, tea • Dried beans • Nuts	• Toxicity is related to overexposure to manganese dust in miners.
• Kidney and skin damage	• Whole grains • Wheat germ • Liver, meat • Beer, wine • Oysters	• Toxicity usually results from exposure in chrome-making industries or overuse of supplements. • Supplements do not build muscle mass or increase endurance.
• Loss of copper from the body • Joint pain • Growth failure • Anemia • Gout	• Dried beans • Grains • Dark green vegetables • Liver • Milk and milk products	• Deficiency is extraordinarily rare.

(continued)

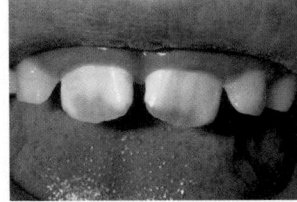

"Mottled teeth" result from an excessive intake of fluoride in children.

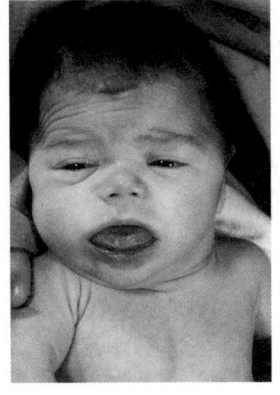

Iodine deficiency during pregnancy produces cretinism in the offspring.

Table 23.2

An intensive course on essential minerals (*continued*)

			PRIMARY FUNCTIONS	CONSEQUENCES OF DEFICIENCY
Sodium			• Needed to maintain the right acid–base balance in body fluids	• Weakness
	AI women:	1.5g		• Apathy
	men:	1.5g		• Poor appetite
	UL:	2.3g	• Helps maintain an appropriate amount of water in blood and body tissues	• Muscle cramps
				• Headache
			• Needed for muscle and nerve activity	• Swelling
Potassium			• Same as for sodium	• Weakness
	AI women:	4.7g		• Irritability, mental confusion
	men:	4.7g		• Irregular heartbeat
				• Paralysis
Chloride			• Component of hydrochloric acid secreted by the stomach (used in digestion)	• Muscle cramps
	AI women:	2.3g		• Apathy
	men:	2.3g		• Poor appetite
	UL:	3.5g	• Needed to maintain the right acid–base balance of body fluids	• Long-term mental retardation in infants
			• Helps maintain an appropriate water balance in the body	

intensifying as the U.S. population ages. It currently appears, however, that a large percentage of the cases of osteoporosis can be prevented. The key to prevention is to build dense bones during childhood and the early adult years and then keep bones dense as you age.[7]

The Timing of Bone Formation   Bones develop and mineralize throughout the first three decades of life. Even after the growth spurt occurs during adolescence and people think they are as tall as they will ever be, bones continue to increase in width and mineral content for 10 to 15 more years. Peak bone density, or the maximal level of mineral content in bones, is reached somewhere between the ages of 30 and 40 years. After that, bone mineral content no longer increases. The higher the peak bone mass, the less likely it is that osteoporosis will develop. People with higher peak bone mass simply have more calcium to lose before bones become weak and fracture easily. That also is the reason males experience osteoporosis less often than females do: They have more bone mass to lose.[11]

Bone size and density often remain fairly stable from age 30 to the mid-40s, but then bones tend to demineralize with increasing age. By the time women are 70, for example, their bones are 30 to 40% less dense on average than they once were.[12] A woman may lose an inch or more in height with age and develop the "dowager's hump" that is characteristic of osteoporosis in the spine (Illustration 23.4).

What has just been described is the way things are in the United States, but this situation is neither normal nor inevitable. The incidence of osteoporosis is far less in many other countries than in the United States.[13] Although genetic and other factors play a role, the causes of osteoporosis are mainly related to diet and other lifestyle behaviors. That means that osteoporosis can largely be prevented.

© Robert Brenner/PhotoEdit

Illustration 23.4   This woman's stooped appearance is due to osteoporosis.

Table 23.2

An intensive course on essential minerals (*continued*)

CONSEQUENCES OF OVERDOSE	PRIMARY FOOD SOURCES	HIGHLIGHTS AND COMMENTS
• High blood pressure in susceptible people • Kidney disease • Heart problems	• Foods processed with salt • Cured foods (corned beef, ham, bacon, pickles, sauerkraut) • Table and sea salt • Bread • Milk, cheese • Salad dressing	• Very few foods naturally contain much sodium. • Processed foods are the leading source of dietary sodium. • High-sodium diets are associated with the development of hypertension in "salt-sensitive" people.
• Irregular heartbeat, heart attack	• Plant foods (potatoes, squash, lima beans, tomatoes, plantains, bananas, oranges, avocados) • Meats • Milk and milk products • Coffee	• Content of vegetables is often reduced in processed foods. • Diuretics (water pills) and other antihypertension drugs may deplete potassium. • Salt substitutes often contain potassium.
• Vomiting	• Same as for sodium. (Most of the chloride in our diets comes from salt.)	• Excessive vomiting and diarrhea may cause chloride deficiency. • Legislation regulating the composition of infant formulas was enacted in response to formula-related chloride deficiency and subsequent mental retardation in infants.

How Do You Build and Maintain Dense Bones?   Since the late 1980s, a rash of studies have examined the relationships among dietary calcium intake, vitamin D status, and bone density. Results have been quite consistent: Bone mass before the age of 30 is increased by calcium intakes of 1000 to 1200 milligrams per day along with adequate vitamin D status. In adults over the age of 50 years, bone density tends to be preserved with calcium intakes around 1200 milligrams per day and 400 to 800 IU of vitamin D.[10, 14] In addition, calcium intakes of 1200 milligrams or more per day with adequate vitamin D status decrease the incidence of stress fractures in athletes.[15] Low bone mineral density can be increased somewhat by switching from a low to an adequate intake of calcium and improved vitamin D status. The earlier in life this occurs, the more improvement in bone density is noted. High levels of bone density once achieved do not last forever; bone must be continually renewed by an adequate supply of calcium and vitamin D.[16]

Vitamin D is important for building dense bones because it increases calcium absorption and the deposition of calcium into bone. Our need for vitamin D can be met from vitamin D–fortified milk, yogurt, soy beverages, breakfast cereals, and seafoods; or by exposing the skin to the sun. The nearby "Health Action" summarizes behaviors that help build dense bones and prevent osteoporosis.

■ **The Current Vitamin D Controversy.**   Sun exposure produces much more vitamin D than is consumed in most diets (see the "On the Side" feature), and levels of vitamin D higher than 800 IU per day may be needed for optimal protection against osteoporosis. Currently scientific thinking around vitamin D requirements and the prevention and treatment of osteoporosis are heading in the "we need more vitamin D" direction. Recommendations for daily doses of vitamin D may increase in the future to around 800 IU (20 mcg) per day.[10]

On the Side

Our bodies produce up to 20,000 IU (500 mcg) of vitamin D with 10 to 15 minutes of whole-body exposure to peak summer sun. Vitamin D toxicity does not occur from sun exposure, but you can overdose on sun and get burned.[18]

© Chris Cole/Getty Images/The ImageBank

We Are Consuming Too Little Calcium   Unfortunately, about half of U.S. women belong to the "600 Club"; they consume 600 milligrams of calcium per day or less.[11] Only 14% of girls and 36% of boys aged 12 to 19 years consume the recommended amount of calcium during these critical years for building bone density. Low calcium intake during the growing years increases the probability that fractures and osteoporosis will occur at younger ages than is the norm. Men tend to consume more calcium than women but on average still fall short of the recommended intake.[17]

■ **Why Do Women Consume Too Little Calcium?**   Women consume too little calcium for several reasons. Although dairy products such as milk, cheese, and yogurt are our richest sources of calcium, some women consider them "fattening" and give them up during the teen years. Skipping breakfast and eating out are common among teens and tend to reduce calcium intake. Many think that milk is a food that only children need, while others stop drinking milk because it "gives them gas." (Such people may be lactose intolerant.)

Increasing intakes of soft drinks, combined with reduced consumption of milk, also appear to be contributing to bone fractures and future osteoporosis in today's youth and women.[21, 22] The wide availability of soft-drinks is thought to be a factor related to its high consumption by teens. According to some college students, one reason their intake of calcium is low is that they frequently eat at fast-food restaurants that feature soft drinks rather than milk on the menu. But when asked if they would choose specials that included fresh-tasting milk (not the kind that tastes like the milk container) as part of the meal, some students reported they would select the specials.[23]

■ **What Else Influences Osteoporosis?**   Table 23.3 summarizes the risk factors identified for osteoporosis. If you were to find all of the various risk factors combined in one person, that individual would be a thin woman with light skin who has consumed too little calcium, has poor vitamin D status, is physically inactive, an excessive alcohol drinker, a smoker, and genetically "small-boned." Additionally, she would have had her ovaries surgically removed for medical reasons before the age of 45.[9] Few people meet every aspect of that description, but people who have several of these characteristics are more likely to develop osteoporosis than those who don't.

At one time, researchers thought that a relatively high intake of protein might intensify the effects of low intakes of calcium on the development of osteoporosis. Studies in humans, however, have failed to demonstrate harmful effects of high protein intake on the development of osteoporosis given adequate calcium intake.[20]

■ **How Is Osteoporosis Treated?**   Calcium supplements or adequate calcium intakes (around 1000 to 1200 milligrams per day), vitamin D supplements

---

Table 23.3

Risk factors for osteoporosis[11]

- Female
- Menopause
- Deficient calcium intake
- Caucasian or Asian heritage
- Thinness ("small bones")
- Cigarette smoking
- Excessive alcohol intake
- Ovarectomy (ovaries removed) before age 45
- Physical inactivity
- Deficient vitamin D status
- Genetic factors

Health *Action*   Behaviors That Help Prevent Osteoporosis[10, 19, 20]

- Consuming adequate amounts of Vitamin D (400 to 800 IU); exposing your hands and arms to direct sunlight for five to 15 minutes daily from early childhood onward
- Moderate alcohol intake
- Consuming recommended amounts of calcium (500 to 800 mg for children, 1000 to 1300 mg per day for adults)

- Following MyPyramid Food Intake Guide or DASH Eating Plan
- Getting regular physical activity
- Not smoking

Table 23.4

Approaches to the treatment of osteoporosis[8, 13]

- Daily intake of 1000–1200 milligrams of calcium
- Regular weight-bearing physical activity
- Adequate vitamin D intake (800 IU daily) or sun exposure
- Five to nine servings of fruits and vegetables daily
- Medication (if needed)

Table 23.5

Caloric content, calcium level, and percentage of available calcium from different foods

Food	Amount	Calories	Calcium (mg)	Calcium Absorbed (%)	Available Calcium (mg)
Yogurt, low fat	1 cup	143	413	32	132
Skim milk	1 cup	85	301	32	96
Soy milk (fortified)	1 cup	79	300	31	93
1% milk	1 cup	163	274	32	88
Tofu	1 cup	188	260	31	81
Cheese	1 ounce	114	204	32	65
Kale, cooked	1 cup	42	94	49	46
Broccoli, cooked	1 cup	44	72	61	44
Bok choy, raw	1 cup	9	73	54	39
Dried beans	1 cup	209	120	24	29
Spinach, cooked	1 cup	42	244	5	12

(800 IU) or sunshine, five to nine servings of fruits and vegetables daily, and weight-bearing exercise like walking and tennis decrease the progression of osteoporosis in many people.[8, 13] Table 23.4 summarizes current recommendations for the treatment of osteoporosis.

Calcium: Where to Find It

Over half of the calcium supplied by the diets of Americans comes from milk and milk products. Milk (including chocolate milk), cheese, and yogurt are all good sources of calcium (see Table 23.5). Some plants such as kale, broccoli, and bok choy provide appreciable amounts of calcium, too. On average, 32% of the calcium content of milk and milk products, calcium-fortified orange juice, and calcium supplements is absorbed, compared to approximately five to 60% of the calcium from different plants (Table 23.5).[24] Calcium absorption decreases with age, vitamin D inadequacy, and ingestion of supplemental iron and zinc.[11] The best way to meet the recommended level of intake (1000 milligrams per day for most adults) is to choose rich sources of absorbable calcium.

Many foods rich in calcium aren't loaded with calories or fat. As Table 23.5 shows, low-fat yogurt, skim and soy milk, kale, and broccoli, for example, provide good amounts of calcium at a low cost in calories.

Calcium is increasingly appearing in unexpected places such as in candy, snack bars, waffles, and bread (Illustration 23.5). The array of calcium-fortified foods is increasing dramatically in the United States in response to publicity about our need for more of it. Although development of such products is making it easier to consume more calcium, the availability of calcium in some of these foods has not been established.[25]

Watch for Calcium on the Label Health claims relating to the benefits of calcium in reducing the risk of osteoporosis may appear on the labels of food products that qualify as good sources of calcium. Look for health claims on foods such as milk, yogurt, and calcium-fortified foods such as orange juice, breakfast cereals, and grain products (Illustration 23.6).

Richard Anderson

Illustration 23.5 The number of calcium-fortified foods is increasing.

Nutrition Facts

Serving Size: 1 cup (240ml)
Servings per Container: 16

Amount per Serving

Calories 110   Calories from Fat 20

	% Daily Value*
Total Fat 2.5g	4%
Saturated Fat 1.5g	8%
Trans Fat 0g	
Cholesterol 15mg	4%
Sodium 135mg	6%
Total Carbohydrate 13g	4%
Dietary Fiber 0g	0%
Sugars 12g	
Protein 8g	

Vitamin A 10%	•	Vitamin C 4%
Calcium 30%	Iron 0%	Vitamin D 25%
Phosphorus 10%		

* Percent Daily Values are based on a 2,000 calorie diet.

MILK
1% LOWFAT MILK

Illustration 23.6 Food products that are good sources of calcium can be labeled with a health claim.

hemoglobin
The iron-containing protein in red blood cells.

myoglobin
The iron-containing protein in muscle cells.

Can You Consume Too Much Calcium? Yes, indeed, you can consume too much calcium. Supplemental calcium in doses exceeding 2.5 grams per day may produce drowsiness, lead to constipation, and cause calcium to deposit in tissues such as the liver and kidneys.[26]

Selected Minerals: Iron

Most of the body's iron supply is found in **hemoglobin**. Small amounts are present in **myoglobin**, and free iron is involved in processes that capture energy released during the breakdown of proteins and fats.

The Role of Iron in Hemoglobin and Myoglobin

What happens to a car when its paint gets scratched? After a while, the exposed metal rusts. The iron in the metal combines with oxygen in the air, and the result is iron oxide, or "rust." Iron readily combines with oxygen, and that property is put to good use in the body. From its location in hemoglobin in red blood cells, iron loosely attaches to oxygen when blood passes near the inner surface of the lungs. The bright red, oxygenated blood is then delivered to cells throughout the body. When the oxygenated blood passes near cells that need oxygen for energy formation or other reasons, oxygen is released from the iron and diffuses into cells (Illustration 23.7). The free iron in hemoglobin then picks up carbon dioxide, a waste product of energy formation. When carbon dioxide attaches to iron, blood turns from bright red to dark bluish red. Blood then circulates back to the lungs, where carbon dioxide is released from the iron and exhaled into the air. The free iron attaches again to oxygen that enters the lungs, and the cycle continues.

Iron in myoglobin traps oxygen delivered by hemoglobin, stores it, and releases it as needed for energy formation for muscle activity. In effect, myoglobin boosts the supply of oxygen available to muscles.

The functions of iron just described operate smoothly when the body's supply of iron is sufficient. Unfortunately, that is often not the case.

Illustration 23.7 Iron's role in carrying oxygen to cells and carbon dioxide away from them by way of the bloodstream.
Fe++ represents charged particles of iron. The charge is neutralized when iron combines with oxygen or carbon dioxide.

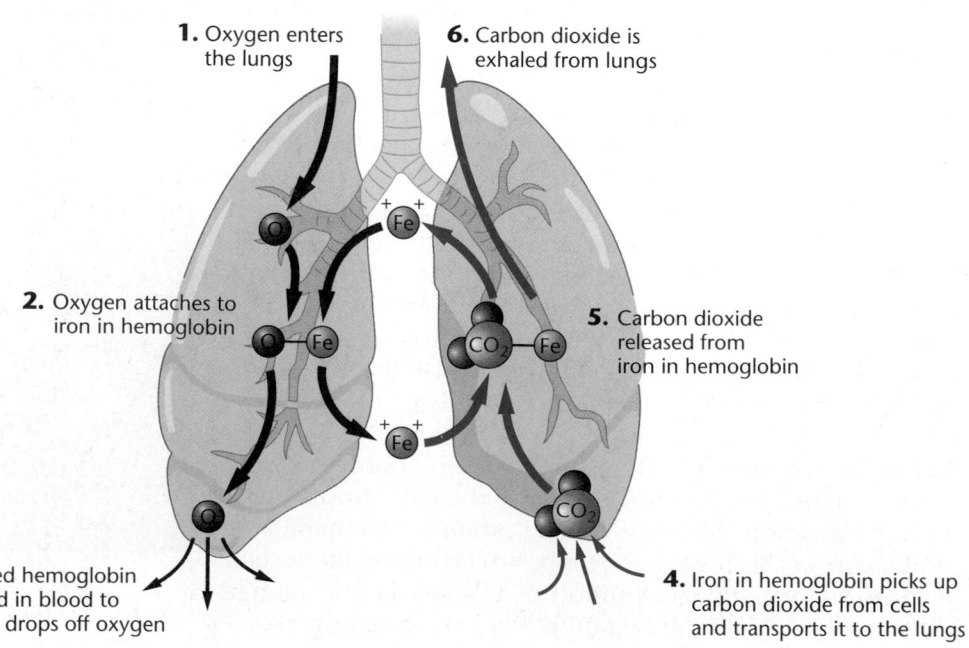

1. Oxygen enters the lungs

6. Carbon dioxide is exhaled from lungs

2. Oxygen attaches to iron in hemoglobin

5. Carbon dioxide released from iron in hemoglobin

3. Oxygenated hemoglobin transported in blood to body cells, drops off oxygen

4. Iron in hemoglobin picks up carbon dioxide from cells and transports it to the lungs

Iron Deficiency Is a Big Problem

The most widespread nutritional deficiency in both developing and developed countries is **iron deficiency** (Table 23.6). It is estimated that one out of every four people in the world is iron deficient.[2] For the most part, iron deficiency affects very young children and women of childbearing age, who have a high need for iron and frequently consume too little of it.[27] Iron deficiency may develop in people who have lost blood due to injury, surgery, or ulcers. Donating blood more than three times a year can also precipitate iron deficiency.[28]

Consequences of Iron Deficiency Many body processes sputter without sufficient oxygen. People with iron deficiency usually feel weak and tired. They have a shortened attention span and a poor appetite, are susceptible to infection, and become irritable easily. If the deficiency is serious enough, **iron-deficiency anemia** develops, and additional symptoms occur. People with iron-deficiency anemia look pale, are easily exhausted, and have rapid heart rates. Iron-deficiency anemia is a particular problem for infants and young children because it is related to lasting retardation in mental development.[30]

Getting Enough Iron in Your Diet

"Enough" iron, according to the RDAs, is eight milligrams for men and 18 milligrams per day for women aged 19 to 50 years. Consuming that much iron can be difficult for women. On average, 1000 calories' worth of food provides about six milligrams of iron. Women would have to consume around 2500 calories per day to obtain even 15 milligrams of iron on average. Selection of good sources of iron has to be done on a better-than-average basis if women are to get enough.

Iron is found in small amounts in many foods, but only a few foods such as liver, beef, and prune juice are rich sources. (A list of food sources of iron is given in Table 23.12.) Foods cooked in iron and stainless steel pans, however, can be a significant source of iron because some of the iron in the pan leaches out during cooking. On average, approximately one milligram of iron is added to each three-ounce serving of food cooked in these pans.[31]

Most of the iron in plants and eggs is tightly bound to substances such as phytates or oxalic acid that limit iron absorption, making these foods relatively poor sources of iron even though they contain a fair amount of it. (Differences in the proportions of iron absorbed from various food sources are shown in Illustration 23.8.) A three-ounce hamburger and a cup of asparagus both contain approximately three milligrams of iron, for example. But 20 times more iron can be absorbed from the hamburger than from the asparagus.[32] Absorption of iron from plants is increased substantially if foods containing vitamin C are included in the same meal.[5]

Iron absorption is also increased by low levels of iron stores in the body (Illustration 23.9). In other words, if you're in need of iron, your body sets off mechanisms that allow more of it to be absorbed from foods or supplements. When iron stores are high, less iron is absorbed.

The body's ability to regulate iron absorption provides considerable protection against iron deficiency and overdose. The protection is not complete, however, as evidenced by widespread iron deficiency and the occurrence of iron toxicity.

Overdosing on Iron Excess iron absorbed into the body cannot be easily excreted. Consequently, it is deposited in various tissues such as the liver, pancreas, and heart. There, the iron reacts with cells, causing damage that can result in liver disease, diabetes, and heart failure. One in 200 people in the United States

Table 23.6

Incidence of iron deficiency[2, 29]

	Population with Iron Deficiency (%)
Worldwide (children under five years):	
Developing countries	51
Developed countries	12
United States:	
Children, 1–3 years	8
Pregnant women	12
Females, 20–49 years	7
Males, 12–49 years	1 or less

iron deficiency
A disorder that results from a depletion of iron stores in the body. It is characterized by weakness, fatigue, short attention span, poor appetite, increased susceptibility to infection, and irritability.

iron-deficiency anemia
A condition that results when the content of hemoglobin in red blood cells is reduced due to a lack of iron. It is characterized by the signs of iron deficiency plus paleness, exhaustion, and a rapid heart rate.

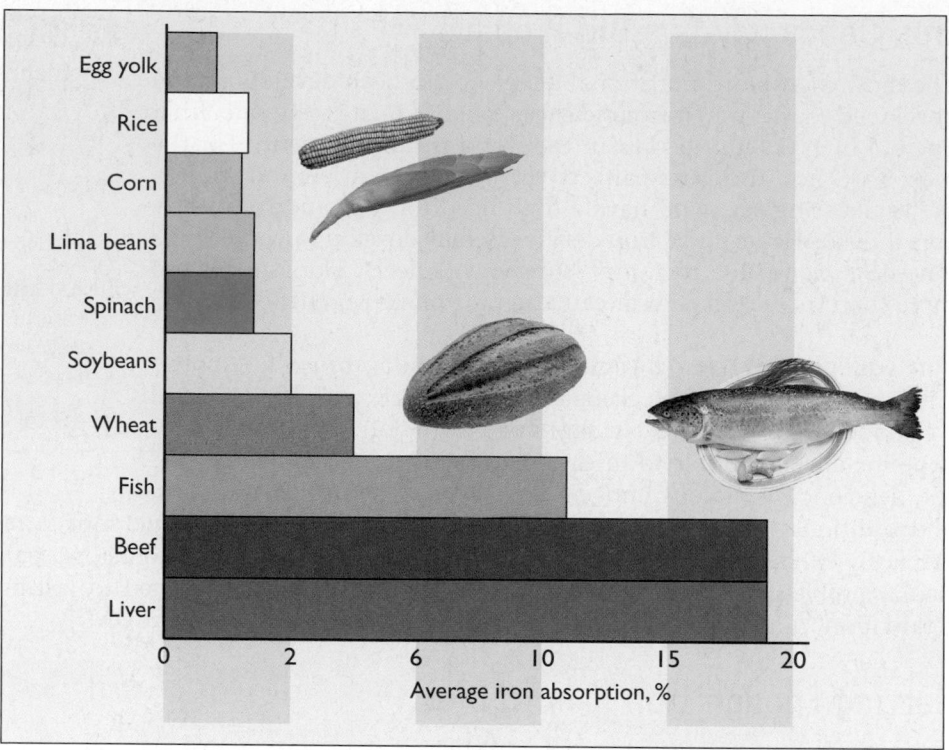

Illustration 23.8 Average percentage of iron absorbed from selected foods by healthy adults.[32]

Egg yolk
Rice
Corn
Lima beans
Spinach
Soybeans
Wheat
Fish
Beef
Liver

Average iron absorption, %
0   2   6   10   15   20

has an inherited tendency to absorb too much iron (a disorder called hemochromatosis and pronounced *hem-oh-chrom-ah-toe-sis*). Other people develop iron toxicity from consuming large amounts of iron with alcohol (alcohol increases iron absorption) or from very high iron intakes, usually due to overdoses of iron supplements.

Each year in the United States, more than 10,000 people accidentally overdose on iron supplements.[33] Commonly, victims are young children who mistakenly think iron pills are candy (Table 23.7 and Illustration 23.10). The lethal dose of iron for a two-year-old child is about three grams,[34] the amount of iron present in 25 pills containing 120 milligrams of iron each. Iron supplements that are not being used should be thrown away or stored in a place where toddlers cannot get to them.

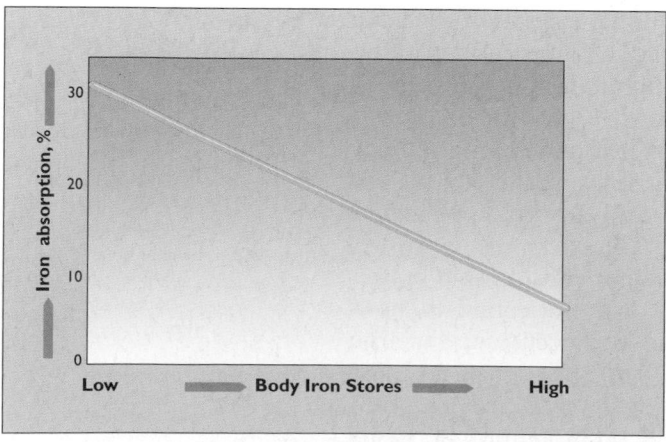

Illustration 23.9 People absorb more iron from foods and supplements when body stores of iron are low than when stores of iron are high.

Iron absorption, %
30
20
10
0
Low ⟶ Body Iron Stores ⟶ High

Table 23.7

Accidental overdoses of chemical substances by children up to age four in the United States[35]

Substance	Cases per 100,000 Children
Aspirin and substitutes	12.7
Solvents/petroleum products	11.2
Tranquilizers	9.9
Iron supplements	8.7
Corrosives and caustics	7.4

Selected Minerals: Sodium

On the one hand, sodium is a gift from the sea (Illustration 23.11); on the other, it is a hazard to health. Sodium's life- and health-sustaining functions are frequently overshadowed by its effects on blood pressure when consumed in excess.

An extremely reactive mineral, sodium occurs in nature combined with other elements. The most common chemical partner of sodium is chloride, and much of the sodium present on this planet is in the form of sodium chloride—table salt. Table salt is 40% sodium by weight; one teaspoon of salt contains about 2300 milligrams of sodium.

Although salt is found in abundance in the oceans, humans have not always had access to enough salt to satisfy their taste for it. During times when salt has been scarce, wars have been fought over it. In such periods, a pocketful of salt was as good as a pocketful of cash. (The word *salary* is derived from the Latin word *salarium*, "salt money.") People were said to be "worth their salt" if they put in a full day's work. Today, salt is widely available in most parts of the world, and the problem is limiting human intakes to levels that do not interfere with health.

What Does Sodium Do in the Body?

Sodium appears directly above potassium in the periodic table, and the two work closely together in the body to maintain normal **water balance**. Both sodium and potassium chemically attract water, and under normal circumstances, each draws sufficient water to the outside or inside of cells to maintain an optimal level of water in both places.

Water balance and cell functions are upset when there's an imbalance in the body's supplies of sodium and potassium. You have probably noticed that you become thirsty when you eat a large amount of salted potato chips or popcorn. Salty foods make you thirsty because your body loses water when the high load of dietary sodium is excreted. The thirst signal indicates that you need water to replace what you have lost.

The loss of body water that accompanies ingestion of large amounts of salt explains why seawater neither quenches thirst nor satisfies the body's need for water. When a person drinks seawater, its high concentration of sodium causes the body to excrete more water than it retains. Rather than increasing the body's supply of water, the ingestion of seawater increases the need for water.

In healthy people, the body's adaptive mechanisms provide a buffer against upsets in water balance due to high sodium intakes. It appears that many people are overwhelming the body's ability to cope with high sodium loads, however. High dietary intakes of sodium appear to play an important role in the development of **hypertension** in many people.[36]

A Bit about Blood Pressure To circulate through the body, blood must exist under pressure in the blood vessels. The amount of pressure exerted on the walls of blood vessels is greatest when pulses of blood are passing through them (that's when "systolic" blood pressure is measured) and least between pulses (that's when "diastolic" blood pressure is taken). Blood pressure measurements note the highest and lowest pressure in blood vessels (Illustration 23.12).

Blood pressure levels less than 120/80 millimeters of mercury (mmHg) are considered normal, whereas levels between 120/80 and 139/89 mmHg are classified as "prehypertension." It is estimated that 28% of U.S. adults qualify as having prehypertension.[3] Values of 140/90 mm Hg and higher qualify as hypertension (Table 23.8). Several blood pressure measurements, taken while a person is relaxed, are needed to obtain an accurate result. (Even going to a clinic or doctor's office to have blood pressure measured can raise it for some people.)

Excess of salty flavor hardens the pulse.

—ABERNATHY, 1000 B.C.

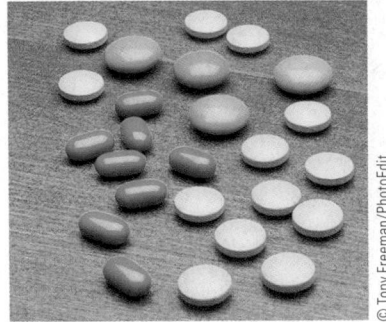

Illustration 23.10 If you were three years old, could you tell which "pills" are candy and which are the iron supplements? Overdoses of iron supplements are a leading cause of accidental poisoning in young children. (The iron supplements are the lightest green in color.)

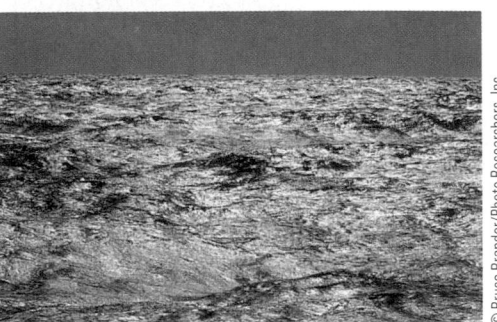

Illustration 23.11 The oceans contain enough salt to cover the surface of the Earth to a depth of more than 400 feet.

water balance
The ratio of the amount of water outside cells to the amount inside cells; a proper balance is needed for normal cell functioning.

hypertension
High blood pressure. It is defined as blood pressure exerted inside blood vessel walls that typically exceeds 140/90 millimeters of mercury.

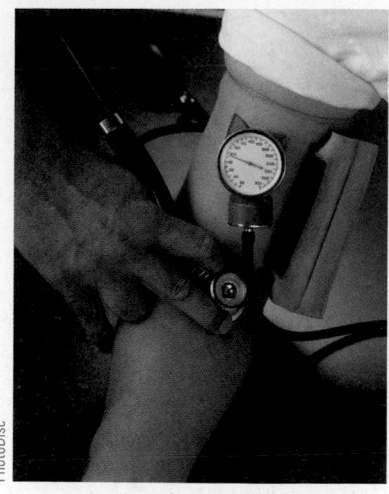

Illustration 23.12 A blood pressure test.

Blood pressure is expressed as two numbers:

$\dfrac{110}{7}$

Systolic pressure measures the force of the blood when the heart contracts.

Diastolic pressure measures the force of the blood when the heart is at rest.

essential hypertension
Hypertension of no known cause; also called primary or idiopathic hypertension, it accounts for 95% of all cases of hypertension.

salt sensitivity
A genetically influenced condition in which a person's blood pressure rises when high amounts of salt or sodium are consumed. Such individuals are sometimes identified by blood pressure increases of 10% or more when switched from a low-salt to a high-salt diet.

Table 23.8

Hypertension categories[40]

Category	Systolic (mm Hg)[a]	Diastolic (mm Hg)
Normal	<120	80
Prehypertension	120–139	80–89
Hypertension		
Stage 1	140–159	90–99
Stage 2	160+	100+

[a]mmHg = millimeters of mercury.

Hypertension is a major public health problem in the United States and other developed countries where Western-type diets are consumed.[37] Although not considered a disease by itself, the presence of hypertension substantially increases the risk that a person will develop heart disease or kidney failure or will experience a heart attack or stroke. Hypertension occurs in 29% of U.S. adults and more than 25% of adults worldwide.[3, 37] It is more likely to develop as people become older. About 40% of all cases of hypertension are classified as mild.[38]

What Causes Hypertension? About 10% of all cases of hypertension can be directly linked to a cause. People who have hypertension with no identifiable cause (95% of all cases) are said to have **essential hypertension.**[37]

A number of risk factors for hypertension have been identified, and dietary factors are among the most important (Table 23.9). Foremost among the evidence linking diet to hypertension are population studies that show a relationship between average salt intake and hypertension. As average salt intake rises, so do rates of hypertension. When people with hypertension reduce their salt intake, their blood pressure tends to drop.[36] Further scrutiny of the data on salt and blood pressure, however, reveals that not everyone is equally susceptible to high-salt diets. Some people are genetically susceptible to salt or have subtle forms of kidney disorders that raise their blood pressure more in response to high-sodium diets than is the case for other people.[36] This condition is known as **salt sensitivity**.

Salt Sensitivity Approximately 51% of people with hypertension and 26% of people with normal blood pressure are salt sensitive. Reduction in salt intake by people who are salt sensitive, along with weight loss if overweight, substantially improves blood pressure in most cases. Because it is currently difficult to identify who in the population is salt sensitive and who is not; because most Americans consume much more sodium than needed; and because high-sodium diets tend to increase blood pressure in the population in general, official advice for Americans is to limit their sodium intake to 2300 milligrams per day (the equivalent of approximately one level teaspoon of salt from all sources).[3, 39]

Other Risk Factors for Hypertension Obesity is a major risk factor for hypertension. For obese people with hypertension, the most effective treatment is weight loss. Excessive alcohol intake can prompt the development of hypertension, and moderation of intake (to two or fewer alcoholic drinks per day) can bring blood pressure back down. Low intakes of vegetables, fruits, whole grains, and potassium all contribute to the development of hypertension. Physically inactive lifestyles also foster the development of hypertension.[40]

How Is Hypertension Treated? The recommended approach to treatment of all cases of hypertension consists of dietary and lifestyle changes and the use of medications if necessary.[40] Weight loss and smoking cessation (if needed),

Table 23.9

Risk factors for hypertension[40]
- Age
- Family history
- High-sodium diet
- Obesity
- Physical inactivity
- Excessive alcohol consumption
- Low vegetable and fruit consumption

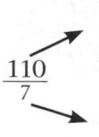

PhotoDisc

a moderate-sodium diet, regular exercise, moderate alcohol consumption (if any), and the DASH diet are basic components of the approach to treatment (Table 23.10). The DASH diet (Illustration 23.13) is based on vegetables, fruits, low-fat dairy products, whole grains, and poultry and fish. Its composition is similar to that of diets recommended for heart disease and cancer prevention. People who adhere to the DASH diet often bring their blood pressure levels back into the normal range.[44] If blood pressure remains elevated after dietary and lifestyle changes have been implemented, or if blood pressure is quite high when diagnosed, antihypertension drugs are prescribed.[42]

Cutting Back on Salt

On average, children and adults in the United States consume twice the recommended amount of sodium.[46] The leading sources of salt (and therefore of sodium) in the U.S. diet are processed foods (Table 23.11). Restricting the use of foods to which salt has been added during processing is the most effective way to lower salt intake and achieve a moderate- to low-salt diet.[47] High-salt processed foods include salad dressings, pickles, canned soups, corned beef, sausages and other luncheon meats, and snack foods such as potato chips and cheese twists. Only a small proportion of our total sodium intake enters our diet from fresh foods. Very few foods naturally contain much sodium—at least not until they are processed (Illustration 23.14).

Table 23.10
Approaches to the treatment of hypertension[42]
• Weight loss (if needed)
• Moderate salt intake (2300 mg of sodium per day or less)
• Moderate alcohol consumption (if any)
• Regular physical activity
• The DASH diet
• Antihypertension drugs (if hypertension not controlled by above measures)

Illustration 23.13
Recommendations for Preventing Hypertension.

The DASH Eating Plan

In the mid-1990s a revolutionary approach to the control of mild and moderate hypertension was tested, and the results have changed health professionals' thinking about high blood pressure prevention and management. Called the DASH diet (Dietary Approaches to Stop Hypertension), it didn't focus on salt restriction and was related to significant reductions in blood pressure within two weeks in most people tested. In some people, reductions in blood pressure were sufficient to erase the need for antihypertension medications, and for others the diet reduced the amount or variety of medications needed. A subsequent study showed that a low-sodium diet boosts the blood pressure lowering effects of the DASH diet, especially in African Americans.[43] This dietary pattern is also associated with reduced risk of heart disease and stroke,[44] and is recommended in MyPyramid as a health-promoting diet for people in general.

The DASH diet consists of eating patterns made up of the following food groups:

	DAILY SERVINGS
Vegetables	4–5
Fruits	4–5
Grain products	7–8
Low-fat milk and dairy products	2–3
Lean meats, fish, poultry	2 or less
Nuts, seeds, dried beans	about 1

Although it does not work for all individuals with hypertension, the DASH diet is a mainstay in clinical efforts to control hypertension. Benefits of the diet last as long as the diet does.[45] A very useful Web site is available that describes refinements in the diet based on research studies and that gives practical tips on how to follow the diet.

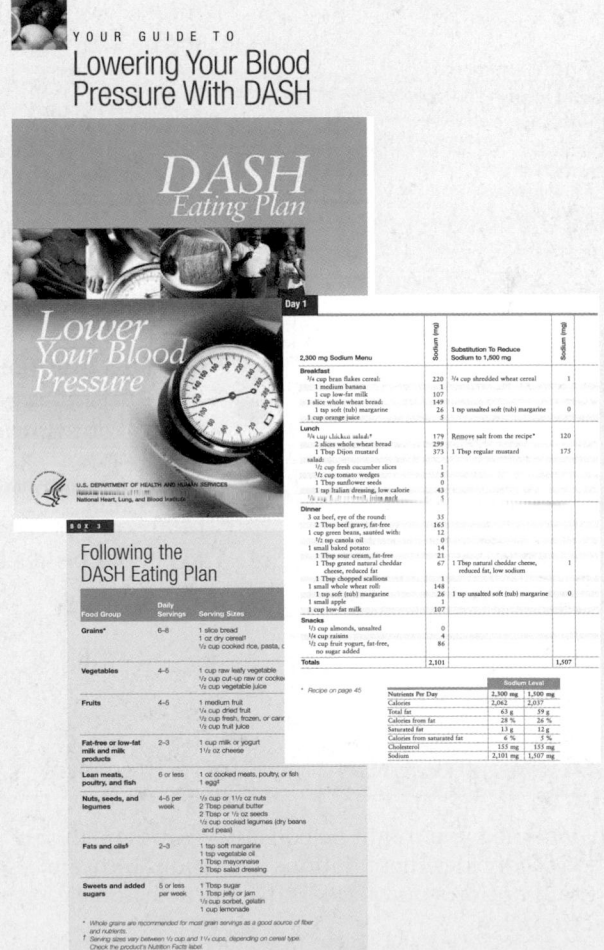

http://www.nhlbi.nih.gov/health/public/heart/hbp/dash

Table 23.11	
Major sources of sodium in the U.S. diet[46]	
Sodium Source	**Contribution to Sodium Intake (%)**
Processed foods	77
Fresh foods	12
Salt added at the table	6
Salt added during cooking and food preparation	5

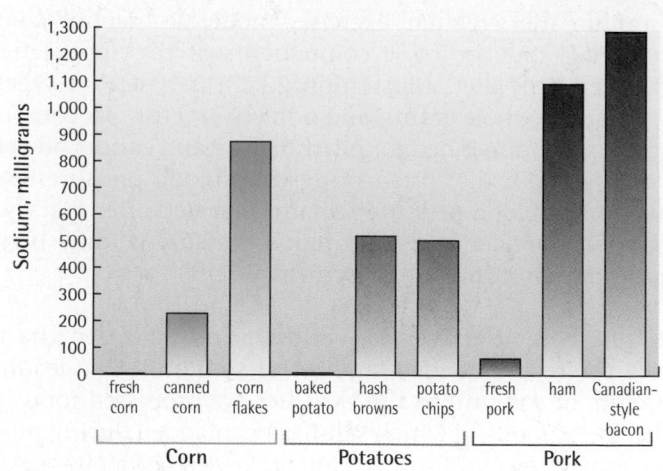

Illustration 23.14 Examples of how processing increases the sodium content of foods.
Sodium values are for a three-ounce serving of each food shown.

Using spices and lemon juice to flavor foods, consuming fresh vegetables (rather than processed ones) and fresh fruits, and checking out the sodium content of foods and then selecting low-sodium ones can also help reduce sodium intake. A few additional steps that can be taken to reduce salt intake are provided in the Take Action feature.

Label Watch Not all processed foods with added sodium taste salty. To find out which processed foods are high in sodium, you have to examine the label. Increasingly, low-salt processed foods are entering the market and can be easily identified by the "low-salt" message on the label (Illustration 23.15). Terms used to identify low-salt (or low-sodium) foods are defined by the Food and Drug Administration. To be considered low-sodium, foods must contain 140 milligrams or less of sodium per serving. Food manufacturers must adhere to the definitions when they make claims about the salt or sodium content of a food on the label.

This unit presented the story behind three of the 15 essential minerals. Much more information could have been relayed about these three, not to mention the wealth of knowledge that could be shared about minerals such as phosphorus, magnesium, zinc, iodine, selenium, and potassium. Critical information about these and other essential minerals is given in Table 23.2, and Table 23.12 lists food sources of many of the essential minerals. For additional information, refer to the Web sites listed at the end of this unit.

Take Action   To reduce high salt intake

Concerned you might be consuming too much salt? If yes, consider these options for action. Indicate which option you would find applicable and easiest to accomplish.

_____ Taste food before you salt it.

_____ Gradually reduce the amount of salt you add to food. (The taste of foods with less salt improves over time.)

_____ Taste test reduced salt snack food (chips, nuts, popcorn) for acceptability.

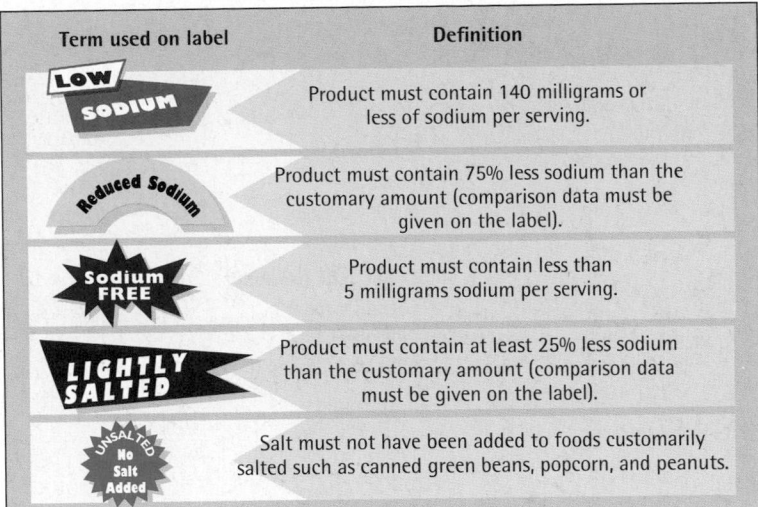

<table>
<tr><th>Term used on label</th><th>Definition</th></tr>
<tr><td>LOW SODIUM</td><td>Product must contain 140 milligrams or less of sodium per serving.</td></tr>
<tr><td>Reduced Sodium</td><td>Product must contain 75% less sodium than the customary amount (comparison data must be given on the label).</td></tr>
<tr><td>Sodium FREE</td><td>Product must contain less than 5 milligrams sodium per serving.</td></tr>
<tr><td>LIGHTLY SALTED</td><td>Product must contain at least 25% less sodium than the customary amount (comparison data must be given on the label).</td></tr>
<tr><td>UNSALTED No Salt Added</td><td>Salt must not have been added to foods customarily salted such as canned green beans, popcorn, and peanuts.</td></tr>
</table>

Illustration 23.15 When a label makes a claim about the sodium content of a food product, the label must list the amount of sodium in a serving and adhere to these definitions.

Table 23.12

Food sources of minerals

Magnesium

FOOD	AMOUNT	MAGNESIUM (mg)
Legumes:		
Lentils, cooked	1/2 cup	134
Split peas, cooked	1/2 cup	134
Tofu	1/2 cup	130
Nuts:		
Peanuts	1/4 cup	247
Cashews	1/4 cup	93
Almonds	1/4 cup	80
Grains:		
Bran buds	1 cup	240
Wild rice, cooked	1/2 cup	119
Breakfast cereal, fortified	1 cup	85
Wheat germ	2 tbs	45
Vegetables:		
Bean sprouts	1/2 cup	98
Black-eyed peas	1/2 cup	58
Spinach, cooked	1/2 cup	48
Lima beans	1/2 cup	32
Milk and milk products:		
Milk	1 cup	30
Cheddar cheese	1 oz	8
American cheese	1 oz	6
Meats:		
Chicken	3 oz	25
Beef	3 oz	20
Pork	3 oz	20

Calcium[a]

FOOD	AMOUNT	CALCIUM (mg)
Milk and milk products:		
Yogurt, low-fat	1 cup	413
Milk shake (low-fat frozen yogurt)	1 1/4 cup	352
Yogurt with fruit, low-fat	1 cup	315
Skim milk	1 cup	301
1% milk	1 cup	300
2% milk	1 cup	298
3.25% milk (whole)	1 cup	288
Swiss cheese	1 oz	270
Milk shake (whole milk)	1 1/4 cup	250
Frozen yogurt, low-fat	1 cup	248
Frappuccino	1 cup	220
Cheddar cheese	1 oz	204
Frozen yogurt	1 cup	200
Cream soup	1 cup	186
Pudding	1/2 cup	185
Ice cream	1 cup	180
Ice milk	1 cup	180
American cheese	1 oz	175
Custard	1/2 cup	150
Cottage cheese	1/2 cup	70
Cottage cheese, low-fat	1/2 cup	69
Vegetables:		
Spinach, cooked	1/2 cup	122
Kale	1/2 cup	47
Broccoli	1/2 cup	36

(continued)

Table 23.12

Food sources of minerals (*continued*)

FOOD	AMOUNT	CALCIUM (mg)
Legumes:		
Tofu	1/2 cup	260
Dried beans, cooked	1/2 cup	60
Foods fortified with calcium:		
Orange juice	1 cup	350
Frozen waffles	2	300
Soy milk	1 cup	200–400
Breakfast cereals	1 cup	150–1000

[a]Actually, the richest source of calcium is alligator meat; 3 1/2 ounces contain about 1231 milligrams of calcium; but just try to find it on your grocer's shelf!

Selenium

FOOD	AMOUNT	SELENIUM (mg)
Seafood:		
Lobster	3 oz	66
Tuna	3 oz	60
Shrimp	3 oz	54
Oysters	3 oz	48
Fish	3 oz	40
Meats/Eggs:		
Liver	3 oz	56
Egg	1 medium	37
Ham	3 oz	29
Beef	3 oz	22
Bacon	3 oz	21
Chicken	3 oz	18
Lamb	3 oz	14
Veal	3 oz	10

Zinc

FOOD	AMOUNT	ZINC (mg)
Meats:		
Liver	3 oz	4.6
Beef	3 oz	4.0
Crab	1/2 cup	3.5
Lamb	3 oz	3.5
Turkey ham	3 oz	2.5
Pork	3 oz	2.4
Chicken	3 oz	2.0
Legumes:		
Dried beans, cooked	1/2 cup	1.0
Split peas, cooked	1/2 cup	0.9
Grains:		
Breakfast cereal, fortified	1 cup	1.5–4.0
Wheat germ	2 tbs	2.4
Oatmeal, cooked	1 cup	1.2
Bran flakes	1 cup	1.0
Brown rice, cooked	1/2 cup	0.6
White rice	1/2 cup	0.4
Nuts and seeds:		
Pecans	1/4 cup	2.0

FOOD	AMOUNT	ZINC (mg)
Cashews	1/4 cup	1.8
Sunflower seeds	1/4 cup	1.7
Peanut butter	2 tbs	0.9
Milk and milk products:		
Cheddar cheese	1 oz	1.1
Whole milk	1 cup	0.9
American cheese	1 oz	0.8

Sodium

FOOD	AMOUNT	SODIUM (mg)
Miscellaneous:		
Salt	1 tsp	2132
Dill pickle	1 (4 1/2 oz)	1930
Sea salt	1 tsp	1716
Ravioli, canned	1 cup	1065
Spaghetti with sauce, canned	1 cup	955
Baking soda	1 tsp	821
Beef broth	1 cup	810
Chicken broth	1 cup	770
Gravy	1/4 cup	720
Italian dressing	2 tbs	720
Pretzels	5 (1 oz)	500
Green olives	5	465
Pizza with cheese	1 wedge	455
Soy sauce	1 tsp	444
Cheese twists	1 cup	329
Bacon	3 slices	303
French dressing	2 tbs	220
Potato chips	1 oz (10 pieces)	200
Catsup	1 tbs	155
Meats:		
Corned beef	3 oz	808
Ham	3 oz	800
Fish, canned	3 oz	735
Meat loaf	3 oz	555
Sausage	3 oz	483
Hot dog	1	477
Fish, smoked	3 oz	444
Bologna	1 oz	370
Milk and milk products:		
Cream soup	1 cup	1070
Cottage cheese	1/2 cup	455
American cheese	1 oz	405
Cheese spread	1 oz	274
Parmesan cheese	1 oz	247
Gouda cheese	1 oz	232
Cheddar cheese	1 oz	175
Skim milk	1 cup	125
Whole milk	1 cup	120
Grains:		
Bran flakes	1 cup	363
Cornflakes	1 cup	325
Croissant	1 medium	270

Table 23.12

Food sources of minerals (*continued*)

FOOD	AMOUNT	SODIUM (mg)
Bagel	1	260
English muffin	1	203
White bread	1 slice	130
Whole-wheat bread	1 slice	130
Saltine crackers	4 squares	125

Iron

FOOD	AMOUNT	IRON (mg)
Meat and meat alternates:		
Liver	3 oz	7.5
Round steak	3 oz	3.0
Hamburger, lean	3 oz	3.0
Baked beans	1/2 cup	3.0
Pork	3 oz	2.7
White beans	1/2 cup	2.7
Soybeans	1/2 cup	2.5
Pork and beans	1/2 cup	2.3
Fish	3 oz	1.0
Chicken	3 oz	1.0
Grains:		
Breakfast cereal, iron-fortified	1 cup	8.0 (4-18)
Oatmeal, fortified, cooked	1 cup	8.0
Bagel	1	1.7
English muffin	1	1.6
Rye bread	1 slice	1.0
Whole-wheat bread	1 slice	0.8
White bread	1 slice	0.6
Fruits:		
Prune juice	1 cup	9.0
Apricots, dried	1/2 cup	2.5
Prunes	5 medium	2.0
Raisins	1/4 cup	1.3
Plums	3 medium	1.1
Vegetables:		
Spinach, cooked	1/2 cup	2.3
Lima beans	1/2 cup	2.2
Black-eyed peas	1/2 cup	1.7
Peas	1/2 cup	1.6
Asparagus	1/2 cup	1.5

Phosphorous

FOOD	AMOUNT	PHOSPHORUS (mg)
Milk and milk products:		
Yogurt	1 cup	327
Skim milk	1 cup	250
Whole milk	1 cup	250
Cottage cheese	1/2 cup	150
American cheese	1 oz	130
Meats:		
Pork	3 oz	275
Hamburger	3 oz	165

FOOD	AMOUNT	PHOSPHORUS (mg)
Tuna	3 oz	162
Lobster	3 oz	125
Chicken	3 oz	120
Nuts and seeds:		
Sunflower seeds	1/4 cup	319
Peanuts	1/4 cup	141
Pine nuts	1/4 cup	106
Peanut butter	1 tbs	61
Grains:		
Bran flakes	1 cup	180
Shredded wheat	2 large biscuits	81
Whole-wheat bread	1 slice	52
Noodles, cooked	1/2 cup	47
Rice, cooked	1/2 cup	29
White bread	1 slice	24
Vegetables:		
Potato	1 medium	101
Corn	1/2 cup	73
Peas	1/2 cup	70
French fries	1/2 cup	61
Broccoli	1/2 cup	54
Other:		
Milk chocolate	1 oz	66
Cola	12 oz	51
Diet cola	12 oz	45

Potassium

FOOD	AMOUNT	POTASSIUM (mg)
Vegetables:		
Potato	1 medium	780
Winter squash	1/2 cup	327
Tomato	1 medium	300
Celery	1 stalk	270
Carrots	1 medium	245
Broccoli	1/2 cup	205
Fruits:		
Avocado	1/2 medium	680
Orange juice	1 cup	469
Banana	1 medium	440
Raisins	1/4 cup	370
Prunes	4 large	300
Watermelon	1 cup	158
Meats:		
Fish	3 oz	500
Hamburger	3 oz	480
Lamb	3 oz	382
Pork	3 oz	335
Chicken	3 oz	208
Grains:		
Bran buds	1 cup	1080
Bran flakes	1 cup	248

(continued)

Table 23.12

Food sources of minerals (*continued*)

FOOD	AMOUNT	POTASSIUM (mg)
Raisin bran	1 cup	242
Wheat flakes	1 cup	96
Milk and milk products:		
Yogurt	1 cup	531
Skim milk	1 cup	400
Whole milk	1 cup	370
Other:		
Salt substitutes	1 tsp	1300–2378

Fluoride

FOOD	AMOUNT	FLUORIDE (mg)
Grape juice, white	6 oz	350
Instant tea	1 cup	335
Raisins	3.5 oz	234
Wine, white	3.5 oz	202
Wine, red	3.5 oz	105

FOOD	AMOUNT	FLUORIDE (mg)
French fries, McDonald's	1 medium	130
Dannon's Fluoride to Go	8 oz	178
Tap water, U.S.	8 oz	59
Municipal water, U.S.	8 oz	186
Bottled water, store brand	8 oz	37

Iodine

FOOD	AMOUNT	IODINE (mg)
Iodized salt	1 tsp	400
Haddock	3 oz	125
Cod	3 oz	87
Shrimp	3 oz	30
Bread	1 oz (1 slice)	35–142
Cottage cheese	$\frac{1}{2}$ cup	50
Egg	1	22
Cheddar cheese	1 oz	17

NUTRITION | Up Close

The Salt of Your Diet

Focal Point: Estimate your sodium intake.

Salt added to foods during processing, during home cooking, and at the table is the major source of sodium in the American diet. To determine if you consume too much sodium, answer the following questions.

How Often Do You Usually:	Daily	4–6 Times per Week	1–3 Times per Week	Less than Weekly
1. Eat processed foods such as luncheon meats, bacon, hot dogs, sausage, canned soups, broths, gravy, TV dinners, or smoked fish?	☐	☐	☐	☐
2. Eat pickles, green olives, potato or tortilla chips, pretzels, or salted crackers?	☐	☐	☐	☐
3. Use soy sauce, garlic salt, steak sauce, or catsup in recipes or on foods?	☐	☐	☐	☐
4. Salt foods at the table?	☐	☐	☐	☐
5. Add salt to foods you are preparing?	☐	☐	☐	☐

FEEDBACK (including scoring) can be found at the end of Unit 23.

Source: Adapted from Limit use of sodium. Washington, DC: USDA Home and Garden Publication 232–4; 1986.

[Key Terms

cofactors, page 23–3
essential hypertension, page 23–18
hemoglobin, page 23–14
hypertension, page 23–17

iron deficiency, page 23–15
iron-deficiency anemia, page 23–15
minerals, page 23–2
myoglobin, page 23–14

osteoporosis, page 23–5
remodeling, page 23–5
salt sensitivity, page 23–18
water balance, page 23–17

Review Questions

TRUE FALSE

1. There are 40 essential minerals that perform specific functions in the body. Each must be obtained from the diet. ☐ ☐

2. Essential minerals are a structural component of teeth; they serve as a source of electrical power that stimulate muscles to contract and help maintain an appropriate balance of fluids in the body. ☐ ☐

3. The amount of certain minerals that is absorbed by the body varies extensively based on what else is consumed along with the minerals. ☐ ☐

4. Calcium is the most abundant minerals found in bones, but other minerals such as magnesium, phosphorus, and carbon are also components of bone. ☐ ☐

5. Bones fully develop and mineralize during the first 18 years of life. ☐ ☐

6. Iodine deficiency and overdose are related to goiter and thyroid disease. ☐ ☐

TRUE FALSE

7. Wilson's disease is related to magnesium deficiency. ☐ ☐

8. An excellent way to develop and maintain a healthy vitamin D status is through brief, daily exposure of the skin to sun. ☐ ☐

9. People with iron deficiency generally feel weak and tired, have a poor appetite, and are susceptible to infection. ☐ ☐

10. Although iron deficiency is a major health problem in America, iron overdose is not. ☐ ☐

11. Blood pressure tends to increase as salt (or sodium) intake increases. ☐ ☐

12. The major source of salt (or sodium) in our diet comes from salt added to food at the table. ☐ ☐

Media Menu

www.ars.usda.gov/Services/docs. htm?docid=9448
This site is the gateway to USDA's Nutrient Retention in Foods tables.

www.nof.org
The National Osteoporosis Foundation is an advocacy group supporting osteoporosis research and education. At this site you can learn 25 things you can do to protect your bones, 25 Facts About Your Bones and Osteoporosis, and 25 Ways to Improve Your Bone Health.

www.webmd.com/hypertension-high-blood-pressure/guide/dash-diet
Obtain user-friendly information on hypertension and the DASH Eating Plan at this WebMD site.

www.nhlbi.nih.gov/health/public/heart/ hbp/dash
This is the National Institutes of Health's DASH Eating Plan site. Obtain a copy of the booklet "Your Guide to Lowering Your Blood Pressure with DASH" here. The booklet includes a week's work of sample menus.

fnic.nal.usda.gov
It's a quick trip from this site to information on minerals, osteoporosis, hypertension, iron deficiency, and other topics covered in this unit.

http://lancaster.unl.edu/food/osteoporosis. shtml
The University of Nebraska Cooperative Extension Service in Lancaster County offers a "Nutrition and Osteoporosis" online Power Point slide show. It is available for viewing from this address.

www.nal.usda.gov/fnic/foodcomp/search
Select "Nutrient lists" from USDA's National Nutrient Database homepage to find lists of food sources of minerals and other nutrients.

fnic.nal.usda.gov/interactiveDRI
Use this terrific tool to identify daily nutrient intake recommendations and Upper Limits for nutrient intake based on the Dietary Reference Intakes (DRIs).

[Notes

1. Mitchell SJ. Changes after taking a college basic nutrition course. J Am Diet Assoc 1990;90:955–61.

2. Mahoney DH Jr. Anemia in at-risk populations—what should be our focus? Am J Clin Nutr 2008;88:1457–8.

3. Ayala C et al. Lower sodium intake recommended for most U.S. adults, MMWR 2009;58:281–3.

4. Prasad AS. Discovery and importance of zinc in human nutrition. Federation Proc 1984;43:2829–34.

5. Thankachan P et al. Iron absorption in young Indian women: the interaction of iron status with the influence of tea and ascorbic acid. Am J Clin Nutr 2008;87:881–6.

6. USDA Tables of Nutrient Retention in Foods, Release 6 (2007), www.ars.usda.gov/Services/docs.htm?docid=9448.

7. Raisz LG. Screening for osteoporosis. N Engl J Med 2005;353:164–5.

8. Sweet MG et al. Review of osteoporosis management, Am Fam Physician 2009;79:193–200.

9. Prestwood KM, et al. Prevention and treatment of osteoporosis, Clin Cornerstone 2002;4:31–41.

10. Khosla S et al. Osteopenia, N Engl J Med 2007;356:2293–300.

11. Follin SL, Hansen LB. Current approaches to the prevention and treatment of postmenopausal osteoporosis, Am J Health—Syst Pharm 2003;60:883–901.

12. Maximizing peak bone mass: calcium supplementation increases bone mineral density in children. Nutr Rev 1992;50:335–7.

13. Lanham-New SA. Fruits and vegetables: the unexpected natural answer to the question of osteoporosis? Am J Clin Nutr 2006;83:1254–5.

14. Reid IR et l. Calcium supplementation and bone mineral density in men, Arch Intern Med 2008;169:2276–82.

15. Harris S. Calcium trials. Experimental Biology Annual Meeting, San Diego, Calif. 2000 Apr. 4.

16. Bischoff-Ferrari HA et al. Effect of calcium supplementation on fracture risk: a double-blind randomized controlled trial, Am J Clin Nutr 2008;87:1945–52.

17. What we eat in America. www.ars.usda.gov/services/docid/htm?/9098, accessed 3/06.

18. Calvo MS et al. Overview of the Proceedings from Experimental Biology 2004 Symposium: vitamin D insufficiency: a significant risk factor in chronic diseases and potential disease-specific biomarkers of vitamin D sufficiency. J Nutr 2005;135:301–3.

19. Tucker KL et al. Effects of beer, wine, and liquor intakes on bone mineral density in older men and women, Am J Clin Nutr 2009;89:1188–96.

20. Tylavsky FA et al. The importance of calcium, potassium, and acid-base homeostasis in bone health and osteoporosis prevention, J Nutr 2008;138:164S–5S.

21. Libuda L et al. Association between long-term consumption of soft drinks and variables of bone modeling and remodeling in a sample of healthy German children and adolescents, Am J Clin Nutr 2008;88:1670–7.

22. Tucker K et al, Cola consumption and bone mineral density in women, Medscape Medical News. 2003, www.medscape.com, accessed 10/03.

23. Lewis NM et al. Food choices of young college women consuming low- or moderate-calcium diets. Nutr Res 1992;12:843–8.

24. New SA. Osteoporosis—ask the experts: dairy and bone health. Medscape Ob/Gyn & Women's Health 2003;8(2).

25. Klausner A. EN's guide to calcium in unexpected places. Envir Nutr 1999;May:5.

26. Looker AC. Interaction of science, consumer practices, and policy: calcium and bone health as a case study. J Nutr 2003;133:1987S–91S.

27. National Report on Biochemical Indicators of Diet and Nutrition in the U.S. Population, www.cdc.gov/nutritionreport, accessed 9/08.

28. Fairbanks VF. Iron in medicine and nutrition. In: Modern nutrition in health and disease, Shils ME et al, eds. 9th ed. Philadelphia: Lippincott Williams & Wilkins, 1999: pp. 193–221.

29. Parvanta I. The hidden hunger: micronutrient malnutrition. Centers for Disease Control, www.cdc.gov/epo/mmwr/preview/mmwrhtml/00051880.htm, March 2000.

30. Lutter CK. Iron deficiency in young children in low-income countries and new approaches for its prevention, J Nutr 2008;138:2523–8.

31. Park J et al. Increased iron content of food due to stainless steel cookware. J Am Diet Assoc 1997;97:659–61.

32. Monsen ER et al. Estimation of available dietary iron. Am J Clin Nutr 1978;31:134–41.

33. Shannon M. Ingestion of toxic substances by children. N Engl J Med 2000;342:186–89.

34. National Research Council Subcommittee on Iron (Committee on Medical and Biological Effects of Environmental Pollutants). Iron. Washington, DC: Division of Medical Sciences, Assembly of Life Sciences, National Academy of Sciences; 1979.

35. Trinkoff AM et al. Poisoning hospitalizations and deaths from solids and liquids among children and teenagers. Am J Public Health 1986;76:657–60.

36. Norat T et al. Blood pressure and interactions between the angiotensin polymorphism AGT M235T and sodium intake: a cross-sectional population study, Am J Clin Nutr 2008;88:392–7.

37. Adrogue HJ et al. Sodium and potassium in the pathogenesis of hypertension, N Engl J Med 2007;356:1966–78.

38. Health, United States, 2005. www.cdc.gov, accessed 9/06.

39. Dietary Guidelines Executive Summary, http://www.health.gov/dietaryguidelines/dga2005/document/html/executivesummary.htm, accessed 7/06.

40. NHLBI issues new high blood pressure clinical practice guidelines. NHLBI Communications Office, 5/14/03, www.nhlbi.nih.gov/index.htm.

41. Wang L et al. Whole-and refined grain intakes and the risk of hypertension in women, Am J Clin Nutr 2007;86:472–9.

42. Apple LJ. The science supporting the 2005 Dietary Guidelines for Americans: sodium, potassium, and physical activity. www.healthierus.gov, accessed 9/05.

43. Conlin PR et al., DASH diet can control stage 1 hypertension, Am J Hypertens 2000;13:949–55.

44. Fung T et al. DASH-style diet and the risk of heart disease and stroke, American Heart Association 2007 Scientific Sessions: Abstract 2369. Presented Monday, November 5, 2007, www.medscape.com/viewarticle/566449.

45. Karanja NM et al. Descriptive characteristics of the dietary patterns used in the Dietary Approaches to Stop Hypertension trial. J Am Diet Assoc 1999;99(suppl):S19–S27.

46. Ervin RB et al. Dietary intake of selected minerals for the United States population: 1999–2000. Advanced Data, No. 341, April 27, 2004.

47. Mattes RD et al. Relative contribution of dietary sodium sources. J Am Coll Nutr 1991;10:383–93.

**Answers
to Review Questions**

1. False, see page 23-2.
2. True, see pages 23-2 and 23-3.
3. True, see pages 23-3, 23-4.
4. True, see page 23-5.
5. False, see page 23-10.
6. True, see pages 23-8, 23-9.
7. False, see page 23-9.
8. True, see page 23-12.
9. True, see page 23-15.
10. False, see pages 23-15, 23-16.
11. True, see page 23-18.
12. False, see page 23-19.

NUTRITION | Up Close

The Salt of Your Diet

Feedback for Unit 23

People in the United States generally consume two times more sodium than they need, and you may be doing so too, if you answered "daily" or "4–6 times per week" to several of the questions. If you wish to cut back on sodium, here are some suggestions:

- Choose fresh instead of processed foods more often.
- Read labels to check for sodium content.
- Reduce salt use in cooking and during food preparation.
- When you do choose processed foods, try the lower-sodium versions available.
- To enhance the flavor of foods, try herbs and spices instead of salt.

UNIT 24

Dietary Supplements and Functional Foods

NUTRITION SCOREBOARD

	TRUE	FALSE
1 Products classified as "dietary supplements" consist of herbs and vitamins and mineral supplements only.		
2 Dietary supplements must be tested for safety and effectiveness before they can be sold.		
3 Herbal remedies have been used for over 100 years in Germany, so they must be safe and effective.		
4 "Probiotics" are "friendly bacteria" that have positive effects on health.		

Key Concepts and Facts

- Dietary supplements include vitamin and mineral pills, herbal remedies, proteins and amino acids, fish oils, and other products.

- Dietary supplements do not have to be shown to be safe or effective prior to being sold.

- Although food is the preferred source of vitamins and minerals, certain people benefit from judiciously selected vitamin or mineral supplements.

- Prebiotics and probiotics can benefit health.

Dietary Supplements

dietary supplements
Any products intended to supplement the diet, including vitamin and mineral supplements; proteins, enzymes, and amino acids; fish oils and fatty acids; hormones and hormone precursors; and herbs and other plant extracts. Such products must be labeled "Dietary Supplement."

What do vitamin E supplements, amino acid pills, and herbal remedies all have in common? They are members of the increasingly popular group of products called **dietary supplements**—and they are all discussed in this unit. Types of dietary supplements available to consumers are presented in Table 24.1. About half of U.S. adults use one or more of these products.[1] Dietary supplements are supposed to "supplement the diet." Many do that and have beneficial health effects in some people, but others neither supplement the diet nor provide health benefits. These

Table 24.1

Types of dietary supplements

Type	Example
1. Vitamins and minerals	Vitamins C and E, selenium
2. Herbs (botanicals)	Dong quai, ginseng, saw palmetto
3. Proteins and amino acids	Shark cartilage, chondroitin, creatine
4. Hormones, hormone precursors	DHEA, "Andro"
5. Fats	Fish oils, DHA, lecithin
6. Other plant extracts	Garlic capsules, fiber, cranberry concentrate, bee pollen

Vitamin, mineral, protein, and amino acid supplements.

Botanical supplements, such as dong quai, often come in gel caps.

Plant extracts can also be taken in liquid form, as tinctures.

Herbal teas are a familiar source of dietary supplements.

Photo Disc

disparate products are grouped together because they are regulated by a common set of rules.

You will learn from this unit that taking some of the dietary supplements available on the market can be a gamble. This "buyer beware" situation exists because of the loose rules that govern dietary supplements. You will also learn about "functional foods" and some exciting, new developments related to intestinal fertilizers and friendly bacteria (no kidding).

Regulation of Dietary Supplements

In 1994 Congress passed the Dietary Supplement Health and Education Act, which started the explosion in the availability of dietary supplements. Under the act, dietary supplements are minimally regulated by the Food and Drug Administration (FDA); they do not have to be tested prior to marketing or shown to be safe or effective.[2] Although often advertised to relieve certain ailments, they are not considered to be drugs. Consequently, dietary supplements are not subjected to vigorous testing to prove safety and effectiveness, as drugs must be. Responsibility for evaluating the safety of dietary supplements lies with manufacturers and not the FDA. Supplements are deemed unsafe when the FDA has proof they are harmful. Since few dietary supplements have been adequately tested, and because results of studies showing negative effects may never see the light of day,[3] it is difficult to prove them to be unsafe. The FDA largely relies on reports of ill effects from manufacturers, health professionals, and consumers to assess supplement safety. Since 1994, the FDA has received thousands of reports of adverse effects of supplements (primarily for herbs), including several hundred deaths.[4]

According to FDA regulations (Table 24.2), dietary supplements must be labeled with a "Supplemental Facts" panel that lists serving size, ingredients, and percent Daily Value (%DV) of essential nutrient ingredients. Products can be labeled with a health claim, such as "high in calcium" or "low fat," if the product qualifies according to the nutrition labeling regulations. Supplements can also be labeled with "structure/function" claims. These claims cannot refer to disease prevention or treatment effects. Claims such as "improves circulation," "prevents wrinkles," "supports the immune system," and "helps maintain mental health" can be used, whereas "prevents heart disease" or "cures depression" cannot be. If a function claim is made on the label or package inserts, the label or insert must include the FDA disclaimer that states the FDA does not support the claim. (This

You can call anything a dietary supplement, even something you grow in your back yard.
—Donna Porter, RD, PhD, Congressional Research Service

Table 24.2

FDA regulations for dietary supplement labeling

1. Product must be labeled "Dietary Supplement."

2. Product must have a "Supplemental Facts" label that includes serving size, amount of the product per serving, % Daily Value of essential nutrients, a list of other ingredients, and the manufacturer's name and address.

3. Nutrient claims (such as "low in sodium" and "high in fiber") can be made on labels of products that qualify based on nutrition labeling regulations.

4. Structure/function claims about how the product affects normal body structures (such as "helps maintain strong bones") or functions ("enhances normal bowel function") can be made on product labels. If a structure/function claim is made, this FDA disclaimer must appear:

   This statement has not been evaluated by the FDA. This product is not intended to diagnose, treat, cure, or prevent any disease.

Nutrition Facts	
Serving size 1 Tablet	
Amount Per Serving	**% DV**
Melatonin 3 mg	*
*Daily Value (DV) not established	

Other Ingredients: Dicalcium Phosphate, Cellulose (Plant Origin), Vegetable Stearic Acid, Vegetable Magnesium Stearate, Silica, Croscarmellose.

GUARANTEED FREE OF: wheat, yeast, soy, corn, sugar, starch, milk, eggs. No artificial colors, flavors. No chemical additives. No preservatives. No animal derivatives.

Directions: As a dietary supplement for adults, take one (1) tablet, under the direction of a physician, only at bedtime as Melatonin may produce drowsiness. **DO NOT EXCEED 3 MG IN A 24 HOUR PERIOD.**

Warning: For Adults. Use only at bedtime. This product is not to be taken by pregnant or lactating women. If you are taking medication or have a medical condition such as an auto-immune condition or a depressive disorder, consult your physician before using this product. **NOT FOR USE BY CHILDREN 16 YEARS OF AGE OR YOUNGER.** Do not take this product when driving a motor vehicle, operating machinery or consuming alcoholic beverages.

In case of accidental overdose, seek professional assistance or contact a Poison Control Center immediately.

KEEP OUT OF REACH OF CHILDREN

is done to reduce the FDA's liability for problems that may be caused by supplements.) Nonetheless, many people wholeheartedly believe health claims made for supplements.[5]

The Federal Trade Commission (FTC) regulates claims for dietary supplements made in print and broadcast advertisements, including direct marketing, web sites, and infomercials. Claims made for dietary supplements in advertisements are supposed to be truthful, but often are not. Although some companies have been prosecuted for making false and misleading claims, neither the FDA nor the FTC has sufficient resources to fully monitor products and enforce laws related to dietary supplements.[6] Rules and regulations related to dietary supplements are expected to become more stringent, and enforcement actions will be expanded in the future.[7,8] The FDA has developed an "Adverse Events Reporting System" Web site (www.cfsan.fda.gov) that simplifies recording and tracking of adverse effects of dietary supplements.

Vitamin and Mineral Supplements

Multivitamin and mineral supplements—such as vitamins E or C, calcium, or magnesium—are among the wide variety of vitamin and mineral supplements used by consumers. They represent the most popular type of dietary supplement, with 43% of Americans using them.[4] Intake levels of vitamins and minerals below the "Tolerable Upper Limits" of the Dietary Reference Intake values (given on the inside front cover of this book) are safe for the vast majority of people, although lower amounts are related to their optimal functioning in the body.[9] Certain concerns related to vitamin and mineral supplements exist. The **bioavailability** of nutrients contained in some supplements remains uncertain. In general, the wider the assortment of vitamins and minerals in a supplement, the lower the absorption of each. Minerals are particularly prone to forming unabsorbable complexes with each other, reducing the bioavailability of multiple minerals in the supplement.[10]

Vitamin and Mineral Supplements: Who Benefits? Supplements can have positive effects on health, as shown by the examples given in Table 24.3. Vitamin and mineral supplements have come a long way from primarily being used to treat vitamin and mineral deficiency diseases.

■ **Using Vitamin and Mineral Supplements for the Wrong Reason.** People often take supplements as a sort of insurance policy against problems caused by poor diets. Although multivitamin and mineral supplements may help fill in some of the nutrient gaps caused by poor food habits, they can't make a bad diet good. Whether a diet is good or bad is determined by more than its vitamin and mineral content. The "goodness" of a diet also depends on its content of essential fatty acid, fiber, water, and other nutrients. Additionally, supplements do not provide phytochemicals found in food such as flavones and antioxidant pigments that have positive influences on health.

One of the most serious consequences of supplements results when they are used as a remedy for health problems that can be treated, but not by vitamins or minerals. Vitamin and mineral supplements have not been found to be an effective treatment for behavioral problems, sexual dysfunction, hair loss, autism, chronic fatigue syndrome, obesity, or stress, for example.

■ **The Rational Use of Vitamin and Mineral Supplements.** Like all medications, vitamin and mineral supplements should be taken only if there is a need for them. If they are taken, dosages should not be excessive. Guidelines for the selection and use of vitamin and mineral supplements can be found in the "Health Action" below.

bioavailability
The amount of a nutrient consumed that is available for absorption and use by the body.

Table 24.3

Who may benefit from vitamin and mineral supplements? Here are some examples:[9, 11–13]

- People with vitamin and mineral deficiency diseases
- Newborns (vitamin K)
- People living in areas without a fluoridated water supply (fluoride)
- Vegans (vitamins B$_{12}$ and D)
- Pregnant women (iron and folate)
- People experiencing blood loss (iron)
- Elderly persons on limited diets (multiple vitamins and minerals)
- People on restricted diets (multiple vitamins and minerals)
- Adults with rheumatoid arthritis (EPA, DHA)
- People at risk for osteoporosis due to low calcium intake and poor vitamin D status (calcium, vitamin D)
- People with alcoholism (multivitamins and minerals)
- People being treated for depression (folic acid)
- Elderly people diagnosed with vitamin B$_{12}$, vitamin D, and folate deficiency

Herbal Remedies

Herbal remedies have a long history, are used worldwide, and are moving from alternative to mainstream health care in North America. Approximately 21% of U.S. adults use herbal supplements each year.[4]

The herb pharmacopoeia includes over 550 primary herbs known by at least 1800 names. Approximately 30% of all modern drugs are derived from plants,[14] and there's no doubt that many additional plants and plant ingredients benefit health. Plant products known to treat disease are considered drugs, however. Those that have not passed the scientific tests needed to demonstrate safety and effectiveness in disease treatment are often considered herbs. Yet, the truth is that some products sold as herbs in the United States have drug-like effects on body functions. An example of an herb that acts powerfully like a drug is given in Illustration 24.1.

Effects of Herbal Remedies Purported effects and side effects of some herbal and similar remedies are listed in Table 24.4. Herbal remedies, like drugs, have biologically active ingredients that can have positive, negative, and neutral effects on body processes. Basically, an herbal remedy (or a drug) is considered valuable if it has beneficial effects on body processes and if the benefits are not outweighed by the risks. Knowledge of the risks and benefits of many herbal supplements remains incomplete. However, available evidence suggests that some herbal remedies are safe and effective, while others appear to be neither.

Which herbal remedies are likely ineffective or unsafe? Human experimentation with various botanicals over the centuries to the present time has helped to identify herbs that lack beneficial effects or have negative side effects. Table 24.5 on page 24–7 lists some of these herbs. The extent to which the herbs included in the table pose a risk to health depends on the amount taken and the duration of use, the age and health status of the user, and other factors.

> *A weed is what we call a plant whose virtues have not yet been discovered.*
>
> —RALPH WALDO EMERSON

Health *Action*   Guidelines for Choosing and Using Vitamin and Mineral Supplements

1. Purchase supplements labeled "USP." They are tested for purity and dose.
   - Terms such as "release assured," "laboratory tested," "quality tested," and "scientifically blended" on supplement labels guarantee nothing.
2. Check the expiration date on supplements. Use unexpired supplements.
3. Choose supplements containing 100% of the Daily Value or less.

Note: 1 mcg (or µg) vitamin A = 3.33 IU, 1 mcg vitamin D = 40 IU, and 1 mg of vitamin E = 1.49 IU.

4. Take supplements with meals.
5. If you have a diagnosed need for a specific vitamin or mineral, take that individual vitamin or mineral and not a multiple supplement.
   - Avoid calcium supplements made from oyster shells, bone, or coral calcium. They may contain lead or aluminum.

6. Store supplements where small children cannot get at them.

Remember! Consult your health care provider about health problems before you start taking supplements to try to treat the problems.

Illustration 24.1 Herb or drug?

"Alternative medicine becomes standard medicine when it is proven true."

—Paul Okunieff, MD

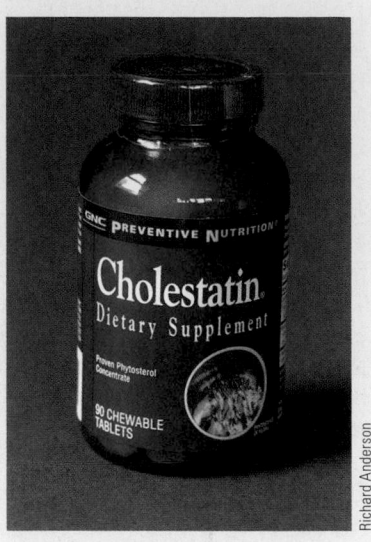

Statins are a popular type of prescription drug that effectively lower LDL cholesterol, raise HDL cholesterol (the good one), and reduce the risk of heart disease. It turns out that statins are also available in herbal form.

Cholestin is an "herb" made from rice grown with red yeast. If subjected to specific growing conditions, the concentration of naturally occurring statins in the rice-yeast mixture can be increased greatly. In 2000, the FDA ruled that cholestin was a drug, and it was taken off the market. Cholestin has been replaced by similar but somewhat different products, such as Cholestatin shown here and Xuezhikang (XZK for short). The FDA recently issued a warning that red yeast rice extracts should not be used.[15]

Richard Anderson

Table 24.4

Proposed effects and potential side effects of herbal and similar remedies[19, 20]

Herb/Other Remedy	Proposed Effects	Potential Side Effects
Glucosamine-chondroitin sulfate	Slows progression of osteoarthritis and its pain	Gastrointestinal upset, fatigue, headache
Ginseng	Anti inflammatory functions, increases energy, normalizes blood glucose, stimulates immune function, relieves impotence in males, cancer prevention	Insomnia, hyperactivity, hypertension, diarrhea, menstrual dysfunction Interacts with blood thinners[a]
SAMe	Relieves mild depression, arthritis pain	May trigger manic excitement, nausea
Garlic	Lowers blood cholesterol, relieves colds and other infections	Heartburn, gas, blood thinner
Cholestin	Maintains desirable blood cholesterol levels	Safety of some ingredients unknown
Echinacea	Prevents and treats colds and sore throat	Allergies to plant components
DHEA	Improves memory, mood, physical well-being	Increases risk of breast cancer
Creatine	Sport supplement (increased performance in short, high-intensity events)	Kidney disease
Saw palmetto	Improves urine flow, reduces urgency of urination in men with prostate enlargement	Nausea, abdominal pain
Ginkgo biloba	Increases mental skills, delays progression of Alzheimer's disease, increases blood flow and sexual performance, decreases depression	Nervousness, headache, diarrhea, nausea. Interacts with blood thinners.
Shark cartilage	Treats lung cancer	Safety unknown
Saint John's wort	Relieves depression	Dry mouth, dizziness, sensitivity to light. Interacts with many drugs and chemotherapy.
Kava (kava-kava)	Relaxation, stress relief, sleep aid, mood enhancer	Liver injury
Black cohosh	Improves menopausal and PMS symptoms	Gastric upset, dizziness, headache, low blood pressure. May increase risk of breast cancer. Interacts with anti-hypertension drugs.
Coenzyme Q10 (ubiquinone)	Remedy for heart disease, aging, Parkinson's disease, cognitive decline	Nausea, diarrhea, rash, low blood glucose Interacts with some drugs

[a]Blood thinners include aspirin, warfarin, coumarin.

Table 24.5

Examples of dietary supplements that may not be effective or safe[18]

Apricot pits (laetrile)	Ephedra	Pennyroyal oil
Androstenedione (Andro)	Eyebright	Poke weed
Aristolochic acid	Ginkgo seed	Sassafras
Belladonna	Kava	Skullcap
Blue cohosh	Licorice root	Star anise
Bitter orange	Liferoot	Vinca
Borage	Lily of the valley	Wild yam
Broom	Lobelia	Wormwood
Chaparral	Ma huang/ephedra	Yohimbe
Chinese yew	Mandrake	
Comfrey	Mistletoe	
Dong quai	Organ/glandular extracts	

■ **Which Herbal Supplements Are Potentially Beneficial?** Results of many studies on the safety and benefits of specfic herbal remedies have been mixed. The bulk of existing evidence, however, indicates that some may be beneficial:

- Ginkgo biloba may decrease symptoms in patients with Alzheimer's disease and increase blood flow to the brain.

- Cranberry may prevent and relieve urinary tract infection.

- Garlic can reduce elevated blood cholesterol levels and blood pressure.

- Saint John's wort may relieve depression.

- Rose hips have mild laxative and diuretic effects (Illustration 24.2).

- SAMe may relieve mild depression and arthritis pain.

- DHEA supplements may be an effective treatment for midlife-onset depression.

- Senna may relieve constipation.

- Ginger may prevent motion sickness.[16–18]

The list of effective herbal remedies may grow, or shrink, with time as additional studies are completed.

■ **Ephedra (Ma Huang).** Ephedra, or Ma huang, is chemically and functionally similar to adrenaline, the "fight-or-flight" hormone. It is traditionally used in China to treat respiratory problems but has gained popularity in the United States as a weight-loss product and athletic performance enhancer. Ephedra delivers limited benefits for these purposes, however, and can produce life-threatening side effects on the heart and nervous system. Side effects and deaths related to ephedra use led to its being banned from the market early in 2004.[18] However, it still appears illegally in some weight loss products.[21]

A Measure of Quality Assurance for Dietary Supplements Not all dietary supplements contain the amounts of herbal and other ingredients declared on the label, and some contain contaminants such as bacteria, mold, and lead. Analyses of the composition of 25 ginseng products, for example, found that concentrations of ginseng compounds in the supplements were up to 36 times different than labeled amounts.[22] Similar studies of echinacea products found that 10% of samples contained no echinacea, and half contained the labeled amount. Some male "enlargement" supplements have been found to be contaminated with

Illustration 24.2 Rose hips in Tuscany.

Judy Brown

Illustration 24.3 These symbols on the labels of dietary supplements certify quality ingredients and accurate labeling but do not address product safety or effectiveness.

Leave no aisle unfortified.
—HEADLINE FOR ARTICLE IN A FOOD INDUSTRY MAGAZINE

functional foods
Generally taken to mean foods, fortified foods, and enhanced food products that may benefit health beyond the effects of essential nutrients they contain.

E. coli, mold, lead, and pesticide residues.[23] Oddly enough, dietary supplements are often labeled as "pure," "natural," or "quality assured."

There is no government body that monitors the contents of herbal supplements. Private groups, such as the U.S. Pharmacopeia (USP), the National Formulary (NF), and Consumer Laboratories (CL), offer testing services to ensure that dietary supplements meet standards for disintegration, purity, potency, and labeling (Illustration 24.3). Products that pass these tests can display "USP," "NF," or the CL symbol boldly on product labels. These letters represent quality ingredients and labeling but do not address product safety or effectiveness.

Due to the lack of studies and potential dangers, the FDA has advised dietary supplement manufacturers not to make claims related to pregnancy for herbs and other products, and to label products truthfully based on scientific evidence.[24,25] Considerations for the use of herbal supplements are summarized in the "Health Action" below.

Functional Foods

Also known as "neutraceuticals," **functional foods** include a variety of foods and products that have in reality or theoretically been modified to enhance their contribution to a healthy diet (Table 24.6).[26] All foods are functional in that they provide nutrients. Foods considered "functional," however, are generally specifically formulated to supply one or more dietary ingredients that may improve health, or they are foods containing high amounts of substances that tend to prevent certain diseases.[27] Because there is no statutory definition for what constitutes functional foods, there are no specific regulations that apply to them. Health claims can be made for functional foods given approval by the FDA.[26] Functional foods containing food additives not on the Generally Accepted As Safe (GRAS) list must be tested and approved by the FDA before being sold.

Functional foods that appear to benefit health are listed in Table 24.7. Increasingly, however, the list of functional foods is becoming infiltrated with sports bars, soups, beverages, and cereals spiked with vitamins, minerals, and herbs. Some of these products carry labels with unsubstantiated health claims and may be of no benefit or are potentially unsafe.[27] For these products, the label "functional food" is a marketing term.

Prebiotics and Probiotics: From "Pharm" to Table

The terms *prebiotics* and *probiotics* were derived from "antibiotics" due to their probable effects on increasing resistance to various diseases. They are in a class

Health Action Considerations for the Use of Herbal Remedies

1. Don't use herbal remedies for serious, self-diagnosed conditions such as depression, persistent headaches, and memory loss. (You might benefit more from a different treatment.)
2. Let your doctor know what herbal remedies you take.
3. If you take prescription medications, clear the use of herbal remedies with your doctor.

4. Don't use herbal remedies without medical advice if you are attempting to become pregnant or if you are pregnant or breast-feeding.
5. Don't mix herbal remedies.
6. If you are allergic to certain plants, make sure herbal remedies are not going to be a problem before you use them.
7. If you have a bad reaction to an herbal remedy, stop using it and

report the reaction to the FDA from the site www.cfsan.fda.gov.
8. Buy herbs labeled with "USP," "NF," or the CL in a beaker symbol.
9. Investigate brands, herb safety, and effectiveness by checking into one or more of the Web sites and resources listed at the end of this unit.

Herbals on the Web
Can you trust information on herbal products you see on the Web?

Who gets thumbs up?
Answers on next page

Sarah
When I'm sick, the first place I go is to the Web to find an herb that will make me feel better.

Pablo
Herbal products I see advertised on the Web look like they'll work for my problem. But I'm conflicted about buying them, because I'm not sure I can trust the information.

---

of functional foods by themselves. **Prebiotics** are fiber-like, nondigestible carbohydrates that are broken down by bacteria in the colon. The breakdown products foster the growth of beneficial bacteria. For this reason they are considered "intestinal fertilizer." **Probiotics** is the term for live, beneficial—or "friendly"—bacteria that enter food through fermentation and aging processes.[25] Table 24.8 lists food and other sources of pre- and probiotics. Availability of foods and other products containing prebiotics and probiotics is much more common in Japan and

prebiotics
Non-digestible food ingredients that beneficially affect a person by selectively stimulating the growth or activity of one or a limited number of bacteria in the colon. Inulin, an extract from chicory root, is a common prebiotic. Also called "intestinal fertilizer."

probiotics
Live microorganisms which, when delivered in adequate amounts, confer a health benefit. Strains of *lactobacillus* (lac-toe-bah-sil-us) and *bifidobacteria* (bif-id-dough bacteria) are the best known probiotics. Also called "friendly bacteria."

Table 24.6

How are foods made to be "functional"?

Foods are made to be "functional" by:

1. Taking out potentially harmful components (e.g., cholesterol in egg yolk and lactose in milk)

2. Increasing the amount of nutrients and beneficial nonnutrients (e.g., fiber-fortified liquid meals, calcium-, and vitamin C-fortified orange juice)

3. Using beneficial substances in food production or products (e.g., using "friendly" bacteria in fermented milk and soy products)

Table 24.7

Example of functional foods with apparent health benefits[27, 31]

Functional Food	Benefit
Stanol and sterol fortified margarine, psyllium fiber, soy protein, whole oat products, garlic, nuts	Reduced blood levels of LDL cholesterol
Omega-3 fatty acids	Reduced heart disease risk
Grapes, grape juice	Decreased blood clots in blood vessels
Cranberry juice	Decreased urinary tract infections
Green tea, cooked tomato products, cruciferous vegetables, conjugated linoleic acid	Decreased risk of certain types of cancer
Folic acid–fortified breads and cereals	Decreased risk of neural tube defects
Probiotics	Decreased risk of infection, lactose intolerance, food allergies, other disorders

Taking the coke out of Coke
Coca Cola used to be fortified with cocaine. During the early part of the 1900s, the "coke" ingredient was removed.

Herbals on the Web Many people use the Web as a source of information about illness remedies. Unfortunately, over half of the sites selling herbal products illegally claim the products prevent or cure specific diseases as if they were really drugs and do not include the required FDA disclaimer.[28]

Take the worry out of decisions about herbal supplements. Check them out using the scientifically reliable Web sites listed at the end of this unit.

Sarah:

Pablo:

European countries than in Canada or the United States.[27] However, availability of such products is increasing as research results shed light on their safety and effectiveness.

The digestive tract, particularly the colon, is home to over 500 species of microorganisms representing 100 trillion bacteria (and billions of viruses and fungi, too). Some species of bacteria such as E. coli may cause disease, whereas others such as lactobacillus and bifidobacteria prevent various diseases.[30] Pre- and probiotics have been credited with important health effects (Table 24.9); the right combination and dose of each fosters the proliferation of healthful bacteria in the colon, nose, and some other internal canals of the body. The concept of the combined benefits of pre- and probiotics has been termed "symbiotics."[30] Because it is difficult to recolonize gut bacteria, the benefits of pre- and probiotics last only as long as dietary intake does.[30]

Prebiotics appear to be safe in general, however, probiotics may be harmful to individuals who may develop blood infections.[34, 35] The primary side-effects associated with prebiotic and probiotic use are flatulence, bloating, and constipation.[35]

Final Thoughts

From dietary supplements to bacteria: The universe of substances considered dietary ingredients is expanding. Knowledge about potential benefits of pre- and probiotics is charging ahead, and advances are catching the attention of consumers and health care professionals. Perhaps you never thought that "intestinal fertilizer" or "friendly bacteria" would ever intentionally pass through your lips. But that may well be the nature of dietary ingredients to come. If you didn't know before, you do now.

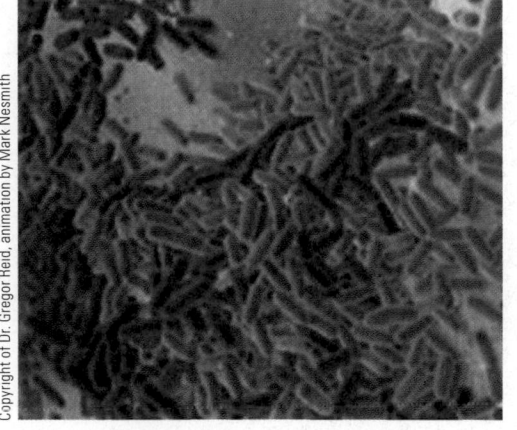

A lactobacillus species (blue) taking over harmful E. coli bacteria (red).

Source: Probiotics: Their tiny worlds... The Scientist July 22, 2002, pp. 20–22; author: Bob Beale.

Table 24.8	
Food and other sources of prebiotics and probiotics[1, 30]	
Prebiotics	**Probiotics**
Chicory	Fermented or aged milk and milk products:
Jerusalem artichokes	• Yogurt with live culture
Wheat	• Buttermilk
Barley	• Kefir
Rye	• Cottage cheese
Onions	• Dairy spreads with added inulin
Garlic	Other fermented products
Leeks	• Soy sauce
Prebiotics tablets and powders and nutritional beverages	• Tempeh
	• Fresh sauerkraut
	• Miso
	Breast milk

Table 24.9

Apparent and potential benefits of prebiotics and probiotics[30, 32, 33]

Prebiotics	Probiotics
• Prevention and treatment of diarrhea and constipation • Prevention of colon cancer • Increased mineral absorption • Decreased blood triglyceride, glucose, and insulin levels	• Treatment for antibiotic-related diarrhea, chronic diarrhea, traveler's diarrhea, and irritable bowel syndrome • Treatment of lactose intolerance and some food allergies • Prevention of infections in urinary tract, gastrointestinal tract, ear canals • Decreased dermatitis in infants • Increased bacterial production of some B vitamins • Decreased blood cholesterol levels • Decreased risk of certain types of cancer • Treatment of irritable bowel syndrome • Decreased dental caries • Decreased high blood pressure • Prevention of wound infections

NUTRITION | Up Close

PhotoDisc

Supplement Use and Misuse

Focal Point: Decide if a dietary supplement is warranted in these situations.

People take dietary supplements for many reasons, but is their use justified? Apply the information from this chapter to determine if you agree with the decisions made in each of the following scenarios.

1. Martha works part-time and takes a full load of classes. Like many college students, she is always on the go, often grabbing something quick to eat at fast-food restaurants or skipping meals altogether. Nevertheless, Martha feels confident her health will not suffer, because she takes a daily vitamin and mineral supplement.

   Is a supplement warranted in this case? _____
   Why or why not? _____

2. Sylvia is a 23-year-old student diagnosed with iron-deficiency anemia. She has learned in her nutrition class that it is preferable to get vitamins and minerals from food instead of supplements. Therefore, instead of taking the iron pills her doctor has prescribed, Sylvia has decided to counteract the anemia by increasing her consumption of iron-rich foods.

   Is a supplement warranted in this case? _____
   Why or why not? _____

3. John is a 21-year-old physical education major involved in collegiate sports. He is very aware that nutrition plays an important role in the way he feels, so he is careful to eat well-balanced meals. In addition, John takes megadoses of vitamins and minerals daily. He is convinced they enhance his physical performance.

   *Is a supplement warranted in this case?*_____
   Why or why not? _____

4. Roberto, a native Californian, is backpacking through Europe when he is slowed down by constipation. He visits a pharmacy where English is spoken and is given senna by the clerk rather than a fast-acting medicine he expected. Roberto has never taken an herb before and is not sure how his body will react to it, or if it will work.

   Should Roberto try the senna, ask for a nonherbal drug, or take another action? (Assume they cost the same.) _____
   What's the rationale for this decision? _____

5. While shopping at the mall, Yuen notices a kiosk selling "Hypermetabolite," a weight-loss product that guarantees you'll lose five pounds a week without dieting. Having gained 10 pounds since she started working full-time, Yuen decides to try it. Her examination of the product's label reveals that an ephedra-derivative and Asian ginseng are major ingredients.

   Should Yuen take Hypermetabolite for weight loss? _____
   Why or why not? _____

FEEDBACK (including answers) can be found at the end of Unit 24.

[Key Terms

bioavailability, page 24-4
dietary supplements, page 24-2

functional foods, page 24-8
prebiotics, page 24-9

probiotics, page 24-9

[Review Questions

	TRUE	FALSE
1. Dietary supplement labels must include a "Supplement Facts" panel and may include qualifying nutrient claims and structure/function claims.	☐	☐
2. Say you read a dietary supplement label that claims the product "helps enhance muscle tone or size." Because the statement appears on the label, it must be true.	☐	☐
3. The Recommended Dietary Allowance (RDA) for vitamin A for breast-feeding women is 1300 mcg (or µg) per day. A supplement containing 3300 IU vitamin A would supply less than the RDA amount.	☐	☐
4. Various vitamin and mineral supplements appears to benefit pregnant women, the elderly, and adults at risk of osteoporosis.	☐	☐

	TRUE	FALSE
5. Vitamin and mineral supplements available over-the-counter can be safely consumed at any dose levels.	☐	☐
6. A "USP" label displayed on dietary supplements indicated the product has been tested for safety and effectiveness.	☐	☐
7. Prebiotics support the growth of beneficial bacteria in the colon.	☐	☐
8. Probiotics have been credited with important health effects.	☐	☐

Media Menu

fnic.nal.usda.gov
Obtain access to the International Bibliographic Information on Dietary Supplements (IBIDS) Database and search tool at this site.

ods.od.nih.gov
The National Institutes of Health maintains an Office of Dietary Supplements that provides facts and background information on dietary supplements, a newsletter, and listserv for email updates on dietary supplements.

ods.od.nih.gov/Health_Information/Health_Information.aspx
Go directly to this address to obtain reliable information on dietary supplement warnings and safety information, false advertising claims, supplement labels.

ods.od.nih.gov/factsheets/botanicalbackground.asp
Looking for basic information about botanicals, their cultivation and preparation for use as herbal supplements? This site addresses these topics.

www.nutrition.gov
Click on dietary supplements in the left column and go to "Questions To Ask Before Taking Vitamin and Mineral Supplements," or "Dietary Supplements for Athletes." Click on "MedlinePlus: Dietary Supplements" and access the National Library of Medicine's treasure trove of information on herbs and dietary supplements.

nccam.nih.gov
The NCCAM Clearinghouse provides information on Complementary and Alternative Medicine, including publications and searches of Federal databases of scientific and medical literature. Toll-free in the United States: 1-888-644-6226, TTY (for deaf and hard-of-hearing callers): 1-866-464-3615.

nccam.nih.gov/camonpubmed
A service of the National Library of Medicine (NLM), PubMed contains publication information and (in most cases) brief summaries of articles from scientific journals.

www.hc-sc.gc.ca/dhp-mps/medeff/report-declaration/index-eng.php
This site leads you to Canada's adverse drug and health product reporting site. It provides a link to adverse reaction report statistics.

www.mayoclinic.com
Search "herbs" and get the skinny on the helpful and harmful ones.

Notes

1. Walker R et al. Dietary Supplements in the US: pitfalls and safety, www.medscape.com/viewarticle/522892, posted 02/07/2006.

2. ADA/APhA Special Report from the Joint Working Group on Dietary Supplements. A healthcare professional's guide to evaluating dietary supplements. American Dietetic Association, American Pharmaceutical Association, 2000:47;1–40.

3. De Smet P. Herbal remedies, N Engl J Med 2002;347:2046–56.

4. Woo J. Vitamin and mineral supplements, presentation at the Office of Dietary Supplements, NIH Consensus Conference, Bethesda, MD, 5/16/06.

5. ADA/APhA Special Report; and Community Nutrition Institute Newsletter, 2000 July 21:2.

6. Herbal remedy sellers on the Web break the rules, www.nlm.nih.gov/medlineplus/news/fullstory14069.html, accessed 10/03.

7. Israelsen L. What Obama means for functional foods and supplements: Part 1, nutraingredients-usa.com/content/view/233507, accessed 1/09.

8. Wasserman I et al. FDA supplement warning letters: 2008 year in review, Jan. 27, 2009, www.nutraingredients-usa.com/content/view/234091, accessed 1/09.

9. Fletcher RH, Fairfield KM, Vitamins for chronic disease prevention in adults: clinical applications, JAMA 2002;287:3127–9.

10. Yetley EA. Multivitamin and multimineral dietary supplements: definitions, characterization, bioavailability, and drug interactions. Am J Clin Nutr 2007, 85:269S–76S.

11. Willett WC, Stampfer MJ. What vitamins should I be taking, doctor? N Engl J Med 2001;345:1819–24.

12. Barringer TA et al., Effect of multivitamin and mineral supplement on infection and quality of life: a randomized, double-blind, placebo-controlled trial, Ann Intern Med 2003;93:365–71.

13. Fairfield KM, Fletcher RH. Vitamins for chronic disease prevention in adults: scientific review. JAMA 2002;9:288:1720–4.

14. Belew C. Herbs and the childbearing woman. Guidelines for midwives. J Nurse Midwifery 1999;44:231–52.

15. Lu Z et al. Effect of Xuezhikang, an extract from red yeast Chinese rice, on coronary events in a Chinese population with previous myocardial infarction, Am J Cardiol 2008; June 9, published online before print, accessed 4/09.

16. Linde K et al. St. John's wort for major depression, Cochrane Systematic Review, 2008, Issue 4:CD000448.

17. Schmidt PJ et al. DHEA may be effective for midlife-onset minor and major depression. Arch Gen Psych 2005;62:154–62.

18. About herbs, botanical, and other products. www.mskcc.org/mskcc/html/11570.cfm, accessed 10/06.

19. Dietary supplement facts, Office of Dietary Supplements, NIH, ods.od.nih.gov, accessed 8/08.

20. Complementary and alternative medicine, NIH, nccam.nih.gov, accessed 8/08.

21. Heller L. Canada targets ephedra and kava, www.nutringredientsusa.com, Aug. 22, 2008, accessed 4/09.

22. Harkey MR et al. 0Variability in commercial ginseng products: an analysis of 25 preparations, Am J Clin Nutr 2001;73:1101–60

23. Gilroy CM et al. Echinacea and truth in labeling, Arch Intern Med 2003;163:699–704.

24. HHS Statement: FDA statement concerning structure/function rule and pregnancy claims. 2000 Jan. 6.

25. Substantiation for dietary supplement claims made under section 403 (r) (6) of the Federal Food, Drug, and Cosmetic Act, Dec. 2008, www.fds/cfsan.gov, accessed 4/09.

26. Hyman P. Claims for functional foods under the current food regulatory scheme, Nutr Today 2002;37:217–9.

27. Position of the American Dietetic Association: functional foods, 2009;109:735–46.

28. Morris et al. Herbal remedy sellers (www.nlm.nih.gov/medline plus/news/ fullstory 14069.html).

29. Sanders ME. Probiotics: considerations for human health, Nutr Rev 2003;61:91–7.

30. Douglas DC et al. Probiotics and prebiotics in dietetics practice, J Am Diet Assoc 2008;108:510–21.

31. Hasler CM. Functional foods: benefits, concerns, and challenges, J Nutr 2002;132:3772–81.

32. Kligler B et al. Probiotics, Am Fam Physician 2008;78:1073–8.

33. Get the facts- an introduction to probiotics, www.nccam.nih.gov, accessed 4/09.

34. Boyle RJ et al. Risks of probiotic treatment. Am J Clin Nutr 2006;83:1254–64.

35. Looijer-van Lagen MAC et al. Probiotics and prebiotics as functional ingredients in inflammatory bowel disease, Nutr Today 2008;43:235–42.

Answers to Review Questions

1. True, see page 24-3.
2. False, see pages 24-3, 24-4.
3. True, see page 24-5.
4. True, see page 24-4.
5. False, see pages 24-4, 24-5.
6. False, see page 24-8.
7. True, see page 24-9.
8. True, see pages 24-10, 24-11.

NUTRITION | # Up Close

Supplement Use and Misuse

Feedback for Unit 24

1. *Is a supplement warranted in this case?* No

   Why or why not? Martha is deceiving herself! She may be getting the vitamins and minerals she needs by taking a supplement, but this cannot make up for her poor food habits. She needs to improve the overall quality of her diet. If Martha follows the MyPyramid recommendations, she should be able to get all the nutrients she needs from what she eats.

2. *Is a supplement warranted in this case?* Yes

   Why or why not? This is one of the times when a supplement is in order. Sylvia needs to follow her doctor's advice and take the prescribed iron preparation to increase her hemoglobin and replete her iron stores. However, she is correct in consuming more iron-rich foods, too, so that after the anemia has been treated, she will not experience a relapse.

3. *Is a supplement warranted in this case?* No

   Why or why not? There is no scientific evidence that megadores of vitamins and minerals enhance physical performance. In fact, John may be setting himself up for toxicity reactions with prolonged intake of supplements at extremely high dosages.

4. *Should Roberto try the senna, ask for a nonherbal drug, or take another action?* If unsure, Roberto should get more information from the pharmacist so he can make a well informed decision about senna use.

5. *Should Yuen take Hypermetabolite for weight loss?* No

   Why or why not? She shouldn't take the product because the ephedra derivative may have serious side effects and might interact with ginseng and because quick weight-loss strategies don't work in the longer run. Plus, there's no guarantee that the product will work or that it isn't mislabeled.

Water Is an Essential Nutrient

NUTRITION SCOREBOARD

		TRUE	FALSE
1	Bottled water is better for health than municipal water from your faucet.		
2	Drinking lots of water helps flush toxins out of your body.		
3	Adequate water intake reduces the risk of some types of cancer and kidney stones.		
4	You can't drink too much water.		

Key Concepts and Facts

- Water is an essential nutrient. It is a required part of the diet. Deficiency symptoms develop when too little is consumed, and toxicity symptoms occur when too much is ingested.

- Functions of water include maintenance of body hydration and temperature, facilitation of digestion, removal of waste products, and participation in energy formation. It is our major source of fluoride.

- Water is a precious resource whose availability and quality are threatened by wasteful use and pollution.

Water: Where Would We be Without It?

Ask any three people you know to name as many essential nutrients as they can. If they mention water, give them a prize. Our need for water is so obvious that it is often taken for granted. Well, that's not going to happen in this text!

Water differs from other essential nutrients in that it is liquid, and our need for it is measured in cups rather than grams or milligrams. Without it, our days are limited to about six. Water is the largest single component of our diet and body. It is a basic requirement of all living things. Now, how could something this important be so easily forgotten?

Water's Roles as an Essential Nutrient

Water qualifies in all respects as an essential nutrient. It is a required part of our diet; it performs specific functions in the body; and deficiency and toxicity signs develop when too little or too much is consumed.

Water is our body's main source of fluoride, an essential mineral needed for the formation and maintenance of enamel and resistance to tooth decay. Fluoride is naturally present in some water supplies and is added to 70% of all municipal water supplies in the United States.[2] Because it is a gas that dissolves in water, fluoride is not removed by water pitcher or faucet filters.

Water is the medium by which many chemical reactions take place within our body. Water plays key roles in digestion and energy formation. It also provides the medium by which digestive enzymes access and break down carbohydrates, proteins, and fats in food. Water is produced as an end product of energy formation from carbohydrates, proteins, and fats. We continue to produce and excrete water even if we quit drinking it for awhile, because energy production is an ongoing process. Water is needed to "carry" nutrients to cells and waste products away from them. Additionally, water acts as the body's cooling system. When our internal temperature gets too high, water transfers heat to the skin and releases it in perspiration. When we're too cool, less water—and heat—is released through the skin. Water's functions are summarized in Table 25.1.

Photo Disc

Table 25.1

Key functions of water in the body

- Provides a medium for chemical reactions involved in digestion and other body processes
- Participates in energy formation
- Transports nutrients and waste products
- Helps regulate body temperature

Water has been given credit for other functions in the body, but undeservedly. Drinking more water than is normally needed does not prevent dry, wrinkled skin, lead to weight loss, or flush toxins out of the body.[3] Nor will it cure chronic fatigue, arthritis, migraines, or hypertension.

Water, Water, Everywhere   The body of a 160-pound person contains about 12 gallons of water (Illustration 25.1). Adults are approximately 60 to 65% water by weight.[4] Water is distributed in the body in blood, the spaces in between cells, and in all cells. The proportion of water in body tissues varies: blood is 83% water, muscle 75%, and bone 22%. Even fat cells are 10% water.[5]

■ **Most Foods Contain Lots of Water, Too.**   Most beverages are more than 85% water, and fruits and vegetables are 75 to 90% (Illustration 25.2). Meats, depending upon their type and how well done they are, contain between 50 and 70% water. Although it is nearly impossible to meet your need for water from solid foods alone, the water content of foods makes an important contribution to our daily intakes. On average, about 81% of water intake comes from plain water and other beverages, and 19% from foods.[6]

Health Benefits of Water   Adequate water consumption may benefit long-term as well as day-to-day health. Consumption of over 10 cups of fluid each day is associated with a decreased risk of bladder, breast, and colon cancer as well as of kidney stone formation. People feel and perform better when they are adequately hydrated.[5]

Illustration 25.1   The body of a 160-pound person contains approximately 12 gallons of water. That's 96 pounds of water!

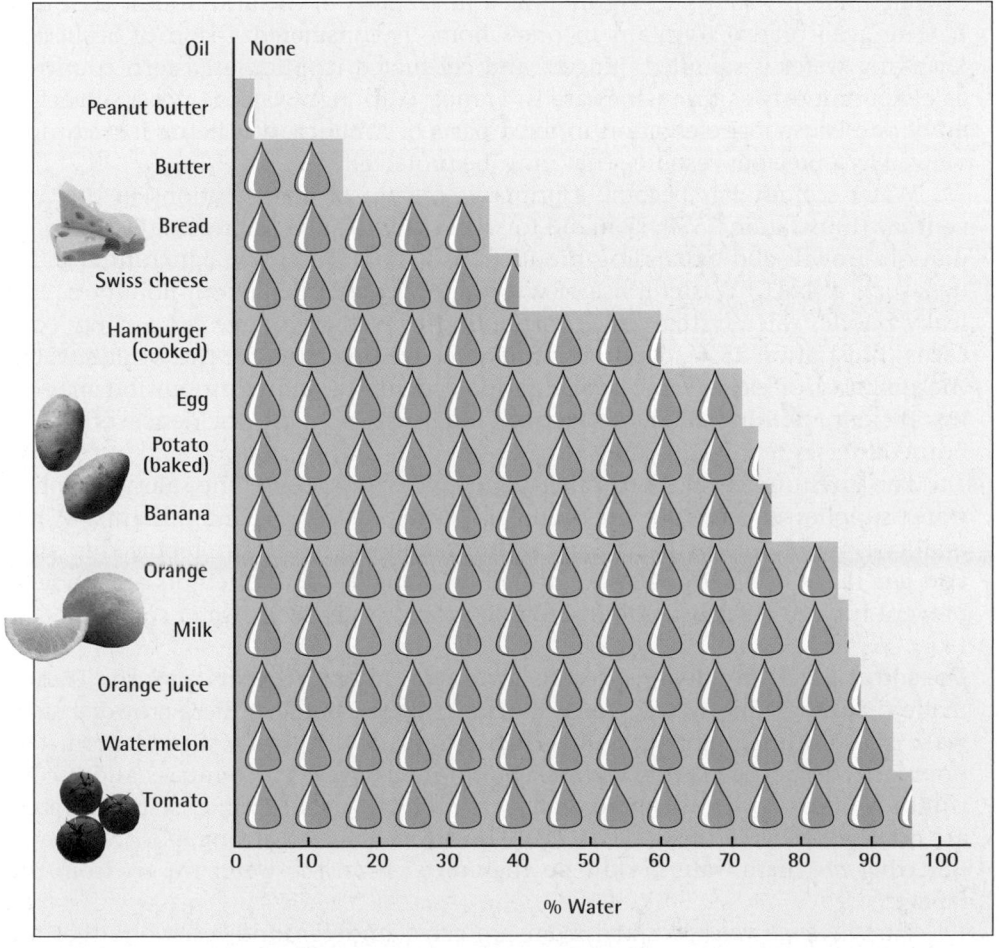

% Water

Photo Disc

Illustration 25.2   The water content of some foods.

To drink, or not to drink water with meals.   According to a health and wellness website, people should not drink water with a meal because it dilutes digestive juices and interferes with digestion. Really?

Barbara Ann
What website made that statement? Why would that happen? Do you know how watery food gets when it's in your stomach?

Carole
I get a bloated feeling when I drink water with food. I think the water is slowing down the digestion of food in my stomach.

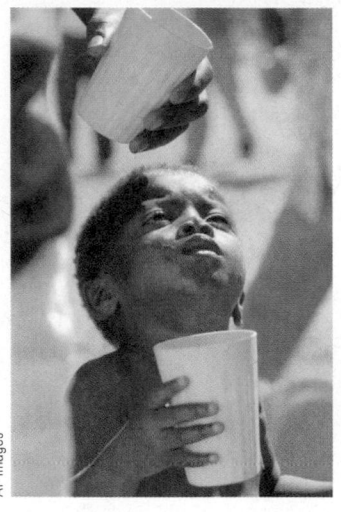

Illustration 25.3   A child waits for water at a refugee camp.

Illustration 25.4   Contaminated water supplies are a threat to the public's health.

The Nature of Our Water Supply

Water covers about three-quarters of the Earth's surface, yet very little of it is drinkable. Nearly 97% of the total supply is salt water, and only 3% is fresh. Of the freshwater supply, only a fourth of the total amount is available for use. The rest is located in polar and glacier ice. Although fresh water is readily available in most locations in the United States, this is not the case in a number of other countries.

Water is so scarce in parts of Russia that drinking-water dispensers are coin-operated. Fresh water is so highly prized in sections of the arid Middle East that having a decorative fountain in one's home is considered a sign of affluence. Drinking water is sampled, judged, and celebrated in Middle Eastern countries as ceremoniously as fine wines are in France. Although water is not a subject of envy or a cause for celebration in most parts of America, it is being increasingly viewed as a precious resource that must be protected.

Water scarcity has become a primary concern of many nations in the 21st century (Illustration 25.3). Demand for water worldwide is increasing due to population growth and expanding production of water-intensive agricultural products such as beef.[8] Wasteful use of water, groundwater depletion, pollution, and leaky public water systems are contributing to water shortages and safety concerns (Illustration 25.4).[9] Without more effective protection of the world's water, the quantity of clean water available for agriculture and consumption may be insufficient, limiting food production and increasing the incidence of water-borne illnesses.[10]

The Environmental Protection Agency is responsible for the safety of public water supplies and has set maximum allowable levels of contaminants. Water quality is monitored by local water utilities, and the results are reported to state and federal officials. Problems identified are remedied and attempts are made to prevent future contamination in order to provide safe public water supplies.[9]

Does the Earth Supply "Gourmet" Water?   About 40 years ago, the French made drinking "fine waters" very stylish. In Paris, boulevardiers crowded sidewalk cafes for hours sipping chilled Perrier served with thinly sliced lemon. The popularity of bottled waters skyrocketed in many Western countries and it continues to grow.[11] In the United States, "mineral," "spring," and "seltzer" waters are now best sellers (Illustration 25.5). These waters have a strong, positive image. But what *are* these waters? How do they differ from the water we get from the faucet?

True *mineral water* is taken from protected underground reservoirs that are lodged between layers of rock. The water dissolves some of the minerals found

in the rocks, and as a result, it contains a higher amount of minerals than most sources of surface water. Actually, most water contains some minerals and could be legitimately considered "mineral water." *Spring water* is taken from freshwater springs that form pools or streams on Earth's surface. True "seltzers" (not the sweetened kind you often find for sale) are *sparkling waters* that are naturally carbonated. Most seltzers, however, become bubbly by the commercial addition of pressurized carbon dioxide.

Bottled waters have their advantages. They are calorie free, are generally sodium free or low in sodium, and quench a thirst better than their major competitor, soft drinks. But they are no "purer" or better for you than tap water.[12] As a matter of fact, some bottled waters contain tap water. The Food and Drug Administration (FDA) estimates that 25% of the bottled water sold in the United States is tap water.

■ **Is Bottled Water Safe?** Every now and then, a story about contaminated bottled water makes the news, raising concerns about the safety of bottled water and the way the industry is monitored. The FDA regulates the bottled water industry. Domestic bottlers must conform to specified standards of water safety (such as allowable levels of chemical contaminants) and labeling requirements. "Mineral" or "spring" water must be just that under FDA regulations. Bottled tap water must be labeled as such.

Bottlers are also subject to unannounced inspections by the FDA. Bottled water is classified as a very low-risk product, however, so domestic bottling plants are inspected, on average, every 5 years.[12] The FDA supplements its monitoring role by requiring that bottlers periodically test their water. Foreign bottling plants are not under the FDA's jurisdiction, but imported waters may be tested when they enter the United States.

Illustration 25.5 Is bottled water better for you than tap water? Consumers often perceive that bottled waters have fewer impurities and are better for health than tap water. Such beliefs are unfounded.

To drink, or not to drink water with meals. Water intake during a meal does not interfere with digestion. On the contrary, it increases the exposure of digestive enzymes to food and facilitates the absorption of nutrients.[7]

Barbara Ann

Carole

Illustration 25.6 "Specialty" bottled water products are of uncertain value to health or fitness.

Safety concerns about bottled water extend to the bottles. Certain types of plastic used in water bottles may contain bisphenol A (BPA) that acts like the hormone estrogen. Current levels of BPA exposure may be related to changes in behavior and the functioning of the brain and prostate in fetuses and young children. Transfer of BPA from plastic to fluids increases with heat.[13]

Some manufacturers of plastic bottles (such as Nalgene) have taken BPA out of consumer plastic bottles and others are in the process of doing so.[14] If concerned, do not heat liquids in clear plastic bottles, look for plastic bottles labeled "without BPA" or "Bisphenol-A free," or use glass bottles.

■ **Water Gimmicks.** New waves of bottled water products are finding their way to grocery shelves throughout America. Tapping into consumer interest in vitamins, minerals, herbs, and fitness, producers are marketing "enhanced" bottled water, fortified with everything from caffeine to oxygen (Illustration 25.6). Although mineral waters provide absorbable calcium and magnesium that contribute to nutrient status and potentially to health,[15] the rationale for many of the products (other than their commercial appeal) is elusive. It makes little sense to fortify water with herbs, oxygen, or with nutrients better obtained from foods. Safety of some of the combinations of herbs put into bottled water is unclear, and they do not appear to supplement the diet in meaningful ways.[16] Adding oxygen to beverages and foods is total nonsense. The oxygen our bodies use for energy formation and metabolic processes comes from the air we breathe, not the beverages we drink or the foods we consume.

Fluoridated Bottled Water Bottled waters have been criticized for their lack of fluoride compared to fluoridated tap water. People of all ages need this mineral for bone health and the prevention of tooth decay. The industry caught on to this shortcoming and responded by offering fluoridated waters (Illustration 25.7). A quick look at the label can tell you if you're getting fluoride along with your water.

Meeting Our Need for Water

In general, individuals require enough water each day to replace water lost in urine, perspiration, stools, and exhaled air. Adequate Intakes (AIs) of water from fluids and food are set at 11 cups per day for women and 15 cups per day for men.[6] No-calorie water is the preferred source for meeting water needs.[17]

Built-in mechanisms that trigger thirst generally protect people from consuming too much or too little water.[18] People who do strenuous work, athletic or otherwise,

Illustration 25.7 Want to help protect your teeth from cavities? Look for word fluoride on the label of bottled water products.

in hot and humid weather need to consume enough water to replace the amount that is lost in sweat, urine, respiration, and evaporation of water from the skin's surface. How much extra water this takes varies from person to person, but it is often within the ballpark of a 50% increase. You know you are drinking enough water if you haven't lost weight after the physical activity and if your urine is pale yellow and produced in normal volume.[3]

Exposure to both cold weather and high altitudes increases water need. Cold air holds little moisture, so you lose more water when you breathe in cold, dry air than warm, moist air. People exposed to cold weather tend to dress in layers and lose body water in the form of sweat that pools next to the skin. Exposure to cold environmental conditions at high altitudes can increase urine production and water loss. High levels of physical activity in cold, high altitude climates further increase water need (Illustration 25.8).[19]

© PICIMPACT/CORBIS

Illustration 25.8 Cold, high altitude conditions increase water need.

Illustration 25.9 This headline is accurate—overdilution of infant formula can lead to water intoxication in infants.

Prolonged bouts of vomiting, diarrhea, and fever increase water need, and that is why you should drink plenty of fluids when you experience these conditions. High-protein and high-fiber diets and alcohol increase your need for water. Losses in body water that accompany high levels of protein consumption are the reason people on high-protein weight-loss diets are encouraged to "drink a lot of water." Adding fiber to your diet augments water need because fiber increases water loss in stools. The increased loss of water that goes along with alcohol intake explains why people get very thirsty after overindulging in spirits.

Are Caffeine-Containing Beverages Hydrating? In the past it was widely believed that beverages containing caffeine were not hydrating because caffeine acted as a diuretic. Recent and better research has demonstrated that this conclusion is incorrect. Caffeine does not increase urine output in people accustomed to drinking coffee, tea, and other beverages that contain caffeine.[18] So, count the coffee or tea you consume as contributing to your overall water intake.

Water Deficiency

A deficiency of water can lead to dehydration. Dehydrated people feel very sick. They are generally nauseated, have a fast heart rate and increased body temperature, feel dizzy, and may find it hard to move. The ingestion of fluids produces quick recovery in all but the most serious cases of dehydration. If it is not resolved, however, dehydration can lead to kidney failure and death.[10]

Water Toxicity

People can overdose on water if they drink too much of it. High intake of water can lead to a condition known as hyponatremia—or low blood sodium level and excessive water accumulation in the brain and lungs. The consequences can be devastating and include confusion, severe headache, nausea, vomiting, and even seizure, coma, and death.[20]

Water intoxication is rare, but it has occurred in marathon runners who consumed too much water during an event, infants given too much water or overdiluted formula (Illustration 25.9), and psychotic patients taking medications that produce cravings for water. The drive for water created by antipsychotic medications can be so strong that access to water (even when showering) has to be limited.[21]

Up Close

© Fotofeeling/Getty Images/Westend61

Foods as a Source of Water

Focal Point: Water is primary component of many foods:

Our requirement for water is met by fluids and solid foods. But how much water is in foods? You may be surprised.

Circle the food in each set that you think contains the highest percentage of water by weight.

1. avocado; potato, boiled; corn, cooked
2. egg; almonds; ripe (black) olives
3. watermelon; celery; pineapple, fresh
4. 2% milk; Coke; cranberry juice, low calorie
5. cheddar cheese; banana; refried beans
6. hot dog; pork sausage, cooked; ham, extra lean
7. apple; mushrooms, raw; orange
8. onions, cooked; lettuce; okra, cooked
9. peanut butter; butter; mayonnaise
10. cake with frosting; bagel; Italian bread

FEEDBACK (answers to questions) can be found at the end of Unit 25.

Review Questions

TRUE FALSE

1. Water plays key roles in the prevention of acne and dry skin and in the dilution of toxins formed during digestion. ☐ ☐

2. Water is our major source of fluoride. ☐ ☐

3. Municipal water supplies are fluoridated in most of the United States. ☐ ☐

4. Adult females living in moderate climates require an average of 8 cups of water a day, and men require twice that amount. ☐ ☐

TRUE FALSE

5. Adequate availability of uncontaminated water is a global health concern. ☐ ☐

6. Americans should limit their consumption of bottled waters because they are often contaminated and therefore unsafe to drink. ☐ ☐

7. The Food and Drug Administration requires that all bottled water be fluoridated. ☐ ☐

8. Beverages containing caffeine are not hydrating. ☐ ☐

Media Menu

www.healthfinder.gov
This is a gateway to reliable information on water. Search "water" and be led to information on the functions of water, water supply concerns, hydration, and public health water standards.

www.cdc.gov
The address for the Centers for Disease Control provides access to water safety information and advice for travelers.

www.ipwr.org
The Institute for Public Health and Water Research (IPWR) is a not-for-profit, independent science and education organization that undertakes research and promotes education about water and health. The site provides updated information on water supplies and water and health.

www.who.int/topics/water/en
This World Health Organization site provides

updates on water-related diseases, benefits of safe water supplies, water scarcity, and guidelines for safe drinking water.

www.cdc.gov/nceh/ehhe/water
The Centers for Disease Control and Prevention's Web site on water-related environmental health addresses water-related environmental hazards, health effects of contaminated water, and provides updates on water-related legislation and policies.

www.epa.gov/safewater/dwh/index.html
The Environmental Protection Agency's Drinking Water and Health: The "What you need to know" page presents facts on tap water safety, health effects of contaminated water, and sanitation rules for well water across the United States.

www.niehs.nih.gov/news/media/questions/sya-bpa.cfm
Got questions about bisphenol A (BPA)? Go to this NIH site to get some answers.

www.iom.edu/fnb
You'll be able to access the 2004 DRI Report on water and recommended intake levels from this site.

www.science.gov
Find out more about water and water quality at this website by searching "water."

www.pueblo.gsa.gov/cic_text/health/house-wells/six.htm
This EPA (Environmental Protection Agency) site lists and explains the six basic steps you should

take to maintain the safety of your drinking water.

www.who.int/water_sanitation_health/en
The World Health Organization's Water, Sanitation, and Health homepage is located at this address. It addresses global health issues related to unsafe water and approaches to improving health by developing safe water supplies.

Notes

1. Kleiner SM. Water: an essential but overlooked nutrient. J Am Diet Assoc 1999;99:200–6.

2. Fluoridation of U.S. municipal water supplies. MMWR 2008;57:737–41.

3. Fiske H. Measuring water's benefits and optimal intake recommendations. Today's Dietitian 2003;Jan:22–4.

4. Kohlstadt I. Safeguarding muscle during weight reduction, Medscape J Med 2008;10:199, www.medscape.com/viewarticle/578880.

5. Askew EW. Water. In: Ziegler EE, Filer LJ Jr., editors. Present knowledge in nutrition. Washington, DC: ILSI Press; 1996: pp. 98–108.

6. Dietary Reference Intakes for Water, Potassium, Sodium Chloride, and Sulfate. Food and Nutrition Board, National Academy of Science. http://books.nap.edu, accessed 5/04.

7. Picco M. Does drinking water during or after a meal disturb digestion? www.

mayoclinic.com/health/digestion/AN01776, accessed 4/29/09.

8. Sin aqua non. The Economist 2009;April 11, p. 59–61.

9. Water. www.intelihealth.com, accessed 6/03.

10. Haskins J. A third of the world population faces water scarcity today. www.eurekalert.com, 8/21/06.

11. Functional drinks show greatest global growth, nutraingredientseurope.com, accessed 12/08.

12. How safe is bottled water? American Institute for Cancer Research Newsletter 1992;36:8.

13. Questions and answers about the National Toxicology Program's Evaluation of Bisphenol A, www.niehs.nih.gov/news/media/questions/sya-bpa.cfm, 4/22/09.

14. Hitti M. Baby bottle makers ditch BPA. www.medscape.com/viewarticle/589362, 3/10/09.

15. Heaney RP. Absorbability and utility of calcium in mineral waters. Am J Clin Nutr 2006; 84;371–4.

16. Welland D. Drink to good health, especially water: Here's why and how much. Environ Nutr 1999;22(Oct):1,6.

17. McBurney MI. Drink fluids to maintain hydration and eat to obtain calories, Nutr Today 2009;44:14–6.

18. Grandjean AC et al. Hydration: issues for the 21st century. Nutr Rev 2003;61:261–71.

19. Wingo JE. Isolated effects of elevated temperatures on sweating in humans, presented at Experimental Biology Annual Meetings, New Orleans, 4/19/09.

20. Noakes TD. Too many fluids as bad as too few. BMJ 2003;327:113–4.

21. Goldman MB et al. Mechanisms of altered water metabolism in psychotic patients with polydipsia and hyponatremia. N Engl J Med 1988;318:397–403.

Answers to Review Questions

1. False, see page 25-3.
2. True, see page 25-2.
3. True, see page 25-2.
4. False, see page 25-6.
5. True, see page 25-4.
6. False, see page 25-5.
7. False, see page 25-6
8. False, see page 25-8.

NUTRITION | Up Close

Foods as a Source of Water

Feedback for Unit 25

The percentage of water is listed after each food.

1. **avocado (80%);** potato, boiled (77%); corn, cooked (73%)
2. egg (75%); almonds (4%); **ripe (black) olives (80%)**
3. watermelon (91%); **celery (95%);** pineapple, fresh (86%)
4. 2% milk (89%); Coke (89%); **cranberry juice, low calorie (95%)**
5. cheddar cheese (37%); **banana (74%);** refried beans (72%)
6. hot dog (53%); pork sausage, cooked (45%); **ham, extra lean (74%)**
7. apple (84%); **mushrooms, raw (92%);** orange (87%)
8. onions, cooked (88%); **lettuce (96%);** okra, cooked (90%)
9. peanut butter (1%); **butter (16%);** mayonnaise (15%)
10. cake with frosting (22%); bagel (33%); **Italian bread (36%)**

Nutrient–Gene Interactions in Health and Disease

NUTRITION SCOREBOARD

		TRUE	FALSE
1	Most chronic diseases are genetically caused.		
2	Individual differences in genetic traits are responsible for large differences in nutrient needs between individuals.		
3	Heart disease, cancer, obesity, and hypertension primarily result from interactions among environmental and genetic factors.		

Key Concepts and Facts

- Nutrients interact in important ways with gene functions and thereby affect health status. Nutrients can turn genes on or off, and nutrient intake can compensate for abnormally functioning genes.

- Health problems related to nutrient–gene interactions originate within cells.

- Advances in knowledge of nutrient–gene interactions are dramatically changing nutritional approaches to disease prevention and treatment.

Answers to **NUTRITION SCOREBOARD**	TRUE	FALSE
1 That's incorrect. Only a very small proportion of diseases are directly caused by genetic traits.[1]		✔
2 Individual genetic traits alter nutrient needs to a small extent in many people, and to a large extent in relatively few. The Dietary Reference Intakes cover the nutrient needs of nearly all healthy people.		✔
3 How true. Common diseases result from interactions between multiple genetic traits and environmental factors such as nutrient intake.[2]	✔	

Nutrition and Genomics

The science of nutritional genomics is in its infancy, but it has the potential to transform the science of nutrition.

—D. Shattuck, Journal of the American Dietetic Association

In the history of the relatively young science of nutrition, at least two breakthroughs have defined the future. The first resulted from experiments in the late 1800s demonstrating that some constituents of foods are essential for life. The second is represented by completion of the draft of the human *genome* and development of the field of nutrigenomics. (Find definitions and explanations of many of the genetic terms used in this unit in Table 26.1.) That breakthrough marked the beginning of a new era in the discovery of underlying causes of a variety of common diseases and the specific roles of components of diets in disease prevention. It is being made possible, in part, by the identification of genetic codes for protein production and nutrients that affect gene activity. Enzymes and other proteins produced as a result of genetic codes are central to life and health because they determine which chemical changes will take place within the body. These changes influence growth, digestion, nutrient absorption, disease resistance, blood pressure control, and many other functions of the body.

Nutrient–Gene Interactions

Research pinpointing nutrient–gene interactions is revolutionizing nutritional approaches to health promotion and to the prevention and treatment of diseases.[3] A number of nutrient-gene interactions identified to date have gotten the revolution well under way. The following are examples of the effects gene types (or genotypes) have on the body's response to nutritional factors. In some people:

- Whole oats lower blood cholesterol levels.

- High folate intake decreases the risk of cancer.

- High polyunsaturated fat, low dietary cholesterol, or low-saturated-fat diets lower blood cholesterol levels.

Table 26.1

Definitions and explanations of genetic terms

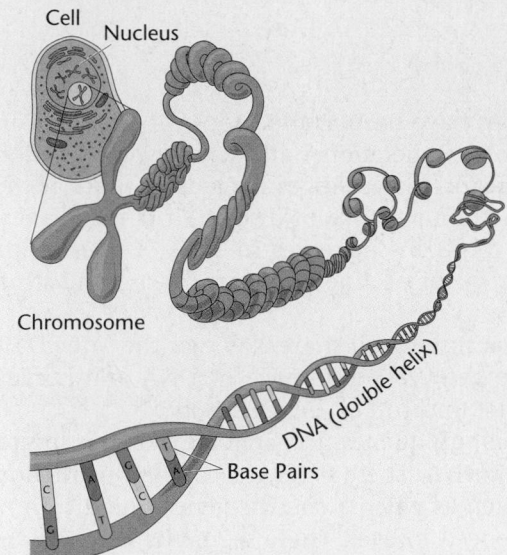

DNA is packed in chromosomes located in cell nuclei

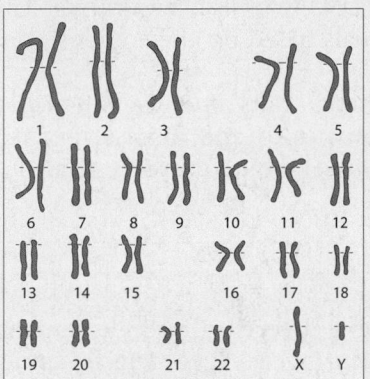

Human chromosomes

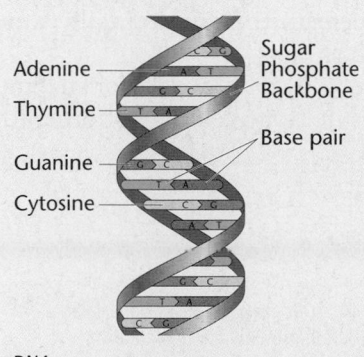

DNA

- **Genome** Combined term for "genes" and "chromosomes." It represents all the genes and DNA contained in an organism, which are principally located in chromosomes. The human genome consists of 20,000 to 24,000 genes.

- **Genes** The basic units of heredity that occupy specific places (loci) on chromosomes. Genes consist of large DNA molecules, each of which contains the code for the manufacture of a specific enzyme or other protein.

- **Genotype** The specific genetic makeup of an individual as coded by DNA.

- **Genomics** The study of the functions and interactions of all genes in the genome. Unlike genetics, it includes the study of genes related to common conditions and their interaction with environmental factors.

- **Nutrigenomics** The study of nutrient-related functions and interactions with genes and their affects on health and disease.

- **Chromosomes** Structures in the nuclei of cells that contain genes. Humans have 23 pairs of chromosomes (shown below); half of each pair comes from the mother and half from the father.

- **DNA (deoxyribonucleic acid)** Segments of genes that provide instructions for the manufacture of enzymes and other proteins by cells. DNA looks something like an immensely long ladder twisted into a helix, or coil. The sides of the ladder structure of DNA are formed by a backbone of sugar-phosphate molecules, and the "rungs" consist of pairs of bases joined by weak chemical bonds. Bases are the "letters" that spell out the genetic code, and there are over 3 billion of them in human DNA. There are two types of base pairs: adenine-guanine and cytosine-thymine. Each sequence of three base pairs on DNA codes for a specific amino acid. A specific enzyme or other protein is formed when coded amino acids are collected and strung together in the sequence dictated by DNA.

  Some sections of DNA do not transmit genetic information because they don't code for protein production. They appear to signal which genes will turn on and how long they will be activated. Characteristics of these segments of DNA vary among individuals, making it possible to identify individuals based on "DNA fingerprinting."

- **DNA fingerprinting** The process of identifying specific individuals by their DNA. This is possible because no two individuals have the same genetic makeup. Differences among individuals are due to variations in the sections of DNA molecules that do not transmit genetic information.

- High-carbohydrate diets increase the risk of type 2 diabetes.

- High alcohol intake during pregnancy produces physical and behavioral abnormalities in the fetus.

- Regular consumption of green tea reduces the risk of prostate cancer.[4-7]

Are you reminded an hour or two later that you ate asparagus? If you said yes, you are among the 1 out of 10 people born with a gene for detecting the odor of a sulfur-containing compound excreted in urine by 4 in 10 people after consumption of asparagus.

single-gene defects
Disorders resulting from one abnormal gene. Also called "inborn errors of metabolism." Over 6,000 single-gene disorders have been cataloged, and most are very rare.[10]

Research into the genetic bases of disease risk is also clarifying which health characteristics have little or nothing to do with genetic traits and everything to do with environmental factors.[1]

Genetic Secrets Unfolded

Our bodies may be new, but our genes have been around for over 40,000 years. Genes replicate themselves exactly over generations, and lasting modifications in them almost never occur. This means that changes in genetic traits do not account for increases or decreases in the incidence of disease. It is why genetic traits cannot be given as the cause of recent increases in rates of obesity and diabetes, nor can they take credit for major declines in heart disease and stroke rates.

Humans are united as a species because we all share 99.9% of the same DNA. We are unique as individuals due to the 0.1% difference in DNA and because environmental exposures leave distinct imprints on gene activity.[8]

Much of the evidence of individual uniqueness in genetic makeup is located in the non-protein coding segments of DNA. These segments contain instructions for the regulation of gene activity, such as when a specific gene should turn on or off, or when a gene should be silenced forever. These segments vary in composition based on environmental factors such as nutrient availability and exposure to infectious agents. Since every individual's environmental exposures are unique, these non-protein coding sections of DNA will also be unique. These differences make it possible to identify individuals based on the composition of their DNA.[8,9]

Many questions about the effects of genetic traits and the interactions among genes and environmental exposures on health and disease remain to be answered. Currently, scientists know the most about disorders related to **single-gene defects**.

Single-Gene Defects

Thousands of rare diseases related to one or more defects in a single gene have been identified, and many of these affect nutrient needs. Such defects can alter the absorption or utilization of nutrients such as amino acids, iron, zinc, and the vitamins B_{12}, B_6, or folate.[10] PKU, celiac disease, lactose intolerance, and hemochromatosis are four examples of single-gene defects that substantially affect nutrient needs. (These are described in Table 26.2.)

PKU and lactose intolerance are caused by defective genetic codes for enzymes, whereas celiac disease and hemochromatosis result from genetic abnormalities

---

Table 26.2

Examples of single-gene disorders that affect nutrient need[10,11]

PKU (phenylketonuria)	A very rare disorder caused by the lack of the enzyme phenylalanine hydroxylase. Lack of this enzyme causes phenylalanine, an essential amino acid, to build up in the blood. High blood levels of phenylalanine during growth lead to mental retardation, poor growth, and other problems. PKU is treated by low-phenylalanine diets.
Celiac disease	An intestinal malabsorption disorder caused by an inherited intolerance to gluten in wheat, rye, and barley. It causes multiple nutrient deficiencies and is treated with gluten-free diets. Celiac disease is also called "nontropical sprue" and "gluten enteropathy."
Lactose intolerance	A common disorder in adults in many countries resulting from lack of the enzyme lactase. Ingestion of lactose in dairy products causes gas, cramps, and nausea due to the presence of undigested lactose in the gut.
Hemochromatosis	A disorder affecting 1 in 200 people that occurs usually due to a genetic deficiency of a protein that helps regulate iron absorption. Individuals with hemochromatosis absorb more iron than normal and have excessive levels of body iron. High levels of body iron have toxic effects on tissues such as the liver and heart. The disorder can also be produced by excessive levels of iron intake over time and frequent iron injections or blood transfusions.

in the formation of other proteins. PKU, celiac disease, and lactose intolerance are treated by diets that limit phenylalanine, gluten, or lactose, respectively. Hemochromatosis is treated by a low iron diet.[1]

Not all single-gene abnormalities produce ill effects. For instance, lack of the gene that codes for an enzyme that helps the body excrete a specific sulfur-containing chemical unique to cruciferous vegetables (such as cabbage, broccoli, and bok choy) may be good. Smokers and nonsmokers who have the gene for the enzyme that helps excrete this beneficial sulfur compound and who consume these vegetables regularly are more likely to develop lung cancer or to have a heart attack than are people who eat the vegetables but lack the gene. Without the gene for the enzyme that causes its quick elimination, the beneficial sulfur-containing compound lingers in the body, extending the time it can play a role in cancer and heart attack prevention.[7,12] Risk of both disorders is decreased by not smoking.

Most diseases related to genetic traits are not as straightforward as are single-gene defects. They are more likely to represent an interwoven mesh of genetic and environmental risk factors.

Chronic Disease: Nurture and Nature

Major health problems of our day, including heart disease, cancer, hypertension, obesity, diabetes, and disorders associated with aging, are not due to single-gene defects but result from interactions among multiple genetic traits and environmental factors. Nutrients, a prominent environmental factor, can turn certain genes on or off and, along with other effects on gene function, can compensate for some of the ill effects certain genotypes have on body processes.[13]

Heart disease primarily stems from plaque buildup in arteries near the heart (Illustration 26.1). Plaque buildup "hardens the arteries," narrows artery openings, and increases the risk of artery blockage and subsequent heart attack.

Some people who consume high-saturated-fat diets build up plaque in their arteries, whereas others don't. For others, low fruit and vegetable diets increase the risk that plaque will accumulate. These differences are due to variations in genotypes. Heart disease is also influenced by blood pressure, body weight, and levels of triglycerides and clotting factors in the blood. Each of these is influenced by environmental factors such as high salt or alcohol intake in people with specific genetic traits. Risk for heart disease is the sum of all the pro and con influences of environmental and genetic factors.[14]

A very large part of our health story is more genetically influenced than genetically caused.

—Elbert Branscomb, Director of the Joint Human Genome Institute

Cancer

Most types of cancer are primarily related to environmental factors such as high fat and alcohol intakes, low vegetable and fruit diets, high levels of body fat,

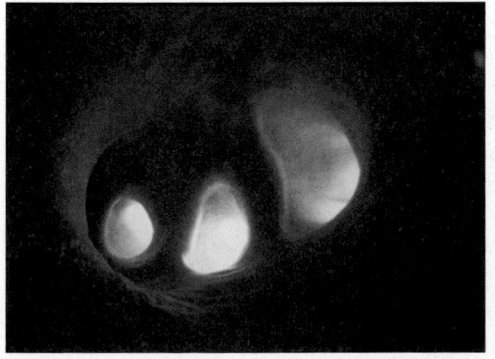
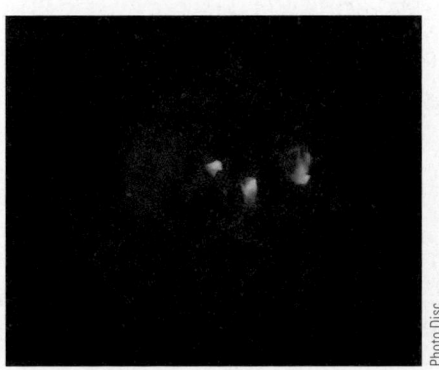

Photo Disc

Illustration 26.1 Healthy (left) versus clogged (right) arteries. How much plaque builds up in arteries is influenced by multiple environmental and genetic factors.

Table 26.3

Estimates of the contribution of environmental and genetic factors to development of some types of cancer

Cancer Site	Environmental Factors (%)	Genetic Factors (%)
Endometrial uterine wall)	100	0
Ovary	78	22
Lung	74	26
Breast	73–90	10–27
Stomach	72	28
Colon	65–95	5–35
Pancreas	64	36
Prostate	58	42

salt sensitivity
A genetically determined condition in which a person's blood pressure rises when high amounts of salt or sodium are consumed. Such individuals are sometimes identified by blood pressure increases of 10% or more when switched from a lower-salt (1 to 3 grams) to a higher-salt (12 to 15 grams) diet.[16]

Illustration 26.2 Obesity is related to complex interactions among nutritional, other environmental, and genetic factors.

Photo Disc

smoking, and toxins in the environment.[15] Some cancers have genetic components that interact with environmental exposures. Studies of the roles played by genetic and environmental factors, to the extent they can currently be separated, show differences by cancer site (Table 26.3). As can be seen by the large contributions of environmental factors to cancer development, the hope for cancer prevention primarily lies in our ability to modify harmful environmental exposures.

Hypertension

The primary causes of most cases of hypertension are unknown, but factors such as high body fat, high alcohol intake, physical inactivity, and genetic traits play a role.[18] High salt (sodium, really) intakes are also related, but not in all studies. Differences in results between studies may be related to the proportion of people in study samples who are genetically predisposed to the effects of salt intake on blood pressure. Such people are **salt sensitive;** their blood pressure increases when they consume high amounts of salt. Approximately 51% of people with hypertension and 26% of people with normal blood pressure are salt sensitive.[19]

Obesity

The current obesity epidemic appears to be driven by a mismatch between multiple components of our 40,000-year old genetic endowment and current food and physical activity environments (Illustration 26.2).[20] Genetic traits that helped our early ancestors survive times of famine and that encouraged food intake, and those that set up metabolic systems around unrefined and unprocessed basic foods are at odds with much of today's food supply and physical activity requirements.[21]

Over 200 genetic traits have been related to obesity development in people exposed to Western-type diets and low levels of physical activity.[22] Many of the traits prompt excess food intake in response to high levels of availability of palatable, energy-dense foods.[20] Other traits relate to genes that ramp-up a feeling of pleasure when energy-dense foods are consumed, and still others promote sensations of hunger when high carbohydrate diets are consumed.[21-24] Gene functions that promote obesity can develop in the fetus and during other periods of rapid growth and development in response to energy and nutrient availability.[22,23]

Reversing the worldwide trend in obesity development will largely depend on decreasing our exposure to environmental triggers of gene responses that encourage excess food intake and physical inactivity. Introducing safe, walkable areas into neighborhoods, expanding access to healthy foods, and expanding the availability of recreational facilities are examples of environmental changes that appear to work.[25]

Genetics of Food Selection

Food preferences are largely learned, but which vegetables we like or don't like may be influenced by genetic traits. There are over 80 genes that help people taste bitter foods, and some people get the set of genes that make them highly sensitive to bitter-tasting foods (which are mainly vegetables). People born with a high sensitivity to bitter tastes tend to dislike cooked cabbage, collard greens, spinach, brussels sprouts, or other vegetables that taste bitter to them. People who tend to like these vegetables generally don't perceive them to be bitter-tasting. A genetic tendency to reject these vegetables is likely to limit intake and therefore may be linked to diseases associated with low vegetable intake.[26]

Changing family history Elena and Alfredo both come from families with a number of relatives who have died from heart disease.

Who gets thumbs up?
Answers on next page

Elena:
Heart disease is in my genes. There's nothing I can do about it.

Photo Disc

Alfredo:
I already know my LDL cholesterol is high and my HDL cholesterol is low. I'm eating and exercising to beat the odds.

Nutrition Tomorrow

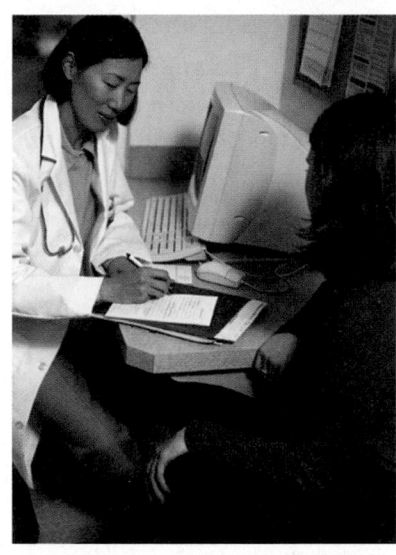

Illustration 26.3 In the future, advice given by registered dietitians may be tailored to the specific genetic traits of individuals.

The promise of advances in knowledge of the genetic bases of disease is longer, healthier lives. No doubt drugs that counter ill effects of genetic traits will continue to be developed, and attempts to fix abnormal genes by gene therapy will broaden. Some of the most meaningful breakthroughs in upcoming decades will be in the area of disease prevention and treatment through nutritional changes.[5]

Although used to some extent now, personalized modifications of dietary intake based on genotypes will become standard practice in clinical dietetics and medicine (Illustration 26.3). It is already clear that lifestyle modifications, such as decreased calorie intake, increased physical activity and intake of vegetables and fruits, for example, will be components of health improvement efforts stemming from knowledge of genetic risk factors.[27]

In our rush to fit medicine with the genetic mantle, we are losing sight of other possibilities for improving public health.

—N. A. Holtzman, Johns Hopkins Medical Institute

Are Gene-Based Designer Diets in Your Future?

Sound far-fetched? It isn't a reality now, but it is coming.[28] In the future, people will have the option of, for example, reducing cancer risk by regularly consuming cruciferous vegetables (cabbage, broccoli, cauliflower, and the like), diabetes risk by switching to a high-protein diet, or heart disease by limiting dietary cholesterol or saturated fat intake (Illustration 26.4). In truth, we have enough knowledge today to implement healthful, disease-preventing dietary changes. The kinds of advances described above will not bring an end to disease or guarantee health, but they will take us a long way down that road.

In less than 10 years, you'll be able to go to a lab and complete a set of genetic tests to identify your personal disease susceptibilities. When you leave you'll be armed with a list of foods to eat and foods to avoid and recommendations of dietary supplements to help prevent your disease.

—D. Evans, CEO of Wellgen, Inc.

Changing family history Alfredo recognizes that a family history of heart disease doesn't mean he is destined to die from it. A small minority of diseases are solely due to genetic factors; many are primarily due to interactions between genetic traits and environmental factors. Disease-promoting traits can be diminished by the right environmental changes.[1]

Elena:

Alfredo:

On the Side

Dogs have a taste for sweet, but cats don't. Somewhere during genetic evolution the gene in cats that enabled them to recognize sweetness was turned off never to be turned on again.[29]

©Royalty-Free/Corbis

Welcome to the Future Café
Today's Specials
Prime Rib au jus
or, for our saturated-fat sensitive guests, try our
Mediterranean Sole
with almonds

Illustration 26.4 A menu of the future.

Up Close

Nature and Nurture

Focal Point: Gaining insight into your family's and your own "dietary" health history.

Families often share dietary factors that interact with genetic traits to influence chronic disease development. Check out your dietary health history and that of your family by completing the following family food tree activity. If you don't know all of your relatives included in the activity or their dietary behaviors, fill out the parts based on what you do know.

Check each dietary behavior that applies:

Relative	Excess Calorie Consumption		Low Vegetable and Fruit Intake		Low Whole Grains Intake		Low Fiber Intake		Low Fish Intake		High Alcohol Intake	
	yes	no	yes	no	yes	no	yes	no	yes	no	yes	no
A. On your mother's side:	—	—	—	—	—	—	—	—	—	—	—	—
Grandmother	—	—	—	—	—	—	—	—	—	—	—	—
Grandfather	—	—	—	—	—	—	—	—	—	—	—	—
B. On your father's side:	—	—	—	—	—	—	—	—	—	—	—	—
Grandmother	—	—	—	—	—	—	—	—	—	—	—	—
Grandfather	—	—	—	—	—	—	—	—	—	—	—	—
C. Your mother	—	—	—	—	—	—	—	—	—	—	—	—
D. Your father	—	—	—	—	—	—	—	—	—	—	—	—
E. Yourself	—	—	—	—	—	—	—	—	—	—	—	—

FEEDBACK (answers to these questions) can be found at the end of Unit 26.

Key Terms

chromosomes, page 26–3
DNA (deoxyribonucleic acid), page 26–3

DNA fingerprinting, page 26–3
genes, page 26–3
genome, page 26–3

genomics, page 26–3
salt sensitivity, page 26–6
single-gene defects, page 26–4

Review Questions

TRUE FALSE

1. Genomics is the study of the functions and interactions of genes. ☐ ☐

2. Nutrigenomics is the study of the interaction of body weight and genes. ☐ ☐

3. Genes provide the codes for the production of enzymes and other proteins in cells. ☐ ☐

4. Unfavorable changes in genetic traits are the primary cause of the current, global epidemics of obesity and diabetes. ☐ ☐

5. Most chronic diseases result from single-gene defects. ☐ ☐

TRUE FALSE

6. Most chronic diseases result from interactions among multiple genetic traits and environmental factors. ☐ ☐

7. How genes function, such as when they are prompted to turn on or off, can be programmed early in life by dietary exposures. ☐ ☐

8. Some of the most meaningful breakthroughs in the prevention and management of chronic diseases in the future will be due to advances in nutrigenomics. ☐ ☐

Media Menu

www.genome.gov
This great site from the National Human Genome Research Institute provides news of the latest developments in genetic research, educational resources, and a written and spoken glossary of genetic terms and useful illustrations.

www.ornl.gov/sci/techresources/Human_Genome/home.shtml
The home page of the Human Genome Project provides background information on human

genome discoveries, ethical, legal, and social issues surrounding genomic information. The site also include reports of breakthroughs in human genome research.

www.healthfinder.gov
Search "genes" and get the facts on topics like how genes work and how genetic conditions and genes are named. Or, you can access news on recent developments in nutrient-gene interaction in health and disease from this site.

www.cdc.gov/genomics/public.htm
Click on the "Student Resources" link and enter "The Gene School." It provides experiments, quizzes, and historical information about genes and health.

Notes

1. Guttmacher AE et al. Welcome to the genomic era. N Engl J Med 2003;349:996–8.

2. Stover PJ. Influence of human genetic variation on nutritional requirements. Am J Clin Nutr 2006;83(suppl): 436S–42S.

3. Zeisel SH. Nutrigenomic and metabolomics will change clinical nutrition and public health practice: insights from studies on dietary requirements for choline, Am J Clin Nutr 2007;86:542–8.

4. Simopoulos AP. Genetic variation and nutrition, Experimental Biology Annual Meeting, San Diego, CA, 4/12/03.

5. Afman L et al. Nutrigenomics: from molecular nutrition to prevention of disease. J Am Diet Assoc 2006;106: 569–78.

6. Guerreriro et al. Risk of colorectal cancer associated with the C677T polymorphism in 5,10-methylenetetrahyrdofolate reductase in Portuguese patients depends on the intake of methyl-donor nutrients. Am J Clin Nutr 2008;88:1413–8.

7. Cornelis MC et al. GSTT1 genotype modifies the association between cruciferous vegetable intake and the risk of myocardial infarction, Am J Clin Nutr 2007;86:752–8.

8. Lee C et al. Structural genomic variation and personalized medicine, N Engl J Med 2008;358:740–1.

9. Barnes S. Nutritional genomics, polyphenols, diets, and their impact on dietetics, J Am Diet Assoc 2008;108:1888–95.

10. Isaacs JS et al. Single-gene autosomal recessive disorders and Prader-Willi Syndrome: an update for food and nutritional professionals, J Am Diet Assoc 2007;107:466–78.

11. Khoury MJ et al. Population screening in the age of genomic medicine, N Engl J Med 2003;348:50–8.

12. Hunter DJ et al. Nutrition and breast cancer, Cancer Causes and Control 1996;7:56–68.

13. London SJ et al. CYP1A1 1462V genetic polymorphism and lung cancer risk in a cohort of men in Shanghai, China. Cancer Epidemiol Biomarkers Prev 2000;9:987–91.

14. Shattuck D. Nutritional genomics. J Am Diet Assoc 2003;103:16–18.

15. Ordovas JM. Genetic interactions with diet influence the risk of cardiovascular disease. Am J Clin Nutr 2006;83(suppl):443S–6S.

16. Lichtenstein P et al. Environmental and heritable factors in the causation of cancer—analyses of cohorts of twins from Sweden, Denmark, and Finland. N Engl J Med 2000;343:78–85.

17. Morimoto A et al. Sodium sensitivity and cardiovascular events in patients with essential hypertension. Lancet 1997;350:1734–7.

18. Foulkes WD. Inherited susceptibility to common cancers. N Engl J Med 2008;359:2143–54.

19. Wong ZY et al. Genetic linkage of beta and gamma subunits of epithelial sodium channel to systolic blood pressure. Lancet 1999;353:1222–5.

20. Kaplan NM. The dietary guideline for sodium: should we shake it up? No. Am J Clin Nutr 2000;71:1020–6.

21. Froguel P et al. The power of the extreme in elucidating obesity. N Engl J Med 2008;359:891–3.

22. Lee YS. The role of genes in the current obesity epidemic, Ann Acad Med Singapore. 2009;38:45–3.

23. Butte NF et al. Viva la Familia Study: genetic and environmental contributions to childhood obesity and its comorbidities in the Hispanic population. Am J Clin Nutr 2006;84:646–54.

24. Edelson E. Overeating? Blame Your Genes Certain DNA may cause people to eat more to get the same pleasure from food, www.healthfinder.gov/news/newsstory.aspx?docID=620371, accessed 4/30/09.

25. Cecil JE et al. An obesity-associated FTO gene variant and increased energy intake in children, N Engl J Med 2008;359:2558–66.

26. Sallis JF et al. Physical activity and food environments: solutions to the obesity epidemic. Milbank Q. 2009;87:123–54.

27. Drewnowski A et al. Genetic taste markers and preferences for vegetables and fruit of female breast care patients. J Am Diet Assoc 2000;100:191–7.

28. Hunter DJ et al. Letting the genome out of the bottle—will we get our wish? N Engl J Med 2008;358:105–7.

29. Kauwell GPA. Epigenetics: what is it and how it can affect dietetic practice, J Am Diet Assoc 2008;108:1056–9.

30. Gorman J. Dogs may laugh, but only cats get the joke. New York Times, 9/5/06.

NUTRITION | # Up Close

Nature and Nurture

Feedback for Unit 26

The more *yes* responses, the better the odds that some of your family's shared dietary traits may influence disease risk. The most important responses are those you gave yourself.

UNIT 27 | Nutrition and Physical Fitness for Everyone

NUTRITION SCOREBOARD

	TRUE	FALSE
1 Physical fitness means being very muscular.		
2 Overweight people can be physically fit.		
3 Exercise is more effective than diet in preventing heart attacks.		
4 Physical fitness can be achieved only by exercising intensively for at least an hour every day of the week.		

Key Concepts and Facts

- Physical fitness, along with a good diet, confers a number of physical and mental health benefits.

- You don't have to be an athlete or be lean to be physically fit. Fitness depends primarily on muscular strength, endurance, and flexibility.

- People who are physically fit have respiratory and circulatory systems capable of delivering large amounts of oxygen to muscles and muscular systems that can utilize large amounts of oxygen for prolonged periods of time.

- Physical fitness can be achieved by resistance training, aerobic exercises, and stretching.

- The fitness level of most people in the United States is poor.

Answers to

NUTRITION SCOREBOARD

		TRUE	FALSE
1	Muscle strength is one part of physical fitness. The other primary components are endurance and flexibility.		✔
2	You can be fit and overweight. You don't have to be thin to be physically fit.	✔	
3	Regular exercise does help reduce the risk of heart attack—but not as much as a combined program of exercise, weight loss, and smoking cessation (if needed), and consumption of a healthy diet.[1]		✔
4	You don't have to exercise to that extent to become physically fit.[2]		✔

I would not wish to imagine a world in which there were no games to play and no chance to satisfy the natural human impulse to run, to jump, to throw, to swim, to dance. Sport and recreation are, in themselves, eminently worthwhile and desirable....Our aim should be, I suggest, to inspire everyone to become involved in sport and recreation by making the choice irresistible in its scope and variety....I see health as a happy consequence of sporting activity.

—Sir Roger Bannister, 1989

AP Images/The Denver Post, Cyrus McCrimmon

Physical Fitness: It Offers Something for Everyone

Physical activity has much to offer the athlete and nonathlete alike. It provides recreation good for the body and soul, it doesn't have to cost anything, and it benefits almost everyone. At its best, physical activity is play with happy consequences for health and well-being. As long as there are activities people enjoy doing, there's an "athlete" in everyone.

The "Happy Consequences" of Physical Activity

Ask people who are in good physical condition what they get from exercise, and you're likely to hear a variety of responses. Someone who is 20 years old may say he wants to stay in shape and improve his stamina. A 40-year-old individual may say that exercise helps keep her cholesterol and weight from getting too high. People who are 80 might tell you they exercise so that they won't have to use a walker and will be able to maintain their independence. Ask children why they exercise, however, and they may not understand the question. Children engage in active play; they don't exercise. For them, fitness is truly an unintended consequence of play.

Regular physical activity benefits both physical and psychological health in people of all ages (Table 27.1). By improving a person's physical health, regular exercise may help ward off heart disease, some types of cancer, hypertension and stroke, osteoporosis, back injury, and diabetes. It tends to increase a person's feeling of well-being and helps relieve depression, anxiety, and stress.[3-6]

Table 27.1

Benefits of regular physical activity

Reduced Risk of Certain Diseases and Disorders	Improved Sense of Well-Being	
• Heart disease • Colon and breast cancer • Hypertension • Stroke • Osteoporosis • Back and other injuries • Obesity, excess abdominal fat • Type 2 diabetes • Bone and joint diseases • Alzheimer's disease	• Increases feeling of well-being • Decreases depression and anxiety • Helps relieve stress • Decreases risk of dementia	

Photo Disc

The Bonus Pack: Exercise plus a Good Diet Exercise benefits health most when combined with a good diet and other healthy behaviors. The risk of developing heart disease, for example, is meaningfully lower when people combine a diet low in animal fat and high in vegetables, fruits, whole grains, and fiber with regular exercise than when exercise alone is relied upon for protection against heart disease.[7] Regular physical activity helps build bone mass and reduces the risk of osteoporosis. But the risk is lowered to a greater extent if exercise is combined with a diet that supplies adequate amounts of calcium and vitamin D. Some other benefits of exercise, such as a reduced chance of developing colon and breast cancer, may be related to the effect of exercise on body fat content. Regular exercise is one of the few factors yet identified that helps people achieve long-term weight control.[8]

■ **Exercise and Body Weight.** Combined with a moderate decrease in usual caloric intake (on the order of 200 calories per day), exercise helps people lose fat, build muscle mass, and become physically fit.[9] Because the body needs more calories to maintain muscle than fat, exercise that results in an increase in muscle mass leads to an increase in caloric requirements. For some people, this increase in caloric requirement makes it easier to maintain weight within the normal range and to lose weight and keep it off.[8]

Exercise combined with a stable caloric intake can lead to weight loss, but the loss is generally smaller than can be achieved by reducing caloric intake.[10] Even if weight loss is not intended or is achieved at a slow pace, the happy consequences of exercise come to all basically healthy individuals who undertake it on a regular basis.

Physical Activity and Fitness

Many of the benefits of physical activity are related to the **physical fitness** it can produce. Physical fitness is not defined by bulging muscles, thin waistlines, or amount of physical activity. Overweight as well as thin people can be physically fit or not. According to the Physical Activity Guidelines for Americans, physical fitness is a state of health primarily measured by strength, endurance, and flexibility.[11]

The strength component of physical fitness relates to the level of maximum force that muscles can produce. Endurance refers to the length of time muscles can perform physical activities, and flexibility refers to a person's range of motion. Physical fitness exists when all three are present at health-promoting levels.

physical fitness
The health of the body as measured by muscular strength, endurance, and flexibility in the conduct of physical activity.

Muscle Strength   In adults, muscles grow in size and strength when muscle cells increase in size. Muscle cell size—and correspondingly, the strength of muscles—increases in response to weight-bearing or resistance exercise. Activities that require muscles to work harder than usual increase muscular strength.

To increase muscle strength, people should lift or push against a heavy weight; to increase muscle endurance, they should lift lighter weights and push against them repeatedly. Resistance training should include exercises that build both muscle strength and endurance.

Endurance: A Measure of Aerobic Fitness   How long a person can perform an activity depends on inherited traits and conditioning.[12] Both factors contribute to a person's stamina—or the extent of one's ability to deliver oxygen to muscles and the ability of muscles to use the oxygen for work. The amount of oxygen an individual is able to deliver to muscles that can be used by muscles for physical activity corresponds to the level of **aerobic fitness.** The more aerobically fit a person is, the longer and harder she or he is able to exercise.

Aerobic activities include jogging, basketball, swimming, soccer, and other low- and moderate-intensity activities. They give the whole body, or most of it, a continuous workout.

■   **How Is Aerobic Fitness Determined?**   Aerobic fitness is classically assessed by measuring **maximal oxygen consumption** (abbreviated as VO_2 max) in a specially equipped laboratory (Illustration 27.1). In the lab, individuals are exercised at increasingly higher intensities, for example, by elevating the grade or speed of a treadmill. The individual performing the exercise breathes through a tube that delivers room air. Monitoring equipment measures the amount of oxygen from the air that is breathed and used during the exercise. The maximal amount of oxygen a person delivers to working muscles is the amount used when the intensity of exercise can no longer be increased. The higher the level of oxygen used at the peak level of activity, the higher the level of aerobic fitness and the longer physical activity can be performed.

People can perform physical activity at 100% of VO_2 max for only a few minutes. Consequently, aerobic fitness goals are set below that level. In general, it is recommended that beginners start an aerobic fitness program with a goal of exercising at 40 to 60% of VO_2 max and working up to a higher level.[11] Aerobically fit people may train at 70 to 85% of VO_2 max.

aerobic fitness
A state of respiratory and circulatory health as measured by the ability to deliver oxygen to muscles and the capacity of muscles to use the oxygen for physical activity.

maximal oxygen consumption
The highest amount of oxygen that can be delivered to, and used by, muscles for physical activity. Also called VO_2 max and maximal volume of oxygen.

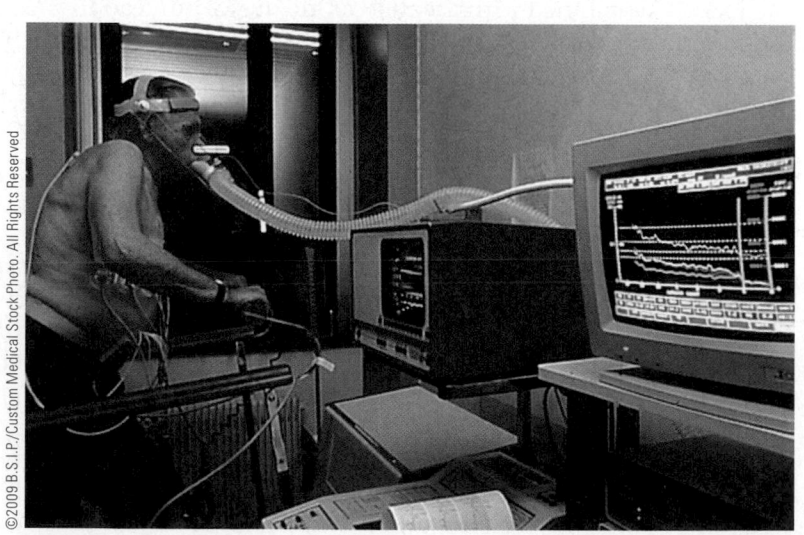

Illustration 27.1   Determining VO_2 max.
This man is running on a treadmill. His nose is plugged, so he must breathe through a tube in his mouth. The tube is attached to an apparatus that measures the total amount of air breathed in and out and the difference between the amount of oxygen inhaled and exhaled. Every few minutes the slope or speed of the treadmill is increased, making the exercise more intense. The amount of oxygen he uses at the point when his exercise intensity can go no higher is considered "maximal oxygen consumption," or VO_2 max.

■ **How Do You Know Your VO$_2$ Max?** There are several ways to estimate VO$_2$ max that don't require a supervised laboratory test. One commonly employed method uses the percentage of maximal heart rate. This formula uses heart rate as a proxy measure of oxygen consumption, and it provides a fairly good estimate of VO$_2$ max because the amount of blood circulated to muscles is roughly equivalent to the number of times the heart sends out a pulse of blood each minute. The more oxygen muscles need to perform an activity, the more oxygen the lungs have to deliver and the faster the heart has to beat (that is, up to the point where the lungs and heart can no longer send an additional supply of oxygen to muscles). The heart rate at which the highest level of oxygen consumption by the body occurs is considered "maximal heart rate." It roughly corresponds to VO$_2$ max.

The formula for estimating maximal heart rate is quite simple: Subtract your age from 220. To obtain a target heart rate for exercise, multiply the result by the desired percentage of maximal heart rate. (Table 27.2 shows an example.) The resulting figure is your "target heart rate" for aerobic exercise. Table 27.3 lists maximal heart rate, 60% of the maximum, and 75% of the maximum rate for different ages.

■ **How Do You Know Your Heart Rate?** You can determine if you are exercising at the appropriate level for aerobic fitness by measuring your pulse rate during a break or as soon as you're done exercising. To take your pulse, place a fingertip gently on an artery in your wrist or neck (Illustration 27.2). Count the number of pulses you feel in a 10-second period and multiply that number by 6. That's your heart rate, measured as heartbeats (or pulses) per minute. As you become aerobically fit your heart rate and blood pressure will decrease because you are better able to deliver oxygen to muscles.[13]

For many years it was assumed that stretching prior to exercise improved flexibility and range of motion and helped prevent injuries. Recent evidence casts doubt on its benefits, however. Stretching muscles to the point of discomfort does not appear to improve flexibility or range of motion, prevent injuries, or enhance performance.[14] Strength exercises that involve major muscle groups, a gentle warm-up of muscle movements prior to and after exercise, and regular exercise may foster flexibility.[14,15]

Table 27.2

Applying the percentage of maximal heart rate (MHR) formula to a 22-year-old individual who will exercise at 60% of MHR

- Target heart rate = (220 − age) × %MHR
- Target heart rate = (220 − 22) × 0.60
  = 198 × 0.60
  = 119

Table 27.3

Target heart rates estimated at 60% and 75% of maximum for people at different ages using the percentage of maximal heart rate formula

Age (Years)	Average Maximal Heart Rate (MHR)	60% of MHR	75% of MHR
20	200	120	150
25	195	117	146
30	190	114	142
35	185	111	138
40	180	108	135
45	175	105	131
50	170	102	127
55	165	99	123
60	160	96	120
65	155	93	116
70	150	90	113

Photo Disc

Illustration 27.2 To find out if you're exercising at the right level, gently take your pulse during a break or right after you finish exercising.

Nutrition and Fitness

Your diet can make a strong contribution to your level of physical fitness. That's because good diets contribute to health, and it is easier to become physically fit if you are in good health. Diet also makes a difference to physical fitness because the foods you eat serve as the source of energy for physical activity. For some activities, the body needs glucose as its primary energy source. For others, the fuel of choice is fat.

Muscle Fuel

Muscles use fat, glucose, and amino acids for energy. The proportion of each that is used, as well as the amount, depends on the intensity of activity (Illustration 27.3). When we're inactive, fat supplies between 85 and 90% of the total amount of energy needed by muscles. The rest is provided by glucose (about 10%) and amino acids (5% at most).[16] Fat is also the primary source of fuel for activities of low-to-moderate intensity such as walking, running, and swimming. As can be seen from Illustration 27.3, amino acids make a relatively small contribution to meeting energy needs for physical activity.

Oxygen is required to convert fat into energy. That makes activities of low-to-moderate intensity "aerobic"—or "oxygen requiring." Consequently, fat-burning, oxygen-requiring exercises are the type used to increase aerobic fitness. Jogging, long-distance running, swimming, and aerobic dance are all aerobic activities.

High-intensity, short-duration activities like spiking a volleyball, running down a deep fly ball, or sprinting down the block to catch a bus are fueled primarily by glucose. Our supply of glucose for intense activities comes principally from glycogen, the storage form of glucose. Glycogen is stored in muscles and the

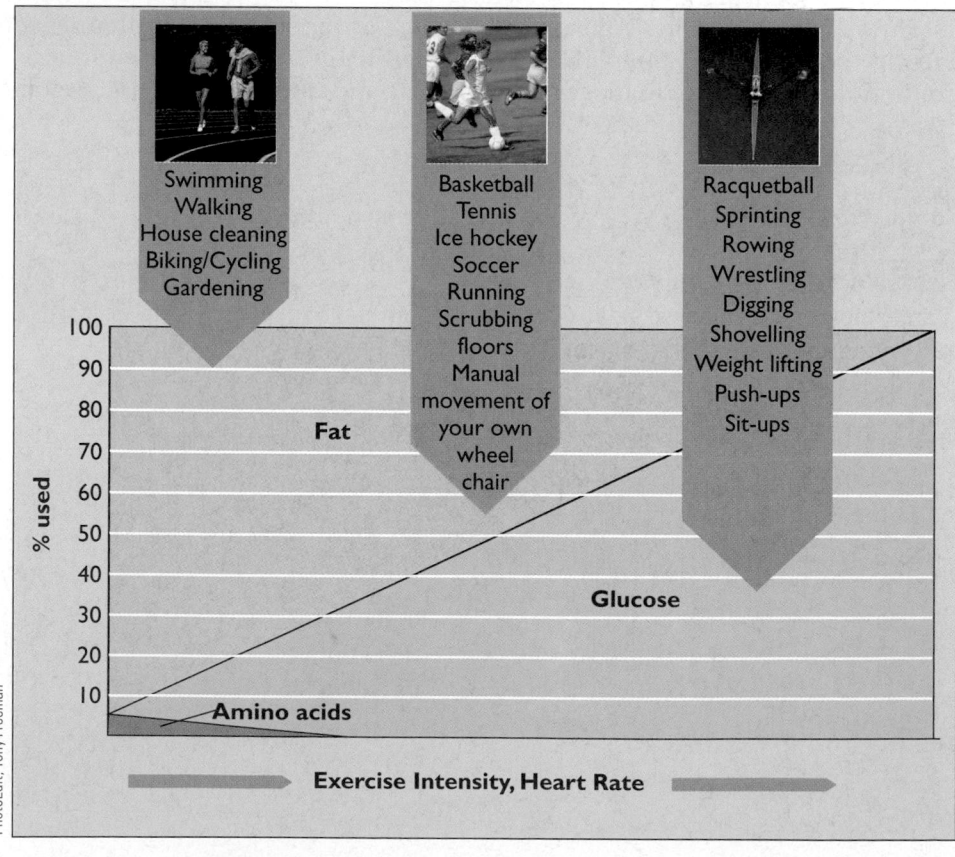

Illustration 27.3 Schematic representation showing the proportionate use of sources of energy for physical activities of various intensities.
Most activities are fueled by both fat and glucose.

liver and can be rapidly converted to glucose when needed by working muscles. The conversion of glucose to energy for intense activity doesn't require oxygen (it's *an*aerobic). People can undertake very intensive activity only as long as their stores of glucose last.[17]

Activities such as basketball, hockey, tennis, and soccer that involve walking, running, sprinting, and the occasional high-intensity, quick move use mainly fat or glucose as the energy source, depending on the intensity of the physical effort.

Stiff Muscles in the Out of Shape   People who are out of shape and exercise intensely, or who change the type of exercise performed, may experience stiff muscles 24 to 48 hours after exercise (Illustration 27.4). Although the cause is not clear, the stiffness usually lasts a day or two. The probability of this happening decreases as people get into shape.

Diet and Aerobic Fitness

The best type of diet for physical fitness is the same one that is recommended for people in general. It includes a variety of vegetables and fruits, whole-grain products, lean meats and fish, and low-fat dairy products. Healthy diets promote physical fitness because they facilitate maintenance of normal weight; reduce plaque buildup in arteries; and supply adequate amounts of vitamins, minerals, essential fatty acids, fiber, and other substances that keep the body running like a well-tuned machine. Diets that promote physical fitness most certainly include well-planned vegetarian diets.[18]

Actually, people who exercise tend to have better diets than people who don't. They are often health conscious, select foods with care and, because they generally eat more than sedentary people, have a better chance of getting all the nutrients they need.[19]

A Reminder about Water

Physical activity increases the body's need for water, and if the climate is hot and humid, this need increases even more. In general, people should drink in response to thirst and, overall, consume enough water to replace the amount lost in sweat, respiration, and urine during exercise.[20] (The amount of water lost during exercise is equivalent to the amount of weight that is lost during the exercise.) You are consuming the right amount of water if your urine is pale yellow and

Illustration 27.4   She's going to feel this tomorrow. People may get stiff muscles if they exercise too hard when they're out of shape.

To drink or not to drink?   Drinking water during exercise upsets your stomach and slows you down. That's what some people say. Are they right?

Who gets thumbs up?
Answers on next page

Tex:
When I played basketball in school, the coach would never let us drink until half-time. Now I know why he did that.

Rose:
I get thirsty when I exercise, so I drink water. It never gave me a stomachache.

ANSWERS TO
Reality Check

To drink or not to drink? Rose is onto something. Thirst in healthy people signals the need for water. Drinking enough water or other dilute fluids during exercise keeps you hydrated. Not drinking enough slows you down.[20]

Tex:

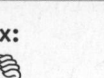

Rose:

People in Washington, DC, and Wyoming tend to be the most physically active, whereas physical activity levels are lowest in Kentucky and Louisiana.[22]

normal in volume. For exercise that lasts over an hour, consumption of a sports drink appears to improve hydration.[20] It's important to keep up fluid intake when exercising in cold weather. Cold air holds less water vapor than hot air, so you lose more water through breathing in cold weather.

A Personal Fitness Program

Commenting on physical activity for the long run, Sir Roger Bannister has said:

> The best advice to those who are unfit is to take exercise unobtrusively. I am not an enthusiast for fitness and exercise schemes that are boring because they tend to fizzle out.... The vast majority of people, I think, can only be attracted for any length of time towards recreational activities if these are rewarding, enjoyable, and satisfying in themselves.

Physical fitness is not something that, once achieved, lasts a lifetime. The beneficial effects of training diminish within 2 weeks after training stops and disappear altogether in 2 to 8 months.[21] That makes it critical to undertake physical activities that "wear well"—those that a person finds rewarding and will reserve a place for in a busy schedule. Too many well-intentioned attempts to get into shape have been abandoned because they were overly ambitious or no fun. As the guidelines for becoming physically fit are presented in the next section, keep firmly in mind that you should seek a realistic and achievable plan—one that feels right and will last because it's good to you and for you.

Becoming Physically Fit: What It Takes

Physical fitness results from regular physical activity that leads to muscle strength, endurance, and flexibility (Table 27.4). These components of physical fitness can be achieved by resistance training and aerobic exercise.

A Resistance Training Plan The resistance training program for physical fitness doesn't prepare people for bodybuilding contests. It increases muscle strength, flexibility, and endurance and improves the performance of aerobic exercises.

The American College of Sports Medicine recommends two to three resistance training sessions per week. Two sessions are enough for those who are not interested in "bulking up." Strength training twice weekly will give you 70 to 80% of the benefits of training three times a week.

Each workout session includes about 10 different exercises, repeated 8 to 12 times each. Each set of exercises should leave the muscles exercised and feeling a bit tired. Exercises selected should involve each of the major muscle groups in the legs and buttocks, arms and shoulders, and abdomen and back. Table 27.5

Table 27.4

An overview of the components of physical fitness training

A. Resistance training:
  - Goal: To improve muscle strength, flexibility, and endurance
  - Components: Two to three resistance training sessions per week
    → Each session includes 10 exercises repeated in sets of 8 to 12.
    → Exercises should involve all major muscle groups.

B. Aerobic training:
  - Goal: To increase the ability of the respiratory and circulatory systems to deliver oxygen to working muscles (to build "cardiorespiratory" fitness)
  - Components: Three exercise sessions per week (lasting 20 to 30 minutes each) at 60 to 75% of maximal heart rate
    → Aerobic exercise should be a continuous activity (not stop-and-go) and involve all or most major muscle groups.

C. Flexibility:
  - Goal: Increase range of motion of muscles and joints.
  - Components:
    → Strength training
    → Regular exercise
    → Gentle warm-up exercise
    → Gentle cool down exercise

(Top) Moving muscles against resistance for strength.
(Middle) Aerobic exercise for oxygen delivery.
(Bottom) Regular exercise for flexibility.

Photo Disc

Photo Disc

©Jason Horowitz/Corbis

lists examples of exercises for the different muscle groups. Muscle strength is increased by the use of progressively heavier weights in workouts.

The Aerobic Fitness Plan

Aerobic fitness can be achieved by exercising at 60 to 75% of maximal heart rate for at least 30 minutes most of the week.[23] The best kind of aerobic exercise is one that involves all, or most, of the body's major muscle groups in the workout. Beginners should start at 40 to 60% of maximum with low-intensity activities, such as alternating walking with jogging, and increase to a higher level as exercise gets easier. Exercises such as swimming, jogging, running, aerobic dance, and cycling are good choices for subsequent aerobic exercise. Activities such as mowing the lawn, shoveling dirt or snow, and household tasks that elevate heart rate to the target level for 30 minutes are also aerobic exercise. Actually, the most common physical activities of American adults are walking, gardening, and yard work.[21]

Some Exercise Is Better than None Physical activity levels that are lower than the recommended levels still benefit health.[22] Sedentary people who take up walking, dancing, gardening, biking, golfing, or similar exercises tend to experience improvement in aerobic fitness, strength, and energy level.[11] For most health outcomes, additional benefits occur as physical activities increase in intensity, duration, or frequency.

One way to improve levels of physical fitness without spending more time on exercising or exercising more frequently is to increase the intensity of physical activities performed.[23] In general, 15 minutes of vigorous activity provides the same benefits as 30 minutes of moderate activity.[11] Vigorous activity burns more calories per minute and more effectively reduces body fat (including abdominal fat) than do lower intensity activities.[24] The Take Action feature for this unit provides some choices for increasing exercise intensity.

Table 27.5

The match between resistance exercises and the body's major muscle groups

Muscle Group Exercised	Activity
Legs and buttocks	Stair climbing, resistance leg pushing, cycling
Arms and shoulders	Free weight lifts, pull-ups, push-ups
Abdomen and back	Abdominal curls, sit-ups, leg lifts (front and side)

I really don't think I need buns of steel. I'd be happy with buns of cinnamon.

—ELLEN DeGeneres, 2006

Here are some options for increasing exercise intensity. Do you see an option or two that would work for you?

I can do that / **No way**

1. Play "Beat your time." Trim seconds off the amount of time it takes you to walk a specific distance each time out for a month. Then maintain that level of walking intensity.

2. Gradually increase your cycling speed until you're safely traveling at least 1 mile in 6 minutes (10 mph).

3. Take the stairs two steps at a time. (Hold the handrail! Start with one floor of steps.)

4. Jog or run, rather than walk, whenever possible.

5. Carry light free weights with you when you walk or jog.

On the Side

What's the largest muscle in your body? If your body is like most, its largest muscle is the gluteus maximus (buttocks).

Photo Disc

You don't have to be a high-performance athlete to benefit from exercise. Do the things you enjoy—walk, ride a bike, garden, swim, play.

The Caloric Value of Exercise   Some people prefer to base their aerobic exercise workout on a particular level of calories (such as 200), while others simply like to know how many calories they're burning off. Table 27.6 lists the caloric value of various exercises in case you would like to know.

Population-Based Physical Activity Recommendations

Population-wide, physical activity recommendations aimed at improving weight and health status of the United State's population are presented in the 2008 Physical Activity Guidelines for Americans. This new set of physical activity recommendations urge Americans to undertake:

- at least 2 hours and 30 minutes each week of moderate intensity aerobic activity. The activity can be performed in bouts of 10 minutes each if desired. Moderate intensity activities are those that increase heart and

Photo Disc

Table 27.6

Average energy output per pound of body weight for selected types of exercise

Exercise	Intensity	Calories (pound/hour)	Exercise	Intensity	Calories (pound/hour)
Walking	3 mph (20 min/mi)	1.6	Weight lifting		2.9
	3½ mph (17 min/mi)	1.8	Wrestling		6.2
	4 mph (15 min/mi)	2.7	Handball	Moderate	4.8
	4½ mph (13 min/mi)	2.9		Vigorous	6.2
Jogging	5 mph (12 min/mi)	4.1	Swimming	Resting strokes	1.4
	5½ mph (11 min/mi)	4.5		20 yd/min (mod.)	2.9
	6 mph (10 min/mi)	4.9		40 yd/min (vig.)	4.8
	6½ mph (9 min/mi)	5.2	Rowing (sculling		
	7 mph (8½ min/mi)	5.6	or machine)		4.8
	7½ mph (8 min/mi)	6.0			
Running	8 mph (7½ min/mi)	6.3	Downhill skiing		3.8
	8½ mph (7 min/mi)	6.7	Cross-country skiing	4 mph (15 min/mi)	4.3
	9 mph (6⅔ min/mi)	7.1	(level)	6 mph (10 min/mi)	5.7
	9½ mph (6⅓ min/mi)	7.4		8 mph (7½ min/mi)	6.7
	10 mph (6 min/mi)	7.8		10 mph (6 min/mi)	7.6
	11 mph (5½ min/mi)	8.5	Aerobic dancing	Moderate	3.4
	12 mph (5 min/mi)	9.5		Vigorous	4.3
Rebound trampoline	50–60 steps/min	4.1	Racquetball/squash	Moderate	4.3
				Vigorous	4.8
Cycling (stationary)	Mild effort	2.9	Tennis	Moderate	3.4
	Moderate effort	3.4		Vigorous	4.3
	Vigorous effort	4.3	Volleyball	Moderate	3.4
Cycling (level)	6 mph (10 min/mi)	1.5		Vigorous	3.8
	8 mph (7½ min/mi)	1.8	Basketball	Moderate	3.8
	10 mph (6 min/mi)	2.0		Vigorous	4.8
	12 mph (5 min/mi)	2.8			
	15 mph (4 min/mi)	3.9			
	20 mph (3 min/mi)	5.7	Football	Moderate	3.8
				Vigorous	4.3
Skating		2.9			
Calisthenics	Moderate	2.4	Baseball/golf/woodcutting/horseback riding/ badminton/canoeing		2.4
	Vigorous	2.9			
Rope skipping	Moderate	4.8	Soccer/hill climbing/fencing/judo/snowshoeing		5.3
Bench stepping	12″ high, 24 steps/min	3.2	Bowling/archery/pool		1.2

Source: Values calculated from Guidelines for graded exercise testing and exercise prescription, 2d ed. Philadelphia: American College of Sports Medicine, Lea & Febiger, 1984.

breathing rates and cause you to sweat a bit. They include brisk walking and jogging, scrubbing floors, operating a wheel chair manually or playing a sport from a wheel chair, running around with children, fast dancing, tennis (doubles), and basketball.

- at least two sessions of strengthening activities per week. Strengthening activities should include push-ups, sit-ups, weight lifting, and other resistance exercises. Exercises should work legs, hips, back, chest, stomach, shoulders, and arms. Each exercise should be repeated 8 to 12 times per session.[11] Higher weekly levels of moderate-intensity physical activity (60 minutes or more daily) and of vigorous-intensity activities promote weight loss and the prevention of weight regain.[25] Vigorous-intensity activities should be added to exercise programs a little at a time in order to reduce the risk of injuries.[11]

A Cautionary Note People who have a chronic disease such as diabetes or hypertension, or are out of shape and over 40, should consult a physician before starting an exercise program. (It may well turn out that exercise will be just what the doctor orders.) Moderate exercise programs (those undertaken at 40 to 60% of VO_2 max) are beneficial and safe for almost everyone.[21]

■ **The Paradox of Death during Exercise.** "Famous athlete dies while jogging" reads the headline. Why would this happen if exercise is good for you? That's the exercise paradox.

Most people who die while shoveling snow, playing basketball, or jogging are those with existing heart disease either due to hardened arteries (atherosclerosis) or an inborn heart problem. It is a very rare event (because exercise really is good for the heart), and the risk can be reduced by regular exercise versus sporadic and intense physical activity.[27]

U.S. Fitness: America Needs to Shape Up

The benefits of regular exercise are sufficient enough to make it a matter of public health policy in the United States. Americans have much to gain by getting in better shape.

Studies are undertaken periodically to see how well Americans are doing at increasing their physical activity level. Not all of the results are encouraging (Illustration 27.5). Approximately 47% of adults do not achieve recommended levels of physical activity, 40% are inactive and only 20% participate in strength training twice weekly.[26] The encouraging news is that recent data indicate

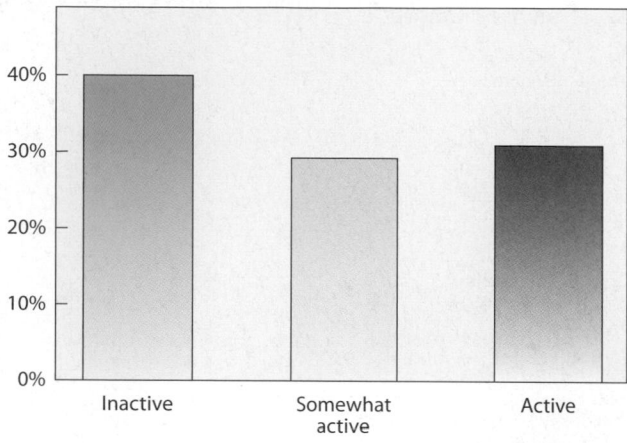

Illustration 27.5 Percent of U.S. who adults are inactive, somewhat active, and active in 2006.[26]

Photo Disc

that rates of physical activity are beginning to increase.[26] If maintained, this trend would help lower rates of overweight, obesity, type 2 diabetes, and other disorders.[28]

A Focus on Fitness in Children

Physical fitness advocates are especially concerned about the lack of opportunity for physical activity in schools. They point out that children who are not physically active are more likely to be overweight and that a lifelong love for physical activity and a physically active lifestyle often develop during childhood. Regular exercise benefits obese children by lowering body fat, increasing bone density, and improving aerobic fitness. Fewer than one in ten public schools require students to participate in physical activity daily—and, on average, children undertake vigorous physical activity for only 25 minutes per week.[29]

Schools are being encouraged to include physical activity as a daily part of the curriculum for all students in all grades. It is recommended that children and adolescents engage in moderate and vigorous-intensity physical activities for an hour or more each day.[11] Physical fitness goals should be stressed above competitive or sport performance goals.[30] Tests of physical fitness, such as the muscle strength and endurance assessment shown in Illustration 27.6, should be performed periodically on all children and the results used to modify instruction and activities.

Physical activity programs for schoolchildren are most likely to be successful when they offer recreational and sporting activities that are irresistible and fulfill the human desire to run, jump, and play. The happy consequences of physical activity come easiest to those who do it for play.

Illustration 27.6 This child is performing modified pull-ups as part of a fitness assessment.

Source: Greg Merhar, reprinted with permission of the Journal of Physical Education, Recreation and Dance, 59 (9), 1987.

© Michael Blann / Getty Images / Riser

Up Close

Exercise: Your Options

Focal Point: Assess your level of physical activity.

About 40% of U.S. adults are physically inactive, and most Americans do not exercise regularly. How much physical activity are you getting?

Physical Activity

1. Check the category that best describes your usual overall daily activity level. Do not consider activities that are part of an exercise program.

- *Inactive:* Sitting most of the day, with less than 2 hours moving about slowly or standing. ☐

- *Average:* Sitting most of the day, walking or standing 2 to 4 hours each day, but not engaging in strenuous activity. ☐

- *Active:* Physically active 4 or more hours each day. Little sitting or standing, engaging in some physically strenuous activities. ☐

Exercise

2. Do you exercise at moderate intensity a total of 150 minutes per week? ☐ Yes ☐ No

3. Do you preform strength-building exercises twice a week? ☐ Yes ☐ No

FEEDBACK (including scoring) can be found at the end of Unit 27.

[Key Terms

aerobic fitness, page 27-4   maximal oxygen consumption, page 27-4   physical fitness, page 27-3

[Review Questions

	TRUE	FALSE
1. Regular physical activity benefits psychological and physical health.	☐	☐
2. The primary components of physical fitness are body fat composition, strength, and heart rate.	☐	☐

	TRUE	FALSE
3. Aerobic fitness is related to a person's ability to deliver oxygen to muscles and the capacity of muscles to use the oxygen for physical activity.	☐	☐
4. Endurance is related to aerobic fitness.	☐	☐

5. Maximal oxygen consumption is related to endurance. ☐ ☐

6. A 44-year-old individual who wants to exercise at 70% of maximal heart rate would aim for a target heart rate of 123 beats per minute. ☐ ☐

7. Glycogen supplies the fuel to meet most of our energy needs while we are at rest. ☐ ☐

8. Fat is the major fuel for low-to-moderate intensity physical activities. ☐ ☐

9. Resistance training primarily benefits endurance. ☐ ☐

10. To build and maintain a state of physical fitness it is recommended that adults engage in vigorous physical activity for 60 minutes seven days each week. ☐ ☐

11. It is recommended that children and adolescents engage in light-intensity physical activity for about 60 minutes daily. ☐ ☐

[Media Menu

www.health.gov/paguidelines
Welcome to the 2008 Physical Activity Guidelines for Americans, the first such set of guidelines to be published by the federal government. The site contains information and tools for increasing physical activity based on individual preferences. Get the "Keeping track of what you do each week" physical activity recording form here.

www.mayoclinic.com/health/weight-training/SM00041&slide=1
Go to this site to view a how-to slide show on weight training exercises for major muscle groups. The show is produced by Mayo Clinic staff. Also access how-to information on fitness ball exercises and other strength training techniques.

http://www.healthfinder.gov/getactive
This site provides help for meeting the physical activity recommendations of 2 hours and 30 minutes per week of aerobic activity and twice weekly strengthening activities.

win.niddk.nih.gov/publications/active.htm
The National Institutes of Health offers a booklet on fun and rewarding physical activities for large people.

www.shapeup.org/shape/steps.php
Shape Up America! is a not-profit organization that offers responsible information on healthy weight management and fitness. You will find a description of the 10,000 Steps Program, fitness tips, a special section on physical activity for children, and a resource center.

www.healthfinder.gov/getactive
Presents practical advice and resources for becoming physically active and for enhancing your physical activity levels.

mypyramidtracker.gov
This My Pyramid link can be used to estimate calorie expenditure for physical activities and help you plan your physical fitness program.

www.adultfitnesstest.org
Take the President's Challenge, Adult Fitness Test here. Complete a one-page questionnaire, submit it, and receive an estimate of your level of aerobic fitness, muscular strength and endurance, flexibility, and body composition. Use the test results to measure your progress toward achieving physical fitness.

www.mypyramidtracker.gov
Click on this link, enter some basic data about yourself and your physical activities, and receive a score and a report of the number of calories expended for the activities.

www.healthierus.gov/exercise.html
Scroll down this page and find an extensive list of Web sites related to physical fitness, physical activity, and health.

www.nlm.nih.gov/medlineplus/exerciseforseniors.html
This site provides pictures and diagrams of exercises that improve balance as well as additional information on exercise for seniors.

www.nlm.nih.gov/medlineplus/exercisephysicalfitness.html
This site covers what's new, exercise recommendations, fitness tips, and public health policies related to physical activity.

www.kidnetic.com
This popular, highly interactive Web site for children focuses on healthy eating and regular physical activity.

[Notes

1. Dwyer JD et al. Leisure time physical activity and atherosclerosis. Am J Med 2003;115:19–25.

2. McCaffree J. Physical activity: how much is enough? J Am Diet Assoc 2003; 103:153–4.

3. Kivipelto M et al. Midlife physical activity and rates of dementia and Alzheimer disease. Lancet Neurol,

10/4/2005, www.medscape.com/viewarticle/514204.

4. Bartholomew JB et al. Brief aerobic exercise may improve mood, well-being in major depression. Med Sci Sports Exerc 2005;37:2032–7.

5. Roux L et al. The cost-effectiveness of community-based interventions aimed at increasing physical activity, Am J Prev

Med, Dec. 2008, www.medscape.com/viewarticle/584921, accessed 12/08.

6. Williams PT. Reduced diabetic, hypertensive, and cholesterol medication use with walking, Med Sci Sports Exerc 2008;40:433–43.

7. Risk factors and coronary heart disease. AHA Scientific Position. www.medscape.com, accessed 7/03.

8. Gorden-Larsen P et al. Fifteen-year longitudinal trends in walking patterns and their impact on weight change, Am J Clin Nutr 2009;89:19–26.

9. Broeder C et al., The effects of aerobic fitness on resting metabolic rate, Am J Clin Nutr 1992;55:795–801.

10. Nicklas BJ et al. Effect of exercise intensity on abdominal fat loss during calorie restriction in overweight and obese postmenopausal women: a randomized, controlled trial, Am J Clin Nutr 2009;89:1043–52.

11. Physical Activity Guidelines for Americans, 2008, U.S. Department of Health & Human Service, http://www.health.gov/PAGuidelines, accessed 10/08.

12. Bouchard C et al. Heredity and trainability of aerobic and anaerobic performances. Sports Med 1988;5:69–73.

13. Gromley SE et al. Effect of intensity of aerobic training on VO2max, Med Sci Sports Exerc 2008;40:1336–43.

14. Behm DG et al. Stretching, balance, and movement times. Med Sci Sports Exerc 2004;36:1397–1402.

15. Hart L. Effect of stretching on sport injury risk: a review. Clin J Sport Med. 2005 Mar;15:113.

16. Horton ES. Metabolic fuels, utilization, and exercise. Am J Clin Nutr 1989; 49(suppl):931–2.

17. Jones NL et al. Exercise limitations in health and disease, N Engl J Med 2000;343:632–41.

18. Barr Sl et al. Nutritional considerations for vegetarian athletes. Nutrition 2004;20:696–703.

19. Nieman DC et al. Nutrient intake of marathon runners. J Am Diet Assoc 1989;89:1273–8.

20. Position of the American Dietetic Association, Dietitians of Canada, and the American College of Sports Medicine: Nutrition and Physical Performance, J Am Diet Assoc 2009;109:509–27.

22. Most US adults not meeting physical activity recommendations. MMWR 2003;53:764–8.

23. Lie D. Moderate intensity exercise effective for weight loss. www.medscape.com, accessed 10/03.

24. Irving BA et al. Effect of exercise training intensity in abdominal visceral fat and body composition. Med Sci Sports Exer 2008;49:1863–72.

25. Phelan S et al. Empirical evaluation of physical activity recommendations for weight control in women, Med Sci Sports Exerc 2007;39:1832–6.

26. Health United States 2008, www.cdc.gov/fastats, accessed 5/09.

27. Foster C et al. The risk in exercise training, Am J Lifestyle Med 2008;2:279–84.

28. Trends in leisure-time physical inactivity by age, sex, and race/ethnicity—United States, 1994–2004. MMWR 2005;54:991–4.

29. School study seeks to promote effective interventions, Chronic Disease News and Notes 2002;15:9–10.

30. Physical activity for everyone: recommendations. www.cdc.gov/nccdphp/dnpa/physical/recommendations/index.htm, accessed 10/9/06.

Answers to Review Questions

1. True, see Table 27.1, page 27-3.
2. False, see page 27-4.
3. True, see page 27-4.
4. True, see page 27-4.
5. True, see page 27-4.
6. True, see page 27-5.
7. False, see pages 27-6, 27-7.
8. True, see pages 27-6, 27-7.
9. False, see pages 27-4, 27-8.
10. False, see pages 27-10, 27-11.
11. False, see page 27-13.

NUTRITION | # Up Close

Exercise: Your Options

Feedback for Unit 27

For those of you who answered "Active" to the first question and "Yes" to the second and third questions, congratulations! You qualify as a mover and shaker. Keep up the good work!

If you answered "Inactive" or "Average" to the first question or "No" to the second or third questions, now is a good time to think about increasing your level of physical fitness. Besides promoting a feeling of well-being, regular exercise can help beat stress, weight gain, heart disease, and high blood pressure. Maybe it's time you exercised your options by incorporating more physical activity into your daily routine.

NUTRITION SCOREBOARD

	TRUE	FALSE
1 The diet an athlete consumes affects energy substrate availability to muscles during exercise.		
2 Carbohydrate loading is a waste of time.		
3 Supplementation with protein and amino acids increases muscle strength to about the same extent as does resistance training.		
4 Scientists can't explain why, but bee pollen actually improves physical performance.		

Key Concepts and Facts

- Endurance is affected by genetics, training, and nutrition.

- Glycogen stores and endurance can be increased by diet and training.

- Abnormal or absent menstrual cycles in female athletes should not be dismissed but corrected with increased caloric intake.

- Glucose and fat are used as energy sources during exercise but in different proportions. Intense physical activity is primarily fueled by glucose, and fats are the main source of energy for low- to moderate-intensity exercise.

- A few ergogenic aids that claim to improve performance work to some extent, but most do not.

Answers to
NUTRITION SCOREBOARD

		TRUE	FALSE
1	Carbohydrate and fat availability to muscles during exercise varies based on usual diet.	✔	
2	Carbohydrate loading increases endurance in trained athletes.[1]		✔
3	Resistance exercise increases muscle strength. Protein and amino acids supplements have not been shown to improve strength in healthy athletes.[1]		✔
4	In bees, maybe—but not in humans.[1]		✔

Sports Nutrition

BAM! Nobody heard it, but Lou felt it. She had "hit the wall." She was ahead of her planned pace, but now her legs felt like lead. She would have to finish the last two miles of the marathon at the slow pace her legs would allow.

From her carefully crafted and scrupulously followed training program to her refined shaping of mental attitude, Lou thought she had done everything right. She had left one thing out, however, and that may have cost her the race. Lou failed to pay attention to her diet while training and ran out of **glycogen** too soon (Illustration 28.1).

Three major factors affect physical performance: genetics, training, and nutrition.[2] The first gives some people an innate edge in sprinting or endurance, and nothing can be done about it. The second is acknowledged as a basic truth. Most athletes know a good bit about proper training, and the trick is to follow the right plan. The third is often ignored or, when taken seriously, misunderstood.

Nutrition has important effects on physical performance, but the legitimate role of nutrition is often poorly understood by athletes and coaches alike.[3] Athletes' lack of knowledge may make them vulnerable to phony nutrition claims and bad dietary advice. It leaves some coaches wishing they had taken a college course on nutrition. Only one in ten coaches has had a course in nutrition, but nearly all of them regularly dispense nutritional advice.[4] (To the coaches and future coaches who are taking this nutrition course: Terrific!)

Basic Components of Energy Formation during Exercise

Understanding the role of nutrition and performance, and the potential effects of some **ergogenic aids** can be fostered by basic knowledge of how energy is formed within muscle cells. Illustration 28.2 summarizes the processes by which energy for muscle movement is formed.

glycogen
The storage form of glucose. Glycogen is stored in muscles and the liver.

ergogenic aids
(ergo = work; genic = producing)
Substances that increase the capacity for muscular work.

There are two main substrates for energy formation in muscles: glucose from muscle and liver glycogen, and fatty acids released from fat stores. How much of each is used depends on the intensity and duration of the exercise, as well as the body's ability to deliver each along with oxygen to muscle cells. Each substrate is used to form **ATP** from **ADP**. ATP serves as the source of energy for muscle contraction.

Anaerobic Energy Formation   Glucose from the liver and muscle glycogen (which is converted to glucose) form ATP without oxygen. This route of energy formation is "anaerobic," or "without oxygen," and it generates most of the energy used for intense muscular work (70% VO_2 max or higher).[1] Creatine phosphate (abbreviated CrP in Illustration 28.2), an amino acid containing a high-energy phosphate molecule, also converts ADP to ATP to a limited extent. Creatine phosphate stores are limited and decrease rapidly during intensive exercise.

Glucose is converted to "pyruvate" during anaerobic energy formation. In the absence of oxygen, pyruvate is converted to lactate. Lactate can build up in muscles and blood if not reconverted to pyruvate by the addition of oxygen. Pyruvate yields additional energy when it enters "aerobic" energy formation pathways along with fatty acids from fat stores.

Aerobic Energy Formation   The conversion of pyruvate and fatty acids to ATP requires oxygen. Much more ATP is delivered by the breakdown of fatty acids than glucose (fats have 9 calories per gram; glucose has only 4). The rate of energy formation from fatty acids is four times slower than that from glucose, however. It's the reason fatty acids are used to fuel low- and moderate-intensity exercise, or those below 60% VO_2 max.[4] Unlike glucose, energy formation from fatty acids is not limited by availability. Muscle cells can continue to produce energy from fatty acids as long as delivery of oxygen from the lungs and the circulation is sufficient.[5]

Illustration 28.1   Hitting the wall. This runner ran out of muscle glycogen before the finish line.

ATP, ADP

Adenosine triphosphate *(ah-den-o-scene tri-phos-fate)* and adenosine diphosphate. Molecules containing a form of phosphorous that can trap energy obtained from the macronutrients. ADP becomes ATP when it traps energy and returns to being ADP when it releases energy for muscular and other work.

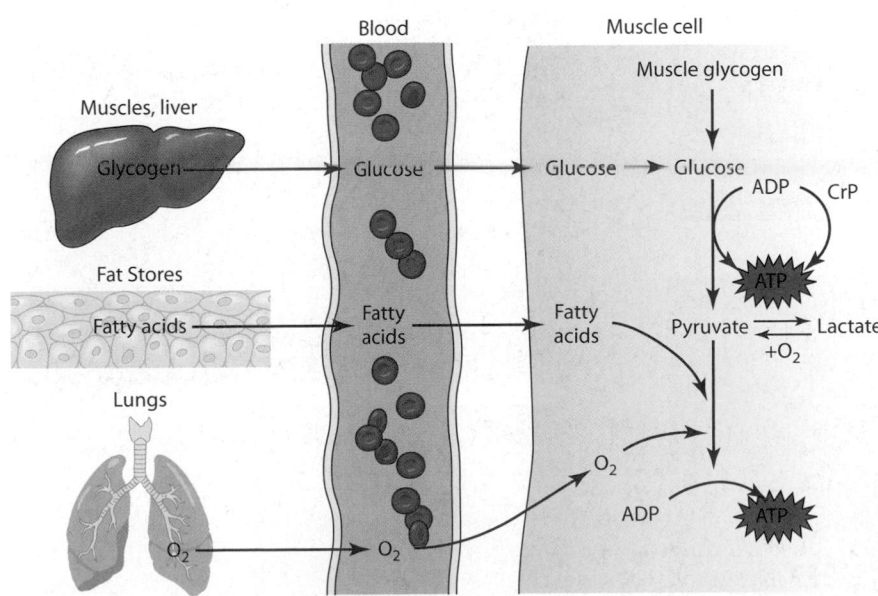

Illustration 28.2   Schematic representation of how ATP is formed for muscular movement.

Nutrition and Physical Performance

Glycogen stores, foods eaten before, during, and after exercise events, fluid intake, and **electrolyte** balance are all related to energy formation and physical performance.

When everything else is equal, nutrition can make the difference between winning and losing.
—R. MAUGHAN, "THE ATHLETE'S DIET"[6]

electrolytes
Minerals such as sodium and potassium that carry a charge when in solution. Many electrolytes help the body maintain an appropriate amount of fluid.

Glycogen Stores and Performance

The vast majority of energy for muscle activity comes from fat and glucose. For most activities, both types of fuel are used; and the more intense the activity, the higher the amount of fat and glucose used for it. Fats, however, are the principal source of fuel for activities of low to moderate intensity, and glucose is the main source of energy for activities of high intensity (Illustration 28.3).[1]

Glycogen stores in muscles and the liver can deliver about 2000 calories worth of energy, whereas adults have access to over 100,000 calories from fat. Consequently, a person's ability to perform continuous, intense physical activity is limited by the amount of glycogen stored.[7] People who run out of muscle glycogen during an event "hit the wall"—they have to slow down their pace substantially because they can no longer use muscle glycogen as a fuel. The pace they are able to maintain will be dictated by the body's ability to use fat and liver glycogen as fuel for muscular work. If athletes keep pushing themselves after muscle glycogen runs out, they may end up "bonking"—using up the liver's supply of glycogen. That's worse than hitting the wall because hypoglycemia (low blood sugar) develops, and the person becomes dizzy and shaky and may pass out.[6] Obviously, endurance athletes do not want to exhaust their glycogen stores before an event is over.

Athletes consuming a typical U.S. diet normally have enough glycogen stores to fuel continuous, intense exercise for about an hour or two.[8] Both glycogen stores and endurance can be increased, however, by loading up muscles with glycogen prior to an event (Table 28.1).

Carbohydrate Loading Carbohydrate loading benefits performance in athletes who participate in endurance events.[1] Such events include activities like running,

Illustration 28.3 Fat is the main source of energy for low- and moderate-intensity activities, whereas glycogen is the primary fuel for high-intensity events.

Activities primarily fueled by fat.

Activities primarily fueled by glycogen.

Table 28.1

Effects of diet on muscle glycogen stores during training[9]

Carbohydrate Intake Level for Previous 3 Days	Muscle Glycogen (Grams per 100 Grams of Muscle)	Time to Exhaustion at 75% Vo_2 Max
Low	0.6	60 minutes
Average	1.8	126 minutes
High	3.5	189 minutes
(65–70% of total calories)		

Table 28.2

How many grams of carbohydrate are in a diet that provides 60 or 70% of total calories from carbohydrates? It depends on your caloric intake.

Usual Caloric Intake	Grams of Carbohydrate	
	60% Carb Diet	70% Carb Diet
2200	330 g	385 g
2400	360 g	420 g
2600	390 g	455 g
2800	420 g	490 g
3000	450 g	525 g
3200	480 g	560 g
3400	510 g	595 g
3600	540 g	630 g
3800	570 g	665 g
4000	600 g	700 g
4200[a]	630 g	735 g

[a]For each additional 200 calories of usual intake, add 30 grams carbohydrate for a 60% carbohydrate diet and 35 grams for a 70% carbohydrate diet.

cross-country skiing, swimming, or cycling and last over 90 minutes. It does not improve performance in high-intensity, short-duration events such as sprints, short races, weight lifting, high jump, and baseball.[10] These activities don't require unusually large glycogen stores.

The currently recommended method for carbohydrate loading is much less stringent than in the past and effectively increases muscle glycogen stores. It calls for increasing carbohydrate intake to 60 to 70% of calories for one-to 4 days prior to an endurance event. Glycogen stores accumulated are maintained by the tapering-off of exercise levels during the period of carbohydrate loading.[11] Table 28.2 shows how many grams of carbohydrate correspond to 60 or 70% of total calories for people with different caloric needs. Average diets provide about 50% of calories from carbohydrates. See Illustration 28.4 for an example of a one-day diet that provides 68% of total calories from carbohydrate.

■ **Carbohydrate-Loading Caveats.** There is a limit on glycogen storage in muscle. Each gram of glycogen binds with 2 grams of water, limiting the total amount that can "fit" into muscle tissue. Once the capacity of muscles to store glycogen is reached, any additional carbohydrate is largely converted to fat and stored.[12]

Not all athletes tolerate the relatively high-carbohydrate, low-fat, carbohydrate-loading diet equally well. Some athletes dislike the stiff, heavy feeling that may occur when muscles become "loaded" with glycogen.[11] In addition, it should be remembered that carbohydrate loading does not improve performance in athletes participating in physical activities that last less than an hour or two.

Illustration 28.4 A high-carbohydrate diet. Carbohydrates provide 68% of the 2420 calories in this day's menus, protein provides 11%, and fats provide 21%. Grams of carbohydrate are shown in parentheses.

BREAKFAST
Orange juice _____ l cup (25 g)
Oatmeal _____ l cup (25 g)
Brown sugar _____ l tablespoon (13 g)
Skim milk _____ l cup (12 g)
Banana _____ l (27 g)

SNACK
Granola _____ 1/2 cup (29 g)
Apple juice _____ l cup (30 g)

LUNCH
Tossed salad _____ 2 cups (10 g)
Salad dressing _____ 2 tablespoons (4 g)
Macaroni & cheese _____ l cup (40 g)
Crackers _____ 4 squares (9 g)
Sliced tomato _____ 1/2 (3 g)
Skim milk _____ l cup (12 g)

SNACK
Cornflakes _____ l cup (20 g)
Canned peaches _____ 1/2 cup (25 g)
Skim milk _____ l cup (12 g)

DINNER
Spanish rice _____ l cup (41 g)
Baked potato _____ l (35 g)
Corn _____ 1/2 cup (17 g)
Apple crisp _____ 1/2 cup (43 g)
Iced tea _____ 2 cups (0 g)

On the Side

Protein Need

Many athletes require no more than their RDA of protein. Individuals undertaking strength or endurance training, however, may need 20 to 40 grams of additional protein daily to support muscle cell growth or repair.[1] For most athletes, this additional amount of protein will already be included in the diet. On average, young adult females in the United States consume 75 grams of protein daily (the RDA is 46 grams) and males 111 grams (versus the RDA of 56 grams).[13] Protein intake generally increases with calorie intake, so athletes tend to have higher protein intakes than non-athletes due to their higher calorie intake levels. Diets providing up to 35% of total calories from protein are compatible with health, but may provide too little carbohydrate and lead to early fatigue in athletes.[3]

■ **Protein and Amino Acid Supplements.** Food sources of high quality protein support muscle growth during strength training, but do not appear to increase muscle strength or performance.[14] Supplemental protein and amino acids marketed to athletes as muscle builders have not been shown to increase muscle strength or performance, either.[15]

Athletes interested in supplementing their diets with protein or amino acids should use high quality protein foods. You can get, for example, higher amounts of the popular branch chain amino acids from a chicken breast than you can from seven branch chain amino acid tablets.[2] A cup of peanuts supplies 11 tablets' worth of these amino acids.

Pre-Event, Event, and Recovery Foods Recommendations about what athletes should eat and drink before, during, and after events in the past were largely based on untested assumptions. As more research has been done, previous recommendations are being replaced by those backed by research results on nutrition and performance. The evidence shows that consuming the right types and amounts of foods and fluids before, during, and after an endurance event benefits performance. Illustration 28.5 gives examples of foods and fluids that are good choices for pre-event, event and post-event recovery meals and snacks. Athletes are encouraged to experiment with foods before using them during an event.

■ **Pre-Event Foods and Fluids.** Foods and fluids consumed 3 to 4 hours prior to an endurance event should be high in carbohydrates (200 to 300 grams) to help fuel the up-coming exercise, low in fat and fiber to prevent digestive upsets, moderate in protein, and provide about two cups of water.[1] Food sources of carbohydrate that release glucose slowly into the blood stream (or low glycemic index foods) are generally preferred over high glycemic foods.[16,17] Honey, oatmeal, corn, and bananas are examples of low glycemic index foods, while white bread, pretzels, cornflakes, and Gatorade have high glycemic index values.

■ **Event Foods and Fluids.** Endurance athletes should consume enough fluids during the event to replace that loss in sweat. (Additional information on

Illustration 28.5
Examples of pre-event, event, and recovery foods and fluids

PhotoDisc

Pre-event	Event	Post-Event	
Oatmeal	Bananas, other fruits	Non-fat milk	Bananas
Honey or brown sugar	Energy bars	Non-fat chocolate	Bagel
Muffin	Sports gels	milk	Protein bars
Unsweetened fruit juice	Sports drinks	Water	Meat sandwich
Non-fat milk, coffee	Water	Fruit juice	Crackers and low-fat
2 cups water		Sports drink	cheese
		Eggs	Soup with pasta
		Soy products	or rice, vegetables,
		Low-fat yogurt	meat or beans

hydration is presented later in this unit.) Adequate fluid and electrolyte consumption during events prevents dehydration and muscle fatigue, and helps prevent muscle cramps. Sports drinks containing 6 to 8% carbohydrate and sodium should be consumed to replace some of the fluid loss.[1] Such drinks lead to faster and more sustained rehydration than water alone.[18]

Carbohydrate should be consumed at the rate of 30 to 60 grams per hour after the first hour of the event. It can be obtained from solid foods or beverage, depending on the preferences of the athlete. Most sports drinks contain 15–25 grams of carbohydrate per cup, energy bars provide about 45 grams per serving, and sports gels 17 to 42 grams per packet. Both energy bars and sports gels should be consumed with about 4 ounces of water. Bananas and other fruits can also be used as supplemental sources of carbohydrate. The Reality Check feature for this unit addresses the use of foods and fluids containing high fructose corn syrup as a carbohydrate fuel for athletes.

Protein needs can be met by consuming 20 grams of high quality protein from low-fat dairy products, lean meats, soy products, and food combinations such as grains and legumes. Protein intakes above 20 grams do not benefit muscle recovery because the excess protein is used for energy or converted to fat.[19]

■ **Post-Event Foods and Fluids.** Food sources of carbohydrate and high quality protein, and fluids should be consumed within 30 minutes after completion of an endurance event. Carbohydrates help rebuild muscle glycogen stores; protein increases protein synthesis and protein tissue repair in muscle cells.[20] Depending on weight, athletes should consume 100 to 200 grams of carbohydrate and up to 20 grams of protein after events. Enough water and other fluids should be consumed to bring body weight back up to its pre-event level.[1]

Hydration

Hydration status is a major factor affecting physical performance and health. Adequate hydration during training and competition enhances performance, prevents excessive body temperatures, delays fatigue, and helps prevent injuries. Dehydration has opposite effects on physical performance and health. Hyperhydration, caused by overdrinking water before, during, and after endurance events, can lead to a loss of sodium from blood and body tissues and **hyponatremia.**[21]

hydration status
The state of the adequacy of fluid in the body tissues. Dehydration indicates the presence of inadequate fluid in body tissues, and hyperhydration means there is too much.

hyponatremia
A deficiency of sodium in the blood (135 mmol/L sodium or less).

Hydration status during exercise is primarily affected by how much a person sweats—or by how much water he or she loses through the skin while exercising. Muscular activity produces heat that must be eliminated to prevent the body from becoming overheated. To keep the body cool internally, the heat produced by muscles is collected in the blood and then released both through blood that circulates near the surface of the skin and in sweat. Sweating cools the body because heat is released when water evaporates on the skin. People sweat more during physical activity in hot, humid weather because it is harder to add moisture and heat to warm, moist air than to dry, cool air. To stay cool during exercise in hot, humid conditions, the body must release more heat and water than when exercise is undertaken in drier, cooler conditions.[18]

Maintaining Hydration Status during Exercise   It is recommended that athletes engaged in events that last an hour or less drink about two cups of water one to two hours before an event and continue to consume water at regular intervals during the event. Athletes undertaking long events (over one hour in duration) should consume beverages that provide sodium (160 to 200 mg sodium per 8 ounces) and carbohydrate (6 to 8% by weight) during the exercise.[22] The sodium is needed to replace that lost in sweat and to prevent hyponatremia, and the carbohydrate helps maintain blood glucose levels and provides muscles with glucose. Hydration status is maintained when athletes do not lose weight during an event and when their urine remains pale yellow and is normal in volume.[1]

sweat rate
Fluid loss per hour of exercise. It equals the sum of body weight loss plus fluid intake.

■ **Fluids That Don't Hydrate.**   Not all fluids help maintain hydration status. Fluids containing over 8% sugar don't quench a thirst and should not be consumed for fluid replacement. Their high-sugar content may draw fluid from the blood into the intestines, thereby increasing the risk of dehydration, nausea, and bloating. Alcohol-containing beverages such as beer, wine, and gin and tonics are not hydrating, nor is water when consumed by a sodium-depleted person.[1] Table 28.3 lists the carbohydrate and sodium content of various beverages.

Table 28.3

Carbohydrate (based on weight) and sodium content of fluids.

	Carbohydrate	Sodium
Gatorade, 1 cup	6%	96 mg
PowerAde, 1 cup	8%	28 mg
Cola, 1 cup	11%	15 mg
Fruit punch, 1 cup	12%	94 mg
Koolaid, 1 cup	10%	37 mg
Apple juice, 1 cup	12%	7 mg

■ **Estimating Fluid Needs: Sweat Rate.**   The amount of fluid athletes need during an event can be estimated by calculating an hourly **sweat rate** for a specific activity (Table 28.4). A person preparing for a marathon who loses a pound in an hour of training and who drinks nothing during the hour would have a sweat rate of one pound (16 ounces). If that athlete drank 8 ounces of fluid in the hour and lost a pound—or 16 ounces—his or her sweat rate would be 24 ounces (16 ounces + 8 ounces). The sweat rate amount of fluid should be consumed per hour of the marathon event. Athletes who gain weight during an event have consumed too much water.[22]

Junior

Lakisha

Dehydration: The Consequences   Loss of more than 2% of body weight (2 to 4 pounds) during an event indicates that the body is becoming dehydrated. Effects of dehydration range from mild to severe, depending on how much body water is lost (Illustration 28.6). Any amount of dehydration impairs physical performance. At the extreme, dehydration can lead to heat exhaustion or heat stroke (Table 28.5). People who overexercise in hot weather when they are out of condition are most likely to suffer heat exhaustion or heat stroke, but these conditions occasionally occur among seasoned athletes, too. Heat exhaustion can be remedied by fluids and electrolytes, but heat stroke requires emergency medical care.[23]

The Health Action feature for this unit highlights dehydration. The information may help you recognize and avoid it.

Hyponatremia and Excess Water   Hyponatremia is an important cause of death and life-threatening illness in marathon runners. In a recent Boston Marathon, 13% of runners developed blood sodium levels that qualified as hyponatremia, and 0.6% became very ill due to extremely low levels of sodium.[25] The condition needs to be treated promptly. The major signs of hyponatremia are difficulty breathing, bloating, confusion, nausea, vomiting, and swelling around the brain.[26]

Hyponatremia is most likely to occur in athletes who gain weight during an event, consume more than 3 liters (12 cups) of low- or no-sodium fluids during the race, and engage in running for four or more hours. Individuals with a low Body Mass Index (under 20 kg/m²) and women are at higher risk of developing hyponatremia than others.[25] Regular consumption of a beverage containing about 200 mg sodium per 8 ounces during a race or eating salty foods decreases the risk of hyponatremia.[1]

Table 28.4

Calculating sweat rate: an example

1. Determine your body weight one hour before and one hour after exercise.
2. Subtract your postexercise weight from your preexercise weight.
3. Convert the number of pounds lost to ounces. (One pound equals 16 ounces.)
4. Add the number of ounces lost or gained to the number of ounces of fluid you consumed during the hour of exercise. The result is your sweat rate, and that's an approximation of the amount of fluid you need to consume during one hour of that exercise.

Example

1. Terrell weighed 172 pounds an hour before an hour-long bout of exercise and 171 pounds an hour after the exercise:
   172 pounds − 171 pounds = 1 pound lost
   1 pound × 16 ounce per pound = 16 ounces
2. He drank 16 ounces of fluid during the hour of exercise.
   16 ounces + 16 ounces = 32 ounces, or "sweat rate."

Terrell would need to consume about 32 ounces of fluid per hour to remain hydrated.

AP Images/The Journal, Doug Lindley

Illustration 28.6   Effects of dehydration.
Even a low level of dehydration impairs physical performance.[24]

Table 28.5

A primer on heat exhaustion and heat stroke[23]

Heat exhaustion: A condition caused by low body water and sodium content due to excessive loss of water through sweat in hot weather. Symptoms include intense thirst, weakness, paleness, dizziness, nausea, fainting, and confusion. Fluids with electrolytes and a cool place are the remedy. Also called "heat prostration" and "heat collapse."

Heat stroke: A condition requiring emergency medical care. It is characterized by hot, dry skin, labored and rapid breathing, a rapid pulse, nausea, blurred vision, irrational behavior, and, often, coma. Internal body temperature exceeds 105°F due to a breakdown of the mechanisms for regulating body temperature. Heat stroke is caused by prolonged exposure to environmental heat or strenuous physical activity. The person affected by heat stroke should be kept cool by any means possible, such as removing clothing and soaking the person in ice-cold water. If conscious, the person should be given fluids. Also called "sunstroke."

Failure to replace lost body water

↓ Blood volume declines

↓ Volume of water in and around cells declines

↓ Sweat, flushing, dry mouth
↑ Body temperature rises
↓ Physical work capacity drops
↑ Electrolyte concentration in muscles increases (causes muscle cramps)

Heat exhaustion

↑↑ Body temperature rises
↑ Heart rate increases
↑ Hot, dry skin

Heat stroke

The signs and symptoms of dehydration can include:
- Thirst
- Lightheadedness or fatigue
- Headache
- Inability to concentrate
- Feeing cold or having chills when it is hot out and you are sweating
- Collapse due to heat exhaustion or stroke

Another sign of dehydration is the amount and color of urine. Large amounts and lighter-color urine indicate good hydration versus urine that is dark in color. One should drink until the urine color becomes lighter.

Body Fat and Weight: Heavy Issues for Athletes

As a group, athletes tend to be very concerned about their weight and body shape, and they are leaner than the U.S. population in general (Table 28.6).[26] But the advantages of low body weight and body fat disappear when they become too low. Body fat levels of less than 5% in men and less than 12% in women can seriously interfere with health. These percentages represent the obligatory body fat content that people need for the functions of fat unrelated to energy production.[6] Everybody needs some fat to use for hormone production, to maintain normal body temperature in cold weather, and to cushion internal organs.

Exercise, Body Fat, and Health in Women   It is not uncommon for female athletes to experience irregular or absent menstrual periods or the late onset of periods during adolescence (Table 28.7).[27] These aberrations in menstrual periods appear to be related to specific effects of high levels of physical activity and caloric intake deficits on hormone production.[28] Female athletes with missing and irregular periods are at risk for developing low bone density, osteoporosis, and bone fractures. Women at particular risk of bone fractures are those with the "female-athlete triad" of disordered eating, amenorrhea (pronounced *a-men-or-re-ah* and meaning "no menstrual periods"), and osteoporosis. Abnormal menstrual cycles should not be dismissed as a normal part of training. Normal menstrual periods should be reinstated through increased caloric intake.[1] Suppressed testosterone levels or other metabolic changes due to caloric deficits in male athletes may be the parallel to menstrual irregularities in female athletes.[29]

Wrestling: The Sport of Weight Cycling   "Making weight" is a common and recurring practice among wrestlers (Illustration 28.7). Most wrestlers will "cut" 1

Table 28.6

Average body fat content of various athletes[28, 29]

	Body Fat Content (As Percentage of Body Weight)	
	Women	Men
Bodybuilders	10%	6%
Long-distance runners	17	6–13
Baseball players	—	12–14
Basketball players	21–27	7–10
Football players	—	9–19
Gymnasts	10	5
Soccer players	—	10
Tennis players	20	16
Wrestlers	—	9

Table 28.7

Incidence of irregular or absent menstrual cycles in female athletes and sedentary women[27, 28]

Joggers (5 to 30 miles per week)	23%
Runners (over 30 miles per week)	34
Long-distance runners (over 70 miles per week)	43
Competitive bodybuilders	86
Noncompetitive bodybuilders	30
Volleyball players	48
Ballet dancers	44
Sedentary women	13

to 20 pounds over a period of days between 50 and 100 times during a high school or college career.[30] That makes wrestling more than a test of strength and agility. It makes it a contest of rapid weight loss.

For competitive reasons, wrestlers often want to stay in the lowest weight class possible and may go to great lengths to achieve it. They may fast, "sweat the weight off" in saunas or rubber suits, or vomit after eating to lose weight before the weigh-in. These practices can be dangerous, even life threatening, if taken too far. After the match is over, the wrestlers may binge and regain the weight they lost.

Wrestlers, like other athletes involved in intense exercise, perform better if they have a good supply of glycogen and a normal amount of body water. Fasting before a weigh-in dramatically reduces glycogen stores, and withholding fluids or losing water by sweating puts the wrestler at risk of becoming dehydrated. Trying to stay within a particular weight class too long may also stunt or delay a young wrestler's growth.

The American Medical Association and the Association for Sports Medicine recommend that wrestling weight be determined after six weeks of training and normal eating. In addition, a minimum of 7% body fat should be used as a qualifier for assigning wrestlers to a particular weight class. In addition, weight classes are now based on normal weight for height and age, and weigh-ins are scheduled close to event times.[31]

Illustration 28.7 Dropping weight quickly in the days before a match can wreck a wrestler's chances and harm his health.

Iron Status of Athletes

Iron status is an important topic in sports nutrition because iron deficiency (or low iron stores) and iron-deficiency anemia (or low blood hemoglobin level) decrease endurance. Iron is a component of hemoglobin, a protein in blood that carries oxygen to cells throughout the body, and it works with enzymes involved in energy production. When iron stores or hemoglobin levels are low, less oxygen is delivered to cells, and less energy is produced than normal.[1]

Female athletes are at higher risk of iron-deficiency anemia than other females. One study of aerobically fit athletes found that 36% of women and 6% of males were iron deficient.[32] Consequently, it is recommended that female athletes especially pay attention to the amount of iron consumed.[1]

Ergogenic Aids: The Athlete's Dilemma

The quest for ownership of a competitive edge has drawn athletes to ergogenic aids throughout much of history. (See Table 28.8 for a historical review of ergogenic aids use.) Relatively few of the hundreds of available products work, and most are sold as "dietary supplements" so they do not have to be tested for safety. Those found to increase muscle mass, strength, or endurance are usually banned for use by competitive athletes. Several of the aids clearly pose serious risks to health. Some contain banned substances, and others simply represent misguided hopes and misspent money.[33] Table 28.9 summarizes claims made for a variety of ergogenic aids, research results, and what is known about adverse effects.

The World Anti-Doping Agency sets the rules for ergogenic aid use by athletes. Its work includes the regular updating of the "Prohibited List" of banned substances that, if used, can disqualify athletes from competing. The 2009 Prohibited List includes androgenic steroids, insulin, stimulants (excluding caffeine), and gene doping.[34]

Table 28.8

Faster, higher, stronger, longer: a brief history of performance-enhancing substances used by athletes

B.C.	Large quantities of beef consumed by athletes in Greece to obtain "the strength of 10 men"; deer liver and lion heart consumed for stamina
1880s	Morphine used to increase performance in (painful) endurance events
1910s	Strychnine consumed for the same reason as morphine
1930s	Amphetamines used to increase energy levels and endurance; testosterone taken to increase muscle mass
1980s	Blood doping, EPO used to increase endurance Ephedra to increase energy
2009	HGH; gene doping to increase strength and endurance

Table 28.9

Ergogenic aids: claims and evidence[1,33-35,37]

Ergogenic Aid–Normal Functions	Claims	Evidence
Amphetamines • No normal function.	Reduce fatigue, improve concentration, increase aggressiveness, decrease appetite and weight.	Reduce fatigue, appetite, and weight; increase aggressiveness. **Adverse effects**: Many, including dizziness, tremors, confusion, paranoia, increased heart rate, chest pain, cardiovascular collapse.
Androgenic anabolic steroids • Testosterone and related substances. • Promote protein synthesis, decrease protein tissue breakdown, increase blood volume and red blood cell production.	Gradually increasing doses of several forms of steroids increase muscle mass.	Increase strength, muscle mass; decrease recovery time. **Adverse effects**: Liver, heart, psychiatric disorders, testicle atrophy, reduced HDL and elevated LDL cholesterol.
Bee pollen • Pollen collected by bees from flowers and mixed with bee secretions. Provides food for bee larva.	Detoxification of blood, improves performance.	No such effects found. **Adverse effects**: Allergic reaction in people who are allergic to bee stings and honey and who have asthma.
Boron • A mineral that influences calcium, magnesium, and phosphorus utilization.	Increases testosterone and muscle mass.	No such effects found. May improve bone density, arthritis.
Branched chain amino acids • Isoleucine, leucine, valine. Are essential amino acids. May lower brain serotonin levels.	Improve performance, delay fatigue, build muscle, increase mental concentration.	May delay muscle breakdown, spare glycogen, increase fat utilization and time to exhaustion. **Adverse effects**: High doses (>20 g) may impair performance.
Caffeine • Central nervous system stimulant, diuretic, circulatory and respiratory stimulant.	Decreases fatigue in endurance events.	May reduce fatigue and improve endurance at low to moderate doses (3–6 mg/kg). Increases insulin resistance **Adverse effects**: Nervousness, sleeplessness, irregular heartbeat; fatal dose 3–10 g. Olympic illegal dose equivalent to 6–8 cups coffee 2–3 hours preevent.
Carnitine • A compound required for energy formation from fatty acids.	Enhances fatty acids utilization for energy.	Most studies show no effect on fatty acid utilization during exercise in healthy adults. **Adverse effects**: diarrhea
Cocaine • No normal functions.	Same as for stimulants but much shorter acting.	
Chromium Picolinate • Involved in energy metabolism and insulin utilization.	Increases lean body mass, decreases fat mass.	No effect on body composition. No evidence of ergogenic effect. **Adverse effects**: May increase oxidative damage to muscles.

Table 28.9

Ergogenic aids: claims and evidence (*continued*)

Ergogenic Aid–Normal Functions	Claims	Evidence
Creatine • Component of muscle; generates ATP from ADP in muscle.	Increases performance in high-intensity, short-duration exercises.	May slightly increase peak muscle force in short-term, high-intensity exercises, especially in vegetarians with low reserves of creatine. No increase in strength observed, no improvement in aerobic exercise performance. May increase energy available for training. **Adverse effects:** Headache, water retention, muscle cramps.
DHEA, Andro (Dehydroepinandrosterone, Androstendione) • Precursors of testosterone.	Promote testosterone production and increase muscle mass.	Promote testosterone production and increase muscle mass. **Adverse effects:** The same as for anabolic steroids.
Energy drinks	Increased energy, attention, alertness	Caffeine increases heart rate, blood pressure; taurine, ginseng, guarana offer no benefit to physical performance. **Adverse effects:** Several deaths, seizure cases related to caffeine in energy drinks.
Ephedra (ma huang) • No normal function.	Central nervous system stimulant; enhances endurance, strength, and body fat loss.	Increases aerobic capacity when combined with caffeine. **Adverse effects:** Nervousness, anxiety, rapid heart rate, headache, hypertension, death. Not recommended for use; no longer sold.
EPO (Erythropoietin) • A hormone that increases red blood cell production by bone marrow.	Replaces blood doping, or injection of red blood cells for increasing oxygen availability and endurance. Is being replaced by altitude tents and other hypoxic devices.	Increases endurance. **Adverse effects:** Increases blood viscosity, blood pressure, blood clot formation; headache. Banned by the National Olympic Committee.
GBL (Gammabutyrolactone, furanodehydro) • No normal function.	Increases muscle mass and muscle recovery from exercise.	No such effects **Adverse effects:** Vomiting, seizures, death. FDA has called for its voluntary removal from the market.
Ginseng, Siberian • No normal function.	Enhances endurance and strength.	Increases muscular strength and aerobic work capacity.
Glycerol • A clear, sweet fluid attached to fatty acids in fat molecules.	Increases hydration.	Conflicting results. **Adverse effects:** Bloating, nausea.
HGH (Human growth hormone) • Stimulates muscle and bone growth; converted to glucose.	Increases muscle mass and strength; similar effects as anabolic steroids but safer.	Increases muscle mass but probably not strength. **Adverse effects:** Growth of facial bones, heart disease, stroke, hypertension, diabetes. Effects generally irreversible.
HMB (ß-hydroxy-ß–methylobutyrate) • Derivative of the amino acid leucine.	Improves muscle size and strength.	No performance improvement, may help muscle recovery. **Adverse effects:** Appears safe at doses of 3 grams per day.

(continued)

Table 28.9

Ergogenic aids: claims and evidence (*continued*)

Ergogenic Aid–Normal Functions	Claims	Evidence
Insulin • Speeds entry of glucose into cells, increases production of glycogen.	Increases muscle mass.	Increases glucose passage into cells and glycogen and protein synthesis. **Adverse effects**: Low blood sugar, seizure, brain damage, death.
Low–GI carbs (low glycemic index carbohydrates)	Increase rate of fat utilization versus high GI carbs; improve endurance.	Low-GI carbs, such as pasta, dried beans, oranges, and peaches, appear to increase fat utilization and endurance compared to high-GI carbs (white bread, corn flakes, candy, and french fries, for example).
Oxygen, oxygen drinks and bars	Increase oxygen availability for muscular work.	No effect on oxygen availability in the body. **Adverse effects**: Burping if consumed.
Protein powders • Can contribute to meeting the body's need for protein.	Increase muscle mass and strength.	No such effects found. **Adverse effects**: Dehydration, liver and kidney problems.
Vitamin and mineral supplements • May contribute to meeting need for vitamins and minerals.	Improve energy and stamina.	Such supplements improve energy and stamina in deficient individuals only. **Adverse effects**: Toxicity reactions.

The Path to Improved Performance Genetics, training, and nutrition: These are the real keys to physical performance. Although other aids will be sought, those that exceed the boundaries of what is considered fair and safe will not be approved for use by athletes. After all, athletic competition is not a test of drugs or performance aids. It's a test of an individual's ability to excel. Anything less wouldn't be sporting.

NUTRITION | Up Close

Testing Performance Aids

Focal Point: The critical examination of studies on performance aids.

Read the following summary of a study (imaginary) of the effects of a phosphorous supplement on strength and then answer the critical thinking questions.

- *Purpose:* To assess the effect of a phosphorous supplement on strength.
- *Methods:* Twenty volunteers from the crew team were given the phosphorous supplement for a week. Strength, assessed as the maximum number of push-ups a study participant could do in one session, was assessed before and after supplementation. Participants recorded any supplement side effects.
- *Results:* The number of push-ups increased by an average of 5% after supplementation. Diarrhea was the only side effect consistently noted by participants.

Critical Thinking Questions

1. Does the study demonstrate that the supplement is safe?

   Yes _____ No _____ Give reasons for your answer.

   _____

2. Does the study demonstrate that the supplement increases strength?

   Yes _____ No _____ Give reasons for your answer.

   _____

FEEDBACK (answers to these questions) can be found at the end of Unit 28.

[Key Terms

ATP, ADP, page 28–3
electrolytes, page 28–4
ergogenic aids, page 28–2

glycogen, page 28–2
hydration status, page 28–7

hyponatremia, page 28–7
sweat rate, page 28–8

Review Questions

1. The two main substrates for energy formation in muscles are amino acids and fatty acids. ☐ ☐

2. Glucose is converted to energy by muscles anaerobically. ☐ ☐

3. Much more ATP is produced from the breakdown of fatty acids than from amino acids or glucose. ☐ ☐

4. Hypoglycemia results when athletes use up the liver's supply of glycogen. ☐ ☐

5. Glucose is the major source of energy for low- and moderate-intensity activities. ☐ ☐

6. Carbohydrate loading is rarely used in practice because it takes six days of a special diet and training regime to prepare for it. ☐ ☐

7. Postevent recovery meals should include rich sources of fat because fats facilitate muscle repair. ☐ ☐

8. Hyponatremia is primarily caused by drinking excessive amounts of water during prolonged exercise. ☐ ☐

9. An athlete who runs for an hour, loses two pounds in that hour, and drinks 12 ounces of fluid during the hour has a sweat rate of 60 ounces. ☐ ☐

10. Signs of dehydration include thirst and lightheadedness or fatigue. ☐ ☐

11. A few ergogenic aids improve performance somewhat, but most do not improve performance. ☐ ☐

12. Athletes attempting to build muscle mass and strength should take protein and amino acid supplements. ☐ ☐

Media Menu

www.mayohealth.com
The Mayo Clinic's Web site offers reliable advice on exercise and sports nutrition.

www.acsm.org
The American College of Sports Medicine Web site offers information on sports nutrition, readily accessible articles on sports nutrition, and the opportunity to join the national Coalition for Promoting Physical Activity.

www.nutrition.gov
Search "physical performance" and find a wealth of information on study results, evaluation systems, and guidelines.

www.gssiweb.com
This connects you to the Gatorade Sports Science Institute. The site provides valuable resources for sports nutrition professionals through "Focus on" sections about ergogenic aids, hydration assessment, and other hot sports topics.

www.steroidabuse.gov
This site links to research information about steroid use, its side effects and reasons why they should not be used.

www.wada-ama.org
The World Anti-Doping Agency's (WADA) home page provides a link to current Prohibited List

of substances banned from use by athletes. The site also provides a 2-minute video about the Montreal-based WADA.

www.fda.gov/medwatch/report.htm
Voluntarily report serious adverse events associated with the use of dietary supplements, medications, and medical devices through this address.

Notes

1. Position of the American Dietetic Association, Dietitians of Canada, and the American College of Sports Medicine: Nutrition and Physical Performance, J Am Diet Assoc 2009;109:509–27.

2. Gleason M. Interrelationships between physical activity and branched-chain amino acids. J Nutr 2005;135:1591S–5S.

3. Rosenbloom C. Athletes (and coaches) say the darndest things. Nutr Today 2009;44:77–80.

4. Corley G et al. Nutrition knowledge and dietary practices of college coaches. J Am Diet Assoc 1990;90:705–9.

5. Brown RC. Nutrition for optimal performance during exercise: carbohydrate and fat. Curr Sports Med 2002;1:222–9.

6. Coleman E. Eating for endurance. Palo Alto (CA): Bull Publishing Co., 1988.

7. Jones NL et al. Exercise limitations in health and disease, N Engl J Med 2000;343:632–41.

8. Maughan R. The athlete's diet: Nutritional goals and dietary strategies. Proc Nutr Soc 2002;61:87–96.

9. Foster C et al. Effects of preexercise feeding on endurance performance. Med Sci Sports Exer 1979;11:1–5.

10. Carbo-loading: Boost your endurance during high-intensity workouts. www.mayoclinic.com/health/carb-loading/HQ00385, 10/8/06.

11. Minehan M for the Australian Sports Commission. Carbohydrate Loading, 2004, www.ausport.gov.au/ais/nutrition/factsheets/competition_and_training2/carbohydrate_loading, accessed 5/16/09.

12. Brouns F et al. Utilization of lipids during exercise in human subjects: metabolic and dietary constraints. Brit J Nutr 1998;79:117–28.

13. What we eat in America, NHANES, 2005–2006, USDA, 2008, available at www.are.usda.gov/ba/bhnrc/fsrg.

14. Verdijk LB et al. Protein supplementation before and after exercise does not further augment skeletal muscle hypertrophy after resistance training in elderly men, Am J Clin Nutr. 2009;89:608–16.

15. Eliot KA et al. The effects of creatine and whey protein supplementation on body composition in men aged 48 to 72 years during resistance training, J Nutr Health Aging, 2008;12:208–12.

16. Coleman, Fat loading; and Suzuki M. Glycemic carbohydrates consumed with amino acids or protein right after exercise enhance muscle formation, Nutr Rev 2003;61:S88–S94.

17. Stevenson EM et al. Influence of high-carbohydrate mixed meals with different glycemic indexes on substrate utilization during subsequent exercise in women. Am J Clin Nutr 2006;84:354–60.

18. Schneid S. Thermoregulation, presented at the Experimental Biology Annual Meeting, New Orleans, 4/19/09.

19. Moore DR et al. Ingested protein dose response of muscle and albumin protein synthesis after resistance exercise in young men, Am J Clin Nutr 2009;8:161–8.

20. Beelan M et al. Coingestion of carbohydrate and protein hydrolysate stimulates muscle protein synthesis during exercise in young men, with no further increase during subsequent overnight recovery, J Nutr 2008;138:2198–2204.

21. Von Duvillard SP et al. Fluids and hydration in prolonged endurance performance. Nutrition. 2004;20:651–6.

22. Stout A. Fueling and weight management strategies in sports nutrition, J Am Diet Assoc 2007;107:1475–9.

23. Casa DJ et al. Cold water immersion: the gold standard for exertional heatstroke treatment, Exerc Sport Sci Re 2007;35:141–9.

24. Noakes TD. Too many fluids as bad as too few. BMJ 2003;327:113–4.

25. Almond CSD et al. Hyponatremia in nonelite marathon runners. N Engl J Med 2005;352:1516–8,1550–6.

26. Wilmore JH. Body composition in sports and exercise: directions for future research. Med Sci Sports Exer 1983;15:21–31.

27. Ireland ML, et al. The female athlete, Philadelphia: W.B. Saunders; 2003.

28. Lloyd T et al. Interrelationships of diet, athletic activity, menstrual status, and bone density in collegiate women, Am J Clin Nutr 1987;46:681–4.

29. Brownell KP et al. Weight regulation practices in athletes: analysis of metabolic and health effects. Med Sci Sports Exer 1987;19:546–56.

30. Oppliger RA et al. The Wisconsin wrestling minimum weight project: a model for weight control among high school wrestlers. Med Sci Sports Exer 1995;27:1220–24.

31. Cunningham E. Where can I find resources to address weight practices for wrestlers? J Am Diet Assoc 2006;106:761.

32. Sinclair LM et al. Prevalence of iron deficiency with and without anemia in recreationally active men and women. J Am Diet Assoc 2005;105:975–8.

33. FDA announces Hydroxycut product recalls, www.medscape.com.viewarticle/702283?src=mp&spon=42&uac=61213 SX, accessed 5/12/09.

34. The World Anti-Doping Agency Prohibited List, www.wada-ama.org.

35. Heller L. FDA orders destruction of supplements worth $1.3 m, www.nutraingredients-usa.com, 5/12/09.

36. Atkinson ML. Panama Canal's locomotive "mules" lock ships dead-center in waterway. Newhouse News Service, Sept. 1955.

37. Steroid abuse, www.steroidabuse.gov, 11/08.

NUTRITION | Up Close

Testing Performance Aids

Feedback for Unit 28

Feedback for Unit 28

1. Answer: No. Side effects, such as blood pressure changes, diarrhea, abnormalities in kidney function, and weight loss, were not assessed. Safety of long-term use was not addressed.

2. Answer: No. A 5% increase in strength is small and may have been due to usual differences in the number of push-ups a person can from time to time. Any increase in strength may have been due to increased training during the week the supplement was taken. There was no control group. All participants knew they were taking phosphorous, a supplement that some may believe increases strength. This belief may have changed performance on the push-up test.

UNIT 29

Good Nutrition for Life: Pregnancy, Breast-Feeding, and Infancy

NUTRITION SCOREBOARD

	TRUE	FALSE
1 The fetus is a parasite.		
2 The United States has the lowest rate of infant mortality in the world.		
3 Normal health, growth, and development of infants are defined by the experience of breast-fed infants.		
4 Drinking alcohol-containing beverages is safe during breast-feeding because the alcohol doesn't pass into the milk.		

Key Concepts and Facts

- There are no "maternal instincts" that draw women to select and consume a good diet during pregnancy.

- The fetus is not a parasite.

- Fetal and infant growth are characterized by "critical periods" during which all essential nutrients needed for growth and development must be available or growth and development will not proceed normally.

- An adult's risk of certain chronic diseases may be partially determined by maternal nutrition during pregnancy and the person's own nutrition early in life.

- Breast-feeding is the optimal method of feeding young infants.

- Rates of growth and development are higher during infancy than at any other time in life.

A Healthy Start

The day has finally arrived, the one Crystal and Tyrone have anticipated for a year. Crystal's pregnancy test is positive!

This is a planned pregnancy, and both Crystal and Tyrone have done all they could to prepare. Crystal's diet has been the picture of perfection, she has kept up her regular exercise schedule, and she has quit drinking alcohol entirely. Now they are ready to have the baby of their dreams. They are convinced that not only will this baby be healthy and strong, but it will be above average in every respect.

All parents want to give their children every advantage in life that they can. Although no one can guarantee that a baby will be born healthy and strong—no matter what the parents do—there are steps parents can take to make the best baby possible.

Photo Disc

Unhealthy Starts on Life

Unfortunately, many infants born in the United States do not have the advantage of good health at birth. Among countries of the world, the United States ranks twenty-ninth in terms of **infant mortality rate** (Table 29.1). The relatively high rate of infant deaths in the United States is due primarily to the proportion of infants born with **low birthweights** or **preterm** (Illustration 29.1). One in 12 U.S. infants is born too small, and over 1 in 8 is born too early[1]. Rates of low birthweight and preterm differ substantially among population groups, with the highest rates occurring in African Americans.[1] These infants are at particular risk for requiring intensive and continuing care and of dying within the first year of life. In contrast, infants weighing between 7 pounds, 11 ounces and 8 pounds,

infant morality rate
Deaths that occur within the first year of life per 1,000 live births.

low-birthweight infants
Infants weighting less than 2500 grams (5.5 pounds) at birth.

preterm infants
Infants born at or before 37 weeks of gestation (pregnancy).

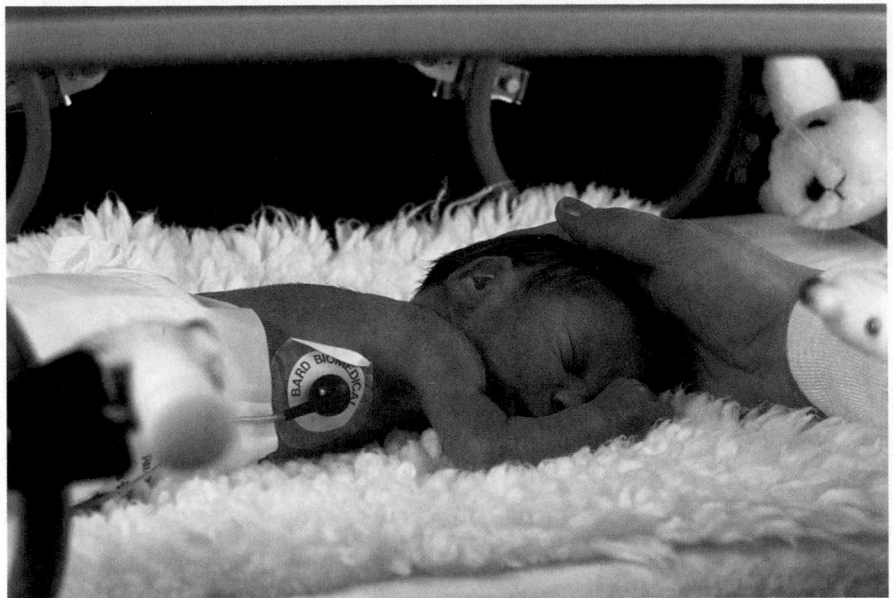

Illustration 29.1 In 2006 8.2% of infants born in the United States had a low birthweight, and 13% were born preterm.

Table 29.1	
Infant mortality rates (deaths per 1,000 live births) for 31 countries listed from lowest to highest, 2004.[2]	
Singapore	2.0
Hong Kong	2.5
Japan	2.8
Sweden	3.1
Norway	3.2
Finland	3.3
Spain	3.5
Czech Republic	3.7
France	3.9
Portugal	4.0
Germany	4.1
Greece	4.1
Italy	4.1
Netherlands	4.1
Switzerland	4.2
Belgium	4.3
Denmark	4.4
Austria	4.5
Israel	4.5
Australia	4.7
Ireland	4.9
Scotland	4.9
England and Wales	5.0
Canada	5.3
Northern Ireland	5.5
New Zealand	5.7
Cuba	5.8
Hungary	6.6
Poland	6.9
Slovakia	6.9
United States	6.9

13 ounces (or 3500 to 4000 grams) are least likely to die within the first year. The average weight of U.S. infants is 7 pounds, 7 ounces (3350 grams).[3]

Improving the Health of U.S. Infants

A high proportion of poor infant outcomes in the United States is attributed to a combination of factors including poverty, poor nutrition, limited access to health care, and a maternal lifestyle that includes the use of illicit drugs, cigarettes, and excessive amounts of alcohol. Many infant deaths, preterm infants, and low-birthweight infants can be prevented, however, by optimizing maternal behaviors that influence health. Of the various behaviors that affect maternal health, none offers potentially greater advantage to both pregnant women and their infants than good nutrition.[4]

Nutrition and Pregnancy

Nutrition is important during pregnancy because the **fetus** depends on it. Energy and every nutrient the fetus needs for **growth** and **development** must be supplied by the mother's diet (preferably) or by supplements. If either insufficient or excessive amounts of nutrients are supplied, fetal growth and development may be compromised. The nature and extent of the impairment depend on which organ or tissue is growing most rapidly when the nutritional deficiencies or overdoses occur.[5]

fetus
A baby in the womb from the eighth week of pregnancy until birth. (Before then, it is referred to as an *embryo*.)

growth
A process characterized by increases in cell number and size.

development
Processes involved in enhancing functional capabilities. For example, the brain grows, but the ability to reason develops.

Critical Periods

Fetal growth and development proceed in a series of critical periods. A **critical period** is an interval of time during which cells of a tissue or organ are genetically programmed to multiply. The period is considered critical because if the cells do not multiply as programmed during this set time interval, they cannot make up the deficiency later. The level of nutrients required for cell multiplication to occur normally must be available during this specific time interval. If the nutrients are not available, the developing tissue or organ will contain fewer cells than normal, will form abnormally, or will function less than optimally.

critical period
A specific interval of time during which cells of a tissue or organ are genetically programmed to multiply. If the supply of nutrients needed for cell multiplication is not available during the specific time interval, the growth and development of the tissue or organ are permanently impaired.

Illustration 29.2 Baby's first picture.
An ultrasound image of a rapidly growing and developing 16-week-old fetus.

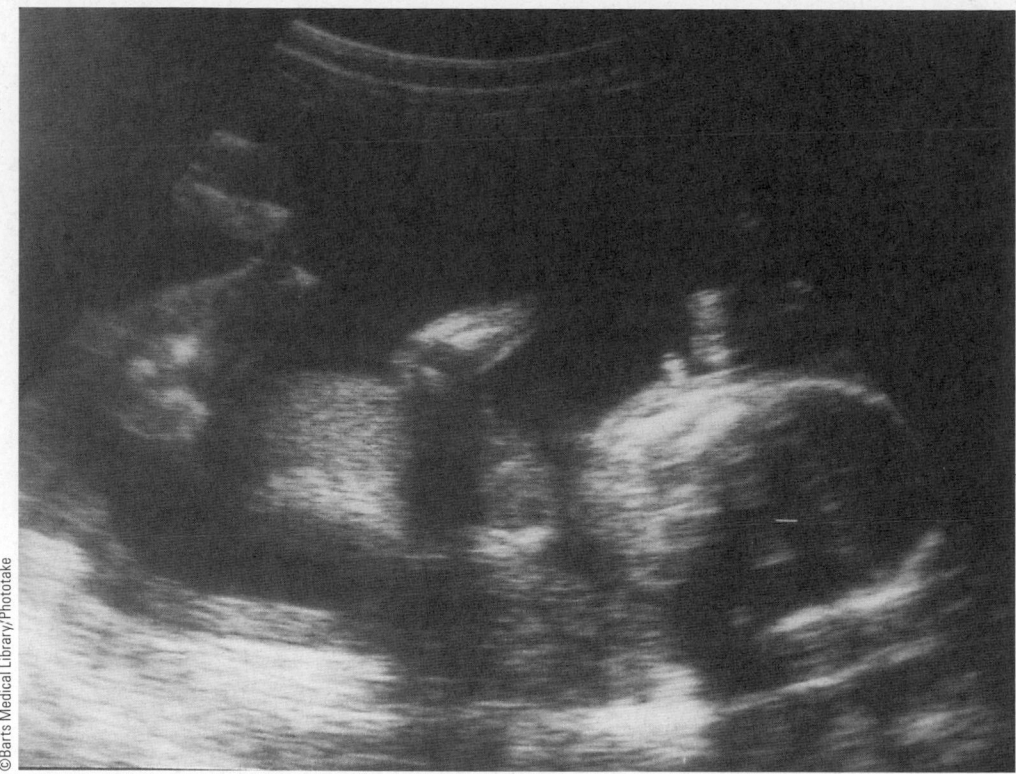

©Barts Medical Library/Phototake

The roof of the mouth (or the hard palate), for example, is formed early in the third month of pregnancy when two developing plates fuse together. This process can only occur early in the third month. If excessive amounts of vitamin A are present in fetal tissues during this period, the two plates may fail to combine, resulting in a cleft palate.[6] (The divided hard palate can be surgically corrected after birth.)

Critical periods of cell multiplication are most intensive in the first few months of pregnancy, when fetal tissues and organs are forming rapidly (Illustration 29.2)—hence the importance of the nutritional status of women at the time of conception and very early in pregnancy. For some organs and tissues, cell multiplication even continues through the first two years after birth. Most of the growth that occurs in the fetus late in pregnancy and throughout the rest of the growing years, however, is due to increases in the size of cells within tissues and organs.

The Fetal Origins Hypothesis

One of the most striking advances in research on pregnancy concerns the potential effects of maternal nutrition on the baby's risk of developing certain chronic diseases later in life. It appears that an important element of increased susceptibility to heart disease, stroke, diabetes, obesity, hypertension, and other disorders may be "programmed," or gene functions modified, by inadequate or excessive supplies of energy or nutrients during pregnancy.[7]

Given optimal conditions, fetal growth and development proceed according to the genetic blueprint established at conception. Organs and tissues are well developed and ready to function optimally after birth. Under less than optimal growing conditions, such as those introduced by maternal weight loss, poor nutrient intakes, or diseases in the mother that alter her ability to supply the fetus with energy or nutrients, fetal growth and development are modified. Fetal tissues undergoing critical phases of development at that time have to make

The ancient Romans believed that if a woman wanted a child with brown eyes, she should eat mice often during pregnancy.

On the Side

Photo Disc

adaptations to cope with the under- or oversupply of nutrients. These adaptations may produce long-term changes in the structure and function of tissues. Low maternal energy intake in the last few months of pregnancy, for example, may hinder the development of cells that produce insulin and limit the body's ability to control blood glucose levels later in life. This could increase the risk of developing diabetes later in life.[8]

A large body of evidence supports the fetal origins hypothesis. Nevertheless, much remains to be discovered about the effects of diet during pregnancy on the risk of later disease.

The Fetus Is Not a Parasite   From early history to current times, every culture has had myths about diet in pregnancy. According to one of the more common pregnancy myths, the fetus is a parasite, capable of drawing whatever nutrients it needs from the mother at the expense of her health. If this were true, there would be no such thing as a small or poorly nourished newborn. Obviously, it is not true.

The fetus is not a parasite because, with few exceptions, it receives an adequate supply of nutrients only if the mother's intake is sufficient to maintain her own health.[9] The situation makes sense in relation to survival of the species. By meeting the mother's needs first, nature protects the reproducer.

Rather striking evidence that the fetus is not a parasite comes from studies that have identified deficiencies of vitamin B_{12}, thiamin, iodine, folate, zinc, and other nutrients in newborns but not in their mothers.[10] Similarly, infants born to women who consumed excessive levels of vitamin or mineral supplements during pregnancy are more likely to display signs of nutrient overdose than are the mothers.[11]

Prepregnancy Weight Status and Prenatal Weight Gain Are Important

Women who enter pregnancy underweight or who fail to gain a certain minimum of weight during pregnancy are much more likely to deliver low-birthweight and preterm infants than are women who enter pregnancy at normal weight or above and gain an appropriate amount of weight.[12] The risk of low birthweight can be reduced substantially by good diets that lead to a desired rate and amount of weight gain during pregnancy. Along with the duration of pregnancy and smoking, prepregnancy weight status and weight gain in pregnancy are the major factors known to influence an infant's birthweight (Table 29.2).

What's the Right Amount of Weight to Gain during Pregnancy?   Weight-gain recommendations for pregnancy vary depending on whether a woman enters pregnancy underweight, normal weight, overweight, or obese. A woman who is obese when she becomes pregnant will need to gain less weight than a woman who enters pregnancy underweight. It is recommended that underweight women gain 28 to 40 pounds, normal-weight women 25 to 35 pounds, over-weight women 15 to 25 pounds, and obese women 11 to 20 pounds during pregnancy. Normal weight women carrying twins should gain 37 to 54 pounds during pregnancy.[13] Table 29.3 can be used to identify prepregnancy weight status by a woman's body mass index (BMI) category and the appropriate recommended weight gain during pregnancy. After the goal is identified, a woman may want to plot her prenatal weight gain on a chart like the one in Illustration 29.3. The weight gained should be the result of a high-quality diet that leads to gradual and consistent gains in weight throughout pregnancy.

Where Does the Weight Gain Go?   If healthy infants tend to weigh about 8 pounds at birth, why do women need to gain more than that? Where does

Table 29.2
Major factors that directly influence birthweight[13]
• Duration of pregnancy
• Prenatal weight gain
• Prepregnancy weight status
• Smoking

Table 29.3
Prepregnancy weight status and recommended weight gain during pregnancy[13]

Prepregnancy weight status, Body Mass Index[a]	Recommended weight gain
Underweight, <18.5 kg/m²	28–40 pounds
Normal weight, 18.5–24.9 kg/m²	25–35 pounds
Overweight, 25–29.9 kg/m²	15–25 pounds
Obese, 30 kg/m² or higher	11–20 pounds
Twin pregnancy, normal weight status	37–54 pounds

[a]Body Mass Index categories modified based on 1997 changes from the National Institutes of Health.

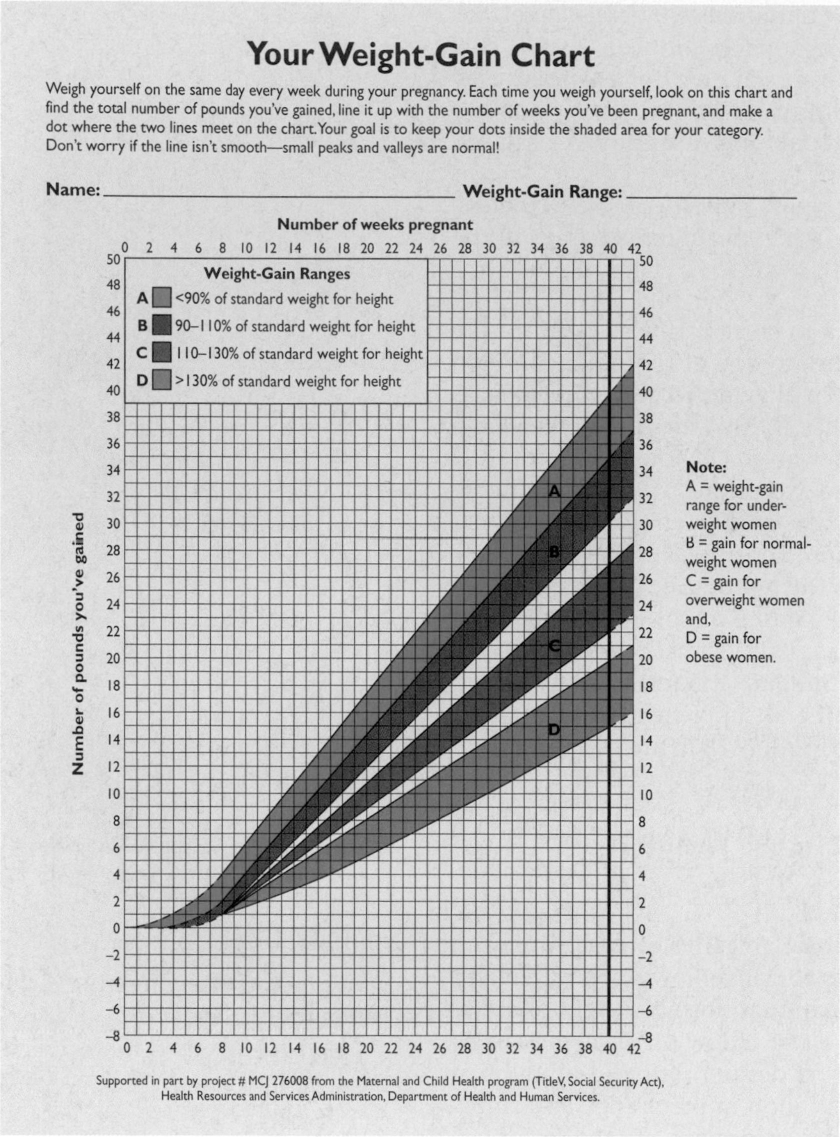

Your Weight-Gain Chart

Weigh yourself on the same day every week during your pregnancy. Each time you weigh yourself, look on this chart and find the total number of pounds you've gained, line it up with the number of weeks you've been pregnant, and make a dot where the two lines meet on the chart. Your goal is to keep your dots inside the shaded area for your category. Don't worry if the line isn't smooth—small peaks and valleys are normal!

Name: _____   Weight-Gain Range: _____

Number of weeks pregnant

Weight-Gain Ranges

A <90% of standard weight for height
B 90–110% of standard weight for height
C 110–130% of standard weight for height
D >130% of standard weight for height

Number of pounds you've gained

Note:
A = weight-gain range for underweight women
B = gain for normal-weight women
C = gain for overweight women and,
D = gain for obese women.

Supported in part by project # MCJ 276008 from the Maternal and Child Health program (TitleV, Social Security Act), Health Resources and Services Administration, Department of Health and Human Services.

Illustration 29.3 An example of a prenatal weight-gain chart for pregnant women.

trimester
One-third of the normal duration of pregnancy. The first trimester is 0 to 13 weeks, the second is 13 to 26 weeks, and the third is 26 to 40 weeks.

the rest of the weight go? Many pregnant women ask themselves these questions, especially when they don't relish the thought of gaining weight. Illustration 29.4 shows where the weight gain goes.

Fetal growth is accompanied by marked increases in maternal blood volume, fat stores, and breast and uterus size, all of which contribute to weight gain. In addition, water accumulates in the amniotic fluid, which cushions and protects the fetus, and the volume of fluid that exists outside cells increases. The placenta (the tissue that transfers nutrients in the mother's blood supply to the fetus) also accounts for some of the weight that is gained during pregnancy.

Where Does the Weight Gain Go—after Pregnancy? Delivery is one of the greatest weight-loss plans known to humankind. On average, women lose 15 pounds within the first week after delivery, and weight loss generally continues for up to a year.[14] Women who gain more than the recommended amount of weight, however, and those who gain weight after delivery may have extra pounds to lose to get back to prepregnancy weight.

The Need for Calories and Key Nutrients during Pregnancy

Pregnant women need more calories, protein, and essential nutrients than nonpregnant women need. All of these are important; but calories, folate, vitamin B_6, vitamin A, calcium, vitamin D, iron, iodine, and EPA and DPH are of paramount concern. Although many pregnant women are concerned about protein, low protein intakes are rarely a problem in pregnancy. Most women consume 10 to 30 grams more protein daily than the recommended intake level of 71 grams.[15]

According to dietary intake recommendations, pregnant women need approximately 15% more calories and up to 50% more of various nutrients than do nonpregnant women (Illustration 29.5).[15] Relatively high nutrient requirements mean that pregnant women should increase their intake of nutrient-dense foods more than their consumption of calorie-rich foods.

Calories On average, women need an additional 340 calories a day in the second **trimester**, and 450 calories in the third trimester of pregnancy.[15] Women entering pregnancy underweight will need more calories than this, and those entering overweight will need fewer. In addition, physically active pregnant women require higher caloric intakes than average. Rather than counting calories, however, it is generally easier and more accurate to monitor the adequacy of caloric intake by tracking weight gain.

Folate Folate is required for protein tissue construction and therefore is in high demand during pregnancy. It is considered an "at risk" nutrient because some pregnant women do not consume enough to meet their daily need for folate.[16]

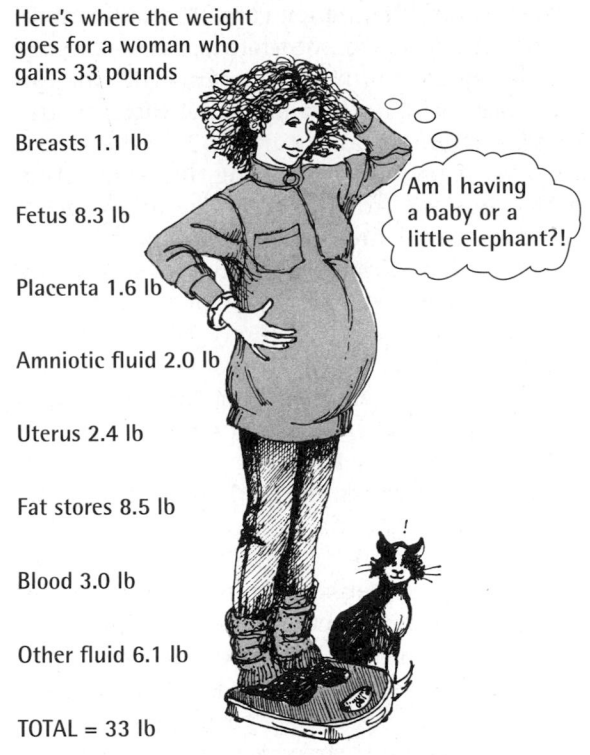

Here's where the weight goes for a woman who gains 33 pounds

Breasts 1.1 lb

Fetus 8.3 lb

Placenta 1.6 lb

Amniotic fluid 2.0 lb

Uterus 2.4 lb

Fat stores 8.5 lb

Blood 3.0 lb

Other fluid 6.1 lb

TOTAL = 33 lb

Am I having a baby or a little elephant?!

Illustration 29.4   Where does all the weight go?

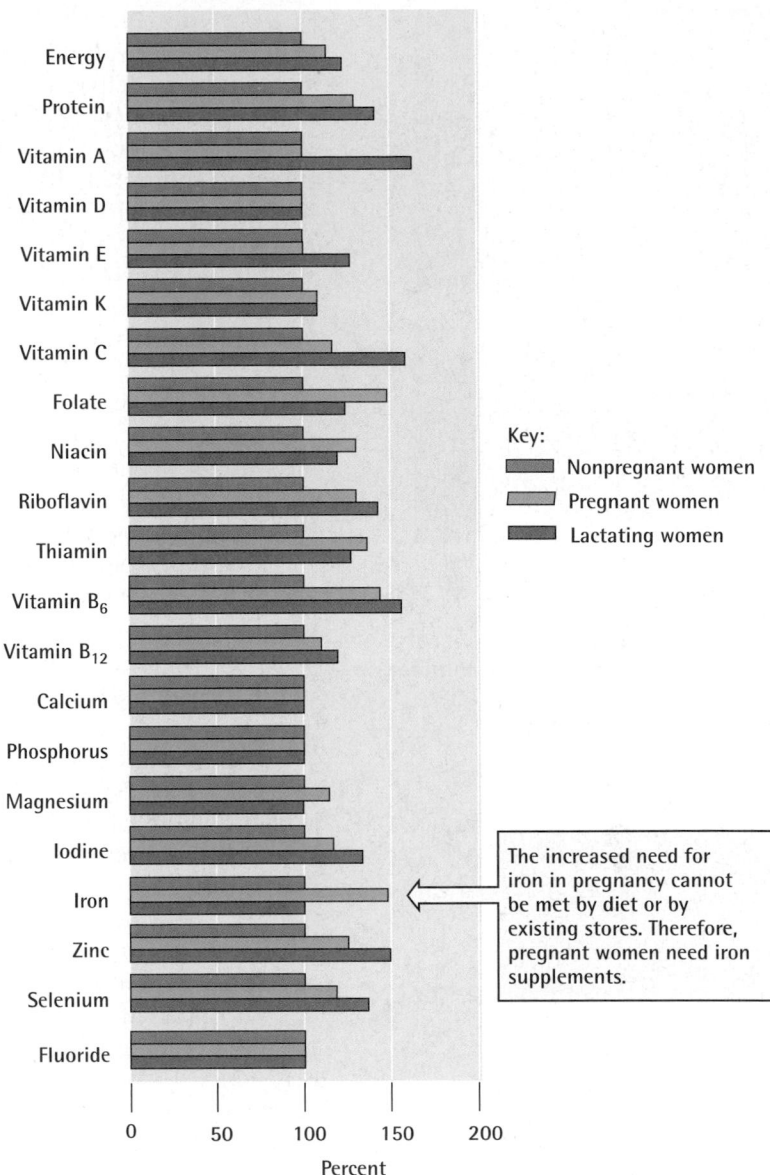

Key:
- Nonpregnant women
- Pregnant women
- Lactating women

Energy, Protein, Vitamin A, Vitamin D, Vitamin E, Vitamin K, Vitamin C, Folate, Niacin, Riboflavin, Thiamin, Vitamin B$_6$, Vitamin B$_{12}$, Calcium, Phosphorus, Magnesium, Iodine, Iron, Zinc, Selenium, Fluoride

0   50   100   150   200
Percent

The increased need for iron in pregnancy cannot be met by diet or by existing stores. Therefore, pregnant women need iron supplements.

Illustration 29.5   Percentage increases in the RDAs or AIs for pregnant and breast-feeding women compared to other women.

Folate deficiency has long been associated with fetal growth failure and malformations, but its link to **neural tube defects** such as spina bifida (Illustration 29.6) is fairly recent. In clinical trials, adequate folate status very early in pregnancy is associated with a 50 to 70% reduction in the occurrence of neural tube defects.[17] This finding, along with the knowledge that many women fail to consume enough folate, led to the fortification of refined grain products with folic acid. (Examples of fortified grain products are shown in Illustration 29.7.) Folic acid is a form of folate used in fortified foods and supplements. It is absorbed from foods about twice as completely as folate is. The vast majority of ready-to-eat breakfast cereals are fortified with at least 100 micrograms (0.1 milligram) of folic acid per serving.

Neural tube defects form before 30 days after conception, so it is important that women are consuming enough folate when they enter pregnancy. "Enough" is 600 micrograms (or 0.6 milligram) daily before and during pregnancy. Women can generally obtain this level of folate by consuming two servings of ready-to-eat breakfast cereal, six servings of grain products, and three servings of vegetables daily.[18] If the need for folate is not met by a diet that includes fortified foods, a 400-microgram folic acid supplement should be taken. Fortification of refined grain products with folic acid is improving folate status in Americans and is decreasing the incidence of neural tube defects.[19]

Vitamin A   Both low and high intakes of vitamin A may cause problems during pregnancy. Too little vitamin A is associated with poor fetal growth. Too much vitamin A in the form of retinol from supplements can cause fetal malformations.[13]

neural tube defects
Malformations of the spinal cord and brain. They are among the most common and severe fetal malformations, occurring in approximately one in every 1000 pregnancies. Neural tube defects include spina bifida (spinal cord fluid protrudes through a gap in the spinal cord; shown in Illustration 29.6), anencephaly (absence of the brain or spinal cord), and encephalocele (protrusion of the brain through the skull).

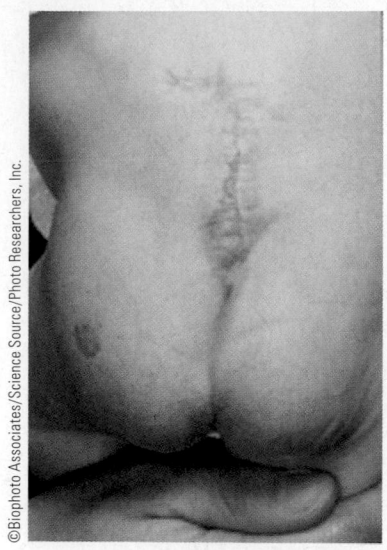

Illustration 29.6 Spina bifida–a neural tube defect.
The interruption in the spinal cord results in paralysis below the injury. This photo shows the back after surgery to close the interruption.

The effects of vitamin A overdoses during pregnancy came to public attention in the early 1990s, when women taking accutane or retinoic acid for acne were found to deliver far more than the expected number of infants with malformations. Use of these drugs before and very early in pregnancy increases the risk that babies will be born with malformations of facial features and the heart; the intake of more than 10,000 to 15,000 IU of retinol daily during the same period may have the same effect. As a precaution, the American College of Obstetrics and Gynecology recommends that women who are or could become pregnant should limit their intake of retinol to less than 5000 IU per day and should not use vitamin A-containing medications. Beta-carotene, a precursor to vitamin A, is not harmful, however.[20]

Calcium   Calcium used to support the mineralization of bones in the fetus is supplied by the mother's diet and, if needed, by the calcium contained in the long bones of the mother's body. Consequently, the fetus has access to as much calcium as it needs. Uptake of calcium by the fetus is especially high during the third trimester, when the fetus's bones are mineralizing.

Pregnant women who regularly consume low-calcium diets lose calcium from their bones during pregnancy but usually regain it after delivery. Because any calcium losses are from the long bones—not from the teeth—the old saying, "for every baby a tooth" has no basis in fact. Several studies have shown that teeth do not demineralize as a result of low-calcium diets during pregnancy.[5]

Vitamin D   Vitamin D during pregnancy supports fetal growth and the addition of calcium to bone and programs genes in ways that can influence the development of chronic diseases such as rheumatoid arthritis and cancer later in life.[21] Lack of it compromises fetal growth and bone development, and this may be happening during many pregnancies. About 42% of African American and 4% of Caucasian women have low blood levels of vitamin D.[22] Vegan women may also be at risk for poor vitamin D status because vitamin D is naturally present only in animal products.

An intake of 5 mcg (200 IU) vitamin D daily is officially recommended for pregnancy. Some credible experts, however, assert that more vitamin D than that is needed. It would not be surprising to see recommended intake levels of vitamin D at least double.[23] Intakes of vitamin D from foods and supplements should not exceed 50 mcg (2000 IU) daily.

Illustration 29.7   Foods fortified with folic acid.

Iron   Iron deficiency is the most common nutrient deficiency in pregnant women.[24] It develops when a woman enters pregnancy with low iron stores and fails to consume enough iron during her pregnancy. Iron requirements increase during pregnancy due to increases in hemoglobin production and storage of iron by the fetus.

Because of the large increase in need, it is difficult for pregnant women to get enough iron from foods. Consequently, 30 milligrams of supplemental iron per day are recommended for the second and third trimesters of pregnancy. Women who do not take supplemental iron are more likely to develop iron deficiency. They are also more likely to deliver infants who are small—and are at risk of developing iron deficiency in their first year—than are women who do take supplements.[25]

Iodine   Iodine is needed for normal thyroid function and plays important roles in protein tissue construction and maintenance. A lack of iodine during pregnancy can interfere with fetal development and, in extreme cases, cause mental and growth retardation and malformation in children.[26]

About half of U.S. pregnant women consume less than the recommended 220 mcg of iodine daily, and 7% have moderate iodine deficiency.[27] The most reliable source of iodine is iodized salt. One teaspoon contains 400 mcg iodine. Salt with added iodine is clearly labeled as "iodized." Fish, shellfish, seaweed, and some types of tea also provide iodine. Women who consume iodized salt are not likely to need supplemental iodine.[28] Usual iodine intake should not exceed 1100 mcg daily during pregnancy.

EPA and DHA   These two omega-3 fatty acids are becoming the folic acid of the recent past. New research is quickly informing us of their roles in fetal and infant development and of the benefits of consuming adequate amounts of them. EPA (eicosapentaenoic acid) and DHA (docosahexaenoic acid) are long-chain, highly unsaturated fatty acids that promote maternal health and support the optimal development of vision and the central nervous system of the fetus and infant. Women who have adequate EPA and DHA intake during pregnancy and breast-feeding tend to deliver infants that develop a somewhat higher level of intelligence, better vision, and otherwise more mature central nervous system functioning than do infants born to women with low intake of EPA and DHA. Rates of preterm delivery are significantly lower among women with adequate EPA and DHA status.[29]

It is currently recommended that women consume 300 mg of EPA and DHA daily during pregnancy and breast-feeding.[30] Most pregnant and breast-feeding women in the United States consume less than half, of this amount. EPA and DHA are found together in fish, fish oils, and seafood (it turns out that fish really is brain food…), and DHA is available in omega-3 fatty acid fortified fortified eggs, orange juice, soy products, margarines, and other products. Consumption of 12 ounces per week of low-mercury fish during pregnancy and breast-feeding is encouraged. Safe fish that provide high levels of EPA and DHA include salmon, herring, anchovies, shrimp, and halibut.

What's a Good Diet for Pregnancy?

Contrary to folklore, women do not instinctively select and consume a healthy diet during pregnancy. A good diet takes planning.

Regardless of a pregnant woman's age and whether or not she is a vegetarian, a healthy diet provides all the nutrients she needs with the possible exception of iron. Foods are recommended over supplements because foods provide fiber, antioxidants, and other beneficial substances that supplements do not. Pregnancy diets should include sufficient fluid and fiber and should consist of regular meals and snacks. A pregnant woman should not consume alcohol and should drink

Table 29.4

Dietary recommendations for pregnancy

- Consume sufficient calories for adequate weight gain.
- Eat a variety of foods from each food group.
- Eat regular meals and snacks.
- Consume sufficient dietary fiber (about 28 grams per day).
- Consume 11 to 12 cups of water each day from fluids and foods.
- Use salt to taste (within reason).
- Do not drink alcoholic beverages.
- Limit coffee to four or fewer cups per day.
- Eat foods you enjoy at pleasant meal times.

Illustration 29.8   An example of food group recommendations for pregnant and breast-feeding women based on MyPyramid Plan for Moms.

coffee only in moderation. Table 29.4 lists these and other recommendations for the diet during pregnancy. Planning a diet for pregnancy around the basic food groups is the most straightforward approach to meeting nutrient needs (Illustration 29.8). These recommendations also apply to breast-feeding women.

Why Alcohol and Pregnancy Don't Mix   As early as the 1800s, maternal consumption of alcohol during pregnancy was said to cause the birth of "sickly" infants. The ill effects of alcohol on babies were not fully acknowledged, however, until the 1970s, when several research reports described a condition called fetal alcohol syndrome (Illustration 29.9). Women who drank heavily or frequently binged on alcohol during pregnancy were found to be at high risk of delivering infants with specific malformations and retarded physical and mental development. The effect of maternal alcohol consumption on the fetus worsens as intake increases. Heavy drinking in the first half of pregnancy is closely associated with the birth of malformed, small, mentally impaired infants. When excessive drinking occurs only in the second half of the pregnancy, infants are less likely to be malformed but are still likely to be small and to suffer abnormal mental

Food group	Ounces/cups recommended per day	Examples of equivalent measures
Grains	8 ounces	• 1 slice bread = 1 ounce • 1 cup cold cereal = 1 ounce • 1 cup cooked rice, pasta, or cereal = 2 ounces
Vegetables	3 cups	• 2 cups tossed salad = 1 cup
Fruits	2 cups	• 1 cup fruit juice = 1 cup
Milk	3 cups	• $^1/_3$ cup shredded cheese = 1 cup • 2 slices American cheese = 1 cup • $1^1/_2$ ounces hard cheese = 1 cup • $1^1/_2$ cups ice cream = 1 cup
Meat and beans	6 1/2 ounces	• 1 small egg = 1 ounce • 1 Tbsp peanut butter = 1 ounce • $^1/_4$ cup dried beans = 1 ounce • $^1/_2$ ounce nuts = 1 ounce • 1 Tbsp mayonnaise = $2^1/_2$ teaspoons oil • 1 Tbsp salad dressing = 1 teaspoon oil

NOTE: Based on MyPyramid food guide for women consuming 2400 calories a day and adjusted for pregnancy nutrient needs. Serving numbers will vary based on actual caloric need.
Source: Judith E. Brown, 2006.

development.[31] These conditions are permanent; they cannot be fully corrected with special treatment, and the child does not outgrow them.

No amount of alcohol has been found to be absolutely safe during pregnancy. When only an occasional drink is consumed, however, the adverse effects of alcohol on fetal development appear to be small and rare. To exclude the possibility of even small impairments in fetal growth and development, it is recommended that women not drink alcohol at all during pregnancy or when they are attempting to become pregnant.[13]

Vitamin and Mineral Supplements   Iron is the only supplement recommended for all pregnant women. About 83% of pregnant women take multiple vitamin and mineral supplements, however. Apparently, supplements are being prescribed as insurance against the possibility of poor diets. Women should be given supplements the same way they are given medications—when they are indicated. Under certain conditions, supplementation with nutrients besides iron is indicated. A multivitamin–mineral preparation is recommended for pregnant women who do not ordinarily consume an adequate diet and for those in high-risk categories, such as women carrying more than one fetus, heavy cigarette smokers, and alcohol and drug abusers.[4,13]

Teen Pregnancy

Approximately 4 of every 100 females between the ages of 15 and 19 years in the United States deliver babies each year.[11] Although teens who do not smoke, drink alcohol, or use drugs and who enter pregnancy in good physical and nutritional health tend to deliver healthy infants, many teens do not enter pregnancy in such good condition. Poor lifestyle habits are thought to be primarily responsible for the high rate of low-birthweight and preterm infants born to teens. Few teens (1%) meet all of the food group recommendations. Individualized attention, nutrition intervention, and support all foster healthy outcomes of teen pregnancy.[32]

Breast-Feeding

A woman's capacity to nourish a growing infant does not end at birth; it continues in the form of breast-feeding (Illustration 29.10). Breast milk from healthy, well-nourished women is ideally suited for infant nutrition and health.[33,34]

What's So Special about Breast Milk?

Breast milk is like a bonus pack. In addition to serving as a complete source of nutrition for infants for the first 4 to 6 months of life, breast milk contains substances that convey a significant degree of protection against a variety of illnesses (Table 29.5)—including infectious diseases such as polio and the "flu" and ear, respiratory tract, and gastrointestinal tract infections. Evidence indicates that breast milk may confer a degree of protection against the development of cancer of the lymph system (lymphoma), asthma, type 2 diabetes, and asthma during childhood.[35,36] In addition, breast milk contains essential and nonessential fats and other substances that appear to promote optimal growth and development of the nervous system and eyes. Intelligence, as measured by IQ, tends to be higher in babies receiving breast milk than in those given formula.[37] The disease-preventing and development-promoting components of human milk are lifesaving assets in many developing countries, where a safe water supply and medical care may be unavailable. Although breast-feeding doesn't protect infants from all infectious diseases and food allergies, it's the ounce of prevention that's worth a pound of cure.

Illustration 29.9   Children with fetal alcohol syndrome (FAS) have characteristic facial features, mental retardation, heart defects, and other problems. Less severe cases can include small size and slow development.

Food is the first enjoyment of life.
—LIN YUTANG

The breastfed infant is the reference model against which all alternative feeding methods must be assessed with regard to growth, health, and development; and other short- and long-term health outcomes.
—HHS BLUEPRINT FOR ACTION ON BREASTFEEDING, 2000

Photo Disc

Illustration 29.10 Babies are born to be breast-fed.

Breast-feeding also offers other benefits—ones that may mean a lot to parents. Table 29.6 discusses them.

Is Breast-Feeding Best for All New Mothers and Infants?

Over 96% of women are biologically capable of breast-feeding, and the vast majority of infants thrive on breast milk.[38] In the United States, about 77% of new mothers breast-feed their infants to some extent (Illustration 29.11).[39] Successful breast-feeding involves more than biology; it is heavily influenced by environmental and psychological conditions.

The increase in women returning to work soon after delivery, a lack of health care provider and emotional support for breast-feeding, embarrassment, early hospital discharge, and inadequate knowledge about how to breast-feed all appear to be deterring U.S. women from breast-feeding. If more women are to have the opportunity to breast-feed, these and other barriers must be broken down. Perhaps a good place to start is with the value our society places on breast-feeding. A number of European countries that actively promote breast-feeding allow women to stay in the hospital after delivery until breast-feeding is going well. The usual practice in some African countries is to relieve a breast-feeding mother's workload so that she can devote nearly full time to feeding and caring for her young infant. A relative may move in with the family and take over household chores, or the mother and baby may live with her parents for a time.

Table 29.6	

Fifteen reasons to breast-feed

1. The milk container is easy to clean.
2. Breast milk is a renewable resource.
3. There's no packaging to discard.
4. Breast milk comes in an attractive container.
5. The temperature of breast milk is always perfect right out of the container.
6. Breast milk tastes really good.
7. There are no leftovers.
8. You don't have to go to the kitchen in the middle of the night to get breast milk ready.
9. It takes just seconds to get a meal ready.
10. There's no bottle to repeatedly pick up off the floor.
11. The price is right.
12. The meal comes in a perfect serving size.
13. Meals and snacks are easy to bring along on a trip or outing.
14. Feeding units come in an assortment of beautiful colors and sizes.
15. One food makes a complete meal.

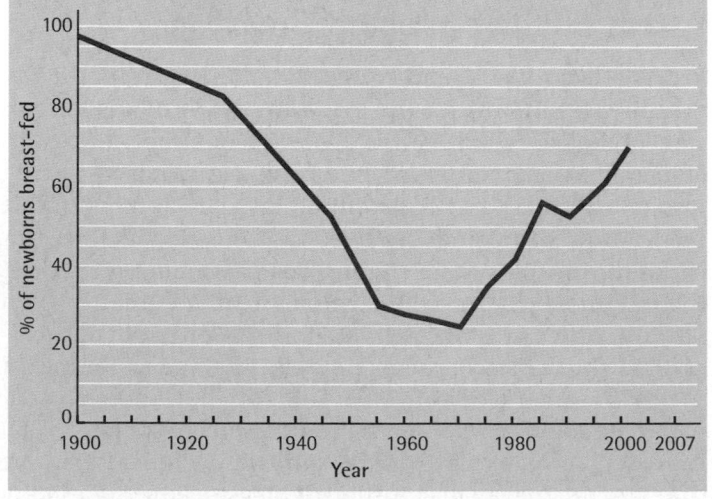

Illustration 29.11 Percentage of breast-fed newborns in the United States.[39,40]

Public health initiatives are in place to facilitate breast-feeding. To meet national health objectives for improving rates of breastfeeding initiation and continuation in the United States (see Table 29.7), it is recommended that:

- Health care workers encourage and facilitate breastfeeding

- "How to" advice and problem solving guidance be available for breast-feeding women from qualified health care staff

- Breast-feeding women returning to work have access to on-site child care facilities and private rooms for breast-feeding

Breast-feeding is at its best when both the mother and infant benefit from the experience. If mutual benefit is not possible, then formula feeding may be necessary. Infant growth and development are well supported by commercially available infant formulas.[42]

Table 29.7

Actual versus national goals for breast-feeding in the United States[39–41]

	Newborns	6 Months	12 Months
National goal	75%	50%	25%
Actual (all women)	77%	39%	20%

How Breast-Feeding Works

The mother's body prepares for breast-feeding during pregnancy. Fat is deposited in breast tissue, and networks of blood vessels and nerves infiltrate the breasts. Ducts that will channel milk from the milk-producing cells forward to the nipple—the milk collection ducts—also mature (Illustration 29.12). Hormonal changes that occur at delivery signal milk production to begin.

Breast milk produced during the first three days or so after delivery is different from the milk produced later. Called **colostrum**, this early milk contains higher levels of protein, minerals, and antibodies than does "mature" milk.

colostrum
The milk produced during the first few days after delivery. It contains more antibodies, protein, and certain minerals than the mature milk that is produced later. It is thicker than mature milk and has a yellowish color.

Breast Milk Production While an infant is consuming one meal, she or he is "ordering" the next. The pressure produced inside the breast by the infant's sucking and the emptying of the breasts during a feeding cause a hormone to be released from specific cells in the brain. The hormone stimulates the production of milk so that more milk is produced for the next feeding. It generally takes about 2 hours for the milk-producing cells to manufacture enough milk for the next feeding. An important exception to this occurs when an infant enters a growth spurt and consumes more milk than usual. Then milk production takes longer, perhaps a day, to catch up with demand.

Only very rarely is a breast-feeding woman unable to produce enough milk. As long as an infant is allowed to satisfy her or his appetite by breast-feeding as often as desired, milk production will catch up with the baby's need.

Nutrition and Breast-Feeding

A breast-feeding woman needs an adequate and balanced diet to replenish her body's nutrient stores, maintain her health, and produce sufficient milk for her baby. Increases in recommended dietary intakes for breast-feeding are generally higher than those for pregnancy. As during pregnancy, proportionately

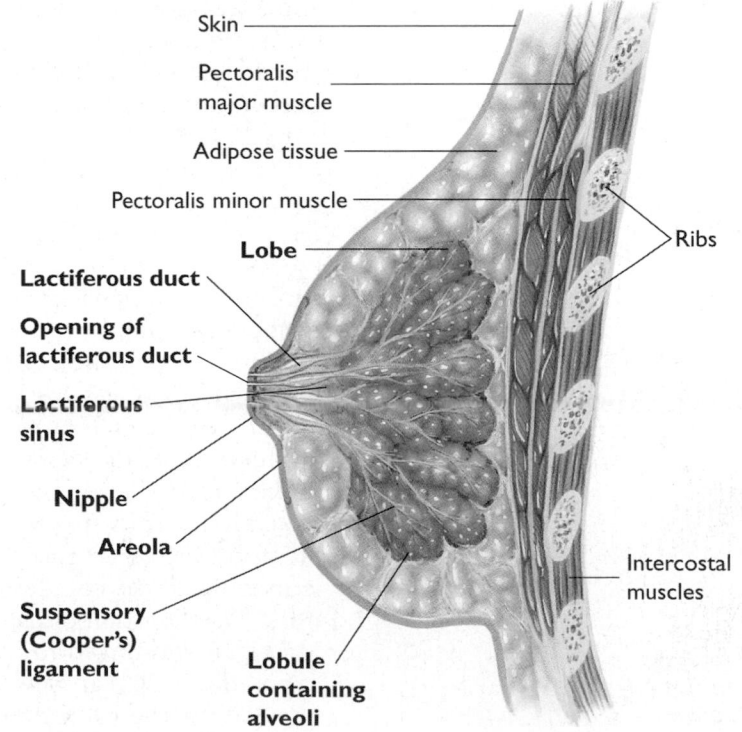

Skin
Pectoralis major muscle
Adipose tissue
Pectoralis minor muscle
Lobe
Lactiferous duct
Opening of lactiferous duct
Lactiferous sinus
Nipple
Areola
Suspensory (Cooper's) ligament
Lobule containing alveoli
Ribs
Intercostal muscles

Illustration 29.12 A view of the interior of the breast.

Can adopting moms breast-feed their infants? Yes, they can. It takes hormones and motivation, however. Breast-feeding success is higher the younger infants are when they begin to breast-feed with their adoptive moms.[43]

Photo Disc

Table 29.8

Dietary recommendations for breast-feeding women[44]

- Diets should supply all of the nutrients breast-feeding women need. (The routine use of vitamin and mineral supplements is not recommended.)
- Fluids should be consumed to thirst (14 cups of fluid each day).
- Weight loss should not exceed 6 to 8 pounds per month after the first month after delivery; caloric intakes should not fall below 1500.
- Alcohol intake should be avoided.

---

higher amounts of nutrients than calories are required, indicating the need for a nutrient-dense diet (see Illustration 29.5 on p. 29–7).

Calorie and Nutrient Needs The RDA for calories is about 15% higher for breast-feeding women than for other women. Actually, a breast-feeding woman needs about 30% more calories than the RDA for women who are not breast-feeding, but she does not have to consume that level of calories from food. Energy supplied from fat stores that normally accumulate during pregnancy contributes to meeting these needs during breast-feeding, so not all of the calories must come from the mother's diet.[44]

A breast-feeding woman has a high requirement for DHA (docohexaenoic acid), an omega-3 fatty acid, because her infant used a good deal of it for central nervous system development. The DHA content of maternal milk directly reflects maternal intake. Adequate intake of DHA by women during breast-feeding is related to gains in intelligence and vision development in infants.[29]

What's a Good Diet for Breast-Feeding Women? The calories and nutrients needed by the breast-feeding woman can be obtained from a varied diet that includes foods from each of the basic food groups. The MyPyramid for Moms recommended numbers of servings from each food group are shown in Illustration 29.8; they are the same as for pregnant women. Dietary recommendations for breast-feeding women are summarized in Table 29.8.

Increases in hunger and food intake that accompany breast-feeding generally take care of meeting caloric needs. Failure to consume enough calories from food can decrease milk production, however. Low-calorie diets (those providing less than 1500 calories per day) and weight loss that exceeds 1.5 to 2 pounds per week—even in women with a good supply of fat stores—can reduce the amount of milk women produce. Weight loss of about a pound a week starting a month after delivery appears to be safe and helpful in women's attempts to return to pre-pregnancy weight.[45]

Are Supplements Recommended for Breast-Feeding Women? Supplements are not recommended for breast-feeding women. Instead, breast-feeding women should try to get all of the nutrients they need from food.[44] Supplements, if needed, should be prescribed on an individual basis.

Dietary Cautions for Breast-Feeding Women Almost anything a woman consumes may end up in her breast milk. When a breast-feeding woman drinks coffee, her infant receives a small dose of caffeine. Breast-fed infants of women who are heavy coffee drinkers (10 or more cups per day) may develop "caffeine jitters." Alcohol also is transferred from a woman's body to breast milk. The development of the brain and nervous system of infants breast-fed by chronic, heavy drinkers appears to be retarded. There is also concern that alcohol ingestion by breast-feeding women may decrease milk production.[46]

Environmental contaminants, such as lead, DDT (dichloro-diphenyl-trichloro-ethane), chlordane, PCBs (polybrominated biphenyls), and PBBs (polychlorinated biphenyls), for example, are transferred into breast milk. Many environmental contaminants may be stored in a woman's fat or bone tissues. When the fat stores are later broken down for use in breast milk or the calcium in bone is mobilized, the contaminants stored in the fat or bone may enter the breast milk. The ingestion of fish from contaminated waters in Lake Ontario and Lake Michigan was linked to abnormally high levels of PCBs in breast milk. Infants exposed to PCBs can develop rashes, digestive upsets, and nervous system problems. With the exception of breast-feeding women known to be exposed to excess levels of environmental toxins, however, it is concluded that the benefits of breast-feeding outweigh the risks to infants from harmful substances in the environment.[47]

Infants are much smaller than women, and it takes a smaller dose of caffeine, alcohol, drugs, or environmental contaminants to have an effect on them than on an adult. Whereas breast-feeding mothers may show no adverse effects from these substances, their infants may.

Breast-feeding women are advised to limit their consumption of regular coffee to 4 or fewer cups a day and to avoid alcohol or limit it to one drink with a meal per day.[44] Drugs or medications should be taken only on the advice of a health care provider.

Infant Nutrition

At no other time during life outside the womb do growth and development proceed at a faster pace than during infancy. Infants grow out of clothes within weeks, long before they wear them out. Each day infants learn new behaviors, and their minds absorb large chunks of information that will serve them well in the future (Illustration 29.13). The rate at which growth and development proceed in the first year of life is truly amazing. Just as infants need security, love, and attention to flourish, so too do they need calories and essential nutrients.

Infant Growth

During the first week or two of life, infants generally lose 5 to 10% of their birthweight while adjusting to the new surroundings. After that, infants grow rapidly. Most infants double their birthweight by 4 months and triple it by 1 year. Length usually increases by 50% during the first year. If this rate of growth were to continue, 10-year-old children would be about 10 stories high and weigh over 220 tons! (Look at Illustration 29.14 to get a clear idea of how tall that is.) After infancy, the growth rate declines and remains at a fairly low level until the adolescent growth spurt begins. Development proceeds in parallel with growth during the first year of life (Illustration 29.15).

Growth Charts for Infants   In 2000 the Centers for Disease Control and Prevention released growth charts for female and male infants, children, and adolescents. Separate charts were not developed by ethnic group or race, because growth potential is a shared human trait that varies primarily due to environmental factors such as nutritional status and disease. An example of the growth charts that include infants is shown in Illustration 29.16 on page 29–17. The charts should be carefully plotted based on accurate, periodic measures of an infant's size. They are best used for screening growth problems; confirmation of underweight or obesity requires assessment of body fat by skinfold thickness and other measures.[48]

Body Composition Changes with Growth   Humans grow up and out, and their body composition and proportions change as growth progresses (Illustration 29.17 on page 29–17). Although generally measured by gains in pounds and inches, growth also reflects changes in bone mass, organ size, body proportions, and composition (the proportionate amounts of water, muscle, and fat). Infants' heads are very large in relation to the rest of their bodies, for example. Brain growth takes precedence over trunk and limb growth early in life, and the body eventually grows to "fit" the head (Illustration 29.18). During the first ten years of life, a child who initially has the shape of a loaf of bread may normally come to resemble a string bean.

Nutrition and Mental Development

Malnutrition has the greatest impact on mental development when it is severe and occurs during the critical period for brain cell multiplication. For humans, this vulnerable period begins during pregnancy and ends after the first year of life. Impairment in mental development is less severe when malnutrition occurs

He who possesses virtue in abundance may be compared to an infant.
—Lao Tzu, SIXTEENTH CENTURY B.C.

Illustration 29.13   Infants develop at a remarkable rate.

Illustration 29.14   This is how tall you would have been as a 10-year-old if you had kept growing at the rate you did during infancy.

Illustration 29.15 Growth and developmental characteristics from birth through one year.

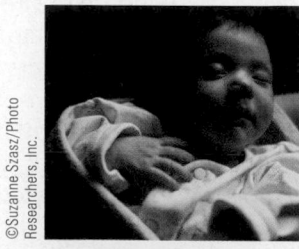

First Month
Generally weighs from 8 to 11 pounds; length is 20 to 23 inches. Head is relatively large and has soft spot on top. Startles and sneezes easily. Jaw may tremble. May hiccup and spit up. Eats every few hours.

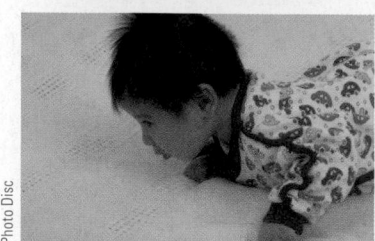

One to Three Months
Lifts head briefly when placed on stomach; smiles, coos, and gurgles. Whole body moves when infant is touched or lifted. Eats every three to four hours.

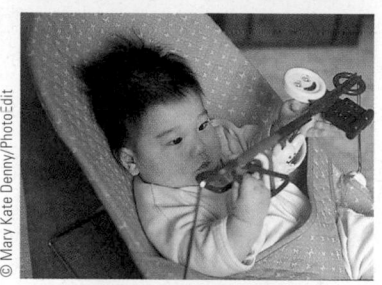

Four to Six Months
Weight nearly doubled. Has grown three to four inches. Follows objects with eyes. Reaches toward objects with both hands; puts fingers and objects into mouth. Turns over, sits unassisted. Awake longer at feeding time. Eats six to seven times per day. Sleeps six to seven hours at night.

Seven to Eleven Months
Gains in weight and height are less rapid, appetite has decreased. Stands up with help, hitches self along the floor. Reaches for, grasps, and examines objects with hands, eyes, and mouth. Has one or two teeth. Takes two naps a day.

Twelve Months
Usually has tripled birthweight and increased length by 50%. Grasps and releases objects with fingers. Holds spoon, but uses it poorly. Begins to walk unassisted.

only during pregnancy or only during infancy than if it occurs throughout both pregnancy and infancy.[49]

Children's mental development is also greatly influenced by the social and psychological environment in which they are raised. Because malnutrition is generally accompanied by both social and psychological deprivation, these factors often contribute jointly to poor mental development. For children in the United States, poverty, neglect, illnesses, and psychological problems appear to be the main cause of undernutrition and poor mental development.[50]

Infant Feeding Recommendations

Current dietary recommendations for infants call, preferably, for breast-feeding or, alternatively, for formula feeding for the first 12 months of life (Table 29.9 on p. 29–18).[53] Infants should be fed "on demand," that is, when they indicate

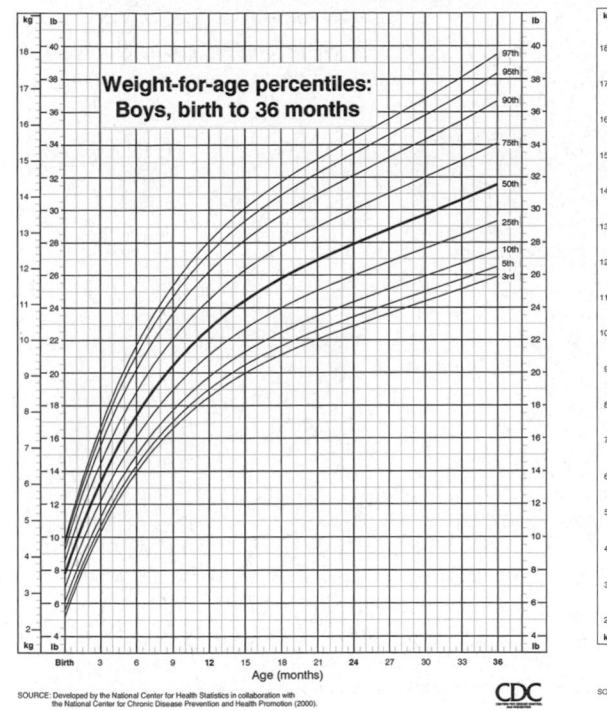

CDC Growth Charts: United States

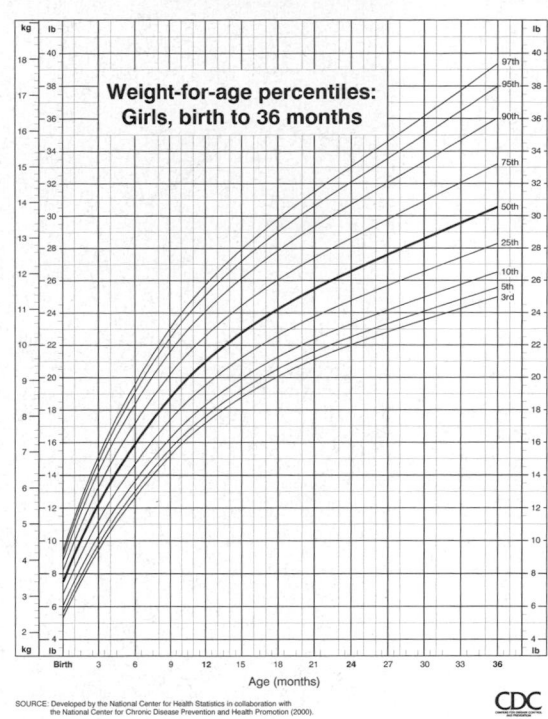

Illustration 29.16 An example of the CDC's growth charts.
The charts can be copied from the Internet using the address www.cdc.gov/growthcharts.

Source: National Center for Chronic Disease Prevention and Health Promotion.

they are hungry, rather than on a rigid schedule set by the clock. In the first few months of life, most infants will get hungry every 3 or 4 hours.

Soft, solid foods should be added to an infant's diet of breast milk or infant formula around 6 months. At one time, it was thought that infants should be offered solid foods within the first few months of life. Although such young infants are unable to swallow much of the food (most of it ends up on their face and bib) or

Photo courtesy of Max Brown and Alyson Sklar

Illustration 29.17 When you hold your arm straight up, is it longer than your head? It's about the same for this 5-month old.

Table 29.9

Overview of infant feeding recommendations[51,52]

- Exclusive breast-feeding for the first 6 months of life and continuation of breast-feeding through the first year of life are preferred. Iron-fortified infant formulas may be used as a secondary option to breastfeeding.
- Introduce solid foods around 6 months of age.
- An infant's first food should be iron-fortified, such as iron-fortified baby cereal.
- After 6 months, offer a variety of healthy foods appropriate for baby. Pay attention to the textures of baby's food, going from smooth to mashed to chopped to tiny pieces.
- Keep trying new foods—babies may need to try a new food ten times or more before they like it!
- Use a small baby toothbrush with a tiny dot of fluoride toothpaste when the teeth start to come in.
- Infants should be fed "on demand" and not by a set schedule. Feeding should stop when the baby loses interest in eating.
- Breast-fed infants receiving little sun exposure should be given a vitamin D supplement (200 IU or 5 micrograms per day).
- Infants not receiving fluoridated water should be given fluoride drops (dose depends on age).

Photo courtesy of Max Brown and Alyson Sklar

Illustration 29.18 Infants' heads are proportionately large for the rest of the body. But, as you can see, there's much more growth to come.

to digest completely what they do swallow, solid foods were thought to help the baby grow and sleep through the night. This belief was incorrect. Infants who receive solids before the age of 6 months are no more likely to grow normally or sleep through the night than are infants who start to receive solid foods around 6 months of age.[54] The age at which an infant begins to sleep for 6 or more hours during the night depends on other factors, including the infant's developmental level and how much he or she has slept during the day. Neither the infant nor the infant's parents are likely to get a full night's sleep for at least 4 months.

Introducing Solid Foods It is recommended that iron-fortified rice cereal be given to all breast-fed infants and to bottle-fed infants who are not receiving iron-fortified formula (Illustration 29.19). The iron provided by iron-fortified rice cereal or formula helps to restock the infant's iron stores, which have been drawn on since birth. Generally, solid foods prepared for infants should consist of single, basic foods such as strained vegetables, fruits, or meats.

Solid foods for infants can be purchased as commercial baby food or prepared at home. When baby foods are prepared at home, care should be taken to avoid contamination and to achieve the right consistency. Infants should be offered new foods one at a time. The new offerings should be separated by several days, so that any allergic reactions to a food can be identified. Variety is the key to achieving a healthy diet for infants in their second 6 months of life, and an assortment of basic, textured foods should be given.[51,55]

By 9 months of age, infants are ready for mashed foods and foods such as yogurt, applesauce, ripe banana pieces, and grits. Most infants have several teeth by this time, and they are able to bite into and chew soft foods.

Infants graduate to adult-type foods after the age of 12 months. Although most foods still need to be mashed or cut up into small pieces for them, 1-year-olds are able to eat the same types of food as the rest of the family. They can drink from a cup and nearly feed themselves with a spoon. Infants have come a long way in 12 months.

Norbert Schaefer/Corbis

Illustration 29.19 Solid foods should be introduced around 6 months.

Table 29.10

Foods to avoid in the first year of life

Foods that May Cause Allergic Reactions	Foods that May Cause Other Problems	Foods that May Lead to Choking
Cow's milk	Blueberries	Grapes
Egg white	Coffee	Frankfurter pieces
Fish, seafood	Corn	Hard candy
Nuts, peanuts	Fruit drinks	Hard pieces of vegetables
Peanut butter	Honey (unpasteurized)	Meat chunks
Soy protein	Prune juice	Nuts and seeds
Wheat products	Tea	Popcorn
		Raw vegetables

Foods to Avoid Not all foods can be considered "baby foods." Some foods should be omitted from an infant's diet because they are apt to cause allergic reactions or are too difficult for infants to chew into small pieces and swallow. Table 29.10 provides a list of foods that should not be offered to infants.

Reduced-fat products are not recommended for infants. Infants need a relatively high-fat diet for brain and nervous system development. Unpasteurized honey is not recommended either, because it may cause botulism in infants due to their still maturing gastrointestinal defenses against bacteria. Beverages containing a lot of sugar, such as soft drinks and sweetened fruit juices, should not be given to infants because they promote tooth decay. To help prevent "baby-bottle tooth decay," shown in Illustration 29.20, infants should not be put to sleep with a bottle containing sweet fluids or formula.

Do Infants Need Supplements? Two situations call for the use of supplements during infancy (see Table 29.11). Breast-fed infants and infants receiving formula from a concentrate that is not diluted with fluoridated water need fluoride supplements after 6 months of age. Since breast milk contains a low amount of vitamin D, and since cases of rickets are increasing among breast-fed infants in the United States, breast-fed infants not exposed regularly to sunshine should receive a vitamin D supplement.[57]

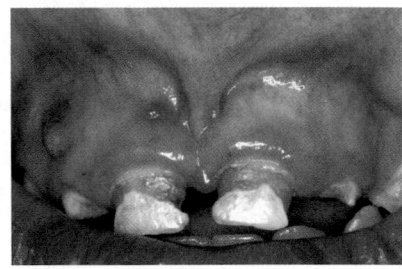

Illustration 29.20 Baby-bottle tooth decay.

© Dietmar A. J. Kennel

The Development of Healthy Eating Habits Begins in Infancy

Older infants and young children should be offered a wide variety of nutritious foods in a positive eating environment. They alone should make the decision about how much to eat of any food offered. Food preferences are primarily learned and are unique to every individual. The learning process begins early in life and is affected by a variety of influences. "Instincts" that draw infants to select certain, nutritious foods are not one of the influences, however. Infants do not instinctively know what to eat. Rather, they are born with mechanisms that help regulate how much they eat. These basic characteristics of infant food intake were beautifully demonstrated by Clara Davis in experiments with 7- to 9-month-old infants in the 1920s and 1930s.[59] One of her most famous experiments is described in Illustration 29.21.

Subsequent research reinforced earlier conclusions that infants and young children should be offered a wide variety of nutritious foods and that decisions about how much to eat should be left up to the infant or child.[61] Beginning at birth, infants who are hungry eat enthusiastically. They stop eating when they are full. Coaxing, cajoling, and pleading by parents or caregivers can override

Table 29.11

Indications for supplementing infants

1. Fluoride supplementation is recommended for infants older than 6 months who live in areas with no or low fluoride in the household water supply (0.25 mg of fluoride/day from 6 months to 3 years of age).[26]

2. Vitamin D supplements (200 IU/day in the United States, 400 IU per day in Canada) should be given to breast-fed infants receiving little exposure to sunshine.[48]

    • Breast-fed infants exposed to about 30 minutes of sunshine each week with only diapers on, or those receiving 2 hours of sun exposure while wearing clothes but no hat, are likely meeting their needs for vitamin D by its manufacture in their skin.[58]

Illustration 29.21   Infant food preferences.

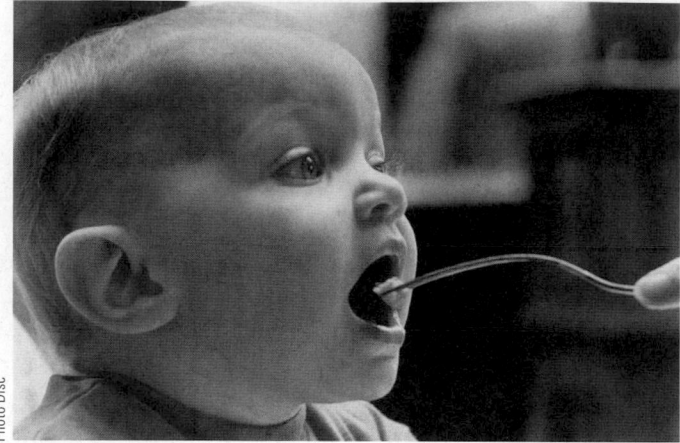

Photo Disc

Clara Davis conducted her most quoted feeding experiment with older infants (7 to 9 months old) living in an orphanage connected to a large hospital in Chicago.

Three times a day a nurse would offer the infants a tray of various foods. The infants would pick up or grab for the foods they wanted, and the nurse would bring the food selected to the infant's hands or mouth.

Altogether 32 different foods were offered, consisting of fresh, unprocessed, unseasoned, and simply prepared basic foods. Foods included milk, beef, kidney, bone marrow, liver, brain, thymus, fish, whole-wheat cereals, raw eggs, sea salt, 15 fresh fruits, and 10 fresh vegetables. Desserts, sugars, syrups, and other sweetened foods were not offered.

Infants quickly formed preferences and narrowed their choices to 14 of the 32 foods. Vegetables were the least liked, whereas milk, bone marrow, eggs, bananas, apples, oranges, and oatmeal were the best liked. They ate enough to grow normally and to remain in good health.

Dr. Davis concluded that appetite is the best guide to how much food an infant or young child needs, and that self-selection will have doubtful value only if the diet is chosen from inferior foods.[60]

infants' decisions about how much food to eat. That, unfortunately, may upset infants' built-in food intake regulating mechanisms and lead to over- or under-eating, and to "fussy" eaters.

Teaching Infants the Right Lessons about Food   Recommendations for feeding infants are primarily based on energy and nutrient needs, the developmental readiness of infants for solid foods, and the prevention of food allergies. But infant feeding recommendations also include a large educational component. Many of the lessons infants learn about food and eating make an impression that lasts a lifetime. Later food habits and preferences, appetite, and food intake regulation are all influenced by early learning experiences.[62] Table 29.12 provides a lesson plan for teaching your infant the right lessons about food and eating.

Making Feeding Time Pleasurable

As Dr. Brazelton, an expert on child development, points out, there is much more to the feeding of infants than meets the eye. Infants learn about communication, relationships, and independence during pleasurable eating experiences. If undertaken in positive and supportive circumstances, mealtimes for infants can provide lessons that will support their physical and psychological health well into the future.

More than any other area of development, feeding is an arena in which parents and baby work out the continuing struggle between dependence (being fed) and independence (feeding oneself). Independence must win. Pushing a child to eat is the surest way to create problems. Feeding has got to be pleasurable.

—T. Berry Brazelton, 1993

Table 29.12

Tips for teaching infants the right lessons about food and eating

1. Infants learn to eat a variety of healthy foods by being offered an assortment of nutritious food choices. There are no inborn mechanisms that direct babies to select a nutritious diet.

2. Infants must be allowed to eat when they are hungry and to stop eating when they are full. Infants, not parents, know when they are hungry and have had enough to eat.

3. Food should be offered in a pleasant environment with positive adult attention.

4. Food should not be used as a reward, punishment, or pacifier.

5. Infants or children should never be coerced into eating anything.

6. Food preferences change throughout infancy. Because an infant rejects a food one time doesn't mean she or he will not accept the food if offered later. Offering a food on a number of occasions often improves acceptance of the food. Babies still may not like strong-flavored vegetables until they are older, however.

© Golden Pixels LLC/Alamy

You Be the Judge!

Focal Point: Apply your nutrition knowledge to the diet of a pregnant woman.

Gloria's baby is due in 5 months. Before becoming pregnant, she never gave much thought to what she ate, but now she is trying to eat healthy foods for her baby and

- *Breakfast:* Bran muffin (2 ounces), cold cereal (1 cup) with strawberries (¹/₂ cup), and a glass of milk (1 cup).
- *Lunch:* Vegetable soup (1 cup vegetables), carrot sticks (¹/₂ cup), and a glass of orange juice (1 cup).

Categorize Gloria's choices into the following food groups to determine if she ate the recommended

herself. Compare her choices with the MyPyramid recommendations in Illustration 29.8 to find out how well she is doing.

- *Afternoon snack:* Yogurt (1 cup).
- *Dinner:* Chicken (3 ounces), turnip greens (¹/₂ cup), a tossed salad (2 cups), and iced tea (2 cups).

servings from MyPyramid. The grains group has been done for you.

Grains
bran muffin, 2 ounces
cold cereal, 1 cup

Gloria's intake: **3 ounces**
Recommended intake: **8 ounces**

Stan Maddock

Vegetables

Gloria's intake:
Recommended intake: **3 cups**

Stan Maddock

Fruits

Gloria's intake:
Recommended intake: **2 cups**

Stan Maddock

Milk and Milk Products

Gloria's intake:
Recommended intake: **3 cups**

Star Maddock

Meat and Beans

Gloria's intake:
Recommended intake: **6 ¹/₂ ounces**

Stan Maddock

Oils

Gloria's intake:
Recommended intake: **7 teaspoons**

Stan Maddock

FEEDBACK (including answers) can be found at the end of Unit 29.

[Key Terms

colostrum, page 29-13
critical period, page 29-3
development, page 29-3
fetus, page 29-3

growth, page 29-3
infant mortality rate, page 29-2
low-birthweight infants, page 29-2
neural tube defects, page 29-7

preterm infants, page 29-2
trimester, page 29-6

Review Questions

TRUE FALSE

1. Relatively high rates of infant mortality, low-birthweight infants, and preterm delivery represent major public health problems in the United States. ☐ ☐

2. "Critical periods" of growth and development are marked by large increases in the size of cells within developing organs. ☐ ☐

3. Inadequate or excessive levels of nutrient availability during pregnancy can modify the function of fetal genes in ways that increase the risk of chronic disease later in life. ☐ ☐

4. A women entering pregnancy with a BMI (Body Mass Index) of 27 should gain 25 to 35 pounds during pregnancy. ☐ ☐

5. Women need 100% more calories during than before pregnancy because they are "eating for two." ☐ ☐

6. Iodine, EPA and DHA, and vitamin D are considered "risk nutrients" during pregnancy because a sizeable portion of women fail to consume enough of them. ☐ ☐

TRUE FALSE

7. Breast-feeding is better for an infant's health than formulas in part because breast milk contains substances that prevent infections and reduce the risk of development of certain chronic disease later in life. ☐ ☐

8. Approximately one in four women is biologically unable to breast-feed successfully. ☐ ☐

9. A woman's milk supply automatically increases as an infant consumes more breast milk. ☐ ☐

10. A daily multivitamin and mineral supplement is recommended for all breast-feeding women. ☐ ☐

11. Infants should be given their first solid food around six months of age and the food should be iron-fortified rice cereal. ☐ ☐

12. Infants are born knowing which foods are good for them. ☐ ☐

Media Menu

www.aap.org
The American Academy of Pediatrics's Web site is loaded with information on breast-feeding and infant and child nutrition.

http://www.americanpregnancy.org
The American Pregnancy Association offers information and resources here that includes a pregnancy calendar, fertility treatment updates, ovulation calendar, and where to find a health care professional.

www.cdc.gov/breastfeeding/
The Centers for Disease Control Web site on breastfeeding can be found here. It includes information on where to go for help, the benefits of breast feeding, and contraindications for breast-feeding.

www.cdc.gov/growthcharts
Need more information about the growth charts, or want copies of them? Get it here.

http://www.cfsan.fda.gov/~pregnant/pregnant.html
Food safety advice for moms-to-be is provided at this Web site.

www.fns.usda.gov/WIC/Breastfeeding/breastfeedingmainpage.htm
Find out about breast-feeding and WIC support for breast-feeding moms at this site.

www.mypyramid.gov/mypyramidmoms/pyramidmoms_plan.aspx
Fill out a few questions about yourself and your pregnancy at My Pyramid for Moms and receive a My Pyramid food plan for a healthy pregnancy diet.

http://mchlibrary.info
The Maternal and Child Health Bureau's site provides information on resources, hot topics, and announcements of new publications

www.nlm.nih.gov/medlineplus/breastfeeding.html
This site is a good source for updates and advice on breast-feeding, problem solving, alternative therapies, and nutrition.

.www.4woman.gov/Breastfeeding
This site offers useful information about breast-feeding and support resources.

Notes

1. Natality statistics, www.cdc.gov/nchs/births.htm. accessed 5/09.

2. DeNoon DJ. U.S. Ranks in Lowest Third for Infant Mortality, 10/15/08, http://www.medscape.com/viewarticle/582117.

3. MacDorman MF et al. Annual summary of vital statistics. Pediatrics 2002;110:1037–50.

4. Position of the American Dietetic Association: Nutrition and lifestyle for a healthy pregnancy outcome, J Am Diet Assoc 2008;108:553–60.

5. Rosso P. Nutrition and metabolism in pregnancy. New York: Oxford University Press; 1990.

6. Johansen AMW et al. Maternal dietary intake of vitamin A and risk of orafacial clefts: a population-based case-control study in Norway, Am J Epidemiol 2008;167:1164–70.

7. Bruce K. Nutritional experiences in early life as determinants of the adult metabolic phenotype. Presented at the Experimental Biology Annual Meeting, New Orleans, April 20, 2009.

8. Langley-Evans SC. Developmental programming of health and disease. Proc Nutr Soc. 2006;65:97–105.

9. Zeisel SH. Is maternal diet supplementation beneficial? Optimal development of infant depends on mother's diet, Am J Clin Nutr 2009;89:685S–7S.

10. Winick M. Nutrition and pregnancy, White Plains (NY): March of Dimes Birth Defects Foundation; 1986.

11. Miller DR, Hayes KC. Vitamin excess and toxicity, In: Hathcock JN, editor. Nutritional toxicology, New York: Academic Press; 1982: pp 81–131.

12. Frederick IO et al. Pre-pregnancy body mass index, gestational weight gain, and other maternal characteristics in relation to infant birth weight, Matern Child Health J 2008;12:557–67.

13. National Academy of Science (Institute of Medicine), Nutrition during pregnancy: I. Weight gain, II. Nutrient supplements, Washington, DC: National Academy Press; 1990: p 48 and the revised guidelines issued by the IOM in May, 2009.

14. Ohlin A, Rossner S. Maternal body weight development after pregnancy. Int J Obesity 1990;14:159–73.

15. Dietary Reference Intakes: Energy, carbohydrate, fiber, fatty acids, cholesterol, protein, and amino acids. Washington, DC: National Academies Press; 2002.

16. Lawrence JM et al. Racial and ethnic disparities in folate intake. Am J Obstet Gynecol 2006;194:520–6.

17. Daly LE et al. Folate levels and neural tube defects. JAMA 1995;274:1698–1702.

18. Brown JE et al. Predictors of red cell folate level in women attempting pregnancy. JAMA 1997;277:548–52.

19. Tamura T et al. Folate and human reproduction. Am J Clin Nutr 2006; 83:993–1016.

20. Brown JE, Kahn ESB. Maternal nutrition and the outcome of pregnancy, Clin Perinatol 1997;24:433–49

21. Hollis BW et al. Vitamin D deficiency during pregnancy: an ongoing epidemic. Am J Clin Nutr 2006;84:273.

22. Nesby-O'Dell et al. Hypovitaminosis D prevalence and determinants among African American and white women of reproductive age: third National Health and Nutrition Examination Survey, 1998–1994. Am J Clin Nutr. 2007;76:187–92.

23. Kovacs CS. Vitamin D in pregnancy and lactation: maternal, fetal, and neonatal outcomes from human and animal studies, Am J Clin Nutr 2008;88(suppl):520S–8S.

24. Murray-Kolb LE et al. Iron deficiency and child and maternal health, Am J Clin Nutr 2009;89(suppl):946S–50S.

25. Preziosi P et al. Effect of iron supplementation on the iron status of pregnant women: consequences for newborns Am J Clin Nutr 1997;66:1178–82.

26. Zimmermann MB. Iodine deficiency in pregnancy and the effects of maternal iodine supplementation on the offspring: a review. Am J Clin Nutr 2009;89:668S–72S.

27. Utiger RD. Iodine nutrition: more is better. N Engl J Med 2006;354:2819–21.

28. Vermiglio F et al. Thyroid function in pregnant women from a mildly iodine deficient area, J Clin Endocrinol Metab 2008;93:2466–8, 2616–21.

29. Carlson SE. DHA supplementation in pregnancy and lactation, Am J Clin Nutr 2009;89:678S–84S.

30. Harris WS et al. Towards establishing dietary reference intakes for eicosapentaenoic and docosahexaenoic acids, J Nutr 2009;139:804S–19S.

31. Mattson SN et al. Heavy prenatal alcohol exposure with or without physical features of fetal alcohol syndrome leads to IQ deficits. J Pediatr 1997; 131:718–21.

32. Dubois S et al. Ability of the Higgins Nutrition Intervention Program to improve adolescent pregnancy outcome, J Am Diet Assoc 1997;97:871–8.

33. Schwarz EB et al. Duration of lactation and risk factors for maternal cardiovascular disease, Obstet Gynecol 2009;113:974–82.

34. Cummings AS. Morbidity in breast-feed and artificially fed infants, J Pediatr 1979;90:726–9.

35. Mayer-Davis EJ et al. Breast-feeding and type 2 diabetes in the youth of three ethnic groups: the SEARCH for diabetes in youth case-control study, Diabetes Care. 2008 Mar;31:470–5.

36. Keuhne CE et al. Breast-feeding and the risk of childhood asthma, 18th Annual Congress of the European Respiratory Society, 10/9/08, www.medscape.com/viewarticle/581836, accessed 5/09.

37. Kramer MS et al. Prolonged and exclusive breast-feeding and cognitive development in children age 6.5 y. Arch Gen Psychiatr 2008;65:578–84.

38. Simopoulos AP, Grave GD. Factors associated with the choice and duration of infant-feeding practices. Pediatrics 1984;74(suppl):603–14.

39. McDowell MM et al. Breast-feeding in the United States: findings from NHANES, 1999–2006, NCHS Data Brief, no. 5, April, 2008.

40. Health, United States, 2003. www.cdc.gov/nchs/hus.htm, accessed 10/06.

41. Hill PD et al. Assessment of breast-feeding and infant growth, J Midwifery Womens Health 2007;52:571–8.

42. Committee on Nutrition, American Academy of Pediatrics. Breastfeeding and the use of human milk. Pediatrics 1997;100:1035–9.

43. Lakhkar BB. Breastfeeding in adoptive babies. Indian Pediatr 2000;37:1114–16.

44. National Academy of Sciences (Institute of Medicine). Nutrition during lactation. Washington, DC: National Academy Press; 1991.

45. Lovelady CA et al. The effect of weight loss in overweight, lactating women on the growth of their infants. N Engl J Med 2000;342:449–53.

46. Menella JA, Beauchamp GK. The transfer of alcohol to human milk, N Engl J Med 1991;325:981–85.

47. Rogan WJ, Pollutants in breast milk, Arch Pediatr Adol Med 1996;150:981–90.

48. Thompson D et al. New CDC growth charts. WIC National Meeting, Salt Lake City, Utah, 9/8/00.

49. Winick M. Malnutrition and brain development. New York: Oxford University Press; 1976.

50. Lloyd-Still JD et al. Intellectual development after severe malnutrition in infancy. Pediatrics 1974;54:306–11.

51. New infant feeding guidelines, accessed 10/06.www.wicworks.ca.gov/education/nutrition/Infant_Feeditn/InfantFeeding.htm

52. Briefel R, Reidy K, Karwe V, Devaney B. Feeding Infants and Toddlers Study: Improvements needed in meeting infant feeding recommendations. J Am Diet Assoc. 2004;101(suppl 1):S31–S37.

53. Rappo PD et al. Pediatrician's responsibility for infant nutrition. Pediatrics 1997;99:749–50.

54. Beal VA. Termination of night feeding in infancy. J Pediatr 1969;75:690–2.

55. Fiocchi A et al. Consensus document for introducing solid foods into an infant's diet to avoid food allergies. Am Allergy Asthma Immunol 2006;97:10–21.

56. American Academy of Pediatric Dentistry. Guidelines for fluoride therapy. www.aapd.org/media/Policies_Guidelines/G_FluorideTherapy.pdf, accessed 10/06.

57. Ziegler EE. Vitamin D deficiency in breastfed infants in Iowa. Pediatrics. 2006;118:603–10.

58. Specker BL et al. Sunshine exposure and serum 25-hydroxyvitamin D concentrations in exclusively breast-fed infants. J Pediatr 1985;107:372–6.

59. Davis C. Self-selection of diets by newly weaned infants: an experimental study. Am J Dis Child 1928;36:651–79.

60. Story M, Brown JE. Do young children instinctively know what to eat? The studies of Clara Davis revisited, N Engl J Med 1987;316:103–6.

61. Birch LL, Deysher M. Caloric compensation and sensory specific satiety: evidence for self-regulation of food intake by young children. Appetite 1986;7:323–31.

62. Skinner JD et al. Do food-related experiences in the first 2 years of life predict variety in school-aged children? J Nutr Ed Behav 2002;34:310–5.

Answers to Review Questions

1. True, see page 29-2.
2. False, see page 29-3.
3. True, see page 29-4.
4. False, see page 29-5.
5. False, see page 29-6.
6. True, see page 29-6.
7. True, see page 29-11.
8. False, see page 29-12.
9. True, see page 29-13.
10. False, see page 29-14.
11. True, see page 29-18.
12. False, see page 29-19.

NUTRITION | # Up Close

You Be the Judge!

Feedback for Unit 29

Gloria's choices:

Grains	3 ounces (bran muffin, cereal)
Vegetables	3 cups (vegetable soup, carrot sticks, turnip greens, tossed salad)
Fruits	$2\frac{1}{2}$ cups (strawberries, apple, orange juice)
Milk and milk products	2 cups (milk, yogurt)
Meat and beans	3 ounces (chicken)
Oils	0 teaspoons

All of Gloria's choices are healthy ones, but she is missing some important foods and sufficient calories. While concentrating hard on eating enough fruits and vegetables, she has neglected to consume enough nutrient-dense foods from the remaining food groups. She needs to include in her diet more dairy products, add more meat or protein alternates, and more grains, especially whole-grain products. With a few additions to her existing choices, the quality and quantity of her diet can be greatly improved. Here is a more balanced version of Gloria's menu for a day:

- *Breakfast:* Bran muffin, cereal with strawberries, and a glass of milk.
- *Lunch:* Vegetable soup, carrot sticks, an apple, **burrito**, and a glass of orange juice.
- *Afternoon snack:* Yogurt, **cheese**, and **whole-grain crackers**.
- *Dinner:* Chicken, grits, turnip greens, tossed salad, **with 2 Tbsp ranch dressing**, **whole grain roll** and iced tea.

Nutrition for the Growing Years: Childhood through Adolescence

NUTRITION SCOREBOARD

		TRUE	FALSE
1	Parents should not allow their children to eat "junk" foods.		
2	It is up to parents and caretakers to decide how much children should eat.		
3	The incidence of overweight among 6- to 11-year-olds has more than quadrupled since the 1960s.		
4	Childhood and adolescence represent "grace periods" during which the diet consumed does not influence future health.		

Key Concepts and Facts

- There is no evidence to support the notion that children are born knowing what foods they should eat.

- Children are born with regulatory processes that help them decide how much to eat.

- Parents and caretakers should decide *what* foods to offer children. Children should decide *how much* to eat.

- Diet and other behaviors of children and adolescents affect health before and during the adult years.

Answers to
NUTRITION SCOREBOARD

		TRUE	FALSE
1	Withholding sweets and other favorite foods makes kids value and want them more. Such foods should not be given special emphasis.		✔
2	Children, not parents or caregivers, should decide how much food is consumed.		✔
3	About 4% of 6- to 11-year-olds were overweight in the 1960s. Now 15% are overweight.[1]	✔	
4	Childhood and adolescent diets can affect future health. Heart disease and diabetes, for example, can have "pediatric" origins that are related to dietary intake patterns.		✔

You are the most important influence on your child. You can do many things to help your children develop healthy eating habits for life.
—MICHELLE OBAMA, 2009

As the twig is bent, so grows the bough.

Juneberg Clark/Photo Researchers, Inc.

The Span of Growth and Development

Physical and mental development proceed at a high rate from infancy through adolescence (Illustration 30.1). Young children generally enter this phase of life able to take a step or two and to guide a spoon into their mouths sideways. They will likely leave adolescence with the ability to drive, work for pay, and solve complex problems. These are the formative years that lay the foundation for the rest of life.

The Nutritional Foundation

Good nutrition takes on particular importance during the growing years, for many reasons. Growth is an energy- and nutrient-requiring process. It will not proceed normally unless the diet supplies enough of both.

Children learn about food and its importance to health and well-being during these early years of life, and they also establish food preferences and physical activity patterns that may endure into the adult years. Food intake regulatory mechanisms are affected by the lessons children learn early in life. Given control over decisions about how much to eat, children generally become responsive to internal cues that signal when they should eat and when they should stop eating.[2] If the lessons go well, they learn to eat for the right reason.

Eating well and learning the right lessons about food and health have implications that transcend the growing years. Early diets may have long-term effects on the risk of developing a number of diseases later in life.

Characteristics of Growth in Children

Growth slows substantially after the first year of life. Between the ages of 2 and 10 years, children normally gain somewhere around 5 pounds and 2 to 3 inches in

One to Two Years

Gains in height and weight continue at a lower rate; appetite is less. Uses finger and thumb to pick up things. Soft spot grows smaller and then disappears. Baby teeth continue to appear. Usually takes one long nap a day. Drinks from a cup, attempts to feed self with a spoon. Likes to eat with hands. Pulls self up to standing position. Walks alone.

Illustration 30.1 Growth and development characteristics from 1 through 16 years.

Two to Three Years

Slower and more irregular gains in height and weight. Has all 20 teeth. Runs and climbs, pushes, pulls, lugs, walks upstairs one step at a time. Feeds self using fingers, spoon, and cup; spills a lot. At times has one favorite food. Associates the sensation of hunger with a need for food.

Three to Four Years

Gains 4–6 pounds and grows about 2–3 inches. Feeds self and drinks from a cup quite neatly, carries things without spilling. May give up sleep at nap time, substituting quiet play.

Four to Five Years

Gains in height and weight about same as previous year. Hops and skips, throws ball. Has increasingly good coordination, masters buttons and shoelaces. Can use knife and fork and is a good self-feeder.

Five to Six Years

Growth continues at about the same rate. Legs lengthen. Six-year permanent molars usually appear (new teeth, not replacing baby teeth). Begins to lose front baby teeth. Prefers plain, bland, and unmixed foods.

(continued)

Illustration 30.1 *(continued)*
Growth and development characteristics from 1 through 16 years.

Six to Nine Years

Slow gains in height and weight (2–3 inches and 4–6 pounds a year). Some additional permanent teeth appear. Likely to have the childhood communicable diseases. Sleeps 11 to 13 hours.

Nine to Twelve Years

May be long legged and rangy, but health is generally sturdy. Permanent teeth continue to appear. Appetite good. Needs about 10 hours sleep. Growth spurt in girls usually begins. May get very irritable when hungry.

Twelve to Fourteen Years

Wide differences in height and weight in children of either sex of same age. Menstruation usually begins (sometimes earlier). Girls develop breasts, are usually taller and heavier than boys of the same age. Growth spurt of boys begins. Muscle growth rapid. Appetite increases. Likes to spend time with friends.

Fourteen to Sixteen Years

Boys are in period of growth spurt. Voice deepens. Girls have usually achieved maximum growth. Menstruation is established though may still be irregular. Enormous appetite. Pimples a common problem. Sleep reaches its adult pattern. All permanent teeth except wisdom teeth.

height per year (Illustration 30.2). Gains in weight and height occur in "spurts" rather than continuously and gradually. Prior to a growth spurt, appetite and food intake increase (given an adequate food supply), and the child puts on a few pounds of fat stores (Illustration 30.3 on p. 30–6). During the growth spurt, these fat stores are used to supply energy needed for growth in height.

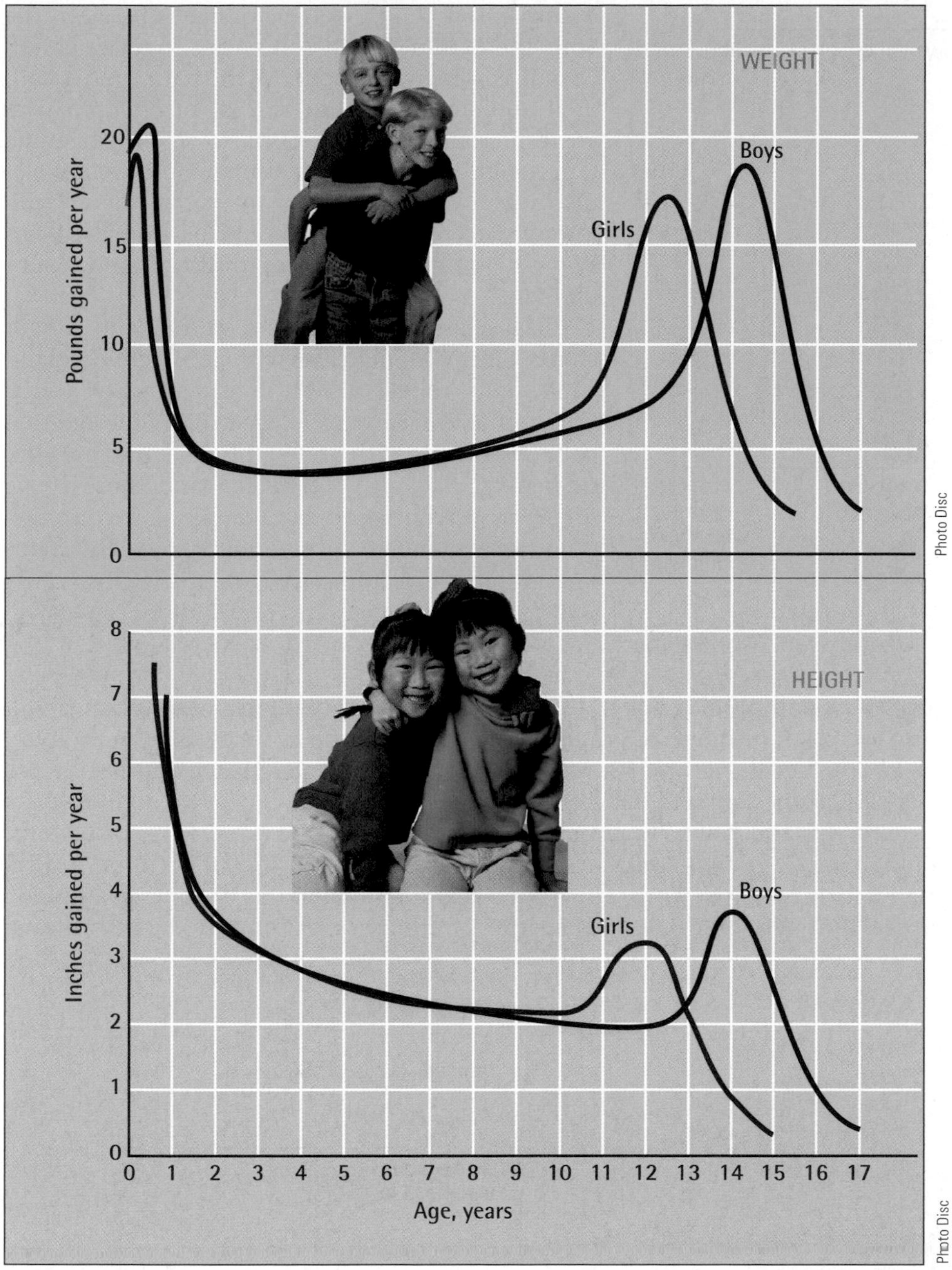

Illustration 30.2 Average yearly growth in weight and height during childhood and adolescence.

CDC's Growth Charts for Children and Adolescents

Growth progress during childhood and adolescence is generally monitored with the use of the Centers for Disease Control (CDC) growth charts. Charts are available for females and males from 0 to 36 months old and from 2 to 20 years. Growth charts for 2- to 20-year-olds consist of graphs of:

- Weight for age (see Illustration 30.4)
- Height (stature) for age
- Weight for height
- Body mass index (BMI) for age

Each graph provides percentiles that reflect the distribution of these measures in a representative sample of 2- to 20-year-olds in the United States. So, for

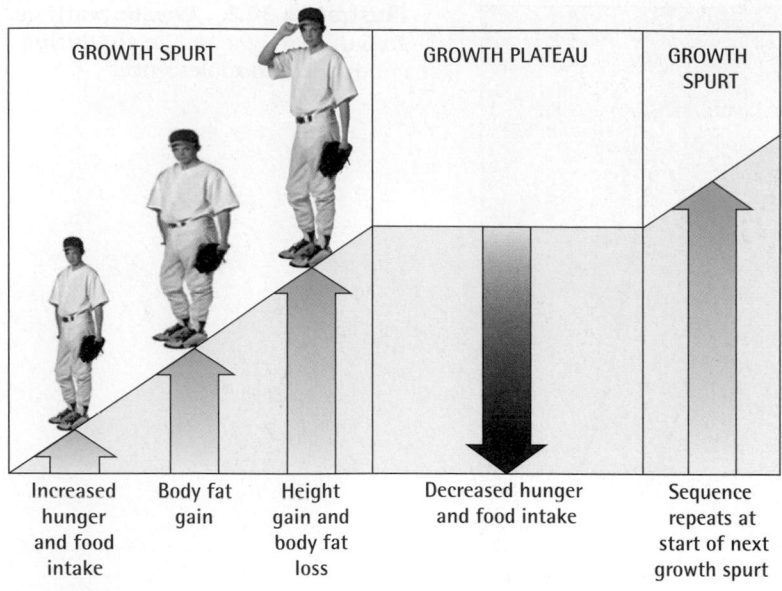

GROWTH SPURT | GROWTH PLATEAU | GROWTH SPURT

Increased hunger and food intake | Body fat gain | Height gain and body fat loss | Decreased hunger and food intake | Sequence repeats at start of next growth spurt

Illustration 30.3 Growth occurs in a series of "spurts." Each spurt follows this sequence.

Illustration 30.4 An example of the CDC's growth charts: Weight for age for boys and girls. The charts can be printed off the Internet using the address www.cdc.gov/growthcharts.

example, if a child's weight for age is between the 50th and 75th percentiles, it means that this child's weight is somewhat higher than that of most children and similar to that of 25% of children who are represented in the 50th to 75th percentile range. Children and adolescents whose weight status measurements place them in the highest and lowest percentile ranges should be evaluated further for potential underlying nutrition and health problems.

BMI charts for age are a feature of the growth charts. Since BMI increases with age in children and adolescents, ranges of BMI used to classify weight status in adults cannot be used. BMIs for age that fall between the 85th and 95th percentiles are considered "at risk" for overweight, and those above the 95th percentile at risk for obesity.[3] Although now recommended for determining weight status in children and adolescents, use of the CDC's new BMI for age charts requires that BMI be calculated. Illustration 30.5 shows you how to calculate BMI.

Food Jags and Normal Appetite Changes When growth is not occurring, children are often disinterested in food and eat very little at times. They may go on "food jags," insisting on eating only a few favorite foods like peanut butter and jelly sandwiches or breakfast cereal.

The ups and downs of children's food intake can be unnerving for parents. But as long as growth continues normally and children remain in good health, there's little reason to worry about food jags and fluctuations in appetite and food intake. Only children know when they are hungry or full.[4] (Actually, children

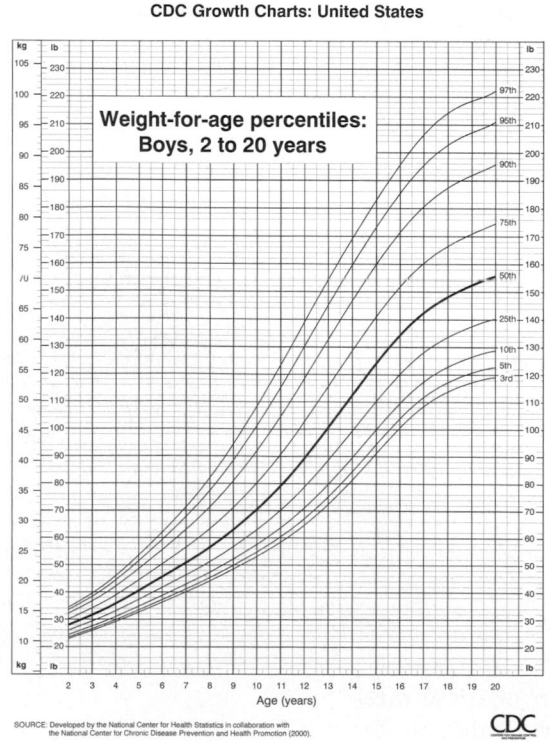

CDC Growth Charts: United States

Weight-for-age percentiles: Boys, 2 to 20 years

SOURCE: Developed by the National Center for Health Statistics in collaboration with the National Center for Chronic Disease Prevention and Health Promotion (2000).

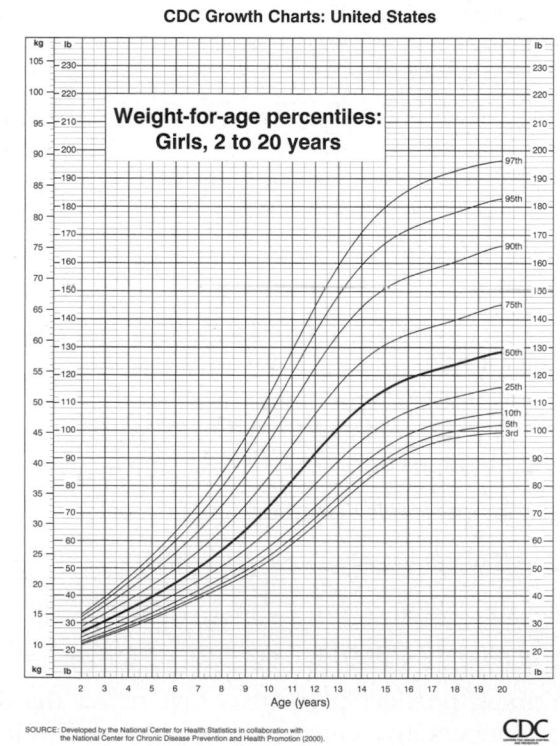

CDC Growth Charts: United States

Weight-for-age percentiles: Girls, 2 to 20 years

SOURCE: Developed by the National Center for Health Statistics in collaboration with the National Center for Chronic Disease Prevention and Health Promotion (2000).

Illustration 30.5   How to calculate BMI.

How to calculate BMI

Lucy just turned 7 years old and has had her weight and height carefully measured. She weighs 57 pounds (lb), and her height (stature) is 4 feet (ft), 2 inches (in.). What's her BMI?

1. Convert her height of 4 ft 2 in. to inches:

$$4 \text{ ft} \times 12 \text{ in. per foot} = 48 \text{ in.}$$
$$\underline{+2 \text{ in.}}$$
$$50 \text{ in.}$$

2. Square the inches:

$$50 \text{ in.} \times 50 \text{ in.} = 2500 \text{ in.}^2$$

3. Apply formula:

$$BMI = \frac{kg}{m^2} \quad or \quad \frac{weight \text{ in. lbs} \times 700}{height \text{ in in.}^2}$$

$$BMI = \frac{57 \text{ lbs} \times 700}{2500 \text{ in.}^2} = \frac{39{,}900}{2500 \text{ in.}^2} = 15.96 \text{ kg/m}^2$$

In this example, Lucy's BMI for age of 15.96 kg/m² falls between the 50th and 75th percentiles, or well within the normal range.

appear to regulate caloric intake meal by meal better than adults do. Read about it in Illustration 30.6.)

Hunger Can Make Kids IrritGable!

It was a hot summer day, and 10-year-old Kim and her friends had spent it in the pool. The fun went on and on ... until five o'clock. That's when the argument started and Kim left the pool in a huff.

When Kim arrived home, still in a huff, her mom knew what was going on. Kim was cranky because she was hungry. She was so cranky, in fact, that suggesting she'd feel better if she ate something would only make her more irritated! Kim was ravenously hungry, but she didn't want anybody telling her what her problem was!

Kim's mom had the solution to her daughter's grumpiness on the table. "Dinner's ready," her mom called softly. Kim stomped to the table and grudgingly started to eat. Within minutes, she did feel better, but it wasn't because she was no longer hungry!

Illustration 30.6   The pudding study.

Lunch at the day care center the day of the study was a bit different than usual—it started with pudding. Preschoolers were offered a serving of pudding that contained 150 calories or another with 40 calories. Both looked and tasted the same. The amount of pudding consumed was secretly recorded, as was the children's food intake during the rest of lunch. The experiment was repeated on adults.

Children compensated almost perfectly for the calories in the different puddings. Those given the higher-calorie pudding ate less at lunch, and those receiving the low-calorie pudding ate more. The adults, however, didn't fare so well. Their caloric intakes for the rest of lunch bore no relationship to calories consumed in the pudding. The study suggests that children are more sensitive to caloric intake on a short-term basis than adults.[5]

Only children know when they are hungry, but sometimes when they stay hungry too long, the signal to eat is overpowered by fun or the effects of not eating. Gently offering food to an irritable child who has skipped a meal or played too long can provide the cure for irritability.

The Adolescent Growth Spurt

The adolescent growth spurt usually occurs in girls between the ages of 9 and 12 years. For boys, this period of growth generally begins around the age of 12 to 14. Actually, though, the age when adolescents start their growth spurt normally varies considerably. Pictured in Illustration 30.7 are three friends, all aged 12. It's not hard to tell which one of them has experienced a growth spurt! The difference was also clear at the dinner table. By the time all three were 19 years old, Max (the one in the middle) had caught up with Ben (on the left), and David had grown taller but less than Max or Ben.

During these years of growth, teenagers gain approximately 50% of their adult weight, 20 to 25% of adult height, and 45% of their total bone mass. In the year of peak growth, girls gain 18 pounds on average, and boys 20 pounds.[6]

Can You Predict or Influence Adult Height? Ultimate height is difficult to predict. On average, children tend to achieve adult heights that are between the heights of their biological parents.[7] There are many exceptions to this general finding, however, and that means that heredity is not the only influence on height. The dramatic increases in height of Japanese youth since World War II provide clear evidence that nongenetic factors have the strongest influence on height. Since the late 1940s, Japanese youth have grown an average of 2 inches taller each generation. (The increase in height has meant that everything from

Illustration 30.7 This photo (on the left) was taken when these boys, born 2 months apart, were 12 years old. Ben (on the left) started his growth spurt at age 11, while Max (in the middle) began his at age 13. David (on the right) began to spring up in height after he turned 14. The photo (on the right) shows the same boys at age 19.

Judith Brown

Judy Brown

shoes to beds must be produced in larger sizes.) The increase in size of Japanese people is largely attributed to higher calorie and protein intakes and more nutrient-dense diets. Indeed, people in most economically developed countries continue to grow taller; a maximal genetically determined height has not yet been reached. If children are less well nourished than their parents, though, they tend to be shorter than their parents as adults.[8]

A healthy birthweight and diet during the growing years, and freedom from frequent bouts of illness, support growth in height.[9] There are no supplements, powders, or special diets that can be used to increase growth rate. Growth hormone injections can increase height somewhat, but the side effects of growth hormone are numerous and its use is limited.

Overweight and Type 2 Diabetes: Growing Problems

Rates of overweight in children and adolescents in the United States have risen remarkably since the 1960s (Illustration 30.8), and overweight-related disorders are also on the rise. Conditions such as type 2 diabetes, bone and joint disorders, abnormal blood lipid levels, and elevated blood pressure that were very rarely observed in children and adolescents in the past are being diagnosed with increasing frequency.[10,11] Approximately 60% of overweight children have risk factors for cardiovascular disease.[12] The emergence of type 2 diabetes as a problem of the early years is of particular concern because diabetes generally worsens with time and causes long-term health impairments. It is estimated that 4% of children and adolescents have impaired glucose tolerance (a strong risk factor for type 2 diabetes), and that 6 to 17% of overweight and obese children and adolescents have type 2 diabetes.[13]

Causes of Overweight in Young People   Jean Mayer, a noted professor of nutrition, said we shouldn't be surprised that so many children are overweight: "A society where most children sit and watch TV, where nobody walks, and where no domestic chores are required is a society where we may expect to have obese children."[14]

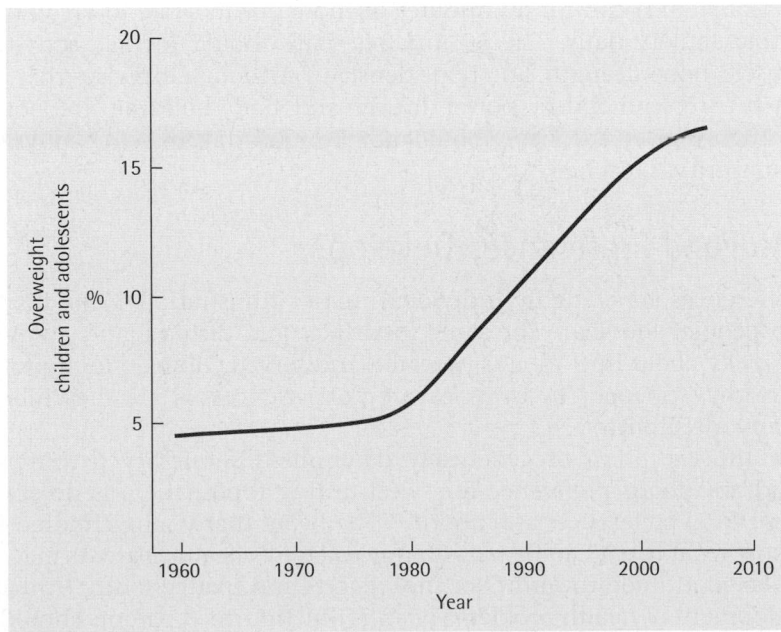

Source: Developed by Judith Brown using data from Chartbook/Health, United States, 2008.

Illustration 30.8   Changes in the percent of overweight children and adolescents in the United States.

This fast-food beverage container holds over a quart of soft drink. It's meant for adolescent males and reads: "Do a whole river of icy-cold refreshment. And don't just quench your thirst, stop it in its tracks and make it run for cover. Because showing your thirst who's boss is doing what tastes right."

A number of "obesigenic" trends have developed over the past several decades that likely account for rising rates of overweight. Compared to 20 or so years ago, children and adolescents now have fewer opportunities for physical activity and are exposed to a generally plentiful supply of energy-dense foods. Most schools currently offer too few physical activity classes and opportunities for exercise.[15] School vending machines in many districts are loaded with energy-dense, empty calorie snack foods, and portion sizes of foods served to children in fast food and other restaurants are often excessive.[16] Many communities lack bike paths or lanes, sidewalks, and safe playgrounds that invite physical activity.[17] Genetically based, inborn processes that regulate appetite and satiety are being overwhelmed by environmental conditions that influence decisions about food intake.

Medicalization of Obesity and Treatment   Stomach stapling and diet drugs are being increasingly used to treat overweight and obesity in children and adolescents, but this trend is not viewed by some as a healthy one.[18] Stomach stapling is forever, and little is known about the psychological and health effects of this type of surgery in children and adolescents. Diet drug use in adults is accompanied by potentially serious side effects and the requirements for long-term use for continued effectiveness. Given that such drastic measures as surgery and diet drugs may not be attractive, how is the problem of the obesity epidemic in children and adolescents to be solved? That's right. The answer is prevention.

Prevention of Overweight   If environmental factors play the predominant role in the overweight epidemic, then the path to prevention is paved with environmental changes. Children and adolescents need more opportunities for physical activity. A wider array of healthy food options should be available at schools and in fast-food restaurants, as well as at home.[16,19] Children who regulate their food intake based on appetite and satiety rather than on environmental cues will have a jump start on overweight prevention.

■ **Physical Activity Guidelines for Children and Adolescents.** Regular exercise during childhood and adolescence helps prevent overweight and improves one's chances of having a healthy adulthood.[20] How much is enough? Recommendations for healthy levels of physical activity for children and adolescents (Table 30.1) call for 60 minutes or more of moderate-to-vigorous intensity aerobic activity daily. The 60 minutes daily should include activities that build muscle, bone strength, and bone density.[20] Although exercises that increase strength are recommended, power lifting and body-building are not. These intense, high-impact exercises should not be undertaken before physical and skeletal maturity is reached.[21]

How Do Food Preferences Develop?

Food preferences are a highly individual matter (Illustration 30.9). It's hard to find two people who share the same food likes and dislikes, and many people are very picky about how food is prepared and served. How do food preferences form? Are they "shaped" by early learning experiences, or are they inborn and beyond anyone's control?

With the exception of genetically determined sensitivity to bitter-tasting foods and an inborn preference for sweet-tasting foods, there is no direct evidence that food preferences are inborn.[22] The belief that young children instinctively know what to eat can be hazardous to a child's health. Parents may take an overly relaxed attitude toward poor food habits and inadvertently contribute to the development of health problems later in life. Informed parents should decide what types of food their child should be offered, but the decision about how much to eat should be left to the child.[23]

I do not like broccoli, and I haven't liked it since I was a little kid and my mother made me eat it. And I'm President of the United States, and I'm not going to eat any more broccoli.

—GEORGE H. W. BUSH, 1990

Table 30.1

Physical activity guidelines for children and adolescents.[20]

Activity	Duration	Examples
Moderate-to-vigorous aerobic exercise	60 minutes daily	Running, hopping, skipping, rope jumping, dance line, cheerleading
Strength building exercise	Included in the 60 minutes daily • 3 day a week	Tree climbing, playing on playground equipment, hill climbing, weight lifting
Bone strengthening	Included in the 60 minutes daily • 3 day a week	Tennis, basketball, rope jumping, skiing, push-ups, soccer

Food likes and dislikes appear to be almost totally shaped by the environment in which children learn about food. Which foods are offered, the way they are offered, and how frequently particular foods are offered all influence whether a child will like a given food or not.[24]

Humans are born with a tendency to be cautious about accepting new things, including foods. They may need to get used to a new food before they trust it. Parents may have to offer a new food on five or more occasions before the child makes a decision.[24] Sometimes children decide they like the food, and sometimes they really don't like it. Often infants and young children do not like strong-flavored vegetables, spicy foods, and mixed foods (Illustration 30.10).

Illustration 30.9 Why select the apple?
Food likes and dislikes appear to be almost totally shaped by the environment in which children learn about food. Which foods are offered, the way they are offered, and how frequently particular foods are offered all influence whether a child will like a given food or not.

Illustration 30.10 MyPyramid for Kids

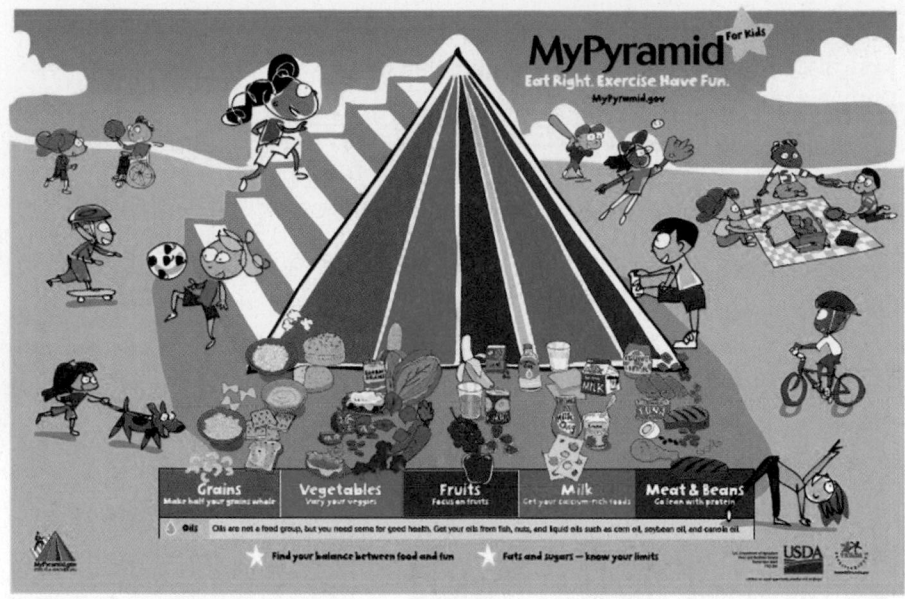

Forcing a young child to eat foods she or he does not like, or totally restricting access to favorite foods, can have lifelong negative effects on food preferences and health. When attempts to get a child to eat a particular food turn the dinner table into a battleground for control, nobody wins. Foods should be offered in an objective, nonthreatening way so that the child has a fair chance to try the food and make a decision about it. Restricting access to, or prohibiting intake of, children's favorite "junk" foods tends to strengthen their interest in the foods and consumption of those foods when they get a chance. Such prohibitions have the opposite effect of that intended because they make kids want the foods even more.[25]

Children today are tyrants. They contradict their parents, gobble their food, and tyrannize their teachers.
—SOCRATES, 470–399 B.C.

What's a Good Diet for Children and Adolescents?

Good diets for the growing years can be achieved by following the food group recommendations shown in Table 30.2. The type of foods selected from the food groups are important. Whole-grain breads and cereals should be represented in diets; and boiled, broiled, and baked foods should be consumed more often than fried foods.

Table 30.2

MyPyramid food group recommendations for preschoolers, children, and adolescents[a]

Age in Years	GRAINS (OUNCES)	VEGETABLES (CUPS)	FRUITS (CUPS)	MILK (CUPS)	MEAT AND BEANS (OUNCES)	OIL (TEASPOONS)
			Food Group			
2	3	1.0	1.0	2	2	2
6	5	1.5	1.5	2	4	4
12	6–7	2.5–3	2	3	5.5–6	6
16	6–10	2.5–3.5	2–2.5	3	5.5–7	6–8

[a] Higher numbers of ounces and cups apply to males. Recommendations vary based on physical activity levels and growth rate.

©Blair Seitz/Photo Researchers, Inc.

Illustration 30.11 Snacks are an important source of calories and nutrition in the diets of children and teenagers.

Foods from the "miscellaneous" group that contribute mainly calories may be needed by children and teenagers. Although cakes, candy, cookies, potato chips, and so on are generally considered "sometimes" foods rather than "always" foods, they can serve as sources of needed calories. Snacks also are an important source of calories in the diets of children and teenagers (Illustration 30.11). Good choices for snacks include:

Yogurt	Bananas	Carrots
Cheese	Oranges	Cucumbers
Low-fat milk	Apples	Popcorn
Nuts, seeds	Dried fruit	Peanuts
Pears	Mangos	Cherry tomatoes
Melons	Grapes	Peanut butter

Recommendation for Fat Intake Fat intakes of 25 to 35% of total calories are compatible with normal growth and health in children and adolescents.[26] The upper limit for fat intake is set at 35% of calories because diets containing more fat than that go downhill in quality. High fat diets are associated with excessive calorie and saturated fat intake, and low intake of vegetables and fruits.[27, 28] As is the case for adults, it is recommended that children and adolescents consume as little saturated and *trans* fat as possible.[26]

Diets of children and adolescents in the United States and Canada tend to provide very low amounts of the omega-3 fatty acids EPA and DHA.[29] Twice a week consumption of non-fried fish and seafood would bring EPA and DHA intakes up to recommended levels.

Is Milk Bad for Children? No! Children certainly should drink milk (Illustration 30.12). If this sounds like a silly question, good. It is, but the notion that children (and adults, for that matter) should not drink milk or consume other animal products erupts in the media from time to time.

Children don't have to consume milk in order to have an adequate diet, because a variety of other foods could provide the nutrients in milk. However, children who fail to drink milk tend to have low calcium intakes, lower bone density, and more fractures than children who drink milk regularly.[30, 31]

Illustration 30.12 Children should drink milk in order to meet their needs for calcium and vitamin D. Virtually all school lunch programs include milk as a beverage.

Photo Disc

Status of Children's and Adolescents' Diets

How well are children and teenagers doing at meeting the MyPyramid recommendations? Children and adolescents tend to consume too few vegetables, fruits, and whole grain products. On average, only 15% of children and 30% of adolescents meet intake recommendation, and 13% of 4- to 13-year-olds consume enough whole grain products.[19,32] Low levels of consumption of these foods contributes to insufficient intake of iron, magnesium, potassium, calcium, vitamin E, folate, and fiber.[19]

Approximately half of all children fail to consume the recommended intake of calcium daily. Low calcium intake and poor vitamin D status, along with limited consumption of vegetables and fruits, during childhood and adolescence decrease peak bone mass and increase the risk of osteoporosis.[33] Effects of inadequate vitamin D status during adolescence extend beyond bone health. Vitamin D insufficiency is also related to metabolic alterations that increase the risk for type 2 diabetes, hypertension, and heart disease later in life.[34] Part of the reason children and teens fail to get enough calcium and vitamin D is that they tend to substitute soft drinks for milk during these years.[25] Children and teens may also have poor vitamin D status due to inadequate intake of milk and exposure to sun.[35]

Diets of children and adolescents generally provide about half of the recommended intake of dietary fiber of 19 to 38 grams per day.[19] This makes constipation a problem for many children and teens.

Early Diet and Later Disease

Children and adolescents generally feel vigorous and healthy even when they are accumulating diet-related risk factors that will influence disease development later in life. Many studies indicate that in populations with a low incidence of heart disease, blood cholesterol values of both children and adults are lower than in countries with high rates of heart disease. Typical cholesterol levels of children in the United States are higher than those of children in countries where the rate of heart disease is lower. In addition, obesity during adolescence is associated with the presence of risk factors for subsequent heart disease: high LDL cholesterol, low HDL cholesterol, and increased blood pressure.[11] High-sodium diets tend to increase blood pressure in children and adolescents who have a hypertensive parent. U.S. children tend to consume far more sodium than is recommended. Adolescents who fail to consume sufficient calcium and to obtain adequate levels of vitamin D are at later risk of osteoporosis.[36] Children and adolescents who become obese are at higher risk not only for obesity later in life but also for disorders related to obesity such as diabetes, heart disease, some cancers, and hypertension.[12]

Healthy eating and activity levels during childhood and adolescence are clearly important to long-term health and quality of life. Improvements in both areas are needed. Families, schools, and communities—as well as children and adolescents themselves—can do many things to help improve eating and physical activity patterns.[20] The rewards of successful efforts would be for life.

NUTRITION | Up Close

Overweight prevention close to home

Focal Point: Tips for parents on preventing overweight in their children.

Debra and Dale have a 1-year-old daughter, and their second adopted infant will arrive in a few weeks. Both are acutely aware of the growing problem of obesity and type 2 diabetes in children and don't want their children to become overweight.

Referring back to the information presented on pages 30-10 and 30-11, list four actions Debra and Dale can take to help their children develop healthful food and activity habits:

1. _____

2. _____

3. _____

4. _____

FEEDBACK (including answers) can be found at the end of Unit 30.

Review Questions

TRUE   FALSE

1. Children generally become good "self-feeders" between the ages of four to five years. ☐ ☐

2. Gains in weight and height in children and adolescents occur in "spurts" rather than as smooth progressions. ☐ ☐

3. Rates of overweight in children and adolescents have finally stopped increasing in the United States. ☐ ☐

4. Food likes and dislikes appear to be almost entirely shaped by the environment in which children learn about food. ☐ ☐

TRUE   FALSE

5. Young children instinctively know what to eat. ☐ ☐

6. According to MyPyramid food intake recommendations, 6-year-old children should consume 1.5 cups of vegetables and fruits daily. ☐ ☐

7. Children have a high need for calories from fat. Therefore, it is not recommended that children limit their fat intake. ☐ ☐

8. A number of chronic diseases such as heart disease, osteoporosis, and type 2 diabetes may begin to develop during childhood and adolescence. ☐ ☐

Media Menu

www.kidnetic.com
This site provides a bunch of fun activities for kids on subjects such as digestion; food, fun, and fitness; healthy and easy recipes; and dancing.

www.nhlbi.nih.gov/health/public/heart/obesity/wecan
We Can!™ is a national program designed for families and communities to help children maintain a healthy weight. The program focuses on three important behaviors: improved food choices, increased physical activity and reduced screen time. The We Can program provides families and communities with resources that can help prevent childhood overweight.

www.hc-sc.gc.ca/fn-an/nutrition/child-enfant/index-eng.php
Health Canada has developed nutrition guidelines for children through key periods of growth, from infancy to adolescence. These resources will be especially useful for health professionals and educators who work with children.

www.mypyramid.gov/preschoolers/index.html
You can use MyPyramid to help preschoolers eat well, be active, and stay healthy. MyPyramid for Preschoolers offers an option for generating a MyPyramid Plan for a preschooler, growth charts, and guidance on physical activity and developing healthy eating habits in preschoolers.

www.mypyramid.gov/kids
Designed specifically for children aged 6 to 11, MyPyramid for kids offers games, worksheets, tips for families, and classroom materials for nutrition education.

www.nichd.nih.gov/msy
The "Media-Smart Youth" curriculum, developed the National Institutes of Child Health and Human Development, offers ten interactive lesson plans that help kids make healthier choices about food and physical activity.

www.nlm.nih.gov/medlineplus/childnutrition.html
The National Library of Medicine shares its collection of information on child nutrition through this Web site.

www.nlm.nih.gov/medlineplus/obesityinchildren.html
Get your news and information on childhood obesity and prevention and management tips here.

www.cdc.gov/growthcharts
Need more information about the growth charts or printouts of them? Get it here.

Notes

1. Health, United States 2008, www.cdc.gov/nchs/hus.htm, accessed 5/09.

2. My Pyramid for Preschoolers, www.mypyramid/preschoolers, accessed 5/09.

3. CDC's growth charts, www.cdc.gov/growthcharts, accessed 5/09.

4. Birch LL. Child feeding practices and the etiology of obesity. Obesity (Silver Spring). 2006;14:343–4.

5. Birch LL. Presentation at the National Conference on Nutrition Education Research, Chicago, September 1986.

6. Tanner JM. Fetus into man: physical growth from conception to maturity. Cambridge (MA): Harvard University Press; 1978.

7. Smith GD. Growth and its disorders. Philadelphia: W. B. Saunders; 1979.

8. Bronner F. Adaptation and nutritional needs. (letter) Am J Clin Nutr 1997;65:1570.

9. Gigante DP et al. Early life factors are determinants of female height at age 19 years in a population-based birth cohort. J Nutr 2006;136:473–8.

10. Denney-Wilson E et al. Measures of adiposity and metabolic risk factors in adolescents, Arch Pediatr Adolesc Med 2008;162:566–73.

11. Raghuveer G et al. Vascular age of carotid arteries in obese children, presented at the American Heart Association's Scientific Sessions,11/7/09, www.medscape.com/viewarticle/583700.

12. Dietz WH et al. Overweight children and adolescents. N Engl J Med 2005; 352:2100–9.

13. Goran MI. Impaired glucose tolerance in obese children and adolescents (letter), N Engl J Med 2002;347:290.

14. Mayer J. Obesity during childhood. In: Winick M, ed. Childhood obesity. New York: John Wiley & Sons; 1975: pp. 73–80.

15. Brewer JD et al. Increasing student physical activity during the school day: opportunities for the physical educator, Strategies: A Journal for Physical and Sport Educators, January 1 2009, www.articlearchives.com/

medicine-health/diseases-disorders-cancer-breast/2310574-1.html, accessed 5/09.

16. Story M. The Third School Nutrition Dietary Assessment Study: findings and policy implications or improving the health of U.S. children, J Am Diet Assoc 2009;109:S7–13.

17. American Public Health Association. Surrounding our kids with the opportunity for physical activity. www.medscape.com/viewarticle/527767, 3/23/06.

18. Treadwell JR et al. Systematic review and meta-analysis of bariatric surgery for pediatric obesity, Ann Surg 2008;248:763–76.

19. Position of the American Dietetic Association: nutrition guidance for healthy children ages 2 to 11 years, J Am Diet Assoc 2008;108;1038–47.

20. Physical Activity Guidelines for Americans, 2008, U.S. Department of Health & Human Service, http://www.health.gov/PAGuidelines.

21. Small EW et al. Guidelines on strength training for children, Pediatrics 2008;121:835–40.

22. Story M, Brown JE. Do young children instinctively know what to eat? The studies of Clara Davis revisited. N Engl J Med 1987;316:103–6.

23. Satter E. The feeding relationship: problems and interventions. J Pediatr 1990;117:S181–S9.

24. Birch LL. The role of experience in children's food acceptance patterns. J Am Diet Assoc 1987;87(suppl):S36–S40.

25. Fisher JO, Birch LL. Restricting access to palatable foods affects children's behavioral response, food selection, and intake. Am J Clin Nutr 1999;69:1264–72.

26. Dietary Reference Intakes: Energy, carbohydrate, fiber, fat, fatty acids, cholesterol, protein, and amino acids, Ch. 8, 11. Washington, DC: National Academies Press, 2002.

27. Sebastian RS et al. U.S. adolescents and My Pyramid: associations between fast food consumption and lower likelihood of meeting recommendations, J Am Diet Assoc 2009;109:226–35.

28. Johnson L et al. Energy-dense, low-fiber, high-fat dietary pattern is associated with increased fatness in childhood, Am J Clin Nutr 2008;87:846–54.

29. Madden SMM et al. Direct diet quantification indicates low intakes of (n-3) fatty acids in children 4 to 8 years old, J Nutr 2009;139:528–32.

30. Moore LL et al. Childhood dairy consumption and bone health, J Pediatr Aug. 13, 2008, published online, www.medscape.com/viewarticle/579032, accessed 9/08.

31. Black RE et al. Children who avoid drinking cow's milk have low calcium intakes and poor bone health, Am J Clin Nutr 2002;76:675–80.

32. Stang J et al., Relationship between vitamin and mineral supplement, dietary intake, and dietary adequacy among adolescents, J Am Diet Assoc 2000; 100:905–10.

33. Vatanparast H et al. Fruit and vegetable intake and bone strength in children. Am J Clin Nutr 2005;82:700–6.

34. Reis JP et al. Vitamin D status and cardiovascular disease risk factors in the U.S. adolescent population, AHA 49th Annual Conference on Cardiovascular Disease Epidemiology and Prevention, 3/11/09, abs. no. P54

35. Maguire J et al. Vitamin D status of Canadian toddlers, Pediatric Academic Societies Annual Meeting, 5/4/09, abs. no. 4545.2.

36. Goulding A et al. Children who avoid drinking cow's milk are at increased risk for prepubertal fractures. J Am Diet Assoc 2004;10:250–3.

NUTRITION | # Up Close

Overweight Prevention Close to Home

Feedback for Unit 30

To help their children develop healthful food and activity habits, Debra and Dale can:

1. Provide lots of opportunities for fun physical activities

2. Provide a variety of healthful food choices

3. Limit meals at fast-food restaurants that promote energy-dense, low-nutrient foods to children

4. Work with other parents and school staff to increase physical activity opportunities at school

Other:

- Work with other parents and school staff to increase availability of healthful food choices at school
- Allow their children to eat when they are hungry and to stop eating when they are full
- Never force their children to eat something or totally restrict access to their favorite foods
- Offer foods in an objective, nonthreatening way

Nutrition and Health Maintenance for Adults of All Ages

NUTRITION SCOREBOARD

	TRUE	FALSE
1 Healthy eating contributes to longevity in part by delaying the age at which chronic disease develops.		
2 Aging is a kind of disease, and people can do little to prevent their health and quality of life from declining as they age.		
3 Calorie restriction and underweight increase longevity.		

Key Concepts and Facts

- Age does not necessarily predict health status. Healthy adults come in all ages.

- Dietary intake, body weight, and physical activity influence changes in health status with age.

- Aging processes begin at the cellular level.

- Medications, diseases, and biological processes associated with aging influence adults' requirements for certain essential nutrients.

Answers to **NUTRITION** SCOREBOARD	TRUE	FALSE
1 Healthy eating pays off.	✔	
2 Aging is not a disease process! Many things can be done to extend life and its quality. (This unit discusses some of those things.)		✔
3 Calorie restriction has not been shown to increase longevity in humans; underweight is associated with shorter life expectancy than normal weight.[1]		✔

One of the most important types of health changes that occurs with age is no change at all.

—Daphne Roe, 1985

You Never Outgrow Your Need for a Good Diet

Eating right during the adult years is a wise practice. Good nutrition helps adults feel healthy and vigorous as they age and improves their overall sense of well-being. Adults who eat healthfully tend to develop heart disease, cancer, hypertension, and diabetes at older ages and may have more life in their years and years of life than adults who do not.[2,3]

Aging is a normal process, not a disease. Although the incidence of many diseases increases with age, the causes of the diseases are often unrelated to aging. They may fully or partially result from the cumulative effects of diets high in saturated fat and low in vegetables and fruits, obesity, smoking, physical inactivity, excessive stress, or other habits that insidiously influence health on a day-to-day basis.[2,3] Aging cannot be prevented, but how healthy we are during aging can be influenced by what we do to our bodies.

Maintaining health as we age is becoming an increasingly important concern. There are more older adults in the United States than ever before, and the numbers are growing.

© Tomas Rodriguez/Corbis

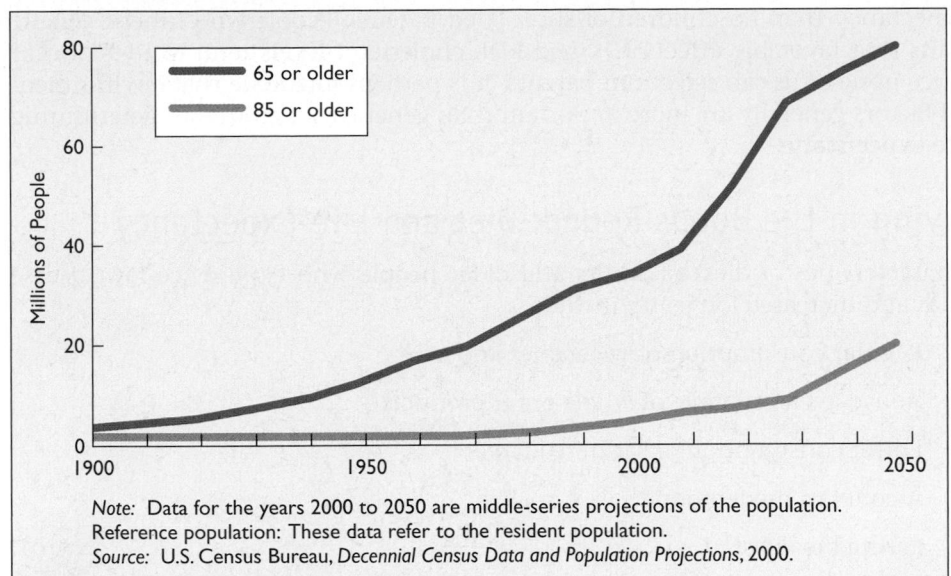

Illustration 31.1 Expected growth in the U.S. population of people aged 65 and 85 years and older.

Note: Data for the years 2000 to 2050 are middle-series projections of the population.
Reference population: These data refer to the resident population.
Source: U.S. Census Bureau, *Decennial Census Data and Population Projections*, 2000.

The Age Wave

The proportion of people in the United States aged 65 years and older is steadily increasing (Illustration 31.1). By the year 2050, approximately 20% of the U.S. population will be 65 years of age or older. Since 1900, **life expectancy** at birth has increased by 63%, from 47.3 to 77.9 years. Although advances in life expectancy are welcomed, there is more progress to be made. The United States ranks thirty-fourth in life expectancy worldwide, behind countries such as Japan, Sweden, Canada, and Greece.[4] Theoretically, the human life span could reach 130 to 135 years.[5]

All groups in the U.S. population are not benefiting equally from the increase in life expectancy. Life expectancy varies a good deal by sex and race. Table 31.1 shows overall life expectancy for Caucasian and African American men and women in 1900, 1960, and 2006. African American men have the shortest life expectancy, followed by Caucasian men. As in other industrialized countries, life expectancy for females exceeds that for males. In the United States, the gender difference in average life span appears to be largely related to behaviors. U.S. males tend to smoke more, consume more alcoholic beverages, pay less attention to what they eat, and seek medical care less often than do females.[6]

Why the Gains in Life Expectancy?

Most of the gains in life expectancy in the United States are attributable to decreased infant deaths and deaths from infectious disease, improved nutrition, and medical advances.[9] In 1900, 1 out of 10 newborns died in the first year of life; today the proportion is less than 1 in 100.[10] Since about 1930, the development and mass use of vaccines that prevent common infectious diseases such as pertussis ("whooping cough"), diphtheria, and tetanus have contributed to reductions in infant and child deaths. Advances in medical treatments and surgery have improved the quality of life for many people and have added approximately 5 years to overall life expectancy in the United States since 1900.[11]

A person's genetic background affects her or his longevity. Life insurance company statistics reveal that children of long-lived parents have a 3-year longer life

Table 31.1

Average length of life for Caucasian and African American males and females in the United States in 1900, 1960, and 2006[7]

	Life Expectancy (Years)		
	1900	1960	2006
Caucasian Americans			
Males	46.6	67.4	75.7
Females	48.7	74.1	80.6
African Americans			
Males	32.5	61.1	69.7
Females	33.5	63.3	76.5

(Data for other ethnic groups were unavailable.)

life expectancy
The average length of life of people of a given age.

"The older an individual gets, the healthier he or she has been."

—T. T. Perls[8]

expectancy than do children of short-lived parents. People who inherit genetic traits that favorably affect HDL and LDL cholesterol levels tend to live longer.[8] Since none of us can select our parents, it is perhaps fortunate that environmental factors generally are more important than genetic background in determining life expectancy.[13]

Living in the Bonus Round: Diet and Life Expectancy

Characteristics of diets of adults and older people who experience low disease rates and increased longevity include

- Regular consumption of vegetables and fruits
- Above-average intake of whole grain products
- Lower consumption of saturated fats
- Alcohol in moderation
- Eating breakfast[3,14]

Taking time to socialize and celebrate with friends and engaging in regular physical activity also appear to contribute bonus years to life.[15]

*All adults should avoid physical in*activity.[16]

Physical Activity and Longevity   Regular physical activity is a core part of healthy lifestyles that maintain health and delay aging processes.[16,17] Regular exercise increases oxygen delivery to organs and tissues, decreases biologic aging, raises HDL-cholesterol levels, and lowers body fat.[17] Substantial health benefits result from performing moderate-intensity exercise for 150 minutes per week, or from 75 minutes of vigorous-intensity exercise weekly. Strengthening exercises performed twice a week further enhance the benefits of regular exercise to health.[16]

All adults, regardless of their age or physical condition, benefit from physical activity.[16] With proper training and spirit, physical fitness can be maintained during aging, or increased to the level of "over 80" event winners (Illustration 31.2).

Illustration 31.2   Over-80 athletes.

Jonathan Sprague/ Redux

Photo Courtesy of brightroom.com

Calorie Restriction and Longevity

For decades it has been known that laboratory animals fed diets providing 30% or less than normal levels of calories and adequate levels of essential nutrients have increased life expectancies. Why this happens isn't clear, but the reasons could be related to modifications in nutrient utilization, reduced metabolic rate, and decreased aging processes.[13] To date, human studies have not shown that caloric restriction increases longevity.[1]

There are some who take the leap and extrapolate results of animal studies to themselves. The number of people practicing calorie restriction is increasing, although their numbers are still small. It's too soon to be adopting a calorie-restricted diet for the purpose of longevity. People who tend to live the longest have normal weights. Life expectancy decreases as body weight decreases below, or increases above, normal.[18]

Nutrition Issues for Adults of All Ages

Researchers and others interested in nutrition and health during middle age and beyond have tended to focus their attention on the cumulative effects of diet on chronic disease. Nutrition exerts its effects on chronic disease development over time, and therefore diseases related to poor diets are most likely to express themselves during the older adult years. The occurrence of diseases related to behavioral traits such as smoking and physical inactivity also increases among adults as they age.

In addition to the health effects of behaviors, people may age biologically.[13] The combined effects of poor diets, other risky lifestyle behaviors, and biological aging increase the rates of serious illness during adulthood. How soon a disease develops largely depends on the intensity and duration of exposure to behavioral risks that contribute to disease development. Behavioral factors also affect the progress of biological aging processes.

Breaking the Chains of Chronic Disease Development

For the most part, the development of chronic disease in middle-aged and older adults can be viewed as a chain that represents the accumulation over time of problems that impair the functions of cells. Each link that is added to the chain, or each additional insult to cellular function, increases the risk that a chronic disease will develop. The presence of a disease indicates that the chain has gotten too long—that the accumulation of problems is sufficient to interfere noticeably with the normal functions of cells and tissues.

It appears that the chain can be shortened by healthful dietary and other behaviors. For example:

- Correcting obesity and stabilizing weight during the adult years may lengthen life expectancy.[19]

- Dietary intakes that correspond to the Dietary Guidelines for Americans are related to longer life expectancy.[3]

- Adequate consumption of the omega-3 fatty acids EPA and DHA are protective against the development of Alzheimer's disease, cognitive and muscular strength decline, dementia, and depression.[20,21] Two to three fish meals a week generally provide adequate amounts of EPA and DHA.[26]

- Maintaining adequate calcium and vitamin D intake and engaging in regular physical activity during the adult years may prevent or postpone the development of osteoporosis.[22]

Table 31.2

Examples of biological changes during aging and nutritional consequences[27–29]

Biological Change	Nutritional Consequences
• Lowered stomach acidity • Decreased lean muscle mass • Reduced production of vitamin D in the skin • Decreased sensation of thirst	• Decreased absorption of vitamin B_{12} and C • Reduced caloric need • Increased dietary requirement for vitamin D, increased risk for inadequate intake of vitamin D and calcium • Dehydration risk

- Above-average intakes of fruit and vegetables may delay or prevent the development of a number of types of cancer, heart disease, and stroke, and the formation of cataracts that "cloud over" the lens of the eye.[3,23,24]

The health status of adults is not necessarily "fixed" by age; it can change for the better or the worse, or not much at all.[25]

Nutrient Needs of Middle-Aged and Older Adults

Biological processes and lifestyle changes that generally accompany aging affect caloric and nutrient needs (Table 31.2). A person's need for calories generally declines with age as physical activity, muscle mass, and basal metabolic rate decrease. However, people who remain physically active into their older years maintain muscle mass, experience less muscle and bone pain, and gain less body fat than people who are inactive.[16]

While caloric need decreases, the requirements for certain nutrients such as protein, vitamin C, vitamin D, vitamin B_{12}, and calcium may increase with aging. The increased need for protein appears to be a result of decreased efficiency of protein utilization. Dietary requirements for vitamin C and calcium may increase due to lower levels of stomach acidity that often occur with advancing age. Decreased stomach acidity reduces the body's ability to absorb vitamins C and B_{12} and calcium. Vitamin D status may become a problem for older adults for two reasons: Their intake of milk and milk products is often low, and exposure of the skin to sunlight produces less than half of the amount of vitamin D as it does in young adults.[30] An estimated 10 to 37% of older adults may be deficient in vitamin D.[31]

The MyPyramid food guide, constructed to reflect the nutrient needs of adults and elderly who exercise 30 to 60 minutes daily is presented in Table 31.3. The food intake plan takes into account the caloric need of most adults and older people. It highlights nutrient-dense foods that are most likely to provide adequate nutrient intakes in this population. An illustrated version of the MyPyramid Food Plan for seniors has been developed at Tuft's University. Called the "Modified MyPyramid for Older Adults" (Illustration 31.3), it shows food options within the basic food groups, emphasizes physical activity, and highlights nutrients most likely to be lacking in the diets of older Americans.[32]

Table 31.3

MyPyramid food group recommendations for older adults[a]

	Food Group					
Age in Years	Grains (ounces)	Vegetables (cups)	Fruits (cups)	Milk (cups)	Meat and Beans (ounces)	Oil (teaspoons)
45	6–9	2.5–3.5	2	3	5.5–6.5	6–8
75	6–7	2.5–3	1.5–2	3	5–6	5–6

[a] Higher numbers of ounces and cups apply to males.

Illustration 31.3 The Modified MyPyramid for Older Adults emphasizes fluid intake, regular physical activity, vitamins B_{12} and D, and calcium.[32]

Fluid Needs For reasons that aren't entirely clear, many older people don't get thirsty when their bodies are running low on water. The implication of this change is that older people may become dehydrated and need medical assistance. Approximately 1 million elderly men and women are admitted to hospitals each year due to dehydration.[33] Fluid needs can be met through consumption of water, juices, teas, and other beverages. Women of all ages need about 11 cups of water a day from fluids and foods, and men need 15 cups.[34]

Does Taste Change with Age? Poor diets observed in some middle-aged and older adults have been ascribed to "declining taste" with age. Besides being a dreadful thought, it's not true that taste declines with age to the extent that it makes eating less pleasurable than before. Although taste sensitivity does diminish somewhat with age, chances are your favorite foods will taste as good to you in your older years as they did in your youth (Illustration 31.4). Sight, smell, and hearing senses usually decline with age a good deal more than taste does.[35]

Illustration 31.4 Taste is less affected by age than are some other senses.

The senses of taste and smell are affected by medications such as antibiotics, antihistamines, some lipid-lowering drugs, and cancer treatments; diseases including Alzheimer's disease, cancer, and allergies; and surgeries that affect parts of the brain and nasal passages. These treatments and conditions are more prevalent in older adults than in middle-aged populations and are primarily responsible for declines in the senses of taste and smell. Changes in these senses increase the likelihood that the person will experience food poisoning or consume an inadequate diet and a low intake of food.[36]

Psychological and Social Aspects of Nutrition in Older Adults

Preparing meals and eating right may not be as simple as it sounds for many older adults. Consuming an adequate and balanced diet may not be easy when you depend on someone else to take you shopping, when mealtimes involve little social life, or when you "don't feel up" to making a meal. Isolation, loneliness, depression, and poor health can be major contributors to poor diets in older adults. The diets of older people are often lacking in nutrients, and because many older adults do not, or cannot, consume enough nutrients to meet their increased need for them, supplementation may be required.

Eating Right during Middle Age and the Older Years

The best diet for middle-aged adults is one that contains a wide variety of basic foods. A balanced and adequate diet can be obtained by judiciously selecting foods from the basic food groups. It is important to select judiciously because not all foods within the respective food groups are equally desirable. Food choices should emphasize the members of the food groups that are low in saturated fat, trans fat, and sodium, and high in fiber. Diets should highlight vegetables, fruits, low-fat dairy products, and whole grains due to their health benefits.

Altering food habits in favor of healthy diets appears to be a common practice among adults of all ages. A study of 100 women and men over the age of 60 found that 100% had made some change in their food choices in the recent past. The changes had been for health, taste, convenience, and social reasons.[37] Clearly, a person's food choices and intake change throughout life, and the changes can be for the better.

I have a theory that chocolate slows down the aging process. It may not be true, but do I dare take the chance?

—AUTHOR UNKNOWN

Up Close

Does He Who Laughs, Last?

Focal Point: Critically thinking about factors that influence longevity.

The following is an adaptation of a letter printed in the "Dear Abby" newspaper column:

> Dear Abby,
>   Since I've reached my 80s, my mail is full of ads for health products to help me live longer.
>   I once had many friends, all of whom were health vigilantes. They shook their heads knowingly as I avoided all health food fads and exercise. They made liquid out of good vegetables and spent fortunes buying all the latest supplements. They argued that "organic" was better and "natural" was best. I would tell them that snake venom, poison ivy and manure were "natural." But they wouldn't listen and they didn't laugh.
>   Now my friends are all dead and I have no one left to argue with.

Critically think about the contents of this letter and identify three alternate explanations for why this individual outlived his friends:

_____

_____

_____

FEEDBACK can be found at the end of Unit 31.

[Key Term

life expectancy, page 31-3

Review Questions

TRUE FALSE

1. One of the main reasons the United States ranks first among all countries in life expectancy is the high-quality diet consumed by adults in general. ☐ ☐

2. Three characteristics of dietary intake related to increased longevity are regular consumption of fruits and vegetables, moderate alcohol consumption, and above-average consumption of whole-grain products. ☐ ☐

3. Obesity is associated with decreased life expectancy. ☐ ☐

TRUE FALSE

4. Elderly persons tend to have a lowered sensation of thirst. ☐ ☐

5. The nutrient needs of older adults are the same as those for young adults. ☐ ☐

6. Adults and the elderly have a higher need for water than younger people. ☐ ☐

7. People often lose weight as they age due to normal declines in the sense of taste. ☐ ☐

8. Food habits rarely change among older individuals. ☐ ☐

Media Menu

www.4woman.gov
The National Women's Health Information Center puts the spotlight on women's health. It provides information on diabetes, diet and heart health, and other conditions of concern to women. Or call 1-800-994-9662, TDD: 1-888-220-5446.

nihseniorhealth.gov
The National Institute on Aging provides health information, exercise videos, and news related to senior health topics.

www.nutrition.gov
Identify federally funded nutrition programs and resources for the elderly by searching the term "elderly nutrition" at this site.

www.nlm.nih.gov/medlineplus/exercise forseniors.html
This site provides pictures and diagrams of exercises that improve balance and additional information on exercise for seniors.

www.cdc.gov/men and www.cdc.gov/women
These CDC sites cover contemporary health information aimed at health concerns of men or women.

Notes

1. Dillin A. Aging processes and calorie restriction, presented at the Experimental Biology Annual Meeting, New Orleans, 4/18/09.

2. Rivlin RS. Keeping the young-elderly healthy: is it too late to improve our health through nutrition? Am J Clin Nutr 2007;86(supl):1572S–6S.

3. Heidemann C et al. Dietary patterns and risk of mortality from cardiovascular disease, cancer, and all causes in a prospective cohort of women, Circulation 2008; 118:230–7.

4. Central Intelligence Agency World Fact Book, www.cia.gov/library/publications/the-world-factbook/rankorder/2102rank.html, accessed 5/09.

5. US Census Bureau's International Data Base. www.about.com, accessed 10/03.

6. Health United States, 2003. www.cdc.gov/nchs/, accessed 10/03.

7. Health, United States 2008, www.cdc.gov/nchs/hus.htm, accessed 5/09.

8. Perls TT. The different paths to 100. Am J Clin Nutr 2006:83(suppl):484S–7S.

9. Bernarducci MP et al. Is there a fountain of youth? A review of current life extension strategies. Pharmacotherapy 1996;16:183–200.

10. Natality statistics, www.cdc.gov/nchs/births.htm, accessed 5/09.

11. Guyer B et al. Annual summary of vital statistics: trends in the health of Americans during the 20th century, Pediatrics 2000;106:307–17.

12. Mariani SM. Magic potions and the quest for eternal youth. Medscape Molecular Medicine 2003;5:1–7.

13. Olshansky SJ et al. What if humans were designed to last? The Scientist, 2007, March, pp. 28–35.

14. Tomey KM et al. Dietary intake is related to prevalent functional limitations in midlife. Am J Epidemiol 2008;167:935–43.

15. Rodriguez-Laso A et al. The effect of social relationships on survival in elderly residents of a Southern European community: a cohort study, BMC Geriatr. 2007;7:19.

16. Physical Activity Guidelines for Americans, 2008, U.S. Department of Health & Human Service, http://www.health.gov/PAGuidelines.

17. Shepard RJ et al. Regular exercise through middle age delays biological aging, Brit J Sports Med Aug. 10, 2008, www.medscape.com/viewarticle/573636.

18. Schneider EL. Weighing your longevity part II: should you restrict your calories? www.healthandage.com, accessed 8/03.

19. Stevens J et al. The effect of age on the association between body-mass index and mortality. New Engl J Med 1998, 338:1–7.

20. Cole GM. Docosahexaenoic acid and Alzheimer's disease, presented at the Experimental Biology Annual Meeting, New Orleans, 4/18/09.

21. Whelan J. (n-6) and (n-3) polyunsaturated fatty acids and the aging brain: food for thought, J Nutr 2009;138:2521–2.

22. Dawson-Hughes B. Serum 25-hydroxyvitamin D and functional outcomes in the elderly, Am J Clin Nutr 2008;88(suppl):537S–40S.

23. Wolfe KL et al. Cellular antioxidant activity of common fruits, J Agric Food Chem 2008;56:8418–26.

24. Tan AG et al. Antioxidant nutrient intake and the long-term incidence of age-related cataract: the Blue Mountain Eye Study, Am J Clin Nutr 2008;87:1899–905.

25. Rowe JW, Kahn RL. Human aging: usual and successful, Clin Nutr 1990; 9:26–33.

26. Connor WE et al. The importance of fish and docosahexaenoic acid in Alzheimer's disease, Am J Clin Nutr 2008;2007;85:929–30.

27. Garry PJ, Vellas BJ. Aging and nutrition. In: Present knowledge in nutrition, Ziegler EE, Filer, LJ eds., ISLI Press: Washington, DC; 1998: pp. 404–19.

28. Bossingham MJ et al. Water balance, hydration status, and fat-free mass hydration in younger and older adults. Am J Clin Nutr 2005;81:1342–50.

29. Francesco VD et al. Unbalanced serum leptin and ghrelin dynamics prolong postprandial satiety and inhibit hunger in healthy elderly: another reason for the "anorexia of aging." Am J Clin Nutr 2006:83:1149–52.

30. North American Menopause Society. Osteoporosis guidelines update. Menopause 2006;13:340–67.

31. Visser M et al. Low serum concentrations of 25-hydroxyvitamin D in older persons and the risk of nursing home admission Am J Clin Nutr 2006;84:616–22.

32. Lichentenstein AH et al. Modified MyPyramid for older adults, J Nutr 2008;138:5–11.

33. Vogelzang JL et al. Overview of fluid maintenance/prevention of dehydration. J Am Diet Assoc 1999;99:605–9.

34. Dietary Reference Intakes for Water, Potassium, Sodium, Chloride, and Sulfate (2004), Food and Nutrition Board, available at: books.nap.edu/openbook/0309091691/gifmid/74.gif.

35. Rolls BJ. Do chemosensory changes influence food intake in the elderly? Physiol Behav 1999;66:193–7.

36. Schiffman SS. Taste and smell losses in normal aging and disease. JAMA 1997;278:1357–62.

37. Bilderbeck N et al., Changing food habits among 100 elderly men and women in the United Kingdom. J Hum Nutr 1981;35:448–55.

NUTRITION | # Up Close

Does He Who Laughs, Last?

Feedback for Unit 31

Alternate explanations:

1. The letter writer's friends may have switched to perceived health foods when they found out they were sick.
2. The letter writer may have consumed a lifelong healthy diet without choosing organic or "natural" foods.
3. The letter writer may come from a family whose members tend to have long lives.

Other possible explanations:

- A sense of humor may help extend life.
- Many factors in addition to diet, exercise, and genetic background influence longevity.
- One individual's experience may not apply to the masses.

UNIT 32 | The Multiple Dimensions of Food Safety

NUTRITION SCOREBOARD

		TRUE	FALSE
1	The ingestion of pesticides on foods is a major contributor to the high incidence of cancer in humans.		
2	Freezing kills bacteria in food.		
3	Alaska has the highest incidence of botulism in the United States.		
4.	You can tell if a food has gone bad by smelling it.		

Key Concepts and Facts

- The leading food safety problem in America is contamination of food by bacteria and viruses.

- Foodborne illnesses primarily result from unsafe methods of producing, storing, and handling food.

- Foodborne illnesses are linked to hundreds of foods but most commonly to raw and undercooked meats and eggs, shellfish, and unpasteurized milk.

- Most cases of foodborne illness are preventable.

More than 50% of consumers do not realize that foodborne diseases can be transmitted by fruits and vegetables.

Photo Disc

On the Side

Threats to the Safety of the Food Supply

Hamburgers contaminated with *E. coli*, chicken tainted with *Salmonella*, fruit coated with organophosphate, mad cow disease—it's enough to give you food fright. Actually, Americans have good reason to be concerned about the safety of some foods (Illustration 32.1). The Centers for Disease Control and Prevention (CDC) estimates that each year in the United States foodborne illnesses cause:

- Sickness in 76 million people

- 325,000 hospitalizations

- Over 5000 deaths[4]

Foodborne illnesses can be caused by bacteria, viruses, marine organisms, fungi, and the toxins they produce, as well as by chemical contaminants in foods or water. They are spread by a wide assortment of foods, including meats, eggs, unpasteurized milk, shellfish, raspberries, sprouts, breakfast cereals, and hummus. Foods most commonly associated with foodborne illness, however, are raw and undercooked meat and eggs, shellfish, and unpasteurized milk.[5]

How Good Foods Go Bad

foodborne illnesses
An illness related to consumption of foods or beverages containing disease-causing bacteria, viruses, parasites, toxins, or other contaminants.

Bacteria and viruses are the most common causes of **foodborne illnesses** and largely enter the food supply during food processing, storage, or preparation (Illustration 32.2). They are transferred to humans in foods through many different routes, a major one being the contamination of food with animal feces.

Illustration 32.1 Problems with food safety often make the headlines.

The lower intestines of many healthy farm animals are colonized by bacteria that may be harmful to humans. These bacteria contaminate food when unsanitary practices are used to prepare and process meats, and when vegetables and fruits are fertilized with animal manure. They can also be transferred to foods by humans who have colonies of harmful bacteria in their gastrointestinal tracts through the use of human sewage on crops and food handling by people carrying the bacteria on their hands (Illustration, 32.2). Humans also transfer certain harmful microorganisms to food when fluids from infected injuries or body secretions contact food.[7] These types of foodborne illnesses are mainly caused by bacteria that are "on" rather than "in" foods.

Although less common, bacteria can be present on the inside of foods. They can enter vegetables and fruits, for example, if the protective skin coatings are

Illustration 32.2 Can you find three potential opportunities for the spread of foodborne illness in this photograph?

broken, allowing an entrance for bacteria. *Salmonella* bacteria can infect the ovaries of hens and cause them to lay normal-looking but infected eggs. Shellfish can concentrate microorganisms present in surrounding water, and although the bacteria may be harmless to the shellfish, they can provide large doses of harmful bacteria and viruses to humans.[8]

Cross-Contamination of Foods

The source of many cases of foodborne illness is food that has come into contact with contaminated food. This situation, referred to as "cross-contamination," increases the reach of foodborne illnesses.

Microorganisms, including bacteria, viruses, toxins, and other harmful substances, can contaminate safe foods during processing, shipping, preparation, or storage. Opportunities for cross-contamination of foods during processing, for example, are plentiful. The hamburger you eat may contain meat from hundreds of different cows, an omelet in a restaurant may contain the eggs of hundreds of different chickens, and the chicken you baked may have bathed with hundreds of others when they were all washed in the same vat of water at the meat-processing plant. Cross-contamination also occurs on cutting boards across America. The failure to routinely wash cutting boards in between the preparation of different raw foods is a major route to the spread of foodborne illness.[9]

Antibiotics, Hormones, and Other Substances in Foods

Potential sources of food contamination include antibiotics and hormones given to animals, pesticides, and industrial pollutants. Effects of these substances on health vary from potentially life threatening to uncertain.

Antibiotic Resistance   Chickens, cattle, and other farm-raised animals are commonly given antibiotics in feed to prevent infectious disease. In many cases, the antibiotics used are the same ones given to humans to treat infections. Unfortunately, microorganisms are very clever: They can transform themselves to become resistant to antibiotics used to kill them. People can become infected with these new strains of antibiotic-resistant microorganisms when they are consumed in foods. These infections can be difficult to treat because the disease-causing microorganisms are not sensitive to the antibiotics used. Many common forms of foodborne illnesses today are represented by these new forms of bacteria.[10]

Hormones   Hormones are commonly given to farm-raised animals to promote growth or improve milk production. Many consumers are concerned about the effects, however, and would prefer such hormones not be present in food. Hormones given to animals do show up in foods along with other hormones naturally present in animal tissues. Whether animal products free of additional hormones are safer than those without them is unknown but under investigation.[11]

Pesticides and PCBs   Pesticides containing organophosphates, mercury-containing fungicides, and DDT remain causes of foodborne illnesses, as do PCBs. Use of DDT on insects and PCBs in transformers was phased out over 20 years ago due to links with cancer, but these long-lasting chemicals still contaminate some land, lakes, and streams. Fewer than 1 in 10,000 foods on the market contains an excessive level of pesticide, and two out of three contain no trace of agricultural chemicals (Illustration 32.3). Agriculture workers in developing countries are most likely to experience illnesses associated with exposure to agricultural chemicals

© Ed Kashi Photography

Illustration 32.3   Gathering broccoli samples for pesticide testing. Pesticide residues do not appear to be a major cause of foodborne illnesses.

(Illustration 32.4). Health problems due to agricultural chemicals appear to be rare in other groups of people.[1]

Causes and Consequences of Foodborne Illness

During the American Civil War, more men died of disease caused by bad food and contaminated water than from all other battle losses.[12]

Over 250 types of foodborne illnesses caused by infectious agents (bacteria, viruses, and parasites) and noninfectious agents (toxins and chemical contaminants) have been identified. Their impact on health ranges from a day or two of nausea and diarrhea to death within minutes. Effects of foodborne illnesses are generally most severe in people with weakened immune systems or certain chronic illnesses, pregnant women, young children, and older persons (Table 32.1). Symptoms of foodborne illness most commonly consist of nausea, vomiting, abdominal cramps, and diarrhea.[7] Many cases go unreported, making it difficult to identify the most common causes of foodborne illness. Estimates of the causes are usually based on cases reported by health care professionals and health departments.[2]

The most prevalent reported causes of foodborne illness result from *Salmonella, Campylobacter,* and *E. coli* 0157:H7 bacteria and Noroviruses (previously called Norwalk, or Norwalk-like viruses).[13] Table 32.2 summarizes facts about these top four leading causes of foodborne illness.

Salmonella Over 37,000 cases of *Salmonella* infection are reported to the CDC yearly, and it is estimated that 38 times that number, or 1,412,498 cases, actually occurred.[2] (A photograph of *Salmonella* bacteria is presented in Illustration 32.5.) An outbreak of Salmonella infection that affected approximately 224,000 people was linked to the transport of an ice cream mixture in a tanker that had previously transported *Salmonella*-infected eggs. The infected ice cream was distributed in 41 states. It is estimated that 5% of the U.S. population experiences a *Salmonella* infection every year.[2]

©Lee Snider/The Image Works

Illustration 32.4 Protective gear—including a chemical-resistant suit, respirator, and goggles—is required for use with some of the most toxic pesticides but is not always worn.

Table 32.1

High-risk groups for severe effects of foodborne illness[2]

- People with weakened immune systems due to HIV/AIDS and others with weakened defenses against infections
- People with certain chronic illnesses such as diabetes and cancer
- Pregnant women
- Young children
- Older persons

Table 32.2

The top four causes of foodborne illness[13]

Bacteria	Onset	Illness Duration	Symptoms	Foods Most Commonly Affected	Usual Source of Contamination
Salmonella	1–3 days	4–7 days	Diarrhea, abdominal pain, chills, fever, vomiting, dehydration	Uncooked or undercooked eggs, unpasteurized milk, raw meat and poultry, vegetables and fruits	Infected animals, human feces on food, contaminated water
Campylobacter	2–5 days	2–5 days	Diarrhea (may be bloody), abdominal cramps, fever, vomiting	Undercooked poultry, unpasteurized milk	Infected poultry and other animals
E. coli 0157:H7	1–8 days	5–10 days	Watery, bloody diarrhea, abdominal cramps, little or no fever	Raw or undercooked beef, unpasteurized milk, raw vegetables and fruits, contaminated water	Infected cattle
Noroviruses	1–2 days	1–3 days	Nausea, vomiting, diarrhea	Undercooked seafood	Human feces contamination of oysters and other shellfish beds

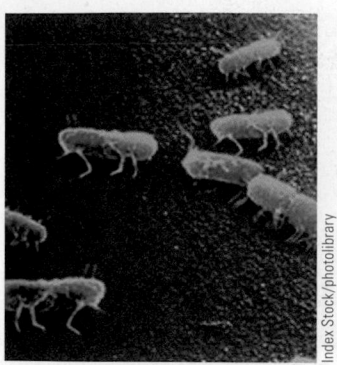

Illustration 32.5 A magnified view of *Salmonella* bacteria.

Campylobacter This bacterium is a common contaminant on chicken and is the cause of the greatest number of cases of bacteria-related foodborne illness each year. That figure is a startling 2,453,926.[2] The bacteria are found on up to 80% of chickens, and 20% represent a strain of *Campylobacter* resistant to antibiotics. In addition to poultry, *Campylobacter* infection has been traced to unpasteurized milk and contaminated water.[14]

E. Coli *E. coli* 0157:H7 is a relatively new strain of E. coli that has evolved into a potential killer. As few as 10 of these bacteria can lead to death by kidney failure in vulnerable people. Nearly 80,000 cases of E. coli 0157:H7 occur each year in the United States, and 8% of infected people die from it.[2] E. coli infections are particularly associated with consumption of undercooked ground beef. A major outbreak of E. coli 0157:H7 infection occurred when contaminated ground beef left over from one day's production was added to the next day's batch of beef. Use of the leftover, contaminated meat kept recontaminating subsequent batches.[15] Some 25 million pounds of ground beef had to be recalled as a result.

Noroviruses Noroviruses are an extremely common but underreported cause of foodborne illnesses. The viruses go underreported because laboratory tests required for diagnosis are not widely performed. Noroviruses usually cause an acute bout of vomiting that resolves within two days. They are thought to be primarily spread by infected kitchen workers and fishermen who have dumped sewage waste into waters above oyster beds. An estimated 23,000,000 cases of illnesses related to noroviruses occur in the United States each year.[2]

Other Causes of Foodborne Illnesses

Of the hundreds of other causes of foodborne illness, seven have been selected for brief review here. They represent a sampling of foodborne illnesses stemming from seafood consumption, and others that can result from a bacterial toxin, a parasite, and prions (a type of protein).

Foodborne Illnesses Related to Seafoods Many types of seafood (Illustration 32.6) may become contaminated due to industrial or human pollution of fresh and ocean waters. Examples presented relate to mercury contamination, ciguatera fish poisoning, and the neurotoxin (or nervous system toxin) responsible for "red tide."

Mercury Contamination Seafoods have come under fire as a potential source of foodborne illnesses due to mercury contamination of waters and fish by fungicides, fossil fuel exhaust, smelting plants, pulp and paper mills, leather-tanning facilities, and chemical manufacturing plants. High levels of mercury are most likely to be present in large, long-lived fish such as shark, swordfish, and tuna. Because mercury can interfere with fetal brain development, pregnant women should limit their consumption of these types of fish. The Food and Drug Administration (FDA) advises that women who are pregnant or may become pregnant not eat shark, swordfish, king mackerel, or tilefish. Pregnant women should limit consumption of fish caught by family or friends to 6 ounces a week, and they may eat an average of 12 ounces of fish from stores and restaurants weekly. Consumption of fish by adults in general does not appear to pose a health risk.[16]

Ciguatera Ciguatera poisoning from fish is caused by a neurotoxin (ciguatoxin) present in microorganisms called dinoflagellates that live in reefs (Illustration 32.7). The toxin is transferred through herbivorous reef fish to carnivorous fish and then to humans who eat these fish. Over 200 types of fish may cause ciguatera poisoning, the most common being grouper, red snapper, and barracuda. Primary areas affected by ciguatera poisoning include the Caribbean

Illustration 32.6 Seafoods are an important potential cause of foodborne illness.

and South Pacific Islands. Symptoms develop within 1 to 30 hours after ingestion of poisoned fish and cause nausea, vomiting, abdominal cramps, watery diarrhea, and then numbness, shooting pains in the legs, and other symptoms. The toxin causing the poisoning is not destroyed by cooking, freezing, or digestive enzymes, and there is no effective treatment.[17]

Red Tide "Red tide" may occur between June and October on Pacific and Atlantic coasts. It is due to the accumulation of a microorganism that produces a nerve toxin. Oysters and other shellfish that consume the microorganism become contaminated, as do humans who ingest the shellfish. Resistant to cooking, the toxin produced causes a burning or prickling sensation in the mouth from 5 to 30 minutes after contaminated shellfish are consumed. This symptom is followed by nausea, vomiting, muscle weakness, and a loss of feeling in the hands and feet. Recovery is usually complete, but it's easy to understand that the warning not to eat mussels, clams, oysters, scallops, or other shellfish from red-tide waters is for real.[18]

Botulism *C. botulinum* bacteria produce a toxin that is one of the deadliest known. It can cause nerve damage and respiratory failure. (An antidote is available but must be given soon after the infection begins.) These bacteria are commonly present in soil and ocean and lake sediment, and they may contaminate crops, honey, animals, and seafood. In humans, botulism usually results from eating underheated, contaminated foods stored in airtight containers. The bacteria thrive without oxygen and produce gases as they grow. The gases expand the food container. (For a real example of a can that exploded due to gas produced by the growth of bacteria, see Illustration 32.8.) Consequently, foods in cans, plastic bags and wraps, and other airtight containers that have bulged-out areas should *not* be eaten.

Alaska has the highest incidence of botulism in the United States. Why does the 49th state rank first? The reason is that some Native Alaskans place uncooked or partially cooked salmon eggs, whale blubber, and other seafoods that harbor *C. botulinum* in plastic bags to ferment. The bags are squeezed to expel the air before sealing, and the no-oxygen environment fosters the growth of botulinum.[19]

Parasites Various parasitic worms, such as tapeworms, flatworms, and roundworms, may enter food and water through fecal material and soil. They are generally killed by freezing and always killed by high temperature. Once consumed, roundworms may attach to the lining of the intestine and feed on the person's blood. This can lead to anemia. One type of roundworm can bore a hole through the stomach within an hour after the worm's source—raw fish—is eaten. The severe pain that results sends people to their doctors on the double. Hundreds of Japanese people experience this foodborne illness every year.[20]

It's a good thing most parasites are destroyed by freezing. One study in Seattle assessed the parasite content of raw fish used in sushi. About 40% of the raw fish samples contained roundworms, but since the fish had been deep frozen, all the worms were dead.[21]

Mad Cow Disease Technically called bovine spongiform encephalopathy (BSE), this rare disease in cattle is suspected of causing at least 150 human deaths in Europe. Only one case of the disease in humans—caused by consumption of affected beef—has been diagnosed in the United States, and it originated in England. The disease appears to have started when cows, who are herbivores by nature, were given sheep intestines and parts of the spinal cord in their feed. Some of the sheep harbored a protein, called a prion, that caused a deadly disease when consumed by cows. A prion is not a bacterium or other microorganism; it's a small protein that can transmit disease when consumed by a similar species. Only one other known foodborne illness is spread by the consumption

Illustration 32.7 Magnified view of dinoflagellates that cause ciguatera poisoning.

©Biophoto Associates/Photo Researchers, Inc.(ALL)

Illustration 32.8 Pressure caused by bacterial gases in this can of glucose drink made the can explode.

Richard Anderson

On the Side

of otherwise healthy body parts. That disease is kuru, and it is transmitted by cannibalism.[22]

Researchers concluded that mad cow disease was transferred to humans who ate the meat of prion-infected cows. The disease represents a new form of Creutzfeldt-Jakob disease (variant Creutzfeldt-Jakob disease, or VCJD) previously identified in humans. It inevitably leads to death in humans, after many years, due to brain damage. Needless to say, it is no longer legal to feed cows animal parts that may transmit the disease. Although the risk of consuming beef from cattle with mad cow disease is extremely small, the possibility that animals may develop this or a similar disease exists. The possibility that mad cow disease may affect cattle in the United States became a reality late in 2003. Stringent monitoring efforts are in place in the United States and many European countries to prevent the occurrence and spread of prions. Rates of BSE are declining due to improved detection and control methods.[23]

Because the disease was first recognized in 1994 and may take 20 or more years to develop, we may not know the full impact of the exposure to contaminated beef for years to come.

Preventing Foodborne Illnesses

There are two major approaches to the prevention of foodborne illnesses. The first relies on food safety regulations that control food processing and handling practices, and the second involves consumer behaviors that lessen the risk of consuming contaminated foods. (Read what can happen when both of these principles are violated in the nearby "Health Action.")

Food Safety Regulations

According to the Federal Food, Drug, and Cosmetic Act, it is illegal to produce or dispense foods that are contaminated with substances that cause illness in humans. Foods are considered "safe" if there is a reasonable certainty that no harm will result from repeated exposure to any substance added to foods. The act governs all substances that are added intentionally or accidentally to foods—except pesticides. According to legislation passed in 1996, pesticides are permitted if their consumption is associated with a "negligible risk" of cancer or other health problems.[26]

Irradiation of Foods   Increasing rates of foodborne illness have led some health officials to recommend the expanded use of food irradiation. Food irradiation is

Health Action   A Double Whammy

From the true story department of *Nutrition Now* comes the tale of a man who took a night off with his wife to dine at their favorite restaurant. The man, a physician, ordered his favorite item: the luscious chicken burritos. He was served more than he could eat, so he took the leftovers home and promptly put them in the refrigerator. About 8 hours later, he felt like his intestines were exploding; he thought he'd rather die than be so sick. Then the light bulb went off—he had probably consumed *C. perfingens* toxin somehow.

By the third day he began to feel better and noticed he was ravenously hungry. Then he remembered there was a leftover burrito in the refrigerator.... Happily, the second round of symptoms wasn't as bad as the first.

The message? It's difficult for consumers to prevent every case of foodborne illness, but we can prevent more cases than we do. Throw out spoiled food.

a safe process when conducted under specified conditions. It destroys bacteria, parasites, and viruses present in or on foods. Availability of irradiated foods is growing but is still rather limited.[27]

Although the proposal has merits, food irradiation is not the silver bullet that will prevent all cases of foodborne illness. Prions, toxins, pesticides, mercury, and PCBs are resistant to radiation, so radiation is unlikely to affect the safety of foods with respect to these contaminants. Once irradiated, foods can later become contaminated in such places as a packing plant, grocery store, restaurant, or home. The absence of all microorganisms in irradiated food may enable individual types of bacteria that contact the food after it is sterilized to grow at unusually high rates. Irradiation does not work well on all foods— some fruits and vegetables develop a poor texture and change flavor when irradiated.[28]

The Consumer's Role in Preventing Foodborne Illnesses

Due in large measure to the demands of individuals and consumer groups, many mechanisms are in place to help ensure the safety of the food supply. However, existing safeguards in food production and processing, and government regulations and enforcement efforts, are insufficient to guarantee that all foods purchased by consumers will be free from contamination. This situation, along with the possibility that food can become contaminated in unanticipated ways, means that responsibility for food safety is shared by consumers.

Food safety education will never be a substitute for a safe food supply.
—Caroline Smith DeWall, 1997

Food Safety Basics The first rule of food safety is to wash your hands thoroughly with soap and water before and after handling food (Illustration 32.9). According to the Centers for Disease Control, this is the single most important means of preventing the spread of foodborne illness caused by bacteria.

■ **There's a Right Way to Wash Your Hands.** Ever watch a TV dramatic series about doctors and hospitals and wondered why they show surgeons scrubbing their hands and upper arms before surgery for so long? TV got it right. It takes about 20 seconds (or the time it takes to repeat the word "mississippi" 20 times) to sanitize your hands. First you need to lather up with soap and very warm water. The soap makes germs lose their grip on your skin. Next you have to make sure to scrub all the crevices between your fingers and under your fingernails. Rinse your hands thoroughly with really warm water and dry them with a paper towel. (If you use a dishcloth, you may reinfect your hands.)[29]

■ **Keep Hot Foods Hot and Cold Foods Cold.** Good foods can go bad if stored improperly. One of the most effective ways to prevent foods from spoiling is to store them at, and heat them to, the right temperatures (see Table 32.3 and Illustrations 32.10 and 32.11 on p. 32–11). When holding prepared foods before serving, hot foods should be kept hot, and cold foods cold. Freezing foods halts the growth of all of the main types of bacteria that contaminate foods. Once the foods are thawed, however, bacteria growth resumes, and the foods may become contaminated with new bacteria while thawing.

Bacteria grow best at temperatures between 40° and 135°F (see Illustration 32.10). The room temperature of homes is within this range, and that's why foods that spoil should not be left outside the refrigerator for more than an hour. If a food has been improperly stored or kept past the expiration date on the label, it should not be eaten even if it tastes, smells, and looks all right. Bacteria that most commonly cause food borne illnesses usually do not change the taste, smell, or appearance of foods.[3] This general rule should apply: When in doubt, throw it out!

■ **Don't Eat Raw Milk, Eggs, or Meats.** Unpasteurized milk and milk products, raw and partially cooked eggs, raw and undercooked meats and fish, and

Illustration 32.9 There's a right way to wash your hands.

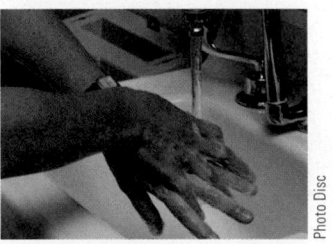

Photo Disc

Table 32.3

A guide for cooking foods to safe temperatures

Raw Food	Internal Temperature
Ground Meats	
Hamburger	160°F
Beef, veal, lamb, pork	160°F
Chicken, turkey	165°F
Beef, Veal, Lamb	
Roasts and steaks	
Medium rare	145°F
Medium	160°F
Well done	170°F
Pork	
Chops, roasts, ribs	
Medium	160°F
Well done	170°F
Ham, fresh	160°F
Sausage, fresh	160°F
Poultry	
Chicken, whole and pieces	180°F
Duck	180°F
Turkey (unstuffed)	180°F
Whole	180°F
Breast	170°F
Dark meat	180°F
Stuffing (cooked separately)	165°F
Eggs	
Fried, poached	Yolk and white are firm
Casseroles	160°F
Sauces, custards	160°F

Fish should be cooked until the flesh is not clear-looking and flakes easily with a fork; shrimp, lobster, and crab until shells are red and the flesh is opaque and not clear- or raw-looking inside; and clams, oysters, and mussels until the shells open.

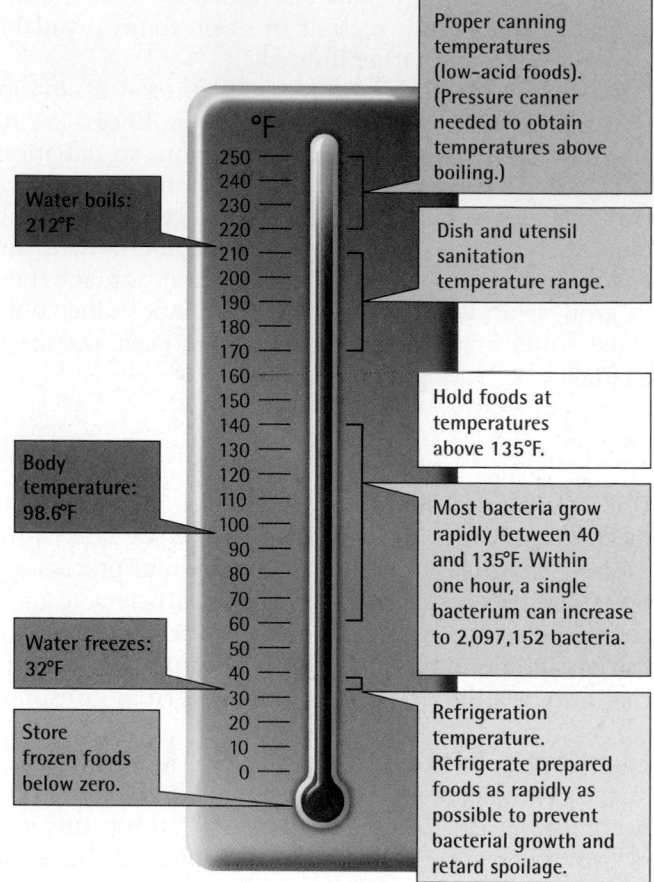

Illustration 32.10 Temperature guide for safe handling of food in the home.

Source: Adapted from Minnesota Extension Service, University of Minnesota; updated in 2004, based on data from USDA.

raw sprouts may be contaminated with microorganisms that cause foodborne illness. Certain raw oysters may be contaminated, too, and should be avoided.[6]

Food Handling and Storage Particular care needs to be taken when handling and storing foods. Dairy products such as milk and cheese may spoil in about a week or sooner if contaminated with bacteria from the air or hands. If you hold cheese by the wrapper when you take it out of the refrigerator to cut a slice off, it will likely last longer than if you touch it with your hands. Handling foods with clean hands or plastic gloves and using clean utensils on washed surface areas helps to reduce the number of bacteria that come in contact with the food. Raw meats should be separated from other foods, and utensils and surface areas used to prepare meats should be thoroughly cleaned after each use.

Illustration 32.12 on page 32–13 shows the USDA's "Safe Handling Instructions" placed on meat and poultry packaging to help consumers protect themselves from potential foodborne illness.

Because consumers have so many questions and concerns about meat and poultry, the USDA established a 24-hour meat and poultry hotline. When in doubt, contact the hotline at 1.888.674.6854, or by email: mphotlne.fsis@usda/gov.

The Safety of Canned Foods Commercially canned foods are heated to the point where all bacteria are killed after the can is sealed. Consequently, the contents of canned foods are sterile and will not spoil if left on the shelf for years.

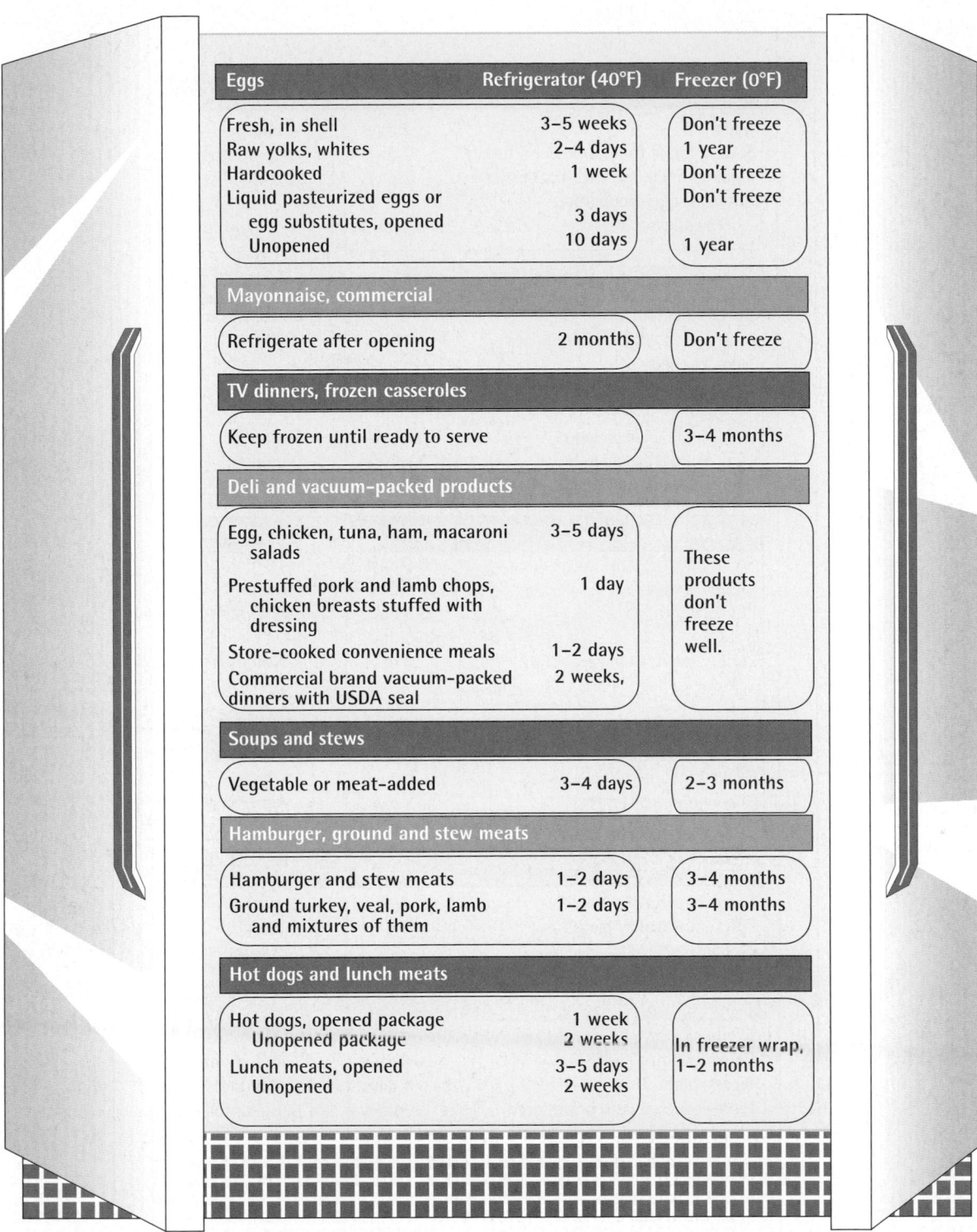

Eggs	Refrigerator (40°F)	Freezer (0°F)
Fresh, in shell	3–5 weeks	Don't freeze
Raw yolks, whites	2–4 days	1 year
Hardcooked	1 week	Don't freeze
Liquid pasteurized eggs or		Don't freeze
egg substitutes, opened	3 days	
Unopened	10 days	1 year

Mayonnaise, commercial		
Refrigerate after opening	2 months	Don't freeze

TV dinners, frozen casseroles		
Keep frozen until ready to serve		3–4 months

Deli and vacuum-packed products		
Egg, chicken, tuna, ham, macaroni salads	3–5 days	These products don't freeze well.
Prestuffed pork and lamb chops, chicken breasts stuffed with dressing	1 day	
Store-cooked convenience meals	1–2 days	
Commercial brand vacuum-packed dinners with USDA seal	2 weeks,	

Soups and stews		
Vegetable or meat-added	3–4 days	2–3 months

Hamburger, ground and stew meats		
Hamburger and stew meats	1–2 days	3–4 months
Ground turkey, veal, pork, lamb and mixtures of them	1–2 days	3–4 months

Hot dogs and lunch meats		
Hot dogs, opened package	1 week	In freezer wrap, 1–2 months
Unopened package	2 weeks	
Lunch meats, opened	3–5 days	
Unopened	2 weeks	

Illustration 32.11 Cold storage.

These safe time limits will help keep refrigerated food from spoiling or becoming dangerous to eat. These time limits will keep frozen food at top quality.

Source: U.S. Department of Agriculture, Food Safety and Inspection Service, September 1990 and 1997.

(continued on next page)

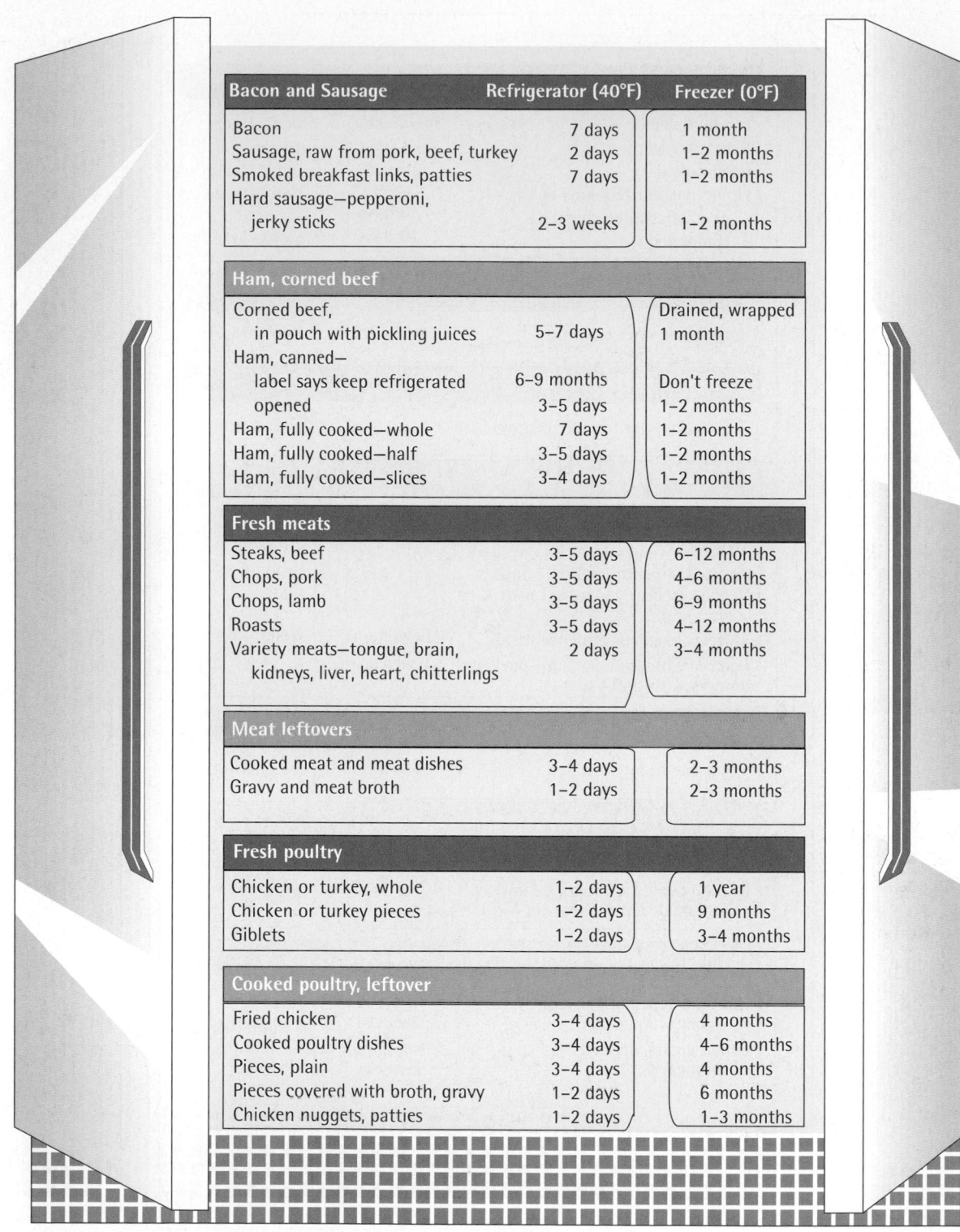

Bacon and Sausage	Refrigerator (40°F)	Freezer (0°F)
Bacon	7 days	1 month
Sausage, raw from pork, beef, turkey	2 days	1–2 months
Smoked breakfast links, patties	7 days	1–2 months
Hard sausage—pepperoni, jerky sticks	2–3 weeks	1–2 months

Ham, corned beef		
Corned beef, in pouch with pickling juices	5–7 days	Drained, wrapped 1 month
Ham, canned— label says keep refrigerated	6–9 months	Don't freeze
opened	3–5 days	1–2 months
Ham, fully cooked—whole	7 days	1–2 months
Ham, fully cooked—half	3–5 days	1–2 months
Ham, fully cooked—slices	3–4 days	1–2 months

Fresh meats		
Steaks, beef	3–5 days	6–12 months
Chops, pork	3–5 days	4–6 months
Chops, lamb	3–5 days	6–9 months
Roasts	3–5 days	4–12 months
Variety meats—tongue, brain, kidneys, liver, heart, chitterlings	2 days	3–4 months

Meat leftovers		
Cooked meat and meat dishes	3–4 days	2–3 months
Gravy and meat broth	1–2 days	2–3 months

Fresh poultry		
Chicken or turkey, whole	1–2 days	1 year
Chicken or turkey pieces	1–2 days	9 months
Giblets	1–2 days	3–4 months

Cooked poultry, leftover		
Fried chicken	3–4 days	4 months
Cooked poultry dishes	3–4 days	4–6 months
Pieces, plain	3–4 days	4 months
Pieces covered with broth, gravy	1–2 days	6 months
Chicken nuggets, patties	1–2 days	1–3 months

Illustration 32.11 (*continued*)
Cold storage: safe time limits

Illustration 32.12   The USDA Safe Handling Instructions label.

Safe Handling Instructions

THIS PRODUCT WAS PREPARED FROM INSPECTED AND PASSED MEAT AND/OR POULTRY. SOME FOOD PRODUCTS MAY CONTAIN BACTERIA THAT CAN CAUSE ILLNESS IF THE PRODUCT IS MISHANDLED OR COOKED IMPROPERLY. FOR YOUR PROTECTION, FOLLOW THESE SAFE HANDLING INSTRUCTIONS.

KEEP REFRIGERATED OR FROZEN.
THAW IN REFRIGERATOR OR MICROWAVE.

KEEP RAW MEAT AND POULTRY SEPARATE FROM OTHER FOODS. WASH WORKING SURFACES (INCLUDING CUTTING BOARDS), UTENSILS, AND HANDS AFTER TOUCHING RAW MEAT OR POULTRY.

COOK THOROUGHLY.

KEEP HOT FOODS HOT. REFRIGERATE LEFTOVERS IMMEDIATELY OR DISCARD.

Nevertheless, canned foods may spoil if the can develops a pinhole leak or if errors made during the canning process allow bacteria to enter the food. A cardinal sign that canned foods have spoiled is a buildup of pressure inside the can. The pressure will cause the top of the can to bulge out instead of curving in. Because a buildup of pressure inside a can may be due to bacteria that cause botulism, the food should be thrown away.

There Is a Limit to What the Consumer Can Do

Keeping food safe from the farm, lake, or ocean to the table represents a major challenge to industry and government. None of the Health Objectives for the Nation for the year 2010 for declines in the incidence of foodborne diseases have been achieved.[30] Until contamination of food is prevented, consumers will continue to play an important role in ensuring food's safety. The Take Action feature offered below presents some options for reducing foodborne illness risk in today's food supply environment.

In the end, it is up to consumers to demand a safe food supply, up to industry to produce it, up to researchers to develop better ways of doing so, and up to government to see that it happens, to make sure it works, and to identify problems still in need of resolution.

—CDC's REPORT ON FOODBORNE ILLNESSES (WWW. CDC.GOV)

Take Action

To limit your exposure to potential sources of foodborne illness[25]

Pick one option from this list you have done, or believe you could do, to reduce your risk of developing foodborne illness.

_____ Buy locally grown produce

_____ Plant a vegetable garden

_____ Buy irradiated raw meats when available

_____ Buy pasteurized dairy products only

© Daniel Pepper/Getty Images

Up Close

Food Safety Detective

Focal Point: Investigating paths to foodborne illness.

Assume an outbreak of foodborne illness occurred among people who ate foods from the onion and lettuce platter shown being prepared in Illustration 32.2. Review the food safety tips in the "Take Action" on page 32–13 and identify three potential causes of the outbreak related to the preparation of the foods as shown in the illustration:

1. _____
2. _____
3. _____

FEEDBACK (answers to these questions) can be found at the end of Unit 32.

[Key Terms

foodborne illnesses, page 32-2

[Review Questions

	TRUE	FALSE
1. All foodborne illnesses are caused by bacteria.	☐	☐
2. Foodborne illnesses often begin with the transfer of harmful bacteria or viruses from feces of an infected animal or person to food.	☐	☐
3. Cross-contamination can result from a failure to wash cutting boards after being used to cut raw meat but not raw vegetables.	☐	☐
4. Antibiotics given to livestock can result in the development of strains of bacteria that resist treatment with antibiotics.	☐	☐

	TRUE	FALSE
5. Symptoms of foodborne illness most commonly consist of headache, blurred vision, and loss of balance.	☐	☐
6. Infection with Norovirus is an extremely common cause of foodborne illness. The illness is marked by episodes of vomiting that resolve with two days.	☐	☐
7. The first rule of food safety is to thoroughly wash your hands before and after food preparation.	☐	☐
8. Cheese can spoil faster if you touch it with your hands.	☐	☐

Media Menu

www.fightbac.org
The Partnership for Food Safety provides materials for use by the media and consumers. The site gives you access to workbooks, posters about food safety, and "mug shots" of the ten least-wanted foodborne pathogens. This site provides a complete cold storage chart and lists the "Core Four" consumer practices for preventing foodborne illness.

www.foodsafety.gov
This site is the national gateway to food safety information, resources, outbreak alerts, and consumer advice. Or you can call the Food Safety 24-Hour Hotline at 1-888-SAFEFOOD.

www.cdc.gov
Search foodborne illnesses, food safety, or international travel and find a plethora of information.

www.epa.gov/ost/fish
The place to go for local fish advisories.

www.cdc.gov/mmwr
This URL sends you to the Centers for Disease Control's Morbidity and Mortality Monthly Report. The site includes updates on foodborne illnesses.

www.cdc.gov/foodnet
This address leads you to FoodNet-Foodborne Disease Active Surveillance Network. It belongs to the Centers for Disease Control and provides updates on foodborne illness outbreaks and advances in understanding them.

Notes

1. Woodall J. Poisoning the poison, The Scientist, Oct. 2007, p. 77.

2. Mead PS et al. Food-related illness and death in the United States. www.cdc.gov, accessed 10/03.

3. Safe food handling: Myth busters. www.fightbac.org/content/view/151/2, accessed 10/06.

4. Food Safety Office, www.cdc.gov/foodsafety, accessed 5/09.

5. What foods are most commonly associated with foodborne illness? www.cdc.gov/ncidod/dbmd/diseaseinfo/foodborneinfections_g.htm#riskiestfoods, accessed 5/09.

6. www.fightbac.org, accessed 10/06, 5/09.

7. Foodborne illness, www.cdc.gov/ncidod/dbmd/diseaseinfo/foodborneinfections_g.htm, accessed 5/09.

8. Foodborne illnesses. www.cdc.gov, accessed 1/01, 5/09.

9. Zhao P et al. Development of a model for evaluation of microbial cross-contamination in the kitchen. J Food Protection 1998;61:960–3.

10. Mathews AW. FDA announces policy designed to curb animal-antibiotic use, Wall Street Journal 2003;Oct. 24:A6.

11. Torisky DM et al. Quantity Feeding During the American Civil War, Binghamton, NY:Haworth Press, Inc., 1998:69-28.

12. Dietary Guideline for Americans. U.S. Dept. of HHS and USDA, 2005.

13. What are the most common foodborne diseases? www.cdc.gov/ncidod/dbmd/diseaseinfo/foodborneinfections_g.htm#mostcommon, accessed 5/09.

14. Wegener HC. The consequences for food safety of the use of fluoroquinolones in food animals, N Engl J Med 1999;340:1581–2.

15. Welland D. E. coli ... salmonella ... Listeria ... unwelcome house guests. Envir Nutr 1997;20(Oct.):1, 6.

16. Clarkson TW et al. The toxicology of mercury—current exposures and clinical manifestations, N Engl J Med 2003;349:1731–8.

17. Treatment of ciguatera poisoning with gabapentin. N Engl J Med 2001;344:692–3.

18. The Merck Manual, 17th ed. Whitehouse Station (NJ): Merck Research Laboratories; 1999, p. 262.

19. Lewis R. Botulism from blubber. The Scientist 2003;Mar. 10:11.

20. Tighter reins sought for imported produce. Envir Nutr 1997;20(Nov.):1, 7.

21. Weir E. Sushi, nemotodes and allergies CMAJ. 2005;1;172:329.

22. Mariani SM. Prions—are they viruses, nucleic acids, or "infectious" proteins? Medscape General Medicine 2003;5:1–8.

23. AMI Foundation Quarterly Report, May 2006. Latest CDC data on foodborne illnesses. www.amif.org/newsletter/AMIFMay2006.pdf, accessed 10/06.

24. Tapeworm, en.wikipedia.org/wiki/Tapeworm, accessed 10/06.

25. There is only so much the consumer can do. How can food be made safer in the first place? http://www.cdc.gov/ncidod/dbmd/diseaseinfo/foodborneinfections_g.htm, accessed 5/09.

26. Madigan D. After Delaney: how will the new pesticide law work? CNI Weekly Report 1996;Oct. 11:4–5.

27. Byrne J. USDA may allow beef carcass irradiation as "processing aid," www.nutraingredientsusa.com, accessed 10/08.

28. Shea KM. Technical report: irradiation of food. Pediatrics 2000;106:1505–9.

29. Henneman A. Don't mess with food safety myths. Nutr Today 1999;34:23–8.

30. Preliminary FoodNet data on the incidence of infection with pathogens transmitted commonly through food—10 states, 2008, MMWR 2009;51(3):333–7.

Answers to Review Questions

1. False, see page 32-2.
2. True, see pages 32-2, 32-3.
3. False, see page 32-4.
4. True, see page 32-4.
5. False, see page 32-5.
6. True, see page 32-6.
7. True, see page 32-9.
8. True, see page 32-10.

NUTRITION | Up Close

Food Safety Detective

Feedback for Unit 32

Potential causes of the foodborne illness outbreak:

1. Disease spread from hands to foods.
2. Use of contaminated cutting board, knife, or food platter.
3. Cross-contamination between onions and lettuce or between other foods stored together.

UNIT 33

Aspects of Global Nutrition

NUTRITION SCOREBOARD

		TRUE	FALSE
1	Approximately 30% of the world's population does not have access to a safe supply of water.		
2	At most, 25% of deaths of young children in developing countries are related to malnutrition and infection.		
3	Many developing countries now face the dual public health problems of increasing rates of overweight as well as underweight.		

Key Concepts and Facts

- The world produces enough food for all its people.

- Poverty, corrupt governments, the HIV/AIDS epidemic, low rates of breast-feeding, unsafe water supplies, and discrimination against females all contribute to malnutrition.

- Malnutrition early in life has long-term effects on mental and physical development.

- Rates of diseases such as heart disease, diabetes, and hypertension increase in developing countries that adopt Western lifestyles and eating habits.

State of the World's Health

The global community consists of countries grouped into the categories of "industrialized nations," "developing nations," and "least-developed nations." (Countries within each category are listed in Table 33.1.) Only 31 countries are considered industrialized, most are developing (113), and, of the developing countries, 49 are least developed. All countries in the global community have interests and issues in common, such as the adequacy of the food supply, availability of health care and education, and safe water. But the countries of the world are also dissimilar in many ways. Differences in financial resources, population growth, and political and other systems translate into large disparities in health status and life expectancy (Table 33.2). People in the least-developed and developing countries are more likely to have substantially shorter life expectancies, to die from infectious diseases, and to experience malnutrition than individuals living in industrialized countries.[5]

The general state of health of populations in various countries is monitored by tracking key environmental, health, and behavioral characteristics. Pregnancy outcome and child growth, rates of breast-feeding, and access to safe drinking water, for example, are key indicators of the health status of a population.[2] As can be seen from Illustration 33.1 and Table 33.3 both on page 33–4, rates of these key indicators vary substantially worldwide. When percentages of low birthweight in newborns and underweight in young children are high, and rates of breast-feeding and access to a safe water supply are low, one can rightfully assume that malnutrition, infection, and other health problems are common. When these key indicators show improvement, the health status of the population and longevity improve.[6]

The health and nutritional status of populations in developing countries is monitored by the World Health Organization (WHO), the Food and Agriculture Organization (FAO), and the United Nations International Children's Emergency Fund (UNICEF). Recent reports have identified leading problem areas related to

Photo Disc

Table 33.1

Countries classified by the U.N. International Children's Emergency Fund (Unicef) as industrialized, developing, and least developed[a]

Industrialized Countries	Developing Countries	Least Developed Countries
Andorra; Australia; Austria; Belgium; Canada; Denmark; Finland; France; Germany; Greece; Holy See; Iceland; Ireland; Israel; Italy; Japan; Liechtenstein; Luxembourg; Malta; Monaco; Netherlands; New Zealand; Norway; Portugal; San Marino; Slovenia; Spain; Sweden; Switzerland; United Kingdom; United States.	Afghanistan; Algeria; Angola; Antigua and Barbuda; Argentina; Armenia; Azerbaijan; Bahamas; Bahrain; Bangladesh; Barbados; Belize; Benin; Bhutan; Bolivia; Botswana; Brazil; Brunei Darussalam; Burkina Faso; Burundi; Cambodia; Cameroon; Cape Verde; Central African Rep.; Chad; Chile; China; Colombia; Comoros; Congo; Congo, Dem. Rep.; Cook Islands; Costa Rica; Cote d'Ivoire; Cuba; Cyprus; Djibouti; Dominica; Dominican Rep.; Ecuador; Egypt; El Salvador; Equatorial Guinea; Eritrea; Ethiopia; Fiji; Gabon; Gambia; Georgia; Ghana; Grenada; Guatemala; Guinea; Guinea-Bissau; Guyana; Haiti; Honduras; India; Indonesia; Iran; Iraq; Jamaica; Jordan; Kazakhstan; Kenya; Kiribati; Korea, Dem. People's Rep.; Korea, Rep. of; Kuwait; Kyrgyzstan; Lao People's Dem. Rep.; Lebanon; Lesotho; Liberia; Libya; Madagascar; Malawi; Malaysia; Maldives; Mali; Marshall Islands; Mauritania; Mauritius; Mexico; Micronesia, Fed. States of; Mongolia; Morocco; Mozambique; Myanmar; Namibia; Nauru; Nepal; Nicaragua; Niger; Nigeria; Niue; Oman; Pakistan; Palau; Panama; Papua New Guinea; Paraguay; Peru; Philippines; Qatar; Rwanda; Saint Kitts and Nevis; Saint Lucia; Saint Vincent/ Grenadines; Samoa; Sao Tome and Principe; Saudi Arabia; Senegal; Seychelles; Sierra Leone; Singapore; Solomon Islands; Somalia; South Africa; Sri Lanka; Sudan; Suriname; Swaziland; Syria; Tajikistan; Tanzania; Thailand; Togo; Tonga; Trinidad and Tobago; Tunisia; Turkey; Turkmenistan; Tuvalu; Uganda; United Arab Emirates; Uruguay; Uzbekistan; Vanuatu; Venezuela; Viet Nam; Yemen; Zambia; Zimbabwe.	Afghanistan; Angola; Bangladesh; Benin; Bhutan; Burkina Faso; Burundi; Cambodia; Cape Verde; Central African Rep.; Chad; Comoros; Congo, Dem. Rep.; Djibouti; Equatorial Guinea; Eritrea; Ethiopia; Gambia; Guinea; Guinea-Bissau; Haiti; Kiribati; Lao People's Dem. Rep.; Lesotho; Liberia; Madagascar; Malawi; Maldives; Mali; Mauritania; Mozambique; Myanmar; Nepal; Niger; Rwanda; Samoa; Sao Tome and Principe; Sierra Leone; Solomon Islands; Somalia; Sudan; Tanzania; Togo; Tuvalu; Uganda; Vanuatu; Yemen; Zambia.

Photo Disc

Photo Disc

[a]Data for some eastern European countries, former Soviet republics, and the Baltic states are not available. Least-developed countries represent a subgroup of the developing countries.

Table 33.2

Life expectancy in selected countries from lowest to highest[4]

	Life Expectancy (Years)
Swaziland	31.9
Zambia	38.6
Mozambique	41.2
Malawi	43.8
Afghanistan	44.6
Chad	47.7
Sudan	51.4
Ethiopia	55.4
Botswana	61.9
Russia	66.0
Philippines	71.1
Iran	71.1
Cuba	77.5
Costa Rica	77.6
United States	78.1
Ireland	78.2
Greece	79.7
Spain	80.0
Italy	80.2
Sweden	80.9
France	81.0
Australia	81.6
Singapore	82.0
Japan	82.1

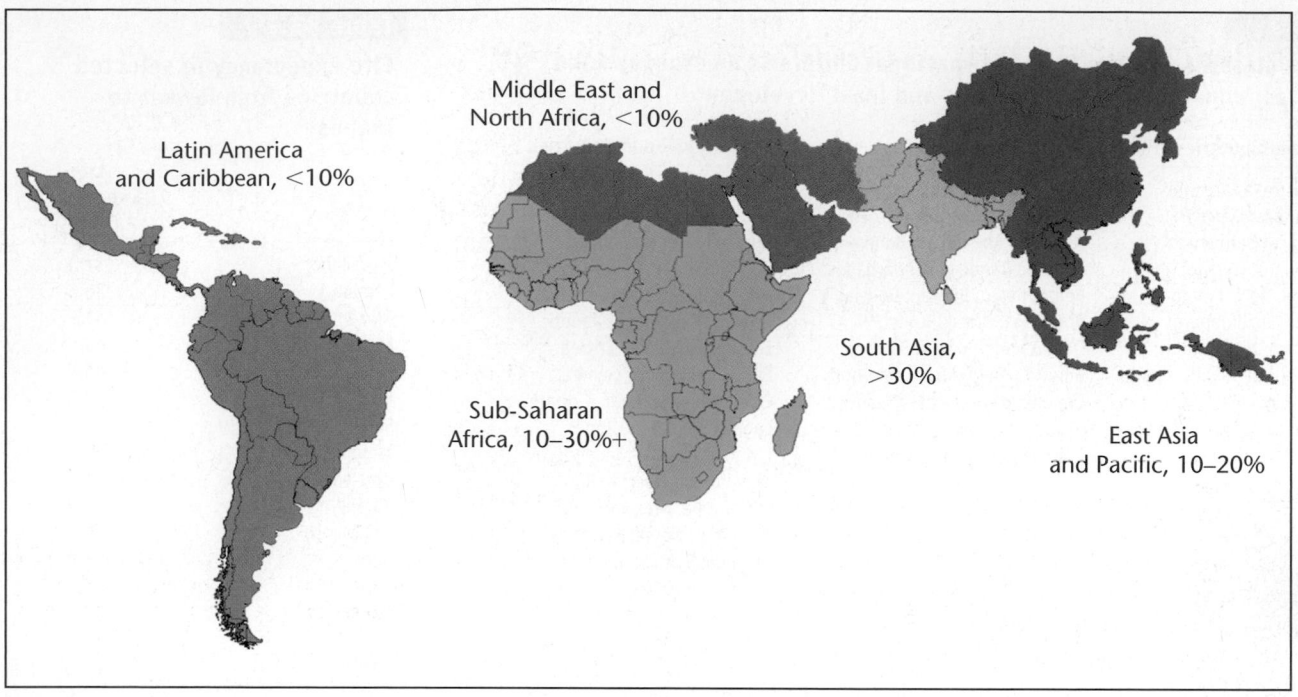

Illustration 33.1 Underweight children are an indicator of malnutrition. This map shows the percentage of children who were underweight in developing regions of the world.

Source: WHO global database on child growth and nutrition, 2004, www.who.org.

malnutrition that must be addressed as part of global strategies aimed at improving health and well-being (Table 33.4). Important progress is being made, and rates of malnutrition, poor growth, and infectious disease are decreasing slowly worldwide.[8] Much progress remains to be made, however.

Table 33.3

Percentage of low-birthweight infants, exclusive breast-feeding, and access to safe water in different areas of the world.[4]

Region[a]	Low-Birthweight Infants (%)	Exclusive Breast-Feeding 0–3 Months (%)	Access To Safe Water (%)
Sub-Saharan Africa	16	32	49
Middle East and North Africa	11	43	81
South Asia	33	46	80
East Asia and Pacific	11	56	67
Latin America and Caribbean	10	38	77
Developing countries	18	44	70
Least developed countries	23	46	54
Industrialized countries[b]	6	—	100
World	17	44	71

[a]Data for the former Soviet republics and the Baltic states are not available.
[b]Data for exclusive breast-feeding are not available; the figure is approximately 50%.

Table 33.4

Priority problem areas related to malnutrition in developing countries[7]

1. Childhood protein-calorie malnutrition
2. Vitamin deficiencies: vitamin A, folate
3. Mineral deficiencies: iodine, iron, zinc
4. Lack of breastfeeding
5. Alcohol abuse
6. Overweight and obesity
7. Insufficient physical activity
8. Low vegetable and fruit intake
9. Malnutrition and increased complications from HIV/AIDS
10. Poor nutritional status of women of child-bearing age

Food and Nutrition: The Global Challenge

We have all heard about starvation in Somalia, Ethiopia, Sudan, and Bangladesh. The pictures of starving children with desperation in their big eyes and heads too large for their frail bodies burn an image in our minds. What we don't hear about in the news, however, is the extent of starvation and malnutrition in the world. These are not just problems of isolated areas experiencing civil war or crop failures. In the developing regions, malnutrition is an ongoing problem for 10 to 51% of children under the age of 5, depending on the region. Approximately 50 to 70% of women worldwide are iron deficient, and 2 billion children suffer from iodine deficiency. Vitamin A deficiency exists among 21% of children in developing countries. These children are at risk of frequent and severe infectious disease, decreased growth, and blindness because of the deficiency. It is estimated that one-third of the world's population is mildly to moderately zinc deficient.[7]

The children we usually see in pictures from famine areas are victims of **marasmus,** a disease caused primarily by a lack of calories and protein. These children look starved (Illustration 33.2) and consume far fewer calories and far less protein and other essential nutrients than they need. In victims of marasmus, the body uses its own muscle and other tissues as an energy source.

Protein-energy malnutrition produces a disease called "**kwashiorkor**" in some young children. It is similar to marasmus in that it is due to a chronic lack of protein and calories. Kwashiorkor is unlike marasmus, however, in some ways. Young children with kwashiorkor appear to be genetically susceptible to developing faulty protein utilization mechanisms when chronically exposed to protein and calorie shortages.[9] These children may appear fat due to massive swelling related to abnormal protein utilization that occurs with the disease (Illustration 33.3). Most victims of kwashiorkor show some of the symptoms of marasmus as well. Children with either type of severe malnutrition are generally deficient in multiple vitamins and minerals and at high risk of dying from infectious diseases due to weakened immune systems.[10,11]

Survivors of Malnutrition

Malnutrition that occurs before a child's brain has completely grown has lasting effects on development.[6] Brain growth occurs primarily during pregnancy through the age of 2 years. If children suffer severe malnutrition during these years, they will experience permanent delays in mental development even if refeeding improves their physical growth and health. The severity of the mental delays depends on the timing and duration of the malnutrition. The longer malnutrition exists, the harder it becomes for young children to achieve a brighter future.[12] In addition, the state of hunger and starvation leads people to become self-centered and to lose any sense of well-being. The need for food for survival may prompt unethical behaviors such as stealing and injuring others to obtain food. The devastating psychological effects

He who is healthy has hope; and he who has hope has everything.

—ARAB PROVERB

marasmus
A severe form of malnutrition primarily due to a chronic lack of calories and protein. Also called protein-energy malnutrition.

kwashiorkor
A severe form of protein-energy malnutrition in young children. It is characterized by swelling, fatty liver, susceptibility to infection, profound apathy, and poor appetite. The cause of kwashiorkor is unclear.

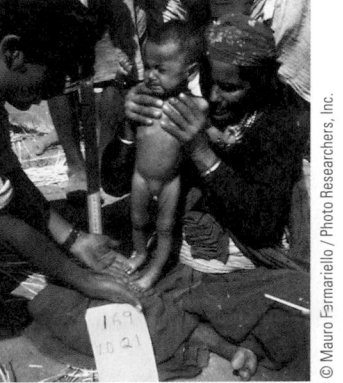

Illustration 33.2 The look of marasmus.
Children with marasmus look like "skin and bones." They don't get enough calories and protein in their diet. This child is suffering from the extreme emaciation of marasmus.

Illustration 33.3 The look of kwashiorkor.
This child has the characteristic "moon face" (edema), swollen belly, and patchy dermatitis often seen with kwashiorkor. Children with kwashiorkor may not look starved because of the massive swelling.

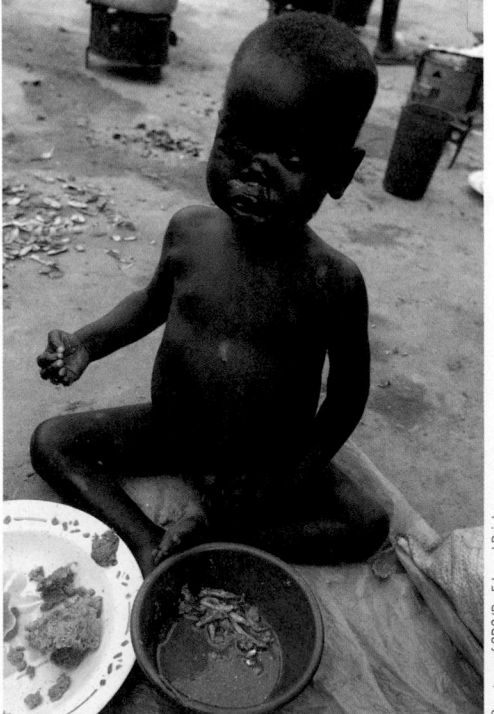

Courtesy of CDC/Dr. Edward Brink

© Mauro Farmariello / Photo Researchers, Inc.

of starvation may follow children into adulthood and have a lasting effect on behavior throughout life. Persistent malnutrition saps the physical and mental energy of people and, ultimately, compromises the economic and social progress of nations.[5]

Malnutrition and Infection   There is a very close relationship between malnutrition and infection: Malnutrition weakens the immune system and increases the likelihood and severity of infection. Simultaneously, repeated infection and the bouts of diarrhea that often accompany it can produce undernutrition. Poor sanitary conditions, contaminated water supplies, and lack of refrigeration contribute to the spread of infectious diseases and thus to the malnutrition the diseases promote. Most deaths of children under the age of 5 years in developing countries are related to the presence of both malnutrition and infection.[13]

Low rates of breast-feeding promote the spread of infection, especially in countries where contaminated water is used to prepare formula. In addition to providing optimal nutrition, breast milk contains a number of substances that protect babies from infection.[7]

Vitamin A deficiency and malnutrition in children play important roles in the spread and severity of infections. Children who are deficient in vitamin A are much more likely to die from a measles infection, for example, than are children whose vitamin A status is sufficient.[12] HIV infection progresses more rapidly into AIDS in malnourished individuals and shortens the time parents can work and support their families. Life expectancy in Sudan, for example, has decreased from 53.0 years in 1996 to 51.4 years in 2009.[4] Over 33 million people in the world have HIV infection.[14]

Why Do Starvation and Malnutrition Happen?

It is very much an oversimplification to assert, without qualification, that people starve because they do not have enough to eat.

—M. Pyke, 1972

Malnutrition and starvation don't have to happen anywhere. The world's agricultural systems have the capacity to produce enough food to feed everyone on Earth.[15] People become malnourished or starve primarily because of poverty. Human-made disasters, including discrimination against women, the HIV/AIDS epidemic, racism, corrupt governance, and other factors (Table 33.5) heavily contribute to malnutrition. In some instances, starvation and malnutrition are due to natural disasters that lead to crop failures or the inability to distribute food to those in need.[12]

The famine in Somalia that led to the death of over 1.5 million people was due to the collapse of the government, fighting among rival clans, and general lawlessness. Farmers and their families were driven off their land by bandits who stole their crops and possessions. Many had nowhere to go for help. Mass starvation in Ethiopia and Sudan was similarly initiated by violent conflict within the country. Seizing the opposing side's food and preventing them from obtaining food aid are among the primary weapons used to fight these civil wars. In Bangladesh, where poor people have no choice but to live on floodplains, cyclones wipe out crops and regularly result in the death of thousands due to floods, starvation, and infectious disease. None of these cases of mass famine was

Table 33.5

Root causes of malnutrition in developing countries[6]

- Poverty
- Discrimination against females
- HIV/AIDS epidemic
- Racism, ethnocentrism
- Poor and corrupt governance
- Unsafe water
- Low levels of education
- Unequitable distribution of the food supply
- Lack of economic opportunities
- Low agricultural productivity

due primarily to the ravages of nature. Starvation and malnutrition are largely initiated by humans, and it is within the power of humans to end them.

Women and female children are at particular risk for malnutrition in some societies because cultural practices call for food to be allocated to men and boys first (Illustration 33.4). If too little food is available, what remains after meals for women and daughters may be too little to support health and growth. In many developing countries, discrimination against women in education and employment, sanctions against the use of birth control, and violence toward women place women at high risk of developing malnutrition and having a low quality of life.[6]

Ending Malnutrition

The long-term solution to malnutrition will depend on the ability of humans to work together to achieve educational and economic development, peace, population growth control, improved sanitation, social equity for women and children, and environmentally sound and productive agricultural policies and practices in developing countries (Illustration 33.5). Bottom-up approaches to problem solving work better than top-down approaches. The people most affected by malnutrition must be active participants in the planning and implementation of improvement programs. Efforts to improve the nutritional status of populations can and have paid off.

Success Stories   Examples of strikingly successful efforts to reduce malnutrition in various countries around the world are available for telling. A repeated theme is the reduction in malnutrition as economic, educational, nutritional, and sanitary conditions improve. Improvements in these same areas were responsible for the dramatic declines in malnutrition and infectious diseases and the increases in life expectancy experienced in the United States, Canada, and many

Illustration 33.4   Women and children in some developing countries are particularly vulnerable to malnutrition because they, and their nutritional needs, may be considered less important than men and their needs. Similarly, boys' needs may take precedence over girls' needs. These children are 2-year-old twins: The child on the left is a girl, and the child on the right is a boy.

Adequate food is the cradle of normal resistance, the playground of normal immunity, the workshop of good health, and the laboratory of long life.

—CHARLES MAYO

Illustration 33.5   Safe water supplies represent a major advance in quality of life for people in developing countries.

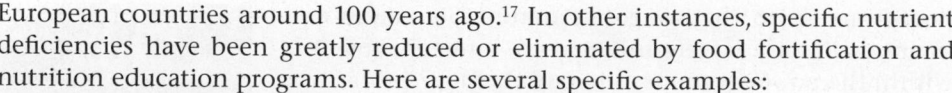

Approximately 12% of the U.S. population lives in poverty, but there are sizable differences in poverty rates by racial group. Twenty-seven percent of Hispanics, 37% of African Americans, and 10% of whites live in poverty.[16]

PhotoDisc

European countries around 100 years ago.[17] In other instances, specific nutrient deficiencies have been greatly reduced or eliminated by food fortification and nutrition education programs. Here are several specific examples:

- Vitamin A supplements and education on vitamin A-rich foods were related to a dramatic decrease in cases of severe and moderate vitamin A deficiency and infection in children in Indonesia.

- Food supplements given to groups of infants in Russia, Brazil, South Africa, and China were associated with higher IQ scores at age 8.

- Iodization of salt has eliminated iodine deficiency in Bolivia and Ecuador. Worldwide, iodization of salt is associated with a drop in the number of iodine-deficient children from 48 million in 1990 to 28 million in 1997.

- Fortification of flour with iron has led to a decrease in iron-deficiency anemia in the Philippines.[18]

In Thailand, a country with a serious iodine-deficiency problem, a highly creative approach to increasing iodine intake was implemented by the king. To celebrate his birthday, salt producers gave the king 10,000 tons of iodized salt. The king had the salt packaged in small plastic bags and, with the help of the Red Cross and the army, delivered a bag to every household in Thailand. A message from the king about the importance of using iodized salt was attached to each bag.[19]

Deaths of young children from malnutrition and its related diseases have dropped substantially in countries experiencing a resurgence of breast-feeding (Illustration 33.6). Breast milk protects infants and young children from a variety of infectious diseases, supports the growth and health of infants, and protects them from the hazards of formulas reconstituted with contaminated water. Worldwide health initiatives have led to the development of the International Code of Marketing Breast Milk Substitutes and to "Baby Friendly Hospitals." The international marketing code calls for the prohibition of free formula samples as well as the promotion of infant formulas by health care professionals. More than 12,700 hospitals have adopted Baby Friendly Hospital policies that effectively promote and facilitate breast-feeding.[12,20]

Health status of people in developing countries has also been improved by the widespread use of oral rehydration fluids that protect children with diarrhea from dehydration. Broadly based vaccination programs have protected many from various diseases.[21]

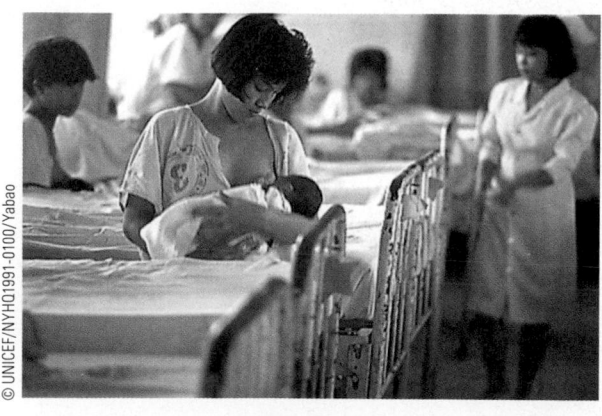

© UNICEF/NYHQ1991-0100/Yabao

Illustration 33.6 After a rooming-in policy that enabled mothers to breastfeed their infants on demand was introduced in one hospital in the Philippines, the incidence of early infant deaths due to infection dropped by 95%.

Source: First Call for Children, A UNICEF Quarterly 1992 (January/ March):1.

The World Food Summit The 1996 World Food Summit ended with a pledge that by 2015, countries would reduce by half the number of people in the world who are hungry and undernourished.[12] Progress is being made toward meeting the goal. Access to food and rates of undernutrition are improving, but at a slower pace than envisioned. Attention is now being placed on expanding agriculture and food production, and decreasing food costs in developing countries.[22] It is anticipated that it will take until the year 2030 to cut rates of hunger and undernutrition by half.[13]

The "Nutrition Transition": Rising Rates of Obesity and Chronic Diseases As countries develop economically and food supplies increase, the incidence of undernutrition usually decreases, and chronic health problems such as heart disease, diabetes, and hypertension tend to increase. Elevated rates of chronic disease appear to be closely related to obesity, high saturated fat and sugar intakes, low vegetable and fruit intakes, low levels of physical activity, and other characteristics of Western lifestyles and dietary habits.[23]

Countries undergoing the nutrition transition often experience rising rates of obesity on top of existing high rates of underweight and malnutrition.[2,24] Children born to poorly nourished women may be biologically programmed while in the womb to conserve energy. When exposed to a generous food supply later in life, they may gain fat more readily than height.[25] Excess accumulation of fat later in life may place the children at increased risk for hypertension, diabetes, heart disease, and a number of other "diseases of civilization." Although infections such as malaria and tuberculosis will likely predominate through the year 2025, the incidence of diseases of "Western civilization" will continue to expand.[2]

The Future

Benefits derived from improving the nutritional status of the world's population are immense. Adequate nutrition is the bedrock on which present generations secure a future for themselves and the next generation. People who are well nourished are more productive, happier, require less medical care, and are more likely to be self-sufficient than malnourished people. When the malnutrition problem is solved, the world will likely wonder why it didn't happen sooner.[6]

If you would like to be part of the solution to food insecurity, hunger, or malnutrition, consider volunteering at a food or meals program, in community courses for language or self-sufficiency skills, or in other public service programs. The experience will likely enrich your life as well as the lives of those you serve.

Knowing is not enough; we must apply. Willingness is not enough; we must do.

—GOETHE

© EyesWideOpen/Getty Images

Up Close

Ethnic Foods Treasure Hunt

FOCAL POINT: Values placed on foods are culturally specific.

The following questions deal with ethnic food preferences. Answer any three of the questions. Resources that will help you find the answers include knowledgeable individuals, the Internet, travel and anthropology books, and encyclopedias.

1. Name three foods commonly served at celebrations by the Igbo in Nigeria (Igbos were formerly referred to as Biafrans):
   a. _____
   b. _____
   c. _____

2. List two special foods given to Hmong women during pregnancy:
   a. _____
   b. _____

3. List the five foods that make up the traditional Costa Rican breakfast:
   a. _____
   b. _____
   c. _____
   d. _____
   e. _____

4. List three foods commonly sold at soccer matches in Italy:
   a. _____
   b. _____
   c. _____

5. Name two traditional foods of the Zulu in South Africa:
   a. _____
   b. _____

6. Name one of the staple foods of people in Nentsi (it's near Siberia):

7. List two foods considered to be "yin" and two considered to be "yang" in Chinese culture:

Yin	Yang
a. _____	a. _____
b. _____	b. _____

FEEDBACK (answers to these questions) can be found at the end of Unit 33.

[Key Terms

kwashiorkor, page 33-5   marasmus, page 33-5

Review Questions

1. Malnutrition and infection are often common health problems in countries with low life expectancy. ☐ ☐

2. Rates of malnutrition are increasing worldwide. ☐ ☐

3. Deficiencies of iodine, iron, and vitamin A remain important public health problems in many developing countries. ☐ ☐

4. The need for food for survival may prompt behaviors such as stealing and injuring others to obtain food. ☐ ☐

5. In many developing countries, discrimination against women places them at high risk of developing malnutrition. ☐ ☐

6. Death rates for young children are decreasing in countries with increasing rates of breast-feeding. ☐ ☐

7. Due to the expanding adoption of Western lifestyles and diet, many developing countries are experiencing increasing rates of overweight and chronic diseases. ☐ ☐

Media Menu

www.who.int
This URL leads to causes of illness and death in industrialized and developing countries, information about disease outbreaks, WHO programs, and travelers' health.

www.fao.org
The United Nations' Food and Agriculture Organization address provides webcasts and photo updates on worldwide malnutrition and meeting the World Food Summit's goal of reducing hunger by half by 2015.

www.ers.usda.gov/AmberWaves/September08/Features/Obesitycountries.htm
Amber Waves, a USDA informational website, offers a podcast (9 minutes, 24 seconds in length) titled "Obesity in the Midst of Unyielding Food Insecurity in Developing Countries." Tune in at this site or download the recording to learn more about this topic.

www.fao.org/ag/agn/nutrition/profiles_en.stm
The Food and Agricultural Organization provides the "Nutrition Country Profiles" here. The Nutrition Country Profiles are concise reviews describing the food and nutrition situation in individual countries, with background statistics on food-related factors.

www.travel.state.gov
Traveling abroad? This site offers travel alerts on health conditions, political instability, crime statistics, and public announcements.

www.census.gov/ipc/www/idb
The U.S. Census Bureau International Data Base (IDB) is a computerized data bank containing statistical tables of demographic data for 228 countries and areas of the world. The site includes mortality causes, life expectancy, family planning, demographic data, and more.

www.cia.gov/library/publications/the-world-factbook/index.html
The CIA World Factbook home page provides an unbelievable amount of information on the world and each country in it. Get data on infant mortality rates, life expectancy, industries, climate, time zones, economies, agriculture, and much from this huge resource.

www.who.int/nutrition
The Word Health Organization's main page for international nutrition is at this address.

Notes

1. Nutrition. www.who.int/nutrition, accessed 10/06.

2. Byers T et al. Public health response to the obesity epidemic: too soon or too late? J Nutr 2007;137:488–92.

3. Meadows DH. The population map: State of the village report, www. odt. org/pop.htm, accessed 6/09.

4. U.S. Census Bureau, International Data Base, www.census.gov/cgi-bin/ipc/agggen, accessed 6/09.

5. The state of food insecurity in the world in 2000. FAO, United Nations Report, Nov. 2000, www.fao.org, accessed 2/01.

6. Position of American Dietetic Association: addressing world hunger, malnutrition, and food insecurity. J Am Diet Assoc 2003;103:1046–56.

7. Quantifying selected major risks to health, World Health Organization, www.who.int/whr/2002/chapter4/en/print.html, accessed 11/02.

8. Christan P. Micronutrients and reproductive health issues: an

international perspective. J Nutr 2003;133:1969S–73S.

9. Amadi B et al. Reduced production of sulfated glycosaminoglycans occurs in Zambian children with kwashiorkor but not for marasmus, Am J Clin Nutr 2009;89:592–600.

10. Neumann CG. Background, Symposium: Food-Based Approaches to Combating Micronutrient Deficiencies in Children of Developing Countries, J Nutr 2007;137:1091–2.

11. Scrimshaw NS. Historical concepts of interactions, synergism and antagonism between nutrition and infection. J Nutr 2003;133:316S–21S.

12. United Nations International Children's Emergency Fund. The state of the world's children, 1998. New York: United Nations; 1998.

13. Underwood BA. Scientific research: essential, but is it enough to combat world food insecurity? J Nutr 2003;133:1434S–7S.

14. AIDS epidemic update : December 2007, data.unaids.org/pub/EPISlides/2007/2007_epiupdate_en.pdf, accessed 6/09.

15. Haddad L et al. Feeding the world in the coming decades requires improvements in investment, technology and institutions. J Nutr 2002;132:3435S–36S.

16. Health, United States, 2008, www.cdc.gov/nchs, accessed 5/09.

17. Gold MR. Presentation at the Tenth Chronic Disease Conference, Washington, DC; 1996.

18. Decimosexta Conferencia Internacional de Nutricion, Montreal. Nutr View 1997; fall:1–36.

19. Food fortification to end micronutrient malnutrition: state of the art. Nutr View 1997; special issue:1–8.

20. Perez-Escamilla R. Evidence based breast-feeding promotion: the Baby-Friendly Hospital Initiative, J Nutr 2007;137:484–7.

21. Pelletier DL, Frongillo EA. Changes in child survival are strongly associated with changes in malnutrition in developing countries. J Nutr 2003; 133:107–19.

22. End Poverty 2015 Campaign, www.endpoverty2015.org/faofoodsummit, accessed 6/09.

23. Obesity in the midst of unyielding food insecurity in developing countries, www.ers.usda.gov/AmberWaves/September08/Features/Obesitycountries.htm, accessed 6/09.

24. Popkin BM. Global nutrition dynamics: the world is shifting rapidly toward a diet linked with noncommunicable diseases. Am J Clin Nutr 2006;84:289–98.

25. De Onis M et al. Prevalence and trends of overweight among preschool children in developing countries. Am J Clin Nutr 2000;72:1032–9.

Answers to Review Questions

1. True, see page 33-2.
2. False, see page 33-4.
3. True, see pages 33-5, 33-6.
4. True, see page 33-5.
5. True, see page 33-7.
6. True, see page 33-8.
7. True, see pages 33-8, 33-9.

NUTRITION | # Up Close

Ethnic Foods Treasure Hunt

Feedback for Unit 33

1. a. yams; b. kola nuts; c. breadfruit with corn and yams; d. cassava.

2. a. chicken; b. rice; c. ripe mango; d. grapes; e. ginger f. milk; g. many more.

3. "Gallo pinto" includes the following:
a. rice and black beans; b. eggs; c. tortillas; d. coffee; e. fruit; f. sometimes steak.

4. a. Italian ices; b. calzones; c. Italian sodas; d. pizza; e. pasta; f. hot dogs.

5. a. beef; b. cassava; c. milk; d. yams; e. pumpkin; f. millet; g. corn; h. dried beans; i. beer; j. honey; k. yogurt; l. cow's blood.

6. a. reindeer; b. polar bear; c. sour milk; d. caribou; e. tea.

7. Yin: a. raw kelp (seaweed); b. boiled rice; c. raw vegetables; d. fruit; e. dairy products.
Yang: a. cooked kelp; b. cooked vegetables; c. fried rice; d. fish; e. eggs; f. chicken.

Appendix

Appendix A

Table of Food Composition

Appendix B

Reliable Sources of Nutrition Information

Appendix C

The U.S. Food Exchange System

Appendix D

Table of Intentional Food Additives

Appendix E

Cells

Appendix F

Canadian Choice System

Appendix A

Table of Food Composition

This edition of the table of food composition includes a wide variety of foods from all food groups. It is updated with each edition to reflect nutrient changes for current foods, remove outdated foods, and add foods that are new to the marketplace.*

The nutrient database for this appendix is compiled from a variety of sources, including the USDA Standard Reference database (Release 16), literature sources, and manufacturers' data. The USDA database provides data for a wider variety of foods and nutrients than other sources. Because laboratory analysis for each nutrient can be quite costly, manufacturers tend to provide data only for those nutrients mandated on food labels. Consequently, data for their foods are often incomplete; any missing information is designated in this table as a blank space. Keep in mind that a blank space means only that the information is unknown and should not be interpreted as a zero.

Whenever using nutrient data, remember that many factors influence the nutrient contents of foods, including the mineral content of the soil, the diet of the animal or the fertilizer of the plant, the season of harvest, the method of processing, the length and method of storage, the method of cooking, the method of analysis, and the moisture content of the sample analyzed. With so many factors involved, users must view nutrient data as a close approximation of the actual amount.

For updates, corrections, and a list of 6000 foods and codes found in the diet analysis software that accompanies this text, visit **www.wadsworth.com/nutrition** and click on *Diet Analysis* in the right-hand column.

- **Fats** Total fats, as well as the breakdown of total fats to saturated, monounsaturated, and polyunsaturated fats, are listed in the table. The fatty acids seldom add up to the total due to rounding and to other fatty acid components that are not included in these basic categories, such as *trans*-fatty acids and glycerol. *Trans*-fatty acids can comprise a large share of the total fat in margarine and shortening (hydrogenated oils) and in any foods that include them as ingredients.

- **Vitamin A and Vitamin E** In keeping with the 2001 RDA for vitamin A, which established a new measure of vitamin A activity—retinol activity equivalents (RAE)—this appendix presents data for vitamin A in micrograms (μg) RAE. Similarly, because the 2000 RDA for vitamin E is based only on the alpha-tocopherol form of vitamin E, this appendix reports vitamin E data in milligrams (mg) alpha-tocopherol (listed in the table as mg α).

- **Bioavailability** Keep in mind that the availability of nutrients from foods depends not only on the quantity provided by a food, but also on the amount absorbed and used by the body—the bioavailability. The bioavailability of folate from fortified foods, for example, is greater than from naturally occurring sources. Similarly, the body can make niacin from the amino acid tryptophan, but niacin values in this table (and most databases) report preformed niacin only. Chapter 10 provides conversion factors and additional details.

- **Using the Table** The items in this table have been organized into several categories, which are listed at the head of each right-hand page. Page numbers have been provided, and each group has been color-coded to make it easier to find individual items.

---

*This food composition table has been prepared for Wadsworth Publishing Company and is copyrighted by ESHA Research in Salem, Oregon—the developer and publisher of the Food Processor and Genesis nutritional software programs. The nutritional data are supported by over 1300 references. Because the list of sources is so extensive, it is not provided here, but is available from the publisher.

In an effort to conserve space, the following abbreviations have been used in the food descriptions and nutrient breakdowns:

- diam = diameter
- ea = each
- enr = enriched
- f/ = from
- frzn = frozen
- g = grams
- liq = liquid
- pce = piece
- pkg = package
- w/ = with
- w/o = without
- t = trace
- 0 = zero (no nutrient value)
- blank space = information not available

Index to Appendix A

Food Composition   (Computer code is for Cengage Diet Analysis program)   (For purposes of calculations, use "0" for t, <1, <.1, <.01, etc.)

DA+ Code	Food Description	Quantity	Measure	Wt (g)	H₂O (g)	Ener (kcal)	Prot (g)	Carb (g)	Fiber (g)	Fat (g)	Sat	Mono	Poly	Trans
											Fat Breakdown (g)			

BREADS, BAKED GOODS, CAKES, COOKIES, CRACKERS, CHIPS, PIES

DA+ Code	Food Description	Quantity	Measure	Wt (g)	H₂O (g)	Ener (kcal)	Prot (g)	Carb (g)	Fiber (g)	Fat (g)	Sat	Mono	Poly	Trans
	Bagels													
8534	Cinnamon and raisin	1	item(s)	71	22.7	194	7.0	39.2	1.6	1.2	0.2	0.1	0.5	—
14395	Multi-grain	1	item(s)	61	—	170	6.0	35.0	1.0	1.5	0.5	0.1	0.4	—
8538	Oat bran	1	item(s)	71	23.4	181	7.6	37.8	2.6	0.9	0.1	0.2	0.3	—
4910	Plain, enriched	1	item(s)	71	25.8	182	7.1	35.9	1.6	1.2	0.3	0.4	0.5	0
4911	Plain, enriched, toasted	1	item(s)	66	18.7	190	7.4	37.7	1.7	1.1	0.2	0.3	0.6	0
	Biscuits													
25008	Biscuits	1	item(s)	41	15.8	121	2.6	16.4	0.5	4.9	1.4	1.4	1.8	—
16729	Scone	1	item(s)	42	11.5	148	3.8	19.1	0.6	6.2	2.0	2.5	1.3	—
25166	Wheat biscuits	1	item(s)	55	21.0	162	3.6	21.9	1.4	6.7	1.9	1.9	2.5	—
	Bread													
325	Boston brown, canned	1	slice(s)	45	21.2	88	2.3	19.5	2.1	0.7	0.1	0.1	0.3	—
8716	Bread sticks, plain	4	item(s)	24	1.5	99	2.9	16.4	0.7	2.3	0.3	0.9	0.9	—
25176	Cornbread	1	piece(s)	55	25.9	141	4.7	18.3	0.9	5.4	2.1	1.4	1.5	0
327	Cracked wheat	1	slice(s)	25	9.0	65	2.2	12.4	1.4	1.0	0.2	0.5	0.2	—
9079	Croutons, plain	¼	cup(s)	8	0.4	31	0.9	5.5	0.4	0.5	0.1	0.2	0.1	—
8582	Egg	1	slice(s)	40	13.9	113	3.8	19.1	0.9	2.4	0.6	0.9	0.4	—
8585	Egg, toasted	1	slice(s)	37	10.5	117	3.9	19.5	0.9	2.4	0.6	1.1	0.4	—
329	French	1	slice(s)	32	8.9	92	3.8	18.1	0.8	0.6	0.2	0.1	0.3	—
8591	French, toasted	1	slice(s)	23	4.7	73	3.0	14.2	0.7	0.5	0.1	0.1	0.2	—
42096	Indian fry, made with lard (Navajo)	3	ounce(s)	85	26.9	281	5.7	41.0	—	10.4	3.9	3.8	0.9	—
332	Italian	1	slice(s)	30	10.7	81	2.6	15.0	0.8	1.1	0.3	0.2	0.4	—
1393	Mixed grain	1	slice(s)	26	9.6	69	3.5	11.3	1.9	1.1	0.2	0.2	0.5	0
8604	Mixed grain, toasted	1	slice(s)	24	7.6	69	3.5	11.3	1.9	1.1	0.2	0.2	0.5	0
8605	Oat bran	1	slice(s)	30	13.2	71	3.1	11.9	1.4	1.3	0.2	0.5	0.5	—
8608	Oat bran, toasted	1	slice(s)	27	10.4	70	3.1	11.8	1.3	1.3	0.2	0.5	0.5	—
8609	Oatmeal	1	slice(s)	27	9.9	73	2.3	13.1	1.1	1.2	0.2	0.4	0.5	—
8613	Oatmeal, toasted	1	slice(s)	25	7.8	73	2.3	13.2	1.1	1.2	0.2	0.4	0.5	—
1409	Pita	1	item(s)	60	19.3	165	5.5	33.4	1.3	0.7	0.1	0.1	0.3	—
7905	Pita, whole wheat	1	item(s)	64	19.6	170	6.3	35.2	4.7	1.7	0.3	0.2	0.7	—
338	Pumpernickel	1	slice(s)	32	12.1	80	2.8	15.2	2.1	1.0	0.1	0.3	0.4	—
334	Raisin, enriched	1	slice(s)	26	8.7	71	2.1	13.6	1.1	1.1	0.3	0.6	0.2	—
8625	Raisin, toasted	1	slice(s)	24	6.7	71	2.1	13.7	1.1	1.2	0.3	0.6	0.2	—
10168	Rice, white, gluten free, wheat free	1	slice(s)	38	—	130	1.0	18.0	0.5	6.0	0	—	—	0
8653	Rye	1	slice(s)	32	11.9	83	2.7	15.5	1.9	1.1	0.2	0.4	0.3	—
8654	Rye, toasted	1	slice(s)	29	9.0	82	2.7	15.4	1.9	1.0	0.2	0.4	0.3	—
336	Rye, light	1	slice(s)	25	9.3	65	2.0	12.0	1.6	1.0	0.2	0.3	0.3	—
8588	Sourdough	1	slice(s)	25	7.0	72	2.9	14.1	0.6	0.5	0.1	0.1	0.2	—
8592	Sourdough, toasted	1	slice(s)	23	4.7	73	3.0	14.2	0.7	0.5	0.1	0.1	0.2	—
491	Submarine or hoagie roll	1	item(s)	135	40.6	400	11.0	72.0	3.8	8.0	1.8	3.0	2.2	—
8596	Vienna, toasted	1	slice(s)	23	4.7	73	3.0	14.2	0.7	0.5	0.1	0.1	0.2	—
8670	Wheat	1	slice(s)	25	8.9	67	2.7	11.9	0.9	0.9	0.2	0.2	0.4	—
8671	Wheat, toasted	1	slice(s)	23	5.6	72	3.0	12.8	1.1	1.0	0.2	0.2	0.4	—
340	White	1	slice(s)	25	9.1	67	1.9	12.7	0.6	0.8	0.2	0.2	0.3	—
1395	Whole wheat	1	slice(s)	46	15.0	128	3.9	23.6	2.8	2.5	0.4	0.5	1.4	—
	Cakes													
386	Angel food, prepared from mix	1	piece(s)	50	16.5	129	3.1	29.4	0.1	0.2	0	0	0.1	—
8772	Butter pound, ready to eat, commercially prepared	1	slice(s)	75	18.5	291	4.1	36.6	0.4	14.9	8.7	4.4	0.8	—
28517	Carrot	1	slice(s)	131	56.6	339	4.8	56.5	1.9	11.1	1.0	5.7	3.8	—
4931	Chocolate with chocolate icing, commercially prepared	1	slice(s)	64	14.7	235	2.6	34.9	1.8	10.5	3.1	5.6	1.2	—
8756	Chocolate, prepared from mix	1	slice(s)	95	23.2	352	5.0	50.7	1.5	14.3	5.2	5.7	2.6	—
393	Devil's food cupcake with chocolate frosting	1	item(s)	35	8.4	120	2.0	20.0	0.7	4.0	1.8	1.6	0.6	—
8757	Fruitcake, ready to eat, commercially prepared	1	piece(s)	43	10.9	139	1.2	26.5	1.6	3.9	0.5	1.8	1.4	—
1397	Pineapple upside down, prepared from mix	1	slice(s)	115	37.1	367	4.0	58.1	0.9	13.9	3.4	6.0	3.8	—
411	Sponge, prepared from mix	1	slice(s)	63	18.5	187	4.6	36.4	0.3	2.7	0.8	1.0	0.4	—
8817	White with coconut frosting, prepared from mix	1	slice(s)	112	23.2	399	4.9	70.8	1.1	11.5	4.4	4.1	2.4	—
8819	Yellow with chocolate frosting, ready to eat, commercially prepared	1	slice(s)	64	14.0	243	2.4	35.5	1.2	11.1	3.0	6.1	1.4	—
8822	Yellow with vanilla frosting, ready to eat, commercially prepared	1	slice(s)	64	14.1	239	2.2	37.6	0.2	9.3	1.5	3.9	3.3	—
	Snack cakes													
8791	Chocolate snack cake, creme filled, with frosting	1	item(s)	50	9.3	200	1.8	30.2	1.6	8.0	2.4	4.3	0.9	—
25010	Cinnamon coffee cake	1	piece(s)	72	22.6	231	3.6	35.8	0.7	8.3	2.2	2.6	3.0	—
16777	Funnel cake	1	item(s)	90	37.6	276	7.3	29.1	0.9	14.4	2.7	4.7	6.1	—
8794	Sponge snack cake, creme filled	1	item(s)	43	8.6	155	1.3	27.2	0.2	4.8	1.1	1.7	1.4	—
	Snacks, chips, pretzels													
29428	Bagel chips, plain	3	item(s)	29	—	130	3.0	19.0	1.0	4.5	0.5	—	—	—
29429	Bagel chips, toasted onion	3	item(s)	29	—	130	4.0	20.0	1.0	4.5	0.5	—	—	—

BREADS, BAKED GOODS, CAKES, COOKIES, CRACKERS, CHIPS, PIES —Continued

Chol (mg)	Calc (mg)	Iron (mg)	Magn (mg)	Pota (mg)	Sodi (mg)	Zinc (mg)	Vit A (µg)	Thia (mg)	Vit E (mg α)	Ribo (mg)	Niac (mg)	Vit B$_6$ (mg)	Fola (µg)	Vit C (mg)	Vit B$_{12}$ (µg)	Sele (µg)
0	13	2.69	19.9	105.1	228.6	0.80	14.9	0.27	0.22	0.19	2.18	0.04	78.8	0.5	0	22.0
0	60	1.08	—	—	310.0	—	0	—	—	—	—	—	—	0	—	—
0	9	2.18	22.0	81.7	360.0	0.63	0.7	0.23	0.23	0.24	2.10	0.03	69.6	0.1	0	24.3
0	63	4.29	15.6	53.3	318.1	1.34	0	0.42	0.07	0.18	2.82	0.04	103.0	0.7	0	16.2
0	65	2.97	15.8	56.1	316.8	0.85	0	0.39	0.07	0.17	2.88	0.04	86.5	0	0	16.6
0	38	0.94	6.0	47.4	206.0	0.20	—	0.16	0.01	0.12	1.20	0.01	31.7	0.1	0.1	7.1
49	79	1.35	7.1	48.7	277.2	0.29	64.7	0.14	0.42	0.15	1.19	0.02	32.3	0	0.1	10.9
0	57	1.21	16.1	81.0	321.1	0.42	—	0.19	0.01	0.14	1.65	0.03	35.3	0.1	0.1	0
0	32	0.94	28.4	143.1	284.0	0.22	11.3	0.01	0.14	0.05	0.50	0.03	5.0	0	0	9.9
0	5	1.02	7.7	29.8	157.7	0.21	0	0.14	0.24	0.13	1.26	0.01	38.9	0	0	9.0
21	94	0.91	10.5	71.5	209.8	0.48	—	0.14	0.32	0.15	1.03	0.04	34.6	1.7	0.2	6.2
0	11	0.70	13.0	44.3	134.5	0.31	0	0.09	—	0.06	0.92	0.08	15.3	0	0	6.3
0	6	0.30	2.3	9.3	52.4	0.06	0	0.04	—	0.02	0.40	0.00	9.9	0	0	2.8
20	37	1.21	7.6	46.0	196.8	0.31	25.2	0.17	0.10	0.17	1.93	0.02	42.0	0	0	12.0
21	38	1.23	7.8	46.6	199.8	0.31	25.5	0.14	0.10	0.16	1.77	0.02	36.3	0	0	12.2
0	14	1.16	9.0	41.0	208.0	0.29	0	0.13	0.05	0.09	1.52	0.03	47.4	0.1	0	8.7
0	11	0.89	7.1	32.2	165.6	0.24	0	0.10	0.04	0.09	1.24	0.02	32.2	0	0	6.8
6	48	3.43	15.3	65.5	279.8	0.29	0	0.36	0.00	0.18	3.91	0.03	103.8	—	0	15.8
0	23	0.88	8.1	33.0	175.2	0.25	0	0.14	0.08	0.08	1.31	0.01	57.3	0	0	8.2
0	27	0.65	20.3	59.8	109.2	0.44	0	0.07	0.09	0.03	1.05	0.06	19.5	0	0	8.6
0	27	0.65	20.4	60.0	109.7	0.44	0	0.06	0.10	0.03	1.05	0.07	16.8	0	0	8.6
0	20	0.93	10.5	44.1	122.1	0.26	0.6	0.15	0.13	0.10	1.44	0.02	24.3	0	0	9.0
0	19	0.92	9.2	33.2	121.0	0.28	0.5	0.12	0.13	0.09	1.29	0.01	18.6	0	0	8.9
0	18	0.72	10.0	38.3	161.7	0.27	1.4	0.10	0.13	0.06	0.84	0.01	16.7	0	0	6.6
0	18	0.74	10.3	38.5	162.8	0.28	1.3	0.09	0.13	0.06	0.77	0.02	13.3	0.1	0	6.7
0	52	1.57	15.6	72.0	321.6	0.50	0	0.35	0.18	0.19	2.77	0.02	64.2	0	0	16.3
0	10	1.95	44.2	108.8	340.5	0.97	0	0.21	0.39	0.05	1.81	0.17	22.4	0	0	28.2
0	22	0.91	17.3	66.6	214.7	0.47	0	0.10	0.13	0.09	0.98	0.04	29.8	0	0	7.8
0	17	0.75	6.8	59.0	101.4	0.18	0	0.08	0.07	0.10	0.90	0.01	27.6	0	0	5.2
0	17	0.76	6.7	59.0	101.8	0.19	0	0.07	0.07	0.09	0.81	0.02	23.5	0.1	0	5.2
0	100	1.08	—	—	140	—	—	0.15	—	0.10	1.20	—	32.0	0	—	—
0	23	0.90	12.8	53.1	211.2	0.36	0	0.13	0.10	0.10	1.21	0.02	35.2	0.1	0	9.9
0	23	0.89	12.5	53.1	210.3	0.36	0	0.11	0.10	0.09	1.09	0.02	29.9	0.1	0	9.9
0	20	0.70	3.9	51.0	175.0	0.18	0	0.10	—	0.08	0.80	0.01	5.3	0	0	8.0
0	11	0.91	7.0	32.0	162.5	0.23	0	0.11	0.05	0.07	1.19	0.03	37.0	0.1	0	6.8
0	11	0.89	7.1	32.2	165.6	0.24	0	0.10	0.04	0.09	1.24	0.02	32.2	0	0	6.8
0	100	3.80	—	128.0	683.0	—	0	0.54	—	0.33	4.50	0.04	—	0	—	42.0
0	11	0.89	7.1	32.2	165.6	0.24	0	0.10	0.04	0.09	1.24	0.02	32.2	0	0	6.8
0	36	0.87	12.0	46.0	130.3	0.30	0	0.09	0.05	0.08	1.30	0.03	21.3	0.1	0	7.2
0	38	0.94	13.6	51.3	140.5	0.34	0	0.10	0.06	0.09	1.44	0.04	19.8	0	0	7.7
0	38	0.94	5.8	25.0	170.3	0.19	0	0.11	0.06	0.08	1.10	0.02	27.8	0	0	4.3
0	15	1.42	37.3	144.4	159.2	0.69	0	0.13	0.35	0.10	1.83	0.09	29.9	0	0	17.8
0	42	0.11	4.0	67.5	254.5	0.06	0	0.04	0.01	0.10	0.08	0.00	9.5	0	0	7.7
166	26	1.03	8.3	89.3	298.5	0.34	111.8	0.10	—	0.17	0.98	0.03	30.8	0	0.2	6.6
0	65	2.18	23.0	279.6	367.7	0.44	—	0.25	0.01	0.19	1.73	0.10	43.4	4.6	0	14.7
27	28	1.40	21.8	128.0	213.8	0.44	16.6	0.01	0.62	0.08	0.36	0.02	10.9	0.1	0.1	2.1
55	57	1.53	30.4	133.0	299.3	0.65	38.0	0.13	—	0.20	1.08	0.03	25.7	0.2	0.2	11.3
19	21	0.70	—	46.0	92.0	—	—	0.04	—	0.05	0.30	—	2.1	0	—	2.0
2	14	0.89	6.9	65.8	116.1	0.11	3.0	0.02	0.38	0.04	0.34	0.02	8.6	0.2	0	0.9
25	138	1.70	15.0	128.8	366.9	0.35	71.3	0.17	—	0.17	1.36	0.03	29.9	1.4	0.1	10.8
107	26	0.99	5.7	88.8	143.6	0.37	48.5	0.10	—	0.19	0.75	0.03	24.6	0	0.2	11.7
1	101	1.29	13.4	110.9	318.1	0.37	13.4	0.14	0.13	0.21	1.19	0.03	34.7	0.1	0.1	12.0
35	24	1.33	19.2	113.9	215.7	0.39	21.1	0.07	—	0.10	0.79	0.02	14.1	0	0.1	2.2
35	40	0.68	3.8	33.9	220.2	0.16	12.2	0.06	—	0.04	0.32	0.01	17.3	0	0.1	3.5
0	58	1.80	18.0	88.0	194.5	0.52	0.5	0.01	0.54	0.03	0.46	0.07	13.0	1.0	0	1.7
26	55	1.36	9.9	91.9	277.6	0.30	—	0.17	0.23	0.16	1.29	0.02	36.1	0.3	0.1	9.6
62	126	1.90	16.2	152.1	269.1	0.65	49.5	0.23	1.54	0.32	1.86	0.04	50.4	0	0.3	17.7
7	19	0.54	3.4	37.0	155.1	0.12	2.1	0.06	0.50	0.05	0.52	0.01	17.0	0	0	1.3
0	0	0.72	—	45.0	70.0	—	0	—	—	—	—	—	—	0	0	—
0	0	0.72	—	50.0	300.0	—	0	—	—	—	—	—	—	0	0	—

Food Composition (Computer code is for Cengage Diet Analysis program) (For purposes of calculations, use "0" for t, <1, <.1, <.01, etc.)

DA+ Code	Food Description	Quantity	Measure	Wt (g)	H₂O (g)	Ener (kcal)	Prot (g)	Carb (g)	Fiber (g)	Fat (g)	Fat Breakdown (g)			
											Sat	Mono	Poly	Trans
38192	Chex traditional snack mix	1	cup(s)	45	—	197	3.0	33.3	1.5	6.1	0.8	—	—	—
654	Potato chips, salted	1	ounce(s)	28	0.6	155	1.9	14.1	1.2	10.6	3.1	2.8	3.5	—
8816	Potato chips, unsalted	1	ounce(s)	28	0.5	152	2.0	15.0	1.4	9.8	3.1	2.8	3.5	—
5096	Pretzels, plain, hard, twists	5	item(s)	30	1.0	114	2.7	23.8	1.0	1.1	0.2	0.4	0.4	—
4632	Pretzels, whole wheat	1	ounce(s)	28	1.1	103	3.1	23.0	2.2	0.7	0.2	0.3	0.2	—
4641	Tortilla chips, plain	6	item(s)	11	0.2	53	0.8	7.1	0.6	2.5	0.3	0.8	0.5	0.3
	Cookies													
8859	Animal crackers	12	item(s)	30	1.2	134	2.1	22.2	0.3	4.1	1.0	2.3	0.6	—
8876	Brownie, prepared from mix	1	item(s)	24	3.0	112	1.5	12.0	0.5	7.0	1.8	2.6	2.3	—
25207	Chocolate chip cookies	1	item(s)	30	3.7	140	2.0	16.2	0.6	7.9	2.1	3.3	2.1	—
8915	Chocolate sandwich cookie with extra creme filling	1	item(s)	13	0.2	65	0.6	8.9	0.4	3.2	0.7	2.1	0.3	1.1
14145	Fig Newtons cookies	1	item(s)	16	—	55	0.5	11.0	0.5	1.3	0	—	—	0
8920	Fortune cookie	1	item(s)	8	0.6	30	0.3	6.7	0.1	0.2	0.1	0.1	0	—
25208	Oatmeal cookies	1	item(s)	69	12.3	234	5.7	45.1	3.1	4.2	0.7	1.3	1.8	—
25213	Peanut butter cookies	1	item(s)	35	4.1	163	4.2	16.9	0.9	9.2	1.7	4.7	2.3	—
33095	Sugar cookies	1	item(s)	16	4.1	61	1.1	7.4	0.1	3.0	0.6	1.3	0.9	—
9002	Vanilla sandwich cookie with creme filling	1	item(s)	10	0.2	48	0.5	7.2	0.2	2.0	0.3	0.8	0.8	—
	Crackers													
9012	Cheese cracker sandwich with peanut butter	4	item(s)	28	0.9	139	3.5	15.9	1.0	7.0	1.2	3.6	1.4	—
9008	Cheese crackers (mini)	30	item(s)	30	0.9	151	3.0	17.5	0.7	7.6	2.8	3.6	0.7	—
33362	Cheese crackers, low sodium	1	serving(s)	30	0.9	151	3.0	17.5	0.7	7.6	2.9	3.6	0.7	—
8928	Honey graham crackers	4	item(s)	28	1.2	118	1.9	21.5	0.8	2.8	0.4	1.1	1.1	—
9016	Matzo crackers, plain	1	item(s)	28	1.2	112	2.8	23.8	0.9	0.4	0.1	0	0.2	—
9024	Melba toast	3	item(s)	15	0.8	59	1.8	11.5	0.9	0.5	0.1	0.1	0.2	—
9028	Melba toast, rye	3	item(s)	15	0.7	58	1.7	11.6	1.2	0.5	0.1	0.1	0.2	—
14189	Ritz crackers	5	item(s)	16	0.5	80	1.0	10.0	0	4.0	1.0	—	—	0
9014	Rye crispbread crackers	1	item(s)	10	0.6	37	0.8	8.2	1.7	0.1	0	0	0.1	—
9040	Rye wafer	1	item(s)	11	0.6	37	1.1	8.8	2.5	0.1	0	0	0	—
432	Saltine crackers	5	item(s)	15	0.8	64	1.4	10.6	0.5	1.7	0.2	1.1	0.2	0.5
9046	Saltine crackers, low salt	5	item(s)	15	0.6	65	1.4	10.7	0.5	1.8	0.4	1.0	0.3	—
9052	Snack cracker sandwich with cheese filling	4	item(s)	28	1.1	134	2.6	17.3	0.5	5.9	1.7	3.2	0.7	—
9054	Snack cracker sandwich with peanut butter filling	4	item(s)	28	0.8	138	3.2	16.3	0.6	6.9	1.4	3.9	1.3	—
9048	Snack crackers, round	10	item(s)	30	1.1	151	2.2	18.3	0.5	7.6	1.1	3.2	2.9	—
9050	Snack crackers, round, low salt	10	item(s)	30	1.1	151	2.2	18.3	0.5	7.6	1.1	3.2	2.9	—
9044	Soda crackers	5	tem(s)	15	0.8	64	1.4	10.6	0.5	1.7	0.2	1.1	0.2	0.5
9059	Wheat cracker sandwich with cheese filling	4	item(s)	28	0.9	139	2.7	16.3	0.9	7.0	1.2	2.9	2.6	—
9061	Wheat cracker sandwich with peanut butter filling	4	item(s)	28	1.0	139	3.8	15.1	1.2	7.5	1.3	3.3	2.5	—
9055	Wheat crackers	10	item(s)	30	0.9	142	2.6	19.5	1.4	6.2	1.6	3.4	0.8	—
9057	Wheat crackers, low salt	10	item(s)	30	0.9	142	2.6	19.5	1.4	6.2	1.6	3.4	0.8	—
9022	Whole wheat crackers	7	item(s)	28	0.8	124	2.5	19.2	2.9	4.8	1.0	1.6	1.8	—
	Pastry													
16754	Apple fritter	1	item(s)	17	6.4	61	1.0	5.5	0.2	3.9	0.9	1.7	1.1	—
41565	Cinnamon rolls with icing, refrigerated dough	1	serving(s)	44	12.3	145	2.0	23.0	0.5	5.0	1.5	—	—	2.0
4945	Croissant, butter	1	item(s)	57	13.2	231	4.7	26.1	1.5	12.0	6.6	3.1	0.6	—
9096	Danish, nut	1	item(s)	65	13.3	280	4.6	29.7	1.3	16.4	3.8	8.9	2.8	—
9115	Doughnut with creme filling	1	item(s)	85	32.5	307	5.4	25.5	0.7	20.8	4.6	10.3	2.6	—
9117	Doughnut with jelly filling	1	item(s)	85	30.3	289	5.0	33.2	0.8	15.9	4.1	8.7	2.0	—
4947	Doughnut, cake	1	item(s)	47	9.8	198	2.4	23.4	0.7	10.8	1.7	4.4	3.7	—
9105	Doughnut, cake, chocolate glazed	1	item(s)	42	6.8	175	1.9	24.1	0.9	8.4	2.2	4.7	1.0	—
437	Doughnut, glazed	1	item(s)	60	15.2	242	3.8	26.6	0.7	13.7	3.5	7.7	1.7	—
10617	Toaster pastry, brown sugar cinnamon	1	item(s)	50	5.3	210	3.0	35.0	1.0	6.0	1.0	4.0	1.0	—
30928	Toaster pastry, cream cheese	1	item(s)	54	—	200	3.0	23.0	0	11.0	4.5	—	—	1.5
	Muffins													
25015	Blueberry	1	item(s)	63	29.7	160	3.4	23.0	0.8	6.0	0.9	1.5	3.3	—
9189	Corn, ready to eat	1	item(s)	57	18.6	174	3.4	29.0	1.9	4.8	0.8	1.2	1.8	—
9121	English muffin, plain, enriched	1	item(s)	57	24.0	134	4.4	26.2	1.5	1.0	0.1	0.2	0.5	—
29582	English muffin, toasted	1	item(s)	50	18.6	128	4.2	25.0	1.5	1.0	0.1	0.2	0.5	—
9145	English muffin, wheat	1	item(s)	57	24.1	127	5.0	25.5	2.6	1.1	0.2	0.2	0.5	—
8894	Oat bran	1	item(s)	57	20.0	154	4.0	27.5	2.6	4.2	0.6	1.0	2.4	—
	Granola bars													
38161	Kudos milk chocolate granola bars w/fruit & nuts	1	item(s)	28	—	90	2.0	15.0	1.0	3.0	1.0	—	—	—
38196	Nature Valley banana nut crunchy granola bars	2	item(s)	42	—	190	4.0	28.0	2.0	7.0	1.0	—	—	—
38187	Nature Valley fruit 'n' nut trail mix bar	1	item(s)	35	—	140	3.0	25.0	2.0	4.0	0.5	—	—	—
1383	Plain, hard	1	item(s)	25	1.0	115	2.5	15.8	1.3	4.9	0.6	1.1	3.0	—
4606	Plain, soft	1	item(s)	28	1.8	126	2.1	19.1	1.3	4.9	2.1	1.1	1.5	—
	Pies													
454	Apple pie, prepared from home recipe	1	slice(s)	155	73.3	411	3.7	57.5	2.3	19.4	4.7	8.4	5.2	—
470	Pecan pie, prepared from home recipe	1	slice(s)	122	23.8	503	6.0	63.7	—	27.1	4.9	13.6	7.0	—

BREADS, BAKED GOODS, CAKES, COOKIES, CRACKERS, CHIPS, PIES—Continued

Chol (mg)	Calc (mg)	Iron (mg)	Magn (mg)	Pota (mg)	Sodi (mg)	Zinc (mg)	Vit A (µg)	Thia (mg)	Vit E (mg α)	Ribo (mg)	Niac (mg)	Vit B$_6$ (mg)	Fola (µg)	Vit C (mg)	Vit B$_{12}$ (µg)	Sele (µg)
0	0	0.55	—	75.8	621.2	—	0	0.09	—	0.05	1.21	—	12.1	0	—	—
0	7	0.45	19.8	465.5	148.8	0.67	0	0.01	1.91	0.06	1.18	0.20	21.3	5.3	0	2.3
0	7	0.46	19.0	361.5	2.3	0.30	0	0.04	2.58	0.05	1.08	0.18	12.8	8.8	0	2.3
0	11	1.29	10.5	43.8	514.5	0.25	0	0.13	0.10	0.18	1.57	0.03	51.3	0	0	1.7
0	8	0.76	8.5	121.9	57.6	0.17	0	0.12	—	0.08	1.85	0.07	15.3	0.3	0	—
0	19	0.25	15.8	23.2	45.5	0.26	0	0.00	0.46	0.01	0.13	0.02	2.2	0	0	0.7
0	13	0.82	5.4	30.0	117.9	0.19	0	0.10	0.03	0.09	1.04	0.01	30.9	0	0	2.1
18	14	0.44	12.7	42.2	82.3	0.23	42.2	0.03	—	0.05	0.24	0.02	7.0	0.1	0	2.8
13	11	0.69	12.4	62.1	108.8	0.24	—	0.08	0.54	0.06	0.87	0.01	17.8	0	0	4.1
0	2	1.01	4.7	17.8	45.6	0.10	0	0.02	0.25	0.02	0.25	0.00	6.0	0	0	1.1
0	10	0.36	—	—	57.5	—	0	—	—	—	—	—	—	0	—	—
0	1	0.12	0.6	3.3	21.9	0.01	0.1	0.01	0.00	0.01	0.15	0.00	5.3	0	0	0.2
0	26	1.93	48.8	176.7	311.1	1.42	—	0.26	0.23	0.13	1.35	0.09	34.9	0.3	0	17.4
13	27	0.65	21.1	112.8	154.1	0.46	—	0.08	0.73	0.09	1.85	0.05	23.5	0.1	0.1	4.8
18	5	0.30	1.7	12.2	49.4	0.08	—	0.04	0.28	0.05	0.31	0.01	9.5	0	0	3.1
0	3	0.22	1.4	9.1	34.9	0.04	0	0.02	0.16	0.02	0.27	0.00	5.0	0	0	0.3
0	14	0.76	15.7	61.0	198.8	0.29	0.3	0.15	0.66	0.08	1.63	0.04	26.3	0	0.1	2.3
4	45	1.43	10.8	43.5	298.5	0.33	8.7	0.17	0.01	0.12	1.40	0.16	45.6	0	0.1	2.6
4	45	1.43	10.8	31.8	137.4	0.33	5.1	0.17	0.09	0.12	1.40	0.16	26.7	0	0.1	2.6
0	7	1.04	8.4	37.8	169.4	0.22	0	0.06	0.09	0.08	1.15	0.01	12.9	0	0	2.9
0	4	0.89	7.1	31.8	0.6	0.19	0	0.11	0.01	0.08	1.10	0.03	4.8	0	0	10.5
0	14	0.55	8.9	30.3	124.4	0.30	0	0.06	0.06	0.04	0.61	0.01	18.6	0	0	5.2
0	12	0.55	5.9	29.0	134.9	0.20	0	0.07	—	0.04	0.70	0.01	12.8	0	0	5.8
0	20	0.72	—	10.0	135.0	—	—	—	—	—	—	—	—	0	—	—
0	3	0.24	7.8	31.9	26.4	0.23	0	0.02	0.08	0.01	0.10	0.02	4.7	0	0	3.7
0	4	0.65	13.3	54.5	87.3	0.30	0	0.04	0.08	0.03	0.17	0.03	5.0	0	0	2.6
0	10	0.84	3.3	23.1	160.8	0.12	0	0.01	0.14	0.06	0.78	0.01	20.9	0	0	1.5
0	18	0.81	4.1	108.6	95.4	0.11	0	0.08	0.01	0.06	0.78	0.01	18.6	0	0	2.9
1	72	0.66	10.1	120.1	392.3	0.17	4.8	0.12	0.06	0.19	1.05	0.01	28.0	0	0	6.0
0	23	0.77	15.4	60.2	201.0	0.31	0.3	0.13	0.57	0.07	1.71	0.04	24.1	0	0	3.0
0	36	1.08	8.1	39.9	254.1	0.20	0	0.12	0.60	0.10	1.21	0.01	27.0	0	0	2.0
0	36	1.08	8.1	106.5	111.9	0.20	0	0.12	0.60	0.10	1.21	0.01	27.0	0	0	2.0
0	10	0.84	3.3	23.1	160.8	0.12	0	0.01	0.14	0.06	0.78	0.01	20.9	0	0	1.5
2	57	0.73	15.1	85.7	255.6	0.24	4.8	0.10	—	0.12	0.89	0.07	17.9	0.4	0	6.8
0	48	0.74	10.6	83.2	226.0	0.23	0	0.10	—	0.08	1.64	0.03	19.6	0	0	6.1
0	15	1.32	18.6	54.9	238.5	0.48	0	0.15	0.15	0.09	1.48	0.04	35.1	0	0	1.9
0	15	1.32	18.6	60.9	84.9	0.48	0	0.15	0.15	0.09	1.48	0.04	15.0	0	0	10.1
0	14	0.86	27.7	83.2	184.5	0.60	0	0.05	0.24	0.02	1.26	0.05	7.8	0	0	4.1
14	9	0.26	2.2	22.4	6.8	0.09	7.1	0.03	0.07	0.04	0.23	0.01	6.3	0.2	0.1	2.6
0	—	0.72	—	—	340.1	—	—	—	—	—	—	—	—	—	—	—
38	21	1.15	9.1	67.3	424.1	0.42	117.4	0.22	0.47	0.13	1.24	0.03	50.2	0.1	0.1	12.9
30	61	1.17	20.8	61.8	236.0	0.56	5.9	0.14	0.53	0.15	1.49	0.06	54.0	1.1	0.1	9.2
20	21	1.55	17.0	68.0	262.7	0.68	9.4	0.28	0.24	0.12	1.90	0.05	59.5	0	0.1	9.2
22	21	1.49	17.0	67.2	249.1	0.63	14.5	0.26	0.36	0.12	1.81	0.08	57.8	0	0.2	10.6
17	21	0.91	9.4	59.7	256.6	0.25	17.9	0.10	0.90	0.11	0.87	0.02	24.4	0.1	0.1	4.4
24	89	0.95	14.3	44.5	142.8	0.23	5.0	0.01	0.08	0.02	0.19	0.01	18.9	0	0	1.7
4	26	0.36	13.2	64.8	205.2	0.46	2.4	0.53	—	0.04	0.39	0.03	13.2	0.1	0.1	5.0
0	0	1.80	—	70.0	190.0	—	—	0.15	—	0.17	2.00	0.20	40.0	0	0	—
10	100	1.80	—	—	220.0	—	—	0.15	—	0.17	2.00	—	40.0	0	0.6	—
20	56	1.02	7.8	70.2	289.4	0.28	—	0.17	0.75	0.15	1.25	0.02	34.0	0.4	0.1	8.8
15	42	1.60	18.2	39.3	297.0	0.30	29.6	0.15	0.45	0.18	1.16	0.04	45.6	0	0.1	8.7
0	30	1.42	12.0	74.7	264.5	0.39	0	0.25	—	0.16	2.21	0.02	42.2	0	0	—
0	95	1.36	11.0	71.5	252.0	0.38	0	0.19	0.16	0.14	1.90	0.02	43.5	0.1	0	13.5
0	101	1.63	21.1	106.0	217.7	0.61	0	0.24	0.25	0.16	1.91	0.05	36.5	0	0	16.6
0	36	2.39	89.5	289.0	224.0	1.04	0	0.14	0.37	0.05	0.23	0.09	50.7	0	0	6.3
0	200	0.36	—	—	60.0	—	0	—	—	—	—	—	—	0	0	—
0	20	1.08	—	120.0	160.0	—	0	—	—	—	—	—	—	0	—	—
0	0	0.00	—	—	95.0	—	0	—	—	—	—	—	—	0	—	—
0	15	0.72	23.8	82.3	72.0	0.50	0	0.06	—	0.03	0.39	0.02	5.6	0.2	0	4.0
0	30	0.72	21.0	92.3	79.0	0.42	0	0.08	—	0.04	0.14	0.02	6.8	0	0.1	4.6
0	11	1.73	10.9	122.5	327.1	0.29	17.1	0.22	—	0.16	1.90	0.05	37.2	2.6	0	12.1
106	39	1.80	31.7	162.3	319.6	1.24	100.0	0.22	—	0.22	1.03	0.07	31.7	0.2	0.2	14.6

DA+ Code	Food Description	Quantity	Measure	Wt (g)	H₂O (g)	Ener (kcal)	Prot (g)	Carb (g)	Fiber (g)	Fat (g)	Fat Breakdown (g) Sat	Mono	Poly	Trans
33356	Pie crust mix, prepared, baked	1	slice(s)	20	2.1	100	1.3	10.1	0.4	6.1	1.5	3.5	0.8	—
9007	Pie crust, ready to bake, frozen, enriched, baked	1	slice(s)	16	1.8	82	0.7	7.9	0.2	5.2	1.7	2.5	0.6	—
472	Pumpkin pie, prepared from home recipe	1	slice(s)	155	90.7	316	7.0	40.9	—	14.4	4.9	5.7	2.8	—
	Rolls													
8555	Crescent dinner roll	1	item(s)	28	9.7	78	2.7	13.8	0.6	1.2	0.3	0.3	0.6	—
489	Hamburger roll or bun, plain	1	item(s)	43	14.9	120	4.1	21.3	0.9	1.9	0.5	0.5	0.8	—
490	Hard roll	1	item(s)	57	17.7	167	5.6	30.0	1.3	2.5	0.3	0.6	1.0	—
5127	Kaiser roll	1	item(s)	57	17.7	167	5.6	30.0	1.3	2.5	0.3	0.6	1.0	—
5130	Whole wheat roll or bun	1	item(s)	28	9.4	75	2.5	14.5	2.1	1.3	0.2	0.3	0.6	—
	Sport bars													
37026	Balance original chocolate bar	1	item(s)	50	—	200	14.0	22.0	0.5	6.0	3.5	—	—	—
37024	Balance original peanut butter bar	1	item(s)	50	—	200	14.0	22.0	1.0	6.0	2.5	—	—	—
36580	Clif Bar chocolate brownie energy bar	1	item(s)	68	—	240	10.0	45.0	5.0	4.5	1.5	—	—	0
36583	Clif Bar crunchy peanut butter energy bar	1	item(s)	68	—	250	12.0	40.0	5.0	6.0	1.5	—	—	0
36589	Clif Luna Nutz over Chocolate energy bar	1	item(s)	48	—	180	10.0	25.0	3.0	4.5	2.5	—	—	0
12005	PowerBar apple cinnamon	1	item(s)	65	—	230	9.0	45.0	3.0	2.5	0.5	1.5	0.5	0
16078	PowerBar banana	1	item(s)	65	—	230	9.0	45.0	3.0	2.5	0.5	1.0	0.5	0
16080	PowerBar chocolate	1	item(s)	65	6.4	230	10.0	45.0	3.0	2.0	0.5	0.5	1.0	0
29092	PowerBar peanut butter	1	item(s)	65	—	240	10.0	45.0	3.0	3.5	0.5	—	—	0
	Tortillas													
1391	Corn tortillas, soft	1	item(s)	26	11.9	57	1.5	11.6	1.6	0.7	0.1	0.2	0.4	—
1669	Flour tortilla	1	item(s)	32	9.7	100	2.7	16.4	1.0	2.5	0.6	1.2	0.5	—
	Pancakes, waffles													
8926	Pancakes, blueberry, prepared from recipe	3	item(s)	114	60.6	253	7.0	33.1	0.8	10.5	2.3	2.6	4.7	—
5037	Pancakes, prepared from mix with egg and milk	3	item(s)	114	60.3	249	8.9	32.9	2.1	8.8	2.3	2.4	3.3	—
1390	Taco shells, hard	1	item(s)	13	1.0	62	0.9	8.3	0.6	2.8	0.6	1.6	0.5	0.6
30311	Waffle, 100% whole grain	1	item(s)	75	32.3	200	6.9	25.0	1.9	8.4	2.3	3.3	2.1	—
9219	Waffle, plain, frozen, toasted	2	item(s)	66	20.2	206	4.7	32.5	1.6	6.3	1.1	3.2	1.5	—
500	Waffle, plain, prepared from recipe	1	item(s)	75	31.5	218	5.9	24.7	1.7	10.6	2.1	2.6	5.1	—
	CEREAL, FLOUR, GRAIN, PASTA, NOODLES, POPCORN													
	Grain													
2861	Amaranth, dry	½	cup(s)	98	9.6	365	14.1	64.5	9.1	6.3	1.6	1.4	2.8	—
1953	Barley, pearled, cooked	½	cup(s)	79	54.0	97	1.8	22.2	3.0	0.3	0.1	0	0.2	—
1956	Buckwheat groats, cooked, roasted	½	cup(s)	84	63.5	77	2.8	16.8	2.3	0.5	0.1	0.2	0.2	—
1957	Bulgur, cooked	½	cup(s)	91	70.8	76	2.8	16.9	4.1	0.2	0	0	0.1	—
1963	Couscous, cooked	½	cup(s)	79	57.0	88	3.0	18.2	1.1	0.1	0	0	0.1	—
1967	Millet, cooked	½	cup(s)	120	85.7	143	4.2	28.4	1.6	1.2	0.2	0.2	0.6	—
1969	Oat bran, dry	½	cup(s)	47	3.1	116	8.1	31.1	7.2	3.3	0.6	1.1	1.3	—
1972	Quinoa, dry	½	cup(s)	85	11.3	313	12.0	54.5	5.9	5.2	0.6	1.4	2.8	—
	Rice													
129	Brown, long grain, cooked	½	cup(s)	98	71.3	108	2.5	22.4	1.8	0.9	0.2	0.3	0.3	—
2863	Brown, medium grain, cooked	½	cup(s)	98	71.1	109	2.3	22.9	1.8	0.8	0.2	0.3	0.3	—
37488	Jasmine, saffroned, cooked	½	cup(s)	280	—	340	8.0	78.0	0	0	0	0	0	0
30280	Pilaf, cooked	½	cup(s)	103	74.0	129	2.1	22.2	0.6	3.3	0.6	1.5	1.0	—
28066	Spanish, cooked	½	cup(s)	244	184.2	241	5.7	50.2	3.3	1.9	0.4	0.6	0.7	0
2867	White glutinous, cooked	½	cup(s)	87	66.7	84	1.8	18.3	0.9	0.2	0	0.1	0.1	—
484	White, long grain, boiled	½	cup(s)	79	54.1	103	2.1	22.3	0.3	0.2	0.1	0.1	0.1	—
482	White, long grain, enriched, instant, boiled	½	cup(s)	83	59.4	97	1.8	20.7	0.5	0.4	0	0.1	0	—
486	White, long grain, enriched, parboiled, cooked	½	cup(s)	79	55.6	97	2.3	20.6	0.7	0.3	0.1	0.1	0.1	—
1194	Wild brown, cooked	½	cup(s)	82	60.6	83	3.3	17.5	1.5	0.3	0	0	0.2	—
	Flour & grain fractions													
505	All purpose flour, self rising, enriched	½	cup(s)	63	6.6	221	6.2	46.4	1.7	0.6	0.1	0	0.2	—
503	All purpose flour, white, bleached, enriched	½	cup(s)	63	7.4	228	6.4	47.7	1.7	0.6	0.1	0	0.2	—
1643	Barley flour	½	cup(s)	56	5.5	198	4.2	44.7	2.1	0.8	0.2	0.1	0.4	—
383	Buckwheat flour, whole groat	½	cup(s)	60	6.7	201	7.6	42.3	6.0	1.9	0.4	0.6	0.6	—
504	Cake wheat flour, enriched	½	cup(s)	69	8.6	248	5.6	53.5	1.2	0.6	0.1	0	0.3	—
426	Cornmeal, degermed, enriched	½	cup(s)	69	7.8	255	5.0	54.6	2.8	1.2	0.1	0.2	0.5	0
424	Cornmeal, yellow whole grain	½	cup(s)	61	6.2	221	4.9	46.9	4.4	2.2	0.3	0.6	1.0	—
1978	Dark rye flour	½	cup(s)	64	7.1	207	9.0	44.0	14.5	1.7	0.2	0.2	0.8	—
1644	Masa corn flour, enriched	½	cup(s)	57	5.1	208	5.3	43.5	5.5	2.1	0.3	0.6	1.0	—
1976	Rice flour, brown	½	cup(s)	79	9.4	287	5.7	60.4	3.6	2.2	0.4	0.8	0.8	—
1645	Rice flour, white	½	cup(s)	79	9.4	289	4.7	63.3	1.9	1.1	0.3	0.3	0.3	—
1980	Semolina, enriched	½	cup(s)	84	10.6	301	10.6	60.8	3.2	0.9	0.1	0.1	0.4	—
2827	Soy flour, raw	½	cup(s)	42	2.2	185	14.7	14.9	4.1	8.8	1.3	1.9	4.9	—
1990	Wheat germ, crude	2	tablespoon(s)	14	1.6	52	3.3	7.4	1.9	1.4	0.2	0.2	0.9	—
506	Whole wheat flour	½	cup(s)	60	6.2	203	8.2	43.5	7.3	1.1	0.2	0.1	0.5	—
	Breakfast bars													
39230	Atkins Morning Start apple crisp breakfast bar	1	item(s)	37	—	170	11.0	12.0	6.0	9.0	4.0	—	—	—
10571	Nutri-Grain apple cinnamon cereal bar	1	item(s)	37	—	140	2.0	27.0	1.0	3.0	0.5	2.0	0.5	—

CEREAL, FLOUR, GRAIN, PASTA, NOODLES, POPCORN—Continued

Chol (mg)	Calc (mg)	Iron (mg)	Magn (mg)	Pota (mg)	Sodi (mg)	Zinc (mg)	Vit A (µg)	Thia (mg)	Vit E (mg α)	Ribo (mg)	Niac (mg)	Vit B$_6$ (mg)	Fola (µg)	Vit C (mg)	Vit B$_{12}$ (µg)	Sele (µg)
0	12	0.43	3.0	12.4	145.8	0.07	0	0.06	—	0.03	0.47	0.01	14.0	0	0	4.4
0	3	0.36	2.9	17.6	103.5	0.05	0	0.04	0.42	0.06	0.39	0.01	8.8	0	0	0.5
65	146	1.96	29.5	288.3	348.8	0.71	660.3	0.14	—	0.31	1.21	0.07	32.6	2.6	0.1	11.0
0	39	0.93	5.9	26.3	134.1	0.18	0	0.11	0.02	0.09	1.16	0.02	31.1	0	0.1	5.5
0	59	1.42	9.0	40.4	206.0	0.28	0	0.17	0.03	0.13	1.78	0.03	47.7	0	0.1	8.4
0	54	1.87	15.4	61.6	310.1	0.53	0	0.27	0.23	0.19	2.41	0.02	54.2	0	0	22.3
0	54	1.86	15.4	61.6	310.1	0.53	0	0.27	0.23	0.19	2.41	0.01	54.2	0	0	22.3
0	30	0.69	24.1	77.1	135.5	0.57	0	0.07	0.26	0.04	1.04	0.06	8.5	0	0	14.0
3	100	4.50	40.0	160.0	180.0	3.75	—	0.37	—	0.42	5.00	0.50	100.0	60.0	1.5	17.5
3	100	4.50	40.0	130.0	230.0	3.75	—	0.37	—	0.42	5.00	0.50	100.0	60.0	1.5	17.5
0	250	4.50	100.0	370.0	150.0	3.00	—	0.37	—	0.25	3.00	0.40	80.0	60.0	0.9	14.0
0	250	4.50	100.0	230.0	250.0	3.00	—	0.37	—	0.25	3.00	0.40	80.0	60.0	0.9	14.0
0	350	5.40	80.0	190.0	190.0	5.25	—	1.20	—	1.36	16.00	2.00	400.0	60.0	6.0	24.5
0	300	6.30	140.0	125.0	100.0	5.25	—	1.50	—	1.70	20.00	2.00	400.0	60.0	6.0	—
0	300	6.30	140.0	190.0	100.0	5.25	0	1.50	—	1.70	20.00	2.00	400.0	60.0	6.0	—
0	300	6.30	140.0	200.0	95.0	5.25	0	1.50	—	1.70	20.00	2.00	400.0	60.0	6.0	5.1
0	300	6.30	140.0	130.0	120.0	5.25	0	1.50	—	1.70	20.00	2.00	400.0	60.0	6.0	—
0	21	0.32	18.7	48.4	11.7	0.34	0	0.02	0.07	0.02	0.39	0.06	1.3	0	0	1.6
0	41	1.06	7.0	49.6	203.5	0.17	0	0.17	0.06	0.08	1.14	0.01	33.3	0	0	7.1
64	235	1.96	18.2	157.3	469.7	0.61	57.0	0.22	—	0.31	1.73	0.05	41.0	2.5	0.2	16.0
81	245	1.48	25.1	226.9	575.7	0.85	82.1	0.22	—	0.35	1.40	0.12	104.6	0.7	0.4	—
0	13	0.25	11.3	29.7	51.7	0.21	0.1	0.03	0.09	0.01	0.25	0.03	9.2	0	0	0.6
71	194	1.60	28.5	171.0	371.3	0.87	48.8	0.15	0.32	0.25	1.47	0.08	28.5	0	0.4	20.0
10	203	4.56	15.8	95.0	481.8	0.35	262.7	0.34	0.64	0.46	5.86	0.68	49.5	0	1.9	8.3
52	191	1.73	14.3	119.3	383.3	0.51	48.8	0.19	—	0.26	1.55	0.04	34.5	0.3	0.2	34.7
0	149	7.40	259.3	356.8	20.5	3.10	0	0.06	—	0.20	1.24	0.20	47.8	4.1	0	—
0	9	1.04	17.3	73.0	2.4	0.64	0	0.06	0.01	0.04	1.61	0.09	12.6	0	0	6.8
0	6	0.67	42.8	73.9	3.4	0.51	0	0.03	0.07	0.03	0.79	0.06	11.8	0	0	1.8
0	9	0.87	29.1	61.9	4.6	0.51	0	0.05	0.01	0.02	0.91	0.07	16.4	0	0	0.5
0	6	0.30	6.3	45.5	3.9	0.20	0	0.05	0.10	0.02	0.77	0.04	11.8	0	0	21.6
0	4	0.75	52.8	74.4	2.4	1.09	0	0.12	0.02	0.09	1.59	0.13	22.8	0	0	1.1
0	27	2.54	110.5	266.0	1.9	1.46	0	0.55	0.47	0.10	0.43	0.07	24.4	0	0	21.2
0	40	3.88	167.4	478.5	4.2	2.62	0.8	0.30	2.06	0.26	1.28	0.40	156.4	0	0	7.2
0	10	0.41	41.9	41.9	4.9	0.61	0	0.09	0.02	0.02	1.49	0.14	3.9	0	0	9.6
0	10	0.51	42.9	77.0	1.0	0.60	0	0.09	—	0.01	1.29	0.14	3.9	0	0	38.0
0	—	2.16	—	—	780.0	—	—	—	—	—	—	—	—	—	—	—
0	11	1.16	9.3	54.6	390.4	0.37	33.0	0.13	0.28	0.02	1.23	0.06	44.3	0.4	0	4.3
0	37	1.52	95.4	330.5	97.1	1.40	—	0.27	0.12	0.05	3.24	0.38	19.0	22.6	0	14.3
0	2	0.12	4.4	8.7	4.4	0.35	0	0.01	0.03	0.01	0.25	0.02	0.9	0	0	4.9
0	8	0.94	9.5	27.7	0.8	0.38	0	0.12	0.03	0.01	1.16	0.07	45.8	0	0	5.9
0	7	1.46	4.1	7.4	3.3	0.40	0	0.06	0.01	0.01	1.43	0.04	57.8	0	0	4.0
0	15	1.43	7.1	44.2	1.6	0.29	0	0.16	0.01	0.01	1.82	0.12	64.0	0	0	7.3
0	2	0.49	26.2	82.8	2.5	1.09	0	0.04	0.19	0.07	1.05	0.11	21.3	0	0	0.7
0	211	2.90	11.9	77.5	793.7	0.38	0	0.42	0.02	0.24	3.64	0.02	122.5	0	0	21.5
0	9	2.90	13.7	66.9	1.2	0.42	0	0.48	0.02	0.30	3.68	0.02	114.4	0	0	21.2
0	16	0.70	45.4	185.9	4.5	1.04	0	0.06	—	0.02	2.56	0.14	12.9	0	0	2.0
0	25	2.42	150.6	346.2	6.6	1.86	0	0.24	0.18	0.10	3.68	0.34	32.4	0	0	3.4
0	10	5.01	11.0	71.9	1.4	0.42	0	0.61	0.01	0.29	4.65	0.02	127.4	0	0	3.4
0	2	2.98	24.1	104.9	4.8	0.48	7.6	0.42	0.10	0.28	3.66	0.12	148.3	0	0	8.0
0	4	2.10	77.5	175.1	21.3	1.10	6.7	0.22	0.24	0.12	2.20	0.18	15.2	0	0	9.4
0	36	4.12	158.7	467.2	0.6	3.58	0.6	0.20	0.90	0.16	2.72	0.28	38.4	0	0	22.8
0	80	4.10	62.7	169.9	2.8	1.00	0	0.80	0.08	0.42	5.60	0.20	132.8	0	0	8.5
0	9	1.56	88.5	228.3	6.3	1.92	0	0.34	0.94	0.06	5.00	0.58	12.6	0	0	—
0	8	0.26	27.6	60.0	0	0.62	0	0.10	0.08	0.02	2.04	0.34	3.2	0	0	11.9
0	14	3.64	39.2	155.3	0.8	0.86	0	0.66	0.20	0.46	5.00	0.08	152.8	0	0	74.6
0	87	2.70	182.0	1067.0	5.5	1.65	2.5	0.24	0.82	0.48	1.83	0.18	146.4	0	0	3.2
0	6	0.90	34.4	128.2	1.7	1.76	0	0.27	—	0.07	0.97	0.18	40.4	0	0	11.4
0	20	2.32	82.8	243.0	3.0	1.74	0	0.26	0.48	0.12	3.82	0.20	26.4	0	0	42.4
0	200	—	—	90.0	70.0	—	—	0.22	—	0.25	3.00	—	—	9.0	—	—
0	200	1.80	8.0	75.0	110.0	1.50	—	0.37	—	0.42	5.00	0.50	40.0	0	—	—

DA+ Code	Food Description	Quantity	Measure	Wt (g)	H₂O (g)	Ener (kcal)	Prot (g)	Carb (g)	Fiber (g)	Fat (g)	Sat	Mono	Poly	Trans
10647	Nutri-Grain blueberry cereal bar	1	item(s)	37	5.4	140	2.0	27.0	1.0	3.0	0.5	2.0	0.5	—
10648	Nutri-Grain raspberry cereal bar	1	item(s)	37	5.4	140	2.0	27.0	1.0	3.0	0.5	2.0	0.5	—
10649	Nutri-Grain strawberry cereal bar	1	item(s)	37	5.4	140	2.0	27.0	1.0	3.0	0.5	2.0	0.5	—
	Breakfast cereals, hot													
1260	Cream of Wheat, instant, prepared	½	cup(s)	121	—	388	12.9	73.3	4.3	0	0	0	0	0
365	Farina, enriched, cooked with water and salt	½	cup(s)	117	102.4	56	1.7	12.2	0.3	0.1	0	0	0	—
363	Grits, white corn, regular and quick, enriched, cooked with water and salt	½	cup(s)	121	103.3	71	1.7	15.6	0.4	0.2	0	0.1	0.1	—
8636	Grits, yellow corn, regular and quick, enriched, cooked with salt	½	cup(s)	121	103.3	71	1.7	15.6	0.4	0.2	0	0.1	0.1	—
8657	Oatmeal, cooked with water	½	cup(s)	117	97.8	83	3.0	14.0	2.0	1.8	0.4	0.5	0.7	0
5500	Oatmeal, maple and brown sugar, instant, prepared	1	item(s)	198	150.2	200	4.8	40.4	2.4	2.2	0.4	0.7	0.8	—
5510	Oatmeal, ready to serve, packet, prepared	1	item(s)	186	158.7	112	4.1	19.8	2.7	2.0	0.4	0.7	0.8	—
	Breakfast cereals, ready to eat													
1197	All-Bran	1	cup(s)	62	1.3	160	8.1	46.0	18.2	2.0	0.4	0.4	1.3	0
1200	All-Bran Buds	1	cup(s)	91	2.7	212	6.4	72.7	39.1	1.9	0.4	0.5	1.2	0
1199	Apple Jacks	1	cup(s)	33	0.9	130	1.0	30.0	0.5	0.5	0	—	—	0
1204	Cap'n Crunch	1	cup(s)	36	0.9	147	1.3	30.7	1.3	2.0	0.5	0.4	0.3	—
1205	Cap'n Crunch Crunchberries	1	cup(s)	35	0.9	133	1.3	29.3	1.3	2.0	0.5	0.4	0.3	—
1206	Cheerios	1	cup(s)	30	1.0	110	3.0	22.0	3.0	2.0	0	0.5	0.5	—
3415	Cocoa Puffs	1	cup(s)	30	0.6	120	1.0	26.0	0.2	1.0	—	—	—	—
1207	Cocoa Rice Krispies	1	cup(s)	41	1.0	160	1.3	36.0	1.3	1.3	0.7	0	0	—
5522	Complete wheat bran flakes	1	cup(s)	39	1.4	120	4.0	30.7	6.7	0.7	—	—	—	0
1211	Corn Flakes	1	cup(s)	28	0.9	100	2.0	24.0	1.0	0	0	0	0	0
1247	Corn Pops	1	cup(s)	31	0.9	120	1.0	28.0	0.3	0	0	0	0	0
1937	Cracklin' Oat Bran	1	cup(s)	65	2.3	267	5.3	46.7	8.0	9.3	4.0	4.7	1.3	0
1220	Froot Loops	1	cup(s)	32	0.8	120	1.0	28.0	1.0	1.0	0.5	0	0	—
38214	Frosted Cheerios	1	cup(s)	37	—	149	2.5	31.1	1.2	1.2	—	—	—	—
372	Frosted Flakes	1	cup(s)	41	1.1	160	1.3	37.3	1.3	0	0	0	0	0
38215	Frosted Mini Chex	1	cup(s)	40	—	147	1.3	36.0	0	0	0	0	0	0
10268	Frosted Mini-Wheats	1	cup(s)	59	3.1	208	5.8	47.4	5.8	1.2	0	0	0.6	0
38216	Frosted Wheaties	1	cup(s)	40	—	147	1.3	36.0	0.3	0	0	0	0	0
1223	Granola, prepared	½	cup(s)	61	3.3	298	9.1	32.5	5.5	14.7	2.5	5.8	5.6	0
2415	Honey Bunches of Oats honey roasted	1	cup(s)	40	0.9	160	2.7	33.3	1.3	2.0	0.7	1.2	0.1	—
1227	Honey Nut Cheerios	1	cup(s)	37	0.9	149	3.7	29.9	2.5	1.9	0	0.6	0.6	—
2424	Honeycomb	1	cup(s)	22	0.3	83	1.5	19.5	0.8	0.4	0	—	—	—
10286	Kashi whole grain puffs	1	cup(s)	19	—	70	2.0	15.0	1.0	0.5	0	—	—	0
41142	Kellogg's Mueslix	1	cup(s)	83	7.2	298	7.6	60.8	6.1	4.6	0.7	2.4	1.5	0
1231	Kix	1	cup(s)	24	0.5	96	1.6	20.8	0.8	0.4	—	—	—	—
30569	Life	1	cup(s)	43	1.7	160	4.0	33.3	2.7	2.0	0.3	0.6	0.6	—
1233	Lucky Charms	1	cup(s)	24	0.6	96	1.6	20.0	0.8	0.8	—	—	—	—
38220	Multi Grain Cheerios	1	cup(s)	30	—	110	3.0	24.0	3.0	1.0	—	—	—	—
1201	Multi-Bran Chex	1	cup(s)	63	1.3	216	4.3	52.9	8.6	1.6	0	0	0.5	0
13633	Post Bran Flakes	1	cup(s)	40	1.5	133	4.0	32.0	6.7	0.7	0	—	—	—
1241	Product 19	1	cup(s)	30	1.0	100	2.0	25.0	1.0	0	0	0	0	0
32432	Puffed rice, fortified	1	cup(s)	14	0.4	56	0.9	12.6	0.2	0.1	0	—	—	—
32433	Puffed wheat, fortified	1	cup(s)	12	0.4	44	1.8	9.6	0.5	0.1	0	—	—	—
13334	Quaker 100% natural granola oats & honey	½	cup(s)	48	—	220	5.0	31.0	3.0	9.0	3.8	4.1	1.2	—
13335	Quaker 100% natural granola oats, honey & raisins	½	cup(s)	51	—	230	5.0	34.0	3.0	9.0	3.6	3.8	1.1	—
2420	Raisin Bran	1	cup(s)	59	5.0	190	4.0	46.0	8.0	1.0	0	0.1	0.4	—
1244	Rice Chex	1	cup(s)	31	0.8	120	2.0	27.0	0.3	0	0	0	0	0
1245	Rice Krispies	1	cup(s)	26	0.8	96	1.6	23.2	0	0	0	0	0	0
5593	Shredded Wheat	1	cup(s)	49	0.4	177	5.8	40.9	6.9	1.1	0.1	0	0.2	0
1248	Smacks	1	cup(s)	36	1.1	133	2.7	32.0	1.3	0.7	—	—	—	—
1246	Special K	1	cup(s)	31	0.9	110	7.0	22.0	0.5	0	0	0	0	0
3428	Total corn flakes	1	cup(s)	23	0.6	83	1.5	18.0	0.6	0	0	0	0	0
1253	Total whole grain	1	cup(s)	40	1.1	147	2.7	30.7	4.0	1.3	—	—	—	—
1254	Trix	1	cup(s)	30	0.6	120	1.0	27.0	1.0	1.0	—	—	—	—
382	Wheat germ, toasted	2	tablespoon(s)	14	0.8	54	4.1	7.0	2.1	1.5	0.3	0.2	0.9	—
1257	Wheaties	1	cup(s)	36	1.2	132	3.6	28.8	3.6	1.2	—	—	—	—
	Pasta, noodles													
449	Chinese chow mein noodles, cooked	½	cup(s)	23	0.2	119	1.9	12.9	0.9	6.9	1.0	1.7	3.9	—
1995	Corn pasta, cooked	½	cup(s)	70	47.8	88	1.8	19.5	3.4	0.5	0.1	0.1	0.2	—
448	Egg noodles, enriched, cooked	½	cup(s)	80	54.2	110	3.6	20.1	1.0	1.7	0.3	0.5	0.4	0
1563	Egg noodles, spinach, enriched, cooked	½	cup(s)	80	54.8	106	4.0	19.4	1.8	1.3	0.3	0.4	0.3	—
440	Macaroni, enriched, cooked	½	cup(s)	70	43.5	111	4.1	21.6	1.3	0.7	0.1	0.1	0.2	0
2000	Macaroni, tricolor vegetable, enriched, cooked	½	cup(s)	67	45.8	86	3.0	17.8	2.9	0.1	0	0	0	—
1996	Plain pasta, fresh-refrigerated, cooked	½	cup(s)	64	43.9	84	3.3	16.0	—	0.7	0.1	0.1	0.3	—

CEREAL, FLOUR, GRAIN, PASTA, NOODLES, POPCORN—Continued

Chol (mg)	Calc (mg)	Iron (mg)	Magn (mg)	Pota (mg)	Sodi (mg)	Zinc (mg)	Vit A (µg)	Thia (mg)	Vit E (mg α)	Ribo (mg)	Niac (mg)	Vit B_6 (mg)	Fola (µg)	Vit C (mg)	Vit B_{12} (µg)	Sele (µg)
0	200	1.80	8.0	75.0	110.0	1.50	—	0.37	—	0.42	5.00	0.50	40.0	0	0	—
0	200	1.80	8.0	70.0	110.0	1.50	—	0.37	—	0.42	5.00	0.50	40.0	0	0	—
0	200	1.80	8.0	55.0	110.0	1.50	—	0.37	—	0.42	5.00	0.50	40.0	0	0	—
0	862	34.91	21.4	150.8	732.6	0.86	—	1.59	—	1.47	21.55	2.15	431.0	0	0	—
0	5	0.58	2.3	15.1	383.3	0.09	0	0.07	0.01	0.05	0.57	0.01	39.6	0	0	10.6
0	4	0.73	6.1	25.4	269.8	0.08	0	0.10	0.02	0.07	0.87	0.03	39.9	0	0	3.8
0	4	0.73	6.1	25.4	269.8	0.08	2.4	0.10	0.02	0.07	0.87	0.03	39.9	0	0	3.3
0	11	1.05	31.6	81.9	4.7	1.17	0	0.09	0.09	0.02	0.26	0.01	7.0	0	0	6.3
0	26	6.83	49.9	126.4	403.5	1.03	0	1.02	—	0.05	1.56	0.30	42.2	0	0	11.1
0	21	3.96	44.7	112.4	240.9	0.92	0	0.60	—	0.04	0.77	0.18	18.7	0	0	3.8
0	241	10.90	224.4	632.4	150.0	3.00	300.1	1.40	—	1.68	9.16	7.44	800.0	12.4	12.0	5.8
0	57	13.64	186.4	909.1	614.5	4.55	464.5	1.09	1.42	1.27	15.45	6.09	1222.7	18.2	18.2	26.3
0	0	4.50	8.0	30.0	130.0	1.50	150.2	0.37	—	0.42	5.00	0.50	100.0	15.0	1.5	2.4
0	5	6.80	20.0	73.3	266.7	5.00	2.5	0.51	—	0.57	6.68	0.67	133.5	0	0	6.7
0	7	6.53	18.7	73.3	240.0	5.13	2.4	0.51	—	0.57	6.68	0.67	133.7	0	0	6.7
0	100	8.10	40.0	95.0	280.0	3.75	150.3	0.37	—	0.42	5.00	0.50	200.0	6.0	1.5	11.3
0	100	4.50	8.0	50.0	170.0	3.75	0	0.37	—	0.42	5.00	0.50	100.0	6.0	1.5	2.0
0	53	6.00	10.7	66.7	253.3	2.00	200.1	0.49	—	0.56	6.67	0.67	133.3	20.0	2.0	5.8
0	0	24.00	53.3	226.7	280.0	20.00	300.1	2.00	—	2.27	26.67	2.67	533.3	80.0	8.0	4.1
0	0	8.10	3.4	25.0	200.0	0.16	149.8	0.37	—	0.42	5.00	0.50	100.0	6.0	1.5	1.4
0	0	1.80	2.5	25.0	120.0	1.50	150.0	0.37	—	0.42	5.00	0.50	100.0	6.0	1.5	2.0
0	27	2.40	80.0	293.3	200.0	2.00	299.9	0.49	—	0.56	6.67	0.67	133.3	20.0	2.0	14.4
0	0	4.50	8.0	35.0	150.0	1.50	150.1	0.37	—	0.42	5.00	0.50	100.0	15.0	1.5	2.3
0	124	5.60	19.9	68.4	261.3	4.67	—	0.46	—	0.52	6.22	0.62	124.4	7.5	1.9	—
0	0	6.00	3.7	26.7	200.0	0.20	200.1	0.49	—	0.56	6.67	0.67	133.3	8.0	2.0	1.8
0	133	12.00	—	33.3	266.7	4.00	—	0.49	—	0.56	6.67	0.67	266.7	8.0	2.0	—
0	0	16.66	69.4	196.7	5.8	1.74	0	0.43	—	0.49	5.78	0.58	115.7	0	1.7	2.4
0	133	10.80	0	46.7	266.7	10.00	—	1.00	—	1.13	13.33	1.33	533.3	8.0	4.0	—
0	48	2.58	106.8	329.4	15.3	2.45	0.6	0.44	6.77	0.17	1.30	0.17	50.0	0.7	0	17.0
0	0	10.80	21.3	0	253.3	0.40	—	0.49	—	0.56	6.67	0.67	133.0	0	2.0	—
0	124	5.60	39.8	112.0	336.0	4.67	—	0.46	—	0.52	6.22	0.62	248.9	7.5	1.9	8.8
0	0	2.03	6.0	26.3	165.4	1.13	—	0.28	—	0.32	3.74	0.37	75.0	0	1.1	—
0	0	0.36	—	60.0	0	—	0	0.03	—	0.03	0.80	0.00	—	0	0	—
0	48	6.83	74.2	363.3	257.5	5.67	136.7	0.67	6.00	0.67	8.33	3.08	615.0	0.3	9.2	14.4
0	120	6.48	6.4	28.0	216.0	3.00	120.2	0.30	—	0.34	4.00	0.40	160.0	4.8	1.2	4.8
0	149	11.87	41.3	120.0	213.3	5.33	0.9	0.53	—	0.60	7.12	0.71	142.4	0	0	10.7
0	80	3.60	12.8	48.0	168.0	3.00	—	0.30	—	0.34	4.00	0.40	160.0	4.8	1.2	4.8
0	100	18.00	24.0	85.0	200.0	15.00	—	1.50	—	1.70	20.00	2.00	400.0	15.0	6.0	—
0	108	17.50	64.8	237.7	410.6	4.05	171.1	0.40	—	0.45	5.40	0.54	432.2	6.5	1.6	4.9
0	0	10.80	80.0	266.7	280.0	2.00	—	0.49	—	0.56	6.67	0.67	133.3	0	2.0	—
0	0	18.00	16.0	50.0	210.0	15.00	225.3	1.50	—	1.70	20.00	2.00	400.0	60.0	6.0	3.6
0	1	4.43	3.5	15.8	0.4	0.14	0	0.36	—	0.25	4.94	0.01	2.7	0	0	1.5
0	3	3.80	17.4	41.8	0.5	0.28	0	0.31	—	0.21	4.23	0.02	3.8	0	0	14.8
0	61	1.20	51.0	220.0	20.0	1.05	0.5	0.13	—	0.12	0.82	0.07	15.0	0.2	0.1	8.3
0	59	1.20	49.0	250.0	20.0	0.99	0.5	0.13	—	0.12	0.80	0.08	14.1	0.4	0.1	8.8
0	20	10.80	80.0	360.0	360.0	2.25	—	0.37	—	0.42	5.00	0.50	100.0	0	2.1	—
0	100	9.00	9.3	35.0	290.0	3.75	—	0.37	—	0.42	5.00	0.50	200.0	6.0	1.5	1.2
0	0	1.44	12.8	32.0	256.0	0.48	120.1	0.30	—	0.34	4.80	0.40	80.0	4.8	1.2	4.1
0	18	2.90	60.3	179.3	1.1	1.37	0	0.14	—	0.12	3.47	0.18	21.1	0	0	2.0
0	0	0.48	10.7	53.3	66.7	0.40	200.2	0.49	—	0.56	6.67	0.67	133.3	8.0	2.0	17.5
0	0	8.10	16.0	60.0	220.0	0.90	225.1	0.52	—	0.59	7.00	2.00	400.0	21.0	6.0	7.0
0	752	13.53	0	22.6	157.9	11.28	112.8	1.13	22.56	1.28	15.04	1.50	300.8	45.1	4.5	1.2
0	1333	24.00	32.0	120.0	253.3	20.00	200.4	2.00	31.32	2.27	26.67	2.67	533.3	80.0	8.0	1.9
0	100	4.50	0	15.0	190.0	3.75	150.3	0.37	—	0.42	5.00	0.50	100.0	6.0	1.5	6.0
0	6	1.28	45.2	133.8	0.6	2.35	0.7	0.23	2.25	0.11	0.79	0.13	49.7	0.8	0	9.2
0	24	9.72	38.4	126.0	264.0	9.00	180.4	0.90	—	1.02	12.00	1.20	240.0	7.2	3.6	1.7
0	5	1.06	11.7	27.0	98.8	0.31	0	0.13	0.78	0.09	1.33	0.02	20.3	0	0	9.7
0	1	0.18	25.2	21.7	0	0.44	2.1	0.04	—	0.02	0.39	0.04	4.2	0	0	2.0
23	10	1.17	16.8	30.4	4.0	0.52	4.8	0.23	0.13	0.11	1.66	0.03	67.2	0	0.1	19.1
26	15	0.87	19.2	29.6	9.6	0.50	8.0	0.19	0.46	0.09	1.17	0.09	51.2	0	0.1	17.4
0	5	0.90	12.6	30.8	0.7	0.36	0	0.19	0.04	0.10	1.18	0.03	51.1	0	0	18.5
0	7	0.33	12.7	20.8	4.0	0.30	3.4	0.08	0.14	0.04	0.72	0.02	43.6	0	0	13.3
21	4	0.73	11.5	15.4	3.8	0.36	3.8	0.13	—	0.10	0.64	0.02	41.0	0	0.1	—

DA+ Code	Food Description	Quantity	Measure	Wt (g)	H₂O (g)	Ener (kcal)	Prot (g)	Carb (g)	Fiber (g)	Fat (g)	Fat Breakdown (g)			
											Sat	Mono	Poly	Trans
1725	Ramen noodles, cooked	½	cup(s)	114	94.5	104	3.0	15.4	1.0	4.3	0.2	0.2	0.2	—
2878	Soba noodles, cooked	½	cup(s)	95	69.4	94	4.8	20.4	—	0.1	0	0	0	—
2879	Somen noodles, cooked	½	cup(s)	88	59.8	115	3.5	24.2	—	0.2	0	0	0.1	—
493	Spaghetti, al dente, cooked	½	cup(s)	65	41.6	95	3.5	19.5	1.0	0.5	0.1	0.1	0.2	—
2884	Spaghetti, whole wheat, cooked	½	cup(s)	70	47.0	87	3.7	18.6	3.2	0.4	0.1	0.1	0.1	—
	Popcorn													
476	Air popped	1	cup(s)	8	0.3	31	1.0	6.2	1.2	0.4	0	0.1	0.2	—
4619	Caramel	1	cup(s)	35	1.0	152	1.3	27.8	1.8	4.5	1.3	1.0	1.6	—
4620	Cheese flavored	1	cup(s)	36	0.9	188	3.3	18.4	3.5	11.8	2.3	3.5	5.5	—
477	Popped in oil	1	cup(s)	11	0.1	64	0.8	5.0	0.9	4.8	0.8	1.1	2.6	—
	FRUIT AND FRUIT JUICES													
	Apples													
952	Juice, prepared from frozen concentrate	½	cup(s)	120	105.0	56	0.2	13.8	0.1	0.1	0	0	0	—
225	Juice, unsweetened, canned	½	cup(s)	124	109.0	58	0.1	14.5	0.1	0.1	0	0	0	—
224	Slices	½	cup(s)	55	47.1	29	0.1	7.6	1.3	0.1	0	0	0	—
946	Slices without skin, boiled	½	cup(s)	86	73.1	45	0.2	11.7	2.1	0.3	0	0	0.1	—
223	Raw medium, with peel	1	item(s)	138	118.1	72	0.4	19.1	3.3	0.2	0	0	0.1	—
948	Dried, sulfured	¼	cup(s)	22	6.8	52	0.2	14.2	1.9	0.1	0	0	0	—
226	Applesauce, sweetened, canned	½	cup(s)	128	101.5	97	0.2	25.4	1.5	0.2	0	0	0.1	—
227	Applesauce, unsweetened, canned	½	cup(s)	122	107.8	52	0.2	13.8	1.5	0.1	0	0	0	—
38492	Crabapples	1	item(s)	35	27.6	27	0.1	7.0	0.9	0.1	0	0	0	—
	Apricot													
228	Fresh without pits	4	item(s)	140	120.9	67	2.0	15.6	2.8	0.5	0	0.2	0.1	—
229	Halves with skin, canned in heavy syrup	½	cup(s)	129	100.1	107	0.7	27.7	2.1	0.1	0	0	0	—
230	Halves, dried, sulfured	¼	cup(s)	33	10.1	79	1.1	20.6	2.4	0.2	0	0	0	—
	Avocado													
233	California, whole, without skin or pit	½	cup(s)	115	83.2	192	2.2	9.9	7.8	17.7	2.4	11.3	2.1	—
234	Florida, whole, without skin or pit	½	cup(s)	115	90.6	138	2.5	9.0	6.4	11.5	2.2	6.3	1.9	—
2998	Pureed	⅛	cup(s)	28	20.2	44	0.5	2.4	1.8	4.0	0.6	2.7	0.5	—
	Banana													
4580	Dried chips	¼	cup(s)	55	2.4	285	1.3	32.1	4.2	18.5	15.9	1.1	0.3	—
235	Fresh whole, without peel	1	item(s)	118	88.4	105	1.3	27.0	3.1	0.4	0.1	0	0.1	—
	Blackberries													
237	Raw	½	cup(s)	72	63.5	31	1.0	6.9	3.8	0.4	0	0	0.2	—
958	Unsweetened, frozen	½	cup(s)	76	62.1	48	0.9	11.8	3.8	0.3	0	0	0.2	—
	Blueberries													
959	Canned in Heavy Syrup	½	cup(s)	128	98.3	113	0.8	28.2	2.0	0.4	0	0.1	0.2	—
238	Raw	½	cup(s)	73	61.1	41	0.5	10.5	1.7	0.2	0	0	0.1	—
960	Unsweetened, frozen	½	cup(s)	78	67.1	40	0.3	9.4	2.1	0.5	0	0.1	0.2	—
	Boysenberries													
961	Canned in heavy syrup	½	cup(s)	128	97.6	113	1.3	28.6	3.3	0.2	0	0	0.1	—
962	Unsweetened, frozen	½	cup(s)	66	56.7	33	0.7	8.0	3.5	0.2	0	0	0.1	—
35576	**Breadfruit**	1	item(s)	384	271.3	396	4.1	104.1	18.8	0.9	0.2	0.1	0.3	—
	Cherries													
967	Sour red, canned in water	½	cup(s)	122	109.7	44	0.9	10.9	1.3	0.1	0	0	0	—
3000	Sour red, raw	½	cup(s)	78	66.8	39	0.8	9.4	1.2	0.2	0.1	0.1	0.1	—
3004	Sweet, canned in heavy syrup	½	cup(s)	127	98.2	105	0.8	26.9	1.9	0.2	0	0.1	0.1	—
969	Sweet, canned in water	½	cup(s)	124	107.9	57	1.0	14.6	1.9	0.2	0	0	0	—
240	Sweet, raw	½	cup(s)	73	59.6	46	0.8	11.6	1.5	0.1	0	0	0	—
	Cranberries													
3007	Chopped, raw	½	cup(s)	55	47.9	25	0.2	6.7	2.5	0.1	0	0	0	—
1717	Cranberry apple juice drink	½	cup(s)	123	102.6	77	0	19.4	0	0.1	0	0	0.1	—
1638	Cranberry juice cocktail	½	cup(s)	127	109.0	68	0	17.1	0	0.1	0	0	0.1	—
241	Cranberry juice cocktail, low calorie, with saccharin	½	cup(s)	119	112.8	23	0	5.5	0	0	0	0	0	—
242	Cranberry sauce, sweetened, canned	¼	cup(s)	69	42.0	105	0.1	26.9	0.7	0.1	0	0	0	—
	Dates													
244	Domestic, chopped	¼	cup(s)	45	9.1	125	1.1	33.4	3.6	0.2	0	0	0	—
243	Domestic, whole	¼	cup(s)	45	9.1	125	1.1	33.4	3.6	0.2	0	0	0	—
	Figs													
975	Canned in heavy syrup	½	cup(s)	130	98.8	114	0.5	29.7	2.8	0.1	0	0	0.1	—
974	Canned in water	½	cup(s)	124	105.7	66	0.5	17.3	2.7	0.1	0	0	0.1	—
973	Raw , medium	2	item(s)	100	79.1	74	0.7	19.2	2.9	0.3	0.1	0.1	0.1	—
	Fruit cocktail & salad													
245	Fruit cocktail, canned in heavy syrup	½	cup(s)	124	99.7	91	0.5	23.4	1.2	0.1	0	0	0	—
978	Fruit cocktail, canned in juice	½	cup(s)	119	103.6	55	0.5	14.1	1.2	0	0	0	0	—
977	Fruit cocktail, canned in water	½	cup(s)	119	107.6	38	0.5	10.1	1.2	0.1	0	0	0	—
979	Fruit salad, canned in water	½	cup(s)	123	112.1	37	0.4	9.6	1.2	0.1	0	0	0	—
	Gooseberries													
982	Canned in light syrup	½	cup(s)	126	100.9	92	0.8	23.6	3.0	0.3	0	0	0.1	—

FRUIT AND FRUIT JUICES —Continued

Chol (mg)	Calc (mg)	Iron (mg)	Magn (mg)	Pota (mg)	Sodi (mg)	Zinc (mg)	Vit A (µg)	Thia (mg)	Vit E (mg α)	Ribo (mg)	Niac (mg)	Vit B$_6$ (mg)	Fola (µg)	Vit C (mg)	Vit B$_{12}$ (µg)	Sele (µg)
18	9	0.89	8.5	34.5	414.5	0.30	—	0.08	—	0.04	0.71	0.03	4.0	0.1	0	—
0	4	0.45	8.5	33.2	57.0	0.11	0	0.09	—	0.02	0.48	0.03	6.6	0	0	—
0	7	0.45	1.8	25.5	141.7	0.19	0	0.01	—	0.03	0.08	0.01	1.8	0	0	—
0	7	1.00	12.4	51.5	0.5	0.35	0	0.12	0.04	0.07	0.90	0.04	7.8	0	0	40.0
0	11	0.74	21.0	30.8	2.1	0.57	0	0.08	0.21	0.03	0.50	0.06	3.5	0	0	18.1
0	1	0.25	11.5	26.3	0.6	0.25	0.8	0.01	0.02	0.01	0.18	0.01	2.5	0	0	0
2	15	0.61	12.3	38.4	72.5	0.20	0.7	0.02	0.42	0.02	0.77	0.01	1.8	0	0	1.3
4	40	0.79	32.5	93.2	317.4	0.71	13.6	0.04	—	0.08	0.52	0.08	3.9	0.2	0.2	4.3
0	0	0.22	8.7	20.0	116.4	0.34	0.9	0.01	0.27	0.00	0.13	0.01	2.8	0	0	0.2
0	7	0.31	6.0	150.6	8.4	0.05	0	0.00	0.01	0.02	0.05	0.04	0	0.7	0	0.1
0	9	0.46	3.7	147.6	3.7	0.04	0	0.03	0.01	0.02	0.12	0.04	0	1.1	0	0.1
0	3	0.06	2.7	58.8	0.5	0.02	1.6	0.01	0.10	0.01	0.05	0.02	1.6	2.5	0	0
0	4	0.16	2.6	75.2	0.9	0.03	1.7	0.01	0.04	0.01	0.08	0.04	0.9	0.2	0	0.3
0	8	0.16	6.9	147.7	1.4	0.05	4.1	0.02	0.24	0.03	0.12	0.05	4.1	6.3	0	0
0	3	0.30	3.4	96.8	18.7	0.04	0	0.00	0.11	0.03	0.20	0.03	0	0.8	0	0.3
0	5	0.44	3.8	77.8	3.8	0.05	1.3	0.01	0.26	0.03	0.24	0.03	1.3	2.2	0	0.4
0	4	0.14	3.7	91.5	2.4	0.03	1.2	0.01	0.25	0.03	0.22	0.03	1.2	1.5	0	0.4
0	6	0.12	2.5	67.9	0.4	—	0.7	0.01	0.20	0.01	0.03	—	2.0	2.8	0	—
0	18	0.54	14.0	362.6	1.4	0.28	134.4	0.04	1.24	0.05	0.84	0.07	12.6	14.0	0	0.1
0	12	0.38	9.0	180.6	5.2	0.14	80.0	0.02	0.77	0.02	0.48	0.07	2.6	4.0	0	0.1
0	18	0.87	10.5	381.5	3.3	0.12	59.1	0.00	1.42	0.02	0.85	0.05	3.3	0.3	0	0.7
0	15	0.66	33.3	583.0	9.2	0.78	8.0	0.08	2.23	0.16	2.19	0.31	102.3	10.1	0	0.4
0	12	0.19	27.6	403.6	2.3	0.45	8.0	0.02	3.03	0.04	0.76	0.08	40.3	20.0	0	—
0	3	0.14	8.0	134.1	1.9	0.17	1.9	0.01	0.57	0.03	0.48	0.07	22.4	2.8	0	0.1
0	10	0.69	41.8	294.8	3.3	0.40	2.2	0.04	0.13	0.01	0.39	0.14	7.7	3.5	0	0.8
0	6	0.30	31.9	422.4	1.2	0.17	3.5	0.03	0.11	0.08	0.78	0.43	23.6	10.3	0	1.2
0	21	0.45	14.4	116.6	0.7	0.38	7.9	0.01	0.84	0.02	0.47	0.02	18.0	15.1	0	0.3
0	22	0.60	16.6	105.7	0.8	0.19	4.5	0.02	0.88	0.03	0.91	0.05	25.7	2.3	0	0.3
0	6	0.42	5.1	51.2	3.8	0.09	2.6	0.04	0.49	0.07	0.14	0.05	2.6	1.4	0	0.1
0	4	0.20	4.4	55.8	0.7	0.12	2.2	0.03	0.41	0.03	0.30	0.04	4.4	7.0	0	0.1
0	6	0.14	3.9	41.9	0.8	0.05	1.6	0.03	0.37	0.03	0.40	0.05	5.4	1.9	0	0.1
0	23	0.55	14.1	115.2	3.8	0.24	2.6	0.03	—	0.04	0.29	0.05	43.5	7.9	0	0.5
0	18	0.56	10.6	91.7	0.7	0.15	2.0	0.04	0.57	0.02	0.51	0.04	41.6	2.0	0	0.1
0	65	2.07	96.0	1881.6	7.7	0.46	0	0.42	0.38	0.11	3.45	0.38	53.8	111.4	0	2.3
0	13	1.67	7.3	119.6	8.5	0.09	46.4	0.02	0.28	0.05	0.22	0.05	9.8	2.6	0	0
0	12	0.25	7.0	134.1	2.3	0.08	49.6	0.02	0.05	0.03	0.31	0.03	6.2	7.8	0	0
0	11	0.44	11.4	183.4	3.8	0.12	10.1	0.02	0.29	0.05	0.50	0.03	5.1	4.6	0	0
0	14	0.45	11.2	162.4	1.2	0.10	9.9	0.03	0.29	0.05	0.51	0.04	5.0	2.7	0	0
0	9	0.26	8.0	161.0	0	0.05	2.2	0.02	0.05	0.02	0.11	0.04	2.9	5.1	0	0
0	4	0.13	3.3	46.8	1.1	0.05	1.7	0.01	0.66	0.01	0.05	0.03	0.6	7.3	0	0.1
0	4	0.09	1.2	20.8	2.5	0.02	0	0.00	0.15	0.00	0.00	0.00	0	48.4	0	0
0	4	0.13	1.3	17.7	2.5	0.04	0	0.00	0.28	0.00	0.05	0.00	0	53.5	0	0.3
0	11	0.05	2.4	29.6	3.6	0.02	0	0.00	0.06	0.00	0.00	0.00	0	38.2	0	0
0	3	0.15	2.1	18.0	20.1	0.03	1.4	0.01	0.57	0.01	0.06	0.01	0.7	1.4	0	0.2
0	17	0.45	19.1	291.9	0.9	0.12	0	0.02	0.02	0.02	0.56	0.07	8.5	0.2	0	1.3
0	17	0.45	19.1	291.9	0.9	0.12	0	0.02	0.02	0.02	0.56	0.07	8.5	0.2	0	1.3
0	35	0.36	13.0	128.2	1.3	0.14	2.6	0.03	0.16	0.05	0.55	0.09	2.6	1.3	0	0.3
0	35	0.36	12.4	127.7	1.2	0.15	2.5	0.03	0.10	0.05	0.55	0.09	2.5	1.2	0	0.1
0	35	0.36	17.0	232.0	1.0	0.14	7.0	0.06	0.10	0.04	0.40	0.10	6.0	2.0	0	0.2
0	7	0.36	6.2	109.1	7.4	0.09	12.4	0.02	0.49	0.02	0.46	0.06	3.7	2.4	0	0.6
0	9	0.25	8.3	112.6	4.7	0.11	17.8	0.01	0.47	0.02	0.48	0.06	3.6	3.2	0	0.6
0	6	0.30	8.3	111.4	4.7	0.11	15.4	0.02	0.47	0.01	0.43	0.06	3.6	2.5	0	0.6
0	9	0.37	6.1	95.6	3.7	0.10	27.0	0.02	—	0.03	0.46	0.04	3.7	2.3	0	1.0
0	20	0.42	7.6	97.0	2.5	0.14	8.8	0.03	—	0.07	0.19	0.02	3.8	12.6	0	0.5

DA+ Code	Food Description	Quantity	Measure	Wt (g)	H₂O (g)	Ener (kcal)	Prot (g)	Carb (g)	Fiber (g)	Fat (g)	Fat Breakdown (g)			
											Sat	Mono	Poly	Trans
981	Raw	½	cup(s)	75	65.9	33	0.7	7.6	3.2	0.4	0	0	0.2	—
	Grapefruit													
251	Juice, pink, sweetened, canned	½	cup(s)	125	109.1	57	0.7	13.9	0.1	0.1	0	0	0	—
249	Juice, white	½	cup(s)	124	111.2	48	0.6	11.4	0.1	0.1	0	0	0	—
3022	Pink or red, raw	½	cup(s)	114	100.8	48	0.9	12.2	1.8	0.2	0	0	0	—
248	Sections, canned in light syrup	½	cup(s)	127	106.2	76	0.7	19.6	0.5	0.1	0	0	0	—
983	Sections, canned in water	½	cup(s)	122	109.6	44	0.7	11.2	0.5	0.1	0	0	0	—
247	White, raw	½	cup(s)	115	104.0	38	0.8	9.7	1.3	0.1	0	0	0	—
	Grapes													
255	American, slip skin	½	cup(s)	46	37.4	31	0.3	7.9	0.4	0.2	0.1	0	0	—
256	European, red or green, adherent skin	½	cup(s)	76	60.8	52	0.5	13.7	0.7	0.1	0	0	0	—
3159	Grape Juice drink, canned	½	cup(s)	125	106.6	71	0	18.2	0.1	0	0	0	0	0
259	Grape Juice, sweetened, with added vitamin C, prepared from frozen concentrate	½	cup(s)	125	108.6	64	0.2	15.9	0.1	0.1	0	0	0	—
3060	Raisins, seeded, packed	¼	cup(s)	41	6.8	122	1.0	32.4	2.8	0.2	0.1	0	0.1	—
987	Guava, raw	1	item(s)	55	44.4	37	1.4	7.9	3.0	0.5	0.2	0	0.2	—
35593	Guavas, strawberry	1	item(s)	6	4.8	4	0	1.0	0.3	0	0	0	0	—
3027	Jackfruit	½	cup(s)	83	60.4	78	1.2	19.8	1.3	0.2	0	0	0.1	—
990	Kiwi fruit or Chinese gooseberries	1	item(s)	76	63.1	46	0.9	11.1	2.3	0.4	0	0	0.2	—
	Lemon													
262	Juice	1	tablespoon(s)	15	13.8	4	0.1	1.3	0.1	0	0	0	0	0
993	Peel	1	teaspoon(s)	2	1.6	1	0	0.3	0.2	0	0	0	0	—
992	Raw	1	item(s)	108	94.4	22	1.3	11.6	5.1	0.3	0	0	0.1	—
	Lime													
269	Juice	1	tablespoon(s)	15	14.0	4	0.1	1.3	0.1	0	0	0	0	—
994	Raw	1	item(s)	67	59.1	20	0.5	7.1	1.9	0.1	0	0	0	—
995	Loganberries, frozen	½	cup(s)	74	62.2	40	1.1	9.6	3.9	0.2	0	0	0.1	—
	Mandarin orange													
1038	Canned in juice	½	cup(s)	125	111.4	46	0.8	11.9	0.9	0	0	0	0	—
1039	Canned in light syrup	½	cup(s)	126	104.7	77	0.6	20.4	0.9	0.1	0	0	0	—
999	Mango	½	cup(s)	83	67.4	54	0.4	14.0	1.5	0.2	0.1	0.1	0	—
1005	Nectarine, raw, sliced	½	cup(s)	69	60.4	30	0.7	7.3	1.2	0.2	0	0.1	0.1	—
	Melons													
271	Cantaloupe	½	cup(s)	80	72.1	27	0.7	6.5	0.7	0.1	0	0	0.1	—
1000	Casaba melon	½	cup(s)	85	78.1	24	0.9	5.6	0.8	0.1	0	0	0	—
272	Honeydew	½	cup(s)	89	79.5	32	0.5	8.0	0.7	0.1	0	0	0	—
318	Watermelon	½	cup(s)	76	69.5	23	0.5	5.7	0.3	0.1	0	0	0	—
	Orange													
14412	Juice with calcium and vitamin D	½	cup(s)	120	—	55	1.0	13.0	0	0	0	0	0	0
29630	Juice, fresh squeezed	½	cup(s)	124	109.5	56	0.9	12.9	0.2	0.2	0	0	0	—
14411	Juice, not from concentrate	½	cup(s)	120	—	55	1.0	13.0	0	0	0	0	0	0
278	Juice, unsweetened, prepared from frozen concentrate	½	cup(s)	125	109.7	56	0.8	13.4	0.2	0.1	0	0	0	—
3040	Peel	1	teaspoon(s)	2	1.5	2	0	0.5	0.2	0	0	0	0	—
273	Raw	1	item(s)	131	113.6	62	1.2	15.4	3.1	0.2	0	0	0	—
274	Sections	½	cup(s)	90	78.1	42	0.8	10.6	2.2	0.1	0	0	0	—
	Papaya, Raw													
16830	Dried, strips	2	item(s)	46	12.0	119	1.9	29.9	5.5	0.4	0.1	0.1	0.1	—
282	Papaya, Raw	½	cup(s)	70	62.2	27	0.4	6.9	1.3	0.1	0	0	0	—
35640	Passion fruit, purple	1	item(s)	18	13.1	17	0.4	4.2	1.9	0.1	0	0	0.1	—
	Peach													
285	Halves, canned in heavy syrup	½	cup(s)	131	103.9	97	0.6	26.1	1.7	0.1	0	0	0.1	—
286	Halves, canned in water	½	cup(s)	122	113.6	29	0.5	7.5	1.6	0.1	0	0	0	—
290	Slices, sweetened, frozen	½	cup(s)	125	93.4	118	0.8	30.0	2.3	0.2	0	0.1	0.1	—
283	Raw, medium	1	item(s)	150	133.3	59	1.4	14.3	2.3	0.4	0	0.1	0.1	—
	Pear													
8672	Asian	1	item(s)	122	107.7	51	0.6	13.0	4.4	0.3	0	0.1	0.1	—
293	D'Anjou	1	item(s)	200	168.0	120	1.0	30.0	5.2	1.0	0	0.2	0.2	—
294	Halves, canned in heavy syrup	½	cup(s)	133	106.9	98	0.3	25.5	2.1	0.2	0	0	0	—
1012	Halves, canned in juice	½	cup(s)	124	107.2	62	0.4	16.0	2.0	0.1	0	0	0	—
291	Raw	1	item(s)	166	139.0	96	0.6	25.7	5.1	0.2	0	0	0	—
1017	Persimmon	1	item(s)	25	16.1	32	0.2	8.4	—	0.1	0	0	0	—
	Pineapple													
3053	Canned in extra heavy syrup	½	cup(s)	130	101.0	108	0.4	28.0	1.0	0.1	0	0	0	—
1019	Canned in juice	½	cup(s)	125	104.0	75	0.5	19.5	1.0	0.1	0	0	0	—
296	Canned in light syrup	½	cup(s)	126	108.0	66	0.5	16.9	1.0	0.2	0	0	0.1	—
1018	Canned in water	½	cup(s)	123	111.7	39	0.5	10.2	1.0	0.1	0	0	0	—
299	Juice, unsweetened, canned	½	cup(s)	125	108.0	66	0.5	16.1	0.3	0.2	0	0	0.1	—
295	Raw, diced	½	cup(s)	78	66.7	39	0.4	10.2	1.1	0.1	0	0	0	—
	FRUIT AND FRUIT JUICES —Continued													
1024	**Plantain, cooked**	½	cup(s)	77	51.8	89	0.6	24.0	1.8	0.1	0.1	0	0	—

Chol (mg)	Calc (mg)	Iron (mg)	Magn (mg)	Pota (mg)	Sodi (mg)	Zinc (mg)	Vit A (µg)	Thia (mg)	Vit E (mg α)	Ribo (mg)	Niac (mg)	Vit B$_6$ (mg)	Fola (µg)	Vit C (mg)	Vit B$_{12}$ (µg)	Sele (µg)
0	19	0.23	7.5	148.5	0.8	0.09	11.3	0.03	0.28	0.02	0.23	0.06	4.5	20.8	0	0.5
0	10	0.45	12.5	202.2	2.5	0.08	0	0.05	0.05	0.03	0.40	0.03	12.5	33.6	0	0.1
0	11	0.25	14.8	200.1	1.2	0.06	1.2	0.05	0.27	0.02	0.25	0.05	12.4	46.9	0	0.1
0	25	0.09	10.3	154.5	0	0.07	66.4	0.04	0.14	0.03	0.23	0.06	14.9	35.7	0	0.1
0	18	0.50	12.7	163.8	2.5	0.10	0	0.04	0.11	0.02	0.30	0.02	11.4	27.1	0	1.1
0	18	0.50	12.2	161.0	2.4	0.11	0	0.05	0.11	0.03	0.30	0.02	11.0	26.6	0	1.1
0	14	0.07	10.4	170.2	0	0.08	2.3	0.04	0.15	0.02	0.30	0.05	11.5	38.3	0	1.6
0	6	0.13	2.3	87.9	0.9	0.02	2.3	0.04	0.09	0.02	0.14	0.05	1.8	1.8	0	0
0	8	0.27	5.3	144.2	1.5	0.05	2.3	0.05	0.14	0.05	0.14	0.07	1.5	8.2	0	0.1
0	9	0.16	7.5	41.3	11.3	0.04	0	0.28	0.00	0.44	0.18	0.04	1.3	33.1	0	0.1
0	5	0.13	5.0	26.3	2.5	0.05	0	0.02	0.00	0.03	0.16	0.05	1.3	29.9	0	0.1
0	12	1.06	12.4	340.3	11.6	0.07	0	0.04	—	0.07	0.46	0.07	1.2	2.2	0	0.2
0	10	0.14	12.1	229.4	1.1	0.12	17.1	0.03	0.40	0.02	0.59	0.06	27.0	125.6	0	0.3
0	1	0.01	1.0	17.5	2.2	—	0.3	0.00	—	0.00	0.03	0.00	—	2.2	0	
0	28	0.49	30.5	250.0	2.5	0.35	12.4	0.02	—	0.09	0.33	0.09	11.5	5.5	0	0.5
0	26	0.23	12.9	237.1	2.3	0.10	3.0	0.02	1.11	0.01	0.25	0.04	19.0	70.5	0	0.2
0	1	0.00	0.9	18.9	0.2	0.01	0.2	0.00	0.02	0.00	0.02	0.01	2.0	7.0	0	0
0	3	0.01	0.3	3.2	0.1	0.01	0.1	0.00	0.01	0.00	0.01	0.00	0.3	2.6	0	0
0	66	0.75	13.0	156.6	3.2	0.10	2.2	0.05	—	0.04	0.21	0.11	—	83.2	0	1.0
0	2	0.02	1.2	18.0	0.3	0.01	0.3	0.00	0.03	0.00	0.02	0.01	1.5	4.6	0	0
0	22	0.40	4.0	68.3	1.3	0.07	1.3	0.02	0.14	0.01	0.13	0.02	5.4	19.5	0	0.3
0	19	0.47	15.4	106.6	0.7	0.25	1.5	0.04	0.64	0.03	0.62	0.05	19.1	11.2	0	0.1
0	14	0.34	13.7	165.6	6.2	0.64	53.5	0.10	0.12	0.04	0.55	0.05	6.2	42.6	0	0.5
0	9	0.47	10.1	98.3	7.6	0.30	52.9	0.07	0.13	0.06	0.56	0.05	6.3	24.9	0	0.5
0	8	0.10	7.4	128.7	1.7	0.03	31.4	0.05	0.92	0.04	0.48	0.11	11.6	22.8	0	0.5
0	4	0.19	6.2	138.7	0	0.12	11.7	0.02	0.53	0.02	0.78	0.02	3.5	3.7	0	0
0	7	0.17	9.6	213.6	12.8	0.14	135.2	0.03	0.04	0.01	0.59	0.05	16.8	29.4	0	0.3
0	9	0.29	9.4	154.7	7.7	0.06	0	0.01	0.04	0.03	0.20	0.14	6.8	18.5	0	0.3
0	5	0.15	8.8	201.8	15.9	0.07	2.7	0.03	0.01	0.01	0.37	0.07	16.8	15.9	0	0.6
0	5	0.18	7.6	85.1	0.8	0.07	21.3	0.02	0.04	0.01	0.13	0.03	2.3	6.2	0	0.3
0	175	0.00	12.0	225.0	0	—	0	0.08	—	0.03	0.40	0.06	30.0	36.0	0	—
0	14	0.25	13.6	248.0	1.2	0.06	12.4	0.11	0.05	0.04	0.50	0.05	37.2	62.0	0	0.1
0	10	0.00	12.5	225.0	0	0.06	0	0.08	—	0.03	0.40	0.06	30.0	36.0	0	0.1
0	11	0.12	12.5	236.6	1.2	0.06	6.2	0.10	0.25	0.02	0.25	0.06	54.8	48.4	0	0.1
0	3	0.01	0.4	4.2	0.1	0.01	0.4	0.00	0.01	0.00	0.01	0.00	0.6	2.7	0	0
0	52	0.13	13.1	237.1	0	0.09	14.4	0.11	0.23	0.05	0.36	0.07	39.3	69.7	0	0.7
0	36	0.09	9.0	162.9	0	0.06	9.9	0.07	0.16	0.03	0.25	0.05	27.0	47.9	0	0.4
0	73	0.30	30.4	782.9	9.2	0.21	83.7	0.06	2.22	0.08	0.93	0.05	58.0	37.7	0	1.8
0	17	0.07	7.0	179.9	2.1	0.05	38.5	0.02	0.51	0.02	0.24	0.01	26.6	43.3	0	0.4
0	2	0.28	5.2	62.6	5.0	0.01	11.5	0.00	0.00	0.02	0.27	0.01	2.5	5.4	0	0.1
0	4	0.35	6.6	120.5	7.9	0.11	22.3	0.01	0.64	0.03	0.80	0.02	3.9	3.7	0	0.4
0	2	0.39	6.1	120.8	3.7	0.11	32.9	0.01	0.59	0.02	0.63	0.02	3.7	3.5	0	0.4
0	4	0.46	6.3	162.5	7.5	0.06	17.5	0.01	0.77	0.04	0.81	0.02	3.8	117.8	0	0.5
0	9	0.37	13.5	285.0	0	0.25	24.0	0.03	1.09	0.04	1.20	0.03	6.0	9.9	0	0.2
0	5	0.00	9.8	147.6	0	0.02	0	0.01	0.14	0.01	0.26	0.02	9.8	4.6	0	0.1
0	22	0.50	12.0	250.0	0	0.24	—	0.04	1.00	0.08	0.20	0.03	14.6	8.0	0	1.0
0	7	0.29	5.3	86.5	6.7	0.10	0	0.01	0.10	0.02	0.32	0.01	1.3	1.5	0	0
0	11	0.36	8.7	119.0	5.0	0.11	0	0.01	0.10	0.01	0.25	0.02	1.2	2.0	0	0
0	15	0.28	11.6	197.5	1.7	0.16	1.7	0.02	0.19	0.04	0.26	0.04	11.6	7.0	0	0.2
0	7	0.62	—	77.5	0.3	—	0	—	—	—	—	—	—	16.5	0	—
0	18	0.49	19.5	132.6	1.3	0.14	1.3	0.11	—	0.03	0.36	0.09	6.5	9.5	0	—
0	17	0.35	17.4	151.9	1.2	0.12	2.5	0.12	0.01	0.02	0.35	0.09	6.2	11.8	0	0.5
0	18	0.49	20.2	132.3	1.3	0.15	2.5	0.11	0.01	0.03	0.36	0.09	6.3	9.5	0	0.5
0	18	0.49	22.1	156.2	1.2	0.15	2.5	0.11	0.01	0.03	0.37	0.09	6.2	9.5	0	0.5
0	16	0.39	15.0	162.5	2.5	0.14	0	0.07	0.03	0.03	0.25	0.13	22.5	12.5	0	0.1
0	10	0.22	9.3	84.5	0.8	0.09	2.3	0.06	0.02	0.03	0.39	0.09	14.0	37.0	0	0.1
0	2	0.45	24.6	358.1	3.9	0.10	34.7	0.04	0.10	0.04	0.58	0.19	20.0	8.4	0	1.1

TABLE A-1   Food Composition

(Computer code is for Cengage Diet Analysis program)   (For purposes of calculations, use "0" for t, <1, <.1, <.01, etc.)

DA+ Code	Food Description	Quantity	Measure	Wt (g)	H₂O (g)	Ener (kcal)	Prot (g)	Carb (g)	Fiber (g)	Fat (g)	Sat	Mono	Poly	Trans
											Fat Breakdown (g)			
300	**Plum, raw, large**	1	item(s)	66	57.6	30	0.5	7.5	0.9	0.2	0	0.1	0	—
1027	**Pomegranate**	1	item(s)	154	124.7	105	1.5	26.4	0.9	0.5	0.1	0.1	0.1	—
	Prunes, Dried													
5644	Dried	2	item(s)	17	5.2	40	0.4	10.7	1.2	0.1	0	0	0	—
305	Dried, stewed	½	cup(s)	124	86.5	133	1.2	34.8	3.8	0.2	0	0.1	0	—
306	Juice, canned	1	cup(s)	256	208.0	182	1.6	44.7	2.6	0.1	0	0.1	0	—
	Raspberries													
309	Raw	½	cup(s)	62	52.7	32	0.7	7.3	4.0	0.4	0	0	0.2	—
310	Red, sweetened, frozen	½	cup(s)	125	90.9	129	0.9	32.7	5.5	0.2	0	0	0.1	—
311	**Rhubarb, cooked with sugar**	½	cup(s)	120	81.5	140	0.5	37.5	2.7	0.1	0	0	0.1	—
	Strawberries													
313	Raw	½	cup(s)	72	65.5	23	0.5	5.5	1.4	0.2	0	0	0.1	—
315	Sweetened, frozen, thawed	½	cup(s)	128	99.5	99	0.7	26.8	2.4	0.2	0	0	0.1	—
16828	**Tangelo**	1	item(s)	95	82.4	45	0.9	11.2	2.3	0.1	0	0	0	—
	Tangerine													
1040	Juice	½	cup(s)	124	109.8	53	0.6	12.5	0.2	0.2	0	0	0	—
316	Raw	1	item(s)	88	74.9	47	0.7	11.7	1.6	0.3	0	0.1	0.1	—
	VEGETABLES, LEGUMES													
	Amaranth													
1043	Leaves, boiled, drained	½	cup(s)	66	60.4	14	1.4	2.7	—	0.1	0	0	0.1	—
1042	Leaves, raw	1	cup(s)	28	25.7	6	0.7	1.1	—	0.1	0	0	0	—
8683	**Arugula leaves, raw**	1	cup(s)	20	18.3	5	0.5	0.7	0.3	0.1	0	0	0.1	—
	Artichoke													
1044	Boiled, drained	1	item(s)	120	100.9	64	3.5	14.3	10.3	0.4	0.1	0	0.2	—
2885	Hearts, boiled, drained	½	cup(s)	84	70.6	45	2.4	10.0	7.2	0.3	0.1	0	0.1	—
	Asparagus													
566	Boiled, drained	½	cup(s)	90	83.4	20	2.2	3.7	1.8	0.2	0	0	0.1	—
568	Canned, drained	½	cup(s)	121	113.7	23	2.6	3.0	1.9	0.8	0.2	0	0.3	—
565	Tips, frozen, boiled, drained	½	cup(s)	90	84.7	16	2.7	1.7	1.4	0.4	0.1	0	0.2	—
	Bamboo shoots													
1048	Boiled, drained	½	cup(s)	60	57.6	7	0.9	1.2	0.6	0.1	0	0	0.1	—
1049	Canned, drained	½	cup(s)	66	61.8	12	1.1	2.1	0.9	0.3	0.1	0	0.1	—
	Beans													
1801	Adzuki beans, boiled	½	cup(s)	115	76.2	147	8.6	28.5	8.4	0.1	0	—	—	—
511	Baked beans with franks, canned	½	cup(s)	130	89.8	184	8.7	19.9	8.9	8.5	3.0	3.7	1.1	—
513	Baked beans with pork in sweet sauce, canned	½	cup(s)	127	89.3	142	6.7	26.7	5.3	1.8	0.6	0.6	0.5	0
512	Baked beans with pork in tomato sauce, canned	½	cup(s)	127	93.0	119	6.5	23.6	5.1	1.2	0.5	0.7	0.3	—
1805	Black beans, boiled	½	cup(s)	86	56.5	114	7.6	20.4	7.5	0.5	0.1	0	0.2	—
14597	Chickpeas, garbanzo beans or bengal gram, boiled	½	cup(s)	82	49.4	134	7.3	22.5	6.2	2.1	0.2	0.5	0.9	—
569	Fordhook lima beans, frozen, boiled, drained	½	cup(s)	85	62.0	88	5.2	16.4	4.9	0.3	0.1	0	0.1	—
1806	French beans, boiled	½	cup(s)	89	58.9	114	6.2	21.3	8.3	0.7	0.1	0	0.4	—
2773	Great northern beans, boiled	½	cup(s)	89	61.1	104	7.4	18.7	6.2	0.4	0.1	0	0.2	—
2736	Hyacinth beans, boiled, drained	½	cup(s)	44	37.8	22	1.3	4.0	—	0.1	0.1	0.1	0	—
570	Lima beans, baby, frozen, boiled, drained	½	cup(s)	90	65.1	95	6.0	17.5	5.4	0.3	0.1	0	0.1	—
515	Lima beans, boiled, drained	½	cup(s)	85	57.1	105	5.8	20.1	4.5	0.3	0.1	0	0.1	—
579	Mung beans, sprouted, boiled, drained	½	cup(s)	62	57.9	13	1.3	2.6	0.5	0.1	0	0	0	—
510	Navy beans, boiled	½	cup(s)	91	58.1	127	7.5	23.7	9.6	0.6	0.1	0.1	0.4	0
32816	Pinto beans, boiled, drained, no salt added	½	cup(s)	63	58.8	14	1.2	2.6	—	0.2	0	0	0.1	—
1052	Pinto beans, frozen, boiled, drained	½	cup(s)	47	27.3	76	4.4	14.5	4.0	0.2	0	0	0.1	—
514	Red kidney beans, canned	½	cup(s)	128	99.0	108	6.7	19.9	6.9	0.5	0.1	0.2	0.2	—
1810	Refried beans, canned	½	cup(s)	127	96.1	119	6.9	19.6	6.7	1.6	0.6	0.7	0.2	—
1053	Shell beans, canned	½	cup(s)	123	111.1	37	2.2	7.6	4.2	0.2	0	0	0.1	—
1670	Soybeans, boiled	½	cup(s)	86	53.8	149	14.3	8.5	5.2	7.7	1.1	1.7	4.4	—
1108	Soybeans, green, boiled, drained	½	cup(s)	90	61.7	127	11.1	9.9	3.8	5.8	0.7	1.1	2.7	—
1807	White beans, small, boiled	½	cup(s)	90	56.6	127	8.0	23.1	9.3	0.6	0.1	0.1	0.2	—
575	Yellow snap, string or wax beans, boiled, drained	½	cup(s)	63	55.8	22	1.2	4.9	2.1	0.2	0	0	0.1	—
576	Yellow snap, string or wax beans, frozen, boiled, drained	½	cup(s)	68	61.7	19	1.0	4.4	2.0	0.1	0	0	0.1	—
	Beets													
584	Beet greens, boiled, drained	½	cup(s)	72	64.2	19	1.9	3.9	2.1	0.1	0	0	0.1	—
2730	Pickled, canned with liquid	½	cup(s)	114	92.9	74	0.9	18.5	3.0	0.1	0	0	0	—
581	Sliced, boiled, drained	½	cup(s)	85	74.0	37	1.4	8.5	1.7	0.2	0	0	0.1	—
583	Sliced, canned, drained	½	cup(s)	85	77.3	26	0.8	6.1	1.5	0.1	0	0	0	—
580	Whole, boiled, drained	2	item(s)	100	87.1	44	1.7	10.0	2.0	0.2	0	0	0.1	—
585	**Cowpeas or black-eyed peas, boiled, drained**	½	cup(s)	83	62.3	80	2.6	16.8	4.1	0.3	0.1	0	0.1	—
	Broccoli													
588	Chopped, boiled, drained	½	cup(s)	78	69.6	27	1.9	5.6	2.6	0.3	0.1	0	0.1	—
	VEGETABLES, LEGUMES—Continued													

Chol (mg)	Calc (mg)	Iron (mg)	Magn (mg)	Pota (mg)	Sodi (mg)	Zinc (mg)	Vit A (µg)	Thia (mg)	Vit E (mg α)	Ribo (mg)	Niac (mg)	Vit B_6 (mg)	Fola (µg)	Vit C (mg)	Vit B_{12} (µg)	Sele (µg)
0	4	0.11	4.6	103.6	0	0.06	11.2	0.02	0.17	0.02	0.27	0.02	3.3	6.3	0	0
0	5	0.46	4.6	398.9	4.6	0.18	7.7	0.04	0.92	0.04	0.46	0.16	9.2	9.4	0	0.9
0	7	0.16	6.9	123.0	0.3	0.07	6.6	0.01	0.07	0.03	0.32	0.03	0.7	0.1	0	0
0	24	0.51	22.3	398.0	1.2	0.24	21.1	0.03	0.24	0.12	0.90	0.27	0	3.6	0	0.1
0	31	3.02	35.8	706.6	10.2	0.53	0	0.04	0.30	0.17	2.01	0.55	0	10.5	0	1.5
0	15	0.42	13.5	92.9	0.6	0.26	1.2	0.02	0.54	0.02	0.37	0.03	12.9	16.1	0	0.1
0	19	0.81	16.3	142.5	1.3	0.22	3.8	0.02	0.90	0.05	0.28	0.04	32.5	20.6	0	0.4
0	174	0.25	16.2	115.0	1.0	—	—	0.02	—	0.03	0.25	—	—	4.0	0	—
0	12	0.30	9.4	110.2	0.7	0.10	0.7	0.02	0.21	0.02	0.28	0.03	17.3	42.3	0	0.3
0	14	0.59	7.7	125.0	1.3	0.06	1.3	0.01	0.30	0.09	0.37	0.03	5.1	50.4	0	0.9
0	38	0.09	9.5	172.0	0	0.06	10.5	0.08	0.17	0.03	0.26	0.05	28.5	50.5	0	0.5
0	22	0.25	9.9	219.8	1.2	0.04	16.1	0.07	0.16	0.02	0.12	0.05	6.2	38.3	0	0.1
0	33	0.13	10.6	146.1	1.8	0.06	29.9	0.05	0.18	0.03	0.33	0.07	14.1	23.5	0	0.1
0	138	1.49	36.3	423.1	13.9	0.58	91.7	0.01	—	0.09	0.37	0.12	37.6	27.1	0	0.6
0	60	0.65	15.4	171.1	5.6	0.25	40.9	0.01	—	0.04	0.18	0.05	23.8	12.1	0	0.3
0	32	0.29	9.4	73.8	5.4	0.09	23.8	0.01	0.09	0.02	0.06	0.01	19.4	3.0	0	0.1
0	25	0.73	50.4	343.2	72.0	0.48	1.2	0.06	0.22	0.10	1.33	0.09	106.8	8.9	0	0.2
0	18	0.51	35.3	240.2	50.4	0.33	0.8	0.04	0.16	0.07	0.93	0.06	74.8	6.2	0	0.2
0	21	0.81	12.6	201.6	12.6	0.54	45.0	0.14	1.35	0.12	0.97	0.07	134.1	6.9	0	5.5
0	19	2.21	12.1	208.1	347.3	0.48	49.6	0.07	1.47	0.12	1.15	0.13	116.2	22.3	0	2.1
0	16	0.50	9.0	154.8	2.7	0.36	36.0	0.05	1.08	0.09	0.93	0.01	121.5	22.0	0	3.5
0	7	0.14	1.8	319.8	2.4	0.28	0	0.01	—	0.03	0.18	0.06	1.2	0	0	0.2
0	5	0.21	2.6	52.4	4.6	0.43	0.7	0.02	0.41	0.02	0.09	0.09	2.0	0.7	0	0.3
0	32	2.30	59.8	611.8	9.2	2.03	0	0.13	—	0.07	0.82	0.11	139.2	0	0	1.4
8	62	2.24	36.3	304.3	556.9	2.42	5.2	0.08	0.21	0.07	1.17	0.06	38.9	3.0	0.4	8.4
9	75	2.08	41.7	326.4	422.5	1.73	0	0.05	0.03	0.07	0.44	0.07	10.1	3.5	0	6.3
9	71	4.09	43.0	373.2	552.8	6.93	5.1	0.06	0.12	0.05	0.62	0.08	19.0	3.8	0	5.9
0	23	1.80	60.2	305.3	0.9	0.96	0	0.21	—	0.05	0.43	0.05	128.1	0	0	1.0
0	40	2.36	39.4	238.6	5.7	1.25	0.8	0.09	0.28	0.05	0.43	0.11	141.0	1.1	0	3.0
0	26	1.54	35.7	258.4	58.7	0.62	8.5	0.06	0.24	0.05	0.90	0.10	17.9	10.9	0	0.5
0	56	0.95	49.6	327.5	5.3	0.56	0	0.11	—	0.05	0.48	0.09	66.4	1.1	0	1.1
0	60	1.88	44.3	346.0	1.8	0.77	0	0.14	—	0.05	0.60	0.10	90.3	1.2	0	3.6
0	18	0.33	18.3	114.0	0.9	0.16	3.0	0.02	—	0.03	0.20	0.01	20.4	2.2	0	0.7
0	25	1.76	50.4	369.9	26.1	0.49	7.2	0.06	0.57	0.04	0.69	0.10	14.4	5.2	0	1.5
0	27	2.08	62.9	484.5	14.5	0.67	12.8	0.11	0.11	0.08	0.88	0.16	22.1	8.6	0	1.7
0	7	0.40	8.7	62.6	6.2	0.29	0.6	0.03	0.04	0.06	0.51	0.03	18.0	7.1	0	0.4
—	63	2.14	48.2	354.0	0	0.93	0	0.21	0.01	0.06	0.59	0.12	127.4	0.8	0	2.6
0	9	0.41	11.3	61.7	32.1	0.10	0	0.04	—	0.03	0.45	0.03	18.3	3.8	0	0.4
0	24	1.27	25.4	303.6	39.0	0.32	0	0.12	—	0.05	0.29	0.09	16.0	0.3	0	0.7
0	32	1.62	35.8	327.7	330.2	2.09	0	0.13	0.02	0.11	0.57	0.10	25.6	1.4	0	0.6
10	44	2.10	41.7	337.8	378.2	1.48	0	0.03	0.00	0.02	0.39	0.18	13.9	7.6	0	1.6
0	36	1.21	18.4	133.5	409.2	0.33	13.5	0.04	0.04	0.07	0.25	0.06	22.1	3.8	0	2.6
0	88	4.42	74.0	442.9	0.9	0.98	0	0.13	0.30	0.24	0.34	0.20	46.4	1.5	0	6.3
0	131	2.25	54.0	485.1	12.6	0.82	7.2	0.23	—	0.14	1.13	0.05	99.9	15.3	0	1.3
0	65	2.54	60.9	414.4	1.8	0.97	0	0.21	—	0.05	0.24	0.11	122.6	0	0	1.2
0	29	0.80	15.6	186.9	1.9	0.23	2.5	0.05	0.28	0.06	0.38	0.04	20.6	6.1	0	0.3
0	33	0.59	16.2	85.1	6.1	0.32	4.1	0.02	0.03	0.06	0.26	0.04	15.5	2.8	0	0.3
0	82	1.36	49.0	654.5	173.5	0.36	275.8	0.08	1.30	0.20	0.35	0.09	10.1	17.9	0	0.6
0	12	0.46	17.0	168.0	299.6	0.29	1.1	0.01	—	0.05	0.28	0.05	30.6	2.6	0	1.1
0	14	0.67	19.6	259.3	65.5	0.30	1.7	0.02	0.03	0.03	0.28	0.05	68.0	3.1	0	0.6
0	13	1.54	14.5	125.8	164.9	0.17	0.9	0.01	0.02	0.03	0.13	0.04	25.5	3.5	0	0.4
0	16	0.79	23.0	305.0	77.0	0.35	2.0	0.02	0.04	0.04	0.33	0.06	80.0	3.6	0	0.7
0	106	0.92	42.9	344.9	3.3	0.85	33.0	0.08	0.18	0.12	1.15	0.05	104.8	1.8	0	2.1
0	31	0.52	16.4	228.5	32.0	0.35	60.1	0.04	1.13	0.09	0.43	0.15	84.2	50.6	0	1.2

DA+ Code	Food Description	Quantity	Measure	Wt (g)	H₂O (g)	Ener (kcal)	Prot (g)	Carb (g)	Fiber (g)	Fat (g)	Fat Breakdown (g) Sat	Mono	Poly	Trans
590	Frozen, chopped, boiled, drained	½	cup(s)	92	83.5	26	2.9	4.9	2.8	0.1	0	0	0.1	—
587	Raw, chopped	½	cup(s)	46	40.6	15	1.3	3.0	1.2	0.2	0	0	0	—
16848	**Broccoflower, raw, chopped**	½	cup(s)	32	28.7	10	0.9	1.9	1.0	0.1	0	0	0	—
	Brussels sprouts													
591	Boiled, drained	½	cup(s)	78	69.3	28	2.0	5.5	2.0	0.4	0.1	0	0.2	—
592	Frozen, boiled, drained	½	cup(s)	78	67.2	33	2.8	6.4	3.2	0.3	0.1	0	0.2	—
	Cabbage													
595	Boiled, drained, no salt added	1	cup(s)	150	138.8	35	1.9	8.3	2.8	0.1	0	0	0	—
35611	Chinese (pak choi or bok choy), boiled with salt, drained	1	cup(s)	170	162.4	20	2.6	3.0	1.7	0.3	0	0	0.1	—
16869	Kim chee	1	cup(s)	150	137.5	32	2.5	6.1	1.8	0.3	0	0	0.2	—
594	Raw, shredded	1	cup(s)	70	64.5	17	0.9	4.1	1.7	0.1	0	0	0	—
596	Red, shredded, raw	1	cup(s)	70	63.3	22	1.0	5.2	1.5	0.1	0	0	0.1	—
597	Savoy, shredded, raw	1	cup(s)	70	63.7	19	1.4	4.3	2.2	0.1	0	0	0	—
35417	**Capers**	1	teaspoon(s)	4	—	2	0	0	0	0	0	0	0	0
	Carrots													
8691	Baby, raw	8	item(s)	80	72.3	28	0.5	6.6	2.3	0.1	0	0	0.1	—
601	Grated	½	cup(s)	55	48.6	23	0.5	5.3	1.5	0.1	0	0	0.1	0
1055	Juice, canned	½	cup(s)	118	104.9	47	1.1	11.0	0.9	0.2	0	0	0.1	—
600	Raw	½	cup(s)	61	53.9	25	0.6	5.8	1.7	0.1	0	0	0.1	0
602	Sliced, boiled, drained	½	cup(s)	78	70.3	27	0.6	6.4	2.3	0.1	0	0	0.1	—
32725	**Cassava or manioc**	½	cup(s)	103	61.5	165	1.4	39.2	1.9	0.3	0.1	0.1	0	—
	Cauliflower													
606	Boiled, drained	½	cup(s)	62	57.7	14	1.1	2.5	1.4	0.3	0	0	0.1	—
607	Frozen, boiled, drained	½	cup(s)	90	84.6	17	1.4	3.4	2.4	0.2	0	0	0.1	—
605	Raw, chopped	½	cup(s)	50	46.0	13	1.0	2.6	1.2	0	0	0	0	—
	Celery													
609	Diced	½	cup(s)	51	48.2	8	0.3	1.5	0.8	0.1	0	0	0	—
608	Stalk	2	item(s)	80	76.3	13	0.6	2.4	1.3	0.1	0	0	0.1	—
	Chard													
1057	Swiss chard, boiled, drained	½	cup(s)	88	81.1	18	1.6	3.6	1.8	0.1	0	0	0	—
1056	Swiss chard, raw	1	cup(s)	36	33.4	7	0.6	1.3	0.6	0.1	0	0	0	—
	Collard greens													
610	Boiled, drained	½	cup(s)	95	87.3	25	2.0	4.7	2.7	0.3	0	0	0.2	—
611	Frozen, chopped, boiled, drained	½	cup(s)	85	75.2	31	2.5	6.0	2.4	0.3	0.1	0	0.2	—
	Corn													
29614	Yellow corn, fresh, cooked	1	item(s)	100	69.2	107	3.3	25.0	2.8	1.3	0.2	0.4	0.6	—
615	Yellow creamed sweet corn, canned	½	cup(s)	128	100.8	92	2.2	23.2	1.5	0.5	0.1	0.2	0.3	—
612	Yellow sweet corn, boiled, drained	½	cup(s)	82	57.0	89	2.7	20.6	2.3	1.1	0.2	0.3	0.5	—
614	Yellow sweet corn, frozen, boiled, drained	½	cup(s)	82	63.2	66	2.1	15.8	2.0	0.5	0.1	0.2	0.3	—
618	**Cucumber**	¼	item(s)	75	71.7	11	0.5	2.7	0.4	0.1	0	0	0	—
16870	**Cucumber, kim chee**	½	cup(s)	75	68.1	16	0.8	3.6	1.1	0.1	0	0	0	—
	Dandelion greens													
620	Chopped, boiled, drained	½	cup(s)	53	47.1	17	1.1	3.4	1.5	0.3	0.1	0	0.1	—
2734	Raw	1	cup(s)	55	47.1	25	1.5	5.1	1.9	0.4	0.1	0	0.2	—
1066	**Eggplant, boiled, drained**	½	cup(s)	50	44.4	17	0.4	4.3	1.2	0.1	0	0	0	—
621	**Endive or escarole, chopped, raw**	1	cup(s)	50	46.9	8	0.6	1.7	1.5	0.1	0	0	0	—
8784	**Jicama or yambean**	½	cup(s)	65	116.5	49	0.9	11.4	6.3	0.1	0	0	0.1	—
	Kale													
623	Frozen, chopped, boiled, drained	½	cup(s)	65	58.8	20	1.8	3.4	1.3	0.3	0	0	0.2	—
29313	Raw	1	cup(s)	67	56.6	33	2.2	6.7	1.3	0.5	0.1	0	0.2	—
	Kohlrabi													
1072	Boiled, drained	½	cup(s)	83	74.5	24	1.5	5.5	0.9	0.1	0	0	0	—
1071	Raw	1	cup(s)	135	122.9	36	2.3	8.4	4.9	0.1	0	0	0.1	—
	Leeks													
1074	Boiled, drained	½	cup(s)	52	47.2	16	0.4	4.0	0.5	0.1	0	0	0	—
1073	Raw	1	cup(s)	89	73.9	54	1.3	12.6	1.6	0.3	0	0	0.1	—
	Lentils													
522	Boiled	¼	cup(s)	50	34.5	57	4.5	10.0	3.9	0.2	0	0	0.1	—
1075	Sprouted	1	cup(s)	77	51.9	82	6.9	17.0	—	0.4	0	0.1	0.2	—
	Lettuce													
625	Butterhead leaves	11	piece(s)	83	78.9	11	1.1	1.8	0.9	0.2	0	0	0.1	—
624	Butterhead, Boston or Bibb	1	cup(s)	55	52.6	7	0.7	1.2	0.6	0.1	0	0	0.1	—
626	Iceberg	1	cup(s)	55	52.6	8	0.5	1.6	0.7	0.1	0	0	0	—
628	Iceberg, chopped	1	cup(s)	55	52.6	8	0.5	1.6	0.7	0.1	0	0	0	—
629	Looseleaf	1	cup(s)	36	34.2	5	0.5	1.0	0.5	0.1	0	0	0	—
1665	Romaine, shredded	1	cup(s)	56	53.0	10	0.7	1.8	1.2	0.2	0	0	0.1	—
	Mushrooms													
15585	Crimini (about 6)	3	ounce(s)	85	—	28	3.7	2.8	1.9	0	0	0	0	0
	VEGETABLES, LEGUMES—Continued													

Chol (mg)	Calc (mg)	Iron (mg)	Magn (mg)	Pota (mg)	Sodi (mg)	Zinc (mg)	Vit A (µg)	Thia (mg)	Vit E (mg α)	Ribo (mg)	Niac (mg)	Vit B$_6$ (mg)	Fola (µg)	Vit C (mg)	Vit B$_{12}$ (µg)	Sele (µg)
0	30	0.56	12.0	130.6	10.1	0.25	46.9	0.05	1.21	0.07	0.42	0.12	51.5	36.9	0	0.6
0	21	0.33	9.6	143.8	15.0	0.19	14.1	0.03	0.36	0.05	0.29	0.08	28.7	40.6	0	1.1
0	11	0.23	6.4	96.0	7.4	0.20	2.6	0.02	0.01	0.03	0.23	0.07	18.2	28.2	0	0.2
0	28	0.93	15.6	247.3	16.4	0.25	30.4	0.08	0.33	0.06	0.47	0.13	46.8	48.4	0	1.2
0	20	0.37	14.0	224.8	11.6	0.18	35.7	0.08	0.39	0.08	0.41	0.22	78.3	35.4	0	0.5
0	72	0.24	22.5	294.0	12.0	0.30	6.0	0.08	0.20	0.04	0.36	0.16	45.0	56.2	0	0.9
0	158	1.76	18.7	630.7	459.0	0.28	360.4	0.04	0.14	0.10	0.72	0.28	69.7	44.2	0	0.7
0	144	1.26	27.0	379.5	996.0	0.36	288.0	0.06	0.36	0.10	0.80	0.32	88.5	79.6	0	1.5
0	28	0.33	8.4	119.0	12.6	0.12	3.5	0.04	0.10	0.02	0.16	0.08	30.1	25.6	0	0.2
0	31	0.56	11.2	170.1	18.9	0.15	39.2	0.04	0.07	0.05	0.29	0.14	12.6	39.9	0	0.4
0	24	0.28	19.6	161.0	19.6	0.18	35.0	0.05	0.11	0.02	0.21	0.13	56.0	21.7	0	0.6
0	0	0.00	—	—	140	—	0	—	—	—	—	—	—	0	—	—
0	26	0.71	8.0	189.6	62.4	0.13	552.0	0.02	—	0.02	0.44	0.08	21.6	2.1	0	0.7
0	18	0.16	6.6	176.0	37.9	0.13	459.2	0.03	0.36	0.03	0.54	0.07	10.4	3.2	0	0.1
0	28	0.54	16.5	344.6	34.2	0.21	1128.1	0.11	1.37	0.07	0.46	0.26	4.7	10.0	0	0.7
0	20	0.18	7.3	195.2	42.1	0.15	509.4	0.04	0.40	0.04	0.60	0.08	11.6	3.6	0	0.1
0	23	0.26	7.8	183.3	45.2	0.15	664.6	0.05	0.80	0.03	0.50	0.11	10.9	2.8	0	0.5
0	16	0.27	21.6	279.1	14.4	0.35	1.0	0.08	0.19	0.04	0.87	0.09	27.8	21.2	0	0.7
0	10	0.19	5.6	88.0	9.3	0.10	0.6	0.02	0.04	0.03	0.25	0.10	27.3	27.5	0	0.4
0	15	0.36	8.1	125.1	16.2	0.11	0	0.03	0.05	0.04	0.27	0.07	36.9	28.2	0	0.5
0	11	0.22	7.5	151.5	15.0	0.14	0.5	0.03	0.04	0.03	0.26	0.11	28.5	23.2	0	0.3
0	20	0.10	5.6	131.3	40.4	0.07	11.1	0.01	0.14	0.03	0.16	0.04	18.2	1.6	0	0.2
0	32	0.16	8.8	208.0	64.0	0.10	17.6	0.01	0.21	0.04	0.25	0.05	28.8	2.5	0	0.3
0	51	1.98	75.3	480.4	156.6	0.29	267.8	0.03	1.65	0.08	0.32	0.07	7.9	15.8	0	0.8
0	18	0.64	29.2	136.4	76.7	0.13	110.2	0.01	0.68	0.03	0.14	0.03	5.0	10.8	0	0.3
0	133	1.10	19.0	110.2	15.2	0.21	385.7	0.03	0.83	0.10	0.54	0.12	88.4	17.3	0	0.5
0	179	0.95	25.5	213.4	42.5	0.22	488.8	0.04	1.06	0.09	0.54	0.09	64.6	22.4	0	1.3
0	2	0.61	32.0	248.0	242.0	0.48	13.0	0.20	0.09	0.07	1.60	0.06	46.0	6.2	0	0.2
0	4	0.48	21.8	171.5	364.8	0.67	5.1	0.03	0.09	0.06	1.22	0.08	57.6	5.9	0	0.5
0	2	0.36	21.3	173.8	0	0.50	10.7	0.17	0.07	0.05	1.32	0.04	37.7	5.1	0	0.2
0	2	0.38	23.0	191.1	0.8	0.51	8.2	0.02	0.05	0.05	1.07	0.08	28.7	2.9	0	0.6
0	12	0.20	9.8	110.6	1.5	0.14	3.8	0.01	0.01	0.01	0.07	0.03	5.3	2.1	0	0.2
0	7	3.61	6.0	87.8	765.8	0.38	—	0.02	—	0.02	0.34	0.08	17.3	2.6	0	—
0	74	0.95	12.6	121.8	23.1	0.15	179.6	0.07	1.28	0.09	0.27	0.08	6.8	9.5	0	0.2
0	103	1.70	19.8	218.3	41.8	0.22	279.4	0.10	1.89	0.14	0.44	0.13	14.8	19.2	0	0.3
0	3	0.12	5.4	60.9	0.5	0.06	1.0	0.04	0.20	0.01	0.30	0.04	6.9	0.6	0	0
0	26	0.41	7.5	157.0	11.0	0.39	54.0	0.04	0.22	0.03	0.20	0.01	71.0	3.2	0	0.1
0	16	0.78	15.5	194.0	5.2	0.20	1.3	0.02	0.59	0.04	0.25	0.05	15.5	26.1	0	0.9
0	90	0.61	11.7	208.7	9.8	0.11	477.8	0.02	0.59	0.07	0.43	0.05	9.1	16.4	0	0.6
0	90	1.14	22.8	299.5	28.8	0.29	515.2	0.07	—	0.08	0.66	0.18	19.4	80.4	0	0.6
0	21	0.33	15.7	280.5	17.3	0.26	1.7	0.03	0.43	0.02	0.32	0.13	9.9	44.6	0	0.7
0	32	0.54	25.7	472.5	27.0	0.04	2.7	0.06	0.64	0.02	0.54	0.20	21.6	83.7	0	0.9
0	16	0.56	7.3	45.2	5.2	0.02	1.0	0.01	—	0.01	0.10	0.04	12.5	2.2	0	0.3
0	53	1.86	24.9	160.2	17.8	0.10	73.9	0.05	0.81	0.02	0.35	0.20	57.0	10.7	0	0.9
0	9	1.65	17.8	182.7	1.0	0.63	0	0.08	0.05	0.04	0.52	0.09	89.6	0.7	0	1.4
0	19	2.47	28.5	247.9	8.5	1.16	1.5	0.17	—	0.09	0.86	0.14	77.0	12.7	0	0.5
0	29	1.02	10.7	196.4	4.1	0.16	137.0	0.04	0.14	0.05	0.29	0.06	60.2	3.1	0	0.5
0	19	0.68	7.1	130.9	2.7	0.11	91.3	0.03	0.09	0.03	0.19	0.04	40.1	2.0	0	0.3
0	10	0.22	3.8	77.5	5.5	0.08	13.7	0.02	0.09	0.01	0.07	0.02	15.9	1.5	0	0.1
0	10	0.22	3.8	77.5	5.5	0.08	13.7	0.02	0.09	0.01	0.07	0.02	15.9	1.5	0	0.1
0	13	0.31	4.7	69.8	10.1	0.06	133.2	0.02	0.10	0.02	0.13	0.03	13.7	6.5	0	0.2
0	18	0.54	7.8	138.3	4.5	0.13	162.4	0.04	0.07	0.03	0.17	0.04	76.2	13.4	0	0.2
0	0	0.67	—	—	32.6	—	0	—	—	—	—	—	—	0	0	—

DA+ Code	Food Description	Quantity	Measure	Wt (g)	H₂O (g)	Ener (kcal)	Prot (g)	Carb (g)	Fiber (g)	Fat (g)	Fat Breakdown (g)			
											Sat	Mono	Poly	Trans
8700	Enoki	30	item(s)	90	79.7	40	2.3	6.9	2.4	0.3	0	0	0.1	—
1079	Mushrooms, boiled, drained	½	cup(s)	78	71.0	22	1.7	4.1	1.7	0.4	0	0	0.1	—
1080	Mushrooms, canned, drained	½	cup(s)	78	71.0	20	1.5	4.0	1.9	0.2	0	0	0.1	—
630	Mushrooms, raw	½	cup(s)	48	44.4	11	1.5	1.6	0.5	0.2	0	0	0.1	—
15587	Portabella, raw	1	item(s)	84	—	30	3.0	3.9	3.0	0	0	0	0	0
2743	Shiitake, cooked	½	cup(s)	73	60.5	41	1.1	10.4	1.5	0.2	0	0.1	0	—
	Mustard greens													
2744	Frozen, boiled, drained	½	cup(s)	75	70.4	14	1.7	2.3	2.1	0.2	0	0.1	0	—
29319	Raw	1	cup(s)	56	50.8	15	1.5	2.7	1.8	0.1	0	0	0	—
	Okra													
16866	Batter coated, fried	11	piece(s)	83	55.6	156	2.1	12.7	2.0	11.2	1.5	3.7	5.5	—
32742	Frozen, boiled, drained, no salt added	½	cup(s)	92	83.8	26	1.9	5.3	2.6	0.3	0.1	0	0.1	—
632	Sliced, boiled, drained	½	cup(s)	80	74.1	18	1.5	3.6	2.0	0.2	0	0	0	—
	Onions													
635	Chopped, boiled, drained	½	cup(s)	105	92.2	46	1.4	10.7	1.5	0.2	0	0	0.1	—
2748	Frozen, boiled, drained	½	cup(s)	106	97.8	30	0.8	7.0	1.9	0.1	0	0	0	—
1081	Onion rings, breaded and pan fried, frozen, heated	10	piece(s)	71	20.2	289	3.8	27.1	0.9	19.0	6.1	7.7	3.6	—
633	Raw, chopped	½	cup(s)	80	71.3	32	0.9	7.5	1.4	0.1	0	0	0	—
16850	Red onions, sliced, raw	½	cup(s)	57	50.7	24	0.5	5.8	0.8	0	0	0	0	—
636	Scallions, green or spring onions	2	item(s)	30	26.9	10	0.5	2.2	0.8	0.1	0	0	0	—
16860	**Palm hearts, cooked**	½	cup(s)	73	50.7	84	2.0	18.7	1.1	0.1	0	0	0.1	—
637	**Parsley, chopped**	1	tablespoon(s)	4	3.3	1	0.1	0.2	0.1	0	0	0	0	—
638	**Parsnips, sliced, boiled, drained**	½	cup(s)	78	62.6	55	1.0	13.3	2.8	0.2	0	0.1	0	—
	Peas													
639	Green peas, canned, drained	½	cup(s)	85	69.4	59	3.8	10.7	3.5	0.3	0.1	0	0.1	—
641	Green peas, frozen, boiled, drained	½	cup(s)	80	63.6	62	4.1	11.4	4.4	0.2	0	0	0.1	—
35694	Pea pods, boiled with salt, drained	½	cup(s)	80	71.1	32	2.6	5.2	2.2	0.2	0	0	0.1	—
1082	Peas and carrots, canned with liquid	½	cup(s)	128	112.4	48	2.8	10.8	2.6	0.3	0.1	0	0.2	—
1083	Peas and carrots, frozen, boiled, drained	½	cup(s)	80	68.6	38	2.5	8.1	2.5	0.3	0.1	0	0.2	—
2750	Snow or sugar peas, frozen, boiled, drained	½	cup(s)	80	69.3	42	2.8	7.2	2.5	0.3	0.1	0	0.1	—
640	Snow or sugar peas, raw	½	cup(s)	32	28.0	13	0.9	2.4	0.8	0.1	0	0	0	—
29324	Split peas, sprouted	½	cup(s)	60	37.4	77	5.3	16.9	—	0.4	0.1	0	0.2	—
	Peppers													
644	Green bell or sweet, boiled, drained	½	cup(s)	68	62.5	19	0.6	4.6	0.8	0.1	0	0	0.1	—
643	Green bell or sweet, raw	½	cup(s)	75	69.9	15	0.6	3.5	1.3	0.1	0	0	0	—
1664	Green hot chili	1	item(s)	45	39.5	18	0.9	4.3	0.7	0.1	0	0	0	—
1663	Green hot chili, canned with liquid	½	cup(s)	68	62.9	14	0.6	3.5	0.9	0.1	0	0	0	—
1086	Jalapeno, canned with liquid	½	cup(s)	68	60.4	18	0.6	3.2	1.8	0.6	0.1	0	0.3	—
8703	Yellow bell or sweet	1	item(s)	186	171.2	50	1.9	11.8	1.7	0.4	0.1	0	0.2	—
1087	**Poi**	½	cup(s)	120	86.0	134	0.5	32.7	0.5	0.2	0	0	0.1	—
	Potatoes													
1090	Au gratin mix, prepared with water, whole milk and butter	½	cup(s)	124	97.7	115	2.8	15.9	1.1	5.1	3.2	1.5	0.2	—
1089	Au gratin, prepared with butter	½	cup(s)	123	90.7	162	6.2	13.8	2.2	9.3	5.8	2.6	0.3	—
5791	Baked, flesh and skin	1	item(s)	202	151.3	188	5.1	42.7	4.4	0.3	0.1	0	0.1	—
645	Baked, flesh only	½	cup(s)	61	46.0	57	1.2	13.1	0.9	0.1	0	0	0	—
1088	Baked, skin only	1	item(s)	58	27.4	115	2.5	26.7	4.6	0.1	0	0	0	—
5795	Boiled in skin, flesh only, drained	1	item(s)	136	104.7	118	2.5	27.4	2.1	0.1	0	0	0.1	—
5794	Boiled, drained, skin and flesh	1	item(s)	150	115.9	129	2.9	29.8	2.5	0.2	0	0	0.1	—
647	Boiled, flesh only	½	cup(s)	78	60.4	67	1.3	15.6	1.4	0.1	0	0	0	—
648	French fried, deep fried, prepared from raw	14	item(s)	70	32.8	187	2.7	23.5	2.9	9.5	1.9	4.2	3.0	—
649	French fried, frozen, heated	14	item(s)	70	43.7	94	1.9	19.4	2.0	3.7	0.7	2.3	0.2	—
1091	Hashed brown	½	cup(s)	78	36.9	207	2.3	27.4	2.5	9.8	1.5	4.1	3.7	—
652	Mashed with margarine and whole milk	½	cup(s)	105	79.0	119	2.1	17.7	1.6	4.4	1.0	2.0	1.2	0.7
653	Mashed, prepared from dehydrated granules with milk, water, and margarine	½	cup(s)	105	79.8	122	2.3	16.9	1.4	5.0	1.3	2.1	1.4	—
2759	Microwaved	1	item(s)	202	145.5	212	4.9	49.0	4.6	0.2	0.1	0	0.1	—
2760	Microwaved in skin, flesh only	½	cup(s)	78	57.1	78	1.6	18.1	1.2	0.1	0	0	0	—
5804	Microwaved, skin only	1	item(s)	58	36.8	77	2.5	17.2	4.2	0.1	0	0	0	—
1097	Potato puffs, frozen, heated	½	cup(s)	64	38.2	122	1.3	17.8	1.6	5.5	1.2	3.9	0.3	—
1094	Scalloped mix, prepared with water, whole milk and butter	½	cup(s)	124	98.4	116	2.6	15.9	1.4	5.3	3.3	1.5	0.2	—
1093	Scalloped, prepared with butter	½	cup(s)	123	99.2	108	3.5	13.2	2.3	4.5	2.8	1.3	0.2	—
	Pumpkin													
1773	Boiled, drained	½	cup(s)	123	114.8	25	0.9	6.0	1.3	0.1	0	0	0	—
656	Canned	½	cup(s)	123	110.2	42	1.3	9.9	3.6	0.3	0.2	0	0	—
	Radicchio													
8731	Leaves, raw	1	cup(s)	40	37.3	9	0.6	1.8	0.4	0.1	0	0	0	—
2498	Raw	1	cup(s)	40	37.3	9	0.6	1.8	0.4	0.1	0	0	0	—

VEGETABLES, LEGUMES—Continued

Chol (mg)	Calc (mg)	Iron (mg)	Magn (mg)	Pota (mg)	Sodi (mg)	Zinc (mg)	Vit A (µg)	Thia (mg)	Vit E (mg α)	Ribo (mg)	Niac (mg)	Vit B_6 (mg)	Fola (µg)	Vit C (mg)	Vit B_{12} (µg)	Sele (µg)
0	1	0.98	14.4	331.2	2.7	0.54	0	0.16	0.01	0.14	5.31	0.07	46.8	0	0	2.0
0	5	1.35	9.4	277.7	1.6	0.67	0	0.05	0.01	0.23	3.47	0.07	14.0	3.1	0	9.3
0	9	0.61	11.7	100.6	331.5	0.56	0	0.06	0.01	0.01	1.24	0.04	9.4	0	0	3.2
0	1	0.24	4.3	152.6	2.4	0.25	0	0.04	0.01	0.19	1.73	0.05	7.7	1.0	0	4.5
0	39	0.35	—	—	9.9	—	0	—	—	—	—	—	—	0	0	—
0	2	0.31	10.2	84.8	2.9	0.96	0	0.02	0.02	0.12	1.08	0.11	15.2	0.2	0	18.0
0	76	0.84	9.8	104.3	18.8	0.15	265.5	0.03	1.01	0.04	0.19	0.08	52.5	10.4	0	0.5
0	58	0.81	17.9	198.2	14.0	0.11	294.0	0.04	1.12	0.06	0.45	0.10	104.7	39.2	0	0.5
2	54	1.13	32.2	170.8	109.7	0.44	14.0	0.16	1.50	0.12	1.29	0.11	39.6	9.2	0	3.6
0	88	0.61	46.9	215.3	2.8	0.57	15.6	0.09	0.29	0.11	0.72	0.04	134.3	11.2	0	0.6
0	62	0.22	28.8	108.0	4.8	0.34	11.2	0.10	0.21	0.04	0.69	0.15	36.8	13.0	0	0.3
0	23	0.24	11.5	174.3	3.1	0.21	0	0.03	0.02	0.02	0.17	0.12	15.7	5.5	0	0.6
0	17	0.32	6.4	114.5	12.7	0.06	0	0.02	0.01	0.02	0.14	0.06	13.8	2.8	0	0.4
0	22	1.20	13.5	91.6	266.3	0.29	7.8	0.19	—	0.09	2.56	0.05	46.9	1.0	0	2.5
0	18	0.16	8.0	116.8	3.2	0.13	0	0.03	0.01	0.02	0.09	0.09	15.2	5.9	0	0.4
0	13	0.10	5.7	82.4	1.7	0.09	0	0.02	0.01	0.01	0.04	0.08	10.9	3.7	0	0.3
0	22	0.44	6.0	82.8	4.8	0.11	15.0	0.01	0.16	0.02	0.15	0.01	19.2	5.6	0	0.2
0	13	1.23	7.3	1318.4	10.2	2.72	2.2	0.03	0.36	0.12	0.62	0.53	14.6	5.0	0	0.5
0	5	0.23	1.9	21.1	2.1	0.04	16.0	0.00	0.02	0.00	0.05	0.00	5.8	5.1	0	0
0	29	0.45	22.6	286.3	7.8	0.20	0	0.06	0.78	0.04	0.56	0.07	45.2	10.1	0	1.3
0	17	0.80	14.5	147.1	214.2	0.60	23.0	0.10	0.02	0.06	0.62	0.05	37.4	8.2	0	1.4
0	19	1.21	17.6	88.0	57.6	0.53	84.0	0.22	0.02	0.08	1.18	0.09	47.2	7.9	0	0.8
0	34	1.57	20.8	192.0	192.0	0.29	41.6	0.10	0.31	0.06	0.43	0.11	23.2	38.3	0	0.6
0	29	0.96	17.9	127.5	331.5	0.74	368.5	0.09	—	0.07	0.74	0.11	23.0	8.4	0	1.1
0	18	0.75	12.8	126.4	54.4	0.36	380.8	0.18	0.41	0.05	0.92	0.07	20.8	6.5	0	0.9
0	47	1.92	22.4	173.6	4.0	0.39	52.8	0.05	0.37	0.09	0.45	0.13	28.0	17.6	0	0.6
0	14	0.65	7.6	63.0	1.3	0.08	17.0	0.04	0.12	0.02	0.19	0.05	13.2	18.9	0	0.2
0	22	1.34	33.6	228.6	12.0	0.62	4.8	0.12	—	0.08	1.84	0.14	86.4	6.2	0	0.4
0	6	0.31	6.8	112.9	1.4	0.08	15.6	0.04	0.34	0.02	0.32	0.15	10.9	50.6	0	0.2
0	7	0.25	7.5	130.4	2.2	0.09	13.4	0.04	0.27	0.02	0.35	0.16	7.5	59.9	0	0
0	8	0.54	11.3	153.0	3.2	0.13	26.6	0.04	0.31	0.04	0.42	0.12	10.4	109.1	0	0.2
0	5	0.34	9.5	127.2	797.6	0.10	24.5	0.01	0.46	0.02	0.54	0.10	6.8	46.2	0	0.2
0	16	1.28	10.2	131.2	1136.3	0.23	57.8	0.03	0.47	0.03	0.27	0.13	9.5	6.8	0	0.3
0	20	0.85	22.3	394.3	3.7	0.31	18.6	0.05	—	0.04	1.65	0.31	48.4	341.3	0	0.6
0	19	1.06	28.8	219.6	14.4	0.26	3.6	0.16	2.76	0.05	1.32	0.33	25.2	4.8	0	0.8
19	103	0.39	18.6	271.0	543.3	0.29	64.4	0.02	—	0.10	1.16	0.05	8.7	3.8	0	3.3
28	146	0.78	24.5	485.1	530.4	0.85	78.4	0.08	—	0.14	1.22	0.21	13.5	12.1	0	3.3
0	30	2.18	56.6	1080.7	20.2	0.72	2.0	0.12	0.08	0.09	2.84	0.62	56.6	19.4	0	0.8
0	3	0.21	15.3	238.5	3.1	0.18	0	0.06	0.02	0.01	0.85	0.18	5.5	7.8	0	0.2
0	20	4.08	24.9	332.3	12.2	0.28	0.6	0.07	0.02	0.06	1.77	0.35	12.8	7.8	0	0.4
0	7	0.42	29.9	515.4	5.4	0.40	0	0.14	0.01	0.02	1.95	0.40	13.6	17.7	0	0.4
0	13	1.27	34.1	572.0	7.4	0.46	0	0.14	0.01	0.03	2.13	0.44	15.0	18.4	0	—
0	6	0.24	15.6	255.8	3.9	0.21	0	0.07	0.01	0.01	1.02	0.21	7.0	5.8	0	0.2
0	16	1.05	30.8	567.0	8.4	0.39	0	0.08	0.09	0.03	1.34	0.37	16.1	21.2	0	0.4
0	8	0.51	18.2	315.7	271.6	0.26	0	0.09	0.07	0.02	1.55	0.12	19.6	9.3	0	0.1
0	11	0.43	27.3	449.3	266.8	0.37	0	0.13	0.01	0.03	1.80	0.37	12.5	10.1	0	0.4
1	23	0.27	19.9	344.4	349.6	0.31	43.0	0.09	0.44	0.04	1.23	0.25	9.4	11.0	0.1	0.8
2	36	0.21	21.0	164.8	179.5	0.26	49.3	0.09	0.53	0.09	0.90	0.16	8.4	6.8	0.1	5.9
0	22	2.50	54.5	902.9	16.2	0.72	0	0.24	—	0.06	3.46	0.69	24.2	30.5	0	0.8
0	4	0.31	19.4	319.0	5.4	0.25	0	0.10	—	0.01	1.26	0.25	9.3	11.7	0	0.3
0	27	3.44	21.5	377.0	9.3	0.29	0	0.04	0.01	0.04	1.28	0.28	9.9	8.9	0	0.3
0	9	0.41	10.9	199.7	307.2	0.21	0	0.08	0.15	0.02	0.97	0.08	9.0	4.0	0	0.4
14	45	0.47	17.4	252.2	423.7	0.31	43.5	0.02	—	0.06	1.28	0.05	12.4	4.1	0	2.0
15	70	0.70	23.3	463.1	410.4	0.49	0	0.08	—	0.11	1.29	0.22	13.5	13.0	0	2.0
0	18	0.69	11.0	281.8	1.2	0.28	306.3	0.03	0.98	0.09	0.50	0.05	11.0	5.8	0	0.2
0	32	1.70	28.2	252.4	6.1	0.20	953.1	0.02	1.29	0.06	0.45	0.06	14.7	5.1	0	0.5
0	8	0.23	5.2	120.8	8.8	0.25	0.4	0.01	0.90	0.01	0.10	0.02	24.0	3.2	0	0.4
0	8	0.23	5.2	120.8	8.8	0.25	0.4	0.01	0.90	0.01	0.10	0.02	24.0	3.2	0	0.4

DA+ Code	Food Description	Quantity	Measure	Wt (g)	H₂O (g)	Ener (kcal)	Prot (g)	Carb (g)	Fiber (g)	Fat (g)	Fat Breakdown (g)			
											Sat	Mono	Poly	Trans
657	**Radishes**	6	item(s)	27	25.7	4	0.2	0.9	0.4	0	0	0	0	—
1099	**Rutabaga, boiled, drained**	½	cup(s)	85	75.5	33	1.1	7.4	1.5	0.2	0	0	0.1	—
658	**Sauerkraut, canned**	½	cup(s)	118	109.2	22	1.1	5.1	3.4	0.2	0	0	0.1	—
	Seaweed													
1102	Kelp	½	cup(s)	40	32.6	17	0.6	3.8	0.5	0.2	0.1	0	0	—
1104	Spirulina, dried	½	cup(s)	8	0.4	22	4.3	1.8	0.3	0.6	0.2	0.1	0.2	—
1106	**Shallots**	3	tablespoon(s)	30	23.9	22	0.8	5.0	—	0	0	0	0	—
	Soybeans													
1670	Boiled	½	cup(s)	86	53.8	149	14.3	8.5	5.2	7.7	1.1	1.7	4.4	—
2825	Dry roasted	½	cup(s)	86	0.7	388	34.0	28.1	7.0	18.6	2.7	4.1	10.5	—
2824	Roasted, salted	½	cup(s)	86	1.7	405	30.3	28.9	15.2	21.8	3.2	4.8	12.3	—
8739	Sprouted, stir fried	½	cup(s)	63	42.3	79	8.2	5.9	0.5	4.5	0.6	1.0	2.5	0
	Soy products													
1813	Soy milk	1	cup(s)	240	211.3	130	7.8	15.1	1.4	4.2	0.5	1.0	2.3	0
2838	Tofu, dried, frozen (koyadofu)	3	ounce(s)	85	4.9	408	40.8	12.4	6.1	25.8	3.7	5.7	14.6	—
13844	Tofu, extra firm	3	ounce(s)	85	—	86	8.6	2.2	1.1	4.3	0.5	0.9	2.8	—
13843	Tofu, firm	3	ounce(s)	85	—	75	7.5	2.2	0.5	3.2	0	0.9	2.3	—
1816	Tofu, firm, with calcium sulfate and magnesium chloride (nigari)	3	ounce(s)	85	72.2	60	7.0	1.4	0.8	3.5	0.7	1.0	1.5	—
1817	Tofu, fried	3	ounce(s)	85	43.0	230	14.6	8.9	3.3	17.2	2.5	3.8	9.7	—
13841	Tofu, silken	3	ounce(s)	85	—	42	3.7	1.9	0	2.3	0.5	—	—	—
13842	Tofu, soft	3	ounce(s)	85	—	65	6.5	1.1	0.5	3.2	0.5	1.1	2.2	—
1671	Tofu, soft, with calcium sulfate and magnesium chloride (nigari)	3	ounce(s)	85	74.2	52	5.6	1.5	0.2	3.1	0.5	0.7	1.8	—
	Spinach													
663	Canned, drained	½	cup(s)	107	98.2	25	3.0	3.6	2.6	0.5	0.1	0	0.2	—
660	Chopped, boiled, drained	½	cup(s)	90	82.1	21	2.7	3.4	2.2	0.2	0	0	0.1	—
661	Chopped, frozen, boiled, drained	½	cup(s)	95	84.5	32	3.8	4.6	3.5	0.8	0.1	0	0.4	—
662	Leaf, frozen, boiled, drained	½	cup(s)	95	84.5	32	3.8	4.6	3.5	0.8	0.1	0	0.4	—
659	Raw, chopped	1	cup(s)	30	27.4	7	0.9	1.1	0.7	0.1	0	0	0	—
8470	Trimmed leaves	1	cup(s)	32	27.5	3	0.9	0	2.8	0.1	—	—	—	—
	Squash													
1662	Acorn winter, baked	½	cup(s)	103	85.0	57	1.1	14.9	4.5	0.1	0	0	0.1	—
29702	Acorn winter, boiled, mashed	½	cup(s)	123	109.9	42	0.8	10.8	3.2	0.1	0	0	0	—
29451	Butternut, frozen, boiled	½	cup(s)	122	106.9	47	1.5	12.2	1.8	0.1	0	0	0	—
1661	Butternut winter, baked	½	cup(s)	102	89.5	41	0.9	10.7	3.4	0.1	0	0	0	—
32773	Butternut winter, frozen, boiled, mashed, no salt added	½	cup(s)	121	106.4	47	1.5	12.2	—	0.1	0	0	0	—
29700	Crookneck and straightneck summer, boiled, drained	½	cup(s)	65	60.9	12	0.6	2.6	1.2	0.1	0	0	0.1	—
29703	Hubbard winter, baked	½	cup(s)	102	86.8	51	2.5	11.0	—	0.6	0.1	0	0.3	—
1660	Hubbard winter, boiled, mashed	½	cup(s)	118	107.5	35	1.7	7.6	3.4	0.4	0.1	0	0.2	—
29704	Spaghetti winter, boiled, drained, or baked	½	cup(s)	78	71.5	21	0.5	5.0	1.1	0.2	0	0	0.1	—
664	Summer, all varieties, sliced, boiled, drained	½	cup(s)	90	84.3	18	0.8	3.9	1.3	0.3	0.1	0	0.1	—
665	Winter, all varieties, baked, mashed	½	cup(s)	103	91.4	38	0.9	9.1	2.9	0.4	0.1	0	0.2	—
1112	Zucchini summer, boiled, drained	½	cup(s)	90	85.3	14	0.6	3.5	1.3	0	0	0	0	—
1113	Zucchini summer, frozen, boiled, drained	½	cup(s)	112	105.6	19	1.3	4.0	1.4	0.1	0	0	0.1	—
	Sweet potatoes													
666	Baked, peeled	½	cup(s)	100	75.8	90	2.0	20.7	3.3	0.2	0	0	0.1	—
667	Boiled, mashed	½	cup(s)	164	131.4	125	2.2	29.1	4.1	0.2	0.1	0	0.1	—
668	Candied, home recipe	½	cup(s)	91	61.1	132	0.8	25.4	2.2	3.0	1.2	0.6	0.1	—
670	Canned, vacuum pack	½	cup(s)	100	76.0	91	1.7	21.1	1.8	0.2	0	0	0.1	—
2765	Frozen, baked	½	cup(s)	88	64.5	88	1.5	20.5	1.6	0.1	0	0	0	—
1136	Yams, baked or boiled, drained	½	cup(s)	68	47.7	79	1.0	18.7	2.7	0.1	0	0	0	—
32785	**Taro shoots, cooked, no salt added**	½	cup(s)	70	66.7	10	0.5	2.2	—	0.1	0	0	0	—
	Tomatillo													
8774	Raw	2	item(s)	68	62.3	22	0.7	4.0	1.3	0.7	0.1	0.1	0.3	—
8777	Raw, chopped	½	cup(s)	66	60.5	21	0.6	3.9	1.3	0.7	0.1	0.1	0.3	—
	Tomato													
16846	Cherry, fresh	5	item(s)	85	80.3	15	0.7	3.3	1.0	0.2	0	0	0.1	—
671	Fresh, ripe, red	1	item(s)	123	116.2	22	1.1	4.8	1.5	0.2	0	0	0.1	—
675	Juice, canned	½	cup(s)	122	114.1	21	0.9	5.2	0.5	0.1	0	0	0.1	—
75	Juice, no salt added	½	cup(s)	122	114.1	21	0.9	5.2	0.5	0.1	0	0	0	—
1699	Paste, canned	2	tablespoon(s)	33	24.1	27	1.4	6.2	1.3	0.2	0	0	0.1	—
1700	Puree, canned	¼	cup(s)	63	54.9	24	1.0	5.6	1.2	0.1	0	0	0.1	—
1118	Red, boiled	½	cup(s)	120	113.2	22	1.1	4.8	0.8	0.1	0	0	0.1	—
3952	Red, diced	½	cup(s)	90	85.1	16	0.8	3.5	1.1	0.2	0	0	0.1	—
1120	Red, stewed, canned	½	cup(s)	128	116.7	33	1.2	7.9	1.3	0.2	0	0	0.1	—
1125	Sauce, canned	¼	cup(s)	61	55.6	15	0.8	3.3	0.9	0.1	0	0	0	—
	VEGETABLES, LEGUMES—Continued													
8778	Sun dried	½	cup(s)	27	3.9	70	3.8	15.1	3.3	0.8	0.1	0.1	0.3	—

Chol (mg)	Calc (mg)	Iron (mg)	Magn (mg)	Pota (mg)	Sodi (mg)	Zinc (mg)	Vit A (µg)	Thia (mg)	Vit E (mg α)	Ribo (mg)	Niac (mg)	Vit B_6 (mg)	Fola (µg)	Vit C (mg)	Vit B_{12} (µg)	Sele (µg)
0	7	0.09	2.7	62.9	10.5	0.07	0	0.00	0.00	0.01	0.06	0.01	6.8	4.0	0	0.2
0	41	0.45	19.6	277.1	17.0	0.30	0	0.07	0.27	0.04	0.61	0.09	12.8	16.0	0	0.6
0	35	1.73	15.3	200.6	780.0	0.22	1.2	0.03	0.17	0.03	0.17	0.15	28.3	17.3	0	0.7
0	67	1.12	48.4	35.6	93.2	0.48	2.4	0.02	0.32	0.04	0.16	0.00	72.0	1.2	0	0.3
0	9	2.14	14.6	102.2	78.6	0.15	2.2	0.18	0.38	0.28	0.96	0.03	7.1	0.8	0	0.5
0	11	0.36	6.3	100.2	3.6	0.12	18.0	0.02	—	0.01	0.06	0.09	10.2	2.4	—	0.4
0	88	4.42	74.0	442.9	0.9	0.98	0	0.13	0.30	0.24	0.34	0.20	46.4	1.5	0	6.3
0	120	3.39	196.1	1173.0	1.7	4.10	0	0.36	—	0.64	0.90	0.19	176.3	4.0	0	16.6
0	119	3.35	124.7	1264.2	140.2	2.70	8.6	0.08	0.78	0.12	1.21	0.17	181.5	1.9	0	16.4
0	52	0.25	60.4	356.6	8.8	1.32	0.6	0.26	—	0.12	0.69	0.10	79.9	7.5	0	0.4
0	60	1.53	60.0	283.2	122.4	0.28	0	0.14	0.26	0.16	1.23	0.18	43.2	0	0	11.5
0	310	8.27	50.2	17.0	5.1	4.16	22.1	0.42	—	0.27	1.01	0.24	78.2	0.6	0	46.2
0	65	1.16	84.1	—	0	—	0	—	—	—	—	—	—	0	0	—
0	108	1.16	56.1	—	0	—	0	—	—	—	—	—	—	0	0	—
0	171	1.36	31.5	125.9	10.2	0.70	0	0.05	0.01	0.05	0.08	0.06	16.2	0.2	0	8.4
0	316	4.14	51.0	124.2	13.6	1.69	0.9	0.14	0.03	0.04	0.08	0.08	23.0	0	0	24.2
0	56	0.34	33.1	—	4.7	—	0	—	—	—	—	—	—	0	1.7	—
0	108	1.16	35.5	—	0	—	0	—	—	—	—	—	—	0	1.9	—
0	94	0.94	23.0	102.1	6.8	0.54	0	0.04	0.01	0.03	0.45	0.04	37.4	0.2	0	7.6
0	136	2.45	81.3	370.2	28.9	0.48	524.3	0.02	2.08	0.14	0.41	0.11	104.8	15.3	0	1.5
0	122	3.21	78.3	419.4	63.0	0.68	471.6	0.08	1.87	0.21	0.44	0.21	131.4	8.8	0	1.4
0	145	1.86	77.9	286.9	92.2	0.46	572.9	0.07	3.36	0.16	0.41	0.12	115.0	2.1	0	5.2
0	145	1.86	77.9	286.9	92.2	0.46	572.9	0.07	3.36	0.16	0.41	0.12	115.0	2.1	0	5.2
0	30	0.81	23.7	167.4	23.7	0.16	140.7	0.02	0.61	0.06	0.22	0.06	58.2	8.4	0	0.3
0	25	2.13	25.5	134.1	38.0	0.18	—	0.03	—	0.05	0.18	0.07	0	7.5	0	—
0	45	0.95	44.1	447.9	4.1	0.17	21.5	0.17	—	0.01	0.90	0.19	19.5	11.1	0	0.7
0	32	0.68	31.9	322.2	3.7	0.13	50.2	0.12	—	0.01	0.65	0.14	13.5	8.0	0	0.5
0	23	0.70	10.9	161.9	2.4	0.14	203.3	0.06	0.14	0.05	0.56	0.08	19.5	4.3	0	0.6
0	42	0.61	29.6	289.6	4.1	0.13	569.1	0.07	1.31	0.01	0.99	0.12	19.4	15.4	0	0.5
0	23	0.70	10.9	161.2	2.4	0.14	202.4	0.06	—	0.05	0.56	0.08	19.4	4.2	0	0.6
0	14	0.31	13.6	137.1	1.3	0.19	5.2	0.03	—	0.02	0.29	0.07	14.9	5.4	0	0.1
0	17	0.48	22.4	365.1	8.2	0.15	308.0	0.07	—	0.04	0.57	0.17	16.3	9.7	0	0.6
0	12	0.33	15.3	252.5	5.9	0.11	236.0	0.05	0.14	0.03	0.39	0.12	11.8	7.7	0	0.4
0	16	0.26	8.5	90.7	14.0	0.15	4.7	0.02	0.09	0.01	0.62	0.07	6.2	2.7	0	0.2
0	24	0.32	21.6	172.8	0.9	0.35	9.9	0.04	0.12	0.03	0.46	0.05	18.0	5.0	0	0.2
0	23	0.45	13.3	247.0	1.0	0.23	267.5	0.02	0.12	0.07	0.51	0.17	20.5	9.8	0	0.4
0	12	0.32	19.8	227.7	2.7	0.16	50.4	0.04	0.11	0.04	0.39	0.07	15.3	4.1	0	0.2
0	19	0.54	14.5	216.3	2.2	0.22	10	0.05	0.13	0.04	0.43	0.05	8.9	4.1	0	0.2
0	38	0.69	27.0	475.0	36.0	0.32	961.0	0.10	0.71	0.10	1.48	0.28	6.0	19.6	0	0.2
0	44	1.18	29.5	377.2	44.3	0.33	1290.7	0.09	1.54	0.08	0.88	0.27	9.8	21.0	0	0.3
7	24	1.03	10.0	172.6	63.9	0.13	0	0.01	—	0.03	0.36	0.03	10.0	6.1	0	0.7
0	22	0.89	22.0	312.0	53.0	0.18	399.0	0.04	1.00	0.06	0.74	0.19	17.0	26.4	0	0.7
0	31	0.47	18.4	330.1	7.0	0.26	913.3	0.05	0.67	0.04	0.49	0.16	19.3	8.0	0	0.5
0	10	0.35	12.2	455.6	5.4	0.13	4.1	0.06	0.23	0.01	0.37	0.15	10.9	8.2	0	0.5
0	10	0.28	5.6	240.8	1.4	0.37	2.1	0.02	—	0.03	0.56	0.07	2.1	13.2	0	0.7
0	5	0.42	13.6	182.2	0.7	0.15	4.1	0.03	0.25	0.02	1.25	0.03	4.8	8.0	0	0.3
0	5	0.41	13.2	176.9	0.7	0.15	4.0	0.03	0.25	0.02	1.22	0.04	4.6	7.7	0	0.3
0	9	0.22	9.4	201.5	4.3	0.14	35.7	0.03	0.45	0.01	0.50	0.06	12.8	10.8	0	0
0	12	0.33	13.5	291.5	6.2	0.20	51.7	0.04	0.66	0.02	0.73	0.09	18.5	15.6	0	0
0	12	0.52	13.4	278.2	326.8	0.18	27.9	0.06	0.39	0.04	0.82	0.14	24.3	22.2	0	0.4
0	12	0.52	13.4	278.2	12.2	0.18	27.9	0.06	0.39	0.04	0.82	0.14	24.3	22.2	0	0.4
0	12	0.97	13.8	332.6	259.1	0.20	24.9	0.02	1.41	0.05	1.00	0.07	3.9	7.2	0	1.7
0	11	1.11	14.4	274.4	249.4	0.22	16.3	0.01	1.23	0.05	0.91	0.07	6.9	6.6	0	0.6
0	13	0.82	10.8	261.6	13.2	0.17	28.8	0.04	0.67	0.03	0.64	0.10	15.6	27.4	0	0.6
0	9	0.24	9.9	213.3	4.5	0.15	37.8	0.03	0.48	0.01	0.53	0.07	13.5	11.4	0	0
0	43	1.70	15.3	263.9	281.8	0.22	11.5	0.06	1.06	0.04	0.91	0.02	6.4	10.1	0	0.8
0	8	0.62	9.8	201.9	319.6	0.12	10.4	0.01	0.87	0.04	0.59	0.06	6.7	4.3	0	0.1
0	30	2.45	52.4	925.3	565.7	0.53	11.9	0.14	0.00	0.13	2.44	0.09	18.4	10.6	0	1.5

DA+ Code	Food Description	Quantity	Measure	Wt (g)	H₂O (g)	Ener (kcal)	Prot (g)	Carb (g)	Fiber (g)	Fat (g)	Sat	Mono	Poly	Trans
												Fat Breakdown (g)		
8783	Sun dried in oil, drained	¼	cup(s)	28	14.8	59	1.4	6.4	1.6	3.9	0.5	2.4	0.6	—
	Turnips													
678	Turnip greens, chopped, boiled, drained	½	cup(s)	72	67.1	14	0.8	3.1	2.5	0.2	0	0	0.1	—
679	Turnip greens, frozen, chopped, boiled, drained	½	cup(s)	82	74.1	24	2.7	4.1	2.8	0.3	0.1	0	0.1	—
677	Turnips, cubed, boiled, drained	½	cup(s)	78	73.0	17	0.6	3.9	1.6	0.1	0	0	0	—
	Vegetables, mixed													
1132	Canned, drained	½	cup(s)	82	70.9	40	2.1	7.5	2.4	0.2	0	0	0.1	—
680	Frozen, boiled, drained	½	cup(s)	91	75.7	59	2.6	11.9	4.0	0.1	0	0	0.1	—
7489	V8 100% vegetable juice	½	cup(s)	120	—	25	1.0	5.0	1.0	0	0	0	0	0
7490	V8 low sodium vegetable juice	½	cup(s)	120	—	25	0	6.5	1.0	0	0	0	0	0
7491	V8 spicy hot vegetable juice	½	cup(s)	120	—	25	1.0	5.0	0.5	0	0	0	0	0
	Water chestnuts													
31073	Sliced, drained	½	cup(s)	75	70.0	20	0	5.0	1.0	0	0	0	0	0
31087	Whole	½	cup(s)	75	70.0	20	0	5.0	1.0	0	0	0	0	0
1135	**Watercress**	1	cup(s)	34	32.3	4	0.8	0.4	0.2	0	0	0	0	—
	NUTS, SEEDS, AND PRODUCTS													
	Almonds													
32940	Almond butter with salt added	1	tablespoon(s)	16	0.2	101	2.4	3.4	0.6	9.5	0.9	6.1	2.0	—
1137	Almond butter, no salt added	1	tablespoon(s)	16	0.2	101	2.4	3.4	0.6	9.5	0.9	6.1	2.0	—
32886	Blanched	¼	cup(s)	36	1.6	211	8.0	7.2	3.8	18.3	1.4	11.7	4.4	—
32887	Dry roasted, no salt added	¼	cup(s)	35	0.9	206	7.6	6.7	4.1	18.2	1.4	11.6	4.4	—
29724	Dry roasted, salted	¼	cup(s)	35	0.9	206	7.6	6.7	4.1	18.2	1.4	11.6	4.4	—
29725	Oil roasted, salted	¼	cup(s)	39	1.1	238	8.3	6.9	4.1	21.7	1.7	13.7	5.3	—
508	Slivered	¼	cup(s)	27	1.3	155	5.7	5.9	3.3	13.3	1.0	8.3	3.3	0
1138	**Beechnuts, dried**	¼	cup(s)	57	3.8	328	3.5	19.1	5.3	28.5	3.3	12.5	11.4	—
517	**Brazil nuts, dried, unblanched**	¼	cup(s)	35	1.2	230	5.0	4.3	2.6	23.3	5.3	8.6	7.2	—
1166	**Breadfruit seeds, roasted**	¼	cup(s)	57	28.3	118	3.5	22.8	3.4	1.5	0.4	0.2	0.8	—
1139	**Butternuts, dried**	¼	cup(s)	30	1.0	184	7.5	3.6	1.4	17.1	0.4	3.1	12.8	—
	Cashews													
32931	Cashew butter with salt added	1	tablespoon(s)	16	0.5	94	2.8	4.4	0.3	7.9	1.6	4.7	1.3	—
32889	Cashew butter, no salt added	1	tablespoon(s)	16	0.5	94	2.8	4.4	0.3	7.9	1.6	4.7	1.3	—
1140	Dry roasted	¼	cup(s)	34	0.6	197	5.2	11.2	1.0	15.9	3.1	9.4	2.7	—
518	Oil roasted	¼	cup(s)	32	1.1	187	5.4	9.6	1.1	15.4	2.7	8.4	2.8	—
	Coconut, Shredded													
32896	Dried, not sweetened	¼	cup(s)	23	0.7	152	1.6	5.4	3.8	14.9	13.2	0.6	0.2	—
1153	Dried, shredded, sweetened	¼	cup(s)	23	2.9	116	0.7	11.1	1.0	8.3	7.3	0.4	0.1	—
520	Shredded	¼	cup(s)	20	9.4	71	0.7	3.0	1.8	6.7	5.9	0.3	0.1	—
	Chestnuts													
1152	Chinese, roasted	¼	cup(s)	36	14.6	87	1.6	19.0	—	0.4	0.1	0.2	0.1	—
32895	European, boiled and steamed	¼	cup(s)	46	31.3	60	0.9	12.8	—	0.6	0.1	0.2	0.2	—
32911	European, roasted	¼	cup(s)	36	14.5	88	1.1	18.9	1.8	0.8	0.1	0.3	0.3	—
32922	Japanese, boiled and steamed	¼	cup(s)	36	31.0	20	0.3	4.5	—	0.1	0	0	0	—
32923	Japanese, roasted	¼	cup(s)	36	18.1	73	1.1	16.4	—	0.3	0	0.1	0.1	—
4958	**Flax seeds or linseeds**	¼	cup(s)	43	3.3	225	8.4	12.3	11.9	17.7	1.7	3.2	12.6	0
32904	**Ginkgo nuts, dried**	¼	cup(s)	39	4.8	136	4.0	28.3	—	0.8	0.1	0.3	0.3	—
	Hazelnuts or filberts													
32901	Blanched	¼	cup(s)	30	1.7	189	4.1	5.1	3.3	18.3	1.4	14.5	1.7	—
32902	Dry roasted, no salt added	¼	cup(s)	30	0.8	194	4.5	5.3	2.8	18.7	1.3	14.0	2.5	—
1156	**Hickorynuts, dried**	¼	cup(s)	30	0.8	197	3.8	5.5	1.9	19.3	2.1	9.8	6.6	—
	Macadamias													
32905	Dry roasted, no salt added	¼	cup(s)	34	0.5	241	2.6	4.5	2.7	25.5	4.0	19.9	0.5	—
32932	Dry roasted, with salt added	¼	cup(s)	34	0.5	240	2.6	4.3	2.7	25.5	4.0	19.9	0.5	—
1157	Raw	¼	cup(s)	34	0.5	241	2.6	4.6	2.9	25.4	4.0	19.7	0.5	—
	Mixed nuts													
1159	With peanuts, dry roasted	¼	cup(s)	34	0.6	203	5.9	8.7	3.1	17.6	2.4	10.8	3.7	—
32933	With peanuts, dry roasted, with salt added	¼	cup(s)	34	0.6	203	5.9	8.7	3.1	17.6	2.4	10.8	3.7	—
32906	Without peanuts, oil roasted, no salt added	¼	cup(s)	36	1.1	221	5.6	8.0	2.0	20.2	3.3	11.9	4.1	—
	Peanuts													
2807	Dry roasted	¼	cup(s)	37	0.6	214	8.6	7.9	2.9	18.1	2.5	9.0	5.7	—
2806	Dry roasted, salted	¼	cup(s)	37	0.6	214	8.6	7.9	2.9	18.1	2.5	9.0	5.7	—
1763	Oil roasted, salted	¼	cup(s)	36	0.5	216	10.1	5.5	3.4	18.9	3.1	9.4	5.5	—
1884	Peanut butter, chunky	1	tablespoon(s)	16	0.2	94	3.8	3.5	1.3	8.0	1.3	3.9	2.4	—
30303	Peanut butter, low sodium	1	tablespoon(s)	16	0.2	95	4.0	3.1	0.9	8.2	1.8	3.9	2.2	—
30305	Peanut butter, reduced fat	1	tablespoon(s)	18	0.2	94	4.7	6.4	0.9	6.1	1.3	2.9	1.8	—
524	Peanut butter, smooth	1	tablespoon(s)	16	0.3	94	4.0	3.1	1.0	8.1	1.7	3.9	2.3	—
2804	Raw	¼	cup(s)	37	2.4	207	9.4	5.9	3.1	18.0	2.5	8.9	5.7	—
	Pecans													
32907	Dry roasted, no salt added	¼	cup(s)	28	0.3	198	2.6	3.8	2.6	20.7	1.8	12.3	5.7	—
	NUTS, SEEDS, AND PRODUCTS —Continued													

Chol (mg)	Calc (mg)	Iron (mg)	Magn (mg)	Pota (mg)	Sodi (mg)	Zinc (mg)	Vit A (µg)	Thia (mg)	Vit E (mg α)	Ribo (mg)	Niac (mg)	Vit B$_6$ (mg)	Fola (µg)	Vit C (mg)	Vit B$_{12}$ (µg)	Sele (µg)
0	13	0.73	22.3	430.4	73.2	0.21	17.6	0.05	—	0.10	0.99	0.08	6.3	28.0	0	0.8
0	99	0.58	15.8	146.2	20.9	0.10	274.3	0.03	1.35	0.05	0.30	0.13	85.0	19.7	0	0.6
0	125	1.59	21.3	183.7	12.3	0.34	441.2	0.04	2.18	0.06	0.38	0.06	32.0	17.9	0	1.0
0	26	0.14	7.0	138.1	12.5	0.09	0	0.02	0.02	0.02	0.23	0.05	7.0	9.0	0	0.2
0	22	0.86	13.0	237.2	121.4	0.33	475.1	0.04	0.24	0.04	0.47	0.06	19.6	4.1	0	0.2
0	23	0.74	20.0	153.8	31.9	0.44	194.7	0.06	0.34	0.10	0.77	0.06	17.3	2.9	0	0.3
0	20	0.36	12.9	260.0	310.0	0.24	100.0	0.05	—	0.03	0.87	0.17	—	30.0	0	—
0	20	0.36	—	450.0	70.0	—	100.0	0.02	—	0.02	0.75	—	—	30.0	0	—
0	20	0.36	12.9	240.0	360.0	0.24	50.0	0.05	—	0.03	0.88	0.17	—	15.0	0	—
0	7	0.00	—	—	5.0	—	0	—	—	—	—	—	—	2.0	—	—
0	7	0.00	—	—	5.0	—	0	—	—	—	—	—	—	2.0	—	—
0	41	0.06	7.1	112.2	13.9	0.03	54.4	0.03	0.34	0.04	0.06	0.04	3.1	14.6	0	0.3
0	43	0.59	48.5	121.3	72.0	0.49	0	0.02	4.16	0.10	0.46	0.01	10.4	0.1	0	0.8
0	43	0.59	48.5	121.3	1.8	0.48	0	0.02	—	0.09	0.46	0.01	10.4	0.1	0	—
0	78	1.34	99.7	249.0	10.2	1.13	0	0.07	8.95	0.20	1.32	0.04	10.9	0	0	1.0
0	92	1.55	98.7	257.4	0.3	1.22	0	0.02	8.97	0.29	1.32	0.04	11.4	0	0	1.0
0	92	1.55	98.7	257.4	117.0	1.22	0	0.02	8.97	0.29	1.32	0.04	11.4	0	0	1.0
0	114	1.44	107.5	274.4	133.1	1.20	0	0.03	10.19	0.30	1.43	0.04	10.6	0	0	1.1
0	71	1.00	72.4	190.4	0.3	0.83	0	0.05	7.07	0.27	0.91	0.03	13.5	0	0	0.7
0	1	1.39	0	579.7	21.7	0.20	0	0.16	—	0.20	0.48	0.38	64.4	8.8	0	4.0
0	56	0.85	131.6	230.7	1.1	1.42	0	0.21	2.00	0.01	0.10	0.03	7.7	0.2	0	671.0
0	49	0.50	35.3	616.7	15.9	0.58	8.5	0.22	—	0.12	4.20	0.22	33.6	4.3	0	8.0
0	16	1.21	71.1	126.3	0.3	0.94	1.8	0.12	—	0.04	0.31	0.17	19.8	1.0	0	5.2
0	7	0.81	41.3	87.4	98.2	0.83	0	0.05	0.15	0.03	0.26	0.04	10.9	0	0	1.8
0	7	0.81	41.3	87.4	2.4	0.83	0	0.05	—	0.03	0.26	0.04	10.9	0	0	1.8
0	15	2.06	89.1	193.5	5.5	1.92	0	0.07	0.32	0.07	0.48	0.09	23.6	0	0	4.0
0	14	1.95	88.0	203.8	4.2	1.73	0	0.12	0.30	0.07	0.56	0.10	8.1	0.1	0	6.5
0	6	0.76	20.7	125.2	8.5	0.46	0	0.01	0.10	0.02	0.13	0.07	2.1	0.3	0	4.3
0	3	0.45	11.6	78.4	60.9	0.42	0	0.01	0.09	0.00	0.11	0.06	1.9	0.2	0	3.9
0	3	0.48	6.4	71.2	4.0	0.21	0	0.01	0.04	0.00	0.11	0.01	5.2	0.7	0	2.0
0	7	0.54	32.6	173.0	1.4	0.33	0	0.05	—	0.03	0.54	0.15	26.1	13.9	0	2.6
0	21	0.80	24.8	328.9	12.4	0.11	0.5	0.06	—	0.03	0.32	0.10	17.5	12.3	0	—
0	10	0.32	11.8	211.6	0.7	0.20	0.4	0.08	0.18	0.05	0.48	0.18	25.0	9.3	0	0.4
0	4	0.19	6.5	42.8	1.8	0.14	0.4	0.04	—	0.01	0.19	0.03	6.1	3.4	0	—
0	13	0.75	23.2	154.8	6.9	0.51	1.4	0.15	—	—	0.24	0.14	21.4	10.1	0	—
0	142	2.13	156.1	354.0	11.9	1.83	0	0.06	0.14	0.06	0.59	0.39	118.4	0.5	0	2.3
0	8	0.62	20.7	390.2	5.1	0.26	21.5	0.17	—	0.07	4.58	0.25	41.4	11.4	0	—
0	45	0.98	48.0	197.4	0	0.66	0.6	0.14	5.25	0.03	0.46	0.17	23.4	0.6	0	1.2
0	37	1.31	51.9	226.5	0	0.74	0.9	0.10	4.58	0.03	0.61	0.18	26.4	1.1	0	1.2
0	18	0.64	51.9	130.8	0.3	1.29	2.1	0.26	—	0.04	0.27	0.06	12.0	0.6	0	2.4
0	23	0.88	39.5	121.6	1.3	0.43	0	0.23	0.19	0.02	0.76	0.12	3.4	0.2	0	3.9
0	23	0.88	39.5	121.6	88.8	0.43	0	0.23	0.19	0.02	0.76	0.12	3.4	0.2	0	3.9
0	28	1.24	43.6	123.3	1.7	0.44	0	0.40	0.18	0.05	0.83	0.09	3.7	0.4	0	1.2
0	24	1.27	77.1	204.5	4.1	1.30	0.3	0.07	—	0.07	1.61	0.10	17.1	0.1	0	1.0
0	24	1.26	77.1	204.5	229.1	1.30	0	0.06	3.74	0.06	1.61	0.10	17.1	0.1	0	2.6
0	38	0.92	90.4	195.8	4.0	1.67	0.4	0.18	—	0.17	0.70	0.06	20.2	0.2	0	—
0	20	0.82	64.2	240.2	2.2	1.20	0	0.16	2.52	0.03	4.93	0.09	52.9	0	0	2.7
0	20	0.82	64.2	240.2	296.7	1.20	0	0.16	2.84	0.03	4.93	0.09	52.9	0	0	2.7
0	22	0.54	63.4	261.4	115.2	1.18	0	0.03	2.49	0.03	4.97	0.16	43.2	0.3	0	1.2
0	7	0.30	25.6	119.2	77.8	0.45	0	0.02	1.01	0.02	2.19	0.07	14.7	0	0	1.3
0	6	0.29	25.4	107.0	2.7	0.47	0	0.01	1.23	0.02	2.14	0.07	11.8	0	0	1.2
0	6	0.34	30.6	120.4	97.2	0.50	0	0.05	1.20	0.01	2.63	0.06	10.8	0	0	1.4
0	7	0.30	24.6	103.8	73.4	0.47	0	0.01	1.44	0.02	2.14	0.09	11.8	0	0	0.9
0	34	1.67	61.3	257.3	6.6	1.19	0	0.23	3.04	0.04	4.40	0.12	87.6	0	0	2.6
0	20	0.78	36.8	118.3	0.3	1.41	1.9	0.12	0.35	0.03	0.32	0.05	4.5	0.2	0	1.1

DA+ Code	Food Description	Quantity	Measure	Wt (g)	H₂O (g)	Ener (kcal)	Prot (g)	Carb (g)	Fiber (g)	Fat (g)	Sat	Mono	Poly	Trans
32936	Dry roasted, with salt added	¼	cup(s)	27	0.3	192	2.6	3.7	2.5	20.0	1.7	11.9	5.6	—
1162	Oil roasted	¼	cup(s)	28	0.3	197	2.5	3.6	2.6	20.7	2.0	11.3	6.5	—
526	Raw	¼	cup(s)	27	1.0	188	2.5	3.8	2.6	19.6	1.7	11.1	5.9	—
12973	Pine nuts or pignolia, dried	1	tablespoon(s)	9	0.2	58	1.2	1.1	0.3	5.9	0.4	1.6	2.9	—
	Pistachios													
1164	Dry roasted	¼	cup(s)	31	0.6	176	6.6	8.5	3.2	14.1	1.7	7.4	4.3	—
32938	Dry roasted, with salt added	¼	cup(s)	32	0.6	182	6.8	8.6	3.3	14.7	1.8	7.7	4.4	—
1167	Pumpkin or squash seeds, roasted	¼	cup(s)	57	4.0	296	18.7	7.6	2.2	23.9	4.5	7.4	10.9	—
	Sesame													
32912	Sesame butter paste	1	tablespoon(s)	16	0.3	94	2.9	3.8	0.9	8.1	1.1	3.1	3.6	—
32941	Tahini or sesame butter	1	tablespoon(s)	15	0.5	89	2.6	3.2	0.7	8.0	1.1	3.0	3.5	—
1169	Whole, roasted, toasted	3	tablespoon(s)	10	0.3	54	1.6	2.4	1.3	4.6	0.6	1.7	2.0	—
	Soy nuts													
34173	Deep sea salted	¼	cup(s)	28	—	119	11.9	8.9	4.9	4.0	1.0	—	—	—
34174	Unsalted	¼	cup(s)	28	—	119	11.9	8.9	4.9	4.0	0	—	—	—
	Sunflower seeds													
528	Kernels, dried	1	tablespoon(s)	9	0.4	53	1.9	1.8	0.8	4.6	0.4	1.7	2.1	—
29721	Kernels, dry roasted, salted	1	tablespoon(s)	8	0.1	47	1.5	1.9	0.7	4.0	0.4	0.8	2.6	—
29723	Kernels, toasted, salted	1	tablespoon(s)	8	0.1	52	1.4	1.7	1.0	4.8	0.5	0.9	3.1	—
32928	Sunflower seed butter with salt added	1	tablespoon(s)	16	0.2	93	3.1	4.4	—	7.6	0.8	1.5	5.0	—
	Trail mix													
4646	Trail mix	¼	cup(s)	38	3.5	173	5.2	16.8	2.0	11.0	2.1	4.7	3.6	—
4647	Trail mix with chocolate chips	¼	cup(s)	38	2.5	182	5.3	16.8	—	12.0	2.3	5.1	4.2	—
4648	Tropical trail mix	¼	cup(s)	35	3.2	142	2.2	23.0	—	6.0	3.0	0.9	1.8	—
	Walnuts													
529	Dried black, chopped	¼	cup(s)	31	1.4	193	7.5	3.1	2.1	18.4	1.1	4.7	11.0	—
531	English or Persian	¼	cup(s)	29	1.2	191	4.5	4.0	2.0	19.1	1.8	2.6	13.8	—
	VEGETARIAN FOODS													
	Prepared													
34222	Brown rice & tofu stir-fry (vegan)	8	ounce(s)	227	244.4	302	16.5	18.0	3.2	21.0	1.7	4.7	13.4	0
34368	Cheese enchilada casserole (lacto)	8	ounce(s)	227	80.3	385	16.6	38.4	4.1	17.8	9.5	6.1	1.1	—
34247	Five bean casserole (vegan)	8	ounce(s)	227	175.8	178	5.9	26.6	6.0	5.8	1.1	2.5	1.9	0
34261	Lentil stew (vegan)	8	ounce(s)	227	227.9	188	11.5	35.9	11.0	0.1	0.1	0.1	0.3	0
34397	Macaroni and cheese (lacto)	8	ounce(s)	227	352.1	391	18.1	37.1	1.0	18.7	9.8	6.0	1.8	0
34238	Steamed rice and vegetables (vegan)	8	ounce(s)	227	222.9	587	11.2	87.9	5.8	23.1	4.1	8.7	9.1	0
34308	Tofu rice burgers (ovo-lacto)	1	piece(s)	218	77.6	435	22.4	68.6	5.6	8.4	1.7	2.4	3.5	—
34276	Vegan spinach enchiladas (vegan)	1	piece(s)	82	59.2	93	4.9	14.5	1.8	2.4	0.3	0.6	1.3	—
34243	Vegetable chow mein (vegan)	8	ounce(s)	227	163.3	166	6.5	22.1	2.0	6.4	0.7	2.7	2.5	0
34454	Vegetable lasagna (lacto)	8	ounce(s)	227	178.9	208	13.7	29.9	2.6	4.1	2.3	1.1	0.3	—
34339	Vegetable marinara (vegan)	8	ounce(s)	252	200.7	104	3.0	16.7	1.4	3.1	0.4	1.4	1.0	0
34356	Vegetable rice casserole (lacto)	8	ounce(s)	227	178.9	238	9.7	24.4	4.0	12.5	4.9	3.5	3.1	—
34311	Vegetable strudel (ovo-lacto)	8	ounce(s)	227	63.1	478	12.0	32.4	2.5	33.8	11.5	16.7	3.9	0
34371	Vegetable taco (lacto)	1	item(s)	85	46.5	117	4.2	13.6	2.9	5.6	2.1	1.9	1.3	—
34282	Vegetarian chili (vegan)	8	ounce(s)	227	191.4	115	5.6	21.4	7.1	1.5	0.2	0.3	0.7	0
34367	Vegetarian vegetable soup (vegan)	8	ounce(s)	227	257.9	111	3.2	16.0	3.2	5.0	1.0	2.1	1.6	0
	Boca burger													
32067	All American flamed grilled patty	1	item(s)	71	—	90	14.0	4.0	3.0	3.0	1.0	—	—	0
32074	Boca Chik n nuggets	4	item(s)	87	—	180	14.0	17.0	3.0	7.0	1.0	—	—	0
32075	Boca meatless ground burger	½	cup(s)	57	—	60	13.0	6.0	3.0	0.5	0	—	—	0
32072	Breakfast links	2	item(s)	45	—	70	8.0	5.0	2.0	3.0	0.5	—	—	0
32071	Breakfast patties	1	item(s)	38	—	60	7.0	5.0	2.0	2.5	0	—	—	0
35780	Cheeseburger meatless burger patty	1	item(s)	71	—	100	12.0	5.0	3.0	5.0	1.5	—	—	0
33958	Original meatless Chik 'n patties	1	item(s)	71	—	160	11.0	15.0	2.0	6.0	1.0	—	—	0
32066	Original patty	1	item(s)	71	—	70	13.0	6.0	4.0	0.5	0	—	—	0
32068	Roasted garlic patty	1	item(s)	71	—	70	12.0	6.0	4.0	1.5	0	—	—	0
37814	Roasted onion meatless burger patty	1	item(s)	71	—	70	11.0	7.0	4.0	1.0	0	—	—	0
	Gardenburger													
37810	BBQ chik'n with sauce	1	item(s)	142	—	250	14.0	30.0	5.0	8.0	1.0	—	—	0
39661	Black bean burger	1	item(s)	71	—	80	8.0	11.0	4.0	2.0	0	—	—	0
39666	Buffalo chick'n wing	3	item(s)	95	—	180	9.0	8.0	5.0	12.0	1.5	—	—	0
39665	Country fried chicken with creamy pepper gravy	1	item(s)	142	—	190	9.0	16.0	2.0	9.0	1.0	—	—	0
37808	Flamed grilled Chik'n	1	item(s)	71	—	100	13.0	5.0	3.0	2.5	0	—	—	0
37803	Garden vegan	1	item(s)	71	—	100	10.0	12.0	2.0	1.0	—	—	—	0
39663	Homestyle classic burger	1	item(s)	71	—	110	12.0	6.0	4.0	5.0	0.5	—	—	0
37807	Meatless breakfast sausage	1	item(s)	43	—	50	5.0	2.0	2.0	3.5	0	—	—	0
37809	Meatless meatballs	6	item(s)	85	—	110	12.0	8.0	4.0	4.5	1.0	—	—	0
37806	Meatless riblets with sauce	1	item(s)	142	—	160	17.0	11.0	4.0	5.0	0	—	—	0
29913	Original	1	item(s)	71	—	90	10.0	8.0	3.0	2.0	0.5	—	—	0
39662	Sun-dried tomato basil burger	1	item(s)	71	—	80	10.0	11.0	3.0	1.5	0.5	—	—	0

VEGETARIAN FOODS —Continued

Chol (mg)	Calc (mg)	Iron (mg)	Magn (mg)	Pota (mg)	Sodi (mg)	Zinc (mg)	Vit A (µg)	Thia (mg)	Vit E (mg α)	Ribo (mg)	Niac (mg)	Vit B_6 (mg)	Fola (µg)	Vit C (mg)	Vit B_{12} (µg)	Sele (µg)
0	19	0.75	35.6	114.5	103.4	1.36	1.9	0.11	0.34	0.03	0.31	0.05	4.3	0.2	0	1.1
0	18	0.68	33.3	107.8	0.3	1.23	1.4	0.13	0.70	0.03	0.33	0.05	4.1	0.2	0	1.7
0	19	0.69	33.0	111.7	0	1.23	0.8	0.18	0.38	0.04	0.32	0.06	6.0	0.3	0	1.0
0	1	0.47	21.6	51.3	0.2	0.55	0.1	0.03	0.80	0.02	0.37	0.01	2.9	0.1	0	0.1
0	34	1.29	36.9	320.4	3.1	0.71	4.0	0.26	0.59	0.05	0.44	0.39	15.4	0.7	0	2.9
0	35	1.34	38.4	333.4	129.6	0.73	4.2	0.26	0.61	0.05	0.45	0.40	16.0	0.7	0	3.0
0	24	8.48	303.0	457.4	10.2	4.22	10.8	0.12	0.00	0.18	0.99	0.05	32.3	1.0	0	3.2
0	154	3.07	57.9	93.1	1.9	1.17	0.5	0.04	—	0.03	1.07	0.13	16.0	0	0	0.9
0	21	0.66	14.3	68.9	5.3	0.69	0.5	0.24	—	0.02	0.85	0.02	14.7	0.6	0	0.3
0	94	1.40	33.8	45.1	1.0	0.68	0	0.07	—	0.02	0.43	0.07	9.3	0	0	0.5
0	59	1.07	—	—	148.1	—	0	—	—	—	—	—	—	0	—	—
0	59	1.07	—	—	9.9	—	0	—	—	—	—	—	—	0	—	—
0	7	0.47	29.3	58.1	0.8	0.45	0.3	0.13	2.99	0.03	0.75	0.12	20.4	0.1	0	4.8
0	6	0.30	10.3	68.0	32.8	0.42	0	0.01	2.09	0.02	0.56	0.06	19.0	0.1	0	6.3
0	5	0.57	10.8	41.1	51.3	0.44	0	0.03	—	0.02	0.35	0.07	19.9	0.1	0	5.2
0	20	0.76	59.0	11.5	83.2	0.85	0.5	0.05	—	0.05	0.85	0.13	37.9	0.4	0	—
0	29	1.14	59.3	256.9	85.9	1.20	0.4	0.17	—	0.07	1.76	0.11	26.6	0.5	0	—
2	41	1.27	60.4	243.0	45.4	1.17	0.8	0.15	—	0.08	1.65	0.09	24.4	0.5	0	—
0	20	0.92	33.6	248.2	3.5	0.41	0.7	0.15	—	0.04	0.51	0.11	14.7	2.7	0	—
0	19	0.97	62.8	163.4	0.6	1.05	0.6	0.01	0.56	0.04	0.14	0.18	9.7	0.5	0	5.3
0	29	0.85	46.2	129.0	0.6	0.90	0.3	0.10	0.20	0.04	0.32	0.15	28.7	0.4	0	1.4
0	353	6.34	118.3	501.4	142.2	2.03	—	0.23	0.07	0.14	1.49	0.36	51.8	24.8	0	14.8
39	441	2.44	34.6	191.2	1139.7	1.84	—	0.31	0.05	0.35	2.23	0.11	87.1	20.4	0.4	20.0
0	48	1.78	40.8	364.1	613.6	0.61	—	0.10	0.52	0.07	0.93	0.11	39.4	8.3	0	3.3
0	34	3.23	50.0	548.8	436.5	1.42	—	0.24	0.14	0.16	2.31	0.29	91.4	26.4	0	12.1
43	415	1.71	45.4	267.8	1641.0	2.32	—	0.32	0.27	0.48	2.18	0.13	74.2	0.9	0.8	33.3
0	91	3.31	153.1	810.1	3117.8	2.04	—	0.37	3.03	0.21	6.16	0.64	61.7	35.2	0	18.8
52	467	9.01	89.7	455.6	2449.5	2.06	—	0.27	0.12	0.26	3.43	0.29	106.6	2.0	0.1	43.0
0	117	1.13	40.4	170.5	134.2	0.68	—	0.07	—	0.07	0.53	0.10	52.1	1.8	0	5.1
0	189	3.70	28.0	310.3	372.7	0.76	—	0.13	0.05	0.11	1.43	0.14	45.6	8.0	0	6.5
10	176	1.86	41.9	470.0	759.4	1.14	—	0.26	0.05	0.25	2.49	0.22	64.7	19.0	0.4	21.8
0	17	0.94	19.1	189.9	439.6	0.42	—	0.15	0.55	0.08	1.36	0.12	40.6	23.5	0	10.8
17	190	1.28	29.3	414.2	626.0	1.24	—	0.16	0.35	0.29	2.00	0.19	72.6	56.0	0.2	5.8
29	200	2.15	24.5	181.0	512.1	1.24	—	0.28	0.20	0.31	2.88	0.11	88.3	17.4	0.2	19.7
7	77	0.88	26.3	174.1	280.7	0.59	—	0.08	0.04	0.06	0.49	0.08	38.7	4.6	0	3.0
0	65	1.98	41.0	543.1	390.7	0.74	—	0.14	0.15	0.10	1.31	0.18	59.3	20.3	0	4.4
0	46	1.87	34.9	550.3	729.5	0.56	—	0.13	0.55	0.09	1.99	0.27	47.8	29.9	0	1.4
5	150	1.00	—	—	280.0	—	0	—	—	—	—	—	—	0	—	—
0	40	1.44	—	—	500.0	—	—	—	—	—	—	—	—	0	—	—
0	60	1.80	—	—	270.0	—	0	—	—	—	—	—	—	0	—	—
0	20	1.44	—	—	330.0	—	0	—	—	—	—	—	—	0	—	—
0	20	1.08	—	—	280.0	—	0	—	—	—	—	—	—	0	—	—
5	80	1.80	—	—	360.0	—	—	—	—	—	—	—	—	0	—	—
0	40	1.80	—	—	430.0	—	—	—	—	—	—	—	—	0	—	—
0	60	1.80	—	—	280.0	—	0	—	—	—	—	—	—	0	—	—
0	60	1.80	—	—	370.0	—	0	—	—	—	—	—	—	0	—	—
0	100	2.70	—	—	300.0	—	—	—	—	—	—	—	—	0	—	—
0	150	1.08	—	—	890.0	—	—	—	—	—	—	—	—	0	—	—
0	40	1.44	—	—	330.0	—	—	—	—	—	—	—	—	0	—	—
0	40	0.72	—	—	1000.0	—	—	—	—	—	—	—	—	0	—	—
5	40	1.44	—	—	550.0	—	—	—	—	—	—	—	—	0	—	—
0	60	3.60	—	—	360.0	—	—	—	—	—	—	—	—	0	—	—
0	40	4.50	—	—	230.0	—	—	—	—	—	—	—	—	0	—	—
0	80	1.44	—	—	380.0	—	—	—	—	—	—	—	—	0	—	—
0	20	0.72	—	—	120.0	—	—	—	—	—	—	—	—	0	—	—
0	60	1.80	—	—	400.0	—	—	—	—	—	—	—	—	0	—	—
0	60	1.80	—	—	720.0	—	—	—	—	—	—	—	—	3.6	—	—
0	80	1.08	30.4	193.4	490.0	0.89	—	0.10	—	0.15	1.08	0.08	10.1	1.2	0.1	7.0
5	60	1.44	—	—	260.0	—	—	—	—	—	—	—	—	3.6	—	—

DA+ Code	Food Description	Quantity	Measure	Wt (g)	H₂O (g)	Ener (kcal)	Prot (g)	Carb (g)	Fiber (g)	Fat (g)	Sat	Mono	Poly	Trans
											Fat Breakdown (g)			
29915	Veggie medley	1	item(s)	71	—	90	9.0	11.0	4.0	2.0	0	—	—	0
	Loma Linda													
9311	Big franks, canned	1	item(s)	51	—	110	11.0	3.0	2.0	6.0	1.0	1.5	3.5	0
9323	Fried Chik'n with gravy	2	piece(s)	80	45.9	150	12.0	5.0	2.0	10	1.5	2.5	5.0	0
9326	Linketts, canned	1	item(s)	35	21.0	70	7.0	1.0	1.0	4.0	0.5	1.0	2.5	0
9336	Redi-Burger patties, canned	1	slice(s)	85	50.5	120	18.0	7.0	4.0	2.5	0.5	0.5	1.5	0
9350	Swiss Stake pattie with gravy, frozen	1	piece(s)	92	65.7	130	9.0	9.0	3.0	6.0	1.0	1.5	3.5	0
9354	Tender Rounds meatball substitute, canned in gravy	6	piece(s)	80	53.9	120	13.0	6.0	1.0	4.5	0.5	1.5	2.5	0
	Morningstar Farms													
33707	America's Original Veggie Dog links	1	item(s)	57	—	80	11.0	6.0	1.0	0.5	0	—	—	0
9362	Better 'n Eggs egg substitute	¼	cup(s)	57	50.3	20	5.0	0	0	0	0	0	0	0
9371	Breakfast bacon strips	2	item(s)	16	6.8	60	2.0	2.0	0.5	4.5	0.5	1.0	3.0	0
9368	Breakfast sausage links	2	item(s)	45	26.8	80	9.0	3.0	2.0	3.0	0.5	1.5	1.0	0
33705	Chik 'n nuggets	4	piece(s)	86	—	190	12.0	18.0	2.0	7.0	1.0	2.0	4.0	0
11587	Chik patties	1	item(s)	71	36.3	150	9.0	16.0	2.0	6.0	1.0	1.5	2.5	0
2531	Garden veggie patties	1	item(s)	67	40.1	100	10.0	9.0	4.0	2.5	0.5	0.5	1.5	0
33702	Spicy black bean veggie burger	1	item(s)	78	—	140	12.0	15.0	3.0	4.0	0.5	1.0	2.5	0
9412	Vegetarian chili, canned	1	cup(s)	230	172.6	180	16.0	25.0	10.0	1.5	0.5	0.5	0.5	0
	Worthington													
9424	Chili, canned	1	cup(s)	230	167.0	280	24.0	25.0	8.0	10.0	1.5	1.5	7.0	0
9436	Diced Chik, canned	¼	cup(s)	55	42.7	50	9.0	2.0	1.0	0	0	0	0	0
9440	Dinner roast, frozen	1	slice(s)	85	53.2	180	14.0	6.0	3.0	11.0	1.5	4.5	5.0	0
9420	Meatless chicken slices, frozen	3	slice(s)	57	38.9	90	9.0	2.0	0.5	4.5	1.0	1.0	2.5	0
36702	Meatless chicken style roll, frozen	1	slice(s)	55	—	90	9.0	2.0	1.0	4.5	1.0	1.0	2.5	0
9428	Meatless corned beef, sliced, frozen	3	slice(s)	57	31.2	140	10.0	5.0	0	9.0	1.0	2.0	5.0	0
9470	Meatless salami, sliced, frozen	3	slice(s)	57	32.4	120	12.0	3.0	2.0	7.0	1.0	1.0	5.0	0
9480	Meatless smoked turkey, sliced	3	slice(s)	57	—	140	10.0	4.0	0	9.0	1.5	2.0	5.0	0
9462	Prosage links	2	item(s)	45	26.8	80	9.0	3.0	2.0	3.0	0.5	0.5	2.0	0
9484	Stakelets patty beef steak substitute, frozen	1	piece(s)	71	41.5	150	14.0	7.0	2.0	7.0	1.0	2.5	3.5	0
9486	Stripples bacon substitute	2	item(s)	16	6.8	60	2.0	2.0	0.5	4.5	0.5	1.0	3.0	0
9496	Vegetable Skallops meat substitute, canned	½	cup(s)	85	—	90	17.0	4.0	3.0	1.0	0	0	0.5	0
	DAIRY													
	Cheese													
1433	Blue, crumbled	1	ounce(s)	28	12.0	100	6.1	0.7	0	8.1	5.3	2.2	0.2	—
884	Brick	1	ounce(s)	28	11.7	105	6.6	0.8	0	8.4	5.3	2.4	0.2	—
885	Brie	1	ounce(s)	28	13.7	95	5.9	0.1	0	7.8	4.9	2.3	0.2	—
34821	Camembert	1	ounce(s)	28	14.7	85	5.6	0.1	0	6.9	4.3	2.0	0.2	—
5	Cheddar, shredded	¼	cup(s)	28	10.4	114	7.0	0.4	0	9.4	6.0	2.7	0.3	—
888	Cheddar or colby	1	ounce(s)	28	10.8	112	6.7	0.7	0	9.1	5.7	2.6	0.3	—
32096	Cheddar or colby, low fat	1	ounce(s)	28	17.9	49	6.9	0.5	0	2.0	1.2	0.6	0.1	—
889	Edam	1	ounce(s)	28	11.8	101	7.1	0.4	0	7.9	5.0	2.3	0.2	—
890	Feta	1	ounce(s)	28	15.7	75	4.0	1.2	0	6.0	4.2	1.3	0.2	—
891	Fontina	1	ounce(s)	28	10.8	110	7.3	0.4	0	8.8	5.4	2.5	0.5	—
8527	Goat cheese, soft	1	ounce(s)	28	17.2	76	5.3	0.3	0	6.0	4.1	1.4	0.1	—
893	Gouda	1	ounce(s)	28	11.8	101	7.1	0.6	0	7.8	5.0	2.2	0.2	—
894	Gruyere	1	ounce(s)	28	9.4	117	8.5	0.1	0	9.2	5.4	2.8	0.5	—
895	Limburger	1	ounce(s)	28	13.7	93	5.7	0.1	0	7.7	4.7	2.4	0.1	—
896	Monterey jack	1	ounce(s)	28	11.6	106	6.9	0.2	0	8.6	5.4	2.5	0.3	—
13	Mozzarella, part skim milk	1	ounce(s)	28	15.2	72	6.9	0.8	0	4.5	2.9	1.3	0.1	—
12	Mozzarella, whole milk	1	ounce(s)	28	14.2	85	6.3	0.6	0	6.3	3.7	1.9	0.2	—
897	Muenster	1	ounce(s)	28	11.8	104	6.6	0.3	0	8.5	5.4	2.5	0.2	—
898	Neufchatel	1	ounce(s)	28	17.6	74	2.8	0.8	0	6.6	4.2	1.9	0.2	—
14	Parmesan, grated	1	tablespoon(s)	5	1.0	22	1.9	0.2	0	1.4	0.9	0.4	0.1	—
17	Provolone	1	ounce(s)	28	11.6	100	7.3	0.6	0	7.5	4.8	2.1	0.2	—
19	Ricotta, part skim milk	¼	cup(s)	62	45.8	85	7.0	3.2	0	4.9	3.0	1.4	0.2	—
18	Ricotta, whole milk	¼	cup(s)	62	44.1	107	6.9	1.9	0	8.0	5.1	2.2	0.2	—
20	Romano	1	tablespoon(s)	5	1.5	19	1.6	0.2	0	1.3	0.9	0.4	0	—
900	Roquefort	1	ounce(s)	28	11.2	105	6.1	0.6	0	8.7	5.5	2.4	0.4	—
21	Swiss	1	ounce(s)	28	10.5	108	7.6	1.5	0	7.9	5.0	2.1	0.3	—
	Imitation cheese													
42245	Imitation American cheddar cheese	1	ounce(s)	28	15.1	68	4.7	3.3	0	4.0	2.5	1.2	0.1	—
53914	Imitation cheddar	1	ounce(s)	28	15.1	68	4.7	3.3	0	4.0	2.5	1.2	0.1	—
	Cottage cheese													
9	Low fat, 1% fat	½	cup(s)	113	93.2	81	14.0	3.1	0	1.2	0.7	0.3	0	—
8	Low fat, 2% fat	½	cup(s)	113	89.6	102	15.5	4.1	0	2.2	1.4	0.6	0.1	—
	Cream cheese													
11	Cream cheese	2	tablespoon(s)	29	15.6	101	2.2	0.8	0	10.1	6.4	2.9	0.4	—
17366	Fat free cream cheese	2	tablespoon(s)	30	22.7	29	4.3	1.7	0	0.4	0.3	0.1	0	—
10438	Tofutti Better than Cream Cheese	2	tablespoon(s)	30	—	80	1.0	1.0	0	8.0	2.0	—	6.0	
	DAIRY —Continued													

Chol (mg)	Calc (mg)	Iron (mg)	Magn (mg)	Pota (mg)	Sodi (mg)	Zinc (mg)	Vit A (µg)	Thia (mg)	Vit E (mg α)	Ribo (mg)	Niac (mg)	Vit B6 (mg)	Fola (µg)	Vit C (mg)	Vit B12 (µg)	Sele (µg)
0	40	1.44	27.0	182.0	290.0	0.46	—	0.07	—	0.08	0.90	0.09	10.6	9.0	0	4.0
0	0	0.77	—	50.0	220.0	—	0	0.22	—	0.10	2.00	0.70	—	0	2.4	—
0	20	1.80	—	70.0	430.0	0.33	0	1.05	—	0.34	4.00	0.30	—	0	2.4	—
0	0	0.36	—	20.0	160.0	0.46	0	0.12	—	0.20	0.80	0.16	—	0	0.9	—
0	0	1.06	—	140.0	450.0	—	0	0.15	—	0.25	4.00	0.40	—	0	1.2	—
0	0	0.72	—	200.0	430.0	—	0	0.45	—	0.25	10.00	1.00	—	0	5.4	—
0	20	1.08	—	80.0	340.0	0.66	0	0.75	—	0.17	2.00	0.16	—	0	1.2	—
0	0	0.72	—	60.0	580.0	—	0	—	—	—	—	—	—	0	—	—
0	20	0.72	—	75.0	90.0	0.60	37.5	0.03	—	0.34	0.00	0.08	24.0	—	0.6	—
0	0	0.36	—	15.0	220.0	0.05	0	0.75	—	0.04	0.40	0.07	—	0	0.2	—
0	0	1.80	—	50.0	300.0	—	0	0.37	—	0.17	7.00	0.50	—	0	3.0	—
0	20	2.70	—	320.0	490.0	—	0	0.52	—	0.25	5.00	0.30	—	0	1.5	—
0	0	1.80	—	210.0	540.0	—	0	1.80	—	0.17	2.00	0.20	—	0	1.2	—
0	40	0.72	—	180.0	350.0	—	—	—	—	—	—	—	—	0	—	—
0	40	1.80	—	320.0	470.0	—	0	—	—	—	0.00	—	—	0	—	—
0	40	3.60	—	660.0	900.0	—	—	—	—	—	—	—	—	0	—	—
0	40	3.60	—	330.0	1130.0	—	0	0.30	—	0.13	2.00	0.70	—	0	1.5	—
0	0	1.08	—	100.0	220.0	0.24	0	0.06	—	0.10	4.00	0.08	—	0	0.2	—
0	20	1.80	—	120.0	580.0	0.64	0	1.80	—	0.25	6.00	0.60	—	0	1.5	—
0	250	1.80	—	250.0	250.0	0.26	0	0.37	—	0.13	4.00	0.30	—	0	1.8	—
0	100	1.08	—	240.0	240.0	—	0	0.37	—	0.13	4.00	0.30	—	0	1.8	—
0	0	1.80	—	130.0	460.0	0.26	0	0.45	—	0.17	5.00	0.30	—	0	1.8	—
0	0	1.08	—	95.0	800.0	0.30	0	0.75	—	0.17	4.00	0.20	—	0	0.6	—
0	60	2.70	—	60.0	450.0	0.23	0	1.80	—	0.17	6.00	0.40	—	0	3.0	—
0	0	1.44	—	50.0	320.0	0.36	0	1.80	—	0.17	2.00	0.30	—	0	3.0	—
0	40	1.08	—	130.0	480.0	0.50	0	1.20	—	0.13	3.00	0.30	—	0	1.5	—
0	0	0.36	—	15.0	220.0	0.05	0	0.75	—	0.03	0.40	0.08	—	0	0.2	—
0	0	0.36	—	10.0	390.0	0.67	0	0.03	—	0.03	0.00	0.01	—	0	0	—
21	150	0.08	6.5	72.6	395.5	0.75	56.1	0.01	0.07	0.10	0.28	0.04	10.2	0	0.3	4.1
27	191	0.12	6.8	38.6	158.8	0.73	82.8	0.00	0.07	0.10	0.03	0.01	5.7	0	0.4	4.1
28	52	0.14	5.7	43.1	178.3	0.67	49.3	0.02	0.06	0.14	0.10	0.06	18.4	0	0.5	4.1
20	110	0.09	5.7	53.0	238.7	0.67	68.3	0.01	0.06	0.14	0.18	0.06	17.6	0	0.4	4.1
30	204	0.19	7.9	27.7	175.4	0.87	74.9	0.01	0.08	0.10	0.02	0.02	5.1	0	0.2	3.9
27	194	0.21	7.4	36.0	171.2	0.87	74.8	0.00	0.07	0.10	0.02	0.02	5.1	0	0.2	4.1
6	118	0.11	4.5	18.7	173.5	0.51	17.0	0.00	0.01	0.06	0.01	0.01	3.1	0	0.1	4.1
25	207	0.12	8.5	53.3	273.6	1.06	68.9	0.01	0.06	0.11	0.02	0.02	4.5	0	0.4	4.1
25	140	0.18	5.4	17.6	316.4	0.81	35.4	0.04	0.05	0.23	0.28	0.12	9.1	0	0.5	4.3
33	156	0.06	4.0	18.1	226.8	0.99	74.0	0.01	0.07	0.05	0.04	0.02	1.7	0	0.5	4.1
13	40	0.53	4.5	7.4	104.3	0.26	81.6	0.02	0.05	0.10	0.12	0.07	3.4	0	0.1	0.8
32	198	0.06	8.2	34.3	232.2	1.10	46.8	0.01	0.06	0.09	0.01	0.02	6.0	0	0.4	4.1
31	287	0.04	10.7	23.0	95.3	1.10	76.8	0.01	0.07	0.07	0.03	0.02	2.8	0	0.5	4.1
26	141	0.03	6.0	36.3	226.8	0.59	96.4	0.02	0.06	0.14	0.04	0.02	16.4	0	0.3	4.1
25	211	0.20	7.7	23.0	152.0	0.85	56.1	0.00	0.07	0.11	0.02	0.02	5.1	0	0.2	4.1
18	222	0.06	6.5	23.8	175.5	0.78	36.0	0.01	0.04	0.08	0.03	0.02	2.6	0	0.2	4.1
22	143	0.12	5.7	21.5	177.8	0.82	50.7	0.01	0.05	0.08	0.02	0.01	2.0	0	0.6	4.8
27	203	0.11	7.7	38.0	178.0	0.79	84.5	0.00	0.07	0.09	0.02	0.01	3.4	0	0.4	4.1
22	21	0.07	2.3	32.3	113.1	0.14	84.5	0.00	—	0.05	0.03	0.01	3.1	0	0.1	0.9
4	55	0.04	1.9	6.3	76.5	0.19	6.0	0.00	0.01	0.02	0.01	0.00	0.5	0	0.1	0.9
20	214	0.14	7.9	39.1	248.3	0.91	66.9	0.01	0.06	0.09	0.04	0.02	2.8	0	0.4	4.1
19	167	0.27	9.2	76.9	76.9	0.82	65.8	0.01	0.04	0.11	0.04	0.01	8.0	0	0.2	10.3
31	127	0.23	6.8	64.6	51.7	0.71	73.8	0.01	0.06	0.12	0.06	0.02	7.4	0	0.2	8.9
5	53	0.03	2.1	4.3	60	0.12	4.8	0.00	0.01	0.01	0.00	0.00	0.5	0	0.1	0.7
26	188	0.15	8.5	25.8	512.9	0.59	83.3	0.01	—	0.16	0.20	0.03	13.9	0	0.2	4.1
26	224	0.05	10.8	21.8	54.4	1.23	62.4	0.01	0.10	0.08	0.02	0.02	1.7	0	0.9	5.2
10	159	0.08	8.2	68.6	381.3	0.73	32.3	0.01	0.07	0.12	0.03	0.03	2.0	0	0.1	4.3
10	159	0.09	8.2	68.6	381.3	0.73	32.3	0.01	0.07	0.12	0.04	0.03	2.0	0	0.1	4.3
5	69	0.15	5.7	97.2	458.8	0.42	12.4	0.02	0.01	0.18	0.14	0.07	13.6	0	0.7	10.2
9	78	0.18	6.8	108.5	458.8	0.47	23.7	0.02	0.02	0.20	0.16	0.08	14.7	0	0.8	11.5
32	23	0.34	1.7	34.5	85.8	0.15	106.1	0.01	0.08	0.05	0.02	0.01	3.8	0	0.1	0.7
2	56	0.05	4.2	48.9	163.5	0.26	83.7	0.01	0.00	0.05	0.04	0.01	11.1	0	0.2	1.5
0	0	0.00	—	—	135.0	—	0	—	—	—	—	—	—	0	—	—

Food Composition   (Computer code is for Cengage Diet Analysis program)   (For purposes of calculations, use "0" for t, <1, <.1, <.01, etc.)

DA+ Code	Food Description	Quantity	Measure	Wt (g)	H₂O (g)	Ener (kcal)	Prot (g)	Carb (g)	Fiber (g)	Fat (g)	Fat Breakdown (g)			
											Sat	Mono	Poly	Trans
	Processed cheese													
24	American cheese food, processed	1	ounce(s)	28	12.3	94	5.2	2.2	0	7.1	4.2	2.0	0.3	—
25	American cheese spread, processed	1	ounce(s)	28	13.5	82	4.7	2.5	0	6.0	3.8	1.8	0.2	—
22	American cheese, processed	1	ounce(s)	28	11.1	106	6.3	0.5	0	8.9	5.6	2.5	0.3	—
9110	Kraft deluxe singles pasteurized process American cheese	1	ounce(s)	28	—	108	5.4	0	0	9.5	5.4	—	—	—
23	Swiss cheese, processed	1	ounce(s)	28	12.0	95	7.0	0.6	0	7.1	4.5	2.0	0.2	—
	Soy cheese													
10437	Galaxy Foods vegan grated parmesan cheese alternative	1	tablespoon(s)	8	—	23	3.0	1.5	0	0	0	0	0	0
10430	Nu Tofu cheddar flavored cheese alternative	1	ounce(s)	28	—	70	6.0	1.0	0	4.0	0.5	2.5	1.0	—
	Cream													
26	Half and half cream	1	tablespoon(s)	15	12.1	20	0.4	0.6	0	1.7	1.1	0.5	0.1	—
32	Heavy whipping cream, liquid	1	tablespoon(s)	15	8.7	52	0.3	0.4	0	5.6	3.5	1.6	0.2	—
28	Light coffee or table cream, liquid	1	tablespoon(s)	15	11.1	29	0.4	0.5	0	2.9	1.8	0.8	0.1	—
30	Light whipping cream, liquid	1	tablespoon(s)	15	9.5	44	0.3	0.4	0	4.6	2.9	1.4	0.1	—
34	Whipped cream topping, pressurized	1	tablespoon(s)	3	1.8	8	0.1	0.4	0	0.7	0.4	0.2	0	—
	Sour cream													
30556	Fat free sour cream	2	tablespoon(s)	32	25.8	24	1.0	5.0	0	0	0	0	0	0
36	Sour cream	2	tablespoon(s)	24	17.0	51	0.8	1.0	0	5.0	3.1	1.5	0.2	—
	Imitation cream													
3659	Coffeemate non dairy creamer, liquid	1	tablespoon(s)	15	—	20	0	2.0	0	1.0	0	0.5	0	—
40	Cream substitute, powder	1	teaspoon(s)	2	0	11	0.1	1.1	0	0.7	0.7	0	0	—
904	Imitation sour cream	2	tablespoon(s)	29	20.5	60	0.7	1.9	0	5.6	5.1	0.2	0	—
35972	Non dairy coffee whitener, liquid, frozen	1	tablespoon(s)	15	11.7	21	0.2	1.7	0	1.5	0.3	1.1	0	—
35976	Non dairy dessert topping, frozen	1	tablespoon(s)	5	2.4	15	0.1	1.1	0	1.2	1.0	0.1	0	—
35975	Non dairy dessert topping, pressurized	1	tablespoon(s)	4	2.7	12	0	0.7	0	1.0	0.8	0.1	0	—
	Fluid milk													
60	Buttermilk, low fat	1	cup(s)	245	220.8	98	8.1	11.7	0	2.2	1.3	0.6	0.1	—
54	Lowfat, 1%	1	cup(s)	244	219.4	102	8.2	12.2	0	2.4	1.5	0.7	0.1	—
55	Lowfat, 1%, with nonfat milk solids	1	cup(s)	245	220.0	105	8.5	12.2	0	2.4	1.5	0.7	0.1	—
57	Nonfat, skim or fat free	1	cup(s)	245	222.6	83	8.3	12.2	0	0.2	0.1	0.1	0	—
58	Nonfat, skim or fat free with nonfat milk solids	1	cup(s)	245	221.4	91	8.7	12.3	0	0.6	0.4	0.2	0	—
51	Reduced fat, 2%	1	cup(s)	244	218.0	122	8.1	11.4	0	4.8	3.1	1.4	0.2	—
52	Reduced fat, 2%, with nonfat milk solids	1	cup(s)	245	217.7	125	8.5	12.2	0	4.7	2.9	1.4	0.2	—
50	Whole, 3.3%	1	cup(s)	244	215.5	146	7.9	11.0	0	7.9	4.6	2.0	0.5	—
	Canned milk													
62	Nonfat or skim evaporated	2	tablespoon(s)	32	25.3	25	2.4	3.6	0	0.1	0	0	0	—
63	Sweetened condensed	2	tablespoon(s)	38	10.4	123	3.0	20.8	0	3.3	2.1	0.9	0.1	—
61	Whole evaporated	2	tablespoon(s)	32	23.3	42	2.1	3.2	0	2.4	1.4	0.7	0.1	—
	Dried milk													
64	Buttermilk	¼	cup(s)	30	0.9	117	10.4	14.9	0	1.8	1.1	0.5	0.1	—
65	Instant nonfat with added vitamin A	¼	cup(s)	17	0.7	61	6.0	8.9	0	0.1	0.1	0	0	—
5234	Skim milk powder	¼	cup(s)	17	0.7	62	6.1	9.1	0	0.1	0.1	0	0	—
907	Whole dry milk	¼	cup(s)	32	0.8	159	8.4	12.3	0	8.5	5.4	2.5	0.2	—
909	**Goat milk**	1	cup(s)	244	212.4	168	8.7	10.9	0	10.1	6.5	2.7	0.4	—
	Chocolate milk													
33155	Chocolate syrup, prepared with milk	1	cup(s)	282	227.0	254	8.7	36.0	0.8	8.3	4.7	2.1	0.5	—
33184	Cocoa mix with aspartame, added sodium and vitamin A, no added calcium or phosphorus, prepared with water	1	cup(s)	192	177.4	56	2.3	10.8	1.2	0.4	0.3	0.1	0	—
908	Hot cocoa, prepared with milk	1	cup(s)	250	206.4	193	8.8	26.6	2.5	5.8	3.6	1.7	0.1	0.2
69	Low fat	1	cup(s)	250	211.3	158	8.1	26.1	1.3	2.5	1.5	0.8	0.1	—
68	Reduced fat	1	cup(s)	250	205.4	190	7.5	30.3	1.8	4.8	2.9	1.1	0.2	—
67	Whole	1	cup(s)	250	205.8	208	7.9	25.9	2.0	8.5	5.3	2.5	0.3	—
70	**Eggnog**	1	cup(s)	254	188.9	343	9.7	34.4	0	19.0	11.3	5.7	0.9	—
	Breakfast drinks													
10093	Carnation Instant Breakfast classic chocolate malt, prepared with skim milk, no sugar added	1	cup(s)	243	—	142	11.1	21.3	0.7	1.3	0.7	—	—	—
10092	Carnation Instant Breakfast classic French vanilla, prepared with skim milk, no sugar added	1	cup(s)	273	—	150	12.9	24.0	0	0.4	0.4	—	—	—
10094	Carnation Instant Breakfast stawberry sensation, prepared with skim milk, no sugar added	1	cup(s)	243	—	142	11.1	21.3	0	0.4	0.4	—	—	—
10091	Carnation Instant Breakfast strawberry sensation, prepared with skim milk	1	cup(s)	273	—	220	12.5	38.8	0	0.4	0.4	—	—	—
1417	Ovaltine rich chocolate flavor, prepared with skim milk	1	cup(s)	258	—	170	8.5	31.0	0	0	0	0	0	0
8539	**Malted milk, chocolate mix, fortified, prepared with milk**	1	cup(s)	265	215.8	223	8.9	28.9	1.1	8.6	5.0	2.2	0.5	—

DAIRY —Continued

Chol (mg)	Calc (mg)	Iron (mg)	Magn (mg)	Pota (mg)	Sodi (mg)	Zinc (mg)	Vit A (µg)	Thia (mg)	Vit E (mg α)	Ribo (mg)	Niac (mg)	Vit B$_6$ (mg)	Fola (µg)	Vit C (mg)	Vit B$_{12}$ (µg)	Sele (µg)
23	162	0.16	8.8	82.5	358.6	0.90	57.0	0.01	0.06	0.14	0.04	0.02	2.0	0	0.4	4.6
16	159	0.09	8.2	68.6	381.3	0.73	49.0	0.01	0.05	0.12	0.03	0.03	2.0	0	0.1	3.2
27	156	0.05	7.7	47.9	422.1	0.80	72.0	0.01	0.07	0.10	0.02	0.02	2.3	0	0.2	4.1
27	338	0.00	0	33.8	459.0	1.22	114.0	—	—	0.14	—	—	—	0	0.2	—
24	219	0.17	8.2	61.2	388.4	1.02	56.1	0.00	0.09	0.07	0.01	0.01	1.7	0	0.3	4.5
0	60	0.00	—	75.0	97.5	—	—	—	—	—	—	—	—	—	—	—
0	200	0.36	—	—	190.0	—	—	—	—	—	—	—	—	0	—	—
6	16	0.01	1.5	19.5	6.2	0.08	14.6	0.01	0.05	0.02	0.01	0.01	0.5	0.1	0	0.3
21	10	0.00	1.1	11.3	5.7	0.03	61.7	0.00	0.15	0.01	0.01	0.00	0.6	0.1	0	0.1
10	14	0.01	1.4	18.3	6.0	0.04	27.2	0.01	0.08	0.02	0.01	0.01	0.3	0.1	0	0.1
17	10	0.00	1.1	14.6	5.1	0.03	41.9	0.00	0.13	0.01	0.01	0.00	0.6	0.1	0	0.1
2	3	0.00	0.3	4.4	3.9	0.01	5.6	0.00	0.01	0.00	0.00	0.00	0.1	0	0	0
3	40	0.00	3.2	41.3	45.1	0.16	23.4	0.01	0.00	0.04	0.02	0.01	3.5	0	0.1	1.7
11	28	0.01	2.6	34.6	12.7	0.06	42.5	0.01	0.14	0.03	0.01	0.00	2.6	0.2	0.1	0.5
0	0	0.00	—	30.0	0	—	0	0.01	—	0.01	0.20	—	—	0	—	—
0	0	0.02	0.1	16.2	3.6	0.01	0	0.00	0.01	0.00	0.00	0.00	0	0	0	0
0	1	0.11	1.7	46.3	29.3	0.34	0	0.00	0.21	0.00	0.00	0.00	0	0	0	0.7
0	1	0.00	0	28.9	12.0	0.00	0.2	0.00	0.12	0.00	0.00	0.00	0	0	0	0.2
0	0	0.00	0.1	0.9	1.2	0.00	0.3	0.00	0.05	0.00	0.00	0.00	0	0	0	0.1
0	0	0.00	0	0.8	2.8	0.00	0.2	0.00	0.04	0.00	0.00	0.00	0	0	0	0.1
10	284	0.12	27.0	370.0	257.3	1.02	17.2	0.08	0.12	0.37	0.14	0.08	12.3	2.5	0.5	4.9
12	290	0.07	26.8	366.0	107.4	1.02	141.5	0.04	0.02	0.45	0.22	0.09	12.2	0	1.1	8.1
10	314	0.12	34.3	396.9	127.4	0.98	144.6	0.09	—	0.42	0.22	0.11	12.3	2.5	0.9	5.6
5	306	0.07	27.0	382.2	102.9	1.02	149.5	0.11	0.02	0.44	0.23	0.09	12.3	0	1.3	7.6
5	316	0.12	36.8	419.0	129.9	1.00	149.5	0.10	0.00	0.42	0.22	0.11	12.3	2.5	1.0	5.4
20	285	0.07	26.8	366.0	100.0	1.04	134.2	0.09	0.07	0.45	0.22	0.09	12.2	0.5	1.1	6.1
20	314	0.12	34.3	396.9	127.4	0.98	137.2	0.09	—	0.42	0.22	0.11	12.3	2.5	0.9	5.6
24	276	0.07	24.4	348.9	97.6	0.97	68.3	0.10	0.14	0.44	0.26	0.08	12.2	0	1.1	9.0
1	93	0.09	8.6	105.9	36.7	0.28	37.6	0.01	0.00	0.09	0.05	0.01	2.9	0.4	0.1	0.8
13	109	0.07	9.9	141.9	48.6	0.36	28.3	0.03	0.06	0.16	0.08	0.02	4.2	1.0	0.2	5.7
9	82	0.06	7.6	95.4	33.4	0.24	20.5	0.01	0.04	0.10	0.06	0.01	2.5	0.6	0.1	0.7
21	359	0.09	33.3	482.5	156.7	1.21	14.9	0.11	0.03	0.48	0.27	0.10	14.2	1.7	1.2	6.2
3	209	0.05	19.9	289.9	93.3	0.75	120.5	0.07	0.00	0.30	0.15	0.06	8.5	1.0	0.7	4.6
3	214	0.05	20.3	296.0	95.3	0.76	123.1	0.07	0.00	0.30	0.15	0.06	8.7	1.0	0.7	4.7
31	292	0.15	27.2	425.6	118.7	1.06	82.2	0.09	0.15	0.38	0.20	0.09	11.8	2.8	1.0	5.2
27	327	0.12	34.2	497.8	122.0	0.73	139.1	0.11	0.17	0.33	0.67	0.11	2.4	3.2	0.2	3.4
25	251	0.90	50.8	408.9	132.5	1.21	70.5	0.11	0.14	0.46	0.38	0.09	14.1	0	1.1	9.6
0	92	0.74	32.6	405.1	138.2	0.51	0	0.04	0.00	0.20	0.16	0.04	1.9	0	0.2	2.5
20	263	1.20	57.5	492.5	110.0	1.57	127.5	0.09	0.07	0.45	0.33	0.10	12.5	0.5	1.1	6.8
8	288	0.60	32.5	425.0	152.5	1.02	145.0	0.09	0.05	0.41	0.31	0.10	12.5	2.3	0.9	4.8
20	273	0.60	35.0	422.5	165.0	0.97	160.0	0.11	0.10	0.45	0.41	0.06	5.0	0	0.8	8.5
30	280	0.60	32.5	417.5	150.0	1.02	65.0	0.09	0.15	0.40	0.31	0.10	12.5	2.3	0.8	4.8
150	330	0.50	48.3	419.1	137.2	1.16	116.8	0.08	0.50	0.48	0.26	0.12	2.5	3.8	1.1	10.7
9	444	4.00	88.9	631.1	195.6	3.38	—	0.33	—	0.45	4.44	0.44	4.0	26.7	1.3	8.0
9	500	4.50	100.0	665.0	192.0	3.75	—	0.37	—	0.51	5.00	0.49	100.0	30.0	1.5	9.0
9	444	4.00	88.9	568.9	186.7	3.38	—	0.33	—	0.45	4.44	0.44	88.9	26.7	1.3	8.0
9	500	4.47	100.0	665.0	288.0	3.75	—	0.37	—	0.51	5.07	0.50	100.0	30.0	1.5	8.8
5	350	3.60	100.0	—	270.0	3.75	—	0.37	—	—	4.00	0.40	—	12.0	1.2	—
27	339	3.76	45.1	577.7	230.6	1.16	903.7	0.75	0.15	1.31	11.08	1.01	18.6	31.8	1.1	12.5

DA+ Code	Food Description	Quantity	Measure	Wt (g)	H₂O (g)	Ener (kcal)	Prot (g)	Carb (g)	Fiber (g)	Fat (g)	Fat Breakdown (g) Sat	Mono	Poly	Trans
	Milkshakes													
73	Chocolate	1	cup(s)	227	164.0	270	6.9	48.1	0.7	6.1	3.8	1.8	0.2	—
3163	Strawberry	1	cup(s)	226	167.8	256	7.7	42.8	0.9	6.3	3.9	—	—	—
74	Vanilla	1	cup(s)	227	169.2	254	8.8	40.3	0	6.9	4.3	2.0	0.3	—
	Ice cream													
4776	Chocolate	½	cup(s)	66	36.8	143	2.5	18.6	0.8	7.3	4.5	2.1	0.3	—
12137	Chocolate fudge, no sugar added	½	cup(s)	71	—	100	3.0	16.0	2.0	3.0	1.5	—	—	0
16514	Chocolate, soft serve	½	cup(s)	87	49.9	177	3.2	24.1	0.7	8.4	5.2	2.4	0.3	—
16523	Sherbet, all flavors	½	cup(s)	97	63.8	139	1.1	29.3	3.2	1.9	1.1	0.5	0.1	—
4778	Strawberry	½	cup(s)	66	39.6	127	2.1	18.2	0.6	5.5	3.4	—	—	—
76	Vanilla	½	cup(s)	72	43.9	145	2.5	17.0	0.5	7.9	4.9	2.1	0.3	—
12146	Vanilla chocolate swirl, fat free, no sugar added	½	cup(s)	71	—	100	3.0	14.0	2.0	3.0	2.0	—	—	0
82	Vanilla, light	½	cup(s)	76	48.3	125	3.6	19.6	0.2	3.7	2.2	1.0	0.2	—
78	Vanilla, light, soft serve	½	cup(s)	88	61.2	111	4.3	19.2	0	2.3	1.4	0.7	0.1	—
	Soy desserts													
10694	Tofutti low fat vanilla fudge non dairy frozen dessert	½	cup(s)	70	—	140	2.0	24.0	0	4.0	1.0	—	—	—
15721	Tofutti premium chocolate supreme non dairy frozen dessert	½	cup(s)	70	—	180	3.0	18.0	0	11.0	2.0	—	—	—
15720	Tofutti premium vanilla non dairy frozen dessert	½	cup(s)	70	—	190	2.0	20.0	0	11.0	2.0	—	—	—
	Ice milk													
16517	Chocolate	½	cup(s)	66	42.9	94	2.8	16.9	0.3	2.1	1.3	0.6	0.1	—
16516	Flavored, not chocolate	½	cup(s)	66	41.4	108	3.5	17.5	0.2	2.6	1.7	0.6	0.1	—
	Pudding													
25032	Chocolate	½	cup(s)	144	109.7	155	5.1	22.7	0.7	5.4	3.1	1.7	0.2	0
1923	Chocolate, sugar free, prepared with 2% milk	½	cup(s)	133	—	100	5.0	14.0	0.3	3.0	1.5	—	—	—
1722	Rice	½	cup(s)	113	75.6	151	4.1	29.9	0.5	1.9	1.1	0.5	0.1	—
4747	Tapioca, ready to eat	1	item(s)	142	102.0	185	2.8	30.8	0	5.5	1.4	3.6	0.1	—
25031	Vanilla	½	cup(s)	136	109.7	116	4.7	17.6	0	2.8	1.6	0.9	0.2	0
1924	Vanilla, sugar free, prepared with 2% milk	½	cup(s)	133	—	90	4.0	12.0	0.2	2.0	1.5	—	—	—
	Frozen yogurt													
4785	Chocolate, soft serve	½	cup(s)	72	45.9	115	2.9	17.9	1.6	4.3	2.6	1.3	0.2	—
1747	Fruit varieties	½	cup(s)	113	80.5	144	3.4	24.4	0	4.1	2.6	1.1	0.1	—
4786	Vanilla, soft serve	½	cup(s)	72	47.0	117	2.9	17.4	0	4.0	2.5	1.1	0.2	—
	Milk substitutes													
	Lactose free													
16081	Fat free, calcium fortified [milk]	1	cup(s)	240	—	80	8.0	13.0	0	0	0	0	0	0
36486	Low fat milk	1	cup(s)	240	—	110	8.0	13.0	0	2.5	1.5	—	—	—
36487	Reduced fat milk	1	cup(s)	240	—	130	8.0	12.0	0	5.0	3.0	—	—	—
36488	Whole milk	1	cup(s)	240	—	150	8.0	12.0	0	8.0	5.0	—	—	—
	Rice													
10083	Rice Dream carob rice beverage	1	cup(s)	240	—	150	1.0	32.0	0	2.5	0	—	—	—
17089	Rice Dream original rice beverage, enriched	1	cup(s)	240	—	120	1.0	25.0	0	2.0	0	—	—	—
10087	Rice Dream vanilla enriched rice beverage	1	cup(s)	240	—	130	1.0	28.0	0	2.0	0	—	—	—
	Soy													
34750	Soy Dream chocolate enriched soy beverage	1	cup(s)	240	—	210	7.0	37.0	1.0	3.5	0.5	—	—	—
34749	Soy Dream vanilla enriched soy beverage	1	cup(s)	240	—	150	7.0	22.0	0	4.0	0.5	—	—	—
13840	Vitasoy light chocolate soymilk	1	cup(s)	240	—	100	4.0	17.0	0	2.0	0.5	0.5	1.0	—
13839	Vitasoy light vanilla soymilk	1	cup(s)	240	—	70	4.0	10.0	0	2.0	0.5	0.5	1.0	—
13836	Vitasoy rich chocolate soymilk	1	cup(s)	240	—	160	7.0	24.0	1.0	4.0	0.5	1.0	2.5	—
13835	Vitasoy vanilla delite soymilk	1	cup(s)	240	—	120	7.0	13.0	1.0	4.0	0.5	1.0	2.5	—
	Yogurt													
3615	Custard style, fruit flavors	6	ounce(s)	170	127.1	190	7.0	32.0	0	3.5	2.0	—	—	—
3617	Custard style, vanilla	6	ounce(s)	170	134.1	190	7.0	32.0	0	3.5	2.0	0.9	0.1	—
32101	Fruit, low fat	1	cup(s)	245	184.5	243	9.8	45.7	0	2.8	1.8	0.8	0.1	—
29638	Fruit, non fat, sweetened with low calorie sweetener	1	cup(s)	241	208.3	123	10.6	19.4	1.2	0.4	0.2	0.1	0	—
93	Plain, low fat	1	cup(s)	245	208.4	154	12.9	17.2	0	3.8	2.5	1.0	0.1	—
94	Plain, nonfat	1	cup(s)	245	208.8	137	14.0	18.8	0	0.4	0.3	0.1	0	—
32100	Vanilla, low fat	1	cup(s)	245	193.6	208	12.1	33.8	0	3.1	2.0	0.8	0.1	—
5242	Yogurt beverage	1	cup(s)	245	199.8	172	6.2	32.8	0	2.2	1.4	0.6	0.1	—
38202	Yogurt smoothie, nonfat, all flavors	1	item(s)	325	—	290	10.0	60.0	6.0	0	0	0	0	0
	Soy yogurt													
34617	Stonyfield Farm O'Soy strawberry-peach pack organic cultured soy yogurt	1	item(s)	113	—	100	5.0	16.0	3.0	2.0	0	—	—	0
34616	Stonyfield Farm O'Soy vanilla organic cultured soy yogurt	1	item(s)	170	—	150	7.0	26.0	4.0	2.0	0	—	—	0
10453	White Wave plain silk cultured soy yogurt	8	ounce(s)	227	—	140	5.0	22.0	1.0	3.0	0.5	—	—	0
	EGGS													
	Eggs													
99	Fried	1	item(s)	46	31.8	90	6.3	0.4	0	7.0	2.0	2.9	1.2	—
	EGGS —Continued													

Chol (mg)	Calc (mg)	Iron (mg)	Magn (mg)	Pota (mg)	Sodi (mg)	Zinc (mg)	Vit A (µg)	Thia (mg)	Vit E (mg α)	Ribo (mg)	Niac (mg)	Vit B_6 (mg)	Fola (µg)	Vit C (mg)	Vit B_{12} (µg)	Sele (µg)
25	300	0.70	36.4	508.9	252.2	1.09	40.9	0.10	0.11	0.50	0.28	0.05	11.4	0	0.7	4.3
25	256	0.24	29.4	412.0	187.9	0.81	58.9	0.10	—	0.44	0.39	0.10	6.8	1.8	0.7	4.8
27	332	0.22	27.3	415.8	215.8	0.88	56.8	0.06	0.11	0.44	0.33	0.09	15.9	0	1.2	5.2
22	72	0.61	19.1	164.3	50.2	0.38	77.9	0.02	0.19	0.12	0.14	0.03	10.6	0.5	0.2	1.7
10	100	0.36	—	—	65.0	—	—	—	—	—	—	—	—	0	—	—
22	103	0.32	19.0	192.0	43.3	0.45	66.6	0.03	0.22	0.13	0.11	0.03	4.3	0.5	0.3	2.5
0	52	0.13	7.7	92.6	44.4	0.46	9.7	0.02	0.02	0.08	0.07	0.02	6.8	5.6	0.1	1.3
19	79	0.13	9.2	124.1	39.6	0.22	63.4	0.03	—	0.16	0.11	0.03	7.9	5.1	0.2	1.3
32	92	0.06	10.1	143.3	57.6	0.49	85.0	0.03	0.21	0.17	0.08	0.03	3.6	0.4	0.3	1.3
10	100	0.00	—	—	65.0	—	—	—	—	—	—	—	—	0	—	—
21	122	0.14	10.6	158.1	56.2	0.55	97.3	0.04	0.09	0.19	0.10	0.03	4.6	0.9	0.4	1.5
11	138	0.05	12.3	194.5	61.6	0.46	25.5	0.04	0.05	0.17	0.10	0.04	4.4	0.8	0.4	3.2
0	0	0.00	—	8.0	90.0	—	0	—	—	—	—	—	—	0	—	—
0	0	0.00	—	7.0	180.0	—	0	—	—	—	—	—	—	0	—	—
0	0	0.00	—	2.0	210.0	—	0	—	—	—	—	—	—	0	—	—
6	94	0.15	13.1	155.2	40.6	0.36	15.7	0.03	0.05	0.11	0.08	0.02	3.9	0.5	0.3	2.2
16	76	0.05	9.2	136.2	48.5	0.47	90.4	0.02	0.05	0.11	0.06	0.01	3.3	0.1	0.2	1.3
35	149	0.46	31.3	226.7	137.0	0.71	—	0.05	0.00	0.22	0.15	0.06	8.3	1.2	0.5	4.9
10	150	0.72	—	330.0	310.0	—	—	0.06	—	0.26	—	—	—	0	—	—
7	113	0.28	15.8	201.4	66.4	0.52	41.6	0.03	0.05	0.17	0.34	0.06	4.5	0.2	0.2	4.8
1	101	0.15	8.5	130.6	205.9	0.31	0	0.03	0.21	0.13	0.09	0.03	4.3	0.4	0.3	0
35	146	0.17	17.2	188.9	136.4	0.52	—	0.04	0.00	0.22	0.10	0.05	8.0	1.2	0.5	4.6
10	150	0.00	—	190.0	380.0	—	—	0.03	—	0.17	—	—	—	0	—	—
4	106	0.90	19.4	187.9	70.6	0.35	31.7	0.02	—	0.15	0.22	0.05	7.9	0.2	0.2	1.7
15	113	0.52	11.3	176.3	71.2	0.31	55.4	0.04	0.10	0.20	0.07	0.04	4.5	0.8	0.1	2.1
1	103	0.21	10.1	151.9	62.6	0.30	42.5	0.02	0.07	0.16	0.20	0.05	4.3	0.6	0.2	2.4
3	500	0.00	—	—	125.0	—	100.0	—	—	—	—	—	—	0	0	—
10	300	0.00	—	—	125.0	—	100.0	—	—	—	—	—	—	0	—	—
20	300	0.00	—	—	125.0	—	98.2	—	—	—	—	—	—	0	—	—
35	300	0.00	—	—	125.0	—	58.1	—	—	—	—	—	—	0	—	—
0	20	0.72	—	82.5	100.0	—	—	—	—	—	—	—	—	1.2	—	—
0	300	0.00	13.3	60.0	90.0	0.24	—	0.06	—	0.00	0.84	0.07	—	0	1.5	—
0	300	0.00	—	53.0	90.0	—	—	—	—	—	—	—	—	0	1.5	—
0	300	1.80	60.0	350.0	160.0	0.60	33.3	0.15	—	0.06	0.80	0.12	60.0	0	3.0	—
U	300	1.80	40.0	260.0	140.0	0.60	33.3	0.15	—	0.06	0.80	0.12	60.0	0	3.0	—
0	300	0.72	24.0	200.0	140.0	0.90	—	0.09	—	0.34	—	—	24.0	0	0.9	—
0	300	0.72	24.0	200.0	120.0	0.90	—	0.09	—	0.34	—	—	24.0	0	0.9	—
0	300	1.08	40.0	320.0	150.0	0.90	—	0.15	—	0.34	—	—	60.0	0	0.9	—
0	40	0.72	—	320.0	115.0	—	0	—	—	—	—	—	—	0	—	—
15	300	0.00	16.0	310.0	100.0	—	—	—	—	0.25	—	—	—	0	—	—
15	300	0.00	16.0	310.0	100.0	—	—	—	—	0.25	—	—	—	0	—	—
12	338	0.14	31.9	433.7	129.9	1.64	27.0	0.08	0.04	0.39	0.21	0.09	22.1	1.5	1.1	6.9
5	369	0.62	41.0	549.5	139.8	1.83	4.8	0.10	0.16	0.44	0.49	0.10	31.3	26.5	1.1	7.0
15	448	0.19	41.7	573.3	171.5	2.18	34.3	0.10	0.07	0.52	0.27	0.12	27.0	2.0	1.4	8.1
5	488	0.22	46.6	624.8	188.7	2.37	4.9	0.11	0.00	0.57	0.30	0.13	29.4	2.2	1.5	8.8
12	419	0.17	39.2	536.6	161.7	2.03	29.4	0.10	0.04	0.49	0.26	0.11	27.0	2.0	1.3	12.0
13	260	0.22	39.2	399.4	98.0	1.10	14.7	0.11	0.00	0.51	0.30	0.14	29.4	2.1	1.5	—
5	300	2.70	100.0	580.0	290.0	2.25	—	0.37	—	0.42	5.00	0.50	100.0	15.0	1.5	—
0	100	1.08	24.0	5.0	20.0	—	0	0.22	—	0.10	—	0.04	—	0	0	—
0	150	1.44	40.0	15.0	40.0	—	—	0.30	—	0.13	—	0.08	—	0	0	—
0	400	1.44	—	0	30.0	—	0	—	—	—	—	—	—	0	—	—
210	27	0.91	6.0	67.6	93.8	0.55	91.1	0.03	0.56	0.23	0.03	0.07	23.5	0	0.6	15.7

(Computer code is for Cengage Diet Analysis program) (For purposes of calculations, use "0" for t, <1, <.1, <.01, etc.)

DA+ Code	Food Description	Quantity	Measure	Wt (g)	H₂O (g)	Ener (kcal)	Prot (g)	Carb (g)	Fiber (g)	Fat (g)	Fat Breakdown (g) Sat	Mono	Poly	Trans
100	Hard boiled	1	item(s)	50	37.3	78	6.3	0.6	0	5.3	1.6	2.0	0.7	—
101	Poached	1	item(s)	50	37.8	71	6.3	0.4	0	5.0	1.5	1.9	0.7	—
97	Raw, white	1	item(s)	33	28.9	16	3.6	0.2	0	0.1	0	0	0	—
96	Raw, whole	1	item(s)	50	37.9	72	6.3	0.4	0	5.0	1.5	1.9	0.7	—
98	Raw, yolk	1	item(s)	17	8.9	54	2.7	0.6	0	4.5	1.6	2.0	0.7	—
102	Scrambled, prepared with milk and butter	2	item(s)	122	89.2	204	13.5	2.7	0	14.9	4.5	5.8	2.6	0.7
	Egg substitute													
4028	Egg Beaters	¼	cup(s)	61	—	30	6.0	1.0	0	0	0	0	0	0
920	Frozen	¼	cup(s)	60	43.9	96	6.8	1.9	0	6.7	1.2	1.5	3.7	—
918	Liquid	¼	cup(s)	63	51.9	53	7.5	0.4	0	2.1	0.4	0.6	1.0	—
	SEAFOOD													
	Cod													
6040	Atlantic cod or scrod, baked or broiled	3	ounce(s)	85	64.6	89	19.4	0	0	0.7	0.1	0.1	0.2	—
1573	Atlantic cod, cooked, dry heat	3	ounce(s)	85	64.6	89	19.4	0	0	0.7	0.1	0.1	0.2	—
2905	**Eel, raw**	3	ounce(s)	85	58.0	156	15.7	0	0	9.9	2.0	6.1	0.8	—
	Fish fillets													
25079	Baked	3	ounce(s)	84	79.9	99	21.7	0	0	0.7	0.1	0.1	0.3	—
8615	Batter coated or breaded, fried	3	ounce(s)	85	45.6	197	12.5	14.4	0.4	10.5	2.4	2.2	5.3	—
25082	Broiled fish steaks	3	ounce(s)	85	68.1	128	24.2	0	0	2.6	0.4	0.9	0.8	—
25083	Poached fish steaks	3	ounce(s)	85	67.1	111	21.1	0	0	2.3	0.3	0.8	0.7	—
25084	Steamed	3	ounce(s)	85	72.2	79	17.2	0	0	0.6	0.1	0.1	0.2	—
25089	**Flounder, baked**	3	ounce(s)	85	64.4	113	14.8	0.4	0.1	5.5	1.1	2.2	1.4	—
1825	**Grouper, cooked, dry heat**	3	ounce(s)	85	62.4	100	21.1	0	0	1.1	0.3	0.2	0.3	—
	Haddock													
6049	Baked or broiled	3	ounce(s)	85	63.2	95	20.6	0	0	0.8	0.1	0.1	0.3	—
1578	Cooked, dry heat	3	ounce(s)	85	63.1	95	20.6	0	0	0.8	0.1	0.1	0.3	—
1886	**Halibut, Atlantic and Pacific, cooked, dry heat**	3	ounce(s)	85	61.0	119	22.7	0	0	2.5	0.4	0.8	0.8	—
1582	**Herring, Atlantic, pickled**	4	piece(s)	60	33.1	157	8.5	5.8	0	10.8	1.4	7.2	1.0	—
1587	**Jack mackerel, solids, canned, drained**	2	ounce(s)	57	39.2	88	13.1	0	0	3.6	1.1	1.3	0.9	—
8580	**Octopus, common, cooked, moist heat**	3	ounce(s)	85	51.5	139	25.4	3.7	0	1.8	0.4	0.3	0.4	—
1831	**Perch, mixed species, cooked, dry heat**	3	ounce(s)	85	62.3	100	21.1	0	0	1.0	0.2	0.2	0.4	—
1592	**Pacific rockfish, cooked, dry heat**	3	ounce(s)	85	62.4	103	20.4	0	0	1.7	0.4	0.4	0.5	—
	Salmon													
2938	Coho, farmed, raw	3	ounce(s)	85	59.9	136	18.1	0	0	6.5	1.5	2.8	1.6	—
1594	Broiled or baked with butter	3	ounce(s)	85	53.9	155	23.0	0	0	6.3	1.2	2.3	2.3	—
29727	Smoked chinook (lox)	2	ounce(s)	57	40.8	66	10.4	0	0	2.4	0.5	1.1	0.6	—
154	**Sardine, Atlantic with bones, canned in oil**	3	ounce(s)	85	50.7	177	20.9	0	0	9.7	1.3	3.3	4.4	—
	Scallops													
155	Mixed species, breaded, fried	3	item(s)	47	27.2	100	8.4	4.7	—	5.1	1.2	2.1	1.3	—
1599	Steamed	3	ounce(s)	85	64.8	90	13.8	2.0	0	2.6	0.4	1.0	0.8	—
1839	**Snapper, mixed species, cooked, dry heat**	3	ounce(s)	85	59.8	109	22.4	0	0	1.5	0.3	0.3	0.5	—
	Squid													
1868	Mixed species, fried	3	ounce(s)	85	54.9	149	15.3	6.6	0	6.4	1.6	2.3	1.8	—
16617	Steamed or boiled	3	ounce(s)	85	63.3	89	15.2	3.0	0	1.3	0.4	0.1	0.5	—
1570	**Striped bass, cooked, dry heat**	3	ounce(s)	85	62.4	105	19.3	0	0	2.5	0.6	0.7	0.9	—
1601	**Sturgeon, steamed**	3	ounce(s)	85	59.4	111	17.0	0	0	4.3	1.0	2.0	0.7	—
1840	**Surimi, formed**	3	ounce(s)	85	64.9	84	12.9	5.8	0	0.8	0.2	0.1	0.4	—
1842	**Swordfish, cooked, dry heat**	3	ounce(s)	85	58.5	132	21.6	0	0	4.4	1.2	1.7	1.0	—
1846	**Tuna, yellowfin or ahi, raw**	3	ounce(s)	85	60.4	92	19.9	0	0	0.8	0.2	0.1	0.2	—
	Tuna, canned													
159	Light, canned in oil, drained	2	ounce(s)	57	33.9	112	16.5	0	0	4.6	0.9	1.7	1.6	—
355	Light, canned in water, drained	2	ounce(s)	57	42.2	66	14.5	0	0	0.5	0.1	0.1	0.2	—
33211	Light, no salt, canned in oil, drained	2	ounce(s)	57	33.9	112	16.5	0	0	4.7	0.9	1.7	1.6	—
33212	Light, no salt, canned in water, drained	2	ounce(s)	57	42.6	66	14.5	0	0	0.5	0.1	0.1	0.2	—
2961	White, canned in oil, drained	2	ounce(s)	57	36.3	105	15.0	0	0	4.6	0.7	1.8	1.7	—
351	White, canned in water, drained	2	ounce(s)	57	41.5	73	13.4	0	0	1.7	0.4	0.4	0.6	—
33213	White, no salt, canned in oil, drained	2	ounce(s)	57	36.3	105	15.0	0	0	4.6	0.9	1.4	1.9	—
33214	White, no salt, canned in water, drained	2	ounce(s)	57	42.0	73	13.4	0	0	1.7	0.4	0.4	0.6	—
	Yellowtail													
8548	Mixed species, cooked, dry heat	3	ounce(s)	85	57.3	159	25.2	0	0	5.7	1.4	2.2	1.5	—
2970	Mixed species, raw	2	ounce(s)	57	42.2	83	13.1	0	0	3.0	0.7	1.1	0.8	—
	Shellfish, meat only													
1857	Abalone, mixed species, fried	3	ounce(s)	85	51.1	161	16.7	9.4	0	5.8	1.4	2.3	1.4	—
16618	Abalone, steamed or poached	3	ounce(s)	85	40.7	177	28.8	10.1	0	1.3	0.3	0.2	0.2	—
	Crab													
1851	Blue crab, canned	2	ounce(s)	57	43.2	56	11.6	0	0	0.7	0.1	0.1	0.2	—
1852	Blue crab, cooked, moist heat	3	ounce(s)	85	65.9	87	17.2	0	0	1.5	0.2	0.2	0.6	—
8562	Dungeness crab, cooked, moist heat	3	ounce(s)	85	62.3	94	19.0	0.8	0	1.1	0.1	0.2	0.3	—
1860	**Clams, cooked, moist heat**	3	ounce(s)	85	54.1	126	21.7	4.4	0	1.7	0.2	0.1	0.5	—
	SEAFOOD —Continued													

Chol (mg)	Calc (mg)	Iron (mg)	Magn (mg)	Pota (mg)	Sodi (mg)	Zinc (mg)	Vit A (μg)	Thia (mg)	Vit E (mg α)	Ribo (mg)	Niac (mg)	Vit B_6 (mg)	Fola (μg)	Vit C (mg)	Vit B_{12} (μg)	Sele (μg)
212	25	0.59	5.0	63.0	62.0	0.52	84.5	0.03	0.51	0.25	0.03	0.06	22.0	0	0.6	15.4
211	27	0.91	6.0	66.5	147.0	0.55	69.5	0.02	0.48	0.20	0.03	0.06	17.5	0	0.6	15.8
0	2	0.02	3.6	53.8	54.8	0.01	0	0.00	0.00	0.14	0.03	0.00	1.3	0	0	6.6
212	27	0.91	6.0	67.0	70.0	0.55	70.0	0.03	0.48	0.23	0.03	0.07	23.5	0	0.6	15.9
210	22	0.46	0.9	18.5	8.2	0.39	64.8	0.03	0.43	0.09	0.00	0.06	24.8	0	0.3	9.5
429	87	1.46	14.6	168.4	341.6	1.22	174.5	0.06	1.33	0.53	0.09	0.14	36.6	0.2	0.9	27.5
0	20	1.08	4.0	85.0	115.0	0.60	112.5	0.15	—	0.85	0.20	0.08	60.0	0	1.2	—
1	44	1.18	9.0	127.8	119.4	0.58	6.6	0.07	0.95	0.23	0.08	0.08	9.6	0.3	0.2	24.8
1	33	1.32	5.6	207.1	111.1	0.82	11.3	0.07	0.17	0.19	0.07	0.00	9.4	0	0.2	15.6
47	12	0.41	35.7	207.5	66.3	0.49	11.9	0.07	0.68	0.06	2.13	0.24	6.8	0.8	0.9	32.0
47	12	0.41	35.7	207.5	66.3	0.49	11.9	0.07	0.68	0.06	2.13	0.24	6.8	0.9	0.9	32.0
107	17	0.42	17.0	231.3	43.4	1.37	887.0	0.13	3.40	0.03	2.97	0.05	12.8	1.5	2.6	5.5
44	8	0.31	29.1	489.0	86.1	0.48	—	0.02	—	0.05	2.47	0.46	8.1	3.0	1.0	44.3
29	15	1.79	20.4	272.2	452.5	0.37	9.4	0.09	—	0.09	1.78	0.08	14.5	0	0.9	7.7
37	55	0.97	96.7	524.3	62.9	0.49	—	0.05	—	0.08	6.47	0.36	12.6	0	1.2	42.5
32	48	0.85	84.0	455.6	54.7	0.42	—	0.05	—	0.07	5.92	0.33	11.5	0	1.1	37.0
41	12	0.29	24.7	319.3	41.7	0.34	—	0.06	—	0.06	1.89	0.21	6.1	0.8	0.8	32.0
44	19	0.34	47.3	224.7	280.2	0.20	—	0.06	0.40	0.07	2.02	0.18	7.4	2.8	1.6	33.5
40	18	0.96	31.5	404.0	45.1	0.43	42.5	0.06	—	0.01	0.32	0.29	8.5	0	0.6	39.8
63	36	1.15	42.5	339.4	74.0	0.40	16.2	0.03	0.42	0.03	3.94	0.29	6.8	0	1.2	34.4
63	36	1.14	42.5	339.3	74.0	0.40	16.2	0.03	—	0.03	3.93	0.29	11.1	0	1.2	34.4
35	51	0.91	91.0	489.9	58.7	0.45	45.9	0.05	—	0.07	6.05	0.33	11.9	0	1.2	39.8
8	46	0.73	4.8	41.4	522.0	0.31	154.8	0.02	1.02	0.08	1.98	0.10	1.2	0	2.6	35.1
45	137	1.15	21.0	110.0	214.9	0.57	73.7	0.02	0.58	0.12	3.50	0.11	2.8	0.5	3.9	21.4
82	90	8.11	51.0	535.8	391.2	2.85	76.5	0.04	1.02	0.06	3.21	0.55	20.4	6.8	30.6	76.2
98	87	0.98	32.3	292.6	67.2	1.21	8.5	0.06	—	0.10	1.61	0.11	5.1	1.4	1.9	13.7
37	10	0.45	28.9	442.3	65.5	0.45	60.4	0.03	1.32	0.07	3.33	0.22	8.5	0	1.0	39.8
43	10	0.29	26.4	382.7	40.0	0.36	47.6	0.08	—	0.09	5.79	0.56	11.1	0.9	2.3	10.7
40	15	1.02	26.9	376.6	98.6	0.56	—	0.13	1.14	0.05	8.33	0.18	4.2	1.8	2.3	41.0
13	6	0.48	10.2	99.2	1134.0	0.17	14.7	0.01	—	0.05	2.67	0.15	1.1	0	1.8	21.6
121	325	2.48	33.2	337.6	429.5	1.10	27.2	0.04	1.70	0.18	4.43	0.14	10.2	0	7.6	44.8
28	20	0.38	27.4	154.8	215.8	0.49	10.7	0.02	—	0.05	0.70	0.06	17.2	1.1	0.6	12.5
27	20	0.22	45.9	238.0	358.7	0.78	32.3	0.01	0.16	0.05	0.84	0.11	10.2	2.0	1.1	18.2
40	34	0.20	31.5	444.0	48.5	0.37	29.8	0.04	—	0.00	0.29	0.39	5.1	1.4	3.0	41.7
221	33	0.85	32.3	237.3	260.3	1.48	9.4	0.04	—	0.39	2.21	0.04	11.9	3.6	1.0	44.1
227	31	0.62	28.9	192.1	356.2	1.49	8.5	0.01	1.17	0.32	1.69	0.04	3.4	3.2	1.0	43.7
88	16	0.91	43.4	279.0	74.8	0.43	26.4	0.09	—	0.03	2.17	0.29	8.5	0	3.8	39.8
63	11	0.59	29.8	239.7	388.5	0.35	198.9	0.06	0.52	0.07	8.30	0.19	14.5	0	2.2	13.3
26	8	0.22	36.6	95.3	121.6	0.28	17.0	0.01	0.53	0.01	0.18	0.02	1.7	0	1.4	23.9
43	5	0.88	28.9	313.8	97.8	1.25	34.9	0.03	—	0.09	10.02	0.32	1.7	0.9	1.7	52.5
38	14	0.62	42.5	377.6	31.5	0.44	15.3	0.37	0.42	0.04	8.33	0.77	1.7	0.8	0.4	31.0
10	7	0.79	17.6	117.3	200.6	0.51	13.0	0.02	0.49	0.07	7.03	0.06	2.8	0	1.2	43.1
17	6	0.87	15.3	134.3	191.5	0.43	9.6	0.01	0.19	0.04	7.52	0.19	2.3	0	1.7	45.6
10	7	0.78	17.6	117.4	28.3	0.51	0	0.02	—	0.06	7.03	0.06	2.8	0	1.2	43.1
17	6	0.86	15.3	134.4	28.3	0.43	0	0.01	—	0.04	7.52	0.19	2.3	0	1.7	45.6
18	2	0.36	19.3	188.8	224.5	0.26	2.8	0.01	1.30	0.04	6.63	0.24	2.8	0	1.2	34.1
24	8	0.55	18.7	134.3	213.6	0.27	3.4	0.00	0.48	0.02	3.28	0.12	1.1	0	0.7	37.2
18	2	0.36	19.3	188.8	28.3	0.26	0	0.01	—	0.04	6.63	0.24	2.8	0	1.2	34.1
24	8	0.54	18.7	134.4	28.3	0.27	3.4	0.00	—	0.02	3.28	0.12	1.1	0	0.7	37.3
60	25	0.53	32.3	457.6	42.5	0.56	26.4	0.14	—	0.04	7.41	0.15	3.4	2.5	1.1	39.8
31	13	0.28	17.0	238.1	22.1	0.29	16.4	0.08	—	0.02	3.86	0.09	2.3	1.6	0.7	20.7
80	31	3.23	47.6	241.5	502.6	0.80	1.7	0.18	—	0.11	1.61	0.12	11.9	1.5	0.6	44.1
144	50	4.84	68.9	295.0	980.1	1.38	3.4	0.28	6.74	0.12	1.89	0.21	6.0	2.6	0.7	75.6
50	57	0.47	22.1	212.1	188.8	2.27	1.1	0.04	1.04	0.04	0.77	0.08	24.4	1.5	0.3	18.0
85	88	0.77	28.1	275.6	237.3	3.58	1.7	0.08	1.56	0.04	2.80	0.15	43.4	2.8	6.2	34.2
65	50	0.36	49.3	347.0	321.5	4.65	26.4	0.04	—	0.17	3.08	0.14	35.7	3.1	8.8	40.5
57	78	23.78	15.3	534.1	95.3	2.32	145.4	0.12	—	0.36	2.85	0.09	24.7	18.8	84.1	54.4

Food Composition — (Computer code is for Cengage Diet Analysis program)   (For purposes of calculations, use "0" for t, <1, <.1, <.01, etc.)

DA+ Code	Food Description	Quantity	Measure	Wt (g)	H₂O (g)	Ener (kcal)	Prot (g)	Carb (g)	Fiber (g)	Fat (g)	Sat	Mono	Poly	Trans
											Fat Breakdown (g)			
1853	**Crayfish, farmed, cooked, moist heat**	3	ounce(s)	85	68.7	74	14.9	0	0	1.1	0.2	0.2	0.4	—
	Oysters													
8720	Baked or broiled	3	ounce(s)	85	68.6	89	5.6	3.2	0	5.8	1.3	2.1	1.9	—
152	Eastern, farmed, raw	3	ounce(s)	85	73.3	50	4.4	4.7	0	1.3	0.4	0.1	0.5	—
8715	Eastern, wild, cooked, moist heat	3	ounce(s)	85	59.8	117	12.0	6.7	0	4.2	1.3	0.5	1.6	—
8584	Pacific, cooked, moist heat	3	ounce(s)	85	54.5	139	16.1	8.4	0	3.9	0.9	0.7	1.5	—
1865	Pacific, raw	3	ounce(s)	85	69.8	69	8.0	4.2	0	2.0	0.4	0.3	0.8	—
1854	**Lobster, northern, cooked, moist heat**	3	ounce(s)	85	64.7	83	17.4	1.1	0	0.5	0.1	0.1	0.1	—
1862	**Mussel, blue, cooked, moist heat**	3	ounce(s)	85	52.0	146	20.2	6.3	0	3.8	0.7	0.9	1.0	—
	Shrimp													
158	Mixed species, breaded, fried	3	ounce(s)	85	44.9	206	18.2	9.8	0.3	10.4	1.8	3.2	4.3	—
1855	Mixed species, cooked, moist heat	3	ounce(s)	85	65.7	84	17.8	0	0	0.9	0.2	0.2	0.4	—
	BEEF, LAMB, PORK													
	Beef													
4450	Breakfast strips, cooked	2	slice(s)	23	5.9	101	7.1	0.3	0	7.8	3.2	3.8	0.4	—
174	Corned beef, canned	3	ounce(s)	85	49.1	213	23.0	0	0	12.7	5.3	5.1	0.5	—
33147	Cured, thin siced	2	ounce(s)	57	32.9	100	15.9	3.2	0	2.2	0.9	1.0	0.1	—
4581	Jerky	1	ounce(s)	28	6.6	116	9.4	3.1	0.5	7.3	3.1	3.2	0.3	—
	Ground beef													
5898	Lean, broiled, medium	3	ounce(s)	85	50.4	202	21.6	0	0	12.2	4.8	5.3	0.4	—
5899	Lean, broiled, well done	3	ounce(s)	85	48.4	214	23.8	0	0	12.5	5.0	5.7	0.3	0.4
5914	Regular, broiled, medium	3	ounce(s)	85	46.1	246	20.5	0	0	17.6	6.9	7.7	0.6	—
5915	Regular, broiled, well done	3	ounce(s)	85	43.8	259	21.6	0	0	18.4	7.5	8.5	0.5	0.6
	Beef rib													
4241	Rib, small end, separable lean, 0" fat, broiled	3	ounce(s)	85	53.2	164	25.0	0	0	6.4	2.4	2.6	0.2	—
4183	Rib, whole, lean and fat, ¼" fat, roasted	3	ounce(s)	85	39.0	320	18.9	0	0	26.6	10.7	11.4	0.9	—
	Beef roast													
16981	Bottom round, choice, separable lean and fat, ⅛" fat, braised	3	ounce(s)	85	46.2	216	27.9	0	0	10.7	4.1	4.6	0.4	—
16979	Bottom round, separable lean and fat, ⅛" fat, roasted	3	ounce(s)	85	52.4	185	22.5	0	0	9.9	3.8	4.2	0.4	—
16924	Chuck, arm pot roast, separable lean and fat, ⅛" fat, braised	3	ounce(s)	85	42.9	257	25.6	0	0	16.3	6.5	7.0	0.6	—
16930	Chuck, blade roast, separable lean and fat, ⅛" fat, braised	3	ounce(s)	85	40.5	290	22.8	0	0	21.4	8.5	9.2	0.8	—
5853	Chuck, blade roast, separable lean, 0" trim, pot roasted	3	ounce(s)	85	47.4	202	26.4	0	0	9.9	3.9	4.3	0.3	—
4296	Eye of round, choice, separable lean, 0" fat, roasted	3	ounce(s)	85	56.5	138	24.4	0	0	3.7	1.3	1.5	0.1	—
16989	Eye of round, separable lean and fat, ⅛" fat, roasted	3	ounce(s)	85	52.2	180	24.2	0	0	8.5	3.2	3.6	0.3	—
	Beef steak													
4348	Short loin, t-bone steak, lean and fat, ¼" fat, broiled	3	ounce(s)	85	43.2	274	19.4	0	0	21.2	8.3	9.6	0.8	—
4349	Short loin, t-bone steak, lean, ¼" fat, broiled	3	ounce(s)	85	52.3	174	22.8	0	0	8.5	3.1	4.2	0.3	—
4360	Top loin, prime, lean and fat, ¼" fat, broiled	3	ounce(s)	85	42.7	275	21.6	0	0	20.3	8.2	8.6	0.7	—
	Beef variety													
188	Liver, pan fried	3	ounce(s)	85	52.7	149	22.6	4.4	0	4.0	1.3	0.5	0.5	0.2
4447	Tongue, simmered	3	ounce(s)	85	49.2	242	16.4	0	0	19.0	6.9	8.6	0.6	0.7
	Lamb chop													
3275	Loin, domestic, lean and fat, ¼" fat, broiled	3	ounce(s)	85	43.9	269	21.4	0	0	19.6	8.4	8.3	1.4	—
	Lamb leg													
3264	Domestic, lean and fat, ¼" fat, cooked	3	ounce(s)	85	45.7	250	20.9	0	0	17.8	7.5	7.5	1.3	—
	Lamb rib													
182	Domestic, lean and fat, ¼" fat, broiled	3	ounce(s)	85	40.0	307	18.8	0	0	25.2	10.8	10.3	2.0	—
183	Domestic, lean, ¼" fat, broiled	3	ounce(s)	85	50.0	200	23.6	0	0	11.0	4.0	4.4	1.0	—
	Lamb shoulder													
186	Shoulder, arm and blade, domestic, choice, lean and fat, ¼" fat, roasted	3	ounce(s)	85	47.8	235	19.1	0	0	17.0	7.2	6.9	1.4	—
187	Shoulder, arm and blade, domestic, choice, lean, ¼" fat, roasted	3	ounce(s)	85	53.8	173	21.2	0	0	9.2	3.5	3.7	0.8	—
3287	Shoulder, arm, domestic, lean and fat, ¼" fat, braised	3	ounce(s)	85	37.6	294	25.8	0	0	20.4	8.4	8.7	1.5	—
3290	Shoulder, arm, domestic, lean, ¼" fat, braised	3	ounce(s)	85	41.9	237	30.2	0	0	12.0	4.3	5.2	0.8	—
	Lamb variety													
3375	Brain, pan fried	3	ounce(s)	85	51.6	232	14.4	0	0	18.9	4.8	3.4	1.9	—
3406	Tongue, braised	3	ounce(s)	85	49.2	234	18.3	0	0	17.2	6.7	8.5	1.1	—
	Pork, cured													
29229	Bacon, Canadian style, cured	2	ounce(s)	57	37.9	89	11.7	1.0	0	4.0	1.3	1.8	0.4	—
161	Bacon, cured, broiled, pan fried or roasted	2	slice(s)	16	2.0	87	5.9	0.2	0	6.7	2.2	3.0	0.7	0
	BEEF, LAMB, PORK —Continued													

Chol (mg)	Calc (mg)	Iron (mg)	Magn (mg)	Pota (mg)	Sodi (mg)	Zinc (mg)	Vit A (µg)	Thia (mg)	Vit E (mg α)	Ribo (mg)	Niac (mg)	Vit B_6 (mg)	Fola (µg)	Vit C (mg)	Vit B_{12} (µg)	Sele (µg)
117	43	0.94	28.1	202.4	82.5	1.25	12.8	0.03	—	0.06	1.41	0.11	9.4	0.4	2.6	29.1
43	36	5.30	37.4	125.0	403.8	72.22	60.4	0.07	0.98	0.06	1.04	0.04	7.7	2.8	14.7	50.7
21	37	4.91	28.1	105.4	151.3	32.23	6.8	0.08	—	0.05	1.07	0.05	15.3	4.0	13.8	54.1
89	77	10.19	80.8	239.0	358.9	154.45	45.9	0.16	—	0.15	2.11	0.10	11.9	5.1	29.8	60.9
85	14	7.82	37.4	256.8	180.3	28.27	124.2	0.10	0.72	0.37	3.07	0.07	12.8	10.9	24.5	131.0
43	7	4.34	18.7	142.9	90.1	14.13	68.9	0.05	—	0.20	1.70	0.04	8.5	6.8	13.6	65.5
61	52	0.33	29.8	299.4	323.2	2.48	22.1	0.01	0.85	0.05	0.91	0.06	9.4	0	2.6	36.3
48	28	5.71	31.5	227.9	313.8	2.27	77.4	0.25	—	0.35	2.55	0.08	64.6	11.6	20.4	76.2
150	57	1.07	34.0	191.3	292.4	1.17	0	0.11	—	0.11	2.60	0.08	15.3	1.3	1.6	35.4
166	33	2.62	28.9	154.8	190.5	1.32	57.8	0.02	1.17	0.02	2.20	0.10	3.4	1.9	1.3	33.7
27	2	0.71	6.1	93.1	509.2	1.44	0	0.02	0.06	0.05	1.46	0.07	1.8	0	0.8	6.1
73	10	1.76	11.9	115.7	855.6	3.03	0	0.01	0.12	0.12	2.06	0.11	7.7	0	1.4	36.5
23	6	1.53	10.8	243.2	815.9	2.25	0	0.04	0.00	0.10	2.98	0.19	6.2	0	1.5	16.0
14	6	1.53	14.5	169.2	627.4	2.29	0	0.04	0.13	0.04	0.49	0.05	38.0	0	0.3	3.0
58	6	2.00	17.9	266.2	59.5	4.63	0	0.05	—	0.23	4.21	0.23	7.6	0	1.8	16.0
69	12	2.21	18.4	250.0	62.4	5.86	0	0.08	—	0.23	5.10	0.16	9.4	0	1.7	19.0
62	9	2.07	17.0	248.3	70.6	4.40	0	0.02	—	0.16	4.90	0.23	7.6	0	2.5	16.2
71	12	2.30	18.5	242.4	72.4	5.18	0	0.08	—	0.23	4.93	0.17	8.5	0	1.6	18.0
65	16	1.59	21.3	319.8	51.9	4.64	0	0.06	0.34	0.12	7.15	0.53	8.5	0	1.4	29.2
72	9	1.96	16.2	251.7	53.6	4.45	0	0.06	—	0.14	2.85	0.19	6.0	0	2.1	18.7
68	6	2.29	17.9	223.7	35.7	4.59	0	0.05	0.41	0.15	5.05	0.36	8.5	0	1.7	29.3
64	5	1.83	14.5	182.0	29.8	3.76	0	0.05	0.34	0.12	3.92	0.29	6.8	0	1.3	23.0
67	14	2.15	17.0	205.8	42.5	5.93	0	0.05	0.45	0.15	3.63	0.25	7.7	0	1.9	24.1
88	11	2.66	16.2	198.2	55.3	7.15	0	0.06	0.17	0.20	2.06	0.22	4.3	0	1.9	20.9
73	11	3.12	19.6	223.7	60.4	8.73	0	0.06	—	0.23	2.27	0.24	5.1	0	2.1	22.7
49	5	2.16	16.2	200.7	32.3	4.28	0	0.05	0.30	0.15	4.69	0.34	8.5	0	1.4	28.0
54	5	1.98	15.3	193.1	31.5	3.95	0	0.05	0.34	0.13	4.37	0.31	7.7	0	1.5	25.2
58	7	2.56	17.9	233.9	57.8	3.56	0	0.07	0.18	0.17	3.29	0.27	6.0	0	1.8	10.0
50	5	3.11	22.1	278.1	65.5	4.34	0	0.09	0.11	0.21	3.93	0.33	6.8	0	1.9	8.5
67	8	1.88	19.6	294.3	53.6	3.85	0	0.06	—	0.15	3.96	0.31	6.0	0	1.6	19.5
324	5	5.24	18.7	298.5	65.5	4.44	6586.3	0.15	0.39	2.91	14.86	0.87	221.1	0.6	70.7	27.9
112	4	2.22	12.8	156.5	55.3	3.47	0	0.01	0.25	0.25	2.96	0.13	6.0	1.1	2.7	11.2
85	17	1.53	20.4	278.1	65.5	2.96	0	0.08	0.11	0.21	6.03	0.11	15.3	0	2.1	23.3
82	14	1.59	19.6	263.7	61.2	3.79	0	0.08	0.11	0.21	5.66	0.11	15.3	0	2.2	22.5
84	16	1.59	19.6	229.5	64.6	3.40	0	0.07	0.10	0.18	5.95	0.09	11.9	0	2.2	20.3
77	14	1.87	24.7	266.1	72.3	4.47	0	0.08	0.15	0.21	5.56	0.12	17.9	0	2.2	26.4
78	17	1.67	19.6	213.4	56.1	4.44	0	0.07	0.11	0.20	5.22	0.11	17.9	0	2.2	22.3
74	16	1.81	21.3	225.3	57.8	5.13	0	0.07	0.15	0.22	4.89	0.12	21.3	0	2.3	24.2
102	21	2.03	22.1	260.3	61.2	5.17	0	0.06	0.12	0.21	5.66	0.09	15.3	0	2.2	31.6
103	22	2.29	24.7	287.5	64.6	6.20	0	0.06	0.15	0.23	5.38	0.11	18.7	0	2.3	32.1
2130	18	1.73	18.7	304.5	133.5	1.70	0	0.14	—	0.31	3.87	0.19	6.0	19.6	20.5	10.2
161	9	2.23	13.6	134.4	57.0	2.54	0	0.06	—	0.35	3.13	0.14	2.6	6.0	5.4	23.8
28	5	0.38	9.6	195.0	798.9	0.78	0	0.42	0.11	0.09	3.53	0.22	2.3	0	0.4	14.2
18	2	0.22	5.3	90.4	369.6	0.56	1.8	0.06	0.04	0.04	1.76	0.04	0.3	0	0.2	9.9

DA+ Code	Food Description	Quantity	Measure	Wt (g)	H₂O (g)	Ener (kcal)	Prot (g)	Carb (g)	Fiber (g)	Fat (g)	Fat Breakdown (g)			
											Sat	Mono	Poly	Trans
35422	Breakfast strips, cured, cooked	3	slice(s)	34	9.2	156	9.8	0.4	0	12.5	4.3	5.6	1.9	—
189	Ham, cured, boneless, 11% fat, roasted	3	ounce(s)	85	54.9	151	19.2	0	0	7.7	2.7	3.8	1.2	—
29215	Ham, cured, extra lean, 4% fat, canned	2	2 ounce(s)	57	41.7	68	10.5	0	0	2.6	0.9	1.3	0.2	—
1316	Ham, cured, extra lean, 5% fat, roasted	3	ounce(s)	85	57.6	123	17.8	1.3	0	4.7	1.5	2.2	0.5	—
16561	Ham, smoked or cured, lean, cooked	1	slice(s)	42	27.6	66	10.5	0	0	2.3	0.8	1.1	0.3	—
	Pork chop													
32671	Loin, blade, chops, lean and fat, pan fried	3	ounce(s)	85	42.5	291	18.3	0	0	23.6	8.6	10	2.6	—
32672	Loin, center cut, chops, lean and fat, pan fried	3	ounce(s)	85	45.1	236	25.4	0	0	14.1	5.1	6.0	1.6	—
32682	Loin, center rib, chops, boneless, lean and fat, braised	3	ounce(s)	85	49.5	217	22.4	0	0	13.4	5.2	6.1	1.1	—
32603	Loin, center rib, chops, lean, broiled	3	ounce(s)	85	55.4	158	21.9	0	0	7.1	2.4	3.0	0.8	0.1
32478	Loin, whole, lean and fat, braised	3	ounce(s)	85	49.6	203	23.2	0	0	11.6	4.3	5.2	1.0	—
32481	Loin, whole, lean, braised	3	ounce(s)	85	52.2	174	24.3	0	0	7.8	2.9	3.5	0.6	—
	Pork leg or ham													
32471	Pork leg or ham, Rump portion, lean and fat, roasted	3	ounce(s)	85	48.3	214	24.6	0	0	12.1	4.5	5.4	1.2	—
32468	Pork leg or ham, Whole, lean and fat, roasted	3	ounce(s)	85	46.8	232	22.8	0	0	15.0	5.5	6.7	1.4	—
	Pork ribs													
32693	Loin, country style, lean and fat, roasted	3	ounce(s)	85	43.3	279	19.9	0	0	21.6	7.8	9.4	1.7	—
32696	Loin, country style, lean, roasted	3	ounce(s)	85	49.5	210	22.6	0	0	12.6	4.5	5.5	0.9	—
	Pork shoulder													
32626	Shoulder, arm picnic, lean and fat, roasted	3	ounce(s)	85	44.3	270	20.0	0	0	20.4	7.5	9.1	2.0	—
32629	Shoulder, arm picnic, lean, roasted	3	ounce(s)	85	51.3	194	22.7	0	0	10.7	3.7	5.1	1.0	—
	Rabbit													
3366	Domesticated, roasted	3	ounce(s)	85	51.5	168	24.7	0	0	6.8	2.0	1.8	1.3	—
3367	Domesticated, stewed	3	ounce(s)	85	50.0	175	25.8	0	0	7.2	2.1	1.9	1.4	—
	Veal													
3391	Liver, braised	3	ounce(s)	85	50.9	163	24.2	3.2	0	5.3	1.7	1.0	0.9	0.3
3319	Rib, lean only, roasted	3	ounce(s)	85	55.0	151	21.9	0	0	6.3	1.8	2.3	0.6	—
1732	Deer or venison, roasted	3	ounce(s)	85	55.5	134	25.7	0	0	2.7	1.1	0.7	0.5	—
	POULTRY													
	Chicken													
29562	Flaked, canned	2	ounce(s)	57	39.3	97	10.3	0.1	0	5.8	1.6	2.3	1.3	—
	Chicken, fried													
29632	Breast, meat only, breaded, baked or fried	3	ounce(s)	85	44.3	193	25.3	6.9	0.2	6.6	1.6	2.7	1.7	—
35327	Broiler breast, meat only, fried	3	ounce(s)	85	51.2	159	28.4	0.4	0	4.0	1.1	1.5	0.9	—
36413	Broiler breast, meat and skin, flour coated, fried	3	ounce(s)	85	48.1	189	27.1	1.4	0.1	7.5	2.1	3.0	1.7	—
36414	Broiler drumstick, meat and skin, flour coated, fried	3	ounce(s)	85	48.2	208	22.9	1.4	0.1	11.7	3.1	4.6	2.7	—
35389	Broiler drumstick, meat only, fried	3	ounce(s)	85	52.9	166	24.3	0	0	6.9	1.8	2.5	1.7	—
35406	Broiler leg, meat only, fried	3	ounce(s)	85	51.5	177	24.1	0.6	0	7.9	2.1	2.9	1.9	—
35484	Broiler wing, meat only, fried	3	ounce(s)	85	50.9	179	25.6	0	0	7.8	2.1	2.6	1.8	—
29580	Patty, fillet or tenders, breaded, cooked	3	ounce(s)	85	40.2	256	14.5	12.2	0	16.5	3.7	8.4	3.7	—
	Chicken, roasted, meat only													
35409	Broiler leg, meat only, roasted	3	ounce(s)	85	55.0	162	23.0	0	0	7.2	1.9	2.6	1.7	—
35486	Broiler wing, meat only, roasted	3	ounce(s)	85	53.4	173	25.9	0	0	6.9	1.9	2.2	1.5	—
35138	Roasting chicken, dark meat, meat only, roasted	3	ounce(s)	85	57.0	151	19.8	0	0	7.4	2.1	2.8	1.7	—
35136	Roasting chicken, light meat, meat only, roasted	3	ounce(s)	85	57.7	130	23.1	0	0	3.5	0.9	1.3	0.8	—
35132	Roasting chicken, meat only, roasted	3	ounce(s)	85	57.3	142	21.3	0	0	5.6	1.5	2.1	1.3	—
	Chicken, stewed													
1268	Gizzard, simmered	3	ounce(s)	85	57.8	124	25.8	0	0	2.3	0.6	0.4	0.3	0.1
1270	Liver, simmered	3	ounce(s)	85	56.8	142	20.8	0.7	0	5.5	1.8	1.2	1.7	0.1
3174	Meat only, stewed	3	ounce(s)	85	56.8	151	23.2	0	0	5.7	1.6	2.0	1.3	—
	Duck													
1286	Domesticated, meat and skin, roasted	3	ounce(s)	85	44.1	287	16.2	0	0	24.1	8.2	11.0	3.1	—
1287	Domesticated, meat only, roasted	3	ounce(s)	85	54.6	171	20	0	0	9.5	3.5	3.1	1.2	—
	Goose													
35507	Domesticated, meat and skin, roasted	3	ounce(s)	85	44.2	259	21.4	0	0	18.6	5.8	8.7	2.1	—
35524	Domesticated, meat only, roasted	3	ounce(s)	85	48.7	202	24.6	0	0	10.8	3.9	3.7	1.3	—
1297	Liver pate, smoked, canned	4	tablespoon(s)	52	19.3	240	5.9	2.4	0	22.8	7.5	13.3	0.4	—
	Turkey													
3256	Ground turkey, cooked	3	ounce(s)	85	50.5	200	23.3	0	0	11.2	2.9	4.2	2.7	—
3263	Patty, batter coated, breaded, fried	1	item(s)	94	46.7	266	13.2	14.8	0.5	16.9	4.4	7.0	4.4	—
219	Roasted, dark meat, meat only	3	ounce(s)	85	53.7	159	24.3	0	0	6.1	2.1	1.4	1.8	—
222	Roasted, fryer roaster breast, meat only	3	ounce(s)	85	58.2	115	25.6	0	0	0.6	0.2	0.1	0.2	—
220	Roasted, light meat, meat only	3	ounce(s)	85	56.4	134	25.4	0	0	2.7	0.9	0.5	0.7	—
1303	Turkey roll, light and dark meat	2	slice(s)	57	39.8	84	10.3	1.2	0	4.0	1.2	1.3	1.0	—
1302	Turkey roll, light meat	2	slice(s)	57	42.5	56	8.4	2.9	0	0.9	0.2	0.2	0.1	0
	PROCESSED MEATS													
	Beef													
1331	Corned beef loaf, jellied, sliced	2	slice(s)	57	39.2	87	13.0	0	0	3.5	1.5	1.5	0.2	—
	PROCESSED MEATS —Continued													

Chol (mg)	Calc (mg)	Iron (mg)	Magn (mg)	Pota (mg)	Sodi (mg)	Zinc (mg)	Vit A (µg)	Thia (mg)	Vit E (mg α)	Ribo (mg)	Niac (mg)	Vit B6 (mg)	Fola (µg)	Vit C (mg)	Vit B12 (µg)	Sele (µg)
36	5	0.67	8.8	158.4	713.7	1.25	0	0.25	0.08	0.12	2.58	0.11	1.4	0	0.6	8.4
50	7	1.13	18.7	347.7	1275.0	2.09	0	0.62	0.26	0.28	5.22	0.26	2.6	0	0.6	16.8
22	3	0.53	9.6	206.4	711.6	1.09	0	0.47	0.09	0.13	3.00	0.25	3.4	0	0.5	8.2
45	7	1.25	11.9	244.1	1023.1	2.44	0	0.64	0.21	0.17	3.42	0.34	2.6	0	0.6	16.6
23	3	0.39	9.2	132.7	557.3	1.07	0	0.28	0.10	0.10	2.10	0.19	1.7	0	0.3	10.7
72	26	0.74	17.9	282.4	57.0	2.71	1.7	0.52	0.17	0.25	3.35	0.28	3.4	0.5	0.7	29.7
78	23	0.77	24.7	361.5	68.0	1.96	1.7	0.96	0.21	0.25	4.76	0.39	5.1	0.9	0.6	33.2
62	4	0.78	14.5	329.1	34.0	1.76	1.7	0.44	—	0.20	3.66	0.26	3.4	0.3	0.4	28.4
56	22	0.57	21.3	291.7	48.5	1.91	0	0.48	0.08	0.18	6.68	0.57	0	0	0.4	38.6
68	18	0.91	16.2	318.1	40.8	2.02	1.7	0.53	0.20	0.21	3.75	0.31	2.6	0.5	0.5	38.5
67	15	0.96	17.0	329.1	42.5	2.10	1.7	0.56	0.17	0.22	3.90	0.32	3.4	0.5	0.5	41.0
82	10	0.89	23.0	318.1	52.7	2.39	2.6	0.63	0.18	0.28	3.95	0.26	2.6	0.2	0.6	39.8
80	12	0.85	18.7	299.4	51.0	2.51	2.6	0.54	0.18	0.26	3.89	0.34	8.5	0.3	0.6	38.5
78	21	0.90	19.6	292.6	44.2	2.00	2.6	0.75	—	0.29	3.67	0.37	4.3	0.3	0.7	31.6
79	25	1.09	20.4	296.8	24.7	3.24	1.7	0.48	—	0.29	3.96	0.37	4.3	0.3	0.7	36.0
80	16	1.00	14.5	276.4	59.5	2.93	1.7	0.44	—	0.25	3.33	0.29	3.4	0.2	0.6	28.6
81	8	1.20	17.0	298.5	68.0	3.46	1.7	0.49	—	0.30	3.66	0.34	4.3	0.3	0.7	32.7
70	16	1.93	17.9	325.7	40.0	1.93	0	0.07	—	0.17	7.17	0.40	9.4	0	7.1	32.7
73	17	2.01	17.0	255.1	31.5	2.01	0	0.05	0.37	0.14	6.09	0.28	7.7	0	5.5	32.7
435	5	4.34	17.0	279.8	66.3	9.55	17983.6	0.15	0.57	2.43	11.18	0.78	281.5	0.9	72.0	16.4
98	10	0.81	20.4	264.5	82.5	3.81	0	0.05	0.30	0.24	6.37	0.23	11.9	0	1.3	9.4
95	6	3.80	20.4	284.9	45.9	2.33	0	0.15	—	0.51	5.70	—	—	0	—	11.0
35	8	0.89	6.8	147.4	408.2	0.79	19.3	0.01	—	0.07	3.58	0.19	2.3	0	0.2	—
67	19	1.05	24.7	222.6	450.2	0.84	—	0.08	—	0.09	10.97	0.46	4.3	0	0.3	—
77	14	0.96	26.4	234.7	67.2	0.91	6.0	0.06	0.35	0.10	12.57	0.54	3.4	0	0.3	22.3
76	14	1.01	25.5	220.3	64.6	0.93	12.8	0.06	0.39	0.11	11.68	0.49	5.1	0	0.3	20.3
77	10	1.13	19.6	194.8	75.7	2.45	21.3	0.06	0.65	0.19	5.13	0.29	8.5	0	0.3	15.6
80	10	1.12	20.4	211.8	81.6	2.73	15.3	0.06	—	0.20	5.22	0.33	7.7	0	0.3	16.7
84	11	1.19	21.3	216.0	81.6	2.53	17.0	0.07	0.38	0.21	5.68	0.33	7.7	0	0.3	16.0
71	13	0.96	17.9	176.9	77.4	1.80	15.3	0.03	0.40	0.10	6.15	0.50	3.4	0	0.3	21.6
49	11	0.75	19.6	244.8	411.4	0.79	4.3	0.09	1.04	0.12	5.99	0.24	24.7	0	0.2	13.9
80	10	1.11	20.4	205.8	77.4	2.43	16.2	0.06	0.22	0.19	5.37	0.31	6.8	0	0.3	18.8
72	14	0.98	17.9	178.6	78.2	1.82	15.3	0.03	0.22	0.10	6.21	0.50	3.4	0	0.3	21.0
64	9	1.13	17.0	190.5	80.8	1.81	13.6	0.05	—	0.16	4.87	0.26	6.0	0	0.2	16.7
64	11	0.91	19.6	200.7	43.4	0.66	6.8	0.05	0.22	0.07	8.90	0.45	2.6	0	0.3	21.9
64	10	1.02	17.9	194.8	63.8	1.29	10.2	0.05	—	0.12	6.70	0.34	4.3	0	0.2	20.9
315	14	2.71	2.6	152.2	47.6	3.75	0	0.02	0.17	0.17	2.65	0.06	4.3	0	0.9	35.0
479	9	9.89	21.3	223.7	64.6	3.38	3385.8	0.24	0.69	1.69	9.39	0.64	491.6	23.7	14.3	70.1
71	12	0.99	17.9	153.1	59.5	1.69	12.8	0.04	0.22	0.13	5.20	0.22	5.1	0	0.2	17.8
71	9	2.29	13.6	173.5	50.2	1.58	53.6	0.14	0.59	0.22	4.10	0.15	5.1	0	0.3	17.0
76	10	2.29	17.0	214.3	55.3	2.21	19.6	0.22	0.59	0.39	4.33	0.21	8.5	0	0.3	19.1
77	11	2.40	18.7	279.8	59.5	2.22	17.9	0.06	1.47	0.27	3.54	0.31	1.7	0	0.3	18.5
82	12	2.44	21.3	330.0	64.6	2.69	10.2	0.07	—	0.33	3.47	0.39	10.2	0	0.4	21.7
78	36	2.86	6.8	71.8	362.4	0.47	520.5	0.04	—	0.15	1.30	0.03	31.2	0	4.9	22.9
87	21	1.64	20.4	229.6	91.0	2.43	0	0.04	0.28	0.14	4.09	0.33	6.0	0	0.3	31.6
71	13	2.06	14.1	258.5	752.0	1.35	9.4	0.09	0.87	0.17	2.16	0.18	38.5	0	0.2	20.8
72	27	1.98	20.4	246.6	67.2	3.79	0	0.05	0.54	0.21	3.10	0.30	7.7	0	0.3	34.8
71	10	1.30	24.7	248.3	44.2	1.48	0	0.03	0.07	0.11	6.37	0.47	5.1	0	0.3	27.3
59	16	1.14	23.8	259.4	54.4	1.73	0	0.05	0.07	0.11	5.81	0.45	5.1	0	0.3	27.3
31	18	0.76	10.2	153.1	332.3	1.13	0	0.05	0.19	0.16	2.72	0.15	2.8	0	0.1	16.6
19	4	0.21	10.8	242.1	590.8	0.50	0	0.01	0.07	0.08	4.05	0.23	2.3	0	0.2	7.4
27	6	1.15	6.2	57.3	540.4	2.31	0	0.00	—	0.06	0.99	0.06	4.5	0	0.7	9.8

DA+ Code	Food Description	Quantity	Measure	Wt (g)	H₂O (g)	Ener (kcal)	Prot (g)	Carb (g)	Fiber (g)	Fat (g)	Fat Breakdown (g) Sat	Mono	Poly	Trans
	Bologna													
13459	Beef	1	slice(s)	28	15.1	90	3.0	1.0	0	8.0	3.5	4.3	0.3	—
13461	Light, made with pork and chicken	1	slice(s)	28	18.2	60	3.0	2.0	0	4.0	1.0	2.0	0.4	—
13458	Made with chicken and pork	1	slice(s)	28	15.0	90	3.0	1.0	0	8.0	3.0	4.1	1.1	—
13565	Turkey bologna	1	slice(s)	28	19.0	50	3.0	1.0	0	4.0	1.0	1.1	1.0	0
	Chicken													
7125	Breast, smoked	1	slice(s)	10	—	10	1.8	0.3	0	0.2	0	—	—	—
	Ham													
7127	Deli-sliced, honey	1	slice(s)	10	—	10	1.7	0.3	0	0.3	0.1	—	—	—
7126	Deli-sliced, smoked	1	slice(s)	10	—	10	1.7	0.2	0	0.3	0.1	—	—	—
8614	Beef & pork mortadella, sliced	2	slice(s)	46	24.1	143	7.5	1.4	0	11.7	4.4	5.2	1.4	—
1323	Pork olive loaf	2	slice(s)	57	33.1	133	6.7	5.2	0	9.4	3.3	4.5	1.1	—
1324	Pork pickle & pimento loaf	2	slice(s)	57	34.2	128	6.4	4.8	0.9	9.1	3.0	4.0	1.6	—
	Sausages & frankfurters													
37296	Beerwurst beef, beer salami (bierwurst)	1	slice(s)	29	16.6	74	4.1	1.2	0	5.7	2.5	2.7	0.2	—
37257	Beerwurst pork beer salami	1	slice(s)	21	12.9	50	3.0	0.4	0	4.0	1.3	1.9	0.5	—
35338	Berliner, pork & beef	1	ounce(s)	28	17.3	65	4.3	0.7	0	4.9	1.7	2.3	0.4	—
37298	Bratwurst pork, cooked	1	piece(s)	74	42.3	181	10.4	1.9	0	14.3	5.1	6.7	1.5	—
37299	Braunschweiger pork liver sausage	1	slice(s)	15	8.2	51	2.0	0.3	0	4.5	1.5	2.1	0.5	—
1329	Cheesefurter or cheese smokie, beef & pork	1	item(s)	43	22.6	141	6.1	0.6	0	12.5	4.5	5.9	1.3	—
1330	Chorizo, beef & pork	2	ounce(s)	57	18.1	258	13.7	1.1	0	21.7	8.2	10.4	2.0	—
8600	Frankfurter, beef	1	item(s)	45	23.4	149	5.1	1.8	0	13.3	5.3	6.4	0.5	—
202	Frankfurter, beef & pork	1	item(s)	45	25.2	137	5.2	0.8	0	12.4	4.8	6.2	1.2	—
1293	Frankfurter, chicken	1	item(s)	45	28.1	100	7.0	1.2	0.2	7.3	1.7	2.7	1.7	0.1
3261	Frankfurter, turkey	1	item(s)	45	28.3	100	5.5	1.7	0	7.8	1.8	2.6	1.8	0.4
37275	Italian sausage, pork, cooked	1	item(s)	68	32.0	234	13.0	2.9	0.1	18.6	6.5	8.1	2.2	—
37307	Kielbasa, kolbassa, pork & beef	1	slice(s)	30	18.5	61	5.0	1.0	0	4.7	1.7	2.2	0.5	—
1333	Knockwurst or knackwurst, beef & pork	2	ounce(s)	57	31.4	174	6.3	1.8	0	15.7	5.8	7.3	1.7	—
37285	Pepperoni, beef & pork	1	slice(s)	11	3.4	51	2.2	0.4	0.2	4.4	1.8	2.1	0.3	—
37313	Polish sausage, pork	1	slice(s)	21	11.4	60	2.8	0.7	0	5.0	1.8	2.3	0.5	—
206	Salami, beef, cooked, sliced	2	slice(s)	52	31.2	136	6.5	1.0	0	11.5	5.1	5.5	0.5	—
37272	Salami, pork, dry or hard	1	slice(s)	13	4.6	52	2.9	0.2	0	4.3	1.5	2.0	0.5	—
40987	Sausage, turkey, cooked	2	ounce(s)	57	36.9	111	13.5	0	0	5.9	1.3	1.7	1.5	0.2
8620	Smoked sausage, beef & pork	2	ounce(s)	57	30.6	181	6.8	1.4	0	16.3	5.5	6.9	2.2	0
8619	Smoked sausage, pork	2	ounce(s)	57	32.0	178	6.8	1.2	0	16.0	5.3	6.4	2.1	0.1
37273	Smoked sausage, pork link	1	piece(s)	76	29.8	295	16.8	1.6	0	24.0	8.6	11.1	2.8	—
1336	Summer sausage, thuringer, or cervelat, beef & pork	2	ounce(s)	57	25.6	205	9.9	1.9	0	17.3	6.5	7.4	0.7	—
37294	Vienna sausage, cocktail, beef & pork, canned	1	piece(s)	16	10.4	37	1.7	0.4	0	3.1	1.1	1.5	0.2	—
	Spreads													
1318	Ham salad spread	¼	cup(s)	60	37.6	130	5.2	6.4	0	9.3	3.0	4.3	1.6	—
32419	Pork and beef sandwich spread	4	tablespoon(s)	60	36.2	141	4.6	7.2	0.1	10.4	3.6	4.6	1.5	—
	Turkey													
13604	Breast, fat free, oven roasted	1	slice(s)	28	—	25	4.0	1.0	0	0	0	0	0	0
13606	Breast, hickory smoked fat free	1	slice(s)	28	—	25	4.0	1.0	0	0	0	0	0	0
16049	Breast, hickory smoked slices	1	slice(s)	56	—	50	11.0	1.0	0	0	0	0	0	0
16047	Breast, honey roasted slices	1	slice(s)	56	—	60	11.0	3.0	0	0	0	0	0	0
16048	Breast, oven roasted slices	1	slice(s)	56	—	50	11.0	1.0	0	0	0	0	0	0
7124	Breast, oven-roasted	1	slice(s)	10	—	10	1.8	0.3	0	0.1	0	0	0	—
13567	Turkey ham, 10% water added	2	slice(s)	56	40.9	70	10.0	2.0	0	3.0	0	0.4	0.6	0
37270	Turkey pastrami	1	slice(s)	28	20.3	35	4.6	1.0	0	1.2	0.3	0.4	0.3	—
3262	Turkey salami	2	slice(s)	57	39.1	98	10.9	0.9	0.1	5.2	1.6	1.8	1.4	0
37318	Turkey salami, cooked	1	slice(s)	28	20.4	43	4.3	0.1	0	2.7	0.8	0.9	0.7	—
	BEVERAGES													
	Beer													
866	Ale, mild	12	fluid ounce(s)	360	332.3	148	1.1	13.3	0.4	0	0	0	0	0
686	Beer	12	fluid ounce(s)	356	327.7	153	1.6	12.7	0	0	0	0	0	0
16886	Beer, non alcoholic	12	fluid ounce(s)	360	328.1	133	0.8	29.0	0	0.4	0.1	0	0.2	0
31609	Bud Light beer	12	fluid ounce(s)	355	335.5	110	0.9	6.6	0	0	0	0	0	0
31608	Budweiser beer	12	fluid ounce(s)	355	327.7	145	1.3	10.6	0	0	0	0	0	0
869	Light beer	12	fluid ounce(s)	354	335.9	103	0.9	5.8	0	0	0	0	0	0
31613	Michelob Beer	12	fluid ounce(s)	355	323.4	155	1.3	13.3	0	0	0	0	0	0
31614	Michelob Light beer	12	fluid ounce(s)	355	329.8	134	1.1	11.7	0	0	0	0	0	0
	Gin, rum, vodka, whiskey													
857	Distilled alcohol, 100 proof	1	fluid ounce(s)	28	16.0	82	0	0	0	0	0	0	0	0
687	Distilled alcohol, 80 proof	1	fluid ounce(s)	28	18.5	64	0	0	0	0	0	0	0	0
688	Distilled alcohol, 86 proof	1	fluid ounce(s)	28	17.8	70	0	0	0	0	0	0	0	0
689	Distilled alcohol, 90 proof	1	fluid ounce(s)	28	17.3	73	0	0	0	0	0	0	0	0
856	Distilled alcohol, 94 proof	1	fluid ounce(s)	28	16.8	76	0	0	0	0	0	0	0	0
	BEVERAGES —Continued													

Chol (mg)	Calc (mg)	Iron (mg)	Magn (mg)	Pota (mg)	Sodi (mg)	Zinc (mg)	Vit A (µg)	Thia (mg)	Vit E (mg α)	Ribo (mg)	Niac (mg)	Vit B$_6$ (mg)	Fola (µg)	Vit C (mg)	Vit B$_{12}$ (µg)	Sele (µg)
20	0	0.36	3.9	47.0	310.0	0.56	0	0.01	—	0.03	0.67	0.04	3.6	0	0.4	—
20	40	0.36	5.6	45.6	300.0	0.45	0	—	—	—	—	—	—	0	—	—
30	20	0.36	5.9	43.1	300.0	0.39	0	—	—	—	—	—	—	0	—	—
20	40	0.36	6.2	42.6	270.0	0.51	0	—	—	—	—	—	—	0	—	—
4	0	0.00	—	—	100.0	—	0	—	—	—	—	—	—	0	—	—
4	0	0.12	—	—	100.0	—	0	—	—	—	—	—	—	0.6	—	—
4	0	0.12	—	—	103.3	—	0	—	—	—	—	—	—	0.6	—	—
26	8	0.64	5.1	75.0	573.2	0.96	0	0.05	0.10	0.07	1.23	0.06	1.4	0	0.7	10.4
22	62	0.30	10.8	168.7	842.9	0.78	34.1	0.16	0.14	0.14	1.04	0.13	1.1	0	0.7	9.3
33	62	0.75	19.3	210.7	740.7	0.95	44.3	0.22	0.22	0.06	1.41	0.23	21.0	4.4	0.3	4.5
18	3	0.44	3.5	66.5	264.9	0.71	0	0.02	0.05	0.03	0.98	0.04	0.9	0	0.6	4.7
12	2	0.15	2.7	53.3	261.0	0.36	0	0.11	0.03	0.04	0.68	0.07	0.6	0	0.2	4.4
13	3	0.32	4.3	80.2	367.7	0.70	0	0.10	—	0.06	0.88	0.05	1.4	0	0.8	4.0
44	33	0.95	11.1	156.9	412.2	1.70	0	0.37	0.01	0.13	2.36	0.15	1.5	0.7	0.7	15.7
24	1	1.42	1.7	27.5	131.5	0.42	641.0	0.03	0.05	0.23	1.27	0.05	6.7	0	3.1	8.8
29	25	0.46	5.6	88.6	465.3	0.96	20.2	0.10	0.10	0.06	1.24	0.05	1.3	0	0.7	6.8
50	5	0.90	10.2	225.7	700.2	1.93	0	0.35	0.12	0.17	2.90	0.30	1.1	0	1.1	12.0
24	6	0.67	6.3	70.2	513.0	1.10	0	0.01	0.09	0.06	1.06	0.04	2.3	0	0.8	3.7
23	5	0.51	4.5	75.2	504.0	0.82	8.1	0.09	0.11	0.05	1.18	0.05	1.8	0	0.6	6.2
43	33	0.52	9.0	90.9	379.8	0.50	0	0.02	0.09	0.11	2.10	0.14	3.2	0	0.2	10.4
35	67	0.66	6.3	176.4	485.1	0.82	0	0.01	0.27	0.08	1.65	0.06	4.1	0	0.4	6.8
39	14	0.97	12.2	206.7	820.8	1.62	6.8	0.42	0.17	0.15	2.83	0.22	3.4	0.1	0.9	15.0
20	13	0.44	4.9	84.4	283.0	0.61	0	0.06	0.06	0.06	0.87	0.05	1.5	0	0.5	5.4
34	6	0.37	6.2	112.8	527.3	0.94	0	0.19	0.32	0.07	1.55	0.09	1.1	0	0.7	7.7
13	2	0.15	2.0	34.7	196.7	0.30	0	0.05	0.00	0.02	0.59	0.04	0.7	0.1	0.2	2.4
15	2	0.29	2.9	37.3	199.3	0.40	0	0.10	0.04	0.03	0.71	0.03	0.4	0.2	0.2	3.7
37	3	1.14	6.8	97.8	592.8	0.92	0	0.04	0.08	0.08	1.68	0.08	1.0	0	1.6	7.6
10	2	0.16	2.8	48.4	289.3	0.53	0	0.11	0.02	0.04	0.71	0.07	0.3	0	0.4	3.3
52	12	0.84	11.9	169.0	377.1	2.19	7.4	0.04	0.10	0.14	3.24	0.18	3.4	0.4	0.7	0
33	7	0.42	7.4	101.5	516.5	0.71	7.4	0.10	0.07	0.06	1.66	0.09	1.1	0	0.3	0
35	6	0.33	6.2	273.9	468.9	0.74	0	0.12	0.14	0.10	1.59	0.10	0.6	0	0.4	10.4
52	23	0.87	14.4	254.6	1136.6	2.13	0	0.53	0.18	0.19	3.43	0.26	3.8	1.5	1.2	16.4
42	5	1.15	7.9	147.4	737.1	1.45	0	0.08	0.12	0.18	2.44	0.14	1.1	9.4	3.1	11.5
14	2	0.14	1.1	16.2	155.0	0.25	0	0.01	0.03	0.01	0.25	0.01	0.6	0	0.2	2.7
22	5	0.35	6.0	90.0	547.2	0.66	0	0.26	1.04	0.07	1.25	0.09	0.6	0	0.5	10.7
23	7	0.47	4.8	66.0	607.8	0.61	15.6	0.10	1.04	0.08	1.03	0.07	1.2	0	0.7	5.8
10	0	0.00	—	—	340.0	—	0	—	—	—	—	—	—	0	—	—
10	0	0.00	—	—	300.0	—	0	—	—	—	—	—	—	0	—	—
25	0	0.72	—	—	720.0	—	0	—	—	—	—	—	—	0	—	—
20	0	0.72	—	—	660.0	—	0	—	—	—	—	—	—	0	—	—
20	0	0.72	—	—	660.0	—	0	—	—	—	—	—	—	0	—	—
4	0	0.06	—	—	103.3	—	0	—	—	—	—	—	—	0	—	—
40	0	0.72	12.3	162.4	700.0	1.44	0	—	—	—	—	—	—	0	—	—
19	3	1.19	4.0	97.8	278.1	0.61	1.1	0.01	0.06	0.07	1.00	0.07	1.4	4.6	0.1	4.6
43	23	0.70	12.5	122.5	569.3	1.31	1.1	0.24	0.13	0.17	2.25	0.24	5.7	0	0.6	15.0
22	11	0.35	6.2	61.2	284.6	0.65	0.6	0.12	0.06	0.08	1.12	0.12	2.8	0	0.3	7.5
0	18	0.07	21.6	90.0	14.4	0.03	0	0.03	0.00	0.10	1.62	0.18	21.6	0	0.1	2.5
0	14	0.07	21.4	96.2	14.3	0.03	0	0.01	0.00	0.08	1.82	0.16	21.4	0	0.1	2.1
0	25	0.21	25.2	28.8	46.8	0.07	—	0.07	0.00	0.18	3.99	0.10	50.4	1.8	0.1	4.3
0	18	0.14	17.8	63.9	9.0	0.10	0	0.03	—	0.10	1.39	0.12	14.6	0	0	4.0
0	18	0.10	21.3	88.8	9.0	0.07	0	0.02	—	0.09	1.60	0.17	21.3	0	0.1	4.0
0	14	0.10	17.7	74.3	14.2	0.03	0	0.01	0.00	0.05	1.38	0.12	21.2	0	0.1	1.4
0	18	0.10	21.3	88.8	9.0	0.07	0	0.02	—	0.09	1.60	0.17	21.3	0	0.1	4.0
0	18	0.14	17.8	63.9	9.0	0.10	0	0.03	—	0.10	1.39	0.12	14.6	0	0	4.0
0	0	0.01	0	0.6	0.3	0.01	0	0.00	—	0.00	0.00	0.00	0	0	0	0
0	0	0.01	0	0.6	0.3	0.01	0	0.00	0.00	0.00	0.00	0.00	0	0	0	0
0	0	0.01	0	0.6	0.3	0.01	0	0.00	0.00	0.00	0.00	0.00	0	0	0	0
0	0	0.01	0	0.6	0.3	0.01	0	0.00	0.00	0.00	0.00	0.00	0	0	0	0
0	0	0.01	0	0.6	0.3	0.01	0	0.00	—	0.00	0.00	0.00	0	0	0	0

Food Composition   (Computer code is for Cengage Diet Analysis program)   (For purposes of calculations, use "0" for t, <1, <.1, <.01, etc.)

DA+ Code	Food Description	Quantity	Measure	Wt (g)	H₂O (g)	Ener (kcal)	Prot (g)	Carb (g)	Fiber (g)	Fat (g)	Fat Breakdown (g)			
											Sat	Mono	Poly	Trans
	Liqueurs													
33187	Coffee liqueur, 53 proof	1	fluid ounce(s)	35	10.8	113	0	16.3	0	0.1	0	0	0	—
3142	Coffee liqueur, 63 proof	1	fluid ounce(s)	35	14.4	107	0	11.2	0	0.1	0	0	0	—
736	Cordials, 54 proof	1	fluid ounce(s)	30	8.9	106	0	13.3	0	0.1	0	0	0	—
	Wine													
861	California red wine	5	fluid ounce(s)	150	133.4	125	0.3	3.7	0	0	0	0	0	0
858	Domestic champagne	5	fluid ounce(s)	150	—	105	0.3	3.8	0	0	0	0	0	0
690	Sweet dessert wine	5	fluid ounce(s)	147	103.7	235	0.3	20.1	0	0	0	0	0	0
1481	White wine	5	fluid ounce(s)	148	128.1	121	0.1	3.8	0	0	0	0	0	0
1811	Wine cooler	10	fluid ounce(s)	300	267.4	159	0.3	20.2	0	0.1	0	0	0	—
	Carbonated													
31898	7 Up	12	fluid ounce(s)	360	321.0	140	0	39.0	0	0	0	0	0	0
692	Club soda	12	fluid ounce(s)	355	354.8	0	0	0	0	0	0	0	0	0
12010	Coca-Cola Classic cola soda	12	fluid ounce(s)	360	319.4	146	0	40.5	0	0	0	0	0	0
693	Cola	12	fluid ounce(s)	368	332.7	136	0.3	35.2	0	0.1	0	0	0	—
2391	Cola or pepper-type soda, low calorie with saccharin	12	fluid ounce(s)	355	354.5	0	0	0.3	0	0	0	0	0	0
9522	Cola soda, decaffeinated	12	fluid ounce(s)	372	333.4	153	0	39.3	0	0	0	0	0	0
9524	Cola, decaffeinated, low calorie with aspartame	12	fluid ounce(s)	355	354.3	4	0.4	0.5	0	0	0	0	0	0
1415	Cola, low calorie with aspartame	12	fluid ounce(s)	355	353.6	7	0.4	1.0	0	0.1	0	0	0	—
1412	Cream soda	12	fluid ounce(s)	371	321.5	189	0	49.3	0	0	0	0	0	0
31899	Diet 7 Up	12	fluid ounce(s)	360	—	0	0	0	0	0	0	0	0	0
12031	Diet Coke cola soda	12	fluid ounce(s)	360	—	2	0	0.2	0	0	0	0	0	0
29392	Diet Mountain Dew soda	12	fluid ounce(s)	360	—	0	0	0	0	0	0	0	0	0
29389	Diet Pepsi cola soda	12	fluid ounce(s)	360	—	0	0	0	0	0	0	0	0	0
12034	Diet Sprite soda	12	fluid ounce(s)	360	—	4	0	0	0	0	0	0	0	0
695	Ginger ale	12	fluid ounce(s)	366	333.9	124	0	32.1	0	0	0	0	0	0
694	Grape soda	12	fluid ounce(s)	372	330.3	160	0	41.7	0	0	0	0	0	0
1876	Lemon lime soda	12	fluid ounce(s)	368	330.8	147	0.2	37.4	0	0.1	0	0	0	—
29391	Mountain Dew soda	12	fluid ounce(s)	360	314.0	170	0	46.0	0	0	0	0	0	0
3145	Orange soda	12	fluid ounce(s)	372	325.9	179	0	45.8	0	0	0	0	0	0
1414	Pepper-type soda	12	fluid ounce(s)	368	329.3	151	0	38.3	0	0.4	0.3	0	0	—
29388	Pepsi regular cola soda	12	fluid ounce(s)	360	318.9	150	0	41.0	0	0	0	0	0	0
696	Root beer	12	fluid ounce(s)	370	330.0	152	0	39.2	0	0	0	0	0	0
12044	Sprite soda	12	fluid ounce(s)	360	321.0	144	0	39.0	0	0	0	0	0	0
	Coffee													
731	Brewed	8	fluid ounce(s)	237	235.6	2	0.3	0	0	0	0	0	0	0
9520	Brewed, decaffeinated	8	fluid ounce(s)	237	234.3	5	0.3	1.0	0	0	0	0	0	0
16882	Cappuccino	8	fluid ounce(s)	240	224.8	79	4.1	5.8	0.2	4.9	2.3	1.0	0.2	—
16883	Cappuccino, decaffeinated	8	fluid ounce(s)	240	224.8	79	4.1	5.8	0.2	4.9	2.3	1.0	0.2	—
16880	Espresso	8	fluid ounce(s)	237	231.8	21	0	3.6	0	0.4	0.2	0	0.2	0
16881	Espresso, decaffeinated	8	fluid ounce(s)	237	231.8	21	0	3.6	0	0.4	0.2	0	0.2	0
732	Instant, prepared	8	fluid ounce(s)	239	236.5	5	0.2	0.8	0	0	0	0	0	0
	Fruit drinks													
29357	Crystal Light sugar free lemonade drink	8	fluid ounce(s)	240	—	5	0	0	0	0	0	0	0	0
6012	Fruit punch drink with added vitamin C, canned	8	fluid ounce(s)	248	218.2	117	0	29.7	0.5	0	0	0	0	0
31143	Gatorade Thirst Quencher, all flavors	8	fluid ounce(s)	240	—	50	0	14.0	0	0	0	0	0	0
260	Grape drink, canned	8	fluid ounce(s)	250	210.5	153	0	39.4	0	0	0	0	0	0
17372	Kool-Aid (lemonade/punch/fruit drink)	8	fluid ounce(s)	248	220.0	108	0.1	27.8	0.2	0	0	0	0	—
17225	Kool-Aid sugar free, low calorie tropical punch drink mix, prepared	8	fluid ounce(s)	240	—	5	0	0	0	0	0	0	0	0
266	Lemonade, prepared from frozen concentrate	8	fluid ounce(s)	248	221.6	99	0.2	25.8	0	0.1	0	0	0	—
268	Limeade, prepared from frozen concentrate	8	fluid ounce(s)	247	212.6	128	0	34.1	0	0	0	0	0	—
14266	Odwalla strawberry 'C' monster smoothie blend	8	fluid ounce(s)	240	—	160	2.0	38.0	0	0	0	0	0	0
10080	Odwalla strawberry lemonade quencher	8	fluid ounce(s)	240	—	110	0	28.0	0	0	0	0	0	0
10099	Snapple fruit punch fruit drink	8	fluid ounce(s)	240	—	110	0	29.0	0	0	0	0	0	0
10096	Snapple kiwi strawberry fruit drink	8	fluid ounce(s)	240	211.2	110	0	28.0	0	0	0	0	0	0
	Slim Fast ready-to-drink shake													
16054	French vanilla ready to drink shake	11	fluid ounce(s)	325	—	220	10.0	40.0	5.0	2.5	0.5	1.5	0.5	—
40447	Optima rich chocolate royal ready-to-drink shake	11	fluid ounce(s)	330	—	180	10.0	24.0	5.0	5.0	1.0	3.5	0.5	0
16055	Strawberries n cream ready to drink shake	11	fluid ounce(s)	325	—	220	10.0	40.0	5.0	2.5	0.5	1.5	0.5	—
	Tea													
33179	Decaffeinated, prepared	8	fluid ounce(s)	237	236.3	2	0	0.7	0	0	0	0	0	0
1877	Herbal, prepared	8	fluid ounce(s)	237	236.1	2	0	0.5	0	0	0	0	0	0
735	Instant tea mix, lemon flavored with sugar, prepared	8	fluid ounce(s)	259	236.2	91	0	22.3	0.3	0.2	0	0	0	—
734	Instant tea mix, unsweetened, prepared	8	fluid ounce(s)	237	236.1	2	0.1	0.4	0	0	0	0	0	0
733	Tea, prepared	8	fluid ounce(s)	237	236.3	2	0	0.7	0	0	0	0	0	0
	Water													
1413	Mineral water, carbonated	8	fluid ounce(s)	237	236.8	0	0	0	0	0	0	0	0	0
	BEVERAGES —Continued													

Chol (mg)	Calc (mg)	Iron (mg)	Magn (mg)	Pota (mg)	Sodi (mg)	Zinc (mg)	Vit A (µg)	Thia (mg)	Vit E (mg α)	Ribo (mg)	Niac (mg)	Vit B₆ (mg)	Fola (µg)	Vit C (mg)	Vit B₁₂ (µg)	Sele (µg)
0	0	0.02	1.0	10.4	2.8	0.01	0	0.00	0.00	0.00	0.05	0.00	0	0	0	0.1
0	0	0.02	1.0	10.4	2.8	0.01	0	0.00	—	0.00	0.05	0.00	0	0	0	0.1
0	0	0.02	0.6	4.5	2.1	0.01	0	0.00	0.00	0.00	0.02	0.00	0	0	0	0.1
0	12	1.43	16.2	170.6	15.0	0.14	0	0.01	0.00	0.04	0.11	0.05	1.5	0	0	—
0	—	—	—	—	—	—	—	—	—	—	—	—	—	—	0	—
0	12	0.34	13.2	135.4	13.2	0.10	0	0.01	0.00	0.01	0.30	0.00	0	0	0	0.7
0	13	0.39	14.8	104.7	7.4	0.18	0	0.01	0.00	0.01	0.15	0.06	1.5	0	0	0.1
0	18	0.75	15.0	129.0	24.0	0.18	—	0.01	0.03	0.03	0.13	0.03	3.0	5.4	0	0.6
0	—	—	—	0.6	75.0	—	—	—	—	—	—	—	—	—	—	—
0	18	0.03	3.5	7.1	74.6	0.35	0	0.00	0.00	0.00	0.00	0.00	0	0	0	0
0	—	—	—	0	49.5	—	0	—	—	—	—	—	—	0	—	—
0	7	0.41	0	7.4	14.7	0.06	0	0.00	0.00	0.00	0.00	0.00	0	0	0	0.4
0	14	0.06	3.5	14.2	56.8	0.11	0	0.00	0.00	0.00	0.00	0.00	0	0	0	0.3
0	7	0.08	0	11.2	14.9	0.03	0	0.00	0.00	0.00	0.00	0.00	0	0	0	0.4
0	11	0.06	0	24.9	14.2	0.03	0	0.02	0.00	0.08	0.00	0.00	0	0	0	0.3
0	11	0.39	3.5	28.4	28.4	0.03	0	0.02	0.00	0.08	0.00	0.00	0	0	0	0
0	19	0.18	3.7	3.7	44.5	0.26	0	0.00	0.00	0.00	0.00	0.00	0	0	0	0
0	—	—	—	77.0	45.0	—	—	—	—	—	—	—	—	—	—	—
0	—	—	—	18.0	42.0	—	0	—	—	—	—	—	—	0	—	—
0	—	—	—	70.0	35.0	—	—	—	—	—	—	—	—	—	—	—
0	—	—	—	30.0	35.0	—	—	—	—	—	—	—	—	—	—	—
0	—	—	—	109.5	36.0	—	0	—	—	—	—	—	—	0	—	—
0	11	0.65	3.7	3.7	25.6	0.18	0	0.00	0.00	0.00	0.00	0.00	0	0	0	0.4
0	11	0.29	3.7	3.7	55.8	0.26	0	0.00	—	0.00	0.00	0.00	0	0	0	0
0	7	0.41	3.7	3.7	33.2	0.14	0	0.00	0.00	0.00	0.05	0.00	0	0	0	0
0	—	—	—	0	70.0	—	—	—	—	—	—	—	—	—	—	—
0	19	0.21	3.7	7.4	44.6	0.36	0	0.00	—	0.00	0.00	0.00	0	0	0	0
0	11	0.14	0	3.7	36.8	0.14	0	0.00	—	0.00	0.00	0.00	0	0	0	0.4
0	—	—	—	0	35.0	—	—	—	—	—	—	—	—	—	—	—
0	18	0.18	3.7	3.7	48.0	0.26	0	0.00	0.00	0.00	0.00	0.00	0	0	0	0.4
0	—	—	—	0	70.5	—	0	—	—	—	—	—	—	0	—	—
0	5	0.02	7.1	116.1	4.7	0.04	0	0.03	0.02	0.18	0.45	0.00	4.7	0	0	0
0	7	0.14	11.8	108.9	4.7	0.00	0	0.00	0.00	0.03	0.66	0.00	0	0	0	0.5
12	144	0.19	14.4	232.8	50.4	0.50	33.6	0.04	0.09	0.27	0.13	0.04	7.2	0	0.4	4.6
12	144	0.19	14.4	232.8	50.4	0.50	33.6	0.04	0.09	0.27	0.13	0.04	7.2	0	0.4	4.6
0	5	0.30	189.6	272.6	33.2	0.11	0	0.00	0.04	0.42	12.34	0.00	2.4	0.5	0	0
0	5	0.30	189.6	272.6	33.2	0.11	0	0.00	0.04	0.42	12.34	0.00	2.4	0.5	0	0
0	10	0.09	9.5	71.6	9.5	0.01	0	0.00	0.00	0.00	0.56	0.00	0	0	0	0.2
0	0	0.00	—	160.0	40.0	—	0	—	—	—	—	—	—	0	—	—
0	20	0.22	7.4	62.0	94.2	0.02	5.0	0.05	0.04	0.05	0.05	0.02	9.9	89.3	0	0.5
0	0	0.00	—	30.0	110.0	—	0	—	—	—	—	—	—	0	—	—
0	130	0.17	2.5	30.0	40.0	0.30	0	0.00	0.00	0.01	0.07	0.01	0	78.5	0	0.3
0	14	0.45	5.0	49.6	31.0	0.19	—	0.03	—	0.05	0.04	0.01	4.3	41.6	0	1.0
0	0	0.00	—	10.1	10.1	—	0	—	—	—	—	—	—	6.0	—	—
0	10	0.39	5.0	37.2	9.9	0.05	0	0.01	0.02	0.05	0.04	0.01	2.5	9.7	0	0.2
0	5	0.00	4.9	24.7	7.4	0.02	0	0.01	0.00	0.01	0.02	0.01	2.5	7.7	0	0.2
0	20	0.72	—	0	20.0	—	0	—	—	—	—	—	—	600.0	0	—
0	0	0.00	—	70.0	10.0	—	0	—	—	—	—	—	—	54.0	0	—
0	0	0.00	—	20.0	10.0	—	0	—	—	—	—	—	—	0	0	—
0	0	0.00	—	40.0	10.0	—	0	—	—	—	—	—	—	0	0	—
5	400	2.70	140.0	600.0	220.0	2.25	—	0.52	—	0.59	7.00	0.70	120.0	60.0	2.1	17.5
5	1000	2.70	140.0	600.0	220.0	2.25	—	0.52	—	0.59	7.00	0.70	120.0	30.0	2.1	17.5
5	400	2.70	140.0	600.0	220.0	2.25	—	0.52	—	0.59	7.00	0.70	120.0	60.0	2.1	17.5
0	0	0.04	7.1	87.7	7.1	0.04	0	0.00	0.00	0.03	0.00	0.00	11.9	0	0	0
0	5	0.18	2.4	21.3	2.4	0.09	0	0.02	0.00	0.01	0.00	0.00	2.4	0	0	0
0	5	0.05	2.6	38.9	5.2	0.02	0	0.00	0.00	0.02	0.00	0.00	0	0	0	0.3
0	7	0.02	4.7	42.7	9.5	0.02	0	0.00	0.00	0.01	0.07	0.00	0	0	0	0
0	0	0.04	7.1	87.7	7.1	0.04	0	0.00	0.00	0.03	0.00	0.00	11.9	0	0	0
0	33	0.00	0	0	2.4	0.00	0	0.00	—	0.00	0.00	0.00	0	0	0	0

Food Composition (Computer code is for Cengage Diet Analysis program)   (For purposes of calculations, use "0" for t, <1, <.1, <.01, etc.)

DA+ Code	Food Description	Quantity	Measure	Wt (g)	H₂O (g)	Ener (kcal)	Prot (g)	Carb (g)	Fiber (g)	Fat (g)	Sat	Mono	Poly	Trans
											Fat Breakdown (g)			
33183	Poland spring water, bottled	8	fluid ounce(s)	237	237.0	0	0	0	0	0	0	0	0	0
1821	Tap water	8	fluid ounce(s)	237	236.8	0	0	0	0	0	0	0	0	0
1879	Tonic water	8	fluid ounce(s)	244	222.3	83	0	21.5	0	0	0	0	0	0
	FATS AND OILS													
	Butter													
104	Butter	1	tablespoon(s)	14	2.3	102	0.1	0	0	11.5	7.3	3.0	0.4	—
2522	Butter Buds, dry butter substitute	1	teaspoon(s)	2	—	5	0	2.0	0	0	0	0	0	0
921	Unsalted	1	tablespoon(s)	14	2.5	102	0.1	0	0	11.5	7.3	3.0	0.4	—
107	Whipped	1	tablespoon(s)	9	1.5	67	0.1	0	0	7.6	4.7	2.2	0.3	—
944	Whipped, unsalted	1	tablespoon(s)	11	2.0	82	0.1	0	0	9.2	5.9	2.4	0.3	—
	Fats, cooking													
2671	Beef tallow, semisolid	1	tablespoon(s)	13	0	115	0	0	0	12.8	6.4	5.4	0.5	—
922	Chicken fat	1	tablespoon(s)	13	0	115	0	0	0	12.8	3.8	5.7	2.7	—
5454	Household shortening with vegetable oil	1	tablespoon(s)	13	0	115	0	0	0	13.0	3.4	5.5	2.7	2.2
111	Lard	1	tablespoon(s)	13	0	115	0	0	0	12.8	5.0	5.8	1.4	—
	Margarine													
114	Margarine	1	tablespoon(s)	14	2.3	101	0	0.1	0	11.4	2.1	5.5	3.4	2.1
5439	Soft	1	tablespoon(s)	14	2.3	103	0.1	0.1	0	11.6	1.7	4.4	2.1	3.0
32329	Soft, unsalted, with hydrogenated soybean and cottonseed oils	1	tablespoon(s)	14	2.5	101	0.1	0.1	0	11.3	2.0	5.4	3.5	—
928	Unsalted	1	tablespoon(s)	14	2.6	101	0.1	0.1	0	11.3	2.1	5.2	3.5	—
119	Whipped	1	tablespoon(s)	9	1.5	64	0.1	0.1	0	7.2	1.2	3.2	2.5	—
	Spreads													
54657	I Can't Believe It's Not Butter!, tub, soya oil (non-hydrogenated)	1	tablespoon(s)	14	2.3	103	0.1	0.1	0	11.6	2.8	2.0	5.1	0.1
2708	Mayonnaise with soybean and safflower oils	1	tablespoon(s)	14	2.1	99	0.2	0.4	0	11.0	1.2	1.8	7.6	—
16157	Promise vegetable oil spread, stick	1	tablespoon(s)	14	4.2	90	0	0	0	10.0	2.5	2.0	4.0	—
	Oils													
2681	Canola	1	tablespoon(s)	14	0	120	0	0	0	13.6	1.0	8.6	3.8	0.1
120	Corn	1	tablespoon(s)	14	0	120	0	0	0	13.6	1.8	3.8	7.4	0
122	Olive	1	tablespoon(s)	14	0	119	0	0	0	13.5	1.9	9.9	1.4	—
124	Peanut	1	tablespoon(s)	14	0	119	0	0	0	13.5	2.3	6.2	4.3	—
2693	Safflower	1	tablespoon(s)	14	0	120	0	0	0	13.6	0.8	10.2	2.0	—
923	Sesame	1	tablespoon(s)	14	0	120	0	0	0	13.6	1.9	5.4	5.7	—
128	Soybean, hydrogenated	1	tablespoon(s)	14	0	120	0	0	0	13.6	2.0	5.8	5.1	—
130	Soybean, with soybean and cottonseed oil	1	tablespoon(s)	14	0	120	0	0	0	13.6	2.4	4.0	6.5	—
2700	Sunflower	1	tablespoon(s)	14	0	120	0	0	0	13.6	1.8	6.3	5.0	—
357	**Pam original no stick cooking spray**	1	serving(s)	0	0.2	0	0	0	0	0	0	0	0	—
	Salad dressing													
132	Blue cheese	2	tablespoon(s)	30	9.7	151	1.4	2.2	0	15.7	3.0	3.7	8.3	—
133	Blue cheese, low calorie	2	tablespoon(s)	32	25.4	32	1.6	0.9	0	2.3	0.8	0.6	0.8	—
1764	Caesar	2	tablespoon(s)	30	10.3	158	0.4	0.9	0	17.3	2.6	4.1	9.9	—
29654	Creamy, reduced calorie, fat free, cholesterol free, sour cream and/or buttermilk and oil	2	tablespoon(s)	32	23.9	34	0.4	6.4	0	0.9	0.2	0.2	0.5	—
29617	Creamy, reduced calorie, sour cream and/or buttermilk and oil	2	tablespoon(s)	30	22.2	48	0.5	2.1	0	4.2	0.6	1.0	2.4	—
134	French	2	tablespoon(s)	32	11.7	146	0.2	5.0	0	14.3	1.8	2.7	6.7	—
135	French, low fat	2	tablespoon(s)	32	17.4	74	0.2	9.4	0.4	4.3	0.4	1.9	1.6	—
136	Italian	2	tablespoon(s)	29	16.6	86	0.1	3.1	0	8.3	1.3	1.9	3.8	—
137	Italian, diet	2	tablespoon(s)	30	25.4	23	0.1	1.4	0	1.9	0.1	0.7	0.5	—
139	Mayonnaise-type	2	tablespoon(s)	29	11.7	115	0.3	7.0	0	9.8	1.4	2.6	5.3	—
942	Oil and vinegar	2	tablespoon(s)	32	15.2	144	0	0.8	0	16.0	2.9	4.7	7.7	—
1765	Ranch	2	tablespoon(s)	30	11.6	146	0.1	1.6	0	15.8	2.3	5.2	7.6	—
3666	Ranch, reduced calorie	2	tablespoon(s)	30	20.5	62	0.1	2.2	0	6.1	1.1	1.8	2.9	—
940	Russian	2	tablespoon(s)	30	11.6	107	0.5	9.3	0.7	7.8	1.2	1.8	4.4	—
939	Russian, low calorie	2	tablespoon(s)	32	20.8	45	0.2	8.8	0.1	1.3	0.2	0.3	0.7	—
941	Sesame seed	2	tablespoon(s)	30	11.8	133	0.9	2.6	0.3	13.6	1.9	3.6	7.5	—
142	Thousand Island	2	tablespoon(s)	32	14.9	118	0.3	4.7	0.3	11.2	1.6	2.5	5.8	—
143	Thousand Island, low calorie	2	tablespoon(s)	30	18.2	61	0.3	6.7	0.4	3.9	0.2	1.9	0.8	—
	Sandwich spreads													
138	Mayonnaise with soybean oil	1	tablespoon(s)	14	2.1	99	0.1	0.4	0	11.0	1.6	2.7	5.8	0
140	Mayonnaise, low calorie	1	tablespoon(s)	16	10.0	37	0	2.6	0	3.1	0.5	0.7	1.7	—
141	Tartar sauce	2	tablespoon(s)	28	8.7	144	0.3	4.1	0.1	14.4	2.2	3.8	7.7	—
	SWEETS													
4799	**Butterscotch or caramel topping**	2	tablespoon(s)	41	13.1	103	0.6	27.0	0.4	0	0	0	0	—
	Candy													
1786	Almond Joy candy bar	1	item(s)	45	4.3	220	2.0	27.0	2.0	12.0	8.0	3.3	0.7	0
1785	Bit-O-Honey candy	6	item(s)	40	—	190	1.0	39.0	0	3.5	2.5	—	—	—
33375	Butterscotch candy	2	piece(s)	12	0.6	47	0	10.8	0	0.4	0.2	0.1	0	—
	SWEETS —Continued													

Chol (mg)	Calc (mg)	Iron (mg)	Magn (mg)	Pota (mg)	Sodi (mg)	Zinc (mg)	Vit A (µg)	Thia (mg)	Vit E (mg α)	Ribo (mg)	Niac (mg)	Vit B$_6$ (mg)	Fola (µg)	Vit C (mg)	Vit B$_{12}$ (µg)	Sele (µg)
0	2	0.02	2.4	0	2.4	0.00	0	0.00	—	0.00	0.00	0.00	0	0	0	0
0	7	0.00	2.4	2.4	7.1	0.00	0	0.00	0.00	0.00	0.00	0.00	0	0	0	0
0	2	0.02	0	0	29.3	0.24	0	0.00	0.00	0.00	0.00	0.00	0	0	0	0
31	3	0.00	0.3	3.4	81.8	0.01	97.1	0.00	0.32	0.01	0.01	0.00	0.4	0	0	0.1
0	0	0.00	0	1.6	120.0	0.00	0	0.00	0.00	0.00	0.00	0.00	0	0	0	—
31	3	0.00	0.3	3.4	1.6	0.01	97.1	0.00	0.32	0.01	0.01	0.00	0.4	0	0	0.1
21	2	0.01	0.2	2.4	77.7	0.01	64.3	0.00	0.21	0.00	0.00	0.00	0.3	0	0	0.1
25	3	0.00	0.2	2.7	1.3	0.01	78.0	0.00	0.26	0.00	0.00	0.00	0.3	0	0	0.1
14	0	0.00	0	0	0	0.00	0	0.00	0.34	0.00	0.00	0.00	0	0	0	0
11	0	0.00	0	0	0	0.00	0	0.00	0.34	0.00	0.00	0.00	0	0	0	0
0	0	0.00	0	0	0	0.00	0	0.00	—	0.00	0.00	0.00	0	0	0	—
12	0	0.00	0	0	0	0.01	0	0.00	0.07	0.00	0.00	0.00	0	0	0	0
0	4	0.01	0.4	5.9	133.0	0.00	115.5	0.00	1.26	0.01	0.00	0.00	0.1	0	0	0
0	4	0.00	0.3	5.5	155.4	0.00	142.7	0.00	1.00	0.00	0.00	0.00	0.1	0	0	0
0	4	0.00	0.3	5.4	3.9	0.00	103.1	0.00	0.98	0.00	0.00	0.00	0.1	0	0	0
0	2	0.00	0.3	3.5	0.3	0.00	115.5	0.00	1.80	0.00	0.00	0.00	0.1	0	0	0
0	2	0.00	0.2	3.4	97.1	0.00	73.7	0.00	0.45	0.00	0.00	0.00	0.1	0	0	0
0	4	0.00	0.3	5.5	155.3	0.00	142.6	0.00	0.72	0.00	0.00	0.00	0.1	0	0	0
8	2	0.06	0.1	4.7	78.4	0.01	11.6	0.00	3.03	0.00	0.00	0.08	1.1	0	0	0.2
0	10	0.18	—	8.7	90.0	—	—	0.00	—	0.00	0.00	0.00	—	0.6	—	—
0	0	0.00	0	0	0	0.00	0	0.00	2.37	0.00	0.00	0.00	0	0	0	0
0	0	0.00	0	0	0	0.00	0	0.00	1.94	0.00	0.00	0.00	0	0	0	0
0	0	0.07	0	0.1	0.3	0.00	0	0.00	1.93	0.00	0.00	0.00	0	0	0	0
0	0	0.00	0	0	0	0.00	0	0.00	2.11	0.00	0.00	0.00	0	0	0	0
0	0	0.00	0	0	0	0.00	0	0.00	4.63	0.00	0.00	0.00	0	0	0	0
0	0	0.00	0	0	0	0.00	0	0.00	0.19	0.00	0.00	0.00	0	0	0	0
0	0	0.00	0	0	0	0.00	0	0.00	1.10	0.00	0.00	0.00	0	0	0	0
0	0	0.00	0	0	0	0.00	0	0.00	1.64	0.00	0.00	0.00	0	0	0	0
0	0	0.00	0	0	0	0.00	0	0.00	5.58	0.00	0.00	0.00	0	0	0	0
0	0	0.00	0	0.3	1.5	0.01	0.1	0.00	0.00	0.00	0.00	0.00	0	0	0	0
5	24	0.06	0	11.1	328.2	0.08	20.1	0.00	1.80	0.03	0.03	0.01	7.8	0.6	0.1	0.3
0	28	0.16	2.2	1.6	384.0	0.08	—	0.01	0.08	0.03	0.01	0.01	1.0	0.1	0.1	0.5
1	7	0.05	0.6	8.7	323.4	0.03	0.6	0.00	1.56	0.00	0.01	0.00	0.9	0	0	0.5
0	12	0.08	1.6	42.6	320.0	0.05	0.3	0.00	0.21	0.01	0.01	0.01	1.9	0	0	0.5
0	2	0.03	0.6	10.8	306.9	0.01	—	0.00	0.71	0.00	0.01	0.01	0	0.1	0	0.5
0	8	0.25	1.6	21.4	267.5	0.09	7.4	0.01	1.60	0.01	0.06	0.00	0	0	0	0
0	4	0.27	2.6	34.2	257.3	0.06	8.6	0.01	0.09	0.01	0.14	0.01	0.6	0	0	0.5
0	2	0.18	0.9	14.1	486.3	0.03	0.6	0.00	1.47	0.01	0.00	0.00	0	0	0	0.6
2	3	0.19	1.2	25.5	409.8	0.05	0.3	0.00	0.06	0.00	0.00	0.02	0	0	0	2.4
8	4	0.05	0.6	2.6	209.0	0.05	6.2	0.00	0.60	0.01	0.00	0.01	1.8	0	0.1	0.5
0	0	0.00	0	2.6	0.3	0.00	0	0.00	1.46	0.00	0.00	0.00	0	0	0	0.5
1	4	0.03	1.2	8.4	354.0	0.01	5.4	0.00	1.84	0.01	0.00	0.00	0.3	0.1	0	0.1
0	5	0.01	1.5	8.4	413.7	0.01	0.9	0.00	0.72	0.01	0.00	0.00	0.3	0.1	0	0.1
0	6	0.20	3.0	51.9	282.3	0.06	13.2	0.01	0.98	0.01	0.16	0.02	1.5	1.4	0	0.5
2	6	0.18	0	50.2	277.8	0.02	0.6	0.00	0.12	0.00	0.00	0.00	1.0	1.9	0	0.5
0	6	0.18	0	47.1	300.0	0.02	0.6	0.00	1.50	0.00	0.00	0.00	0	0	0	0.5
8	5	0.37	2.6	34.2	276.2	0.08	4.5	0.46	1.28	0.01	0.13	0.00	0	0	0	0.5
0	5	0.27	2.1	60.6	249.3	0.05	4.8	0.01	0.30	0.01	0.13	0.00	0	0	0	0
5	1	0.03	0.1	1.7	78.4	0.02	11.2	0.01	0.72	0.01	0.00	0.08	0.7	0	0	0.2
4	0	0.00	0	1.6	79.5	0.01	0	0.00	0.32	0.00	0.00	0.00	0	0	0	0.3
8	6	0.20	0.8	10.1	191.5	0.05	20.2	0.00	0.97	0.00	0.01	0.07	2.0	0.1	0.1	0.5
0	22	0.08	2.9	34.4	143.1	0.07	11.1	0.01	—	0.03	0.01	0.01	0.8	0.1	0	0
0	18	0.33	30.3	126.5	65.0	0.36	0	0.01	—	0.06	0.21	—	—	0	—	—
0	20	0.00	—	—	150.0	—	0	—	—	—	—	—	—	0	—	—
1	0	0.00	0	0.4	46.9	0.01	3.4	0.00	0.01	0.00	0.00	0.00	0	0	0	0.1

DA+ Code	Food Description	Quantity	Measure	Wt (g)	H₂O (g)	Ener (kcal)	Prot (g)	Carb (g)	Fiber (g)	Fat (g)	Sat	Mono	Poly	Trans
											Fat Breakdown (g)			
1701	Chewing gum, stick	1	item(s)	3	0.1	7	0	2.0	0.1	0	0	0	0	—
33378	Chocolate fudge with nuts, prepared	2	piece(s)	38	2.9	175	1.7	25.8	1.0	7.2	2.5	1.5	2.9	0.1
1787	Jelly beans	15	item(s)	43	2.7	159	0	39.8	0.1	0	0	0	0	—
1784	Kit Kat wafer bar	1	item(s)	42	0.8	210	3.0	27.0	0.5	11.0	7.0	3.5	0.3	0
4674	Krackel candy bar	1	item(s)	41	0.6	210	2.0	28.0	0.5	10.0	6.0	3.9	0.4	0
4934	Licorice	4	piece(s)	44	7.3	154	1.1	35.1	0	1.0	0	0.1	0	—
1780	Life Savers candy	1	item(s)	2	—	8	0	2.0	0	0	0	0	0	0
1790	Lollipop	1	item(s)	28	—	108	0	28.0	0	0	0	0	0	0
4679	M & Ms peanut chocolate candy, small bag	1	item(s)	49	0.9	250	5.0	30.0	2.0	13.0	5.0	5.4	2.1	—
1781	M & Ms plain chocolate candy, small bag	1	item(s)	48	0.8	240	2.0	34.0	1.0	10.0	6.0	3.3	0.3	—
4673	Milk chocolate bar, Symphony	1	item(s)	91	0.9	483	7.7	52.8	1.5	27.8	16.7	7.2	0.6	—
1783	Milky Way bar	1	item(s)	58	3.7	270	2.0	41.0	1.0	10.0	5.0	3.5	0.3	—
1788	Peanut brittle	1 ½	ounce(s)	43	0.3	207	3.2	30.3	1.1	8.1	1.8	3.4	1.9	—
1789	Reese's peanut butter cups	2	piece(s)	51	0.8	280	6.0	19.0	2.0	15.5	6.0	7.2	2.7	0
4689	Reese's pieces candy, small bag	1	item(s)	43	1.1	220	5.0	26.0	1.0	11.0	7.0	0.9	0.4	0
33399	Semisweet chocolate candy, made with butter	½	ounce(s)	14	0.1	68	0.6	9.0	0.8	4.2	2.5	1.4	0.1	—
1782	Snickers bar	1	item(s)	59	3.2	280	4.0	35.0	1.0	14.0	5.0	6.1	2.9	—
4694	Special Dark chocolate bar	1	item(s)	41	0.4	220	2.0	25.0	3.0	12.0	8.0	4.6	0.4	0
4695	Starburst fruit chews, original fruits	1	package(s)	59	3.9	240	0	48.0	0	5.0	1.0	2.1	1.8	—
4698	Taffy	3	piece(s)	45	2.2	179	0	41.2	0	1.5	0.9	0.4	0.1	0.1
4699	Three Musketeers bar	1	item(s)	60	3.5	260	2.0	46.0	1.0	8.0	4.5	2.6	0.3	—
4702	Twix caramel cookie bars	2	item(s)	58	2.4	280	3.0	37.0	1.0	14.0	5.0	7.7	0.5	—
4705	York peppermint pattie	1	item(s)	39	3.9	160	0.5	32.0	0.5	3.0	1.5	1.2	0.1	0
	Frosting, icing													
4760	Chocolate frosting, ready to eat	2	tablespoon(s)	31	5.2	122	0.3	19.4	0.3	5.4	1.7	2.8	0.6	—
4771	Creamy vanilla frosting, ready to eat	2	tablespoon(s)	28	4.2	117	0	19.0	0	4.5	0.8	1.4	2.2	0
17291	Dec-A-Cake variety pack candy decoration	1	teaspoon(s)	4	—	15	0	3.0	0	0.5	0	—	—	—
536	White icing	2	tablespoon(s)	40	3.6	162	0.1	31.8	0	4.2	0.8	2.0	1.2	—
	Gelatin													
13697	Gelatin snack, all flavors	1	item(s)	99	96.8	70	1.0	17.0	0	0	0	0	0	0
2616	Sugar free, low calorie mixed fruit gelatin mix, prepared	½	cup(s)	121	—	10	1.0	0	0	0	0	0	0	0
548	**Honey**	1	tablespoon(s)	21	3.6	64	0.1	17.3	0	0	0	0	0	0
	Jams, jellies													
550	Jam or preserves	1	tablespoon(s)	20	6.1	56	0.1	13.8	0.2	0	0	0	0	—
42199	Jams, preserves, dietetic, all flavors, w/ sodium saccarin	1	tablespoon(s)	14	6.4	18	0	7.5	0.4	0	0	0	0	—
552	Jelly	1	tablespoon(s)	21	6.3	56	0	14.7	0.2	0	0	0	0	—
545	**Marshmallows**	4	item(s)	29	4.7	92	0.5	23.4	0	0.1	0	0	0	—
4800	**Marshmallow cream topping**	2	tablespoon(s)	40	7.9	129	0.3	31.6	0	0.1	0	0	0	—
555	**Molasses**	1	tablespoon(s)	20	4.4	58	0	14.9	0	0	0	0	0	—
4780	**Popsicle or ice pop**	1	item(s)	59	47.5	47	0	11.3	0	0.1	0	0	0	—
	Sugar													
559	Brown sugar, packed	1	teaspoon(s)	5	0.1	17	0	4.5	0	0	0	0	0	0
563	Powdered sugar, sifted	⅓	cup(s)	33	0.1	130	0	33.2	0	0	0	0	0	—
561	White granulated sugar	1	teaspoon(s)	4	0	16	0	4.2	0	0	0	0	0	—
	Sugar substitute													
1760	Equal sweetener, packet size	1	item(s)	1	—	0	0	0.9	0	0	0	0	0	0
13029	Splenda granular no calorie sweetener	1	teaspoon(s)	1	—	0	0	0.5	0	0	0	0	0	0
1759	Sweet N Low sugar substitute, packet	1	item(s)	1	0.1	4	0	0.5	0	0	0	0	0	0
	Syrup													
3148	Chocolate syrup	2	tablespoon(s)	38	11.6	105	0.8	24.4	1.0	0.4	0.2	0.1	0	—
29676	Maple syrup	¼	cup(s)	80	25.7	209	0	53.7	0	0.2	0	0.1	0.1	—
4795	Pancake syrup	¼	cup(s)	80	30.4	187	0	49.2	0	0	0	0	0	0
	SPICES, CONDIMENTS, SAUCES													
	Spices													
807	Allspice, ground	1	teaspoon(s)	2	0.2	5	0.1	1.4	0.4	0.2	0	0	0	—
1171	Anise seeds	1	teaspoon(s)	2	0.2	7	0.4	1.1	0.3	0.3	0	0.2	0.1	—
729	Bakers' yeast, active	1	teaspoon(s)	4	0.3	12	1.5	1.5	0.8	0.2	0	0.1	0	—
683	Baking powder, double acting with phosphate	1	teaspoon(s)	5	0.2	2	0	1.1	0	0	0	0	0	0
1611	Baking soda	1	teaspoon(s)	5	0	0	0	0	0	0	0	0	0	—
8552	Basil	1	teaspoon(s)	1	0.8	0	0	0	0	0	0	0	0	—
34959	Basil, fresh	1	piece(s)	1	0.5	0	0	0	0	0	0	0	0	—
808	Basil, ground	1	teaspoon(s)	1	0.1	4	0.2	0.9	0.6	0.1	0	0	0	—
809	Bay leaf	1	teaspoon(s)	1	0	2	0	0.5	0.2	0.1	0	0	0	—
11720	Betel leaves	1	ounce(s)	28	—	17	1.8	2.4	0	0	—	—	—	—
730	Brewers' yeast	1	teaspoon(s)	3	0.1	8	1.0	1.0	0.8	0	0	0	0	0
11710	Capers	1	teaspoon(s)	5	—	0	0	0	0	0	0	0	0	—
1172	Caraway seeds	1	teaspoon(s)	2	0.2	7	0.4	1.0	0.8	0.3	0	0.2	0.1	—
	SPICES, CONDIMENTS, SAUCES —Continued													

Chol (mg)	Calc (mg)	Iron (mg)	Magn (mg)	Pota (mg)	Sodi (mg)	Zinc (mg)	Vit A (µg)	Thia (mg)	Vit E (mg α)	Ribo (mg)	Niac (mg)	Vit B_6 (mg)	Fola (µg)	Vit C (mg)	Vit B_{12} (µg)	Sele (µg)
0	0	0.00	0	0.1	0	0.00	0	0.00	0.00	0.00	0.00	0.00	0	0	0	0
5	22	0.74	20.9	69.5	14.8	0.54	14.4	0.02	0.09	0.03	0.12	0.03	6.1	0.1	0	1.1
0	1	0.05	0.9	15.7	21.3	0.02	0	0.00	0.00	0.01	0.00	0.00	0	0	0	0.5
3	60	0.36	16.4	126.0	30.0	0.51	0	0.07	—	0.22	1.07	0.05	59.6	0	0.1	2.0
3	40	0.36	—	168.8	50.0	—	0	—	—	—	—	—	—	0	—	—
0	0	0.22	2.6	28.2	126.3	0.07	0	0.01	0.07	0.01	0.04	0.00	0	0	0	—
0	0	0.00	—	0		—	0	0.00	—	0.00	0.00	—	—	0	—	0
0	0	0.00	—	—	10.8	—	0	0.00	—	0.00	0.00	—	—	0	—	1.0
5	40	0.36	36.5	170.6	25.0	1.13	14.8	0.03	—	0.06	1.60	0.04	17.3	0.6	0.1	1.9
5	40	0.36	19.6	127.4	30.0	0.46	14.8	0.02	—	0.06	0.10	0.01	2.9	0.6	0.1	1.4
22	228	0.82	61.0	398.6	91.9	1.00	0	0.06	—	0.25	0.14	0.10	10.9	2.0	0.4	—
5	60	0.18	19.8	140.1	95.0	0.41	15.1	0.02	—	0.06	0.20	0.02	5.8	0.6	0.2	3.3
5	11	0.51	17.9	71.4	189.2	0.37	16.6	0.05	1.08	0.01	1.12	0.03	19.6	0	0	1.1
3	40	0.72	45.4	217.4	180.0	0.93	0	0.12	—	0.08	2.35	0.07	28.1	0	0.1	2.3
0	20	0.00	18.9	169.9	80.0	0.32	0	0.04	—	0.06	1.22	0.03	12.0	0	0.1	0.8
3	5	0.44	16.3	51.7	1.6	0.23	0.4	0.01	—	0.01	0.06	0.01	0.4	0	0	0.5
5	40	0.36	42.3	—	140.0	1.37	15.3	0.03	—	0.06	1.60	0.05	23.5	0.6	0.1	2.7
0	0	1.80	45.5	136.0	50.0	0.59	0	0.01	—	0.02	0.16	0.01	0.8	0	0	1.2
0	10	0.18	0.6	1.2	0	0.00	—	0.00	—	0.00	0.00	0.00	0	30	0	0.5
4	4	0.00	0	1.4	23.4	0.09	12.2	0.01	0.04	0.01	0.00	0.00	0	0	0	0.3
5	20	0.36	17.5	80.3	110.0	0.33	14.5	0.01	—	0.03	0.20	0.01	0	0.6	0.1	1.5
5	40	0.36	18.5	116.8	115.0	0.45	15.0	0.09	—	0.13	0.69	0.01	13.9	0.6	0.1	1.2
0	0	0.33	23.4	66.1	10.0	0.28	0	0.01	—	0.03	0.31	0.01	1.5	0	0	—
0	2	0.44	6.4	60.0	56.1	0.09	0	0.00	0.48	0.00	0.03	0.00	0.3	0	0	0.2
0	1	0.04	0.3	9.5	51.5	0.01	0	0.00	0.43	0.08	0.06	0.00	2.2	0	0	0
0	0	0.00	—	—	15.0	—	0	—	—	—	—	—	—	0	—	—
0	4	0.01	0.4	5.6	76.4	0.01	44.4	0.00	0.32	0.00	0.00	0.00	0	0	0	0.3
0	0	0.00	—	0	40.0	—	0	—	—	—	—	—	—	0	—	—
0	0	0.00	0	0	50.0	0.00	0	0.00	0.00	0.00	0.00	0.00	0	0	0	—
0	1	0.08	0.4	10.9	0.8	0.04	0	0.00	0.00	0.01	0.02	0.01	0.4	0.1	0	0.2
0	4	0.10	0.8	15.4	6.4	0.01	0	0.00	0.02	0.02	0.01	0.01	2.2	1.8	0	0.4
0	1	0.56	0.7	9.7	0	0.01	0	0.00	0.01	0.00	0.00	0.00	1.3	0	0	0.2
0	1	0.04	1.3	11.3	6.3	0.01	0	0.00	0.00	0.01	0.01	0.00	0.4	0.2	0	0.1
0	1	0.06	0.6	1.4	23.0	0.01	0	0.00	0.00	0.00	0.02	0.00	0.3	0	0	0.5
0	1	0.08	0.8	2.0	32.0	0.01	0	0.00	0.00	0.00	0.03	0.00	0.4	0	0	0.7
0	41	0.94	48.4	292.8	7.4	0.05	0	0.01	0.00	0.00	0.18	0.13	0	0	0	3.6
0	0	0.31	0.6	8.9	4.1	0.08	0	0.00	0.00	0.00	0.00	0.00	0	0.4	0	0.1
0	4	0.03	0.4	6.1	1.3	0.00	0	0.00	0.00	0.00	0.01	0.00	0	0	0	0.1
0	0	0.01	0	0.7	0.3	0.00	0	0.00	0.00	0.00	0.00	0.00	0	0	0	0.2
0	0	0.00	0	0.1	0	0.00	0	0.00	0.00	0.00	0.00	0.00	0	0	0	0
0	0	0.00	0	0	0	0.00	0	0.00	0.00	0.00	0.00	0.00	0	0	0	0
0	0	0.00	—	—	0	—	—	0.00	—	0.00	0.00	—	—	0	0	—
0	0	0.00	—	—	0	—	0	—	0.00	—	—	—	—	0	—	—
0	5	0.79	24.4	84.0	27.0	0.27	0	0.00	0.01	0.01	0.12	0.00	0.8	0.1	0	0.5
0	54	0.96	11.2	163.2	7.2	3.32	0	0.01	0.00	0.01	0.02	0.00	0	0	0	0.5
0	2	0.02	1.6	12.0	65.6	0.06	0	0.01	0.00	0.01	0.00	0.00	0	0	0	0
0	13	0.13	2.6	19.8	1.5	0.01	0.5	0.00	—	0.00	0.05	0.00	0.7	0.7	0	0.1
0	14	0.77	3.6	30.3	0.3	0.11	0.3	0.01	—	0.01	0.06	0.01	0.2	0.4	0	0.1
0	3	0.66	3.9	80.0	2.0	0.25	0	0.09	0.00	0.21	1.59	0.06	93.6	0	0	1.0
0	339	0.51	1.8	0.2	363.1	0.00	0	0.00	0.00	0.00	0.00	0.00	0	0	0	0
0	0	0.00	0	0	1258.6	0.00	0	0.00	0.00	0.00	0.00	0.00	0	0	0	0
0	2	0.02	0.6	2.6	0	0.01	2.3	0.00	0.01	0.00	0.01	0.00	0.6	0.2	0	0
0	1	0.01	0.4	2.3	0	0.00	1.3	0.00	—	0.00	0.00	0.00	0.3	0.1	0	0
0	30	0.58	5.9	48.1	0.5	0.08	6.6	0.00	0.10	0.00	0.09	0.03	3.8	0.9	0	0
0	5	0.25	0.7	3.2	0.1	0.02	1.9	0.00	—	0.00	0.01	0.01	1.1	0.3	0	0
0	110	2.29	—	155.9	2.0	—	—	0.04	—	0.07	0.19	—	—	0.9	0	—
0	6	0.46	6.1	50.7	3.3	0.21	0	0.41	—	0.11	1.00	0.06	104.3	0	0	0
0	—	—	—	—	105.0	—	—	—	—	—	—	—	—	—	0	—
0	14	0.34	5.4	28.4	0.4	0.11	0.4	0.01	0.05	0.01	0.07	0.01	0.2	0.4	0	0.3

Food Composition (Computer code is for Cengage Diet Analysis program) (For purposes of calculations, use "0" for t, <1, <.1, <.01, etc.)

DA+ Code	Food Description	Quantity	Measure	Wt (g)	H₂O (g)	Ener (kcal)	Prot (g)	Carb (g)	Fiber (g)	Fat (g)	Fat Breakdown (g)			
											Sat	Mono	Poly	Trans
1173	Celery seeds	1	teaspoon(s)	2	0.1	8	0.4	0.8	0.2	0.5	0	0.3	0.1	—
1174	Chervil, dried	1	teaspoon(s)	1	0	1	0.1	0.3	0.1	0	0	0	0	—
810	Chili powder	1	teaspoon(s)	3	0.2	8	0.3	1.4	0.9	0.4	0.1	0.1	0.2	—
8553	Chives, chopped	1	teaspoon(s)	1	0.9	0	0	0	0	0	0	0	0	—
51420	Cilantro (coriander)	1	teaspoon(s)	0	0.3	0	0	0	0	0	0	0	0	—
811	Cinnamon, ground	1	teaspoon(s)	2	0.2	6	0.1	1.9	1.2	0	0	0	0	0
812	Cloves, ground	1	teaspoon(s)	2	0.1	7	0.1	1.3	0.7	0.4	0.1	0	0.1	—
1175	Coriander leaf, dried	1	teaspoon(s)	1	0	2	0.1	0.3	0.1	0	0	0	0	—
1176	Coriander seeds	1	teaspoon(s)	2	0.2	5	0.2	1.0	0.8	0.3	0	0.2	0	—
1706	Cornstarch	1	tablespoon(s)	8	0.7	30	0	7.3	0.1	0	0	0	0	—
1177	Cumin seeds	1	teaspoon(s)	2	0.2	8	0.4	0.9	0.2	0.5	0	0.3	0.1	—
11729	Cumin, ground	1	teaspoon(s)	5	—	11	0.4	0.8	0.8	0.4	—	—	—	—
1178	Curry powder	1	teaspoon(s)	2	0.2	7	0.3	1.2	0.7	0.3	0	0.1	0.1	—
1179	Dill seeds	1	teaspoon(s)	2	0.2	6	0.3	1.2	0.4	0.3	0	0.2	0	—
1180	Dill weed, dried	1	teaspoon(s)	1	0.1	3	0.2	0.6	0.1	0	0	0	0	—
34949	Dill weed, fresh	5	piece(s)	1	0.9	0	0	0.1	0	0	0	0	0	—
4949	Fennel leaves, fresh	1	teaspoon(s)	1	0.9	0	0	0.1	0	0				
1181	Fennel seeds	1	teaspoon(s)	2	0.2	7	0.3	1.0	0.8	0.3	0	0.2	0	—
1182	Fenugreek seeds	1	teaspoon(s)	4	0.3	12	0.9	2.2	0.9	0.2	0.1	—	—	—
11733	Garam masala, powder	1	ounce(s)	28	—	107	4.4	12.8	0	4.3	—	—	—	—
1067	Garlic clove	1	item(s)	3	1.8	4	0.2	1.0	0.1	0	0	0	0	—
813	Garlic powder	1	teaspoon(s)	3	0.2	9	0.5	2.0	0.3	0	0	0	0	—
1068	Ginger root	2	teaspoon(s)	4	3.1	3	0.1	0.7	0.1	0	0	0	0	—
1183	Ginger, ground	1	teaspoon(s)	2	0.2	6	0.2	1.3	0.2	0.1	0	0	0	—
35497	Leeks, bulb and lower-leaf, freeze-dried	¼	cup(s)	1	0	3	0.1	0.6	0.1	0	0	0	0	—
1184	Mace, ground	1	teaspoon(s)	2	0.1	8	0.1	0.9	0.3	0.6	0.2	0.2	0.1	—
1185	Marjoram, dried	1	teaspoon(s)	1	0	2	0.1	0.4	0.2	0	0	0	0	—
1186	Mustard seeds, yellow	1	teaspoon(s)	3	0.2	15	0.8	1.2	0.5	0.9	0	0.7	0.2	—
814	Nutmeg, ground	1	teaspoon(s)	2	0.1	12	0.1	1.1	0.5	0.8	0.6	0.1	0	—
2747	Onion flakes, dehydrated	1	teaspoon(s)	2	0.1	6	0.1	1.4	0.2	0	0	0	0	—
1187	Onion powder	1	teaspoon(s)	2	0.1	7	0.2	1.7	0.1	0	0	0	0	—
815	Oregano, ground	1	teaspoon(s)	2	0.1	5	0.2	1.0	0.6	0.2	0	0	0.1	—
816	Paprika	1	teaspoon(s)	2	0.2	6	0.3	1.2	0.8	0.3	0	0	0.2	—
817	Parsley, dried	1	teaspoon(s)	0	0	1	0.1	0.2	0.1	0	0	0	0	—
818	Pepper, black	1	teaspoon(s)	2	0.2	5	0.2	1.4	0.6	0.1	0	0	0	—
819	Pepper, cayenne	1	teaspoon(s)	2	0.1	6	0.2	1.0	0.5	0.3	0.1	0	0.2	—
1188	Pepper, white	1	teaspoon(s)	2	0.3	7	0.3	1.6	0.6	0.1	0	0	0	—
1189	Poppy seeds	1	teaspoon(s)	3	0.2	15	0.5	0.7	0.3	1.3	0.1	0.2	0.9	—
1190	Poultry seasoning	1	teaspoon(s)	2	0.1	5	0.1	1.0	0.2	0.1	0	0	0	—
1191	Pumpkin pie spice, powder	1	teaspoon(s)	2	0.1	6	0.1	1.2	0.3	0.2	0.1	0	0	—
1192	Rosemary, dried	1	teaspoon(s)	1	0.1	4	0.1	0.8	0.5	0.2	0.1	0	0	—
11723	Rosemary, fresh	1	teaspoon(s)	1	0.5	1	0	0.1	0.1	0	0	0	0	—
2722	Saffron powder	1	teaspoon(s)	1	0.1	2	0.1	0.5	0	0	0	0	0	—
11724	Sage	1	teaspoon(s)	1	—	1	0	0.1	0	0	—	—	—	—
1193	Sage, ground	1	teaspoon(s)	1	0.1	2	0.1	0.4	0.3	0.1	0	0	0	—
30189	Salt substitute	¼	teaspoon(s)	1	—	0	0	0	0	0	0	0	0	0
30190	Salt substitute, seasoned	¼	teaspoon(s)	1	—	1	0	0.1	0	0	0	—	—	—
822	Salt, table	¼	teaspoon(s)	2	0	0	0	0	0	0	0	0	0	0
1194	Savory, ground	1	teaspoon(s)	1	0.1	4	0.1	1.0	0.6	0.1	0	—	—	—
820	Sesame seed kernels, toasted	1	teaspoon(s)	3	0.1	15	0.5	0.7	0.5	1.3	0.2	0.5	0.6	—
11725	Sorrel	1	teaspoon(s)	3	—	1	0.1	0.1	0	0	0	—	—	—
11721	Spearmint	1	teaspoon(s)	2	1.6	1	0.1	0.2	0.1	0	0	0	0	—
35498	Sweet green peppers, freeze-dried	¼	cup(s)	2	0	5	0.3	1.1	0.3	0	0	0	0	—
11726	Tamarind leaves	1	ounce(s)	28	—	33	1.6	5.2	0	0.6	—	—	—	—
11727	Tarragon	1	ounce(s)	28	—	14	1.0	1.8	0	0.3	—	—	—	—
1195	Tarragon, ground	1	teaspoon(s)	2	0.1	5	0.4	0.8	0.1	0.1	0	0	0.1	—
11728	Thyme, fresh	1	teaspoon(s)	1	0.5	1	0	0.2	0.1	0	0	0	0	—
821	Thyme, ground	1	teaspoon(s)	1	0.1	4	0.1	0.9	0.5	0.1	0	0	0	—
1196	Turmeric, ground	1	teaspoon(s)	2	0.3	8	0.2	1.4	0.5	0.2	0.1	0	0	—
11995	Wasabi	1	tablespoon(s)	14	10.7	10	0.7	2.3	0.2	0	—	—	—	—
	Condiments													
674	Catsup or ketchup	1	tablespoon(s)	15	10.4	15	0.3	3.8	0	0	0	0	0	—
703	Dill pickle	1	ounce(s)	28	26.7	3	0.2	0.7	0.3	0	0	0	0	—
138	Mayonnaise with soybean oil	1	tablespoon(s)	14	2.1	99	0.1	0.4	0	11.0	1.6	2.7	5.8	0
140	Mayonnaise, low calorie	1	tablespoon(s)	16	10.0	37	0	2.6	0	3.1	0.5	0.7	1.7	—
1682	Mustard, brown	1	teaspoon(s)	5	4.1	5	0.3	0.3	0	0.3	—	—	—	—
700	Mustard, yellow	1	teaspoon(s)	5	4.1	3	0.2	0.3	0.2	0.2	0	0.1	0	0
706	Sweet pickle relish	1	tablespoon(s)	15	9.3	20	0.1	5.3	0.2	0.1	0	0	0	—
141	Tartar sauce	2	tablespoon(s)	28	8.7	144	0.3	4.1	0.1	14.4	2.2	3.8	7.7	—

SPICES, CONDIMENTS, SAUCES —Continued

Chol (mg)	Calc (mg)	Iron (mg)	Magn (mg)	Pota (mg)	Sodi (mg)	Zinc (mg)	Vit A (µg)	Thia (mg)	Vit E (mg α)	Ribo (mg)	Niac (mg)	Vit B6 (mg)	Fola (µg)	Vit C (mg)	Vit B12 (µg)	Sele (µg)
0	35	0.89	8.8	28.0	3.2	0.13	0.1	0.01	0.02	0.01	0.06	0.01	0.2	0.3	0	0.2
0	8	0.19	0.8	28.4	0.5	0.05	1.8	0.00	—	0.00	0.03	0.01	1.6	0.3	0	0.2
0	7	0.37	4.4	49.8	26.3	0.07	38.6	0.01	0.75	0.02	0.20	0.09	2.6	1.7	0	0.2
0	1	0.01	0.4	3.0	0	0.01	2.2	0.00	0.00	0.00	0.01	0.00	1.1	0.6	0	0
0	0	0.01	0.1	1.7	0.2	0.00	1.1	0.00	0.01	0.00	0.00	0.00	0.2	0.1	0	0
0	23	0.19	1.4	9.9	0.2	0.04	0.3	0.00	0.05	0.00	0.03	0.00	0.1	0.1	0	0.1
0	14	0.18	5.5	23.1	5.1	0.02	0.6	0.00	0.17	0.01	0.03	0.01	2.0	1.7	0	0.1
0	7	0.25	4.2	26.8	1.3	0.02	1.8	0.01	0.01	0.01	0.06	0.00	1.6	3.4	0	0.2
0	13	0.29	5.9	22.8	0.6	0.08	0	0.00	—	0.01	0.03	—	0	0.4	0	0.5
0	0	0.03	0.2	0.2	0.7	0.01	0	0.00	0.00	0.00	0.00	0.00	0	0	0	0.2
0	20	1.39	7.7	37.5	3.5	0.10	1.3	0.01	0.07	0.01	0.09	0.01	0.2	0.2	0	0.1
0	20	—	—	43.6	4.8	—	—	—	—	—	—	—	—	—	—	—
0	10	0.59	5.1	30.9	1.0	0.08	1.0	0.01	0.44	0.01	0.06	0.02	3.1	0.2	0	0.3
0	32	0.34	5.4	24.9	0.4	0.10	0.1	0.01	—	0.01	0.05	0.01	0.2	0.4	0	0.3
0	18	0.48	4.5	33.1	2.1	0.03	2.9	0.00	—	0.00	0.02	0.01	1.5	0.5	0	—
0	2	0.06	0.6	7.4	0.6	0.01	3.9	0.00	0.01	0.00	0.01	0.00	1.5	0.9	0	—
0	1	0.02	—	4.0	0.1	—	—	0.00	—	0.00	0.01	0.00	—	0.3	0	—
0	24	0.37	7.7	33.9	1.8	0.07	0.1	0.01	—	0.01	0.12	0.01	—	0.4	0	—
0	7	1.24	7.1	28.5	2.5	0.09	0.1	0.01	—	0.01	0.06	0.02	2.1	0.1	0	0.2
0	215	9.24	93.6	411.1	27.5	1.07	—	0.09	—	0.09	0.70	—	0	0	0	—
0	5	0.05	0.8	12.0	0.5	0.03	0	0.01	0.00	0.00	0.02	0.03	0.1	0.9	0	0.4
0	2	0.07	1.6	30.8	0.7	0.07	0	0.01	0.01	0.00	0.01	0.08	0.1	0.5	0	1.1
0	1	0.02	1.7	16.6	0.5	0.01	0	0.00	0.01	0.00	0.02	0.01	0.4	0.2	0	0
0	2	0.20	3.3	24.2	0.6	0.08	0.1	0.00	0.32	0.00	0.09	0.01	0.7	0.1	0	0.7
0	3	0.06	1.3	19.2	0.3	0.01	0.1	0.01	—	0.00	0.02	0.01	2.9	0.9	0	0
0	4	0.23	2.8	7.9	1.4	0.03	0.7	0.01	—	0.01	0.02	0.00	1.3	0.4	0	0
0	12	0.49	2.1	9.1	0.5	0.02	2.4	0.00	0.01	0.00	0.01	0.01	1.6	0.3	0	0
0	17	0.32	9.8	22.5	0.2	0.18	0.1	0.01	0.09	0.01	0.26	0.01	2.5	0.1	0	4.4
0	4	0.06	4.0	7.7	0.4	0.04	0.1	0.01	0.00	0.00	0.02	0.00	1.7	0.1	0	0
0	4	0.02	1.5	27.1	0.4	0.03	0	0.01	0.00	0.00	0.01	0.02	2.8	1.3	0	0.1
0	8	0.05	2.6	19.8	1.1	0.04	0	0.01	0.01	0.00	0.01	0.02	3.5	0.3	0	0
0	24	0.66	4.1	25.0	0.2	0.06	5.2	0.01	0.28	0.01	0.09	0.01	4.1	0.8	0	0.1
0	4	0.49	3.9	49.2	0.7	0.08	55.4	0.01	0.62	0.03	0.32	0.08	2.2	1.5	0	0.1
0	4	0.29	0.7	11.4	1.4	0.01	1.5	0.00	0.02	0.00	0.02	0.00	0.5	0.4	0	0.1
0	9	0.60	4.1	26.4	0.9	0.03	0.3	0.00	0.01	0.01	0.02	0.01	0.2	0.4	0	0.1
0	3	0.14	2.7	36.3	0.5	0.04	37.5	0.01	0.53	0.01	0.15	0.04	1.9	1.4	0	0.2
0	6	0.34	2.2	1.8	0.1	0.02	0	0.00	—	0.00	0.01	0.00	0.2	0.5	0	0.1
0	41	0.26	9.3	19.6	0.6	0.28	0	0.02	0.03	0.01	0.02	0.01	1.6	0.1	0	0
0	15	0.53	3.4	10.3	0.4	0.04	2.0	0.00	0.02	0.00	0.04	0.02	2.1	0.2	0	0.1
0	12	0.33	2.3	11.3	0.9	0.04	0.2	0.00	0.01	0.00	0.03	0.01	0.9	0.4	0	0.2
0	15	0.35	2.6	11.5	0.6	0.03	1.9	0.01	—	0.01	0.01	0.02	3.7	0.7	0	0.1
0	2	0.04	0.6	4.7	0.2	0.01	1.0	0.00	—	0.00	0.01	0.00	0.8	0.2	0	—
0	1	0.07	1.8	12.1	1.0	0.01	0.2	0.00	—	0.00	0.01	0.01	0.7	0.6	0	0
0	4	—	1.1	2.7	0	0.01	—	0.00	—	—	—	—	—	—	0	—
0	12	0.19	3.0	7.5	0.1	0.03	2.1	0.01	0.05	0.00	0.04	0.01	1.9	0.2	0	0
0	7	0.00	0	603.6	0.1	—	0	—	—	—	—	—	—	0	—	—
0	0	0.00	—	476.3	0.1	—	—	—	—	—	—	—	—	0	—	—
0	0	0.01	0	0.1	581.4	0.00	0	0.00	0.00	0.00	0.00	0.00	0	0	0	0
0	30	0.53	5.3	14.7	0.3	0.06	3.6	0.01	—	—	0.05	0.02	—	0.7	0	0.1
0	3	0.21	9.2	10.8	1.0	0.27	0.1	0.03	0.01	0.01	0.15	0.00	2.6	0	0	0
0	—	—	—	—	0.1	—	—	—	—	—	—	—	—	—	0	—
0	4	0.22	1.2	8.7	0.6	0.02	3.9	0.00	—	0.00	0.01	0.00	2.0	0.3	0	—
0	2	0.16	3.0	50.7	3.1	0.03	4.5	0.01	0.06	0.01	0.11	0.03	3.7	30.4	0	0.1
0	85	1.48	20.2	—	—	—	—	0.06	—	0.02	1.16	—	—	0.9	0	—
0	48	—	14.5	128.1	2.6	0.17	—	0.04	—	—	0.14	0.03	4.4	0.6	0	0.1
0	18	0.51	5.6	48.3	1.0	0.06	3.4	0.00	—	0.02	0.14	0.03	4.4	0.8	0	0.1
0	3	0.14	1.3	4.9	0.1	0.01	1.9	0.00	—	0.00	0.01	0.00	0.4	1.3	0	—
0	26	1.73	3.1	11.4	0.8	0.08	2.7	0.01	0.10	0.01	0.06	0.01	3.8	0.7	0	0.1
0	4	0.91	4.2	55.6	0.8	0.09	0	0.00	0.06	0.01	0.11	0.04	0.9	0.6	0	0.1
0	13	0.11	—	—	—	—	—	0.02	—	0.01	0.07	—	—	11.2	0	—
0	3	0.07	2.9	57.3	167.1	0.03	7.1	0.00	0.21	0.02	0.21	0.02	1.5	2.3	0	0
0	12	0.10	2.0	26.1	248.1	0.03	2.6	0.01	0.02	0.01	0.03	0.01	0.3	0.2	0	0
5	1	0.03	0.1	1.7	78.4	0.02	11.2	0.01	0.72	0.01	0.00	0.08	0.7	0	0	0.2
4	0	0.00	0	1.6	79.5	0.01	0	0.00	0.32	0.00	0.00	0.00	0	0	0	0.3
0	6	0.09	1.0	6.8	68.1	0.01	0	0.00	0.09	0.00	0.01	0.00	0.2	0.1	0	—
0	3	0.07	2.5	6.9	56.8	0.03	0.2	0.01	0.01	0.00	0.02	0.00	0.4	0.1	0	1.6
0	0	0.13	0.8	3.8	121.7	0.02	9.2	0.00	0.08	0.01	0.03	0.00	0.2	0.2	0	0
8	6	0.20	0.8	10.1	191.5	0.05	20.2	0.00	0.97	0.00	0.01	0.07	2.0	0.1	0.1	0.5

DA+ Code	Food Description	Quantity	Measure	Wt (g)	H₂O (g)	Ener (kcal)	Prot (g)	Carb (g)	Fiber (g)	Fat (g)	Sat	Mono	Poly	Trans
	Sauces													
685	Barbecue sauce	2	tablespoon(s)	31	18.9	47	0	11.3	0.2	0.1	0	0	0.1	0
834	Cheese sauce	¼	cup(s)	63	44.4	110	4.2	4.3	0.3	8.4	3.8	2.4	1.6	—
32123	Chili enchilada sauce, green	2	tablespoon(s)	57	53.0	15	0.6	3.1	0.7	0.3	0	0	0.1	0
32122	Chili enchilada sauce, red	2	tablespoon(s)	32	24.5	27	1.1	5.0	2.1	0.8	0.1	0	0.4	0
29688	Hoisin sauce	1	tablespoon(s)	16	7.1	35	0.5	7.1	0.4	0.5	0.1	0.2	0.3	—
1641	Horseradish sauce, prepared	1	teaspoon(s)	5	3.3	10	0.1	0.2	0	1.0	0.6	0.3	0	—
16670	Mole poblano sauce	½	cup(s)	133	102.7	156	5.3	11.4	2.7	11.3	2.6	5.1	3.0	—
29689	Oyster sauce	1	tablespoon(s)	16	12.8	8	0.2	1.7	0	0	0	0	0	—
1655	Pepper sauce or Tabasco	1	teaspoon(s)	5	4.8	1	0.1	0	0	0	0	0	0	—
347	Salsa	2	tablespoon(s)	32	28.8	9	0.5	2.0	0.5	0.1	0	0	0	—
52206	Soy sauce, tamari	1	tablespoon(s)	18	12.0	11	1.9	1.0	0.1	0	0	0	0	—
839	Sweet and sour sauce	2	tablespoon(s)	39	29.8	37	0.1	9.1	0.1	0	0	0	0	—
1613	Teriyaki sauce	1	tablespoon(s)	18	12.2	16	1.1	2.8	0	0	0	0	0	0
25294	Tomato sauce	½	cup(s)	150	132.8	63	2.2	11.9	2.6	1.8	0.2	0.4	0.9	0
728	White sauce, medium	¼	cup(s)	63	46.8	92	2.4	5.7	0.1	6.7	1.8	2.8	1.8	—
1654	Worcestershire sauce	1	teaspoon(s)	6	4.5	4	0	1.1	0	0	0	0	0	0
	Vinegar													
30853	Balsamic	1	tablespoon(s)	15	—	10	0	2.0	0	0	0	0	0	0
727	Cider	1	tablespoon(s)	15	14.0	3	0	0.1	0	0	0	0	0	0
1673	Distilled	1	tablespoon(s)	15	14.3	2	0	0.8	0	0	0	0	0	0
12948	Tarragon	1	tablespoon(s)	15	13.8	2	0	0.1	0	0	0	0	0	0
	MIXED FOODS, SOUPS, SANDWICHES													
	Mixed dishes													
16652	Almond chicken	1	cup(s)	242	186.8	281	21.8	15.8	3.4	14.7	1.8	6.3	5.6	—
25224	Barbecued chicken	1	serving(s)	177	99.3	327	27.1	15.7	0.5	17.1	4.8	6.8	3.8	0
25227	Bean burrito	1	item(s)	149	81.8	326	16.1	33.0	5.6	14.8	8.3	4.7	0.9	—
9516	Beef and vegetable fajita	1	item(s)	223	143.9	397	22.4	35.3	3.1	18.0	5.9	8.0	2.5	—
16796	Beef or pork egg roll	2	item(s)	128	85.2	225	9.9	18.4	1.4	12.4	2.9	6.0	2.6	—
177	Beef stew with vegetables, prepared	1	cup(s)	245	201.0	220	16.0	15.0	3.2	11.0	4.4	4.5	0.5	—
30233	Beef stroganoff with noodles	1	cup(s)	256	190.1	343	19.7	22.8	1.5	19.1	7.4	5.7	4.4	—
16651	Cashew chicken	1	cup(s)	242	186.8	281	21.8	15.8	3.4	14.7	1.8	6.3	5.6	—
30274	Cheese pizza with vegetables, thin crust	2	slice(s)	140	76.6	298	12.7	35.4	2.5	12.0	4.9	4.7	1.6	—
30330	Cheese quesadilla	1	item(s)	54	18.3	190	7.7	15.3	1.0	10.8	5.2	3.6	1.3	—
215	Chicken and noodles, prepared	1	cup(s)	240	170.0	365	22.0	26.0	1.3	18.0	5.1	7.1	3.9	—
30239	Chicken and vegetables with broccoli, onion, bamboo shoots in soy based sauce	1	cup(s)	162	125.5	180	15.8	9.3	1.8	8.6	1.7	3.0	3.1	—
25093	Chicken cacciatore	1	cup(s)	244	175.7	284	29.9	5.7	1.3	15.3	4.3	6.2	3.3	0
28020	Chicken fried turkey steak	3	ounce(s)	492	276.2	706	77.1	68.7	3.6	12.0	3.4	2.9	3.9	—
218	Chicken pot pie	1	cup(s)	252	154.6	542	22.6	41.4	3.5	31.3	9.8	12.5	7.1	—
30240	Chicken teriyaki	1	cup(s)	244	158.3	364	51.0	15.2	0.7	7.0	1.8	2.0	1.7	—
25119	Chicken waldorf salad	½	cup(s)	100	67.2	179	14.0	6.8	1.0	10.8	1.8	3.1	5.2	—
25099	Chili con carne	¾	cup(s)	215	174.4	198	13.7	21.4	7.5	6.9	2.5	2.8	0.5	0
1062	Coleslaw	¾	cup(s)	90	73.4	70	1.2	11.2	1.4	2.3	0.3	0.6	1.2	—
1574	Crab cakes, from blue crab	1	item(s)	60	42.6	93	12.1	0.3	0	4.5	0.9	1.7	1.4	—
32144	Enchiladas with green chili sauce (enchiladas verdes)	1	item(s)	144	103.8	207	9.3	17.6	2.6	11.7	6.4	3.6	1.0	0
2793	Falafel patty	3	item(s)	51	17.7	170	6.8	16.2	—	9.1	1.2	5.2	2.1	—
28546	Fettuccine alfredo	1	cup(s)	244	88.7	279	13.1	46.1	1.4	4.2	2.2	1.0	0.4	0
32146	Flautas	3	item(s)	162	78.0	438	24.9	36.3	4.1	21.6	8.2	8.8	2.3	—
29629	Fried rice with meat or poultry	1	cup(s)	198	128.5	333	12.3	41.8	1.4	12.3	2.2	3.5	5.7	—
16649	General Tso chicken	1	cup(s)	146	91.0	296	18.7	16.4	0.9	17.0	4.0	6.3	5.3	—
1826	Green salad	¾	cup(s)	104	98.9	17	1.3	3.3	2.2	0.1	0	0	0	—
1814	Hummus	½	cup(s)	123	79.8	218	6.0	24.7	4.9	10.6	1.4	6.0	2.6	—
16650	Kung pao chicken	1	cup(s)	162	87.2	434	28.8	11.7	2.3	30.6	5.2	13.9	9.7	—
16622	Lamb curry	1	cup(s)	236	187.9	257	28.2	3.7	0.9	13.8	3.9	4.9	3.3	—
25253	Lasagna with ground beef	1	cup(s)	237	158.4	284	16.9	22.3	2.4	14.5	7.5	4.9	0.8	—
442	Macaroni and cheese, prepared	1	cup(s)	200	122.3	390	14.9	40.6	1.6	18.6	7.9	6.4	2.9	—
29637	Meat filled ravioli with tomato or meat sauce, canned	1	cup(s)	251	198.7	208	7.8	36.5	1.3	3.7	1.5	1.4	0.3	—
25105	Meat loaf	1	slice(s)	115	84.5	245	17.0	6.6	0.4	16.0	6.1	6.9	0.9	0
16646	Moo shi pork	1	cup(s)	151	76.8	512	18.9	5.3	0.6	46.4	6.9	15.8	21.2	—
16788	Nachos with beef, beans, cheese, tomatoes and onions	1	serving(s)	551	253.5	1576	59.1	137.5	20.4	90.8	32.6	41.9	9.4	—
6116	Pepperoni pizza	2	slice(s)	142	66.1	362	20.2	39.7	2.9	13.9	4.5	6.3	2.3	—
29601	Pizza with meat and vegetables, thin crust	2	slice(s)	158	81.4	386	16.5	36.8	2.7	19.1	7.7	8.1	2.2	—
655	Potato salad	½	cup(s)	125	95.0	179	3.4	14.0	1.6	10.3	1.8	3.1	4.7	—
25109	Salisbury steaks with mushroom sauce	1	serving(s)	135	101.8	251	17.1	9.3	0.5	15.5	6.0	6.7	0.8	0
16637	Shrimp creole with rice	1	cup(s)	243	176.6	309	27.0	27.7	1.2	9.2	1.7	3.6	2.9	—
	MIXED FOODS, SOUPS, SANDWICHES —Continued													
497	Spaghetti and meat balls with tomato sauce, prepared	1	cup(s)	248	174.0	330	19.0	39.0	2.7	12.0	3.9	4.4	2.2	—

Chol (mg)	Calc (mg)	Iron (mg)	Magn (mg)	Pota (mg)	Sodi (mg)	Zinc (mg)	Vit A (µg)	Thia (mg)	Vit E (mg α)	Ribo (mg)	Niac (mg)	Vit B6 (mg)	Fola (µg)	Vit C (mg)	Vit B12 (µg)	Sele (µg)
37	31	1.56	49.4	181.3	591.6	1.05	—	0.18	0.35	0.13	1.82	0.17	44.6	22.6	0.1	3.2
46	38	2.71	33.0	396.9	569.5	2.08	—	0.51	0.38	0.20	5.07	0.30	103.0	18.8	0.4	22.8
30	188	2.26	49.4	403.0	471.5	1.41	—	0.26	0.00	0.26	3.83	0.24	80.0	17.9	0.2	28.9
214	45	1.84	18.7	135.7	463.3	0.98	106.1	0.13	0.67	0.28	1.35	0.13	62.4	1.9	0.7	20.3
11	23	2.15	25.0	202.8	340.1	0.78	45.2	0.26	0.24	0.07	2.76	0.14	76.4	3.6	0.3	13.9
0	20	1.54	18.7	96.7	152.9	0.68	25.0	0.19	0.12	0.03	1.86	0.13	73.3	2.3	0	9.8
74	40	1.76	35.6	619.9	621.8	2.53	—	0.81	0.20	0.37	6.69	0.66	14.7	11.9	0.7	49.6
0	30	1.21	35.2	249.6	796.8	0.48	54.4	0.07	2.43	0.04	1.11	0.11	30.4	26.1	0	0.5
0	26	0.96	15.5	144.8	224.2	0.30	—	0.02	0.88	0.04	0.26	0.04	32.2	10.0	0	2.7
13	17	1.02	19.5	182.5	412.1	0.57	24.6	0.03	—	0.07	6.86	0.08	8.2	2.3	1.2	42.2
77	69	2.56	33.2	400.8	577.1	2.51	—	0.23	0.28	0.30	6.41	0.29	61.4	1.4	1.1	33.4
60	29	1.65	17.9	193.3	549.1	0.48	25.6	0.15	1.28	0.20	1.59	0.09	46.1	5.5	0.2	11.3
0	23	2.38	21.8	157.6	369.7	0.82	48.4	0.28	0.15	0.05	2.44	0.12	85.8	3.7	0	8.1
21	79	2.36	27.9	351.0	944.6	1.08	44.3	0.32	1.16	0.24	4.36	0.21	73.8	9.7	0.2	27.1
40	258	2.38	24.4	215.3	941.3	1.88	102.1	0.31	0.55	0.35	2.97	0.14	59.9	0.2	0.8	20.4
17	100	2.24	16.6	138.6	579.3	1.02	44.8	0.29	0.49	0.21	2.92	0.12	58.1	0.2	0.5	16.0
22	233	2.04	19.9	127.0	733.7	1.24	97.1	0.25	0.47	0.30	2.25	0.05	58.1	0	0.3	13.1
104	309	4.47	44.4	401.5	986.1	5.75	0	0.35	—	0.77	8.26	0.49	109.2	0	2.8	38.9
98	267	4.03	44.9	464.1	1314.3	5.20	0	0.33	—	0.67	8.25	0.47	95.6	1.4	2.4	6.6
71	157	4.57	46.7	464.9	1087.3	1.82	41.8	0.54	1.52	0.40	12.82	0.61	118.1	6.4	0.4	42.3
36	189	2.50	68.4	394.4	1650.7	2.57	70.7	1.00	—	0.79	5.49	0.13	86.6	12.3	1.1	30.8
46	81	3.04	19.5	127.4	1206.4	2.26	2.6	0.23	0.20	0.24	3.42	0.11	59.8	0.3	0.9	31.2
219	85	2.25	18.8	159.0	423.1	0.87	—	0.27	0.12	0.43	2.06	0.15	73.6	0.9	0.6	29.9
206	104	2.79	17.3	117.1	438.7	0.92	89.3	0.26	0.66	0.40	2.26	0.10	79.7	0	0.6	24.2
22	235	2.05	19.9	128.7	763.6	1.26	129.5	0.19	0.72	0.28	2.05	0.05	38.2	0	0.2	13.2
28	47	1.77	22.1	218.4	235.2	2.33	9.5	0.23	0.26	0.20	3.12	0.13	44.1	3.2	0.9	18.3
58	130	3.24	16.1	290.5	770.9	1.37	96.4	0.30	0.29	0.48	2.68	0.20	75.9	2.8	0.5	23.1
34	91	2.47	23.5	210.6	1097.6	1.13	5.6	0.57	0.50	0.26	3.80	0.25	60.5	2.2	0.2	20.2
122	102	5.85	49.7	569.5	791.0	5.67	0	0.36	—	0.38	7.57	0.54	76.8	1.1	4.1	25.5
71	74	3.57	27.4	267.2	474.0	4.11	0	0.28	—	0.28	6.24	0.23	60.3	0	2.1	27.1
87	96	4.92	43.6	479.6	824.0	4.88	0	0.41	—	0.37	7.28	0.32	82.8	2.6	2.4	33.6
39	115	1.88	20.0	172.4	505.1	1.19	—	0.23	0.28	0.22	4.84	0.19	46.4	0.5	0.2	19.9
37	114	1.99	21.3	189.0	494.5	1.07	—	0.22	0.28	0.20	4.26	0.22	46.4	0.5	0.2	23.4
44	24	2.31	12.7	143.1	670.3	1.98	0	0.23	—	0.27	3.64	0.04	48.0	0.1	0.5	26.0
51	80	3.02	22.8	182.2	1364.1	2.70	2.7	0.28	0.26	0.26	4.97	0.14	60.3	0.3	1.0	14.3
0	110	2.92	66.0	226.0	580.3	1.32	0	0.31	2.39	0.23	6.72	0.18	92.1	0	0	13.2
0	94	2.50	56.7	198.1	492.9	1.12	—	0.26	2.01	0.20	5.66	0.15	78.1	0.1	0	11.2
73	41	2.80	67.0	330.5	844.6	4.38	30.2	0.41	—	0.41	5.96	0.32	71.3	5.6	1.8	25.7
51	54	4.22	30.6	315.5	792.3	3.39	11.1	0.37	—	0.30	5.86	0.26	57.0	2.1	1.2	29.2
73	92	5.16	49.0	524.3	797.6	4.52	0	0.40	—	0.36	7.30	0.36	89.8	5.5	1.6	42.0
59	78	2.97	35.9	316.3	724.7	1.02	—	0.27	0.34	0.26	12.07	0.46	60.8	1.8	2.4	76.9
64	307	4.59	49.9	534.6	1759.0	2.74	74.8	0.49	1.19	0.65	3.91	0.31	121.9	10.5	0.6	42.1
67	100	3.46	34.3	304.6	564.9	3.00	5.7	0.28	0.74	0.32	6.84	0.46	64.4	0	0.3	40
5	79	3.05	61.8	588.8	689.0	1.41	—	0.27	0.02	0.15	3.63	0.23	140.1	3.6	0.2	7.9
3	78	1.89	43.0	371.9	883.0	0.96	43.0	0.08	1.08	0.03	0.52	0.03	30.4	1.5	0	7.8
5	20	1.07	7.3	97.6	929.6	1.51	12.2	0.06	1.22	0.05	1.03	0.03	19.5	0.5	0.2	7.3
48	289	0.80	20.1	341.4	1019.1	0.67	358.9	0.06	—	0.33	0.50	0.07	10.0	1.3	0.4	7.0
0	10	0.51	2.4	209.8	775.9	0.24	0	0.01	0.04	0.07	3.34	0.02	4.9	0	0.2	0
24	25	1.38	16.4	340.1	774.5	0.77	—	0.15	0.02	0.16	5.57	0.13	37.2	1.8	0.3	10.2
12	14	1.59	9.6	53.0	638.7	0.38	26.5	0.13	0.07	0.10	1.30	0.04	19.3	0	0.1	11.6
10	5	0.50	7.6	32.8	577.1	0.20	2.5	0.20	0.12	0.07	1.08	0.02	17.6	0	0.1	9.6
22	174	0.86	19.8	359.6	1041.6	0.91	62.0	0.10	—	0.27	0.88	0.06	29.8	4.0	0.5	8.0
32	186	0.69	22.3	310.0	1009.4	0.19	114.1	0.07	—	0.24	0.43	0.06	7.4	1.5	0.5	4.7
27	181	0.67	17.4	272.8	1046.6	0.67	178.6	0.07	—	0.25	0.92	0.06	7.4	1.2	0.5	8.0
10	34	0.61	2.4	87.8	985.8	0.63	163.5	0.02	—	0.06	0.82	0.01	2.4	0.2	0.1	7.0
10	164	1.36	19.8	267.8	823.4	0.79	81.8	0.10	1.01	0.29	0.62	0.05	7.4	0	0.6	6.0
0	17	1.31	4.9	73.2	775.9	0.24	9.8	0.05	0.97	0.05	0.50	0.00	2.4	0	0	2.9
1	80	1.20	17.5	340.7	787.9	0.56	—	0.12	1.05	0.18	3.32	0.12	39.5	10.7	0.3	4.3
102	22	0.75	4.9	219.6	729.6	0.48	41.5	0.02	0.29	0.19	3.02	0.05	14.6	0	0.5	7.6
4	262	0.78	42.4	542.2	515.6	0.88	—	0.16	0.52	0.30	1.14	0.16	32.7	12.5	0.7	6.0
34	29	1.24	19.5	373.3	1561.6	1.43	—	0.26	0.12	0.24	4.96	0.20	14.6	0.5	0.4	19.3

DA+ Code	Food Description	Quantity	Measure	Wt (g)	H₂O (g)	Ener (kcal)	Prot (g)	Carb (g)	Fiber (g)	Fat (g)	Sat	Mono	Poly	Trans
												Fat Breakdown (g)		
28054	Lentil chowder	1	cup(s)	244	202.8	153	11.4	27.7	12.6	0.5	0.1	0.1	0.2	0
28560	Macaroni and bean	1	cup(s)	246	138.8	146	5.8	22.9	5.1	3.7	0.5	2.2	0.6	0
714	Manhattan clam chowder, condensed, prepared with water	1	cup(s)	244	225.1	73	2.1	11.6	1.5	2.1	0.4	0.4	1.2	—
28561	Minestrone	1	cup(s)	241	185.4	103	4.5	16.8	4.8	2.3	0.3	1.4	0.4	0
717	Minestrone, condensed, prepared with water	1	cup(s)	241	220.1	82	4.3	11.2	1.0	2.5	0.6	0.7	1.1	—
28038	Mushroom & wild rice	1	cup(s)	244	199.7	86	4.7	13.2	1.7	0.3	0	0	0.2	0
828	New England clam chowder, condensed, prepared with milk	1	cup(s)	248	212.2	151	8.0	18.4	0.7	5.0	2.1	0.7	0.6	0
28036	New England style clam chowder	1	cup(s)	244	227.5	61	3.8	8.8	1.8	0.2	0.1	0	0	0
28566	Old country pasta	1	cup(s)	252	183.3	146	6.5	18.3	3.6	4.5	2.0	2.4	0.9	0
725	Onion, dehydrated, prepared with water	1	cup(s)	246	235.7	30	0.8	6.8	0.7	0	0	0	0	—
16667	Shrimp gumbo	1	cup(s)	244	207.2	166	9.5	18.2	2.4	6.7	1.3	2.9	2.0	
28037	Southwestern corn chowder	1	cup(s)	244	217.8	98	4.9	17.0	2.4	0.5	0.1	0.1	0.2	0
30282	Soybean (miso)	1	cup(s)	240	218.6	84	6.0	8.0	1.9	3.4	0.6	1.1	1.4	—
25140	Split pea	1	cup(s)	165	119.7	72	4.5	16.0	1.6	0.3	0.1	0	0.2	0
718	Split pea with ham, condensed, prepared with water	1	cup(s)	253	206.9	190	10.3	28.0	2.3	4.4	1.8	1.8	0.6	—
726	Tomato vegetable, dehydrated, prepared with water	1	cup(s)	253	238.4	56	2.0	10.2	0.8	0.9	0.4	0.3	0.1	0
710	Tomato, condensed, prepared with milk	1	cup(s)	248	213.4	136	6.2	22.0	1.5	3.2	1.8	0.9	0.3	—
719	Tomato, condensed, prepared with water	1	cup(s)	244	223.0	73	1.9	16.0	1.5	0.7	0.2	0.2	0.2	—
28595	Turkey noodle	1	cup(s)	244	216.9	114	8.1	15.1	1.9	2.4	0.3	1.1	0.7	—
28051	Turkey vegetable	1	cup(s)	244	220.8	96	12.2	6.6	2.0	1.1	0.3	0.2	0.3	0
25141	Vegetable	1	cup(s)	252	228.1	82	5.2	16.5	4.5	0.3	0	0	0.1	0
720	Vegetable beef, condensed, prepared with water	1	cup(s)	244	224.0	76	5.4	9.9	2.0	1.9	0.8	0.8	0.1	—
28598	Vegetable gumbo	1	cup(s)	252	184.5	170	4.4	28.9	3.6	4.7	0.7	3.2	0.5	0
721	Vegetarian vegetable, condensed, prepared with water	1	cup(s)	241	222.7	67	2.1	11.8	0.7	1.9	0.3	0.8	0.7	—
	FAST FOOD													
	Arby's													
36094	Au jus sauce	1	serving(s)	85	—	43	1.0	7.0	0	1.3	0.4	—	—	0.4
751	Beef 'n cheddar sandwich	1	item(s)	195	—	445	22.0	44.0	2.0	21.0	6.0	—	—	1.0
9279	Cheddar curly fries	1	serving(s)	198	—	631	8.0	73.0	7.0	37.4	6.8	—	—	5.7
34770	Chicken breast fillet sandwich, grilled	1	item(s)	233	—	414	32.0	36.0	3.0	17.0	3.0	—	—	0
36131	Chocolate shake, regular	1	serving(s)	397	—	507	13.0	83.0	0	13.0	8.0	—	—	0
36045	Curly fries, large size	1	serving(s)	198	—	631	8.0	73.0	7.0	37.0	7.0	—	—	6.0
36044	Curly fries, medium size	1	serving(s)	128	—	406	5.0	47.0	5.0	24.0	4.0	—	—	4.0
752	Ham 'n cheese sandwich	1	item(s)	167	—	304	23.0	35.0	1.0	7.0	2.0	—	—	0
36048	Homestyle fries, large size	1	serving(s)	213	—	566	6.0	82.0	6.0	37.0	7.0	—	—	5.0
36047	Homestyle fries, medium size	1	serving(s)	142	—	377	4.0	55.0	4.0	25.0	4.0	—	—	4.0
33465	Homestyle fries, small size	1	serving(s)	113	—	302	3.0	44.0	3.0	20.0	4.0	—	—	3.0
9249	Junior roast beef sandwich	1	item(s)	125	—	272	16.0	34.0	2.0	10.0	4.0	—	—	0
9251	Large roast beef sandwich	1	item(s)	281	—	547	42.0	41.0	3.0	28.0	12.0	—	—	2.0
39640	Market Fresh chicken salad with pecans sandwich	1	item(s)	322	—	769	30.0	79.0	9.0	39.0	10.0	—	—	0
39641	Market Fresh Martha's Vineyard salad, without dressing	1	serving(s)	330	—	277	26.0	24.0	5.0	8.0	4.0	—	—	0
34769	Market Fresh roast turkey & Swiss sandwich	1	serving(s)	359	—	725	45.0	75.0	5.0	30.0	8.0	—	—	1.0
9267	Market Fresh roast turkey ranch & bacon sandwich	1	serving(s)	382	—	834	49.0	75.0	5.0	38.0	11.0	—	—	1.0
39642	Market Fresh Santa Fe salad, without dressing	1	serving(s)	372	—	499	30.0	42.0	7.0	23.0	8.0	—	—	2.0
39650	Market Fresh Southwest chicken wrap	1	serving(s)	251	—	567	36.0	42.0	4.0	29.0	9.0	—	—	1.0
37021	Market Fresh Ultimate BLT sandwich	1	item(s)	294	—	779	23.0	75.0	6.0	45.0	11.0	—	—	1.0
750	Roast beef sandwich, regular	1	item(s)	154	—	320	21.0	34.0	2.0	14.0	5.0	—	—	1.0
36132	Strawberry shake, regular	1	serving(s)	397	—	498	13.0	81.0	0	13.0	8.0	—	—	0
2009	Super roast beef sandwich	1	item(s)	198	—	398	21.0	40.0	2.0	19.0	6.0	—	—	1.0
36130	Vanilla shake, regular	1	serving(s)	369	—	437	13.0	66.0	0	13.0	8.0	—	—	0
	Auntie Anne's													
35371	Cheese dipping sauce	1	serving(s)	35	—	100	3.0	4.0	0	8.0	4.0	—	—	0
35353	Cinnamon sugar soft pretzel	1	item(s)	120	—	350	9.0	74.0	2.0	2.0	0	—	—	0
35354	Cinnamon sugar soft pretzel with butter	1	item(s)	120	—	450	8.0	83.0	3.0	9.0	5.0	—	—	0
35372	Marinara dipping sauce	1	serving(s)	35	—	10	0	4.0	0	0	0	0	0	0
35357	Original soft pretzel	1	serving(s)	120	—	340	10.0	72.0	3.0	1.0	0	—	—	0
35358	Original soft pretzel with butter	1	item(s)	120	—	370	10.0	72.0	3.0	4.0	2.0	—	—	0
35359	Parmesan herb soft pretzel	1	item(s)	120	—	390	11.0	74.0	4.0	5.0	2.5	—	—	—
35360	Parmesan herb soft pretzel with butter	1	item(s)	120	—	440	10.0	72.0	9.0	13.0	7.0	—	—	—
35361	Sesame soft pretzel	1	item(s)	120	—	350	11.0	63.0	3.0	6.0	1.0	—	—	0
35362	Sesame soft pretzel with butter	1	item(s)	120	—	410	12.0	64.0	7.0	12.0	4.0	—	—	0
	FAST FOOD —Continued													
35364	Sour cream & onion soft pretzel	1	item(s)	120	—	310	9.0	66.0	2.0	1.0	0	—	—	0
35366	Sour cream & onion soft pretzel with butter	1	item(s)	120	—	340	9.0	66.0	2.0	5.0	3.0	—	—	0

Chol (mg)	Calc (mg)	Iron (mg)	Magn (mg)	Pota (mg)	Sodi (mg)	Zinc (mg)	Vit A (µg)	Thia (mg)	Vit E (mg α)	Ribo (mg)	Niac (mg)	Vit B6 (mg)	Fola (µg)	Vit C (mg)	Vit B12 (µg)	Sele (µg)
0	48	4.38	59.3	626.2	26.7	1.57	—	0.24	0.06	0.12	1.87	0.32	176.3	16.1	0	3.5
0	59	1.90	32.4	275.9	531.0	0.51	—	0.16	0.37	0.12	1.44	0.10	58.2	9.1	0	8.8
2	27	1.56	9.8	180.6	551.4	0.87	48.8	0.02	1.22	0.03	0.77	0.09	9.8	3.9	3.9	9.0
0	62	1.70	29.9	287.5	442.7	0.42	—	0.09	0.23	0.09	0.70	0.07	47.3	13.3	0	3.5
2	34	0.91	7.2	313.3	911.0	0.74	118.1	0.05	—	0.04	0.94	0.09	36.2	1.2	0	8.0
0	30	1.38	27.3	376.9	283.8	1.00	—	0.06	0.07	0.23	3.21	0.14	18.1	3.7	0.1	4.7
17	169	3.00	29.8	456.3	887.8	0.99	91.8	0.20	0.54	0.43	1.96	0.17	22.3	5.2	11.9	10.9
3	89	1.30	29.2	503.7	256.8	0.52	—	0.06	0.02	0.10	1.30	0.15	24.2	11.8	3.0	3.8
5	57	2.43	51.9	500.4	355.7	0.78	—	0.21	0.01	0.14	2.64	0.20	80.0	22.0	0.1	10.3
0	22	0.12	9.8	76.3	851.2	0.12	0	0.03	0.02	0.03	0.15	0.06	0	0.2	0	0.5
51	105	2.85	48.8	461.2	441.6	0.90	80.5	0.18	1.90	0.12	2.52	0.19	85.4	17.6	0.3	14.9
1	83	1.03	26.3	434.4	217.4	0.57	—	0.08	0.09	0.13	1.81	0.21	32.1	39.6	0.2	1.7
0	65	1.87	36.0	362.4	988.8	0.86	232.8	0.06	0.96	0.16	2.61	0.15	57.6	4.6	0.2	1.0
0	28	1.26	29.8	328.1	602.4	0.52	—	0.10	0.00	0.07	1.50	0.16	49.5	8.1	0	0.7
8	23	2.27	48.1	399.7	1006.9	1.31	22.8	0.14	—	0.07	1.47	0.06	2.5	1.5	0.3	8.0
0	20	0.60	10.1	169.5	334.0	0.20	10.1	0.06	0.43	0.09	1.26	0.06	12.7	3.0	0.1	2.0
10	166	1.36	29.8	466.2	711.8	0.84	94.2	0.09	0.44	0.31	1.35	0.15	5.0	15.6	0.6	9.2
0	20	1.31	17.1	273.3	663.7	0.29	24.4	0.04	0.41	0.07	1.23	0.10	0	15.4	0	6.1
26	28	1.40	24.1	223.8	395.9	0.75	—	0.21	0.01	0.12	2.85	0.15	45.0	7.1	0.1	13.7
21	38	1.48	25.0	423.4	348.7	0.99	—	0.09	0.01	0.09	3.64	0.27	23.9	10.6	0.2	10.3
0	40	2.08	39.1	681.5	670.3	0.67	—	0.16	0.00	0.09	2.70	0.26	36.3	22.4	0	2.2
5	20	1.09	7.3	168.4	773.5	1.51	190.3	0.03	0.58	0.04	1.00	0.07	9.8	2.4	0.3	2.7
0	56	1.85	38.9	360.2	518.4	0.64	—	0.18	0.64	0.08	1.77	0.17	57.6	21.9	0	4.1
0	24	1.06	7.2	207.3	814.6	0.45	171.1	0.05	1.39	0.04	0.90	0.05	9.6	1.4	0	4.3
0	0	—	—	—	1510.0	—	—	—	—	—	—	—	—	—	—	—
51	80	3.96	—	—	1274.0	—	—	—	—	—	—	—	—	1.8	—	—
0	80	3.24	—	—	1476.0	—	—	—	—	—	—	—	—	9.6	—	—
9	90	3.06	—	—	913.0	—	—	—	—	—	—	—	—	10.8	—	—
34	510	0.54	—	—	357.0	—	—	—	—	—	—	—	—	5.4	—	—
0	80	3.24	—	—	1476.0	—	—	—	—	—	—	—	—	9.6	—	—
0	50	1.98	—	—	949.0	—	—	—	—	—	—	—	—	6.0	—	—
35	160	2.70	—	—	1420.0	—	—	—	—	—	—	—	—	1.2	—	—
0	50	1.62	—	—	1029.0	—	—	—	—	—	—	—	—	12.6	—	—
0	30	1.08	—	—	686.0	—	—	—	—	—	—	—	—	8.4	—	—
0	30	0.90	—	—	549.0	—	—	—	—	—	—	—	—	6.6	—	—
29	60	3.06	—	—	740.0	—	0	—	—	—	—	—	—	0	—	—
102	70	6.30	—	—	1869.0	—	0	—	—	—	—	—	—	0.6	—	—
74	180	4.32	—	—	1240.0	—	—	—	—	—	—	—	—	30.0	—	—
72	200	1.62	—	—	454.0	—	—	—	—	—	—	—	—	33.6	—	—
91	360	5.22	—	—	1788.0	—	—	—	—	—	—	—	—	10.2	—	—
109	330	5.40	—	—	2258.0	—	—	—	—	—	—	—	—	11.4	—	—
59	420	3.60	—	—	1231.0	—	—	—	—	—	—	—	—	36.6	—	—
88	240	4.50	—	—	1451.0	—	—	—	—	—	—	—	—	7.8	—	—
51	170	4.68	—	—	1571.0	—	—	—	—	—	—	—	—	16.8	—	—
44	60	3.60	—	—	953.0	—	0	—	—	—	—	—	—	0	—	—
34	510	0.72	—	—	363.0	—	—	—	—	—	—	—	—	6.6	—	—
44	70	3.78	—	—	1060.0	—	—	—	—	—	—	—	—	6.0	—	—
34	510	0.36	—	—	350.0	—	—	—	—	—	—	—	—	5.4	—	—
10	100	0.00	—	—	510.0	—	—	—	—	—	—	—	—	0	—	—
0	20	1.98	—	—	410.0	—	0	—	—	—	—	—	—	0	—	—
25	30	2.34	—	—	430.0	—	—	—	—	—	—	—	—	0	—	—
0	0	0.00	—	—	180.0	—	0	—	—	—	—	—	—	0	—	—
0	30	2.34	—	—	900.0	—	0	—	—	—	—	—	—	0	—	—
10	30	2.16	—	—	930.0	—	—	—	—	—	—	—	—	0	—	—
10	80	1.80	—	—	780.0	—	—	—	—	—	—	—	—	1.2	—	—
30	60	1.80	—	—	660.0	—	—	—	—	—	—	—	—	1.2	—	—
0	20	2.88	—	—	840.0	—	0	—	—	—	—	—	—	0	—	—
15	20	2.70	—	—	860.0	—	—	—	—	—	—	—	—	0	—	—
0	30	1.98	—	—	920.0	—	—	—	—	—	—	—	—	0	—	—
10	40	2.16	—	—	930.0	—	—	—	—	—	—	—	—	0	—	—

DA+ Code	Food Description	Quantity	Measure	Wt (g)	H₂O (g)	Ener (kcal)	Prot (g)	Carb (g)	Fiber (g)	Fat (g)	Fat Breakdown (g)			
											Sat	Mono	Poly	Trans
35373	Sweet mustard dipping sauce	1	serving(s)	35	—	60	0.5	8.0	0	1.5	1.0	—	—	0
35367	Whole wheat soft pretzel	1	item(s)	120	—	350	11.0	72.0	7.0	1.5	0	—	—	0
35368	Whole wheat soft pretzel with butter	1	item(s)	120	—	370	11.0	72.0	7.0	4.5	1.5	—	—	0
	Boston Market													
34978	Butternut squash	¾	cup(s)	143	—	140	2.0	25.0	2.0	4.5	3.0	—	—	0
35006	Caesar side salad	1	serving(s)	71	—	40	3.0	3.0	1.0	20.0	2.0	—	—	1.5
35013	Chicken Carver sandwich with cheese and sauce	1	item(s)	321	—	700	44.0	68.0	3.0	29.0	7.0	—	—	0
34979	Chicken gravy	4	ounce(s)	113	—	15	1.0	4.0	0	0.5	0	—	—	0
35053	Chicken noodle soup	¾	cup(s)	283	—	180	13.0	16.0	1.0	7.0	2.0	—	—	0
34973	Chicken pot pie	1	item(s)	425	—	800	29.0	59.0	4.0	49.0	18.0	—	—	7.0
35054	Chicken tortilla soup with toppings	¾	cup(s)	227	—	340	12.0	24.0	1.0	22.0	7.0	—	—	0
35007	Cole slaw	¾	cup(s)	125	—	170	2.0	21.0	2.0	9.0	2.0	—	—	0
35057	Cornbread	1	item(s)	45	—	130	1.0	21.0	0	3.5	1.0	—	—	1.0
34980	Creamed spinach	¾	cup(s)	191	—	280	9.0	12.0	4.0	23.0	15.0	—	—	0
34998	Fresh vegetable stuffing	1	cup(s)	136	—	190	3.0	25.0	2.0	8.0	1.0	—	—	0
34991	Garlic dill new potatoes	¾	cup(s)	156	—	140	3.0	24.0	3.0	3.0	1.0	—	—	0
34983	Green bean casserole	¾	cup(s)	170	—	60	2.0	9.0	2.0	2.0	1.0	—	—	0
34982	Green beans	¾	cup(s)	91	—	60	2.0	7.0	3.0	3.5	1.5	—	—	0
34984	Homestyle mashed potatoes	¾	cup(s)	221	—	210	4.0	29.0	3.0	9.0	6.0	—	—	0
34985	Homestyle mashed potatoes and gravy	1	cup(s)	334	—	225	5.0	33.0	3.0	9.5	6.0	—	—	0
34988	Hot cinnamon apples	¾	cup(s)	145	—	210	0	47.0	3.0	3.0	0	—	—	0
34989	Macaroni and cheese	¾	cup(s)	221	—	330	14.0	39.0	1.0	12.0	7.0	—	—	0.5
51193	Market chopped salad with dressing	1	item(s)	563	—	580	11.0	31.0	9.0	48.0	9.0	—	—	1.0
34970	Meatloaf	1	serving(s)	218	—	480	29.0	23.0	2.0	33.0	13.0	—	—	0
39383	Nestle Toll House chocolate chip cookie	1	item(s)	78	—	370	4.0	49.0	2.0	19.0	9.0	—	—	0
34965	Quarter chicken, dark meat, no skin	1	item(s)	134	—	260	30.0	2.0	0	13.0	4.0	—	—	0
34966	Quarter chicken, dark meat, with skin	1	item(s)	149	—	280	31.0	3.0	0	15.0	4.5	—	—	0
34963	Quarter chicken, white meat, no skin or wing	1	item(s)	173	—	250	41.0	4.0	0	8.0	2.5	—	—	0
34964	Quarter chicken, white meat, with skin and wing	1	item(s)	110	—	330	50.0	3.0	0	12.0	4.0	—	—	0
34968	Roasted turkey breast	5	ounce(s)	142	—	180	38.0	0	0	3.0	1.0	—	—	0
35011	Seasonal fresh fruit salad	1	serving(s)	142	—	60	1.0	15.0	1.0	0	0	0	0	0
51192	Spinach with garlic butter sauce	1	serving(s)	170	—	130	5.0	9.0	5.0	9.0	6.0	—	—	0
34969	Spiral sliced holiday ham	8	ounce(s)	227	—	450	40.0	13.0	0	26.0	10.0	—	—	0
35003	Steamed vegetables	1	cup(s)	136	—	50	2.0	8.0	3.0	2.0	0	—	—	0
35005	Sweet corn	¾	cup(s)	176	—	170	6.0	37.0	2.0	4.0	1.0	—	—	0
35004	Sweet potato casserole	¾	cup(s)	198	—	460	4.0	77.0	3.0	17.0	6.0	—	—	0
	Burger King													
29731	Biscuit with sausage, egg & cheese	1	item(s)	191	—	610	20.0	33.0	1.0	45.0	15.0	—	—	1.0
14249	Cheeseburger	1	item(s)	133	—	330	17.0	31.0	1.0	16.0	7.0	—	—	0.5
14251	Chicken sandwich	1	item(s)	219	—	660	24.0	52.0	4.0	40.0	8.0	—	—	2.5
3808	Chicken Tenders, 8 pieces	1	serving(s)	123	—	340	19.0	21.0	0.5	20.0	5.0	—	—	3.0
14259	Chocolate shake, small	1	item(s)	315	—	470	8.0	75.0	1.0	14.0	9.0	—	—	0
29732	Croissanwich with sausage & cheese	1	item(s)	106	37.2	370	14.0	23.0	0.5	25.0	9.0	12.7	3.3	2.0
14261	Croissanwich with sausage, egg & cheese	1	item(s)	159	71.4	470	19.0	26.0	0.5	32.0	11.0	15.8	6.1	2.5
3809	Double cheeseburger	1	item(s)	189	—	500	30.0	31.0	1.0	29.0	14.0	—	—	1.5
14244	Double Whopper sandwich	1	item(s)	373	—	900	47.0	51.0	3.0	57.0	19.0	—	—	2.0
14245	Double Whopper with cheese sandwich	1	item(s)	398	—	990	52.0	52.0	3.0	64.0	24.0	—	—	2.5
14250	Fish Filet sandwich	1	item(s)	250	—	630	24.0	67.0	4.0	30.0	6.0	—	—	2.5
14255	French fries, medium, salted	1	serving(s)	116	—	360	4.0	41.0	4.0	20.0	4.5	—	—	4.5
14262	French toast sticks (5)	1	serving(s)	112	37.6	390	6.0	46.0	2.0	20.0	4.5	10.6	2.9	4.5
14248	Hamburger	1	item(s)	121	—	290	15.0	30.0	1.0	12.0	4.5	—	—	0
14263	Hash brown rounds, small	1	serving(s)	75	27.1	230	2.0	23.0	2.0	15.0	4.0	—	—	5.0
14256	Onion rings, medium	1	serving(s)	91	—	320	4.0	40.0	3.0	16.0	4.0	—	—	3.5
39000	Tendercrisp chicken sandwich	1	item(s)	286	—	780	25.0	73.0	4.0	43.0	8.0	—	—	4.0
37514	TenderGrill chicken sandwich	1	item(s)	258	—	450	37.0	53.0	4.0	10.0	2.0	—	—	0
14258	Vanilla shake, small	1	item(s)	296	—	400	8.0	57.0	0	15.0	9.0	—	—	0
1736	Whopper sandwich	1	item(s)	290	—	670	28.0	51.0	3.0	39.0	11.0	—	—	1.5
14243	Whopper with cheese sandwich	1	item(s)	315	—	760	33.0	52.0	3.0	47.0	16.0	—	—	1.5
	Carl's Jr													
33962	Carl's bacon Swiss crispy chicken sandwich	1	item(s)	268	—	750	31.0	91.0	—	28.0	28.0	—	—	—
10801	Carl's Catch Fish sandwich	1	item(s)	215	—	560	19.0	58.0	2.0	27.0	7.0	—	1.9	—
10862	Carl's Famous Star hamburger	1	item(s)	254	—	590	24.0	50.0	3.0	32.0	9.0	—	—	—
10785	Charbroiled chicken club sandwich	1	item(s)	270	—	550	42.0	43.0	4.0	23.0	7.0	—	2.9	—
10866	Charbroiled chicken salad	1	item(s)	437	—	330	34.0	17.0	5.0	7.0	4.0	—	1.0	—
10855	Charbroiled Santa Fe Chicken sandwich	1	item(s)	266	—	610	38.0	43.0	4.0	32.0	8.0	—	—	—
10790	Chicken stars (6 pieces)	6	item(s)	85	—	260	13.0	14.0	1.0	16.0	4.0	—	1.6	—
34864	Chocolate shake, small	1	serving(s)	595	—	540	15.0	98.0	0	11.0	7.0	—	—	—
	FAST FOOD —Continued													
10797	Crisscut fries	1	serving(s)	139	—	410	5.0	43.0	4.0	24.0	5.0	—	—	—
10799	Double Western Bacon cheeseburger	1	item(s)	308	—	920	51.0	65.0	2.0	50	21.0	—	6.6	—

Chol (mg)	Calc (mg)	Iron (mg)	Magn (mg)	Pota (mg)	Sodi (mg)	Zinc (mg)	Vit A (µg)	Thia (mg)	Vit E (mg α)	Ribo (mg)	Niac (mg)	Vit B$_6$ (mg)	Fola (µg)	Vit C (mg)	Vit B$_{12}$ (µg)	Sele (µg)
40	0	0.00	—	—	120.0	—	0	—	—	—	—	—	—	0	—	—
0	30	1.98	—	—	1100.0	—	0	—	—	—	—	—	—	0	—	—
10	30	2.34	—	—	1120.0	—	—	—	—	—	—	—	—	0	—	—
10	59	0.80	—	—	35.0	—	—	—	—	—	—	—	—	22.2	—	—
0	60	0.43	—	—	75.0	—	—	—	—	—	—	—	—	5.4	—	—
90	211	2.85	—	—	1560.0	—	—	—	—	—	—	—	—	15.8	—	—
0	0	0.00	—	—	570.0	—	0	—	—	—	—	—	—	0	—	—
55	0	1.07	—	—	220.0	—	—	—	—	—	—	—	—	1.8	—	—
115	40	4.50	—	—	800.0	—	—	—	—	—	—	—	—	1.2	—	—
45	123	1.32	—	—	1310.0	—	—	—	—	—	—	—	—	18.4	—	—
10	41	0.48	—	—	270.0	—	—	—	—	—	—	—	—	24.5	—	—
5	0	0.71	—	—	220.0	—	0	—	—	—	—	—	—	0	—	—
70	264	2.84	—	—	580.0	—	—	—	—	—	—	—	—	9.5	—	—
0	41	1.48	—	—	580.0	—	—	—	—	—	—	—	—	2.5	—	—
0	0	0.85	—	—	120.0	—	0	—	—	—	—	—	—	14.3	—	—
5	20	0.72	—	—	620.0	—	—	—	—	—	—	—	—	2.4	—	—
0	43	0.38	—	—	180.0	—	—	—	—	—	—	—	—	5.1	—	—
25	51	0.46	—	—	660.0	—	—	—	—	—	—	—	—	19.2	—	—
25	100	0.59	—	—	1230.0	—	—	—	—	—	—	—	—	24.9	—	—
0	16	0.28	—	—	15.0	—	—	—	—	—	—	—	—	0	—	—
30	345	1.65	—	—	1290.0	—	—	—	—	—	—	—	—	0	—	—
10	—	—	—	—	2010.0	—	—	—	—	—	—	—	—	—	—	—
125	140	3.77	—	—	970.0	—	—	—	—	—	—	—	—	1.8	—	—
20	0	1.32	—	—	340.0	—	—	—	—	—	—	—	—	0	—	—
155	0	1.52	—	—	260.0	—	0	—	—	—	—	—	—	0	—	—
155	0	2.14	—	—	660.0	—	0	—	—	—	—	—	—	0	—	—
125	0	0.89	—	—	480.0	—	0	—	—	—	—	—	—	0	—	—
165	0	0.78	—	—	960.0	—	0	—	—	—	—	—	—	0	—	—
70	20	1.80	—	—	620.0	—	0	—	—	—	—	—	—	0	—	—
0	16	0.29	—	—	20.0	—	—	—	—	—	—	—	—	29.5	—	—
20	—	—	—	—	200.0	—	—	—	—	—	—	—	—	—	—	—
140	0	1.73	—	—	2230.0	—	0	—	—	—	—	—	—	0	—	—
0	53	0.46	—	—	45.0	—	—	—	—	—	—	—	—	24.0	—	—
0	0	0.43	—	—	95.0	—	—	—	—	—	—	—	—	5.8	—	—
20	44	1.18	—	—	210.0	—	—	—	—	—	—	—	—	9.8	—	—
210	250	2.70	—	—	1620.0	—	89.9	—	—	—	—	—	—	0	—	—
55	150	2.70	—	—	780.0	—	—	0.24	—	0.31	4.17	—	—	1.2	—	—
70	64	2.89	—	—	1440.0	—	—	0.50	—	0.32	10.29	—	—	0	—	—
55	20	0.72	—	—	960.0	—	—	0.14	—	0.11	10.93	—	—	0	—	—
55	333	0.79	—	—	350.0	—	—	0.11	—	0.61	0.26	—	—	2.7	0	—
50	99	1.78	20.1	217.3	810.0	1.51	—	0.34	1.03	0.33	4.33	—	—	0	0.6	22.2
180	146	2.63	28.6	313.2	1060.0	2.08	—	0.38	1.66	0.51	4.72	0.28	—	0	1.1	38.0
105	250	4.50	—	—	1030.0	—	—	0.26	—	0.44	6.37	—	—	1.2	—	—
175	150	8.07	—	—	1090.0	—	—	0.39	—	0.59	11.05	—	—	9.0	—	—
195	299	8.08	—	—	1520.0	—	—	0.39	—	0.66	11.03	—	—	9.0	—	—
60	101	3.62	—	—	1380.0	—	—	—	—	—	—	—	—	3.6	—	—
0	20	0.71	—	—	590.0	—	0	0.15	—	0.48	2.30	—	—	8.9	—	—
0	60	1.80	21.3	124.3	440.0	0.57	—	0.31	0.98	0.19	2.88	0.05	—	0	0	13.7
40	80	2.70	—	—	560.0	—	—	0.25	—	0.28	4.25	—	—	1.2	—	—
0	0	0.36	—	—	450.0	—	0	0.11	0.83	0.06	1.35	0.17	—	1.2	—	—
0	100	0.00	—	—	460.0	—	0	0.14	—	0.09	2.32	—	—	0	—	—
75	79	4.43	—	—	1730.0	—	—	—	—	—	—	—	—	8.9	—	—
75	57	6.82	—	—	1210.0	—	—	—	—	—	—	—	—	5.7	—	—
60	348	0.00	—	—	240.0	—	—	0.11	—	0.63	0.21	—	—	2.4	0	—
51	100	5.38	—	—	1020.0	—	—	0.38	—	0.43	7.30	—	—	9.0	—	—
115	249	5.38	—	—	1450.0	—	—	0.38	—	0.51	7.28	—	—	9.0	—	—
80	200	5.40	—	—	1900.0	—	—	—	—	—	—	—	—	2.4	—	—
80	150	2.70	—	—	990.0	—	60.0	—	—	—	—	—	—	2.4	—	—
70	100	4.50	—	—	910.0	—	—	—	—	—	—	—	—	6.0	—	—
95	200	3.60	—	—	1330.0	—	—	—	—	—	—	—	—	9.0	—	—
75	200	1.80	—	—	880.0	—	—	—	—	—	—	—	—	30.0	—	—
100	200	3.60	—	—	1440.0	—	—	—	—	—	—	—	—	9.0	—	—
35	19	1.02	—	—	470.0	—	0	—	—	—	—	—	—	0	—	—
45	600	1.08	—	—	360.0	—	0	—	—	—	—	—	—	0	—	—
0	20	1.80	—	—	950.0	—	0	—	—	—	—	—	—	12.0	—	—
155	300	7.20	—	—	1730.0	—	—	—	—	—	—	—	—	1.2	—	—

DA+ Code	Food Description	Quantity	Measure	Wt (g)	H₂O (g)	Ener (kcal)	Prot (g)	Carb (g)	Fiber (g)	Fat (g)	Fat Breakdown (g) Sat	Mono	Poly	Trans
14238	French fries, small	1	serving(s)	92	—	290	5.0	37.0	3.0	14.0	3.0	—	—	—
10798	French toast dips without syrup, 5 pieces	1	serving(s)	155	—	370	8.0	49.0	0	17.0	5.0	—	1.4	—
10802	Onion rings	1	serving(s)	128	—	440	7.0	53.0	3.0	22.0	5.0	—	0.8	—
34858	Spicy chicken sandwich	1	item(s)	198	—	480	14.0	48.0	2.0	26.0	5.0	—	—	—
34867	Strawberry shake, small	1	serving(s)	595	—	520	14.0	93.0	0	11.0	7.0	—	—	—
10865	Super Star hamburger	1	item(s)	348	—	790	41.0	52.0	3.0	47.0	14.0	—	—	—
38925	The Six Dollar burger	1	item(s)	429	—	1010	40.0	60.0	3.0	66.0	26.0	—	—	—
10818	Vanilla shake, small	1	item(s)	398	—	314	10.0	51.5	0	7.4	4.7	—	—	—
10770	Western Bacon cheeseburger	1	item(s)	225	—	660	32.0	64.0	2.0	30.0	12.0	—	4.8	—
	Chick Fil-A													
38746	Biscuit with bacon, egg & cheese	1	item(s)	163	—	470	18.0	39.0	1.0	26.0	9.0	—	—	3.0
38747	Biscuit with egg	1	item(s)	135	—	350	11.0	38.0	1.0	16.0	4.5	—	—	3.0
38748	Biscuit with egg & cheese	1	item(s)	149	—	400	14.0	38.0	1.0	21.0	7.0	—	—	3.0
38753	Biscuit with gravy	1	item(s)	192	—	330	5.0	43.0	1.0	15.0	4.0	—	—	4.0
38752	Biscuit with sausage, egg & cheese	1	item(s)	212	—	620	22.0	39.0	2.0	42.0	14.0	—	—	3.0
38771	Carrot & raisin salad	1	item(s)	113	—	170	1.0	28.0	2.0	6.0	1.0	—	—	0
38761	Chargrilled chicken Cool Wrap	1	item(s)	245	—	390	29.0	54.0	3.0	7.0	3.0	—	—	0
38766	Chargrilled chicken garden salad	1	item(s)	275	—	180	22.0	9.0	3.0	6.0	3.0	—	—	0
38758	Chargrilled chicken sandwich	1	item(s)	193	—	270	28.0	33.0	3.0	3.5	1.0	—	—	0
38742	Chicken biscuit	1	item(s)	145	—	420	18.0	44.0	2.0	19.0	4.5	—	—	3.0
38743	Chicken biscuit with cheese	1	item(s)	159	—	470	21.0	45.0	2.0	23.0	8.0	—	—	3.0
38762	Chicken Caesar Cool Wrap	1	item(s)	227	—	460	36.0	52.0	3.0	10.0	6.0	—	—	0
38757	Chicken deluxe sandwich	1	item(s)	208	—	420	28.0	39.0	2.0	16.0	3.5	—	—	0
38764	Chicken salad sandwich on wheat bun	1	item(s)	153	—	350	20.0	32.0	5.0	15.0	3.0	—	—	0
38756	Chicken sandwich	1	item(s)	170	—	410	28.0	38.0	1.0	16.0	3.5	—	—	0
38768	Chick-n-Strip salad	1	item(s)	327	—	400	34.0	21.0	4.0	20.0	6.0	—	—	0
38763	Chick-n-Strips	4	item(s)	127	—	300	28.0	14.0	1.0	15.0	2.5	—	—	0
38770	Cole slaw	1	item(s)	128	—	260	2.0	17.0	2.0	21.0	3.5	—	—	0
38776	Diet lemonade, small	1	cup(s)	255	—	25	0	5.0	0	0	0	0	0	0
38755	Hashbrowns	1	serving(s)	84	—	260	2.0	25.0	3.0	17.0	3.5	—	—	1.0
38765	Hearty breast of chicken soup	1	cup(s)	241	—	140	8.0	18.0	1.0	3.5	1.0	—	—	0
38741	Hot buttered biscuit	1	item(s)	79	—	270	4.0	38.0	1.0	12.0	3.0	—	—	3.0
38778	IceDream, small cone	1	item(s)	135	—	160	4.0	28.0	0	4.0	2.0	—	—	0
38774	IceDream, small cup	1	serving(s)	227	—	240	6.0	41.0	0	6.0	3.5	—	—	0
38775	Lemonade, small	1	cup(s)	255	—	170	0	41.0	0	0.5	0	—	—	0
38777	Nuggets	8	item(s)	113	—	260	26.0	12.0	0.5	12.0	2.5	—	—	0
38769	Side salad	1	item(s)	108	—	60	3.0	4.0	2.0	3.0	1.5	—	—	0
38767	Southwest chargrilled salad	1	item(s)	303	—	240	25.0	17.0	5.0	8.0	3.5	—	—	0
40481	Spicy chicken cool wrap	1	serving(s)	230	—	380	30.0	52.0	3.0	6.0	3.0	—	—	0
38772	Waffle potato fries, small, salted	1	serving(s)	85	—	270	3.0	34.0	4.0	13.0	3.0	—	—	1.5
	Cinnabon													
39572	Caramellata Chill with whipped cream	16	fluid ounce(s)	480	—	406	10.0	61.0	0	14.0	8.0	—	—	0
39571	Cinnabon Bites	1	serving(s)	149	—	510	8.0	77.0	2.0	19.0	5.0	—	—	5.0
39570	Cinnabon Stix	5	item(s)	85	—	379	6.0	41.0	1.0	21.0	6.0	—	—	4.0
39567	Classic roll	1	item(s)	221	—	813	15.0	117.0	4.0	32.0	8.0	—	—	5.0
39568	Minibon	1	item(s)	92	—	339	6.0	49.0	2.0	13.0	3.0	—	—	2.0
39573	Mochalatta Chill with whipped cream	16	fluid ounce(s)	480	—	362	9.0	55.0	0	13.0	8.0	—	—	0
39569	Pecanbon	1	item(s)	272	—	1100	16.0	141.0	8.0	56.0	10.0	—	—	5.0
	Dairy Queen													
1466	Banana split	1	item(s)	369	—	510	8.0	96.0	3.0	12.0	8.0	—	—	0
38552	Brownie Earthquake®	1	serving(s)	304	—	740	10.0	112.0	0	27.0	16.0	—	—	0.5
38561	Chocolate chip cookie dough blizzard®, small	1	item(s)	319	—	720	12.0	105.0	0	28.0	14.0	—	—	2.5
1464	Chocolate malt, small	1	item(s)	418	—	640	15.0	111.0	1.0	16.0	11.0	—	—	0.5
38541	Chocolate shake, small	1	item(s)	397	—	560	13.0	93.0	1.0	15.0	10.0	—	—	0.5
17257	Chocolate soft serve	½	cup(s)	94	—	150	4.0	22.0	0	5.0	3.5	—	—	0
1463	Chocolate sundae, small	1	item(s)	163	—	280	5.0	49.0	0	7.0	4.5	—	—	0
1462	Dipped cone, small	1	item(s)	156	—	340	6.0	42.0	1.0	17.0	9.0	4.0	3.0	1.0
38555	Oreo cookies blizzard, small	1	item(s)	283	—	570	11.0	83.0	0.5	21.0	10.0	—	—	2.5
38547	Royal Treats Peanut Buster® Parfait	1	item(s)	305	—	730	16.0	99.0	2.0	31.0	17.0	—	—	0
17256	Vanilla soft serve	½	cup(s)	94	—	140	3.0	22.0	0	4.5	3.0	—	—	0
	Domino's													
31606	Barbeque buffalo wings	1	item(s)	25	—	50	6.0	2.0	0	2.5	0.5	—	—	—
31604	Breadsticks	1	item(s)	30	—	115	2.0	12.0	0	6.3	1.1	—	—	—
37551	Buffalo Chicken Kickers	1	item(s)	24	—	47	4.0	3.0	0	2.0	0.5	—	—	—
37548	CinnaStix	1	item(s)	30	—	123	2.0	15.0	1.0	6.1	1.1	—	—	—
37549	Dot, cinnamon	1	item(s)	28	7.6	99	1.9	14.9	0.7	3.7	0.7	—	—	—
31605	Double cheesy bread	1	item(s)	35	—	123	4.0	13.0	0	6.5	1.9	—	—	—
	FAST FOOD —Continued													
31607	Hot buffalo wings	1	item(s)	25	—	45	5.0	1.0	0	2.5	0.5	—	—	—
	Domino's Classic hand tossed pizza													

Chol (mg)	Calc (mg)	Iron (mg)	Magn (mg)	Pota (mg)	Sodi (mg)	Zinc (mg)	Vit A (µg)	Thia (mg)	Vit E (mg α)	Ribo (mg)	Niac (mg)	Vit B$_6$ (mg)	Fola (µg)	Vit C (mg)	Vit B$_{12}$ (µg)	Sele (µg)
0	0	1.08	—	—	170.0	—	0	—	—	—	—	—	—	21.0	—	—
3	0	0.00	—	—	470.0	—	0	0.25	—	0.23	2.00	—	—	0	—	—
0	20	0.72	—	—	700.0	—	0	—	—	—	—	—	—	3.6	—	—
40	100	3.60	—	—	1220.0	—	—	—	—	—	—	—	—	6.0	—	—
45	600	0.00	—	—	340.0	—	0	—	—	—	—	—	—	0	—	—
130	100	7.20	—	—	980.0	—	—	—	—	—	—	—	—	9.0	—	—
145	279	4.29	—	—	1960.0	—	—	—	—	—	—	—	—	16.7	—	—
30	401	0.00	—	—	234.0	—	0	—	—	—	—	—	—	0	—	—
85	200	5.40	—	—	1410.0	—	60.0	—	—	—	—	—	—	1.2	—	—
270	150	2.70	—	—	1190.0	—	—	—	—	—	—	—	—	0	—	—
240	80	2.70	—	—	740.0	—	—	—	—	—	—	—	—	0	—	—
255	150	2.70	—	—	970.0	—	—	—	—	—	—	—	—	0	—	—
5	60	1.80	—	—	930.0	—	0	—	—	—	—	—	—	0	—	—
300	200	3.60	—	—	1360.0	—	—	—	—	—	—	—	—	0	—	—
10	40	0.36	—	—	110.0	—	—	—	—	—	—	—	—	4.8	—	—
65	200	3.60	—	—	1020.0	—	—	—	—	—	—	—	—	6.0	—	—
65	150	0.72	—	—	620.0	—	—	—	—	—	—	—	—	30	—	—
65	80	2.70	—	—	940.0	—	—	—	—	—	—	—	—	6.0	—	—
35	60	2.70	—	—	1270.0	—	0	—	—	—	—	—	—	0	—	—
50	150	2.70	—	—	1500.0	—	—	—	—	—	—	—	—	0	—	—
80	500	3.60	—	—	1350.0	—	—	—	—	—	—	—	—	1.2	—	—
60	100	2.70	—	—	1300.0	—	—	—	—	—	—	—	—	2.4	—	—
65	150	1.80	—	—	880.0	—	—	—	—	—	—	—	—	0	—	—
60	100	2.70	—	—	1300.0	—	—	—	—	—	—	—	—	0	—	—
80	150	1.44	—	—	1070.0	—	—	—	—	—	—	—	—	6.0	—	—
65	40	1.44	—	—	940.0	—	—	—	—	—	—	—	—	0	—	—
25	60	0.36	—	—	220.0	—	—	—	—	—	—	—	—	36.0	—	—
0	0	0.36	—	—	5.0	—	0	—	—	—	—	—	—	15.0	—	—
5	20	0.72	—	—	380.0	—	—	—	—	—	—	—	—	0	—	—
25	40	1.08	—	—	900.0	—	—	—	—	—	—	—	—	0	—	—
0	60	1.80	—	—	660.0	—	0	—	—	—	—	—	—	0	—	—
15	100	0.36	—	—	80.0	—	—	—	—	—	—	—	—	0	—	—
25	200	0.36	—	—	105.0	—	—	—	—	—	—	—	—	0	—	—
0	0	0.36	—	—	10.0	—	0	—	—	—	—	—	—	15.0	—	—
70	40	1.08	—	—	1090.0	—	0	—	—	—	—	—	—	0	—	—
10	100	0.00	—	—	75.0	—	—	—	—	—	—	—	—	15.0	—	—
60	200	1.08	—	—	770.0	—	—	—	—	—	—	—	—	24.0	—	—
60	200	3.60	—	—	1090.0	—	—	—	—	—	—	—	—	3.6	—	—
0	20	1.08	—	—	115.0	—	0	—	—	—	—	—	—	1.2	—	—
46	—	—	—	—	187.0	—	—	—	—	—	—	—	—	—	—	—
35	—	—	—	—	530.0	—	—	—	—	—	—	—	—	—	—	—
16	—	—	—	—	413.0	—	—	—	—	—	—	—	—	—	—	—
67	—	—	—	—	801.0	—	—	—	—	—	—	—	—	—	—	—
27	—	—	—	—	337.0	—	—	—	—	—	—	—	—	—	—	—
46	—	—	—	—	252.0	—	—	—	—	—	—	—	—	—	—	—
63	—	—	—	—	600.0	—	—	—	—	—	—	—	—	—	—	—
30	250	1.80	—	—	180.0	—	—	—	—	—	—	—	—	15.0	—	—
50	250	1.80	—	—	350.0	—	—	—	—	—	—	—	—	0	—	—
50	350	2.70	—	—	370.0	—	—	—	—	—	—	—	—	1.2	—	—
55	450	1.80	—	—	340.0	—	—	—	—	—	—	—	—	2.4	—	—
50	450	1.44	—	—	280.0	—	—	—	—	—	—	—	—	2.4	—	—
15	100	0.72	—	—	75.0	—	—	—	—	—	—	—	—	0	—	—
20	200	1.08	—	—	140.0	—	—	—	—	—	—	—	—	0	—	—
20	200	1.08	—	—	130.0	—	—	—	—	—	—	—	—	1.2	—	—
40	350	2.70	—	—	430.0	—	—	—	—	—	—	—	—	1.2	—	—
35	300	1.80	—	—	400.0	—	—	—	—	—	—	—	—	1.2	—	—
15	150	0.72	—	—	70.0	—	—	—	—	—	—	—	—	0	—	—
26	10	0.36	—	—	175.5	—	—	—	—	—	—	—	—	0	—	—
0	0	0.72	—	—	122.1	—	—	—	—	—	—	—	—	0	—	—
9	0	0.00	—	—	162.5	—	—	—	—	—	—	—	—	0	—	—
0	0	0.72	—	—	111.4	—	—	—	—	—	—	—	—	0	—	—
0	6	0.59	—	—	85.7	—	—	—	—	—	—	—	—	0	—	—
6	40	0.72	—	—	162.3	—	—	—	—	—	—	—	—	0	—	—
26	10	0.36	—	—	254.5	—	—	—	—	—	—	—	—	1.2	—	—
																—

DA+ Code	Food Description	Quantity	Measure	Wt (g)	H₂O (g)	Ener (kcal)	Prot (g)	Carb (g)	Fiber (g)	Fat (g)	Fat Breakdown (g)			
											Sat	Mono	Poly	Trans
31573	America's favorite feast, 12"	1	slice(s)	102	—	257	10.0	29.0	2.0	11.5	4.5	—	—	—
31574	America's favorite feast, 14"	1	slice(s)	141	—	353	14.0	39.0	2.0	16.0	6.0	—	—	—
37543	Bacon cheeseburger feast, 12"	1	slice(s)	99	—	273	12.0	28.0	2.0	13.0	5.5	—	—	—
37545	Bacon cheeseburger feast, 14"	1	slice(s)	137	—	379	17.0	38.0	2.0	18.0	8.0	—	—	—
37546	Barbeque feast, 12"	1	slice(s)	96	—	252	11.0	31.0	1.0	10.0	4.5	—	—	—
37547	Barbeque feast, 14"	1	slice(s)	131	—	344	14.0	43.0	2.0	13.5	6.0	—	—	—
31569	Cheese, 12"	1	slice(s)	55	—	160	6.0	28.0	1.0	3.0	1.0	—	—	0
31570	Cheese, 14"	1	slice(s)	75	—	220	8.0	38.0	2.0	4.0	1.0	—	—	0
37538	Deluxe feast, 12"	1	slice(s)	201	101.8	465	19.5	57.4	3.5	18.2	7.7	—	—	—
37540	Deluxe feast, 14"	1	slice(s)	273	138.4	627	26.4	78.3	4.7	24.1	10.2	—	—	—
31685	Deluxe, 12"	1	slice(s)	100	—	234	9.0	29.0	2.0	9.5	3.5	—	—	—
31694	Deluxe, 14"	1	slice(s)	136	—	316	13.0	39.0	2.0	12.5	5.0	—	—	—
31686	Extravaganzza, 12"	1	slice(s)	122	—	289	13.0	30.0	2.0	14.0	5.5	—	—	—
31695	Extravaganzza, 14"	1	slice(s)	165	—	388	17.0	40.0	3.0	18.5	7.5	—	—	—
31575	Hawaiian feast, 12"	1	slice(s)	102	—	223	10.0	30.0	2.0	8.0	3.5	—	—	—
31576	Hawaiian feast, 14"	1	slice(s)	141	—	309	14.0	41.0	2.0	11.0	4.5	—	—	—
31687	Meatzza, 12"	1	slice(s)	108	—	281	13.0	29.0	2.0	13.5	5.5	—	—	—
31696	Meatzza, 14"	1	slice(s)	146	—	378	17.0	39.0	2.0	18.0	7.5	—	—	—
31571	Pepperoni feast, extra pepperoni & cheese, 12"	1	slice(s)	98	—	265	11.0	28.0	2.0	12.5	5.0	—	—	—
31572	Pepperoni feast, extra pepperoni & cheese, 14"	1	slice(s)	135	—	363	16.0	39.0	2.0	17.0	7.0	—	—	—
31577	Vegi feast, 12"	1	slice(s)	102	—	218	9.0	29.0	2.0	8.0	3.5	—	—	—
31578	Vegi feast, 14"	1	slice(s)	139	—	300	13.0	40.0	3.0	11.0	4.5	—	—	—
	Domino's thin crust pizza													
31583	America's favorite, 12"	1	slice(s)	72	—	208	8.0	15.0	1.0	13.5	5.0	—	—	—
31584	America's favorite, 14"	1	slice(s)	100	—	285	11.0	20.0	2.0	18.5	7.0	—	—	—
31579	Cheese, 12"	1	slice(s)	49	—	137	5.0	14.0	1.0	7.0	2.5	—	—	—
31580	Cheese, 14"	1	slice(s)	68	27.0	214	8.8	19.0	1.4	11.4	4.6	2.9	2.5	—
31688	Deluxe, 12"	1	slice(s)	70	—	185	7.0	15.0	1.0	11.5	4.0	—	—	—
31697	Deluxe, 14"	1	slice(s)	94	—	248	10.0	20.0	2.0	15.0	5.5	—	—	—
31689	Extravaganzza, 12"	1	slice(s)	92	—	240	11.0	16.0	1.0	15.5	6.0	—	—	—
31698	Extravaganzza, 14"	1	slice(s)	123	—	320	14.0	21.0	2.0	20.5	8.0	—	—	—
31585	Hawaiian, 12"	1	slice(s)	71	—	174	8.0	16.0	1.0	9.5	3.5	—	—	—
31586	Hawaiian, 14"	1	slice(s)	100	—	240	11.0	21.0	2.0	13.0	5.0	—	—	—
31690	Meatzza, 12"	1	slice(s)	78	—	232	11.0	15.0	1.0	15.0	6.0	—	—	—
31699	Meatzza, 14"	1	slice(s)	104	—	310	14.0	20.0	2.0	20	8.0	—	—	—
31581	Pepperoni, extra pepperoni & cheese, 12"	1	slice(s)	68	—	216	9.0	14.0	1.0	14.0	5.5	—	—	—
31582	Pepperoni, extra pepperoni & cheese, 14"	1	slice(s)	93	—	295	13.0	20.0	1.0	19.0	7.5	—	—	—
31587	Vegi, 12"	1	slice(s)	71	—	168	7.0	15.0	1.0	9.5	3.5	—	—	—
31588	Vegi, 14"	1	slice(s)	97	—	231	10.0	21.0	2.0	13.5	5.0	—	—	—
	Domino's Ultimate deep dish pizza													
31596	America's favorite, 12"	1	slice(s)	115	—	309	12.0	29.0	2.0	17.0	6.0	—	—	—
31702	America's favorite, 14"	1	slice(s)	162	—	433	17.0	42.0	3.0	23.5	8.0	—	—	—
31590	Cheese, 12"	1	slice(s)	90	—	238	9.0	28.0	2.0	11.0	3.5	—	—	—
31591	Cheese, 14"	1	slice(s)	128	53.9	351	14.5	41.0	2.9	13.2	5.2	3.8	2.5	—
31589	Cheese, 6"	1	item(s)	215	—	598	22.9	68.4	3.9	27.6	9.9	—	—	—
31691	Deluxe, 12"	1	slice(s)	122	—	287	11.0	29.0	2.0	15.0	5.0	—	—	—
31700	Deluxe, 14"	1	slice(s)	156	—	396	15.0	42.0	3.0	20.0	7.0	—	—	—
31692	Extravaganzza, 12"	1	slice(s)	136	—	341	14.0	30.0	2.0	19.0	7.0	—	—	—
31701	Extravaganzza, 14"	1	slice(s)	186	—	468	20.0	43.0	3.0	25.5	9.5	—	—	—
31599	Hawaiian, 12"	1	slice(s)	114	—	275	12.0	30.0	2.0	13.0	5.0	—	—	—
31600	Hawaiian, 14"	1	slice(s)	162	—	389	17.0	43.0	3.0	18.0	6.5	—	—	—
31693	Meatzza, 12"	1	slice(s)	121	—	333	14.0	29.0	2.0	19.0	7.0	—	—	—
31703	Meatzza, 14"	1	slice(s)	167	—	458	19.0	42.0	3.0	25.0	9.5	—	—	—
31593	Pepperoni, extra pepperoni & cheese, 12"	1	slice(s)	110	—	317	13.0	29.0	2.0	17.5	6.5	—	—	—
31594	Pepperoni, extra pepperoni & cheese, 14"	1	slice(s)	155	—	443	18.0	42.0	3.0	24.0	9.0	—	—	—
31602	Vegi, 12"	1	slice(s)	114	—	270	11.0	30.0	2.0	13.5	5.0	—	—	—
31603	Vegi, 14"	1	slice(s)	159	—	380	15.0	43.0	3.0	18.0	6.5	—	—	—
31598	With ham and pineapple tidbits, 6"	1	item(s)	430	—	619	25.2	69.9	4.0	28.3	10.2	—	—	—
31595	With Italian sausage, 6"	1	item(s)	430	—	642	24.8	69.6	4.2	31.1	11.3	—	—	—
31592	With pepperoni, 6"	1	item(s)	430	—	647	25.1	68.5	3.9	32.0	11.7	—	—	—
31601	With vegetables, 6"	1	item(s)	430	—	619	23.4	70.8	4.6	28.7	10.1	—	—	—
	In-n-Out Burger													
34391	Cheeseburger with mustard & ketchup	1	serving(s)	268	—	400	22.0	41.0	3.0	18.0	9.0	—	—	0.5
34374	Cheeseburger	1	serving(s)	268	—	480	22.0	39.0	3.0	27.0	10.0	—	—	0.5
34390	Cheeseburger, lettuce leaves instead of buns	1	serving(s)	300	—	330	18.0	11.0	3.0	25.0	9.0	—	—	0
34377	Chocolate shake	1	serving(s)	425	—	690	9.0	83.0	0	36.0	24.0	—	—	1.0
34375	Double-Double cheeseburger	1	serving(s)	330	—	670	37.0	39.0	3.0	41.0	18.0	—	—	1.0
	FAST FOOD —Continued													
34393	Double-Double cheeseburger with mustard & ketchup	1	serving(s)	330	—	590	37.0	41.0	3.0	32.0	17.0	—	—	1.0

Chol (mg)	Calc (mg)	Iron (mg)	Magn (mg)	Pota (mg)	Sodi (mg)	Zinc (mg)	Vit A (µg)	Thia (mg)	Vit E (mg α)	Ribo (mg)	Niac (mg)	Vit B$_6$ (mg)	Fola (µg)	Vit C (mg)	Vit B$_{12}$ (µg)	Sele (µg)
22	100	1.80	—	—	625.5	—	—	—	—	—	—	—	—	0.6	—	—
31	140	2.52	—	—	865.5	—	—	—	—	—	—	—	—	0.6	—	—
27	140	1.80	—	—	634.0	—	—	—	—	—	—	—	—	0	—	—
38	190	2.52	—	—	900.0	—	—	—	—	—	—	—	—	0	—	—
20	140	1.62	—	—	600.0	—	—	—	—	—	—	—	—	0.6	—	—
27	190	2.16	—	—	831.5	—	—	—	—	—	—	—	—	0.6	—	—
0	0	1.80	—	—	110.0	—	0	—	—	—	—	—	—	0	—	—
0	0	2.70	—	—	150.0	—	0	—	—	—	—	—	—	0	—	—
40	199	3.56	—	—	1063.1	—	—	—	—	—	—	—	—	1.4	—	—
53	276	4.84	—	—	1432.2	—	—	—	—	—	—	—	—	1.8	—	—
17	100	1.80	—	—	541.5	—	—	—	—	—	—	—	—	0.6	—	—
23	130	2.34	—	—	728.5	—	—	—	—	—	—	—	—	1.2	—	—
28	140	1.98	—	—	764.0	—	—	—	—	—	—	—	—	0.6	—	—
37	190	2.70	—	—	1014.0	—	—	—	—	—	—	—	—	1.2	—	—
16	130	1.62	—	—	546.5	—	—	—	—	—	—	—	—	1.2	—	—
23	180	2.34	—	—	765.0	—	—	—	—	—	—	—	—	1.2	—	—
28	130	1.80	—	—	739.5	—	—	—	—	—	—	—	—	0	—	—
37	190	2.52	—	—	983.5	—	—	—	—	—	—	—	—	0	—	—
24	130	1.62	—	—	670.0	—	70.9	—	—	—	—	—	—	0	—	—
33	180	2.34	—	—	920.0	—	104.7	—	—	—	—	—	—	0	—	—
13	130	1.62	—	—	489.0	—	—	—	—	—	—	—	—	0.6	—	—
18	180	2.34	—	—	678.0	—	—	—	—	—	—	—	—	0.6	—	—
23	100	0.90	—	—	533.0	—	—	—	—	—	—	—	—	2.4	—	—
32	140	1.26	—	—	736.5	—	—	—	—	—	—	—	—	3.0	—	—
10	90	0.54	—	—	292.5	—	60.0	—	—	—	—	—	—	1.8	—	—
14	151	0.48	17.7	125.1	338.0	0.02	64.6	0.05	1.01	0.07	0.69	—	—	2.4	0.5	24.1
19	100	0.90	—	—	449.0	—	—	—	—	—	—	—	—	2.4	—	—
24	130	1.08	—	—	601.0	—	—	—	—	—	—	—	—	3.6	—	—
29	140	1.08	—	—	671.5	—	—	—	—	—	—	—	—	2.4	—	—
38	190	1.44	—	—	886.5	—	—	—	—	—	—	—	—	3.6	—	—
17	130	0.72	—	—	454.0	—	—	—	—	—	—	—	—	3.0	—	—
24	180	0.90	—	—	637.5	—	—	—	—	—	—	—	—	3.6	—	—
29	140	0.90	—	—	647.0	—	—	—	—	—	—	—	—	1.8	—	—
38	190	1.26	—	—	865.5	—	—	—	—	—	—	—	—	2.4	—	—
26	130	0.72	—	—	577.0	—	80.0	—	—	—	—	—	—	1.8	—	—
35	80	1.08	—	—	792.5	—	105.8	—	—	—	—	—	—	2.4	—	—
14	130	0.72	—	—	396.5	—	—	—	—	—	—	—	—	2.4	—	—
19	180	1.08	—	—	550.5	—	—	—	—	—	—	—	—	3.0	—	—
25	120	2.34	—	—	796.5	—	—	—	—	—	—	—	—	0.6	—	—
34	170	3.24	—	—	1110.0	—	—	—	—	—	—	—	—	0.6	—	—
11	110	1.98	—		555.5	—	70.0	—	—	—	—	—	—	0	—	—
18	189	3.78	32.0	209.9	718.1	1.75	99.8	0.29	1.13	0.31	5.44	—	—	0	0.6	15.6
36	295	4.67	—	—	1341.4	—	174.0	—	—	—	—	—	—	0.5	—	—
20	120	2.16	—	—	712.0	—	—	—	—	—	—	—	—	1.2	—	—
26	170	3.06	—	—	974.5	—	—	—	—	—	—	—	—	1.2	—	—
31	160	2.52	—	—	934.5	—	—	—	—	—	—	—	—	1.2	—	—
40	220	3.42	—	—	1260.0	—	—	—	—	—	—	—	—	1.2	—	—
19	150	1.98	—	—	717.0	—	—	—	—	—	—	—	—	1.2	—	—
26	210	2.88	—	—	1011.0	—	—	—	—	—	—	—	—	1.8	—	—
31	160	2.34	—	—	910.5	—	—	—	—	—	—	—	—	0	—	—
40	220	3.24	—	—	1230.0	—	—	—	—	—	—	—	—	0.6	—	—
27	150	2.16	—	—	840.5	—	86.5	—	—	—	—	—	—	0	—	—
37	220	3.06	—	—	1166.0	—	115.4	—	—	—	—	—	—	0.6	—	—
15	150	2.16	—	—	659.5	—	—	—	—	—	—	—	—	0.6	—	—
21	220	3.06	—	—	924.0	—	—	—	—	—	—	—	—	1.2	—	—
43	298	4.84	—	—	1497.8	—	—	—	—	—	—	—	—	1.5	—	—
45	302	4.89	—	—	1478.1	—	—	—	—	—	—	—	—	0.6	—	—
47	299	4.81	—	—	1523.7	—	167.9	—	—	—	—	—	—	0.6	—	—
36	307	5.10	—	—	1472.5	—	—	—	—	—	—	—	—	4.7	—	—
60	200	3.60	—	—	1080.0	—	—	—	—	—	—	—	—	12.0	—	—
60	200	3.60	—	—	1000.0	—	—	—	—	—	—	—	—	9.0	—	—
60	200	2.70	—	—	720.0	—	—	—	—	—	—	—	—	12.0	—	—
95	300	0.72	—	—	350.0	—	—	—	—	—	—	—	—	0	—	—
120	350	5.40	—	—	1440.0	—	—	—	—	—	—	—	—	9.0	—	—
115	350	5.40	—	—	1520.0	—	—	—	—	—	—	—	—	12.0	—	—

DA+ Code	Food Description	Quantity	Measure	Wt (g)	H₂O (g)	Ener (kcal)	Prot (g)	Carb (g)	Fiber (g)	Fat (g)	Fat Breakdown (g) Sat	Mono	Poly	Trans
	Double-Double cheeseburger, lettuce leaves instead													
34392	of buns	1	serving(s)	362	—	520	33.0	11.0	3.0	39.0	17.0	—	—	1.0
34376	French fries	1	serving(s)	125	—	400	7.0	54.0	2.0	18.0	5.0	—	—	0
34373	Hamburger	1	item(s)	243	—	390	16.0	39.0	3.0	19.0	5.0	—	—	0
34389	Hamburger with mustard & ketchup	1	serving(s)	243	—	310	16.0	41.0	3.0	10.0	4.0	—	—	0
34388	Hamburger, lettuce leaves instead of buns	1	serving(s)	275	—	240	13.0	11.0	3.0	17.0	4.0	—	—	0
34379	Strawberry shake	1	serving(s)	425	—	690	9.0	91.0	0	33.0	22.0	—	—	0.5
34378	Vanilla shake	1	serving(s)	425	—	680	9.0	78.0	0	37.0	25.0	—	—	1.0
	Jack in the Box													
30392	Bacon ultimate cheeseburger	1	item(s)	338	—	1090	46.0	53.0	2.0	77.0	30.0	—	—	3.0
1740	Breakfast Jack	1	item(s)	125	—	290	17.0	29.0	1.0	12.0	4.5	—	—	0
14074	Cheeseburger	1	item(s)	131	—	350	18.0	31.0	1.0	17.0	8.0	—	—	1.0
14106	Chicken breast strips, 4 piece	4	piece(s)	201	—	500	35.0	36.0	3.0	25.0	6.0	—	—	6.0
37241	Chicken club salad, plain, without salad dressing	1	serving(s)	431	—	300	27.0	13.0	4.0	15.0	6.0	—	—	0
14064	Chicken sandwich	1	item(s)	145	—	400	15.0	38.0	2.0	21.0	4.5	—	—	2.5
14111	Chocolate ice cream shake, small	1	serving(s)	414	—	880	14.0	107.0	1.0	45.0	31.0	—	—	2.0
14073	Hamburger	1	item(s)	118	—	310	16.0	30.0	1.0	14.0	6.0	—	—	1.0
14090	Hash browns	1	serving(s)	57	—	150	1.0	13.0	2.0	10.0	2.5	—	—	3.0
14072	Jack's Spicy Chicken sandwich	1	item(s)	270	—	620	25.0	61.0	4.0	31.0	6.0	—	—	3.0
1468	Jumbo Jack hamburger	1	item(s)	261	—	600	21.0	51.0	3.0	35.0	12.0	—	—	1.5
1469	Jumbo Jack hamburger with cheese	1	item(s)	286	—	690	25.0	54.0	3.0	42.0	16.0	—	—	1.5
14099	Natural cut french fries, large	1	serving(s)	196	—	530	8.0	69.0	5.0	25.0	6.0	—	—	7.0
14098	Natural cut french fries, medium	1	serving(s)	133	—	360	5.0	47.0	4.0	17.0	4.0	—	—	5.0
1470	Onion rings	1	serving(s)	119	—	500	6.0	51.0	3.0	30.0	6.0	—	—	10
33141	Sausage, egg & cheese biscuit	1	item(s)	234	—	740	27.0	35.0	2.0	55.0	17.0	—	—	6.0
14095	Seasoned curly fries, medium	1	serving(s)	125	—	400	6.0	45.0	5.0	23.0	5.0	—	—	7.0
14077	Sourdough Jack	1	item(s)	245	—	710	27.0	36.0	3.0	51.0	18.0	—	—	3.0
37249	Southwest chicken salad, plain, without salad dressing	1	serving(s)	488	—	300	24.0	29.0	7.0	11.0	5.0	—	—	0
14112	Strawberry ice cream shake, small	1	serving(s)	417	—	880	13.0	105.0	0	44.0	31.0	—	—	2.0
14078	Ultimate cheeseburger	1	item(s)	323	—	1010	40.0	53.0	2.0	71.0	28.0	—	—	3.0
14110	Vanilla ice cream shake, small	1	serving(s)	379	—	790	13.0	83.0	0	44.0	31.0	—	—	2.0
	Jamba Juice													
31645	Aloha Pineapple smoothie	24	fluid ounce(s)	730	—	500	8.0	117.0	4.0	1.5	1.0	—	—	—
31646	Banana Berry smoothie	24	fluid ounce(s)	719	—	480	5.0	112.0	4.0	1.0	0	—	—	—
31656	Berry Lime Sublime smoothie	24	fluid ounce(s)	728	—	460	3.0	106.0	5.0	2.0	1.0	—	—	—
31647	Carribean Passion smoothie	24	fluid ounce(s)	730	—	440	4.0	102.0	4.0	2.0	1.0	—	—	—
38422	Carrot juice	16	fluid ounce(s)	472	—	100	3.0	23.0	0	0.5	0	—	—	—
31648	Chocolate Moo'd smoothie	24	fluid ounce(s)	634	—	720	17.0	148.0	3.0	8.0	5.0	—	—	—
31649	Citrus Squeeze smoothie	24	fluid ounce(s)	727	—	470	5.0	110.0	4.0	2.0	1.0	—	—	—
31651	Coldbuster smoothie	24	fluid ounce(s)	724	—	430	5.0	100.0	5.0	2.5	1.0	—	—	—
31652	Cranberry Craze smoothie	24	fluid ounce(s)	793	—	460	6.0	104.0	4.0	0.5	0	—	—	—
31654	Jamba Powerboost smoothie	24	fluid ounce(s)	738	—	440	6.0	105.0	6.0	1.0	0	—	—	—
38423	Lemonade	16	fluid ounce(s)	483	—	300	1.0	75.0	0	0	0	0	0	0
31657	Mango-a-go-go smoothie	24	fluid ounce(s)	690	—	440	3.0	104.0	4.0	1.5	0.5	—	—	—
38424	Orange juice, freshly squeezed	16	fluid ounce(s)	496	—	220	3.0	52.0	0.5	1.0	0	—	—	—
38426	Orange/carrot juice	16	fluid ounce(s)	484	—	160	3.0	37.0	0	1.0	0	—	—	—
31660	Orange-a-peel smoothie	24	fluid ounce(s)	726	—	440	8.0	102.0	5.0	1.5	0	—	—	—
31662	Peach Pleasure smoothie	24	fluid ounce(s)	720	—	460	4.0	108.0	4.0	2.0	1.0	—	—	—
31665	Protein Berry Pizzaz smoothie	24	fluid ounce(s)	710	—	440	20.0	92.0	5.0	1.5	0	—	—	—
31668	Razzmatazz smoothie	24	fluid ounce(s)	730	—	480	3.0	112.0	4.0	2.0	1.0	—	—	—
31669	Strawberries Wild smoothie	24	fluid ounce(s)	725	—	450	6.0	105.0	4.0	0.5	0	—	—	—
38421	Strawberry Tsunami smoothie	24	fluid ounce(s)	740	—	530	4.0	128.0	4.0	2.0	1.0	—	—	—
38427	Vibrant C juice	16	fluid ounce(s)	448	—	210	2.0	50.0	1.0	0	0	0	0	0
38428	Wheatgrass juice, freshly squeezed	1	ounce(s)	28	—	5	0.5	1.0	0	0	0	0	0	0
	Kentucky Fried Chicken (KFC)													
31850	BBQ baked beans	1	serving(s)	136	—	220	8.0	45.0	7.0	1.0	0	—	—	0
31853	Biscuit	1	item(s)	57	—	220	4.0	24.0	1.0	11.0	2.5	—	—	3.5
51223	Boneless Fiery Buffalo Wings	6	item(s)	211	—	530	30.0	44.0	3.0	26.0	5.0	—	—	2.5
39386	Boneless Honey BBQ Wings	6	item(s)	213	—	570	30.0	54.0	5.0	26.0	5.0	—	—	2.5
51224	Boneless Sweet & Spicy Wings	6	item(s)	203	—	550	30.0	50.0	3.0	26.0	5.0	—	—	2.5
31851	Cole slaw	1	serving(s)	130	—	180	1.0	22.0	3.0	10.0	1.5	—	—	0
31842	Colonel's Crispy Strips	3	item(s)	151	—	370	28.0	17.0	1.0	20.0	4.0	—	—	2.5
31849	Corn on the cob	1	item(s)	162	—	150	5.0	26.0	7.0	3.0	1.0	—	—	0
51221	Double Crunch sandwich	1	item(s)	213	—	520	27.0	39.0	3.0	29.0	5.0	—	—	1.5
3761	Extra Crispy chicken, breast	1	item(s)	162	—	370	33.0	10.0	2.0	22.0	5.0	—	—	1.5
3762	Extra Crispy chicken, drumstick	1	item(s)	60	—	150	12.0	4.0	0	10.0	2.5	—	—	1.0
	FAST FOOD —Continued													
3763	Extra Crispy chicken, thigh	1	item(s)	114	—	290	17.0	16.0	1.0	18.0	4.0	—	—	1.5
3764	Extra Crispy chicken, whole wing	1	item(s)	52	—	150	11.0	11.0	1.0	7.0	1.5	—	—	0
51218	Famous Bowls mashed potatoes with gravy	1	serving(s)	531	—	720	26.0	79.0	6.0	34.0	9.0	—	—	3.5

Chol (mg)	Calc (mg)	Iron (mg)	Magn (mg)	Pota (mg)	Sodi (mg)	Zinc (mg)	Vit A (µg)	Thia (mg)	Vit E (mg α)	Ribo (mg)	Niac (mg)	Vit B$_6$ (mg)	Fola (µg)	Vit C (mg)	Vit B$_{12}$ (µg)	Sele (µg)
120	350	4.50	—	—	1160.0	—	—	—	—	—	—	—	—	12.0	—	—
0	20	1.80	—	—	245.0	—	0	—	—	—	—	—	—	0	—	—
40	40	3.60	—	—	650.0	—	—	—	—	—	—	—	—	9.0	—	—
35	40	3.60	—	—	730.0	—	—	—	—	—	—	—	—	12.0	—	—
40	40	2.70	—	—	370.0	—	—	—	—	—	—	—	—	12.0	—	—
85	300	0.00	—	—	280.0	—	—	—	—	—	—	—	—	0	—	—
90	300	0.00	—	—	390.0	—	—	—	—	—	—	—	—	0	—	—
140	308	7.38	—	540.0	2040.0	—	—	—	—	—	—	—	—	0.6	—	—
220	145	3.48	—	210.0	760.0	—	—	—	—	—	—	—	—	3.5	—	—
50	151	3.61	—	270.0	790.0	—	40.2	—	—	—	—	—	—	0	—	—
80	18	1.60	—	530.0	1260.0	—	—	—	—	—	—	—	—	1.1	—	—
65	280	3.35	—	560.0	880.0	—	—	—	—	—	—	—	—	50.4	—	—
35	100	2.70	—	240.0	730.0	—	—	—	—	—	—	—	—	4.8	—	—
135	460	0.47	—	840.0	330.0	—	—	—	—	—	—	—	—	0	—	—
40	100	3.60	—	250.0	600.0	—	0	—	—	—	—	—	—	0	—	—
0	10	0.18	—	190.0	230.0	—	0	—	—	—	—	—	—	0	—	—
50	150	1.80	—	450.0	1100.0	—	—	—	—	—	—	—	—	9.0	—	—
45	164	4.92	—	380.0	940.0	—	—	—	—	—	—	—	—	9.8	—	—
70	234	4.20	—	410.0	1310.0	—	—	—	—	—	—	—	—	8.4	—	—
0	20	1.42	—	1240.0	870.0	—	0	—	—	—	—	—	—	8.9	—	—
0	19	1.01	—	840.0	590.0	—	0	—	—	—	—	—	—	5.6	—	—
0	40	2.70	—	140.0	420.0	—	40.0	—	—	—	—	—	—	18.0	—	—
280	88	2.36	—	310.0	1430.0	—	—	—	—	—	—	—	—	0	—	—
0	40	1.80	—	580.0	890.0	—	—	—	—	—	—	—	—	0	—	—
75	200	4.50	—	430.0	1230.0	—	—	—	—	—	—	—	—	9.0	—	—
55	274	4.10	—	670.0	860.0	—	—	—	—	—	—	—	—	43.8	—	—
135	466	0.00	—	750.0	290.0	—	—	—	—	—	—	—	—	0	—	—
125	308	7.39	—	480.0	1580.0	—	—	—	—	—	—	—	—	0.6	—	—
135	532	0.00	—	750.0	280.0	—	—	—	—	—	—	—	—	0	—	—
5	200	1.80	60.0	1000.0	30.0	0.30	—	0.37	—	0.34	2.00	0.60	60.0	102.0	0	1.4
0	200	1.44	40.0	1010.0	115.0	0.60	—	0.09	—	0.25	0.80	0.70	24.0	15.0	0.2	1.4
5	200	1.80	16.0	510.0	35.0	0.30	—	0.06	—	0.25	6.00	0.70	140.0	54.0	0	1.4
5	100	1.80	24.0	810.0	60.0	0.30	—	0.09	—	0.25	5.00	0.50	100.0	78.0	0	1.4
0	150	2.70	80.0	1030.0	250.0	0.90	—	0.52	—	0.25	5.00	0.70	80.0	18.0	0	5.6
30	500	1.08	60.0	810.0	380.0	1.50	—	0.22	—	0.76	0.40	0.16	16.0	6.0	1.5	4.2
5	100	1.80	80.0	1170.0	35.0	0.30	—	0.37	—	0.34	1.90	0.60	100.0	180.0	0	1.4
5	100	1.08	60.0	1260.0	35.0	16.50	—	0.37	—	0.34	3.00	0.40	121.5	1302.0	0	1.4
0	250	1.44	16.0	500.0	50.0	0.30	—	0.03	—	0.25	5.00	0.60	120.0	54.0	0	1.4
0	1200	1.80	480.0	1070.0	45.0	16.50	—	5.55	—	6.12	68.00	7.40	640.0	288.0	10.8	77.0
0	20	0.00	8.0	200.0	10.0	0.00	—	0.03	—	0.17	14.00	1.80	320.0	36.0	0	0
5	100	1.08	24.0	780.0	50.0	0.30	—	0.15	—	0.25	5.00	0.70	120.0	72.0	0	1.4
0	60	1.08	60.0	990.0	0	0.30	—	0.45	—	0.13	2.00	0.20	160.0	246.0	0	0
0	100	1.80	60.0	1010.0	125.0	0.60	—	0.45	—	0.25	3.00	0.50	120.0	132.0	0	2.8
0	250	1.80	80.0	1380.0	160.0	0.90	—	0.45	—	0.42	2.00	0.50	140.0	240.0	0.6	1.4
5	100	0.72	32.0	740.0	60.0	0.30	—	0.06	—	0.25	4.00	0.60	80.0	18.0	0	1.4
0	1100	2.62	60.0	650.0	240.0	0.58	—	0.08	—	0.17	1.20	0.70	58.3	60.0	0	5.6
5	150	1.80	32.0	810.0	70.0	0.30	—	0.09	—	0.34	6.00	1.00	160.0	60.0	0	1.4
5	250	1.80	40.0	1050.0	180.0	0.90	—	0.12	—	0.34	0.80	0.40	40.0	60.0	0.6	1.4
5	100	1.08	24.0	480.0	10.0	0.30	—	0.06	—	0.34	14.00	1.80	320.0	90.0	0	1.4
0	20	1.08	40.0	720.0	0	0.30	—	0.30	—	0.10	1.60	0.40	80.0	678.0	0	0
0	0	1.80	8.0	80.0	0	0.00	0	0.03	—	0.03	0.40	0.04	16.0	3.6	0	2.8
0	100	2.70	—	—	730.0	—	—	—	—	—	—	—	—	1.2	—	—
0	40	1.80	—	—	640.0	—	—	—	—	—	—	—	—	0	—	—
65	40	1.80	—	—	2670.0	—	—	—	—	—	—	—	—	1.2	—	—
65	40	1.80	—	—	2210.0	—	—	—	—	—	—	—	—	1.2	—	—
65	60	1.80	—	—	2000.0	—	—	—	—	—	—	—	—	1.2	—	—
5	40	0.72	—	—	270.0	—	—	—	—	—	—	—	—	12.0	—	—
65	40	1.44	—	—	1220.0	—	0	—	—	—	—	—	—	1.2	—	—
0	60	1.08	—	—	10.0	—	—	—	—	—	—	—	—	6.0	—	—
55	100	2.70	—	—	1220.0	—	—	—	—	—	—	—	—	6.0	—	—
85	20	2.70	—	—	1020.0	—	—	—	—	—	—	—	—	1.2	—	—
55	0	1.44	—	—	300.0	—	0	—	—	—	—	—	—	0	—	—
95	20	2.70	—	—	700.0	—	—	—	—	—	—	—	—	—	—	—
45	20	1.08	—	—	340.0	—	—	—	—	—	—	—	—	0	—	—
35	200	5.40	—	—	2330.0	—	—	—	—	—	—	—	—	6.0	—	—

DA+ Code	Food Description	Quantity	Measure	Wt (g)	H₂O (g)	Ener (kcal)	Prot (g)	Carb (g)	Fiber (g)	Fat (g)	Fat Breakdown (g)			
											Sat	Mono	Poly	Trans
51219	Famous Bowls rice with gravy	1	serving(s)	384	—	610	25.0	67.0	5.0	27.0	8.0	—	—	2.5
31841	Honey BBQ chicken sandwich	1	item(s)	147	—	290	23.0	40.0	2.0	4.0	1.0	—	—	0
31833	Honey BBQ wing pieces	6	item(s)	157	—	460	27.0	26.0	3.0	27.0	6.0	—	—	2.0
10859	Hot wings pieces	6	piece(s)	134	—	450	26.0	19.0	2.0	30.0	7.0	—	—	2.0
42382	KFC Snacker sandwich	1	serving(s)	119	—	320	14.0	29.0	2.0	17.0	3.0	—	—	1.0
31848	Macaroni & cheese	1	serving(s)	136	—	180	8.0	18.0	0	8.0	3.5	—	—	1.0
31847	Mashed potatoes with gravy	1	serving(s)	151	—	140	2.0	20.0	1.0	5.0	1.0	—	—	0.5
10825	Original Recipe chicken, breast	1	item(s)	161	—	340	38.0	9.0	2.0	17.0	4.0	—	—	1.0
10826	Original Recipe chicken, drumstick	1	item(s)	59	—	140	13.0	3.0	0	8.0	2.0	—	—	0.5
10827	Original Recipe chicken, thigh	1	item(s)	126	—	350	19.0	7.0	1.0	27.0	7.0	—	—	1.0
10828	Original Recipe chicken, whole wing	1	item(s)	47	—	140	10.0	4.0	0	9.0	2.0	—	—	0.5
51222	Oven roasted Twister chicken wrap	1	item(s)	269	—	520	30.0	46.0	4.0	23.0	3.5	—	—	0
31844	Popcorn chicken, small or individual	1	serving(s)	114	—	370	19.0	21.0	2.0	24.0	4.5	—	—	2.5
31852	Potato salad	1	serving(s)	128	—	180	2.0	22.0	2.0	9.0	1.5	—	—	0
10845	Potato wedges, small	1	serving(s)	102	—	250	4.0	32.0	3.0	12.0	2.0	—	—	1.5
31839	Tender Roast chicken sandwich with sauce	1	item(s)	236	—	430	37.0	29.0	2.0	18.0	3.5	—	—	0
	Long John Silver													
39392	Baked cod	1	serving(s)	101	—	120	22.0	1.0	0	4.5	1.0	—	—	0
3777	Batter dipped fish sandwich	1	item(s)	177	—	470	18.0	48.0	3.0	23.0	5.0	—	—	4.5
37568	Battered fish	1	item(s)	92	—	260	12.0	17.0	0.5	16.0	4.0	—	—	4.5
37569	Breaded clams	1	serving(s)	85	—	240	8.0	22.0	1.0	13.0	2.0	—	—	2.5
37566	Chicken plank	1	item(s)	52	—	140	8.0	9.0	0.5	8.0	2.0	—	—	2.5
39404	Clam chowder	1	item(s)	227	—	220	9.0	23.0	0	10.0	4.0	—	—	1.0
39398	Cocktail sauce	1	ounce(s)	28	—	25	0	6.0	0	0	0	0	0	0
3770	Coleslaw	1	serving(s)	113	—	200	1.0	15.0	3.0	15.0	2.5	1.8	4.1	0
39400	French fries, large	1	item(s)	142	—	390	4.0	56.0	5.0	17.0	4.0	—	—	5.0
3774	Fries, regular	1	serving(s)	85	—	230	3.0	34.0	3.0	10.0	2.5	—	—	3.0
3779	Hushpuppy	1	piece(s)	23	—	60	1.0	9.0	1.0	2.5	0.5	—	—	1.0
3781	Shrimp, batter-dipped, 1 piece	1	piece(s)	14	—	45	2.0	3.0	0	3.0	1.0	—	—	1.0
39399	Tartar sauce	1	ounce(s)	28	—	100	0	4.0	0	9.0	1.5	—	—	—
39395	Ultimate Fish sandwich	1	item(s)	199	—	530	21.0	49.0	3.0	28.0	8.0	—	—	5.0
	McDonald's													
50828	Asian salad with grilled chicken	1	item(s)	362	—	290	31.0	23.0	6.0	10.0	1.0	—	—	0
2247	Barbecue sauce	1	item(s)	28	—	45	0	11.0	0	0	0	0	0	0
737	Big Mac hamburger	1	item(s)	219	—	560	25.0	47.0	3.0	30.0	10.0	—	—	1.5
29777	Caesar salad dressing	1	package(s)	44	—	150	1.0	5.0	0	13.0	2.5	—	—	—
38391	Caesar salad with grilled chicken, no dressing	1	serving(s)	278	230.6	181	26.4	10.5	3.1	6.0	2.9	1.7	0.8	0.2
38393	Caesar salad without chicken, no dressing	1	serving(s)	190	170.4	84	6.0	8.1	3.0	3.9	2.2	0.9	0.3	0.1
738	Cheeseburger	1	item(s)	119	—	310	15.0	35.0	1.0	12.0	6.0	—	—	1.0
29775	Chicken McGrill sandwich	1	item(s)	213	—	400	27.0	38.0	3.0	16.0	3.0	—	—	0
1873	Chicken McNuggets, 6 piece	6	item(s)	96	—	250	15.0	15.0	0	15.0	3.0	—	—	1.5
3792	Chicken McNuggets, 4 piece	4	item(s)	64	—	170	10.0	10.0	0	10.0	2.0	—	—	1.0
29774	Crispy chicken sandwich	1	item(s)	232	121.8	500	27.0	63.0	3.0	16.0	3.0	5.7	7.4	1.5
743	Egg McMuffin	1	item(s)	139	76.8	300	17.0	30.0	2.0	12.0	4.5	3.8	2.5	—
742	Filet-O-Fish sandwich	1	item(s)	141	—	400	14.0	42.0	1.0	18.0	4.0	—	—	1.0
2257	French fries, large	1	serving(s)	170	—	570	6.0	70.0	7.0	30.0	6.0	—	—	8.0
1872	French fries, small	1	serving(s)	74	—	250	2.0	30.0	3.0	13.0	2.5	—	—	3.5
33822	Fruit 'n Yogurt Parfait	1	item(s)	149	111.2	160	4.0	31.0	1.0	2.0	1.0	0.2	0.1	0
739	Hamburger	1	item(s)	105	—	260	13.0	33.0	1.0	9.0	3.5	—	—	0.5
2003	Hash browns	1	item(s)	53	—	140	1.0	15.0	2.0	8.0	1.5	—	—	2.0
2249	Honey sauce	1	item(s)	14	—	50	0	12.0	0	0	0	0	0	0
38397	Newman's Own creamy caesar salad dressing	1	item(s)	59	32.5	190	2.0	4.0	0	18.0	3.5	4.6	9.6	0
38398	Newman's Own low fat balsamic vinaigrette salad dressing	1	item(s)	44	29.1	40	0	4.0	0	3.0	0	1.0	1.2	0
38399	Newman's Own ranch salad dressing	1	item(s)	59	30.1	170	1.0	9.0	0	15.0	2.5	9.0	3.7	0
1874	Plain Hotcakes with syrup and margarine	3	item(s)	221	—	600	9.0	102.0	2.0	17.0	4.0	—	—	4.0
740	Quarter Pounder hamburger	1	item(s)	171	—	420	24.0	40.0	3.0	18.0	7.0	—	—	1.0
741	Quarter Pounder hamburger with cheese	1	item(s)	199	—	510	29.0	43.0	3.0	25.0	12.0	—	—	1.5
2005	Sausage McMuffin with egg	1	item(s)	165	82.4	450	20.0	31.0	2.0	27.0	10.0	10.9	4.6	0.5
50831	Side salad	1	item(s)	87	—	20	1.0	4.0	1.0	0	0	0	0	0
	Pizza Hut													
39009	Hot chicken wings	2	item(s)	57	—	110	11.0	1.0	0	6.0	2.0	—	—	0.3
14025	Meat Lovers hand tossed pizza	1	slice(s)	118	—	300	15.0	29.0	2.0	13.0	6.0	—	—	0.5
14026	Meat Lovers pan pizza	1	slice(s)	123	—	340	15.0	29.0	2.0	19.0	7.0	—	—	0.5
31009	Meat Lovers stuffed crust pizza	1	slice(s)	169	—	450	21.0	43.0	3.0	21.0	10.0	—	—	1.0
14024	Meat Lovers thin 'n crispy pizza	1	slice(s)	98	—	270	13.0	21.0	2.0	14.0	6.0	—	—	0.5
	FAST FOOD —Continued													
14031	Pepperoni Lovers hand tossed pizza	1	slice(s)	113	—	300	15.0	30.0	2.0	13.0	7.0	—	—	0.5
14032	Pepperoni Lovers pan pizza	1	slice(s)	118	—	340	15.0	29.0	2.0	19.0	7.0	—	—	0.5
31011	Pepperoni Lovers stuffed crust pizza	1	slice(s)	163	—	420	21.0	43.0	3.0	19.0	10.0	—	—	1.0

Chol (mg)	Calc (mg)	Iron (mg)	Magn (mg)	Pota (mg)	Sodi (mg)	Zinc (mg)	Vit A (µg)	Thia (mg)	Vit E (mg α)	Ribo (mg)	Niac (mg)	Vit B$_6$ (mg)	Fola (µg)	Vit C (mg)	Vit B$_{12}$ (µg)	Sele (µg)
35	200	4.50	—	—	2130.0	—	—	—	—	—	—	—	—	6.0	—	—
60	80	2.70	—	—	710.0	—	—	—	—	—	—	—	—	2.4	—	—
140	40	1.80	—	—	970.0	—	—	—	—	—	—	—	—	21.0	—	—
115	40	1.44	—	—	990.0	—	—	—	—	—	—	—	—	1.2	—	—
25	60	2.70	—	—	690.0	—	—	—	—	—	—	—	—	2.4	—	—
15	150	0.72	—	—	800.0	—	—	—	—	—	—	—	—	1.2	—	—
0	40	1.44	—	—	560.0	—	—	—	—	—	—	—	—	1.2	—	—
135	20	2.70	—	—	960.0	—	—	—	—	—	—	—	—	6.0	—	—
70	20	1.08	—	—	340.0	—	—	—	—	—	—	—	—	0	—	—
110	20	2.70	—	—	870.0	—	—	—	—	—	—	—	—	1.2	—	—
50	20	1.44	—	—	350.0	—	0	—	—	—	—	—	—	1.2	—	—
60	40	6.30	—	—	1380.0	—	—	—	—	—	—	—	—	15.0	—	—
25	40	1.80	—	—	1110.0	—	0	—	—	—	—	—	—	0	—	—
5	0	0.36	—	—	470.0	—	—	—	—	—	—	—	—	6.0	—	—
0	20	1.08	—	—	700.0	—	0	—	—	—	—	—	—	0	—	—
80	80	2.70	—	—	1180.0	—	—	—	—	—	—	—	—	9.0	—	—
90	20	0.72	—	—	240.0	—	—	—	—	—	—	—	—	0	—	—
45	60	2.70	—	—	1210.0	—	—	—	—	—	—	—	—	2.4	—	—
35	20	0.72	—	—	790.0	—	—	—	—	—	—	—	—	4.8	—	—
10	20	1.08	—	—	1110.0	—	0	—	—	—	—	—	—	0	—	—
20	0	0.72	—	—	480.0	—	0	—	—	—	—	—	—	2.4	—	—
25	150	0.72	—	—	810.0	—	—	—	—	—	—	—	—	0	—	—
0	0	0.00	—	—	250.0	—	—	—	—	—	—	—	—	0	—	—
20	40	0.36	—	222.7	340.0	0.70	—	0.07	—	0.08	2.34	—	—	18.0	—	—
0	0	0.00	—	—	580.0	—	0	—	—	—	—	—	—	24.0	—	—
0	0	0.00	—	370.0	350.0	0.30	0	0.09	—	0.01	1.60	—	—	15.0	—	—
0	20	0.36	—	—	200.0	—	0	—	—	—	—	—	—	0	—	—
15	0	0.00	—	—	160.0	—	0	—	—	—	—	—	—	1.2	—	—
15	0	0.00	—	—	250.0	—	0	—	—	—	—	—	—	0	—	—
60	150	2.70	—	—	1400.0	—	—	—	—	—	—	—	—	4.8	—	—
65	150	3.60	—	—	890.0	—	—	—	—	—	—	—	—	54.0	—	—
0	0	0.00	—	55.0	260.0	—	—	—	—	—	—	—	—	0	—	—
80	250	4.50	—	400.0	1010.0	—	—	—	—	—	—	—	—	1.2	—	—
10	40	0.18	—	30.0	400.0	—	—	—	—	—	—	—	—	0.6	—	—
67	178	1.77	—	708.9	767.3	—	—	0.15	—	0.19	10.62	—	127.9	29.2	0.2	—
10	163	1.15	17.1	410.4	157.7	—	—	0.08	—	0.07	0.40	—	102.6	26.8	0	0.4
40	200	2.70	—	240.0	740.0	—	60.0	—	—	—	—	—	—	1.2	—	—
70	150	2.70	—	510.0	1010.0	—	—	—	—	—	—	—	—	6.0	—	—
35	20	0.72	—	240.0	670.0	—	—	—	—	—	—	—	—	1.2	—	—
75	0	0.36	—	160.0	450.0	—	—	—	—	—	—	—	—	1.2	—	—
60	80	3.60	62.6	526.6	1380.0	1.53	41.8	0.46	2.77	0.39	12.85	0.20	104.4	6.0	0.4	—
230	300	2.70	26.4	218.2	860.0	1.59	—	0.36	0.82	0.51	4.31	0.20	109.8	1.2	0.9	—
40	150	1.80	—	250.0	640.0	—	36.2	—	—	—	—	—	—	0	—	—
0	20	1.80	—	—	330.0	—	0	—	—	—	—	—	—	9.0	—	—
0	20	0.72	—	—	140.0	—	0	—	—	—	—	—	—	3.6	—	—
5	150	0.67	20.9	248.8	85.0	0.53	0	0.06	—	0.17	0.35	—	19.4	9.0	0.3	—
30	150	2.70	—	210.0	530.0	—	5.0	—	—	—	—	—	—	1.2	—	—
0	0	0.36	—	210.0	290.0	—	0	—	—	—	—	—	—	1.2	—	—
0	0	0.00	—	0	0	—	0	—	—	—	—	—	—	0	—	—
20	61	0.00	3.0	16.0	500.0	0.20	—	0.01	15.43	0.02	0.01	0.64	2.4	0	0.1	0.1
0	4	0.00	1.3	8.8	730.0	0.01	—	0.00	0.00	0.00	0.00	0.00	0	2.4	0	0
0	40	0.00	1.8	70.4	530.0	0.03	0	0.01	—	0.08	0.01	0.02	0.6	0	0	0.2
20	150	2.70	—	280.0	620.0	—	—	—	—	—	—	—	—	0	—	—
70	150	4.50	—	390.0	730.0	—	10.0	—	—	—	—	—	—	1.2	—	—
95	300	4.50	—	440.0	1150.0	—	100.0	—	—	—	—	—	—	1.2	—	—
255	300	3.60	29.7	282.2	950.0	2.01	—	0.43	0.82	0.56	4.83	0.24	—	0	1.2	—
0	20	0.72	—	—	10.0	—	—	—	—	—	—	—	—	15.0	—	—
70	0	0.36	—	—	450.0	—	—	—	—	—	—	—	—	0	—	—
35	150	1.80	—	—	760.0	—	—	—	—	—	—	—	—	6.0	—	—
35	150	2.70	—	—	750.0	—	—	—	—	—	—	—	—	6.0	—	—
55	250	2.70	—	—	1250.0	—	—	—	—	—	—	—	—	9.0	—	—
35	150	1.44	—	—	740.0	—	—	—	—	—	—	—	—	6.0	—	—
40	200	1.80	—	—	710.0	—	57.7	—	—	—	—	—	—	2.4	—	—
40	200	2.70	—	—	700.0	—	57.7	—	—	—	—	—	—	2.4	—	—
55	300	2.70	—	—	1120.0	—	—	—	—	—	—	—	—	3.6	—	—

Food Composition   (Computer code is for Cengage Diet Analysis program)   (For purposes of calculations, use "0" for t, <1, <.1, <.01, etc.)

DA+ Code	Food Description	Quantity	Measure	Wt (g)	H₂O (g)	Ener (kcal)	Prot (g)	Carb (g)	Fiber (g)	Fat (g)	Fat Breakdown (g)			
											Sat	Mono	Poly	Trans
14030	Pepperoni Lovers thin 'n crispy pizza	1	slice(s)	92	—	260	13.0	21.0	2.0	14.0	7.0	—	—	0.5
10834	Personal Pan pepperoni pizza	1	slice(s)	61	—	170	7.0	18.0	0.5	8.0	3.0	—	—	1.0
10842	Personal Pan supreme pizza	1	slice(s)	77	—	190	8.0	19.0	1.0	9.0	3.5	—	—	1.0
39013	Personal Pan Veggie Lovers pizza	1	slice(s)	69	—	150	6.0	19.0	1.0	6.0	2.0	—	—	0.5
14028	Veggie Lovers hand tossed pizza	1	slice(s)	118	—	220	10.0	31.0	2.0	6.0	3.0	—	—	0.3
14029	Veggie Lovers pan pizza	1	slice(s)	119	—	260	10.0	30.0	2.0	12.0	4.0	—	—	0.3
31010	Veggie Lovers stuffed crust pizza	1	slice(s)	172	—	360	16.0	45.0	3.0	14.0	7.0	—	—	0.5
14027	Veggie Lovers thin 'n crispy pizza	1	slice(s)	101	—	180	8.0	23.0	2.0	7.0	3.0	—	—	0.5
39012	Wing blue cheese dipping sauce	1	item(s)	43	—	230	2.0	2.0	0	24.0	5.0	—	—	1.0
39011	Wing ranch dipping sauce	1	item(s)	43	—	210	0.5	4.0	0	22.0	3.5	—	—	0.5
	Starbucks													
38052	Cappuccino, tall	12	fluid ounce(s)	360	—	120	7.0	10.0	0	6.0	4.0	—	—	—
38053	Cappuccino, tall nonfat	12	fluid ounce(s)	360	—	80	7.0	11.0	0	0	0	0	0	0
38054	Cappuccino, tall soymilk	12	fluid ounce(s)	360	—	100	5.0	13.0	0.5	2.5	0	—	—	—
38059	Cinnamon spice mocha, tall nonfat w/o whipped cream	12	fluid ounce(s)	360	—	170	11.0	32.0	0	0.5	—	—	—	—
38057	Cinnamon spice mocha, tall w/ whipped cream	12	fluid ounce(s)	360	—	320	10.0	31.0	0	17.0	11.0	—	—	—
38051	Espresso, single shot	1	fluid ounce(s)	30	—	5	0	1.0	0	0	0	0	0	0
38088	Flavored syrup, 1 pump	1	serving(s)	10	—	20	0	5.0	0	0	0	0	0	0
32562	Frappuccino bottled coffee drink, mocha	9 ½	fluid ounce(s)	298	—	190	6.0	39.0	3.0	3.0	2.0	—	—	—
32561	Frappuccino coffee drink, all bottled flavors	9 ½	fluid ounce(s)	281	—	190	7.0	35.0	0	3.5	2.5	—	—	—
38073	Frappuccino, mocha	12	fluid ounce(s)	360	—	220	5.0	44.0	0	3.0	1.5	—	—	—
38067	Frappuccino, tall caramel w/o whipped cream	12	fluid ounce(s)	360	—	210	4.0	43.0	0	2.5	1.5	—	—	—
38070	Frappuccino, tall coffee	12	fluid ounce(s)	360	—	190	4.0	38.0	0	2.5	1.5	—	—	—
39894	Frappuccino, tall coffee, light blend	12	fluid ounce(s)	360	—	110	5.0	22.0	2.0	1.0	0	—	—	—
38071	Frappuccino, tall espresso	12	fluid ounce(s)	360	—	160	4.0	33.0	0	2.0	1.5	—	—	—
39897	Frappuccino, tall mocha, light blend	12	fluid ounce(s)	360	—	140	5.0	28.0	3.0	1.5	0	—	—	—
39887	Frappuccino, tall Strawberries & Creme, w/o whipped cream	12	fluid ounce(s)	360	—	330	10.0	65.0	0	3.5	1.0	—	—	—
38063	Frappuccino, tall Tazo chai creme w/o whipped cream	12	fluid ounce(s)	360	—	280	10.0	52.0	0	3.5	1.0	—	—	—
38066	Frappuccino, tall Tazoberry	12	fluid ounce(s)	360	—	140	0.5	36.0	0.5	0	0	0	0	0
38065	Frappuccino, tall Tazoberry Crème	12	fluid ounce(s)	360	—	240	4.0	54.0	0.5	1.0	0	—	—	—
38080	Frappuccino, tall vanilla w/o whipped cream	12	fluid ounce(s)	360	—	270	10.0	51.0	0	3.5	1.0	—	—	—
39898	Frappuccino, tall white chocolate mocha, light blend	12	fluid ounce(s)	360	—	160	6.0	32.0	2.0	2.0	1.0	—	—	—
38074	Frappuccino, tall white chocolate w/o whipped cream	12	fluid ounce(s)	360	—	240	5.0	48.0	0	3.5	2.5	—	—	—
39883	Java Chip Frappuccino, tall w/o whipped cream	12	fluid ounce(s)	360	—	270	5.0	51.0	1.0	7.0	4.5	—	—	—
33111	Latte, tall w/ nonfat milk	12	fluid ounce(s)	360	335.3	120	12.0	18.0	0	0	0	0	0	0
33112	Latte, tall w/ whole milk	12	fluid ounce(s)	360	—	200	11.0	16.0	0	11.0	7.0	—	—	—
33109	Macchiato, tall caramel w/ nonfat milk	12	fluid ounce(s)	360	—	170	11.0	30.0	0	1.0	0	—	—	—
33110	Macchiato, tall caramel w/ whole milk	12	fluid ounce(s)	360	—	240	10.0	28.0	0	10.0	6.0	—	—	—
33107	Mocha coffee drink, tall nonfat, w/o whip cream	12	fluid ounce(s)	360	—	170	11.0	33.0	1.0	1.5	0	—	—	—
38089	Mocha syrup	1	serving(s)	17	—	25	1.0	6.0	0	0.5	0	—	—	—
33108	Mocha, tall mocha w/ whole milk	12	fluid ounce(s)	360	—	310	10.0	32.0	1.0	17.0	10.0	—	—	—
38042	Steamed apple cider, tall	12	fluid ounce(s)	360	—	180	0	45.0	0	0	0	0	0	0
38087	Tazo chai black tea, soymilk, tall	12	fluid ounce(s)	360	—	190	4.0	39.0	0.5	2.0	0	—	—	—
38084	Tazo chai black tea, tall	12	fluid ounce(s)	360	—	210	6.0	36.0	0	5.0	3.5	—	—	—
38083	Tazo chai black tea, tall nonfat	12	fluid ounce(s)	360	—	170	6.0	37.0	0	0	0	0	0	0
38076	Tazo iced tea, tall	12	fluid ounce(s)	360	—	60	0	16.0	0	0	0	0	0	0
38077	Tazo tea, grande lemonade	16	fluid ounce(s)	480	—	120	0	31.0	0	0	0	0	0	0
38045	Vanilla creme steamed nonfat milk, tall w/whipped cream	12	fluid ounce(s)	360	—	260	11.0	33.0	0	8.0	5.0	—	—	—
38046	Vanilla crème steamed soymilk, tall w/whipped cream	12	fluid ounce(s)	360	—	300	8.0	37.0	1.0	12.0	6.0	—	—	—
38044	Vanilla creme steamed whole milk, tall w/whipped cream	12	fluid ounce(s)	360	—	330	10.0	31.0	0	18.0	11.0	—	—	—
38090	Whipped cream	1	serving(s)	27	—	100	0	2.0	0	9.0	6.0	—	—	—
38062	White chocolate mocha, tall nonfat w/o whipped cream	12	fluid ounce(s)	360	—	260	12.0	45.0	0	4.0	3.0	—	—	—
38061	White chocolate mocha, tall w/ whipped cream	12	fluid ounce(s)	360	—	410	11.0	44.0	0	20.0	13.0	—	—	—
38048	White hot chocolate, tall nonfat w/o whipped cream	12	fluid ounce(s)	360	—	300	15.0	51.0	0	4.5	3.5	—	—	—
38050	White hot chocolate, tall soymilk w/whipped cream	12	fluid ounce(s)	360	—	420	11.0	56.0	1.0	16.0	9.0	—	—	—
38047	White hot chocolate, tall w/whipped cream	12	fluid ounce(s)	360	—	460	13.0	50.0	0	22.0	15.0	—	—	—
	FAST FOOD —Continued													
	Subway													
15842	Cheese steak sandwich, 6", wheat bread	1	item(s)	250	—	360	24.0	47.0	5.0	10.0	4.5	—	—	0
40478	Chicken & bacon ranch sandwich, 6", white or wheat bread	1	serving(s)	297	—	540	36.0	47.0	5.0	25.0	10.0	—	—	0.5
38622	Chicken & bacon ranch wrap with cheese	1	item(s)	257	—	440	41.0	18.0	9.0	27.0	10.0	—	—	0.5
32045	Chocolate chip cookie	1	item(s)	45	—	210	2.0	30.0	1.0	10.0	6.0	—	—	0

Chol (mg)	Calc (mg)	Iron (mg)	Magn (mg)	Pota (mg)	Sodi (mg)	Zinc (mg)	Vit A (µg)	Thia (mg)	Vit E (mg α)	Ribo (mg)	Niac (mg)	Vit B₆ (mg)	Fola (µg)	Vit C (mg)	Vit B₁₂ (µg)	Sele (µg)
40	200	1.44	—	—	690.0	—	58.0	—	—	—	—	—	—	2.4	—	—
15	80	1.44	—	—	340.0	—	38.5	—	—	—	—	—	—	1.4	—	—
20	80	1.86	—	—	420.0	—	—	—	—	—	—	—	—	3.6	—	—
10	80	1.80	—	—	280.0	—	—	—	—	—	—	—	—	3.6	—	—
15	150	1.80	—	—	490.0	—	—	—	—	—	—	—	—	9.0	—	—
15	150	2.70	—	—	470.0	—	—	—	—	—	—	—	—	9.0	—	—
35	250	2.70	—	—	980.0	—	—	—	—	—	—	—	—	9.0	—	—
15	150	1.44	—	—	480.0	—	—	—	—	—	—	—	—	9.0	—	—
25	20	0.00	—	—	550.0	—	0	—	—	—	—	—	—	0	—	—
10	0	0.00	—	—	340.0	—	0	—	—	—	—	—	—	0	—	—
25	250	0.00	—	—	95.0	—	—	—	—	—	—	—	—	1.2	0	—
3	200	0.00	—	—	100.0	—	—	—	—	—	—	—	—	0	0	—
0	250	0.72	—	—	75.0	—	—	—	—	—	—	—	—	0	0	—
5	300	0.72	—	—	150.0	—	—	—	—	—	—	—	—	0	0	—
70	350	1.08	—	—	140.0	—	—	—	—	—	—	—	—	2.4	0	—
0	0	0.00	—	—	0	—	0	—	—	—	—	—	—	0	0	—
0	0	0.00	—	—	0	—	0	—	—	—	—	—	—	0	0	—
12	219	1.08	—	530.0	110.0	—	—	—	—	—	—	—	—	0	—	—
15	250	0.36	—	510.0	105.0	—	—	—	—	—	—	—	—	0	—	—
10	150	0.72	—	—	180.0	—	—	—	—	—	—	—	—	0	0	—
10	150	0.00	—	—	180.0	—	—	—	—	—	—	—	—	0	0	—
10	150	0.00	—	—	180.0	—	—	—	—	—	—	—	—	0	0	—
0	150	0.00	—	—	220.0	—	—	—	—	—	—	—	—	0	—	—
10	100	0.00	—	—	160.0	—	—	—	—	—	—	—	—	0	0	—
0	150	0.72	—	—	220.0	—	—	—	—	—	—	—	—	0	—	—
3	350	0.00	—	—	270.0	—	—	—	—	—	—	—	—	21.0	—	—
3	350	0.00	—	—	270.0	—	—	—	—	—	—	—	—	3.6	0	—
0	0	0.00	—	—	30.0	—	0	—	—	—	—	—	—	0	0	—
0	150	0.00	—	—	125.0	—	0	—	—	—	—	—	—	1.2	0	—
3	350	0.00	—	—	370.0	—	—	—	—	—	—	—	—	3.6	0	—
3	150	0.00	—	—	250.0	—	—	—	—	—	—	—	—	0	—	—
10	150	0.00	—	—	210.0	—	—	—	—	—	—	—	—	0	0	—
10	150	1.44	—	—	220.0	—	—	—	—	—	—	—	—	0	—	—
5	350	0.00	39.8	—	170.0	1.35	—	0.12	—	0.47	0.36	0.13	17.5	0	1.3	—
45	400	0.00	46.6	—	160.0	1.28	—	0.12	—	0.54	0.34	0.14	16.8	2.4	1.2	—
5	300	0.00	—	—	160.0	—	—	—	—	—	—	—	—	1.2	—	—
30	300	0.00	—	—	135.0	—	—	—	—	—	—	—	—	2.4	—	—
5	300	2.70	—	—	135.0	—	—	—	—	—	—	—	—	0	—	—
0	0	0.72	—	—	0	—	0	—	—	—	—	—	—	0	0	—
55	300	2.70	—	—	115.0	—	—	—	—	—	—	—	—	0	—	—
0	0	1.08	—	—	15.0	—	0	—	—	—	—	—	—	0	0	—
0	200	0.72	—	—	70.0	—	—	—	—	—	—	—	—	0	0	—
20	200	0.36	—	—	85.0	—	—	—	—	—	—	—	—	1.2	0	—
5	200	0.36	—	—	95.0	—	—	—	—	—	—	—	—	0	0	—
0	0	0.00	—	—	0	—	0	—	—	—	—	—	—	0	0	—
0	0	0.00	—	—	15.0	—	—	—	—	—	—	—	—	4.8	0	—
35	350	0.00	—	—	170.0	—	—	—	—	—	—	—	—	0	0	—
30	400	1.44	—	—	130.0	—	—	—	—	—	—	—	—	0	0	—
65	350	0.00	—	—	140.0	—	—	—	—	—	—	—	—	0	0	—
40	0	0.00	—	—	10.0	—	—	—	—	—	—	—	—	0	0	—
5	400	0.00	—	—	210.0	—	—	—	—	—	—	—	—	0	0	—
70	400	0.00	—	—	210.0	—	—	—	—	—	—	—	—	2.4	0	—
10	450	0.00	—	—	250.0	—	—	—	—	—	—	—	—	0	0	—
35	500	1.44	—	—	210.0	—	—	—	—	—	—	—	—	0	0	—
75	500	0.00	—	—	250.0	—	—	—	—	—	—	—	—	3.6	0	—
35	150	8.10	—	—	1090.0	—	—	—	—	—	—	—	—	18.0	—	—
90	250	4.50	—	—	1400.0	—	—	—	—	—	—	—	—	21.0	—	—
90	300	2.70	—	—	1680.0	—	—	—	—	—	—	—	—	9.0	—	—
15	0	1.08	—	—	150.0	—	—	—	—	—	—	—	—	0		

DA+ Code	Food Description	Quantity	Measure	Wt (g)	H₂O (g)	Ener (kcal)	Prot (g)	Carb (g)	Fiber (g)	Fat (g)	Fat Breakdown (g)			
											Sat	Mono	Poly	Trans
32048	Chocolate chip M&M cookie	1	item(s)	45	—	210	2.0	32.0	0.5	10.0	5.0	—	—	0
32049	Chocolate chunk cookie	1	item(s)	45	—	220	2.0	30.0	0.5	10.0	5.0	—	—	0
4024	Classic Italian B.M.T. sandwich, 6", white bread	1	item(s)	236	—	440	22.0	45.0	2.0	21.0	8.5	—	—	0
15838	Classic tuna sandwich, 6", wheat bread	1	item(s)	250	—	530	22.0	45.0	4.0	31.0	7.0	—	—	0.5
15837	Classic tuna sandwich, 6", white bread	1	item(s)	243	—	520	21.0	43.0	2.0	31.0	7.5	—	—	0.5
16397	Club salad, no dressing and croutons	1	item(s)	412	—	160	18.0	15.0	4.0	4.0	1.5	—	—	0
3422	Club sandwich, 6", white bread	1	item(s)	250	—	310	23.0	45.0	2.0	6.0	2.5	—	—	0
4030	Cold cut combo sandwich, 6", white bread	1	item(s)	242	—	400	20.0	45.0	2.0	17.0	7.5	—	—	0.5
34030	Ham & egg breakfast sandwich	1	item(s)	142	—	310	16.0	35.0	3.0	13.0	3.5	—	—	0
3885	Ham sandwich, 6", white bread	1	item(s)	238	—	310	17.0	52.0	2.0	5.0	2.0	—	—	0
3888	Meatball marinara sandwich, 6", wheat bread	1	item(s)	377	—	560	24.0	63.0	7.0	24.0	11.0	—	—	1.0
4651	Meatball sandwich, 6", white bread	1	item(s)	370	—	550	23.0	61.0	5.0	24.0	11.5	—	—	1.0
15839	Melt sandwich, 6", white bread	1	item(s)	260	—	410	25.0	47.0	4.0	15.0	5.0	—	—	—
32046	Oatmeal raisin cookie	1	item(s)	45	—	200	3.0	30.0	1.0	8.0	4.0	—	—	0
16379	Oven-roasted chicken breast sandwich, 6", wheat bread	1	item(s)	238	—	330	24.0	48.0	5.0	5.0	1.5	—	—	0
32047	Peanut butter cookie	1	item(s)	45	—	220	4.0	26.0	1.0	12.0	5.0	—	—	0
4655	Roast beef sandwich, 6", wheat bread	1	item(s)	224	—	290	19.0	45.0	4.0	5.0	2.0	—	—	0
3957	Roast beef sandwich, 6", white bread	1	item(s)	217	—	280	18.0	43.0	2.0	5.0	2.5	—	—	0
16378	Roasted chicken breast, 6", white bread	1	item(s)	231	—	320	23.0	46.0	3.0	5.0	2.0	—	—	0
34028	Southwest steak & cheese sandwich, 6", Italian bread	1	item(s)	271	—	450	24.0	48.0	6.0	20.0	6.0	—	—	0
4032	Spicy Italian sandwich, 6", white bread	1	item(s)	220	—	470	20.0	43.0	2.0	25.0	9.5	—	—	0
4031	Steak & cheese sandwich, 6", white bread	1	item(s)	243	—	350	23.0	45.0	3.0	10.0	5.0	—	—	0
32050	Sugar cookie	1	item(s)	45	—	220	2.0	28.0	0.5	12.0	6.0	—	—	0
40477	Sweet onion chicken teriyaki sandwich, 6", white or wheat bread	1	serving(s)	281	—	370	26.0	59.0	4.0	5.0	1.5	—	—	0
38623	Turkey breast & bacon melt wrap with chipotle sauce	1	item(s)	228	—	380	31.0	20.0	9.0	24.0	7.0	—	—	0
15834	Turkey breast & ham sandwich, 6", white bread	1	item(s)	227	—	280	19.0	45.0	2.0	5.0	2.0	—	—	0
16376	Turkey breast sandwich, 6", white bread	1	item(s)	217	—	270	17.0	44.0	2.0	4.5	2.0	—	—	0
15841	Veggie Delite sandwich, 6", wheat bread	1	item(s)	167	—	230	9.0	44.0	4.0	3.0	1.0	—	—	0
16375	Veggie Delite, 6", white bread	1	item(s)	160	—	220	8.0	42.0	2.0	3.0	1.5	—	—	0
32051	White chip macadamia nut cookie	1	item(s)	45	—	220	2.0	29.0	0.5	11.0	5.0	—	—	0
	Taco Bell													
29906	7-Layer burrito	1	item(s)	283	—	490	17.0	65.0	9.0	18.0	7.0	—	—	1.0
744	Bean burrito	1	item(s)	198	—	340	13.0	54.0	8.0	9.0	3.5	—	—	0.5
749	Beef burrito supreme	1	item(s)	248	—	410	17.0	51.0	7.0	17.0	8.0	—	—	1.0
33417	Beef Chalupa Supreme	1	item(s)	153	—	380	14.0	30.0	3.0	23.0	7.0	—	—	0.5
34474	Beef Gordita Baja	1	item(s)	153	—	340	13.0	29.0	4.0	19.0	5.0	—	—	0
29910	Beef Gordita Supreme	1	item(s)	153	—	310	14.0	29.0	3.0	16.0	6.0	—	—	0.5
2014	Beef soft taco	1	item(s)	99	—	200	10.0	21.0	3.0	9.0	4.0	—	—	0
10860	Beef soft taco supreme	1	item(s)	135	—	250	11.0	23.0	3.0	13.0	6.0	—	—	0.5
34472	Chicken burrito supreme	1	item(s)	248	—	390	20.0	49.0	6.0	13.0	6.0	—	—	0.5
33418	Chicken Chalupa Supreme	1	item(s)	153	—	360	17.0	29.0	2.0	20.0	5.0	—	—	0
34475	Chicken Gordita Baja	1	item(s)	153	—	320	17.0	28.0	3.0	16.0	3.5	—	—	0
29909	Chicken quesadilla	1	item(s)	184	—	520	28.0	40.0	3.0	28.0	12.0	—	—	0.5
29907	Chili cheese burrito	1	item(s)	156	—	390	16.0	40.0	3.0	18.0	9.0	—	—	1.5
10794	Cinnamon twists	1	serving(s)	35	—	170	1.0	26.0	1.0	7.0	0	—	—	0
29911	Grilled chicken Gordita Supreme	1	item(s)	153	—	290	17.0	28.0	2.0	12.0	5.0	—	—	0
14463	Grilled chicken soft taco	1	item(s)	99	—	190	14.0	19.0	1.0	6.0	2.5	—	—	—
29912	Grilled Steak Gordita Supreme	1	item(s)	153	—	290	15.0	28.0	2.0	13.0	5.0	—	—	0
29904	Grilled steak soft taco	1	item(s)	128	—	270	12.0	20.0	2.0	16.0	4.5	—	—	0
29905	Grilled steak soft taco supreme	1	item(s)	135	—	235	13.0	21.0	1.0	11.0	6.0	—	—	—
2021	Mexican pizza	1	serving(s)	216	—	530	20.0	42.0	7.0	30.0	8.0	—	—	1.0
29894	Mexican rice	1	serving(s)	131	—	170	6.0	23.0	1.0	11.0	3.0	—	—	0
10772	Meximelt	1	serving(s)	128	—	280	15.0	22.0	3.0	14.0	7.0	—	—	0.5
2011	Nachos	1	serving(s)	99	—	330	4.0	32.0	2.0	21.0	3.5	—	—	2.0
2012	Nachos Bellgrande	1	serving(s)	308	—	770	19.0	77.0	12.0	44.0	9.0	—	—	3.0
2023	Pintos 'n cheese	1	serving(s)	128	—	150	9.0	19.0	7.0	6.0	3.0	—	—	0.5
34473	Steak burrito supreme	1	item(s)	248	—	380	18.0	49.0	6.0	14.0	7.0	—	—	0.5
33419	Steak Chalupa Supreme	1	item(s)	153	—	360	15.0	28.0	2.0	21.0	6.0	—	—	0
747	Taco	1	item(s)	78	—	170	8.0	13.0	3.0	10.0	3.5	—	—	0
2015	Taco salad with salsa, with shell	1	serving(s)	548	—	840	30.0	80.0	15.0	45.0	11.0	—	—	1.5
	FAST FOOD —Continued													
14459	Taco supreme	1	item(s)	113	—	210	9.0	15.0	3.0	13.0	6.0	—	—	0
748	Tostada	1	item(s)	170	—	230	11.0	27.0	7.0	10.0	3.5	—	—	0.5
	CONVENIENCE MEALS													
	Banquet													
29961	Barbeque chicken meal	1	item(s)	281	—	330	16.0	37.0	2.0	13.0	3.0	—	—	—
14788	Boneless white fried chicken meal	1	item(s)	286	—	310	10.0	21.0	4.0	20.0	5.0	—	—	—
29960	Fish sticks meal	1	item(s)	207	—	470	13.0	58.0	1.0	20.0	3.5	—	—	—

Chol (mg)	Calc (mg)	Iron (mg)	Magn (mg)	Pota (mg)	Sodi (mg)	Zinc (mg)	Vit A (µg)	Thia (mg)	Vit E (mg α)	Ribo (mg)	Niac (mg)	Vit B$_6$ (mg)	Fola (µg)	Vit C (mg)	Vit B$_{12}$ (µg)	Sele (µg)
10	20	1.00	—	—	100.0	—	—	—	—	—	—	—	—	0	—	—
10	0	1.00	—	—	100.0	—	—	—	—	—	—	—	—	0	—	—
55	150	2.70	—	—	1770.0	—	—	—	—	—	—	—	—	16.8	—	—
45	100	5.40	—	—	1030.0	—	—	—	—	—	—	—	—	21.0	—	—
45	100	3.60	—	—	1010.0	—	—	—	—	—	—	—	—	16.8	—	—
35	60	3.60	—	—	880.0	—	—	—	—	—	—	—	—	30.0	—	—
35	60	3.60	—	—	1290.0	—	—	—	—	—	—	—	—	13.8	—	—
60	150	3.60	—	—	1530.0	—	—	—	—	—	—	—	—	16.8	—	—
190	80	4.50	—	—	720.0	—	66.7	—	—	—	—	—	—	3.6	—	—
25	60	2.70	—	—	1375.0	—	—	—	—	—	—	—	—	13.8	—	—
45	200	7.20	—	—	1610.0	—	—	—	—	—	—	—	—	36.0	—	—
45	200	5.40	—	—	1590.0	—	—	—	—	—	—	—	—	31.8	—	—
45	150	5.40	—	—	1720.0	—	—	—	—	—	—	—	—	24.0	—	—
15	20	1.08	—	—	170.0	—	—	—	—	—	—	—	—	0	—	—
45	60	4.50	—	—	1020.0	—	—	—	—	—	—	—	—	18.0	—	—
15	20	0.72	—	—	200.0	—	—	—	—	—	—	—	—	0	—	—
20	60	6.30	—	—	920.0	—	—	—	—	—	—	—	—	18.0	—	—
20	60	4.50	—	—	900.0	—	—	—	—	—	—	—	—	13.8	—	—
45	60	2.70	—	—	1000.0	—	—	—	—	—	—	—	—	13.8	—	—
45	150	8.10	—	—	1310.0	—	—	—	—	—	—	—	—	21.0	—	—
55	60	2.70	—	—	1650.0	—	—	—	—	—	—	—	—	16.8	—	—
35	150	6.30	—	—	1070.0	—	—	—	—	—	—	—	—	13.8	—	—
15	0	0.72	—	—	140.0	—	—	—	—	—	—	—	—	0	—	—
50	80	4.50	—	—	1220.0	—	—	—	—	—	—	—	—	24.0	—	—
50	200	2.70	—	—	1780.0	—	—	—	—	—	—	—	—	6.0	—	—
25	60	2.70	—	—	1210.0	—	—	—	—	—	—	—	—	13.8	—	—
20	60	2.70	—	—	1000.0	—	—	—	—	—	—	—	—	13.8	—	—
0	60	4.50	—	—	520.0	—	—	—	—	—	—	—	—	18.0	—	—
0	60	2.70	—	—	500.0	—	—	—	—	—	—	—	—	13.8	—	—
15	20	0.72	—	—	160.0	—	—	—	—	—	—	—	—	0	—	—
25	250	5.40	—	—	1350.0	—	—	—	—	—	—	—	—	15.0	—	—
5	200	4.50	—	—	1190.0	—	5.9	—	—	—	—	—	—	4.8	—	—
40	200	4.50	—	—	1340.0	—	9.9	—	—	—	—	—	—	6.0	—	—
40	150	2.70	—	—	620.0	—	—	—	—	—	—	—	—	3.6	—	—
35	100	2.70	—	—	780.0	—	—	—	—	—	—	—	—	2.4	—	—
40	150	2.70	—	—	620.0	—	—	—	—	—	—	—	—	3.6	—	—
25	100	1.80	—	—	630.0	—	—	—	—	—	—	—	—	1.2	—	—
40	150	2.70	—	—	650.0	—	—	—	—	—	—	—	—	3.6	—	—
45	200	4.50	—	—	1360.0	—	—	—	—	—	—	—	—	9.0	—	—
45	100	2.70	—	—	650.0	—	—	—	—	—	—	—	—	4.8	—	—
40	100	1.80	—	—	800.0	—	—	—	—	—	—	—	—	3.6	—	—
75	450	3.60	—	—	1420.0	—	—	—	—	—	—	—	—	1.2	—	—
40	300	1.80	—	—	1080.0	—	—	—	—	—	—	—	—	0	—	—
0	0	0.37	—	—	200.0	—	0	—	—	—	—	—	—	0	—	—
45	150	1.80	—	—	650.0	—	—	—	—	—	—	—	—	4.8	—	—
30	100	1.08	—	—	550.0	—	14.6	—	—	—	—	—	—	1.2	—	—
40	100	2.70	—	—	530.0	—	—	—	—	—	—	—	—	3.6	—	—
35	100	2.70	—	—	660.0	—	—	—	—	—	—	—	—	3.6	—	—
35	120	1.44	—	—	565.0	—	29.2	—	—	—	—	—	—	3.6	—	—
40	350	3.60	—	—	1000.0	—	—	—	—	—	—	—	—	4.8	—	—
15	100	1.44	—	—	790.0	—	—	—	—	—	—	—	—	3.6	—	—
40	250	2.70	—	—	880.0	—	—	—	—	—	—	—	—	2.4	—	—
3	80	0.71	—	—	530.0	—	0	—	—	—	—	—	—	0	—	—
35	200	3.60	—	—	1280.0	—	—	—	—	—	—	—	—	4.8	—	—
15	150	1.44	—	—	670.0	—	—	—	—	—	—	—	—	3.6	—	—
35	200	4.50	—	—	1250.0	—	9.9	—	—	—	—	—	—	9.0	—	—
40	100	2.70	—	—	530.0	—	—	—	—	—	—	—	—	3.6	—	—
25	80	1.08	—	—	350.0	—	—	—	—	—	—	—	—	1.2	—	—
65	450	7.20	—	—	1780.0	—	—	—	—	—	—	—	—	12.0	—	—
40	100	1.08	—	—	370.0	—	—	—	—	—	—	—	—	3.6	—	—
15	200	1.80	—	—	730.0	—	—	—	—	—	—	—	—	4.8	—	—
50	40	1.08	—	—	1210.0	—	0	—	—	—	—	—	—	4.8	—	—
45	80	1.44	—	—	1200.0	—	—	—	—	—	—	—	—	18.0	—	—
55	20	1.44	—	—	710.0	—	—	—	—	—	—	—	—	0	—	—

DA+ Code	Food Description	Quantity	Measure	Wt (g)	H₂O (g)	Ener (kcal)	Prot (g)	Carb (g)	Fiber (g)	Fat (g)	Fat Breakdown (g)			
											Sat	Mono	Poly	Trans
29957	Lasagna with meat sauce meal	1	item(s)	312	—	320	15.0	46.0	7.0	9.0	4.0	—	—	—
14777	Macaroni and cheese meal	1	item(s)	340	—	420	15.0	57.0	5.0	14.0	8.0	—	—	—
1741	Meatloaf meal	1	item(s)	269	—	240	14.0	20.0	4.0	11.0	4.0	—	—	—
39418	Pepperoni pizza meal	1	item(s)	191	—	480	11.0	56.0	5.0	23.0	8.0	—	—	—
33759	Roasted white turkey meal	1	item(s)	255	—	230	14.0	30.0	5.0	6.0	2.0	—	—	—
1743	Salisbury steak meal	1	item(s)	269	196.9	380	12.0	28.0	3.0	24.0	12.0	—	—	—
	Budget Gourmet													
1914	Cheese manicotti with meat sauce entree	1	item(s)	284	194.0	420	18.0	38.0	4.0	22.0	11.0	6.0	1.3	—
1915	Chicken with fettucini entree	1	item(s)	284	—	380	20.0	33.0	3.0	19.0	10.0	—	—	—
3986	Light beef stroganoff entrée	1	item(s)	248	177.0	290	20.0	32.0	3.0	7.0	4.0	—	—	—
3996	Light sirloin of beef in herb sauce entrée	1	item(s)	269	214.0	260	19.0	30.0	5.0	7.0	4.0	2.3	0.3	—
3987	Light vegetable lasagna entrée	1	item(s)	298	227.0	290	15.0	36.0	4.8	9.0	1.8	0.9	0.6	—
	Healthy Choice													
9425	Cheese French bread pizza	1	item(s)	170	—	340	22.0	51.0	5.0	5.0	1.5	—	—	—
9306	Chicken enchilada suprema meal	1	item(s)	320	251.5	360	13.0	59.0	8.0	7.0	3.0	2.0	2.0	—
3821	Familiar Favorites lasagna bake with meat sauce entrée	1	item(s)	255	—	270	13.0	38.0	4.0	7.0	2.5	—	—	—
13744	Familiar Favorites sesame chicken with vegetables and rice entrée	1	item(s)	255	—	260	17.0	34.0	4.0	6.0	2.0	2.0	2.0	—
9316	Lemon pepper fish meal	1	item(s)	303	—	280	11.0	49.0	5.0	5.0	2.0	1.0	2.0	—
9322	Traditional salisbury steak meal	1	item(s)	354	250.3	360	23.0	45.0	5.0	9.0	3.5	4.0	1.0	—
9359	Traditional turkey breasts meal	1	item(s)	298	—	330	21.0	50.0	4.0	5.0	2.0	1.5	1.5	—
	Stouffers													
2313	Cheese French bread pizza	1	serving(s)	294	—	380	15.0	43.0	3.0	16.0	6.0	—	—	—
11138	Cheese manicotti with tomato sauce entrée	1	item(s)	255	—	360	18.0	41.0	2.0	14.0	6.0	—	—	—
2366	Chicken pot pie entrée	1	item(s)	284	—	740	23.0	56.0	4.0	47.0	18.0	12.4	10.5	—
11116	Homestyle baked chicken breast with mashed potatoes and gravy entrée	1	item(s)	252	—	270	21.0	21.0	2.0	11.0	3.5	—	—	—
11146	Homestyle beef pot roast and potatoes entrée	1	item(s)	252	—	260	16.0	24.0	3.0	11.0	4.0	—	—	—
11152	Homestyle roast turkey breast with stuffing and mashed potatoes entrée	1	item(s)	273	—	290	16.0	30.0	2.0	12.0	3.5	—	—	—
11043	Lean Cuisine Comfort Classics baked chicken and whipped potatoes and stuffing entrée	1	item(s)	245	—	240	15.0	34.0	3.0	4.5	1.0	2.0	1.0	0
11046	Lean Cuisine Comfort Classics honey mustard chicken with rice pilaf entrée	1	item(s)	227	—	250	17.0	37.0	1.0	4.0	1.0	1.0	1.0	0
9479	Lean Cuisine Deluxe French bread pizza	1	item(s)	174	—	310	16.0	44.0	3.0	9.0	3.5	0.5	0.5	0
360	Lean Cuisine One Dish Favorites chicken chow mein with rice	1	item(s)	255	—	190	13.0	29.0	2.0	2.5	0.5	1.0	0.5	0
11054	Lean Cuisine One Dish Favorites chicken enchilada Suiza with Mexican-style rice	1	serving(s)	255	—	270	10.0	47.0	3.0	4.5	2.0	1.5	1.0	0
9467	Lean Cuisine One Dish Favorites fettucini alfredo entree	1	item(s)	262	—	270	13.0	39.0	2.0	7.0	3.5	2.0	1.0	0
11055	Lean Cuisine One Dish Favorites lasagna with meat sauce entrée	1	item(s)	298	—	320	19.0	44.0	4.0	7.0	3.0	2.0	0.5	0
	Weight Watchers													
11164	Smart Ones chicken enchiladas suiza entrée	1	item(s)	255	—	340	12.0	38.0	3.0	10.0	4.5	—	—	—
39763	Smart Ones chicken oriental entrée	1	item(s)	255	—	230	15.0	34.0	3.0	4.5	1.0	—	—	—
11187	Smart Ones pepperoni pizza	1	item(s)	198	—	400	22.0	58.0	4.0	9.0	3.0	—	—	—
39765	Smart Ones spaghetti bolgnese entrée	1	item(s)	326	—	280	17.0	43.0	5.0	5.0	2.0	—	—	—
31512	Smart Ones spicy szechuan style vegetables & chicken	1	item(s)	255	—	220	11.0	34.0	4.0	5.0	1.0	—	—	—
	BABY FOODS													
787	Apple juice	4	fluid ounce(s)	127	111.6	60	0	14.8	0.1	0.1	0	0	0	—
778	Applesauce, strained	4	tablespoon(s)	64	56.7	26	0.1	6.9	1.1	0.1	0	0	0	—
779	Bananas with tapioca, strained	4	tablespoon(s)	60	50.4	34	0.2	9.2	1.0	0	0	0	0	—
604	Carrots, strained	4	tablespoon(s)	56	51.7	15	0.4	3.4	1.0	0.1	0	0	0	—
770	Chicken noodle dinner, strained	4	tablespoon(s)	64	54.8	42	1.7	5.8	1.3	1.3	0.4	0.5	0.3	—
801	Green beans, strained	4	tablespoon(s)	60	55.1	16	0.7	3.8	1.3	0.1	0	0	0	—
910	Human milk, mature	2	fluid ounce(s)	62	53.9	43	0.6	4.2	0	2.7	1.2	1.0	0.3	—
760	Mixed cereal, prepared with whole milk	4	ounce(s)	113	84.6	128	5.4	18.0	1.5	4.0	2.2	1.2	0.4	—
772	Mixed vegetable dinner, strained	2	ounce(s)	57	50.3	23	0.7	5.4	0.8	0	—	—	0	—
762	Rice cereal, prepared with whole milk	4	ounce(s)	113	84.6	130	4.4	18.9	0.1	4.1	2.6	1.0	0.2	—
758	Teething biscuits	1	item(s)	11	0.7	44	1.0	8.6	0.2	0.6	0.2	0.2	0.1	—

Chol (mg)	Calc (mg)	Iron (mg)	Magn (mg)	Pota (mg)	Sodi (mg)	Zinc (mg)	Vit A (µg)	Thia (mg)	Vit E (mg α)	Ribo (mg)	Niac (mg)	Vit B6 (mg)	Fola (µg)	Vit C (mg)	Vit B12 (µg)	Sele (µg)
20	100	2.70	—	—	1170.0	—	—	—	—	—	—	—	—	0	—	—
20	150	1.44	—	—	1330.0	—	0	—	—	—	—	—	—	0	—	—
30	0	1.80	—	—	1040.0	—	0	—	—	—	—	—	—	0	—	—
35	150	1.80	—	—	870.0	—	0	—	—	—	—	—	—	0	—	—
25	60	1.80	—	—	1070.0	—	—	—	—	—	—	—	—	3.6	—	—
60	40	1.44	—	—	1140.0	—	0	—	—	—	—	—	—	0	—	—
85	300	2.70	45.4	484.0	810.0	2.29	—	0.45	—	0.51	4.00	0.22	30.7	0	0.7	—
85	100	2.70	—	—	810.0	—	—	0.15	—	0.42	6.00	—	—	0	—	—
35	40	1.80	38.9	280.0	580.0	4.71	—	0.17	—	0.36	4.28	0.27	18.9	2.4	2.5	—
30	40	1.80	57.7	540.0	850.0	4.81	—	0.15	—	0.29	5.53	0.37	38.4	6.0	1.6	—
15	283	3.03	78.5	420.0	780.0	1.39	—	0.22	—	0.45	3.13	0.32	74.8	59.1	0.2	—
10	350	3.60	—	—	600.0	—	—	—	—	—	—	—	—	0	—	—
30	40	1.44	—	—	580.0	—	—	—	—	—	—	—	—	3.6	—	—
20	100	1.80	—	—	600.0	—	—	—	—	—	—	—	—	0	—	—
35	18	0.72	—	—	580.0	—	—	—	—	—	—	—	—	12.0	—	—
35	20	0.36	—	—	580.0	—	—	—	—	—	—	—	—	30.0	—	—
45	80	2.70	—	—	580.0	—	—	—	—	—	—	—	—	21.0	—	—
35	40	1.80	—	—	600.0	—	—	—	—	—	—	—	—	0	—	—
30	200	1.80	—	230.0	660.0	—	—	—	—	—	—	—	—	2.4	—	—
70	250	1.44	—	550.0	920.0	—	—	—	—	—	—	—	—	6.0	—	—
65	150	2.70	—	—	1170.0	—	—	—	—	—	—	—	—	2.4	—	—
55	20	0.72	—	490.0	770.0	—	0	—	—	—	—	—	—	0	—	—
35	20	1.80	—	800.0	960.0	—	—	—	—	—	—	—	—	6.0	—	—
45	40	1.08	—	490.0	970.0	—	—	—	—	—	—	—	—	3.6	—	—
25	40	1.16	—	500.0	650.0	—	—	—	—	—	—	—	—	3.6	—	—
30	64	0.38	—	370.0	650.0	—	—	—	—	—	—	—	—	0	—	—
20	150	2.70	—	300.0	700.0	—	—	—	—	—	—	—	—	15.0	—	—
25	40	0.72	—	380.0	650.0	—	—	—	—	—	—	—	—	2.4	—	—
20	150	0.72	—	350.0	510.0	—	—	—	—	—	—	—	—	2.4	—	—
15	200	0.72	—	290.0	690.0	—	0	—	—	—	—	—	—	0	—	—
30	250	1.47	—	610.0	690.0	—	—	—	—	—	—	—	—	2.4	—	—
40	200	0.72	—	—	800.0	—	—	—	—	—	—	—	—	2.4	—	—
35	40	0.72	—	—	790.0	—	—	—	—	—	—	—	—	6.0	—	—
15	200	1.08	—	401.0	700.0	—	69.1	—	—	—	—	—	—	4.8	—	—
15	150	3.60	—	—	670.0	—	—	—	—	—	—	—	—	9.0	—	—
10	40	1.44	—	—	890.0	—	—	—	—	—	—	—	—	0	—	—
0	5	0.72	3.8	115.4	3.8	0.03	1.3	0.01	0.76	0.02	0.10	0.03	0	73.4	0	0.1
0	3	0.12	1.9	45.4	1.3	0.01	0.6	0.01	0.36	0.02	0.04	0.02	1.3	24.5	0	0.2
0	3	0.12	6.0	52.8	5.4	0.04	1.2	0.01	0.36	0.02	0.08	0.04	3.6	10.0	0	0.4
0	12	0.20	5.0	109.8	20.7	0.08	320.9	0.01	0.29	0.02	0.25	0.04	8.4	3.2	0	0.1
10	17	0.40	9.0	89.0	14.7	0.32	70.4	0.03	0.12	0.04	0.44	0.04	7.0	0	0	2.4
0	23	0.40	12.0	87.6	3.0	0.12	10.8	0.02	0.04	0.04	0.20	0.02	14.4	0.2	0	0
9	20	0.02	1.8	31.4	10.5	0.10	37.6	0.01	0.04	0.02	0.10	0.01	3.1	3.1	0	1.1
12	249	11.82	30.6	225.7	53.3	0.80	28.4	0.49	—	0.65	6.54	0.07	12.5	1.4	0.3	
—	12	0.18	6.2	68.6	4.5	0.08	77.1	0.01	—	0.02	0.28	0.04	4.5	1.6	0	0.4
12	271	13.82	51.0	215.5	52.2	0.72	24.9	0.52	—	0.56	5.90	0.12	9.1	1.4	0.3	4.0
0	11	0.39	3.9	35.5	28.4	0.10	3.1	0.02	0.02	0.05	0.47	0.01	5.4	1.0	0	2.6

Appendix B

Reliable Sources of Nutrition Information

Many sources of nutrition information are available to consumers, but the quality of the information they provide varies widely. All of the sources listed here provide scientifically based information.

Expert Advice

Registered dietitians (hospitals and the yellow pages) Public health nutritionists (public health departments) College nutrition instructors/professors (colleges and universities) Extension Service home economists (state and county U.S. Department of Agriculture Extension Service offices) Consumer affairs staff of the Food and Drug Administration (national, regional, and state FDA offices)

You can find hundreds of toll-free telephone numbers for health information through the following Website: www.healthfinder.gov. After you are connected, search the term toll-free numbers.

U.S. Government

- **Federal Trade Commission (FTC)**
  Public Reference Branch
  (202) 326-2222
  www.ftc.gov

- **Food and Drug Administration (FDA)**
  Office of Consumer Affairs, HFE 1
  Room 16-85
  5600 Fishers Lane
  Rockville, MD 20857
  (301) 443-1544
  www.fda.gov

- **FDA Consumer Information Line**
  (301) 827-4420

- **FDA Office of Food Labeling, HFS 150**
  Washington, DC 20204
  (202) 205-4561; fax (202) 205-4564
  www.cfsan.fda.gov

- **FDA Office of Plant and Dairy Foods and Beverages**
  HFS 300
  200 C Street SW
  Washington, DC 20204
  (202) 205-4064; fax (202) 205-4422

- **FDA Office of Special Nutritionals,**
  HFS 450
  200 C Street SW
  Washington, DC 20204
  (202) 205-4168; fax (202) 205-5295

- **Food and Nutrition Information Center**
  National Agricultural Library,
  Room 304
  10301 Baltimore Avenue
  Beltsville, MD 20705-2351
  (301) 504-5719; fax (301) 504-6409
  www.nal.usda.gov/fnic

- **Food Research Action Center (FRAC)**
  1875 Connecticut Avenue NW,
  Suite 540
  Washington, DC 20009
  (202) 986-2200; fax (202) 986-2525

- **Superintendent of Documents**
  U.S. Government Printing Office
  Washington, DC 20402
  (202) 512-1071
  www.access.gpo.gov/su_docs

- **U.S. Department of Agriculture (USDA)**
  14th Street SW and Independence Avenue
  Washington, DC 20250
  (202) 720-2791
  www.usda.gov/fcs

- **USDA Center for Nutrition Policy and Promotion**
  1120 20th Street NW, Suite 200
  North Lobby
  Washington, DC 20036
  (202) 208-2417
  www.usda.gov/fcs/cnpp.htm

- **USDA Food Safety and Inspection Service**
  Food Safety Education Office,
  Room 1180-S
  Washington, DC 20250
  (202) 690-0351
  www.usda.gov/fsis

- **U.S. Department of Education (DOE)**
  Accreditation Agency Evaluation Branch
  7th and D Street SW
  ROB 3, Room 3915
  Washington, DC 20202-5244
  (202) 708-7417

- **U.S. Department of Health and Human Services**
  200 Independence Avenue SW
  Washington, DC 20201
  (202) 619-0257
  www.os.dhhs.gov

- **U.S. Environmental Protection Agency (EPA)**
  401 Main Street SW
  Washington, DC 20460
  (202) 260-2090
  www.epa.gov

- **U.S. Public Health Service**
  Assistant Secretary of Health
  Humphrey Building, Room 725-H
  200 Independence Avenue SW
  Washington, DC 20201
  (202) 690-7694

Health Canada

Headquarters

- **Health Canada**
  A.L. 0900C2
  Ottawa, Canada
  K1A 0K9
  Telephone: (613) 957-2991
  TTY: 1-800-267-1245
  http://www.hc-sc.gc.ca/

Regional Headquarters

- **British Columbia/Yukon**
  Suite 405, Winch Building
  757 West Hastings Street
  Vancouver, BC
  V6C 1A1
  Tel: (604) 666-2083
  Fax: (604) 666-2258

- **Alberta/NWT**
  Suite 710, Canada Place
  9700 Jasper Avenue
  Edmonton, AB
  T5J 4C3
  Tel: (780) 495-2651
  Fax: (780) 495-3285

- **Manitoba/Saskatchewan**
  391 York Avenue, Suite 425
  Winnipeg, MB
  R3C 0P4
  Tel: (204) 983-2508
  Fax: (204) 983-3972

- **Ontario/Nunavut**
  25 St. Clair Avenue East, 4th Floor
  Toronto, ON
  M4T 1M2
  Tel: (416) 973-4389
  Toll free: 1-866-999-7612
  Fax: (416) 973-1423

- **Quebec**
  Room 218, Complexe Guy-Favreau
  East Tower
  200 René Lévesque Blvd. West
  Montreal, QC
  H2Z 1X4
  Tel: (514) 283-2306
  Fax: (514) 283-6739

- **Atlantic**
  Suite 1525, 15th Floor,
  Maritime Centre
  1505 Barrington Street
  Halifax, NS B3J 3Y6
  Tel: (902) 426-2700
  Fax: (902) 426-9689

International Agencies

- **Food and Agriculture Organization of the United Nations (FAO)**
  Liaison Office for North America
  2175 K Street, Suite 300
  Washington, DC 20437
  (202) 653-2400
  www.fao.org

- **International Food Information Council Foundation**
  1100 Connecticut Avenue NW,
  Suite 430
  Washington, DC 20036
  (202) 296-6540
  ificinfo.health.org

- **UNICEF**
  3 United Nations Plaza
  New York, NY 10017
  (212) 326-7000
  www.unicef.com

- **World Health Organization (WHO)**
  Regional Office
  525 23rd Street NW
  Washington, DC 20037
  (202) 974-3000
  www.who.org

Professional Nutrition Organizations

- **American Dietetic Association (ADA)**
  216 West Jackson Boulevard,
  Suite 800
  Chicago, IL 60606-6995
  (800) 877-1600; (312) 899-0040
  www.eatright.org

- **ADA, The Nutrition Hotline**
  (800) 366-1655

- **American Society for Clinical Nutrition**
  9650 Rockville Pike
  Bethesda, MD 20814-3998
  (301) 530-7110; fax (301) 571-1863
  www.faseb.org/ascn

- **Dietitians of Canada**
  480 University Avenue, Suite 604
  Toronto, Ontario M5G 1V2, Canada
  (416) 596-0857; fax (416) 596-0603
  www.dietitians.ca

- **Human Nutrition Institute (INACG)**
  1126 Sixteenth Street NW
  Washington, DC 20036
  (202) 659-0789
  www.ilsi.org

- **National Academy of Sciences/ National Research Council (NAS/ NRC)**
  2101 Constitution Avenue, NW
  Washington, DC 20418
  (202) 334-2000
  www.nas.edu

- **National Institute of Nutrition**
  265 Carling Avenue, Suite 302
  Ottawa, Ontario K1S 2E1
  (613) 235-3355; fax (613) 235-7032
  www.nin.ca

- **Society for Nutrition Education**
  7101 Wisconsin Avenue, Suite 901
  Bethesda, MD 20814-4805
  (301) 656-4938

Aging

- **Administration on Aging**
  330 Independence Avenue SW
  Washington, DC 20201
  (202) 619-0724
  www.aoa.dhhs.gov

- **American Association of Retired Persons (AARP)**
  601 E Street NW
  Washington, DC 20049
  (202) 434-2277
  www.aarp.org

- **National Aging Information Center**
  330 Independence Avenue SW
  Washington, DC 20201
  (202) 619-7501
  www.aoa.dhhs.gov/naic

- **National Institute on Aging**
  Public Information Office
  31 Center Drive, MSC 2292
  Bethesda, MD 20892
  (301) 496-1752
  www.nih.gov/nia

Alcohol and Drug Abuse

- **Al-Anon Family Group Headquarters, Inc.**
  1600 Corporate Landing Parkway
  Virginia Beach, VA 23454-5617
  (800) 356-9996
  www.al-anon.alateen.org

- **Alateen**
  1600 Corporate Landing Parkway
  Virginia Beach, VA 23454-5617
  (800) 356-9996
  www.al-anon.alateen.org

- **Alcoholics Anonymous (AA)**
  General Service Office
  475 Riverside Drive
  New York, NY 10115
  (212) 870-3400
  www.aa.org

- **Narcotics Anonymous (NA)**
  P.O. Box 9999
  Van Nuys, CA 91409
  (818) 773-9999; fax (818) 700-0700
  www.wsoinc.com

- **National Clearinghouse for Alcohol and Drug Information (NCADI)**
  P.O. Box 2345
  Rockville, MD 20847-2345
  (800) 729-6686
  www.health.org

- **National Council on Alcoholism and Drug Dependence (NCADD)**
  12 West 21st Street
  New York, NY 10010
  (800) NCA-CALL or (800) 622-2255
  (212) 206-6770; fax (212) 645-1690
  www.ncadd.org

- **U.S. Center for Substance Abuse Prevention**
  1010 Wayne Avenue, Suite 850
  Silver Spring, MD 20910
  (301) 459-1591 ext. 244;
  fax (301) 495-2919
  www.covesoft.com/csap.html

Consumer Organizations

- **Center for Science in the Public Interest (CSPI)**
  1875 Connecticut Avenue NW,
  Suite 300
  Washington, DC 20009-5728
  (202) 332-9110; fax (202) 265-4954
  www.cspinet.org

- **Choice in Dying, Inc.**
  1035 30th Street NW
  Washington, DC 20007
  (202) 338-9790; fax (202) 338-0242
  www.choices.org

- **Consumer Information Center**
  Pueblo, CO 81009
  (888) 8 PUEBLO or (888) 878-3256
  www.pueblo.gsa.gov

- **Consumers Union of US Inc.**
  101 Truman Avenue
  Yonkers, NY 10703-1057
  (914) 378-2000
  www.consunion.org

- **National Council Against Health Fraud, Inc. (NCAHF)**
  P.O. Box 1276
  Loma Linda, CA 92354
  (909) 824-4690
  www.ncahf.org

Fitness

- **American College of Sports Medicine**
  P.O. Box 1440
  Indianapolis, IN 46206 1440
  (317) 637-9200
  _www.acsm.org/sportsmed

- **American Council on Exercise (ACE)**
  5820 Oberlin Drive, Suite 102
  San Diego, CA 92121
  (800) 529-8227
  www.acefitness.org

- **President's Council on Physical Fitness and Sports**
  Humphrey Building, Room 738
  200 Independence Avenue SW
  Washington, DC 20201
  (202) 690-9000; fax (202) 690-5211
  _www.indiana.edu/~preschal

- **Shape Up America!**
  6707 Democracy Boulevard,
  Suite 306
  Bethesda, MD 20817
  (301) 493-5368
  www.shapeup.org

- **Sport Medicine and Science Council of Canada**
  1600 James Naismith Drive, Suite 314
  Gloucester, Ontario K1B 5N4, Canada
  (613) 748-5671; fax (613) 748-5729
  www.smscc.ca

Food Safety

- **Alliance for Food & Fiber Food Safety Hotline**
  (800) 266-0200

- **FDA Center for Food Safety and Applied Nutrition**
  200 C Street SW
  Washington, DC 20204
  (800) FDA-4010 or (800) 332-4010
  vm.cfsan.fda.gov

- **National Lead Information Center**
  (800) LEAD-FYI or (800) 532-3394
  (800) 424-LEAD or (800) 424-5323

- **National Pesticide Telecommunications Network (NPTN)**
  Oregon State University
  333 Weniger Hall
  Corvallis, OR 97331-6502
  (541) 737-6091
  _www.ace.orst.edu/info/nptn

- **USDA Meat and Poultry Hotline**
  (800) 535-4555

- **U.S. EPA Safe Drinking Water Hotline**
  (800) 426-4791

Health and Disease

- **Alzheimer's Disease Education and Referral Center**
  P. O. Box 8250
  Silver Spring, MD 20907-8250
  (800) 438-4380
  www.alzheimers.org

- **Alzheimer's Disease Information and Referral Service**
  919 North Michigan Avenue,
  Suite 1000
  Chicago, IL 60611
  (800) 272-3900
  www.alz.org

- **American Academy of Allergy, Asthma, and Immunology**
  611 East Wells Street
  Milwaukee, WI 53202
  (414) 272-6071; fax (414) 276-3349
  www.aaaai.org

- **American Cancer Society National Home Office**
  1599 Clifton Road NE
  Atlanta, GA 30329-4251
  (800) ACS-2345 or (800) 227-2345
  www.cancer.org

- **American Council on Science and Health**
  1995 Broadway, 2nd Floor
  New York, NY 10023-5860
  (212) 362-7044; fax (212) 362-4919
  www.acsh.org

- **American Dental Association**
  211 East Chicago Avenue
  Chicago, IL 60611
  (312) 440-2800
  www.ada.org

- **American Diabetes Association**
  1660 Duke Street
  Alexandria, VA 22314
  (800) 232-3472 or (703) 549-1500
  www.diabetes.org

- **American Heart Association**
  Box BHG, National Center
  7320 Greenville Avenue
  Dallas, TX 75231
  (800) 275-0448 or (214) 373-6300
  www.amhrt.org

- **American Institute for Cancer Research**
  1759 R Street NW
  Washington, DC 20009
  (800) 843-8114 or (202) 328-7744;
  fax (202) 328-7226
  www.aicr.org

- **American Medical Association**
  515 North State Street
  Chicago, IL 60610
  (312) 464-5000
  www.ama-assn.org

- **American Public Health Association (APHA)**
  1015 Fifteenth Street NW, Suite 300
  Washington, DC 20005
  (202) 789-5600
  www.apha.org

- **American Red Cross**
  National Headquarters
  8111 Gatehouse Road
  Falls Church, VA 22042
  (703) 206-7180
  www.redcross.org

- **Canadian Diabetes Association**
  15 Toronto Street, Suite 800
  Toronto, ON M5C 2E3
  (800) BANTING or (800) 226-8464
  (416) 363-3373
  www.diabetes.ca

- **Canadian Public Health Association**
  400-1565 Carling Avenue
  Ottawa, Ontario K1Z 8R1
  (613) 725-3769; fax (613) 725-9826
  www.cpha.ca

- **Centers for Disease Control and Prevention (CDC)**
  1600 Clifton Road NE
  Atlanta, GA 30333
  (404) 639-3311
  www.cdc.gov

- **The Food Allergy Network**
  10400 Eaton Place, Suite 107
  Fairfax, VA 22030-2208
  (800) 929-4040 or (703) 691-3179
  www.foodallergy.org
- **Internet Health Resources**
  www.ihr.com
- **National AIDS Hotline (CDC)**
  (800) 342-AIDS (English)
  (800) 344-SIDA (Spanish)
  (800) 2437-TTY (Deaf)
  (900) 820-2437
- **National Cancer Institute**
  Office of Cancer Communications
  Building 31, Room 10824
  Bethesda, MD 20892
  (800) 4-CANCER or (800) 422-6237
  www.nci.nih.gov
- **National Diabetes Information Clearinghouse**
  1 Information Way
  Bethesda, MD 20892-3560
  (301) 654-3327
  www.niddk.nih.gov
- **National Digestive Disease Information Clearinghouse (NDDIC)**
  2 Information Way
  Bethesda, MD 20892-3570
  (301) 654-3810
  www.niddk.nih.gov
- **National Health Information Center (NHIC)**
  Office of Disease Prevention and Health Promotion
  (800) 336-4797
  nhic-nt.health.org
- **National Heart, Lung, and Blood Institute**
  Information Center
  P.O. Box 30105
  Bethesda, MD 20824-0105
  (301) 251-1222
  _www.nhlbi.nih.gov/nhlbi/nhlbi.htm
- **National Institute of Allergy and Infectious Diseases**
  Office of Communications
  Building 31, Room 7A50
  31 Center Drive, MSC2520
  Bethesda, MD 20892-2520
  (301) 496-5717
  www.niaid.nih.gov
- **National Institute of Dental Research (NIDR)**
  National Institute of Health
  Bethesda, MD 20892-2190
  (301) 496-4261
  www.nidr.nih.gov

- **National Institutes of Health (NIH)**
  9000 Rockville Pike
  Bethesda, MD 20892
  (301) 496-2433
  www.nih.gov
- **National Osteoporosis Foundation**
  1150 17th Street NW, Suite 500
  Washington, DC 20036
  (202) 223-2226
  www.nof.org
- **Office of Disease Prevention and Health Promotion**
  odphp.osophs.dhhs.gov
- **Office on Smoking and Health (OSH)**
  _www.americanheart.org/heart.org/
  Heart_and_stroke_A_Z_Guide/osh
  .html

Infancy and Childhood

- **American Academy of Pediatrics**
  141 Northwest Point Boulevard
  Elk Grove Village, IL 60007-1098
  (847) 228-5005
  www.aap.org
- **Association of Birth Defect Children, Inc.**
  930 Woodcock Road, Suite 225
  Orlando, FL 32803
  (407) 245-7035
  www.birthdefects.org
- **Canadian Paediatric Society**
  100-2204 Walkley Road
  Ottawa, ON K1G 4G8
  (613) 526-9397; fax (613) 526-3332
  www.cps.ca
- **National Center for Education in Maternal & Child Health**
  2000 15th Street North, Suite 701
  Arlington, VA 22201-2617
  (703) 524-7802
  www.ncemch.org

Pregnancy and Lactation

- **American College of Obstetricians and Gynecologists Resource Center**
  409 12th Street SW
  Washington, DC 20024-2188
  (202) 638-5577
  www.acog.org
- **La Leche International, Inc.**
  1400 N. Meacham Road
  Schaumburg, IL 60173
  (847) 519-7730
  www.lalecheleague.org
- **March of Dimes Birth Defects Foundation**
  1275 Mamaroneck Avenue
  White Plains, NY 10605
  (914) 428-7100
  www.sunkist.com

World Hunger

- **Bread for the World**
  1100 Wayne Avenue, Suite 1000
  Silver Spring, MD 20910
  (301) 608-2400
  www.bread.org
- **Center on Hunger, Poverty and Nutrition Policy**
  Tufts University School of Nutrition
  11 Curtis Avenue
  Medford, MA 02155
  (617) 627-3956
- **Freedom from Hunger**
  P.O. Box 2000
  1644 DaVinci Court
  Davis, CA 95617
  (530) 758-6200
  www.freefromhunger.org
- **Oxfam America**
  26 West Street
  Boston, MA 02111
  (617) 482-1211
  www.oxfamamerica.org
- **SEEDS Magazine**
  P.O. Box 6170
  Waco, TX 76706
  (254) 755-7745
  _www.helwys.com/seedhome.htm
- **Worldwatch Institute**
  1776 Massachusetts Avenue NW, Suite 800
  Washington, DC 20036
  (202) 452-1999
  www.worldwatch.org

Scientific Literature

Nutrition Journals

American Journal of Clinical Nutrition
British Journal of Nutrition
Human Nutrition, Applied Nutrition
Journal of the American College of Nutrition
Journal of the American Dietetic Association
Journal of the Canadian Dietetic Association
Journal of Food Composition and Analysis
Journal of Nutrition
Journal of Nutrition Education
Nutrition Abstracts and Reviews
Nutrition and Metabolism
Nutrition Reports International
Nutrition Research
Nutrition Reviews
Nutrition Today

Other Journals

American Journal of Epidemiology
American Journal of Nursing
American Journal of Public Health
Annals of Internal Medicine
Annals of Surgery
Canadian Journal of Public Health
Caries Research

Food Technology
Gastroenterology
International Journal of Obesity
Journal of the American Dental Association
Journal of the American Medical Association
Journal of Clinical Investigation
Journal of Food Science
Journal of Home Economics

Journal of Pediatrics
Lancet
New England Journal of Medicine
Nutrition Today
Pediatrics
Science
The Scientist

Appendix C

The U.S. Food Exchange System

The U.S. exchange system divides the foods suitable for use in planning a healthy diet into six lists—the starch/bread, meat/meat alternate, vegetable, fruit, milk, and fat lists.a These lists are shown in Tables C.1 through C.6. Following these lists are three other sets of foods: free foods, combination foods, and foods for occasional use (Tables C.7, C.8, and C.9).

aThe Exchange Lists are the basis of a meal planning system designed by a committee of the American Diabetes Association and The American Dietetic Association. While designed primarily for people with diabetes and others who must follow special diets, the Exchange Lists are based on principles of good nutrition that apply to everyone. © 2008 American Dietetic Association, The American Diabates Association.

Table C.1

The U.S.. Exchange System: Starch/Bread List

15 g carbohydrate, 0-3 g protein, 0-15, 80 cal

Amount	Food	Amount	Food
Cereals/Grains/Pasta		**Bread**	
$^1/_2$ c	Bran cereals,	$^1/_4$ (1 oz)	Bagel, large (4 oz)
$^1/_2$ c	Bulgur, cooked	$^1/_2$	English muffins
$^1/_3$ c	Cooked cereals	$^1/_2$ (1 oz)	Frankfurter or hamburger buns
$^1/_2$ c	Grits, cooked	$^1/_2$ piece	Pita, 6" across
$^3/_4$ c	Other ready-to-eat unsweetened cereals	1 (1 oz)	Plain rolls, small
$^1/_3$ c	Pasta, cooked	1 slice (1 oz)	Raisin, unfrosted
$1^1/_2$ c	Puffed cereals	1 slice (1 oz)	Rye, pumpernickel ◢
$^1/_3$ c	Rice, white or brown, cooked	1	Tortillas, 6" across
$^1/_2$ c	Shredded wheat	1 slice (1 oz)	White (including French, Italian)
3 tbs	Wheat germ ◢	1 slice (1 oz)	Whole-wheat
Dried Beans/Peas/Lentils		2 slices ($1^1/_2$ oz)	Reduced-calorie
$^1/_3$ c	Baked beans ◢	**Crackers/Snacks**	
$^1/_2$ c	Beans and peas, cooked, such as kidney, white, split, black-eyed ◢	8	Animal crackers
		3	Graham crackers, $2^1/_2$" square
$^1/_2$ c	Lentils, cooked ◢	$^3/_4$ oz	Matzoh
Starchy Vegetables		4 pieces	Melba toast, 2" × 4" piece
		20	Oyster crackers
$^1/_2$ c	Corn ◢	3 c	Popcorn, popped, no fat added
$^1/_2$ cob (5 oz)	Corn on the cob, large ◢	$^3/_4$ oz	Pretzels
$^1/_2$ c	Lima beans ◢	6	Saltine-type crackers
$^1/_2$ c	Peas, green, canned or frozen ◢	2 to 5 ($^3/_4$ oz)	Whole-wheat crackers, no fat added (crisp breads)
$^1/_3$ c	Plantains ◢	**Starch Foods Prepared with Fat**	
1 small (3 oz)	Potatoes, baked	(Count as 1 starch/bread serving, plus 1 fat serving.)	
$^1/_2$ c	Potatoes, mashed	1	Biscuits, $2^1/_2$" across
1 c	Squash, winter (acorn, butternut)	1 ($1^1/_2$ oz)	Cornbread, $1^3/_4$" cube
$^1/_2$ c	Yams, sweet potatoes, plain	6	Crackers, round butter type
		1 cup (2 oz)	French fries, oven-baked
		1	Muffins, plain, small
		1	Pancakes, 4" across
		$^1/_3$ c	Stuffing, bread, prepared
		2	Taco shells, 5" across
		1	Waffles, 4" square
		2 to 5 ($^3/_4$ oz)	Whole-wheat crackers, fat added

◢ 3 grams or more dietary fiber per serving. Average fiber contents of whole-grain products is 2 grams per serving. For starch foods not on this list, the general rule is that ½ cup cereal, grain, or pasta is 1 serving: 1 ounce of a bread product is 1 serving.

Table C.2

U.S. Exchange System: Meat/Meat Alternate Lists

Lean meat = 7 g protein, 3 g fat, 55 cal; medium-fat meat = 7 g protein, 5 g fat, 75 cal; high-fat meat = 7 g protein, 8 g fat, 100 cal.

Category	Amount	Food
Lean Meat and Alternates		
Beef	1 oz	USDA Select or Choice grades of lean beef, trimmed of fat, such as round, sirloin, and flank steak; tenderloin;
Pork	1 oz	Lean pork, such as Canadian bacon 🖉, tenderloin, riborloin chop/roast, warm
Veal	1 oz	Chops, roasts
Poultry	1 oz	Chicken, turkey, Cornish hen (without skin)
Fish	1 oz	All fresh and frozen fish
	1 oz	Crab, lobster, scallops, shrimp, clams (fresh or canned in water 🖉)
	6 medium	Oysters
	1 oz	Tuna, canned in water or oil
	1 oz	Herring uncreamed or smoked
	2 medium	Sardines, canned
Wild game	1 oz	Venison, rabbit, buffalo, ostrich
Cheese	1/4 c	Any cottage cheese
	1 oz	Cheeses with 3 grams of fat or less per oz
Other	1 oz	Lunch meats with 3 grams of fat or less per oz
	2 whites	Egg whites
	1/4 c	Egg substitutes, plain
Medium-Fat Meat and Alternates		
Beef	1 oz	Most beef products fall into this category; examples: all ground beef, roasts (rib, chuck, rump), steak (cubed, porterhouse, T-bone), meatloaf
Pork	1 oz	Most pork products fall into this category; examples: cutlet, loin roast, shoulder roast
Lamb	1 oz	Most lamb products fall into this category; examples: chops, leg, roast, ground

Category	Amount	Food
Medium-Fat Meat and Alternates (*continued*)		
Veal	1 oz	Cutlet, ground or cubed, unbreaded
Poultry	1 oz	Chicken (with skin), wild duck or goose (well-drained of fat), ground turkey, fried chicken
Fish	1 oz	Any Fried Product
Cheese		Skim or part-skim milk cheeses, such as:
	1/4 c	Ricotta
	1 oz	Mozzarella
	1 oz	Cheeses with 4-7 grams of fat per oz
Other	1	Eggs (high in cholesterol, limit to 3 per week)
	4 oz	Tofu, 2½" × 2¾" × 1"
	1 oz	Sausage with 4-7 grams of fat per oz
High-Fat Meat and Alternates[a]		
Pork	1 oz	Spareribs, ground pork, pork sausages (patties or links)
Cheese	1 oz	All regular cheeses, such as American, blue, Cheddar, Monterey, Swiss, brie
Other	1 oz	Lunch meats 🖉, such as bologna, salami, pastrami pimento loaf, with 8 grams of fat or more per oz
	1 oz	Sausage 🖉, such as Polish, Italian
	1 oz	Knockwurst, smoked 🖉
	1 oz	Bratwurst 🖉
	1 (10/lb)	Frankfurters 🖉 (turkey or chicken)
Count as 1 high-fat meat plus 1 fat exchange:		
1 frank	(10/lb)	Frankfurters 🖉 (beef, pork, or combination)

🖉 400 milligrams or more sodium per exchange. Meats contribute no fiber to the diet.

[a]These items are high in saturated fat, cholesterol, and calories and should be used no more than three times per week.

Table C.3

U.S. Exchange System: Vegetable List

5 g carbohydrate, 2 g protein, 25 cal
All portion sizes, except as otherwise noted, are $^1/_2$ c of any cooked vegetable or vegetable juice, 1 c of any raw vegetable.

Artichokes, $^1/_2$ medium	Cabbage, cooked (green, bok chog, Chinese)	Leeks	Spinach, cooked
Asparagus	Carrots	Mushrooms, fresh	Summer squash (crookneck)
Bean sprouts	Cauliflower	Okra	Tomatoes, 1 large
Beans (green, wax, Italian)	Eggplant	Onions	Tomato/vegetable juice 🖉
Beets	Peppers	Pea pods	Turnips
Broccoli	Greens (collard, mustard,	Rutabagas	Water chestnuts
Brussels sprouts	turnip)	Sauerkraut 🖉	
	Kohlrabi		

Starchy vegetables such as corn, peas, and potatoes are found on the Starch/Bread List.
For free vegetables, see the Free Food List (Table D.7).

🖉 400 milligrams or more sodium per serving. Most vegetable servings contain 2 to 3 grams dietary fiber.

Table C.4

U.S. Exchange System: Fruit List

15 g carbohydrate, 60 cal
All portion sizes, unless otherwise noted, are $^1/_2$ c fresh fruit or fruit juice, $^1/_4$ c dried fruit.

Amount	Food	Amount	Food
Fresh, Frozen, and Unsweetened Canned Fruit		$^1/_2$ c (2 halves)	Pears, canned
		$^3/_4$ c	Pineapple, raw
1	Apples, raw, 2" across	$^1/_2$ c	Pineapple, canned
$^1/_2$ c	Applesauce, unsweetened	2	Plums, raw, 2" across
4	Apricots, medium, raw	1 c	Raspberries, raw 🌿
$^1/_2$ c (4 halves)	Apricots, canned	$1^1/_4$ c	Strawberries, raw, whole
1	Bananas, extra small	2	Tangerines, $2^1/_2$" across
$^3/_4$ c	Blackberries, raw 🌿	$1^1/_4$ c	Watermelon, cubes
$^3/_4$ c	Blueberries, raw 🌿		
$^1/_3$	Cantaloupe, 5" across	**Dried Fruit**	
1 c	Cantaloupe, cubes	4 rings	Apples 🌿
12	Cherries, large, raw	8 halves	Apricots 🌿
$^1/_2$ c	Cherries, canned	3 medium	Dates
2	Figs, raw, 2" across	$1^1/_2$	Figs 🌿
$^1/_2$ c	Fruit cocktail, canned	3 medium	Prunes 🌿
$^1/_2$	Grapefruit, large	**Fruit Juice**	
$^3/_4$ c	Grapefruit, segments	$^1/_2$ c	Apple juice/cider
17	Grapes, small	$^1/_3$ c	Cranberry juice cocktail
1 slice	Honeydew melon, medium	$^1/_3$ c	Grape juice
1 c	Honeydew melon, cubes	$^1/_2$ c	Grapefruit juice
1	Kiwis, large	$^1/_2$ c	Orange juice
$^3/_4$ c	Mandarin oranges	$^1/_2$ c	Pineapple juice
$^1/_2$	Mangoes, small	$^1/_3$ c	Prune juice
1	Nectarines, $1^1/_2$" across 🌿		
1	Oranges, $2^1/_2$" across		
1 c	Papayas, cubes		
1 ($^3/_4$ c)	Peaches, $2^3/_4$" across		
$^1/_2$ c (2 halves)	Peaches, canned		
$^1/_2$ large or 1 small	Pears		

🌿 3 grams or more dietary fiber per serving. Average fiber contents of fresh, frozen, and dry fruits: 2 grams per serving.

Table C.5

U.S. Exchange System: Milk List

Nonfat and very low-fat milk = 12 g carbohydrate, 8 g protein, trace fat, 100 cal; low-fat milk = 12 g carbohydrate, 8 g protein, = g fat, 120 cal; whole milk = 12 g carbohydrate, 8 g protein, 8 g fat, 160 cal.

Amount	Food	Amount	Food
Nonfat and Very Low-Fat Milk		**Low-Fat Milk**	
1 c	Nonfat milk	1 c fluid	2% milk
1 c	1% milk	6 oz	Plain low-fat yogurt, with added nonfat milk solids
½ c	Evaporated nonfat milk		
1 c	Low-fat buttermilk	**Whole Milk**	
6 oz	Plain nonfat yogurt		
		1 c	Whole milk, buttermilk, goat's milk
		½ c	Evaporated whole milk
		8 oz	Whole plain yogurt

Table C.6

U.S. Exchange System: Fat List

5 g fat, 45 cal

Amount	Food	Amount	Food
Unsaturated Fats		**Saturated Fats**	
2 tbs	Avocados (*trans* fat free)	1 slice	Bacon[a]
1 tsp	Margarine	1 tsp	Butter
1 tbs	Margarine, diet[a] (30%-50% vegetable oil, trans-fatfree)	½ oz (2 tbs)	Chitterlings
		2 tbs	Coconut, shredded
2 tsp	Mayonnaise	1 tbs	Cream (heavy, whipping)
1 tbs	Mayonnaise, reduced fat[a]	1½ tbs	Cream (light, coffee, table)
	Nuts and seeds:	2 tbs	Cream (sour)
6 whole	Almonds, dry roasted	1 tbs	Cream cheese
6 whole	Cashews, dry roasted	¼ oz	Salt pork[a]
10 whole	Peanuts		
2 whole	Pecans		
1 tbp	Pumpkin seeds		
1 tbs	Other nuts		
1 tbs	Seeds, pine nuts, sunflower seeds (without shells)		
2 whole	Walnuts		
1 tbp	Oil (canola, olive, peanut)		
8 large	Olives[a], black		
10 large	Olives, green		
1 tbs	Salad dressing, all varieties[a]		
1 tbp	Salad dressing, mayonnaise type		
1 tbs	Salad dressing, mayonnaise type, reduced calorie		
2 tbs	Salad dressing, reduced calorie		

Two tablespoons of low-calorie salad dressing is a free food.

Table C.7

U.S. Exchange System: Free Foods

A free food is any food or drink that contains less than 20 cal/serving. People with diabetes are advised to eat as much as they want of those items that have no serving size specified. They may eat two or three servings per day of those items that have a specific serving size. It is suggested that they spread the servings out through the day.

Amount	Food	Amount	Food
Drinks	Bouillon, low-sodium	**Condiments**	
	Bouillon 🖊 or broth without fat	1 tbs	Catsup
	Carbonated drinks, sugar-free		Horseradish
	Carbonated water		Mustard
	Club soda	1½ medium	Pickles 🖊, dill, unsweetened
1 tbs	Cocoa powder, unsweetened	1 tbs	Taco sauce
	Coffee/tea		Vinegar
	Drink mixes, sugar-free	**Seasonings**	Basil, fresh
	Tonic water, sugar-free		Celery seeds
			Chili powder
Nonstick Pan Spray			Chives
			Cinnamon
Fruit			Curry
½ c	Cranberries, unsweetened		Dill
½ c	Rhubarb, unsweetened		Flavoring extracts (almond, butter, lemon,
Vegetables (raw, 1 c)	Cabbage		peppermint, vanilla,
	Celery		walnut, etc.)
	Chinese cabbage 🌿		Garlic
	Cucumbers		Garlic powder
	Green onions		Herbs
	Hot peppers		Hot pepper sauce
	Mushrooms		Lemon
	Radishes		Lemon juice
	Arugula 🌿		Lemon pepper
Salad Greens	Endive		Lime
	Escarole		Lime juice
	Lettuce		Mint
	Romaine		Onion powder
	Spinach		Oregano
	Watercress		Paprika
Sweet Substitutes	Candy, hard, sugar-free		Pepper
	Gelatin, sugar-free		Pimento
	Gum, sugar-free		Spices
2 tsp	Jam/jelly, sugar-free	¼ c	Wine, used in cooking
2 tbs	Pancake syrup, sugar-free		Worcestershire sauce
	Sugar substitutes (saccharin, aspartame, acesulfame-K)		
2 tbs	Whipped topping, light or fat-free		

🌿 3 grams or more dietary fiber per serving.

🖊 400 milligrams or more sodium per serving.

Table C.8

U.S. Exchange Combination Foods

Much of the food we eat is mixed together in various combinations. These combination foods do not fit into only one exchange list. It can be quite hard to tell what is in a certain casserole dish or baked food item. This is a list of average values for some typical combination foods. This list will help you fit these foods into your meal plan. Ask your dietitian for information about any other foods you'd like to eat. The *American Diabetes Association/American Dietetic Association Family Cookbooks* and the *American Diabetes Associates Holiday Cookbook* have many recipes and further information about many foods, including combination foods. Check your library or local bookstore.

Food	Amount	Exchanges	Food	Amount	Exchanges
Casseroles, homemade	1 c (8 oz)	2 carbohydrates + 2 medium-fat meat	Spaghetti and meatballs 🖉, canned	1 c (8 oz)	2 carbohydrates + 2 medium-fat meats
Cheese pizza 🖉, thin crust	¼ of 12"	2 carbohydrates + 2 medium-fat meats	Sugar-free pudding, made with nonfat milk	½ c	1 carbohydrate
Chili with beans 🌾🖉, commercial	1 c (8 oz)	2 starch, 2 medium-fat meat, 2 fat			
Chow mein 🌾🖉, with noodles and vegetables in source	1 c	2 carbohydrate + 1 fat	**If beans are used as a meat substitute:**		
Macaroni and cheese 🖉 meat, 2 fat	1 c (8 oz)	2 carbohydrates + 2 medium-fat meats	Dried beans 🌾, peas 🌾, lentils 🌾	1 c (cooked)	2 starch, 1 lean meat
Soups:					
Bean 🌾🖉	1 c (8 oz)	1 carbohydrate + 1 lean meat			
Cream 🖉 made with water	1 c (8 oz)	1 carbohydrate + 1 fat			
Vegetable 🖉 or broth 🖉	1 c (8 oz)	1 carbohydrate			

🖉 400 milligrams or more sodium per serving.   🌾 3 grams or more dietary fiber per serving.

Table C.9

U.S. Exchange Foods for Occasional Use

The following list includes average exchange values for some foods high in sugar and fat. People are advised to use them only occasionally and in moderate amounts.

Food	Amount	Exchanges	Food	Amount	Exchanges
Angel food cake	$1/_{12}$ cake	2 carbohydrates	Granola bars	1 small	$1^1/_2$ carbohydrate
Cake, no icing	$1/_{12}$ cake or a 2" square	1 carbohydrate + 1 fat	Ice cream, any flavor, fat-free	$1/_2$ c	$1^1/_2$ carbohydrate
Cookies	2 small, $2^1/_4$" across	1 carbohydrate + 2 fats	Sherbet, any flavor	$1/_2$ c	2 carbohydrates
			Vanilla wafers	5 small	1 carbohydrate + 1 fat
Frozen fruit yogurt, fat free	$1/_3$ c	1 carbohydrate			
Gingersnaps	3	1 carbohydrate			

If more than one serving is eaten, these foods have 400 milligrams or more sodium.

Appendix D

Table of Intentional Food Additives

Table D.1

A guide to Intentional Food Additives

Abbreviation	Type of Additive	Uses
AA	Anticaking agents	Keeps dry powders and crystals from clumping together (e.g., salt, powdered sugar).
C	Colors	Synthetic (laboratory-made) vegetable and fruit concentrates and other substances used to color foods (e.g., soft drinks, frosting).
E	Emulsifiers	Used to make oil and water mix (e.g., salad dressings, sauces).
EX	Extenders	"Fillers" such as fruit pulp and texturized protein (e.g., fruit drinks, hamburger).
FA	Flavoring agents	Used to add particular flavors to food (e.g., pudding, rye bread).
N	Nutrients	Used to add vitamins or minerals to foods (e.g., breakfast cereals, skim milk).
P	Preservatives	Used to keep food from spoiling (e.g., breads, breakfast cereals).
S	Sweeteners	Used to sweeten foods. Some are "artificial," such as aspartame and saccharin, and some are extracted from plants, such as sugar cane and sugar beets (e.g., soft drinks, catsup).
T	Texturizers	Used to improve the texture of food by stabilizing moisture content, dryness, volume, tenderness, or hardness (e.g., cakes, breads).
TH	Thickeners	Used to improve the consistency of foods (e.g., low-fat salad dressings, low-calorie jams)

Common Food Additives and Their Primary Function

Alginates (T)
Alpha tocopherol (vitamin E) (N)
Alpha tocopheryl acetate (vitamin E) (N)
Ascorbic acid (vitamin C) (P)
Aspartame (NutraSweet™) (S)
Baking powder (T)
Beet juice (C)
Beet sugar (S)
Beta-carotene (N)
BHA (P)
BHT (P)
Calcium carbonate (calcium) (N)
Calcium pantothenate (pantothenic acid, a B vitamin) (N)
Calcium propionate (P)
Calcium silicate (AA)
Cane sugar (S)

Carotene (C)
Carrageenan (E, TH)
Cellulose gum (TH)
Chromium chloride (N)
Citric acid (FA, P)
Corn syrup (S)
Cupric oxide (copper) (N)
Cyanocobalamin (vitamin B12) (N)
Dextrin (TH)
Dextrose (S)
Dibasic calcium phosphate (calcium phosphorus) (N)
Diglycerides (E)
EDTA (P)
Extracts (FA)
FD&C blue no. 1 (C)
FD&C red no. 3 (use is being phased out) (C)

(continued)

Table D.1

A guide to Intentional Food Additives

Common Food Additives and Their Primary Function

FD&C yellow no. 5 (C)
Ferrous fumarate (iron) (N)
Ferrous sulfate (iron) (N)
Fructose (S)
Fruit pulp (EX)
Gelatin (TH)
Glycerol (T)
Glyceryl abietate (E, T)
Guar gum (T, TH)
High-fructose corn syrup (S)
Honey (S)
Hydrolyzed protein (EX)
Lecithin (E)
Magnesium oxide (magnesium) N
Maltodextrin (S, TH)
Manganese sulfate (manganese) (N)
Modified food starch (TH)
Monoglycerides (E)
Monosodium glutamate (MSG) (FA)
Natural flavorings (FA)
Niacinamide (niacin, vitamin B3) (N)
Nitrates (P)
Nitrites (P)
Paprika (C)
Pectin (TH)
Phosphoric acid (P)
Phytonadione (vitamin K) (N)
Polysorbates (P)
Potassium benzoate (P)
Potassium bicarbonate (T)
Potassium chloride (N)
Potassium iodite (iodine) (N)
Potassium metabisulfite (P)
Potassium sorbate (P)
Propyl gallate (P)
Propyleneglycol (T)
Pyridoxine hydrochloride (vitamin B6) (N)
Reduced iron (iron) (N)
Retinol (vitamin A) (N)
Riboflavin (vitamin B2) (N)
Saccharin (S)

Salt (FA, P)
Silicondioxide (AA)
Sodium aluminum phosphate (T)
Sodium ascorbite (vitamin C) (N)
Sodium benzoate (P)
Sodium bicarbonate (baking soda) (T)
Sodium bisulfite (P)
Sodium chloride (P, FA)
Sodium citrate (P)
Sodium erythorbate (P)
Sodium hexametaphosphate (P, T)
Sodium metabisulfite (P)
Sodium molydate (molybdenum) (N)
Sodium nitrate (P)
Sodium nitrite (P)
Sodium propionate (P)
Sodium saccharin (S)
Sodium selenate (selenium) (N)
Sodium stearolyn-2-lactylate (P, E)
Sodium sulfite (P)
Sorbitan monostearate (E)
Sorbitol (S, T)
Sorghum (S, P)
Starches (TH)
Sucrose (S)
Sugar (P)
Sulfur dioxide (P)
Sweeteners (FA)
Texturized protein (EX)
Thiamin hydrochloride (thiamin, B1) (N)
Thiamin mononitrate (thiamin, B1) (N)
Tocopherols (P, N)
Turmeric (C)
Vitamin A palmitate (vitamin A) (N)
Vitamin C (ascorbic acid) (N, P)
Vitamin E (N, P)
Xanthan gum (T)
Xylitol (S)
Yeast (T)
Zinc oxide (zinc) (N)

Appendix E

Cells

This appendix presents an overview of the basic structure and functions of cells in the human body. Cell structure and function are central to discussions of nutrition for it is within cells that nutrients are utilized to sustain life and health. The life-sustaining processes that take place within each of the more than one hundred trillion cells in the human body are maintained by the nutrients we consume in our diet. Chemical reactions within the cells produce energy from carbohydrates, proteins, and fats; cause proteins to be broken down or built up; and in thousands of other ways keep us in a state of health.

A cell is a basic unit of life. Any substance that does not consist of one or more cells cannot be alive. Cells are the "building blocks" of tissues (such as muscles and bones), organs (such as the kidneys and liver), and systems (the respiratory and digestive systems, for example). Normal cell health and functioning are maintained when a state of nutritional and environmental utopia exists within and around the cells. A disruption in the availability of nutrients or the presence of harmful substances in the cell's environment can initiate disorders that eventually affect our health or growth. Health problems in general begin with disruptions in the normal activity of cells.

The types and amounts of food and supplements people consume affect the cells' environment and their ability to function normally. Excessive or inadequate supplies of nutrients and other chemical substances disrupt cell functions and result in health problems. Humans remain in a state of health as long as their cells do.

Cells
The basic unit of life, of which all living things are composed. Every cell is surrounded by a membrane and contains cytoplasm, within which are organelles and a nucleus; the cell nucleus contains chromosomes.

Cell Structures and Functions

A generalized diagram of a human cell is shown in Illustration E.1. Although all cells have some structures and functions in common, the specific functions performed, and the structures that support those functions, can vary a good deal from cell to cell. Cells lining the esophagus, stomach, and intestines, for example, are specialized to produce and secrete mucus that helps food pass through the digestive tract. Red blood cells are specially formed to transport oxygen and carbon dioxide.

Every cell is surrounded by a cell membrane that helps move nutrients into and out of the cell. Inside the cell membrane lies the cytoplasm, a fluid material that contains many organelles, tubes, and particles. Among the organelles in the cytoplasm are ribosomes, mitochondria, and lysosomes. Each of these "little organs" is encased in a cell membrane and performs specific functions. Ribosomes assemble amino acids into proteins following the instructions of DNA and its messenger RNA. The mitochondria are made of intricately folded membranes that bear thousands of highly organized sets of enzymes on their surfaces. These enzymes are actively involved in the production of energy and are found in particularly dense quantities in muscle cells. The lysosomes are like packets of enzymes. The enzymes are used to break down old cell particles that are being recycled and to destroy substances that are harmful to the body.

Cytoplasm also contains a highly organized system of membranes called the endoplasmic reticulum. When these membranes are dotted with ribosomes, they are called "rough endoplasmic reticulum." When ribosomes are absent, the endoplasmic reticulum is referred to as "smooth." Some membranes within the cytoplasm form tubes that collect certain types of cellular material and transport it out of the cell. These membranous tubes are called "Golgi apparatus." The rough and smooth endoplasmic reticulum are continuous with the Golgi apparatus, so secretions produced throughout the cell can be collected and transported to its exterior.

Within each cell is a nucleus covered by a two-layer membrane. The nucleus contains chromosomes and the genetic material DNA. DNA encodes all of the instructions a cell needs to conduct protein synthesis and to replicate life.

All cells within the body are part of a complex communication system that uses hormones, electrical impulses, and other chemical messengers to link each cell to the others. No cell is an island that operates independently from the others.

Illustration E.1 Generalized structure of a human cell.

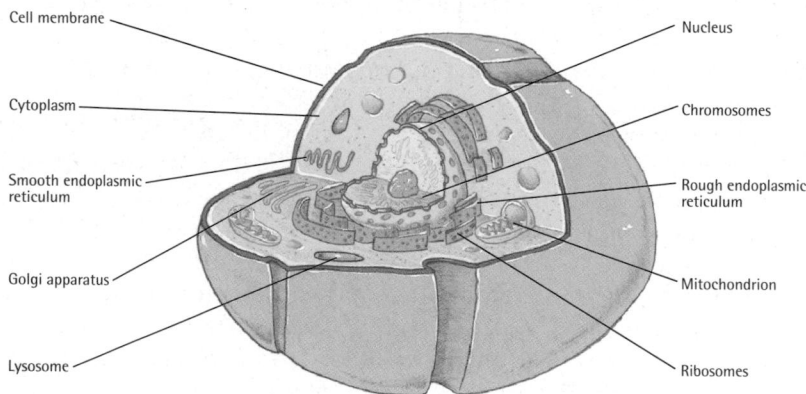

cell membrane: the membrane that surrounds the cell and encloses its contents; made primarily of lipid and protein.

chromosomes: a set of structures within the nucleus of every cell that contain the cell's genetic material, DNA, associated with other materials (primarily proteins).

cytoplasm (SIGH-toe-plazm): the cell contents, except for the nucleus.
cyto = cell
plasm = a form

Golgi (GOAL-gee) apparatus: a set of membranes within the cell where secretory materials are packaged for export.

lysosomes: cellular organelles; membrane-enclosed sacs of degradative enzymes.
lysis = dissolution

mitochondria (my-toe-KON-dree-uh; *singular* mitochondrion): the cellular organelles responsible for producing ATP aerobically; made of membranes (lipid and protein) with enzymes mounted on them.

mitos = thread (referring to their slender shape)
chondros = cartilage (referring to their external appearance)

nucleus: a major membrane-enclosed body within every cell, which contains the cell's genetic material, DNA, embedded in chromosomes.
nucleus = a kernel

organelles: membrane-bound subcellular structures such as ribosomes, mitochondria, and lysosomes.
organelle = little organ

ribosomes: protein-making organelles in cells; composed of RNA and protein.
ribo = containing the sugar ribose
some = body

rough endoplasmic reticulum (en-doh-PLAZ-mic reh-TIC-you-lum): intracellular membranes dotted with ribosomes, where protein synthesis takes place.
endo = inside
plasm = the cytoplasm

smooth endoplasmic reticulum: smooth intracellular membranes bearing no ribosomes.

Appendix F

WHO: Nutrition Recommendations
Canada: Choice System and Guidelines

This appendix first presents nutrition recommendations from the World Health Organization (WHO) and then provides details for Canadians on Canada's *Food Guide to Healthy Eating* and on the exchange system (called the choice system).

Nutrition Recommendations from WHO

The World Health Organization (WHO) has assessed the relationships between diet and the development of chronic diseases. Its recommendations include:

- Total energy: sufficient to support normal growth, physical activity, and healthy body weight (body mass index = 20 to 22).

- Total fat: 15 to 30 percent of total energy.

- Saturated fat: less than 10 percent of total energy.

- Total carbohydrate: 55 to 75 percent of total energy.

- Added sugars: less than 10 percent of total energy.

- Protein: 10 to 15 percent of total energy.

- Salt: less than 5 grams/day, preferably iodized.

- Fruit and vegetables: at least 400 grams (almost 1 pound) daily.

- Physical activity: one hour per day of moderate intensity on most days of the week.

Canada's *Food Guide to Healthy Eating*

Figure F-1 presents the 2007 Canada's *Food Guide to Healthy Eating,* which interprets Canada's *Guidelines for Healthy Eating* for consumers and recommends a range of servings to consume daily from each of the four food groups. Publications available from Health Canada through its website explain how to use the *Guide.* Figure F-2 presents Canada's Physical Activity Guide. *Food Guide Basics, Choosing Foods, Using the Food Guide, Maintaining Healthy Habits* and *Create My Food Guide* features are available on the home page (www.hc-sc.gc.ca/fn-an/food-guide-aliment/index-eng.php). The site includes information for educators, First Nation populations, and about physical activity Figure F-2 presents Canada's Physical Activity Guide.

Illustration F-1 Canadian Food Guide to Healthy Eating

Recommended Number of Food Guide Servings per Day

Age in Years	Children			Teens		Adults				
	2-3	4-8	9-13	14-18		19-50		51+		
Sex	Girls and Boys			Females	Males	Females	Males	Females	Males	
Vegetables and Fruit	4	5	6	7	8	7-8	8-10	7	7	
Grain Products	3	4	6	6	7	6-7	8	6	7	
Milk and Alternatives	2	2	3-4	3-4	3-4	2	2	3	3	
Meat and Alternatives	1	1	1-2	2	3	2	3	2	3	

The chart above shows how many Food Guide Servings you need from each of the four food groups every day.

Having the amount and type of food recommended and following the tips in *Canada's Food Guide* will help:
- Meet your needs for vitamins, minerals and other nutrients.
- Reduce your risk of obesity, type 2 diabetes, heart disease, certain types of cancer and osteoporosis.
- Contribute to your overall health and vitality.

What is One Food Guide Serving?
Look at the examples below.

Vegetables and Fruit

Fresh, frozen or canned vegetables
125 mL (½ cup)

Leafy vegetables
Cooked: 125 mL (½ cup)
Raw: 250 mL (1 cup)

Fresh, frozen or canned fruits
1 fruit or 125 mL (½ cup)

100% Juice
125 mL (½ cup)

Grain Products

Bread
1 slice (35 g)

Bagel
½ bagel (45 g)

Flat breads
½ pita or ½ tortilla (35 g)

Cooked rice, bulgur or quinoa
125 mL (½ cup)

Cereal
Cold: 30 g
Hot: 175 mL (¾ cup)

Cooked pasta or couscous
125 mL (½ cup)

Milk and Alternatives

Milk or powdered milk (reconstituted)
250 mL (1 cup)

Canned milk (evaporated)
125 mL (½ cup)

Fortified soy beverage
250 mL (1 cup)

Yogurt
175 g (¾ cup)

Kefir
175 g (¾ cup)

Cheese
50 g (1 ½ oz.)

Meat and Alternatives

Cooked fish, shellfish, poultry, lean meat
75 g (2 ½ oz.)/125 mL (½ cup)

Cooked legumes
175 mL (¾ cup)

Tofu
150 g / 175 mL (¾ cup)

Eggs
2 eggs

Peanut or nut butters
30 mL (2 Tbsp)

Shelled nuts and seeds
60 mL (¼ cup)

Oils and Fats
- Include a small amount – 30 to 45 mL (2 to 3 Tbsp) – of unsaturated fat each day. This includes oil used for cooking, salad dressings, margarine and mayonnaise.
- Use vegetable oils such as canola, olive and soybean.
- Choose soft margarines that are low in saturated and trans fats.
- Limit butter, hard margarine, lard and shortening.

Make each Food Guide Serving count...
wherever you are – at home, at school, at work or when eating out!

► **Eat at least one dark green and one orange vegetable each day.**
- Go for dark green vegetables such as broccoli, romaine lettuce and spinach.
- Go for orange vegetables such as carrots, sweet potatoes and winter squash.
- **Choose vegetables and fruit prepared with little or no added fat, sugar or salt.**
- Enjoy vegetables steamed, baked or stir-fried instead of deep-fried.
- **Have vegetables and fruit more often than juice.**

► **Make at least half of your grain products whole grain each day.**
- Eat a variety of whole grains such as barley, brown rice, oats, quinoa and wild rice.
- Enjoy whole grain breads, oatmeal or whole wheat pasta.
- **Choose grain products that are lower in fat, sugar or salt.**
- Compare the Nutrition Facts table on labels to make wise choices.
- Enjoy the true taste of grain products. When adding sauces or spreads, use small amounts.

► **Drink skim, 1%, or 2% milk each day.**
- Have 500 mL (2 cups) of milk every day for adequate vitamin D.
- Drink fortified soy beverages if you do not drink milk.
- **Select lower fat milk alternatives.**
- Compare the Nutrition Facts table on yogurts or cheeses to make wise choices.

► **Have meat alternatives such as beans, lentils and tofu often.**
► **Eat at least two Food Guide Servings of fish each week.**
- Choose fish such as char, herring, mackerel, salmon, sardines and trout.
- **Select lean meat and alternatives prepared with little or no added fat or salt.**
- Trim the visible fat from meats. Remove the skin on poultry.
- Use cooking methods such as roasting, baking or poaching that require little or no added fat.
- If you eat luncheon meats, sausages or prepackaged meats, choose those lower in salt (sodium) and fat.

Enjoy a variety of foods from the four food groups.

Satisfy your thirst with water!

Drink water regularly. It's a calorie-free way to quench your thirst. Drink more water in hot weather or when you are very active.

* Health Canada provides advice for limiting exposure to mercury from certain types of fish. Refer to www.healthcanada.gc.ca for the latest information.

Illustration F-1 Canadian Food Guide to Healthy Eating (continued)

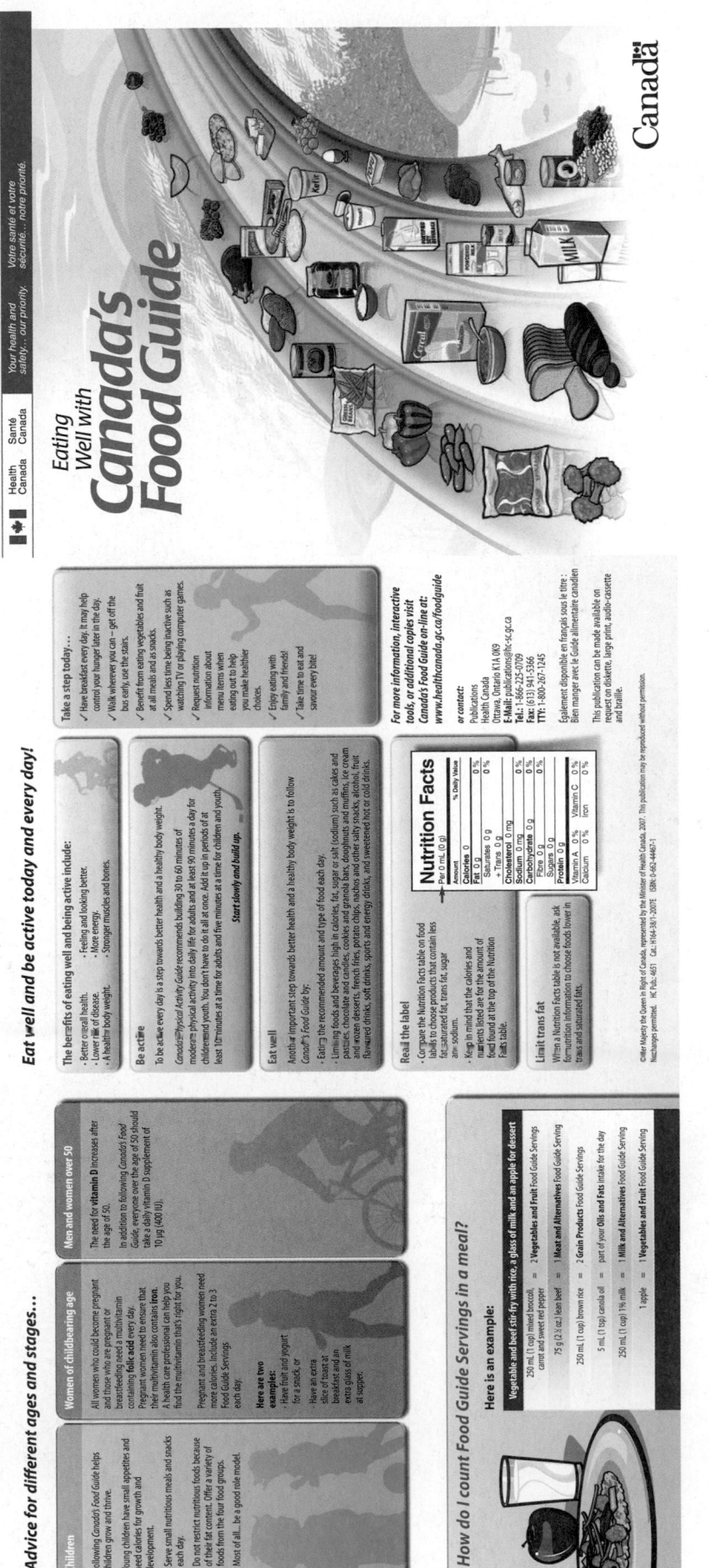

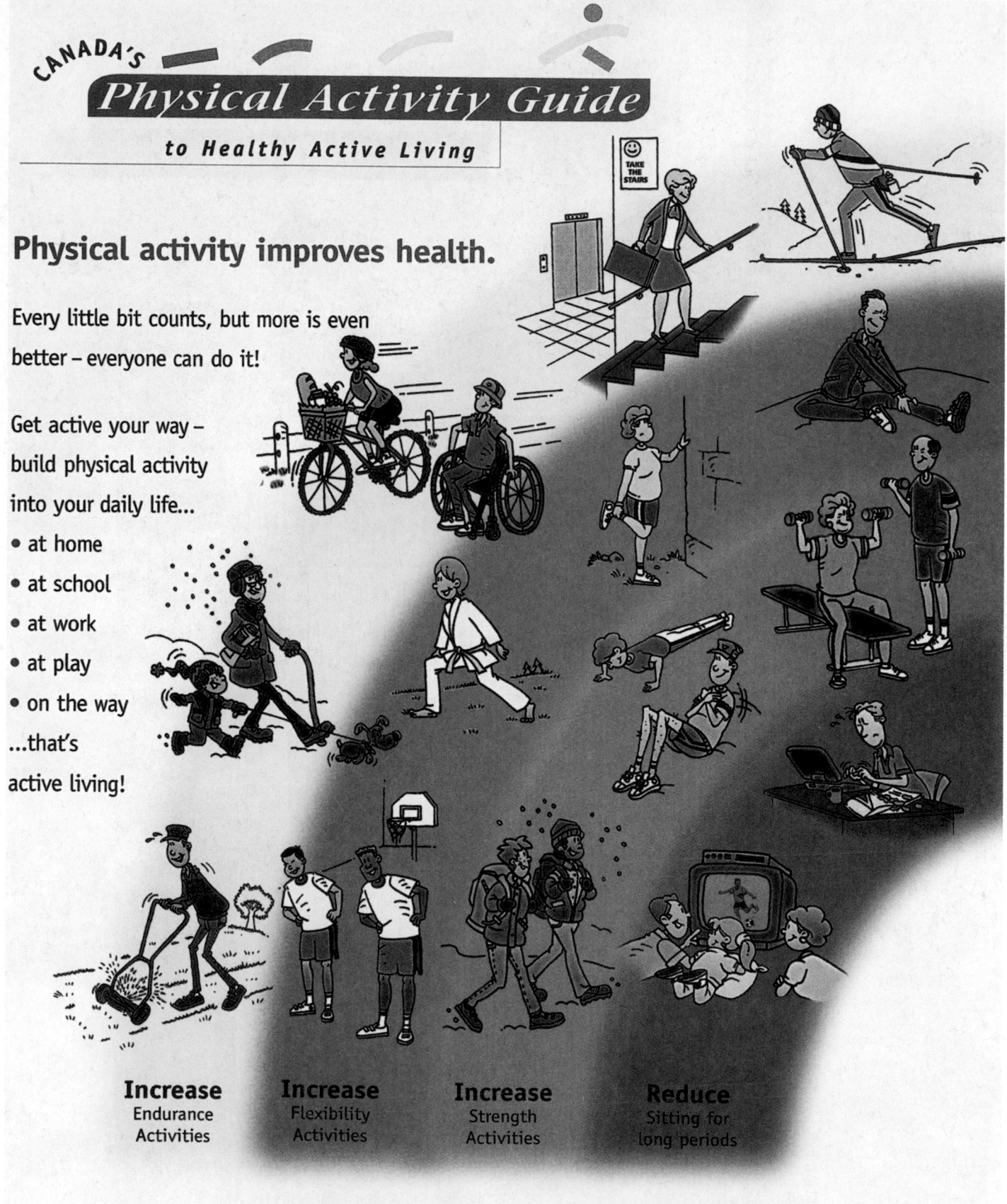

Choose a variety of activities from these three groups:

Endurance

4-7 days a week
Continuous activities for your heart, lungs and circulatory system.

Flexibility

4-7 days a week
Gentle reaching, bending and stretching activities to keep your muscles relaxed and joints mobile.

Strength

2-4 days a week
Activities against resistance to strengthen muscles and bones and improve posture.

Starting slowly is very safe for most people. Not sure? Consult your health professional.

For a copy of the *Guide Handbook* and more information: **1-888-334-9769**, or **www.paguide.com**

Eating well is also important. Follow *Canada's Food Guide to Healthy Eating* to make wise food choices.

Get Active Your Way, Every Day—For Life!

Scientists say accumulate 60 minutes of physical activity every day to stay healthy or improve your health. As you progress to moderate activities you can cut down to 30 minutes, 4 days a week. Add-up your activities in periods of at least 10 minutes each. Start slowly... and build up.

Time needed depends on effort

Very Light Effort	Light Effort *60 minutes*	Moderate Effort *30-60 minutes*	Vigorous Effort *20-30 minutes*	Maximum Effort
• Strolling • Dusting	• Light walking • Volleyball • Easy gardening • Stretching	• Brisk walking • Biking • Raking leaves • Swimming • Dancing • Water aerobics	• Aerobics • Jogging • Hockey • Basketball • Fast swimming • Fast dancing	• Sprinting • Racing

Range needed to stay healthy

You Can Do It – Getting started is easier than you think

Physical activity doesn't have to be very hard. Build physical activities into your daily routine.

- Walk whenever you can – get off the bus early, use the stairs instead of the elevator.
- Reduce inactivity for long periods, like watching TV.
- Get up from the couch and stretch and bend for a few minutes every hour.
- Play actively with your kids.
- Choose to walk, wheel or cycle for short trips.

- Start with a 10 minute walk – gradually increase the time.
- Find out about walking and cycling paths nearby and use them.
- Observe a physical activity class to see if you want to try it
- Try one class to start, you don't have to make a long-term commitment.
- Do the activities you are doing now, more often.

Benefits of regular activity:

- better health
- improved fitness
- better posture and balance
- better self-esteem
- weight control
- stronger muscles and bones
- feeling more energetic
- relaxation and reduced stress
- continued independent living in later life

Health risks of inactivity:

- premature death
- heart disease
- obesity
- high blood pressure
- adult-onset diabetes
- osteoporosis
- stroke
- depression
- colon cancer

No changes permitted. Permission to photocopy this document in its entirety not required.
Cat. No. H39-429/1998-1E ISBN 0-662-86627-7

CANADA'S *Physical Activity Guide*
to Healthy Active Living

Glossary

Note to the reader: If you have a customized book, this glossary may include some entries that will not be found in your book.

absorption
The process by which nutrients and other substances are transferred from the digestive system into body fluids for transport throughout the body.

adequate diet
A diet consisting of foods that together supply sufficient protein, vitamins, and minerals and enough calories to meet a person's need for energy.

Adequate Intakes (AIs)
Provisional RDAs developed when there is insufficient evidence to support a specific level of intake.

aerobic fitness
A state of respiratory and circulatory health as measured by the ability to deliver oxygen to muscles, and the capacity of muscles to use the oxygen for physical activity.

age-related macular degeneration
Eye damage caused by oxidation of the macula, the central portion of the eye that allows you to see details clearly. It is the leading cause of blindness in U.S. adults over the age of 65. Antioxidants provided by the carotenoids in dark green, leafy vegetables such as kale, collard greens, spinach, and Swiss chard may help prevent macular degeneration.

alcohol sugars
Simple sugars containing an alcohol group in their molecular structure. The most common are xylitol, mannitol, and sorbitol.

alcoholism
An illness characterized by a dependence on alcohol and by a level of alcohol intake that interferes with health, family and social relations, and job performance.

Alzheimer's disease
A brain disease that represents the most common form of dementia. It is characterized by memory loss for recent events that expands to more distant memories over the course of five to ten years. It eventually produces profound intellectual decline characterized by dementia and personal helplessness.

amylophagia (am-e-low-phag-ah)
Laundry starch or cornstarch eating.

anaphylactic shock (an-ah-fa-lac-tic)
A condition due to reduced oxygen supply to the heart and other tissues. It is caused by the body's reaction to an allergen in food or other substance. Symptoms of anaphylactic shock (or "anaphylaxis") may include abdominal cramps, vomiting, chest tightness, paleness, weak and rapid pulse, and difficulty breathing.

anorexia nervosa
An eating disorder characterized by extreme weight loss, poor body image, and irrational fears of weight gain and obesity.

antibodies
Blood proteins that help the body fight particular diseases. They help the body develop an immunity, or resistance, to many diseases.

antioxidants
Chemical substances that prevent or repair damage to cells caused by oxidizing agents such as pollutants, ozone, smoke, and reactive oxygen. Oxidation reactions are a normal part of cellular processes. Vitamins C and E and certain phytochemicals function as antioxidants.

appetite
The desire to eat; a pleasant sensation that is aroused by thoughts of the taste and enjoyment of food.

association
The finding that one condition is correlated with, or related to another condition, such as a disease or disorder. For example, diets low in vegetables are associated with breast cancer. Associations do *not* prove that one condition (such as a diet low in vegetables) *causes* an event (such as breast cancer). They indicate that a statistically significant relationship between a condition and an event exists.

atherosclerosis
"Hardening of the arteries" due to a build-up of plaque.

ATP, ADP
Adenosine triphosphate (ah-den-o-scene tri-phos-fate) and adenosine diphosphate. Molecules containing a form of phosphorous that can trap energy obtained from the macronutrients. ADP becomes ATP when it traps energy, and returns to being ADP when it releases energy for muscular and other work.

autoimmune disease
A disease related to the destruction of the body's own cells by substances produced by the immune system that mistakenly recognize certain cell components as harmful.

balanced diet
A diet that provides neither too much nor too little of nutrients and other components of food such as fat and fiber.

bariatrics
The field of medicine concerned with weight loss.

basal metabolism
Energy used to support body processes such as growth, health, tissue repair and maintenance, and other functions. Assessed while at rest, basal metabolism includes energy the body expends for breathing, the pumping of the heart, the maintenance of body temperature, and other life-sustaining, ongoing functions. Also called resting metabolism.

bile
A yellowish-brown or green fluid produced by the liver, stored in the gallbladder, and secreted into the small intestine. It acts like a detergent, breaking down globs of fat entering the small intestine to droplets, making the fats more accessible to the action of lipase.

binge eating
The consumption of a large amount of food in a small amount of time.

binge-eating disorder
An eating disorder characterized by periodic binge eating, which normally is not followed by vomiting or the use of laxatives. People must experience eating binges twice a week on average over a period of 6 months to qualify for the diagnosis.

bioavailability
The amount of a nutrient consumed that is available for absorption and use by the body.

biotechnology
As applied to food products, the process of modifying the composition of foods by biologically altering their genetic makeup. Also called *genetic engineering* of foods. The food products produced are sometimes referred to as "GM" and GMOs (genetically modified organisms).

body mass index (BMI)
An indicator of the appropriateness of a person's weight for their height. It is calculated by dividing weight in kilograms by height in meters. It can also be calculated using the method shown in Table 9.2.

bulimia nervosa
An eating disorder characterized by recurrent episodes of rapid, uncontrolled eating of large amounts of food in a short period of time. Episodes of binge eating are often followed by purging.

calorie (*calor = heat*)
A unit of measure used to express the amount of energy produced by foods in the form of heat. The calorie used in nutrition is the large "Calorie," or the "kilocalorie" (kcal). It equals the amount of energy needed to raise the temperature of 1 kilogram of water (about 4 cups) from 15 to 16°C (59 to 61°F). The term *kilocalorie*, or "calorie" as used in this text, is gradually being replaced by the "kilojoule" (kJ) in the United States; 1 kcal = 4.2 kJ.

cancer
A group of diseases in which abnormal cells grow out of control and can spread throughout the body. Cancer is not contagious and has many causes.

carbohydrates
Chemical substances in foods that consist of a simple sugar molecule or multiples of them in various forms.

cardiovascular disease
Disorders related to plaque build-up in arteries of the heart, brain, and other organs and tissues.

cataracts
Complete or partial clouding over the lens of the eye.

cause and effect
A finding that demonstrates that a condition causes a particular event. For example, vitamin C deficiency causes the deficiency disease scurvy.

celiac disease
An autoimmune disease characterized by inflammation of the small intestine lining resulting from a genetically based intolerance to gluten. The inflammation produces diarrhea, fatty stools, weight loss, and vitamin and mineral deficiencies. (Also called *celiac sprue* and *gluten-sensitive enteropathy*.)

cholesterol
A fat-soluble, colorless liquid found primarily in animal products. Cholesterol is used by the body to form hormones such as testosterone and estrogen and is a component of cell membranes. Cholesterol is present in plant cell membranes but the quantity is small and plants are not considered to be a significant dietary source of cholesterol.

chronic diseases
Slow-developing, long-lasting diseases that are not contagious (e.g., heart disease, cancer, diabetes). They can be treated but not always cured.

chronic inflammation
Low-grade inflammation that lasts weeks, months, or years. Inflammation is the first response of the body's immune system to infectious agents, toxins, or irritants. It triggers the release of biologically active substances that promote oxidation and other reactions to counteract the infection, toxin, or irritant. A side-effect of chronic inflammation is that it also damages lipids, cells, and tissues.

circulatory system
The heart, arteries, capillaries, and veins responsible for circulating blood throughout the body.

cirrhosis
Cirrhosis (pronounced sear-row-sis) is a disease of the liver characterized by widespread fibrous tissue buildup and disruption of normal liver structure and function. It can be caused by a number of chronic conditions that affect the liver such as alcoholism, diabetes, and obesity.

clinical trial
A study design in which one group of randomly assigned subjects (or subjects selected by the "luck of the draw") receives an active treatment and another group receives an inactive treatment, or "sugar pill," called the placebo.

coenzymes
Chemical substances, including many vitamins, that activate specific enzymes. Activated enzymes increase the rate at which reactions take place in the body, such as the breakdown of fats or carbohydrates in the small intestine and the conversion of glucose and fatty acids into energy within cells.

cofactors
Individual minerals required for the activity of certain proteins. For example: • iron is needed for hemoglobin's function in oxygen and carbon dioxide transport, • zinc is needed to activate or is a structural component of over 200 enzymes, and • magnesium activates over 300 enzymes involved in the formation of energy and proteins.

colostrum
The milk produced during the first few days after delivery. It contains more antibodies, protein, and certain minerals than the mature milk that is produced later. It is thicker than mature milk and has a yellowish color.

complementary protein sources
Plant sources of protein that together provide sufficient quantities of the nine essential amino acids.

complete proteins
Proteins that contain all of the essential amino acids in amounts needed to support growth and tissue maintenance.

complex carbohydrates
The form of carbohydrate found in starchy vegetables, grains, and dried beans and in many types of dietary fiber. The most common form of starch is made of long chains of interconnected glucose units.

control group
Subjects in a study who do not receive the active treatment or who do *not* have the condition under investigation. Control periods, or times when subjects are not receiving the treatment, are sometimes used instead of a control group.

critical period
A specific interval of time during which cells of a tissue or organ are genetically programmed to multiply. If the supply of nutrients needed for cell multiplication is not available during the specific time interval, the growth and development of the tissue or organ are permanently impaired.

cruciferous vegetables
Sulfur-containing vegetables whose outer leaves form a cross (or crucifix). Vegetables in this family include broccoli, cabbage, cauliflower, brussels sprouts, mustard and collard greens, kale, bok choy, kohlrabi, rutabaga, turnips, broccoflower, and watercress.

Daily Values (DVs)
Scientifically agreed-upon daily dietary intake standards for fat, saturated fat, cholesterol, carbohydrate, dietary fiber, and protein intake compatible with health. DVs are intended for use on nutrition labels only and are listed in the Nutrition Facts panel. "% Daily Value" on Nutrition Facts panels is calculated as the percentage of each DV supplied by a serving of the labeled food.

development
Processes involved in enhancing functional capabilities. For example, the brain grows, but the ability to reason develops.

diabetes
A disease characterized by abnormal utilization of glucose by the body and elevated blood glucose levels. There are three main types of diabetes: type 1, type 2, and gestational diabetes. The word *diabetes* in this text refers to type 2 diabetes, by far the most common. Diabetes is short for the term "diabetes mellitus."

diarrhea
The presence of three or more liquid stools in a 24-hour period.

dietary fiber
Naturally occurring, intact forms of nondigestible carbohydrates in plants and "woody" plant cell walls. Oat and wheat bran, and raffinose in dried beans, are examples of this type of fiber.

dietary supplements
Any products intended to supplement the diet, including vitamin and mineral supplements; proteins, enzymes, and amino acids; fish oils and fatty acids; hormones and hormone precursors; and herbs and other plant extracts. Such products must be labeled "Dietary Supplement."

dietary thermogenesis
Thermogenesis means "the production of heat." Dietary thermogenesis is the energy expended during food ingestion, the digestion of food, and the absorption and utilization of nutrients. Some of the energy escapes as heat. It accounts for approximately 10% of the body's total energy need. Also called *diet-induced thermogenesis* and *thermic effect of foods or feeding*.

digestion
The mechanical and chemical processes whereby ingested food is converted into substances that can be absorbed by the intestinal tract and utilized by the body.

disaccharides (di = two, saccharide = sugar)
Simple sugars consisting of two molecules of monosaccharides linked together. Sucrose, maltose, and lactose are disaccharides.

DNA (deoxyribonucleic acid)
Genetic material contained in cells that initiates and directs the production of proteins in the body.

double blind
A study in which neither the subjects participating in the research nor the scientists performing the research know which subjects are receiving the treatment and which are getting the placebo. Both subjects and investigators are "blind" to the treatment administered.

double-blind, placebo-controlled food challenge
A test used to determine the presence of a food allergy or other adverse reaction to a food. In this test, neither the patient nor the care provider knows whether a suspected offending food or a placebo is being tested.

duodenal (do-odd-en-all) and stomach ulcers
Open sores in the lining of the duodenum (the uppermost part of the small intestine) or the stomach.

edema
Swelling due to an accumulation of fluid in body tissues.

electrolytes
Minerals such as sodium and potassium that carry a charge when in solution. Many electrolytes help the body maintain an appropriate amount of fluid.

empty-calorie foods
Foods that provide an excess of calories in relation to nutrients. Soft drinks, candy, sugar, alcohol, and fats are considered empty-calorie foods.

endothelium
(Pronounced *n-dough-theil-e-um*) The layer of cells lining the inside of blood vessels.

energy-dense foods
Food that provide relatively high levels of calories per unit weight of the food. Fried chicken, cheeseburgers, a biscuit, egg, sausage sandwich, and potato chips are energy-dense foods.

energy density
The number of calories in a gram of food. It is calculated by dividing the number of calories in a portion of food by the food's weight in grams.

enrichment
The replacement of thiamin, riboflavin, niacin, and iron lost when grains are refined.

environmental trigger
An environmental factor, such as inactivity, a high-fat diet, or a high sodium intake, that causes a genetic tendency toward a disorder to be expressed.

enzymes
Protein substances that speed up chemical reactions. Enzymes are found throughout the body but are present in particularly large amounts in the digestive system.

epidemiological studies
Research that seeks to identify conditions related to particular events within a population. This type of research does *not* identify cause-and-effect relationships. For example, much of the information known about diet and cancer is based on epidemiological studies that have found that diets low in vegetables and fruits are associated with the development of heart disease.

ergogenic aids (ergo = work; genic = producing)
Substances that increase the capacity formuscular work.

essential amino acids
Amino acids that cannot be synthesized in adequate amounts by humans and therefore must be obtained from the diet. They are sometimes referred to as "indispensable amino acids."

essential fatty acids
Components of fats (linoleic acid–pronounced lynn-oh-lay-ick and alpha-linolenic acid–lynn-oh-len-ick) required in the diet.

essential hypertension
Hypertension of no known cause; also called primary or idiopathic hypertension, it accounts for 95% of all cases of hypertension.

essential nutrients
Substances required for normal growth and health that the body cannot generally produce, or produce in sufficient amounts; they must be obtained in the diet.

experimental group
Subjects in a study who receive the treatment being tested or have the condition that is being investigated.

fatty liver disease
A reversible condition characterized by fat infiltration of the liver (10% or more by weight). If not corrected, fatty liver disease can produce liver damage and other disorders. The condition is primarily associated with obesity, diabetes, and excess alcohol consumption. The disease is called "steatohepatitis" when accompanied by inflammation.

fermentation
The process by which carbohydrates are converted to ethanol by the action of the enzymes in yeast.

fetus
A baby in the womb from the eighth week of pregnancy until birth. (Before then, it is referred to as an embryo.)

flatulence (flat-u-lens)
Presence of excess gas in the stomach and intestines.

food additives
Any substances added to food that become part of the food or affect the characteristics of the food. The term applies to substances added both intentionally and unintentionally to food.

food allergen
A substance in food (almost always a protein) that is identified as harmful by the body and elicits an allergic reaction from the immune system.

food allergy
Adverse reaction to a normally harmless substance in food that involves the body's immune system. (Also called food hypersensitivity.)

food insecurity
Limited or uncertain availability of safe, nutritious foods or the ability to acquire them in socially acceptable ways.

food intolerance
Adverse reaction to a normally harmless substance in food that does not involve the body's immune system.

food security
Access at all times to a sufficient supply of safe, nutritious foods.

foodborne illnesses
An illness related to consumption of foods or beverages containing disease-causing bacteria, viruses, parasites, toxins, or other contaminants.

fortification
The addition of one or more vitamins and/or minerals to a food product.

free radicals
Chemical substances (often oxygen-based) that are missing electrons. The absence of electrons make the chemical substance reactive and prone to oxidizing nearby molecules by stealing an electron from them. Free radicals can damage

lipids, proteins, DNA (genetic material contained in cells), cells, and tissues by altering their chemical structure and functions.

fruitarian
A form of vegetarian diet in which fruits are the major ingredient. Such diets provide inadequate amounts of a variety of nutrients.

functional fiber
Specific types of nondigestible carbohydrates that have beneficial effects on health. Two examples of functional fibers are psyllium and pectin.

functional foods
Generally taken to mean foods, fortified foods, and enhanced food products that may benefit health beyond the effects of essential nutrients they contain.

genotype
The specific genetic makeup of an individual as coded by DNA.

geophagia (ge-oh-phag-ah)
Clay or dirt eating.

gestational diabetes
Diabetes first discovered during pregnancy.

glycemic index (GI)
A measure of the extent to which blood glucose level is raised by a 50-gram portion of a carbohydrate-containing food compared to 50 grams of glucose or white bread.

glycemic load (GL)
A measure of the extent to which blood glucose level is raised by a given amount of a carbohydrate-containing food. GL is calculated by multiplying a food's GI by its carbohydrate content.

glycogen
The body's storage form of glucose. Glycogen is stored in the liver and muscles.

growth
A process characterized by increases in cell number and size.

health
The WHO defines health as state of complete physical, mental, and social well-being and not merely the absence of disease or infirmity.

heart disease
One of a number of disorders that result when circulation of blood to parts of the heart is inadequate. Also called coronary heart disease. ("Coronary" refers to the blood vessels at the top of the heart. They look somewhat like a crown.)

heartburn
A condition that results when acidic stomach contents are released into the esophagus, usually causing a burning sensation.

hemoglobin
The iron-containing protein in red blood cells.

hemorrhoids (hem-or-oids)
Swelling of veins in the anus or rectum.

histamine (hiss-tah-mean)
A substance released in allergic reactions. It causes dilation of blood vessels, itching, hives, and a drop in blood pressure and stimulates the release of stomach acids and other fluids. Antihistamines neutralize the effects of histamine and are used in the treatment of some cases of allergies.

homocysteine
A compound produced when the amino acid methionine is converted to another amino acid, cysteine. High blood levels of homocysteine increase the risk of hardening of the arteries, heart attack, and stroke.

hormone
A substance, usually a protein or steroid (a cholesterol-derived chemical), produced by one tissue and conveyed by the bloodstream to another. Hormones affect the body's metabolic processes such as glucose utilization and fat deposition.

hunger
Unpleasant physical and psychological sensations (weakness, stomach pains, irritability) that lead people to acquire and ingest food.

hydrogenation
The addition of hydrogen to unsaturated fatty acids.

hypertension
High blood pressure. It is defined as blood pressure exerted inside of blood vessel walls that typically exceeds 140/90 mm Hg (or, millimeters of mercury).

hypoglycemia
A disorder resulting from abnormally low blood glucose levels. Symptoms of hypoglycemia include irritability, nervousness, weakness, sweating, and hunger. These symptoms are relieved by consuming glucose or foods that provide carbohydrate.

hypothesis
A statement made prior to initiating a study of the relationship sought to be tested by the research.

immunoproteins
Blood proteins such as antibodies that play a role in the functioning of the immune system (the body's disease defense system). Antibodies attack foreign proteins.

immune system
Body tissues that provide protection against bacteria, viruses, and other substances identified by cells as harmful.

incomplete proteins
Proteins that are deficient in one or more essential amino acids.

infant mortality rate
Deaths that occur within the first year of life per 1,000 live births.

initiation
The start of the cancer process; it begins with the alteration of DNA within cells.

insulin resistance
A condition in which cell membrane have reduced sensitivity to insulin so that more insulin than normal is required to transport a given amount of glucose into cells. It is characterized by elevated levels of serum insulin, glucose, and triglycerides, and increased blood pressure.

iron deficiency
A disorder that results from a depletion of iron stores in the body. It is characterized by weakness, fatigue, short attention span, poor appetite, increased susceptibility to infection, and irritability.

iron-deficiency anemia
A condition that results when the content of hemoglobin in red blood cells is reduced due to a lack of iron. It is characterized by the signs of iron deficiency plus paleness, exhaustion, and a rapid heart rate.

irritable bowel syndrome (IBS)
A disorder of bowel function characterized by chronic or episodic gas, abdominal pain, diarrhea or constipation, or both.

kwashiorkor (kwa-she-or-kor)
A severe form of protein-energy malnutrition in young children. It is characterized by swelling, fatty liver, susceptibility to infection, profound apathy, and poor appetite. The cause of kwashiorkor is unclear.

lactose intolerance
The term for gastrointestinal symptoms (flatulence, bloating, abdominal pain, diarrhea, and "rumbling in the bowel") resulting from the consumption of more lactose than can be digested with available lactase.

lactose maldigestion
A disorder characterized by reduced digestion of lactose due to the low availability of the enzyme lactase.

life expectancy
The average length of life of people of a given age.

lipids
Compounds that are insoluble in water and soluble in fat. Triglycerides, saturated and unsaturated fats, and essential fatty acids are examples of lipids, or "fats."

low-birthweight infants
Infants weighing less than 2,500 grams (5.5 pounds) at birth.

lymphatic system
A network of vessels that absorb some of the products of digestion and transport them to the heart, where they are mixed with the substances contained in blood.

macronutrients
The group name for the energy-yielding nutrients of carbohydrate, protein, and fat. They are called macronutrients because we need relatively large amounts of them in our daily diet.

malnutrition
Poor nutrition resulting from an excess or lack of calories or nutrients.

marasmus
A severe form of malnutrition primarily due to a chronic lack of calories and protein. Also called protein-energy malnutrition.

maximal oxygen consumption
The highest amount of oxygen that can be delivered to, and utilized by, muscles for physical activity. Also called VO$_2$ max and maximal volume of oxygen.

meta-analysis
An analysis of data from multiple studies. Results are based on larger samples than the individual studies and are therefore more reliable. Differences in methods and subjects among the studies may bias the results of meta-analyses.

metabolic syndrome
A constellation of metabolic abnormalities that increase the risk of heart disease, hypertension, type 2 diabetes, and other disorders. Metabolic syndrome is characterized by insulin resistance, abdominal obesity, high blood pressure and triglycerides levels, low levels of HDL cholesterol, and impaired glucose tolerance. It is also called *Syndrome X* and *insulin resistance syndrome*.

metabolism
The chemical changes that take place in the body. The conversion of glucose to energy or to body fat is an example of a metabolic process.

minerals
In the context of nutrition, minerals are specific, single atoms that perform particular functions in the body. There are 15 essential minerals—or minerals required in the diet.

monosaccharides
(mono=one, saccharide=sugar): Simple sugars consisting of one sugar molecule. Glucose, fructose, and galactose are common monosaccharides.

myoglobin
The iron-containing protein in muscle cells.

neural tube defects
Malformations of the spinal cord and brain. They are among the most common and severe fetal malformations, occurring in approximately

1 in every 1000 pregnancies. Neural tube defects include spina bifida (spinal cord fluid protrudes through a gap in the spinal cord; shown in Illustration 29.6), anencephaly (absence of the brain or spinal cord), and encephalocele (protrusion of the brain through the skull).

nonessential amino acids
Amino acids that can be readily produced by humans from components of the diet. Also referred to as "dispensable amino acids."

nonessential nutrients
Nutrients required for normal growth and health that the body can manufacture in sufficient quantities from other components of the diet. We do not require a dietary source of nonessential nutrients.

nutrigenomics
The study of diet- and nutrient-related functions and interactions of genes and their affects on health and disease.

nutrient-dense foods
Foods that contain relatively high amounts of nutrients compared to their calorie value. Broccoli, collards, bread, cantaloupe, and lean meats are examples of nutrient-dense foods.

nutrients
Chemical substances found in food that are used by the body for growth and health. The six categories of nutrients are carbohydrates, proteins, fats, vitamins, minerals, and water.

nutrition
The study of foods, their nutrients and other chemical constituents, and the effects of food constituents on health.

obesity
A condition characterized by excess body fat.

osteoporosis (osteo=bones; poro=porous, osis=abnormal condition)
A condition in which bones become fragile and susceptible to fracture due to a loss of calcium and other minerals.

overweight
A high weight-for-height.

oxidative stress
A condition that occurs when cells are exposed to more oxidizing molecules (such as free radicals) than to antioxidant molecules that neutralize them. Over time, oxidative stress causes damage to lipids, DNA, cells and tissues. It increases the risk of heart disease, type 2 diabetes, cancer, and other diseases.

pagophagia (pa-go-phag-ah)
Ice eating.

peer review
Evaluation of the scientific merit of research or scientific reports by experts in the area under

review. Studies published in scientific journals have gone through peer review prior to being accepted for publication.

% Daily Value (%DV)
Scientifically agreed-upon standards of daily intake of nutrients from the diet developed for use on nutrition labels. The "% Daily Values" listed in nutrition labels represent the percentages of the standards obtained from one serving of the food product.

phenylketonuria (feen-ol-key-tone-u-re-ah), PKU
A rare genetic disorder related to the lack of the enzyme phenylalanine hydroxylase. Lack of this enzyme causes the essential amino acid phenylalanine to build up in blood.

physical fitness
The health of the body as measured by muscular strength, endurance, and flexibility in the conduct of physical activity.

phytochemicals (phyto=plant)
Chemical substances in plants, some of which perform important functions in the human body. Phytochemicals give plants color and flavor, participate in processes that enable plants to grow, and protect plants against insects and diseases.

pica (pike-eh)
The regular consumption of nonfood substances such as clay or laundry starch.

placebo
A "sugar pill," an imitation treatment given to subjects in research.

placebo effect
Changes in health or perceived health that result from expectations that a "treatment" will produce an effect on health.

plant stanols or sterols
Substances in corn, wheat, oats, rye, olives, wood, and some other plants that are similar in structure to cholesterol but that are not absorbed by the body. They decrease cholesterol absorption.

plaque (dental)
A soft, sticky, white material on teeth; formed by bacteria.

plaque (arterial)
Deposits of cholesterol, other fats, calcium, and cell materials in the lining of the inner wall of arteries.

plumbism
Lead (primarily from old paint flakes) eating.

polysaccharides (poly=many, saccharide=sugar)
Carbohydrates containing many molecules of monosaccharides linked together. Starch, glycogen, and dietary fiber are

the three major types of polysaccharides. Polysaccharides consisting of three to 10 monosaccharides may be referred to as "oligosaccharides."

prebiotics
Non-digestible food ingredients that beneficially affect a person by selectively stimulating the growth or activity of one or a limited number of bacteria in the colon. Inulin, an extract from chicory root, is a common prebiotic. Also called "intestinal fertilizer."

precursor
In nutrition, a nutrient that can be converted into another nutrient. (Also called provitamin.) Beta-carotene is a precursor of vitamin A.

prediabetes
A condition in which blood glucose levels are higher than normal but not high enough for the diagnosis of diabetes. It is characterized by impaired glucose tolerance, or fasting blood glucose levels between 110 and 126 mg/dl.

preterm
Infants born at or before 37 weeks of gestation (pregnancy).

probiotics
Live microorganisms which when delivered in adequate amounts confer a health benefit. Strains of *lactobacillus* (lac-toe-bah-sil-us) and *bifidobacteria* (bif-id-dough bacteria) are the best known probiotics. Also called "friendly bacteria."

progression
The uncontrolled growth of abnormal cells.

promotion
The period in cancer development when the number of cells with altered DNA increases.

prostate
A gland located above the testicles in males. The prostate secretes a fluid that surrounds sperm.

protein
Chemical substance in foods made up of chains of amino acids.

purging
The use of self-induced vomiting, laxatives, or diuretics (water pills) to prevent weight gain.

Recommended Dietary Allowances (RDAs)
Intake levels of essential nutrients that meet the nutritional needs of practically all healthy people while decreasing the risk of certain chronic diseases.

remodeling
The breakdown and buildup of bone tissue.

restrained eating
The purposeful restriction of food intake below desired amounts in order to control body weight.

salt sensitivity
A genetically determined condition in which a person's blood pressure rises when high amounts of salt or sodium are consumed. Such individuals are sometimes identified by blood pressure increases of 10% or more when switched from a low-salt (1–3 grams) to a high-salt (12–15 grams) diet.

satiety
A feeling of fullness or of having had enough to eat.

saturated fats
The type of fat that tends to raise blood cholesterol levels and the risk for heart disease. They are solid at room temperature and are found primarily in animal products such as meat, butter, and cheese.

serotonin (pronounced sare-uh-tone-in)
A neurotransmitter, or chemical messenger, for nerve cell activities that excite or inhibit various behaviors and body functions. It plays a role in mood, appetite regulation, food intake, respiration, pain transmission, blood vessel constriction, and other body processes.

simple sugars
Carbohydrates that consist of a glucose, fructose, or galactose molecule; or a combination of glucose and either fructose or galactose. High-fructose corn syrup and alcohol sugars are also considered simple sugars. Simple sugars are often referred to as "sugars."

single-gene defects
Disorders resulting from one abnormal gene. Also called "inborn errors of metabolism." Over 800 single-gene defects have been cataloged, and most are very rare.

starch
Complex carbohydrates made up of complex chains of glucose molecules. Starch is the primary storage form of carbohydrate in plants. The vast majority of carbohydrate in our diet consists of starch, monosaccharides, and disaccharides.

statistically significant
Research findings that likely represent a true or actual result and not one due to chance.

steatohepatitis
Steatohepatitis (pronounced ste-at-oh-hep-ah-tie-tis) is a disease characterized by inflammation of, and fat accumulation in the liver. It is associated with alcoholism and may occur in obesity and diabetes. Steatohepatitis may progress to cirrhosis.

stroke
The event that occurs when a blood vessel in the brain suddenly ruptures or becomes blocked, cutting off blood supply to a portion of the brain. Stroke is often associated with "hardening of the arteries" in the brain. (Also called a *cerebral vascular accident*.)

subcutaneous fat
(Pronounced *sub-q-tain-e-ous*) Fat located under the skin.

tooth decay
The disintegration of teeth due to acids produced by bacteria in the mouth that feed on sugar. Also called dental caries or cavities.

total fiber
The sum of functional and dietary fiber.

trans fats
A type of unsaturated fat present in hydrogenated oils, margarine, shortening, pastries, and some cooking oils that increase the risk of heart disease. Fats containing fatty acids in the trans form are generally referred to as trans fats.

trimester
One-third of the normal duration of pregnancy. The first trimester is 0 to 13 weeks, the second is 13 to 26 weeks, and the third is 26 to 40 weeks.

tryptophan (pronounced trip-tuh-fan)
An essential amino acid that is used to form the chemical messenger serotonin (among other functions). Tryptophan is generally present in lower amounts in food protein than most other essential amino acids. It can be produced in the body from niacin, a B vitamin.

type 1 diabetes
A disease characterized by high blood glucose levels resulting from destruction of the insulin-producing cells of the pancreas. This type of diabetes was called juvenile-onset diabetes and insulin-dependent diabetes in the past, and its official medical name is type 1 diabetes mellitus.

type 2 diabetes
A disease characterized by high blood glucose levels due to the body's inability to use insulin normally, or to produce enough insulin. This type of diabetes was called adult-onset diabetes and non-insulin-dependent diabetes in the past, and its official medical name is type 2 diabetes mellitus.

underweight
Usually defined as a low weight-for-height. May also represent a deficit of body fat.

unsaturated fats
The type of fat that tends to lower blood cholesterol level and the risk of heart disease. They are liquid at room temperature and found in foods such as nuts, seeds, fish, shellfish, and vegetable oils.

visceral fat
(Pronounced *vis-sir-el*) Fat located under the skin and muscle of the abdomen.

vitamins
Chemical substances that perform specific functions in the body.

water balance
The ratio of the amount of water outside cells to the amount inside cells; this balance is needed for normal cell functioning.

zoochemicals
Chemical substances in animal foods, some of which likely perform important functions in the body.

Index

Note to the reader: If you have a customized book, this index may include some entries that will not be found in your book.

Credits

This page constitutes an extension of the copyright page. We have made every effort to trace the ownership of all copyrighted material and to secure permission from copyright holders. In the event of any question arising as to the use of any material, we will be pleased to make the necessary corrections in future printings. Thanks are due to the following authors, publishers, and agents for permission to use the material indicated.

Dietary Reference Intakes (DRIs): Recommended Intakes for Individuals, Vitamins

Food and Nutrition Board, Institute of Medicine, National Academies

Life Stage Group	Vit A (µg/d)[a]	Vit C (mg/d)	Vit D (µg/d)[b,c]	Vit E (mg/d)[d]	Vit K (µg/d)	Thiamin (mg/d)	Riboflavin (mg/d)	Niacin (mg/d)[e]	Vit B₆ (mg/d)	Folate (µg/d)[f]	Vit B₁₂ (µg/d)	Pantothenic Acid (mg/d)	Biotin (µg/d)	Choline (mg/d)[g]
Infants														
0–6 mo	400*	40*	5*	4*	0.2*	0.2*	0.3*	2*	0.1*	65*	0.4*	1.7*	5*	125*
7–12 mo	500*	50*	5*	5*	2.5*	0.3*	0.4*	4*	0.3*	80*	0.5*	1.8*	6*	150*
Children														
1–3 y	300	15	5*	6	30*	0.5	0.5	6	0.5	150	0.9	2*	8*	200*
4–8 y	400	25	5*	7	55*	0.6	0.6	8	0.6	200	1.2	3*	12*	250*
Males														
9–13 y	600	45	5*	11	60*	0.9	0.9	12	1.0	300	1.8	4*	20*	375*
14–18 y	900	75	5*	15	75*	1.2	1.3	16	1.3	400	2.4	5*	25*	550*
19–30 y	900	90	5*	15	120*	1.2	1.3	16	1.3	400	2.4	5*	30*	550*
31–50 y	900	90	5*	15	120*	1.2	1.3	16	1.3	400	2.4	5*	30*	550*
51–70 y	900	90	10*	15	120*	1.2	1.3	16	1.7	400	2.4[i]	5*	30*	550*
>70 y	900	90	15*	15	120*	1.2	1.3	16	1.7	400	2.4[i]	5*	30*	550*
Females														
9–13 y	600	45	5*	11	60*	0.9	0.9	12	1.0	300	1.8	4*	20*	375*
14–18 y	700	65	5*	15	75*	1.0	1.0	14	1.2	400[i]	2.4	5*	25*	400*
19–30 y	700	75	5*	15	90*	1.1	1.1	14	1.3	400[i]	2.4	5*	30*	425*
31–50 y	700	75	5*	15	90*	1.1	1.1	14	1.3	400[i]	2.4	5*	30*	425*
51–70 y	700	75	10*	15	90*	1.1	1.1	14	1.5	400	2.4[h]	5*	30*	425*
>70 y	700	75	15*	15	90*	1.1	1.1	14	1.5	400	2.4[h]	5*	30*	425*
Pregnancy														
14–18 y	750	80	5*	15	75*	1.4	1.4	18	1.9	600[j]	2.6	6*	30*	450*
19–30 y	770	85	5*	15	90*	1.4	1.4	18	1.9	600[j]	2.6	6*	30*	450*
31–50 y	770	85	5*	15	90*	1.4	1.4	18	1.9	600[j]	2.6	6*	30*	450*
Lactation														
14–18 y	1,200	115	5*	19	75*	1.4	1.6	17	2.0	500	2.8	7*	35*	550*
19–30 y	1,300	120	5*	19	90*	1.4	1.6	17	2.0	500	2.8	7*	35*	550*
31–50 y	1,300	120	5*	19	90*	1.4	1.6	17	2.0	500	2.8	7*	35*	550*

NOTE: This table (taken from the DRI reports, see www.nap.edu) presents Recommended Dietary Allowances (RDAs) in **bold type** and Adequate Intakes (AIs) in ordinary type followed by an asterisk (*). RDAs and AIs may both be used as goals for individual intake. RDAs are set to meet the needs of almost all (97 to 98 percent) individuals in a group. For healthy breastfed infants, the AI is the mean intake. The AI for other life stage and gender groups is believed to cover needs of all individuals in the group, but lack of data or uncertainty in the data prevent being able to specify with confidence the percentage of individuals covered by this intake.

[a] As retinol activity equivalents (RAEs). 1 RAE = 1 µg retinol, 12 µg β-carotene, 24 µg β-carotene, or 24 µg β-cryptoxanthin. The RAE for dietary provitamin A carotenoids is twofold greater than retinol equivalents (RE), whereas the RAE for preformed vitamin A is the same as RE.

[b] As cholecalciferol. 1 µg cholecalciferol = 40 IU vitamin D.

[c] In the absence of adequate exposure to sunlight.

[d] As α-tocopherol. α-Tocopherol includes RRR-α-tocopherol, the only form of α-tocopherol that occurs naturally in foods, and the 2R-stereoisomeric forms of α-tocopherol (RRR-, RSR-, RRS-, and RSS-α-tocopherol) that occur in fortified foods and supplements. It does not include the 2S-stereoisomeric forms of α-tocopherol (SRR-, SSR-, SRS-, and SSS-α-tocopherol), also found in fortified foods and supplements.

[f] As dietary folate equivalents (DFE). 1 DFE = 1 µg food folate = 0.6 µg of folic acid from fortified food or as a supplement consumed with food = 0.5 µg of a supplement taken on an empty stomach.

[g] Although AIs have been set for choline, there are few data to assess whether a dietary supply of choline is needed at all stages of the life cycle, and it may be that the choline requirement can be met by endogenous synthesis at some of these stages.

[h] Because 10 to 30 percent of older people may malabsorb food-bound B₁₂, it is advisable for those older than 50 years to meet their RDA mainly by consuming foods fortified with B₁₂ or a supplement containing B₁₂.

[i] In view of evidence linking folate intake with neural tube defects in the fetus, it is recommended that all women capable of becoming pregnant consume 400 µg from supplements or fortified foods in addition to intake of food folate from a varied diet.

[j] It is assumed that women will continue consuming 400 µg from supplements or fortified food until their pregnancy is confirmed and they enter prenatal care, which ordinarily occurs after the end of the periconceptional period—the critical time for formation of the neural tube.

Dietary Reference Intakes (DRIs): Recommended Intakes for Individuals, Elements

Food and Nutrition Board, Institute of Medicine, National Academies

LIFE STAGE GROUP	Calcium (mg/d)	Chromium (µg/d)	Copper (µg/d)	Fluoride (mg/d)	Iodine (µg/d)	Iron (mg/d)	Magnesium (mg/d)	Manganese (mg/d)	Molybdenum (µg/d)	Phosphorus (mg/d)	Selenium (µg/d)	Zinc (mg/d)	Potassium (g/d)	Sodium (g/d)	Chloride (g/d)
Infants															
0–6 mo	210*	0.2*	200*	0.01*	110*	0.27*	30*	0.003*	2*	100*	15*	2*	0.4*	0.12*	0.18*
7–12 mo	270*	5.5*	220*	0.5*	130*	11	75*	0.6*	3*	275*	20*	3	0.7*	0.37*	0.57*
Children															
1–3 y	500*	11*	340	0.7*	90	7	80	1.2*	17	460	20	3	0.3*	1.0*	1.5*
4–8 y	800*	15*	440	1*	90	10	130	1.5*	22	500	30	5	3.8*	1.2*	1.9*
Males															
9–13 y	1,300*	25*	700	2*	120	8	240	1.9*	34	1,250	40	8	4.5*	1.5*	2.3*
14–18 y	1,300*	35*	890	3*	150	11	410	2.2*	43	1,250	55	11	4.7*	1.5*	2.3*
19–30 y	1,000*	35*	900	4*	150	8	400	2.3*	45	700	55	11	4.7*	1.5*	2.3*
31–50 y	1,000*	35*	900	4*	150	8	420	2.3*	45	700	55	11	4.7*	1.5*	2.3*
51–70 y	1,200*	30*	900	4*	150	8	420	2.3*	45	700	55	11	4.7*	1.3*	2.0*
>70 y	1,200*	30*	900	4*	150	8	420	2.3*	45	700	55	11	4.7*	1.2*	1.8*
Females															
9–13 y	1,300*	21*	700	2*	120	8	240	1.6*	34	1,250	40	8	4.5*	1.5*	2.3*
14–18 y	1,300*	24*	890	3*	150	15	360	1.6*	43	1,250	55	9	4.7*	1.5*	2.3*
19–30 y	1,000*	25*	900	3*	150	18	310	1.6*	45	700	55	8	4.7*	1.5*	2.3*
31–50 y	1,000*	25*	900	3*	150	18	320	1.6*	45	700	55	8	4.7*	1.5*	2.3*
51–70 y	1,200*	20*	900	3*	150	8	320	1.8*	45	700	55	8	4.7*	1.3*	2.0*
>70 y	1,200*	20*	900	3*	150	8	320	1.8*	45	700	55	8	4.7*	1.2*	1.8*
Pregnancy															
14–18 y	1,300*	29*	1,000	3*	220	27	400	2.0*	50	1,250	60	12	4.7*	1.5*	2.3*
19–30 y	1,000*	30*	1,000	3*	220	27	350	2.0*	50	700	60	11	4.7*	1.5*	2.3*
31–50 y	1,000*	30*	1,000	3*	220	27	360	2.0*	50	700	60	11	4.7*	1.5*	2.3*
Lactation															
14–18 y	1,300*	44*	1,300	3*	290	10	360	2.6*	50	1,250	70	13	5.1*	1.5*	2.3*
19–30 y	1,000*	45*	1,300	3*	290	9	310	2.6*	50	700	70	12	5.1*	1.5*	2.3*
31–50 y	1,000*	45*	1,300	3*	290	9	320	2.6*	50	700	70	12	5.1*	1.5*	2.3*

NOTE: This table presents Recommended Dietary Allowances (RDAs) in **bold type** and Adequate Intakes (AIs) in ordinary type followed by an asterisk (*). RDAs and AIs may both be used as goals for individual intake. RDAs are set to meet the needs of almost all (97 to 98 percent) individuals in a group. For healthy breastfed infants, the AI is the mean intake. The AI for other life stage and gender groups is believed to cover needs of all individuals in the group, but lack of data or uncertainty in the data prevent being able to specify with confidence the percentage of individuals covered by this intake.

SOURCES: *Dietary Reference Intakes for Calcium, Phosphorous, Magnesium, Vitamin D, and Fluoride* (1997); *Dietary Reference Intakes for Thiamin, Riboflavin, Niacin, Vitamin B₆, Folate, Vitamin B₁₂, Pantothenic Acid, Biotin, and Choline* (1998); *Dietary Reference Intakes for Vitamin C, Vitamin E, Selenium, and Carotenoids* (2000); *Dietary Reference Intakes for Vitamin A, Vitamin K, Arsenic, Boron, Chromium, Copper, Iodine, Iron, Manganese, Molybdenum, Nickel, Silicon, Vanadium, and Zinc* (2001); and *Dietary Reference Intakes for Water, Potassium, Sodium, Chloride, and Sulfate* (2004). These reports may be accessed via http://www.nap.edu.

Dietary Reference Intakes (DRIs): Tolerable Upper Intake Levels (UL[a]), Elements

Food and Nutrition Board, Institute of Medicine, National Academies

LIFE STAGE GROUP	Arsenic[b]	Boron (mg/d)	Calcium (g/d)	Chromium	Copper (μg/d)	Fluoride (mg/d)	Iodine (μg/d)	Iron (mg/d)	Magnesium (mg/d)[c]	Manganese (mg/d)	Molybdenum (μg/d)	Nickel (mg/d)	Phosphorus (g/d)	Potassium	Selenium (μg/d)	Silicon[d]	Sulfate	Vanadium (mg/d)[e]	Zinc (mg/d)	Sodium (g/d)	Chloride (g/d)
Infants																					
0–6 mo	ND[f]	ND	ND	ND	ND	0.7	ND	40	ND	ND	ND	ND	ND	ND	45	ND	ND	ND	4	ND	ND
7–12 mo	ND	ND	ND	ND	ND	0.9	ND	40	ND	ND	ND	ND	ND	ND	60	ND	ND	ND	5	ND	ND
Children																					
1–3 y	ND	3	2.5	ND	1,000	1.3	200	40	65	2	300	0.2	3	ND	90	ND	ND	ND	7	1.5	2.3
4–8 y	ND	6	2.5	ND	3,000	2.2	300	40	110	3	600	0.3	3	ND	150	ND	ND	ND	12	1.9	2.9
Males, Females																					
9–13 y	ND	11	2.5	ND	5,000	10	600	40	350	6	1,100	0.6	4	ND	280	ND	ND	ND	23	2.2	3.4
14–18 y	ND	17	2.5	ND	8,000	10	900	45	350	9	1,700	1.0	4	ND	400	ND	ND	ND	34	2.3	3.6
19–70 y	ND	20	2.5	ND	10,000	10	1,100	45	350	11	2,000	1.0	4	ND	400	ND	ND	1.8	40	2.3	3.6
>70 y	ND	20	2.5	ND	10,000	10	1,100	45	350	11	2,000	1.0	3	ND	400	ND	ND	1.8	40	2.3	3.6
Pregnancy																					
14–18 y	ND	17	2.5	ND	8,000	10	900	45	350	9	1,700	1.0	3.5	ND	400	ND	ND	ND	34	2.3	3.6
19–50 y	ND	20	2.5	ND	10,000	10	1,100	45	350	11	2,000	1.0	3.5	ND	400	ND	ND	ND	40	2.3	3.6
Lactation																					
14–18 y	ND	17	2.5	ND	8,000	10	900	45	350	9	1,700	1.0	4	ND	400	ND	ND	ND	34	2.3	3.6
19–50 y	ND	20	2.5	ND	10,000	10	1,100	45	350	11	2,000	1.0	4	ND	400	ND	ND	ND	40	2.3	3.6

[a] UL = The maximum level of daily nutrient intake that is likely to pose no risk of adverse effects. Unless otherwise specified, the UL represents total intake from food, water, and supplements. Due to lack of suitable data, ULs could not be established for arsenic, chromium, silicon, potassium, and sulfate. In the absence of ULs, extra caution may be warranted in consuming levels above recommended intakes.

[b] Although the UL was not determined for arsenic, there is no justification for adding arsenic to food or supplements.

[c] The ULs for magnesium represent intake from a pharmacological agent only and do not include intake from food and water.

[d] Although silicon has not been shown to cause adverse effects in humans, there is no justification for adding silicon to supplements.

[e] Although vanadium in food has not been shown to cause adverse effects in humans, there is no justification for adding vanadium to food and vanadium supplements should be used with caution. The UL is based on adverse effects in laboratory animals and this data could be used to set a UL for adults but not children and adolescents.

[f] ND = Not determinable due to lack of data of adverse effects in this age group and concern with regard to lack of ability to handle excess amounts. Source of intake should be from food only to prevent high levels of intake.

SOURCES: Dietary Reference Intakes for Calcium, Phosphorous, Magnesium, Vitamin D, and Fluoride (1997); Dietary Reference Intakes for Thiamin, Riboflavin, Niacin, Vitamin B_6, Folate, Vitamin B_{12}, Pantothenic Acid, Biotin, and Choline (1998); Dietary Reference Intakes for Vitamin C, Vitamin E, Selenium, and Carotenoids (2000); Dietary Reference Intakes for Vitamin A, Vitamin K, Arsenic, Boron, Chromium, Copper, Iodine, Iron, Manganese, Molybdenum, Nickel, Silicon, Vanadium, and Zinc (2001); and Dietary Reference Intakes for Water, Potassium, Sodium, Chloride, and Sulfate (2004). These reports may be accessed via http://www.nap.edu.